A Specialist Periodical Report

Spectroscopic Properties of Inorganic and Organometallic Compounds
Volume 6

A Review of the Literature Published during 1972

Senior Reporter
N. N. Greenwood, *School of Chemistry, University of Leeds*

Reporters
D. M. Adams, *University of Leicester*
J. H. Carpenter, *University of Newcastle upon Tyne*
G. Davidson, *University of Nottingham*
M. Goldstein, *North London Polytechnic*
R. Greatrex, *University of Leeds*
B. E. Mann, *University of Sheffield*
S. R. Stobart, *Queen's University, Belfast*

© Copyright 1973

The Chemical Society
Burlington House, London, W1V 0BN

ISBN: 0 85186 053 2

Library of Congress Catalog No. 74-6662

Organic formulae composed by Wright's Symbolset method

PRINTED IN GREAT BRITAIN BY JOHN WRIGHT AND SONS LTD., AT THE STONEBRIDGE PRESS, BRISTOL

Foreword

This Report provides a comprehensive review of literature published during 1972 on the n.m.r., n.q.r., microwave, vibrational, and Mössbauer spectra of inorganic and organometallic compounds. The coverage and arrangement closely parallel those adopted last year. The Reporters have made strenuous efforts to contain the growing weight of material within the confines of a readable narrative of similar length to those produced in previous years. Extensive use of tabular material at appropriate points has enabled much factual and bibliographical information to be presented without destroying the flow of textual commentary.

It has been our aim not only to record results, but to indicate the wide variety of ways in which spectroscopic information is being used. Each chapter is self-contained and its length, and the number of references it contains, indicate the extent to which each technique is being applied to the study of inorganic and organometallic compounds. Each chapter states the range of work covered and the areas which have been omitted. The references have been obtained by means of the techniques outlined in previous volumes and we have tried to include all significant information published during 1972, though a few papers of Russian origin have had to be held over until next year.

We have been greatly heartened to read the many warm reviews of earlier volumes—it is always gratifying to have one's work appreciated—but equally we would greatly value suggestions from users on the ways in which the coverage of material or its presentation could be improved.

N. N. G.

Contents

Chapter 1 Nuclear Magnetic Resonance Spectroscopy   1
By B. E. Mann

1 Introduction   1
   Techniques, Coupling Constants, Chemical Shifts,
     and Relaxation Measurements   4
    Techniques   4
    Coupling Constants   6
    ^{1}H and ^{19}F Chemical Shifts   8
    ^{13}C Chemical Shifts   10
    ^{11}B, ^{14}N, ^{15}N, ^{29}Si, ^{31}P, ^{33}S, ^{35}Cl, ^{37}Cl, ^{77}Se, ^{79}Br,
     ^{81}Br, and ^{129}Xe Chemical Shifts   12
    Metal Chemical Shifts   14
    Relaxation Measurements   15

2 Stereochemistry   15
   Complexes of Li, Be, Mg, Sc, Y, La, and U   16
   Complexes of Ti, Zr, Hf, Th, V, Nb, and Ta   18
   Complexes of Cr, Mo, and W   20
   Complexes of Mn and Re   27
   Complexes of Fe, Ru, and Os   29
   Complexes of Co, Rh, and Ir   43
   Complexes of Ni, Pd, and Pt   52
   Complexes of Cu, Ag, and Au   60
   Complexes of Zn, Cd, and Hg   61

3 Dynamic Systems   64
   Fluxional Molecules   64
   Equilibria   74
    Solvation Studies of Ions   74
    Ionic Equilibria   77
    Equilibria among Uncharged Species   80
   Course of Reactions   84

4 Paramagnetic Complexes   87
   Compounds of the *d*-Block Transition Elements   88
   Compounds of the Lanthanides and Actinides   94

5 Solid-state N.M.R.	99
Motion in Solids	100
Structure of Solids	103
6 Group III Compounds	112
Boron Hydrides and Carbaboranes	112
Other Compounds of Boron	118
Complexes of Other Group III Elements	121
7 Compounds of Silicon, Germanium, Tin, and Lead	123
8 Compounds of Group V	138
9 Compounds of Group VI, VII, and Xenon	145
10 Bibliography	148

Chapter 2 Nuclear Quadrupole Resonance Spectroscopy	154
By J. H. Carpenter	
1 Introduction	154
2 Instrumentation and Techniques	155
3 Main Group Elements	156
Group I (Sodium-23, Potassium-39, Potassium-40, Potassium-41, Rubidium-85, and Rubidium-87)	156
Group III (Aluminium-27, Gallium-69, and Gallium-71)	156
Group V (Nitrogen-14, Arsenic-75, Antimony-121, Antimony-123, and Bismuth-209)	157
Group VI (Oxygen-17)	162
Group VII (Chlorine-35, Chlorine-37, Bromine-79, Bromine-81, and Iodine-127)	162
4 Transition Metals	172
Manganese-55	172
Cobalt-59	172
Copper-63 and Copper-65	173
Niobium-93	173
Tantalum-181	174

Chapter 3 Microwave Spectroscopy	175
By J. H. Carpenter	
1 Introduction	175
2 Instrumentation and Technique	176
3 Diatomic Molecules	177

Contents vii

    4 Triatomic Molecules 181

    5 Tetra-atomic Molecules 183

    6 Penta-atomic Molecules 187

    7 Molecules containing Six or More Atoms 191

Chapter 4 Vibrational Spectra of Small Symmetric Species and of Single Crystals 196
    By D. M. Adams

    1 General Introduction 196

    2 Spectra of Small Symmetric Species 198
        Diatomic Species 198
        Triatomic Species 200
        Tetra-atomic Species 207
        Penta-atomic Species 210
        Hexa-atomic Species 218
        Hepta-atomic Species 221
        Larger Symmetric Species 225

    3 Single-crystal and other Solid-state Spectroscopy 227
        'Simple' Lattice Types 227
        Mixed Oxides and Fluorides 230
        Sheet and Chain Structures 231
        Complex Halides 233
        Oxoanion-containing Crystals 235
        Complex Cationic Salts 237
        Complex Anionic Salts 238
        Molecular Crystals 243
        Others 244

Chapter 5 Characteristic Vibrational Frequencies of Compounds containing Main-group Elements 245
    By S. R. Stobart

    1 Group I Elements 245

    2 Group II Elements 246

    3 Group III Elements 248
        Compounds containing B—H Bonds 248
        Compounds containing Al—H or Ga—H Bonds 253

Compounds containing M—C Bonds (M = B, Al, Ga, In, or Tl)   254
Compounds containing M—N Bonds (M = B, Al, or Ga) or B—P Bonds   257
Compounds containing M—O Bonds (M = B, Al, Ga, In, or Tl)   260
Compounds containing M—S Bonds (M = Al or In) or M—Se Bonds (M = B or Al)   262
Compounds containing M—Halogen Bonds (M = B, Al, Ga, In, or Tl)   262

4 Group IV Elements   264
Compounds containing M—H Bonds (M = Si, Ge, or Sn)   265
Compounds containing M—C Bonds (M = Si, Ge, Sn, or Pb)   270
Compounds containing M—M Bonds (M = Si, Ge, or Sn)   273
Compounds containing M—N Bonds (M = Si, Ge, Sn, or Pb), M—P Bonds (M = Si or Sn), or Si—As Bonds   275
Compounds containing M—O Bonds (M = Si, Ge, Sn, or Pb)   277
Compounds containing M—S Bonds (M = Si, Ge, or Sn), M—Se Bonds (M = Si, Ge, or Sn), or Sn—Te Bonds   278
Compounds containing M—Halogen Bonds (M = Si, Sn, or Pb)   280

5 Group V Elements   282
Compounds containing E—H Bonds (E = N or P)   282
Compounds containing E—C Bonds (E = N, P, As, Sb, or Bi)   283
Compounds containing N—N or N—P Bonds   286
Compounds containing As—N, Sb—N, or P—P Bonds   291
Compounds containing E—O Bonds (E = N, P, As, Sb, or Bi)   292
Compounds containing N—S or P—S Bonds   294
Other Compounds containing a Group V Element Bonded to a Group VI Element   297
Compounds containing Group V–Halogen Bonds   297

6 Group VI Elements   299
Compounds containing O—H, S—H, or Te—H Bonds   299
Compounds containing E—C Bonds (E = S, Se, or Te)   300

Compounds containing O—O, S—O, Se—O, Te—O, S—S, S—Se, or Te—Te Bonds	301
Compounds containing Group VI–Halogen Bonds	304
7 Group VII Elements	306
8 Group VIII Elements	307

Chapter 6 Vibrational Spectra of Transition-element Compounds 309
By M. Goldstein

1 Introduction	309
2 General	309
3 Scandium	314
4 Titanium, Zirconium, and Hafnium	315
5 Vanadium, Niobium, and Tantalum	318
6 Chromium, Molybdenum, and Tungsten	321
7 Manganese, Technetium, and Rhenium	327
8 Iron, Ruthenium, and Osmium	330
9 Cobalt, Rhodium, and Iridium	335
10 Nickel, Palladium, and Platinum	341
11 Copper, Silver, and Gold	348
12 Zinc, Cadmium, and Mercury	352
13 Lanthanides	356
14 Actinides	358

Chapter 7 Vibrational Spectra of some Co-ordinated Ligands 361
By G. Davidson

1 Carbon Donors	361
2 Carbonyls	384
3 Nitrogen Donors	394
Molecular Nitrogen, Azido-, and Related Complexes	394
Amines and Related Ligands	399

Oximes	406
Ligands containing $>\!\!C\!=\!N\!\!<$ Groups	407
Cyanides and Isocyanides	411
Nitrosyls	416

4 Phosphorus and Arsenic Donors — 419

5 Oxygen Donors — 423
- Molecular Oxygen, Peroxo-, and Hydroxy-complexes — 423
- Acetylacetonates and Related Complexes — 424
- Carboxylates — 427
- Keto-, Alkoxy-, Phenoxy-, and Ether Ligands — 434
- O-Bonded Amides and Ureas — 436
- Nitrates and Nitrato-complexes — 437
- Ligands containing O—N, O—P, or O—As Bonds — 439
- Ligands containing O—S Bonds — 443

6 Sulphur and Selenium Donors — 445

7 Potentially Ambident Ligands — 454
- Cyanate, Thiocyanate Complexes, *etc.*, and Iso-analogues — 454
- Ligands containing N and O Donor Atoms — 458
- Ligands containing either N and As or N and S Donor Atoms — 463
- Ligands containing O and S Donor Atoms — 463

8 Appendix: Additional References to Metal Carbonyl Complexes — 466
- Vanadium, Niobium, and Tantalum Carbonyl Complexes — 466
- Chromium Carbonyl Complexes — 467
- Molybdenum Carbonyl Complexes — 469
- Tungsten Carbonyl Complexes — 473
- Manganese Carbonyl Complexes — 475
- Technetium and Rhenium Carbonyl Complexes — 477
- Iron Carbonyl Complexes — 478
- Ruthenium Carbonyl Complexes — 485
- Osmium Carbonyl Complexes — 487
- Cobalt Carbonyl Complexes — 488
- Rhodium Carbonyl Complexes — 489
- Iridium Carbonyl Complexes — 490
- Nickel, Palladium, and Platinum Carbonyl Complexes — 491
- Mixed Transition-metal Carbonyls — 491

Contents xi

Chapter 8 Mössbauer Spectroscopy 494
By R. Greatrex

1 Introduction 494
   Books and Reviews 494

2 Theoretical 497

3 Instrumentation and Methodology 500

4 Iron-57 502
   General Topics 502
      Nuclear Parameters, Hyperfine Interactions, and
        New Effects 502
      Pressure-dependence Studies 505
      Lattice Dynamics 506
      Alloy-type Systems 507
      ^{57}Fe Impurity Studies 509
      ^{57}Co Source Experiments and Decay After-effect
        Phenomena 512
   Compounds of Iron 516
      High-spin Iron(II) Compounds 516
      High-spin Iron(III) Compounds 524
      Spin-crossover systems, Unusual Electronic States,
        and Biological Compounds 530
      Low-spin and Covalent Complexes 537
   Oxide and Chalcogenide Systems containing Iron 542
      Binary Oxides 542
      Spinel Oxides and Garnets 547
      Other Oxide Systems 552
      Minerals 558
      Chalcogenides 565

5 Tin-119 567
   General Topics 567
   Tin(II) Compounds 574
   Tin(IV) Compounds 575

6 Other Elements 590
   Main Group Elements 590
      Germanium (^{73}Ge) 590
      Krypton (^{83}Kr) 591
      Antimony (^{121}Sb) 591
      Tellurium (^{125}Te) 594
      Iodine (^{127}I, ^{129}I) 595
      Caesium (^{133}Cs) 600

Transition Elements	601
Nickel (^{61}Ni)	601
Zinc (^{67}Zn)	601
Ruthenium (^{99}Ru)	601
Hafnium (^{178}Hf, ^{180}Hf)	601
Tantalum (^{181}Ta)	602
Tungsten (^{180}W, ^{182}W, ^{183}W, ^{184}W, ^{186}W)	604
Osmium (^{186}Os, ^{188}Os, ^{189}Os, ^{190}Os)	605
Iridium (^{191}Ir)	607
Platinum (^{195}Pt)	607
Gold (^{197}Au)	607
Lanthanide and Actinide Elements	608
Samarium (^{149}Sm)	608
Europium (^{151}Eu, ^{153}Eu)	609
Gadolinium (^{155}Gd, ^{157}Gd)	611
Dysprosium (^{161}Dy)	612
Holmium (^{165}Ho)	612
Erbium (^{166}Er)	612
Ytterbium (^{170}Yb, ^{171}Yb)	612
Uranium (^{236}U, ^{238}U)	613
Neptunium (^{237}Np)	613
Plutonium (^{239}Pu)	616
7 Bibliography	616
Author Index	623

Abbreviations

acac	acetylacetone anion
astp	tris(*o*-diphenylphosphinophenyl)arsine
ata	$N(CH_2CO_2)_3^{3-}$
azb	azobenzene
bipy	bipyridyl
bn	butylenediamine
CNDO	complete neglect of differential overlap
1,7-cth	5,7,7,12,14,14-hexamethyl-1,4,8,11-tetra-azacyclotetradecane
cydta	cyclohexanediaminotetra-acetic acid
dbm	dibenzoylmethane anion
depe	1,2-bisdiethylphosphinoethane
DHMB	Dewar hexamethylbenzene
diars	C$_6$H$_4$(AsMe$_2$)$_2$ (1,2-bis(dimethylarsino)benzene)
diars′	$Ph_2AsCH_2CH_2AsPh_2$
dien	diethylenetriamine
dimetrien	$H_2NCH_2CH_2NHCH(Me)CH(Me)NHCH_2CH_2NH_2$
diphos, Pf-Pf	$Ph_2PCH_2CH_2PPh_2$
dmaq	8-dimethylarsinoquinoline
DMF	dimethylformamide
dmgH	dimethylglyoxime
dmg	monoanion of dimethylglyoxime
DMSO	dimethyl sulphoxide
dpm	dipivaloylmethane
dppa	1,2-bisdiphenylphosphinoacetylene
dpt	dipropylenetriamine
ed3a	$(HO_2CCH_2)_2NCH_2CH_2NHCH_2CO_2H$
edda	ethylenediaminediacetic acid
edpa	HO_2CCH_2\\NCH$_2$CH$_2$N/CH$_2$CO$_2$H ; HO$_2$CCH(Me)/ \\CH(Me)CO$_2$H
edta	$(HO_2CCH_2)_2NCH_2CH_2N(CH_2CO_2H)_2$
eee	$H_2NCH_2CH_2SCH_2CH_2SCH_2CH_2NH_2$
en	ethylenediamine

ffars	$n = 1$, $X = Me_2As$	
ffos	$n = 1$, $X = Ph_2P$	$\begin{array}{c} X X \\ \diagdown \diagup \\ C{=}C \\ \diagup \diagdown \\ (CF_2)_n{-}CF_2 \end{array}$
f$_6$fos	$n = 2$, $X = Ph_2P$	
f$_8$fos	$n = 4$, $X = Ph_2P$	
fod	1,1,1,2,2,3,3-heptafluoro-7,7-dimethyloctane-4,6-dionato	
H$_5$dtpa	$(HO_2CCH_2)_2NCH_2CH_2N(CH_2CO_2H)CH_2CH_2N\text{-}(CH_2CO_2H)_2$	
hfac	hexafluoroacetylacetonato	
nta	$R = CH_2CO_2H$	
mida	$R = Me$	$RN(CH_2CO_2H)_2$
ida	$R = H$	
4-mpdpa	4-methyl-2-pyryldi-(2-pyridyl)amine	
OECy	$-SCH_2CH(NH_2)CO_2Et$	
pdta	propylenediaminetetra-acetic acid	
Pf=Pf	$Ph_2PCH{=}CHPPh_2$	
phen	1,10-phenanthroline	
pic	picoline	
γ-picO	γ-picoline N-oxide	
Pm-Pm	$Me_2PCH_2CH_2PMe_2$	
pn	propylenediamine	
ptas	tris(o-diphenylarsinophenyl)phosphine	
py	pyridine	
pyO	pyridine N-oxide	
qp	tris(o-diphenylphosphinophenyl)phosphine	
R-en	$RNHCH_2CH_2NH_2$	
sal	salicylate anion	
salen	NN'-ethylenebis(salicylaldiminato)	
sbtp	tris(o-diphenylphosphinophenyl)stibine	
SMCy	$MeSCH_2CH(NH_2)CO_2^-$	
sp	$o\text{-}CH_2{=}CHC_6H_4PPh_2$	
tcne	tetracyanoethylene	
terpy	$2,2',2''$-terpyridyl	
terpyO$_3$	$2,2',2''$-terpyridyl $NN'N''$-trioxide	
tetren	tetraethylenepentamine	
THF	tetrahydrofuran	
tmed	$Me_2NCH_2CH_2NMe_2$	
TPP	tetraphenylporphorin	
trien	triethylenetetramine	
tripyam	tri-(2-pyridyl)amine	
tta	anion of thenoyltrifluoroacetone	
ttn	1,1,1-tris(dimethylaminomethyl)ethane	
tu	thiourea	

Conversion factors

	cm^{-1}	J mol^{-1}	eV	kcal mol^{-1}	Mc s^{-1} (MHz)
cm^{-1}	1	11.957	1.2394×10^{-4}	2.8584×10^{-3}	2.9979×10^4
J mol^{-1}	8.3626×10^{-2}	1	1.0364×10^{-5}	2.3904×10^{-4}	2506.2
eV	8068.3	9.6484×10^4	1	2.3063	2.4188×10^8
kcal mol^{-1}	349.83	4183.3	4.3359×10^{-2}	1	1.0487×10^7
Mc s^{-1} (MHz)	3.3356×10^{-5}	3.9903×10^{-4}	4.1344×10^{-9}	9.5345×10^{-8}	1

Mössbauer spectra

For ^{57}Fe ($E\gamma = 14.413$ keV): 1 mm s^{-1} = 3.879×10^{-4} cm^{-1} = 4.638×10^{-3} J mol^{-1}
= 4.809×10^{-8} eV = 1.109×10^{-6} kcal mol^{-1}
= 11.63 Mc s^{-1} (MHz)

For other nuclides multiply the above conversion factors by $E\gamma$ (keV)/14.413.

1
Nuclear Magnetic Resonance Spectroscopy

BY B. E. MANN

1 Introduction

This year has been marked by the welcome appearance of a Specialist Periodical Report devoted to Nuclear Magnetic Resonance.[1] Consequently, the terms of reference of the present Report have been modified to include only inorganic and organometallic compounds, and many developments in instrumentation and techniques have been omitted. In spite of this, the number of references abstracted have increased yet again. In order to reduce the volume of the Report, tables have been incorporated into the text. The use of the Chemical Society's n.m.r. Macroprofile (UKCIS) has been changed. All 1972 references obtained *via* this Macroprofile have been incorporated into the text and only 1971 references which were not discussed in last year's Report are included in the bibliography (pp. 148—153).

As would be expected, ^{1}H n.m.r. spectroscopy continues to provide most of the information reported here. However, interest in other nuclei has been widespread and includes ^{2}H, ^{7}Li, ^{11}B, ^{13}C, ^{14}N, ^{15}N, ^{17}O, ^{19}F, ^{23}Na, ^{25}Mg, ^{27}Al, ^{29}Si, ^{31}P, ^{33}S, ^{35}Cl, ^{37}Cl, ^{39}K, ^{47}Ti, ^{49}Ti, ^{51}V, ^{55}Mn, ^{57}Fe, ^{59}Co, ^{63}Cu, ^{69}Ga, ^{71}Ga, ^{75}As, ^{77}Se, ^{79}Br, ^{81}Br, ^{85}Rb, ^{93}Nb, ^{105}Pd, ^{119}Sn, ^{129}Xe, ^{133}Cs, ^{187}Re, ^{199}Hg, and ^{201}Hg.

The flood of relevant books and reviews has continued. It has proved necessary to split Volume 5 of 'Annual Reports on N.M.R. Spectroscopy' into at least two parts, with Part A occupying 696 pages.[2] Volume 5A contains chapters on 'General review of proton magnetic resonance' by T. N. Huckerby, 'Fluorine-19 N.M.R. spectroscopy' by R. Fields, 'N.M.R. spectroscopy in the study of carbohydrates and related compounds' by T. D. Inch, 'Heteronuclear magnetic double resonance' by W. McFarlane, 'Nitrogen N.M.R. spectroscopy' by G. A. Webb and M. Witanowski, 'N.M.R. spectroscopy in liquids containing compounds of aluminium and gallium' by J. W. Akitt, and 'The application of Fourier transformation to high resolution N.M.R. spectroscopy' by D. G. Gilles and D. Shaw. 'Advances in Magnetic Resonance' has continued with the publication of Volume 5, which contains chapters on 'Pulsed-Fourier-transform nuclear

---

[1] 'Nuclear Magnetic Resonance', ed. R. K. Harris (Specialist Periodical Reports), The Chemical Society, London, 1972, vol. 1, 1973, vol. 2.
[2] 'Annual Reports on N.M.R. Spectroscopy', ed. E. F. Mooney, Academic Press, London and New York, vol. 5A, 1972.

magnetic resonance spectrometer' by A. G. Redfield and R. K. Gupta, 'Spectrometers for multiple-pulse N.M.R.' by J. D. Ellett, jun., M. G. Gibby, U. Haeberlen, L. M. Huber, M. Mehring, A. Pines, and J. S. Waugh, 'N.M.R. and ultraslow motions' by D. C. Ailion, and 'N.M.R. in helium-three' by M. G. Richards.[3]

One volume of a new series of books has been devoted to magnetic resonance and contains chapters on 'Nuclear spin relaxation in gases' by M. Bloom, 'N.M.R. studies of molecular motion in solids' by P. S. Allen, 'Carbon-13 nuclear spin relaxation' by J. R. Lyerla, jun. and D. M. Grant, and 'N.M.R. and E.S.R. in liquid crystals' by J. Bulthuis, C. W. Hilbers, and C. MacLean.[4]

A number of books devoted to aspects of n.m.r. spectroscopy have been published, including 'The Sadtler Guide to N.M.R. Spectra' by W. W. Simons and M. Zanger,[5] 'Magnetic Resonance' by K. A. McLauchlan,[6] 'Theory and Interpretation of Magnetic Resonance Spectra, by W. T. Dixon,[7] 'Introductory Fourier Transform Spectroscopy, by R. J. Bell,[8] 'Carbon-13 Nuclear Magnetic Resonance for Organic Chemists' by G. C. Levy and G. L. Nelson,[9] 'Carbon-13 N.M.R. Spectra' by L. F. Johnson and W. C. Jankowski,[10] 'The Nuclear Magnetic Resonance of Polymers' by I. Ya. Slonin and A. N. Lyubimov,[11] and 'Large-field Nuclear Magnetic Resonance Spectrometer' by P. Thomas and C. Zermati.[12] Also a number of books have been published which contain at least one chapter devoted to n.m.r. spectroscopy,[13–17] and the book 'Computers in Chemical and

[3] 'Advances in Magnetic Resonance', ed. J. S. Waugh, Academic Press, New York and London, vol. 5, 1972.
[4] MTP International Review of Science, Physical Chemistry', Series One, ed. A. D. Buckingham, Volume 4, 'Magnetic Resonance', ed. C. A. McDowell, Butterworths, London, and University Park Press, Baltimore, 1972.
[5] W. W. Simons and M. Zanger, 'The Sadtler Guide to NMR Spectra', Sadtler Research Laboratories, Inc., Philadelphia, 1972.
[6] K. A. McLauchlan, 'Magnetic Resonance', Oxford University Press, Oxford, 1972.
[7] W. T. Dixon, 'Theory and Interpretation of Magnetic Resonance Spectra', Plenum Press, London and New York, 1972.
[8] R. J. Bell, 'Introductory Fourier Transform Spectroscopy', Academic Press, New York and London, 1972.
[9] G. C. Levy and G. L. Nelson, 'Carbon-13 Nuclear Magnetic Resonance for Organic Chemists', Wiley–Interscience, New York, 1972.
[10] L. F. Johnson and W. C. Jankowski, 'Carbon-13 N.M.R. Spectra', Wiley–Interscience, New York, 1972.
[11] 'The Nuclear Magnetic Resonance of Polymers', I. Ya. Slonin and A. N. Lyubimov, Heyden, London, 1972.
[12] P. Thomas and C. Zermati, 'Large-field Nuclear Magnetic Resonance Spectrometer', C.N.R.S., Paris, 1970.
[13] 'Comprehensive Analytical Chemistry', ed. C. L. Wilson and D. W. Wilson, Elsevier, Amsterdam, Volume IIC, 1971.
[14] 'An Introduction to Spectroscopic Methods for the Identification of Organic Compounds', ed. F. Scheinmann, Pergamon Press, Oxford, 1970.
[15] 'Coordination Chemistry', ed. A. E. Martell, A.C.S. Monograph 168, Von Nostrand Reinhold Co., New York, 1972.
[16] L. A. Kazitsyma and N. B. Kupletskaya, 'Uses of Ultraviolet, Infrared, and N.M.R. Spectroscopy in Organic Chemistry', Vyssh. Shkola, Moscow, 1971.
[17] C. J. Hawkins, 'Absolute Configuration of Metal Complexes', Wiley–Interscience, New York, 1971.

Biochemical Research' contains a chapter on Fourier transform spectrometry by S. T. Dunn, C. T. Foskett, P. Curbelo, and P. R. Griffiths.[18]

A number of reviews have also been published on various topics, including a general review of n.m.r. spectroscopy,[19] 'Applications of 220 MHz N.M.R.',[20] 'N.M.R. in solutions',[21] 'Recent developments in high resolution n.m.r. spectroscopy,[22] 'Pulse Fourier transform n.m.r. spectroscopy',[23] 'Application of Fourier transform n.m.r. spectroscopy',[24] 'Signal-to-noise enhancement through instrumental techniques',[25] 'The use of a computer in nuclear magnetic resonance spectroscopy',[26, 27] 'Interference effects in nuclear magnetic resonance',[28] 'Nuclear magnetic double resonance',[29] 'Spin–spin coupling between geminal and vicinal protons',[30] '^{1}H N.m.r. spectra of the AA′XX′ and AA′BB′ type:— analysis and classification',[31] 'The application of n.m.r. and e.s.r. to co-ordination compounds in solution and the solid state',[32] 'Spectroscopy and structure of pentacoordinated molecules',[33] 'The application of proton magnetic resonance to the investigation of organophosphorus compounds',[34] 'The use of ^{31}P n.m.r.',[35] 'N.m.r. studies of carbon–lithium bonding in organo-lithium compounds',[36] 'Chemically induced nuclear polarization, I: Phenomenon, examples and applications,[37] II: The radical-pair model',[38] a review of hydride complexes of the transition metals which contains a useful, but incomplete, listing of many hydride chemical shifts,[39] and 'N.M.R. of molecules orientated by electric fields'.[40]

The attention of the reader is drawn to the I.U.P.A.C. publication 'Recommendations for the Presentation of N.M.R. Data for Publication

---

[18] 'Computers in Chemical and Biochemical Research', ed. C. Klopfenstein and C. Wilkins, Academic Press, New York and London, 1972.
[19] D. L. Rabenstein, *Guide Mod. Methods Instr. Analysis*, 1972, 231.
[20] A. A. Grey, *Canad. J. Spectroscopy*, 1972, **17**, 82.
[21] N. H. Velthorst, *Chem. Tech. (Amsterdam)*, 1972, **27**, 341.
[22] E. D. Becker, *Appl. Spectroscopy*, 1972, **26**, 421.
[23] D. A. Netzel, *Appl. Spectroscopy*, 1972, **26**, 430.
[24] E. D. Becker and T. C. Farrar, *Science*, 1972, **178**, 361.
[25] G. M. Hieftje, *Analyt. Chem.*, 1972, **44**, No. 6, 81A; No. 7, 69A.
[26] E. G. Hoffmann, W. Stempfle, G. Schroth, B. Weimann, E. Ziegler, and J. Brandt, *Angew. Chem. Internat. Edn.*, 1972, **11**, 375.
[27] R. R. Ernst, *Chimia (Switz.)*, 1972, **26**, 53.
[28] J. S. Blicharski, *Inst. Nuclear Phys., Cracow, Rep.*, 1972 792/PL.
[29] R. B. Johannesen and T. D. Coyle, *Endeavour*, 1972, **31**, 112.
[30] V. F. Bystrov, *Russ. Chem. Rev.*, 1972, **41**, 281.
[31] H. Günther, *Angew. Chem. Internat. Edn.*, 1972, **11**, 861.
[32] H. J. Keller, *Ber. Bunsengesellschaft phys. Chem.*, 1972, **76**, 1080.
[33] R. H. Holmes, *Accounts Chem. Res.*, 1972, **5**, 296.
[34] B. I. Ionin and T. N. Timofeeva, *Russ. Chem. Rev.*, 1972, **41**, 390; *Uspekhi Khim.*, 1972, **41**, 758.
[35] E. Fluck, *Chem.-Ztg*, 1972, **96**, 517.
[36] L. D. McKeever, *Ions Ion Pairs Org. React.*, 1972, **1**, 263.
[37] H. R. Ward, *Accounts Chem. Res.*, 1972, **5**, 18.
[38] R. G. Lawler, *Accounts Chem. Res.*, 1972, **5**, 25.
[39] H. D. Kaesz and R. B. Saillant, *Chem. Rev.*, 1972, **72**, 231.
[40] C. W. Hilbers and C. Maclean, *Nuclear Magn. Resonance*, 1972, **7**, 1.

in Chemical Journals'.[41] Few papers reach the high standards recommended. The invited lectures presented at the Fourth International Symposium on Magnetic Resonance, Rehovot/Jerusalem, Israel, have been published.[42] They include 'Multi-site chemical exchange by n.m.r.' by E. A. Allan, M. G. Hogben, L. W. Reeves, and K. N. Shaw, 'Fourier transform studies of nuclear spin relaxation' by W. A. Anderson, R. Freeman, and H. Hill, 'Developments in the motional narrowing of the n.m.r. spectra of solids – microscopic and macroscopic' by E. R. Andrew, '^{13}C N.M.R. spectroscopy: relaxation times of ^{13}C and methods for sensitivity enhancement' by E. D. Becker and R. R. Shoup, 'New spin echo techniques in the earth's magnetic field range' by G. J. Bene, 'Solvent effects on n.m.r. spectra of gases and liquids' by H. J. Bernstein, 'Nuclear magnetic resonance and nuclear spin symmetry in molecular solids' by M. Bloom, 'Study intramolecular motion by n.m.r. of orientated molecules' by P. Diehl, 'N.m.r. satellites as a probe for chemical investigations' by S. Fujiwara, Y. Arata, H. Ozawa, and M. Kunugi, 'N.m.r. studies of paramagnetic molecules' by R. M. Golding, 'Nuclear magnetic ordering' by M. Goldman, 'Intermolecular nuclear relaxation and molecular pair distribution in liquid mixtures' by R. Göller, H. G. Hertz, and R. Tutsch, 'N.m.r. study of induced dielectric alignment, in pure and binary liquids' by C. W. Hilbers, J. Biemond, and C. MacLean, 'Biologically important isotope hybrid compounds in n.m.r.: ^{1}H Fourier transform n.m.r. at unnatural abundance' by J. J. Katz and H. L. Crespi, 'Developments in n.m.r. in liquid crystalline solvents' by S. Meiloom, R. C. Hewitt, and L. C. Snyder, 'Isotropic n.m.r. shifts in the "solid" and "liquid" states' by A. J. Vega and D. Fiat, 'Spin echoes and Loschmidt's paradox' by J. S. Waugh, W.-K. Rhim, and A. Pines, 'N.m.r. in superconductors' by M. Weger, and 'N.m.r. study of the electronic structure of solid and liquid metals' by D. Zamir and U. El-Hanany.

Techniques, Coupling Constants, Chemical Shifts, and Relaxation Measurements.—The papers reviewed in this section are concerned primarily with the development of techniques, or the measurement of coupling constants, chemical shifts, or relaxation times. The section is thence split into six subsections covering (a) techniques, (b) coupling constants, (c) ^{1}H and ^{19}F chemical shifts, (d) ^{13}C chemical shifts, (e) ^{11}B, ^{14}N, ^{15}N, ^{29}Si, ^{31}P, ^{33}S, ^{35}Cl, ^{37}Cl, ^{77}Se, ^{79}Br, ^{81}Br, and ^{129}Xe chemical shifts, and (f) metal chemical shifts.

Techniques. The ASIS (aromatic solvent induced shifts) effect has been exploited for a number of inorganic and organometallic compounds. It has been applied to trisubstituted borazine derivatives to assess aromatic character;[43] for H_3BPR_3 and H_3BNMe_3 it was found that the hydridic protons were shifted upfield in C_6D_6 and downfield in C_6F_6. These results

---

[41] I.U.P.A.C. 'Recommendations for the Presentation of N.M.R. Data for Publication in Chemical Journals', *Pure Appl. Chem.*, 1972, **29**, 625.
[42] *Pure and Applied Chemistry*, 1972, **32**.
[43] M. Pasdeloup, J.-P. Laurent, and G. Commenges, *J. Chim. phys.*, 1972, **69**, 1022.

demonstrate for the first time that the ASIS effect can be reversed when the proton bears a partial negative charge.[44] In order to explain the ASIS effect in some complexes of the types NiL_4, $(OC)_{5-n}FeL_n$, and $(OC)_{6-n}ML_n$ (M = Cr, Mo, or W; n = 1 or 2), it was necessary to postulate that they can be induced by local dipolar regions in absence of a net solute dipole,[45] whereas for complexes $(acac)_2R_2Sn$ and $(acac)R_3Sb$ it was found that the derived charge distribution correlates better with the inductive effect of the substituents at the metal atom than with the molecular dipole moments.[46] ASIS has also been used to assist in interpretation of the ^1H n.m.r. spectrum of 2-ferrocenyl-4-n-butylquinoline and 1,2-dihydro-2-ferrocenyl-4-n-butyl-quinoline.[47] An alternative way to induce changes in the ^1H n.m.r. shifts of diols is to add borate.[48]

An interesting new application of n.m.r. is to follow moderately rapid reactions using flow methods,[49] and this method was applied to the determination of the rate of the reaction

$$[Ni(NH_3)(H_2O)_5]^{2+} + H_3O^+ \longrightarrow [Ni(H_2O)_6]^{2+} + [NH_4]^+$$

Good agreement with optical measurements was found.[50]

It has been reported that for isotopically rare nuclei, e.g. ^{13}C, ^{15}N, ^2H, ^{29}Si, in the presence of other nuclei, e.g. ^1H, it is possible to irradiate the abundant nuclei, get spin polarization of the isotopically rare nuclei, and record free induction decays with greatly increased sensitivity. This technique has been used to obtain the principal components of the chemical shielding tensor of ^{13}C for adamantane,[51] and calcium carbonate,[52] of ^{15}N for $(NH_4)_2SO_4$ and NH_4NO_3,[53] and of ^{29}Si for polycrystalline organosilicon compounds.[54]

Interest in ^{13}C n.m.r. spectroscopy has continued with the report of the ^{13}C chemical shifts of many reference compounds with respect to Me_4Si, and the solvent effects on Me_4Si have been investigated.[55] The problem of sensitivity of ^{13}C n.m.r. signals has attracted attention and it has been found that addition of a paramagnetic compound, e.g. $Cr(acac)_3$, can produce up to a 40-fold increase in sensitivity for $Fe(CO)_5$ with no significant chemical

[44] A. H. Cowley, M. C. Damasco, J. A. Mosbo, and J. G. Verkade, *J. Amer. Chem. Soc.*, 1972, **94**, 6715.
[45] J. A. Mosbo, J. R. Pipal, and J. G. Verkade, *J. Magn. Resonance*, 1972, **8**, 243.
[46] A. Mackor and H. A. Meinema, *Rec. Trav. chim.*, 1972, **91**, 911.
[47] D. J. Booth, B. W. Rockett, and J. Ronayne, *J. Organometallic Chem.*, 1972, **44**, C29.
[48] W. Voelter, C. Bürvenich, and E. Breitmaier, *Angew. Chem. Internat. Edn.*, 1972, **11**, 539.
[49] Y. Ashai and E. Mizuta, *Talanta*, 1972, **19**, 567.
[50] J. Grimaldi, J. Baldo, C. McMurray, and B. D. Sykes, *J. Amer. Chem. Soc.*, 1972, **94**, 7641.
[51] A. Pines, M. G. Gibby, and J. S. Waugh, *J. Chem. Phys.*, 1972, **56**, 1776.
[52] A. Pines, W.-K. Rhim, and J. S. Waugh, *J. Magn. Resonance*, 1972, **6**, 457.
[53] M. G. Gibby, R. G. Griffin, A. Pines, and J. S. Waugh, *Chem. Phys. Letters*, 1972, **17**, 80.
[54] M. G. Gibby, A. Pines, and J. S. Waugh, *J. Amer. Chem. Soc.*, 1972, **94**, 6231.
[55] G. C. Levy and J. D. Cargioli, *J. Magn. Resonance*, 1972, **6**, 143.

shift changes.[56, 57] When this technique was applied to ^{15}N n.m.r., increases up to only six-fold were found.[58]

A calculational procedure for substituent effects has enabled the calculation of chemical shifts in ^{1}H, ^{11}B, ^{13}C, and ^{31}P n.m.r. spectroscopy and the explanation of additivity of substituent effects. An exception was found for ^{119}Sn chemical shifts.[59]

Coupling Constants. $^1J(^{13}C-^1H)$ and $^1J(^{29}Si-^1H)$ of some methylphenylsilanes have been shown to correlate with Hammett σ constants, and the mechanisms of transmission of substituent effects across carbon–silicon and silicon–silicon bonds were discussed.[60, 61] $^1J(^{75}As-^1H)$ and $^1J(^{75}As-^2H)$ in AsH$_3$ and AsD$_3$ have been measured using pulse n.m.r. techniques. It was possible to show by T_1 measurements that the arsine molecule reorientates as a free rotor at high temperatures.[62] It has been found that for [SnH$_3$]$^-$, $^1J(^{119}Sn-^1H)$ is 109.4 Hz, compared with 1933 Hz for SnH$_4$. These coupling constants were discussed in terms of the Pople–Santry treatment.[63]

The ^{11}B n.m.r. spectrum of B$_5$H$_9$ with complete ^{1}H decoupling at $+46$ °C shows a well-resolved $^1J(^{11}B-^{11}B) = 19.4$ Hz, but cooling to -51 °C produces decoupling.[64] ^{1}H, ^{11}B, and ^{19}F n.m.r. and INDOR have been used to determine the relative signs of coupling constants for F$_3$BNMe$_3$.[65] $^1J(^{11}B-^{31}P)$ has been reported for a number of adducts F$_2$XP,BH$_3$ and F$_2$X,PB$_4$H$_8$ (X = F, Cl, Br, or I), and a linear correlation was found between $^1J(^{11}B-^{31}P)$ and ν(B–H).[66]

The ^{13}C n.m.r. spectra of a number of silicon compounds have been reported and it was found that the magnitude of $^1J(^{13}C-^{29}Si)$ is roughly proportional to the *s*-character of the carbon.[67] $^1J(^{195}Pt-^{13}C) = 1035$ Hz has been reported for [Pt(CN)$_4$]$^{2-}$ in a review on the application of n.m.r. and e.s.r. to co-ordination compounds.[68] In a study on the ^{13}C n.m.r. spectra of Bu^{t}CH$_2$HgR, many linear relationships between coupling constants, including $J(^{199}Hg-^{13}C)$, were noted.[69]

The first report of $^1J(Cl-^{19}F)$ has appeared for [ClF$_6$]$^+$, where $^1J(^{35}Cl-^{19}F) = 337$ Hz and $^1J(^{37}Cl-^{19}F) = 281$ Hz.[70] The compound

[56] S. Barcza and N. Engstrom, *J. Amer. Chem. Soc.*, 1972, **94**, 1762.
[57] O. A. Gansow, A. R. Burke, and G. N. LaMar, *J.C.S. Chem. Comm.*, 1972, 456.
[58] L. F. Farnell, E. W. Randall, and A. I. White, *J.C.S. Chem. Comm.*, 1972, 1159.
[59] L. Phillips and V. Wray, *J.C.S. Perkin II*, 1972, 214.
[60] F. K. Cartledge and K. H. Riedel, *J. Organometallic Chem.*, 1972, **34**, 11.
[61] Y. Nagai, M.-A. Ohtsuki, T. Nakano, and H. Watanabe, *J. Organometallic Chem.*, 1972, **35**, 81.
[62] L. J. Burnett and A. H. Zeltmann, *J. Chem. Phys.*, 1972, **56**, 4695.
[63] T. Birchall and A. Pereira, *J.C.S. Chem. Comm.*, 1972, 1150.
[64] D. W. Lowman, P. D. Ellis, and J. D. Odom, *J. Magn. Resonance*, 1972, **8**, 289.
[65] V. V. Negrebetskii, V. S. Bogdanov, and A. V. Kessenikh, *Zhur. strukt. Khim.*, 1972, **13**, 327.
[66] R. T. Paine and R. W. Parry, *Inorg. Chem.*, 1972, **11**, 1237.
[67] G. C. Levy, D. M. White, and J. C. Cargioli, *J. Magn. Resonance*, 1972, **8**, 280.
[68] H. J. Keller, *Ber. Bunsengesellschaft phys. Chem.*, 1972, **76**, 1080.
[69] G. Singh and G. S. Reddy, *J. Organometallic Chem.*, 1972, **42**, 267.
[70] K. O. Christe, *Inorg. Nuclear Chem. Letters*, 1972, **8**, 741.

[IF$_6$]$^+$ shows $^1J(^{19}F-^{127}I)$ = 2730 Hz. It is interesting to note that the linewidth of the outermost line is 110 Hz whereas for the inner lines the linewidth can be up to 170 Hz.[71] A linear relationship has been found between $^1J(^{31}P-^{183}W)$ for W(CO)$_5$L and the electronegativity of the phosphorus ligand. Points due to L = PCl$_3$, PBr$_3$, or PI$_3$ fall off this line, forming another linear relationship, with PF$_3$ falling on both lines.[72]

The ^{1}H n.m.r. spectrum of B$_5$H$_9$ with ^{11}B decoupling shows signals 4:1:4 in intensity with the apical hydrogen split into a nine-line pattern by the other eight hydrogens, $^3J(^1H-^1H)$ = 5.7 Hz.[73] From an analysis of the ^{13}C–H satellite magnetic resonance spectrum of (C$_5$H$_5$)SnMe$_3$, J(H–H) have been determined. It is found that these coupling constants can be useful in differentiating between fluxional σ- and π-cyclopentadienyl groups.[74] $J(^{11}B-^1H)$ has been determined indirectly for some vinyl compounds of boron. It is found that J[B–H(vic)] = J[B–H($trans$)] > J[B–H(cis)] > J[B–H(gem)]. J[B–H(cis)], and as J[B–H(gem)] decreases with increased electronegativity of substituent, the signs are probably positive.[75] The ^{1}H n.m.r. spectra of B$_2$H$_6$, MeB$_2$H$_5$, 1,1-Me$_2$B$_2$H$_4$, 1,2-Me$_2$B$_2$H$_4$, 1,1,2-Me$_3$B$_2$H$_3$, 1,1,2,2-Me$_4$B$_2$H$_2$, and ClB$_2$H$_5$ have been fully analysed. It was found that J(H$_T$–H$_\mu$) is 7.5—8.7 Hz, J(Me–H$_\mu$) is 2.4—3.5 Hz, and J(Me–H$_{T,gem}$) is 5.0—5.3 Hz.[76]

The observation that $^2J(^{119}$Sn–C–^{1}H) increases as r(C—Sn) decreases for a variety of methyl–tin complexes has been explained using the Fermi contact equation,[77] and for Me$_6$W, $^2J(^{183}W-^1H)$ is 3.0 Hz.[78] Double-resonance experiments on compound (1) show $^2J(^{195}$Pt–CH) and $^3J(^{195}$Pt–CH$_3$) to be opposite in sign.[79] Calculations of $J(^{199}$Hg–CH$_3$) for some methylmercury complexes indicate that the Fermi contact mechanism is dominant.[80] $J(^{205}$Tl–^{1}H) coupling constants in various monoarylthallium

(1)

[71] M. Brownstein and H. Selig, *Inorg. Chem.*, 1972, **11**, 656.
[72] E. O. Fischer, L. Knauss, R. L. Keiter, and J. G. Verkade, *J. Organometallic Chem.*, 1972, **36**, C7.
[73] T. Onak, *J.C.S. Chem. Comm.*, 1972, 351.
[74] Yu. K. Grishin, N. M. Sergeyev, and Yu. A. Ustynyuk, *J. Organometallic Chem.*, 1972, **34**, 105.
[75] V. S. Bogdanov, A. V. Kessenikh, and A. Y. A. Shchteinshneider, *Zhur. strukt. Khim.*, 1972, **13**, 226.
[76] J. B. Leach, C. B. Ungermann, and T. P. Onak, *J. Magn. Resonance*, 1972, **6**, 74.
[77] M. K. Das, J. Buckle, and P. G. Harrison, *Inorg. Chim. Acta*, 1972, **6**, 17.
[78] A. Shortland and G. Wilkinson, *J.C.S. Chem. Comm.*, 1972, 318.
[79] D. F. Gill and B. L. Shaw, *J.C.S. Chem. Comm.*, 1972, 65.
[80] H. F. Henneike, *J. Amer. Chem. Soc.*, 1972, **94**, 5945.

dichlorides and substituted monoarylthallium dichlorides have been determined in DMSO solutions in order to compare the couplings with the corresponding $J(^1H-^1H)$ coupling constants, and close resemblances were found.[81]

The measurement of $^2J(^{31}P-^{31}P)$ has once again attracted attention. The ^{19}F and ^{31}P n.m.r. spectra of $\{F_2P(=E)\}_2Z$ have been measured and analysed as $AA'XX'X''X'''$. For $(SPF_2)_2O$ it was found that $^2J(^{31}P-^{31}P)$ shows a marked temperature dependence.[82] Similarly, $^2J(^{31}P-^{31}P)$ has been evaluated from ^{19}F n.m.r. spectra of $cis\text{-}L_2M(CO)_4$ (M = Cr, Mo, or W) and it is found that for M = Mo, L = $(CF_3)_2PR$ or $(CF_3)PR_2$, $^2J(^{31}P-^{31}P)$ decreases in the order R = F > Cl > Br > I > H.[83] $^1H\{^{31}P\}$ INDOR has been used to show that for compound (2), $^2J(P^1-CH_3) = ca. \mp 10.7$ Hz,

$$\begin{array}{c} P^2Ph_2ClP^1Me_2Ph \\ H_2C\diagdown|\diagup \\ C\text{---}Ir \\ H_2C\diagup|\diagdown \\ P^3Ph_2ClX \end{array}$$

(2)

$^4J(P^3-CH_3) = ca. \pm 2.2$ Hz, and $^2J(P^1-P^3) = ca. \pm 440$ Hz, but was significantly different when X = NO_2 (± 408 Hz), py (± 398 Hz), or CO (± 320 Hz).[84] It is often assumed that $^2J(P-M-P)$ for mutually *trans* tertiary phosphines is large, but for *trans*-$SnCl_4(PEt_3)(PEt_2Ph)$, $^2J(P-Sn-P)$ is only 43.2 Hz. Unlike tungsten, rhodium, and platinum complexes, $^1J(^{119}Sn-^{31}P)$ *decreases* as the number of phenyl groups on the tertiary phosphine is increased.[85]

^1H and ^{19}F Chemical Shifts. $Me_3SiCH_2CH_2CH_2SO_3^-Na^+$ is commonly used as a reference in aqueous solution, but it has been found that it will react with $Hg(OAc)_2$ to give $MeHg(OAc)$ and $Me_2Si(OAc)CH_2CH_2CH_2SO_3^-Na^+$. Thus care must be exercised in using $Me_3SiCH_2CH_2CH_2SO_3^-Na^+$ as internal reference.[86]

For the complexes (3; M = Ti, Zr, or Hf), a correlation has been found between the angle between the cyclopentadienyl rings and the chemical

(3)

[81] J. P. Maher, M. Evans, and M. Harrison, *J.C.S. Dalton*, 1972, 188.
[82] T. L. Charlton and R. G. Cavell, *Inorg. Chem.*, 1972, **11**, 1583.
[83] J. F. Nixon and J. R. Swain, *J.C.S. Dalton*, 1972, 1038.
[84] B. E. Mann, C. Masters, and B. L. Shaw, *J.C.S. Dalton*, 1972, 48.
[85] J. F. Malone and B. E. Mann, *Inorg. Nuclear Chem. Letters*, 1972, **8**, 819.
[86] R. E. DeSimone, *J.C.S. Chem. Comm.*, 1972, 780.

shift difference between the α and β protons on the ring.[87] From ^{1}H, ^{19}F, and ^{31}P n.m.r. measurements on $(C_5H_5)Fe(CO)_2X$ and $(C_5H_5)Fe(CO)$-$(PPh_3)C_6H_4R$-p, it was concluded that the effects of X and R substituents upon the cyclopentadienyl-ring chemical shifts are predominantly inductive.[88] The ^{1}H n.m.r. spectra of $MeCo(dmg)_2(base)$ have been measured for a wide variety of bases. For substituted pyridines, the slope of the dependence of the chemical shift of the methyl group on the Hammett σ constant is 2/3 that for Hammett σ constant in the N-methylpyridinium salts. It was therefore concluded that a cobalt atom will conduct about 2/3 of the electronic effect of a substituent.[89]

The Buckingham–Stevens theory has been applied. For $HIr(piperidine)_4$-$(NCX)_2$ (X = O, S, or Se), the theory is only partly in accord with results and the chemical shift is more dependent on the covalent character of the *trans* ligand.[90] $[IrHCl_2(PBu^t{}_2R)_2]$ shows a very high-field hydride signal at τ 60.5 for R = Me. This very large high-field shift was explained as due to a small value for ΔE. ^{31}P N.m.r. data were also reported.[91]

Molecular properties of diborane and decaborane(14), including ^{1}H and ^{11}B chemical shifts, have been calculated from a minimum basis set and extended Slater orbital wavefunctions, but agreement with experiment was poor.[92, 93] Variable-temperature ^{1}H n.m.r. spectra of borazine and [^{10}B]-borazine have been analysed for chemical shifts, coupling constants, and linewidths. It was concluded that the broadening is due to a combination of quadrupolar relaxation resulting from the high-spin nuclei present and long-range spin coupling.[94]

A correlation has been established between the chemical shifts of the silicon methyl groups and the resonance constant of the *para* substituent in p-$XC_6H_4SiMe_3$.[95] A similar relationship has been found for $(MeO)Me_2$-SnC_6H_4X.[96] The ^{1}H chemical shift and $^2J(^{119}Sn-CH_3)$ for $MeSnR_3$ have been determined and interpreted in terms of $(d-p)\pi$ bonding between tin and the π-system of R; $MeHgR$ is analogous.[97] Del Re calculations have been extended to correlate and interpret the n.m.r. data in organotin compounds. The methyl proton chemical shifts have been correlated with the partial charge on the methyl hydrogen atom. Variations in $^1J(^{13}C-^1H)$

[87] M. Hillman and A. J. Weiss, *J. Organometallic Chem.*, 1972, **42**, 123.
[88] A. N. Nesmeyanov, I. F. Leshcheva, I. V. Polovyanyuk, Yu. A. Ustynyuk, and L. G. Makarova, *J. Organometallic Chem.*, 1972, **37**, 159.
[89] J. P. Fox, R. Banninger, R. T. Proffitt, and L. L. Ingraham, *Inorg. Chem.*, 1972, **11**, 2379.
[90] E. R. Birnbaum, *J. Inorg. Nuclear Chem.*, 1972, **34**, 3499.
[91] C. Masters, B. L. Shaw, and R. E. Stainbank, *J.C.S. Dalton*, 1972, 664.
[92] E. A. Laws, R. M. Stevens, and W. N. Lipscomb, *J. Amer. Chem. Soc.*, 1972, **94**, 4461.
[93] E. A. Laws, R. M. Stevens, and W. N. Lipscomb, *J. Amer. Chem. Soc.*, 1972, **94**, 4467.
[94] E. K. Mellon, B. M. Coker, and P. B. Dillon, *Inorg. Chem.*, 1972, **11**, 852.
[95] A. P. Kreshkov, V. F. Andronov, and V. A. Drozdov, *Russ. J. Phys. Chem.*, 1972, **46**, 574.
[96] J. Pijselman and M. Pereyre, *J. Organometallic Chem.*, 1972, **44**, 309.
[97] G. Barbieri and F. Taddei, *J.C.S. Perkin II*, 1972, 1323.

and $^2J(\text{Sn-C-H})$ have been correlated with the calculated Coulomb integrals.[98] It is usual to use the rule that when $\text{Me}_2\text{NCS}_2^-$ is bidentate it gives a signal at τ 6.72 and when it is unidentate at τ 7.24. However, $\text{Me}_3\text{SnS}_2\text{CNMe}_2$ is known to contain $\text{Me}_2\text{NCS}_2^-$ as a unidentate ligand but the signal is at τ 6.55. Thus caution is required.[99] From ^1H and ^{19}F shift measurements on HOF it has been concluded that the hydrogen and fluorine atoms carry charges of $+0.5\,e$ and $-0.5\,e$, respectively.[100] The temperature dependence of the ^1H shielding constant of HCl gas has been measured and the shielding constant factorized. There is an isotope shift between H^{35}Cl and H^{37}Cl of 0.001 p.p.m.[101]

From the ^{19}F chemical shifts of m- and p-$\text{FC}_6\text{H}_4\text{HgX}$, σ_I and σ_R^0 have been determined for the mercury substituents.[102] A number of substituted $\alpha(1)$- and $\beta(2)$-fluoronaphthalenes with metallo-substituents of the type HgX or MR_3 (M = Group IVB metalloid) have been synthesized and their fluorine n.m.r. spectra have been measured. The fluorine chemical shifts have been used to provide experimental evidence for d_π-p_π and p_π-p_π bonding.[103] Similarly, ^{19}F n.m.r. chemical shifts of some aryl silicon compounds have been measured in order to assess π-bonding.[104, 105]

^{13}C **Chemical Shifts.** The use of CNDO/2 calculations to determine ^{13}C chemical shifts, $^1J(^{13}\text{C}-^1\text{H})$, and $^2J(^1\text{H}-\text{C}-^1\text{H})$ for CH_3X compounds has been described.[106] For the complexes $(\text{PhCH}_2)_n\text{X}$ (X = Ti, Zr, Hg, B, Al, Si, or Sn) a linear relationship has been reported between both the ^1H and ^{13}C chemical shifts of the CH_2 group and the electronegativity of X. The deviations found for titanium and zirconium were attributed to electronic interactions of the CH_2 with the metal.[107] The ^{13}C and ^{19}F n.m.r. chemical shifts of $(\text{C}_5\text{H}_5)_2\text{TiX}_2$ have been discussed in terms of bonding.[108] For Grignard reagents, a linear relationship has been reported between the ^{13}C chemical shift and that of the corresponding hydrocarbon.[109]

The effect of changes in the electronic environment of the sp^2 hybridized carbene carbon atom bound to the metal in a series of 18 complexes of the type $(\text{OC})_5\text{MCXY}$ (M = Cr or W; X = OR or NR^1R^2; Y = organic

[98] R. Gupta and B. Majee, *J. Organometallic Chem.*, 1972, **40**, 97.
[99] B. W. Fitzsimmons and A. C. Sawbridge, *J.C.S. Dalton*, 1972, 1678.
[100] J. C. Hindman, A. Svirmickas, and E. H. Appelman, *J. Chem. Phys.*, 1972, **57**, 4542.
[101] W. T. Raynes and B. P. Chadburn, *Mol. Phys.*, 1972, **24**, 853.
[102] O. N. Kravtsov, B. A. Kvasov, L. S. Golovchenko, and E. I. Fedin, *J. Organometallic Chem.*, 1972, **36**, 227.
[103] W. Adcock, S. Q. A. Rizvi, W. Kitching, and A. J. Smith, *J. Amer. Chem. Soc.*, 1972, **94**, 369.
[104] J. Lipowitz, *J. Amer. Chem. Soc.*, 1972, **94**, 1582.
[105] A. P. Kreshkov, V. F. Andronov, and V. A. Drozdov, *Zhur. fiz. Khim.*, 1972, **46**, 977; *Russ. J. Phys. Chem.*, 1972, **46**, 564.
[106] R. Radeglia and E. Gey, *J. prakt. Chem.*, 1971, **313**, 1070.
[107] L. Zelta and G. Gatti, *Org. Magn. Resonance*, 1972, **4**, 585.
[108] A. N. Nesmeyanov, O. V. Nogina, E. I. Fedin, V. A. Dubovitskii, B. A. Kvasov, and P. V. Petrovskii, *Doklady Akad. Nauk S.S.S.R.*, 1972, **205**, 857.
[109] D. Leibfritz, B. O. Wagner, and J. D. Roberts, *Annalen*, 1972, **763**, 173.

group) has been examined with the aid of ^{13}C n.m.r. spectroscopy.[110] The ^{13}C n.m.r. spectra of $(OC)_5MCXY$ have been measured and the carbene ^{13}C n.m.r. signals found at very high frequency.[111] Similarly, the ^{13}C chemical shifts of $(C_5H_5)Fe(CO)_2X$ have been reported. The carbonyl chemical shifts are linearly dependent on the Taft σ_I values of the substituent and on the measured CO stretching frequencies.[112]

It has been shown that for mer-$RhCl_3(CO)(PBu^n{}_2Ph)_2$, $trans$-$RuCl_2$-$(CO)_2(PEt_3)_2$, and $trans$-$PdCl_2(PBu^n{}_2Bu^t)_2$, the ^{13}C signals of the mutual $trans$ tertiary phosphine ligands are either triplets or singlets, thus offering useful stereochemical information. Similarly, for $IrClIMe(CO)(AsEt_2Ph)_2$ the ^{13}C signals of the ethyl group are doubled, showing the absence of a plane of symmetry through the iridium–arsenic bond.[113] The ^{13}C n.m.r. spectra of $[(1\text{-}MeC_3H_4)NiX]_2$ and $[(C_3H_4R)PdX]_2$ have been reported, and a $trans$ influence $1 < Br < Cl$ has been found; $^1J(^{13}C\text{-}^1H) = ca.$ 160 Hz is consistent with sp^2 hybridization.[114, 115] The ^{13}C n.m.r. spectra of (4) and

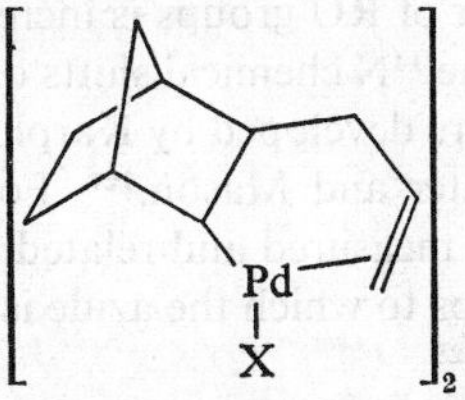

(4)

related compounds have been measured. Linear relationships have been found between ^{13}C chemical shifts of various atoms.[116]

The ^{13}C n.m.r. spectra of some olefinic and acetylenic complexes of platinum have been reported. It was felt that the data (which bear considerable similarity to data on cyclopropane) were best interpreted in terms of a π-bonding mechanism![117] Data have also been reported for olefin–silver complexes.[118]

The ^{13}C nuclear Overhauser enhancement of 1.65 has been measured for Me_4Si.[119] ^{13}C Chemical shifts of Me_4Si have been determined at infinite

---

[110] J. A. Connor, E. M. Jones, E. W. Randall, and E. Rosenberg, *J.C.S. Dalton*, 1972, 2419.

[111] C. G. Kreiter and V. Formáček, *Angew. Chem. Internat. Edn.*, 1972, **11**, 141.

[112] O. A. Gansow, D. A. Schexnayder, and B. Y. Kimura, *J. Amer. Chem. Soc.*, 1972, **94**, 3406.

[113] B. E. Mann, B. L. Shaw, and R. E. Stainbank, *J.C.S. Chem. Comm.*, 1972, 151.

[114] L. A. Churlyaeva, M. I. Lobach, G. P. Kondratenkov, and V. A. Kormer, *J. Organometallic Chem.*, 1972, **39**, C23.

[115] V. N. Sokolov, G. M. Khvostik, I. Ya. Poddubnyi, and G. P. Kondratenkov, *Doklady Akad. Nauk S.S.S.R.*, 1972, **204**, 120.

[116] D. G. Cooper, R. P. Hughes, and J. Powell, *J. Amer. Chem. Soc.*, 1972, **94**, 9244.

[117] M. H. Chisholm, H. C. Clark, L. E. Manzer, and J. B. Stothers, *J. Amer. Chem. Soc.*, 1972, **94**, 5087.

[118] C. D. M. Beverwijk and J. P. C. M. van Dongen, *Tetrahedron Letters*, 1972, 4291.

[119] T. C. Farrar, S. J. Druck, R. R. Shoup, and E. D. Becker, *J. Amer. Chem. Soc.* 1972, **94**, 699.

dilution in sixteen solvents: shifts of up to 4.68 p.p.m. were found. Cyclohexane was also examined and found to be a much better choice of internal reference. External referencing was considered to be even better.[120]

^{11}B, ^{14}N, ^{15}N, ^{29}Si, ^{31}P, ^{33}S, ^{35}Cl, ^{37}Cl, ^{77}Se, ^{79}Br, ^{81}Br, *and* ^{129}Xe *Chemical Shifts*. The chemical shifts of the boron 1s binding energies of some gaseous compounds have been measured but no correlation was found with the ^{11}B n.m.r. chemical shifts.[121] For the pair of compounds [BH$_4$]$^-$ and [BH$_3$D]$^-$ a ^{2}H isotope effect of 4.5 Hz has been observed in the ^{11}B n.m.r. spectra.[122] Use can be made of the different values of T_1 to extract unresolved lines in the ^{11}B n.m.r. spectrum and was applied to n-B$_9$H$_{15}$.[123]

The ^{15}N n.m.r. spectra of [Co(^{15}NH$_3$)$_5$X]$^{n+}$ have been measured but are relatively insensitive to X.[124] For a number of amino-boranes, a decrease in shielding of the ^{14}N nucleus is observed with increasing boron–nitrogen bond order. Neighbour anisotropic effects can also be important.[125] From ^{1}H and ^{14}N n.m.r. measurements on aminosilanes it has been concluded that as the number of RO groups is increased on the silicon, $(p-d)\pi$ bonding is reduced.[126] The ^{14}N chemical shifts of N$_3$$^-$ and NCO$^-$ have been calculated by the procedure developed by Karplus and Pople and compared with the method of Grinter and Mason.[127] For a number of azides, ^{14}N chemical shifts have been measured and related to the electronic characteristics of the atom or groups to which the azide is covalently attached and to the mode of attachment.[128]

The use of ^{29}Si n.m.r. has been examined and applied to a wide variety of silicon compounds.[129, 130] For P(SiMe$_3$)$_3$, $^1J(^{29}$Si–^{31}P) = 25 Hz.

There is a linear relationship between the ^{31}P chemical shift of the free tertiary phosphine and the change in chemical shift on co-ordination for the complexes (C$_6$H$_6$)RuCl$_2$PR$_3$[131] and *mer*-MCl$_3$(PR$_3$)$_3$ (M = Rh or Ir).[132] For IrCl$_3$(PEt$_3$)$_2$(PMe$_2$Ph), ^2J(P–P) (*trans*) is 427 Hz.[132] The ^{31}P chemical shifts of (R$_3$M)$_n$PX$_{3-n}$ (M = Si, Ge, or Sn) have been discussed using the theory of Letcher and Van Wazer. It was shown that, in both the Ge—P and Sn—P bond, π interactions are important.[133] The ^{31}P chemical

[120] D. Ziessow and M. Carroll, *Ber. Bunsengesselschaft phys. Chem.*, 1972, **76**, 61.
[121] P. Finn and W. L. Jolly, *J. Amer. Chem. Soc.*, 1972, **94**, 1540.
[122] M. M. Kreevoy and J. E. C. Hutchins, *J. Amer. Chem. Soc.*, 1972, **94**, 6371.
[123] A. Allerhand, A. O. Clouse, R. R. Rietz, T. Roseberry, and R. Schaeffer, *J. Amer. Chem. Soc.*, 1972, **94**, 2445.
[124] J. W. Lehman and B. M. Fung, *Inorg. Chem.*, 1972, **11**, 214.
[125] W. Beck, W. Becker, H. Nöth, and B. Wrackmeyer, *Chem. Ber.*, 1972, **105**, 2883.
[126] K. A. Andrianov, V. F. Andronov, V. A. Drozdov, D. Ya. Zhinkin, A. P. Kreshkov, and M. M. Morgunova, *Doklady Akad. Nauk S.S.S.R.*, 1972, **202**, 583.
[127] K. F. Chew, W. Derbyshire, and N. Logan, *J.C.S. Faraday II*, 1972, 594.
[128] W. Beck, W. Becker, K. F. Chew, W. Derbyshire, N. Logan, D. M. Revitt, and D. B. Sowerby, *J.C.S. Dalton*, 1972, 245.
[129] G. C. Levy, J. D. Cargioli, P. C. Juliano, and T. D. Mitchell, *J. Magn. Resonance*, 1972, **8**, 399.
[130] H. C. Marsmann, *Chem.-Ztg.*, 1972, **96**, 287.
[131] R. A. Zelonka and M. C. Baird, *J. Organometallic Chem.*, 1972, **44**, 383.
[132] B. E. Mann, C. Masters, and B. L. Shaw, *J.C.S. Dalton*, 1972, 704.
[133] G. Engelhardt, *Z. anorg. Chem.*, 1972, **387**, 52.

shift of phosphorus vapour relative to liquid white phosphorus has been measured. A linear relationship between shift and pressure was found and extrapolated to zero pressure to obtain the shift of free P_4.[134] The ^{31}P n.m.r. chemical shifts of P_4 have been measured in substituted benzenes and halogen-free solvents. The shifts are mainly due to van der Waals interactions.[135] There has been discussion on the application of the Van Wazer–Letcher theory to interpret the ^{31}P chemical shifts and to calculate phosphorus–nitrogen bond orders in phosphazo-derivatives.[136, 137] The deuterium isotope effect in $(MeO)_2PHO$ has been studied by INDOR and isotope shifts of up to 0.49 p.p.m. have been found.[138] The gas-phase and solvent chemical shifts of PBr_3 have been measured and found to be temperature dependent.[139]

The ^{33}S chemical shifts, covering a range of nearly 600 p.p.m., have been reported for 12 compounds and linewidths of up to 16 G (1200 p.p.m.) found. It is interesting to note that when concentrated H_2SO_4 is diluted to 10 mol l^{-1}, the ^{33}S resonance moves 90 p.p.m. and sharpens ten-fold.[140]

^{1}H-{^{77}Se} INDOR has been used to determine ^{77}Se chemical shifts in 80 organoselenium compounds. The shifts cover a range of over 1500 p.p.m. and are relatively insensitive to solvent effects. The behaviour of the chemical shifts parallels that found in similar phosphorus compounds but the shifts are several times larger. Correlations are found with the extent of α-chain branching in alkyl derivatives and with Hammett σ-constants in substituted aryl derivatives.[141] Even for R_3PSe, where ^{77}Se does not couple to ^{1}H, ^{77}Se shifts could be determined by ^{1}H-{^{77}Se} INDOR.[142] Similarly, ^{77}Se chemical shifts have been measured for $(PF_3)_2Se$ and $(GeH_3)_2Se$. ^{1}H, ^{19}F, and ^{31}P n.m.r. spectra were also recorded.[143]

The magnetic screening constants of the halogen nuclei in CuCl and CuBr have been calculated on the basis of the overlap of electron shells of the ions. Satisfactory correspondence with the experimental data has been obtained.[144] ^{35}Cl N.m.r. chemical shifts and linewidths in alkylchlorosilanes have been correlated with the polar constants of the substituents attached to the silicon atom and the quadrupole resonance frequencies. The ^{35}Cl chemical shifts can be used to assess the acid–base properties and reactivity of Si—Cl bonds.[145] The variations of the ^{35}Cl, ^{81}Br, and ^{2}H n.m.r. signals

[134] G. Heckmann and E. Fluck, *Mol. Phys.*, 1972, **23**, 175.
[135] G. Heckmann and E. Fluck, *Z. Naturforsch.*, 1972, **27b**, 764.
[136] E. S. Kozlov and S. N. Gaidamaka, *Teor. i eksp. Khim.*, 1972, **8**, 420.
[137] Yu. P. Egorov and A. S. Tarasevich, *Teor. i eksp. Khim.*, 1972, **8**, 422.
[138] W. McFarlane and D. S. Rycroft, *Mol. Phys.*, 1972, **24**, 893.
[139] G. Heckmann, *Mol. Phys.*, 1972, **23**, 627.
[140] H. L. Retcofsky and R. A. Friedel, *J. Amer. Chem. Soc.*, 1972, **94**, 6579.
[141] W. McFarlane and R. J. Wood, *J.C.S. Dalton*, 1972, 1397.
[142] W. McFarlane and D. S. Rycroft, *J.C.S. Chem. Comm.*, 1972, 902.
[143] D. E. J. Arnold, J. S. Dryburgh, E. A. V. Ebsworth, and D. W. H. Rankin, *J.C.S. Dalton*, 1972, 2518.
[144] V. M. Bouznik and L. G. Falaleeva, *J. Magn. Resonance*, 1972, **6**, 197.
[145] A. P. Kreshkov, V. F. Andronov, and V. A. Drozdov, *Russ. J. Phys. Chem.*, 1972, **46**, 183; *Zhur. fiz. Khim.*, 1972, **46**, 309.

with concentration, the ratios of ^{35}Cl, ^{37}Cl, ^{79}Br, and ^{81}Br relative to ^{2}H, and the solvent isotope effect have been measured in H$_2$O and D$_2$O.[146] Analysis of the large ^{129}Xe n.m.r. shifts induced by gaseous O$_2$ and NO indicates that they arise from a 'contact-overlap' mechanism.[147]

Metal Chemical Shifts. The resonance frequencies of ^{47}Ti and ^{49}Ti are so close that both isotopes are observed in a normal sweep. ^{47}Ti and ^{49}Ti n.m.r. spectroscopy were used to investigate halogen exchange between TiCl$_4$, TiBr$_4$, and TiF$_6^{2-}$.[148]

The paramagnetic component σ_p^A of the magnetic screening constant σ^A has been discussed for the case of the octahedral molecular σ-bonds in AX$_{6-n}$Y$_n$, where A is a transition element with the nd^6 electronic configuration, and applied to ^{59}Co chemical shifts.[149] ^{59}Co Chemical shifts have been reported for a number of complexes of the type Co(NH$_3$)$_{6-n}$X$_n$ and it was possible to explain the chemical shifts using the theory of Griffith and Orgel.[150] It has been concluded from ^{59}Co n.m.r. measurements on Co(NO)$_2$(PR$_3$)X that covalency increases X = Cl < Br < I. The ^{59}Co chemical shifts have also been used to set up an order of π acceptor ability for PR$_3$.[151]

^{119}Sn Chemical shifts have been determined by ^{1}H–{^{119}Sn} INDOR for a wide range of tin complexes. For 32 ethyltin complexes it is found that high-field shifts occur when an unsaturated group is attached to tin, indicating that π-bonding involving the tin 5d orbitals is important.[152] The solvent dependence of the ^{119}Sn chemical shift of Me$_3$SnCl has been attributed to complex formation and both stability constants and ΔH were derived.[153] A linear relationship has been reported between the pK_a of RCO$_2$H and the ^{119}Sn chemical shift of Ar$_3$SnO$_2$CR.[154] For the complexes (YC$_6$H$_4$CH$_2$)$_n$SnCl$_{4-n}$, it was concluded from ^{119}Sn n.m.r. spectra that one conformation predominates. The ^{119}Sn chemical shifts were related to inductive and neighbour anisotropy effects and to $(p_\pi-d_\pi)$ interactions.[155] ^{119}Sn N.m.r. spectroscopy has also been used to investigate association in Bu$^n_2$Sn(OR)$_2$ and Bu$^n_3$SnOR [156] and chemical shifts have been reported for a variety of tin complexes.[157]

---

[146] J. Blazer, O. Lutz, and W. Steinkilberg, *Z. Naturforsch.*, 1972, **27a**, 72.
[147] A. D. Buckingham and P. A. Kollman, *Mol. Phys.*, 1972, **23**, 65.
[148] R. G. Kidd, R. W. Matthews, and H. G. Spinney, *J. Amer. Chem. Soc.*, 1972, **94**, 6686.
[149] S. P. Ionov and V. S. Lyubimov, *Russ. J. Phys. Chem.*, 1971, **45**, 1407.
[150] N. S. Biradar and M. A. Pujar, *Z. anorg. Chem.*, 1972, **391**, 54.
[151] D. Rehder and J. Schmidt, *Z. Naturforsch.*, 1972, **27b**, 625.
[152] W. McFarlane, J. C. Maire, and M. Delmas, *J.C.S. Dalton*, 1972, 1862.
[153] V. N. Torocheshnikov, A. P. Tupčiauskas, N. M. Sergeyev, and Yu. A. Ustynyuk, *J. Organometallic Chem.*, 1972, **35**, C25.
[154] W. McFarlane and R. J. Wood, *J. Organometallic Chem.*, 1972, **40**, C17.
[155] L. Verdonck and G. P. Van der Kelen, *J. Organometallic Chem.*, 1972, **10**, 139.
[156] P. J. Smith, R. F. M. White, and L. Smith, *J. Organometallic Chem.*, 1972, **40**, 341.
[157] A. G. Davies, L. Smith, and P. J. Smith, *J. Organometallic Chem.*, 1972, **39**, 279.

^{1}H–{^{199}Hg} INDOR has been used to derive ^{199}Hg chemical shifts for a number of dialkyl- or diaryl-mercury compounds.[158]

Relaxation Measurements. The ^{13}C nuclear spin–lattice relaxation time T_1 was studied in liquid Ni(CO)$_4$ and Fe(CO)$_5$. Only anisotropic chemical shift and spin–rotation interaction contribute to the relaxation rate. It was possible to derive the anisotropy of the chemical shift and both spin–rotation interaction constants.[159] From T_1 measurements on substituted ferrocenes it has proved possible to measure the correlation time and to show that the unsubstituted ring is spinning up to seven times faster than the substituted ring.[160]

T_1 has been measured for ^{15}N in ^{15}NH$_3$ (186 s) and ^{15}ND$_3$ (413 s).[161] The use of ^{29}Si n.m.r. spectroscopy has been examined. However, as a consequence of the long T_1 and the negative nuclear Overhauser enhancement the total destruction of the signal can result. As Me$_4$Si has a relatively short T_1 at room temperature and a small Overhauser enhancement it makes a good reference.[162]

From the measurement of T_1 for ^{1}H, ^{2}H, and ^{31}P for PhPH$_2$, PhPD$_2$, C$_6$D$_5$PH$_2$, and C$_6$D$_5$PD$_2$ it has proved possible to determine the diffusion coefficient and the rotational diffusion coefficient, and the potential barrier to rotation was also measured.[163] The complete rotational diffusion tensors, and hence average correlation times, of the three symmetric top molecules PCl$_3$, PBr$_3$, and POCl$_3$ have been obtained from viscosity and halogen n.m.r. relaxation time data.[164] Similarly, $^1J(^{31}$P–^{35}Cl) = 127 Hz for PCl$_3$ and $^1J(^{31}$P–^{79}Br) = 296 Hz for PBr$_3$ have been measured.[165]

The relaxation rates of ^{69}Ga and ^{71}Ga have been measured in aqueous solution and compared with the theoretical ratio of relaxation rates of 2.5. This ratio is only approached in concentrated solution.[166] Contrary to a commonly held myth, T_1 and T_2 for ^{119}SnCl$_4$ (T_1 = ca. 2 s, T_2 = ca. 2 ms) and ^{119}SnI$_4$ (T_1 = ca. 0.5 s, T_2 = ca. 10 ms) are short. It was possible to derive $^1J(^{119}$Sn–^{35}Cl) = 470 Hz and $^1J(^{119}$Sn–^{127}I) = 940 Hz.[167]

2 Stereochemistry

This section is subdivided into nine parts, which contain n.m.r. information about lithium, beryllium, magnesium, calcium, and transition-metal complexes, presented by groups according to the Periodic Table. Within each group classification is by ligand type. As far as possible cross-references are

[158] A. P. Tupčiauskas, N. M. Sergeyev, Yu. A. Ustynyuk, and A. N. Kashin, *J. Magn. Resonance*, 1972, **7**, 124.
[159] H. W. Spiess and H. Mahnke, *Ber. Bunsengesellschaft phys. Chem.*, 1972, **76**, 990.
[160] G. C. Levy, *Tetrahedron Letters*, 1972, 3709.
[161] W. M. Litchman and M. Alei, jun., *J. Chem. Phys.*, 1972, **56**, 5818.
[162] G. C. Levy, *J. Amer. Chem. Soc.*, 1972, **94**, 4793.
[163] S. J. Seymour and J. Jonas, *J. Magn. Resonance*, 1972, **8**, 376.
[164] K. T. Gillen, *J. Chem. Phys.*, 1972, **56**, 1573.
[165] A. D. Jordan, R. G. Cavell, and R. B. Jordan, *J. Chem. Phys.*, 1972, **56**, 483.
[166] V. P. Tarasov and Yu. A. Buslaev, *Mol. Phys.*, 1972, **24**, 665.
[167] R. R. Sharp, *J. Chem. Phys.*, 1972, **57**, 5321.

given at the beginning of each subgroup to compounds discussed elsewhere in this chapter. In this cross-referencing it has not proved possible within the space available to include the many compounds that occur within the sections on dynamic systems, paramagnetic systems, and solid-state n.m.r. Thus many more compounds of relevance to this chapter appear in these sections.

Complexes of Li, Be, Mg, Ca, Sc, Y, La, and U.—Information concerning complexes of these elements can be found at the following sources: R_2Mg [109] and $(C_5H_5)Ni(PPh_3)MgBr$.[790]

^1H N.m.r. spectroscopy has been used to show that $Bu^tCH_2CH=CHLi$ is a *cis–trans* mixture,[168] and to investigate the reaction of $(D_3C)_3CLi$ with isoprene to yield oligomers such as (5).[169] Data have also been reported

$$(CD_3)_3CCHMe\diagdown_H C=C \diagup^{CH_2-CHMe}_{Me} \quad \overset{Li}{\underset{CH=CMe}{\underset{|}{CH_2}}}$$

(5)

for $LiCH_2CN$ and $PhCHLiCN$.[170] ^1H N.m.r. spectroscopy is useful in determining the ratio of $LiOBu^n$ to $LiBu^n$ by observation of the α-methylene resonances.[171] The ^1H and ^7Li n.m.r. spectra of a variety of *meta*- and *para*-substituted aryl-lithium and aryl Grignard compounds in ether have been examined. In contrast to covalently substituted benzenes, the inherently large chemical shifts between the ring protons permit a detailed analysis of the spectra, and thus the influence of the substituent on both the chemical shifts and spin–spin coupling were investigated. A correlation between ^7Li and *o*-hydrogen chemical shifts was found.[172] The ^{13}C n.m.r. spectrum of (6; R = H or Me) shows a quartet due to $^1J(^{13}C-^7Li)$. Both the NMe and NCH_2 proton resonance patterns are temperature dependent, e.g. NMe_2 is a singlet at 25 °C and two singlets at −60 °C.[173] The ^1H n.m.r. spectra of (7; M = Li, Na, MgBr, or K) have been reported.[174] The ^1H n.m.r. spectra of $H_2C=CHCH_2CD_2M$ and $\overline{CH_2CH_2CHCH_2CD_2}M$ (M = Li or MgBr) have been examined in an unsuccessful search for rearrangements.[175] The ^7Li n.m.r. spectrum of *p*-t-amylbenzyl-lithium is asymmetric.[176] The ^1H n.m.r. spectrum of the addition product from Bu^tLi and butadiene polymer has been used to show *cis–trans* isomerism.[177]

---

[168] W. H. Glaze, J. E. Hanicak, M. L. Moore, and J. Chaudhuri, *J. Organometallic Chem.*, 1972, **44**, 39.
[169] A. Ulrich, J. Cressely, A. Deluzarche, F. Schué, J. Sledz, and C. Tanielian, *Bull. Soc. chim. France*, 1972, 3556.
[170] R. Das and C. A. Wilkie, *J. Amer. Chem. Soc.*, 1972, **94**, 4555.
[171] P. J. Reed and J. R. Urwin, *J. Organometallic Chem.*, 1972, **39**, 1.
[172] J. A. Ladd and J. Parker, *J.C.S. Dalton*, 1972, 930.
[173] G. van Koten and J. G. Noltes, *J.C.S. Chem. Comm.*, 1972, 940.
[174] G. Fraenkel, C. C. Ho, Y. Liang, and S. Yu, *J. Amer. Chem. Soc.*, 1972, **94**, 4733.
[175] A. Maercker and W. Theysohn, *Annalen*, 1972, **759**, 132.
[176] R. Waack, M. A. Doran, and A. L. Gatzke, *J. Organometallic Chem.*, 1972, **46**, 1.
[177] S. Bywater, D. J. Worsfold, and G. Hollingsworth, *Macromolecules*, 1972, **5**, 389.

The lowest recorded shift for protons bound to carbon which is itself bound to beryllium has been reported for $(PhCH_2)_2BeOEt$ (τ 8.25).[177a] Data have also been reported for $(Bu^tO)_4Be_4Cl_4$, $X_2Be_3(OBu^t)_4$,[177b] $Me_3AlBe_3(OBu^t)_6$, $[Be(OCEt_2Me)_2]_3$,[178] $[Me_4N][(Et_2Be)_2SCN]$,[179] $Be(acac)_2$, and Be(ethyl acetoacetate)$_2$.[180]

^1H N.m.r. spectra of the *cis–trans* stereoisomers of vinylic organomagnesium compounds have been studied and the parameters discussed in terms of structural modifications.[181] When $Me_2NCH_2CH_2NMe_2$ is added to EtMgBr, two sets of resonances are observed due to Et_2Mg-$(Me_2NCH_2CH_2NMe_2)$ and $EtBrMg(Me_2NCH_2CH_2NMe_2)$.[182] ^1H N.m.r. data have also been reported for (8; M = Mg or Hg), $\overline{OCH_2CH_2Mg}$,[183] and (tetraphenylporphyrin)Mg,(py)$_2$.[184] After enrichment to 90% in ^{13}C, the ^{13}C n.m.r. spectra of chlorophyll a and b have been recorded and many assignments made.[185] Data have also been reported for (8)[186] and $Mg\{N(SiMe_3)_2\}_2$.[187]

[177a] G. E. Coates and R. C. Srivastava, *J.C.S. Dalton*, 1972, 1541.
[177b] R. A. Andersen, N. A. Bell, and G. E. Coates, *J.C.S. Dalton*, 1972, 577.
[178] R. A. Andersen and G. E. Coates, *J.C.S. Dalton*, 1972, 2153.
[179] N. Atam, H. Müller, and K. Dehnicke, *J. Organometallic Chem.*, 1972, **37**, 15.
[180] L. Maijs, I. Vevere, and I. Strauss, *Latv. Psr. Zinat. Akad. Vestis, Kim. Ser.*, 1972, **4**, 486.
[181] B. Méchin and N. Naulet, *J. Organometallic Chem.*, 1972, **39**, 229.
[182] J. A. Magnuson and J. D. Roberts, *J. Magn. Resonance*, 1972, **37**, 133.
[183] C. Blomberg, G. Schat, H. H. Grootveld, A. D. Vreugdenhil, and F. Bickelhaupt, *Annalen*, 1972, **763**, 148.
[184] H. Kobayashi, T. Hara, and Y. Kaizu, *Bull. Chem. Soc. Japan*, 1972, **45**, 2148.
[185] C. E. Strouse, V. H. Kollman, and N. A. Matwiyoff, *Biochem. Biophys. Res. Comm.*, 1972, **46**, 328.
[186] M. Kirilov and G. Petrov, *Monatsh.*, 1972, **103**, 1651.
[187] U. Wannagat, H. Autzen, H. Kuckertz, and H.-J. Wismar, *Z. anorg. Chem.*, 1972, **394**, 254.

An ^{1}H n.m.r. investigation has demonstrated the formation of 1:2 complexes of La^{3+} and Y^{3+} with edta,[188] and the presence of a mole of water in aquoglyoxalbis-(2-hydroxyanil)dioxouranium.[189] Data have also been reported for Sc[N(SiMe$_3$)$_2$]$_3$,[190] UO$_2$(salen), and related compounds.[191] The ^{1}H n.m.r. spectrum of (9) shows no evidence for co-ordination of —NR^{1}R^2.[192]

Complexes of Ti, Zr, Hf, Th, V, Nb, and Ta.—Information concerning complexes of these elements can be found at the following sources: TiCl$_4$, TiBr$_4$, [TiF$_6$]$^{2-}$,[148] (CH$_2$)$_3$(C$_5$H$_4$)$_2$MCl$_2$ (3; M = Ti, Zr, or Hf),[87] Ti(CH$_2$Ph)$_4$, Zr(CH$_2$Ph)$_4$,[107] (C$_5$H$_5$)$_2$TiX$_2$,[108] (C$_5$H$_5$)$_2$Ti(OCOPh)$_2$,[483] and Nb(OMe)$_5$.[317]

When (C$_5$Me$_5$)$_2$TiMe$_2$ is heated, one mole of methane is formed. ^{1}H N.m.r. spectroscopy shows one C$_5$Me$_5$ ring to have four methyl singlets and there is a high-field methyl signal. Structures such as (10) were suggested but the CH$_2$ resonance was not found. The Evans method of determining susceptibility was applied to (C$_5$Me$_5$)$_2$Ti and a hydride resonance at δ 0.28 was found in the decomposition product (C$_5$Me$_5$)$_2$TiH$_2$.[193] Data

(10)

(11)

have also been reported for (11),[194] (C$_5$H$_5$)$_2$M^{1}Cl(M^{2}Ph$_3$) (M^1 = Ti, Zr, or Hf; M^2 = Si, Ge, or Sn), (C$_5$H$_5$)$_2$Ti(SnPh$_3$)$_2$,[195] (C$_5$H$_5$)$_2$M(CH$_2$Ph)$_2$ (M = Ti or Zr),[196] and (Et$_2$N)$_3$Ti(vinylic group).[197]

Li$_2$P$_4$Ph$_4$ reacts with (C$_5$H$_5$)$_2$MX$_2$ (M = Ti or Zr) to yield (12; M = Ti or Zr), which gives an AX$_2$ ^{31}P n.m.r. spectrum with $^1J(^{31}P-^{31}P) = 323$ Hz. The ^{31}P chemical shifts are very metal-dependent.[198] It has been concluded from ^{1}H n.m.r. and i.r. spectroscopy measurements on (OC)$_5$CrCMeOTi-(C$_5$H$_5$)$_2$Cl and {(OC)$_5$CrCMeO}$_2$Ti(C$_5$H$_5$) that the electronegativity of the

---

[188] N. A. Kostromina and T. V. Ternovaya, *Zhur. neorg. Khim.*, 1972, **17**, 1596.
[189] G. Bandoli, L. Caltalini, D. A. Clemente, M. Vidali, and P. A. Vigato, *J.C.S. Chem Comm.*, 1972, 344.
[190] E. C. Alyea, D. C. Bradley, and R. G. Copperthwaite, *J.C.S. Dalton*, 1972, 1580.
[191] A. Pasini, M. Gullotti, and E. Cesarotti, *J. Inorg. Nuclear Chem.*, 1972, **34**, 3821.
[192] M. Vidali, P. A. Vigato, G. Bandoli, D. A. Clemente, and U. Casellato, *Inorg. Chim. Acta*, 1972, **6**, 671.
[193] J. E. Bercaw, R. H. Marvich, L. G. Bell, and H. H. Brintzinger, *J. Amer. Chem. Soc.*, 1972, **94**, 1219.
[194] K. Yasufuku and H. Yamazaki, *Bull. Chem. Soc. Japan*, 1972, **45**, 2664.
[195] B. M. Kingston and M. F. Lappert, *J.C.S. Dalton*, 1972, 69.
[196] G. Fachinetti and C. Floriani, *J.C.S. Chem. Comm.*, 1972, 654.
[197] H. Bürger and H.-J. Neese, *J. Organometallic Chem.*, 1972, **36**, 101.
[198] K. Issleib, G. Wille, and F. Krech, *Angew. Chem. Internat. Edn.*, 1972, **11**, 527.

(OC)$_5$CrCMeO group towards Ti is the same as that of Cl.^{199} ^{1}H N.m.r. spectroscopy has been used to determine the extent of deuteriation of (MeC$_5$H$_4$)$_2$Ti(allyl) after treatment with D$_2$ and conversion into (MeC$_5$H$_4$)$_2$TiCl$_2$.^{200} Data have also been reported for (13).^{201} For MeTiCl$_3$(XCH$_2$CH$_2$Y) (X, Y = OMe, NMe$_2$, or SMe), it was necessary to cool to stop the process (14) ↔ (15).^{202} ^{1}H N.m.r. spectroscopy has been used to show that TiCl$_4$,HCO$_2$R and TiCl$_4$,2HCO$_2$R do not dissociate in solution^{203} and data have been reported for (16),^{204} Cl$_3$TiS$_2$PF$_2$,

$$(R_2N)_{4-m}Ti\left(N\genfrac{}{}{0pt}{}{(CH_2)_n}{(CH_2)_n}CH_2\right)_m \qquad (C_5H_5)_2M\genfrac{}{}{0pt}{}{PPh}{PPh}PPh$$

(16)   (12)

(13)   (14)   (15)

Cl$_3$Nb(S$_2$PF$_2$)$_2$,^{205} (C$_5$H$_5$)$_2$ZrMe$_2$, (C$_5$H$_5$)$_2$ZrMeX, (C$_5$H$_5$)$_2$ZrCl$_2$,^{206} Hf(OPr^i)$_4$,Pr^{i}OH, [Hf$_2$(OPr^i)$_9$]$^-$, [Hf$_3$(OPr^i)$_{14}$]$^{2-}$, and HfAl$_2$(OPr^i)$_{10}$.^{207} When (C$_5$H$_5$)$_2$ZrH$_2$ is treated with AlMe$_3$ the hydrides become inequivalent and couple. The structure (17) was suggested.^{208} ^{1}H N.m.r. spectroscopy has been used to investigate the reaction of (C$_8$H$_8$)$_2$Hf with AlHEt$_2$ and data are also reported for (C$_8$H$_8$)Hf(allyl)$_2$,^{209} L$_2$ThCl$_2$ (L = Schiff base),^{210} and (C$_5$H$_5$)$_3$ThBH$_4$.^{211}

^{51}V N.m.r. spectroscopy has been used to help to characterize M[V(CO)$_6$], [Et$_4$N]$_3$[V(CO)$_4$(CN)$_2$]$_2$, Na[V(CO)$_5$NH$_3$], [V(CO)$_5$(CN)]$^-$, and [V(CO)$_5$(PPh$_3$)]$^-$.^{212} The methylene resonance of VO(CH$_2$SiMe$_3$)$_3$ is *ca.* 150 Hz broad and this was attributed to ^{51}V coupling.^{213} Data have also

[199] E. O. Fischer and S. Fontana, *J. Organometallic Chem.*, 1972, **40**, 159.
[200] H. A. Martin, M. Van Gorkom, and R. O. De Jongh, *J. Organometallic Chem.*, 1972, **36**, 93.
[201] C. Floriani and G. Fachinetti, *J.C.S. Chem. Comm.*, 1972, 790.
[202] R. J. H. Clark and A. J. McAlees, *Inorg. Chem.*, 1972, **11**, 342.
[203] M. Basso-Bert, D. Gervais, and J.-P. Laurent, *J. Chim. phys.*, 1972, **69**, 982.
[204] H. Bürger and U. Dämmgen, *Z. anorg. Chem.*, 1972, **394**, 209.
[205] R. G. Cavell and A. R. Sanger, *Inorg. Chem.*, 1972, **11**, 2016.
[206] P. C. Wailes, H. Weigold, and A. P. Bell, *J. Organometallic Chem.*, 1972, **34**, 155.
[207] R. C. Mehrotra and A. Mehrotra, *J.C.S. Dalton*, 1972, 1203.
[208] P. C. Wailes, H. Weigold, and A. P. Bell, *J. Organometallic Chem.*, 1972, **43**, C29.
[209] H.-J. Kablitz, R. Kallweit, and G. Wilke, *J. Organometallic Chem.*, 1972, **44**, C49.
[210] N. S. Biradar and V. H. Kulkarni, *Z. anorg. Chem.*, 1972, **387**, 275.
[211] T. J. Marks, W. J. Kennelly, J. R. Kolb, and L. A. Shimp, *Inorg. Chem.*, 1972, **11**, 2540.
[212] D. Rehder, *J. Organometallic Chem.*, 1972, **37**, 303.
[213] W. Mowat, A. Shortland, G. Yagupsky, N. J. Hill, M. Yagupsky, and G. Wilkinson, *J.C.S. Dalton*, 1972, 533.

$$(C_5H_5)_2ZrHAlMe_3$$
$$\diagup \;\;\; \diagdown$$
$$H \;\;\;\;\; H$$
$$\diagdown \;\;\; \diagup$$
$$(C_5H_5)_2ZrHAlMe_3$$
(17)

$$\left[\left((C_5H_5)_2Nb\begin{array}{c}Me\\S\\S\\Me\end{array}M\right)_2\right]^{2+}$$
(18)

been reported for $(C_5H_5)V(CO)_2(PF_2NC_5H_{10})_2$,[214] $VO(OH)(cupferron)_2$,[215] $(C_5H_5)_2NbH(PMe_2Ph)$, $[(C_5H_5)NbH_2(PMe_2Ph)]^+PF_6^-$,[216] (18; M = Ni, Pd, or Pt),[217] and $Me_nNbCl_{5-n}L$.[218] The adducts of $NbCl_5$ and $TaCl_5$ with some aliphatic and cyclic oxides and sulphides have been studied by ^1H n.m.r. spectroscopy and are found to have 1:1 stoicheiometry at room temperature and lower. For $O(CH_2CH_2)_2STaCl_5$, two species were detected, with one co-ordinated via oxygen and the other via sulphur. The stability of the complexes appeared to be controlled by steric factors.[219]

Complexes of Cr, Mo, and W.—Information concerning complexes of these elements can be found at the following sources: $(OC)_5MCXY$ (M = Cr or W; X = OR or NR^1R^2; Y = organic group),[110, 111] cis-$L_2M(CO)_4$ [M = Cr or Mo; L = $(CF_3)_2PR$ or CF_3PR_2],[83] $(C_4H_6)Cr(CO)_4$,[414] $(OC)_{6-n}ML_n$ (L = phosphorus ligand; M = Cr, Mo, or W),[45] (C_5H_5)-$Mo(CO)(PF_2NMe_2)_2Cl$,[214] $M_2(CH_2SiMe_3)_6$ (M = Mo or W),[213] $W(CO)_5PX_3$,[72] Me_6W,[78] and $Cr(acac)_3$.[57]

^1H and ^{14}N n.m.r. spectroscopies have been used to determine the mode of bonding in (19; M = Cr, Mo, or W; X = O, S, or Se). When the

$$(OC)_5M\left(\begin{array}{c}\\ \end{array}\underset{X}{\overset{N}{\diagup\diagdown}}-Me\right)$$
(19)

ligand is complexed via N, the ^{14}N chemical shift is ca. 80 p.p.m., but when complexed via X is ca. 10 p.p.m.[220] An n.m.r. study of a series of bis-μ-phosphido-dimetallic species has demonstrated that large increases in $J(P-P')$ occur on reduction to the dianion. These increases were attributed to an increase of the metal–metal distance and a marked decrease in the phosphorus–phosphorus distance. Activation parameters were measured for the inversion of $(OC)_nM(\mu-PMe_2)_2M(CO)_n$ (M = Cr, Mo, or W;

[214] R. B. King, W. C. Zipperer, and M. Ishaq, *Inorg. Chem.*, 1972, **11**, 1361.
[215] A. T. Pilipenko, L. L. Shevchenko, V. V. Trachevskii, V. N. Strokan, and L. A. Zyuzya, *Zhur. priklad. Spektroskopii*, 1972, **16**, 290.
[216] C. R. Lucas and M. L. H. Green, *J.C.S. Chem. Comm.*, 1972, 1005.
[217] W. E. Douglas and M. L. H. Green, *J.C.S. Dalton*, 1972, 1796.
[218] G. W. A. Fowles, D. A. Rice, and J. D. Wilkins, *J.C.S. Dalton*, 1972, 2313.
[219] A. Merbach and J. C. Bünzli, *Helv. Chim. Acta*, 1972, **55**, 580.
[220] J. C. Weis and W. Beck, *J. Organometallic Chem.*, 1972, **44**, 325.

$n = 4$; M = Fe or Ru; $n = 3$).[221] The ^{1}H n.m.r. spectra of $(OC)_5$-MPMe$_2$PMe$_2$ (M = Cr, Mo, or W) are complex, and data have been given for all combinations of the type $(OC)_5M^1PMe_2PMe_2M^2(CO)_5$,[222] Mo(CO)$_4$-(PMePh$_2$)$_2$,[223] $(OC)_5M^1PPh_2SnMe_{3-n}X_n$ (including ^{31}P n.m.r. data),[224] $M^1(CO)_5\{(Ar_2Sb)_2CH_2\}$, $\{M^1(CO)_5\}_2(Me_2Sb)_2CH_2$,[225] [Ph$_4$As][M^1(CO)$_5$-N$_4$CR],[226] Me$_3$SnSMeM^1(CO)$_5$, $(C_5H_5)M^3(CO)_3SMe$, $(C_5H_5)(CO)_2$-FeSMeCr(CO)$_5$,[227, 228] $(C_5H_5)M^1(CO)_3HgS_2CNEt_2$,[229] (M^1, M^2 = Cr, Mo, or W; M^3 = Mo or W), and $[(C_5Me_5)Cr(CO)_2]_2$.[230]

The ^{1}H n.m.r. spectrum of (20; M = Cr, Mo, or W) has been analysed as AA'BB'X. As the NH proton was most affected by co-ordination it was postulated that co-ordination via S=C$\langle$ occurred.[231] The ^{1}H n.m.r. spectra of (21; M = Cr, Mo, or W; L = CO or PPh$_3$) have been measured

$$(OC)_5M \leftarrow S = C \begin{matrix} H \\ \diagup \\ N-CH_2 \\ | \\ S-CH_2 \end{matrix}$$
(20)

$$\begin{matrix} & & H \\ & N & \diagup \\ L & \diagdown & C \\ & M & \diagdown \\ OC & | & N-R \\ & C \\ & O \end{matrix}$$
(21)

and used to show that when L = PPh$_3$, L enters the cis position.[232] The ^{1}H and ^{19}F n.m.r. spectra of ML$_6$ and M(CO)L$_5$ [M = Cr, Mo, or W; L = Pr^{n}OPF$_2$, P(OMe)$_2$F, P(OMe)$_3$, MeP(OMe)$_2$, or Me$_2$P(OMe)] are very complex as they are of the [AX]$_6$, [AX$_2$]$_6$, or AX$_3$[BY$_3$]$_4$ type.[233]

^{1}H N.m.r. spectroscopy has been used to investigate the protonation of (arene)Cr(CO)$_2$L (L = CO or PPh$_3$). The observation of a signal at ca. τ 14 led to the suggestion that protonation occurs on the metal.[234–236] The position of base-catalysed deuteriation of (arene)Cr(CO)$_3$ has been

[221] R. E. Dessy, A. L. Rheingold, and G. D. Howard, *J. Amer. Chem. Soc.*, 1972, **94**, 746.
[222] M. Brockhaus, F. Staudacher, and H. Vahrenkamp, *Chem. Ber.*, 1972, **105**, 3716.
[223] P. M. Treichel, W. M. Douglas, and W. K. Dean, *Inorg. Chem.*, 1972, **11**, 1615.
[224] H. Nöth and S. N. Sze, *J. Organometallic Chem.*, 1972, **43**, 249.
[225] T. Fukumoto, Y. Matsumura, and R. Ohawara, *J. Organometallic Chem.*, 1972, **37**, 113.
[226] J. C. Weis and W. Beck, *Chem. Ber.*, 1972, **105**, 3202.
[227] W. Ehrl and H. Vahrenkamp, *Chem. Ber.*, 1972, **105**, 1471.
[228] W. J. Schlientz and J. K. Ruff, *Inorg. Chem.*, 1972, **11**, 2265.
[229] W. K. Glass and T. Shiels, *Inorg. Nuclear Chem. Letters*, 1972, **8**, 257.
[230] R. B. King and A. Efraty, *J. Amer. Chem. Soc.*, 1972, **94**, 3773.
[231] D. De Filippo, F. Devillanova, C. Preti, E. F. Trogu, and P. Viglino, *Inorg. Chim. Acta*, 1972, **6**, 23.
[232] H. Brunner and W. A. Herrmann, *Chem. Ber.*, 1972, **105**, 770.
[233] R. Mathieu and R. Poilblanc, *Inorg. Chem.*, 1972, **11**, 1859.
[234] C. P. Lillya and R. A. Sahatjian, *Inorg. Chem.*, 1972, **11**, 889.
[235] B. V. Lokshin, V. I. Zdanovich, N. K. Baranetskaya, V. N. Setkina, and D. N. Kursanov, *J. Organometallic Chem.*, 1972, **37**, 331.
[236] D. N. Kursanov, V. N. Setkina, P. V. Petrovskii, V. I. Zdanovich, N. K. Baranetskaya, and I. D. Rubin, *J. Organometallic Chem.*, 1972, **37**, 339.

investigated by ^{1}H n.m.r. spectroscopy.[237, 238] The ^{1}H n.m.r. spectra of the $Cr(CO)_3$ and $Mn(CO)_3$ complexes of 2-R-pyrrole (R = Ph or $PhCH_2$) have been interpreted in terms of electronic effects transmitted through the σ-electron system and ring-current effects.[239] ^{1}H N.m.r. spectroscopy has been used to show that $Cr(CO)_3$ co-ordinates to a phenyl group in Zn(tetraphenylporphyrin),[240] and the spectrum of (22) is of the AA'BB' type.[241] Data have also been reported for (arene)$Cr(CO)_3$,[242, 243] (23) (including ^{31}P n.m.r. data),[244] $(C_7H_8)Cr(CO)_2\{P(OPh)_3\}$,[245] (24),[246] $(OC)_5$-$CrCPh(NR^1R^2)$,[247] $(OC)_5CrCR^1(SR^2)$,[248] $(OC)_5Cr(CHNMe_2)$,[249] $(OC)_5$-

(22), (23), (24)

$CrPh(OMe)(PHMe_2)$,[250] $C_4F_7(NO)_2(C_5H_5)Cr$ (^{19}F n.m.r. data),[251] $(C_5H_5)Cr(CO)(NO)$(olefin),[252] $[(C_5H_5)Fe(C_5H_4)]_3PCr(CO)_5$,[253] Me_2PCH_2-$CH_2PMe_2Cr(CO)_5$,[254] and (alkoxynorbornene)$Cr(CO)_4$.[255]

Extensive use has been made of the Karplus equation in order to determine the conformation of the chelate ring in (25).[256] ^{1}H and ^{19}F n.m.r. data have also been given for $Cr(CO)_4(f_4fars)$ [f_4fars = (26)].[257]

When isocyanides co-ordinate to $Cr(CO)_6$ or $Mo(CO)_6$ to form $(RNC)_n$-$M(CO)_{6-n}$, $^2J(^{14}N^{-1}H)$ is lost. The ASIS effect was measured in an attempt

[237] M. Ashraf, *Canad. J. Chem.*, 1972, **50**, 118.
[238] W. S. Trahanosky and R. J. Card, *J. Amer. Chem. Soc.*, 1972, **94**, 2897.
[239] N. J. Gogan and C. S. Davies, *J. Organometallic Chem.*, 1972, **39**, 129.
[240] N. J. Gogan and Z. U. Siddiqui, *Canad. J. Chem.*, 1972, **50**, 720.
[241] K. Öfele and E. Dotzauer, *J. Organometallic Chem.*, 1972, **42**, C87.
[242] G. R. Knox, D. G. Leppard, P. L. Pauson, and W. E. Watts, *J. Organometallic Chem.*, 1972, **34**, 347.
[243] M. Ashraf and W. R. Jackson, *J.C.S. Perkin II*, 1972, 103.
[244] H. Vahrenkamp and H. Nöth, *Chem. Ber.*, 1972, **105**, 1148.
[245] W. P. Anderson, W. G. Blenderman, and K. A. Drews, *J. Organometallic Chem.*, 1972, **42**, 139.
[246] J. A. S. Howell, B. F. G. Johnson, and J. Lewis, *J. Organometallic Chem.*, 1972, **42**, C54.
[247] E. O. Fischer and M. Leupold, *Chem. Ber.*, 1972, **105**, 599.
[248] E. O. Fischer and M. Leupold, C. G. Kreiter, and J. Müller, *Chem. Ber.*, 1972, **105**, 150.
[249] B. Çetinkaya, M. F. Lappert, and K. Turner, *J.C.S. Chem. Comm.*, 1972, 851.
[250] F. R. Kreissl, C. G. Kreiter, and E. O. Fischer, *Angew. Chem. Internat. Edn.*, 1972, **11**, 643.
[251] R. B. King and W. C. Zipperer, *Inorg. Chem.*, 1972, **11**, 2119.
[252] M. Herberhold and H. Alt, *J. Organometallic Chem.*, 1972, **42**, 407.
[253] C. U. Pittman, jun. and G. O. Evans, *J. Organometallic Chem.*, 1972, **43**, 361.
[254] J. A. Connor, E. M. Jones, and G. K. McEwen, *J. Organometallic Chem.*, 1972, **43**, 357.
[255] D. Wege and S. P. Wilkinson, *J.C.S. Chem. Comm.*, 1972, 1335.
[256] W. R. Cullen, L. D. Hall, and J. E. H. Ward, *J. Amer. Chem. Soc.*, 1972, **94**, 5702.
[257] J. P. Crow, W. R. Cullen, and F. L. Hou, *Inorg. Chem.*, 1972, **11**, 2125.

(25) structure: H₂C and RHC bridging two AsMe₂ groups on Cr(CO)₄

(26) Me₂AsC=CAsMe₂ with CF₂–CF₂ bridge

(27) [complex cluster structure with Fe, two Cr(CO)₄ groups, and two P-containing cages]

to place the positive charge.[258] The ^{1}H, ^{11}B, and ^{13}C n.m.r. spectra of (27; M = Cr or Mo) and 7,9- or 7,8-B_9H_9(3-Br-pyridine)CHPM(CO)$_5$ have been reported.[259]

The separation of the α and β carbon protons of the substituted ferrocene ring has been used to estimate the Hammett substituent constant, σ_P, and the resonance constituent constant, σ_R, for (OC)$_5$MCX(C$_5$H$_4$)Fe(C$_5$H$_5$) (M = Cr or W).[260] The ^{1}H n.m.r. spectrum of PhC≡C–C(OEt)Cr(CO)$_5$ is invariant with temperature, implying a low-energy barrier to COEt rotation. Data were also given for PhC≡CC(OEt)W(CO)$_5$, PhC(NMe)=CHC(NMe$_2$)W(CO)$_5$,[261] (OC)$_5$CrC(OEt)NR$_2$, (OC)$_5$WC(SeMe)Me, and cis-(OC)$_4$Cr{C(OEt)PMe$_2$}$_2$.[262] For cis-(OC)$_4$M[C(OEt)PMe$_2$]$_2$ (M = Cr or W), the PMe$_2$ protons form an intermediate [AX$_6$]$_2$ pattern, implying a significant value for $^4J(^{31}P-^{31}P)$.[263]

The methanolysis of (ClR$_2$P)Mo(CO)$_5$ has been followed by ^{1}H n.m.r. and data were reported for (XR$_2$P)(OC)$_5$Mo (X = RO, NH$_2$, NHMe, NMe$_2$, or SH; R = Me or Ph),[264] [Et$_3$NH][Mo(CO)$_5$PPh$_2$O], and (OC)$_5$Mo(μ-PR$^1_2$OPR$^2_2$)Mo(CO)$_5$.[265] It is interesting to note that for the

[258] J. A. Connor, E. M. Jones, G. K. McEwen, M. K. Lloyd, and J. A. McCleverty, J.C.S. Dalton, 1972, 1246.
[259] D. C. Beer and L. J. Todd, J. Organometallic Chem., 1972, 36, 77.
[260] J. A. Connor and J. P. Lloyd, J.C.S. Dalton, 1972, 1470.
[261] E. O. Fischer and F. R. Kreissl, J. Organometallic Chem., 1972, 35, C47.
[262] E. O. Fischer, Pure Appl. Chem., 1972, 30, 353.
[263] E. O. Fischer, F. R. Kreissl, C. G. Kreiter, and E. W. Meineke, Chem. Ber., 1972, 105, 2558.
[264] C. S. Kraihanzel and C. M. Bartish, J. Organometallic Chem., 1972, 43, 343.
[265] C. S. Kraihanzel and C. M. Bartish, J. Amer. Chem. Soc., 1972, 94, 3572.

latter compound, when $R^1 = R^2 = Me$, apparent triplets are observed for the methyl groups. Compounds of the type 1-X-2,4,6-trimethylbenzene have been synthesized in order to detect ring currents in X by measuring the chemical shift separation of the 2,6- and 4-methyl groups. When X = (28), the difference is 0.25 p.p.m., and when X = (29), 0.33 p.p.m., implying the presence of a ring current, but when X = (30), the difference is -0.15 p.p.m.[266] Data were also given for (31; X = O, NR, *etc.*; L = CO or PPh₃).[267] ¹H N.m.r. spectroscopy has been used to provide evidence of donor–acceptor interactions in $[C_5H_5NMe][Mo(CO)_5I]$ and $[N_4P_4Me_9][Mo(CO)_5I]$,[268] and (32; $R^1 = Me, R^2 = H; R^1 = H, R^2 = Et$) show the presence of two conformers on cooling.[269] The ¹H n.m.r. spectrum of (33) shows inequivalent methyl groups owing to restricted rotation about

the carbon–nitrogen bond. Methylation occurs at the sulphur atoms, as shown by the observation of three methyl resonances.[270] Data have also been reported on $[(C_5H_5)Mo(CO)_3PPh_2H]PF_6$, $(C_5H_5)Mo(CO)_2(PPh_2H)Br$, $Cr(CO)_4(PPh_2H)_2$, $Mo(CO)_3(PhPH_2)_3$, $W_2(CO)_6(PPh_2)_2$,[271] $(C_5H_5)Mo(CO)_2(PPh_3)H$, $(C_5H_5)Mo(CO)_2(PPh_3)(C_5F_4N)$ (and related complexes with ¹⁹F n.m.r. data),[272] $(C_5H_5)Mo(CO)_2$(allylic ligand),[273] $(C_5H_5)Mo(CO)_2$-

---

[266] G. Häfelinger, R. G. Weissenhorn, F. Hack, and G. Westermayer, *Angew. Chem. Internat. Edn.*, 1972, **11**, 725.
[267] I. W. Renk and H. tom Dieck, *Chem. Ber.*, 1972, **105**, 1403.
[268] N. L. Paddock, T. N. Ranganathan, and J. N. Wingfield, *J.C.S. Dalton*, 1972, 1578.
[269] J. L. Roustan, C. Charrier, J. Y. Mérour, J. Bénaïm, and C. Giannotti, *J. Organometallic Chem.*, 1972, **38**, C37.
[270] P. M. Treichel and W. K. Dean, *J.C.S. Dalton*, 1972, 804.
[271] P. M. Treichel, W. K. Dean, and W. M. Douglas, *J. Organometallic Chem.*, 1972, **42**, 145.
[272] M. I. Bruce, B. L. Goodall, D. N. Sharrocks, and F. G. A. Stone, *J. Organometallic Chem.*, 1972, **39**, 139.
[273] J. Y. Mérour, C. Charrier, J. Bénaïm, J. L. Roustan, and D. Commereuc, *J. Organometallic Chem.*, 1972, **39**, 321.

LSnMe$_3$,[274, 275] RMo(CO)$_2$(PMe$_2$Ph)(C$_5$H$_5$),[276] (C$_5$H$_5$)Mo(CO)$_3$SnCl$_2$S$_2$-CNMe,[277] (C$_5$H$_5$)Mo(CO)$_2$CN(AsMe$_2$CH$_2$CH=CH$_2$),[278] (C$_5$H$_5$)Mo-(CO)$_2$(2-NR=CH-pyrrole),[279] {(Ph$_2$AsCH$_2$CH$_2$)$_2$PPh}Mo(CO)$_2$(COMe)-(C$_5$H$_5$), [(C$_5$H$_5$)Mo(CO)$_2${(AsPh$_2$CH$_2$CH$_2$)$_2$PPh}]PF$_6$,[280] and (C$_5$H$_5$)$_2$Mo-(CO).[281]

As the cyclopentadienyl resonance of (NC)$_2$C=CClMo(PPh$_3$)$_2$(C$_5$H$_5$) consists of two 1:2:1 triplets in the ratio 4:1, it was suggested that this complex exists as a mixture of *cis*- and *trans*-isomers.[282] The ^{13}C n.m.r. spectrum of (Bu^{t}NC)$_4$Mo(CN)$_4$ has been examined over the temperature range of +40 to −47 °C, but no changes were observed.[283] Data have also been reported for (C$_5$H$_5$)Mo(NO)LX$_2$,[284] (34),[285] OMo(S$_2$PF$_2$)$_2$

(34)

(^{19}F n.m.r. data),[286] (C$_5$H$_5$)$_2$MHR (M = Mo or W),[287] [(C$_5$H$_5$)$_2$MH-(C$_2$H$_4$)]$^+$ (M = Mo or W),[288] [Me$_4$N]$_2$[(C$_5$H$_5$)$_2$M$_2$Fe$_2$(CO)$_{10}$],[289] (Ph$_3$P-CHCO)M(CO)$_5$,[290] (C$_7$H$_7$)M(CO)[P(OR)$_3$]I,[291] K[(C$_5$H$_5$)M(CO)$_2$(COR)-CN],[292] (C$_5$H$_5$)M(CO)$_3$SCF$_3$, [(C$_5$H$_5$)Mo(CO)$_2$SCF$_3$]$_2$,[293] (C$_5$H$_5$)M(CO)$_3$-COPh,[294] Me$_2$Tl[M(C$_5$H$_5$)(CO)$_3$],[295] and (C$_5$H$_5$)(CO)$_3$MSiMe$_2$X,[296] where M = Mo or W.

[274] T. A. George, *Inorg. Chem.*, 1972, **11**, 77.
[275] M. D. Curtis, *Inorg. Chem.*, 1972, **11**, 802.
[276] P. J. Craig and J. Edwards, *J. Organometallic Chem.*, 1972, **46**, 335.
[277] W. K. Glass and T. Shiels, *J. Organometallic Chem.*, 1972, **35**, C65.
[278] K. P. Wainwright and S. B. Wild, *J.C.S. Chem. Comm.*, 1972, 571.
[279] H. Brunner and W. A. Herrmann, *Chem. Ber.*, 1972, **105**, 3600.
[280] R. B. King and P. N. Kapoor, *Inorg. Chim. Acta*, 1972, **6**, 391.
[281] J. L. Thomas and H. H. Brintzinger, *J. Amer. Chem. Soc.*, 1972, **94**, 1386.
[282] R. B. King and M. S. Saran, *J.C.S. Chem. Comm.*, 1972, 1053.
[283] M. Novotny, D. F. Lewis, and S. J. Lippard, *J. Amer. Chem. Soc.*, 1972, **94**, 6961.
[284] J. A. McCleverty and D. Seddon, *J.C.S. Dalton*, 1972, 2526.
[285] J. Chatt and J. R. Dilworth, *J.C.S. Chem. Comm.*, 1972, 549.
[286] R. G. Cavell and A. R. Sanger, *Inorg. Chem.*, 1972, **11**, 2011.
[287] A. Nakamura and S. Otsuka, *J. Amer. Chem. Soc.*, 1972, **94**, 1886.
[288] F. W. S. Benfield, B. R. Francis, and M. L. H. Green, *J. Organometallic Chem.*, 1972, **44**, C13.
[289] A. T. T. Hsieh and M. J. Hays, *J. Organometallic Chem.*, 1972, **39**, 157.
[290] A. Greco, *J. Organometallic Chem.*, 1972, **43**, 351.
[291] T. W. Beall and L. W. Houk, *Inorg. Chem.*, 1972, **11**, 915.
[292] T. Kruck, M. Höfler, and L. Liebig, *Chem. Ber.*, 1972, **105**, 1174.
[293] J. L. Davidson and D. W. A. Sharp, *J.C.S. Dalton*, 1972, 107.
[294] A. N. Nesmeyanov, L. G. Makarova, N. A. Ustynyuk, and L. V. Bogatyreva, *J. Organometallic Chem.*, 1972, **46**, 105.
[295] B. Walther and C. Rockstroh, *J. Organometallic Chem.*, 1972, **44**, C4.
[296] W. Malisch, *J. Organometallic Chem.*, 1972, **39**, C28.

The ^{1}H and ^{19}F n.m.r. spectra of $FC_6H_4M(CO)_2L$ (M = Mo or W; L = CO or PPh$_3$), $FC_6H_4COMo(CO)_2PPh_3$, and $(C_5H_5)W(CO)_3C_6H_4$-COMe-p, have been measured. The σ_R^0 and σ_I constants estimated from the Taft equation had negative values, indicating that metal carbonyl substituents display donor inductive and resonance effects with respect to the phenyl ring.[297] The ^{1}H chemical shifts of $Me_3SnM(C_5H_5)(CO)_3$ (M = Mo or W) have been measured as a function of solvent.[298] The ^{1}H n.m.r. spectra of $(C_5H_5)M(CO)_3CONR^1R^2$ (M = Mo or W) has been measured. When $R^1 = R^2$ = Me, only one methyl signal is observed at room temperature and two at -30 °C.[299] The observation of only one methyl resonance for $(C_5H_5)M(CO)_2N=C(p\text{-tol})_2$ (M = Mo or W) over the temperature range -40 to $+40$ °C has been taken as evidence for a linear M—N—C skeleton. Data were also given for $(C_5H_5)M(CO)_2$-$\{(p\text{-tol})_2CNC(p\text{-tol})_2\}$, $(C_5H_5)M(CO)_2\{(p\text{-tol})_2CNC(p\text{-tol})_2\}(p\text{-tol})_2CO$ (M = Mo or W),[300] $[(C_5H_5)Mo(NO)XY]_2$,[301] $M(NO)_2\{(Ph_2As)_2CH_2\}_nX_2$ (M = Mo or W; n = 1 or 2),[301a] MOF_4, $[M_2O_2F_9]^-$ (M = Mo or W),[302] and $Mo_2Cl_4L_4$.[303]

The reaction of $MoOF_4$ and $[WOF_5]^-$ with ethanol and $MeCOCH_2COMe$ in MeCN has been studied by ^{19}F n.m.r. spectroscopy, and species such as $MoOF_4$,MeCN and $[WOF_4(OEt)]^-$ have been identified.[304]

Data have been reported for $WH_6(PR_2Ph)_3$,[305] $(C_5H_5)_2WH_2$, and $(C_5H_5)_2WH,CHCl_2$.[306] Me_6W reacts with NO to yield compounds with two sharp singlets in the ratio 4:2 and X-ray structure determination shows the structure to be $WMe_4(ONMeNO)_2$.[307] The ^{31}P n.m.r. spectrum of $(OC)_5WPPh_2CH_2PPh_2$ shows $^3J(^{31}P-^{183}W)$ = 6.3 Hz.[308] Data have been reported for $W(CO)_5\{C(OMe)(CH=CMeO_2CMe)\}$,[309] $(OC)_5WPPh_2CH_2$-PPh_2, $[(OC)_5WPPh_2CH_2PPh_2R]^+$,[310] $W(N_2)_2(PMe_2Ph)_4$, mer-$W(CO)_3$-$(PRPh_2)_3$, $W(CO)_4(PBu^nPh_2)_2$ (including ^{31}P n.m.r. data),[311] $(OC)_5WEMe_2$-$XEMe_2W(CO)_5$ (E = P or As; X = O, S, NMe, or NPh) and related

[297] A. N. Nesmeyanov, L. G. Makarova, N. A. Ustynyuk, B. A. Kvasov, and L. V. Bogatyreva, *J. Organometallic Chem.*, 1972, **34**, 185.
[298] R. M. G. Roberts, *J. Organometallic Chem.*, 1972, **40**, 359.
[299] W. Jetz and R. J. Angelici, *J. Amer. Chem. Soc.*, 1972, **94**, 3799.
[300] H. R. Keable and M. Kilner, *J.C.S. Dalton*, 1972, 153.
[301] J. A. McCleverty and D. Seddon, *J.C.S. Dalton*, 1972, 2588.
[301a] J. A. Bowden, R. Colton, and C. J. Commons, *Austral. J. Chem.*, 1972, **25**, 1393.
[302] Yu. A. Buslaev, Yu. V. Kokunov, V. A. Bochkareva, and E. M. Shustorovich, *Zhur. strukt. Khim.*, 1972, **13**, 526.
[303] J. San Filippo, jun., *Inorg. Chem.*, 1972, **11**, 3140.
[304] Yu. A. Buslayev, Yu. V. Kokunov, V. A. Bochkaryova, and E. M. Shustorovich, *J. Inorg. Nuclear Chem.*, 1972, **34**, 2861.
[305] J. R. Moss and B. L. Shaw, *J.C.S. Dalton*, 1972, 1910.
[306] K. S. Chen, J. Kleinberg, and J. A. Landgrebe, *J.C.S. Chem. Comm.*, 1972, 295.
[307] S. R. Fletcher, A. Shortland, A. C. Skapski, and G. Wilkinson, *J.C.S. Chem. Comm.*, 1972, 922.
[308] R. L. Keiter and L. W. Cary, *J. Amer. Chem. Soc.*, 1972, **94**, 9232.
[309] C. P. Casey, R. A. Boggs, and R. L. Anderson, *J. Amer. Chem. Soc.*, 1972, **94**, 8947.
[310] R. L. Keiter and D. P. Shah, *Inorg. Chem.*, 1972, **11**, 191.
[311] B. Bell, J. Chatt, and G. J. Leigh, *J.C.S. Dalton*, 1972, 2492.

complexes,[312] $(C_5H_5)W(CO)_2(PF_3)Me$,[313] $[-OS(=O)FOW(=O)X_3-]_n$ (X = OSO_2F),[314] and $(diphos)_2Cl_2W(N_2HCOR)$.[315]

^1H N.m.r. spectroscopy has been used to show that Me_3Al interacts with complexes such as $W(N_2)_2(diphos)_2$.[316] Similarly, ^1H, ^{19}F, and ^{31}P n.m.r. spectroscopy have been used to investigate the reaction of WF_6 with a wide variety of compounds such as $Nb(OMe)_5$, $B(OMe)_3$, $(MeO)_3P=O$, etc., and the stereochemistry of the resulting complexes has been determined.[317] ^1H and ^{19}F n.m.r. spectroscopy have been used to determine the stereochemistry of $(MeO)_nWX_{6-n}$ (X = F or Cl),[318] and to characterize the product formed between WF_6 and benzene or toluene.[319] ^{19}F N.m.r. spectroscopy has also been used to determine the structure of the products of reaction of WO_3 and HF. Species such as $[WO_2F_4]^{2-}$ and $[WO_2F_3(H_2O)]^-$ were detected.[320]

Complexes of Mn and Re.—Information concerning complexes of these elements can be found at the following sources: $Re(CO)_5CF_2CF_2Me$,[337] $(OC)_4Mn\{CH_2C(CO_2R^1)CHR^2\}$,[269] $Mn(CO)_4(PPhH_2)Br$,[271] $M_2(CO)_8Me_2$-$AsC=C(AsMe_2)CF_2CF_2$, (M = Mn or Re),[257] $C_4F_7Mn(CO)_5$,[251] $(OC)_5Mn$-$Mn(CO)_4CCOCH_2CH_2CHMc$,[309] $(OC)_3Mn(2$-R-pyrrole),[239] $(C_5Me_5)Mn$-$(CO)_3$,[230] $(MeC_5H_4)Mn(CO)_2(PMePh_2)$,[223] $(C_5H_5)Mn(CO)_2SMeSnMe_3$, $(C_5H_5)Mn(CO)_2SMeMo(CO)_3(C_5H_5)$,[227] $(C_5H_5)Mn(CO)(PF_2NMe_2)_2$,[214] $Re(CO)_3(PF_3)_2Br$,[313] and $Re_2Cl_6L_2$.[303]

^1H N.m.r. spectroscopy has been used to follow the reaction of $Me_3SnMn(CO)_5$ with $SnCl_4$.[321] For the complexes $(RC_5H_4)Mn(CO)_3$ [R = asymmetric group, e.g. $-CH(OH)CF_3$] the α-protons of the ring are inequivalent.[322] The ^1H chemical shifts of $(C_5H_5)Mn(CO)_2L$ have been reported and compared with data for $RC_5H_4Mn(CO)_2L$ and EtC_5H_4Mn-$(CO)(diphos)$.[323] The inductive effects deduced from the ^{19}F n.m.r. spectra of $(OC)_3MC_5H_4C_6H_4F$ (M = Mn or Re) agreed with those deduced from redox potentials.[324] The compound (35) shows three ^{11}B n.m.r. signals, consistent with the proposed structure.[325] Data have also been reported

[312] H. Vahrenkamp, *Chem. Ber.*, 1972, **105**, 3574.
[313] R. B. King and A. Efraty, *J. Amer. Chem. Soc.*, 1972, **94**, 3768.
[314] R. Dev and G. H. Cady, *Inorg. Chem.*, 1972, **11**, 1134.
[315] J. Chatt, G. A. Heath, and G. J. Leigh, *J.C.S. Chem. Comm.*, 1972, 444.
[316] J. Chatt, R. H. Crabtree, and R. L. Richards, *J.C.S. Chem. Comm.*, 1972, 534.
[317] D. W. Walker and J. M. Winfield, *J. Inorg. Nuclear Chem.*, 1972, **34**, 759.
[318] L. B. Handy, K. G. Sharp, and F. E. Brinckman, *Inorg. Chem.*, 1972, **11**, 523.
[319] R. R. McLean, D. W. A. Sharp, and J. M. Winfield, *J.C.S. Dalton*, 1972, 676.
[320] Yu. A. Buslaev, S. P. Petrosyants, and V. I. Chagin, *Zhur. neorg. Khim.*, 1972, **17**, 704.
[321] R. A. Burnham, F. Glockling, and S. R. Stobart, *J.C.S. Dalton*, 1972, 1991.
[322] D. N. Kursanov, Z. N. Parnes, N. M. Loim, N. E. Kolobova, I. B. Zlotina, P. V. Petrovskii, and E. I. Fedin, *J. Organometallic Chem.*, 1972, **44**, C15.
[323] A. G. Ginzburg, B. D. Lavrukhin, V. N. Setkina, and P. O. Okulevich, *Zhur. obshchei Khim.*, 1972, **42**, 514.
[324] S. P. Gubin, A. A. Koridze, N. A. Ogorodnikova, and B. A. Kvasov, *Doklady Akad. Nauk S.S.S.R.*, 1972, **205**, 346.
[325] J. W. Howard and R. N. Grimes, *Inorg. Chem.*, 1972, **11**, 263.

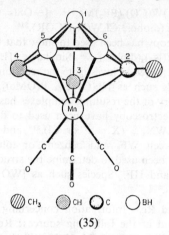

(35)

for $H_3GeMn(CO)_5$,[326] $Me_3SiCF_2CF_2Mn(CO)_5$,[327] $(OC)_9Mn_2PH_3$,[328] $(OC)_9$-$M_2C(OMe)R$ (M = Mn, Tc, or Re).[329] $Mn(CO)_3(PMe_2Bz)_2Me$,[330] $(OC)_4$-$M(C_6H_4CH=NR)$ (M = Mn or Re),[331] $cis\text{-}MnCl(\overline{COCH_2CH_2O})(CO)_4$,[332] $(Et_5C_5)Mn(CO)_2PPh_3$,[333] $(C_5H_4COR)Mn(CO)_3$,[334] $(C_5H_5)Mn(CO)_2$-$NH=NHMn(CO)_2(C_5H_5)$,[335] $Mn(CO)_{5-n}(CNMe)_nBr$, $[Mn(CO)_{6-n}$-$(CNMe)_n]^+$,[336] $[(Bu^tNC)_6Mn]^+$,[337] $(C_5H_5)Re(CO)_2N_2$, $(C_5H_5)Re(CO)_3$,[338] $(OC)_3Re(S_2PEt_2)PPh_2CH_2CH_2PPh_2Re(CO)_3(S_2PEt_2)$,[339] $[ReBr(CO)_3$-$(THF)]$,[340] $Cl_3SiHRe_2(CO)_9$, $PhCl_2SiHRe_2(CO)_9$, $HRe_3(CO)_{14}$,[341] (C_5H_5)-$ReMe(CO)(NO)$, $[(C_5H_5)ReH(CO)]^+$,[342] $cis\text{-}ReCl(CO)_2(PMe_2Ph)_3$,[343] $ReCl_3(N=CH_2)(PMe_2Ph)_3$, and $ReCl_2(N=CHMe)(py)(PMePh_2)_2$.[344]

[326] R. D. George, K. M. Mackay, and S. R. Stobart, *J.C.S. Dalton*, 1972, 1505.
[327] H. C. Clark and T. L. Hauw, *J. Organometallic Chem.*, 1972, **42**, 429.
[328] E. O. Fischer and W. A. Herrmann, *Chem. Ber.*, 1972, **105**, 286.
[329] E. O. Fischer, E. Offhaus, J. Müller, and D. Nöthe, *Chem. Ber.*, 1972, **105**, 3027.
[330] R. L. Bennett, M. I. Bruce, and F. G. A. Stone, *J. Organometallic Chem.*, 1972, **38**, 325.
[331] R. L. Bennett, M. I. Bruce, B. L. Goodall, M. Z. Iqbal, and F. G. A. Stone, *J.C.S. Dalton*, 1972, 1787.
[332] M. Green, J. R. Moss, I. W. Nowell, and F. G. A. Stone, *J.C.S. Chem. Comm.*, 1972, 1339.
[333] B. L. Lokshin, A. G. Ginzburg, V. N. Setkina, D. N. Kursanov, and I. B. Nemirovskaya, *J. Organometallic Chem.*, 1972, **37**, 347.
[334] M. Le Plouzennec and R. Dabard, *Bull. Soc. chim. France*, 1972, 3600.
[335] D. Sellmann, *J. Organometallic Chem.*, 1972, **44**, C46.
[336] P. M. Treichel, G. E. Dirreen, and H. J. Mueh, *J. Organometallic Chem.*, 1972, **44**, 339.
[337] R. B. King and M. H. Saran, *Inorg. Chem.*, 1972, **11**, 2112.
[338] D. Sellmann, *J. Organometallic Chem.*, 1972, **36**, C27.
[339] E. Lindner and H. Berke, *J. Organometallic Chem.*, 1972, **39**, 145.
[340] D. Vitali and F. Calderazzo, *Gazzetta*, 1972, **102**, 587.
[341] J. K. Hoyano and W. A. G. Graham, *Inorg. Chem.*, 1972, **11**, 1265.
[342] R. P. Stewart, N. Okamoto, and W. A. G. Graham, *J. Organometallic Chem.*, 1972, **42**, C32.
[343] D. J. Darensbourg, *Inorg. Nuclear Chem. Letters*, 1972, **8**, 529.
[344] J. Chatt, R. J. Dosser, and G. J. Leigh, *J.C.S. Chem. Comm.*, 1972, 1243.

Complexes of Fe, Ru, and Os.—Information concerning complexes of these elements can be found at the following sources: $Fe(CO)_5$,[57, 159] $Fe(CO)_4(Ar_2SbCH_2SbAr_2)$, $\{Fe(CO)_4\}_2(Me_2SbCH_2SbMe_2)$,[225] $Fe(CO)_{5-n}$-L_n,[45, 330] $Fe(CHNMe_2)(CO)_4$,[249] $(OC)_3Fe(\mu\text{-}PMe_2)_2Fe(CO)_3$,[221] $Fe(CO)_4$-$(PMePh_2)$, $Fe_2(CO)_6(PMe_2Ph)_2$,[223] $Fe(CO)_4(C_4F_7)(C_2F_5)$, $C_4F_7Fe(CO)_2$-(C_5H_5),[251] $Ru_3(CO)_9(PMe_2Bz)_3$, $Ru(CO)_2(PMe_2Bz)_2Cl_2$,[330] $[(Ph_3PCHCO)$-$Fe(CO)_3]^-$,[290] $[Fe(CO)_3SCF_3]_2$, $[Fe(CO)_2PPh_3SCF_3]_2$,[293] $Ru(CO)_2(C_6H_4$-$CH=NR)_2$,[331] $(Me_5C_5)Fe(CO)_2COMe$,[230] $(C_5H_5)Fe(CO)_2SMe$, $(C_5H_5)Fe$-$(CO)_2SMeM(CO)_5$ (M = Cr or W), $(C_5H_5)Fe(CO)_2SMeMn(CO)_2$-$(C_5H_5)$,[227] $(C_5H_5)Fe(CO)_2X$,[88, 112] $[(C_5H_5)Fe(CO)_2(PHPh_2)]^+$, $(C_5H_5)Fe$-$(CO)(PHPh_2)X$, $[\{(C_5H_5)Fe\}_2(P_2Ph_4)_2(PPh_2H)_2]^{2+}$,[271] $(C_5H_5)Fe(CO)$-$(PF_3)I$,[313] $(C_5H_5)_2Fe_2(CO)_3(PF_2NEt_2)$, $(C_5H_5)Fe(PF_2NEt_2)_2I$,[214] $\{(Ph_2As$-$CH_2CH_2)_2PPh\}Fe(CO)(COMe)(C_5H_5)$,[280] $K[(C_5H_5)Fe(CO)_2(COR)(CN)]$, $(C_5H_5)Fe(CO)(COR)(CNMe)$,[292] $(C_5H_5)(CO)_2FeSiMe_2X$,[296] $(C_5H_5)RuR$-$ClPPh_3$,[131] $(OC)_5M-CFcX$ (M = Cr or W),[260] $Fc_3PCr(CO)_5$,[253] FcX,[47, 160] $Fe(NMe=CH-CR=NMe)_3$,[266] $[(1,7\text{-}B_9H_9CHE)_2Fe]^{2-}$, and $[\{1,7\text{-}B_9H_9\text{-}CHEMo(CO)_5\}_2Fe]^{2-}$ (E = P or As).[259]

For $H_2Fe_2(CO)_6\{P(CF_3)_2\}_2$, the hydride resonance is a triplet and one $[AX_6]_2$ pattern is found for the fluorine spectrum, which is consistent with (36). On standing, another species, which is thought to be (37), appears.[345] The ^1H n.m.r. spectrum of $(tol_3P)_3RuLH_2$ shows a hydride resonance as a

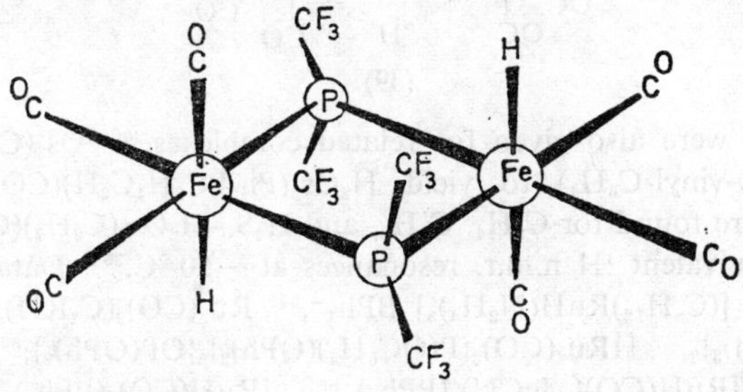

(36)

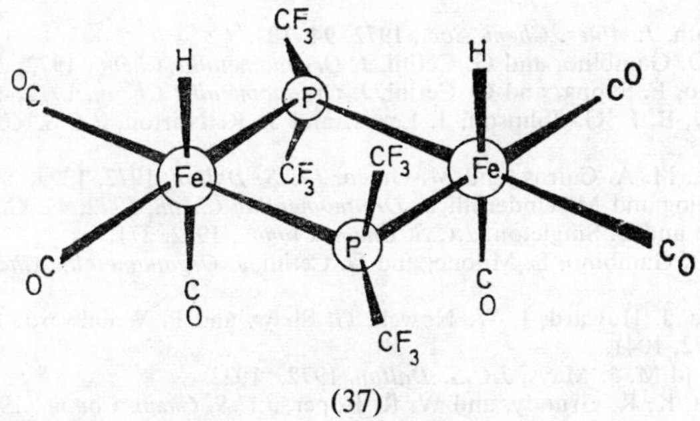

(37)

[345] R. C. Dobbie, M. J. Hopkinson, and D. Whittaker, *J.C.S. Dalton*, 1972, 1030.

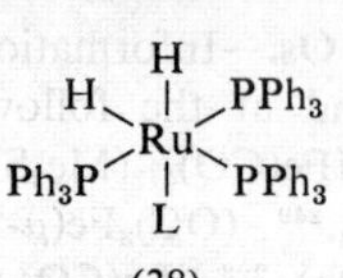

$$\begin{array}{c} H \\ H \diagdown | \diagup PPh_3 \\ Ru \\ Ph_3P \diagup | \diagdown PPh_3 \\ L \end{array}$$
(38)

doublet of triplets with hydrogen coupling, *i.e.* the stereochemistry is (38); $(Ph_3P)_3RuHClN_2B_{10}H_8SMe_2$ and $(Ph_3P)_2RuHCl(NCMe)N_2B_{10}H_8SMe_2$ were also prepared.[346] A linear correlation has been reported between the hydride chemical shift of $H_2Ru_3(CO)_9E$ (E = S, Se, or Te) and the electronegativity of E.[347] The hydride signal of $H_2Ru_3(CO)_9C_2Ph_2$ shows doubling due to two inequivalent protons.[348] The ^1H n.m.r. spectrum of H_3Ru_3-$(CO)_9CMe$ shows two singlets at τ 5.88 and τ 27.63 and the ^{13}C n.m.r. spectrum shows two carbonyl environments.[349] Compound (39) shows coupling between the hydride signal and the central proton of the allyl

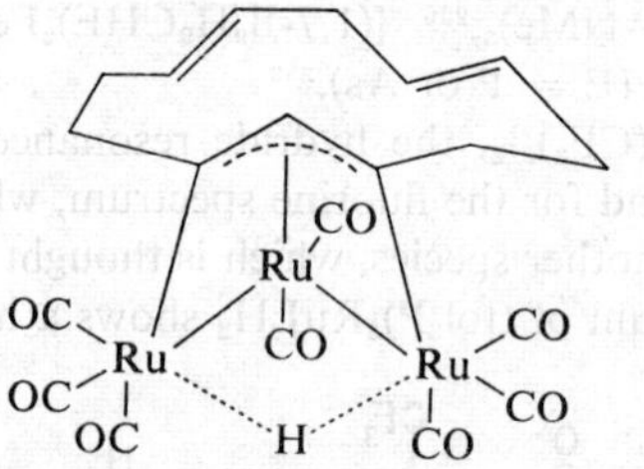

(39)

group; data were also given for related complexes.[350] $Os_3(CO)_{12}$ reacts with $Ph_2P(o\text{-vinyl-}C_6H_4)$ to yield $H_2Os_3(Ph_2PC_6H_4C_2H)(CO)_8$; similar reactions were found for C_2H_4, C_6H_6, and H_2S. $H_2Os_3(C_2H_2)(CO)_9$ shows four non-equivalent ^1H n.m.r. resonances at -50 °C.[351] Data were also reported for $[(C_8H_{12})RuH(N_2H_4)_3]^+BPh_4^-$,[352] $Ru_3(CO)_9(C_2RH)$,[353] H_4Ru_4-$(CO)_9\{P(OPh)_3\}_3$, $HRu_2(CO)_3\{P(OC_6H_4)(OPh)_2\}_2\{OP(OPh)_2\}$,[354] $[HRu_3$-$(CO)_{11}]$,[355] $[RuH(CO)(MeCN)_2(PPh_3)_2]^+$, $[RuH(CO)_2(PPh_3)_3]^+$,[356] and $OsH(CO)(NO)(PPh_3)_2$.[357]

---

[346] W. H. Knoth, *J. Amer. Chem. Soc.*, 1972, **94**, 104.
[347] E. Sappa, O. Gambino, and G. Cetini, *J. Organometallic Chem.*, 1972, **35**, 375.
[348] O. Gambino, E. Sappa, and G. Cetini, *J. Organometallic Chem.*, 1972, **44**, 185.
[349] A. J. Canty, B. F. G. Johnson, J. Lewis, and J. R. Norton, *J.C.S. Chem. Comm.*, 1972, 1331.
[350] M. I. Bruce, M. A. Cairns, and M. Green, *J.C.S. Dalton*, 1972, 1293.
[351] A. J. Deeming and M. Underhill, *J. Organometallic Chem.*, 1972, **42**, C60.
[352] J. J. Hough and E. Singleton, *J.C.S. Chem. Comm.*, 1972, 371.
[353] E. Sappa, O. Gambino, L. Milone, and G. Cetini, *J. Organometallic Chem.*, 1972, **39**, 169.
[354] M. I. Bruce, J. Howard, I. W. Nowell, G. Shaw, and P. Woodward, *J.C.S. Chem. Comm.*, 1972, 1041.
[355] J. Knight and M. J. Mays, *J.C.S. Dalton*, 1972, 1022.
[356] B. E. Cavitt, K. R. Grundy, and W. R. Roper, *J.C.S. Chem. Comm.*, 1972, 60.
[357] G. R. Clark, K. R. Grundy, W. R. Roper, J. M. Waters, and K. R. Whittle, *J.C.S. Chem. Comm.*, 1972, 119.

¹H N.m.r. spectroscopy has been used to show that when a ferrocene is dissolved in BF_3,H_2O, a hydride is formed.[358] Complexes of the type H_2FeL_4 are normally *cis*, but ¹H n.m.r. spectroscopy has shown that $H_2Fe\{PPh(OR)_2\}_4$ are *cis–trans* mixtures and $H_2Fe\{o\text{-}C_6H_4(PEt_2)_2\}_2$ predominates in the *trans* form. $H_2Fe(PRPh_2)_3CO$ is shown to be of conformation (40) at -50 °C but is fluxional at higher temperature. However, $H_2Ru(PMePh_2)_3CO$ is static even at 100 °C.[359] Similarly, $FeH_2(PHPh_2)_4$ is static at room temperature but becomes fluxional at 85 °C. The

```
         PHPh₂                          H
  Ph₂PH \ | / H              OC \  |  / PPh₃
         Fe                         Os
     OC / | \ H              Ph₃P / | \ Cl
         PHPh₂                       PPh₃
         (40)                         (41)
```

ruthenium complex is also static and data were given for *trans*-$MHX(PHPh_2)_4$ (M = Fe or Ru) and $[FeH(PHPh_2)_5]^+$.[360] The ¹H n.m.r. spectrum of (41) is a doublet (87.0 Hz) of triplets (24.5 Hz). Data were also given for related complexes [361] and $FeH_2(PF_3)_4$.[362]

¹¹B N.m.r. spectroscopy has been used to investigate the structure of boron complexes with iron. B_6H_{10} reacts with $Fe(CO)_5$ to yield $\mu\text{-}[Fe(CO)_4]\text{-}B_6H_{10}$, which has an ¹¹B n.m.r. spectrum consistent with C_s symmetry for the 4,5-bridged hexaborane unit.[363] $NaC_2B_4H_7$ reacts with $(C_5H_5)Fe(CO)_2I$ to yield $\mu\text{-}\{(\pi\text{-}C_5H_5)Fe(CO)_2\}C_2B_4H_7$, which gives four ¹¹B n.m.r. signals. U.v. irradiation yields $(C_5H_5)Fe(C_2B_4H_7)$ and paramagnetic $(C_5H_5)Fe\text{-}(C_2B_4H_6)$; the latter compound gives broad signals with marked contact shifts.[364] An AA′BB′ pattern is found for the C_5H_4 ring in $X(CO)_2FeC_5H_4\text{-}(3\text{-}B_{10}H_9C_2H_2)$, and for $(3\text{-}B_{10}H_9C_2H_2Fe)(CO)(C_5H_5)PPh_3$ the CH protons are inequivalent.[365]

A number of carbenes, $(OC)_4FeCRX$, have been prepared. When X = NMe_2, two methyl ¹H n.m.r. signals were observed which coalesce at 14 °C.[366] Similarly, $Fe(CO)_5$ reacts with $C_6H_3(OMe)_2Li$ and $Et_3O^+BF_4^-$ to yield $Fe_2C_{18}H_{14}O_{10}$. The ¹H n.m.r. spectrum was not very informative and an *X*-ray structure determination was necessary to establish the structure as (42).[367] Compound (42) shows three methoxy resonances in

---

[358] T. E. Bitterwolf and A. C. Ling, *J. Organometallic Chem.*, 1972, **40**, 197.
[359] D. H. Gerlach, W. G. Peet, and E. L. Muetterties, *J. Amer. Chem. Soc.*, 1972, **94**, 4545.
[360] J. R. Sanders, *J.C.S. Dalton*, 1972, 1333.
[361] N. Ahmad, S. D. Robinson, and M. F. Uttley, *J.C.S. Dalton*, 1972, 843.
[362] T. Kruck and R. Kobelt, *Chem. Ber.*, 1972, **105**, 3765.
[363] A. Davison, D. D. Traficante, and S. S. Wreford, *J.C.S. Chem. Comm.*, 1972, 1155.
[364] L. G. Sneddon and R. N. Grimes, *J. Amer. Chem. Soc.*, 1972, **94**, 7161.
[365] L. I. Zakharkin, L. V. Orlova, B. V. Lokshin, and L. A. Fedorov, *J. Organometallic Chem.*, 1972, **40**, 15.
[366] E. O. Fischer, H.-J. Beck, C. G. Kreiter, J. Lynch, J. Müller, and E. Winkler, *Chem. Ber.*, 1972, **105**, 162.
[367] E. O. Fischer, E. Winkler, G. Huttner, and D. Regler, *Angew. Chem. Internat. Edn.*, 1972, **11**, 238.

the ratio 1:2:1 and two ethoxy signals, demonstrating the presence of two components in solution.[368] Data have also been given for $(OC)_4FeC(NMe_2)OAl(NMe_2)_2$, $(OC)_4FeC(NMe_2)Ph$,[369] $Fe_2Pt(CO)_8L_2$,[370] (tolyl-NC)$_n$-Fe(CO)$_{5-n}$,[371] $(H_2C=CHOR)Fe(CO)_4$ (R = SiMe$_3$, OCHNMe, or H),[372] (43),[373] (44),[374] $(C_5H_5)Fe(CO)(PPh_3)\{CH(OEt)Me\}$,[375] and (45).[376]

The reaction of hexa-2,4-diene with $Ru_3(CO)_{12}$ has been shown by ^1H and ^{13}C n.m.r. spectroscopy and X-ray crystal structure determination to yield (46). There are five ^{13}C carbonyl resonances observed, some of which couple to the hydride resonance.[377, 378] $HOs_3(CO)_8{}'(PPh_3)_2{}'$ shows a

[368] E. O. Fischer, E. Winkler, G. Huttner, and D. Regler, *Angew. Chem. Internat. Edn.*, 1972, **11**, 238.
[369] W. Petz and G. Schmid, *Angew. Chem. Internat. Edn.*, 1972, **11**, 934.
[370] M. I. Bruce, G. Shaw, and F. G. A. Stone, *J.C.S. Dalton*, 1972, 1082.
[371] H. Alper and R. A. Partio, *J. Organometallic Chem.*, 1972, **35**, C40.
[372] H. Thyret, *Angew. Chem. Internat. Edn.*, 1972, **11**, 520.
[373] D. H. Gibson and R. L. Vonnahme, *J. Amer. Chem. Soc.*, 1972, **94**, 5090.
[374] K. Yasufuku and H. Yamazaki, *J. Organometallic Chem.*, 1972, **35**, 367.
[375] A. Davison and D. L. Reger, *J. Amer. Chem. Soc.*, 1972, **94**, 9237.
[376] A. L. Balch and J. Miller, *J. Amer. Chem. Soc.*, 1972, **94**, 417.
[377] M. Evans, M. Hursthouse, E. W. Randall, E. Rosenberg, L. Milone, and M. Valle, *J.C.S. Chem. Comm.*, 1972, 545.
[378] M. Valle, O. Gambino, L. Milone, G. A. Vaglio, and G. Cetini, *J. Organometallic Chem.*, 1972, **38**, C46.

hydride signal as a 1:1:1:1 quartet, whereas (47) has a 1:1 doublet for the hydride signal.[379]

$H_2C=\overline{CCH_2CHCH}=CH_2$ reacts with $Fe_3(CO)_9$ to yield (48), for which ^1H and ^{13}C n.m.r. data have been reported.[380] For (49), the large values found for $^4J(H^2-H^5)$ (1.9—4.2 Hz) and $^4J(H^3-H^6)$ (1.9—5.4 Hz) have been attributed to the 'W' pathway.[381] ^1H N.m.r. spectroscopy has been used to distinguish between the ψ-endo and ψ-exo forms of $(R^2CHC=CH-CH=CHCHMeOR^1)Fe(CO)_3$.[382] These compounds can exist in two

forms, (50) and (51), and their mirror images. With the aid of $Eu(fod)_3$, the spectra have been completely analysed and the previous assignments shown to be wrong.[383] C_2F_4, C_3F_6, and $(CF_3)_2CO$ react on irradiation with $Fe(\pi\text{-}C_4R_4)(CO)_3$ to yield π-cyclobutenyl complexes. Extensive use was made of ^1H and ^{19}F n.m.r. spectroscopy to determine the stereochemistry of the products.[384] $Eu(dpm)_3$ has been used to simplify the ^1H n.m.r. spectra of complexes such as (52).[385] When (tropone)$Fe(CO)_3$ is protonated, there is no high-field signal produced and a new signal appears with a coupling constant of 8.2 Hz. It is therefore suggested that the α-carbon is

[379] C. W. Bradford, R. S. Nyholm, G. J. Gainsford, J. M. Guss, P. R. Ireland, and R. Mason, *J.C.S. Chem. Comm.*, 1972, 87.
[380] W. E. Billups, L.-P. Lin, and O. A. Gansow, *Angew. Chem. Internat. Edn.*, 1972, **11**, 637.
[381] K. Erlich and G. F. Emerson, *J. Amer. Chem. Soc.*, 1972, **94**, 2464.
[382] D. E. Kuhn and C. P. Lillya, *J. Amer. Chem. Soc.*, 1972, **94**, 1682.
[383] M. I. Foreman, *J. Organometallic Chem.*, 1972, **39**, 161.
[384] A. Bond and M. Green, *J.C.S. Dalton*, 1972, 763.
[385] B. F. G. Johnson, J. Lewis, P. McArdle, and G. L. P. Randall, *J.C.S. Dalton*, 1972, 2076.

protonated. However, (53) protonates on the oxygen.[386] The ^{1}H n.m.r. spectrum of $(C_6H_6)_3Fe_3(CO)_{10}$ shows methyl signals in the ratio 1:2:1:1:1 and structure (54) was suggested.[387]

As conversion of (55) into (56) causes low-field shift and reduces $J(H^1–H^6)$ and $J(H^1–H^2)$, it was suggested that the dihedral angles $\chi_{1,6}$ and $\chi_{1,2}$ increase.[388] The ^{1}H n.m.r. spectrum of (57) has been fully analysed,[389] and the signal of H^4 in (58) comes at δ 1.32, which was considered to be

[386] D. F. Hunt, G. C. Farrant, and G. T. Rodeheaver, *J. Organometallic Chem.*, 1972, **38**, 349.
[387] R. B. King and C. W. Eavenson, *J. Organometallic Chem.*, 1972, **42**, C95.
[388] R. Aumann, *Angew. Chem. Internat. Edn.*, 1972, **11**, 522.
[389] A. Eisenstadt, *Tetrahedron Letters*, 1972, 2005.

typical of such protons.[390] Data have also been reported for (diene)-Fe(CO)$_3$,[391-402] (59),[401] (cyclobutadiene)Fe(CO)$_3$,[402, 403] (RCH=CH-—CH=X)Fe(CO)$_3$,[404] (C$_{10}$H$_{12}$)Fe$_2$(CO)$_6$, (C$_{10}$H$_{12}$)Fe(CO)$_3$,[405] (60),[406] (61),[407] (62) (X = S or Se),[408] (C$_4$H$_4$SO$_2$)Fe(CO)$_3$,[409] (63),[410] (64),[411] (65),[412] (diene)$_2$Fe(CO),[413] (diene)$_2$FePF$_3$,[414, 415] (C$_5$H$_5$)$_2$Fe$_2$(CO)$_3$-{C=C(CN)$_2$},[416] (C$_5$H$_5$)Fe(CO)$_2$CH$_2$CH$_2$OH,[417] (C$_5$H$_5$)Fe(CO)$_2$C≡CPh,-CuCl,[418] (C$_5$H$_5$)Fe(CO)$_2$R,[419, 420] (C$_3$H$_5$)COFe(CO)$_2$(C$_3$H$_3$N$_2$)$_3$BH,[421] (66),[422] (C$_5$H$_6$)Fe(PF$_3$)$_3$,[423] (67),[424] and [Fe(C$_5$H$_5$)(CO)$_n${Ph$_2$P(CH$_2$)$_n$-PPh$_2$}]$^+$.[425]

^{13}C N.m.r. spectroscopy has been used to investigate the binding of propene in [(C$_5$H$_5$)Fe(CO)$_2$(propene)]$^+$, Rh(C$_3$H$_6$)$_2$(acac), and [(C$_3$H$_6$)$_2$Ag]$^+$,

[390] Y. Becker, A. Eisenstadt, and Y. Shvo, *J.C.S. Chem. Comm.*, 1972, 1156.
[391] B. F. G. Johnson, J. Lewis, P. McArdle, and G. L. P. Randall, *J.C.S. Dalton*, 1972, 456.
[392] G. Schröder, S. R. Ramadas, and P. Nickoloff, *Chem. Ber.*, 1972, 105, 1072.
[393] H. Alper and E. C.-H. Keung, *J. Amer. Chem. Soc.*, 1972, 94, 2144.
[394] H. A. Brune, G. Horlbeck, and W. Schwab, *Tetrahedron*, 1972, 28, 4455.
[395] A. E. Hill and H. M. R. Hoffmann, *J.C.S. Chem. Comm.*, 1972, 574.
[396] J. M. Landesberg and L. Katz, *J. Organometallic Chem.*, 1972, 35, 327.
[397] H. A. Brune, G. Horlbeck, and W. Schwab, *Tetrahedron*, 1972, 28, 4455.
[398] R. Victor, R. Ben-Shoshan, and S. Sarel, *J. Org. Chem.*, 1972, 37, 1930.
[399] R. J. H. Cowles, B. F. G. Johnson, J. Lewis, and A. W. Parkins, *J.C.S. Dalton*, 1972, 1768.
[400] J. A. S. Howell, B. F. G. Johnson, P. L. Josty, and J. Lewis, *J. Organometallic Chem.*, 1972, 39, 329.
[401] J. M. Landesberg, L. Katz, and C. Olsen, *J. Org. Chem.*, 1972, 37, 930.
[402] C. H. Maudlin, E. R. Biehl, and P. C. Reeves, *Tetrahedron Letters*, 1972, 2955.
[403] R. B. King and I. Haiduc, *J. Amer. Chem. Soc.*, 1972, 94, 4044.
[404] A. M. Brodie, B. F. G. Johnson, P. L. Josty, and J. Lewis, *J.C.S. Dalton*, 1972, 2031.
[405] F. A. Cotton and G. Deganello, *J. Organometallic Chem.*, 1972, 38, 147.
[406] D. F. Hunt and J. W. Russell, *J. Amer. Chem. Soc.*, 1972, 94, 7198.
[407] D. F. Hunt and J. W. Russell, *J. Organometallic Chem.*, 1972, 46, C22.
[408] T. L. Gilchrist, P. G. Mente, and C. W. Rees, *J.C.S. Perkin I*, 1972, 2165.
[409] Y. L. Chow, J. Fossey, and R. A. Perry, *J.C.S. Chem. Comm.*, 1972, 501.
[410] A. Eisenstadt, *J. Organometallic Chem.*, 1972, 38, C32.
[411] D. J. Ehntholt and R. C. Kerber, *J. Organometallic Chem.*, 1972, 38, 139.
[412] R. P. Ferrari, G. A. Vaglio, O. Gambino, M. Valle, and G. Cetini, *J.C.S. Dalton*, 1972, 1998.
[413] A. Carbonaro and F. Cambisi, *J. Organometallic Chem.*, 1972, 44, 171.
[414] E. K. von Gustorf, O. Jaenicke, and O. E. Polansky, *Chem. Ber.*, 1972, 11, 532.
[415] D. L. Williams-Smith, L. R. Wolf, and P. S. Skell, *J. Amer. Chem. Soc.*, 1972, 94, 4042.
[416] R. B. King and M. S. Saran, *J. Amer. Chem. Soc.*, 1972, 94, 1784.
[417] W. P. Giering, M. Rosenblum, and J. Tancrede, *J. Amer. Chem. Soc.*, 1972, 94, 7170.
[418] M. I. Bruce, R. Clark, J. Howard, and P. Woodward, *J. Organometallic Chem.*, 1972, 42, C107.
[419] W. Jetz and R. J. Angelici, *Inorg. Chem.*, 1972, 11, 1960.
[420] H. G. Ang, W. E. Kow, and K. F. Mok, *Inorg. Nuclear Chem. Letters*, 1972, 8, 829.
[421] R. B. King and A. Bond, *J. Organometallic Chem.*, 1972, 46, C53.
[422] E. K. von Gustorf, I. Fischler, J. Leitich, and H. Dreeskamp, *Angew. Chem. Internat. Edn.*, 1972, 11, 1088.
[423] T. Kruck and L. Knoll, *Chem. Ber.*, 1972, 105, 3783.
[424] R. M. Moriarity, K.-N. Chen, C.-L. Yeh, J. L. Flippen, and J. Karle, *J. Amer. Chem. Soc.*, 1972, 94, 8944.
[425] M. L. Brown, J. L. Cramer, J. A. Ferguson, T. J. Meyer, and N. Winterton, *J. Amer. Chem. Soc.*, 1972, 94, 8707.

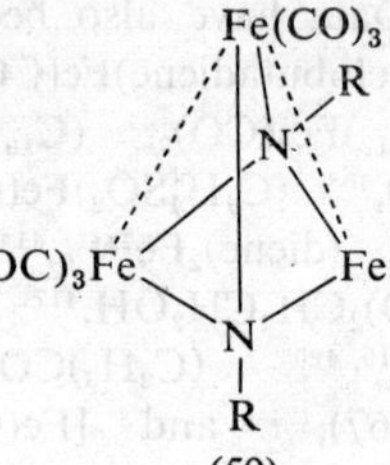

(59)

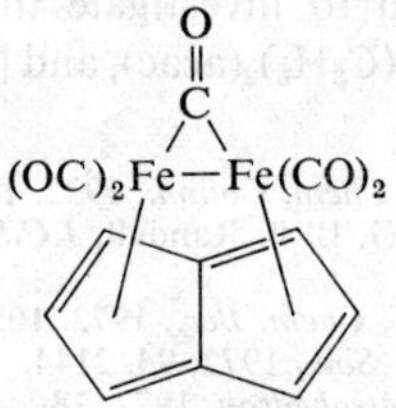

(60)

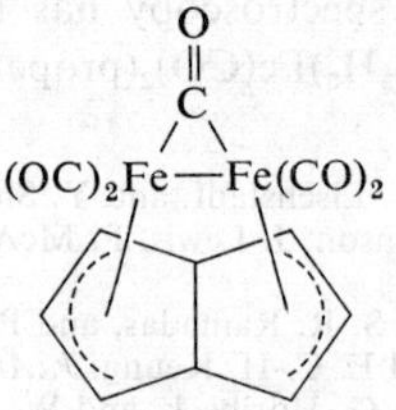

(61)

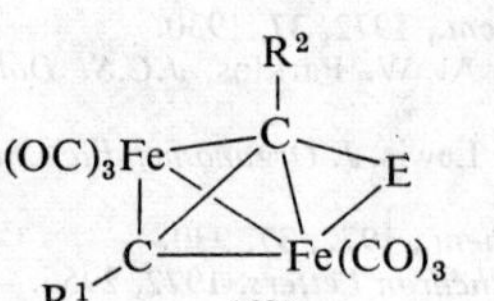

(62)

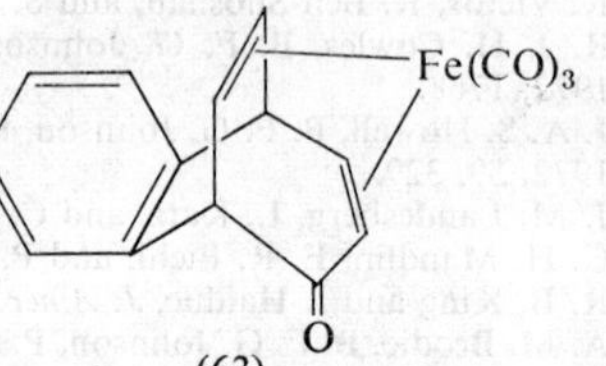

(63)

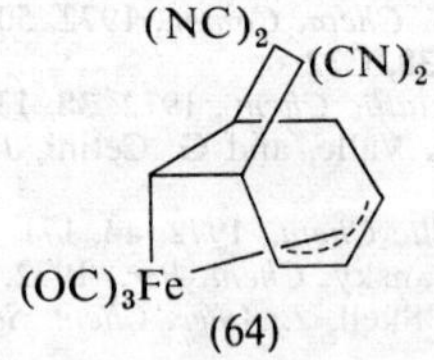

(64)

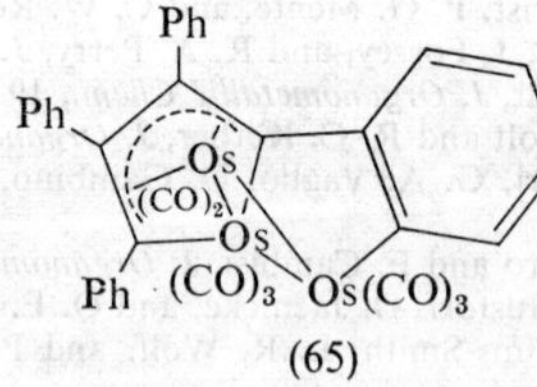

(65)

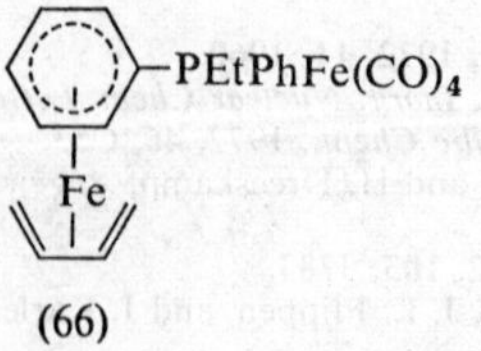

(66)

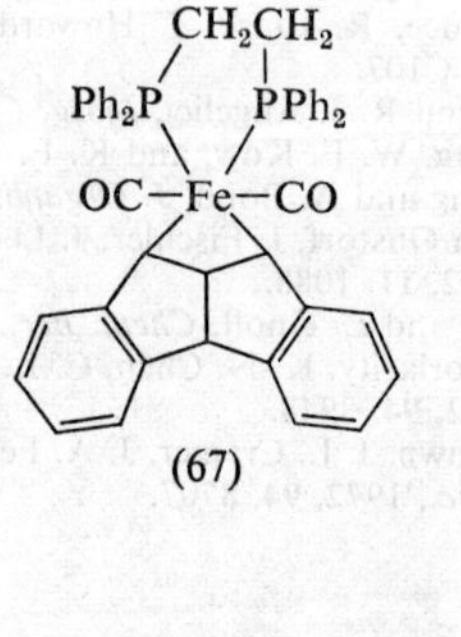

(67)

and $^1J(^{103}\text{Rh}-^{13}\text{C})$ was observed.[426] The observation of a signal at τ 6.21 in $C_8H_6\{Fe(CO)_2(C_5H_5)\}_2$ was taken as evidence for (68) rather than (69).[427] In the ^{1}H n.m.r. spectrum of $(C_5H_5)Fe(CO)_2CHCH_2CH_2$, two sets of proton resonances were observed at τ 9.48 and τ 10.12, due to the protons *trans* and *cis* to the metal.[428] The methylene protons of (C_5H_5)Fe-

$$(C_5H_5)(OC)_2Fe-\overset{}{\underset{H\ H}{\boxed{}}}-Fe(CO)_2(C_5H_5) \qquad (C_5H_5)(OC)_2Fe-\overset{}{\boxed{}}-Fe(CO)_2(C_5H_5)$$
(68)   (69)

$(CO)(C=NC_6H_{11})_3CH_2Ar$ are magnetically inequivalent.[429] The ^{1}H n.m.r. spectrum of $(C_5H_5)_2Fe_2(CO)_3(CNBu^t)$ shows only one Bu^t resonance, although the i.r. spectrum showed the presence of *cis*- and *trans*-isomers.[430] $[(C_5H_5)Fe(CO)_2]^-$ reacts with $S=PF_2Br$ to yield either $(C_5H_5)Fe(CO)_2$-SPF_2 or $(C_5H_5)Fe(CO)_2PF_2S$. The ^{31}P n.m.r. signal was observed at a position consistent with P^V and consequently that latter structure was favoured.[431] ^{1}H N.m.r. spectroscopy has shown $(Me_2Ge)Fe_2(CO)_3(C_5H_5)$ to be a mixture of probably (70) and (71). In the *cis*-form the methyl groups are inequivalent.[432] Similarly, in (72) the bridging methyl groups are inequivalent.[433] By use of ^{19}F n.m.r. spectroscopy it has been concluded that for $(C_5H_5)Fe(CO)L(C_6H_4F)$ the aryl–iron interaction is predominantly inductive.[434] Data have also been reported for $(C_5H_5)Fe(CO)_2CR=CHPh$,[435] (73),[436] $(C_5H_5)Fe(CO)_2P(O)(CF_3)_2$ (^{19}F n.m.r.),[437] $(BzC_5H_5)Fe(CO)_3$,[438] $(C_5H_5)Fe(CO)_2N(SO_2Cl)COCH_2CMeCH_2$,[439] $(C_5H_5)Fe(CO)_2GeH_3$, $(GeH_3)_2Fe(CO)_4$, $(GeH_3)HFe(CO)_3$,[440] $(C_5H_5)_2Fe_2(CO)_2(\text{diphos})$,[441, 442] $(C_5H_5)Fe(CO)_2GeX_3$,[443] $(C_5H_5)Fe(CO)(CNR)I$,[444] (74), $\{(C_5H_5)Fe(CO)_2\}$-

[426] K. R. Aris, V. Aris, and J. M. Brown, *J. Organometallic Chem.*, 1972, **42**, C67.
[427] R. B. King, A. Efraty, and W. C. Zipperer, *J. Organometallic Chem.*, 1972, **38**, 121.
[428] A. Cutler, R. W. Fish, W. P. Giering, and M. Rosenblum, *J. Amer. Chem. Soc.*, 1972, **94**, 4354.
[429] Y. Yamamoto and H. Yamazaki, *Inorg. Chem.*, 1972, **11**, 211.
[430] W. Jetz and R. J. Angelici, *J. Organometallic Chem.*, 1972, **35**, C37.
[431] C. B. Colburn, W. E. Hill, and D. W. A. Sharp, *Inorg. Nuclear Chem. Letters*, 1972, **8**, 625.
[432] M. D. Curtis and R. C. Job, *J. Amer. Chem. Soc.*, 1972, **94**, 2153.
[433] S. A. R. Knox, R. P. Phillips, and F. G. A. Stone, *J.C.S. Chem. Comm.*, 1972, 1227.
[434] A. N. Nesmeyanov, L. G. Makarova, and I. V. Polovyanyuk, *Izvest. Akad. Nauk S.S.S.R., Ser. khim.*, 1972, 607.
[435] D. W. Lichtenberg and A. Wojcicki, *J. Amer. Chem. Soc.*, 1972, **94**, 8271.
[436] W. P. Giering, S. Raghu, M. Rosenblum, A. Cutler, D. Ehntholt, and R. W. Fish, *J. Amer. Chem. Soc.*, 1972, **94**, 8251.
[437] R. C. Dobbie, P. R. Mason, and R. J. Porter, *J.C.S. Chem. Comm.*, 1972, 612.
[438] M. Y. Darensbourg, *J. Organometallic Chem.*, 1972, **38**, 133.
[439] Y. Yamamoto and A. Wojcicki, *J.C.S. Chem. Comm.*, 1972, 1088.
[440] S. R. Stobart, *J.C.S. Dalton*, 1972, 2442.
[441] R. J. Haines and A. L. du Preez, *Inorg. Chem.*, 1972, **11**, 330.
[442] J. A. Ferguson and T. J. Meyer, *Inorg. Chem.*, 1972, **11**, 631.
[443] R. C. Edmondson, E. Eisner, M. J. Newlands, and L. K. Thompson, *J. Organometallic Chem.*, 1972, **35**, 119.
[444] H. Brunner and M. Vogel, *J. Organometallic Chem.*, 1972, **35**, 169.

(70) [Fe-Ge-Fe structure with Me₂, CO, C₅H₅, C=O bridge]

(71) [similar Fe-Ge-Fe structure]

(73) $(C_5H_5)Fe(CO)_2$—[pyrrolinone ring]—Tosyl

(72) [Ru-Ge-Ru structure with Me₂, CO, GeMe₃]

(74) $(C_5H_5)Ni$=$Fe(C_5H_5)L$ with two C=O

$SnCl_2\{Ni(CO)(C_5H_5)\}$,[445] $(C_5H_5)Ru(CO)_2X$,[446] $[Me_3SiFe(COSiMe_3)(CO)_3]$,[447] and $Fe(CO)_4CH_2CH_2CH_2SiMe_2$.[448]

The ¹H n.m.r. spectrum of (75) is complex but can be interpreted as being consistent with this structure.[449] The binding of ¹³CO to a variety of haemoglobins has been investigated by ¹³C n.m.r. spectroscopy and in a number of cases two ¹³C signals were observed.[450] Data have also been reported for $[Fe(CO)_2(NO)L]^+$, $Fe(CO)(NO)(diphos)CO_2Me$,[451] (76),[452] $(C_{23}H_{18}N_2)Fe_2(CO)_6$,[453] and $Fe(CO)(NO)_2C(OEt)NR_2$.[454]

(75) [vinyl chloride Fe complex with ON, CO, PPh₃]

(76) [PhH-N-N bridged bis-Fe(CO)₃ with $(CO_2Me)_2$ groups]

[445] K. Yasufuku and H. Yamazaki, *J. Organometallic Chem.*, 1972, **38**, 367.
[446] R. J. Haines and A. L. du Preez, *J.C.S. Dalton*, 1972, 944.
[447] M. A. Nasta, A. G. MacDiarmid, and F. E. Saalfeld, *J. Amer. Chem. Soc.*, 1972, **94** 2449.
[448] C. S. Cundy and M. F. Lappert, *J.C.S. Chem. Comm.*, 1972, 445.
[449] G. Cadraci, S. M. Murgia, and A. Foffani, *J. Organometallic Chem.*, 1972, **36**, C11.
[450] R. B. Moon and J. H. Richards, *J. Amer. Chem. Soc.*, 1972, **94**, 5093.
[451] B. F. G. Johnson and J. A. Segal, *J.C.S. Dalton*, 1972, 1268.
[452] H. Kisch, *J. Organometallic Chem.*, 1972, **38**, C19.
[453] D. P. Madden, A. J. Carty, and T. Birchall, *Inorg. Chem.*, 1972, **11**, 1453.
[454] E. O. Fischer, F. R. Kreissl, E. Winkler, and C. G. Kreiter, *Chem. Ber.*, 1972, **105**, 588.

The ^{1}H n.m.r. spectrum of $Fe(CO)_3(PMe_2CH_2CH_2SiMe_3)_2$ shows a doublet for the PMe_2 group,[455] whereas for (77), when $R^1 \neq R^2$ there are two isomers found.[456] The ^{1}H and ^{31}P n.m.r. spectra of trans-$P(OCH_2)_3$-$PFe(CO)_3P(OCH_2)_3P$ have been measured and $^2J(^{31}P-^{31}P) = 38.0$ Hz and $^3J(^{31}P-^{31}P) = 9.0$ Hz.[457] Similarly, ^{1}H n.m.r. spectroscopy shows that $[Fe(CO)_2P(OR)_3SPh]_2$ exists in solution as a mixture of two species. When R = Me, the spectrum shows a triplet and a doublet and structures (78) and (79) were suggested.[458] The ^{19}F n.m.r. spectrum of $[F_2PFe(CO)_3]_2$ shows that there are two types of fluorine,[459] and the ^{1}H n.m.r. spectrum of $Fe_3(CO)_6(Ph_2AsC_2CF_3)_2\{P(OMe)_3\}_2$ shows two $P(OMe)_3$ resonances.[460]

$$\begin{array}{c} R^1 \diagdown \quad \diagup R^2 \\ C=C \\ (C_5H_5)Ni \diagdown \quad \diagup Fe(CO)_3 \\ P \\ Ph_2 \end{array}$$

(77)

(78) $(RO)_3P-Fe\underset{CO}{\overset{Ph}{\underset{S}{\diagup}}}\underset{S}{\overset{Ph}{\diagdown}}Fe-P(OR)_3$ with OC, CO ligands

(79) analogous structure with $P(OR)_3$

Data have also been reported for $Fe_2(CO)_5L(SBu^t)_2$, $[Fe(CO)_2P(OMe)_3$-$SBu^t]_2$ (virtually coupled triplet for the OMe resonance),[461] $(C_5H_5)Fe_2$-$SR(CO)_6$, $(C_5H_5)Fe_2PPh_2(CO)_5$ (two C_5H_5 resonances),[462] $Fe_n(CO)_{4n-x}L_x$,[463] and $Fe_2(CO)_5L(COR)_2$.[464]

$(C_5H_5)Ru(PPh_3)_2CH_2Ph$ reacts with $CF_3C\equiv CCF_3$ to yield $(C_5H_5)Ru$-$(PPh_3)\{C_2(CF_3)_2\}CH_2Ph$, and the ^{1}H and ^{19}F n.m.r. spectra were considered to be consistent with (80).[465] In $Ru(CO)(CH=NR)(PPh_3)(S_2CNEt_2)$, the ethyl groups are inequivalent owing to restricted rotation about the

---

[455] J. Grobe and U. Möller, *J. Organometallic Chem.*, 1972, **36**, 335.
[456] K. Yasufuku and H. Yamazaki, *J. Organometallic Chem.*, 1972, **35**, 367.
[457] D. A. Allison, J. Clardy, and J. G. Verkade, *Inorg. Chem.*, 1972, **11**, 2804.
[458] J. A. De Beer and R. J. Haines, *J. Organometallic Chem.*, 1972, **36**, 297.
[459] W. M. Douglas and J. K. Ruff, *Inorg. Chem.*, 1972, **11**, 901.
[460] T. O'Connor, A. J. Carty, M. Mathew, and G. J. Palenik, *J. Organometallic Chem.*, 1972, **38**, C15.
[461] J. A. De Beer and R. J. Haines, *J. Organometallic Chem.*, 1972, **37**, 173.
[462] R. J. Haines and C. R. Nolte, *J. Organometallic Chem.*, 1972, **36**, 163.
[463] P. M. Treichel, W. K. Dean, and W. M. Douglas, *Inorg. Chem.*, 1972, **11**, 1609.
[464] V. Kliener and E. O. Fischer, *J. Organometallic Chem.*, 1972, **42**, 447.
[465] M. I. Bruce, R. C. F. Gardner, and F. G. A. Stone, *J. Organometallic Chem.*, 1972, **40**, C39.

carbon–nitrogen bond.[466] Data have also been reported on (olefin)-$Ru(PPh_3)_3$,[467] $RuCl_2(CO)_2\{PMe(CH_2SiMe_3)_2\}_2$,[468] $[Ru(CO)_2PPh_3I_3]$,[469] $Ru(PF_3)_n(CO)_{5-n}$ (^{19}F n.m.r. data),[470] $RuPt_2(CO)_5(PMe_2Ph)_3$,[471] $Ru_3(CO)_7$-$(PPh)_2(C_6H_4)$,[472] $(C_6H_6)RuCl_2L$,[473–475] $[(C_6H_6)(C_6H_8)RuClHg_3ClRu$-$(C_6H_6)(C_6H_8)]Cl_4$,[475] and $[Fe(CO)_3EMe]_2$ (E = S or Se).[476, 477]

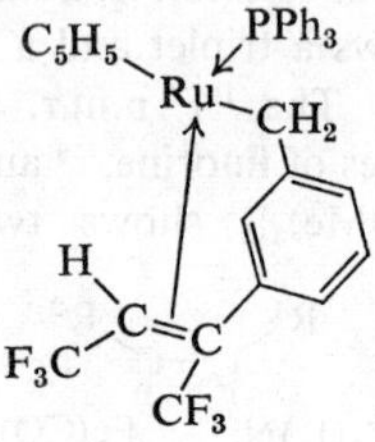

(80)

It has been shown that previous assignments for $(MeOC_5H_4)Fe(C_5H_5)$ are wrong and new assignments are given;[478] it was then possible to follow the deuteriation of this and related compounds.[479] The 220 MHz n.m.r. spectra of $RC_5H_4FeC_5H_5$ (R = Me, Et, Pr, or Bu^t) have been reported. It was found that the predominant shielding effect is due to the 3,4-protons, contrary to previous predictions and assignments.[480] A comparative study of the ^{1}H n.m.r. spectra of some ferrocenes and cobalticinium salts has been carried out. The positive charge on $[(C_5H_5)_2Co]^+$ leads to a decrease in the ring-proton shielding and substituent effects were examined.[481] Comparison of the ^{1}H n.m.r. spectra of epimeric compounds in a series of homoannularly bridged [3](1,2)ferrocenophane derivatives has established that the protons of an *exo* substituent are shielded compared with the protons of the same substituent in an *endo* configuration.[482] Data have also been reported for a

---

[466] D. F. Christian, G. R. Clarke, W. R. Roper, J. M. Waters, and K. R. Whittle, *J.C.S. Chem. Comm.*, 1972, 458.
[467] S. Komiya, A. Yamamoto, and S. Ikeda, *J. Organometallic Chem.*, 1972, 42, C65.
[468] A. T. T. Hsieh, J. D. Ruddick, and G. Wilkinson, *J.C.S. Dalton*, 1972, 1966.
[469] J. Jeffery and R. J. Mawby, *J. Organometallic Chem.*, 1972, 40, C42.
[470] C. A. Udovich and R. J. Clark, *J. Organometallic Chem.*, 1972, 36, 355.
[471] M. I. Bruce, G. Shaw, and F. G. A. Stone, *J.C.S. Dalton*, 1972, 1781.
[472] M. I. Bruce, G. Shaw, and F. G. A. Stone, *J.C.S. Dalton*, 1972, 2094.
[473] R. A. Zelonka and M. C. Baird, *J. Organometallic Chem.*, 1972, 35, C43.
[474] M. A. Bennett, G. B. Robertson, and A. K. Smith, *J. Organometallic Chem.*, 1972, 43, C41.
[475] R. A. Zelonka and M. C. Baird, *Canad. J. Chem.*, 1972, 50, 3063.
[476] H. Büttner and R. D. Feltham, *Inorg. Chem.*, 1972, 11, 971.
[477] P. Rosenbuch and N. Welcman, *J.C.S. Dalton*, 1972, 1963.
[478] D. W. Slocum and C. R. Ernst, *Tetrahedron Letters*, 1972, 5217.
[479] D. W. Slocum, B. P. Koonsvitsky, and C. R. Ernst, *J. Organometallic Chem.*, 1972, 38, 125.
[480] D. W. Slocum, W. E. Jones, and C. R. Ernst, *J. Org. Chem.*, 1972, 37, 4278.
[481] A. N. Nesmeyanov, N. S. Kochetkova, E. V. Leonova, E. I. Fedin, and P. V. Petrovskii, *J. Organometallic Chem.*, 1972, 39, 173.
[482] T. D. Turbitt and W. E. Watts, *Tetrahedron*, 1972, 28, 1227.

wide range of ferrocene derivatives,[483-508] $Fe(C_5H_4PPh_2)_2Pt(CH_2CD_2Et)_2$,[509] $(C_7H_9)_2Fe$,[510] $[(RC_6H_5)Fe(C_5H_5)]^+$,[511] (81),[512] $Me_3GeSiMe_2C_5H_4FeC_5H_4SiMe_2OEt$,[513] and $[(C_5H_5)_2RuHgRu(C_5H_5)_2]^{2+}$.[514]

(81)

Iron(II) phthalocyanine forms 1:2 complexes with amines, which show large shifts of up to τ 19 and hence can be used as shift reagents.[515] ^1H N.m.r. spectroscopy has been used to assign the stereochemistry of cis- and

[483] D. R. Morris and B. W. Rockett, *J. Organometallic Chem.*, 1972, **35**, 179.
[484] E. W. Neuse and R. K. Crossland, *J. Organometallic Chem.*, 1972, **43**, 385.
[485] M. Kumada, T. Kondo, K. Mimura, M. Ishikawa, K. Yamamoto, S. Ikeda, and M. Kondo, *J.C.S. Chem. Comm.*, 1972, **43**, 293.
[486] L. H. Ali, A. Cox, and T. J. Kemp, *J.C.S. Chem. Comm.*, 1972, 265.
[487] P. Dixneuf and R. Dabard, *J. Organometallic Chem.*, 1972, **37**, 167.
[488] R. F. Kovar and M. D. Rausch, *J. Organometallic Chem.*, 1972, **35**, 351.
[489] J. A. Winstead, R. R. McGuire, R. E. Cochoy, A. D. Brown, jun., and G. J. Gauthier, *J. Org. Chem.*, 1972, **37**, 2055.
[490] C. Moise, J.-P. Monin, and J. Tirouflet, *Bull. Soc. chim. France*, 1972, 2048.
[491] P. Dixneuf and R. Dabard, *Bull. Soc. chim. France*, 1972, 2847.
[492] E. W. Neuse, *J. Organometallic Chem.*, 1972, **40**, 387.
[493] D. R. Morris and B. W. Rockett, *J. Organometallic Chem.*, 1972, **40**, C21.
[494] I. Agranat, M. Rabinovitz, M. Weissman, and M. R. Pick, *Tetrahedron Letters*, 1972, 3379.
[495] H. L. Lentzner and W. E. Watts, *Tetrahedron*, 1972, **28**, 121.
[496] S. Allenmark, *Tetrahedron Letters*, 1972, 2885.
[497] T. D. Turbitt and W. E. Watts, *J. Organometallic Chem.*, 1972, **46**, 109.
[498] Y. Omote, T. Komatsu, R. Kobayashi, and N. Sugiyama, *Nippon Kagaku Kaishi*, 1972, 780.
[499] D. Touchard and R. Dabard, *Tetrahedron Letters*, 1972, **49**, 5005.
[500] J. B. Evans and G. Marr, *J.C.S. Perkin I*, 1972, 2502.
[501] W. M. Horspool, S. T. McNeilly, J. A. Miller, and I. M. Young, *J.C.S. Perkin I*, 1972, 1113.
[502] A. Eisenstadt and M. Cais, *J.C.S. Chem. Comm.*, 1972, 216.
[503] F. H. Hon and T. T. Tidwell, *J. Org. Chem.*, 1972, **37**, 1782.
[504] G. Bigam, J. Hooz, S. Linke, R. E. D. McClung, M. W. Mosher, and D. D. Tanner, *Canad. J. Chem.*, 1972, **50**, 1825.
[505] P. Dixneuf and R. Dabard, *Bull. Soc. chim. France*, 1972, 2838.
[506] P. V. Rowling, and M. D. Rausch, *J. Org. Chem.*, 1972, **37**, 729.
[507] J. J. McDonnell and D. J. Pochopien, *J. Org. Chem.*, 1972, **37**, 4064.
[508] U. Mueller-Westerhoff, *Tetrahedron Letters*, 1972, 4639.
[509] G. M. Whitesides, J. F. Gaasch, and E. R. Stedronsky, *J. Amer. Chem. Soc.*, 1972, **94**, 5258.
[510] J. Müller and B. Metrschenk, *Chem. Ber.*, 1972, **105**, 3346.
[511] C. C. Lee, R. G. Sutherland, and B. J. Thomson, *J.C.S. Chem. Comm.*, 1972, 907.
[512] T. J. Katz, N. Acton, and J. McGinnis, *J. Amer. Chem. Soc.*, 1972, **94**, 6205.
[513] M. Kumada, T. Kondo, K. Mimura, K. Yamamoto, and M. Ishikawa, *J. Organometallic Chem.*, 1972, **43**, 307.
[514] D. N. Hendrickson, Y. S. Sohn, W. H. Morrison, jun., and H. B. Gray, *Inorg. Chem.*, 1972, **11**, 808.
[515] J. E. Maskasky, J. R. Mooney, and M. E. Kenney, *J. Amer. Chem. Soc.*, 1972, **94**, 2132.

trans-Ru(py)$_4$Cl$_2$. When *trans*, only one α-hydrogen resonance is observed, but when *cis*, two α-hydrogen resonances appear.[516] The ^{1}H and ^{19}F n.m.r. spectra of [Ru(NH$_3$)$_5$RCN]$^{2+}$ and [Rh(NH$_3$)$_5$RCN]$^{3+}$ have been fully analysed and it was concluded that for the ruthenium complex there is ligand π-back bonding.[517] The ^{1}H n.m.r. spectra of [Fe(terpy)$_2$]$^{2+}$ and [Co(terpy)$_2$]$^{3+}$ have also been analysed.[518] It has been concluded from the ^{1}H n.m.r. spectra of [(NH$_3$)$_5$RuL]$^{2+}$ (L = nitrogen base) that there is a net positive charge on the ligands owing to complexation.[519] Data have also been reported for Na$_2$[Fe(HN=CN—CH=NH)(CN)$_4$],4H$_2$O,[520] [RuClN$_2$-{o-(Me$_2$As)$_2$C$_6$H$_4$}$_2$]$^+$,[521] [{MeC(=NOH)C(=NOH)—R—C(=NOH)C-(=NOH)Me}Fe(py)$_2$]$_n$,[522] $\overline{\text{-(CH}_2\text{CMe}_2\text{NHCH}_2\text{CH}_2\text{NHCHMe)}_2}$ FeX$_2$,[523] [Fe(Me$_6$[14]1,4,8,11-tetraene-N$_4$)(MeCN)$_2$]$^{2+}$,[524] Ru(CO)(octaethylporphyrin)pyridinate,[525] Fe^{II} porphyrins,[526] (di-2-pyridyl ketone)Fe(NO)$_2$,[527] and *cis*-Os(N$_3$)$_2$(PMe$_2$Ph)$_4$.[528]

^{1}H N.m.r. spectroscopy has been used to investigate the structure of modified haemoglobins.[529] 250 MHz ^{1}H n.m.r. spectroscopy provides evidence that an aliphatic amino-acid is situated very close to the distal side of the haem plane in *Chironomus* Hb and *Aplysia* myoglobin,[530] and has been used to demonstrate that the haem environments of the α- and β-chains in carbonmonoxyhaemoglobin A are not equivalent.[531] ^{19}F N.m.r spectroscopy has been used to study the interaction of haemoglobin with ligands and allosteric effectors. Oxidation of the haem iron produced significant changes which were discussed in the light of the crystal structure.[532] ^{13}C N.m.r. spectroscopy has been used to investigate conformations of various haemoglobins.[533]

^{1}H N.m.r. data for a series of complexes, {(XC$_6$H$_4$)EtNCS$_2$}$_3$Fe, have been successfully interpreted by assuming that the iron atom is in an intermediate crystal field of octahedral symmetry. These results have been

[516] D. W. Raichart and H. Taube, *Inorg. Chem.*, 1972, **11**, 999.
[517] R. D. Foust, jun. and P. C. Ford, *J. Amer. Chem. Soc.*, 1972, **94**, 5686.
[518] H. Elsbrand and J. K. Beattie, *J. Inorg. Nuclear Chem.*, 1972, **34**, 771.
[519] D. K. Lavalee and E. B. Fleischer, *J. Amer. Chem. Soc.*, 1972, **94**, 2583.
[520] V. L. Goedken, *J.C.S. Chem. Comm.*, 1972, 207.
[521] P. G. Douglas and R. D. Feltham, *J. Amer. Chem. Soc.*, 1972, **94**, 5254.
[522] J. Backes, I. Masuda, and K. Shinra, *Bull. Chem. Soc. Japan*, 1972, **45**, 1724.
[523] J. C. Dabrowiak, P. H. Merrell, and D. H. Busch, *Inorg. Chem.*, 1972, **11**, 1979.
[524] V. L. Goedken and D. H. Busch, *J. Amer. Chem. Soc.*, 1972, **94**, 7355.
[525] G. W. Sovocool, F. R. Hopf, and D. G. Whitten, *J. Amer. Chem. Soc.*, 1972, **94**, 4350.
[526] R. Bonnett and M. J. Dimsdale, *J.C.S. Perkin I*, 1972, 2540.
[527] R. E. Dessy, J. C. Charkoudian, and A. L. Rheingold, *J. Amer. Chem. Soc.*, 1972, **94**, 738.
[528] B. Bell, J. Chatt, J. R. Dilworth, and G. J. Leigh, *Inorg. Chim. Acta*, 1972, **6**, 635.
[529] K. H. Winterhalter and K. Wuethrich, *J. Mol. Biol.*, 1972, **63**, 477.
[530] K. Wuethrich, R. M. Keller, M. Brunori, G. Giacometti, R. Huber, and H. Formanek, *F.E.B.S. Letters*, 1972, **21**, 63.
[531] T. R. Lindstrom, I. B. E. Noren, S. Charache, H. Lehmann, and C. Ho, *Biochemistry*, 1972, **11**, 1677.
[532] W. H. Huestis and M. A. Raftery, *Biochemistry*, 1972, **11**, 1648.
[533] R. B. Moon and J. H. Richards, *Proc. Nat. Acad. Sci. U.S.A.*, 1972, **69**, 2193.

interpreted as arising from changes in the π character of the nitrogen–phenyl bond owing to the degree of hindered rotation with aromatic substitution.[534] Data have also been reported for $Fe(PF_3)_4X_2$,[535] $[Ru(hfac)_3]^-$,[536] $[Ru_3O(O_2CR)_6L_3]^+$,[537] and $(RO)_2OsO_2L_2$.[538] The ^1H n.m.r. spectrum of $RuCl_2(Ph_2AsCH_2AsPh_2)_3$ shows two CH_2 signals in the ratio 2:1. Therefore there are two unidentate and one bidentate ligands.[539]

Complexes of Co, Rh, and Ir.—Information concerning complexes of these elements can be found at the following sources: $(C_5H_5)Fe(C_2Ph)Co_2(CO)_6$,[194] $(C_8H_6)_2Co_2$,[512] $(C_7H_9)(C_7H_{10})Co$,[510] $[(C_5H_5)_2Co]^+$,[481] $[Rh(CO)Cl(C_6H_4CH=NR)]_2$,[331] $(C_4F_7)Co(CO)(C_5H_5)C_2F_5$,[251] $IrHCl_2(PBu^t_2R)_2$,[91] $Rh(CO)Cl(PMe_2CH_2Ph)_2$,[330] $Rh(C_3H_6)_2(acac)$,[426] $(C_5H_5)M(PF_3)_2$ (M = Rh or Ir),[313] $IrClIMe(CO)(AsEt_2Ph)_2$,[113] $(C_{23}H_{18}N_2)Rh(CO)_2Cl$,[453] $RhCl_3(CHNMe_2)(PEt_3)_2$,[249] $MCl_3\{PMe(CH_2SiMe_3)_2\}_3$ (M = Rh or Ir),[468] $[Co(NH_3)_5X]^{2+}$,[124, 150] $[Co(terpy)_3]^{3+}$,[518] $[Rh(NH_3)_5RCN]^{3+}$,[517] $[Co(tetraphenylprophyrin)]^-$,[184] cobalt porphyrins,[526] $Co(NO)_2(PR_3)X$,[151] MCl_3L_3 (M = Rh or Ir)[132] $Co(S_2CNMe_2)_3$,[476] $IrCl_2X(PMe_2Ph)(Ph_2PCH_2CH_2PPh_2)$,[84] $MeCo(dmg)_2OH_2$,[89] and $HIr(piperidine)_4(NCE)_2$ (E = O, S, or Se).[90]

^1H N.m.r. spectroscopy shows a 1:2:1 triplet for the hydride resonance and a 1:4:6:4:1 quintet for the methyl group in $[Co(bipy)(PEt_3)_2H_2]ClO_4$, demonstrating that the tertiary phosphines are mutually *trans*;[540] $[Rh(bipy)(PPh_3)_2H_2]^+$ is similar.[541] Deuteriation of the *o*-hydrogen in $[(PhO)_3P]_4RhH$ has been followed by ^1H n.m.r. spectroscopy.[542] ^{31}P N.m.r. spectroscopy has been used to show that $Rh(PPh_3)_3Cl$ reacts with H_2 to yield $RhH_2Cl(PPh_3)_2$.[543] Data have also been reported for $[RhHX(cis-Ph_2PCH=CHPPh_2)]^+$,[544] $[RhHCl\{PPh_2(OMe)\}_4]^+$, $[IrH_2(CO)(PPh_3)_3]^+$,[545, 546] $IrH(CO)L(PPh_3)_2$,[546] $IrH_2(PPh_3)_2(S_2CNEt_2)$,[547] and $IrH_2X(p\text{-tol-NC})(Ph_3As)_2$.[548]

trans-$IrCl(CO)(PPh_3)$ adds $SnMe_3H$ to give a complex with 14 lines in the hydride spectrum, and the product was thought to be $IrH_2(SnMe_3)(CO)(PPh_3)_2$.[549] Extensive use has been made of ^1H and ^{31}P n.m.r. spectroscopy to characterize compounds of the type $IrCl_2H(PR_3)_2$ and adducts

---

[534] R. M. Golding, B. D. Lukeman, and E. Sinn, *J. Chem. Phys.*, 1972, **56**, 4147.
[535] T. H. Kruck, R. Kobelt, and A. Prasch, *Z. Naturforsch.*, 1972, **27b**, 344.
[536] G. S. Patterson and R. H. Holm, *Inorg. Chem.*, 1972, **11**, 2285.
[537] A. Spencer and G. Wilkinson, *J.C.S. Dalton*, 1972, 1570.
[538] L. R. Subbaraman, J. Subbaraman, and E. J. Behrman, *Inorg. Chem.*, 1972, **11**, 2621.
[539] J. T. Mague and J. P. Mitchener, *Inorg. Chem.*, 1972, **11**, 2714.
[540] A. Camus, C. Cocevar, and G. Mestroni, *J. Organometallic Chem.*, 1972, **39**, 355.
[541] I. I. Bhayat and W. R. McWhinnie, *J. Organometallic Chem.*, 1972, **46**, 159.
[542] E. K. Barfield and G. W. Parshall, *Inorg. Chem.*, 1972, **11**, 964.
[543] P. Meakin, J. P. Jesson, and C. A. Tolman, *J. Amer. Chem. Soc.*, 1972, **94**, 3240.
[544] J. T. Mague, *Inorg. Chem.*, 1972, **11**, 2558.
[545] P.-C. Kong and D. M. Roundhill, *Inorg. Chem.*, 1972, **11**, 1437.
[546] M. S. Fraser and W. H. Baddley, *J. Organometallic Chem.*, 1972, **36**, 377.
[547] A. Aràneo and T. Napoletano, *Inorg. Chim. Acta*, 1972, **6**, 363.
[548] A. Aràneo, T. Napoletano, and P. Fantucci, *J. Organometallic Chem.*, 1972, **42**, 471.
[549] F. Glocking and J. G. Irwin, *Inorg. Chim. Acta*, 1972, **6**, 355.

with small ligands.[550] Data have also been reported for $IrClH(PPh_3)_2$-$(PMe_2Ph)(CO_2Ph)$,[551] $IrHX(SC_6H_4Y)(CO)(PPh_3)_2$,[552] $IrH_2X(CO)L_2$, and $IrXMeI(CO)L_2$.[553]

^1H N.m.r. spectroscopy has been used to investigate reactions of $IrCl_2(CH_2CH=CH_2)L_2$ and the spectrum of $IrCl_2(CH_2CHOHMe)(CO)$-$(PMe_2Ph)_2$ has been analysed with the aid of $Eu(fod)_3$.[554] Products from the reaction of $Rh_2Cl_2(CO)_4$ with cyclopropanes have been characterized,[555] and products from the metallation of PMe_2(1-naphthyl) and $PMePh$-(1-naphthyl) by rhodium and iridium have been identified by ^1H and ^{31}P n.m.r. spectroscopy.[556] For (82; R = C_6H_4F), if the ^{19}F chemical shifts are plotted against analogous data for $FC_6H_4Pt(PEt_3)_2X$, a linear relationship results from which it was concluded that there is the same *trans* effect.[557] A linear correlation has also been found between the methyl chemical shift and the pK_a of the base for $MeCo(dmg)_2L$.[558] Data have also been reported for (1-adamantyl)pentacyanocobaltate,[559] [MeCo(2,3,9,10-tetramethyl-1,4,8,11-tetra-azacyclotetradeca-1,3,8,10-tetraene)X]$^+$,[560] (83),[561] [RIrCl_3(CO)]_2,[562] MeCORh(tetraphenylporphyrin),[563] MeRh(PPh_3)_3,[564]

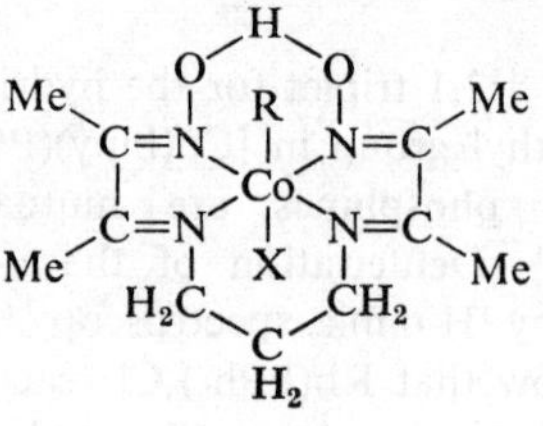

(82)

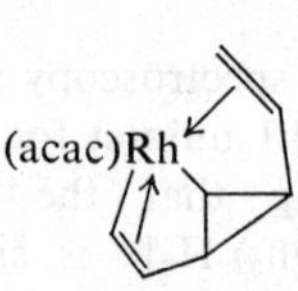

(84)

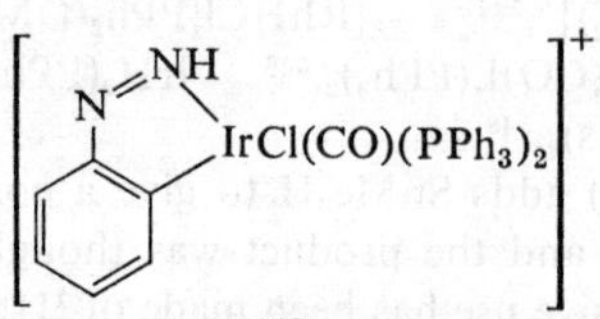

(83)

[550] B. L. Shaw and R. E. Stainbank, *J.C.S. Dalton*, 1972, 2108.
[551] S. A. Smith, D. M. Blake, and M. Kubota, *Inorg. Chem.*, 1972, **11**, 660.
[552] J. R. Gaylor and C. V. Senoff, *Canad. J. Chem.*, 1972, **50**, 1868.
[553] B. L. Shaw and R. E. Stainbank, *J.C.S. Dalton*, 1972, 223.
[554] J. M. Duff, B. E. Mann, E. M. Miller, and B. L. Shaw, *J.C.S. Dalton*, 1972, 2337.
[555] F. J. McQuillin and K. C. Powell, *J.C.S. Dalton*, 1972, 2129.
[556] J. M. Duff and B. L. Shaw, *J.C.S. Dalton*, 1972, 2219.
[557] H. A. O. Hill, K. G. Morallee, F. Cervinez, and G. Pellizer, *J. Amer. Chem. Soc.*, 1972, **94**, 277.
[558] C. Bied-Charreton, L. Alais, and A. Gaudemer, *Bull. Soc. chim. France*, 1972, 861.
[559] S. H. Goh and L.-Y. Goh, *J. Organometallic Chem.*, 1972, **43**, 401.
[560] K. Farmery and D. H. Busch, *Inorg. Chem.*, 1972, **11**, 2901.
[561] F. W. B. Einstein, A. B. Gilchrist, G. W. Rayner-Canham, and D. Sutton, *J. Amer. Chem. Soc.*, 1972, **94**, 645.
[562] M. A. Bennett and R. Charles, *J. Amer. Chem. Soc.*, 1972, **94**, 666.
[563] B. R. James and D. V. Stynes, *J.C.S. Chem. Comm.*, 1972, 1261.
[564] C. S. Cundy, C. Eaborn, and M. F. Lappert, *J. Organometallic Chem.*, 1972, **44**, 291.

(84),[565] Rh(acac)IMe(PPh$_3$)$_2$,[566] [(HON=CRC$_6$H$_4$)$_2$RhX]$_2$,[567] (PhN=CHC$_6$H$_4$)RhCl$_2$(PhNH$_2$)$_2$,[568] (PhN=NC$_6$H$_4$)$_2$Rh(μ-Cl)Rh(CO)$_2$,[569] RCo(dmg)$_2$L,[570-577] and RRh(salen)py.[578]

Cu$_4$Ir$_2$(PMePh$_2$)$_2$(C≡CPh)$_8$ shows an apparent triplet for the methyl resonances, implying that $J(^{31}P-^{31}P) > 50$ Hz, and the structure is thought to be (85).[579] When PhFHCCHXBr reacts with Ir(PMe$_3$)$_2$Cl(CO), racemization occurs.[580] Data have also been reported for Ir(C≡CPh)(CO)(PPh$_3$)$_n$,[581] CF$_3$HC=CCF$_3$Rh(CO)(PPh$_3$)$_2$, (^{1}H and ^{19}F n.m.r.),[582] (86),[583] (Me$_5$C$_5$)Rh(PF$_3$)R$_F$I,[584] and (ON)(Ph$_3$P)Ir(CF$_3$C=CCF$_3$)$_2$Ir(NO)(PPh$_3$).[585]

Although the ^{13}C n.m.r. spectrum of (H$_3$N)$_4$RhXC—NHCR^1=CR^{2}NH has been recorded, the metal-bound carbon was not observed.[586] The ^{1}H

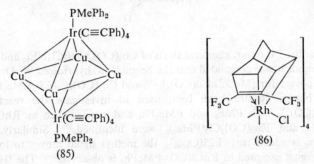

[565] V. Aris, J. M. Brown, and B. T. Golding, *J.C.S. Chem. Comm.*, 1972, 1206.
[566] D. M. Barlex, M. J. Hacker, and R. D. W. Kemmitt, *J. Organometallic Chem.*, 1972, **43**, 425.
[567] H. Onoue and I. Moritani, *J. Organometallic Chem.*, 1972, **44**, 189.
[568] I. Jardine and F. J. McQuillin, *Tetrahedron Letters*, 1972, 459.
[569] M. I. Bruce, M. Z. Iqbal, and F. G. A. Stone, *J. Organometallic Chem.*, 1972, **40**, 393.
[570] B. T. Golding and S. Sakrikar, *J.C.S. Chem. Comm.*, 1972, 1183.
[571] C. Fontaine, K. N. V. Duong, C. Merienne, A. Gaudemer, and C. Giannotti, *J. Organometallic Chem.*, 1972, **38**, 167.
[572] A. Gaudemer, F. Gaudemer, and L. Diep, *Bull. Soc. chim. France*, 1972, 886.
[573] C. Giannotti, C. Fontaine, and A. Gaudemer, *J. Organometallic Chem.*, 1972, **39**, 381.
[574] C. Giannotti, C. Fontaine, B. Septe, and D. Doue, *J. Organometallic Chem.*, 1972, **39**, C74.
[575] S. N. Anderson, D. H. Ballard, and M. D. Johnson, *J.C.S. Perkin II*, 1972, 311.
[576] R. B. Silverman, D. Dolphin, and B. M. Babior, *J. Amer. Chem. Soc.*, 1972, **94**, 4028.
[577] C. Giannotti, B. Septe, and D. Benlian, *J. Organometallic Chem.*, 1972, **39**, C5.
[578] R. J. Cozens, K. S. Murray, and B. O. West, *J. Organometallic Chem.*, 1972, **38**, 391.
[579] O. M. Abu Salah, M. I. Bruce, M. R. Churchill, and S. A. Bezman, *J.C.S. Chem. Comm.*, 1972, 858.
[580] J. S. Bradley, D. E. Connor, D. Dolphin, J. A. Labinger, and J. A. Osborn, *J. Amer. Chem. Soc.*, 1972, **94**, 4043.
[581] R. Nast and L. Dahlenburg, *Chem. Ber.*, 1972, **105**, 1456.
[582] B. L. Booth and A. D. Lloyd, *J. Organometallic Chem.*, 1972, **35**, 195.
[583] J. A. Evans, R. D. W. Kemmitt, B. Y. Kimura, and D. R. Russell, *J.C.S. Chem. Comm.*, 1972, 509.
[584] R. B. King and A. Efraty, *J. Organometallic Chem.*, 1972, **36**, 371.
[585] J. Clemens, M. Green, M.-C. Kuo, C. J. Fritchie, jun., J. T. Mague, and F. G. A. Stone, *J.C.S. Chem. Comm.*, 1972, 53.
[586] R. J. Sundberg, R. E. Shepherd, and H. Taube, *J. Amer. Chem. Soc.*, 1972, **94**, 6558.

n.m.r. spectrum of $[(Bu^tNC)_3RhC(NBu^tH)(NBu^iH)]^+$ shows two NH resonances and three t-butyl resonances. It was suggested that there is isomerism about the C—N bond.[587] The ^1H chemical shift and C—CH$_3$ bond length of MeCCo$_3$(CO)$_9$ have been interpreted in terms of 'much "s" character' in the C—CH$_3$ bond.[588] Data have also been reported for L(Ph$_3$P)RhClC(NRCH$_2$)$_2$,[589] [Rh$_6$(CO)$_{15}$COEt]$^-$,[590, 591] RCCo$_3$(CO)$_9$,[592] R$_3$NBH$_2$OCCo$_3$(CO)$_9$ (^{11}B n.m.r. data),[593] and (87).[594]

$$\begin{array}{c} SiMe_2R \\ | \\ C \\ RMe_2Si-C \diagdown \diagup Co(C_5H_5) \\ (C_5H_5)Co-C \\ \| \\ O \end{array}$$

(87)

From the ^{11}B n.m.r. chemical shifts of Co$_3$(CO)$_9$COBCl$_2$NEt$_3$ and related compounds it was concluded that the boron is sp^3 hybridized.[595] Data were also reported for [MeOZnCo(CO)$_4$]$_4$ [596] and Co$_2$(CO)$_4$C$_6$(CF$_3$)$_2$Me$_2$H$_2$.[597]

^1H N.m.r. spectra have been used to investigate the reaction of [RhCl(CO)$_2$]$_2$ with PMe$_3$ and PMe$_2$Ph, and species such as RhCl(CO)(PMe$_3$)$_3$ and Rh$_2$(CO)$_2$Cl$_2$(PMe$_3$)$_2$ were identified.[598] Similarly, when PMePh$_2$ is added to [RhCl(CO)$_2$]$_2$, the methyl signal moves to low field and a signal assigned to RhCl(CO)$_2$PMePh$_2$ is observed.[599] The ^1H n.m.r. spectrum of RhCl(CO)(PPh$_3$)(PMePh$_2$) shows only a singlet for the methyl group and this was attributed to exchange.[600]

The ^{14}N n.m.r. spectrum of compounds such as (Ph$_3$P)$_2$Rh(CO)N$_3$ shows three signals.[601] Treatment of [RhCl(CO)$_2$]$_2$ with PF$_3$ in 1:4 ratio gives a mixture showing two ^{19}F resonances which were attributed to [(F$_3$P)$_2$RhCl]$_2$ and [(F$_3$P)(CO)RhCl]$_2$.[602] The ^{17}O n.m.r. spectra of Ir(CO)X(PR$_3$)$_2$$^{17}O_2$ shows no resonance attributable to bound ^{17}O. Similarly, the ^{17}O resonance of IrI(S$^{17}O_4$)(CO)(P-p-tol$_3$)$_2$ could not be

[587] P. R. Branson and M. Green, *J.C.S. Dalton*, 1972, 1303.
[588] M. D. Brice and B. R. Penfold, *Inorg. Chem.*, 1972, **11**, 1381.
[589] D. J. Cardin, M. J. Doyle, and M. F. Lappert, *J.C.S. Dalton*, 1972, 927.
[590] P. Chini, S. Martinengo, and G. Garlaschelli, *J.C.S. Chem. Comm.*, 1972, 709.
[591] P. Chini, S. Martinengo, and G. Giordano, *Gazzetta*, 1972, **102**, 330.
[592] R. Dolby and B. H. Robinson, *J.C.S. Dalton*, 1972, 2046.
[593] G. Schmid and B. Stutte, *J. Organometallic Chem.*, 1972, **37**, 375.
[594] H. Sakurai and J. Hayashi, *J. Organometallic Chem.*, 1972, **39**, 365.
[595] G. Schmid and V. Bätzel, *J. Organometallic Chem.*, 1972, **46**, 149.
[596] J. M. Burlitch and S. E. Hayes, *J. Organometallic Chem.*, 1972, **42**, C13.
[597] R. S. Dickson and P. J. Fraser, *Austral. J. Chem.*, 1972, **25**, 1179.
[598] J. Gallay, D. De Montauzon, and R. Poilblanc, *J. Organometallic Chem.*, 1972, **38**, 179.
[599] L. D. Rollmann, *Inorg. Chim. Acta*, 1972, **6**, 137.
[600] D. F. Steele and T. A. Stephenson, *J.C.S. Dalton*, 1972, 2161.
[601] K. von Werner and W. Beck, *Chem. Ber.*, 1972, **105**, 3209.
[602] J. F. Nixon and J. R. Swain, *J.C.S. Dalton*, 1972, 1044.

detected. It was suggested that Vaska-type compounds tumble at a rate which broadens the resonance beyond detection.[603] Data have also been reported for $(CF_3)_n(NC)_{3-n}PCo(NO)(CO)_2$,[604] $Rh(CO)_2(PPh_3)_2Hg(p\text{-tolyl})$, $Ir(CO)_3PPh_3Hg(benzyl)$,[605] $Ir(CO)Cl(PMePh_2)_2(o\text{-chloranil})$,[606] $RhXL\text{-}\{As(o\text{-}CH_2=CHC_6H_4)_3\}$,[607] $(\text{anisyl-NC})_2Rh(CO)MeI_2$,[608] $IrCl(O_2Bu^t)_2\text{-}(CO)L$,[609] and $L_2(CO)ClIr(CF_3NNCF_3)$.[610]

The arene ^1H n.m.r. spectrum of $[L_2Rh(\text{arene})]^+$ is temperature invariant but when L = ethylene, the ethylene resonances broaden on warming.[611] The effects of R on the ^1H chemical shifts of the chelate ring have been discussed for (88).[612] Data have also been reported for $(HC\equiv CR)Co_2\text{-}(CO)_6$,[613] $Co(C_8H_{11})(C_8H_{12})$,[614] $(C_5H_5)Rh(C_2H_4)PPh_3$,[615] $[Rh(C_2H_4)_3\text{-}(MeCN)_2]^+$,[616] (89),[617] and $Ir\{N=C=C(CN)C(CN)_2H\}\{(NC)_2C=C(CN)_2\}\text{-}(PPh_3)_2$.[618]

(88)   (89)

^1H and ^{19}F N.m.r. spectroscopy have been used to demonstrate the formation of a hydride when HCl is added to $(C_3H_5)Rh(PF_3)_3$ at $-75\ °C$.[619] Similarly, 2-acetyl-π-allylcobalt tricarbonyl reacts with conc. H_2SO_4 to yield (90).[620] The ^{19}F n.m.r. spectrum of (91) demonstrates a static structure.[621] Protonation and deuteriation studies have been carried out on (cyclohexa-1,3-diene)(π-cyclopentadienyl)M (M = Rh or Ir),[622] and

[603] A. Lapidot and C. S. Irving, *J.C.S. Dalton*, 1972, 668.
[604] I. H. Sabherwal and A. B. Burg, *Inorg. Chem.*, 1972, **11**, 3138.
[605] G. M. Intille and M. J. Braithwaite, *J.C.S. Dalton*, 1972, 645.
[606] Y. S. Sohn and A. L. Balch, *J. Amer. Chem. Soc.*, 1972, **94**, 1144.
[607] D. I. Hall and R. S. Nyholm, *J.C.S. Dalton*, 1972, 804.
[608] R. V. Parish and P. G. Simms, *J.C.S. Dalton*, 1972, 809.
[609] B. L. Booth, R. N. Haszeldine, and G. R. H. Neuss, *J.C.S. Chem. Comm.*, 1972, 1075.
[610] J. Ashley-Smith, M. Green, and F. G. A. Stone, *J.C.S. Dalton*, 1972, 1805.
[611] M. Green and T. A. Kuc, *J.C.S. Dalton*, 1972, 832.
[612] K. Bouchal, J Škramovská, P. Schmidt, and F. Hrabák, *Coll. Czech. Chem. Comm.*, 1972, **37**, 3081.
[613] K. M. Nicholas and R. Pettit, *J. Organometallic Chem.*, 1972, **44**, C21.
[614] S. Otsuka and T. Taketomi, *J.C.S. Dalton*, 1972, 1879.
[615] R. Cramer, *J. Amer. Chem. Soc.*, 1972, **94**, 5681.
[616] F. Maspero, E. Perrotti, and F. Simonetti, *J. Organometallic Chem.*, 1972, **38**, C43.
[617] B. F. G. Johnson, H. V. P. Jones, and J. Lewis, *J.C.S. Dalton*, 1972, 463.
[618] M. S. Fraser, G. F. Everitt, and W. H. Baddley, *J. Organometallic Chem.*, 1972, **35**, 403.
[619] J. F. Nixon and B. Wilkins, *J. Organometallic Chem.*, 1972, **44**, C25.
[620] S. Otsuka and A. Nakamura, *Inorg. Chem.*, 1972, **11**, 644.
[621] D. M. Barlex, A. C. Jarvis, R. D. W. Kemmitt, and B. Y. Kimura, *J.C.S. Dalton*, 1972, 2549.
[622] B. F. G. Johnson, J. Lewis, and D. J. Yarrow, *J.C.S. Dalton*, 1972, 2084.

data have been reported for $(C_5H_5)_3Co_3(CO)C_{14}H_{20}$,[623] $[Co(C_5H_5)_2]^+$-$[Co(PPh_3)X_3]^-$,[624] $[(R^1R^2R^3CC_5H_3R^4)Co(C_5H_5)]^+$,[625] $[(C_5H_5)Co(C_5H_5\text{-}BR)]^+$,[626] (92),[627] and $[(C_5H_5)M(NO)PPh_3]^+$ (M = Co or Rh).[628]

(90)

(91)

(92)

The thermal rearrangement of π-cyclopentadienyl-π-dicarbollyl derivatives of cobalt has been examined by use of ^1H and ^{11}B n.m.r. spectroscopy,[629] and the 80.5 MHz ^{11}B n.m.r. spectra of species such as (93) have been reported.[630] Data have also been reported for $[(C_5H_5)Co(\pi\text{-}2\text{-}B_7CH_8)]^-$,[631] and $(C_5H_5)Co\{\pi\text{-}(1)\text{-}2,4\text{-}B_8C_2H_{10}\}$.[632] The ^1H resonance of the cyclopentadienyl group of $[(C_5H_5)Co(CNEt)_3]^{2+}$ is at rather low field, τ 3.59.[633]

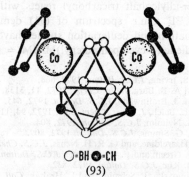

○=BH ●=CH
(93)

[623] R. B. King and A. Efraty, *J. Amer. Chem. Soc.*, 1972, **94**, 3021.
[624] M. Van Den Akker, R. Olthof, F. Van Bolhuis, and F. Jellinek, *Rec. Trav. chim.*, 1972, **91**, 75.
[625] N. U. Murr and R. Dabard, *J. Organometallic Chem.*, 1972, **39**, C82.
[626] G. E. Herberich and G. Greiss, *Chem. Ber.*, 1972, **105**, 3413.
[627] R. S. Dickson and H. P. Kirsch, *Austral. J. Chem.*, 1972, **25**, 2535.
[628] N. G. Connelly and J. D. Davies, *J. Organometallic Chem.*, 1972, **38**, 385.
[629] M. K. Kaloustian, R. J. Wiersema, and M. F. Hawthorne, *J. Amer. Chem. Soc.*, 1972, **94**, 6679.
[630] W. J. Evans and M. F. Hawthorne, *J.C.S. Chem. Comm.*, 1972, 611.
[631] D. F. Dustin and M. F. Hawthorne, *J.C.S. Chem. Comm.*, 1972, 1329.
[632] C. J. Jones, J. N. Francis, and M. F. Hawthorne, *J. Amer. Chem. Soc.*, 1972, **94** 8391.
[633] J. A. Dineen and P. L. Pauson, *J. Organometallic Chem.*, 1972, **43**, 209.

The ^{19}F n.m.r. spectra of $(C_5H_5)Rh(RPF_2)_2$ are of the $[AX_n]_2$ spin type with rhodium coupling.[634]

$[Co(NH_3)_4(NCS)Cl]ClO_4$ shows only one NH_3 resonance, demonstrating the *trans* stereochemistry,[635] whereas $[Rh(NH_3)_5(CH_3CN)]$ shows $^4J(^{103}Rh-H) = 0.4$ Hz.[636] Two series of complexes of *cis*-$[Co(2\text{-picolyl-amine})_2X_2]^+$ have been prepared. One shows a simple singlet for the CH_2 resonance, whereas the other has a complex pattern.[637] Although *trans*-$[Rh(3\text{-methylpyridine})_4Cl_2]^+$ and *mer*-$Rh(3\text{-methylpyridine})_3Cl_3$ should show restricted rotation, the ^1H n.m.r. spectrum is consistent with free rotation.[638] Data have also been reported for $[Ir(NH_3)_5(NH_2Cl)]^{3+}$,[639] and $[RhCl_2(4\text{-aminopyridine})_4]^+$.[640]

Addition of NaOH or KOH to $[Co(en)_3]^{3+}$ causes the CH_2 resonance to move to higher field and become asymmetric. From this and other evidence it was concluded that deprotonation is occurring.[641] The ^1H n.m.r. spectra of *N*-deuteriated $[Co(en)_3]^{3+}$ and $[Co(R\text{-pn})_3]^{3+}$ have been measured at 251 MHz with ^{59}Co decoupling in D_2O as a function of temperature and phosphate concentration. It was shown that the Δ-$\delta\lambda\lambda$ form is more abundant than the Δ-$\lambda\lambda\lambda$ form, but at high phosphate concentration the reverse is true. The ^{59}Co chemical shifts show large deuterium isotope effects.[642] The ^1H n.m.r. spectrum of $[(o\text{-phen})_2CoCO_3]^+$ has been analysed, but the spectrum of $[(bipy)_2CoCO_3]^+$ was insufficiently well resolved to analyse.[643] Data have also been reported for $[Co(MeNH-CH_2CH_2NHMe)_nX_{6-2n}]^-$,[644, 645] *trans*-$[Co(R\text{-MeNHCH}_2CHMeNHMe)_2\text{-}X_2]^+$,[646] $[Co(H_2NCHMeCH_2CHMeNH_2)_3]^{3+}$,[647] $\{(Bu_3P)_2Co(dmg)_2\}$-$\{Co(dmg)_2Cl_2\}$,[648] $Co(dmg)_2(4\text{-t-butylpyridine})(NCS)$,[649] *trans*-$[Co(SCN)(NH_3)(en)_2]^{2+}$,[650] *cis*-$[Rh(bipy)_2(OH_2)_2]^{3+}$, and *cis*-$[Ir(bipy)_2Cl_2]^+$.[651]

^1H N.m.r. spectroscopy has been used to differentiate between three

---

[634] J. F. Nixon and A. A. Pinkerton, *J. Organometallic Chem.*, 1972, **37**, C47.
[635] K. S. Mok, C. K. Poon, and H. W. Tong, *J.C.S. Dalton*, 1972, 1701.
[636] R. D. Foust, jun. and P. C. Ford, *Inorg. Chem.*, 1972, **11**, 899.
[637] K. Michelsen, *Acta Chem. Scand.*, 1972, **26**, 769.
[638] A. W. Addison, K. Dawson, R. D. Gillard, B. T. Heaton, and H. Shaw, *J.C.S. Dalton*, 1972, 589.
[639] B. C. Lane, J. W. McDonald, F. Basolo, and R. G. Pearson, *J. Amer. Chem. Soc.*, 1972, **94**, 3786.
[640] C. McRobbie and H. Frye, *Austral. J. Chem.*, 1972, **25**, 893.
[641] G. Navon, R. Panigel, and D. Meyerstein, *Inorg. Chim. Acta*, 1972, **6**, 299.
[642] J. L. Sudmeier, G. L. Blackmer, C. H. Bradley, and F. A. L. Anet, *J. Amer. Chem. Soc.*, 1972, **94**, 757.
[643] D. J. Francis and R. B. Jordan, *Inorg. Chem.*, 1972, **11**, 461.
[644] K. W. Larson, A. A. Ryan, and J. G. Brushmiller, *Inorg. Nuclear Chem. Letters*, 1972, **8**, 991.
[645] J. A. Tiethof and D. W. Cooke, *Inorg. Chem.*, 1972, **11**, 315.
[646] J. A. Tiethof and D. W. Cooke, *Inorg. Nuclear Chem. Letters*, 1972, **8**, 1013.
[647] F. Mizukami, H. Ito, J. Fujita, and K. Saito, *Bull. Chem. Soc. Japan*, 1972, **45**, 2129.
[648] L. G. Marzilli, J. G. Salerno, and L. A. Epps, *Inorg. Chem.*, 1972, **11**, 2050.
[649] L. A. Epps and L. G. Marzilli, *J.C.S. Chem. Comm.*, 1972, 109.
[650] D. A. Buckingham, I. I. Creaser, W. Marty, and A. M. Sargeson, *Inorg. Chem.*, 1972, **11**, 2738.
[651] P. M. Gidney, R. D. Gillard, and B. T. Heaton, *J.C.S. Dalton*, 1972, 2621.

isomers of $[Co(H_2NCH_2CH_2NHCH_2CH_2NH_2)_2]^{3+}$,[652, 653] and to follow the deuteriation of this compound.[654] The magnetic equivalence of ligand protons has been examined by ^1H n.m.r. spectroscopy for a large number of cobalt(III) complexes of terdentate azo-dyestuffs.[655] Data have also been reported for trans-$[Co\{(R)-H_2N(CH_2)_3NHCHMeCHMeNH(CH_2)_3NH_2\}-Cl_2]^+$,[656] $[CoCl(NH_3)\{N(CH_2CH_2NH_2)_3\}]^+$,[657] $[Co(CF_3COCHCOCH_3)_2-\{N(CH_2CH_2NH_2)_3\}]^+$ (including ^{19}F n.m.r. data),[658] and trans-$[Co(H_2NCH_2CHMeNHCH_2CH_2NHCHMeCH_2NH_2)Cl_2]^+$.[659]

A correlation has been found between the chemical shifts of the methyl groups and the ligand-field strength of X in the complexes $[Co(MAC)X_2]^+$, where MAC is one of four methyl-substituted quadridentate macrocycles.[660] For $[Co\{Me_2C(CH_2NHCH_2CH_2NHCH_2)_2\}X_2]^+$ in the trans-series there is a 0.15 p.p.m. separation of the methyl resonances, whereas in the cis-series the separation is 0.35 p.p.m.[661] ^{13}C N.m.r. spectroscopy has been used to show that uroporphyrinogen III is a precursor of vitamin B_{12} [662] and to determine the biosynthetic pathway for the synthesis of this vitamin.[663] Data have also been reported for cobalt complexes of $CH_2(CMe_2NHCH_2-CH_2CHMe)_2CH_2$,[664] $[Co(N-[14]-4,11-diene-N_4)(OH_2)Me]^+$,[665] $[CoCl_2N_4-C_{10}H_{12}Me_4]^+$,[666] corrins,[667] and rhodium porphins and porphyrins.[668]

The X-ray structure of $Co(acac)(OC_6H_4CH=NCH_2CH_2N=CHC_6H_4O)$ shows that the methine hydrogens of the salen ligand are inequivalent, but they are equivalent in the ^1H n.m.r. spectrum.[669, 670] ^1H N.m.r. spectroscopy has been used to study the ligand-exchange reactions of cobalt(III) complexes containing bisacetylacetone-ethylenedi-imine and an amino-acid. Three sets of signals due to DD-, LL-, and DL-forms were observed.[671] The ^1H n.m.r. spectrum of $[Co\{^-O_2CCH_2CH_2N(CH_2CO_2^-)CH_2CH_2N(CH_2CO_2^-)CH_2CH_2-CO_2^-\}]^-$ shows a well resolved AB pattern for the $>NCH_2CO_2^-$ group but

[652] F. R. Kenne and G. H. Searle, Inorg. Chem., 1972, 11, 148.
[653] Y. Yoshikawa and K. Yamasaki, Bull. Chem. Soc. Japan, 1972, 45, 179.
[654] G. H. Searle and F. R. Keene, Inorg. Chem., 1972, 11, 1006.
[655] G. Schetty and E. Steiner, Helv. Chim. Acta, 1972, 55, 1509.
[656] M. Saburi, C. Hattori, and S. Yoshikawa, Inorg. Chim. Acta, 1972, 6, 427.
[657] C.-H. L. Yang and M. W. Grieb, J.C.S. Chem. Comm., 1972, 656.
[658] S. C. Cummings and R. E. Sievers, Inorg. Chem., 1972, 11, 1483.
[659] M. Saburi and S. Yoshikawa, Bull. Chem. Soc. Japan, 1972, 45, 1443.
[660] E. S. Gore, J. C. Dabrowiak, and D. H. Busch, J.C.S. Dalton, 1972, 923.
[661] N. F. Curtis and G. W. Reader, J.C.S. Dalton, 1972, 1453.
[662] A. I. Scott, C. A. Townsend, K. Okada, M. Kajiwara, and R. J. Cushley, J. Amer. Chem. Soc., 1972, 94, 8271.
[663] A. I. Scott, C. A. Townsend, K. Okada, M. Kajiwara, P. J. Whitman, and R. J. Cushley, J. Amer. Chem. Soc., 1972, 94, 8267.
[664] L. A. P. Kane-Maguire, J. F. Endicott, and D. P. Rillema, Inorg. Chim. Acta, 1972, 6, 443.
[665] T. S. Roche and J. F. Endicott, J. Amer. Chem. Soc., 1972, 94, 8622.
[666] S. C. Jackels, K. Farmery, E. K. Barefield, N. J. Rose, and D. H. Busch, Inorg. Chem., 1972, 11, 2893.
[667] A. W. Johnson and W. R. Overend, J.C.S. Perkin I, 1972, 2681.
[668] R. Grigg, G. Shelton, A. Sweeney, and A. W. Johnson, J.C.S. Perkin I, 1972, 1789.
[669] M. Calligaris, G. Manzini, G. Nardin, and L. Randaccio, J.C.S. Dalton, 1972, 543.
[670] R. J. Cozens and K. S. Murray, Austral. J. Chem., 1972, 25, 911.
[671] Y. Fujii, Bull. Chem. Soc. Japan, 1972, 45, 3084.

overlapping AA'BB' patterns for the rest. $J(A-B) = 18.4$ Hz is consistent with out-of-plane glycinate rings.[672] Before photolysis, $[(-)_{5461}\text{-Rh}^{III}\text{-D-}(-)\text{-pdta}]^-$ shows only one methyl doublet in the ^{1}H n.m.r. spectrum, but on photolysis another doublet appears. It was concluded that the methyl group was equatorial, but on photolysis becomes axial.[673] Data have also been reported for $[\text{Co}(\text{OCMeCHCMe}=\text{NCH}_2\text{CH}_2\text{N}=\text{CMeCHCMeO})\text{-}L_2]^{n+}$,[674] and $[\text{Co}(^-\text{O}_2\text{CCH}_2\text{NHCH}_2\text{CH}_2\text{NHCH}_2\text{CO}_2^-)L_2]^+$.[675]

The ^{1}H n.m.r. spectra of $(+)_{5461}\text{-[Co(acac)}_2(\text{L-val)]}$, $(+)_{5461}\text{-[Co(acac)}_2\text{-(L-phe)]}$, and the laevorotatory isomers show four methyl resonances but only one methine resonance.[676] The ^{1}H n.m.r. resonances of $[\text{Co(NH}_3)_3\text{-(gly-gly)}]^+$ are pH-dependent and this was attributed to protonation of the carboxyl oxygens.[677] Data have also been reported for $\text{Co(L-aspH)}_2(\text{ala})$,[678] $[\text{Co(trien)(gly-L-asp)}]^+$,[679] $[\text{Co(gly)}_2(\text{L-glu)}]^-$,[680] $\text{Co}\{(S)\text{-amino-acid}\}\text{-}(\text{acac})_2$,[681] $trans\text{-(O,X)CoX(amino-acid)(H}_2\text{NCH}_2\text{CH}_2\text{NHCH}_2\text{CH}_2\text{NH}_2)$,[682] $\Lambda\text{-}(+)_{5893}\text{-[Co(en)}_2\{(R)\text{-}(+)\text{-pantoate}\}]^+$,[683] $[\text{Co(en)}_2\text{Cl}\{\text{NH}_2(\text{CH}_2)_5\text{CO}_2\text{-}R\}]^{2+}$,[684] $[\text{Co(L-ala)}_2(\text{acac})]$,[685] $[\text{Co(H}_2\text{NCH}_2\text{CH}_2\text{NHCH}_2\text{CH}_2\text{NH}_2)\text{-}(\text{gly-gly-gly-gly-OEt})]^{2+}$,[686] complexes of cobalt(III) with glycyl-L-histidine,[687] and $trans\text{-[Co(acac)}_2(\text{N}_3)_2]^+$.[688]

$[\text{IrH(MeNC)(dppe)}_2]^{2+}$ shows a doublet of doublets of triplets for the hydride resonance, which is said to be consistent with a cis stereochemistry.[689] The ^{19}F n.m.r. spectrum of $[\text{RhCl(PF}_3)_2]_2$ has been analysed as $[\text{AX}_3]_2$ and the exchange of PF_3 and C_2H_4 studied.[690] The ^{1}H n.m.r. spectrum of $[\text{Ir}\{\text{P(OMe)}_3\}_5]\text{BPh}_4$ shows that all the methyl groups are equivalent but the pattern is a rather odd triplet.[691] ^{1}H N.m.r. spectroscopy has been used to demonstrate that $\text{MX}_3(\text{SR}_2)_3$ (M = Rh or Ir) has the mer configuration and $^3J(^{103}\text{Rh}-^1\text{H})$ was observed.[692] Data have also been

---

[672] W. Byers and B. E. Douglas, *Inorg. Chem.*, 1972, **11**, 1470.
[673] G. L. Blackmer, J. L. Sudmeier, R. N. Thibedeau, and R. M. Wing, *Inorg. Chem.*, 1972, **11**, 189.
[674] Y. Fujii, A. Osawa, Y. Furukoate, F. Ebina, and S. Takahashi, *Bull. Chem. Soc. Japan*, 1972, **45**, 2459.
[675] K. Kuroda, *Bull. Chem. Soc. Japan*, 1972, **45**, 2176.
[676] S. H. Laurie, *J.C.S. Dalton*, 1972, 573.
[677] I. G. Browning, R. D. Gillard, J. R. Lyons, P. R. Mitchell, and D. A. Phillips, *J.C.S. Dalton*, 1972, 1815.
[678] T. Matsuda, Y. Okumoto, and M. Shibata, *Bull. Chem. Soc. Japan*, 1972, **45**, 802.
[679] A. Y. Girgis and J. I. Legg, *J. Amer. Chem. Soc.*, 1972, **94**, 8420.
[680] K. Kawasaki and M. Shibata, *Bull. Chem. Soc. Japan*, 1972, **45**, 3100.
[681] D. J. Seematter and J. G. Brushmiller, *J.C.S. Chem. Comm.*, 1972, 1277.
[682] K. Ohkawa, J. Fujita, and Y. Shimura, *Bull. Chem. Soc. Japan*, 1972, **45**, 161.
[683] E. B. Kipp and R. A. Haines, *Inorg. Chem.*, 1972, **11**, 271.
[684] R. W. Hay, R. Bennett, and D. J. Barnes, *J.C.S. Dalton*, 1972, 1524.
[685] Y. Fujii and T. Ejiri, *Bull. Chem. Soc. Japan*, 1972, **45**, 283.
[686] Y. Wu and D. H. Busch, *J. Amer. Chem. Soc.*, 1972, **94**, 4115.
[687] R. D. Gillard and A. Spencer, *J.C.S. Dalton*, 1972, 902.
[688] L. J. Boucher and D. R. Herrington, *Inorg. Chem.*, 1972, **11**, 1772.
[689] W. M. Bedford and G. Rouschias, *J.C.S. Chem. Comm.*, 1972, 1224.
[690] D. A. Clement and J. F. Nixon, *J.C.S. Dalton*, 1972, 2553.
[691] L. H. Haines and E. Singleton, *J.C.S. Dalton*, 1972, 1891.
[692] E. A. Allen and W. Wilkinson, *J.C.S. Dalton*, 1972, 613.

reported for $RhX(EPh_3)_2(RNC)$ (E = P or As),[693] $RhCl(C_8H_{12})\{o\text{-}CH_2\text{=}CHC_6H_4)PPh_2\}$,[694] $[Rh\{(o\text{-}CH_2\text{=}CHC_6H_4)PPh_2\}_2]^+$,[695] $RhCl_3(PR_3)_3$,[696] $MCl(Ph_2PC_6H_4CH\text{=}CHC_6H_4PPh_2)$ (M = Rh or Ir),[697] $IrCl_3(EtONO)(PPh_3)_2$,[698] $Rh_2(O_2CPh)_4$,[699] and $(Et_2NCS_2)_nM$ (M = Co, Ni, Cu, Zn, or Cd).[700]

Complexes of Ni, Pd, and Pt.—Information concerning complexes of these elements can be found at the following sources: $Fe(C_5H_4PPh_2)_2Pt(CH_2\text{-}CD_2Et)_2$,[470] $(C_5H_5)NiCR^1CR^2(PPh_2)Fe(CO)_3$,[374] $Pt(acac)(MeCHC_6H_4\text{-}PBu^t_2)$,[79] $[PtCl(C_6H_4CHNPh)]_2$,[568] $PtX(C_6H_4F)(PEt_3)_2$,[557] $Ni(CO)_3\text{-}C(OEt)NEt_2$,[262] $PtCl_4(CHNMe_2)(PEt_3)$,[249] $Ni(CO)_4$,[159] $(C_5H_5)Ni(CO)\text{-}SnCl_2Fe(CO)_2(C_5H_5)$,[445] $\{(C_5H_5)Ni(CO)\}_2GeCl_2$,[443] $Fe_2Pt(CO)_8L_2$,[370] $RuPt_2(CO)_5(PMe_2Ph)_3$,[471] $(C_5H_5)_2Ti(C\equiv CPh)_2Ni(CO)$,[194] $[MePt(PMe_2Ph)_2\text{-}(C_2H_4)]^+$,[117] $[(C_3H_4Me)NiX]_2$,[114] $[(C_{10}H_{15})PdX]_2$,[116] $[(C_3H_4R)PdX]_2$,[115] $[(C_5H_5)NiSCF_3]_2$,[293] $[Pt(CN)_4]^{2-}$,[68] $[Ni(NH_3)(H_2O)_5]^{2+}$,[50] $[(diphos)\text{-}Ni(N_3)]_2^{2+}$, $[(Ph_3P)_2PdN_3]_2^{2+}$,[128, 601] $Pd(porphyrin)$,[668] $Ni\{P(OCH_2)_3CEt\}_4$,[45] $MCl_2(PR_3)_2$ (M = Ni, Pd, or Pt),[303, 330, 468] $[\{(C_5H_5)_2Nb(SMe)_2\}_2\text{-}M]^{2+}$ (M = Ni, Pd, or Pt),[217] and $(Et_2NCS_2)_2Ni$.[700]

Protonation of $Pt(PPh_3)_4$ yields $[(Ph_2P)_3PtH]^+$, which gives an AB_2X spectrum where the hydride signal appears as a doublet of doublets of doublets. Addition of excess PPh_3 catalyses *cis*-phosphine exchange;[701] $Pt(ROPPh_2)_4$ reacts similarly.[702] ^1H N.m.r. and i.r. spectroscopic studies of *trans*-$PtHX(PR_3)_2$ has been used to give information about the relative *trans* influence of $-SCN$ and $-NCS$ ligands, the magnetic anisotropic effects of the isoelectronic ligands CN^-, CNR, and CO, and the facile phosphine exchange reactions of *trans*-$PtH(CN)(PMePh_2)_2$ and *trans*-$Pt(CN)_2(PMePh_2)_2$. A linear relationship was found between $\nu(Pt-H)$ and $\tau(Pt-H)$.[703] The hydride signal of *trans*-$Pt(PPh_3)_2HCN$ is broad, but addition of PPh_3 or cooling results in sharpening.[704] Data have also been reported for $[HNiL_4]^+$,[705] $[HNi\{P(OMe)_3\}_4]^+$,[706] and *trans*-$PtHXL_2$.[707–710]

---

[693] A. Nakamura, Y. Tatsuno, and S. Otsuka, *Inorg. Chem.*, 1972, **11**, 2058.
[694] P. R. Brookes, *J. Organometallic Chem.*, 1972, **42**, 459.
[695] P. R. Brookes, *J. Organometallic Chem.*, 1972, **43**, 415.
[696] G. M. Intille, *Inorg. Chem.*, 1972, **11**, 695.
[697] M. A. Bennett, P. W. Clark, G. B. Robertson, and P. O. Whimp, *J.C.S. Chem. Comm.*, 1972, 1011.
[698] C. A. Reed and W. R. Roper, *J.C.S. Dalton*, 1972, 1243.
[699] F. A. Cotton and J. G. Norman, jun., *J. Amer. Chem. Soc.*, 1972, **94**, 5697.
[700] R. M. Golding, C. M. Harris, K. J. Jessop, and W. C. Tennant, *Austral. J. Chem.*, 1972, **25**, 2567.
[701] K. Thomas, J. T. Dumler, B. W. Renoe, C. J. Nyman, and D. M. Roundhill, *Inorg. Chem.*, 1972, **11**, 1795.
[702] P.-C. Kong and D. M. Roundhill, *Inorg. Chem.*, 1972, **11**, 749.
[703] H. C. Clark and H. Kurosawa, *J. Organometallic Chem.*, 1972, **36**, 399.
[704] M. W. Adlard and G. Socrates, *J.C.S. Chem. Comm.*, 1972, 17.
[705] C. A. Tolman, *Inorg. Chem.*, 1972, **11**, 3128.
[706] C. A. Tolman, *J. Amer. Chem. Soc.*, 1972, **94**, 2994.
[707] M. W. Adlard and G. Socrates, *J. Inorg. Nuclear Chem.*, 1972, **34**, 2339.
[708] M. W. Adlard and G. Socrates, *J.C.S. Dalton*, 1972, 797.
[709] H. C. Clark and H. Kurosawa, *Inorg. Chem.*, 1972, **11**, 1275.
[710] M. Lenarda and W. H. Baddley, *J. Organometallic Chem.*, 1972, **39**, 217.

NiMe$_2$(PMe$_3$)$_2$ has a virtual triplet for the PMe$_3$ resonance, demonstrating the *trans* stereochemistry.[711] At room temperature the ^{1}H n.m.r. spectrum of [PtMeL$_2$(CNR)]$^+$ shows no phosphorus or platinum coupling, but is normal at $-20\ °C$. Treatment with amines produces carbene analogues.[712] For these complexes and *trans*-[MePtL(PMe$_2$Ph)$_2$]$^+$, ^2J(Pt–C–H) has been used to evaluate the *trans* influence.[713] The ^{1}H n.m.r. spectrum of the product obtained from the reaction between [PtMe$_3$]$_2$-SO$_4$,4H$_2$O and NaI–NaOH has given evidence of tetramers containing both iodo and hydroxo bridging groups in the same framework.[714] For PtMe$_3$L$_2$X, ^2J(^{195}Pt–C–H) has been used to produce an order of *trans* influence for X.[715] Me$_3$Pt(salicylaldimine) gives three methyl resonances, which is consistent with a dimeric structure.[716] For the complexes [PtMe$_2$(PMe$_2$Ph)$_2$L$_2$]$^{2+}$ and [PtMe$_2$(PMe$_2$Ph)$_2$LI]$^+$, ^2J(^{195}Pt–CH$_3$) has been used to establish an n.m.r. *trans* influence series and the data compared with data for platinum(II) complexes.[717] Data have also been reported for Me$_2$PdL$_2$,[718] *trans*-[PtMe(olefin)L$_2$]$^+$,[719] PtBr$_2$Me$_2$(CNEt)$_2$, and *cis*-Pt(CF$_3$)$_2$L$_2$.[720]

At $-65\ °C$, MeC=CMeC=O displaces ethylene from Pt(PPh$_3$)$_2$(C$_2$H$_4$), and at $-30\ °C$ the product isomerizes to (Ph$_3$P)$_2$PtCMe=CMeC=O.[721] Extensive use has been made of ^{1}H and ^{31}P n.m.r. spectroscopy to determine the stereochemistry of complexes such as PtCl{CH$_2$C$_6$H$_4$PBu^t(*o*-tol)}-{PBu^t(*o*-tol)$_2$}. When the tertiary phosphines are mutually *trans*, ^2J(^{31}P–Pt–^{31}P) = *ca*. 400 Hz is found.[722, 723] For *trans*-PtCl(COPh)-(PPh$_3$)$_2$, ^1J(^{195}Pt–^{31}P) is 3328 Hz, which is rather large.[724] Data have also been reported for PdCl(CH$_2$COCH$_2$CO$_2$R)L$_2$,[725] Pd(CO$_2$Me)(O$_2$CCF$_3$)-(PPh$_3$)$_2$, Pt(CO$_2$Me)$_2$(PPh$_3$)$_2$,[726] [Pt$_3$(PPh$_3$)$_4$(CO)(RC=CHPh)]BF$_4$,[727] *cis*-PtR$^1_2$(PR$^2_3$)$_2$ (R = PhCH$_2$, Me$_3$SiCH$_2$, or Ph$_2$MeSi),[728] Pt(CH$_2$CH$_2$CH$_2$)-

---

[711] H.-F. Klien and H. H. Karsch, *Chem. Ber.*, 1972, **105**, 2628.
[712] H. C. Clark and L. E. Manzer, *Inorg. Chem.*, 1972, **11**, 503.
[713] M. H. Chisholm, H. C. Clark, and L. E. Manzer, *Inorg. Chem.*, 1972, **11**, 1269.
[714] J. R. Hall and G. A. Swile, *J. Organometallic Chem.*, 1972, **44**, 201.
[715] D. E. Clegg, J. R. Hall, and G. A. Swile, *J. Organometallic Chem.*, 1972, **38**, 403.
[716] V. Romano, R. Badalamenti, T. Pizzino, and F. Maggio, *J. Organometallic Chem.*, 1972, **42**, 199.
[717] H. C. Clark and L. E. Manzer, *Inorg. Chem.*, 1972, **11**, 2749.
[718] N. Garty and M. Michman, *J. Organometallic Chem.*, 1972, **36**, 391.
[719] M. H. Chisholm and H. C. Clark, *J. Amer. Chem. Soc.*, 1972, **94**, 1532.
[720] H. C. Clark and L. E. Manzer, *J. Organometallic Chem.*, 1972, **38**, C41.
[721] J. P. Visser and J. E. Ramakers-Blom, *J. Organometallic Chem.*, 1972, **44**, C63.
[722] A. J. Cheney and B. L. Shaw, *J.C.S. Dalton*, 1972, 754.
[723] A. J. Cheney and B. L. Shaw, *J.C.S. Dalton*, 1972, 860.
[724] S. P. Dent, C. Eaborn, A. Pidcock, and B. Ratclif, *J. Organometallic Chem.*, 1972, **46**, C68.
[725] S. Baba, T. Sobata, T. Ogura, and S. Kawaguchi, *Inorg. Nuclear Chem. Letters*, 1972, **8**, 605.
[726] K. von Werner and W. Beck, *Chem. Ber.*, 1972, **105**, 3947.
[727] D. M. Blake and L. M. Leung, *Inorg. Chem.*, 1972, **11**, 2879.
[728] M. R. Collier, C. Eaborn, B. Jovanović, M. F. Lappert, L. Manojlović-Muir, K. W. Muir, and M. M. Truelock, *J.C.S. Chem. Comm.*, 1972, 613.

Cl(py),[729] (Ph$_3$P)$_2$PtC(CN)$_2$CH$_2$C(CN)$_2$,[730] [Pd(8-CHMe-quinoline)-Cl]$_2$,[731] [Pd(C$_6$H$_4$CH=NPh)O$_2$CMe]$_2$,[732] Pt(C$_6$H$_4$CH$_2$NEt$_2$)$_2$,[733] trans-PtCl$_2$(DMSO)(4-{3-[1-ethyl-4(1H)-quinolylidene]propenyl}quinoline),[734] (94),[735] cis-Pt(CCMeB$_{10}$H$_{10}$)$_2$L$_2$ (^{1}H and ^{11}B n.m.r. data),[736] (95),[737] (96),[738] and [Pd{CHR^{1}C(OMe)R^{2}CH$_2$SR^2}Cl]$_2$.[739]

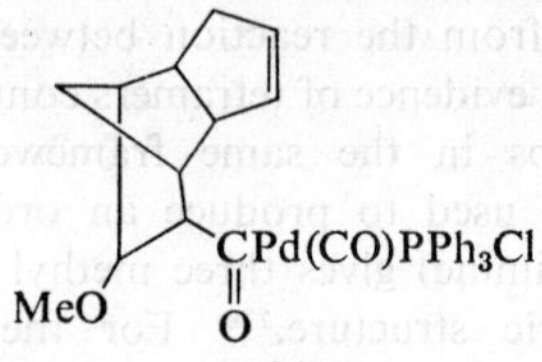

MeO—C(=O)—Pd(CO)PPh$_3$Cl

(94)

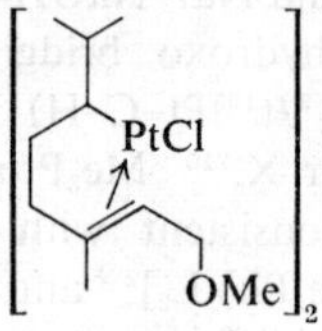

[PtCl / OMe]$_2$

(95)

OMe—Pt(amine)(Cl)—O

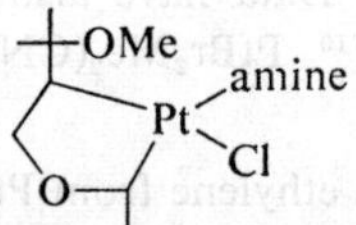

(96)

For Pt(CF=CFCF$_3$)X(PMe$_2$Ph)$_2$, the methyl groups are inequivalent, owing to restricted rotation about the platinum–carbon bond.[740] The ^{1}H n.m.r. spectra of M(EPh$_3$)$_2$(O$_2$CCF$_3$)(CR=CRCR=CRH) (M = Pd or Pt; E = P or As) have been reported. When M = Pt, E = P, ^6J(P–H) = 1.4 Hz was observed for the terminal hydrogen.[741] It has been found that for trans-[Pt(CF$_3$)L(PMe$_2$Ph)$_2$]$^+$ and PtXMe$_2$(CF$_3$)(PMe$_2$Ph)$_2$, a linear relationship exists between ^2J(Pt–CF$_3$) and ^2J(Pt–CH$_3$) for analogous compounds.[742] Both ^{1}H and ^{19}F n.m.r. data have been reported for

[729] F. J. McQuillin and K. G. Powell, J.C.S. Dalton, 1972, 2123.
[730] M. Lenarda, R. Ros, M. Graziani, and U. Belluco, J. Organometallic Chem., 1972, 46, C29.
[731] V. I. Sokolov, T. A. Sorokina, L. L. Troitskaya, L. I. Solovieva, and O. A. Reutov, J. Organometalllic Chem., 1972, 36, 389.
[732] H. Onoue and I. Moritani, J. Organometallic Chem., 1972, 43, 431.
[733] G. Longoni, P. Fantucci, P. Chini, and F. Canziani, J. Organometallic Chem., 1972, 39, 413.
[734] T. Winkler and C. Mayer, Helv. Chim. Acta, 1972, 55, 2351.
[735] G. Carturan, M. Graziani, R. Ros, and U. Belluco, J.C.S. Dalton, 1972, 262.
[736] R. Rogowski and K. Cohn, Inorg. Chem., 1972, 11, 1429.
[737] M. A. Schwartz and T. J. Dunn, J. Amer. Chem. Soc., 1972, 94, 4205.
[738] E. Benedetti, A. de Renzi, A. Panunzi, and C. Pedone, J. Organometallic Chem., 1972, 39, 403.
[739] Y. Takahasi, A. Tokuda, S. Sakai, and Y. Ishii, J. Organometallic Chem., 1972, 35, 415.
[740] T. G. Appleton, H. C. Clark, and R. J. Puddephatt, Inorg. Chem., 1972, 11, 2074.
[741] D. M. Roe, P. M. Bailey, K. Moseley, and P. M. Maitlis, J.C.S. Chem. Comm., 1972, 1273.
[742] T. G. Appleton, M. H. Chisholm, H. C. Clark, and L. E. Manzer, Inorg. Chem., 1972, 11, 1786.

$(Et_3P)_2\overline{Ni(CF_2)_3CF_2}$, $(Et_3P)_2\overline{MOC(CF_3)_2C(CF_3)_2O}$ (M = Pd or Pt),[743] $[CH_2=CMeCH_2C(CF_3)=CCF_3PdCl(PMe_2Ph)]_2$,[744] $\overline{PdOC(CF_3)_2NHC-(CF_3)_2}L_2$,[745] and $Pt(OCOCF_3)(CF_2CF_2H)L_2$.[746]

$(Bu^tNC)PdX_2\{C(NHBu^t)(NMe_2)\}$ gives two NMe signals owing to restricted rotation.[747] Data were also reported for $[M\{C(NHMe)_2\}_4]^{2+}$ (M = Pd or Pt), $[Pt\{C(SEt)NHMe\}_2(CNMe)_2]^{2+}$,[748] $[Pt(CSNHMe)(CNMe)-(PPh_3)_2]^{2+}$,[749] trans-$[MX\{C(NHR^1)NR^2R^3\}L_2]^+$,[750] and $\overline{Pt\{C(NHMe)-NHNHC(NHMe)\}}I_2$.[751]

$Ni(C_2H_4)(PPh_3)_2$ shows fast C_2H_4 exchange, whereas the palladium complex shows the C_2H_4 resonance at the free position. $Pt(C_2H_4)(PPh_3)_2$ shows platinum coupling to the ethylene, and addition of excess ethylene catalyses exchange. Data were also given for $M(PPh_3)_n$ (including ^{31}P n.m.r. data),[752] and $Pd(MeO_2CC\equiv CCO_2Me)\{P(OPh)_3\}_2$.[753]

The ^1H n.m.r. spectra of two isomers of $PtCl_2(C_2H_1)$(isoquinoline) were reported. These isomers were thought to be rotamers but cis–trans isomerism was not discussed.[754] The ^1H n.m.r. parameters for the alkyl protons of some aliphatic α-olefins have been compared with those for the corresponding trans-dichloro(amine)(olefin)platinum(II) complexes. The extent of the downfield shifts were related to the distance between the alkyl protons and the platinum atom.[755] Data have also been reported for $Pt(PPh_3)_2(CH_2=CClCN)$,[756] $[XC_6H_4CH=CH_2PtCl_2]_2$,[757] cis-$(CH_2=CHR)$-$LPtCl_2$,[758] $Pt(PrC\equiv CPr)(PPh_3)_2$,[759] $Pt[MeEtC(OH)C\equiv CC(OH)MeEt]_2$,[760]

$(o-PhC_6H_4O)_3PNi(allene)_2$,[761] and $Pt\{\overline{HC=C=CH(CH_2)_n}\}(PPh_3)_2$.[762]

---

[743] F. G. A. Stone, *Pure Appl. Chem.*, 1972, **30**, 551.
[744] T. G. Appleton, H. C. Clark, R. C. Poller, and R. J. Puddephatt, *J. Organometallic Chem.*, 1972, **39**, C13.
[745] H. D. Empsall, M. Green, and F. G. A. Stone, *J.C.S. Dalton*, 1972, 96.
[746] R. D. W. Kemmitt, B. Y. Kimura, G. W. Littlecott, and R. D. Moore, *J. Organometallic Chem.*, 1972, **44**, 403.
[747] G. A. Larkin, R. P. Scott, and M. G. H. Wallbridge, *J. Organometallic Chem.*, 1972, **37**, C21.
[748] J. S. Miller and A. L. Balch, *Inorg. Chem.*, 1972, **11**, 2069.
[749] P. M. Treichel and W. J. Knebel, *Inorg. Chem.*, 1972, **11**, 1285.
[750] L. Busetto, A. Palazzi, B. Crociani, U. Belluco, E. M. Badley, B. J. L. Kilby, and R. L. Richards, *J.C.S. Dalton*, 1972, 1800.
[751] A. L. Balch, *J. Organometallic Chem.*, 1972, **37**, C19.
[752] C. A. Tolman, W. C. Seidel, and D. H. Gerlach, *J. Amer. Chem. Soc.*, 1972, **94**, 2669.
[753] Ts. Ito, S. Hasegawa, Y. Takahashi, and Y. Ishii, *J.C.S. Chem. Comm.*, 1972, 629.
[754] L. Spaulding and M. Orchin, *J.C.S. Chem. Comm.*, 1972, 1249.
[755] R. Lazzaroni, P. Salvadori, and P. Pino, *J. Organometallic Chem.*, 1972, **43**, 233.
[756] K. Suzuki and H. Okuda, *Bull. Chem. Soc. Japan*, 1972, **45**, 1938.
[757] T. Iwao, A. Saika, and T. Kinugasa, *Inorg. Chem.*, 1972, **11**, 3106.
[758] A. De Renzi, G. Paiaro, and A. Panunzi, *Gazzetta*, 1972, **102**, 413.
[759] G. Wittig and S. Fischer, *Chem. Ber.*, 1972, **105**, 3542.
[760] F. D. Rochon and T. Theophanides, *Canad. J. Chem.*, 1972, **50**, 1325.
[761] M. Englert, P. W. Jolly, and G. Wilke, *Angew. Chem. Internat. Edn.*, 1972, **11**, 136.
[762] J. P. Visser and J. E. Ramakers, *J.C.S. Chem. Comm.*, 1972, 178.

The ^{1}H n.m.r. spectra of the 2,6,10-dodecatriene-1,12-diylnickel–hydrochloric acid reaction product and the bis-(π-allylnickel trifluoroacetate)–butadiene reaction product are not consistent with the previously proposed structures.[763] In the ^{1}H n.m.r. spectrum of (97; X = Cl or Br; n = 4), the central allylic proton is at *higher* field than the *syn* protons.[764] When (98) is dissolved in DCCl$_3$ at $-50\,°$C, only one species is observed, but

$$\left[\begin{array}{c} \overset{\text{PdX}}{\diagup} \\ (CH_2)_n \end{array} \right]_2$$

(97)

(98)   (99)

on warming to room temperature another species appears, which is thought to be (99).[765] It is interesting to note that in (100) $J(^{31}P-^1H^4)$ = 6 Hz, which was attributed to through-space coupling.[766] Extensive use of ^{1}H spectroscopy has been made in an investigation of the mechanism of 1,3-diene insertion into allyl–palladium bonds.[767] For (101), $J(H^3-H^4)$ = 0.71 Hz.[768] Data have also been reported for Ni(CH$_2$CH=CHMeCH$_2$-CH$_2$CHCMeCH$_2$)P(C$_6$H$_{11}$)$_3$,[769] 1,2,3-h^3-(1-acetyl-2,3-dimethylallyl)(pyridine)chloropalladium,[770] Pd(CH$_2$CR^{1}CH$_2$)(OC$_6$H$_4$CH=NR^2),[771] PdCl-(MeCHCHCHCOMe)(C$_5$H$_5$N),[772] PdCl(C$_3$H$_5$)(benzotriazole),[773] (102),[774,775] (103),[776] (104),[777] (105),[778] (106),[779] Bu^t$_3$C$_3$Ni(CO)Br,[780] (Me$_2$AlSPh)$_2$-

[763] V. A. Kormer and M. I. Lobach, *J. Polymer Sci.*, Part B, Polymer Letters, 1972, **10**, 177.
[764] H. A. Quinn, W. R. Jackson, and J. J. Rooney, *J.C.S. Dalton*, 1972, 180.
[765] A. Musco, R. Rampone, P. Ganis, and C. Pedone, *J. Organometallic Chem.*, 1972, **34**, C48.
[766] Y. Takahashi, S. Sakai, and Y. Ishii, *Inorg. Chem.*, 1972, **11**, 1516.
[767] R. P. Hughes and J. P. Powell, *J. Amer. Chem. Soc.*, 1972, **94**, 7723.
[768] P. F. Swanton, G. Zannoni, and R. Rampone, *Spectroscopy Letters*, 1972, **5**, 307.
[769] B. Barnett, B. Büssemeier, P. Heimbach, P. W. Jolly, C. Krüger, I. Tkatchenko, and G. Wilke, *Tetrahedron Letters*, 1972, 1457.
[770] J. W. Faller and M. T. Tully, *J. Amer. Chem. Soc.*, 1972, **94**, 2676.
[771] B. E. Reichert and B. O. West, *J. Organometallic Chem.*, 1972, **36**, C29.
[772] B. T. Heaton and D. J. A. McCaffrey, *J. Organometallic Chem.*, 1972, **43**, 437.
[773] Y. Watanabe, T. Mitsudo, M. Tanaka, K. Yamamoto, and Y. Takegami, *Bull. Chem. Soc. Japan*, 1972, **45**, 925.
[774] W. G. Dauben and A. J. Kielbania, jun., *J. Amer. Chem. Soc.*, 1972, **94**, 3669.
[775] S. Masamune, M. Sakai, and N. Darby, *J.C.S. Chem. Comm.*, 1972, 471.
[776] C. W. Alexander and W. R. Jackson, *J.C.S. Perkin II*, 1972, 1601.
[777] T. Hosokawa and P. M. Maitlis, *J. Amer. Chem. Soc.*, 1972, **94**, 3238.
[778] C. Calvo, T. Hosokawa, H. Reinheimer, and P. M. Maitlis, *J. Amer. Chem. Soc.*, 1972, **94**, 3237.
[779] R. P. Hughes and J. Powell, *J. Organometallic Chem.*, 1972, **34**, C51.
[780] W. K. Olander and T. L. Brown, *J. Amer. Chem. Soc.*, 1972, **94**, 2139.

Ni(C$_8$H$_8$),[781] Pd{C(NHAr^1)NHAr^2}Cl$_2$,[782] and (5-methylenecycloheptene)-PtCl$_2$.[783]

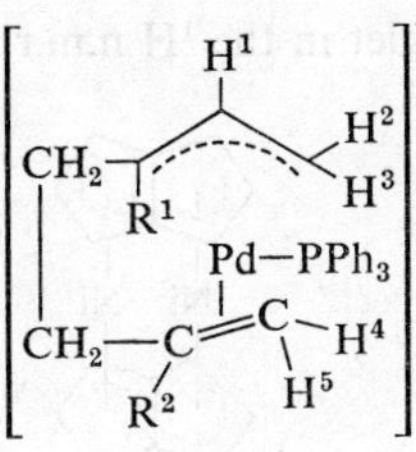

(100)

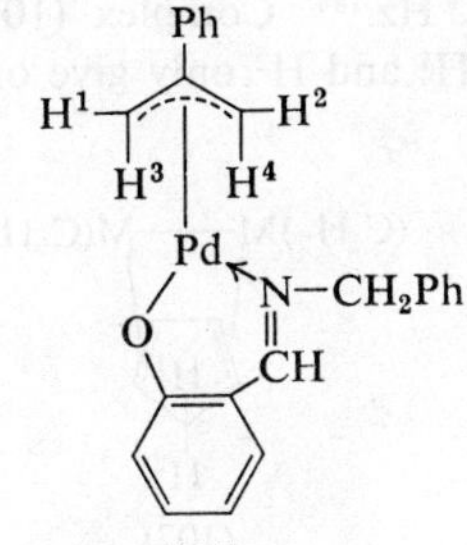

(101)

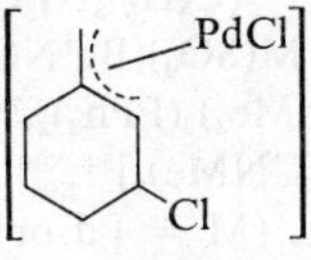

(102)

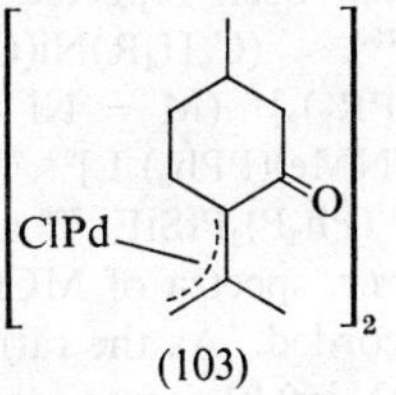

(103)

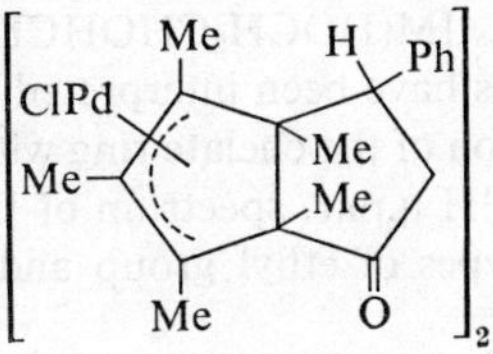

(104)

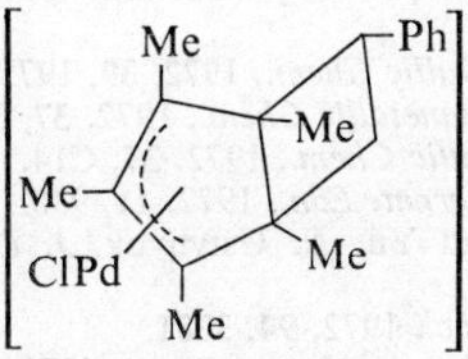

(105)

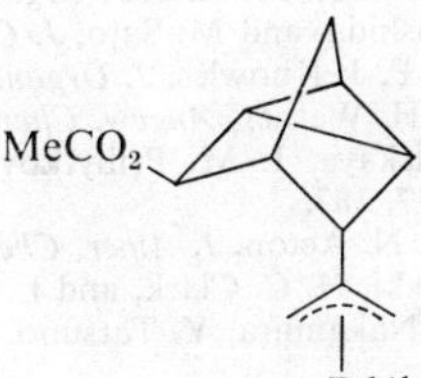

(106)

[781] T. Hirabayashi and Y. Ishii, *J. Organometallic Chem.*, 1972, **39**, C85.
[782] B. Crociani, T. Boschi, G. G. Troilo, and U. Croatto, *Inorg. Chim. Acta*, 1972, **6**, 655.
[783] C. B. Anderson and J. T. Michalowski, *J.C.S. Chem. Comm.*, 1972, 459.

The cyclopentadienyl proton chemical shifts of $(C_5H_5)Ni(PBu^n_3)SR$ show a linear correlation with the Taft σ^* constant of R.[784] The ^{31}P n.m.r. spectrum of $(C_5H_5)NiCN(Ph_2PCH_2PPh_2)$ shows two doublets with $^2J(P-P) = 72$ Hz.[785] Complex (107; M = Ni or Pd) is thought to be fluxional as H^1 and H^2 only give one singlet in the ^1H n.m.r. spectrum.[786]

$(C_5H_5)M$——$M(C_5H_5)$

H^1

H^2

(107)

Ni Ni

(108)

Data have also been reported for $(C_5H_5)Ni(PBu^n_3)SR$,[787-789] $(C_5H_5)Ni(PPh_3)MgBr$,[790] $(C_5H_4R)Ni(C_5H_4R)$,[791] $(C_5H_5)_3Ni_3CPh$,[792] (108),[793] $[MCl(RNC)(PR_3)_2]^+$ (M = Ni or Pd),[794] $M(SO_4)(Bu^tNC)_2$ (M = Ni or Pd),[795] $[Pt(CNMe)(PPh_3)_2L]^{2+}$,[796] $NiCl_2(SnMe_3)_2(PPh_3)_2$,[797] $PtCl_2(HgCl_2)(PMe_2Ph)_2$,[798] $(Ph_3P)_2PtSiF_4$,[799] and $[Pt_2Cl_2(CNMe)_4]^{2+}$.[800]

The ^1H n.m.r. spectra of $MCl_n(NMe_3)_{4-n}$ (M = Pd or Pt; n = 2 or 3) have been recorded. As the ratio of $^3J(Pt-H)$ for trans-$PtCl_2(NMe_3)_2$ and $[PtCl_3(NMe_3)]^-$ is 0.91, it was concluded that the trans influence of nitrogen is comparable to that of chloride.[801] The ^1H n.m.r. spectra of $[Pt(H_2NCH_2CHClCH_2NH_2)(NH_3)_2]^{2+}$, $[M(HOCH_2CHOHCH_2OH)_2]Cl_2$ (M = Pd or Pt), and related complexes have been interpreted as showing preference for the chair-type conformation of the chelate ring with the chloro- or hydroxy-substituent axial.[802] The ^1H n.m.r. spectrum of $[Pt(Et_2NCH_2CH_2NHCH_2CH_2NEt_2)Cl]^+$ has two types of ethyl group and the effect of solvent on

[784] M. Sato and T. Yoshida, *J. Organometallic Chem.*, 1972, **39**, 389.
[785] F. Sato, T. Uemura, and M. Sato, *J. Organometallic Chem.*, 1972, **39**, C25.
[786] E. O. Fischer, P. Meyer, C. G. Kreiter, and J. Müller, *Chem. Ber.*, 1972, **105**, 3014.
[787] M. Sato, F. Sato, N. Takemoto, and K. Iida, *J. Organometallic Chem.*, 1972, **34**, 205.
[788] F. Sato, K. Iida, and M. Sato, *J. Organometallic Chem.*, 1972, **39**, 197.
[789] F. Sato, T. Yoshida, and M. Sato, *J. Organometallic Chem.*, 1972, **37**, 381.
[790] H. Felkin and P. J. Knowles, *J. Organometallic Chem.*, 1972, **37**, C14.
[791] A. Salzer and H. Werner, *Angew. Chem. Internat. Edn.*, 1972, **11**, 930.
[792] T. I. Voyevodskaya, I. M. Pribytkova, and Yu. A. Ustynyuk, *J. Organometallic Chem.*, 1972, **37**, 187.
[793] T. J. Katz and N. Acton, *J. Amer. Chem. Soc.*, 1972, **94**, 3281.
[794] W. J. Cherwinski, H. C. Clark, and L. E. Manzer, *Inorg. Chem.*, 1972, **11**, 1511.
[795] S. Otsuka, A. Nakamura, Y. Tatsuno, and M. Miki, *J. Amer. Chem. Soc.*, 1972, **94**, 3761.
[796] P. M. Treichel and W. J. Knebel, *Inorg. Chem.*, 1972, **11**, 1289.
[797] P. E. Garrou and G. E. Hartwell, *J.C.S. Chem. Comm.*, 1972, 881.
[798] R. W. Baker, M. J. Braithwaite, and R. S. Nyholm, *J.C.S. Dalton*, 1972, 1924.
[799] T. R. Durkin and E. P. Schram, *Inorg. Chem.*, 1972, **11**, 1048.
[800] P. M. Treichel, K. P. Wagner, and W. J. Knebel, *Inorg. Chim. Acta*, 1972, **6**, 674.
[801] P. L. Goggin, R. J. Goodfellow, and F. J. S. Reed, *J.C.S. Dalton*, 1972, 1298.
[802] T. G. Appleton and J. R. Hall, *Inorg. Chem.*, 1972, **11**, 117.

the separation of lines was examined.[803] Data have also been reported for some nickel complexes of nitrogen macrocycle ligands,[804-809] [(PhN=NCMe=CMeNNHPh)PdCl]$_2$,[810] [(MeC$_5$H$_3$NCH$_2$NHCH$_2$C$_5$H$_3$NMe)-PdCl]$^+$,[811] {RN=CMeC(COMe)=N—O}$_2$M (M = Ni or Pd),[812] [Ni(Ph$_2$PCH$_2$CH$_2$N=CMeCMe=NCH$_2$CH$_2$PPh$_2$)]$^{2+}$,[813] Ni(OC$_6$H$_3$Bu^{t}CH=N-CHMeCMe=NCH$_2$C$_6$H$_3$Bu^{t}O),[814] Ni(OCPh=CPhN=NCR^{1}R^{2}N=NC-Ph=CPhO),[815] and [Pd(L-ala-gly-gly)]$^-$.[816]

^{11}B and ^{31}P n.m.r. spectroscopy have been used to show that cis-PtX$_2$(PBu$^n_3$)$_2$ reacts with BX$_3$ to yield [(Bu$^n_3$P)$_2$PtX$_2$Pt(PBu$^n_3$)$_2$][BX$_4$].[817] Data have also been reported for Ni(PR$_2$F)$_4$ (^{19}F and ^{31}P n.m.r.),[818] Ni{P(OMe)$_3$}$_4$,[819] [(CF$_3$)$_2$PS$_2$PtSP(CF$_3$)$_2$]$_2$,[820] Pt(OCOCF$_3$)$_2$(PMe$_2$Ph)$_2$,[821] [Pt$_2$(OH)$_2$(PR$_3$)$_4$]$^{2+}$,[822] [{CH$_2$(PPh$_2$)$_2$}PtCl]$_2$,[823] and trans-PtCl(SO$_2$Ph)L$_2$ (L = Et$_3$P, Et$_3$As, Et$_2$Se, or Et$_2$Te).[824, 825]

For [(Me$_2$SO)$_2$Pt{S(=O)Me$_2$}$_2$]$^{2+}$ the S-bonded Me$_2$SO shows platinum coupling of 23.1 Hz.[826] The ^{1}H n.m.r. signal of PdI$_2$(SMe$_2$)$_2$ is broad at 30 °C but sharp at −20 °C.[827] ^{19}F N.m.r. spectra of [MF(PR$_3$)$_3$]$^+$[BF$_4$]$^-$ (M = Pd or Pt) have been reported with $^2J(^{31}$P–^{19}F)(cis) = ca. 30 Hz, $^2J(^{31}$P–^{19}F)(trans) = ca. 150 Hz, and $^1J(^{195}$Pt–^{19}F) = ca. 200 Hz, cf. [PtF$_6$]$^{2-}$ where $^1J(^{195}$Pt–^{19}F) = 2080 Hz. The low value of $^1J(^{195}$Pt–^{19}F) was attributed to the small s-character of the bond.[828] Data have also been

---

[803] R. Roulet and H. B. Gray, Inorg. Chem., 1972, **11**, 2101.
[804] V. L. Goedken, P. H. Merrell, and D. H. Busch, J. Amer. Chem. Soc., 1972, **94**, 3397.
[805] C. J. Hipp and D. H. Busch, J.C.S. Chem. Comm., 1972, 737.
[806] T. J. Truex and R. H. Holm, J. Amer. Chem. Soc., 1972, **94**, 4529.
[807] E. K. Barefield, F. V. Lovecchio, N. E. Tokel, E. Ochiai, and D. H. Busch, Inorg. Chem., 1972, **11**, 283.
[808] C. J. Hipp, L. F. Lindoy, and D. H. Busch, Inorg. Chem., 1972, **11**, 1988.
[809] B. M. Higson and E. D. McKenzie, J.C.S. Dalton, 1972, 269.
[810] L. Caglioti, L. Cattalini, M. Ghedini, F. Gasparrini, and P. A. Vigato, J.C.S. Dalton, 1972, 154.
[811] M. G. B. Drew, M. J. Riedl, and J. Rodgers, J.C.S. Dalton, 1972, 234.
[812] B. C. Sharma, K. S. Bose, and C. C. Patel, Inorg. Nuclear Chem. Letters, 1972, **8**, 805.
[813] T. D. DuBois, Inorg. Chem., 1972, **11**, 718.
[814] H. Kanatomi and I. Murase, Inorg. Chem., 1972, **11**, 1356.
[815] C. M. Kerwin and G. A. Melson, Inorg. Chem., 1972, **11**, 726.
[816] T. P. Pitner, E. W. Wilson, jun., and R. B. Martin, Inorg. Chem., 1972, **11**, 738.
[817] P. M. Druce, M. F. Lappert, and P. N. K. Riley, J.C.S. Dalton, 1972, 438.
[818] M.-H. Micoud, J.-M. Savariault, and P. Cassoux, Bull. Soc. chim. France, 1972, 3774.
[819] D. F. Bachman, E. D. Stevens, T. A. Lane, and J. T. Yoke, Inorg. Chem., 1972, **11**, 109.
[820] R. G. Cavell, W. Byers, E. D. Day, and P. M. Watkins, Inorg. Chem., 1972, **11**, 1598.
[821] D. M. Barlex and R. D. W. Kemmitt, J.C.S. Dalton, 1972, 1436.
[822] G. W. Bushnell, K. R. Dixon, R. G. Hunter, and J. J. McFarland, Canad. J. Chem., 1972, **50**, 3694.
[823] F. Glockling and R. J. I. Pollock, J.C.S. Chem. Comm., 1972, 467.
[824] F. Faraone, L. Silvestro, S. Sergi, and R. Pietropaolo, J. Organometallic Chem., 1972, **34**, C55.
[825] F. Faraone, L. Silvestro, S. Sergi, and R. Pietropaulo, J. Organometallic Chem., 1972, **46**, 379.
[826] J. H. Price, A. N. Williamson, R. F. Schramm, and B. B. Wayland, Inorg. Chem., 1972, **11**, 1280.
[827] P. L. Goggin, R. J. Goodfellow, S. R. Haddock, F. J. S. Reed, J. G. Smith, and K. M. Thomas, J.C.S. Dalton, 1972, 1904.
[828] K. R. Dixon and J. J. McFarland, J.C.S. Chem. Comm., 1972, 1274.

reported for Ni(EtOCSCH=CMeS)$_2$,[829] Ni(S$_2$CSR)$_2$,[830] Ni(S$_2$CNEt$_2$)$_2$,[831] M(S$_2$CNBu$^n_2$)$_3$X (M = Ni or Pt),[832] [{CH$_2$=CHCH$_2$S(CH$_2$)$_2$S}PdBr]$_2$,[833] and [(C$_7$H$_5$MeS$_2$)$_2$Ni]$^{2-}$.[834]

Complexes of Cu, Ag, and Au.—Information concerning complexes of these elements can be found at the following sources: (RC$_6$H$_3$CH$_2$NMe$_2$)$_4$-Cu$_2$Li$_2$,[173] (cyclo-octyne)MBr (M = Cu or Au),[759] [(C$_5$H$_5$)Fe(CO)$_2$C≡CPhCuCl]$_2$,[418] [(alkene)$_2$Ag]$^+$,[118, 426] Ph$_3$PAuSMeMn(CO)$_2$(C$_5$H$_5$),[227] Cu$_4$Ir$_2$(PMePh$_2$)$_2$(C≡CPh)$_8$,[579] [Au(N$_3$)$_n$]$^-$ (n = 2 or 4),[128] [M{P(CH$_2$-SiMe$_3$)$_3$}$_4$]$^+$ (M = Cu, Ag, or Au),[468] Pt(Bu$^n_2$N=CS$_2$)$_3$MBr$_2$ (M = Cu or Au),[832] Et$_2$NCS$_2$Cu,[700] [MI(PBu$^n_3$)]$_4$ (M = Cu or Ag),[303] and CuCl and CuBr.[144]

^{1}H N.m.r. spectroscopy has been used to investigate the complexation of MeAu and Au$^+$ by PMe$_3$ and P(OMe)$_3$. Rapid exchange was found and only average coupling constants and shifts were observed.[835] MeI reacts with Ph$_3$PAuMe to give Me$_3$AuPPh$_3$ and no Me$_2$AuIPPh$_3$ was detected. If cis-AuIMe$_2$PPh$_3$ is added to MeAuPPh$_3$, AuMe$_3$PPh$_3$ is formed.[836] AuCF$_2$CF$_2$Me(PPh$_3$) has the α-^{19}F resonance at 100.5 p.p.m., whereas in Re(CO)$_5$CF$_2$CF$_2$Me it is at 52.9 p.p.m. The ^{19}F n.m.r. spectrum of Ph$_3$PAuC(CF$_3$)=C(CF$_3$)AuPPh$_3$ is of the [AX$_3$]$_2$ type, with J(P–P) or J(F–F) = 50 Hz.[837] For olefin–copper(I) complexes, chemical shifts have been used to evaluate the relative contributions of σ and π components to

$$\begin{array}{c} \text{Me} \quad \text{L} \\ \diagdown \quad \diagup \\ \text{Au} \\ \text{F}_3\text{C}-\text{C}----\text{C}-\text{CF}_3 \\ \diagup \quad \diagdown \\ \text{Au} \\ \diagup \quad \diagdown \\ \text{Me} \quad \text{L} \end{array}$$
(109)

the bonding.[838] The ^{19}F n.m.r. spectrum of (109) is unsymmetric and therefore the CF$_3$ groups are probably inequivalent.[839] Data have also been reported for [AuC(OMe)=NC$_6$H$_{11}$]$_4$,[840] Ph$_3$PAuC=NAr(OR),[841] Cu(allyl

---

[829] A. R. Hendrickson and R. L. Martin, *Austral. J. Chem.*, 1972, **25**, 257.
[830] J. M. Andrews, D. Coucouvanis, and J. P. Fackler, jun., *Inorg. Chem.*, 1972, **11**, 493.
[831] J. L. K. F. de Vries and R. H. Herber, *Inorg. Chem.*, 1972, **11**, 2458.
[832] J. Willemse and J. A. Cras, *Rec. Trav. chim.*, 1972, **91**, 1309.
[833] L. Cattalini, J. S. Coe, S. Degetto, A. Dondoni, and A. Vigato, *Inorg. Chem.*, 1972, **11**, 1519.
[834] T. Herskovitz, C. E. Forbes, and R. H. Holm, *Inorg. Chem.*, 1972, **11**, 1318.
[835] H. Schmidbaur and R. Franke, *Chem. Ber.*, 1972, **105**, 2985.
[836] A. Tamaki and J. K. Kochi, *J. Organometallic Chem.*, 1972, **40**, C81.
[837] C. M. Mitchell and F. G. A. Stone, *J.C.S. Dalton*, 1972, 102.
[838] R. G. Salomon and J. K. Kochi, *J. Organometallic Chem.*, 1972, **43**, C7.
[839] A. Johnson, R. J. Puddephatt, and J. L. Quirk, *J.C.S. Chem. Comm.*, 1972, 938.
[840] G. Minghetti and F. Bonati, *Angew. Chem. Internat. Edn.*, 1972, **11**, 429.
[841] G. Minghetti and F. Bonati, *Gazzetta*, 1972, **102**, 205.

alcohol)ClO_4,[842] Ag(hfac)(olefin),[843] [(MeC≡CMe)$_2$Au]AuCl$_4$, and Au(CMe=CMeCl)Cl$_2$(py).[844, 845]

The low-temperature ^{31}P n.m.r. spectra of (Ph$_3$P)$_3$Ag(NCBH$_3$) and [{(p-tol)$_3$P}$_4$Ag]$^+$[BH$_3$CN]$^-$ shows ^1J(Ag–^{31}P).[846, 847] Data have also been reported for Cu{Ph$_2$P(CH$_2$)$_n$PPh$_2$}$_m$X,[848] M(PMe$_3$)$_n$Cl (M = Cu or Ag),[849] (Me$_3$P)$_n$MOSiMe$_3$ (M = Cu, Ag, or Au),[850] [Au$_9$(PPh$_3$)$_8$]$^{3+}$[PF$_6$]$^-_3$ (also ^{31}P n.m.r. data),[851] [Cu{(Me$_2$P=S)$_2$O}$_2$][CuBr$_2$] (also ^{31}P n.m.r. data),[852] M(PPh$_3$)$_2$N$_3$,CF$_3$CN (^{19}F n.m.r. data, M = Cu or Ag),[853] Au(SO$_3$F)$_3$,[854] and [C$_{12}$H$_{19}$][AuCl$_4$].[855]

Complexes of Zn, Cd, and Hg.—Information concerning complexes of these elements can be found at the following sources: MeHgX,[80, 86] Bu^{t}CH$_2$HgX,[69] (PhCH$_2$)$_2$Hg,[107] Hg(CH$_2$CH$_2$CH$_2$)$_2$O,[183] FC$_6$H$_4$HgX,[102, 103] C$_8$H$_{12}$HgCl$_2$,[759] (p-t-amylbenzyl)$_2$Hg,[176] (C$_5$H$_5$)Fe(CO)$_2$HgCl,[419] [{(C$_5$H$_5$)$_2$-Ru}$_2$Hg]$^{2+}$,[514] [MeOZnCo(CO)$_4$]$_4$,[596] Zn(C$_{14}$H$_{22}$N$_4$),[806] Zn{(triphenylporphyrin)(C$_6$H$_5$)Cr(CO)$_3$},[240] Zn(porphyrin),[668] MX$_2$\{P(CH$_2$SiMe$_3$)$_3$\}$_2$ (M = Zn, Cd, or Hg),[468] M{(Me$_2$P=S)E}$_2$ (M = Cd or Hg; E = O or S),[852] M(EtOCECHCMeS)$_2$ (M = Zn, Cd, or Hg; E = O or S),[829] and (Et$_2$NCS$_2$)$_2$M (M = Zn or Cd).[700]

The ^{1}H n.m.r. spectra of MeHgX has been interpreted as showing that the electron-acceptor ability of the X systems is in the order p-CHB$_{10}$H$_{10}$-CH ≤ p-AsB$_{10}$H$_{10}$CH < p-PB$_{10}$H$_{10}$CH; and m-CHB$_{10}$H$_{10}$CH < m-AsB$_{10}$-H$_{10}$CH < m-PB$_{10}$H$_{10}$CH.[856] The addition of SCN$^-$ to MeHgSCN causes ^2J(^{199}Hg–^{1}H) to increase from 205.6 to 212.5 Hz, but it was concluded that the MeHgSCN unit is unaffected in the complex ion.[857] The ^{1}H n.m.r. spectrum of (allyl)HgCl is very broad and, when a little HgCl$_2$ is added, the spectrum becomes of the AX$_4$ type.[858] Addition of dipivaloylmethane to Hg(OAc)$_2$ yields (110). The value of ^2J(^{199}Hg–^{1}H) of 333 Hz at −30 °C is the largest in the literature.[859] This structure was subsequently confirmed

---

[842] Y. Ishino, T. Ogura, K. Noda, T. Hirashima, and O. Manabe, *Bull. Chem. Soc. Japan*, 1972, **45**, 150.
[843] W. Partenheimer and E. H. Johnson, *Inorg. Chem.*, 1972, **11**, 2840.
[844] R. Hüttel and H. Forkl, *Chem. Ber.*, 1972, **105**, 2913.
[845] R. Hüttel and H. Forkl, *Chem. Ber.*, 1972, **105**, 1664,
[846] S. J. Lippard and P. S. Welcker, *Inorg. Chem.*, 1972, **11**, 6.
[847] E. L. Muetterties and C. W. Alegranti, *J. Amer. Chem. Soc.*, 1972, **94**, 6386.
[848] N. Marsich, A. Camus, and E. Cebulec, *J. Inorg. Nuclear Chem.*, 1972, **34**, 933.
[849] H. Schmidbaur, J. Adlkofer, and K. Schwirten, *Chem. Ber.*, 1972, **105**, 3382.
[850] H. Schmidbaur, J. Adlkofer, and A. Shiotani, *Chem. Ber.*, 1972, **105**, 3389.
[851] F. Cariati and L. Naldini, *J.C.S. Dalton*, 1972, 2286.
[852] D. A. Wheatland, C. H. Clapp, and R. H. Waldron, *Inorg. Chem.*, 1972, **11**, 2340.
[853] R. F. Ziolo, J. A. Thick, and Z. Dori, *Inorg. Chem.*, 1972, **11**, 626.
[854] W. M. Johnson, R. Dev, and G. H. Cady, *Inorg. Chem.*, 1972, **11**, 2260.
[855] R. Hüttel, P. Tauchner, and H. Forkl, *Chem. Ber.*, 1972, **105**, 1.
[856] L. A. Fedorov, V. I. Kyskin, and L. I. Zakharkin, *Izvest. Akad. Nauk S.S.S.R., Ser. khim.*, 1972, 536.
[857] J. Relf, R. P. Cooney, and H. F. Henneike, *J. Organometallic Chem.*, 1972, **39**, 75.
[858] W. Kitching, M. L. Bullpitt, P. D. Sleezer, S. Winstein, and W. G. Young, *J. Organometallic Chem.*, 1972, **34**, 233.
[859] R. H. Fish, R. E. Lundin, and W. F. Haddon, *Tetrahedron Letters*, 1972, 921.

$$\begin{pmatrix} O=C\diagdown_{But} \\ O=C\diagup^{But} \end{pmatrix}_2 CH-Hg$$

(110)

by an X-ray structure determination.[860] ¹H N.m.r. spectroscopy has been used to determine the *cis–trans* ratio of mercuriated olefins.[861] The metal satellite n.m.r. spectra of Et$_2$Hg, (cyclopropyl)$_2$Hg, Et$_2$Sn, and (cyclopropyl)$_4$M (M = Sn or Pb) have been reported. It was found that there are linear relationships between the coupling constants. The results were correlated with extended Hückel molecular orbital calculations.[862] Data have also been reported for BrZnCH$_2$C≡N, BrZnCH=C=CH$_2$,[863] MeZnCl, MePh$_2$SnZnCl,[864] (Me$_3$SiCHI)$_2$Hg,[865] FC$_6$H$_4$OHgR,[866] (*p*-tol)-HgCCl$_2$Br, PhHgCF$_3$,[867] CF$_3$HgX,[868] (CF$_3$CH$_2$CH$_2$)HgX,[869] PhCH$_2$-HgCH$_2$I,[870] Me$_3$CCH$_2$HgO$_2$CMe,[871] RHgCl,[872-878] and R$_2$Hg.[879]

For (R$_F$)$_2$Hg, it is found that a fluorine in the 1-position produces a low $^2J(^{199}Hg-C-^1H)$ and a high $^2J(^{199}Hg-C-^{19}F)$.[880] The ¹H n.m.r. spectra of (C$_2$H$_3$)$_2$M (M = Zn, Cd, or Hg) have been reported. The change in chemical shifts and coupling constants are discussed as a function of the central metal atom.[881] For some complexes such as (CH$_2$=CHCH$_2$CH$_2$)$_2$Zn, the α-protons are 0.2 p.p.m. to high field of that expected. Complexation of the double bond was suggested.[882] For alkenylmercuric chlorides, the

---

[860] R. Allmann, K. Flatau, and H. Musso, *Chem. Ber.*, 1972, **105**, 3067.
[861] R. F. Richter, J. C. Phillips, and R. D. Bach, *Tetrahedron Letters*, 1972, 4327.
[862] P. A. Scherr and J. P. Oliver, *J. Amer. Chem. Soc.*, 1972, **94**, 8026.
[863] N. Goasdoué and M. Gaudemar, *J. Organometallic Chem.*, 1972, **39**, 17.
[864] F. J. A. Des Tombe, G. J. M. Van der Kerk, and J. G. Noltes, *J. Organometallic Chem.*, 1972, **43**, 323.
[865] D. Seyferth, S. B. Andrews, and R. L. Lambert, jun., *J. Organometallic Chem.*, 1972, **37**, 69.
[866] A. N. Nesmeyanov, D. N. Kravtsov, B. A. Kvasov, E. M. Rokhlina, V. M. Pachevskaya, L. S. Golovchenko, and E. I. Fedin, *J. Organometallic Chem.*, 1972, **38**, 307.
[867] D. Seyferth and S. P. Hopper, *J. Organometallic Chem.*, 1972, **44**, 97.
[868] D. Seyferth, S. P. Hopper, and G. J. Murphy, *J. Organometallic Chem.*, 1972, **46**, 201.
[869] A. K. Prokof'ev and O. Yu. Okhlobystin, *J. Organometallic Chem.*, 1972, **36**, 239.
[870] R. Scheffold and U. Michel, *Angew. Chem. Internat. Edn.*, 1972, **11**, 231.
[871] R. F. Heck, *J. Organometallic Chem.*, 1972, **37**, 389.
[872] H. K. Hall, jun., J. P. Schaefer, and R. J. Spanggord, *J. Org. Chem.*, 1972, **37**, 3069.
[873] M. Matsuo and Y. Saito, *J. Org. Chem.*, 1972, **37**, 3350.
[874] R. D. Bach and R. F. Richter, *J. Amer. Chem. Soc.*, 1972, **94**, 4747.
[875] E. Vedejs and M. F. Salomon, *J. Org. Chem.*, 1972, **37**, 2075.
[876] J. J. Perie, J. P. Laval, J. Roussel, and A. Lattes, *Tetrahedron*, 1972, **28**, 675.
[877] J. E. Galle and A. Hassner, *J. Amer. Chem. Soc.*, 1972, **94**, 3930.
[878] E. Samuel and M. D. Rausch, *J. Organometallic Chem.*, 1972, **37**, 29.
[879] A. J. Bloodworth and R. J. Bunce, *J.C.S. Perkin I*, 1972, 2787.
[880] L. A. Fedorov, Z. Stumbreviciute, B. L. Dyatkin, B. I. Martynov, and S. R. Sterlin, *Doklady Akad. Nauk S.S.S.R.*, 1972, **204**, 1135.
[881] H. D. Visser and J. P. Oliver, *J. Organometallic Chem.*, 1972, **40**, 7.
[882] J. St. Denis, J. P. Oliver, and J. B. Smart, *J. Organometallic Chem.*, 1972, **44**, C32.

spectra are complex and $J(^{199}\text{Hg}-^{1}\text{H})$, due to both *gem* and *cis* coupling of ca. 300 Hz, were observed.[883] Data have also been reported for $\text{Me}_3\text{SiHg-CF}_2\text{CFClSiMe}_3$,[884] p-$\text{HC}_6\text{F}_4\text{HgX}$,[885] $\text{HgCl}(R^1C=CHR^2)$,[886] $\text{Hg(OAc)-}$$(\overline{\text{C}_6\text{H}_3\text{CH}_2\text{CH}_2})$,[887] $\text{PhHgCX}_2\text{CO}_2\text{Me}$,[888] chloromercuriated olefins,[889] acetoxymercuri-sugars,[890] and $\text{RHgSC}_6\text{H}_4\text{F-}p$ (^{19}F n.m.r. data).[891]

The oxidation of zinc octaethylporphin by bromine has been examined by ^{1}H n.m.r. The diamagnetic shift found for the dimer is ten times greater than that found for chlorophyll dimers.[892] The binding of HgCl_2 to S-β-(2-pyridylethyl)-L-cysteine has been investigated by ^{1}H n.m.r. spectroscopy. The mercury binds *via* nitrogen and not *via* sulphur or oxygen.[893] The ^{199}Hg satellites of complexes such as $\text{HgX}_2(\text{EMe}_3)_2$ (E = P or As) are broad. This was explained as due to relatively rapid ^{199}Hg relaxation. Similar effects were found for platinum complexes. $[\text{Hg(PMe}_3)_2]^{2+}$ gives a virtually coupled triplet.[894] The ^{1}H n.m.r. spectrum of $\text{Hg(PBu}^t_2)_2$ shows a doublet.[895] Variable-temperature n.m.r. spectra of $\text{ZnX}_2(NN'\text{-diethylthiourea})_2$ have been recorded in the range +30 to −90 °C. The measurements were interpreted as there being intramolecular —NH···X interactions and hydrogen-bonding to the solvent at low temperatures.[896] Data have

(111)

[883] R. C. Larock, S. K. Gupta, and H. C. Brown, *J. Amer. Chem. Soc.*, 1972, **94**, 4371.
[884] R. Fields, R. N. Haszeldine, and A. F. Hubbard, *J.C.S. Perkin I*, 1972, 847.
[885] H. B. Albrecht and G. B. Deacon, *Austral. J. Chem.*, 1972, **25**, 57.
[886] R. C. Larock and H. C. Brown, *J. Organometallic Chem.*, 1972, **36**, 1.
[887] C. Eaborn, A. A. Najam, and D. R. M. Walton, *J.C.S. Perkin I*, 1972, 2481.
[888] D. Seyferth, R. A. Woodruff, D. C. Mueller, and R. L. Lambert. jun., *J. Organometallic Chem.*, 1972, **43**, 55.
[889] F. R. Jensen, J. R. Miller, S. J. Cristol, and R. S. Beckley, *J. Org. Chem.*, 1972, **37**, 4341.
[890] D. Horton, J. M. Tarelli, and J. D. Wander, *Carbohydrate Res.*, 1972, **23**, 440.
[891] D. N. Kravstov, B. A. Kvasov, L. S. Golovchenko, E. M. Rokhlina, and E. I. Fedin, *J. Organometallic Chem.*, 1972, **39**, 107.
[892] J. H. Fuhrhop, P. Wasser, D. Riesner, and D. Mauzerall, *J. Amer. Chem. Soc.*, 1972, **94**, 7996.
[893] R. H. Fish and M. Friedman, *J.C.S. Chem. Comm.*, 1972, 812.
[894] P. L. Goggin, R. J. Goodfellow, S. R. Haddock, and J. G. Eary, *J.C.S. Dalton*, 1972, 647.
[895] M. Baudler and A. Zarkadas, *Chem. Ber.*, 1972, **105**, 3844.
[896] A. M. Giuliani, *J.C.S. Dalton*, 1972, 497.

also been reported for (111),[897] Zn(salicylaldimino-L-alanine),[898] M{S$_2$P-(CF$_3$)$_2$}$_2$ (M = Zn or Cd; ^{19}F n.m.r. data),[899] Hg{P(CF$_3$)$_2$}$_2$,[900] and HgI$_2$(S$_2$CNR$_2$).[901]

3 Dynamic Systems

The layout of this section has been altered from last year. This section is now in three parts: (a) 'Fluxional Molecules' dealing with rate processes involving no molecular change, (b) 'Equilibria' dealing with the use of n.m.r. to measure the position of equilibria and ligand-exchange reactions, and (c) 'Course of Reactions' dealing with the use of n.m.r. to monitor the course of reactions. Each section is ordered by the Periodic Table.

In a review on the structure of ion-pair solvation complexes, the use of n.m.r. has been discussed.[902] In 'Progress in Inorganic Chemistry', two relevant reviews have appeared: 'Nuclear Magnetic Resonance Cation Solvation Studies' by A. Fratiello, and 'Kinetics and Mechanisms of Isomerization and Racemization Processes of Six-Coordinate Chelate Complexes' by N. Serpone and D. G. Bickley.[903] 'Proton relaxation enhancement probes: – applications and limitations to systems containing macromolecules'[904] and 'The stereochemistry of dynamic transition metal hydrides'[905] have also been reviewed. The use of n.m.r. total lineshape analysis to determine the rates and activation energies of internal rotation has been critically examined.[906] Multisite chemical exchange by n.m.r. has been treated theoretically[907] and a new method of treating chemical exchange in nuclear magnetic relaxation has been proposed.[908]

Fluxional Molecules.—The modes of intramolecular rearrangement in octahedral complexes have been discussed with reference to recent n.m.r. experiments. It is pointed out that experiments cannot distinguish between a mode of rearrangement which involves racemization and one which does not.[909] A generalized treatment has been used to enumerate all distinct permutational isomerization reactions of molecules MH$_n$P$_4$, where n = 1, 2, 3, or 4, M is a transition metal, and P is a trisubstituted

---

[897] C. O. Bender, R. Bonnett, and R. G. Smith, *J.C.S. Perkin I*, 1972, 771.
[898] G. N. Weinstein and R. H. Holm, *Inorg. Chem.*, 1972, **11**, 2553.
[899] R. G. Cavell, E. D. Day, W. Byers, and P. M. Watkins, *Inorg. Chem.*, 1972, **11**, 1759.
[900] J. Grobe and R. Demuth, *Angew. Chem. Internat. Edn.*, 1972, **11**, 1097.
[901] H. C. Brinkhoff and J. M. A. Dautzenberg, *Rec. Trav. chim.*, 1972, **91**, 117.
[902] J. Smid, *Angew. Chem. Internat. Edn.*, 1972, **11**, 112.
[903] 'Progress in Inorganic Chemistry', series ed. S. J. Lippard, 'Inorganic Reaction Mechanisms', Part II, volume ed. J. O. Edwards, Interscience Publishers, New York, London, Sydney, Toronto, 1972.
[904] R. A. Dwek, *Adv. Mol. Relax. Processes*, 1972, **4**, 1.
[905] J. P. Jesson, *Du Pont Innovation*, 1972, **4**, 8.
[906] R. R. Shoup, E. D. Becker, and M. L. McNeel, *J. Phys. Chem.*, 1972, **76**, 71.
[907] L. W. Reeves, *Pulsed Magn. Opt. Resonance, Proc. Ampere Int. Summer Sch. 2, 1971*, 1972, 271.
[908] H. Wennerström, *Mol. Phys.*, 1972, **24**, 69.
[909] J. I. Musher, *Inorg. Chem.*, 1972, **11**, 2335.

phosphorus ligand.[910] Unusual structure and temperature-dependent behaviour of the ^{1}H n.m.r. spectra in a number of diamagnetic complexes, $M(S_2CNR_2)_n$, has been accounted for,[911] Aryl exchange in Cu_4MgAr_6 has been suggested.[912] Slowing of the dynamic interchange of bridge and terminal hydrogens in a metal tetrahydroborate has been observed for the first time in the low-temperature ^{1}H n.m.r. spectra of $(C_5H_5)_3UBH_4$ and $\Delta G^{\ddagger}$ has been estimated.[913]

At $-27\,°C$, $(C_5H_5)_4Ti$ shows the presence of two types of cyclopentadienyl groups, whereas $(C_5H_5)_4M$ (M = Zr or Hf) are still fluxional at $-27\,°C$.[914] $[Me_2Ti_2Cl_6Br]^-$ and $[Me_2Ti_2Cl_6Br_2]^{2-}$ show two methyl resonances at $-90\,°C$, which coalesce at $-82\,°C$.[915] $TiCl_3(OMe)(MeOCH_2CH_2OMe)$ is fluxional at room temperature but at low temperatures the *mer*- and *fac*-isomers were observed.[916] The variable-temperature behaviour of the ^{1}H n.m.r. spectra of $M(BH_4)_4$ (M = Zr or Hf) has been ascribed to ^{10}B and ^{11}B quadrupolar spin–lattice relaxation.[917] $Nb(NCS)_2(OMe)_3bipy$ has one methyl resonance at $40\,°C$, two at $10\,°C$, and three at $-40\,°C$.[918] $MoH_4(PMePh_2)_4$ is fluxional at room temperature and static at $-30\,°C$ whereas $WH_4(PMePh_2)_2$ is static at room temperature but fluxional at $70\,°C$.[919] The ^{1}H n.m.r. spectrum of (112) has been measured and exchange of H_a and H_b established by double resonance.[920] $(C_5H_5)Cr(CO)(NO)(C_2H_2)$ has intramolecular mobility of the π-bonded acetylene ligand in solution and $\Delta G^{\ddagger}$ was estimated.[921] $(C_5H_5)Mo(CO)_3Mo(CO)_2(CNMe)(C_5H_5)$ shows one cyclopentadienyl signal at $60\,°C$ and two at $-43\,°C$, which were attributed to (113) and (114).[922] The ^{1}H n.m.r. spectra of $H_2B(N_2C_3H_3)_2Mo(CO)_2(C_7H_7)$ shows that at $-120\,°C$ the C_7H_7 group is *trihapto*.[923, 924] At $-49\,°C$, $HB(N_2C_3H_3)_3Mo(CO)_2(C_5H_5)$ is static and is not completely fluxional even at $114\,°C$.[925] $B(N_2C_3H_3)_4Mo(CO)_2(C_5H_5)$ is static at low temperature as two conformers which interconvert with an activation energy of $10\,kcal\,mol^{-1}$.[926] The mechanism for fluxional behaviour in $RB(N_2C_3H_3)_3Mo(CO)_2(\pi$-allyl$)$ involves an internal rotation

[910] W. G. Klemperer, *Inorg. Chem.*, 1972, **11**, 2668.
[911] R. M. Golding, P. C. Healy, P. W. G. Newman, E. Sinn, and A. H. White, *Inorg. Chem.*, 1972, **11**, 2435.
[912] L. M. Seitz and R. Madl, *J. Organometallic Chem.*, 1972, **34**, 415.
[913] T. J. Marks and J. R. Kolb, *J.C.S. Chem. Comm.*, 1972, 1019.
[914] E. M. Brainina, N. P. Gambaryan, B. V. Lokshin, P. V. Petrovskii, Yu. T. Struchkov, and E. N. Kharlamova, *Izvest. Akad. Nauk S.S.S.R., Ser. khim.*, 1972, 187.
[915] R. J. H. Clark and M. A. Coles, *J.C.S. Dalton*, 1972, 2454.
[916] R. J. H. Clark and A. J. McAlees, *J.C.S. Dalton*, 1972, 640.
[917] T. J. Marks and L. A. Shimp, *J. Amer. Chem. Soc.*, 1972, **94**, 1542.
[918] N. Vuletić and C. Djordjević, *J.C.S. Dalton*, 1972, 2322.
[919] B. Bell, J. Chatt, G. J. Leigh, and T. Ito, *J.C.S. Chem. Comm.*, 1972, 34.
[920] J. A. Connor and P. D. Rose, *J. Organometallic Chem.*, 1972, **46**, 329.
[921] M. Heberhold, H. Alt, and C. G. Kreiter, *J. Organometallic Chem.*, 1972, **42**, 413.
[922] R. D. Adams and F. A. Cotton, *J. Amer. Chem. Soc.*, 1972, **94**, 6193.
[923] F. A. Cotton, J. L. Calderon, M. Jeremic, and A. Shaver, *J.C.S. Chem. Comm.*, 1972, 777.
[924] J. L. Calderon, F. A. Cotton, and A. Shaver, *J. Organometallic Chem.*, 1972, **42**, 419.
[925] J. L. Calderon, F. A. Cotton, and A. Shaver, *J. Organometallic Chem.*, 1972, **38**, 105.
[926] J. L. Calderon, F. A. Cotton, and A. Shaver, *J. Organometallic Chem.*, 1972, **37**, 127.

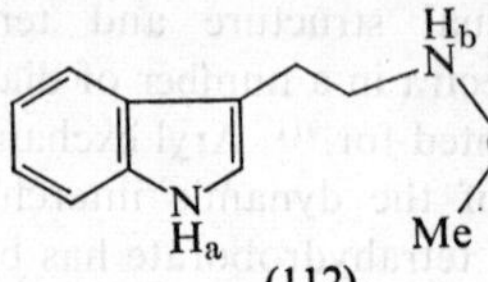

(112)

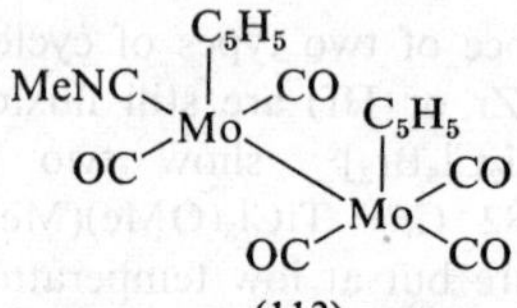

(113) 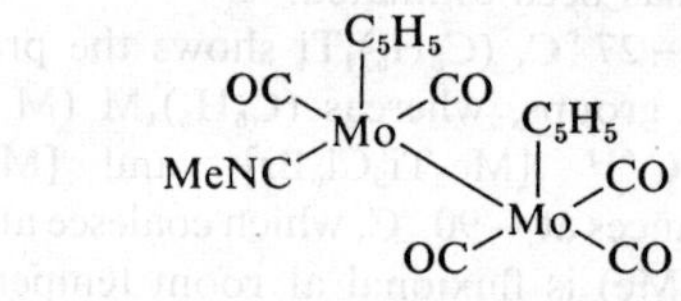 (114)

of the [RB(N$_2$C$_3$H$_3$)$_3$]$^-$ group around the B—Mo axis.[927] The variable-temperature ^{1}H n.m.r. studies of the complexes (C$_5$H$_5$)M(CO)$_2${(p-tol)$_2$-CNC(p-tol)$_2$} (M = Mo or W) showed the molecules to be fluxional, the observed temperature dependence being consistent with the occurrence of two distinct averaging processes within the aza-allyl–allene ligand.[928] Complex (115) is fluxional and activation parameters have been determined from variable-temperature ^{1}H n.m.r. spectroscopy.[929] The ^{1}H n.m.r. spectrum of [(C$_5$H$_5$)Mn(CO)(NO)]$_2$ shows a singlet at 40 °C but two doublets of unequal intensity at −62 °C. This behaviour was attributed to interconversion of (116) and (117).[930]

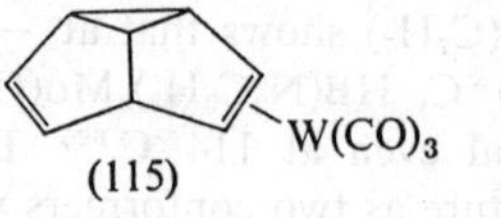

(115)

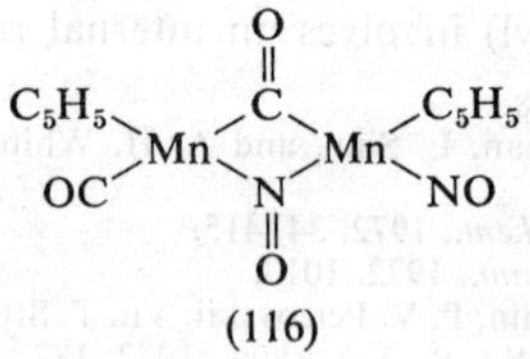

(116) 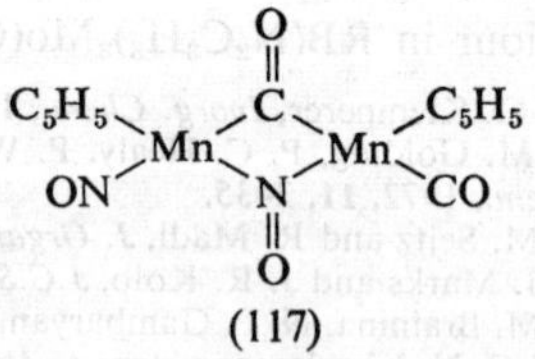 (117)

[HM^1(PF$_3$)$_4$]$^-$, HM^2(PF$_3$)$_4$, HM^3{P(OEt)$_3$}$_4$, and HM^4(Ph$_2$PCH$_2$CH$_2$-PPh$_2$)$_2$ (M^1 = Fe, Ru, or Os; M^2 = Co, Rh, or Ir; M^3 = Co or Rh; M^4 = Rh or Ir) are all stereochemically non-rigid but most give limiting spectra at low temperature. The data for HIr(CO)$_2$(PR$_3$)$_2$ are inconsistent

[927] P. Meakin, S. Trofimenko, and J. P. Jesson, *J. Amer. Chem. Soc.*, 1972, **94**, 5677.
[928] H. R. Keable and M. Kilner, *J.C.S. Dalton*, 1972, 1535.
[929] R. M. Moriarty, C. L. Yeh, E.-L. Yeh, and K. C. Ramey, *J. Amer. Chem. Soc.*, 1972, **94**, 9228.
[930] T. J. Marks and J. S. Kristoff, *J. Organometallic Chem.*, 1972, **42**, C91.

with the idealized Berry rearrangement.[931] $H_2Fe(PF_3)_4$ and $H_2Ru_3(CO)_9$-(C_8H_8) also show fluxional hydride behaviour.[932, 933] Cyclo-octa-1,5-diene reacts with α-$H_4Ru_4(CO)_{12}$ to yield $H_2Ru_3(CO)_9C_8H_{12}$, which shows two hydride signals at $-55\,°C$ which coalesce at $46\,°C$, and $\Delta G^{\neq}$ was measured.[934] $[(C_5H_5)Fe(CO)_2]_2$ has only one cyclopentadienyl resonance at $-44\,°C$ but two at $-66\,°C$: this was attributed to the presence of (118) and (119). $[(C_5H_5)Ru(CO)_2]_2$ is similar but there are four species.[935] ^{13}C N.m.r. spectroscopy has also been applied to $[(C_5H_5)Fe(CO)_2]_2$.[936] Variable-temperature ^{13}C n.m.r. spectroscopy has been used to demonstrate that 6-cyclopentadienyl rearrangement in $(\pi$-$C_5H_5)Fe(CO)_2(\sigma$-$C_5H_5)$ occurs via 1,2-shifts.[937] When cooled to $-120\,°C$, the ^{13}C n.m.r. spectrum of $C_8H_8Fe(CO)_3$ shows four C_8H_8 signals and two carbonyl signals, showing that the carbonyl groups are also rigid.[938] The ^1H n.m.r. spectra of (120) and (121) are temperature dependent, owing to $Fe(CO)_3$ migration. The

(118)

(119)

(120)

(121)

4, 5, 6, and 7 positions were deuteriated and deuterium decoupling was used to simplify the ^1H n.m.r. spectrum.[939] The ^{13}C n.m.r. spectrum of $Fe(CO)_3(Me_2PCH_2CH_2PMe_2)$ shows non-rigid behaviour owing to rapid intramolecular exchange. Thus the temperature dependence of the ^1H n.m.r. spectrum is attributed to solvent effects; $^2J(^{31}P-^{31}P)$ was measured.[940] $Os_3(CO)_7(C_6H_4)(PMe_2)_2$ shows an AA'XX' pattern for the C_6H_4 group at

---

[931] P. Meakin, E. L. Muetterties, and J. P. Jesson, *J. Amer. Chem. Soc.*, 1972, **94**, 5271.
[932] T. Kruck and R. Kobelt, *Chem. Ber.*, 1972, **105**, 3772.
[933] A. J. P. Domingos, B. F. G. Johnson, and J. Lewis, *J. Organometallic Chem.*, 1972, **36**, C43.
[934] A. J. Canty, B. F. G. Johnson, and J. Lewis, *J. Organometallic Chem.*, 1972, **43**, C35.
[935] J. G. Bullitt, F. A. Cotton, and T. J. Marks, *Inorg. Chem.*, 1972, **11**, 671.
[936] O. A. Gansow, A. R. Burke, and W. D. Vernon, *J. Amer. Chem. Soc.*, 1972, **94**, 2550.
[937] D. J. Ciappenelli, F. A. Cotton, and L. Kruczynski, *J. Organometallic Chem.*, 1972, **42**, 159.
[938] G. Rigatti, G. Boccalon, A. Ceccon, and G. Giacometti, *J.C.S. Chem. Comm.*, 1972, 1165.
[939] H. W. Whitlock, jun. and H. Stucki, *J. Amer. Chem. Soc.*, 1972, **94**, 8594.
[940] M. Akhtar, P. D. Ellis, A. G. MacDiarmid, and J. D. Odom, *Inorg. Chem.*, 1972, **11**, 2917.

room temperature and an ABXY pattern at $-60\,°C$, and $\Delta G^{\neq}$ was estimated.[941] $Os(CO)_4(SiMe_3)_2$ shows two methyl resonances at $0\,°C$ which coalesce at $80\,°C$.[942] $(OC)_4FeC\{NHMe\}_2$ shows two methyl resonances which coalesce at $69\,°C$.[943] Compounds of the type $(C_5H_5)Fe(CO)(PPh_3)CH_2R$ exhibit temperature-dependent ^1H n.m.r. spectra because of hindered rotation about the metal–carbon bond.[944] At room temperature the ^1H n.m.r. spectra of $[\{(C_5H_5)FeC_5H_4\}_2CH]^+X^-$ are of the AA′BB′ type, but at $-68\,°C$ the spectra are more complex owing to restricted rotation about the ferrocene–carbon bond.[945] At room temperature the ^{19}F n.m.r. spectrum of $(C_6H_8)Fe(PF_3)_2(CO)$ is of the AA′X$_3$X$_3$′ type, but at low temperature the spectrum changes to the ABX$_3$Y$_3$ type; $C_6H_8Fe(PF_3)(CO)_2$ is similar.[946] When (122) is cooled to $-73\,°C$, three types of proton are observed.[947] The first

$$\begin{array}{c} CH_2 \\ \|\| \\ H_2C \overset{C}{\underset{|}{\diagdown\diagup}} CH_2 \\ Fe(CO)_2PF_3 \end{array}$$

(122)

direct experimental evidence that indicates the axis and mode of rotation of the co-ordinated olefin molecules has been obtained from variable-temperature ^1H and ^{13}C n.m.r. studies on $[Os(CO)(NO)(C_2H_4)(PPh_3)_2][PF_6]$.[948]

Iron(III) tetraphenylporphyrin and its adduct with N-methylimidazole have been investigated. It was found that 20% of the iron is in the high-spin form and rates of exchange were determined.[949] Both inter- and intra-molecular site exchange in complexes of tetra-(p-isopropylphenyl)-porphinatoruthenium carbonyl with nitrogen bases have been studied by total lineshape analysis of the variable-temperature ^1H n.m.r. spectra.[950] In aqueous solution the methylene AB pattern of $[Fe\{(C_5H_4NCH=NCH)_3CMe\}]^{2+}$ collapses to a singlet at $95\,°C$, indicating racemization of the pseudooctahedral structure.[951]

[941] A. J. Deeming, R. S. Nyholm, and M. Underhill, *J.C.S. Chem. Comm.*, 1972, 224.
[942] R. K. Pomeroy and W. A. G. Graham, *J. Amer. Chem. Soc.*, 1972, **94**, 274.
[943] K. Öfele and C. G. Kreiter, *Chem. Ber.*, 1972, **105**, 529.
[944] J. Thomson, W. Keeney, M. C. Baird, and W. F. Reynolds, *J. Organometallic Chem.*, 1972, **40**, 205.
[945] S. Lupin, M. Kapon, M. Cais, and F. H. Herbstein, *Angew. Chem. Internat. Edn.*, 1972, **11**, 1025.
[946] J. D. Warren, M. A. Busch, and R. J. Clark, *Inorg. Chem.*, 1972, **11**, 452.
[947] R. J. Clark, M. R. Abraham, and M. A. Busch, *J. Organometallic Chem.*, 1972, **35**, C33.
[948] B. F. G. Johnson and J. A. Segal, *J.C.S. Chem. Comm.*, 1972, 1312.
[949] G. N. LaMar and F. A. Walker, *J. Amer. Chem. Soc.*, 1972, **94**, 8607.
[950] S. S. Eaton, G. R. Eaton, and R. H. Holm, *J. Organometallic Chem.*, 1972, **39**, 179.
[951] S. O. Wandiga, J. E. Sarneski, and F. L. Urbach, *Inorg. Chem.*, 1972, **11**, 1349.

Below $-80\,°C$, the ^1H n.m.r. spectrum of Fe(NN-disubstituted dithiocarbamato)maleonitriledithiolen corresponds to a chiral molecular configuration and slow carbon–nitrogen bond rotation. On warming, two processes occur: inversion, probably due to twist about the pseudo threefold axis of the complex, and carbon–nitrogen bond rotation.[952] Rearrangements have also been reported for $(MeBzNCS_2)_3Fe$[953] and $Ru(S_2PMe_2)L_2$.[954] At $-81\,°C$, $Rh_2(PF_3)_6(PhC\equiv CMe)$ shows in the ^{19}F n.m.r. spectrum only one type of PF_3, but two are observed at lower temperature.[955] At 30 °C, $IrCl(C_2H_4)_4$ has only one C_2H_4 resonance, but two resonances in the ratio 3:1 are observed at low temperature.[956] Temperature-dependent ^1H n.m.r. spectra of (diene)Rh(PhCOCHCSPh),[957] $Rh(\pi\text{-allyl})(PF_3)_3$,[958] $[(RC_6H_4NC)_4\text{-}Rh\{(NC)_2C=C(CN)_2\}]ClO_4$,[959] and $YCCo_3(CO)_7$(norbornadiene)[960] have been reported. The ^{13}C n.m.r. spectrum of $Rh_4(CO)_{12}$ is a quintet at 63.2 °C, which broadens on cooling.[961] The ^1H n.m.r. spectra of $[Co(CO)_n\text{-}L_{5-n}]^+[BPh_4]^-$ and $Co_4(CO)_{12-n}\{P(OMe)_3\}_n$ are temperature dependent with features such as doublets changing to singlets.[962,963] When (123) is heated

$$\begin{array}{c} C_5H_5 \\ | \\ Co \\ \diagdown\!\!\!\diagup \\ Co\!-\!C_5H_5 \end{array}$$

(123)

the two cyclopentadiene rings interchange.[964] Temperature-dependent ^1H n.m.r. spectra of Rh^I, Pd^{II}, and Pt^{II} complexes of $-NAr-N=NAr$ have been reported.[965] The temperature dependence of the n.m.r. spectra of several alkylcobaloxime dimers, $[RCo(dmg)_2]_2$, have been studied as a function of temperature and dissociation parameters obtained.[966]

At room temperature, the ^1H n.m.r. spectrum of trans-$PtH(SC_6H_4X\text{-}p)$-$(PPh_3)_2$ shows only a singlet with platinum satellites, but on cooling

---

[952] L. H. Pignolet, R. A. Lewis, and R. H. Holm, *Inorg. Chem.*, 1972, **11**, 99.
[953] D. J. Duffy and L. H. Pignolet, *Inorg. Chem.*, 1972, **11**, 2843.
[954] D. J. Cole-Hamilton, P. W. Armit, and T. A. Stephenson, *Inorg. Nuclear Chem. Letters*, 1972, **8**, 917.
[955] M. A. Bennett, R. N. Johnson, G. B. Robertson, T. W. Turney, and P. O. Whimp, *J. Amer. Chem. Soc.*, 1972, **94**, 6540.
[956] A. L. Onderdelinden and A. van der Ent, *Inorg. Chim. Acta*, 1972, **6**, 420.
[957] H. I. Heitner and S. J. Lippard, *Inorg. Chem.*, 1972, **11**, 1447.
[958] D. A. Clement, J. F. Nixon, and B. Wilkins, *J. Organometallic Chem.*, 1972, **37**, C43.
[959] K. Kawakami, T. Kaneshima, and T. Tanaka, *J. Organometallic Chem.*, 1972, **34**, C21.
[960] P. A. Elder and B. H. Robinson, *J. Organometallic Chem.*, 1972, **36**, C45.
[961] F. A. Cotton, L. Kruczynski, B. L. Shapiro, and L. F. Johnson, *J. Amer. Chem. Soc.*, 1972, **94**, 6191.
[962] S. Attali and R. Poilblanc, *Inorg. Chim. Acta*, 1972, **6**, 475.
[963] D. Labroue and R. Poilblanc, *Inorg. Chim. Acta*, 1972, **6**, 387.
[964] M. Rosenblum, B. North, D. Wells, and W. P. Giering, *J. Amer. Chem. Soc.*, 1972, **94**, 1239.
[965] S. D. Robinson and M. F. Uttley, *J.C.S. Chem. Comm.*, 1972, 184.
[966] A. W. Herlinger and T. L. Brown, *J. Amer. Chem. Soc.*, 1972, **94**, 388.

phosphorus coupling is observed.[967] The ^{1}H n.m.r. spectrum of $(Et_3P)_2PtB_3H_7$ gives signals 3:2:2 for B_3H_7, but on cooling the signals become 2:1:2:2 and are broad.[968] $L_2PtCl\{CO(NMe_2)\}$ shows restricted rotation about the carbon–nitrogen bond.[969] The ^{1}H n.m.r. spectrum of (di-t-butylsulphurdi-imine)$Pt(C_2H_4)Cl_2$ shows two isomers at $-60\,^\circ$C but only one set of resonances at 41 $^\circ$C.[970] The barriers to rotation of ethylene in $PtXYL(C_2H_4)$ have been measured.[971] When the free olefin is added to bis-(3,3-dimethylpent-1-ene)$PtCl_2$, exchange is observed in the ^{1}H n.m.r. spectrum.[972]

The ^{1}H n.m.r. spectra of (1,2-disubstituted π-allyl)palladium complexes have been measured to determine the stereochemistry of the 1-substituent. On heating there is interchange of the *syn* and *anti* protons.[973] [(π-1-methyl-1-chloroallyl)PdCl]$_2$ exists as a *syn–anti* mixture; on warming exchange begins.[974] As a consequence of a variable-temperature ^{1}H n.m.r. investigation of some π-crotyl(amine)palladium(II) halide complexes, three pathways of isomerization and epimerization have been suggested.[975] ^{1}H N.m.r. spectroscopy has been used to show that for (allyl)Pd(S_2COMe)(PMe$_2$Ph) the allyl group is π-bonded and the xanthate acts as a unidentate ligand, but that for (2-methylallyl)Pd(S_2CNMe$_2$)(PMe$_2$Ph) the allyl group is σ-bonded. Cooling to low temperature was necessary to stop fluxional

(124)

behaviour and ΔG was derived for *syn–anti* proton exchange.[976] Temperature dependence has also been observed in the ^{1}H n.m.r. spectra of some allylic palladium carboxylates,[977, 978] of (124) on the addition of a pyridine base,[979] and of (C_3H_5)Pd(PhCOCHCSPh).[980]

At 26 $^\circ$C, Ni$\{SC_6H_4CH=N(CH_2)_3\}_2$NH shows only one set of resonances,

---

[967] A. E. Keskinen and C. V. Senoff, *J. Organometallic Chem.*, 1972, **37**, 201.
[968] L. J. Guggenberger, A. R. Kane, and E. L. Muetterties, *J. Amer. Chem. Soc.*, 1972, **94**, 5665.
[969] C. R. Green and R. J. Angelici, *Inorg. Chem.*, 1972, **11**, 2095.
[970] J. Kuyper, K. Vrieze, and A. Oskam, *J. Organometallic Chem.*, 1972, **46**, C25.
[971] J. Ashley-Smith, I. Douek, B. F. G. Johnson, and J. Lewis, *J.C.S. Dalton*, 1972, 1776.
[972] C. Masters, *J.C.S. Chem. Comm.*, 1972, 1258.
[973] J. W. Faller, M. T. Tully, and K. J. Laffey, *J. Organometallic Chem.*, 1972, **37**, 193.
[974] D. J. S. Guthrie and S. M Nelson, *Co-ordination Chem. Rev.*, 1972, **8**, 139.
[975] J. W. Faller and M. J. Maltina, *Inorg. Chem.*, 1972, **11**, 1296.
[976] J. Powell and A. W.-L. Chan, *J. Organometallic Chem.*, 1972, **35**, 203.
[977] P. W. N. M. Van Leeuwen, J. Lukas, A. P. Praat, and M. Appelman, *J. Organometallic Chem.*, 1972, **38**, 199.
[978] J. Powell and T. Jack, *Inorg. Chem.*, 1972, **11**, 1039.
[979] E. Boan, A. Chan, and J. Powell, *J. Organometallic Chem.*, 1972, **34**, 405.
[980] S. J. Lippard and S. M. Morehouse, *J. Amer. Chem. Soc.*, 1972, **94**, 6949.

but at $-47\,°C$ two sets of resonances are observed.[981] Variable-temperature ^1H n.m.r. spectra have been recorded for $(RSC_2H_4SR)MX_2$ (R = Me, Et, Pr^n, Pr^i, or Bu^n; M = Pd or Pt; X = Cl, Br, or I). The pyramidal conformation at sulphur allows the detection of *meso* and racemic isomeric forms at low temperature, but at higher temperature inversion at the sulphur occurs.[982]

At room temperature, $Me_2Au(MeSCH_2CH_2SMe)Cl$ shows one set of resonances, but two sets are observed at low temperature when sulphur inversion is slow.[983] $Me_2AuCl(C_5H_5N)$ shows two methyl signals until excess pyridine is added, when coalescence occurs.[984] $(PhCH_2)_2SAuCl_3$ shows an AB pattern for the CH_2 group which coalesces at $62\,°C$.[985] Dynamic n.m.r. methods have been used to determine the average lifetimes of magnetically non-equivalent acetate methylenic protons before interchange in the zinc and cadmium complexes of 1,3-propenediaminetetra-acetic acid.[986]

The temperature effects in the ^{11}B n.m.r. spectrum of $F_2PPF_2BH_3$ have been ascribed to quadrupolar effects.[987] Intramolecular hydrogen exchange in tetraborane(10) has been studied by n.m.r. and rate data have been obtained; the mechanisms were discussed.[988] Complexes of BF_3 with acetylacetone and related complexes show fast exchange at room temperature but slow exchange at $-50\,°C$.[989] In the ^1H n.m.r. spectra of (2-methylallyl)$_3$B, the CH_2 groups are equivalent at $+80\,°C$ but are static at $-70\,°C$; similar results were found for 2-MeC$_3$H$_4$AlMe$_2$.[990] The rates of nitrogen inversion and topomerization have been determined for $PhNMePr^i$ or $PhCHMeNMe_2$ complexed with $AlCl_3$, $MeAlCl_2$, or Me_3Al.[991]

By use of ^1H n.m.r. spectroscopy it has been shown that in $MeN=NNMeSiR_3$ the R_3Si group migrates from one end of the molecule to the other.[992] Rotational barriers have been determined for $Me_3Si-NR^1CR^2O$,[993] but other workers have explained the non-equivalence of the $SiMe_3$ group as migration from nitrogen to oxygen.[994] The rearrangement of organosilylhydrazine anions has been studied by variable-temperature ^1H n.m.r. spectroscopy, and approximate rates have been determined from coalescence temperatures for non-equivalent organosilyl groups.[995] At

---

[981] I. Bertini, L. Sacconi, and G. P. Speroni, *Inorg. Chem.*, 1972, **11**, 1323.
[982] R. J. Cross, I. G. Dalgleish, G. J. Smith, and R. Wardle, *J.C.S. Dalton*, 1972, 992.
[983] H. Schmidbauer and K. C. Dash, *Chem. Ber.*, 1972, **105**, 3662.
[984] H. Hagnauer, G. C. Stocco, and R. S. Tobias, *J. Organometallic Chem.*, 1972, **46**, 179.
[985] F. Coletta, R. Ettorre, and A. Gambaro, *Inorg. Nuclear Chem. Letters*, 1972, **8**, 667.
[986] D. L. Rabenstein and B. J. Fuhr, *Inorg. Chem.*, 1972, **11**, 2430.
[987] H. L. Hodges and R. W. Rudolph, *Inorg. Chem.*, 1972, **11**, 2845.
[988] R. Schaeffer and L. G. Sneddon, *Inorg. Chem.*, 1972, **11**, 3098.
[989] A. Fratiello and R. E. Schuster, *J. Org. Chem.*, 1972, **37**, 2237.
[990] A. Stefani and P. Pino, *Helv. Chim. Acta*, 1972, **55**, 1110.
[991] Z. Buczkowski, A. Gryff-Keller, and P. Szczecinski, *Roczniki Chem.*, 1972, **46**, 195.
[992] N. Wiberg and H. J. Pracht, *Chem. Ber.*, 1972, **105**, 1388.
[993] A. Komoriya and C. H. Yoder, *J. Amer. Chem. Soc.*, 1972, **94**, 5285.
[994] M. Fukai, K. Itoh, and Y. Ishii, *J.C.S. Perkin II*, 1972, 1043.
[995] R. West and B. Bichlmeir, *J. Amer. Chem. Soc.*, 1972, **94**, 1649.

room temperature, the ^{1}H and ^{19}F n.m.r. spectra of $H_3SiNHPF_2$ are first order but at -73 °C the number of lines is too great to be accounted for by merely postulating inequivalence of the fluorines.[996] The inversion barriers for $PhPr^iM^1M^2Me_3$, $PhM^1(SiHMe_2)_2$ (M^1 = P or As; M^2 = Ge or Sn),[997] $Bu^tSiMe_2P(SiMe_3)Si(OMe)_3$,[998] $Pr^i_2NSiPh_2Cl$, $Pr^i_2NSiCl_2C_6F_5$,[999] Pr^iPhP-$SiMe_3$,[1000] and (125)[1001] have been determined. $(CF_3)_3P(OSiMe_3)_2$, when

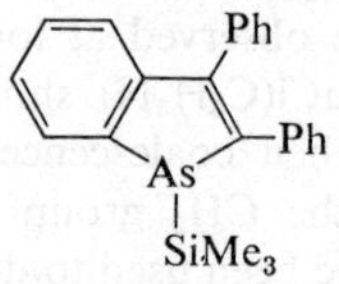

(125)

cooled, shows two sets of ^{19}F resonances in the ratio 6:3. Cooling to -140 °C produces additional splitting due to restricted rotation of the CF_3 groups.[1002]

The variable-temperature ^{1}H n.m.r. spectra of $C_5H_4(MMe_3)_2$ (M = Si, Ge, or Sn) have been measured. Several interconverting isomers were found and activation parameters determined.[1003, 1004] ^{13}C N.m.r. spectra of $C_5H_5SiMe_nCl_{3-n}$, $C_5H_5MMe_3$, $(\pi\text{-}C_5H_5)Fe(CO)_2(\sigma\text{-}C_5H_5)$, C_5H_5HgMe (M = Ge or Sn), and trimethylstannylindene have been measured and activation parameters determined.[1005, 1006] Similarly, the variable-temperature ^{1}H n.m.r. spectra of $(C_5H_5)MMe_3$ (M = Ge or Sn) have been measured and 1,2-shifts demonstrated.[1007] The ^{1}H n.m.r. spectrum of (126) is consistent with a static structure below -40 °C, but on warming migration occurs causing H^1 and H^5 to become equivalent.[1008] On heating, the 3- and 5-positions of (127) become equivalent, but (128) is static.[1009] Low-temperature ^{1}H n.m.r. spectra of $Me_2NCO(OSnMe_3)$ have been

---

[996] D. E. J. Arnold, E. A. V. Ebsworth, H. F. Jessep, and D. W. H. Rankin, *J.C.S. Dalton*, 1972, 1681.
[997] R. D. Baechler, J. P. Casey, R. J. Cook, G. H. Senkler, jun., and K. Mislow, *J. Amer. Chem. Soc.*, 1972, **94**, 2859.
[998] O. J. Scherer and R. Mergner, *J. Organometallic Chem.*, 1972, **40**, C64.
[999] W. R. Jackson and T. G. Kee, *Tetrahedron Letters*, 1972, 5051.
[1000] R. D. Baechler and K. Mislow, *J.C.S. Chem. Comm.*, 1972, 185.
[1001] R. H. Bowman and K. Mislow, *J. Amer. Chem. Soc.*, 1972, **94**, 2861.
[1002] R. G. Cavell, R. D. Leary, and A. J. Tomlinson, *Inorg. Chem.*, 1972, **11**, 2578.
[1003] Yu. A. Ustynyuk, A. V. Kisin, I. M. Pribytkova, A. A. Zenkin, and N. D. Antonova, *J. Organometallic Chem.*, 1972, **42**, 47.
[1004] Yu. A. Ustynyuk, A. V. Kisin, and A. A. Zenkin, *J. Organometallic Chem.*, 1972, **37**, 101.
[1005] Yu. K. Grishin, N. M. Sergeyev, and Yu. A. Ustynyuk, *Org. Magn. Resonance*, 1972, **4**, 377.
[1006] N. M. Sergeyev, Yu. K. Grishin, Yu. N. Luzikov, and Yu. A. Ustynyuk, *J. Organometallic Chem.*, 1972, **38**, C1.
[1007] A. V. Kisin, V. A. Korenevsky, N. M. Sergeyev, and Yu. A. Ustynyuk, *J. Organometallic Chem.*, 1972, **34**, 93.
[1008] M. D. Curtis and R. Fink, *J. Organometallic Chem.*, 1972, **38**, 299.
[1009] F. A. Cotton and D. J. Ciappenelli, *Synth. Inorg. Metal-org. Chem.*, 1972, **2**, 197.

(126) (127) (128)

measured and activation parameters for hindered rotation about the nitrogen–carbon bond were found.[1010]

The ^{1}H n.m.r. spectrum of molten $(NH_4)HSO_4$ indicates that proton scrambling occurs both between cations and between cations and anions, and rate constants have been determined.[1011] The ^{19}F n.m.r. spectrum of $(CF_3)_2C=NC(CF_3)_2NSO$ shows two singlets at 90 °C, but at -60 °C three lines were observed.[1012] The value of $\Delta G^{\ddagger}$ for the barrier to rotation around a phosphorus–nitrogen bond has been determined for some aminophosphines and is dependent on the substituents bound to the phosphorus.[1013] An analysis of the temperature-dependent ^{19}F n.m.r. lineshapes for $PF_3(NH_2)_2$ has established that the internal rotation about the two equivalent P—N bonds is essentially uncorrelated.[1014] The low-temperature ^{19}F n.m.r. spectrum of $(C_5H_5)PF_2$ is a doublet of doublets, but above 25 °C it is a doublet of sextets.[1015] $Me_2Sb(OMe)_3$ shows one MeO resonance but two at low temperature.[1016]

Selenolidine (129) has sharp ^{1}H n.m.r. resonances but the methyl resonance splits into two at -70 °C, which was attributed to two conformers of the ring.[1017] The low-temperature ^{1}H n.m.r. spectrum of $(H_2N)_2CE$ (E = O, S, or Se) shows inner and outer NH_2 protons. ^{14}N Decoupling

(129) (130)

and variable-temperature n.m.r. spectroscopy were used to obtain the barrier to internal rotation.[1018] The ^{19}F n.m.r. spectrum of $C_6F_5SF_3$ can be best explained by (130), with a relatively high barrier to rotation about the carbon–sulphur bond. ^6J(F–F) was observed and activation energies were calculated.[1019] Water catalyses fluorine exchange in R_2NSF_3 and thus

---

[1010] A. E. Lemire and J. C. Thompson, *Canad. J. Chem.*, 1972, **50**, 1386.
[1011] C. Hall, *J.C.S. Chem. Comm.*, 1972, 889.
[1012] R. F. Swindell, D. P. Babb, T. J. Ouellette, and J. M. Shreeve, *Inorg. Chem.*, 1972, **11**, 242.
[1013] M.-P. Simonnin, C. Charrier, and R. Burgada, *Org. Magn. Resonance*, 1972, **4**, 113.
[1014] E. L. Muetterties, P. Meakin, and R. Hoffmann, *J. Amer. Chem. Soc.*, 1972, **94**, 5674.
[1015] J. E. Bentham, E. A. V. Ebsworth, H. Moretto, and D. W. H. Rankin, *Angew. Chem. Internat. Edn.*, 1972, **11**, 640.
[1016] H. A. Meinema and J. G. Noltes, *J. Organometallic Chem.*, 1972, **36**, 313.
[1017] C. Draguet and M. Renson, *Bull. Soc. chim. belges*, 1972, **81**, 279.
[1018] W. Walter, E. Schaumann, and H. Rose, *Tetrahedron*, 1972, **28**, 3233.
[1019] W. A. Sheppard and D. W. Ovenall, *Org. Magn. Resonance*, 1972, **4**, 695.

addition of Me_3SiNEt_2 or silylation of the n.m.r. tube with $(Me_3Si)_2NH$ can reduce exchange.[1020, 1021] The temperature dependence of the ^{19}F n.m.r. spectrum of $ClOF_5$ could not be explained.[1022] At room temperature the ^{19}F n.m.r. spectrum of IF_4OMe is broad but cooling produces sharpening.[1023]

Equilibria.—*Solvation Studies of Ions.* Association constants and hydroxyl ^1H n.m.r. shifts have been measured for one-to-one complexes in acetonitrile of Li^+, Na^+, Mg^{2+}, Ca^{2+}, Sr^{2+}, and Ba^{2+} with water and alcohols. The results have been explained on an electrostatic model.[1024] Chemical shift measurements on $^{35}Cl^-$, $^{79}Br^-$, and $^{127}I^-$ in the solvents water, MeOH, MeCN, Me_2SO, and Me_2NCHO have shown a strong solvent dependence but only a weak cation dependence for the cations Li^+, Na^+, K^+, and Et_4N^+. A close correlation between chemical shifts and the charge-transfer-to-solvent u.v. absorption band energies has been interpreted as showing that the n.m.r. paramagnetic term is dominant in determining shifts.[1025] LiCl in aqueous pyridine forms a solvate Li^+Cl^-,py.[1026] For $NC_8H_8^-$, the dependence of the α-proton chemical shift on the cation has been interpreted as due to ion pairing in acetone.[1027]

The free energies of transfer of NaF from H_2O to H_2O_2–H_2O mixtures have been measured by use of e.m.f. measurements, but are in poor agreement with n.m.r. measurements.[1028] The ^1H n.m.r. spectra of NaX have been measured in alcohols and anion–solvent interactions postulated.[1029] ^{23}Na N.m.r. chemical shifts of aqueous $NaClO_4$ and NaOH have been interpreted by the formation of short-lived ion pairs.[1030] The ^{23}Na chemical shifts of NaI and $NaBPh_4$ in strongly basic solvents are downfield of aqueous saturated NaCl, indicating strong solvation.[1031] The hydration of sodium ions which occurs when a solution of $NaBPh_4$ in THF is titrated with water can be followed by the ^{23}Na chemical shift, and a hydration number of 3—4 is obtained.[1032] ^{23}Na Linewidths of $NaBH_4$ and $NaBPh_4$ in $Me(OCH_2CH_2)_nOMe$ are proportional to viscosity (absolute temperature)$^{-1}$.[1033] ^{23}Na N.m.r. spectra have also been used to study solvation by glycerol acetate.[1034] ^{23}Na N.m.r. of soap solutions

---

[1020] A. F. Janzen, J. A. Gibson, and D. G. Ibbott, *Inorg. Chem.*, 1972, **11**, 2853.
[1021] D. G. Ibbott and A. F. Janzen, *Canad. J. Chem.*, 1972, **50**, 2428.
[1022] K. Züchner and O. Glemser, *Angew. Chem. Internat. Edn.*, 1972, **11**, 1094.
[1023] G. Oates and J. M. Winfield, *Inorg. Nuclear Chem. Letters*, 1972, **8**, 1093.
[1024] G. W. Stockton and J. S. Martin, *J. Amer. Chem. Soc.*, 1972, **94**, 6921.
[1025] T. R. Stengle, Y.-C. E. Pan, and C. H. Langford, *J. Amer. Chem. Soc.*, 1972, **94**, 9037.
[1026] D. W. Larsen, *J. Phys. Chem.*, 1972, **76**, 53.
[1027] A. G. Anastassiou and S. W. Eachus, *J. Amer. Chem. Soc.*, 1972, **94**, 2537.
[1028] A. K. Covington, K. E. Newman, and M. Wood, *J.C.S. Chem. Comm.*, 1972, 1234.
[1029] B. S. Krumgal'z, K. P. Mishchenko, D. G. Traber, and Yu. I. Tsereteli, *Zhur. strukt. Khim.*, 1972, **13**, 396.
[1030] G. J. Templeman and A. L. Van Geet, *J. Amer. Chem. Soc.*, 1972, **94**, 5578.
[1031] M. Herlem and A. I. Popov, *J. Amer. Chem. Soc.*, 1972, **94**, 1431.
[1032] A. L. Van Geet, *J. Amer. Chem. Soc.*, 1972, **94**, 5583.
[1033] G. W. Canters, *J. Amer. Chem. Soc.*, 1972, **94**, 5230.
[1034] E. M. Arnett, H. C. Ko, and R. J. Minasz, *J. Phys. Chem.*, 1972, **76**, 2474.

(lyotropic nematic phase) shows a triplet structure due to incomplete averaging of the quadrupole interaction constant.[1035] Similar results have been obtained for sodium in other liquid crystals.[1036-1038]

The ^{39}K chemical shifts in a number of aqueous electrolyte solutions are thought to arise from overlapping of the outer electron orbitals of anion and cation during random ionic collisions.[1039] By observing the nuclear fluorine–proton Overhauser effect in aqueous solutions of KF it has been shown that the longitudinal components of the ^{19}F magnetic moments relax by dipole–dipole interaction with the protons of the surrounding water molecules and with other fluorine nuclei.[1040] As the ^{55}Mn linewidth of $[C_6H_4\{OCH_2CH_2OCH_2CH_2O\}_2C_6H_4K]^+[MnO_4]^-$ is greater than that for free MnO_4^-, ion pairing has been suggested.[1041] ^{1}H and ^{13}C n.m.r. spectroscopy have been used to show that RbI goes into the cage of $O\{CH_2CH_2N(CH_2CH_2OCH_2CH_2)_2NCH_2CH_2\}_2O$.[1042] The nuclear quadrupole relaxation rate of ^{85}Rb in aqueous solution of rubidium caprylate and caproate have been studied as a function of soap concentration and temperature.[1043]

From ^{1}H n.m.r. measurement on complexes of $BeCl_2$ or $Al(ClO_4)_3$ with $(Me_2N)_3P=O$, $(MeO)_2MeP=O$, $(MeO)_3P$, or H_2O, a total solvation number of 6 for aluminium and 4 for beryllium has been found. For $[Be\{(Me_2N)_3PO\}_4]^{2+}$, $^2J(^9Be-^{31}P) = 6.1$ Hz.[1044] Below -73 °C, in a water–acetone mixture, it is possible to observe bulk and co-ordinated water or acetone on $Mg(ClO_4)_2$.[1045] Similarly, an aquation number of 5 has been found for $Lu(NO_3)_3$ in water–acetone. It was suggested that the NO_3 is bound.[1046]

The ligand-exchange processes of $[UO_2]^{2+}$ complexes which form during an extraction of $UO_2(NO_3)_2$ by organic solutions of $(BuO)_3PO$, $(BuO)Bu_2PO$, or Bu_3PO have been studied by spin-echo n.m.r.[1047] ^{1}H N.m.r. spectroscopy has been used to follow the extraction of uranyl chloride with 0.082M tricaprylmonomethylammonium chloride into CCl_4.[1048] The relaxation time of protons in aqueous solutions of U^{4+} in the presence of F^- was studied by spin-echo n.m.r. The presence of F^-

[1035] D. M. Chen and L. W. Reeves, *J. Amer. Chem. Soc.*, 1972, **94**, 4384.
[1036] G. Lindblom, *Acta Chem. Scand.*, 1972, **26**, 1745.
[1037] G. J. T. Tiddy, *J.C.S. Faraday I*, 1972, **68**, 670.
[1038] G. J. T. Tiddy, *J.C.S. Faraday I*, 1972, **68**, 369.
[1039] E. G. Bloor and R. G. Kidd, *Canad. J. Chem.*, 1972, **50**, 3926.
[1040] G. Keller, *Ber. Bunsengesellschaft phys. Chem.*, 1972, **76**, 24.
[1041] D. J. Sam and H. E. Simmons, *J. Amer. Chem. Soc.*, 1972, **94**, 4024.
[1042] J. Cheney, J. M. Lehn, J. P. Sauvage, and M. E. Stubbs, *J.C.S. Chem. Comm.*, 1972, 1100.
[1043] B. Lindman and I. Danielsson, *J. Colloid Interface Sci.*, 1972, **39**, 349.
[1044] J. J. Delpuech, A. Peguy, and M. R. Khaddar, *J. Magn. Resonance*, 1972, **6**, 325.
[1045] R. D. Green and N. Sheppard, *J.C.S. Faraday II*, 1972, **68**, 821.
[1046] K. B. Yatsimirskii, V. A. Bidzilya, and N. K. Davidenko, *Doklady Akad. Nauk S.S.S.R.*, 1972, **202**, 1379.
[1047] A. A. Vashman, T. Ya. Vereshchagina, and I. S. Pronin, *Zhur. neorg. Khim.*, 1972, **17**, 471.
[1048] T. Sato, *J. Inorg. Nuclear Chem.*, 1972, **34**, 3835.

in the inner co-ordination sphere of U^{4+} increases the bond strength of $U-OH_2$.[1049]

The ligand-exchange reactions of $NbCl_5,NCR$ have been studied by ^1H n.m.r. spectroscopy.[1050] The exchange rates of the solvents methanol and DMF from Mn^{III}(protoporphyrin 1X dimethyl ester) have been determined from the temperature dependence of line-broadenings and chemical shift.[1051] ^1H N.m.r. spectroscopy has been used to measure the methanol exchange rate from the *cis* and *trans* co-ordination sites of $[Co(MeOH)_5-NCS]^+$ by lineshape analysis of the methyl resonance of the co-ordinated methanol.[1052] ^1H and ^{14}N n.m.r. spectroscopy have been used to study acetonitrile exchange between the first co-ordination sphere of cobalt(II) and the bulk solvent.[1053-1055]

From a ^{17}O n.m.r. study of aqueous Ni^{2+} it has been concluded that there are six moles of water co-ordinated, and water in the second co-ordination sphere can just be detected.[1056] The ^1H n.m.r. spectrum of $[Ni(thiourea)_6]-[ClO_4]_2$ shows rapid exchange even at $-90\ °C$, but exchange in $[Ni(dimethylthiourea)_6][ClO_4]_2$ is stopped at $-90\ °C$.[1057] For $Ni^{II}\beta\beta'\beta''$-triaminotriethylamine, two water exchange rates have been found from ^{17}O n.m.r. line-broadening, but the copper analogue shows only one slow exchange rate.[1058] From n.m.r. relaxation measurements on $Cu(ClO_4)_2$ in MeOH, a pseudo-first-order rate constant and activation energy for MeOH exchange have been measured.[1059]

^1H N.m.r. measurements of the exchange of MeCN with BX_3 have shown that the rate and activation energy for dissociation are influenced by solvent effects.[1060] The use of aromatic counter ions to produce shifts has been examined and, using $[BPh_4]^-$, upfield shifts of up to 3 p.p.m. have been found.[1061] The low-temperature ^1H n.m.r. spectra for $AlCl_3$, BCl_3, $GaCl_3$, $GaBr_3$, and $SbCl_5$ in MeCN show signals due to free and bound MeCN. An apparent co-ordination of 1.5 for $AlCl_3$ is attributed to $[Al(MeCN)_6]^{3+}$-$[AlCl_4]_3^-$ and of 1 for $SbCl_5$ to $[SbCl_4(MeCN)_2]^+[SbCl_6]^-$.[1062, 1063] Similarly, ^1H and ^{71}Ga nuclear magnetic resonances have been used to show that solvation of $GaCl_3$ by MeCN produces $[Ga(NCMe)_6][GaCl_4]_3$.[1064] Al^{3+}, Ga^{3+}, and In^{3+} in $(MeO)_3PO-H_2O$-acetone have been shown to have

[1049] V. A. Glebov, *Zhur. neorg. Khim.*, 1972, **17**, 1175.
[1050] A. Merbach and J. C. Bünzli, *Helv. Chim. Acta*, 1972, **55**, 179.
[1051] L. Rusnak and R. B. Jordan, *Inorg. Chem.*, 1972, **11**, 196.
[1052] J. R. Vriesenga, *Inorg. Chem.*, 1972, **11**, 2724.
[1053] R. J. West and S. F. Lincoln, *Inorg. Chem.*, 1972, **11**, 1688.
[1054] G. Beech and K. Miller, *J.C.S. Dalton*, 1972, 801.
[1055] S. F. Lincoln and R. J. West, *Austral. J. Chem.*, 1972, **25**, 469.
[1056] J. W. Neely and R. E. Connick, *J. Amer. Chem. Soc.*, 1972, **94**, 3419.
[1057] D. R. Eaton and K. Zaw, *J. Amer. Chem. Soc.*, 1972, **94**, 4394.
[1058] D. P. Rablen, H. W. Dodgen, and J. P. Hunt, *J. Amer. Chem. Soc.*, 1972, **94**, 1771.
[1059] R. Poupko and Z. Luz, *J. Chem. Phys.*, 1972, **51**, 3311.
[1060] J. Fogelman and J. M. Miller, *Canad. J. Chem.*, 1972, **50**, 1262.
[1061] G. P. Schiemenz, *J. Magn. Resonance*, 1972, **6**, 291.
[1062] I. Y. Ahmed and C. D. Schmulbach, *Inorg. Chem.*, 1972, **11**, 228.
[1063] I. Y. Ahmed, *J. Magn. Resonance*, 1972, **7**, 196.
[1064] S. F. Lincoln, *Austral. J. Chem.*, 1972, **25**, 2705.

a total hydration number of six.[1065, 1066] Separate OH and OCH_2 resonances have been observed for free and bound Pr^nOH in solutions of $AlCl_3$ in Pr^nOH at low temperature with an aluminium solvation number of four.[1067] ^1H and ^{27}Al n.m.r. spectra of Bu^i_3Al have been measured to study solvation.[1068] In co-ordinating solvents, $J(^1H-^{27}Al)$ is observed for $NaAlEt_4$ and it was concluded that, in co-ordinating solvents, solvation occurs to give symmetric solvent-separated ion pairs.[1069]

The ^1H n.m.r. chemical shifts of R_4NX have been discussed in terms of the water structure.[1070] From the ^1H n.m.r. spectra of concentrated aqueous solutions of R_4NBr, association is postulated at higher concentrations to give a quasi-crystalline lattice.[1071] The ^1H n.m.r. chemical shifts of Me_3SX have been interpreted in terms of ion association.[1072]

Ionic Equilibria. N.m.r. has been used to study the binding of Na^+ to the saliva of patients with cystic fibrosis of the pancreas.[1073] ^{23}Na N.m.r. relaxation times of sodium with phosphatidylserine, phosphoserine, and phosphoethanolamine have been studied as a function of pH and ionic association has been concluded.[1074] ^{13}C N.m.r. spectroscopy has been used to investigate complexing of Na^+ and K^+ by valinomycin and beauvericin.[1075, 1076] Low-temperature ^1H n.m.r. spectra of mixtures of $(C_5H_5)_2Mg$ and MgX_2 indicate that cyclopentadienylmagnesium halides consist of Schlenk equilibrium mixtures.[1077] ^{25}Mg N.m.r. spectra of aqueous solutions containing phosphate groups have been used to determine rates of exchange.[1078] The ligand $[MeO_2C(CH_2)_{10}NMeCOCH_2OCH_2]_2$ interacts with Ca^{2+}, as demonstrated by changes in the ^1H and ^{13}C n.m.r. spectra.[1079] Complexes of carbohydrates with metal cations have been examined by ^1H n.m.r. spectroscopy and it is shown that La^{3+}, Ca^{2+}, and Sr^{2+} form the strongest complexes.[1080]

[1065] J. Crea and S. F. Lincoln, *Inorg. Chem.*, 1972, **11**, 1131.
[1066] L. S. Frankel and R. S. Drago, *Inorg. Chem.*, 1972, **11**, 1964.
[1067] H. Grasdalen, *J. Magn. Resonance*, 1972, **6**, 336.
[1068] S. I. Vinogradova, V. M. Denisov, and A. I. Kol'tsov, *Zhur. obshchei Khim.*, 1972, **42**, 1031.
[1069] T. D. Westmoreland, jun., N. S. Bhacca, J. D. Wander, and M. C. Day, *J. Organometallic Chem.*, 1972, **38**, 1.
[1070] M.-M. Marciacq-Rousselot, A. de Trobriand, and M. Lucas, *J. Phys. Chem.*, 1972, **76**, 1455.
[1071] A. LoSurdo and H. E. Wirth, *J. Phys. Chem.*, 1972, **76**, 130.
[1072] A. K. Covington, M. L. Hassall, and I. R. Lantzke, *J.C.S. Faraday II*, 1972, **68**, 1352.
[1073] D. Martinez and A. A. Silvidi, *Arch. Biochem. Biophys.*, 1972, **148**, 224.
[1074] T. L. James and J. H. Noggle, *Analyt. Biochem.*, 1972, **49**, 208.
[1075] V. F. Bystrov, V. T. Ivanov, S. A. Koz'min, I. I. Mikhaleva, K. K. H. Khalilulina, Yu. A. Ovchinnikov, E. I. Fedin, and P. V. Petrovskii, *F.E.B.S. Letters*, 1972, **21**, 34.
[1076] M. Ohnishi, M. C. Fedarko, J. D. Baldeschwieler, and L. F. Johnson, *Biochem. Biophys. Res. Comm.*, 1972, **46**, 312.
[1077] W. T. Ford and J. B. Grutzner, *J. Org. Chem.*, 1972, **37**, 2561.
[1078] R. G. Bryant, *J. Magn. Resonance*, 1972, **6**, 159.
[1079] D. Ammann, E. Pretsch, and W. Simon, *Tetrahedron Letters*, 1972, 2473.
[1080] S. J. Angyal, *Austral. J. Chem.*, 1972, **25**, 1957.

¹H N.m.r. spectroscopy has been used to study complex formation between H₄edta or $(HO_2CCH_2)_2N(CH_2)_2N(CH_2CO_2H)(CH_2CH_2OH)$ and La, Y, or Lu.[1081, 1082] The exchange kinetics of $\beta\beta'\beta''$-triaminotriethylamine (tren) with $[Nd(tren)_2]^{3+}$ in acetonitrile have been measured using ¹H n.m.r. line-broadening.[1083] Complex formation between histidine and Nd^{3+} has also been investigated,[1084] and the rates of H₄edta exchange in complexes with Lu have also been measured.[1085] Complex formation between H₄edta or H₄edda and vanadium(v) has also been examined.[1086]

¹⁷O N.m.r. studies have been made on Mn^{2+} complexes with the ligands H₃nta, H₄egta, and H₄edta, and kinetic parameters measured.[1087] The ¹H spin relaxation in $Mn^{II}(N\text{-Me-}\gamma\text{-butyrolactam})$ complexes has been interpreted as being due to first-solvation-sphere relaxation at high temperature and chemical exchange at low temperature.[1088] ¹H and ¹³C n.m.r. have been used to study the binding of Mn^{2+} to fructose 1,6-diphosphate [1089] and of Mn^{2+}, Eu^{3+}, Gd^{3+}, and $[Fe(CN)_6]^{3-}$ to choline.[1090] Gd^{III} and Mn^{II} have been used as paramagnetic probes to map the bonding site of N-fluoroacetyl-O-glucosamine, and analogues, to henegg lysozyme by ¹H and ¹⁹F n.m.r. spectroscopy.[1091] ¹H N.m.r. spectroscopy has also been used to investigate the systems substrate–Mn^{2+}–pig-heart triphosphopyridine nucleotide-dependent isocitrate dehydrogenase,[1092] Mn^{2+}–alkaline phosphatase,[1093] adenosine triphosphate–Mn^{2+}–activated rabbit muscle phosphofructokinase,[1094] phosphoenolypyruvate–Mn^{2+}–muscle pyruvate kinase,[1095] and guanosine 5′-monophosphate–M–bovine serum albumin (M = Mn^{2+} or Cu^{2+}).[1096]

¹H N.m.r. spectroscopy has been used to investigate human haemoglobin histidine titrations.[1097] ¹⁹F N.m.r. spectroscopy of fluorinated haemoglobin derivatives has been used to determine the pK_a of histidine β146 imidazole,[1098] and the binding of n-butyl isocyanide.[1099] The cyanide and carbon

[1081] E. Bruecher and N. A. Kostromina, *Teor. i eksp. Khim.*, 1972, **8**, 210.
[1082] N. A. Kostromina and T. V. Ternovaya, *Russ. J. Inorg. Chem.*, 1972, **17**, 825.
[1083] M. F. Johnson and J. H. Forsberg, *Inorg. Chim. Acta*, 1972, **11**, 2683.
[1084] A. D. Sherry, E. R. Birnbaum, and D. W. Darnall, *J. Biol. Chem.*, 1972, **247**, 3489.
[1085] K. B. Yatsimirskii, E. D. Romanenko, and L. I. Budarin, *Doklady Akad. Nauk S.S.S.R.*, 1972, **202**, 1140.
[1086] L. W. Amos and D. T. Sawyer, *Inorg. Chem.*, 1972, **11**, 2692.
[1087] M. S. Zetter, M. W. Grant, E. J. Wood, H. W. Dodgen, and J. P. Hunt, *Inorg. Chem.*, 1972, **11**, 2701.
[1088] T.-M. Chen and L. O. Morgan, *J. Phys. Chem.*, 1972, **11**, 1973.
[1089] S. J. Benkovic, J. L. Engle, and A. S. Mildvan, *Biochem. Biophys. Res. Comm.*, 1972, **47**, 852.
[1090] R. J. Kostelnik and S. M. Castellano, *J. Magn. Resonance*, 1972, **7**, 219.
[1091] C. G. Butchard, R. A. Dwek, S. J. Ferguson, P. W. Kent, R. J. P. Williams, and A. V. Xavier, *F.E.B.S. Letters*, 1972, **25**, 91.
[1092] J. J. Villafranca and R. F. Colman, *J. Biol. Chem.*, 1972, **247**, 209.
[1093] G. L. Cottam and B. C. Thompson, *J. Magn. Resonance*, 1972, **6**, 352.
[1094] R. Jones, R. A. Dwek, and I. O. Walker, *European J. Biochem.*, 1972, **28**, 74.
[1095] T. Nowak and A. S. Mildvan, *Biochemistry*, 1972, **11**, 2819.
[1096] L. S. Kan and N. C. Li, *J. Magn. Resonance*, 1972, **7**, 161.
[1097] N. J. Greenfield and M. N. Williams, *Biochem. Biophys. Acta*, 1972, **257**, 187.
[1098] W. H. Huestis and M. A. Raftery, *Proc. Nat. Acad. Sci., U.S.A.*, 1972, **69**, 1887.
[1099] W. H. Huestis and M. A. Raftery, *Biochem. Biophys. Res. Comm.*, 1972, **48**, 678.

monoxide adducts of haemoglobin have been examined and the pK_a values of histidines determined.[1100] Pulsed Fourier transform n.m.r. investigations on mixed solutions of the two oxidation states of cytochrome c reveal that the rate of exchange of oxidation states by electron transfer is sensitive to the conformation of the protein molecule.[1101]

The electron exchange rate between vitamin B_{12a} and vitamin B_{12r} has been measured by ^{1}H n.m.r. line-broadening.[1102] From ^{1}H n.m.r. studies on vitamin B_{12} coenzyme, the ratio of base-on to base-off and protonated forms has been determined.[1103]

The addition of Ni^{2+} to a solution of thiamine pyrophosphate causes shifting and broadening of the ^{1}H n.m.r. signals. The spectra were examined as a function of pH and temperature.[1104] By use of T_1 and T_2 measurements, the rates of NH proton exchange for $[(NH_3)_6Pt]^{4+}$ and $[(NH_3)_5Pt(NH_2)]^{3+}$ have been measured over a wide range of conditions.[1105] The binding of Cu^{2+} to RNase A has been investigated by ^{1}H n.m.r. spectroscopy.[1106] ^{1}H N.m.r. spectroscopy has been used to measure the complex-forming power of Ag^+ with heterocyclic compounds containing oxygen or sulphur.[1107]

From the ^{1}H n.m.r. spectra of zinc, cadmium, and lead complexes of acetylglycine, acid ionization constants and the rates of exchange of the protons have been determined.[1108] The binding of zinc, cadmium, and lead ions by diglycine, triglycine, and tetraglycine in aqueous solution has been investigated by ^{1}H n.m.r. spectroscopy. It was found that the binding site of the metal is pH dependent. Formation constants and rates of proton exchange were determined.[1109] ^{35}Cl N.m.r. has been used to show that, in $RSHgCl_n$, n is always 1 with no evidence for a second chlorine atom co-ordinating.[1110] N.m.r. spectroscopy has been used to study the interaction of mercurials with sulphydryl groups in myosin.[1111] Bromine n.m.r. relaxation measurements give the relative stability constants for organomercurials with various ligands and have been used to define a 'softness scale'.[1112]

^{1}H and ^{19}F resonances have been used to study the interaction of methanol with BF_3 in liquid SO_2, and species such as $[MeOBF_3]^-$ and

[1100] H. Sick, K. Gersonde, J. C. Thompson, W. Maurer, W. Haar, and H. Rueterjans, *European J. Biochem.*, 1972, **29**, 217.
[1101] R. K. Gupta, S. H. Koenig, and A. G. Redfield, *J. Magn. Resonance*, 1972, **7**, 66.
[1102] D. P. Rillema, J. F. Endicott, and L. A. P. Kane-Maguire, *J.C.S. Chem. Comm.*, 1972, 495.
[1103] J. D. Brodie and M. Poe, *Biochemistry*, 1972, **11**, 2534.
[1104] A. A. Gallo, I. L. Hansen, H. Z. Sable, and J. T. Swift, *J. Biol. Chem.*, 1972, **247**, 5913.
[1105] E. Grunwald and D.-W. Fong, *J. Amer. Chem. Soc.*, 1972, **94**, 7371.
[1106] M. Ihnat, *Biochemistry*, 1972, **11**, 3483.
[1107] K. K. Deb, J. E. Bloor, and T. C. Cole, *Inorg. Chem.*, 1972, **11**, 2428.
[1108] D. L. Rabenstein, *Canad. J. Chem.*, 1972, **50**, 1036.
[1109] D. L. Rabenstein and S. Libich, *Inorg. Chem.*, 1972, **11**, 2960.
[1110] R. G. Bryant, *J. Inorg. Nuclear Chem.*, 1972, **34**, 3467.
[1111] R. G. Bryant, Y. Legler, and M. H. Han, *Biochemistry*, 1972, **11**, 3846.
[1112] M. W. Garnett and T. K. Halstead, *J.C.S. Chem. Comm.*, 1972, 587.

[MeOH$_2$]$^+$ were identified; exchange rates were also measured.[1113] ^{19}F and ^{11}B resonances have been used to follow halogen exchange in mixtures of [BF$_4$]$^-$, [BCl$_4$]$^-$, [BBr$_4$]$^-$, and [BI$_4$]$^-$, and species such as [BF$_2$ClBr]$^-$ were identified and chemical shifts discussed.[1114] The system LiBH$_4$–Al(BH$_4$)$_3$–ether has been investigated by ^{11}B n.m.r. spectroscopy and LiAl(BH$_4$)$_4$ found.[1115]

The hydrolysis of aluminium salt solutions has been studied using ^{1}H and ^{27}Al n.m.r. and this has led to reliable values for the concentrations of [Al(H$_2$O)$_6$]$^{3+}$, [Al$_2$(OH)$_2$(H$_2$O)$_8$]$^{4+}$, and [Al$_{13}$O$_4$(OH)$_{24}$(H$_2$O)$_{12}$]$^{7+}$.[1116] A sulphatoaluminium complex and a small amount of a dimeric cation have been detected in aqueous aluminium sulphate solutions by means of ^{27}Al n.m.r. spectroscopy.[1117] The equilibrium in acetonitrile solution among the five ions [AlCl$_n$Br$_{4-n}$] has been examined by ^{27}Al n.m.r. spectroscopy and equilibrium constants have been determined.[1118] Al^{3+} and Ga^{3+} have been added to ferrichromes to assist conformation determination.[1119]

^{19}F N.m.r. spectroscopy has been used to investigate equilibria in aqueous solutions containing Li$_2$SiF$_6$ and HClO$_4$. Equilibrium constants for formation of [SiF$_5$(OH$_2$)]$^-$ and SiF$_4$(OH$_2$)$_2$ were measured as a function of temperature.[1120, 1121]

^{14}N and ^{15}N resonances have been used to show that NO$_2$F is completely dissociated to [NO$_2$]$^+$ and F$^-$ in liquid HF, but that in HNO$_3$–HF there is an equilibrium.[1122] Similarly, ^{14}N chemical shifts in HNO$_3$–H$_2$SO$_4$ mixtures have been used to determine the concentration of [NO$_2$]$^+$.[1123]

Structural changes which occur at the active site of creatine kinase on binding of substrates and inhibitory anions has been examined by ^{1}H n.m.r. relaxation rate studies.[1124] ^{35}Cl N.m.r. spectroscopy shows that the initial identity of the subunits in coenzyme-free horse-liver alcohol dehydrogenase disappears on addition of one mole of coenzyme.[1125] ^{35}Cl N.m.r. spectroscopy has also been used to investigate the binding of dodecyl sulphate to bovine serum albumin.[1126]

Equilibria among Uncharged Species. ^{1}H and ^{7}Li n.m.r. spectra of mixtures of LiMe and LiBr or LiI have been used to identify species such as

[1113] K. L. Servis and L. Jao, *J. Phys. Chem.*, 1972, **76**, 329.
[1114] J. S. Hartman and G. J. Schrobilgen, *Inorg. Chem.*, 1972, **11**, 940.
[1115] M. Ehemann, H. Nöth, and G. Schmidt-Sudhoff, *Z. anorg. Chem.*, 1972, **394**, 33.
[1116] J. W. Akitt, N. N. Greenwood, B. L. Khandelwal, and G. D. Lester, *J.C.S. Dalton*, 1972, 604.
[1117] J. W. Akitt, N. N. Greenwood, and B. L. Khandelwal, *J.C.S. Dalton*, 1972, 1226.
[1118] D. E. H. Jones, *J.C.S. Dalton*, 1972, 567.
[1119] M. Llinas, M. P. Klein, and J. B. Neilands, *J. Mol. Biol.*, 1972, **68**, 265.
[1120] P. M. Borodin and N. K. Zao, *Russ. J. Inorg. Chem.*, 1971, **16**, 1720.
[1121] P. M. Borodin and K. Z. Naguyen, *Vestnik Leningrad. Univ. (Fiz. Khim.)*, 1972, **2**, 56.
[1122] F. Seel and V. Hartmann, *J. Fluorine Chem.*, 1972, **2**, 27.
[1123] F. Seel, V. Hartmann, and W. Gombler, *Z. Naturforsch.*, 1972, **27b**, 325.
[1124] G. H. Reed and M. Cohn, *J. Biol. Chem.*, 1972, **247**, 3073.
[1125] B. Lindman, M. Zeppezauer, and A. Akeson, *Biochim. Biophys. Acta*, 1972, **257**, 173.
[1126] J. A. Magnuson and N. S. Magnuson, *J. Amer. Chem. Soc.*, 1972, **94**, 5461.

Li$_4$Me$_3$I.[1127] ^{13}C N.m.r. spectroscopy has been used to study molecular association of chlorophyll a.[1128] The low-temperature ^{1}H n.m.r. spectrum of MeMgOCPh$_2$Me in ether shows two resonances, which were attributed to (131) and (132).[1129] ^{1}H N.m.r. spectroscopy has been used to investigate exchange in Me$_2$TiCl$_2$ and related species.[1130] ^{19}F N.m.r. spectroscopy has been used to investigate the TiCl$_4$–TiF$_4$–MeOCH$_2$CH$_2$OMe system; species such as TiClF$_3$(MeOCH$_2$CH$_2$OMe) were identified.[1131] Low-temperature ^{19}F n.m.r. spectra of systems such as TiF$_4$–TiBr$_4$–tetrahydrofuran, and TiF$_4$–TiI$_4$–donor have been examined and the positions of equilibria determined.[1132] ^{1}H N.m.r. spectroscopy has been used to measure the stability constant for pyridine complexing to (PhCH$_2$)$_4$M (M = Zr or Hf),[1133] and for the phosphorus ligands complexing to (C$_5$H$_5$)Mo-(CO)$_3$Me.[1134] ^{31}P N.m.r. spectroscopy has been used to demonstrate the

formation of NbCl$_5$,OPCl$_3$ and NbCl$_5$,OPBr$_3$ in CHCl$_3$ and to determine the enthalpy and entropy of activation for exchange with free ligand.[1135]

The equilibrium constants for cis–trans isomerism in [(C$_5$H$_5$)Fe(CO)$_2$]$_2$,-2Et$_3$Al have been measured.[1136] The ^{1}H chemical shifts of (133) are temperature dependent. Isomerization was suggested and an activation energy estimated.[1137] Equilibrium constants for

[MeCo(dmg)$_2$]$_2$ + 2 L ⇌ 2 MeCo(dmg)$_2$L

[1127] D. P. Novak and T. L. Brown, *J. Amer. Chem. Soc.*, 1972, **94**, 3793.
[1128] J. J. Katz, T. R. Janson, A. G. Kostka, R. A. Uphaus, and G. L. Closs, *J. Amer. Chem. Soc.*, 1972, **94**, 2883.
[1129] J. A. Nackashi and E. C. Ashby, *J. Organometallic Chem.*, 1972, **35**, C1.
[1130] J. F. Hanlan and J. D. McCowan, *Canad. J. Chem.*, 1972, **50**, 747.
[1131] R. S. Borden, P. A. Loeffler, and D. S. Dyer, *Inorg. Chem.*, 1972, **11**, 2481.
[1132] R. S. Borden, *U.S. Nat. Tech. Inform. Serv., AD Rep.*, 1972, No. 744496.
[1133] J. J. Felten and W. P. Anderson, *J. Organometallic Chem.*, 1972, **36**, 87.
[1134] K. W. Barnett, T. G. Pollman, and T. W. Solomon, *J. Organometallic Chem.*, 1972, **36**, C23.
[1135] J. C. Bunzli and A. Merbach, *Helv. Chim. Acta*, 1972, **55**, 2867.
[1136] A. Alich, N. J. Nelson, D. Strope, and D. F. Shriver, *Inorg. Chem.*, 1972, **11**, 2976.
[1137] R. Aumann and B. Lohmann, *J. Organometallic Chem.*, 1972, **44**, C51.

have been determined and activation energies calculated for proton exchange between the OHO bridges and water.[1138] ^{19}F N.m.r. spectroscopy has been used to measure the equilibrium position of ligand-exchange reactions of $\{(CF_3)_3P\}_2Ni(CO)_2$.[1139] ^{1}H N.m.r. spectroscopy has been used to show that $(C_5H_5)Ni(PBu^n_3)\{SC(=NPh)SEt\}$ is partially dissociated in solution to $(C_5H_5)Ni(PBu^n_3)(SEt)$ and PhNCS.[1140] From a study of the position of the HCN resonance as a function of added $Ni(CN)_2(PPr^n_3)_2$, the formation of $Ni(CN)_2(PPr^n_3)_2$,HCN has been postulated.[1141] ^{1}H N.m.r. spectroscopy has been used to measure the dissociation energy of [Ni(mesoporphyrin 1X dimethyl ester)]$_2$.[1142] The Evans method of measuring solution susceptibility has been used to study the complexation of pyridine by $\{PhNNN(=O)R\}_2Ni$.[1143] The activation energy for sulphur-atom exchange between dithiolate and perthiolate ligands in nickel and zinc complexes has been measured.[1144] ^{1}H N.m.r. spectroscopy has provided evidence that [PtMe$_3$LI]$_2$ is in equilibrium with PtMe$_3$L$_2$I and [PtMe$_3$I]$_4$,[1145] and has been used to measure the pK_a of (acac)PtCl(CH$_2$=CHOH).[1146]

Stability constants for complexes formed between ZnMe$_2$ and NEt$_3$, OEt$_2$, or SEt$_2$ have been measured.[1147] The equilibria between Me$_3$M^{1}HgM^{2}Me$_3$, (Me$_3$M^1)$_2$Hg, and (Me$_3$M)$_2$Hg (M^1, M^2 = Si, Ge, or Sn) have been investigated by ^{1}H n.m.r. spectroscopy.[1148] A solution containing two methylmercury salts which undergo rapid anion exchange gives intermediate values for $J(^{199}$Hg$-^1$H). This has been used [1149] to follow the equilibrium

$$MeHg(O_2CMe) + PhHgSPh \rightleftharpoons MeHgSPh + PhHg(O_2CMe)$$

Exchange reactions of the type

$$Y^1C_6H_4SO_2NR^1Me + Y^2C_6H_4SO_2NR^2Me \rightleftharpoons$$
$$Y^2C_6H_4SO_2NR^1Me + Y^1C_6H_4SO_2NR^2Me$$
(R^1, R^2 = H or HgMe; Y^1, Y^2 = p-NO$_2$, p-Cl, p-Ph, p-NH$_2$, or o-Cl)

have been followed by ^{1}H n.m.r. spectroscopy and substituent effects discussed.[1150]

[1138] T. L. Brown, L. M. Ludwick, and R. S. Stewart, *J. Amer. Chem. Soc.*, 1972, **94**, 384.
[1139] D. K. Kang and A. B. Burg, *Inorg. Chem.*, 1972, **11**, 902.
[1140] F. Sato and M. Sato, *J. Organometallic Chem.*, 1972, **46**, C63.
[1141] B. Corain, *Co-ordination Chem. Rev.*, 1972, **8**, 159.
[1142] L. Petrakis and F. E. Dickson, *J. Mol. Structure*, 1972, **11**, 361.
[1143] P. S. Zacharias and A. Chakravorty, *Inorg. Chim. Acta*, 1972, **6**, 623.
[1144] J. P. Fackler, jun., J. A. Fetchin, and D. C. Fries, *J. Amer. Chem. Soc.*, 1972, **94**, 7323.
[1145] J. R. Hall and G. A. Swile, *J. Organometallic Chem.*, 1972, **42**, 479.
[1146] M. Tsutsui, M. Ori, and J. Francis, *J. Amer. Chem. Soc.*, 1972, **94**, 1414.
[1147] G. Levy, P. de Loth, and F. Gallais, *J. Chim. phys.*, 1972, **69**, 601.
[1148] T. N. Mitchell, *J. Organometallic Chem.*, 1972, **38**, 17.
[1149] R. J. Kline and L. F. Sytsma, *J. Organometallic Chem.*, 1972, **38**, 1.
[1150] L. A. Fedorov, A. S. Peregudov, and D. N. Kravtsov, *J. Organometallic Chem.*, 1972, **40**, 251.

^{19}F N.m.r. spectroscopy has been used [1151] to measure the equilibria

$$F_2XP + Me_2OB_3H_7 \rightleftharpoons F_2XPB_3H_7 + Me_2O.$$

^{1}H and ^{11}B n.m.r. spectroscopy have been used to show that $R_2^1N(CH_2)_n$-$B(OH)R^2$ exists in solution as an equilibrium between open-chain and cyclically co-ordinated forms, and thermodynamic parameters were estimated.[1152]

^{1}H and ^{19}F n.m.r. spectra have been used to identify all 20 of the possible BXYZ adducts with Me$_3$N.[1153] ^{1}H and ^{11}B n.m.r. spectroscopy have been used to examine ligand scrambling in the B(OMe)$_3$-B(NMe$_2$)$_3$ system and enthalpies determined for the stability constants.[1154] ^{1}H N.m.r. spectra have been used to measure the exchange of methyl groups for Al$_2$Me$_6$-GaMe$_3$, Al$_2$Me$_6$-ZnMe$_2$, Al$_2$Me$_6$-AlMe$_3$L, and GaMe$_3$-AlMe$_3$L. In general, the kinetics do not follow a simple first-order dissociative or simple second-order associative rate law; other mechanisms were discussed.[1155] The parameters for the equilibrium

$$\tfrac{1}{2} Et_6Al_2 + Ph_2O \rightleftharpoons Et_3Al,OPh_2$$

have been estimated by the n.m.r. chemical shift method.[1156] ^{1}H N.m.r.-spectroscopy has been used to examine the adducts of MeIn(toluene-3,4-dithiolate) with nitrogen bases.[1157]

^{1}H N.m.r. spectroscopy has been used to measure *cis–trans* isomerism of PhN=NN(SiMe$_3$)$_2$ as a function of temperature,[1158] and has been used to derive rates, activation energies, and entropies for equilibria between (134), (135), (136), and the Diels–Alder dimer.[1159] Similarly, (germyl)(methyl)-cyclopentadiene is fluxional and is a *ca.* 50:50 mixture of (137) and (138).[1160] Tautomerism between (139) and (140) has been postulated from ^{1}H n.m.r. evidence.[1161] The SiO$_2$–HF–HClO$_4$–H$_2$O system has been studied by

[1151] R. T. Paine and R. W. Parry, *Inorg. Chem.*, 1972, **11**, 268.
[1152] V. S. Bogdanov, V. G. Kiselev, A. D. Naumov, L. S. Vasil'ev, V. P. Dmitrikov, V. A. Dorokhov, and B. M. Mikhailov, *Zhur. obshchei Khim.*, 1972, **42**, 1547.
[1153] B. Benton-Jones, M. E. A. Davidson, J. S. Hartman, J. J. Klassen, and J. M. Miller, *J.C.S. Dalton*, 1972, 2603.
[1154] J.-P. Laurent, G. Cros, G. Copin, and J. Praud, *J. Chim. phys.*, 1972, **69**, 695.
[1155] T. L. Brown and L. L. Murrell, *J. Amer. Chem. Soc.*, 1972, **94**, 378.
[1156] P. E. M. Allen, A. E. Byers, and R. M. Lough, *J.C.S. Dalton*, 1972, 479.
[1157] A. F. Berniaz and D. G. Tuck, *J. Organometallic Chem.*, 1972, **46**, 243.
[1158] N. Wiberg and H. J. Pracht, *Chem. Ber.*, 1972, **105**, 1392.
[1159] G. I. Avramenko, N. M. Sergeyev, and Yu. A. Ustynyuk, *J. Organometallic Chem.*, 1972, **37**, 89.
[1160] S. R. Stobart, *J. Organometallic Chem.*, 1972, **43**, C26.
[1161] B. Dejak and Z. Lasocki, *J. Organometallic Chem.*, 1972, **44**, C39.

(138)   (139)   (140)

^{19}F n.m.r. spectroscopy and equilibrium constants and ΔH have been obtained[1162, 1163] for

$$4\,HF + SiO_2 \rightleftharpoons SiF_4(OH_2)_2$$

A Job plot on the magnitude of J(Pb–H) for $Me_2Pb(acac)_2$ as $(Me_2N)_3PO$ is added has provided evidence for a 1:1 adduct.[1164] ^{1}H N.m.r. spectroscopy has been used to demonstrate that PhPHPHPh is in equilibrium with $(PPh)_5$ and $PhPH_2$. The spectra were analysed to give $^1J(^{31}P-^{31}P)$.[1165] ^{19}F and ^{31}P n.m.r. spectroscopy have been used to follow redistribution reactions in the systems PCl_3–PBr_3, CF_3PCl_2–CF_3Br_2, and PF_3–PCl_3. Contrary to previous results, exchange is slow, but catalysed by traces of water; thermodynamic data were derived.[1166] ^{1}H N.m.r. spectroscopy has been used to determine the position of equilibria for species such as Me_3SbXY.[1167–1169] There are two species in solution for $ArSbCl_2(acac)$ and, in the case of $PhSbF_2(acac)$, the signals coalesce at 116 °C.[1170]

Course of Reactions.—Chemically induced dynamic nuclear polarization (CIDNP) has been used to provide evidence for a radical pathway in the formation of a Grignard reagent.[1171] (See also p. 87.) The isomerization and decomposition of $H_2C=CPhCH_2CD_2MgBr$ has been followed by ^{1}H n.m.r. spectroscopy.[1172–1174] ^{2}H N.m.r. spectroscopy has been used to follow the reaction of $Ti(CD_3)_4$ with $AlMe_3$.[1175]

The extent of deuteriation of aromatic molecules catalysed by transition-metal hydrides has been measured by ^{1}H n.m.r. spectroscopy.[1176] The isomerization of $(1\text{-}R\text{-}C_7H_7)Cr(CO)_3$ to $(2\text{-}R\text{-}C_7H_7)Cr(CO)_3$ etc. has been

[1162] P. M. Borodin and N. K. Zao, *Zhur. neorg. Khim.*, 1972, **17**, 1850.
[1163] P. M. Borodin and N. K. Zao, *Russ. J. Inorg. Chem.*, 1972, **17**, 959.
[1164] M. Aritomi, Y. Kawasaki, and R. Okawara, *Inorg. Nuclear Chem. Letters*, 1972, **8**, 1053.
[1165] J. P. Albrand and D. Gagnaire, *J. Amer. Chem. Soc.*, 1972, **94**, 8630.
[1166] A. D. Jordan and R. G. Cavell, *Inorg. Chem.*, 1972, **11**, 564.
[1167] C. G. Moreland and R. J. Beam, *Inorg. Chem.*, 1972, **11**, 3112.
[1168] C. G. Moreland and G. G. Long, *Inorg. Nuclear Chem. Letters*, 1972, **8**, 347.
[1169] W. A. Kustes, C. G. Moreland, and G. G. Long, *Inorg. Nuclear Chem. Letters*, 1972, **8**, 695.
[1170] N. Nishii and R. Okawara *J. Organometallic Chem.*, 1972, **38**, 335.
[1171] H. W. H. J. Bodewitz, C. Blomberg, and F. Bickelhaupt, *Tetrahedron Letters*, 1972, 281.
[1172] A. Maercker and K. Weber, *Annalen*, 1972, **756**, 20.
[1173] A. Maercker and K. Weber, *Annalen*, 1972, **756**, 33.
[1174] A. Maercker and K. Weber, *Annalen*, 1972, **756**, 43.
[1175] A. S. Khachaturov, L. S. Bresler, and I. Ya. Poddubnyi, *J. Organometallic Chem.*, 1972, **42**, C18.
[1176] U. Klabunde and G. W. Parshall, *J. Amer. Chem. Soc.*, 1972, **94**, 9081.

followed by ^1H n.m.r. spectroscopy and 1,5-migration demonstrated.[1177] The rates of cis–trans isomerism of cis-$Cr(CO)_4(PEt_3)\{C(OMe)Me\}$,[1178] and MeCN exchange in $(MeCN)_3W(CO)_3$ have been followed by ^1H n.m.r. spectroscopy.[1179]

Protonation of (trans,trans-hexa-2,4-dienyl)iron tricarbonyl has been followed by ^1H n.m.r. spectroscopy.[1180] The rearrangement of α-substituted 2-ferrocenyl-2-propyl cations in CF_3CO_2H has been followed by ^1H n.m.r. spectroscopy.[1181, 1182] Similarly, the deuteriation of $[(1,2-Me_2C_3H_3)-Fe(CO)_4]^+[BF_4]^-$ has been followed.[1183] ^1H N.m.r. spectroscopy has also been used to follow the reaction of butadiene or isoprene with $[HCo(CN)_5]^{3-}$,[1184] deuteriation of (cyclohexadiene)$(C_5H_5)Rh$,[1185] and protonation of (cyclo-octatriene)$M(C_5H_5)$ (M = Rh or Ir).[1186]

The reaction of $[Co(NH_3)_6][HSO_4]_3$ in concentrated H_2SO_4 has been followed by ^1H n.m.r. spectroscopy and $[Co(NH_3)_5,HSO_3]^{2+}$, cis-$[Co(NH_3)_4(HSO_3)_2]^+$, and possibly $Co(NH_3)_3(HSO_3)_3$ have been identified as intermediates.[1187] ^{59}Co N.m.r. spectroscopy has been used to follow isomerization reactions of $[Co(en)(NH_3)_2(OH)_2]^+$ and related species.[1188] ^1H N.m.r. spectroscopy shows that conversion of cis-$[Co(en)_2(NH_2-CH_2CN)Cl]^{2+}$ into $[Co(en)\{H_2NCH_2C(NH_2)=NCH_2CH_2NH_2\}Cl]^{2+}$ is 40 times slower than proton exchange.[1189] The protons of the methylene group of co-ordinated malonate in some cobalt(III) complexes have been shown to exchange with deuterium in H_2SO_4–H_2O.[1190] ^1H N.m.r. spectroscopy has been used to follow the conversion of $Co(S_2CSR)_3$ into $[Co(S_2CSR)_2(SR)]_2$, and for similar reactions for iron and nickel complexes.[1191]

syn–anti Isomerism in allylic palladium complexes,[1192] and the reaction of organic compounds such as (141) with $PdCl_2(PhCN)_2$ have been followed by ^1H n.m.r. spectroscopy.[1193] ^1H N.m.r. spectroscopy has also been used

---

[1177] M. I. Foreman, G. R. Knox, P. L. Pauson, K. H. Todd, and W. E. Watts, *J.C.S. Perkin II*, 1972, 1141.
[1178] E. O. Fischer, H. Fischer, and H. Werner, *Angew. Chem. Internat. Edn.*, 1972, **11**, 644.
[1179] M. Green and S. H. Taylor, *J.C.S. Dalton*, 1972, 2629.
[1180] M. Brookhart and D. L. Harris, *J. Organometallic Chem.*, 1972, **42**, 441.
[1181] T. D. Turbitt and W. E. Watts, *J.C.S. Chem. Comm.*, 1972, 947.
[1182] N. M. D. Brown, T. D. Turbitt, and W. E. Watts, *J. Organometallic Chem.*, 1972, **46**, C19.
[1183] D. H. Gibson and R. L. Vonnahme, *J.C.S. Chem. Comm.*, 1972, 1021.
[1184] T. Funabiki, M. Matsumoto, and K. Tarama, *Bull. Chem. Soc. Japan*, 1972, **45**, 2723.
[1185] B. F. G. Johnson, J. Lewis, and D. Yarrow, *J.C.S. Chem. Comm.*, 1972, 235.
[1186] J. Evans, B. F. G. Johnson, and J. Lewis, *J.C.S. Dalton*, 1972, 2668.
[1187] R. A. Sutula and J. B. Hunt, *Inorg. Chem.*, 1972, **11**, 1879.
[1188] F. Yajima, Y. Koike, T. Sakai, and S. Fujiwara, *Inorg. Chem.*, 1972, **11**, 2054.
[1189] D. A. Buckingham, B. M. Foxman, A. M. Sargeson, and A. Zanella, *J. Amer. Chem. Soc.*, 1972, **94**, 1007.
[1190] M. E. Farago and M. A. R. Smith, *J.C.S. Chem. Comm.*, 1972, 2120.
[1191] D. F. Lewis, S. J. Lippard, and J. A. Zubieta, *J. Amer. Chem. Soc.*, 1972, **94**, 1563.
[1192] J. Lukas, J. E. Ramakers-Blom, T. G. Hewitt, and J. J. De Boer, *J. Organometallic Chem.*, 1972, **46**, 167.
[1193] G. Albelo and M. F. Rettig, *J. Organometallic Chem.*, 1972, **42**, 183.

to determine the rate of reaction of 2,2′,2″-terpyridine with Ni^{2+}.[1194] For $[Pt(HMeNCH_2CH_2CH_2NMeH)(NH_3)_2]^{2+}$, the *meso* and racemic forms are in the ratio 2.8:1. The racemic form deuteriates faster than the *meso*-form,

(141)

which is consistent with equatorial NH protons exchanging faster than axial NH protons.[1195] The observation of CIDNP during the decomposition of Bu^tHgSnR_3 in benzene has provided evidence for a radical mechanism.[1196]

^{11}B N.m.r. spectroscopy has been used to examine the decomposition of $[H_2B(NH_3)_2][B_4H_7]$ with rising temperature,[1197] and to investigate the reaction of $C_2B_3H_7$ with Et_3N.[1198] ^1H and ^{11}B resonances have been used to follow the reaction of RY with BX_3 to give RX and BX_2Y *etc.*, and to obtain rates.[1199] ^{11}B N.m.r. spectroscopy has been used to follow the reaction of Me_3NAlH_3 with B_2H_6 to yield Me_3NBH_3, $Al(BH_4)_3$, and $Me_3NAl(BH_4)_3$. Similarly, B_2H_6 reacts with Me_2NAlH_2 to yield species such as $(THF)_2AlH(BH_4)_2$.[1200] The reaction between $Al_4Me_8(NMe_2)_2H_2$ and NMe_3 has been followed by ^1H n.m.r. spectroscopy, and species such as $Al_3Me_6(NMe_2)_2H$ have been identified.[1201] Evidence has been found for complex formation between Bu^i_3Al and Bu^i_2AlH.[1202] The rearrangement of $MeCHBrBEt_2$ to $MeEtCHBBrEt$, catalysed by $AlBr_3$ and related electrophilic catalysts, has been followed by ^1H n.m.r. spectroscopy.[1203]

^1H N.m.r. spectroscopy has been used to follow reactions

$(Me_3Si)_2NM + XSiMe_3 \longrightarrow N(SiMe_3)_3 + MX$

It was found that M = Na reacts faster than M = Li, and X = NC > I ≫ Cl; a complex, $Me_3SiCN,nNaN(SiMe_3)_2$, was detected.[1204] ^1H N.m.r. spectroscopy has been used[1205] to follow reactions such as

$Me_3SiAsMe_2 + SO_2Cl_2 \longrightarrow (Me_3Si)_2O + Me_2AsCl + Me_3SiCl + SO_2$

The diethylamine-catalysed hydrolysis of Me_3SiF in MeCN has been studied by ^1H n.m.r. spectroscopy.[1206] The reaction between Me_3SiO_2-$CCHBr_2$ and Ph_3P has been followed by ^1H n.m.r. spectroscopy: the

---

[1194] P. A. Cock, C. E. Cottrell, and R. K. Boyd, *Canad. J. Chem.*, 1972, **50**, 402.
[1195] T. G. Appleton and J. R. Hall, *Inorg. Chem.*, 1972, **11**, 124.
[1196] T. N. Mitchell, *Tetrahedron Letters*, 1972, 2281.
[1197] G. Kodama, U. Engelhardt, C. Lafrenz, and R. W. Parry, *J. Amer. Chem. Soc.*, 1972, **94**, 407.
[1198] D. A. Franz, V. R. Miller, and R. N. Grimes, *J. Amer. Chem. Soc.*, 1972, **94**, 412.
[1199] M. Goldstein, L. I. B. Haines, and J. A. G. Hemmings, *J.C.S. Dalton*, 1972, 2260.
[1200] P. C. Keller, *Inorg. Chem.*, 1972, **11**, 256.
[1201] J. D. Glore and E. P. Schram, *Inorg. Chem.*, 1972, **11**, 1532.
[1202] J. J. Eisch and S. G. Rhee, *J. Organometallic Chem.*, 1972, **42**, C73.
[1203] H. C. Brown and Y. Yamamoto, *J.C.S. Chem. Comm.*, 1972, 71.
[1204] M. Murray, G. Schirawski, and U. Wannagat, *J.C.S. Dalton*, 1972, 911.
[1205] J. E. Byrne and C. R. Russ, *J. Organometallic Chem.*, 1972, **38**, 319.
[1206] J. A. Gibson and A. F. Janzen, *Canad. J. Chem.*, 1972, **50**, 5087.

reaction proceeds *via* an intermediate which may be [Me$_3$SiOC(OPPh$_3$)=CHBr]Br to give Me$_3$SiBr.[1207] The cleavage of *cis*-1,3,5-trimethyl-1,3,5-triphenylcyclotrisiloxane in CCl$_4$ initiated by Et$_2$NH prior to polymerization was attended by a displacement of the amine proton in the ^{1}H n.m.r. spectrum by 1 p.p.m. downfield.[1208]

The decomposition of Me$_2$PbX$_2$ in D$_2$O and pyridine has also been followed.[1209, 1210]

The reaction

$$[H_2PO_2]^- + H_2O \xrightarrow{\text{catalyst}} [H_2PO_3]^- + H_2$$

has been followed by ^{1}H n.m.r. spectroscopy.[1211] Halogen redistribution reactions of MePX$_2$O[1212] and the reaction of (CF$_3$PCF$_2$)$_2$ to yield CF$_3$P(CF$_2$Br)$_2$Br$_2$, CF$_3$P(CF$_2$Br)Br$_3$, and CF$_3$PBr$_4$ have been followed by ^{1}H, ^{19}F, and ^{31}P n.m.r. spectroscopy.[1213]

^{1}H N.m.r. spectroscopy has been used to follow the reaction of phenylvinylselenide with benzoyl peroxide to yield PhSe(OCH$_2$Ph)$_2$(vinyl) and PhSeCH(OBz)CH$_2$OBz.[1214] The decomposition of Ar$_2$Se$_2$ has been studied by ^{1}H and ^{77}Se n.m.r. spectroscopy. Reaction rates and thermodynamic parameters were determined.[1215]

4 Paramagnetic Complexes

In this section compounds of the *d*-block transition elements will be considered first and then those of the lanthanide and actinide elements. A number of paramegnatic complexes have already been referred to in refs. 325, 327, 479, 895—902, 926, 927, 930—943, 946, and 976.

Chemically induced dynamic nuclear polarization (CIDNP) has been reviewed.[1216, 1217]

Often a limitation in the use of n.m.r. spectroscopy to investigate paramagnetic complexes is the great linewidth encountered. It has been shown that, theoretically, if protons are replaced by deuterium, and the ^{2}H n.m.r. spectrum is measured, the linewidth is reduced by a factor of 40. This theory was tested for a number of paramagnetic transition-metal acetylacetonate complexes, and this reduction was indeed found for Cr(acac)$_3$.[1218]

Formulae have been presented for molecular magnetic multipoles in a uniform field and for the corresponding susceptibilities. The well-known

---

[1207] T. Okada and R. Okawara, *J. Organometallic Chem.*, 1972, **42**, 117.
[1208] K. A. Andrianov, B. G. Zavin, A. A. Zhdanov, A. M. Evdokimov, T. V. Biryukova, and B. D. Lavrukhin, *Vysokomol. Soedineniya, Ser. B.*, 1972, **14**, 327.
[1209] H. J. Haupt, F. Huber, and J. Gmehling, *Z. anorg. Chem.*, 1972, **390**, 31.
[1210] J. Gmehling and F. Huber, *Z. anorg. Chem.*, 1972, **393**, 131.
[1211] K. A. Holbrook and P. J. Twist, *J.C.S. Dalton*, 1972, 1865.
[1212] J. G. Riess and R. Bender, *Bull. Soc. chim. France*, 1972, 3700.
[1213] D.-K. Kang and A. B. Burg, *J.C.S. Chem. Comm.*, 1972, 763.
[1214] Y. Okamoto, R. Homsany, and T. Yano, *Tetrahedron Letters*, 1972, 2529.
[1215] M. A. Lardon, *Ann. New York Acad. Sci.*, 1972, **192**, 132.
[1216] S. H. Pine, *J. Chem. Educ.*, 1972, **49**, 664.
[1217] A. L. Buchachenko and F. M. Zhidomirov, *Russ. Chem. Rev.*, 1971, **40**, 81.
[1218] A. Johnson and G. W. Everett, jun., *J. Amer. Chem. Soc.*, 1972, **94**, 1419.

dipole formula for the neighbour-anisotropy or 'pseudo-contact' contribution to nuclear magnetic shielding was generalized to include higher multipoles.[1219] A semiempirical DODS treatment for the calculation of spin densities has been used to simulate experimental (^{1}H and ^{17}O n.m.r.) trends and corroborate existing experimental values, including negative spin densities in aquo-complexes of the first-row transition-metal series. The spin density is linearly dependent on the number of unpaired t_{2g} electrons.[1220] The theory of paramagnetic relaxation phenomena in liquids has been reviewed.[1221]

Relaxation times and nuclear Overhauser enhancements for protons in ammonia solutions of K, Li, Na, or Ca have been measured and interpreted as showing that more than one paramagnetic species is in solution.[1222] The Knight shift of sodium in ammonia solution has been interpreted using the Becker–Linquist–Alder two-equilibria mode.[1223] Paramagnetic Br_2^+ causes no significant broadening of FSO_3H, but I_2^+ causes significant broadening at the same concentration.[1224]

Compounds of d-Block Transition Elements.—The n.m.r. spectra of nucleic acids and the effect of paramagnetic metal ions have been examined.[1225] The ^{13}C n.m.r. contact shifts for $(MeC_5H_4)_2M$ (M = V, Cr, Co, or Ni) have been reported and mechanisms of spin delocalization discussed. Exchange polarization was found to dominate the spin transfer to the ring carbons for vanadium and chromium although a strongly competitive direct delocalization is indicated by the data. Previous reports of Jahn–Teller effects based on n.m.r. data for the chromium and cobalt species are also brought into question.[1226] ^{17}O and ^{1}H n.m.r. spectroscopy have been used to determine spin densities and kinetic parameters for water exchange in $[U(OH_2)_6]^{3+}$.[1227]

^{1}H N.m.r. spectra of $[CrL_2L^1]^{2+}$, where L = chelating nitrogen base, have been reported. Analysis of the contact shifts shows that there are at least two competing spin delocalization mechanisms.[1228] Analysis of the non-Curie temperature dependence of the ligand contact shifts in a large series of mixed ligand complexes of Cr^{II} with symmetrically substituted o-phenanthrolines has yielded the orbital ground states for each complex.[1229] The data on these complexes have been analysed theoretically and discussed in terms of π-bonding.[1230] WCl_4L_2 shows sharp ^{1}H n.m.r. resonances but

[1219] A. D. Buckingham and P. J. Stiles, *Mol. Phys.*, 1972, **24**, 99.
[1220] A. M. Chmelnick and D. Fiat, *J. Magn. Resonance*, 1972, **7**, 418.
[1221] H. Sillescu, *Adv. Mol. Relax. Processes*, 1972, **3**, 91.
[1222] A. W. Mehner and W. Mueller-Warmuth, *Z. Naturforsch.*, 1972, **27a**, 833.
[1223] A. Demortier and G. Lepoutre, *J. Chim. phys.*, 1972, **69**, 179.
[1224] R. J. Gillespie and M. J. Morton, *Inorg. Chem.*, 1972, **11**, 586.
[1225] H. Fritzche, *Z. Chem.*, 1972, **12**, 1.
[1226] S. E. Anderson, jun. and N. A. Matwiyoff, *Chem. Phys. Letters*, 1972, **13**, 150.
[1227] A. M. Chmelnick and D. Fiat, *J. Magn. Resonance*, 1972, **8**, 325.
[1228] G. N. LaMar and G. R. Van Hecke, *J. Amer. Chem. Soc.*, 1972, **94**, 9042.
[1229] G. N. LaMar and G. R. Van Hecke, *J. Amer. Chem. Soc.*, 1972, **94**, 9049.
[1230] G. N. LaMar, *J. Amer. Chem. Soc.*, 1972, **94**, 9055.

^{31}P coupling is not observed. The chemical shifts vary linearly with (absolute temperature)$^{-1}$.[1231] ^{14}N Spin relaxation rates in solutions of Co^{II}, Ni^{II}, and Mn^{II} ions have been studied as a function of temperature and frequency. The 'excess' ^{14}N spin relaxation is governed by the rate of first-shell relaxation.[1232] Paramagnetic shifts and broadening of N-vinyl-imidazole complexes of manganese, iron, cobalt, nickel, and copper chlorides have been reported.[1233] T_1 and T_2 measurements on carbon-bound protons of citrate and *trans*-aconitate have been made using Mn^{2+}, Fe^{2+}, and aconitase. A H—Mn distance of 4.3 Å was derived in the aconitase–Mn^{2+}–citrate complex.[1234] ^{1}H and ^{31}P n.m.r. spectra of adenosine 5′-monophosphate (AMP) have been recorded as a function of pD in the presence and absence of low molar ratios of the paramagnetic metal ions Mn^{2+}, Ni^{2+}, Co^{2+}, Cr^{2+}, Cu^{2+}, Fe^{2+}, and Fe^{3+}. All of these cations form chelate complexes involving the phosphate moiety and the adenine ring of the AMP, but different binding sites on the adenine ring appear to be used.[1235] The binding of sulphacetamide to manganese carbonic anhydrase has been investigated by ^{1}H n.m.r. spectroscopy.[1236]

Criticism of work on [Fe(dipy)$_3$]$^{3+}$ has been answered, and it has been concluded that there is extensive π-delocalization. The work also notes that other workers have made an error in INDO calculations and that spin in the σ-framework does not put significant spin density on to a methyl proton.[1237] The ^{1}H n.m.r. spectrum of (142) has been examined and various

$$\left[(H_3N)_5RuN \diagup \hspace{-0.5em}=\hspace{-0.5em} \diagdown NRu(NH_3)_5 \right]^{5+}$$

(142)

possible behaviours for the unpaired electron have been considered.[1238] For Fe^{III}(protoporphyrin IX)(pyridine)$_2$ complexes, a linear relationship between the paramagnetic chemical shifts of the porphyrin protons and the basicity of the co-ordinated pyridine has been found. The temperature dependence of the shifts provides evidence for a low-spin–high-spin equilibrium.[1239] The shifts of [Fe(porphyrin)]$_2$O have been attributed solely to contact interaction.[1240] ^{1}H N.m.r. spectroscopy has been used to demonstrate the presence of five-co-ordinate high-spin Fe^{II} in carboxymethylated cytochrome *c*.[1241] ^{13}C Chemical shifts and spin–lattice relaxation

---

[1231] A. V. Butcher, J. Chatt, G. J. Leigh, and P. L. Richards, *J.C.S. Dalton*, 1972, 1064.
[1232] T.-M. Chen, *J. Phys. Chem.*, 1972, **76**, 1968.
[1233] V. K. Voronov, Yu. N. Ivlev, E. S. Domnina, and G. G. Skvortsova, *Khim. geterotsikl. Soedinenii*, 1972, **7**, 994.
[1234] J. J. Villafranca and A. S. Mildvan, *J. Biol. Chem.*, 1972, **247**, 3454.
[1235] A. W. Missen, D. F. S. Natusch, and L. J. Porter, *Austral. J. Chem.*, 1972, **25**, 129.
[1236] A. Lanir and G. Navon, *Biochemistry*, 1972, **11**, 3536.
[1237] R. E. DeSimone and R. S. Drago, *Inorg. Chem.*, 1972, **11**, 668.
[1238] J. H. Elias and R. S. Drago, *Inorg. Chem.*, 1972, **11**, 415.
[1239] H. A. O. Hill and K. G. Morallee, *J. Amer. Chem. Soc.*, 1972, **94**, 731.
[1240] M. Wicholas, R. Mastacich, and D. Jayne, *J. Amer. Chem. Soc.*, 1972, **94**, 4518.
[1241] R. M. Keller, I. Aviram, A. Schejter, and K. Wuethrich, *F.E.B.S. Letters*, 1972, **20**, 90.

times have been measured for cyanoferrimyoglobins of sperm whale and harbour seal and bovine pancreatic ribonuclease A which have been carboxymethylated with enriched bromo[2-^{13}C]acetate.[1242] In a ^{1}H n.m.r. investigation of haemoglobin, signals to -56 p.p.m. were observed.[1243] ^{1}H N.m.r. spectroscopy has been used to show the presence of high-spin iron(II) in ferrous cytochrome P450$_{cam}$.[1244]

If a paramagnetic first-row transition-metal is added to a phosphorus-containing compound, P–H decoupling can occur.[1245] Data have also been reported for RuCl$_3$(AsPh$_3$)(SMe$_2$)$_2$.[1246] T_1, T_2, and chemical shifts of ^{7}Li$^+$ and ^{133}Cs$^+$ in aqueous solution containing Fe^{3+} and various counter-anions have been reported. In general the results for ^{7}Li$^+$ can be understood in terms of dipolar interaction with the Fe^{3+} unpaired electrons. However, for ^{133}Cs$^+$ in the presence of a counter-ion of F$^-$ or Cl$^-$, the relaxation rates are faster than expected and ion-pair formation between Cs$^+$ and ferric halide complexes was postulated.[1247]

^{1}H N.m.r. spectra of octahedral Fe^{II}, Co^{II}, and Ni^{II} complexes with pyridine N-oxide and benzamide in solution with excess ligand have been recorded. The data provide support for the ratio method.[1248] The ^{1}H n.m.r. spectrum of methionine in the presence of Fe^{3+} shows that as the thio-ether group is least affected Fe^{3+} does not co-ordinate to the sulphur.[1249] ^{1}H and ^{2}H n.m.r. spectra of the four diastereoisomers of (143) have been reported.[1250]

(143)

The ^{11}B n.m.r. signals of Fe(S$_2$CNPr$^i_2$)$_2$BF$_3$ and [Fe(S$_2$CNPr$^i_2$)$_3$][BF$_4$] are sharp and 15.8 p.p.m. upfield from that of (MeO)$_3$B.[1251] The variable-temperature ^{1}H n.m.r. spectrum of Fe(S$_2$CNMePh)$_3$ has, at -89 °C, four methyl resonances. On warming, the two outer peaks coalesce first; the

---

[1242] A. M. Nigen, P. Keim, R. C. Marshall, J. S. Morrow, and F. R. N. Gurd, *J. Biol. Chem.*, 1972, **247**, 4100.
[1243] T. R. Lindstrom, C. Ho, and A. V. Pisciotta, *Nature New Biol.*, 1972, **237**, 263.
[1244] R. M. Keller, K. Wuethrich, and P. G. Debrunner, *Proc. Nat. Acad. Sci. U.S.A.*, 1972, **69**, 2073.
[1245] R. Engel and L. Gelbaum, *J.C.S. Perkin I*, 1972, 1233.
[1246] E. S. Switkes, L. Ruiz-Ramirez, T. A. Stephenson, and J. Sinclair, *Inorg. Nuclear Chem. Letters*, 1972, **8**, 593.
[1247] M. Shporer, R. Poupko, and Z. Luz, *Inorg. Chem.*, 1972, **11**, 2441.
[1248] I. Bertini, D. Gatteschi, and A. Scozzafava, *Inorg. Chim. Acta*, 1972, **6**, 185.
[1249] E. J. Halbert and M. J. Rogerson, *Austral. J. Chem.*, 1972, **25**, 421.
[1250] G. W. Everett, jun. and R. M. King, *Inorg. Chem.*, 1972, **11**, 2041.
[1251] E. A. Pasek and D. K. Straub, *Inorg. Chem.*, 1972, **11**, 259.

mechanism was discussed.[1252] The study of the contact-shifted resonances in the ¹H n.m.r. spectra of oxidized and reduced spinach ferrodoxin were extended to lower magnetic field than had been reported by other workers, and two new resonances found.[1253]

The isotropic shifts for cobalt(II) systems undergoing octahedral–tetrahedral equilibria have been analysed and separated into a term due to tetrahedral and a term due to octahedral species.[1254] The ¹H n.m.r. spectrum of hydrotris-(1,2,4-triazol-1-yl)boratocobalt(II) has been measured in solutions of varying D_2O–D_2SO_4 content. Signals were observed for the dissociated ligand, the normal paramagnetic complex, and a third where one (or more) triazole unit is dissociated.[1225] The ¹H n.m.r. chemical shifts of a number of cobalt and nickel complexes such as (144) have been

(144)

recorded and shifts of up to 70 p.p.m. observed.[1256] ¹H N.m.r. spectra of pseudo-tetrahedral bis-(N-alkylsalicylaldiminato)cobalt(II) complexes have been recorded. Both Fermi-contact shifts and dipolar shifts have been found.[1257] Explanation of the ¹H n.m.r. chemical shifts in $Co(acac)_2$(amine) complexes require both contact and pseudo-contact terms, whereas for $Ni(acac)_2$(amine) complexes only contact shifts are found.[1258] The effects of $Co(acac)_2$, $Ni(acac)_2$, and Ni(α-thiopicolinanilide)$_2$ on the ¹H n.m.r. spectra with cumyl hydroperoxide have been studied.[1259] ¹H Chemical shifts of CH_2Cl_2, $CHCl_3$, and H_2O added to a pyridine solution of $Co(acac)_2(py)_2$ have been observed and interpreted as outer-sphere co-ordination.[1260] The ¹H n.m.r. chemical shifts of $[Co(2,6\text{-lutidine } N\text{-oxide})_4]^{2+}$ have been

---

[1252] M. C. Palazzotto and L. H. Pignolet, *J.C.S. Chem. Comm.*, 1972, 6.
[1253] I. Salmeen and G. Palmer, *Arch. Biochem. Biophys.*, 1972, **150**, 767.
[1254] K. W. Jolley, P. D. Buckley, and L. F. Blackwell, *Austral. J. Chem.*, 1972, **25**, 1311.
[1255] T. W. McGaughy and B. M. Fung, *Inorg. Chem.*, 1972, **11**, 2728.
[1256] E. Larsen, G. N. LaMar, B. E. Wagner, J. E. Parks, and R. H. Holm, *Inorg. Chem.*, 1972, **11**, 2652.
[1257] C. Benelli, I. Bertini, and D. Gatteschi, *J.C.S. Dalton*, 1972, 661.
[1258] K. Fricke, *Spectrochim. Acta*, 1972, **28A**, 735.
[1259] G. M. Bulgakova, A. N. Shupik, K. I. Zamaraev, and I. I. Skibida, *Doklady Akad. Nauk. S.S.S.R.*, 1972, **203**, 863.
[1260] A. N. Shupik, G. A. Senyukova, V. M. Nekipelov, and K. I. Zamaraev, *Phys. Letters (A)*, 1972, **41**, 227.

interpreted as delocalization of α spin density on to the ligand *via* the π-system.[1261]

Ion-pairing interaction in [Bu$^n_4$N][Ph$_3$PCoBr$_3$] has been studied by ^{1}H n.m.r. spectroscopy in a number of solvents.[1262] A freshly prepared solution of thiomalic acid and Co^{2+} at pH 2.6 is paramagnetic but on standing it becomes diamagnetic.[1263] The ^{1}H n.m.r. spectra of several unsymmetrically substituted quaternary ammonium cations, ion-paired to [CoBr$_4$]$^{2-}$ and [NiBr$_4$]$^{2-}$, have been studied in CH$_2$Cl$_2$. For the cobalt salts isotropic contact shifts occur, whereas for the nickel salts both isotropic and dipolar shifts occur.[1264]

The theory for the contact shift and pseudo-contact shift in the n.m.r. spectra of tetrahedral and distorted-tetrahedral complexes of nickel(II) has been developed. It was shown that, as the pseudo-contact shift is no longer proportional to T^{-1}, published values of ΔH and ΔS for the 'planar–tetrahedral' equilibrium are in error.[1265] Relative ^{1}H n.m.r. isotropic shifts for NiL$_n$(NO$_3$)$_2$ (L = nitrogen base) were determined and compared with values calculated using the INDO–MO method. It was found that spin transfer is dominated by a σ-mechanism with σ–π polarization at the nitrogen atom.[1266] N.m.r. contact shift measurements have been made on a number of Ni^{II} aminotroponeiminates containing heterocyclic substituents. Competitive π-bonding in mixed Ni^{II} aminotroponeiminates enables a quantitative measurement of the electron-accepting or -donating properties of a heterocyclic group relative to a phenyl group to be obtained.[1267] Complex (145) shows contact shifts.[1268]

(145)

Contact shifts in quinoline and isoquinoline co-ordinated with Ni(acac)$_2$ have been studied by ^{1}H and ^{13}C n.m.r. spectroscopy and compared with electron spin densities calculated using the unrestricted Hartree–Fock INDO–SCF method. The electron spin distributions are dominated by

[1261] D. W. Herlocker, *Inorg. Chim. Acta*, 1972, **6**, 211.
[1262] Y.-Y. Lim and R. S. Drago, *J. Amer. Chem. Soc.*, 1972, **94**, 84.
[1263] S. G. Modak, P. L. Khare, and C. Mande, *Proc. Indian Acad. Sci., Sect. A*, 1972, **75**, 262.
[1264] L. Rosenthal and I. M. Walker, *Inorg. Chem.*, 1972, **11**, 2444.
[1265] B. R. McGarvey, *J. Amer. Chem. Soc.*, 1972, **94**, 1103.
[1266] M. S. Sun, F. Grein, and D. G. Brewer, *Canad. J. Chem.*, 1972, **50**, 2626.
[1267] D. R. Eaton, R. E. Benson, C. G. Bottomley, and A. D. Josey, *J. Amer. Chem. Soc.*, 1972, **94**, 5996.
[1268] P. S. Zacharias and A. Chakravorty, *Inorg. Nuclear Chem. Letters*, 1972, **8**, 203.

σ-electron effects.[1269] DMF Exchange with (146) and (147) has been examined.[1270] The absolute values of n.m.r. contact shifts have been

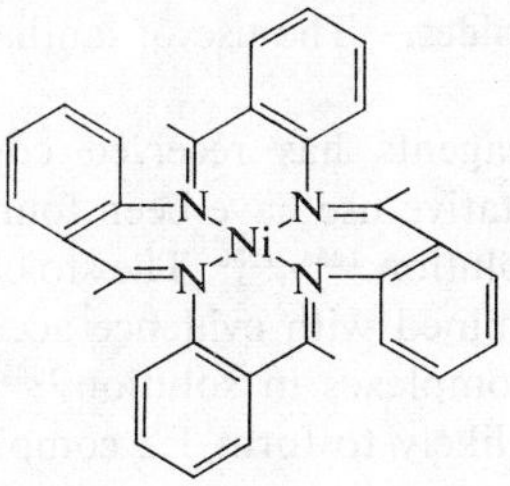

(146) 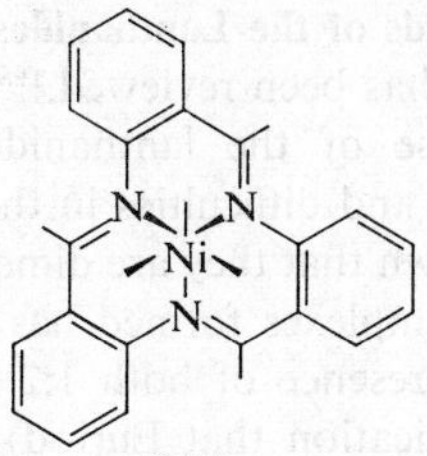   (147)

measured in a number of adducts of $Ni(S_2PEt_2)_2$ with aliphatic and cyclic secondary amines.[1271] $Ni\{S_2P(OEt_2)\}_2$ is paramagnetic and forms a 1:1 complex with PPh_3 which retains ^{31}P splitting of the ethyl group.[1272]

Fermi contact shifts observed in the broad-line ^1H n.m.r. spectrum of CuX_2L (X = halide, L = triazole) indicates diffusion of a positive hole from copper to the heterocycle.[1273] ^1H Chemical shifts have been reported for the 1:1 adduct $Cu(hfac)_2(2,2,6,6$-tetramethylpiperidine N-oxyl) and the Evans method was used to determine the susceptibility.[1274] The solution ^1H n.m.r. spectra of a number of $Cu_2(O_2CAr)_4$ complexes show quite large contact shifts consistent with a superexchange mechanism for the spin exchange.[1275] A singlet–triplet separation of 913 ± 150 cm^{-1} has been determined for $CuCl_2(H_2O)(pyO)$.[1276] $Cu(hfac)_2L$ has a broad ^1H n.m.r. spectrum.[1277]

A new n.m.r. technique for the determination of magnetic susceptibilities has been described.[1278] A modification of the Evans method has been described which is suitable for micro use. Only 5×10^{-7} mol of the paramagnetic material is required.[1279] A practical-course teaching experiment to determine the magnetic susceptibility of $[Mn(CNBu^t)_6][PF_6]_2$ by n.m.r. spectroscopy has been described.[1280] The Evans method has also been used to determine the solution susceptibility of tetraphenylporphine-iron(I) anion,[1281] $FeCl_3,THF$,[1282] μ-superoxo-bis[bis-(L-histidinato)cobalt],[1283]

[1269] I. Morishima, K. Okada, and T. Yonezawa, *J. Amer. Chem. Soc.*, 1972, **94**, 1425.
[1270] L. L. Rusnak, J. E. Letter, jun. and R. B. Jordan, *Inorg. Chem.*, 1972, **11**, 199.
[1271] M. M. Dhingra, G. Govil, and C. R. Kanekar, *J. Magn. Resonance*, 1972, **6**, 577.
[1272] N. Yoon, M. J. Incorvia, and J. I. Zink, *J.C.S. Chem. Comm.*, 1972, 499.
[1273] S. Emori, M. Inoue, and M. Kubo, *Bull. Chem. Soc. Japan*, 1972, **45**, 2259.
[1274] Y. Y. Lim and R. S. Drago, *Inorg. Chem.*, 1972, **11**, 1334.
[1275] R. A. Zelonka and M. C. Baird, *Inorg. Chem.*, 1972, **11**, 134.
[1276] U. Sakaguchi, Y. Arata, and S. Fujiwara, *J. Magn. Resonance*, 1972, **8**, 341.
[1277] R. A. Zelonka and M. C. Baird, *Canad. J. Chem.*, 1972, **50**, 1269.
[1278] J. Homer and P. M. Whitney, *J.C.S. Chem. Comm.*, 1972, 153.
[1279] J. Löliger and R. Scheffold, *J. Chem. Educ.*, 1972, **49**, 646.
[1280] R. A. Bailey, *J. Chem. Educ.*, 1972, **49**, 297.
[1281] I. A. Cohen, D. Ostfeld, and B. Lichtenstein, *J. Amer. Chem. Soc.*, 1972, **94**, 4522.
[1282] L. S. Benner and C. A. Root, *Inorg. Chem.*, 1972, **11**, 652.
[1283] M. Woods, J. A. Weil, and J. K. Kinnaird, *Inorg. Chem.*, 1972, **11**, 1713.

Co{(Me$_2$N)$_2$PF}$_3$I$_2$,[1284] and complexes between Co^{II}(salen) and various imidazoles.[1285]

Compounds of the Lanthanides and Actinides.—The use of lanthanide shift reagents has been reviewed.[1286-1293]

The use of the lanthanide shift reagents has received considerable attention and difficulties in their quantitative use have been found. It has been shown that they are dimerized in solution.[1294, 1295] The stoicheiometry of the complexes formed has been examined with evidence accumulating for the presence of both 1:2 and 1:1 complexes in solution,[1296-1303] with some indication that Eu(fod)$_3$ is more likely to form 1:2 complexes than Eu(dpm)$_3$.

For the pseudo-contact equation to apply, the lanthanide–substrate bond should be axially symmetric, but X-ray structure measurements have shown that this is not true.[1304] Dipolar shifts of Ln(dpm)$_3$(4-picoline)$_2$ have been measured in the solid state. The susceptibility tensors are highly anisotropic and non-axial.[1305] However, if certain conditions occur the equation still holds.[1306-1308] A number of workers have examined procedures by which it is possible to obtain structural data from lanthanide-induced shifts and to

[1284] T. Nowlin and K. Cohn, *Inorg. Chem.*, 1972, **11**, 560.
[1285] L. G. Marzilli and P. A. Marzilli, *Inorg. Chem.*, 1972, **11**, 457.
[1286] M. Holik, *Chem. Listy*, 1972, **66**, 449.
[1287] F. J. Smentowski and R. D. Stipanovic, *J. Amer. Oil Chemists' Soc.*, 1972, **49**, 48.
[1288] K. Tori, *Farumashia*, 1972, **8**, 8.
[1289] J. Grandjean, *Ind. chim. belge*, 1972, **37**, 220.
[1290] J.-P. Bégué, *Bull. Soc. chim. France*, 1972, 2073.
[1291] R. von Ammon and R. D. Fischer, *Angew. Chem. Internat. Edn.*, 1972, **11**, 675.
[1292] M. R. Peterson, jun. and G. H. Wahl, jun., *J. Chem. Educ.*, 1972, **49**, 790.
[1293] F. Lefevre and M. L. Martin, *Org. Magn. Resonance*, 1972, **4**, 737.
[1294] J. F. Desreux, L. E. Fox, and C. N. Reilley, *Analyt. Chem.*, 1972, **44**, 2217.
[1295] B. Feibush, M. F. Richardson, R. E. Sievers, and C. S. Springer, jun., *J. Amer. Chem. Soc.*, 1972, **94**, 6717.
[1296] H. N. Andersen, B. J. Botting, and S. E. Smith, *J.C.S. Chem. Comm.*, 1972, 1193.
[1297] J. K. M. Sanders, S. W. Hanson, and D. H. Williams, *J. Amer. Chem. Soc.*, 1972, **94**, 5325.
[1298] D. F. Evans and M. Wyatt, *J.C.S. Chem. Comm.*, 1972, 312.
[1299] I. Armitage, G. Dunsmore, L. D. Hall, and A. G. Marshall, *Canad. J. Chem.*, 1972, **50**, 2119.
[1300] K. Roth, M. Grosse, and D. Rewicki, *Tetrahedron Letters*, 1972, 435.
[1301] I. Armitage, G. Dunsmore, L. D. Hall, and A. G. Marshall, *Chem. and Ind.*, 1972, 79.
[1302] B. L. Shapiro and M. D. Johnston, jun., *J. Amer. Chem. Soc.*, 1972, **94**, 8183.
[1303] V. G. Gibb, I. M. Armitage, L. D. Hall, and A. G. Marshall, *J. Amer. Chem. Soc.*, 1972, **94**, 8919.
[1304] R. E. Cremer and K. Seff, *J.C.S. Chem. Comm.*, 1972, 400.
[1305] W. D. Horrocks, jun. and J. P. Sipe, tert., *Science*, 1972, **177**, 4053.
[1306] J. M. Briggs, G. P. Moss, E. W. Randall, and K. D. Sales, *J.C.S. Chem. Comm.*, 1972, 1180.
[1307] H. Huber, *Tetrahedron Letters*, 1972, 3559.
[1308] C. L. Honeybourne, *Tetrahedron Letters*, 1972, 1095.

remove experimental errors such as competing species in solution.[1309-1318] Stability constants have also been measured.[1319-1321] A deuterium isotope effect can be observed, e.g. aliphatic alcohols which are deuteriated are more electropositive than those that are not.[1322-1326]

Attention has been drawn to the fact that when an organic substrate is complexed to $Eu(dpm)_3$ or $Eu(fod)_3$, not only the chemical shifts change, but coupling constants can also change, with hydrogen–hydrogen coupling constants changing by up to 2 Hz.[1327-1329] Some new shift reagents have been introduced: $Ln(CF_3COCHCOBu^t)_3$ and $Ln(C_2F_5COCHCOBu^t)_3$, which are intermediate in strength between $Ln(dpm)_3$ and $Ln(fod)_3$,[1330] and $Eu(C_2F_5COCHCOC_2F_5)_3$, which has the advantage that is has no t-butyl signal.[1331] An 'easy' preparation of $Ln(d$-3-trifluoroacetylcamphor$)_3$ has been given.[1332]

The shifting abilities of $Eu(dpm)_3$ and $Eu(fod)_3$ have been compared and, at the same number of molar equivalents, $Eu(dpm)_3$ gives rise to a larger shift than does $Eu(fod)_3$.[1333] CS_2 is a good solvent for $Eu(dpm)_3$ as it is more soluble, and shifts are larger.[1334]

Evidence is accumulating that the induced shifts are not purely pseudo-contact shifts. Contact shifts can also occur, with an increasing

[1309] D. E. Williams, *Tetrahedron Letters*, 1972, 1345.
[1310] D. R. Kelsey, *J. Amer. Chem. Soc.*, 1972, **94**, 1765.
[1311] M. R. Willcott, tert., R. E. Lenkinski, and R. E. Davis, *J. Amer. Chem. Soc.*, 1972, **94**, 1742.
[1312] K. Doerffel, R. Ehrig, H. G. Hauthal, H. Kasper, and G. Zimmermann, *J. prakt. Chem.*, 1972, **314**, 385.
[1313] J. Goodisman and R. S. Matthews, *J.C.S. Chem. Comm.*, 1972, 127.
[1314] B. L. Shapiro, M. L. Shapiro, A. D. Godwin, and M. D. Johnston, jun., 1972, **8**, 402.
[1315] J. W. ApSimon and H. Beierbeck, *J.C.S. Chem. Comm.*, 1972, 172.
[1316] J. W. ApSimon, H. Beierbeck, and A. Fruchier, *Canad. J. Chem.*, 1972, **50**, 2905.
[1317] B. Bleaney, C. M. Dobson, B. A. Levine, R. B. Martin, R. J. P. Williams, and A. V. Xavier, *J.C.S. Chem. Comm.*, 1972, 791.
[1318] P. V. Demarco, B. J. Cerimele, R. W. Crane, and A. L. Thakkar, *Tetrahedron Letters*, 1972, 3539.
[1319] T. A. Wittstruck, *J. Amer. Chem. Soc.*, 1972, **94**, 5130.
[1320] H. Booth and D. V. Griffiths, *J.C.S. Perkin II*, 1972, 2361.
[1321] H. Huber and J. Seelig, *Helv. Chim. Acta*, 1972, **55**, 135.
[1322] A. M. Grotens, J. Smid, and E. de Boer, *J. Magn. Resonance*, 1972, **6**, 612.
[1323] Y. Takagi, S. Teratani, and J. Uzawa, *J.C.S. Chem. Comm.*, 1972, 280.
[1324] J. K. M. Sanders and D. H. Williams, *J.C.S. Chem. Comm.*, 1972, 436.
[1325] C. C. Hinckley, W. A. Boyd, and G. V. Smith, *Tetrahedron Letters*, 1972, 879.
[1326] A. M. Grotens, C. W. Hilbers, E. de Boer, and J. Smid, *Tetrahedron Letters*, 1972, 2067.
[1327] B. L. Shapiro, M. D. Johnston, jun., and R. L. R. Towns, *J. Amer. Chem. Soc.*, 1972, **94**, 4381.
[1328] K.-T. Liu, *Tetrahedron Letters*, 1972, 5039.
[1329] T. B. Patrick and P. H. Patrick, *J. Amer. Chem. Soc.*, 1972, **94**, 6230.
[1330] H. E. Francis and W. F. Wagner, *Org. Magn. Resonance*, 1972, **4**, 189.
[1331] C. A. Burgett and P. Warner, *J. Magn. Resonance*, 1972, **8**, 87.
[1332] V. Schurig, *Tetrahedron Letters*, 1972, 3297.
[1333] B. L. Shapiro, M. D. Johnston, jun., A. D. Godwin, T. W. Proulx, and M. J. Shapiro, *Tetrahedron Letters*, 1972, 3233.
[1334] D. B. Walters, *Analyt. Chim. Acta*, 1972, **60**, 421.

contribution along the series $Pr(fod)_3 < Yb(fod)_3 < Eu(dpm)_3 < Er(fod)_3 < Eu(fod)_3$.[1335–1337] There is a large contact-term contribution to the ^{13}C paramagnetic shifts induced by $Eu(dpm)_3$, $Pr(dpm)_3$, and $Eu(fod)_3$ in pyridine and β-picoline. This contribution could not be explained as due mainly to spin delocalization through π-bonds.[1338] Experimental ^{14}N and ^{17}O n.m.r. results in a series of lanthanide complexes have been interpreted using a second-order perturbation theory. It was found that $4f$ orbitals are not involved in bonding.[1339] However, other workers find that shifts are predominantly of the pseudo-contact type.[1340] $Dy(dpm)_3$ has been suggested as the preferred high-field shift reagent and $Yb(dpm)_3$ as the best low-field shift reagent.[1341]

The pseudo-contact contribution to the n.m.r. shifts for lanthanide complexes in solution has been derived from the anisotropy in the susceptibility. Provided that the molecular geometry is independent of temperature and of lanthanide ion, the shift should vary as T^{-2}. Exceptions should be found for Sm^{3+} and Eu^{3+}, owing to low-lying excited states:[1342] such an exception has been found.[1343] However, the shifts of $Yb(dpm)_3$ are proportional to $T^{-\frac{1}{2}}$.[1344]

The interaction between $Ln(dpm)_3$ (Ln = Eu, Pr, or Yb) and 1,10-phenanthroline and αα′-bipyridyl has been studied. The shifts do not fit the Robertson–McConnell expression. A conformational process involving biphenyl-type rotational isomerism was deduced from the temperature dependence of the $Eu(dpm)_3(bipy)$ n.m.r. spectrum.[1345] The use of shift reagents in exchanging situations has been examined, the chief drawback being the relatively high temperature at which shift reagents have to be used to avoid line-broadening.[1346–1350] The dependence of the lanthanide-induced shift on the geometry of the molecule has been discussed.[1351] $Eu(dpm)_3$ has been used for bulky substituted methanols to investigate

---

[1335] R. K. Mackie and T. M. Shepherd, *Org. Magn. Resonance*, 1972, **4**, 557.
[1336] A. A. Chalmers and K. G. R. Pachler, *Tetrahedron Letters*, 1972, 4033.
[1337] B. F. G. Johnson, J. Lewis, P. McArdle, and J. R. Norton, *J.C.S. Chem. Comm.*, 1972, 535.
[1338] M. Hirayama, E. Edagawa, and Y. Hanyu, *J.C.S. Chem. Comm.*, 1972, 1343.
[1339] R. M. Golding and M. P. Halton, *Austral. J. Chem.*, 1972, **25**, 2577.
[1340] I. Armitage, J. R. Campbell, and L. D. Hall, *Canad. J. Chem.*, 1972, **50**, 2139.
[1341] Z. W. Wolkowski, C. Beauté, and R. Jantzen, *J.C.S. Chem. Comm.*, 1972, 619.
[1342] B. Bleaney, *J. Magn. Resonance*, 1972, **8**, 91.
[1343] U. Sakaguchi, M. Kunugi, T. Fukumi, S. Tadokoro, A. Yamasaki, and S. Fujiwara, *Chem. Letters*, 1972, 177.
[1344] C. Beauté, S. Cornuel, D. Lelandais, N. Thoai, and Z. W. Wolkowski, *Tetrahedron Letters*, 1972, 1099.
[1345] N. S. Bhacca, J. Selbin, and J. D. Wander, *J. Amer. Chem. Soc.*, 1972, **94**, 8719.
[1346] H. N. Cheng and H. S. Gutowsky, *J. Amer. Chem. Soc.*, 1972, **94**, 5505.
[1347] H. N. Cheng and H. S. Gutowsky, *U.S. Nat. Tech. Inform. Serv., AD Rep.*, 1972, No. 739737.
[1348] G. Borgen, *Acta Chem. Scand.*, 1972, **26**, 1740.
[1349] S. Rengaraju and K. D. Berlin, *J. Org. Chem.*, 1972, **37**, 3304.
[1350] R. A. Fletton, G. F. H. Green, and J. E. Page, *J.C.S. Chem. Comm.*, 1972, 1134.
[1351] R. Radeglia, K. Doerffel, and H. Kasper, *J. prakt. Chem.*, 1972, **314**, 266.

steric crowding.[1352] The Evans method has been used to determine the solution susceptibility of Eu(dpm)$_3$ in the presence of substrates.[1353]

Lanthanide shift reagents have also been applied to derivatives of ferrocene, $(C_5H_5)Mn(CO)_3$, and $(arene)Cr(CO)_3$ containing oxygen substituents,[1354] $Me_2PhSnCH_2CH_2CHMeOH$,[1355] and a wide variety of organic compounds.[1356−1434]

[1352] F. H. Hon, H. Matsumura, H. Tanida, and T. T. Tidwell, *J. Org. Chem.*, 1972, **37**, 1778.
[1353] D. Schwendiman and J. T. Zink, *Inorg. Chem.*, 1972, **11**, 3051.
[1354] J. Paul, K. Schlögl, and W. Silhan, *Monatsh.*, 1972, **103**, 243.
[1355] J. Nasielski, *Pure Appl. Chem.*, 1972, **30**, 449.
[1356] P. Joseph-Nathan, J. E. Herz, and V. M. Rodríguez, *Canad. J. Chem.*, 1972, **50**, 2788.
[1357] M. Kainosho, K. Ajisaka, W. H. Pirkle, and S. D. Beare, *J. Amer. Chem. Soc.*, 1972, **94**, 5924.
[1358] R. J. Cushley, D. R. Anderson, and S. R. Lipsky, *J.C.S. Chem. Comm.*, 1972, 636.
[1359] A. K. Bose, B. Dayal, H. P. S. Chawla, and M. S. Manhas, *Tetrahedron Letters*, 1972, 3599.
[1360] R. D. Bennett and R. E. Schuster, *Tetrahedron Letters*, 1972, 673.
[1361] R. R. Fraser, M. A. Petit, and M. Miskow, *J. Amer. Chem. Soc.*, 1972, **94**, 3253.
[1362] R. E. Davis and M. R. Willcott, tert., *J. Amer. Chem. Soc.*, 1972, **94**, 1744.
[1363] R. E. Rondeau, M. A. Berwick, R. N. Steppel, and M. P. Servé, *J. Amer. Chem. Soc.*, 1972, **94**, 1096.
[1364] S.-O. Almqvist, R. Andersson, Y. Shahab, and K. Olsson, *Acta Chem. Scand.*, 1972, **26**, 3378.
[1365] A. Zschunke, J. Tauchnitz, and R. Borsdorf, *Z. Chem.*, 1972, **12**, 425.
[1366] D. D. Tanner and P. Van Bostelen, *J. Amer. Chem. Soc.*, 1972, **94**, 3187.
[1367] W. G. Bentrude, H.-W. Tan, and K. C. Lee, *J. Amer. Chem. Soc.*, 1972, **94**, 3264.
[1368] H. Wamhoff, C. Materne, and F. Knoll, *Chem. Ber.*, 1972, **105**, 753.
[1369] K. von Bredow, G. Helferich, and C. D. Weis, *Helv. Chim. Acta*, 1972, **55**, 553.
[1370] R. Muntwyler and W. Keller-Schierlein, *Helv. Chim. Acta*, 1972, **55**, 460.
[1371] E. Pretsch, M. Vašák, and W. Simon, *Helv. Chim. Acta*, 1972, **55**, 1096.
[1372] W. B. Smith and D. L. Deavenport, *J. Magn. Resonance*, 1972, **6**, 256.
[1373] K. Sisido, M. Naruse, A. Saito, and K. Utimoto, *J. Org. Chem.*, 1972, **37**, 733.
[1374] D. E. U. Ekong, J. I. Okogun, and M. Shok, *J.C.S. Perkin I*, 1972, 653.
[1375] H. Hogeveen, C. F. Roobeeck, and H. C. Volger, *Tetrahedron Letters*, 1972, 221.
[1376] A. Kato and M. Numata, *Tetrahedron Letters*, 1972, 203.
[1377] R. C. Taylor and D. B. Walters, *Tetrahedron Letters*, 1972, 63.
[1378] W. Walter and R. F. Becker, *Annalen*, 1972, **755**, 127.
[1379] G. J. Martin, N. Naulet, F. Lefevre, and M. L. Martin, *Org. Magn. Resonance*, 1972, **4**, 121.
[1380] W. Wiegrebe, J. Fricke, H. Budzikiewicz, and L. Pohl, *Tetrahedron*, 1972, **28**, 2849.
[1381] R. M. Cory and A. Hassner, *Tetrahedron Letters*, 1972, 1245.
[1382] A. Ius, G. Vecchio, and G. Carrea, *Tetrahedron Letters*, 1972, 1543.
[1383] L. J. Luskus and K. N. Houk, *Tetrahedron Letters*, 1972, 1925.
[1384] A. van Bruijnsvoort, C. Kruk, E. R. de Waard, and H. O. Huisman, *Tetrahedron Letters*, 1972, 1737.
[1385] P.-C. Casals and G. Boccaccio, *Tetrahedron Letters*, 1972, 1647.
[1386] J. R. Salaün and J. M. Conia, *Tetrahedron Letters*, 1972, 2849.
[1387] L. W. Morgan and M. C. Bourlas, *Tetrahedron Letters*, 1972, 2631.
[1388] T. J. Leitereg, *Tetrahedron Letters*, 1972, 2617.
[1389] P. E. Pfeffer and H. L. Rothbart, *Tetrahedron Letters*, 1972, 2536.
[1390] A. A. M. Roof, A. van Wageningen, C. Kruk, and H. Cerfontain, *Tetrahedron Letters*, 1972, 367.
[1391] L. H. Keith, A. L. Alford, and J. D. McKinney, *Analyt. Chim. Acta*, 1972, **60**, 1.
[1392] R. Seux, G. Morel, and A. Foucaud, *Tetrahedron Letters*, 1972, 1003.
[1393] F. Bohlmann and C. Zdero, *Tetrahedron Letters*, 1972, 851.
[1394] A. F. Cockerill, N. J. A. Gutteridge, D. M. Rackham, and C. W. Smith, *Tetrahedron Letters*, 1972, 3059.

References continued overleaf

^{1}H N.m.r. spectroscopy has been used to show that Eu^{3+} binds to sodium galacturonate but not to sodium glucoronate.[1435] The solvation of Eu^{3+} in acetonitrile–water solutions has been investigated by ^{1}H n.m.r. spectroscopy.[1436] Tris-terdentate chelate complexes of 2,6-dipicolinate (dpa) and Ln^{3+} yield ^{1}H n.m.r. shifts that increase upfield for Eu < Yb < Er < Tm and downfield for Sm < Nd < Pr < Ho < Tb < Dy.[1437] Contact

References continued from overleaf

[1395] L. A. Paquette, S. A. Lang, jun., S. K. Porter, and J. Clardy, *Tetrahedron Letters*, 1972, 3137.
[1396] M. Ochiai, E. Mizuta, O. Aki, A. Morimoto, and T. Okada, *Tetrahedron Letters*, 1972, 3245.
[1397] A. F. Bramwell and R. D. Wells, *Tetrahedron*, 1972, **28**, 4155.
[1398] A. J. M. Reuvers, A. Sinnema, and H. van Bekkum, *Tetrahedron*, 1972, **28**, 4353.
[1399] N. Platzer and P. Demerseman, *Bull. Soc. chim. France*, 1972, 192.
[1400] G. A. Neville and H. W. Avdovich, *Canad. J. Chem.*, 1972, **50**, 880.
[1401] R. Wasylishen and T. Schaefer, *Canad. J. Chem.*, 1972, **50**, 274.
[1402] G. A. Neville, *Canad. J. Chem.*, 1972, **50**, 1253.
[1403] C. E. Crawforth, O. Meth-Cohn, and C. A. Russell, *J.C.S. Chem. Comm.*, 1972, 259.
[1404] C. P. Casey and R. A. Boggs, *J. Amer. Chem. Soc.*, 1972, **94**, 6457.
[1405] P. E. Manni, G. A. Howie, B. Katz, and J. M. Cassady, *J. Org. Chem.*, 1972, **37**, 2769.
[1406] F. G. Klärner, *Angew. Chem. Internat. Edn.*, 1972, **11**, 832.
[1407] L. E. Legler, S. L. Jindal, and R. W. Murray, *Tetrahedron Letters*, 1972, 3907.
[1408] L. Crombie, D. A. R. Findley, and D. A. Whiting, *Tetrahedron Letters*, 1972, 4027.
[1409] A. J. M. Reuvers, A. Sinnema, and H. van Bekkum, *Tetrahedron*, 1972, **28**, 4353.
[1410] P. Caubere and J. J. Brunet, *Tetrahedron*, 1972, **28**, 4859.
[1411] E. Brown, R. Dhal, and P. F. Casals, *Tetrahedron*, 1972, **28**, 5607.
[1412] M. Hájek, L. Vodička, Z. Ksandr, and S. Landa, *Tetrahedron Letters*, 1972, 4103.
[1413] E. B. Dongala, A. Solladié-Cavallo, and G. Solladié, *Tetrahedron Letters*, 1972, 4233.
[1414] G. P. Schiemenz, *Tetrahedron Letters*, 1972, 4267.
[1415] L. A. Paquette and G. L. Thompson, *J. Amer. Chem. Soc.*, 1972, **94**, 7118.
[1416] N. S. Angerman, S. S. Danyluk, and T. A. Victor, *J. Amer. Chem. Soc.*, 1972, **94**, 7137.
[1417] K. Tsukida, M. Ito, and F. Ikeda, *Internat. J. Vitam. Nutr. Res.*, 1972, **42**, 91.
[1418] M. Matsui and M. Okada, *Chem. and Pharm. Bull. (Japan)*, 1972, **20**, 1033.
[1419] D. Fleischer and R. C. Schulz, *Makromol. Chem.*, 1972, **152**, 311.
[1420] Y. Yoshimura, Y. Mori, and K. Tori, *Chem. Letters*, 1972, 181.
[1421] J. P. Wineburg and D. Swern, *J. Amer. Oil Chemists' Soc.*, 1972, **49**, 267.
[1422] B. S. Perrett and I. A. Stenhouse, *U.K. At. Energy Auth., Res. Group, Rep.*, 1972, AERE-R 7042.
[1423] K. Tsukida, M. Ito, and F. Ikeda, *J. Vitaminol.*, 1972, **18**, 24.
[1424] P. R. Serve, R. E. Rosenberg, and M. Herbert, *J. Heterocyclic Chem.*, 1972, **9**, 721.
[1425] Y. Kobayashi, *Tetrahedron Letters*, 1972, 5093.
[1426] O. Ceder and B. Beijer, *Acta Chem. Scand.*, 1972, **26**, 2977.
[1427] A. J. Dale, *Acta Chem. Scand.*, 1972, **26**, 2985.
[1428] P. S. Mariano and R. McElroy, *Tetrahedron Letters*, 1972, 5305.
[1429] A. K. Bose, B. Dayal, H. P. S. Chawla, and M. S. Manhas, *Tetrahedron*, 1972, **28**, 5977.
[1430] J. J. Uebel and R. M. Wing, *J. Amer. Chem. Soc.*, 1972, **94**, 8910.
[1431] R. Radeglia and A. Weber, *J. prakt. Chem.*, 1972, **314**, 884.
[1432] R. Chujo, K. Koyama, I. Ando, and Y. Inoue, *Brit. Polymer J.*, 1972, **3**, 394.
[1433] A. R. Katritzky and A. Smith, *Brit. Polymer J.*, 1972, **4**, 199.
[1433a] R. B. Wetzel and G. L. Kenyon, *J. Amer. Chem. Soc.*, 1972, **94**, 9230.
[1434] G. A. Neville, *Org. Magn. Resonance*, 1972, **4**, 633.
[1434a] O. Achmatowicz, jun., A. Ejchart, J. Jurczak, L. Kozerski, J. St. Pyrek, and A. Zamojski, *Roczniki Chem.*, 1972, **46**, 903.
[1435] T. Anthonsen, B. Larsen, and O. Smidsrød, *Acta Chem. Scand.*, 1972, **26**, 2988.
[1436] Y. Hass and G. Navon, *J. Phys. Chem.*, 1972, **76**, 1449.
[1437] H. Donato, jun. and R. B. Martin, *J. Amer. Chem. Soc.*, 1972, **94**, 4129.

shifts have been reported for $M(O_2CCH_nCl_{3-n})_3$,[1438] $[Ln(1,8-naphthyl-pyridine)_n]^{3+}$,[1439] and $M\{N(SiMe_3)_2\}_2$.[1440]

The paramagnetic shifts of the C_6H_{11} protons in the low-temperature n.m.r. spectra of $(C_6H_{11}NC)Ln(C_5H_5)_3$ have been analysed.[1441] Large paramagnetic shifts are found for $(C_5H_5)_3UOR$.[1442]

5 Solid-state N.M.R.

A book on n.m.r. spectra of solids has appeared.[1443] The application of n.m.r. in solid-state chemistry has been reviewed.[1444, 1445] A review on nuclear physical crystallography and solid-state physics contains a section on n.m.r. spectroscopy.[1446] The use of n.m.r. and e.s.r. spectroscopy to investigate the surfaces of solids has been reviewed.[1447] Nuclear magnetic resonance and relaxation of molecules adsorbed on solids have been reviewed.[1448] A review has also appeared on the theory and interpretation of n.m.r. spectra and applications to coating chemistry.[1449] N.m.r. relaxation of absorbed molecules with emphasis on adsorbed water has been reviewed.[1450] A review has also appeared on the surface diffusion of benzene adsorbed on modified silica surfaces investigated by n.m.r. relaxation techniques.[1451]

Van Vleck's moment equations have been numerically evaluated to obtain analytical expressions for the second and fourth moments of dipole-broadened magnetic resonance lines for cubic crystals and powders as functions of the spin concentration.[1452] The n.m.r. spectra for spin-1 nuclei in the presence of quadrupole coupling have been calculated for isotropic powders. The method is claimed to be both rapid and accurate.[1453] The role of spin symmetry conservation in nuclear relaxation in solids has been examined.[1454] The Provotorev theory of partial saturation

---

[1438] C. R. Kanekar, M. M. Dhingra, and N. V. Thakur, *J. Inorg. Nuclear Chem.*, 1972, **34**, 3257.
[1439] R. J. Foster, R. L. Bodner, and D. G. Hendricker, *J. Inorg. Nuclear Chem.*, 1972, **34**, 3795.
[1440] D. C. Bradley, J. S. Ghotra, and F. A. Hart, *J.C.S. Chem. Comm.*, 1972, 349.
[1441] R. von Ammon and B. Kanellakopulos, *Ber. Bunsengesellschaft phys. Chem.*, 1972, **76**, 995.
[1442] R. von Ammon, R. D. Fischer, and B. Kanellakopulos, *Chem. Ber.*, 1972, **105**, 45.
[1443] M. Goodman, 'Spin Temperature and Nuclear Magnetic Resonance in Solids', Oxford University Press, New York and Oxford, 1970.
[1444] A. Weiss, *Angew. Chem. Internat. Edn.*, 1972, **11**, 607.
[1445] U. Kh. Kopvillem and V. R. Nagibarov, *Mekh. Relaksatsionnykh Yavlenii Tverd. Telakh, Vses. Nauch. Konf., 1969*, 1972, 245.
[1446] A. Andreeff and M. Schenk, *Krist. Tech.*, 1972, **7**, 317.
[1447] T. Chirulescu, *Stud. Cercet. Chim.*, 1972, **20**, 705.
[1448] H. Pfeifer, *NMR (Nuclear Magn. Resonance)*, 1972, **7**, 53.
[1449] L. C. Afremow, *Amer. Soc. Test. Mater. Spec. Tech. Publ.*, 1972, *STP 500*, 564.
[1450] H. A. Resing, *Adv. Mol. Relax. Processes*, 1972, **3**, 199.
[1451] B. Boddenberg, R. Haul, and G. Oppermann, *Adv. Mol. Relax. Processes*, 1972, **3**, 61.
[1452] G. W. Canters and C. S. Johnson, jun., *J. Magn. Resonance*, 1972, **6**, 1.
[1453] J. L. Colot and F. Grandjean, *J. Magn. Resonance*, 1972, **8**, 60.
[1454] S. Clough and F. Poldy, *J. Chem. Phys.*, 1972, **56**, 1790.

in solids has been used to extract T_1, the Zeeman spin–lattice relaxation from on-resonance absorption-mode second-derivative and dispersion-mode first-derivative continuous wave signals. The results were found to be in agreement with the usual 90–τ–90° pulse method.[1455] It has been found that the permeability, K, of a rock can be estimated from $K = A^{\alpha^2}$ where

$$\alpha = m_{\text{ef}}\{(T_1 . T_{11})/(T_{11} - T_1)\}^2 + m'_{\text{ef}}\{(T'_1 . T_{11})/(T_{11} - T'_1)\}^2$$

T_1 and T'_1 are the spin–lattice relaxation rates of free and bound ligand, T_{11} is that of the liquid in the pores, and m is the effective pore volume.[1456]

Motion in Solids.—^{7}Li and ^{23}Na n.m.r. investigations of orientated DNA fibres show splitting into three lines owing to quadrupolar interactions. The spectra indicate that the alkali-metal ions are in rapid motion around the fibre axis.[1457] ^{19}F N.m.r. spectra of the low-temperature quartz-like and crystobalite modification of BeF_2 and BeF_2,H_2O have been studied to examine fluoride diffusion.[1458]

From T_1 and T_2 measurements on solid binary metal hydrides, the frequency of atomic jumps and the activation energy of self diffusion for the diffusing atom can be calculated. The pulsed-gradient spin-echo technique was outlined for direct measurement of the translational diffusion coefficient.[1459] High-resolution quasi-elastic neutron scattering and n.m.r. have been used to measure the jump rate of protons in β-NbH.[1460] Polycrystalline WOF_4 shows a ^{19}F line motionally narrowed to 200 mG at room temperature. At low temperatures, two broad lines are observed showing two different fluorine sites. This is not consistent with the previously suggested structure.[1461] The ^{1}H n.m.r. spectrum of a single crystal of $CsMnCl_3,2H_2O$ has been measured as a function of temperature and orientation. From the splittings at high temperature, the proton position in the unit cell was confirmed. The temperature dependence conforms with the model where water molecules are rapidly interchanging with an activation energy of 30.5 kJ mol^{-1}.[1462]

'Wide-line' ^{1}H n.m.r. measurements on the 'fluxional' molecules $C_8H_8Fe(CO)_3$ and $C_8H_8Fe_2(CO)_5$ reveal that the cyclo-octatraene rings possess considerable motional freedom in the solid state, reorientating in their approximate molecular planes, but that the ring in $C_8H_8\{Fe(CO)_3\}_2$ is

---

[1455] Z. Trontelj, J. L. Bjorkstam, and R. Johnson, *J. Magn. Resonance*, 1972, **8**, 35.
[1456] S. V. Vedenin, G. R. Bulka, V. M. Nizamutdinov, and V. D. Shchepkin, *Geol. Nefti Gaza*, 1972, **8**, 63.
[1457] H. T. Edzes, A. Rupprecht, and H. J. C. Berendsen, *Biochem. Biophys. Res. Comm.*, 1972, **46**, 790.
[1458] I. P. Aleksandrova and L. R. Batsanova, *Zhur. struckt. Khim.*, 1972, **13**, 232.
[1459] R. M. Colts, *Ber. Bunsengesellschaft phys. Chem.*, 1972, **76**, 760.
[1460] B. Alefeld, H. G. Bohn, and N. Stump, *Ber. Bunsengesellschaft phys. Chem.*, 1972, **76**, 781.
[1461] L. B. Asprey, R. R. Ryan, and E. Fukushima, *Inorg. Chem.*, 1972, **11**, 3122.
[1462] A. J. Vega and D. Fiat, *J. Magn. Resonance*, 1972, **7**, 278.

fixed rigidly in the crystal.[1463] The *monohapto*-cyclopentadienyl rings in the fluxional molecules $(\sigma\text{-}C_5H_5)(\pi\text{-}C_5H_5)Fe(CO)_2$ and $(C_5H_5)HgX$ have been shown by 'wide-line' n.m.r. to reorientate in the solid state, and activation energies were estimated.[1464] The 'wide-line' ^1H n.m.r. spectrum of $(C_8H_8)_2(CO)_4Ru_3$ is markedly temperature dependent, indicating that the C_8H_8 rings are reorientating in the solid state with an activation energy of 21.5 kJ mol^{-1}. Second-moment calculations support 1,2-shifts.[1465] N.m.r. spectroscopy has been used to study ionic motion in $(NH_4)_4Fe(CN)_6,H_2O$ and an activation energy of 19 kJ mol^{-1} obtained.[1466] The activation energies for rotational mobility of $(C_5H_5)_2Co$, $(C_5H_5)_2CoBr$, and $(C_5H_5)_2Ni$ have been determined by ^1H n.m.r. spectroscopy. All are *ca.* 7.5 kJ mol^{-1}.[1467]

Coherent averaging techniques have been used to examine ^{19}F shielding tensors of the CF_3 group in single crystals of AgO_2CCF_3. At room temperature only one line was observed, but at -233 °C six lines were observed, consistent with a static structure.[1468] The ^1H n.m.r. spectrum of $Cd(en)Ni(CN)_4,2C_6D_6$ at -120 °C has a very broad ethylenediamine signal, but on warming to 20 °C the line markedly sharpens owing to motion of the ethylenediamine molecule.[1469]

Broad-line and pulse ^1H n.m.r. spectra of Me_3N,BMe_3 have been recorded. The Me_3B group reorientates even at low temperature, but Me_3N only reorientates when the whole molecule moves; activation energies were determined.[1470] ^{23}Na N.m.r. spectra of samples of sodium beta-alumina have been interpreted in terms of sodium diffusion and water absorption. Intense motional narrowing is observed below room temperature, consistent with a low activation energy for diffusion. The quadrupole interaction decreases markedly with increasing water content, indicating the penetration of water into the beta-alumina structure on a molecular scale.[1471]

The ^{27}Al n.m.r. spectra of $Gd_3Al_2(AlO_4)_3$ has been investigated between 1.5 and 300 K. Below 2.5 K a new kind of magnetic line splitting was observed. It is thought that this is due to the rotation of alternate AlO_6 octahedra and AlO_4 tetrahedra about symmetry axes.[1472] The ^1H n.m.r.

[1463] A. J. Campbell, C. A. Fyfe, and E. Maslowsky, jun., *J. Amer. Chem. Soc.*, 1972, **94**, 2690.
[1464] A. J. Campbell, C. A. Fyfe, R. G. Goel, E. Maslowsky, jun., and C. V. Senoff, *J. Amer. Chem. Soc.*, 1972, **94**, 8387.
[1465] C. E. Cottrell, C. A. Fyfe, and C. V. Senoff, *J. Organometallic Chem.*, 1972, **43**, 203.
[1466] M. S. Whittingham, P. S. Connell, and R. A. Huggins, *J. Solid State Chem.*, 1972, **5**, 321.
[1467] M. K. Makova, Yu. S. Karimov, N. S. Kochetkova, and E. V. Leonova, *Teor. i eksp. Khim.*, 1972, **8**, 259.
[1468] R. G. Griffin, J. D. Ellett, jun., M. Mehring, J. G. Bullitt, and J. S. Waugh, *J. Chem. Phys.*, 1972, **57**, 2147.
[1469] T. Miyoshi, T. Iwamoto, and Y. Sasaki, *Inorg. Chim. Acta*, 1972, **6**, 59.
[1470] T. T. Ang and B. A. Dunell, *J.C.S. Faraday II*, 1972, **68**, 1331.
[1471] O. Kline, H. S. Story, and W. L. Roth, *J. Chem. Phys.*, 1972, **57**, 5180.
[1472] B. Derighetti, E. Zubkovska, and E. Brun, *J. Magn. Resonance*, 1972, **6**, 426.

spectrum of H_3GaNMe_3 has been measured over a temperature range 63—300 K. The activation energy for rotation of GaH_3 is 3.6 ± 0.3 kJ mol^{-1}.[1473]

The β-phase of Me_4Si has been examined. At -183 °C there is a minimum in the T_1 curve owing to methyl reorientation and in the $T_{1\rho}$ curve at -102 °C owing to molecular tumbling.[1474] Broad-line ^1H n.m.r. spectra of crystalline $(Me_3Si)_4C$ show a transitional change at 213 °C and have been compared with calorimetric data.[1475] T_1 and $T_{1\rho}$ measurements have also been carried out on $Me_3SiSiMe_3$. Methyl rotation has an activation energy of 6.5 kJ mol^{-1}, Me_2Si rotation requires 31.4 ± 1.7 kJ mol^{-1}.[1476] Magnetic resonance data at 10 MHz on porous glass powder showed two environmental states of water; Na$^+$ slows down the rate of exchange between the two sites.[1477] The motion of water in glass pores of 15 μm diameter has been investigated by ^1H n.m.r. spectroscopy.[1478] N.m.r. has been used to investigate ion exchange resins swollen in aliphatic alcohols, and the motion of the alcohols examined.[1479] ^1H and ^{19}F nuclear spin-relaxation times have been studied in $MgSiF_6,H_2O$; $[SiF_6]^{2-}$ reorientates with an activation energy 31.4 ± 4.2 kJ mol^{-1}.[1480]

T_1 and $T_{1\rho}$ have been measured in polycrystalline NH_4Cl. The order–disorder transition has been considered dynamically as a result of flips of $[NH_4]^+$.[1481] Calculation of ^1H n.m.r. powder lineshapes of the ammonium ions undergoing quantum mechanical tunnelling have been made and previously unexplained lineshapes explained.[1482] From ^1H and ^2H n.m.r. measurements on Me_2NH_2Cl and Me_2NO_2Cl, rapid reorientation of the methyl groups has been found to have an activation of 13.8 kJ mol^{-1}. At higher temperatures, the whole ion reorientates.[1483] Other workers have found the line of $[Me_2NH_2]^+$ narrowed by tunnelling of the methyl groups even at 4 K.[1484] ^1H N.m.r. lineshape and second-moment data have been reported for the β- and γ-forms of $MeNH_3Cl$, CD_3NH_3Cl, and $MeND_3Cl$ and the two forms show different behaviour.[1485] The ^1H n.m.r. spectrum of Me_4NCl has been measured over the temperature range 77—567 K, and a new phase found. Methyl-group rorientation occurs without tunnelling and at slightly higher temperatures pseudo-isotropic reorientations of the cations occur.[1486]

[1473] M. B. Dunn and C. A. McDowell, *Mol. Phys.*, 1972, **24**, 969.
[1474] S. Albert and J. A. Ripmeester, *J. Chem. Phys.*, 1972, **57**, 2641.
[1475] J. M. Dereppe and J. H. Magill, *J. Phys. Chem.*, 1972, **76**, 4037.
[1476] S. Albert, H. S. Gutowsky, and J. A. Ripmeester, *J. Chem. Phys.*, 1972, **56**, 1332.
[1477] G. Belfort, *Nature Phys. Sci.*, 1972, **237**, 60.
[1478] N. K. Roberts and H. L. Northey, *Nature Phys. Sci.*, 1972, **237**, 144.
[1479] S. Kamata and N. Ishibashi, *Nippon Kagaku Kaishi*, 1972, 1684.
[1480] D. B. Utton and T. Tsang, *J. Chem. Phys.*, 1972, **56**, 116.
[1481] T. Kodama, *J. Magn. Resonance*, 1972, **7**, 137.
[1482] M. B. Dunn, R. Ikeda, and C. A. McDowell, *Chem. Phys. Letters*, 1972, **16**, 226.
[1483] R. Sjöblom and J. Tegenfeldt, *Acta Chem. Scand.*, 1972, **26**, 3068.
[1484] E. R. Andrew and P. C. Canepa, *J. Magn. Resonance*, 1972, **7**, 429.
[1485] J. Tegenfeldt, T. Keousim, and C. Säterkvist, *Acta Chem. Scand.*, 1972, **26**, 3524.
[1486] A. A. V. Gibson and R. E. Raab, *J. Chem. Phys.*, 1972, **57**, 4688.

T_1 in $N_2H_6SO_4$ has been measured. A single relaxation mechanism, probably NH_3 reorientation, and a first-order phase transition were found.[1487] Dipolar relaxation of ^7Li by rotation of ND_nH_{3-n} in $LiN_2H_{5-m}D_mSO_4$ has been examined.[1488] Free induction decays of a single crystal of P_4S_3 and powders of Zn_3P_2, Mg_3P_2, and P_4S_{10} have been measured. P_4S_3 has two ^{31}P shielding tensors, one for the apical phosphorus and the other for the three phosphorus atoms in the base triangle. From the symmetry tensors, rapid rotation about the C_3 axis was concluded.[1489] ^2H N.m.r. has been measured in deuteriated ice. The results indicate that the Bjerrum defect-diffusion mechanism is not dominant and support the Schottky vacancy-diffusion mechanism.[1490] ^{19}F T_1 data for SF_6 chlathrate deuteriate have been measured over a temperature range 40—250 K and analysed in terms of dipole–dipole and spin–rotation interactions. An activation energy of 0.9 kJ mol^{-1} was found for reorientation of the SF_6 molecules.[1491]

Structure of Solids.—Phase diagrams have been determined for a number of liquid-crystal solvent mixtures useful as solvents in n.m.r. spectroscopy.[1492] The ^1H, ^2H, and ^7Li n.m.r. spectra of $Li(NH_4)C_4O_6H_4,H_2O$ have been measured above and below the ferroelectric–paraelectric transition temperature. Only the ^2H n.m.r. signal changes on going through this temperature.[1493] ^1H and ^7Li n.m.r. spectra of dispersions of lithium and sodium montmorillonite reveal that under the action of the energy field formed by the surfaces of the mineral in the coagulation nets of the dispersion, the layers of water are characterized by a lower freezing point, anisotropic motion of separate molecules, and a lower number of hydrogen bonds.[1494] N.m.r. spectroscopy has been used to investigate Pr^i_2O and Bu_2O adsorbed on to LiF.[1495] The ^{23}Na n.m.r. spectra of liquid crystals of sodium linoleate in water have been measured. Only 34% of the total sodium could be accounted for by the central resonance.[1496]

The ^1H n.m.r. spectra of the sodium, potassium, manganese, cobalt, and copper forms of three types of montmorillonite have been detected and linewidths and second moments calculated. The manganese complex shows broadening due to the magnetic moment of that ion.[1497] The

---

[1487] J. W. Harrell, jun. and F. L. Howell, *J. Magn. Resonance*, 1972, **8**, 311.
[1488] R. S. Parker and V. H. Schmidt, *J. Magn. Resonance*, 1972, **6**, 507.
[1489] M. G. Gibby, A. Pines, W.-K. Rhim, and J. S. Waugh, *J. Chem. Phys.*, 1972, **56**, 991.
[1490] V. H. Schmidt, *Pulsed Magn. Opt. Resonance, Proc. Ampere Int. Summer Sch.*, 2nd, 1971, 1972, 232.
[1491] M. B. Dunn, *Chem. Phys. Letters*, 1972, **15**, 508.
[1492] R. A. Bernheim and T. A. Shuhler, *J. Phys. Chem.*, 1972, **76**, 925.
[1493] Z. M. El Saffar, D. E. O'Reilly, E. M. Peterson, and C. Flick, *J. Chem. Phys.*, 1972, **57**, 2372.
[1494] V. V. Mank, Z. E. Suyunova, Yu. I. Tarasevich, and F. D. Ovcharenko, *Doklady Akad. Nauk S.S.S.R.*, 1972, **202**, 117.
[1495] G. Karagounis and J. M. Tsangaris, *Chem. Chron.*, 1972, **1**, 95.
[1496] M. Shporer and M. M. Civan, *Biophys. J.*, 1972, **12**, 114.
[1497] Z. E. Suyunova, V. V. Mank, Yu. I. Tarasevich, and A. G. Brekhunets, *Ukrain. khim. Zhur.*, 1972, **38**, 281.

deuterium quadrupole coupling constant of $KD(CF_3CO_2)_2$ measured by ^2H n.m.r. spectroscopy is in agreement with CNDO/2 calculations.[1498]

The ^1H n.m.r. signal strength in lanthanum magnesium nitrate tetracosahydrate at 1.6 K is 6.5 times the thermal equilibrium value. The dynamic polarization in this diamagnetic crystal was analysed.[1499] The n.m.r. spectra of manganese spinels, $(Mg,Mn)Al_2O_4$, have been reported.[1500] ^1H N.m.r. spectroscopy has been used to determine the stability of water in human dental enamel.[1501] The free precession decay of nuclear pairs in solids and the hydrogen–hydrogen distances in gypsum, cyanamide, and CH_2Cl_2 have been determined.[1502] ^{19}F Chemical shifts have been measured for the solids CaF_2, SrF_2, BaF_2, CdF_2, HgF_2, MgF_2, and ZnF_2. In the latter two cases the lattice is non-cubic and there are three components to the shift tensor. A theoretical calculation of the chemical shift tensor for MgF_2 has been presented and gives good agreement with the experimental data.[1503] The positions of the hydrogen atoms and the OH distances in datolite, $CaBSiO_4(OH)$, have been determined by ^1H n.m.r. spectroscopy using a natural single crystal. The results are in reasonable agreement with X-ray and i.r. studies.[1504] The ^2H n.m.r. spectrum of $Sr(DCO_2)_2,2D_2O$ has been recorded and the deuterium quadrupole coupling constant determined for the two crystallographically independent formate ions. One formate ion is planar and the second is bent out of the plane by 22°.[1505]

N.m.r. measurements of spin–lattice relaxation times and resonance lineshapes have been measured as a function of temperature in the range 77—715 K for protons contained in $YH_{2.61}$—$YH_{2.94}$. Within a limited temperature range, samples with higher hydrogen concentrations displayed complex lineshapes with non-exponential magnetization recovery. These phenomena were associated with the formation of YH_3 within the hydrogen-deficient structure.[1506] The principle of spin-echo spectroscopy has been discussed and a survey made of some of the peculiarities resulting from the application of spin-echo technique to magnetically ordered compounds such as $Gd_{1-n}Y_nAl_2$.[1507] It has been previously shown that the ^{19}F n.m.r. shift is positive in the first half of the group of rare-earth trifluorides and negative in the second half of the group. It has now been shown that for SmF_3 the ^{19}F chemical shift is negative, although samarium is in the first

---

[1498] J. Stepišnik and D. Hadži, *J. Mol. Structure*, 1972, **13**, 307.
[1499] V. Montelatica and G. Tomassetti, *Lett. Nuovo Cimento Soc. Ital. Fis.*, 1972, (2)**3**, 391.
[1500] C. P. Poole, jun., *U.S. Nat. Tech. Inform. Serv.*, AD Rep., 1972, No. 742302.
[1501] G. H. Dibdin, *Arch. Oral Biol.*, 1972, **17**, 433.
[1502] B. A. van Baren, S. Emid, Chr. Steenbergen, and R. A. Wind, *J. Magn. Resonance*, 1972, **6**, 466.
[1503] R. W. Vaughan, D. D. Elleman, W.-K. Rhim, and L. M. Stacey, *J. Chem. Phys.*, 1972, **57**, 5383.
[1504] Y. Sugitani, M. Watanabe, and K. Nagashima, *Acta Cryst.*, 1972, **B28**, 326.
[1505] M. da G. C. Dillon and J. A. S. Smith, *J.C.S. Faraday II*, 1972, **68**, 2183.
[1506] H. T. Weaver, *J. Chem. Phys.*, 1972, **56**, 3193.
[1507] E. Dormann, *Festkoerperprobleme*, 1972, **12**, 487.

half. It was suggested that crystal-field effects may be upsetting the rule.[1508]

The temperature and frequency dependences of the ^{1}H spin relaxation in binary and ternary hydrides of titanium have been measured, in order to study diffusion.[1509, 1510] Measurements have been reported of the nuclear electric quadrupole interaction parameters of the ^{11}B n.m.r. spectra of the transition-metal monoborides MB (M = Ti, V, Cr, Ni, Mo, or W). The measured-^{11}B quadrupole coupling constants and asymmetry parameters are quite small and are considered to be inconsistent with the assumption that the boron atoms are linked by strong covalent sp hybrid bonds.[1511] N.m.r. has been used to study the interaction of TiO_2 with fused $NaHSO_4$.[1512] The ^{11}B n.m.r. spectra of TiB_2 and ZrB_2 and ^{27}Al n.m.r. spectra of $ThAl_2$ have been analysed in terms of nuclear dipole doublet interaction, nuclear electric quadrupole, and anisotropic Knight shift interactions.[1513]

^{1}H N.m.r. spectroscopy has proved useful for zirconium and hafnium hydroxides, basic nitrates, and ferrocyanides to show the presence of both OH and H_2O groups.[1514, 1515] The broad-line n.m.r. spectra of $Na_4Hf(SO_4)_4,3H_2O$, $K_4Hf(SO_4)_4,H_2O$, $(NH_4)_2Hf(SO_4)_3,2H_2O$, and $Hf(SO_4)_2,4H_2O$ have been recorded to investigate the nature of the water.[1516, 1517]

^{1}H, ^{2}H, and ^{51}V n.m.r. spectroscopy have been used to investigate the vanadium–hydrogen and vanadium–deuterium systems. It was deduced that in the tetragonal β-phase, the octahedral c-sites are occupied by hydrogen and deuterium atoms, whereas in the δ-phase the tetrahedral sites are occupied in an arrangement as proposed by Somenkov.[1518, 1519] The proton T_1 has been measured for the niobium–hydrogen system for hydrogen concentrations ranging from 3 to 78 atom %, and from 4 to 500 K. In the β-phase, the hydrogen atoms jump at 10^8 s^{-1} at 300 K

[1508] S. K. Malik, R. Vijayaraghavan, and P. Berniër, *J. Magn. Resonance*, 1972, **8**, 161.
[1509] A. Schmolz, H. Goretzki, and F. Noack, *Ber. Bunsengesellschaft phys. Chem.*, 1972, **76**, 780.
[1510] A. Schmolz, H. Goretzki, and F. Noack, *Ber. Kernforschungsanlage Juelich*, 1972, 263.
[1511] R. B. Creel and R. G. Barnes, *J. Chem. Phys.*, 1972, **56**, 1549.
[1512] L. T. Savcchenko, R. V. Chernov, and A. M. Kalinichenko, *Ukrain. khim. Zhur.*, 1972, **38**, 550.
[1513] D. R. Torgeson, R. G. Barnes, and R. B. Creel, *J. Chem. Phys.*, 1972, **56**, 4178.
[1514] Z. N. Prozorovskaya, V. F. Chuvaev, L. N. Komissarova, N. M. Kosinova, and Z. A. Vladimirova, *Zhur. neorg. Khim.*, 1972, **17**, 1524 (*Russ. J. Inorg. Chem.*, 1972, **17**, 787).
[1515] N. F. Savenko, I. A. Sheka, I. V. Matyash, and A. M. Kalinichenko, *Ukraine khim. Zhur.*, 1972, **38**, 146.
[1516] N. F. Savenko, L. I. Fedoryako, I. A. Sheka, I. V. Matyash, and A. M. Kalinichenko, *Ukrain. khim. Zhur.*, 1972, **38**, 410.
[1517] L. I. Fedoryako, N. F. Savenko, I. A. Sheka, and F. G. Kramarenko, *Ukrain. khim. Zhur.*, 1972, **38**, 519.
[1518] R. R. Arons, H. G. Bohn, and H. Lütgemeier, *Ber. Bunsengesellschaft phys. Chem.*, 1972, **76**, 781.
[1519] R. R. Arons, H. G. Bohn, and H. Lütgemeier, *Ber. Kernforschungsanlage Juelich*, 1972, 272.

and activation energy of 0.222 eV.[1520, 1521] This work has been extended to the niobium–deuterium system.[1522]

The ^{1}H n.m.r. spectrum of $NbVO_4,3H_2O$ shows an intense sharp line due to weakly bound water, whereas $Nb_4(V_2O_7)_3,4H_2O$ and $Nd_2V_{10}O_{28},13H_2O$ show two resonances due to weakly and strongly bound water.[1523] The angle variation of the ^{93}Nb n.m.r. linewidths has been measured for an off-stoicheiometry lithium niobate crystal. It is shown that the widths can be calculated on the basis of a model in which both the magnitude and direction of the electric field gradient are treated as random quantities. A second type of niobium, ca. 6%, was found.[1524] ^{2}H N.m.r. spectra have been measured for the tantalum–deuterium system as a function of composition and temperature and the structure has been discussed.[1525] The broad-line ^{1}H n.m.r. spectrum of $H_3PW_{12}O_{40}$ and $H_{3+n}PW_{12-n}V_nO_{40}$ has been studied and the positions of the protons have been discussed.[1526]

The ^{55}Mn n.m.r. broad-line spectra of single crystals of $RMn(CO)_5$ (R = $Ph_3Ge, Ph_3Sn, Ph_3Pb,$ or Cl_3Sn) have been measured. The quadrupole coupling constants are related to electronegativity and were compared with data on $RCo(CO)_4$.[1527] Single-crystal broad-line n.m.r. spectra of $Mn_2(CO)_{10}$, $ReMn(CO)_{10}$, and $Re_2(CO)_{10}$ have been measured at room temperature. For $Mn_2(CO)_{10}$, the y-axis of the field gradient tensor is within 10° of the Mn—Mn bond, but for $Re_2(CO)_{10}$ the z-axis of the field gradient tensor is within 6° of the Re—Re bond.[1528] The ^{55}Mn and ^{59}Co n.m.r. spectra of a single crystal and solution of $CoMn(CO)_9$ have been measured. Two pairs of independent crystallographic sites related by the two-fold axis and mirror plane of the monoclinic crystal were found. It was concluded that the electrons forming the cobalt–manganese bond are unequally shared by the two metal atoms, with cobalt having the large MO coefficient.[1529]

The temperature dependence of the n.m.r. frequency of ^{57}Fe in $FeBO_3$ has been investigated in detail. The results agree with a prediction of the spin wave theory for an easy plane weak ferromagnet. A new phenomenon, the secondary nuclear spin echo, was discovered and discussed.[1530] ^{59}Co N.m.r. signals in $(Fe_{1-n}Co_n)Si$ solid solutions have been observed. The

---

[1520] H. Lütgemeier, R. R. Arons, and H. G. Bohn, *J. Magn. Resonance*, 1972, **8**, 74.
[1521] B. Alefeld, H. G. Bohn, and N. Stump, *Ber. Kernforschungsanlage Juelich*, 1972, 286.
[1522] H. Lütgemeier, H. G. Bohn, and R. R. Arons, *J. Magn. Resonance*, 1972, **8**, 80.
[1523] A. P. Nakhodnova, T. D. Voloshina, and I. V. Mokhosoeva, *Zhur. neorg. Khim.*, 1972, **17**, 1282; *Russ. J. Inorg. Chem.*, 1972, **17**, 665.
[1524] G. E. Peterson and A. Carnevale, *J. Chem. Phys.*, 1972, **56**, 4848.
[1525] K. Nakamura, *Bull. Chem. Soc. Japan*, 1972, **45**, 3356.
[1526] V. F. Chuvaev, G. N. Shinnik, N. A. Polotebneva, and V. I. Spitsyn, *Doklady Akad. Nauk S.S.S.R.*, 1972, **204**, 1403.
[1527] J. L. Slater, M. Pupp, and R. K. Sheline, *J. Chem. Phys.*, 1972, **57**, 2105.
[1528] E. S. Mooberry, H. W. Spiess, and R. K. Sheline, *J. Chem. Phys.*, 1972, **57**, 813.
[1529] E. S. Mooberry and R. K. Sheline, *J. Chem. Phys.*, 1972, **56**, 1852.
[1530] M. P. Petrov, G. A. Smolenskii, A. P. Paugurt, and S. A. Kizhaev, *Amer. Inst. Phys. Conf. Proc.*, 1972, 379.

corresponding small hyperfine fields can be interpreted as due to the manifestations of a weak itinerant ferromagnet in this system.[1531]

^{59}Co N.m.r. spectra of Co_3S_4 have been measured and the 'apparent' hyperfine field at the cobalt(III) was estimated as -1.5×10^2 Oe spin^{-1}, whereas for Co_3O_4 it is 8.1×10^3 Oe spin^{-1}.[1532] ^{1}H and ^{19} n.m.r. spectroscopy have been used to investigate $MF_2,5HF,6H_2O$ (M = Co or Ni). The second moment is temperature dependent and the energy barrier was determined.[1533] N.m.r. and microwave studies of $CsCoCl_3,2H_2O$ have led to a model of the magnetic structure and exchange interactions.[1534]

A ^{105}Pd n.m.r. signal has been observed in α-PdH$_n$ for the first time A linear relationship between the Knight shift and hydrogen concentration has been found. The hydrogen resonance goes to low field on increasing the hydrogen concentration and is up to 23 p.p.m. to low field of water.[1535] The characteristics of the behaviour of the palladium–hydrogen electrode in acidic and alkaline solution have been investigated by n.m.r. spectroscopy.[1536] ^{1}H N.m.r. spectroscopy has been used to investigate but-1- and -2-enes adsorbed on a 2% Pt–Al_2O_3 catalyst.[1537]

^{1}H N.m.r. spectroscopy can be useful in distinguishing between bound and adsorbed water, e.g. in $CuSO_4,5H_2O + nH_2O$ or molecular sieves.[1538] Ethylene and propene can adsorb on to ZnO as π-complexes and ^{1}H n.m.r. data have been reported.[1539] ^{1}H and ^{19}F n.m.r. spectroscopy have been used to investigate polycrystalline MF(OH) (M = Zn, Cd, or Hg) and the OH···F distance determined. The ^{19}F chemical shift of CdF(OH) of 655 p.p.m. to high field of F_2 has been taken as showing a strong fluorine ionic bond.[1540]

The ^{2}H n.m.r. spectra of several boron deuterides have been studied in the solid state, and deuteron quadrupole coupling constants have been obtained from powder spectra. The occurrence of abnormally high coupling constants in decaborane has been explained by the electron deficiency in the compound.[1541] The lineshape of the ^{1}H n.m.r. signal of borazole is dominated by electric-quadrupole-dependent nuclear spin–lattice relaxation of boron and nitrogen.[1542] The ^{11}B n.m.r. spectra in crystalline boron oxides have been measured. At 16 MHz, two ^{11}B signals

[1531] S. Kawarazaki, H. Yasuoka, and Y. Nakamura, *Solid State Comm.*, 1972, **11**, 81.
[1532] H. Saji and T. Yamadaya, *Phys. Letters (A)*, 1972, **41**, 365.
[1533] K. Muthukrishnan and J. Ramakrishna, *Mol. Phys.*, 1972, **24**, 231.
[1534] W. J. M. De Jonge, A. L. M. Bongaarts, and J. A. Cowen, *Amer. Inst. Phys. Conf. Proc.*, 1972, 441.
[1535] P. Brill and J. Voitländer, *Ber. Bunsengesellschaft phys. Chem.*, 1972, **76**, 847.
[1536] S. I. Berezina and A. N. Gil'manov, *Probl. Org. Fiz. Khim.*, 1971, 228.
[1537] A. Jutand, D. Vivien, and J. Conard, *Surface Sci.*, 1972, **32**, 258.
[1538] M. Drahgicescu and G. Todireanu, *Stud. Cercet. Fiz.*, 1972, **24**, 465.
[1539] A. G. Whitney and I. D. Gay, *J. Catalysis*, 1972, **25**, 176.
[1540] L. M. Avkhutsky, Yu. V. Gargarinsky, V. Saraswati, and R. Vijayaraghavan, *J. Magn. Resonance*, 1972, **8**, 194.
[1541] J. Witschel, jun. and B. M. Fung, *J. Chem. Phys.*, 1972, **56**, 5417.
[1542] G. M. Whitesides, S. L. Regen, J. B. Lisle, and R. Mays, *J. Phys. Chem.*, 1972, **76**, 2871.

were found but at 6 MHz the lower-field signal splits into a doublet. It was not possible to distinguish between two models, one using one boron site and the other using two.[1543] ^{1}H N.m.r. spectroscopy has been used to determine the structure of KBO_3,H_2O, $NaBO_3,H_2O$, $KBO_4,\frac{1}{2}H_2O$, and $NaBO_{3.8},H_2O$.[1544] ^{11}B N.m.r. studies on strontium–borate glasses indicate that oxygens are shared by three borons with four co-ordination, but the fraction of three-co-ordinate boron increases linearly with SrO content.[1545] The ^{11}B n.m.r. spectra of germanium borate glasses and vitreous B_2O_3 are basically identical. The ^{11}B quadrupole coupling constant was measured and the bonding discussed.[1546]

The ^{1}H n.m.r. spectra of $[MeNH_3]^+$, $[H_2NNH_3]^+$, and $[HONH_3]^+$ alums have been studied as a function of temperature and compared with ^{27}Al n.m.r. data. The angular dependence of the spectral splitting of $MeNO_3Al(SO_4)_2,12H_2O$ was also reported.[1547] The n.m.r. spectrum of preheated polycrystalline $Al(OH)_3$ has been studied and found to be different to data found by previous workers.[1548] The second moment of the n.m.r. resonances of laminar aluminosilicates depends linearly on the content of paramagnetic Fe^{3+} ions. The broadening is mainly due to dipole–dipole interactions.[1549] Wide-line ^{1}H n.m.r. spectroscopy has been used to determine an OH$\cdots$Al distance of 2.3 Å; the H—H distances are random. The data were explained in terms of hydrogen diffusion.[1550] ^{27}Al N.m.r. spectra of topaz, $Al_2SiO_4(F,OH)_2$, has been used to obtain the ^{27}Al quadrupole coupling constant, 1.67 ± 0.03 MHz, and asymmetry parameter, $\eta = 0.38 \pm 0.05$.[1551] The ^{1}H spectrum of the low-temperature form of cordierite, $Mg_2Al_4Si_5O_{18},nH_2O$, indicates the presence of isolated water in a channel formed by the $AlSi_5O_{18}$ ring. Two crystallographically non-equivalent aluminium sites were observed. The ^{27}Al n.m.r. spectrum indicates a high degree of aluminium–silicon order.[1552] ^{23}Na, ^{27}Al, and ^{29}Si n.m.r. have been used to investigate aluminium–silicon disorder in nepheline.[1553]

Extensive use has been made of n.m.r. spectroscopy to investigate zeolites and ion exchange resins, and the state of adsorbed water on zeolite 13-X and other materials has been reviewed.[1554] N.m.r. relaxation measure-

---

[1543] C. Rhee and P. J. Bray, *J. Chem. Phys.*, 1972, **56**, 2476.
[1544] U. Sommer and G. Heller, *J. Inorg. Nuclear Chem.*, 1972, **34**, 2713.
[1545] M. J. Park and P. J. Bray, *Phys. and Chem. Glasses*, 1972, **13**, 50.
[1546] J. F. Baugher and P. J. Bray, *Phys. and Chem. Glasses*, 1972, **13**, 63.
[1547] I. S. Vinogradova, *Kristallografiya*, 1972, **17**, 410.
[1548] G. Serfozo, G. Varhegyi, and K. Tompa, *Banyasz. Kohasz. Lapok. Kohasz.*, 1972, **105**, 177.
[1549] Yu. V. Shulepov, V. V. Mank, and Yu. I. Tatasevich, *Teor. i eksp. Khim.*, 1972, **8**, 117.
[1550] D. Freude, D. Müller, and H. Schmiedel, *Z. phys. Chem.* (Leipzig), 1972, **250**, 345.
[1551] T. Tsang and S. Ghose, *J. Chem. Phys.*, 1972, **56**, 261.
[1552] T. Tsang and S. Ghose, *J. Chem. Phys.*, 1972, **56**, 3329.
[1553] D. Brinkmann, S. Ghose, and F. Laces, *Z. Krist.*, 1972, **135**, 208.
[1554] H. A. Resing, *Adv. Mol. Relax. Processes*, 1972, **3**, 199.

ments of water protons in zeolites and in Na-X zeolite have been made.[1555, 1556] The second moment and longitudinal and transverse relaxation times have been determined for decationated Y zeolite between -180 and $+400\,°C$. The room-temperature second moment is due to ^1H–^{27}Al interaction. The variations in relaxation times were explained as being due to proton jumping.[1557] ^{205}Tl N.m.r. spectroscopy has been used to investigate Tl^+ in zeolite.[1558] When benzene is absorbed on zeolites, a ^{13}C resonance shift of 5 p.p.m. has been observed compared with the signal of the free liquid.[1559] Adsorbed pyridine and ammonia on Y zeolite have been investigated by wide-line n.m.r. spectroscopy. For Na–Y zeolites, two to three pyridine molecules are adsorbed at each site. At 77 K pyridine is immobile on Y zeolite but rotates on Na–Y zeolite.[1560]

Pulsed n.m.r. techniques have been used to measure T_1 and T_2 for water protons and sodium in a sulphonic acid ion exchanger in the sodium form. Cross-linking affects T_1 and T_2 for water only slightly. Above six moles of water per exchange site the environment of the water protons is water-like, but below that figure there is structuring.[1561] The ^1H n.m.r. spectra of packed beds of Dowex AG50W in 18 ionic forms have been measured in order to determine effects of counterions on the chemical shifts of the internal water. The counterions studied are H^+, Li^+, Na^+, K^+, Rb^+, Ag^+, NH_4^+, NMe_4^+, NEt_4^+, NBu_4^+, $NPhMe_3^+$, Mg^{2+}, Ca^{2+}, Sr^{2+}, Ba^{2+}, Zn^{2+}, Cd^{2+}, and La^{3+}.[1562] From a similar study the hydration numbers of Mg^{2+} and Al^{3+} have been estimated.[1563]

The hydration numbers of ions in KU-2 and AU-17 ion exchange resins have been studied by ^1H n.m.r. spectroscopy. For univalent cations a hydration number of *ca.* 3 was found.[1564] ^1H and ^7Li n.m.r. spectra of the lithium form of the KU-2 ion exchange resin have been measured. At a low moisture content the lines are broad, owing to tightly bound water, and the diameter of the cavity was estimated.[1565] ^1H N.m.r. spectra of water and NH_4^+ in an aqueous suspension of the ion exchange resin KU-2 over the temperature range -30 to $+90\,°C$ have been recorded. The

---

[1555] A. Gutsze, D. Deininger, H. Pfeifer, G. Finger, W. Schirmer, and H. Stach, *Z. phys. Chem. (Leipzig)*, 1972, **249**, 383.

[1556] A. Gutsze, D. Deininger, H. Pfeifer, G. Finger, W. Schirmer, and H. Stach, *Z. phys. Chem. (Leipzig)*, 1972, **249**, 383.

[1557] M. M. Mestdagh, W. E. Stone, and J. J. Fripiat, *J. Phys. Chem.*, 1972, **76**, 1220.

[1558] D. Freude, A. Hauser, H. Pankau, and H. Schmiedel, *Z. phys. Chem. (Leipzig)*, 1972, **251**, 13.

[1559] D. Geschke, *Z. phys. Chem. (Leipzig)*, 1972, **249**, 125.

[1560] D. Deininger and B. Reïmann, *Z. phys. Chem. (Leipzig)*, 1972, **251**, 353.

[1561] W. J. Blaedel, L. E. Brower, T. L. James, and J. H. Noggle, *Analyt. Chem.*, 1972, **44**, 982.

[1562] D. G. Howery, L. Shore, and B. H. Kohn, *J. Phys. Chem.*, 1972, **76**, 578.

[1563] H. D. Sharma and N. Subramanian, *Canad. J. Chem.*, 1972, **49**, 3948.

[1564] V. V. Mank, V. D. Grebenyuk, and O. D. Kurilenko, *Doklady Akad. Nauk S.S.S.R.*, 1972, **203**, 1115.

[1565] V. V. Mank, V. D. Grevenyuk, N. P. Gnusin, and E. D. Trunov, *Zhur. fiz. Khim.*, 1972, **46**, 344 (*Russ. J. Phys. Chem.*, 1972, **46**, 200).

difference between the signal position of water adsorbed and external water was used to estimate the hydration number of the NH_4^+ cation.[1566]

The ^1H n.m.r. spectrum of water adsorbed on aerosil has been recorded.[1567] The state of methanol on the surface of heat-treated aerosil has been investigated. The OH resonance is broad but the width of the methyl adsorption increases with heat treatment.[1568] The diffusion coefficients of benzene into porous plugs compressed from aerosil have been measured by the spin-echo technique with the use of constant and pulsed field gradients.[1569] N.m.r. relaxation times of benzene or cyclohexane on aerosil have been measured. Two-phase behaviour and restricted rotation were found.[1570] ^{19}F N.m.r. spectroscopy has been used to investigate the adsorption of C_6H_5F on aerosil.[1571]

^1H N.m.r. data have been used to determine the surface hydroxyl concentration on silica.[1572] The state of hydroxy-groups in silica gels has been studied by ^1H n.m.r. spectroscopy and it has been calculated that the surface OH groups are 5.0—5.4 Å apart.[1573] The position of water and hydroxy-groups in hydrosodalite has been determined by ^1H n.m.r. spectroscopy.[1574] The hysteresis observed in the change of ^1H n.m.r. linewidth for water adsorbed on microporous SiO_2 at coverages $1.5 < \theta < 4.0$ during cooling and heating from just below 0 to -40 °C was probably due to the movement, during freezing and thawing, of the phase boundary between the bulk ice structure and the liquid structure adjacent to the adsorbent surface.[1575] The self-diffusion coefficients of water adsorbed on two types of silica for 0.7—5 statistical monolayers have been measured by a pulsed-gradient spin-echo n.m.r. method.[1576] The influence of various surfaces, *e.g.* silica glass and metal, on nuclear spin relaxation of ^{199}Hg and ^{201}Hg vapour has been studied. On silica there is a discontinuity at 250 °C, where there is a silica structure change.[1577] ^1H N.m.r. spectroscopy has been used to investigate the adsorption of benzene on silica gel and cyclohexane on alumina, and activation energies were determined.[1578]

[1566] V. V. Mank, V. D. Grebenyuk, and O. D. Kurilenko, *Dopovidi Akad. Nauk Ukrain. R.S.R., Ser. B*, 1972, **34**, 639.
[1567] F. Reessing and P. Fink, *Z. Chem.*, 1972, **12**, 479.
[1568] A. V. Volkov, A. V. Kiselev, V. I. Lygin, and V. B. Khlebnikov, *Russ. J. Phys. Chem.*, 1972, **46**, 290.
[1569] B. Boddenberg, R. Haul, and G. Oppermann, *J. Colloid Interface Sci.*, 1972, **38**, 210.
[1570] R. Haul and B. Boddenberg, *Porous Struct. Catal. Transp. Processes Heterogeneous Catal., 1968*, 1972, 309.
[1571] F. Reessing and P. Fink, *Z. Chem.*, 1972, **12**, 428.
[1572] V. V. Morariu and R. Mills, *Z. phys. Chem. (Frankfurt)*, 1972, **78**, 298.
[1573] V. V. Mank, A. A. Baran, and G. F. Yankovskaya, *Ukrain. khim. Zhur.*, 1972, **38**, 939.
[1574] V. Yu. Galitskii, V. N. Shcherbakov, and S. P. Gabuda, *Kristallografiya*, 1972, **17**, 788.
[1575] V. V. Morariu and R. Mills, *J. Colloid Interface Sci.*, 1972, **39**, 406.
[1576] V. V. Morariu and R. Mills, *Z. phys. Chem. (Frankfurt)*, 1972, **79**, 1.
[1577] A. M. Bonnot and B. Cagnac, *Compt. rend.*, 1972, **274**, B, 947.
[1578] D. Freude, D. Geschke, H. Pfeifer, and H. Winkler, *Porous Struct. Catal. Transp. Processes Heterogeneous Catal., 1968*, 1972, 281.

¹H N.m.r. spectroscopy has been used to examine salt–gel solutions at room temperature and low temperatures. Salts with a large K_D value displace more water from the gel.[1579]

¹H N.m.r. and i.r. spectroscopy have been used to show that in metamict cyrtolite, water is present both as H_2O and OH^-.[1580] The molecular motions of $SnMe_3F$ in the solid state have been examined by continuous-wave and pulsed n.m.r. spectroscopy. Me_3Sn rotates at high temperature and the methyl groups rotate even at 77 K.[1581]

A simple method has been proposed which enables chemical shifts in solids to be detected where only one chemical shift tensor contributes to the spectrum. The measurements of anisotropic chemical shifts have been made for ¹H and ³¹P in a single crystal of KH_2PO_4 by using this method.[1582] The solid-state ³¹P n.m.r. spectra of some adducts of $POCl_3$ and Lewis acids have been reported and it was concluded from the chemical shifts that in each case co-ordination is *via* oxygen.[1583] The compounds $[PCl_nBr_{4-n}]^+$ have been characterized in the solid state by ³¹P n.m.r. spectroscopy.[1584]

As the ferroelectric Curie temperature of KH_2AsO_4 is approached from the paraelectric phase, additional resonance lines appear in the ⁷⁵As n.m.r. spectrum. These lines were shown to result from polarization fluctuations with lifetimes greater than 1 ms.[1585] The electric-field gradient (E.F.G.) at the caesium sites in CsH_2AsO_4 and the ¹³³Cs spin–lattice relaxation rate in CsD_2AsO_4 have been determined. The results show that the time-averaged value of the E.F.G. tensor at the caesium site in the paraelectric phase is the average of the two ferroelectric E.F.G. tensors for opposite directions.[1586]

Wide-line ¹H n.m.r. spectroscopy has been used to determine water in anodic coatings.[1587] The n.m.r. spectra of ice crystals have been examined.[1588] The ¹H n.m.r. lineshape and second moment have been studied in H_2S and H_2Se at 58 K, and it was concluded that the lattice is rigid. Hydrogen–hydrogen distances of 1.883 ± 0.007 Å for H_2S and 2.014 ± 0.011 Å for H_2Se were found.[1589] T_1 has been measured for solid SF_6.[1590] Rigid-lattice n.m.r. spectra characteristic of intramolecular spin–spin interactions have been measured for SF_6, $\overline{OCH_2CH_2CH_2}$, and cyclobutanone in cages of D_2O clathrate.[1591] Resolved ¹⁹F n.m.r. lineshapes

[1579] J. J. Pesek and R. L. Pecsok, *Analyt. Chem.*, 1972, **44**, 620.
[1580] E. S. Rudnitskaya and I. M. Lipova, *Izvest. V.U.Z., Geol. Razved.*, 1972, **15**, 43.
[1581] S. E. Ulrich and B. A. Dunell, *J.C.S. Faraday II*, 1972, **68**, 680.
[1582] T. Terao and T. Hashi, *J. Magn. Resonance*, 1972, **7**, 238.
[1583] K. B. Dillon and T. C. Waddington, *J. Inorg. Nuclear Chem.*, 1972, **34**, 1825.
[1584] K. B. Dillon and P. N. Gates, *J.C.S. Chem. Comm.*, 1972, 348.
[1585] G. J. Andriaenssens, J. L. Bjorkstam, and J. Aikins, *J. Magn. Resonance*, 1972, **7**, 99.
[1586] R. Blink, M. Mali, J. Slak, J. Stepišnik, and S. Žumer, *J. Chem. Phys.*, 1972, **56**, 3566.
[1587] B. R. Baker and R. M. Pearson, *J. Electrochem. Soc.*, 1972, **119**, 160.
[1588] H. Graenicher, *Pulsed Magn. Opt. Resonance, Proc. Ampere Int. Summer Sch. 2nd, 1971*, 1972, 223.
[1589] Z. M. El. Saffar, and P. Schultz, *J. Chem. Phys.*, 1972, **56**, 2524.
[1590] O. P. Revokatov and S. V. Parfenov, *Pis'ma Zhur. eksp. i teor. Fiz.*, 1972, **15**, 151.
[1591] S. K. Garg and D. W. Davidson, *Chem. Phys. Letters*, 1972, **13**, 73.

have been observed for SF_6 deuterioclathrate hydrate at 4.2 K and lower temperatures. By comparison with computer-simulated spectra, it was postulated that a distribution of correlation times exists in this system and that there is a ^{19}F chemical shift anisotropy of at least 100 p.p.m.[1592]

6 Group III Compounds

Information concerning complexes of these elements can be found at the following sources: $[BH_3D]^-$,[122] $(C_5H_5)_3ThBH_4$,[211] H_3BPR_3,[44, 66] $(PPh_3)_3$-$Ag(NCBH_3)$,[846] $Co_3(CO)_9COBH_2NEt_3$,[595] $\{H_2B(N_2C_3H_2)_2\}_2Fe(CO)_2$,[421] B_2H_6,[92] $Me_nB_2H_{6-n}$,[76] 2,4-$Me_2C_3B_3H_5$, $(\pi$-$MeC_3B_3H_3)Mn(CO)_5$,[325] μ-$\{(\pi$-$C_5H_5)Fe(CO)_2\}C_2B_4H_7$,[364] B_5H_9,[64, 73] μ-$Fe(CO)_4B_6H_{10}$,[363] B_9H_{15},[123] $B_{10}H_{14}$,[93] $[(1,7$-$B_9H_9CHE)_2Fe]^{2-}$, $[\{1,7$-$B_9H_9CHEMo(CO)_5\}_2Fe]^{2-}$,[259] π-cyclopentadienyl-π-dicarbollyl derivatives of cobalt,[629] $(C_5H_5)CoB_8C_2H_{10}Co(C_5H_5)$,[630] $(Ph_3P)_2RuHCl(NCMe)N_2B_{10}H_8SMe_2$,[346] $[(C_5H_5)Co(2$-$B_7CH_8)]^-$,[631] (C_5H_5)-$Co\{\pi$-(1)-$2,4$-$B_8C_2H_{10}\}$,[632] cis-$Pt(C_2MeB_{10}H_{10})_2L_2$,[736] $M(CH_2Ph)_3$ (M = B or Al),[107] $Et_2BN(NO)OEt$,[206] $MeB(\mu$-$NMe_2)_2B(\mu$-$NMe_2)_2BMe$,[1776] $B(vinyl)_3$,[75] $R^1{}_nB(NR^2{}_2)_{3-n}$,[125] borazines,[43, 94] $B(OMe)_3$,[317] F_3B,NMe_3,[65] $[L_2PtX_2PtL_2]^{2+}[BX_4]_2^-$,[817] $Me_3AlBe_3(OBu^t)_6$,[178] $[Me_3AlH_2Zr(C_5H_5)_2]_2$,[208] $[Me_2AlSCH_2]_2$, $(Me_2AlSPh)_2Ni(COD)_2$,[781] $(vinyl)_3MNMe_3$ (M = Al, Ga, or In),[881] $(OC)_4FeC(NMe_2)OAl(NMe_2)_2$,[369] $HfAl_2(OPr^i)_{10}$,[207] $Me_2Tl\{M$-$(C_5H_5)(CO)_3M\}$ (M = Mo or W),[295] and $ArTlCl_2$.[81]

Boron Hydrides and Carbaboranes.—^{11}B N.m.r. spectroscopy has been used to show that acidic hydrolysis of $[BH_4]^-$ produces $[BH_3OH]^-$ and borate,[1593] but if the hydrolysis is carried out in the presence of CN^-, then some $[BH_3CN]^-$ is formed.[1594] ^1H N.m.r. spectroscopy, coupled with i.r. and Raman spectroscopy, has been used to show that in $Me_2NNH_2BH_3$, the BH_3 group is indeed attached to the NH_2.[1595] As HF is added to Me_3N,BH_3, the initial ^1H n.m.r. signal vanishes to give Me_3N,BH_2F, Me_3B,HF_2, Me_3N,BF_3, and, finally, $[Me_3NH]^+BF_4^-$. ^{11}B and ^{19}F n.m.r. data were also reported.[1596] The previously unreported compounds CH_3SPF_2, $(MeS)_2PF$, $MePF_2,BH_3$, $MeSPF_2,BH_3$, and $(MeS)_2PF,BH_3$ have been prepared and characterized by ^1H, ^{11}B, ^{19}F, and ^{31}P n.m.r. spectroscopy. A series of base-displacement reactions established the base strengths towards borane as $MePF_2 > Me_2NPF_2 > MeOPF_2 > MeSPF_2 \geqslant (MeS)_2PF$, while $^1J(^{11}B$–$^{31}P)$ for the fluorophosphine–borane adducts decreases in the series $Me_2NPF_2 > MeOPF_2 > MePF_2 > MeSPF_2 > (MeS)_2PF$.[1597] For Cl_{3-n}-$(MeN)_nPBH_3$, $^1J(^{11}B$–$^1H)$ and $^1J(^{11}B$–$^{31}P)$ increase linearly with n.[1598]

[1592] M. B. Dunn and C. A. McDowell, *Chem. Phys. Letters*, 1972, **13**, 268.
[1593] F. T. Wang and W. L. Jolly, *Inorg. Chem.*, 1972, **11**, 1933.
[1594] L. A. Levine and M. M. Kreevoy, *J. Amer. Chem. Soc.*, 1972, **94**, 3346.
[1595] J. R. Durig, S. Chatterjee, J. M. Casper, and J. D. Odom, *J. Inorg. Nuclear Chem.*, 1972, **34**, 1805.
[1596] J. M. Van Paasschen and R. A. Geanangel, *J. Amer. Chem. Soc.*, 1972, **94**, 2680.
[1597] R. Foester and K. Cohn, *Inorg. Chem.*, 1972, **11**, 2590.
[1598] C. Jouany, G. Jugie, and J.-P. Laurent, *Bull. Soc. chim. France*, 1972, 880.

Data have also been reported for $R_3NAlH_3BH_3$,[1599] $\overline{\text{-(CH}_2)_n\text{NMH}_2}$ (M = B, Al, or Ga),[1600] $(MeO)_3PBH_{3-n}Br_n$,[1601] $R_3BPF_2NMe_2$ (R = H, Me, or F),[1602] $[Me_3NBH_2OR]^+[PF_6]^-$,[1603] S_7NBH_2,[1604] $Et_3NBH_2C_5H_8Me$,[1605] $(Et_2N)_nBH_{3-n}$, and μ-$RNHB_2H_5$.[1606]

^1H and ^{11}B n.m.r. spectra of $Me_3NBH_2PMe_2BH_3$ have been reported. It is interesting that the ^{11}B n.m.r. spectrum of BH_3 shows no phosphorus coupling but BH_2 shows phosphorus coupling of ca. 60 Hz.[1607] The ^{11}B n.m.r. spectrum of $F_4P_2,2BH_3$ is a 1:3:3:1 quartet while the ^{19}F n.m.r. spectrum is of the $[AX_2]_2$ type.[1608] ^1H and ^{11}B n.m.r. spectra of $C_4B_2H_6$, $Me_nC_4B_2H_{6-n}$,[1609] $K[EtS(BH_3)_2]$,[1610] (148), and (149) have been measured.[1611]

(148)   (149)

^{11}B and ^{19}F n.m.r. spectroscopy have been used to characterize 2-$CF_3SPCF_3B_3H_8$. On standing for two days, isomerization to the 1-isomer occurs.[1612] The ^{11}B n.m.r. spectrum of $Me_2MB_3H_8$ (M = Al or Ga) is temperature dependent, showing signals 2:1 at low temperature (M = Ga) and a nine-line multiplet at room temperature. The structure is thought to be (150).[1613] ^1H and ^{11}B n.m.r. spectroscopy have been used to help characterize B-(cis-prop-1-enyl)-closo-1,5-dicarbapentaborane(5) and $[C_2B_3H_4]_2$.[1614] The ^1H n.m.r. spectrum of $Al(BH_4)_3,NHMe_2$ shows a dodecet for the BH_4 resonance, in the ratio 1:1:2:2:3:3:3:3:2:2:1:1, owing to coupling to ^{11}B and ^{27}Al with $J(Al-H) \approx \frac{1}{2}J(B-H)$.[1615] The ^{11}B n.m.r. spectrum shows a 1:2:1 triplet and 1:4:6:4:1 quintet, and structure (151) was suggested.[1616] ^1H and ^{11}B n.m.r. spectroscopy have been used to follow the separation of (152) and (153),[1617] and the ^{11}B n.m.r. spectrum of

[1599] M. Ehemann, N. Davies, and H. Nöth, Z. anorg. Chem., 1972, **389**, 235.
[1600] A. Storr, B. S. Thomas, and A. D. Penland, J.C.S. Dalton, 1972, 326.
[1601] G. Jugie and J. P. Laussac, Compt. rend., 1972, **214**, C, 1668.
[1602] S. Fleming and R. W. Parry, Inorg. Chem., 1972, **11**, 1.
[1603] D. L. Reznicek and N. E. Miller, Inorg. Chem., 1972, **11**, 858.
[1604] M. H. Mendelsohn and W. L. Jolly, Inorg. Chem., 1972, **11**, 1944.
[1605] H. C. Brown, E. Negishi, and J.-J. Katz, J. Amer. Chem. Soc., 1972, **94**, 5893.
[1606] L. D. Schwartz and P. C. Keller, J. Amer. Chem. Soc., 1972, **94**, 3015.
[1607] L. D. Schwartz and P. C. Keller, Inorg. Chem., 1972, **11**, 1931.
[1608] R. T. Paine and R. W. Parry, Inorg. Chem., 1972, **11**, 210.
[1609] V. R. Miller and R. N. Grimes, Inorg. Chem., 1972, **11**, 862.
[1610] J. J. Mielcarek and P. C. Keller, J.C.S. Chem. Comm., 1972, 1090.
[1611] E. Negishi, P. L. Burke, and H. C. Brown, J. Amer. Chem. Soc., 1972, **94**, 7431.
[1612] I. B. Mishra and A. B. Burg, Inorg. Chem., 1972, **11**, 664.
[1613] J. J. Borlin and D. F. Grimes, J. Amer. Chem. Soc., 1972, **94**, 1367.
[1614] A. B. Burg and T. J. Reilly, Inorg. Chem., 1972, **11**, 1962.
[1615] N. Davies and M. G. H. Wallbridge, J.C.S. Dalton, 1972, 1421.
[1616] P. C. Keller, J. Amer. Chem. Soc., 1972, **94**, 4020.
[1617] O. T. Beachley, jun., J. Amer. Chem. Soc., 1972, **94**, 4223.

(154) shows two ^{11}B triplets.[1618] ^{1}H and ^{11}B n.m.r. spectroscopy have been used to assign structures to (155; X = Me$_3$Si, Me$_3$Ge, Me$_2$B, or Me$_2$ClSi),[1619] and (156).[1620] Pyrolysis of (Me$_2$B)$_2$CH$_2$ yields among the products B$_3$C$_5$H$_{11}$, which shows two CH and three CH$_3$ groups in the ^{1}H n.m.r. spectrum.[1621]

C$_2$H$_4$B$_4$H$_8$ reacts with PMe$_3$ to yield a solid C$_2$H$_4$B$_4$H$_8$(PMe$_3$)$_2$. The ^{11}B n.m.r. spectrum shows four ^{11}B resonances, which is consistent with

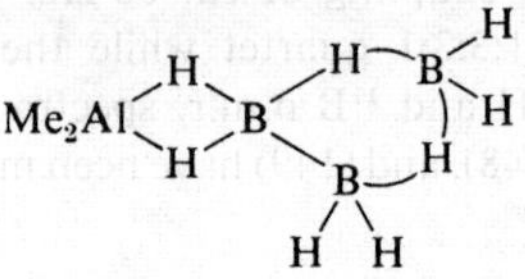

(150)

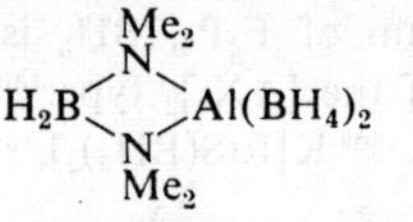

(151)

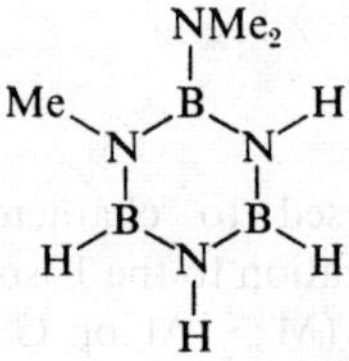

(152)

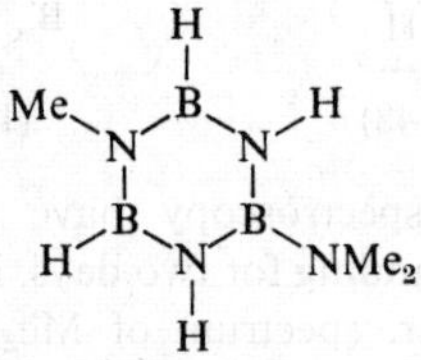

(153)

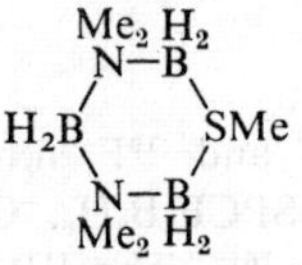

(154)

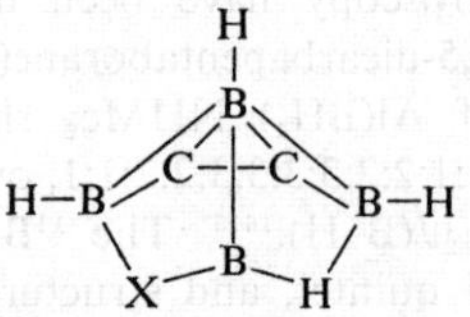

(155)

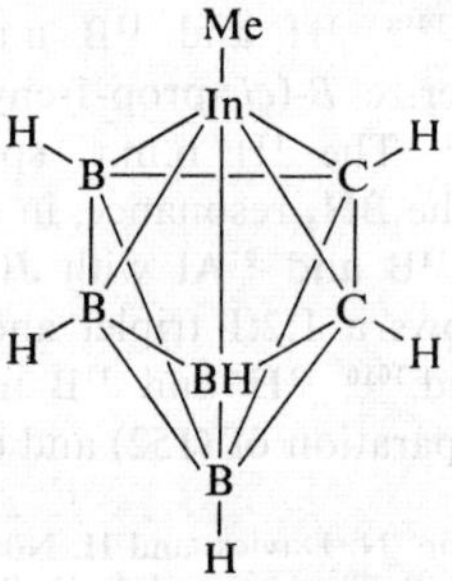

(156)

(157).[1622] ^{1}H and ^{11}B n.m.r. data have also been reported for silyl and germyl derivatives of C$_2$B$_4$H$_8$.[1623]

The ^{1}H and ^{11}B chemical shifts of XB$_5$H$_8$ and p-XC$_6$H$_4$B(OH)$_2$ have been examined and linear correlations found.[1624] The ^{11}B n.m.r. spectrum of decomposition products of [B$_5$H$_8$]$^-$ in solution contains resonances at

[1618] A. B. Burg, *Inorg. Chem.*, 1972, **11**, 2283.
[1619] C. G. Savory and M. G. H. Wallbridge, *J.C.S. Dalton*, 1972, 918.
[1620] R. N. Grimes, W. J. Rademaker, M. E. Denniston, R. F. Bryan, and P. T. Greene, *J. Amer. Chem. Soc.*, 1972, **94**, 1865.
[1621] M. P. Brown, A. K. Holliday, and G. M. Way, *J.C.S. Chem. Comm.*, 1972, 850.
[1622] R. E. Bowen and C. R. Phillips, *J. Inorg. Nuclear Chem.*, 1972, **34**, 382.
[1623] M. L. Thompson and R. N. Grimes, *Inorg. Chem.*, 1972, **11**, 1925.
[1624] A. R. Siedel and G. M. Bodner, *Inorg. Chem.*, 1972, **11**, 3108.

$$\text{Me}_3\text{PBH}_2\text{CH}_2\text{CH}_2\overset{\text{H}-\text{BH}_2}{\underset{\text{H}-\text{B}}{\text{BH}}}\overset{}{\underset{\text{PMe}_3}{\bigg|}}\text{H}$$

(157)

δ −10.3 and 36.7, which cannot be assigned to known species.[1625] The product from the reaction of 1- or 2-MeB$_5$H$_8$ and Me$_3$N gives the same ^{11}B n.m.r. spectrum.[1626] The SiH$_2$ resonance in 1-ClSiH$_2$B$_5$H$_8$ is a 1:1:1:1 quartet at room temperature, but on cooling to −85 °C much of this fine structure is lost.[1627] ^{11}B N.m.r. data have also been reported for (BH$_2$CN)$_5$.[1628]

Protonation of B$_6$H$_{10}$ and 2-MeB$_6$H$_9$ occurs on the basal boron and [BCl$_4$]$^-$ is formed.[1629] Low-temperature ^{1}H and ^{11}B n.m.r. spectra of B$_6$H$_{10}$ and 2-MeB$_6$H$_9$ have been used to demonstrate that (a) the bridge proton exchange rate is slow on the n.m.r. time-scale, (b) the B$_6$H$_{10}$ spectrum is consistent with the solid-state structure, and (c) the static structure of 2-MeB$_6$H$_9$ is without a plane of symmetry.[1630]

SiH$_2$Cl$_2$ reacts with NaC$_2$B$_4$H$_7$ to yield μ,μ'-SiH$_2$(C$_2$B$_4$H$_7$)$_2$. The ^{11}B n.m.r. spectra show four doublets of equal area: therefore there are no boron–silicon terminal bonds. Two possible isomers, (158) or (159), were suggested.[1631] The ^{19}F chemical shifts of 1-(m- or -p-FC$_6$H$_4$)-1,6-B$_8$C$_2$H$_9$ or 1,10-B$_8$C$_2$H$_9$ have been used to calculate resonance substituent constants. Both carbaboranes had weak electron-acceptor ability.[1632] Alkaline ethanolysis of 5,6-dicarba-nido-decaborane(12) produces C$_2$H$_{14}$B$_8$. The ^{11}B n.m.r. spectrum shows three doublets in the ratio 2:4:2 and the structure (160) is suggested.[1633] ^{11}B N.m.r. spectroscopy has been used to characterize the low-temperature product from the reaction of NH$_3$ with n-B$_9$H$_{15}$.[1634] ^{1}H and ^{11}B n.m.r. spectra of B$_8$H$_{12}$, B$_8$H$_{18}$, and 2,2'-(B$_5$H$_8$)$_2$ have been measured at 51.7 kG.[1635]

^{11}B and ^{13}C n.m.r. chemical shifts of a number of carbaboranes have been reported. It is found that six-co-ordinate carbon atoms are at ca. 130 p.p.m., five-co-ordinate carbon atoms ca. 115 p.p.m., and four-co-ordinate carbon atoms ca. 90 p.p.m. from CS$_2$.[1636] The ^{1}H and ^{11}B n.m.r.

[1625] V. T. Brice, H. D. Johnson, sec., D. L. Denton, and S. G. Shore, *Inorg. Chem.*, 1972, **11**, 1135.
[1626] G. Kodama, *J. Amer. Chem. Soc.*, 1972, **94**, 5907.
[1627] T. C. Geisler and A. D. Norman, *Inorg. Chem.*, 1972, **11**, 2549.
[1628] B. F. Spielvogel, R. F. Bratton, and C. G. Moreland, *J. Amer. Chem. Soc.*, 1972, **94**, 8597.
[1629] H. D. Johnson, sec., V. T. Brice, G. L. Brubaker, and S. G. Shore, *J. Amer. Chem. Soc.*, 1972, **94**, 6711.
[1630] V. T. Brice, H. D. Johnson, sec., and S. G. Shore, *J.C.S. Chem. Comm.*, 1972, 1128.
[1631] A. Tabereaux and R. N. Grimes, *J. Amer. Chem. Soc.*, 1972, **94**, 4769.
[1632] L. I. Zakharkin, V. N. Kalinin, E. G. Rys, and B. A. Kvasov, *Izvest. Akad. Nauk S.S.S.R., Ser. khim.*, 1972, 507.
[1633] B. Štibr, J. Plešek, and S. Heřmánek, *Chem. and Ind.*, 1972, 649.
[1634] R. Schaeffer and L. G. Sneddon, *Inorg. Chem.*, 1972, **11**, 3102.
[1635] R. R. Rietz, R. Schaeffer, and L. G. Sneddon, *Inorg. Chem.*, 1972, **11**, 1242.
[1636] J. L. Todd, *Pure Appl. Chem.*, 1972, **30**, 587.

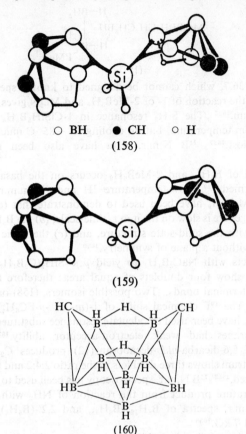

spectra of CsB$_9$H$_{14}$ (161) are consistent with the delocalization of the bridge hydrogen atoms and one hydrogen atom from each BH$_2$ group around the open face of the icosahedral fragment, despite the X-ray structure.[1637] The 70.6 MHz ^{11}B n.m.r. spectrum of [(3)-1,2-B$_9$C$_2$H$_{12}$]$^-$ shows five doublets in intensity ratio 2:3:2:1:1, whereas [{(3)-1,2-C$_9$B$_2$H$_{11}$}$_2$Co]$^-$ shows five doublets in the ratio 1:1:4:2:1.[1638] On the basis of ^{1}H and ^{11}B n.m.r. spectroscopy, the (3)-1,7-structure has been suggested for (1,5-C$_8$H$_{12}$)-Ni(Me$_2$C$_2$B$_9$H$_9$).[1639]

Reduction of 1,2-B$_{10}$C$_2$H$_{12}$ yields [B$_{10}$C$_2$H$_{12}$]$^{2-}$ and protonation yields two isomeric [B$_{10}$C$_2$H$_{13}$]$^-$ anions. For one, the ^{11}B n.m.r. spectrum shows seven doublets with an intensity ratio 1:2:1:2:1:2:1; for the other, the ^{11}B

[1637] N. N. Greenwood, J. A. McGinnety, and J. D. Owen, *J.C.S. Dalton*, 1972, 986.
[1638] A. R. Siedle, G. M. Bodner, and L. T. Todd, *U.S. Nat. Tech. Inform. Serv., AD Rep.*, 1972, No. 735932.
[1639] J. L. Spencer, M. Green, and F. G. A. Stone, *J.C.S. Chem. Comm.*, 1972, 1178.

Nuclear Magnetic Resonance Spectroscopy

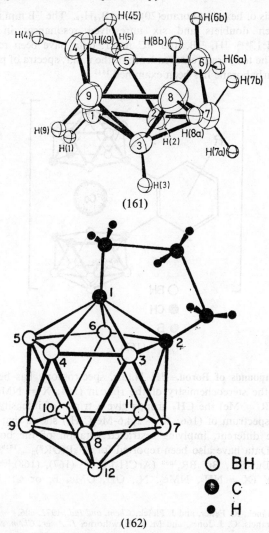

(161)

(162)

n.m.r. spectrum shows six doublets with an intensity ratio 2:1:1:2:2:2.[1640] The 80.5 MHz ^{11}B n.m.r. spectrum of $Rb_2B_{10}H_{14}$ consists of doublet:doublet:triplet:doublet with an intensity ratio 2:4:2:2, which was considered to be consistent with 2632 topology.[1641] Data have also been reported for $B_{10}C_5H_{16}$ (162), $[Me_4N][Co(B_9C_5H_{15})_2]$, and $[Ni(B_9C_5H_{15})_2B]$.[1642]

[1640] G. B. Dunks, R. J. Wiersema, and M. F. Hawthorne, *J.C.S. Chem. Comm.*, 1972, 899.
[1641] W. N. Lipscomb, R. J. Wiersema, and M. F. Hawthorne, *Inorg. Chem.*, 1972, **11**, 651.
[1642] T. E. Paxson, M. K. Kaloustian, G. M. Tom, R. J. Wiersema, and M. F. Hawthorne, *J. Amer. Chem. Soc.*, 1972, **94**, 4882.

Hydrolysis of hexadecaborane(20) gives $B_{14}H_{18}$. The ^{11}B n.m.r. spectrum shows seven doublets and six unresolved resonances in the ratio 2:1:6:1:1:1:1:1.[1643] ^1H, ^{11}B, and ^{13}C n.m.r. data have been reported for (163).[1644] The temperature dependence of the n.m.r. spectra of polymerized 1-vinyl-o-carbaborane has been examined.[1645]

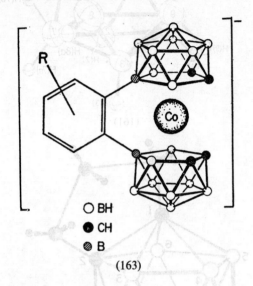

(163)

Other Compounds of Boron.—^1H N.m.r. spectroscopy has been used to determine the stereochemistry of the ring in (164; X = NMe or O).[1646] For (165; R = Me) the CH_2 groups give a triplet of intensity 4.[1647] The ^1H n.m.r. spectrum of (166; R = 2,4,6-$Me_3C_6H_2$) shows that the methyl groups are different, implying restricted rotation of the boron–carbon bond.[1648] Data have also been reported for $R^1{}_nB(OR^2)_{3-n}$,[1649–1651] $Bu^n{}_3B$, $Bu^s{}_3B$,[1652] $Bu^n{}_2BK$, $Bu^n{}_2BR$,[1653] $(ArCH_2)_3B$,[1654] (167), (168),[1655] MeEtBX, Me_2BCH_2Y (X = NH_2, NMe_2, N_3, OH, OMe, F, or Cl; Y = PMe_2,

---

[1643] S. Heřmánek, K. Fetter, and J. Plešek, *Chem. and Ind.*, 1972, 606.
[1644] J. N. Francis, C. J. Jones, and M. F. Hawthorne, *J. Amer. Chem. Soc.*, 1972, **94**, 4878.
[1645] J. R. Wright and T. J. Klingen, *J. Inorg. Nuclear Chem.*, 1972, **34**, 3284.
[1646] F. A. Davis, I. J. Turchi, B. E. Maryanoff, and R. O. Hutchins, *J. Org. Chem.*, 1972, **37**, 1583.
[1647] R. H. Fish, *J. Organometallic Chem.*, 1972, **42**, 345.
[1648] R. Van Veen and F. Bickelhaupt, *J. Organometallic Chem.*, 1972, **43**, 241.
[1649] S. Korcek, G. B. Watts, and K. U. Ingold, *J.C.S. Perkin II*, 1972, 242.
[1650] I. Kronawitter and H. Nöth, *Chem. Ber.*, 1972, **105**, 2423.
[1651] E. Negishi, J.-J. Katz, and H. C. Brown, *J. Amer. Chem. Soc.*, 1972, **94**, 4027.
[1652] A. G. Davies, B. P. Roberts, and J. C. Scaiano, *J.C.S. Perkin II*, 1972, 803.
[1653] D. J. Pasto and P. W. Wojtkowski, *J. Organometallic Chem.*, 1972, **34**, 251.
[1654] B. G. Ramsey and N. K. Das, *J. Amer. Chem. Soc.*, 1972, **94**, 4227.
[1655] P. Jutzi, *Angew. Chem. Internat. Edn.*, 1972, **11**, 53.

AsMe₂, SMe, Cl, or I),[1656] F₂C=CHBX₂,[1657] (S=PF₂NH)₂BPh,[1658] (C₅H₅)B₃N₃Me₃,[1659] (169),[1660] (170) or (171),[1661] (172),[1662] (173), (174)[1663]

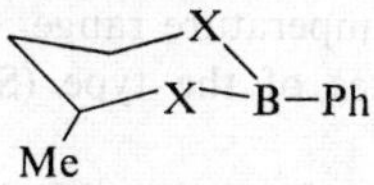

(164)

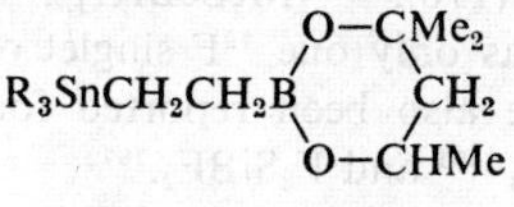

(165)

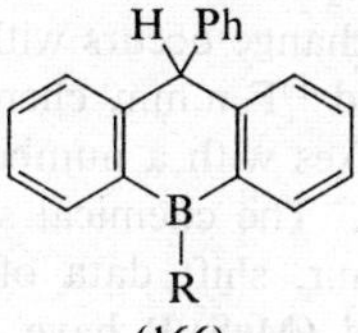

(166)

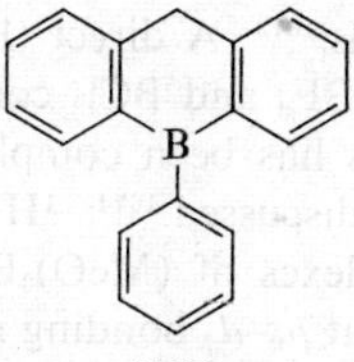

(167)

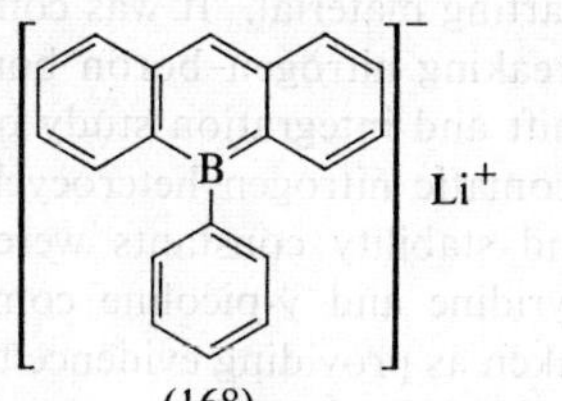

(168)

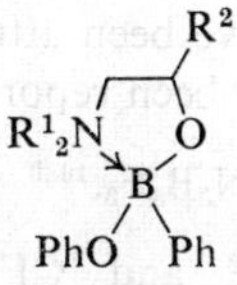

(169)

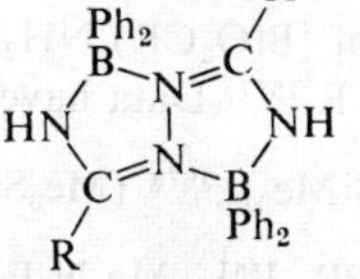

(170)

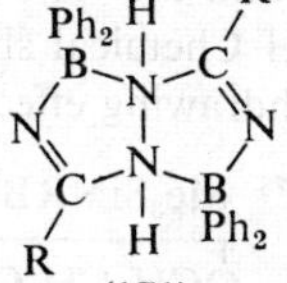

(171)

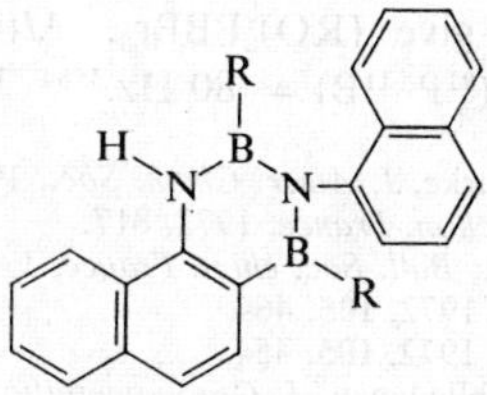

(172)

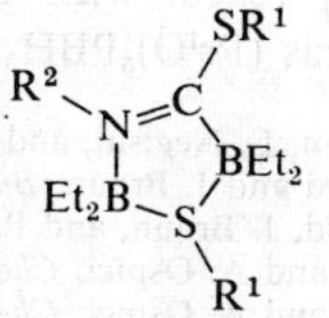

(173)

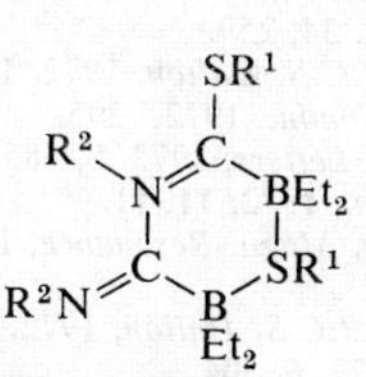

(174)

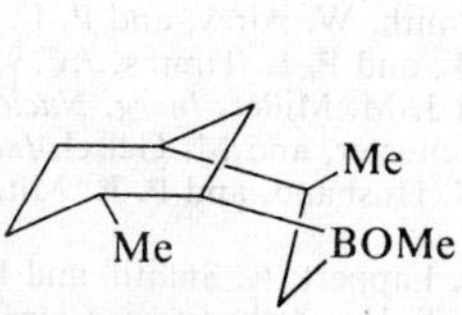

(175)

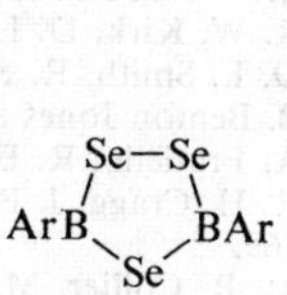

(176)

[1656] J. Rathke and R. Schaeffer, *Inorg. Chem.*, 1972, **11**, 1150.
[1657] J. J. Ritter, T. D. Coyle, and J. M. Bellama, *J. Organometallic Chem.*, 1972, **42**, 25.
[1658] H. W. Roesky, *Chem. Ber.*, 1972, **105**, 1726.
[1659] B. L. Therrell, jun. and E. K. Mellon, *Inorg. Chem.*, 1972, **11**, 1137.
[1660] W. Kliegel, *Annalen*, 1972, **763**, 61.
[1661] T.-T. Wang and K. Niedenzu, *J. Organometallic Chem.*, 1972, **35**, 231.
[1662] J. Cueilleron and B. Frange, *Bull. Soc. chim. France*, 1972, 107.
[1663] A. Grote, A. Haag, and G. Hesse, *Annalen*, 1972, **755**, 67.

(175),[1664] adducts of $C_5H_{6-n}Cl_n$ and $\overline{(vinyl)BOCH_2CH_2O}$,[1665] $\overline{RBOCH_2\text{-}CH_2O}$,[1666] (176),[1667] $[MeSeBRI]_n$,[1668] $PhBX_2$,[1669] and $Cl_2BCH_2CH_2BCl_2$.[1670] B_8F_{12} has only one ^{19}F singlet over the temperature range -80—$0\,°C$. Data have also been reported for complexes of the type $(Si_2F_5)(SiF_3)$-$(BF_2)BPF_3$ [1671] and F_3SiBF_2.[1672]

^1H N.m.r. spectroscopy has been used to show that Me_3N,BF_nCl_{3-n} formed from $Me_3N,^{10}BF_3$ still contained ^{10}B in the same amount as the starting material. It was concluded that halogen exchange occurs without breaking nitrogen–boron bonds.[1673] A direct ^1H and ^{19}F n.m.r. chemical shift and integration study of BF_3 and BCl_3 complexes with a number of aromatic nitrogen heterocycles has been completed. The chemical shifts and stability constants were discussed.[1674] ^1H N.m.r. shift data of the pyridine and γ-picoline complexes of $(MeO)_3B$ and $(MeS)_3B$ have been taken as providing evidence that p_π–d_π bonding in the boron–sulphur bond is much weaker than p_π–p_π bonding in the boron–oxygen bond.[1675] The ^{11}B n.m.r. shift of 32 ± 4 p.p.m. for $Bu^t_2C{=}NBR_2$ is consistent with three co-ordination, and the ^1H signal of Bu^t indicates nitrogen–boron double-bonding.[1676] ^1H Chemical shifts of $[B(O_2CR)_2NH_2]_2$ have been attributed to electron-withdrawing effects of R.[1677] Data have also been reported for $B(OC_2H_4)_3N$,[1678] $Me_3SiNRBFN(SiMe_3)_2$,[1679] $(Me_3Si)_2HN_3B_3F_3$,[1680] $\overline{OCH_2CH_2OBClNR_3}$, $\overline{OCH_2CH_2OBClPR_3}$,[1681] Me_2N_4BX,[1682] and $CF_3CON(NMe_2)BF_2$.[1683]

$(RO)_3PBH_3$ reacts with Br_2 to give $(RO)_3PBBr_3$. $^1J(^{31}P\text{–}^{11}B) = ca.$ 280 Hz whereas $(Pr^i O)_3PBH_3$ has $^1J(^{31}P\text{–}^{11}B) = 80$ Hz.[1684] $^1J(^{11}B\text{–}^{31}P)$ has

---

[1664] H. C. Brown, E. Negishi, and P. L. Burke, *J. Amer. Chem. Soc.*, 1972, **94**, 3561.
[1665] G. Coindard and J. Braun, *Bull. Soc. chim. France*, 1972, 817.
[1666] G. Coindard, J. Braun, and P. Cardiot, *Bull. Soc. chim. France*, 1972, 811.
[1667] W. Siebert and A. Ospici, *Chem. Ber.*, 1972, **105**, 464.
[1668] W. Siebert and A. Ospici, *Chem. Ber.*, 1972, **105**, 454.
[1669] F. C. Nahm, E. F. Rothergy, and K. Niedenzu, *J. Organometallic Chem.*, 1972, **35**, 9.
[1670] M. Zeldin and A. Rosen, *J. Organometallic Chem.*, 1972, **34**, 259.
[1671] R. W. Kirk, D. L. Smith, W. Airey, and P. L. Timms, *J.C.S. Dalton*, 1972, 1392.
[1672] D. L. Smith, R. Kirk, and P. L. Timms, *J.C.S. Chem. Comm.*, 1972, 295.
[1673] B. Benton-Jones and J. M. Miller, *Inorg. Nuclear Chem. Letters*, 1972, **8**, 485.
[1674] A. Fratiello, R. E. Schuster, and M. Geisel, *Inorg. Chem.*, 1972, **11**, 11.
[1675] R. H. Cragg, J. P. N. Husband, and P. R. Mitchell, *Org. Magn. Resonance*, 1972, **4**, 469.
[1676] M. R. Collier, M. F. Lappert, R. Snaith, and K. Wade, *J.C.S. Dalton*, 1972, 370.
[1677] G. J. Barrett and D. T. Haworth, *Inorg. Chim. Acta*, 1972, **6**, 504.
[1678] D. Fenske and H. J. Becker, *Chem. Ber.*, 1972, **105**, 2085.
[1679] G. Elter, O. Glemser, and W. Herzog, *Inorg. Nuclear Chem. Letters*, 1972, **8**, 191.
[1680] G. Elter, O. Glemser, and W. Herzog, *Chem. Ber.*, 1972, **105**, 115.
[1681] S. G. Shore, C. L. Crist, B. Lockman, J. R. Long, and A. D. Coon, *J.C.S. Dalton*, 1972, 1123.
[1682] B. Hessett, J. B. Leach, J. H. Morris, and P. G. Perkins, *J.C.S. Dalton*, 1972, 131.
[1683] G. Czieslik and O. Glemser, *Z. anorg. Chem.*, 1972, **394**, 26.
[1684] T. Reetz, *Inorg. Chem.*, 1972, **11**, 650.

also been measured for $(Me_2N)_2BER_2$ (E = P, As, or Sb).[1685] ^1H N.m.r. spectroscopy has been used to investigate the $PH_3-BX_3-BY_3$ system and species such as H_3PBCl_2Br have been detected.[1686] Data have also been reported for H_3P,BI_3.[1687]

^{11}B N.m.r. spectroscopy has been used to examine borate solutions and signals due to $B(OH)_3$, pentaborate, tetraborate, and metaborate were identified. A linear relationship was found between the Na_2O to B_2O_3 ratio and ^{11}B chemical shift.[1688] From the observation of a sharp ^{11}B n.m.r. signal for $SO_3-K_2B_4O_7$ it was concluded that the boron is tetrahedral. I.r. spectroscopy indicates triangular co-ordination for the boron.[1689] ^1H, ^{11}B, and ^{19}F n.m.r. spectroscopy have been used to study halogen redistribution in MeO_2,BX_3 adducts in solution: the equilibrium is attained rapidly. A linear relationship was found between $^1J(^{11}B-^{19}F)$ and the ^{19}F chemical shift.[1690] The n.m.r. spectra of RSCl in SO_2 and in the presence of BCl_3, BF_3, or SbF_5 indicated an equilibrium involving a dimeric monocationic species.[1691] Data have also been reported for $ArCHOBF_3$,[1692] $X_nB(SCF_3)_{3-n}$,[1693] and $(Me_2N)_2COBF_3$.[1694]

Complexes of Other Group III Elements.—The low-temperature n.m.r. spectrum of a mixture of $Bu^i{}_2AlH$ and $Bu^i{}_2AlCl$ shows the presence of three species which are thought to be (177), (178), and (179).[1695] Data have also been reported for $LiAl(C_3F_7)H_2I$.[1696]

For $Li^+[Me_3MSnMe_3]^-$ (M = Al, Ga, In, or Tl), it is found that $^2J(^{119}Sn-C-^1H)$ increases with the size of M whereas $^3J(^{119}Sn-M-C-^1H)$ is in the order Al < Ga > In > Tl.[1697] The reaction of Me_3Al with PbO to give compounds such as $(Me_2Al)_2O$ has been followed by ^1H n.m.r. spectroscopy.[1698] The ^1H n.m.r. spectrum of $[R^1{}_2AlCH=CHR^2]_2$ shows

---

[1685] W. Becker and H. Nöth, *Chem. Ber.*, 1972, **105**, 1962.
[1686] J. E. Drake and B. Rapp, *J.C.S. Dalton*, 1972, 2341.
[1687] M. Schmidt and H. H. J. Schröder, *Z. anorg. Chem.*, 1972, **394**, 290.
[1688] H. D. Smith, jun. and R. J. Wiersema, *Inorg. Chem.*, 1972, **11**, 1152.
[1689] S. N. Kondrat'ev and S. I. Mel'nikova, *Russ. J. Inorg. Chem.*, 1972, **17**, 489.
[1690] M. J. Bula, D. E. Hamilton, and J. S. Hartman, *J.C.S. Dalton*, 1972, 1405.
[1691] G. Capozzi, V. Lucchini, and G. Modena, *Chimica e Industria*, 1972, **54**, 41.
[1692] M. Rabinovitz and A. Grinvald, *J. Amer. Chem. Soc.*, 1972, **94**, 2724.
[1693] A. Haas and M. Häberlein, *Chem.-Ztg.*, 1972, **96**, 412.
[1694] J. S. Harlman and G. J. Schrobilgen, *Canad. J. Chem.*, 1972, **50**, 713.
[1695] J. J. Eisch and S. G. Rhee, *J. Organometallic Chem.*, 1972, **38**, C25.
[1696] R. S. Dickson and G. D. Sutcliffe, *Austral. J. Chem.*, 1972, **25**, 761.
[1697] A. T. Weibel and J. P. Oliver, *J. Amer. Chem. Soc.*, 1972, **94**, 8590.
[1698] M. Boleslawski and S. Pasynkiewicz, *J. Organometallic Chem.*, 1972, **43**, 81.

that ether co-ordinates and breaks up the dimer.[1699] The ^{1}H n.m.r. spectra of R_3M and R_2MQ (R = Me, Et, or Bu^t; M = Al, Ga, or In; Q = anion of quinolin-8-ol) have been reported. The nature of bonding of quinolin-8-ol has been discussed from the shifts and coupling constants.[1700] The interaction of Me$_3$Al and acetylacetone has been investigated and the presence of Me$_2$Al(acac)(OCMeCH=CMeOH)$_2$ suggested.[1701, 1702] [Me$_2$Al-NHMe]$_3$ can be separated into fractions which readily interconvert in solution.[1703] The cyclopentadienyl resonance of TlMe(C$_5$H$_5$)X is a singlet even at −72 °C.[1704] Data have also been reported for R$_2$AlXSiMe$_3$, PhMeC(OAlMe$_2$)(NMeSiMe$_3$),[1705] R$^1{}_2$AlNR^{2}CO$_2$Me,[1706] (Me$_3$Si)$_2$NAl$_2$-Me$_5$,[1707, 1708] AlMe$_3$(PPh$_3$),[1709] [Me$_2$GaO$_2$PX$_2$]$_2$,[1710] MeGa$_2$X$_5$,[1711] MeIn-(tetraphenylporphyrin),[1712] Me$_2$InO$_3$SMe,[1713] Me$_2$InCl,SSbMe$_3$,[1714] RIn-Br$_2$,[1715] Me$_2$Tl(succinimide),[1716] [Me$_2$TlOPPh$_2$]$_2$,[1717] Me$_n$Tl(OCOPr^i)$_{3-n}$,[1718] and R(ClCH$_2$)TlX.[1719]

The adduct formation of [Et$_3$Al]$_2$ or [Et$_2$AlCl]$_2$ and Ph$_2$P(CH$_2$)$_n$PPh$_2$ in benzene has been investigated by ^{1}H n.m.r. spectroscopy, and species such as (AlEt$_3$)$_2$(diphosphine) have been found.[1720] In mixtures of Et$_2$AlOEt and EtAlCl$_2$, a new species which may be [EtAlCl(OEt)]$_2$ is formed.[1721] For Ar(Me$_2$NCS$_2$CH$_2$)TlX, a coupling of 7 Hz is observed over six bonds.[1722] ^{1}H N.m.r. spectra of cyclohexene and AlBr$_3$ show that, as the concentration of AlBr$_3$ increases, the olefinic hydrogen moves to low field, implying formation of a complex.[1723] Data have also been reported for RC(OMe)-(OAlEt$_2$)(SAlEt$_2$),[1724] Et$_2$IAl,NHMe$_2$,[1725] Et$_2$In(oxinate),[1726] [(cyclopropyl)$_2$-

---

[1699] G. Zweifel and G. M. Clark, *J. Organometallic Chem.*, 1972, **39**, C33.
[1700] B. Sen, G. L. White, and J. D. Wander, *J.C.S. Dalton*, 1972, 447.
[1701] S. Pasynkiewicz and K. Dowbor, *J. Organometallic Chem.*, 1972, **43**, 75.
[1702] S. Pasynkiewicz and K. Dowbor, *J. Organometallic Chem.*, 1972, **39**, C1.
[1703] K. J. Alford, K. Gosling, and J. D. Smith, *J.C.S. Dalton*, 1972, 2203.
[1704] T. Abe and R. Okawara, *J. Organometallic Chem.*, 1972, **35**, 27.
[1705] T. Sakakibara, T. Hirabayashi, and Y. Ishii, *J. Organometallic Chem.*, 1972, **46**, 231.
[1706] T. Hirabayashi, T. Sakakibara, and Y. Ishii, *J. Organometallic Chem.*, 1972, **35**, 19.
[1707] N. Wiberg, W. Baumeister, and P. Zahn, *J. Organometallic Chem.*, 1972, **36**, 267.
[1708] N. Wiberg and W. Baumeister, *J. Organometallic Chem.*, 1972, **36**, 277.
[1709] T. R. Durkin and E. P. Schram, *Inorg. Chem.*, 1972, **11**, 1054.
[1710] B. Schaible and J. Weidlein, *J. Organometallic Chem.*, 1972, **35**, C7.
[1711] W. Lind and I. J. Worrall, *J. Organometallic Chem.*, 1972, **40**, 35.
[1712] M. Bhatti, W. Bhatti, and E. Mast, *Inorg. Nuclear Chem. Letters*, 1972, **8**, 133.
[1713] H. Olapinski and J. Weidlein, *J. Organometallic Chem.*, 1972, **35**, C53.
[1714] T. Maeda, G. Yoshida, and R. Okawara, *J. Organometallic Chem.*, 1972, **44**, 237.
[1715] M. J. S. Gynane, L. G. Waterworth, and I. J. Worrall, *J. Organometallic Chem.*, 1972, **43**, 257.
[1716] B. Walther and C. Rockstroh, *J. Organometallic Chem.*, 1972, **42**, 41.
[1717] B. Walther, *J. Organometallic Chem.*, 1972, **38**, 237.
[1718] R. Okawara, *Pure Appl. Chem.*, 1972, **30**, 499.
[1719] T. Abe and R. Okawara, *J. Organometallic Chem.*, 1972, **43**, 117.
[1720] T. Kagawa and H. Hashimoto, *Bull. Chem. Soc. Japan*, 1972, **45**, 1739.
[1721] A. C. L. Su and J. W. Collette, *J. Organometallic Chem.*, 1972, **36**, 177.
[1722] T. Abe, S. Numata, and R. Okawara, *Inorg. Nuclear Chem. Letters*, 1972, **8**, 909.
[1723] H.-H. Perkampus and G. Prescher, *Z. phys. Chem. (Frankfurt)*, 1972, **77**, 333.
[1724] T. Hirabayashi, H. Imaeda, K. Itoh, and Y. Ishii, *J. Organometallic Chem.*, 1972, **42**, 33.
[1725] K. Gosling and A. L. Bhuiyan, *Inorg. Nuclear Chem. Letters*, 1972, **8**, 329.
[1726] T. Maeda and R. Okawara, *J. Organometallic Chem.*, 1972, **39**, 87.

$M(NCH_2CH_2)]_2$ (M = Al or Ga),[1727] RGa_2Br_5,[1728] $[Tl(CH_2COMe)_n]^{(3-n)+}$,[1729] $In(C_5H_5)_n$ (n = 1 or 3),[1730] $MeC{\equiv}C(CMe_2OMe)Tl(OAc)_2$,[1731] and $[PhTl]^{2+}$.[1732]

^{1}H N.m.r. spectroscopy has been used to show that $M\{Al(OPr^i)_4\}_3$ (M = Sc or In) has a structure similar to that of $[Al(OPr^i)_3]_4$ and shows magnetic inequivalence of the methyl protons in the bridging as well as the terminal isopropoxy-groups.[1733, 1734] A correlation has been found between the position of the C_α-H signal in the ^{1}H n.m.r. spectrum and the reactivity of $Al(N{=}CPhCH_2Me)XCl$ with $H_2C{=}CRCN$.[1735] Data have also been reported for $(p\text{-}Me_2NC_6H_4)_3M^1(M^2Cl_n)_2$ (M^1 = As or Sb; M^2 = Al, n = 4; M^2 = Sb, n = 6),[1736] $(2\text{-}HO\text{-}3,5\text{-}Pr^i_2C_6H_2CO_2)_3M$ (M = Al, Ga, or In),[1737] and $Tl(O_2NPr^i)$.[1738]

7 Compounds of Silicon, Germanium, Tin, and Lead

There have been many references to complexes of these elements in the preceding sections. The previously mentioned complexes are given here in the order of the Periodic Table. Information concerning complexes of these elements with magnesium and the transition metals can be found at the following sources: $Mg\{N(SiMe_3)_2\}_2$,[187] $Sc\{N(SiMe_3)_2\}_3$,[190] $(C_5H_5)_2$-$M^1Cl(M^2Ph_3)$ (M^1 = Zr or Hf; M^2 = Si, Ge, or Sn),[195] $Cr(CO)_4\{Me_2AsCH$-$(SiMe_3)CH_2AsMe_2\}$,[256] $(OC)_5CrPPh_2SnMe_3$,[224] $Me_3SnSMeM(CO)_5$ (M = Cr or W),[227] $Mo_2(CH_2SiMe_3)_6$,[213] $(C_5H_5)(OC)_nMSiMe_2X$ (M = Mo or W, n = 3; M = Fe, n = 2; X = Me, Cl, or Br),[296] $(C_5H_5)Mo(CO)_2(PPh_3)$-(MR_3) (M = Ge or Sn),[275] $(C_5H_5)Mo(CO)_2LSnMe_3$,[274] $(C_5H_5)(OC)_3$-$\overline{MoSnCl_2S_2CNMe_2}$,[277] $W(CO)_5MeSSnMe_3$,[228] $\overline{Me_3MC(CF_3)C(CF_3)}$-${=}C(CF_3)\dot{C}(CF_3)Mn(CO)_5$ (M = Si, Ge, or Sn),[327] $Me_3SnMn(CO)_5$,[321] $Mn(CO)_5GeH_3$,[326] $MeSiCl_2HRe_2(CO)_9$,[341] $(C_5H_5)Fe(CO)_2MR_3$ (M = Si, Ge, or Sn; R = Me or Ph),[112] $(Me_3SiOCH{=}CH_2)Fe(CO)_4$,[372] $[Me_3SiFe$-$(COSiMe_3)(CO)_3]_2$,[447] $Fe(CO)_n(PMe_2CH_2CH_2SiMe_3)_{5-n}$,[455] $\{(C_5H_5)Fe$-$(C_5H_4)SiMe_2\}_2SiMe_2$,[485] $Me_3GeSiMe_2C_5H_4FeC_5H_4SiMe_2OSiMe_3$,[592] $(OC)_4$-$\overline{FeCH_2CH_2CH_2SiMe_2}$,[448] $Fe(CO)_4(GeH_3)_2$,[440] $Me_2GeFe_2(CO)_3(C_5H_5)$,[432] $\{(C_5H_5)Fe(C_5H_3CH_2NMe_2)\}_2SnBu_2$,[483] $(C_5H_5)Ni(CO)SnCl_2Fe(CO)_2$-$(C_5H_5)$,[445] $(C_8H_9)Ru_2(CO)_4(GeMe_2)_2GeMe_3$,[433] $(C_5H_5)_2Co(CO)(Me_3SiCC$-$SiMe_3)$,[594] $IrH_2(CO)(PPh_3)_2SnMe_3$,[549] $(C_5H_5)Ni(CO)MX_3$ (M = Ge or

---

[1727] J. Müller, K. Margiolis, and K. Dehnicke, *J. Organometallic Chem.*, 1972, **46**, 219.
[1728] W. Lind and I. J. Worrall, *J. Organometallic Chem.*, 1972, **36**, 35.
[1729] P. Abley, J. E. Byrd, and J. Halpern, *J. Amer. Chem. Soc.*, 1972, **94**, 1985.
[1730] J. S. Poland and D. G. Tuck, *J. Organometallic Chem.*, 1972, **42**, 307.
[1731] R. K. Sharma and E. D. Martinez, *J.C.S. Chem. Comm.*, 1972, 1129.
[1732] S. Uemura, Y. Ikeda, and K. Ichikawa, *Tetrahedron*, 1972, **28**, 3025.
[1733] A. Mehrotra and R. C. Mehrotra, *J.C.S. Chem. Comm.*, 1972, 189.
[1734] A. Mehrotra and R. C. Mehrotra, *Inorg. Chem.*, 1972, **11**, 2170.
[1735] H. Hoberg and R. Kieffer, *Annalen*, 1972, **760**, 141.
[1736] J. M. Keck and G. Klar, *Z. Naturforsch.*, 1972, **27b**, 596.
[1737] H. F. Eicke, V. Arnold, and F. L'Eplattenier, *Angew. Chem. Internat. Edn.*, 1972, **11**, 1096.
[1738] A. G. Lee, *Spectrochim. Acta*, 1972, **28A**, 133.

Sn),[443] cis-Pt(CH_2SiMe_3)$_2$(PR_3)$_2$,[728] (Ph_3P)$_2$PtSiF_4,[799] R_3PAuOSiMe_3,[850] MePh_2SnZnCl,[864] and Me$_3$SiHgCF_2CFClSiMe_3.[884]

Information concerning complexes of these elements with Group IIIB elements can be found at the following sources: SiH_2($C_2B_4H_7$)$_2$,[1631] R_3MC$_2B_4H_7$ (R = H or Me; M = Si or Ge),[1623] Me$_3$SiNRBFN(SiMe_3)$_2$,[1679] (Me$_3$Si)$_2$NBF(SiMe_3)NH,[1680] Me$_3$MC$_2$Me$_2B_4H_5$ (M = Si or Ge),[1619] F_3SiBF_2,[1672] (Si_2F_5)(SiF_3)(BF_2)BPF_3,[1671] R_3SnCH_2CH_2BOCMeHCH_2C-Me$_2$,[1647] R_2AlXSiMe_3,[1705] [(Me$_3$Si)$_2$N]$_3$Al,[1707] LiMe$_3$SnMMe$_3$ (M = Al, Ga, In, or Tl),[1697] and (Me$_3$Si)$_2$NAl$_2$Me$_5$.[1708]

Information concerning complexes of these elements with Group VB and VIB elements can be found at the following sources: (Me$_3$Si)$_3$N,[126] Me$_3$SiX (X = CN, C_6F_5, P{SiMe_3}$_2$, etc.)[130] (R_3M)$_n$PX$_{3-n}$ (M = Si, Ge, or Sn; R = H, Me, Ph, or Bu; X = H, Me, or Ph),[133] SnCl_4(PEt_3)$_2$,[85] Me$_3$SiO(SiMe$_2$O)$_n$SiMe$_3$,[129] Me$_2$Si(OAc)CH_2CH_2CH_2SO_3^-Na$^+$,[86] Si-(OMe)$_4$,[317] (acac)$_2R_2$Sn,[46] (MeO)Me$_2$SnAr,[96] Bu$^n_2$Sn(OR)$_2$,[156] FC_6H_4OMR_3 (M = Sn or Pb),[866] Ar_3SnO_2CR,[154] (H$_3$M)$_2$E (M = Si or Ge; E = S, Se, or Te),[143] and Me$_3$SnS_2CNMe_2.[99]

Information concerning other complexes of these elements can be found at the following sources: Me$_3$SiC≡CPh,[67] p-XC_6H_4SiMe_3,[95] $C_6H_4C_2H_3$-MMe$_3$ (M = Si or Sn),[887] Me$_3$SiCH_2CHCH_2CCl(CO_2Me),[888] Me$_4$Si,[119, 120] Me$_3$SiCHISnMe_3,[865] phenylmethylsilanes,[61] phenyltetramethyldisilanes,[60] alkylarylfluorosilanes,[105] alkylchlorosilanes,[145] M(CH_2SiMe_3)$_4$ (M = Sn or Pb),[213] (PhCH_2)$_n$M (M = Si or Sn),[107] FC_6H_4SiXYZ,[104] [SnH_3]$^-$,[63] organotin compounds,[98] methyltin compounds,[77] Me$_3$SnC_5H_5,[74] Me$_3$SnCl,[153] RSnCl_3,[157] MeSnR_3,[97] EtSnX_3,[152] SnBu$^n_3$(C_3F_7),[1696] (ArCH_2)$_n$SnCl_{4-n},[155] and SnCl_4.[167]

The effect of d_π–p_π interaction in organic compounds of Group IVB elements has been reviewed.[1739]

^{1}H N.m.r. spectra of [GeH_3]$^-$ and GeH_4 have been reported. The spectra were cation dependent, which was taken to indicate that there is a contact ion pair for Cs$^+$[GeH_3]$^-$ and a solvent-separated ion pair for Li$^+$, Na$^+$, and K$^+$.[1740] For ArSiMe$_n$H$_{3-n}$, quantitative correlations of the chemical shifts with relative anisotropic contributions of the aromatic ring have been examined as a function of the molecular parameters, bond angles, bond lengths, and the dihedral angle. It was suggested that d_π–p_π bonding may be responsible for the high-field chemical shift of the aryl and vinyl silanes.[1741] ^{1}H Chemical shifts of disilanylamines are interpreted as indicating that the basicity decreases in the order Me$_3$N > H$_3$SiSiH_2NMe_2 > (H$_3$SiSiH_2)$_2$NMe > (H$_3$SiSiH_2)$_3$N.[1742] ^{1}H N.m.r. spectroscopy has been

---

[1739] A. N. Egorochkin, N. S. Vyazankin, and S. Ya. Khorshev, *Russ. Chem. Rev.*, 1972, **41**, 425.

[1740] T. Birchall and I. Drummond, *Inorg. Chem.*, 1972, **11**, 250.

[1741] R. J. Ouellette, J. M. Pang, and S. H. Williams, *J. Organometallic Chem.*, 1972, **39**, 267.

[1742] M. Abedini, *Quart. Bull. Fac. Sci., Tehran Univ.*, 1972, **3**, 1.

used to investigate the mechanism of addition of Bu^n_3SnH to acetylenes.[1743] The effect of concentration on the chemical shifts of R_3SiH has been examined.[1744] An equilibrium between $MeCS(OSiH_3)$ and $MeCO(SSiH_3)$ has been postulated and data have been reported for $H_3GeSSiMe_3$ and related compounds.[1745] The Karplus equation has been used to determine the conformation of species such as $H_3SiCH_2SiH_2Me$.[1746] Data have also been reported for $\{(H_3Si)_2N\}_2SiH_2$,[1747] H_3SiAsR_2, Me_3SiAsR_2,[1748] $(GeH_3\text{-}CH_2SiH_2)_2O$,[1749] SiH_3OSiH_2F,[1750] $Si_3H_6Me_2$,[1751] $H_3SiSiHBrSiH_3$,[1752] $MeClSiHSiH_2Cl$,[1753] $(PhH_2GeGeH_2)_2O$,[1754] $Et_2MeSiSiMe_2SiMe_2H$,[1755] (indenyl)-$SiMe_2H$,[1756] $H(SiMe_2)_4H$,[1757] (1-naphthyl)$SiPhBu^nH$,[1758] (180; X = H or Me),[1759] (181),[1760] $Me_2HSiC_6Br_5$,[1761] $Me_2SiHOCH_2PEt_2$, $Me_2M(OCH_2PEt_2)_2$ (M = Si or Ge),[1762] $HSi(SiF_3)_2(Si_2F_5)$, $1,1\text{-}(SiF_2)_nB_2H_4$,[1763] $HMe_2Si(CF_2)_6\text{-}SiMe_2H$,[1764] silacyclobutanes,[1765] $Me_3SiSiIH_2$,[1766] $(GeH_2)_2AsMe$,[1767] $Ph_2SiHCH(OMe)Ph$,[1768] HCl_2SiR,[1769] and (182; R = H or Me).[1770]

(180)

(181)

(182)

[1743] J.-P. Quintard and M. Pereyre, *J. Organometallic Chem.*, 1972, **42**, 75.
[1744] V. O. Reikhsfel'd, V. B. Pukhnarevich, S. P. Sushchinskaya, and A. M. Evdokimov, *Zhur. obshchei Khim.*, 1972, **42**, 163.
[1745] S. Cradock, E. A. V. Ebsworth, and H. F. Jessep, *J.C.S. Dalton*, 1972, 359.
[1746] R. J. Ouellette, D. Barton, J. Stolfo, A. Rosenblum, and P. Weber, *Tetrahedron*, 1972, **28**, 2163.
[1747] W. M. Scantlin and A. D. Norman, *Inorg. Chem.*, 1972, **11**, 3082.
[1748] J. W. Anderson and J. E. Drake, *J. Inorg. Nuclear Chem.*, 1972, **34**, 2455.
[1749] C. H. Van Dyke, E. W. Kifer, and G. A. Gibbon, *Inorg. Chem.*, 1972, **11**, 408.
[1750] E. W. Kifer and C. H. van Dyke, *Inorg. Chem.*, 1972, **11**, 404.
[1751] P. S. Skell and P. W. Owen, *J. Amer. Chem. Soc.*, 1972, **94**, 5434.
[1752] T. C. Geisler, C. G. Cooper, and A. D. Norman, *Inorg. Chem.*, 1972, **11**, 1710.
[1753] A. J. Vanderwielen and M. A. Ring, *Inorg. Chem.*, 1972, **11**, 246.
[1754] P. Rivière and J. Satgé, *Helv. Chim. Acta*, 1972, **55**, 1164.
[1755] P. Ishikawa, T. Takaoka, and M. Kumada, *J. Organometallic Chem.*, 1972, **42**, 333.
[1756] P. E. Rakita and G. A. Taylor, *Inorg. Chem.*, 1972, **11**, 2136.
[1757] M. Ishikawa and M. Kumada, *J. Organometallic Chem.*, 1972, **42**, 325.
[1758] R. J. P. Corriu, G. F. Lanneau, and G. L. Royo, *J. Organometallic Chem.*, 1972, **35**, 35.
[1759] M. R. Smith, jun. and H. Gilman, *J. Organometallic Chem.*, 1972, **42**, 1.
[1760] H. Sakurai and M. Murakami, *J. Amer. Chem. Soc.*, 1972, **94**, 5080.
[1761] C. F. Smith, G. J. Moore, and C. Tamborski, *J. Organometallic Chem.*, 1972, **42**, 257.
[1762] C. Couret, J. Satgé, and F. Couret, *Inorg. Chem.*, 1972, **11**, 2274.
[1763] D. Solan and A. B. Burg, *Inorg. Chem.*, 1972, **11**, 1253.
[1764] M. R. Smith, jun. and H. Gilman, *J. Organometallic Chem.*, 1972, **46**, 251.
[1765] J. Dubac, P. Mazerolles, and B. Serres, *Tetrahedron Letters*, 1972, 3495.
[1766] E. Hengge, G. Bauer, and H. Marketz, *Z. anorg. Chem.*, 1972, **394**, 93.
[1767] J. W. Anderson and J. E. Drake, *J.C.S. Dalton*, 1972, 951.
[1768] E. O. Fischer and K. H. Dötz, *J. Organometallic Chem.*, 1972, **36**, C4.
[1769] R. Nakao, T. Fukumoto, and J. Tsurugi, *J. Org. Chem.*, 1972, **37**, 4349.
[1770] J. V. Scribelli and D. M. Curtis, *J. Organometallic Chem.*, 1972, **40**, 317.

Infinite-dilution chemical shifts of $Me_3M(CH_2)_nOH$ (M = C, Si, or Ge) have been reported and discussed.[1771] ^1H and ^{13}C n.m.r. spectroscopy have been used to determine the position of deuterium in complexes such as $Me_3MCMeDCHMeCHCl_2$ (M = Si or Sn).[1772] The ^1H and ^{119}Sn n.m.r. spectra of $R^1_3M(CH_2)_nCH=CHR^2$ (M = Si, Ge, or Sn; R^1 = Me, Et, Bu^n, or Ph; n = 1 or 2) have been analysed. It was suggested that the data provided no evidence of $d_\pi-p_\pi$ overlap in these compounds.[1773] Data have also been reported for $Me_3SiCClBrSnMe_3$,[1774] $Me_2S(O)=CHMMe_3$ (M = Si or Ge), $Me_2Si\{CH=S(O)Me_3\}_2$,[1775] and (183; M = Si or Ge).[1776]

(183)

The n.m.r. spectra of 18 silanes containing fluorine in the aliphatic and aromatic groups have been reported.[1777] ^1H N.m.r. spectroscopy has been used to find the stereochemistry of the reaction products of Me_3SiLi with alkyl halides.[1778] $Me_3SiC\equiv C-C\equiv CSiMe_3$ reacts with Na_2Te to yield (184); ^1H n.m.r. data were reported.[1779] $(Me_3Si)C_3H_2$ can have either stereochemistry (185) or (186), but as the ^{13}C satellite of the allenic hydrogen shows a hydrogen-hydrogen coupling constant of 7 Hz, the stereochemistry is (186).[1780] The n.m.r. spectra of p-trimethylsilylstyrene and its trichloroplatinic derivatives have been recorded. The chemical shifts observed for the ethylenic protons indicate that the $SiMe_3$ group exerts only a very slight electron-releasing effect on the ethylenic system.[1781] The ^1H n.m.r. spectrum of [187; X = SiR_3 or $Si(hal)_3$] has been measured. The chemical shift

(184)   (185)   (186)   (187)

[1771] J. Dědina, J. Schraml, and V. Chvalovský, *Coll. Czech. Chem. Comm.*, 1972, **37**, 3762.
[1772] D. Seyferth, Y. M. Cheng, and D. D. Traficante, *J. Organometallic Chem.*, 1972, **46**, 9.
[1773] R. G. Jones, P. Partington, W. J. Rennie, and R. M. G. Roberts, *J. Organometallic Chem.*, 1972, **35**, 291.
[1774] D. Seyferth, F. M. Armbrecht, jun., R. L. Lambert, jun., and W. Tronich, *J. Organometallic Chem.*, 1972, **44**, 299.
[1775] H. Schmidbaur and W. Kapp, *Chem. Ber.*, 1972, **105**, 1203.
[1776] M. A. Weiner and P. Schwartz, *J. Organometallic Chem.*, 1972, **35**, 285.
[1777] I. V. Romashkin, G. V. Odabashyan, V. F. Andronov, and V. A. Drozdov, *Zhur. obshchei Khim.*, 1972, **42**, 1060.
[1778] G. S. Koerner, M. L. Hall, and T. G. Traylor, *J. Amer. Chem. Soc.*, 1972, **94**, 7205.
[1779] T. J. Barton and R. W. Roth, *J. Organometallic Chem.*, 1972, **39**, C66.
[1780] P. S. Skell and P. W. Owen, *J. Amer. Chem. Soc.*, 1972, **94**, 1578.
[1781] Y. Limouzin and J. C. Marie, *J. Organometallic Chem.*, 1972, **39**, 255.

of the proton attached to carbon-3 decreases linearly with increasing Hammett σ^* constant of X. The degree of $d_\pi-p_\pi$ conjugation between silicon and carbon was greater for SiR_3 than $SiCl_3$.[1782] Data have also been reported for $Me_3SiCH_2(p\text{-tolyl})$,[1783] $p\text{-}Me_3SiC_6H_4CH(SiMe_3)NPhMe$,[1784] $MeSiCHPhCH_2CO_2Et$, $Et_3SiCHPhCH=C(OEt)OSiMe_3$,[1785] $Me_3SiOCH=\overline{CHCHCH_2CH_2}$, $Me\overline{SiCHCH_2CH}=CHCH_2O$,[1786] $Me_3SiCH=\overline{C(CH_2)_5}$[1787] (188),[1788] (189), (190),[1789] $PhCH(SiMe_3)EH$ (E = O or S),[1790] $SiMe_3R$,[1791, 1792] $Me_3SiCHCl_2$,[1793] $Me_3Si(MeO_2C)\overline{CCH_2CH_2}$, $R^1R^2C=C(OMe)OSiMe_3$,[1794] $(Me_3Si)_4C$,[1795] $(Me_3Si)_nCH_{3-n}CO_2H$,[1796] $Me_3SiCH=CHCl$,[1797] $Ph_3P=C(SiMe_3)_2$,[1798] (191),[1799] (192),[1800] (193),[1801] Me_3SiCH-

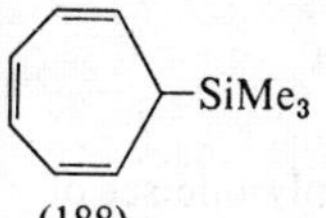

(188) 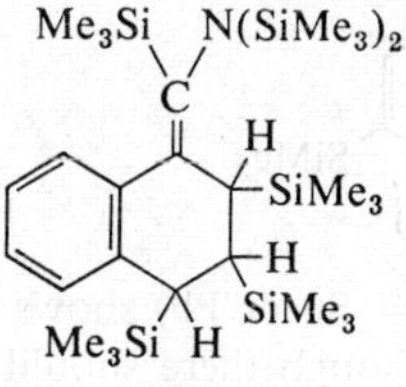 (189) 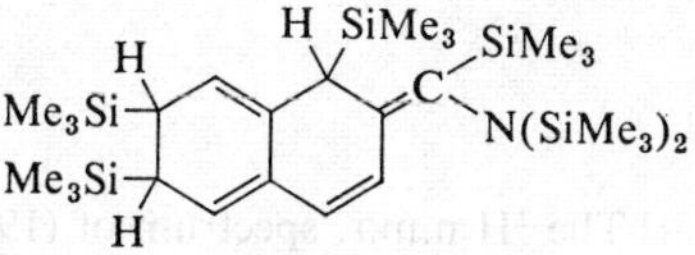 (190)

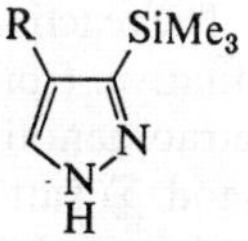

(191) 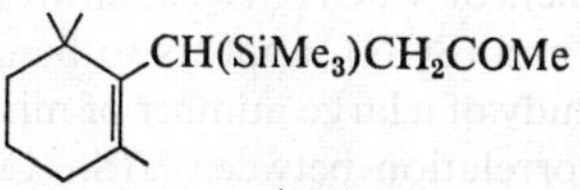 (192) 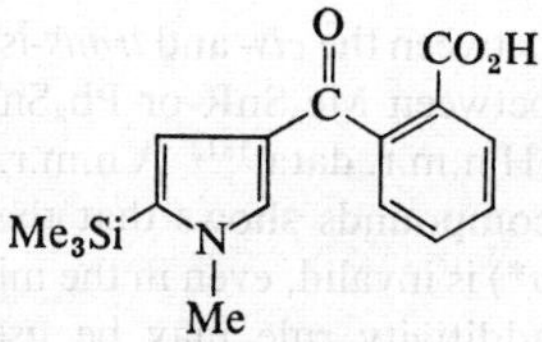 (193)

[1782] A. N. Egorochkin, N. S. Vyazankin, A. I. Burov, E. A. Chernyshev, V. I. Savushkina, and B. M. Tabenko, *Khim. geterotsikl. Soedinenii*, 1972, 911.
[1783] C. Eaborn, A. A. Najam, and D. R. M. Walton, *J. Organometallic Chem.*, 1972, **46**, 255.
[1784] P. Bourgeois and N. Duffaut, *J. Organometallic Chem.*, 1972, **35**, 63.
[1785] J.-P. Picard, *J. Organometallic Chem.*, 1972, **34**, 279.
[1786] V. Rautenstrauch, *Helv. Chim. Acta*, 1972, **55**, 594.
[1787] B. Martel and M. Varache, *J. Organometallic Chem.*, 1972, **40**, C53.
[1788] A. J. Ashe, tert., *J. Org. Chem.*, 1972, **37**, 2053.
[1789] C. Biran, J. Dédier, J. Dunoguès, R. Calas, and N. Duffaut, *J. Organometallic Chem.*, 1972, **35**, 263.
[1790] A. Wright, D. Ling, P. Boudjouk, and R. West, *J. Amer. Chem. Soc.*, 1972, **94**, 4784.
[1791] M. Bolourtchian, P. Bourgeois, J. Dunoguès, N. Duffaut, and R. Calas, *J. Organometallic Chem.*, 1972, **43**, 139.
[1792] G. Märkl and R. Fuchs, *Tetrahedron Letters*, 1972, 4691.
[1793] D. R. Dimmel, C. A. Wilkie, and F. Ramon, *J. Org. Chem.*, 1972, **37**, 2662.
[1794] C. Ainsworth, F. Chen, and Y.-N. Kuo, *J. Organometallic Chem.*, 1972, **46**, 59.
[1795] C. Chung and R. J. Lagow, *J.C.S. Chem. Comm.*, 1972, 1078.
[1796] O. W. Steward, J. S. Johnson, and C. Eaborn, *J. Organometallic Chem.*, 1972, **46**, 97.
[1797] R. F. Cunico and E. M. Dexheimer, *J. Amer. Chem. Soc.*, 1972, **94**, 2868.
[1798] H. Schmidbaur, H. Stühler, and W. Vornberger, *Chem. Ber.*, 1972, **105**, 1084.
[1799] L. Birkofer and M. Franz, *Chem. Ber.*, 1972, **105**, 1759.
[1800] R. Calas, M. Bolourtchian, J. Dunoguès, N. Duffaut, and B. Barbe, *J. Organometallic Chem.*, 1972, **34**, 269.
[1801] J. R. Pratt, F. H. Pinkerton, and S. F. Thames, *J. Organometallic Chem.*, 1972, **38**, 29.

=CHSiMe₃, (194),[1802] CH₂=C=CHSiMe₃, BrCH₂C≡CSiMe₃,[1803] (PhCO₂)PhC=C(SiMe₃)P(O)(OEt)₂,[1804] (195),[1805] (196),[1806] (SiMe₃)₂-C₆H₄,[1807] (197),[1808] (naphthyl)SiMe₃,[1809] and Me₃Si(C≡C)ₙR (R = Ar, SiMe₃, or H).[1810]

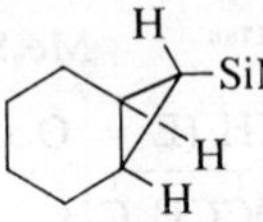

(194)

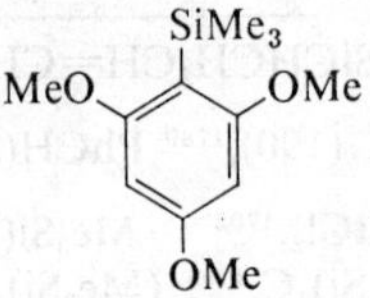

(195)

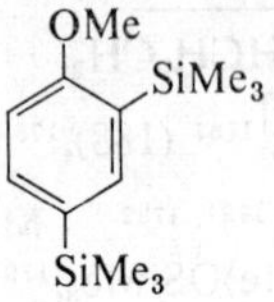

(196)

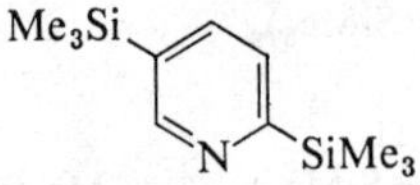

(197)

The ¹H n.m.r. spectrum of (198; M = Sn or Pb) shows only one set of methyl resonances even at −50 °C, although there should be two methyl resonances in the ratio 2:1.[1811] Solvent effects have been used to distinguish between the *cis*- and *trans*-isomers of MeCHCH₂CHSnMe₃.[1812] The reaction between Me₃SnR or Ph₃SnR and (SCN)₂ appears to be quantitative, from ¹H n.m.r. data.[1813] A n.m.r. study of a large number of mixed tetraorganotin compounds shows that the correlation between ²J(Sn–C–H) and ∑(Taft's σ*) is invalid, even in the mixed tetra-alkyltin series. However, Malinovski's additivity rule may be used to predict tin–methyl, tin–benzyl, and tin–t-butyl coupling constants.[1814] Data have also been reported for (199;

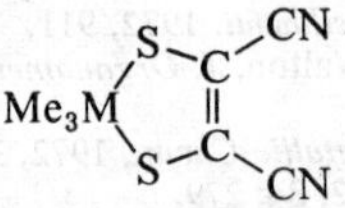

(198)

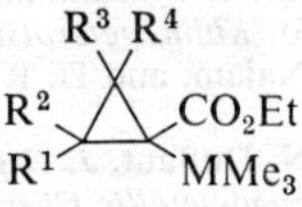

(199)

[1802] D. Seyferth, H. Menzel, A. W. Dow, and T. C. Flood, *J. Organometallic Chem.*, 1972, **44**, 279.
[1803] P. Bourgeois and G. Mérault, *J. Organometallic Chem.*, 1972, **39**, C44.
[1804] F. A. Carey and A. S. Court, *J. Org. Chem.*, 1972, **37**, 939.
[1805] C. Eaborn, Z. S. Salih, and D. R. M. Walton, *J. Organometallic Chem.*, 1972, **36**, 47.
[1806] C. Eaborn, Z. S. Salih, and D. R. M. Walton, *J. Organometallic Chem.*, 1972, **36**, 41.
[1807] D. Seyferth and D. L. White, *J. Amer. Chem. Soc.*, 1972, **94**, 3132.
[1808] E. Heilbronner, V. Hornung, F. H. Pinkerton, and S. F. Thames, *Helv. Chim. Acta*, 1972, **55**, 289.
[1809] L. Birkofer and N. Ramadan, *J. Organometallic Chem.*, 1972, **44**, C41.
[1810] D. R. M. Walton and F. Waugh, *J. Organometallic Chem.*, 1972, **37**, 45.
[1811] E. S. Bretschneider and C. W. Allen, *J. Organometallic Chem.*, 1972, **38**, 43.
[1812] M. Grielen, P. Baekelmans, and J. Nasielski, *J. Organometallic Chem.*, 1972, **34**, 329.
[1813] M. L. Bullpitt and W. Kitching, *J. Organometallic Chem.*, 1972, **34**, 321.
[1814] M. Gielen, M. De Clercq, and B. de Poorter, *J. Organometallic Chem.*, 1972, **34**, 305.

M = Ge, Sn, or Ph), $R^1R^2C=CR^3CH_2CH(GeMe_3)CO_2Et$,[1815] and $Me_3GeCMe(CN)CH_2PEt_2$.[1816]

^{1}H N.m.r. spectroscopy has been used to show that CO_2N_2 reacts with $Me_3SiSiMe_3$ to yield $MeCO_2SiMe_2SiMe_3$ and not $HCO_2CH_2SiMe_2SiMe_3$.[1817] Data have also been reported 1-naphthyl derivatives such as (1-naphthyl)-$\overline{Si_3Me_7}$,[1818] $Me_3PNSi_2Me_4Cl$,[1819] $R_3PCHSi_2Me_5$, $Me_2P=CH-SiMe_2Si-\overline{Me_2CH_2}$,[1820] Me_3SiSiR_3,[1821] (200),[1822] $RGeMe_2GeMe_3$,[1823] (201), and $Cl(CH_2)_4CH=CPhOSiMe_3$.[1824]

$(CF_3)_2C=C(CN)_2$ and $(CF_3)(CF_2Cl)C=C(CN)_2$ undergo reactions with a wide variety of organometallic compounds by a 1,4-addition. ^{1}H and ^{19}F n.m.r. data have been reported for compounds such as $Me_3MNCC-(CN)C(CF_3)_2SMe$ (M = Si or Sn) and $Bu^n_3GeNCC(CN)C(CF_3)_2NMe_2$.[1825] ^{1}H N.m.r. data have also been reported for hydrazines substituted with $SiMe_3$ or $GeMe_3$.[1826]

The observation of two methyl signals for $(Me_3Si)_2NCRO$ or $Me_3SiOCR=NSiMe_3$ could be due to either restricted rotation or fluxional inequivalence. The observation of $^1J(^{15}N-^1H) = 77.5$ Hz for $Me_3Si-NHCOMe$ provides extra evidence for the structure being $Me_3SiOCR-=NSiMe_3$.[1827] The ^{1}H n.m.r. spectra of $Bu^t_2C=NSiMe_nCl_{3-n}$ show only one Bu^t signal even at low temperature. This is in contrast to $Bu^t_2C=NH$, which shows two Bu^t signals at -60 °C.[1828] ^{1}H N.m.r. spectroscopy has been used to determine the ratio of isomers for (202) and (203).[1829] ^{1}H N.m.r. spectroscopic investigations of the silyltriazene $PhN=NN(SiMe_3)_2$ show hindered rotation of the amino-group about the N—N single bond

---

[1815] U. Shöllkopf, B. Bánhidai, and H.-U. Scholz, *Annalen*, 1972, **761**, 137.
[1816] J. Satgé, C. Couret, and J. Escudié, *J. Organometallic Chem.*, 1972, **34**, 83.
[1817] R. T. Conlin, P. P. Gaspar, R. H. Levin, and M. Jones, jun., *J. Amer. Chem. Soc.*, 1972, **94**, 7165.
[1818] C. G. Pitt, R. N. Carey, and E. C. Toren, jun., *J. Amer. Chem. Soc.*, 1972, **94**, 3806.
[1819] H. Schmidbaur and W. Vornberger, *Chem. Ber.*, 1972, **105**, 3187.
[1820] H. Schmidbaur and W. Vornberger, *Chem. Ber.*, 1972, **105**, 3173.
[1821] A. Hosomi and H. Sakurai, *Bull. Chem. Soc. Japan*, 1972, **45**, 248.
[1822] H. Sakurai, S. Tasaka, and M. Kira, *J. Amer. Chem. Soc.*, 1972, **94**, 9285.
[1823] K. Yamamoto and M. Kumada, *J. Organometallic Chem.*, 1972, **35**, 297.
[1824] K. Itoh, S. Kato, and Y. Ishii, *J. Organometallic Chem.*, 1972, **34**, 293.
[1825] E. W. Abel, J. P. Crow, and J. N. Wingfield, *J.C.S. Dalton*, 1972, 787.
[1826] L. K. Peterson and K. I. Thé, *Canad. J. Chem.*, 1972, **50**, 553.
[1827] C. H. Yoder and D. Bonelli, *Inorg. Nuclear Chem. Letters*, 1972, **8**, 1027.
[1828] J. B. Farmer, R. Snaith, and K. Wade, *J.C.S. Dalton*, 1972, 1501.
[1829] K. Itoh, T. Katsuura, I. Matsuda, and Y. Ishii, *J. Organometallic Chem.*, 1972, **34**, 63.

similar to that shown for the alkyltriazene PhN=NNMe$_2$.[1830] ^{1}H and ^{31}P n.m.r. data, including $J(^{31}P-^{31}P)$, have been reported for compounds such as SPCl$_2$N=PCl$_2$N=PCl$_2$N=PCl$_2$NHSiMe$_3$.[1831] Data have also been reported for Me$_3$SiX,[1832, 1833] Me$_3$SiNHR, Me$_3$SiOR,[1834] (204),[1835] (205),[1836]

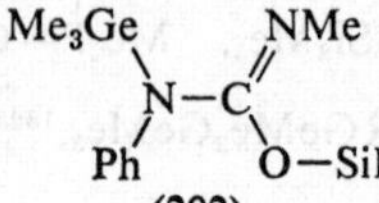

(202)

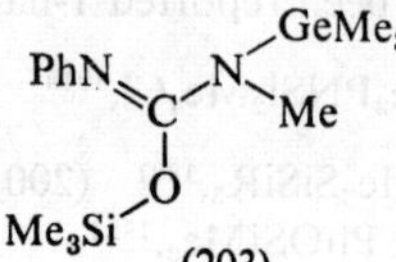

(203)

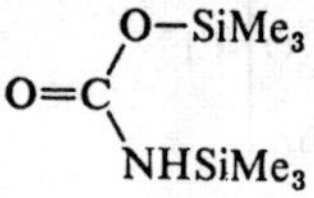

(204)

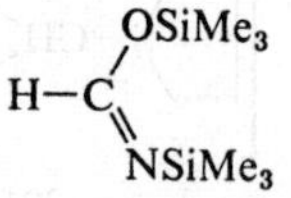

(205)

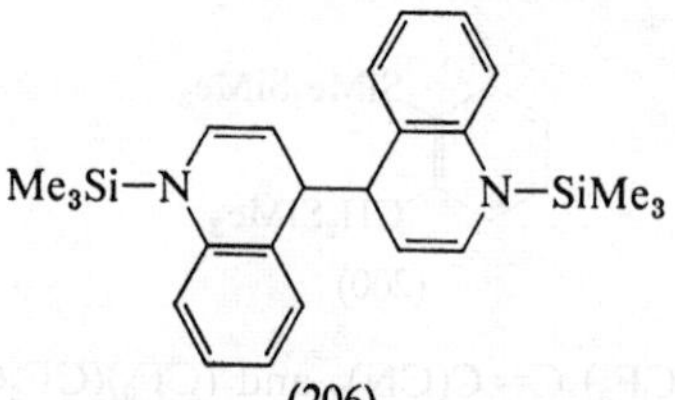

(206)

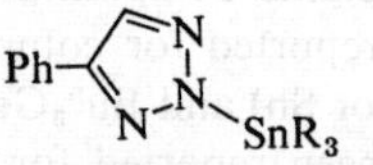

(207)

Me$_3$SiNRBF$_2$,[1837] (206),[1838] R^{1}N=NNR^{2}SiMe$_3$,[1839] Me$_3$SiN=NN(SiMe$_3$)-NHSiMe$_3$,[1840] Me$_3$SiN=S(O)=NSiMe$_3$,[1841] [Me$_2$SNSiMe$_3$]$_2$,[1842] PhN=C(OGeMe$_3$)NMeGeMe$_3$,[1843] Me$_3$SnNR$_2$, Me$_2$SN(MeN$_3$Me)$_2$,[1844] (207),[1845] Me$_3$SnNSO, and Me$_2$Si(NSO)$_2$.[1846]

The ^{1}H n.m.r. spectrum of Me$_3$PbNCS, with $^2J(^{207}Pb-C-^1H) = 79.5$ Hz, is consistent with a five-co-ordinate species where the methyl group is

---

[1830] N. Wiberg and H. J. Pracht, *Chem. Ber.*, 1972, **105**, 1399.
[1831] H. W. Roesky, *Chem. Ber.*, 1972, **105**, 1439.
[1832] G. Neumann and W. P. Neumann, *J. Organometallic Chem.*, 1972, **42**, 293.
[1833] S. S. Washburne, W. R. Peterson, jun., and D. A. Berman, *J. Org. Chem.*, 1972, **37**, 1738.
[1834] F. Chen and C. Ainsworth, *J. Amer. Chem. Soc.*, 1972, **94**, 4037.
[1835] L. Birkofer and P. Sommer, *J. Organometallic Chem.*, 1972, **35**, C15.
[1836] W. Kantlehner, W. Kugel, and H. Bredereck, *Chem. Ber.*, 1972, **105**, 2264.
[1837] G. Elter, O. Glemser, and W. Herzog, *J. Organometallic Chem.*, 1972, **36**, 257.
[1838] H. P. Becker and W. P. Neumann, *J. Organometallic Chem.*, 1972, **37**, 57.
[1839] N. Wiberg and H. J. Pracht, *Chem. Ber.*, 1972, **105**, 1377.
[1840] N. Wiberg and W. Uhlenbrock, *Chem. Ber.*, 1972, **105**, 63.
[1841] O. Glemser, M. F. Feser, S. P. von Halasz, and J. Saran, *Inorg. Nuclear Chem. Letters*, 1972, **8**, 321.
[1842] R. Appel, I. Ruppert, and F. Knoll, *Chem. Ber.*, 1972, **105**, 2492.
[1843] K. Itoh, I. Matsuda, T. Katsuura, S. Kato, and Y. Ishii, *J. Organometallic Chem.*, 1972, **34**, 75.
[1844] J. Hollaender, W. P. Neumann, and G. Alester, *Chem. Ber.*, 1972, **105**, 1540.
[1845] S. Kozima, T. Itano, N. Mihara, K. Sisido, and T. Isida, *J. Organometallic Chem.*, 1972, **44**, 117.
[1846] D. A. Armitage and A. W. Sinden, *J. Organometallic Chem.*, 1972, **44**, C43.

equatorial and the NCS and solvent are axial.[1847] For the compounds Me_3MPHPh (M = C, Si, Ge, or Sn), the magnitude of $^1J(^{31}P-^1H)$ has been explained using the Fermi contact equation.[1848]

^1H and ^{29}Si chemical shifts in $Me_nSi(O_2CR)_{4-n}$ correlate linearly with the electron-withdrawing ability of the carboxylate. The results were explained without invoking $p_\pi-d_\pi$ bonding.[1849] If OH groups in organic compounds are converted into Me_3SiO groups they can be very readily determined by ^1H n.m.r. spectroscopy.[1850] This method is suitable to detect eluates collected using g.l.c.[1851] If flavanoids are converted into $SiMe_3$ ethers and the ^1H n.m.r. spectra measured in CCl_4 or C_6H_6, both the MeO and Me_3Si groups show diagnostic benzene-induced shifts.[1852] The shifts and coupling constants of the anomeric protons of 30 pertrimethyl-silyloligosaccharides have been measured and a method has been presented for the determination of the configuration of the glycosidic bond in oligosaccharides.[1853] The ^{13}C chemical shifts of compounds such as (norbornyl)-$OSiMe_3$ are useful in conformational analysis.[1854] Correlation of the chemical shift, δ, of the silicon-attached methyl groups in $p\text{-}XC_6H_4OSiMe_3$ with the resonance constants σ_R was unsatisfactory, but the equation

$$\delta = 15.18 + 4.27\sigma_R$$

was obtained.[1855] $MOSiMe_3$ (M = Li, Na, or K) forms 1:1 complexes (tetramers) with Me_3PO.[1856] Data have also been reported for Me_3SiOR,[1857-1863] $Me_3SiO(sugar)$,[1864] $R^1R^2C=C(OSiMe_3)_2$,[1865] Me_3Si-(polyphenols),[1866] (208),[1867] $EtCH=CHC(OSiMe_3)=CHEt$,[1868] Me_3Si-

---

[1847] N. Bertazzi, G. Alonzo, A. Silvestri, and G. Consiglio, *J. Organometallic Chem.*, 1972, **37**, 281.
[1848] P. G. Harrison, S. E. Ulrich, and J. J. Zuckerman, *Inorg. Chem.*, 1972, **11**, 25.
[1849] W. McFarlane and J. M. Sealy, *J.C.S. Perkin II*, 1972, 1561.
[1850] A. Hase and T. Hase, *Analyst*, 1972, **97**, 998.
[1851] G. M. Bebault, J. M. Berry, G. G. S. Dutton, and K. B. Gibney, *Analyt. Letters*, 1972, **5**, 413.
[1852] E. Rodriguez, N. J. Carman, and T. J. Mabry, *Phytochemistry*, 1972, **11**, 409.
[1853] J. P. Kamerling, M. J. A. de Bie, and J. F. G. Vliegenthart, *Tetrahedron*, 1972, **28**, 3037.
[1854] H.-J. Schneider, *J. Amer. Chem. Soc.*, 1972, **94**, 3636.
[1855] A. P. Kreshkov, V. F. Andronov, and V. A. Drozdov, *Zhur. fiz. Khim.*, 1972, **46**, 992.
[1856] H. Schmidbaur and J. Adlkofer, *Chem. Ber.*, 1972, **105**, 1956.
[1857] G. G. S. Dutton, N. Funnell, and K. B. Gibney, *Canad. J. Chem.*, 1972, **50**, 3913.
[1858] H. R. Kricheldorf, *Chem. Ber.*, 1972, **105**, 3958.
[1859] P. Bajaj, R. C. Mehrotra, J. C. Maire, and R. Ouaki, *J. Organometallic Chem.*, 1972, **40**, 301.
[1860] T. Murakawa, K. Fujii, S. Murai, and S. Tsutsumi, *Bull. Chem. Soc. Japan*, 1972, **45**, 2520.
[1861] A. P. Kurtz and C. R. Dawson, *J. Org. Chem.*, 1972, **37**, 2767.
[1862] H. R. Kricheldorf, *Annalen*, 1972, **763**, 17.
[1863] G. Neumann and W. P. Neumann, *J. Organometallic Chem.*, 1972, **42**, 277.
[1864] J. Lehmann and H. Schäfer, *Chem. Ber.*, 1972, **105**, 3503.
[1865] C. Ainsworth and Y.-N. Kup, *J. Organometallic Chem.*, 1972, **46**, 73.
[1866] A. Sato, T. Kitamura, and T. Higuchi, *Mokuzai Gakkaishi*, 1972, **18**, 253.
[1867] P. Cazeau and F. Frainnet, *Bull. Soc. chim. France*, 1972, 1658.
[1868] K. Rühlmann, B. Fichte, T. Kiriakidis, C. Michael, G. Michael, and E. Gründemann, *J. Organometallic Chem.*, 1972, **34**, 41.

ONO_2,[1869] (209),[1870] $Me_3SiOCHPhSiR^1R^2R^3$,[1871] $F_2PE^1E^2SiMe_3$ (E^1, E^2 = O or S),[1872] $Me_3MOCOCH_2CH_2COPEt_2$ (M = Si, Ge, or Sn),[1873] Me_2N-$NHCE_2MMe_3$ (M = Si or Ge; E = O or S),[1874] $Me_nPb(NO_3)_{4-n}$,[1875] and

(208) Me_3SiO—(CN)$_2$—(CN)$_2$

(209) $OSiMe_3$ / $OSiMe_3$

$(4-R^1C_6H_4)_2C(OSnR^2_3)C(OSnR^2_3)(C_6H_4R^3-4)_2$.[1876] ^{13}C N.m.r. spectroscopy has been used to assess the charge distribution in $PhSMMe_3$ (M = C, Si, Ge, Sn, or Pb) and compared with ESCA measurements on the sulphur. A linear correlation was found between the ESCA data and the $^{13}C-S$ chemical shift.[1877] ^1H N.m.r. spectroscopy has been used to show that $Me_3SnSCH_2CO_2SnMe_3$ slowly disproportionates to give Me_4Sn and $Me_2Sn(SCH_2CO_2)_2$.[1878] ^1H N.m.r. data have also been reported for $PhC(O)SeMMe_3$ (M = Ge or Sn).[1879] ^1H N.m.r. data for R_nSnX_{4-n} (R = Me or Et; X = halogen) have been measured. J(Sn–H) correlates with the electronegativity of the substituents. The Fermi contact mechanism was assumed and a correlation with tin s-character was drawn.[1880]

^1H and ^{29}Si n.m.r. spectroscopy have been used to demonstrate the stereochemistry of (210) and related compounds.[1881] ^{13}C and ^{29}Si n.m.r. spectra, including $^1J(^{13}C-^{29}Si)$, have been reported for (211).[1882] Data have also been reported for (212; M = Si, Ge, Sn, or Pb; R = Me or Ph),[1883] $Me(Ph)(1-naphthyl)SiR$,[1884, 1885] (213; X = H, Cl, or Me),[1886] (214),[1887] (215),[1888] (216),[1889] $Me(Ph)(1-naphthyl)SiCHX_2$,[1890] (217),[1891] (218),[1892]

[1869] L. Birkofer and M. Franz, *Chem. Ber.*, 1972, **105**, 470.
[1870] J. J. Bloomfield, R. A. Martin, and J. M. Nelke, *J.C.S. Chem. Comm.*, 1972, 96.
[1871] H. Watanabe, T. Kogure, and Y. Nagai, *J. Organometallic Chem.*, 1972, **43**, 285.
[1872] R. G. Cavell, R. D. Leary, and A. J. Tomlinson, *Inorg. Chem.*, 1972, **11**, 2573.
[1873] C. Couret, J. Escudié, and J. Satgé, *Rec. Trav. chim.*, 1972, **91**, 429.
[1874] L. K. Peterson and K. I. Thé, *Canad. J. Chem.*, 1972, **50**, 562.
[1875] K. C. Williams and D. W. Imhoff, *J. Organometallic Chem.*, 1972, **42**, 107.
[1876] H. Hillgärtner, B. Schroeder, and W. P. Neumann, *J. Organometallic Chem.*, 1972, **42**, C83.
[1877] S. Pignataro, L. Lunazzi, C. A. Boicelli, R. Di Marino, A. Ricci, A. Mangini, and R. Danieli, *Tetrahedron Letters*, 1972, 5341.
[1878] M. Wada, S.-I. Sato, M. Aritomi, M. Harakawa, and R. Okawara, *J. Organometallic Chem.*, 1972, **39**, 99.
[1879] H. Ishihara and S. Kato, *Tetrahedron Letters*, 1972, 3751.
[1880] G. Barbieri and F. Taddei, *J.C.S. Perkin II*, 1972, 1327.
[1881] G. Fritz and M. Hähnke, *Z. anorg. Chem.*, 1972, **390**, 137.
[1882] R. L. Lambert, jun. and D. Seyferth, *J. Amer. Chem. Soc.*, 1972, **94**, 9246.
[1883] J. Y. Corey, M. Dueber, and M. Malaidza, *J. Organometallic Chem.*, 1972, **36**, 49.
[1884] R. J. P. Corriu and J. P. R. Massé, *J. Organometallic Chem.*, 1972, **34**, 221.
[1885] R. J. P. Corriu and G. Royo, *Bull. Soc. chim. France*, 1972, 1497.
[1886] L. Birkofer and H. Haddad, *Chem. Ber.*, 1972, **105**, 2101.
[1887] R. A. Felix and W. P. Weber, *J. Org. Chem.*, 1972, **37**, 2323.
[1888] C. L. Frye and J. M. Klosowski, *J. Amer. Chem. Soc.*, 1972, **94**, 7186.
[1889] T. J. Barton, J. L. Witiak, and C. L. McIntosh, *J. Amer. Chem. Soc.*, 1972, **94**, 6229.
[1890] L. H. Sommer, L. A. Ulland, and G. A. Parker, *J. Amer. Chem. Soc.*, 1972, **94**, 3469.
[1891] W. Adcock, S. Q. A. Rizvi, and W. Kitching, *J. Amer. Chem. Soc.*, 1972, **94**, 3657.
[1892] T. J. Barton, A. J. Nelson, and J. Clardy, *J. Org. Chem.*, 1972, **37**, 895.

Me₂SiCH₂SiMe₂CH=CHSiMe₂CH₂ and 44 related compounds,[1893] (219) (including ²⁹Si n.m.r. data) and related compounds,[1894] (220),[1895] Me₂Si-(CH₂SiMeHCH₂)₂SiMe₂,[1896] MeSi(CH₂SiMe₂)₃CH,[1897] and Me(Ph)-(1-naphthyl)germane.[1898]

(210) (211) (212) (213) (214) (215) (216) (217) (218) (219) (220)

[1893] G. Fritz and M. Hähnke, *Z. anorg. Chem.*, 1972, **390**, 104.
[1894] G. Fritz and M. Hähnke, *Z. anorg. Chem.*, 1972, **390**, 157.
[1895] H. Okinoshima, K. Yamamoto, and M. Kumada, *J. Amer. Chem. Soc.*, 1972, **94**, 9263.
[1896] G. Fritz and M. Hähnke, *Z. anorg. Chem.*, 1972, **390**, 185.
[1897] G. Fritz and M. Hähnke, *Z. anorg. Chem.*, 1972, **390**, 191.
[1898] R. J. P. Corriu and J. J. E. Moreau, *J. Organometallic Chem.*, 1972, **40**, 55.

The ^{1}H n.m.r. spectra of some vinyl derivatives of germanium, e.g. Me$_2$Ge(C$_2$H$_3$)$_2$, have been analysed using LAOCN3.[1899] The ^{1}H n.m.r. spectrum of (1-naphthyl)(Ph)Sn(Me)CH(Me)C≡CH shows the presence of two species in solution.[1900] The inequivalence of the methyl groups in RMe$_2$SnCHMeY has been examined as a function of R, solvent, and temperature.[1901] Data have also been reported for (221),[1902] (222),[1903] Me$_{2n}$Ge$_n$,[1904] (EtO)$_2$Si(CH$_2$)$_3$NSiMe$_2$(OEt),[1905] MeRSi(NSiRMeCl)$_2$-SiMeR,[1906] Me$_2$SiXSiMe$_2$NPrSiMe$_2$SiMe$_2$NPrSiMe$_2$X (X = NMe or O),[1907] Me$_2$S(=NSnMe$_3$)$_2$, and [−N=SMe$_2$=N−SnMe$_2$−]$_n$.[1908] For (223),

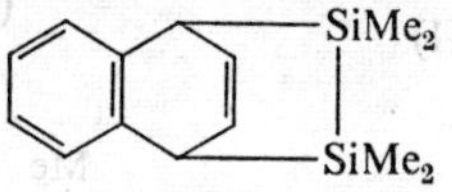

(221)

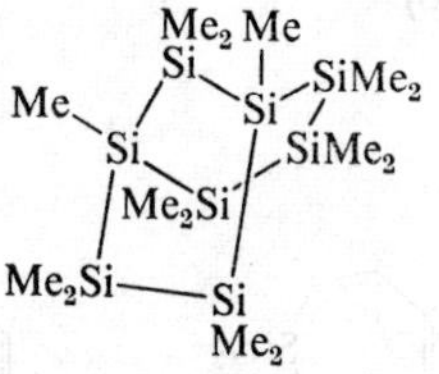

(222)

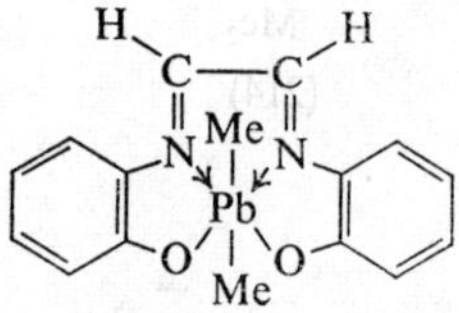

(223)

^2J(Pb–C–H) = 177 Hz, which is the highest such coupling reported for a Me$_2$Pb^{IV} group.[1909]

^{1}H N.m.r. spectroscopy has been used to analyse the cyclotetrasiloxanes formed from treatment of cis- or trans-Me$_3$Ph$_3$Si$_3$O$_3$ with KOH.[1910] T_1 Measurements have been carried out on polysiloxanes and related to the effect of substituents on segmental motion.[1911] For Me$_2$SnClO$_2$CR, $^2J(^{119}$Sn–C–^{1}H) is ca. 75 Hz in CHCl$_3$, indicating five co-ordination, and

---

[1899] R. C. Job and M. D. Curtis, Inorg. Nuclear Chem. Letters, 1972, **8**, 251.
[1900] A. Jean and M. Lequan, J. Organometallic Chem., 1972, **36**, C9.
[1901] M. Gielen, M. R. Barthels, M. de Clercq, C. Dehouck, and G. Mayence, J. Organometallic Chem., 1972, **34**, 315.
[1902] D. N. Roark and G. J. D. Peddle, J. Amer. Chem. Soc., 1972, **94**, 5837.
[1903] R. West and A. Indriksons, J. Amer. Chem. Soc., 1972, **94**, 6110.
[1904] E. Carberry, B. D. Dombek, and S. C. Cohen, J. Organometallic Chem., 1972, **36**, 61.
[1905] T.-T. Tsai and C. J. Marshall, jun., J. Organometallic Chem., 1972, **37**, 596.
[1906] L. W. Breed and J. C. Wiley, jun., Inorg. Chem., 1972, **11**, 1634.
[1907] U. Wannagat and S. Meier, Z. anorg. Chem., 1972, **392**, 179.
[1908] D. Hanssgen and R. Appel, Chem. Ber., 1972, **105**, 3271.
[1909] F. Di Bianco, E. Rivarola, G. C. Stocco, and R. Barbieri, Z. anorg. Chem., 1972, **387**, 126.
[1910] D. Harber, A. Holt, and A. W. P. Jarvie, J. Organometallic Chem., 1972, **38**, 255.
[1911] J. A. Barrie, M. J. Fredrickson, and R. Sheppard, Polymer, 1972, **13**, 431.

Nuclear Magnetic Resonance Spectroscopy

is ca. 90 Hz in $(CD_3)_2CO$, indicating six co-ordination.[1912] The observation of 132 Hz for $^2J(Pb-C-H)$ in $Me_2Pb(OMe)_2$ has been taken as evidence for octahedral co-ordination, whereas 93.5 Hz for $^2J(PbCH)$ in $Me_2Pb(OBu^t)_2$ indicates tetrahedral co-ordination.[1913] For $Me_2Pb(acac)_2$, $^2J(PbCH)$ is larger than for Me_4Pb. This has been attributed to the accommodation of some positive charge on the central lead atom.[1914] Data have also been reported for $(NCCH_2CH_2SiMe_2)_2O$,[1915] $PhEtCHO-SiMeCl_2$,[1916] $(RO)MeSi(CH_2)_2CHMe$,[1917] $Me(N_3)SnOSnMe_2OMe$,[1918] and (224; R = Me or Ph).[1919]

The ^1H n.m.r. spectrum of (225) in liquid SO_2 is sharp but in $CDCl_3$ it vanishes and this has been attributed to radical formation.[1920] The ^1H n.m.r. spectrum of (226) shows two methyl resonances at room temperature, owing to restricted rotation about the carbon–nitrogen bond.[1921] Data have also been reported for $\overline{Cl_2SiCH_2CH_2CH_2}$ [1922] and (227).[1923]

[1912] C. S.-C. Wang and J. M. Shreeve, *J. Organometallic Chem.*, 1972, **38**, 287.
[1913] R. J. Puddephatt and G. H. Thistlethwaite, *J.C.S. Dalton*, 1972, 570.
[1914] M. Aritomi, Y. Kawasaki, and R. Okawara, *Inorg. Nuclear Chem. Letters*, 1972, **8**, 69.
[1915] E. S. Brown, E. A. Rick, and F. D. Mendicino, *J. Organometallic Chem.*, 1972, **38**, 37.
[1916] K. Yamamoto, T. Hayashi, and M. Kumada, *J. Organometallic Chem.*, 1972, **46**, C65.
[1917] J. Dubac, P. Mazerolles, and B. Serres, *Tetrahedron Letters*, 1972, 525.
[1918] H. Matsuda, F. Mori, A. Kashiwa, S. Matsuda, N. Kasai, and K. Jitsumori, *J. Organometallic Chem.*, 1972, **34**, 341.
[1919] R. H. Abu-Samn and H. Latscha, *Chem.-Ztg.*, 1972, **96**, 222.
[1920] Y. Takaya, G. Matsubayashi, and T. Tanaka, *Inorg. Chim. Acta*, 1972, **6**, 339.
[1921] K. Tanaka and T. Tanaka, *Bull. Chem. Soc. Japan*, 1972, **45**, 489.
[1922] R. Damrauer, R. A. Davis, M. T. Burke, R. A. Karn, and G. T. Goodman, *J. Organometallic Chem.*, 1972, **43**, 121.
[1923] N. M. Sergeyev, G. I. Avramenko, V. A. Korenevsky, and Yu. A. Ustynyuk, *Org. Magn. Resonance*, 1972, **4**, 39.

The ^{1}H chemical shifts of Et$_4$M (M = Si, Ge, or Sn) in 13 solvents have been reported and found similar to those reported for Et$_2$Hg. The effects did not correlate with electronegativity.[1924] ^{19}F Chemical shifts have been reported for (F-aryl)CH$_2$MPh$_3$ (M = Si, Ge, Sn, or Pb): it was concluded that hyperconjugation occurs.[1925] The n.m.r. spectra of Ph$_3$MX (M = C, Si, Ge, Sn, or Pb; X = Ph, vinyl, or N$_3$) have been fully analysed. An electronegativity order was proposed using *meta* coupling constants.[1926] The acceptor properties of some alkynyl-lead compounds have been investigated by ^{1}H n.m.r. spectroscopy. ^4J(MeC≡CPb) appears to reflect qualitatively the *s*-character of the lead–carbon bond. The coupling constant decreases when another ligand co-ordinates.[1927] The ^{1}H n.m.r. spectra of (228) and (229) (X = O or S) have been measured and the coupling constants derived. The relative signs of the long-range J(Pb–H) appear to be the same and are thought to arise from the Fermi contact mechanism.[1928] Data have also been reported for $\overline{R^1R^2\dot{M}CH_2S(O)CH_2\text{-}}$ $\overline{CH_2\dot{C}H_2}$ (M = C, Si, or Ge),[1929] (230),[1930] [Ph$_3$SiCH$_2$CH(Me)—]$_2$,[1931]

(228)   (229)   (230)

Et$_3$SiC≡CC≡CH,[1932] *p*-Et$_3$Ge(CH$_2$)$_2$C$_6$H$_4$COMe,[1933] R$_3$GeCH=CHPh,[1934] Et$_3$GeCHFCF$_3$,[1935] Ph$_3$SnCH$_2$CH$_2$CH$_2$Cl,[1936] and (neophyl)$_3$PhSn.[1937]

The ^{1}H n.m.r. spectra of *p*-substituted phenoxysilanes and phenoxygermanes have been interpreted as showing that in these compounds the silicon–oxygen bond has more p_π–d_π character than the germanium–oxygen bond, but that p_π–d_π bonding is not negligible in phenoxygermanes.[1938] A melt of Et$_3$SnOH shows resonances due to (Et$_3$Sn)$_2$O and H$_2$O. The same happens on other attempts to observe the ^{1}H n.m.r. spectrum of

[1924] V. S. Petrosyan, N. S. Yashina, and O. A. Reutov, *Izvest. Akad. Nauk S.S.S.R., Ser. khim.*, 1972, **5**, 1018.
[1925] W. Kitching, A. J. Smith, W. Adcock, and S. Q. A. Rizvi, *J. Organometallic Chem.*, 1972, **42**, 373.
[1926] P. N. Preston, L. H. Sutcliffe, and B. Taylor, *Spectrochim. Acta*, 1972, **28A**, 197.
[1927] R. J. Puddephatt and G. H. Thistlethwaite, *J. Organometallic Chem.*, 1972, **40**, 143.
[1928] G. Barbieri and F. Taddei, *J.C.S. Perkin II*, 1972, 262.
[1929] J. Dubac, P. Mazerolles, M. Joly, W. Kitching, C. W. Fong, and W. H. Atwell, *J. Organometallic Chem.*, 1972, **34**, 17.
[1930] R. Corriu and J. Massé, *J. Organometallic Chem.*, 1972, **35**, 51.
[1931] A. W. P. Jarvie, A. J. Bourne, and R. J. Rowley, *J. Organometallic Chem.*, 1972, **39**, 93.
[1932] R. Eastmond, T. R. Johnson, and D. R. M. Walton, *Tetrahedron*, 1972, **28**, 4601.
[1933] P. Mazerolles and H. Cousse, *Bull. Soc. chim. France*, 1972, 1361.
[1934] R. J. P. Corriu and J. J. E. Moreau, *J. Organometallic Chem.*, 1972, **40**, 73.
[1935] B. I. Petrov, O. A. Kruglaya, N. S. Vyazankin, B. I. Martynov, S. R. Sterlin, and B. L. Dyatkin, *J. Organometallic Chem.*, 1972, **34**, 299.
[1936] M. Gielen and J. Topart, *Bull. Soc. chim. belges*, 1971, **80**, 655.
[1937] H.-J. Götze, *Chem. Ber.*, 1972, **105**, 1775.
[1938] J. R. Chipperfield, D. F. Ewing, and G. E. Gould, *J. Organometallic Chem.*, 1972, **46**, 263.

Et_3SnOH.[1939] Data have also been reported for $ArN=NN(SiR_3)_2$,[1940] $Ph_3SnNCNSnPh_3$,[1941] $(PhCH_2)(Ph_3C)NCNSnBrPh_3$,[1942] $Ph_2Si(OMe)$-CH_2D,[1943] $R_3GeONPh(CH_2Ph)$,[1944] Et_3GeO_2SAr,[1945] $Bu^n_3SnOCR^1R^2$-$N=CPh_2$,[1946] $(tolyl)_3SnO_2SAr$,[1947] and $ArPb(O_2CMe)_3$.[1948]

From ^1H n.m.r. investigations it has been concluded that SiF_2 reacts with allene to yield $H_2C=C=CHSiF_2SiF_2CH_2CH=CH_2$ and $HC\equiv CCH_2$-$SiF_2CH_2CH=CH_2$.[1949] The ^{19}F n.m.r. spectra of $CF_3CF_2SiF_2X$ are second order and have been computer simulated.[1950] The X-ray structure of (231) shows that for (a) the F—Si—F angle is 106.1° and $^2J(F-F)$ is 56 Hz and for (b) the F—Si—F angle is 104° and $^2J(F-F)$ is 31 Hz. Therefore there is no simple relationship between $J(F-F)$ and bond angle.[1951] The ^1H n.m.r. spectrum of $(C_5H_5)SnX$ shows no tin coupling. It was therefore suggested

(231) (232)

that there is rapid cyclopentadienyl exchange.[1952] The neophyl methylene resonance in $Bu^tPh(neophyl)SnI$ is AB with tin satellites $J(Sn-^1H) = 55$ and 35 Hz. Hence the correlation between $^2J(Sn-^1H)$ and percentage s-character is inaccurate.[1953] Data have also been reported for (232),[1954] $SiF_2SiF_2CH_2CH=CHCH_2$,[1955] $PhSiMeX_2$,[1956] $(F_3SiCH_2)_2SiF_2$,[1957] $p-Cl_3Si$-$(CH_2)_3C_6H_4CH_2Cl$,[1958] and fluorovinylsilanes.[1959]

[1939] J. M. Brown, A. C. Chapman, R. Harper, D. J. Mowthorpe, A. G. Davies, and P. J. Smith, *J.C.S. Dalton*, 1972, 338.
[1940] N. Wiberg and H. J. Pracht, *J. Organometallic Chem.*, 1972, **40**, 289.
[1941] R. A. Cardona and E. J. Kupchik, *J. Organometallic Chem.*, 1972, **34**, 129.
[1942] R. A. Cardona and E. J. Kupchik, *J. Organometallic Chem.*, 1972, **43**, 163.
[1943] P. Boudjouk, J. R. Roberts, C. M. Golino, and L. H. Sommer, *J. Amer. Chem. Soc.*, 1972, **94**, 7926.
[1944] J. Satgé, M. Lesbre, P. Rivière, and S. Richelme, *J. Organometallic Chem.*, 1972, **34**, C18.
[1945] E. Lindner and K. Schardt, *J. Organometallic Chem.*, 1972, **44**, 111.
[1946] P. G. Harrison, *J.C.S. Perkin I*, 1972, 130.
[1947] U. Kunze, E. Lindner, and J. Koola, *J. Organometallic Chem.*, 1972, **40**, 327.
[1948] D. de Vos and J. Wolters, *J. Organometallic Chem.*, 1972, **39**, C63.
[1949] C. S. Liu and J. C. Thompson, *J. Organometallic Chem.*, 1972, **38**, 249.
[1950] K. G. Sharp and T. D. Coyle, *Inorg. Chem.*, 1972, **11**, 1259.
[1951] C. S. Liu, S. C. Nyburg, J. T. Szymanski, and J. C. Thompson, *J.C.S. Dalton*, 1972, 1129.
[1952] K. D. Bos, E. J. Bulten, and J. G. Noltes, *J. Organometallic Chem.*, 1972, **39**, C52.
[1953] C. E. Holloway, S. A. Kandil, and I. M. Walker, *J. Amer. Chem. Soc.*, 1972, **94**, 4027.
[1954] C. S. Liu, J. L. Margrave, J. C. Thompson, and P. L. Timms, *Canad. J. Chem.*, 1972, **50**, 459.
[1955] J. C. Thompson and J. L. Margrave, *Inorg. Chem.*, 1972, **11**, 913.
[1956] P. Hencsei, J. Reffy, O. Mestyanek, T. Veszpremi, and J. Nagy, *Periodica Polytech.*, 1972, **16**, 101.
[1957] G. Fritz, M. Berndt, and R. Huber, *Z. anorg. Chem.*, 1972, **391**, 219.
[1958] W. Parr and K. Grohmann, *Angew. Chem. Internat. Edn.*, 1972, **11**, 314.
[1959] A. Orlando, C. S. Liu, and J. C. Thompson, *J. Fluorine Chem.*, 1972, **2**, 103.

A solution of $SnCl_4$ in Me_2NCHO shows ^1H n.m.r. signals for bound and free Me_2NCHO, and the formation of $[SnCl_3(Me_2NCHO)]^+$ and $[Sn_3Cl_{13}]^-$ has been suggested.[1960] Data have also been reported for $(Et_2N)_2SnCl_2$,·$2HCl$,[1961] $[Sn(acac)X_4]^-$,[1962] $(Bu^tO)_2Si(SH)_2$,[1963] and $[Sn(S_2C_6H_3Me)_3]^{2-}$.[1964]

8 Compounds of Group V

Information concerning complexes of these elements can be found at the following sources: P_4,[134, 135] $(MeO)_2PHO$,[135, 138] AsH_3,[62] $Cr(CO)_4\{Me_2As\text{-}\overline{C=CCF_2CF_2AsMe_2}\}$, $M_2(CO)_8\{Me_2As\overline{C=CCF_2CF_2AsMe_2}\}$ (M = Mn or Re),[257] $Cr(CO)_4(Me_2AsCR^1R^2CR^3R^4AsMe_2)$,[156] $\{M(CO)_5\}_2L$ [M = Cr, Mo, or W; L = $(Me_2Sb)_2CH_2$ or $(Ph_2Sb)_2CH_2$],[225] $(C_5H_5)Mo(CO)_2\text{-}(EPh_3)SnMe_3$ (E = As or Sb),[274] $(C_5H_5)Mo(CO)_2(CN)(Me_2AsCH_2CH=CH_2)$,[278] $(OC)_5WEMe_2XEMe_2W(CO)_5$ (E = P or As; X = O, S, NMe, or PPh),[312] $Fe_3(CO)_6(Ph_2AsC_2CF_3)_2\{P(OMe)_3\}_2$,[460] $RuCl_2(CO)_2(Ph_2As\text{-}CH_2Ph_2)_2$,[539] $RhX\{As(o\text{-}vinyl\text{-}C_6H_4)_3\}$,[607] $RX(AsPh_3)_2(RNC)$,[693] $IrH_2\text{-}(AsPh_3)_2(S_2COEt)$,[547] $IrH_2X\{(p\text{-}tolyl)NC\}(AsPh_3)_2$,[548] $trans\text{-}PtCl(O_2SPh)\text{-}(AsEt_3)_2$,[825] $Pd(AsPh_3)_2(O_2CCF_3)(CR=CRCR=CRH)$,[741] $(Me_2N)_2BEEt_2$ (E = As or Sb),[1685] HN_3, $R_2As(N_3)$,[128] $[N_4P_4Me_9]^+[Mo(CO)_5I]^-$,[268] $(acac)R_4Sb$,[46] R_2InX,$5SbMe_3$,[1714] $Me_2BCH_2AsMe_2$,[1656] H_3SiAsR_2,[1748] $R_2GeAsMe_2$,[1767] $[F_2PhPN=SN(tosyl)Me_2]^+$,[1842] $S=PCl_2N=PCl_2N=PCl_2N=PCl_2NHSiMe_3$,[1831] PBr_3,[139, 165] and $[Et_4N]^+[E(S_2C_6H_3Me)_2]^-$ (E = As, Sb, or Bi).[1964]

The use of spectroscopic techniques to investigate phosphorus–fluorine–chalcogen compounds has been reviewed.[1965] A review on recent advances in phosphonitrilic chemistry contains a survey of the use of ^{19}F and ^{31}P n.m.r. spectroscopy to investigate these systems.[1966, 1967] The application of high-resolution n.m.r. spectroscopy to phosphorus compounds[1968] and the structure determination of nitrogen compounds with the aid of n.q.r., n.m.r., and ESCA have been reviewed.[1969]

The ^{13}C n.m.r. spectra of $R^1{}_2PCH=CR^2R^3$ have been reported.[1970] On dilution, the ^1H n.m.r. spectrum of R_4P_2 broadens: the presence of aggregates was suggested.[1971] ^1H N.m.r. spectroscopy has been used to show that in CF_3CO_2H, (233) protonates in the 5-position.[1972] The ^{13}C n.m.r. spectrum of Ph_5Sb showed three resonances, but the carbon attached

---

[1960] W. G. Movius, *J. Inorg. Nuclear Chem.*, 1972, **34**, 3571.
[1961] G. E. Manoussakis and J. A. Tossidis, *J. Inorg. Nuclear Chem.*, 1972, **34**, 2449.
[1962] D. W. Thompson, J. F. Lefelhocz, and K. S. Wong, *Inorg. Chem.*, 1972, **11**, 1139.
[1963] W. Wojnowski and M. Wojnowska, *Z. anorg. Chem.*, 1972, **389**, 302.
[1964] E. Gagliardi and A. Durst, *Monatsh.*, 1972, **103**, 292.
[1965] H.-G. Horn, *Chem.-Ztg.*, 1972, **96**, 666.
[1966] H. R. Allcock, *Chem. Rev.*, 1972, **72**, 315.
[1967] D. B. Denney, *Analyt. Chem. Phosphorus Compounds*, 1972, 611.
[1968] J. R. Van Wazer and T. Glonek, *Analyt. Chem. Phosphorus Compounds*, 1972, 151.
[1969] H. G. Fitzky, D. Wendisch, and R. Holm, *Angew. Chem. Internat. Edn.*, 1972, **11**, 979.
[1970] M.-P. Simonnin, R.-M. Lequan, and F. W. Wehrli, *Tetrahedron Letters*, 1972, 1559.
[1971] H. C. E. McFarlane and W. McFarlane, *J.C.S. Chem. Comm.*, 1972, 1189.
[1972] D. Lloyd and M. I. C. Singer, *Tetrahedron*, 1972, **28**, 353.

to antimony was not detected. The ^{1}H n.m.r. spectrum of (3,4,5-trideuteriophenyl)$_5$Sb showed only a single resonance down to $-142\,°C$, although there was significant broadening below $-100\,°C$.[1973] A general correlation has been found between chemical shift, electronegativity, and the principal quantum number of the valence electron for $(p\text{-Me}_2NC_6H_4)_3MX_2$ (M = P, As, or Sb).[1974] Data have also been reported for Me$_4$POMe,[1975] Me$_4$PF,[1976] [PhCH$_2$(NH$_2$)P(C$_2$H$_4$)$_2$P(NH$_2$)CH$_2$Ph]$^{2+}$Cl$^-_2$,[1977] [PR$_F$]$_n$,[1978] (234),[1979]

(233) (234) (235) (236)

Me$_2$AsC=C(AsMe$_2$)(CF$_2$)$_n$CF$_2$,[1980] (235),[1981] (236),[1982, 1983] As(o-C$_6$H$_4$)$_3$-CH,[1984] Ph$_3$Sb, Ph$_3$SbCl$_2$,[1985] Ph(Me)RSb,[1986] Ph$_2$EtSb,[1987] MeOC$_6$H$_4$-SbMe$_2$,[1988] and (C$_6$F$_5$)$_3$Bi.[1989]

It is interesting that for $(CF_3)_2CFN=S=NC(CF_3)_2N=(CF_3)_2$, $^{10}J(^{19}F-^{19}F) = 1.5$ Hz. It was suggested that the coupling is through space.[1990, 1991] ^{1}H N.m.r. spectroscopy has been used to demonstrate adduct formation between S$_4$N$_4$ and olefins.[1992] Data have also been reported for [(CF$_3$)$_2$C=NS]$_2$,[1993] CF$_2$ClCF$_2$N=S=NMe,[1994] (F$_3$CS)(ClF$_2$CS)NHC$_5$-H$_4$N,[1995] and FN=CFN=NCF=NF.[1996]

[1973] I. R. Beattie, K. M. S. Livingston, G. A. Ozin, and R. Sabine, *J.C.S. Dalton*, 1972, 784.
[1974] J. M. Keck and G. Klar, *Z. Naturforsch.*, 1972, **27b**, 591.
[1975] H. Schmidbaur and H. Stühler, *Angew. Chem. Internat. Edn.*, 1972, **11**, 145.
[1976] H. Schmidbaur, K.-H. Mitschke, and J. Weidlein, *Angew. Chem. Internat. Edn.*, 1972, **11**, 144.
[1977] S. E. Frazier and H. H. Sisler, *Inorg. Chem.*, 1972, **11**, 1431.
[1978] H. G. Ang, M. E. Redwood, and B. O. West, *Austral. J. Chem.*, 1972, **25**, 493.
[1979] G. Märkl, J. Advena, and H. Hauptmann, *Tetrahedron Letters*, 1972, 3961.
[1980] L. S. Chia and W. R. Cullen, *Canad. J. Chem.*, 1972, **50**, 1421.
[1981] G. Märkl, H. Hauptmann, and J. Advena, *Angew. Chem. Internat. Edn.*, 1972, **11**, 441.
[1982] G. Märkl and H. Hauptmann, *Angew. Chem. Internat. Edn.*, 1972, **11**, 439.
[1983] G. Märkl and H. Hauptmann, *Angew. Chem. Internat. Edn.*, 1972, **11**, 441.
[1984] H. Vermeer, P. C. J. Kevenaar, and F. Bickelhaupt, *Annalen*, 1972, **763**, 155.
[1985] D. L. Venezky, C. W. Sink, B. A. Nevett, and W. F. Fortescue, *J. Organometallic Chem.*, 1972, **35**, 131.
[1986] S. Sato, Y. Matsumura, and R. Okawara, *J. Organometallic Chem.*, 1972, **43**, 333.
[1987] S. Sato, Y. Matsumura, and R. Okawara, *Inorg. Nuclear Chem. Letters*, 1972, **8**, 837.
[1988] G. G. de Paoli, B. Zarli, and L. Volponi, *Synth. Inorg. Metal-org. Chem.*, 1972, **2**, 77.
[1989] G. B. Deacon and I. K. Johnson, *Inorg. Nuclear Chem. Letters*, 1972, **8**, 271.
[1990] R. F. Swindell and J. M. Shreeve, *J. Amer. Chem. Soc.*, 1972, **94**, 5713.
[1991] R. R. Swindell and J. M. Shreeve, *Inorg. Nuclear Chem. Letters*, 1972, **8**, 759.
[1992] M. R. Brinkman and C. W. Allen, *J. Amer. Chem. Soc.*, 1972, **94**, 1550.
[1993] S. G. Metcalf and J. M. Shreeve, *Inorg. Chem.*, 1972, **11**, 1631.
[1994] R. Mews and O. Glemser, *Inorg. Chem.*, 1972, **11**, 2521.
[1995] A. Haas and R. Lorenz, *Chem. Ber.*, 1972, **105**, 3161.
[1996] J. B. Hyne, T. E. Austin, and L. A. Bigelow, *Inorg. Chem.*, 1972, **11**, 418.

The ^{31}P n.m.r. spectrum of $R(Me_2N)P(O)OP(O)(NMe_2R)$ shows the presence of two species, presumably the *meso* and racemic forms. The influence of R on the separation of the two resonances was examined.[1997] Variable-temperature ^{19}F n.m.r. measurements have been carried out on hexafluoroacetone adducts of a series of 1-substituted phosphetans, in order to obtain data on the relative apicophilicities of different groups. The data were interpreted in terms of electronegativity and back-bonding factors.[1998] Similarly, ^{19}F n.m.r. measurements on $N_nP_nF_{2n-1}Ar$ (n = 3—8; Ar = C_6F_5 or FC_6H_4) have been used to determine σ_I and σ_R.[1999] The ^1H n.m.r. spectrum of $P(NMeNMe)_3P$ shows a virtual triplet pattern, but the methyl resonance for $ClP(NMeNMe)_2PCl$ is only a doublet; however, the methyl resonance of $Cl_2P(NMeNMe)PCl_2$ is a virtual triplet.[2000] The magnitude of $^4J(P-H)$ for $HClC=CClN=PCl_3$ is temperature dependent and this behaviour has been attributed to the position of the equilibrium changing.[2001] ^{13}C N.m.r. data, including $^2J(P-N-C)$, have been reported for Me_2NPPhX.[2002] ^1H N.m.r. spectroscopy provides a quick and specific way of assaying sodium cacodylate injections.[2003] Data have also been reported for $P_3N_3F_4(NMe_2)_2$,[2004] (237),[2005, 2006] $Cl_3P=NPCl_2=NSO_2Cl$,[2007] (238),[2008] $R^1_2NR^2PC\equiv CPR^2NR^1_2$,[2009] R_2AsNEt_2,[2010] $[R_3AsNMe_2]^+Cl^-$,[2011] and (239).[2012]

The $^3J(^{13}C-^{31}P)$ coupling constants in some 1,3,2-dioxaphosphorinan-2-ones have been shown to be dependent on the POCC dihedral angle and the orientation of the phosphorus–oxygen bond.[2013] The α-methyl

(237) (238) (239)

[1997] M. D. Joesten and Y. T. Chen, *Inorg. Chem.*, 1972, **11**, 429.
[1998] R. K. Oram and S. Trippett, *J.C.S. Chem. Comm.*, 1972, 554.
[1999] T. Chivers and N. L. Paddock, *Inorg. Chem.*, 1972, **11**, 848.
[2000] M. D. Havlicek and J. W. Gilje, *Inorg. Chem.*, 1972, **11**, 1624.
[2001] E. Fluck and W. Steck, *Z. anorg. Chem.*, 1972, **387**, 349.
[2002] M.-P. Simonnin, R.-M. Lequan, and F. W. Wehrli, *J.C.S. Chem. Comm.*, 1972, 1204.
[2003] W. Holak, *J. Pharm. Sci.*, 1972, **61**, 1635.
[2004] E. Niecke, H. Thamm, and D. Böhler, *Inorg. Nuclear Chem. Letters*, 1972, **8**, 261.
[2005] R. Appel, R. Kleinstück, and K.-D. Ziehn, *Chem. Ber.*, 1972, **105**, 2476.
[2006] A. Schmidpeter, J. Ebeling, H. Stary, and C. Weingand, *Z. anorg. Chem.*, 1972, **394**, 171.
[2007] W. Haubold and E. Fluck, *Z. Naturforsch.*, 1972, **27b**, 368.
[2008] M. Bermann and J. R. Van Wazer, *Inorg. Chem.*, 1972, **11**, 2515.
[2009] W. Kuchen and K. Koch, *Z. anorg. Chem.*, 1972, **394**, 74.
[2010] L. S. Sagan, R. A. Zingaro, and K. J. Irgolic, *J. Organometallic Chem.*, 1972, **39**, 301.
[2011] L. K. Krannich and H. H. Sisler, *Inorg. Chem.*, 1972, **11**, 1226.
[2012] O. J. Scherer and R. Wies, *Angew. Chem. Internat. Edn.*, 1972, **11**, 592.
[2013] A. A. Borisenko, N. M. Sergeyev, E. Ye. Nifant'ev, and Yu. A. Ustynyuk, *J.C.S. Chem. Comm.*, 1972, 406.

$^2J(^{31}P-^{13}C)$ in 2,2,3,4,4-pentamethylphosphetans is stereospecific with respect to the exocyclic phosphorus substituent. The coupling is large (27—37 Hz) when the methyl group is *trans* and small (0—5 Hz) when the methyl group is *cis* to the exocyclic phosphorus substituent.[2014] This sort of behaviour appears to be frequently found in a very wide range of phosphorus-containing four-membered heterocyclic rings.[2015, 2016] ^1H, ^{19}F, and ^{31}P n.m.r. spectroscopy have been used to show the presence of two conformers for (240) at $-100\,°C$.[2017] However, the ^1H n.m.r. spectrum of (241) is temperature invariant.[2018] The ^{19}F n.m.r. spectrum of $R^1PF_3(OR^2)$ is almost independent of R^2.[2019] The ^1H n.m.r. spectrum of $(CF_3)_2PO_2H$

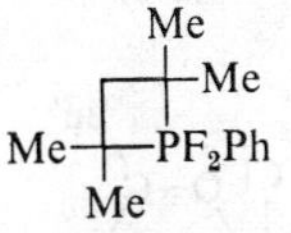

does not show phosphorus coupling even at $-100\,°C$, implying rapid exchange. The ^{19}F and ^{31}P n.m.r. spectra of $(CF_3)_2P(S)OP(S)(CF_3)_2$ are of the $AA'X_6X'_6$ spin type.[2020] Data have also been reported for $F_2ClCSN=CClS_2CClF_2$,[2021] $R_2P(O_2CCF_3)$,[2022] $Me(HOCH_2)PO_2H$,[2023] $(EtO_2POCH=CHOR$,[2024] $Ph_{3-n}(PhC\equiv C)_nP=S$,[2025] Me_2PEMe (E = O or S),[2026] and RPHFO.[2027]

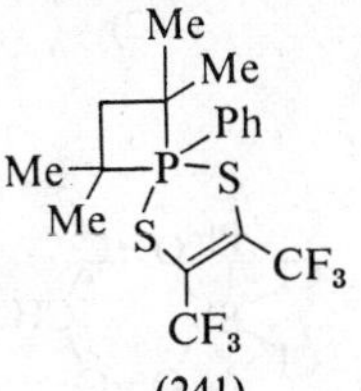

The n.m.r. spectra of (242) have been measured and used to estimate the anisotropic diamagnetic susceptibility of the arsenic–oxygen and arsenic–chlorine bonds.[2028] For $Cl_nR_{4-n}Sb(acac)$ (R = Me or Ph) and $EtCl_2Sb(acac)$, if $\delta(CH)$ and $\delta(CH_3)$ are plotted against $\sum\sigma_I$ for the substituents, a linear relationship is obtained. The stereochemistry was assigned by signal multiplicity and isomers were found.[2029] In order to explain the ^1H chemical

---

[2014] G. A. Gray and S. E. Cremer, *J.C.S. Chem. Comm.*, 1972, 367.
[2015] G. A. Gray and S. E. Cremer, *J. Org. Chem.*, 1972, **37**, 3458.
[2016] G. A. Gray and S. E. Cremer, *J. Org. Chem.*, 1972, **37**, 3470.
[2017] N. J. De'Ath, D. Z. Denney, and D. B. Denney, *J.C.S. Chem. Comm.*, 1972, 272.
[2018] N. J. De'Ath and D. B. Denney, *J.C.S. Chem. Comm.*, 1972, 395.
[2019] D. U. Robert, G. N. Flatau, C. Demay, and J. G. Riess, *J.C.S. Chem. Comm.*, 1972, 1127.
[2020] A. A. Pinkerton and R. G. Cavell, *J. Amer. Chem. Soc.*, 1972, **94**, 1870.
[2021] P. Gielow and A. Haas, *Z. anorg. Chem.*, 1972, **394**, 53.
[2022] P. Sartori and M. Thomzik, *Z. anorg. Chem.*, 1972, **394**, 157.
[2023] L. Maier, *Z. anorg. Chem.*, 1972, **394**, 117.
[2024] L. Maier, *Z. anorg. Chem.*, 1972, **394**, 111.
[2025] E. Fluck and N. Seng, *Z. anorg. Chem.*, 1972, **393**, 126.
[2026] F. Seel, W. Gombler, and K.-D. Velleman, *Annalen*, 1972, **756**, 181.
[2027] U. Ahrens and H. Falius, *Chem. Ber.*, 1972, **105**, 3317.
[2028] G. Kamai, N. A. Chadaeva, I. I. Saidashev, and N. K. Tazeeva, *Vop. Stereokhim.*, 1971, **1**, 127.
[2029] H. A. Meinema, A. Mackor, and J. G. Noltes, *J. Organometallic Chem.*, 1972, **37**, 285.

shifts in (243) it was necessary to postulate that the phenyl group is perpendicular to the acetylacetonate C–H axis. This postulate was confirmed by an X-ray structure determination.[2030] Ph_2SbCl_2(dpm) has two t-butyl and two CH resonances in the ^1H n.m.r. spectrum, due to (244) and (245). When a freshly prepared solution in C_6D_6 is examined, only

(242)

(243)

(244)

(245)

one isomer is observed but, slowly (24 h), the second species appears.[2031] For $Me_3Sb(O_2CR)_2$, a plot of methyl chemical shift against pK_a of the acid produces a straight line.[2032] Data have also been reported for $R^1{}_2NCO_2$-$CH_2CHR^2XMMe_2$ (M = As or Sb) [2033] and Ph_3BiX(oxalate).[2034]

The ^{19}F n.m.r. spectrum of H_2NPF_4 shows two ^{19}F resonances of equal intensity. The axial fluorines and the amino-protons form an AA'XX' spin system and $J(^{15}N-{}^1H)$ = 90.3 Hz. This has been taken to indicate that the nitrogen is sp^2 hybridized.[2035] ^{19}F and ^{31}P n.m.r. spectra have been reported for F_2PN_3, F_2PON_3, and $FPO(N_3)_2$. Such work appears to be dangerous and expensive as F_2PN_3 detonated, destroying the phosphorus probe.[2036] ^1H, ^{19}F, and ^{31}P n.m.r. spectroscopy have been used to show that $\overline{FPNMeCH_2CH_2NMe}$ binds to BH_3 via phosphorus. With BF_3, species such as (246) are formed.[2037] ^{31}P N.m.r. spectroscopy has been used to show that PCl_5, BCl_3, and NH_4Cl react to form $[Cl(Cl_2P=N)_nPCl_3]^+$.[2038] ^{31}P N.m.r. spectroscopy has been used to assist characterization of mixed-substituent poly(aminophosphazenes).[2039] The ^{19}F n.m.r. spectra of a

---

[2030] J. Kroon, J. B. Hulscher, and A. F. Peerdeman, *J. Organometallic Chem.*, 1972, **37**, 297.
[2031] H. A. Meinema and J. G. Noltes, *J. Organometallic Chem.*, 1972, **37**, C31.
[2032] R. G. Goel and D. R. Ridley, *J. Organometallic Chem.*, 1972, **38**, 83.
[2033] J. Koketsu, S. Kokjma, and Y. Ishii, *J. Organometallic Chem.*, 1972, **38**, 69.
[2034] G. Faragalia, E. Rivarola, and F. Di Bianca, *J. Organometallic Chem.*, 1972, **38**, 91.
[2035] A. H. Cowley and J. R. Schweiger, *J.C.S. Chem. Comm.*, 1972, 560.
[2036] S. R. O'Neill and J. M. Shreeve, *Inorg. Chem.*, 1972, **11**, 1629.
[2037] S. Fleming, M. K. Lupton, and K. Jekot, *Inorg. Chem.*, 1972, **11**, 2534.
[2038] K. Niedenzu, I. A. Boenig, and E. B. Bradley, *Z. anorg. Chem.*, 1972, **393**, 88.
[2039] H. R. Allcock, W. J. Cook, and D. P. Mack, *Inorg. Chem.*, 1972, **11**, 2584.

(246) [structure: bicyclic P-F-P cation with Me-N groups and $[B_2F_7]^-$ counterion]

series of complexes such as (247) have been measured. The results were discussed in terms of molecular conformation and of intramolecular exchange.[2040] The ^{19}F n.m.r. resonances for (248) are not equivalent and an ABX n.m.r. spectrum is observed.[2041] Data have also been reported for PF_2N_3,[2042] (249),[2043] $\{Cl_2P(S)\}_2NR$,[2044] $RSPF_nCl_{2-n}=NPX_2O$,[2045] SO_2-$(NPClNEt_2)_2NMe$,[2046] $\{Cl(Me_2N)P(O)\}_2NMe$ and many similar compounds,[2047] $F_2PN=C=NPF_2$,[2048] (250),[2049] (251),[2050] $[-N=PF_2-]_n$,[2051]

(247) OC–N(Me)–N(Me)–PF_2Y

(248) [S_3N_3 ring with OPF_2 substituent]

(249) H_2C-O, H_2C-O P–NHNMe_2·NH_2 Cl$^-$

(250) [cyclic structure with Ph_2, NMe_2, H substituents]

(251) [cyclic phosphazene with (OCH_2CF_3)_2, HN, O=P, CF_3CH_2O, P(OCH_2CF_3)_2]

(252) [S-N ring with Cl, Cl_2P, PCl_2]

[2040] R. K. Harris, J. R. Woplin, R. E. Dunmur, M. Murray, and R. Schmutzler, *Ber. Bunsengesellschaft phys. Chem.*, 1972, **76**, 44.
[2041] H. W. Roesky and L. F. Grimm, *Angew. Chem. Internat. Edn.*, 1972, **11**, 642.
[2042] E. L. Lines and L. F. Centofanti, *Inorg. Chem.*, 1972, **11**, 2269.
[2043] S. E. Frazier and H. H. Sisler, *Inorg. Chem.*, 1972, **11**, 1223.
[2044] R. Keat, *J.C.S. Dalton*, 1972, 2189.
[2045] H. W. Roesky, B. H. Kuhtz, and L. F. Grimm, *Z. anorg. Chem.*, 1972, **389**, 167.
[2046] U. Klingebiel and O. Glemser, *Chem. Ber.*, 1972, **105**, 1510.
[2047] I. Irvine and R. Keat, *J.C.S. Dalton*, 1972, 17.
[2048] D. W. H. Rankin, *J.C.S. Dalton*, 1972, 869.
[2049] M. Bermann and J. R. Van Wazer, *Inorg. Chem.*, 1972, **11**, 209.
[2050] H. R. Allcock and E. J. Walsh, *J. Amer. Chem. Soc.*, 1972, **94**, 119.
[2051] H. R. Allcock, R. L. Kugel, and E. G. Stroh, *Inorg. Chem.*, 1972, **11**, 1120.

(252),[2052] [MeNPF$_3$]$_4$, [(MeN)$_4$P$_3$F$_6$]$^+$[PF$_6$]$^-$,[2053] P$_3$N$_3$F$_{6-n}$Cl$_n$,[2054] N$_4$P$_4$Cl$_n$-(NCS)$_n$,[2055] P$_4$N$_4$Cl$_{6-n}$(NMe$_2$)$_n$,[2056] and Sb(NMe$_2$)$_2$(NR$_2$).[2057]

For MeO$_2$SNSOF$_2$ and OE(NSOF$_2$)$_2$ (E = S or Se) the fluorine atoms are inequivalent, giving rise to an AB pattern.[2058] The CF$_2$ fluorines in CF$_3$SO(NH)CF$_2$CF$_3$ are also inequivalent.[2059] Data have also been reported for Cl$_2$FSCN(SCCl$_{3-n}$F$_n$)$_2$,[2060] O=PF$_2$N=SF$_2$=NPF$_2$O,[2061] and F$_2$S=NC(O)NCO.[2062]

The ^{31}P n.m.r. spectra of some cyclic metaphosphates have been reported. It was found that the order of ^{31}P chemical shifts to high field for [HPO$_3$]$_n$ is 3 < n < 8 < 4 < 5 < 6 < 7 in tetramethylurea as solvent. Other orders are found in water and as a function of pH. The order was discussed in terms of conformers.[2063] ^{31}P N.m.r. spectroscopy has been used to assist the identification of products such as H$_3$PO$_6$ from the reaction of phosphorus compounds, e.g. P$_4$O$_{10}$ or P$_2$O$_3$F$_4$, with H$_2$O$_2$.[2064] The ^{31}P n.m.r. spectra of polymetaphosphates have been measured in order to determine the ratio of middle : end groups and hence to determine the amount of polymerization.[2065] Spin–lattice relaxation times of ^{1}H, ^{2}H, ^{19}F, and ^{31}P in solutions of NaPF$_6$ and NaPO$_3$F in H$_2$O and D$_2$O have been measured. ^{19}F and ^{31}P spin-relaxation times occur mainly by spin–rotation and dipole–dipole interactions.[2066] ^{1}H, ^{19}F, and ^{31}P n.m.r. spectra have been used to examine FPO$_3$H$_2$, F$_2$PO$_2$H$_2$, F$_3$PO, PF$_3$, phosphonic acid, phosphinic acid, pyrophosphoric acid, and polyphosphoric acid with FSO$_3$H or FSO$_3$H–SbF$_5$ in order to determine the amount of protonation.[2067] The ^{19}F and ^{31}P n.m.r. spectra of {RFP(S)}$_2$S have been analysed to obtain ^2J(P–S–P) for the racemic and *meso*-forms.[2068] The ^{31}P n.m.r. spectrum of P$_4$S$_3$ shows a quartet and a doublet which is consistent with the structure (253).[2069] Data have also been reported for (254),[2070] and EtO$_2$PCl-NCCl$_2$.[2071]

[2052] H. W. Roesky, *Angew. Chem. Internat. Edn.*, 1972, **11**, 642.
[2053] K. Utvary and W. Czysch, *Monatsh.*, 1972, **103**, 1048.
[2054] P. Clare, D. B. Sowerby, and B. Green, *J.C.S. Dalton*, 1972, 2374.
[2055] R. L. Dieck and T. Moeller, *Inorg. Nuclear Chem. Letters*, 1972, **8**, 763.
[2056] D. Millington and D. B. Sowerby, *J.C.S. Dalton*, 1972, 2035.
[2057] A. Kiennemann, G. Levy, and C. Tanielian, *J. Organometallic Chem.*, 1972, **46**, 305.
[2058] A. Roland, K. Seppelt, and W. Sundermeyer, *Z. anorg. Chem.*, 1972, **393**, 141.
[2059] D. T. Sauer and J. M. Shreeve, *Inorg. Chem.*, 1972, **11**, 238.
[2060] A. Haas and R. Lorenz, *Chem. Ber.*, 1972, **105**, 237.
[2061] O. Glemser, J. Wegener, and R. Höfer, *Chem. Ber.*, 1972, **105**, 474.
[2062] A. F. Clifford, J. S. Harman, and C. A. McAuliffe, *Inorg. Nuclear Chem. Letters*, 1972, **8**, 567.
[2063] T. Glonek, J. R. Van Wazer, M. Mudgett, and T. C. Myers, *Inorg. Chem.*, 1972, **11**, 567.
[2064] E. Fluck and W. Steck, *Z. anorg. Chem.*, 1972, **388**, 53.
[2065] R. C. Mehrotra, P. C. Vyas, and C. K. Oza, *Indian J. Chem.*, 1972, **10**, 726.
[2066] M. F. Froix and E. Price, *J. Chem. Phys.*, 1972, **56**, 6050.
[2067] G. A. Olah and C. W. McFarland, *Inorg. Chem.*, 1972, **11**, 845.
[2068] R. K. Harris, J. R. Woplin, M. Murray, and R. Schmutzler, *J.C.S. Dalton*, 1972, 1590.
[2069] L. Kolditz and E. Wahner, *Z. Chem.*, 1972, **12**, 389.
[2070] W. G. Bentrude, W. D. Johnson, and W. A. Kahn, *J. Amer. Chem. Soc.*, 1972, **94**, 3058.
[2071] W. Haubold and E. Fluck, *Z. anorg. Chem.*, 1972, **392**, 59.

The ^{1}H n.m.r. spectrum of (255) has been analysed as AA'X$_3$X'$_3$ and J(AA) used to determine which isomer is *cis* and which is *trans*.[2072] ^{15}N N.m.r. spectroscopy has been used to show two isomers for (256) and (257).[2073]

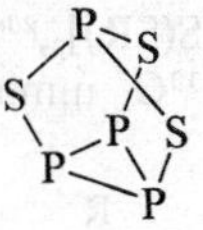

(253)

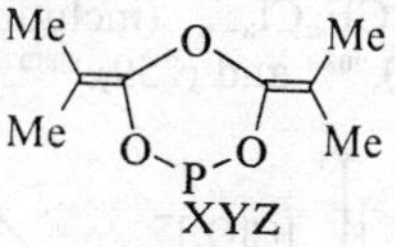

(254) XYZ

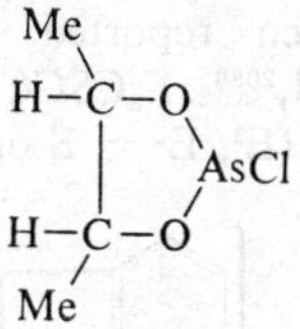

(255)

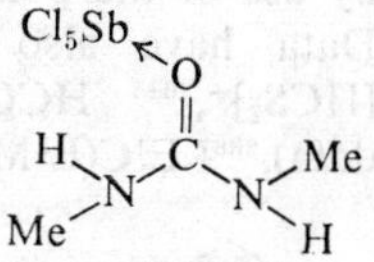

(256)

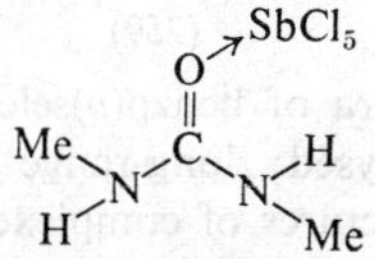

(257)

^{13}C N.m.r. data have been reported for [PhMeI]$^+$[SbF$_6$]$^-$.[2074] ^{19}F, ^{121}Sb, and ^{123}Sb n.m.r. spectra have been observed for [SbF$_6$]$^-$. The ^{19}F n.m.r. spectra show ^1J(^{19}F–^{121}Sb) = 1934 Hz and ^1J(^{19}F–^{123}Sb) = 1047 Hz.[2075] Data have also been reported for MePXY.[2076]

9 Compounds of Groups VI and VII and of Xenon

Information concerning complexes of these elements can be found at the following sources: TeC$_4$H$_4$,[1779] TeC$_4$H$_4$Cr(CO)$_3$, TeC$_4$H$_4$Fe$_2$(CO)$_6$,[241] R^{1}R^{2}C$_2$SeFe$_2$(CO)$_6$,[408] [Fe(CO)$_3$SeMe$_2$]$_2$,[477] *trans*-PtCl(SO$_2$Ph)(EEt$_2$)$_2$ (E = Se or Te),[824] *trans*-PtCl(O$_2$SPh)(EEt$_2$)$_2$ (E = Se or Te),[825] [Me$_2$NCSe$_2$-C$_2$H$_4$]$^+$,[1921] H$_2$Ru$_3$(CO)$_9$E (E = S, Se, or Te),[347] (H$_3$M)$_2$E, (F$_2$P)$_2$E (M = Si or Ge; E = O, S, Se, or Te),[143] H$_3$SiSeMe,[1745] OSe(NSOF$_2$)$_2$,[2058] R$_3$PSe,[142] HOF,[100] HCl,[101] [ClF$_6$]$^+$,[70] and [IF$_6$]$^+$.[71]

A book has appeared which lists nuclear magnetic resonance data of sulphur compounds.[2077]

The ^{1}H n.m.r. spectra of (258; E = O, S, Se, or Te) have been analysed and solvent effects examined.[2078] The ^{17}O n.m.r. spectrum of CF$_3$OOOCF$_3$ shows resonances at −321 p.p.m. (intensity 2) and −479 p.p.m. (intensity 1) relative to H$_2$^{17}O. OF$_2$ reacts with C^{17}OF$_2$ to give the resonance at −321 p.p.m. and ^{17}OF$_2$ reacts with COF$_2$ to give the resonance at

---

[2072] D. W. Aksnes and O. Vikane, *Acta Chem. Scand.*, 1972, **26**, 835.
[2073] P. Stilbs, *Tetrahedron Letters*, 1972, 227.
[2074] G. A. Olah and E. G. Melby, *J. Amer. Chem. Soc.*, 1972, **94**, 6220.
[2075] R. G. Kidd and R. W. Matthews, *Inorg. Chem.*, 1972, **11**, 1156.
[2076] H. W. Schiller and R. W. Rudolph, *Inorg. Chem.*, 1972, **11**, 187.
[2077] N. F. Chamberlain and J. J. R. Reed, in 'The Analytical Chemistry of Sulphur and its Compounds, Part III', ed. J. H. Karchmer, 'Nuclear Magnetic Resonance Data of Sulphur Compounds', Wiley–Interscience, New York, 1971.
[2078] P. Faller and J. Weber, *Bull. Soc. chim. France*, 1972, 3193.

−479 p.p.m.[2079] [19]F N.m.r. data have also been reported for CF_3OOF [2080] and $(CF_3OO)_2CO$.[2081]

[19]F Chemical shifts have been measured for compounds of the type FC_6H_4SOX. The effects of substituents, X, at sulphur have been analysed by use of the polar and resonance substituent parameters σ_I and σ_R.[2082] Data have also been reported for $[HCS(SH)]_3$,[2083] $[HCS(SR)]_3$,[2084] $[HCS_2]^-$,[2085] HCOSH,[2086] $F_3CSCCl_2SSCF_nCl_{3-n}$ (including [13]C n.m.r. data),[2087] $E^1C(E^2Me)_2$ (E^1, E^2 = S or Se),[2088] and (259).[2089]

(258)   (259)   (260)

The [1]H n.m.r. spectra of benzo(b)selenophen and benzo(b)tellurophen have been fully analysed; long-range coupling between the rings is observed.[2090] The structures of complexes between selenane and bromine or iodine have been determined in solution. By analysis of the coupling constants in the ring for the selenane–bromine complex it was possible to show trigonal-bipyramidal stereochemistry.[2091] [1]H N.m.r. spectroscopy has been used to show axial preference of the substituent in (260).[2092] Data have also been reported for R^1R^2SeCXY,[2093] poly-(3,3-dimethyl-selenetan),[2094] $ArECl_3$ (E = Se or Te),[2095] $RN=C=Se$,[2096] PhSeX,[2097] ArSeMe,[2098] (261),[2099] PhCOSeH,[2100] (262),[2101] (263),[2102] $(PhSe)_2CHMe$,

[2079] I. J. Solomon, A. J. Kacmarek, W. K. Sumida, and J. K. Raney, Inorg. Chem., 1972, 11, 195.
[2080] D. D. DesMarteau, Inorg. Chem., 1972, 11, 193.
[2081] D. Pilopovich, C. J. Schack, and R. D. Wilson, Inorg. Chem., 1972, 11, 2531.
[2082] W. A. Sheppard and R. W. Taft, J. Amer. Chem. Soc., 1972, 94, 1919.
[2083] R. Engler and G. Gattow, Z. anorg. Chem., 1972, 389, 145.
[2084] R. Engler, G. Gattow, and M. Dräger, Z. anorg. Chem., 1972, 390, 64.
[2085] R. Engler, G. Gattow, and M. Dräger, Z. anorg. Chem., 1972, 388, 229.
[2086] R. Engler and G. Gattow, Z. anorg. Chem., 1972, 388, 78.
[2087] A. Haas, W. Klug, and H. Marsmann, Chem. Ber., 1972, 105, 820.
[2088] M. Dräger and G. Gattow, Spectrochim. Acta, 1972, 28A, 425.
[2089] A. Tadino, L. Christiaens, M. Renson, and P. Cagniant, Bull. Soc. chim. belges, 1972, 81, 595.
[2090] G. Llabrès, M. Baiwir, J. Denoel, J. L. Piette, and L. Christiaens, Tetrahedron Letters, 1972, 3177.
[2091] J. B. Lambert, D. H. Johnson, R. G. Keske, and C. E. Mixan, J. Amer. Chem. Soc., 1972, 94, 8172.
[2092] J. B. Lambert, C. E. Mixan, and D. H. Johnson, Tetrahedron Letters, 1972, 4335.
[2093] N. N. Magdesieva, R. A. Kandgetcyan, and A. A. Ibragimov, J. Organometallic Chem., 1972, 42, 399.
[2094] E. J. Goethals, E. Schacht, and D. Tack, J. Polymer Sci., Part A-1, Polymer Chem., 1972, 10, 533.
[2095] K. J. Wynne, A. J. Clark, and M. Ber, J.C.S. Dalton, 1972, 2370.
[2096] N. Sonoda, G. Yamamoto, and S. Tsutsumi, Bull. Chem. Soc. Japan, 1972, 45, 2937.
[2097] K. J. Wynne and P. S. Pearson, Inorg. Chem., 1972, 11, 1196.
[2098] F. Mantovani, L. Christiaens, and P. Faller, Bull. Soc. chim. France, 1972, 1595.
[2099] C. Dragnet and M. Renson, Bull. Soc. chim. belges, 1972, 81, 295.
[2100] K. A. Jensen, L. Bøje, and L. Hendriksen, Acta Chem. Scand., 1972, 26, 1465.
[2101] L. Fitjer and W. Lüttke, Chem. Ber., 1972, 105, 919.
[2102] L. Fitjer and W. Lüttke, Chem. Ber., 1972, 105, 907.

$CH_2C(SPh)_2C(SePh)_2$,[2103] (264),[2104] (265),[2105] benzo[b]selenophens,[2106-2108] PhTeOAc,[2109] o-C_6H_4{$CH(OEt)_2$}{$TeCH_2CH(OEt)_2$},[2110] o-$C_6H_4(TeBr)$-(COCH=CHPh),[2111] (266),[2112] and 1-(tellurophen-2-yl)ethyl acetate.[2113]

(261) (262) (263) (264)

(265) (266)

The ^{19}F n.m.r. spectra of ArO_2SF_3 and $(ArO)_2S(O)F_2$ show resonances for equatorial and axial fluorines.[2114] The ^{19}F n.m.r. spectrum of $HOSeF_5$ is of the AB_4 type with selenium satellites.[2115] Cation dependence is found for the ^{19}F n.m.r. spectrum of $[SeOF_5]^-$.[2116] XeF_2 reacts with $HOSeF_5$ to give $Xe(OSeF_5)_2$ and $FXe(OSeF_5)$, which give AB_4 ^{19}F n.m.r. spectra with ^{77}Se and ^{129}Xe satellites.[2117] $[C_5H_5NH]^+[TeF_5O]^-$ and SeF_5Cl also give AB_4 ^{19}F n.m.r. spectra.[2118, 2119] Data have also been reported for $S_3N_3F_3$,[2120] $F_5SeOSO_2OSO_2F$,[2121] $R^1R^2P(OH)Se$,[2122] and F_5TeOR.[2123]

Nuclear spin relaxation and molecular motion in liquid HBr and HCl have been examined.[2124] The ^1H shielding of $[ClHCl]^-$, $[BrHBr]^-$, and $[HIH]^-$ has been derived and a deshielding of ca. 13 p.p.m. found with

---

[2103] D. Seebach and N. Peleties, *Chem. Ber.*, 1972, **105**, 511.
[2104] B. Decroix, J. Morel, C. Paulmer, and P. Pastour, *Bull. Soc. chim. France*, 1972, 1848.
[2105] F. Terrier, A.-P. Chatrousse, R. Schaal, C. Paulmier, and P. Pastour, *Tetrahedron Letters*, 1972, 1961.
[2106] T. Q. Mink, L. Christiaens, and M. Renson, *Tetrahedron*, 1972, **28**, 5397.
[2107] T. Q. Mink, P. Thibaut, L. Christiaens, and M. Renson, *Tetrahedron*, 1972, **28**, 5393.
[2108] N. N. Magdesieva and V. A. Vdovin, *Khim. geterotsikl. Soedinenii*, 1972, 15.
[2109] B. C. Pant, *Tetrahedron Letters*, 1972, 4779.
[2110] J.-L. Piette, R. Lysy, and M. Renson, *Bull. Soc. chim. France*, 1972, 3559.
[2111] J.-L. Piette and M. Renson, *Bull. Soc. chim. belges*, 1971, **80**, 669.
[2112] F. Fringuelli and A. Taticchi, *J.C.S. Perkin I*, 1972, 199.
[2113] F. Fringuelli, G. Marino, and A. Taticchi, *Gazzetta*, 1972, **102**, 534.
[2114] D. S. Ross and D. W. A. Sharp, *J.C.S. Dalton*, 1972, 34.
[2115] K. Seppelt, *Angew. Chem. Internat. Edn.*, 1972, **11**, 630.
[2116] K. Seppelt, *Chem. Ber.*, 1972, **105**, 2431.
[2117] K. Seppelt, *Angew. Chem. Internat. Edn.*, 1972, **11**, 723.
[2118] G. W. Fraser and J. B. Millar, *J.C.S. Chem. Comm.*, 1972, 1113.
[2119] C. J. Schack, R. D. Wilson, and J. F. Hon, *Inorg. Chem.*, 1972, **11**, 208.
[2120] N. J. Maraschin and R. L. Lagow, *J. Amer. Chem. Soc.*, 1972, **94**, 8601.
[2121] K. Seppelt, *Chem. Ber.*, 1972, **105**, 3131.
[2122] M. Mikołajczyk and J. Łuczak, *Tetrahedron*, 1972, **28**, 5411.
[2123] F. Sladky and H. Kropshofer, *Inorg. Nuclear Chem. Letters*, 1972, **8**, 195.

respect to the solvated hydrogen halide. The equilibrium constants are in the order consistent with a simple electrostatic model of ion–molecule association.[2125] The shielding constants were shown to be dominated by hydrogen charge-density, which is less than for the corresponding molecule. This is the charge shift predicted by a simple electrostatic model but not by the independent-electron molecular orbital model.[2126]

The ^{19}F n.m.r. spectrum of $[ClF_2]^+$ has been reported for the first time.[2127] The ^{19}F n.m.r. spectrum of ClF_3O has been measured as a neat liquid and as a gas; no extra resonances were found on cooling.[2128] From ^{19}F n.m.r. spectra, the presence of $I(SO_3F)_3$ in mixtures of I_2 and $S_2O_6F_2$ has been suggested.[2129] The ^{19}F n.m.r. spectrum of CF_3IF_4 is A_3X_4 at room temperature.[2130]

The ^{19}F n.m.r. spectrum of $[F_5Xe]^+$ shows the presence of two types of fluorine of intensity one, $^1J(^{129}Xe-^{19}F) = 170$ Hz, and intensity four, $^1J(^{129}Xe-^{19}F) = 1377$ Hz.[2131] In SbF_5, $OXeF_4,2SbF_5$ and $XeO_2F_2,2SbF_5$ have the structures $[XeOF_3]^+[Sb_2F_{11}]^-$ and $[XeO_2F]^+[Sb_2F_{11}]^-$ respectively.[2132]

10 Bibliography

The following is a list of those references obtainable *via* the Chemical Society's n.m.r. Macroprofile (UKCIS) which are in journals not abstracted from the main text and omitted from Volume 5 of this S.P.R. The *Chemical Abstracts* reference number is given in brackets.

Chemical Abstracts, 1972, Volume 76.

A. G. Yurchenko and S. D. Isaev. ^{1}H n.m.r. spectra of adamantanone and adamantanone oxime in the presence of europium tris(dipivaloylmethanate). *Zhur. org. Khim.*, 1971, **7**, 2628 (*CA* 71 918).

M. Karras and E. Rahkamaa. Determination of moisture in pulp by the n.m.r. method. *Pap. Puu*, 1971, **53**, 653 (*CA* 73 963).

J. R. Campbell. Lanthanide chemical shift reagents. *Aldrichimica Acta*, 1971, **4**, 55 (*CA* 78 585).

H. Pfeiffer. N.m.r. and relaxation of adsorbed molecules. *Paramagn. Rezonans 1944—1969, Vses. Yubileinaya Konf. 1969*, 1971, 255 (*CA* 78 987).

L. Tomic, Z. Majerski, M. Tomic, and D. E. Sunko. Tris(dipivaloylmethanato)holmium-induced n.m.r. shifts. *Croat. Chem. Acta*, 1971, **43**, 267 (*CA* 79 097).

Yu. V. Belov, Yu. L. Kleiman, N. V. Morkovin, and Yu. Yu. Samitov. Observation of double resonance in the rya-2307 n.m.r. spectrometer. *Prib. Tekh. Eksp.*, 1971, 186 (*CA* 79 239).

V. Niculescu, S. Mandache, and I. Pop. Device for temperature regulation in n.m.r. measurements. *Stud. Cercet. Fiz.*, 1971, **23**, 1123 (*CA* 79 248).

[2124] K. Krynicki and J. G. Powles, *J. Magn. Resonance*, 1972, **6**, 539.
[2125] F. Y. Fujiwara and J. S. Martin, *J. Chem. Phys.*, 1972, **56**, 4091.
[2126] J. S. Martin and F. Y. Fujiwara, *J. Chem. Phys.*, 1972, **56**, 4098.
[2127] M. Brownstein and J. Shamir, *Canad. J. Chem.*, 1972, **50**, 3409.
[2128] D. Pilopovich, C. B. Lindahl, C. J. Schack, R. D. Wilson, and K. O. Christe, *Inorg. Chem.*, 1972, **11**, 2189.
[2129] C. Chung and G. H. Cady, *Inorg. Chem.*, 1972, **11**, 2528.
[2130] O. R. Chambers, G. Oates, and J. M. Winfield, *J.C.S. Chem. Comm.*, 1972, 839.
[2131] D. D. Desmarteau and M. Eisenberg, *Inorg. Chem.*, 1972, **11**, 2641.
[2132] R. J. Gillespie, B. Landa, and G. J. Schrobilgen, *J.C.S. Chem. Comm.*, 1972, 607.

Yu. D. Gavrilov and V. F. Bystrov. Suppression of a heteronuclear spin–spin interaction in n.m.r. spectra using a spin generator. *Prib. Tekh. Eksp.*, 1971, 140 (*CA* 79 250).

S. A. Al'tshuler and M. A. Teplov. N.m.r. of paramagnetic ions in singlet electronic states. *Paramagn. Rezonans 1944—1969, Vses. Yubileinaya Konf. 1969*, 1971, 166 (*CA* 92 181).

I. A. Nuretdinov and E. I. Loginova. Phosphorus-31 and selenium-77 spin–spin interaction constants and structure of chlorophosphine selenides. *Izvest. Akad. Nauk S.S.S.R., Ser. khim.*, 1971, 2360 (*CA* 92 662).

A. N. Gil'manov, S. I. Berezina, G. S. Vozdvizhenskii, and V. D. Lapshin. Proton magnetic relaxation study of the state of hydrogen adsorbed by palladium and platinum cathodes. *Elektrokhimiya*, 1971, **7**, 1336 (*CA* 92 663).

T. Kanashiro, T. Ohno, T. Taki, and M. Satoh. Acoustic excitation of n.m.r. in sodium chlorate. *Bull. Fac. Eng., Tokushima Univ.*, 1971, **8**, 19 (*CA* 92 664).

I. A. Nuretdinov, V. V. Negrebetskii, A. Z. Yankelevich, A. V. Kessenikh, L. K. Nikonorova, and E. I. Loginova. Proton n.m.r., phosphorus-31 n.m.r., and proton–phosphorus-31 INDOR-^{31}P spectra of compounds containing the $=PX-N=E-PY=$ group. *Izvest. Akad. Nauk S.S.S.R., Ser. khim.*, 1971, 2589 (*CA* 92 667).

V. S. Lyubimov and S. P. Ionov. Electronic structure and n.m.r. spectra of boron compounds. *Izvest. Akad. Nauk S.S.S.R., Ser. khim.*, 1971, 2584 (*CA* 92 679).

V. I. Chizhik and Yu. A. Ermakov. Quadrupole relaxation of the nuclei of lithium and sodium ions in aqueous solutions of electrolytes. *Yad. Magn. Rezonans*, 1971, 60 (*CA* 92 697).

H. Yokoyama, S. Chiba, and N. Ichinose. N.m.r. and magnetic properties of cobalt chromium sulphide ($CoCr_2S_4$) and its solid solution. *Ferrites, Proc. Internat. Conf. 1970*, 1971, 611 (*CA* 92 710).

S. Albert, H. S. Gutowsky, and J. A. Ripmeester. Spin–lattice relaxation time study of molecular motion and phase transitions in the tetramethylammonium halides. *U.S. Nat. Tech. Inform. Serv., AD Rep.*, 1971, 22 (*CA* 92 721).

V. K. Voronov. Paramagnetic shifts in complexes of nickel with some derivatives of the five-membered N-heterocyles. *Izvest. Sibirsk. Otdel. Akad. Nauk S.S.S.R., Ser. khim. Nauk*, 1971, 62 (*CA* 92 723).

Yu. N. Molin, R. Z. Sagdeev, E. V. Dvornikuv, and V. A. Grigor'ev. N.m.r. of 'other' nuclei in paramagnetic complexes. *Paramagn. Rezonans 1944—1969, Vses. Yubileinaya Konf. 1969*, 1971, 246 (*CA* 92 734).

Z. Kecki. Proton magnetic resonance of electrolyte solutions in methanol and water. *Paramagn. Rezonans 1944—1969, Vses. Yubileinaya Konf. 1969*, 1971, 234 (*CA* 92 742).

R. S. Borden. Fluorine-19 n.m.r. study of some bromofluorotitanate complexes and hydrolysis of some titanium tetrafluoride complexes. *U.S. Nat. Tech. Inform. Serv., AD Rep.*, 1971, 40 (*CA* 92 756).

F. I. Bashirov, Yu. L. Popov, K. S. Saikin, and R. A. Dautov. Apparatus for studying nuclear magnetic relaxation in the laboratory and rotating frames. *Prib. Tekh. Eksp.*, 1971, 137 (*CA* 92 917).

V. R. Nagibarov. Inertial spin–lattice relaxation. *Paramagn. Rezonans 1944—1969, Vses. Yubileinaya Konf., 1969*, 1971, 250 (*CA* 92 183).

V. V. Frolov. Theory of the effect of internal rotation on the spin-lattice relaxation rate. *Yad. Magn. Rezonans*, 1971, 18 (*CA* 92 682).

R. R. Fraser, T. Durst, M. R. McClory, R. Viau, and Y. Y. Wigfield. Comparison of the effects of solvents and Eu(dpm)$_3$ on proton shieldings in sulphoxides. *Internat. J. Sulfur Chem. (A)*, 1971, **1**, 133 (*CA* 98 577).

M. L. Filleux-Blanchard and A. Durand. Hindered rotation around the carbon–nitrogen bond in thioureas and selenoureas. *Compt., rend.* 1971, **273**, C, 1770 (*CA* 98 585).

N. N. Magdesieva, V. A. Vdovin, and N. M. Sergeev. Chemistry of benzo[b]selenophen: ^{1}H n.m.r. spectra of deuterio-derivatives of the benzo[b]selenophen series. *Khim. geterotsikl. Soedinenii*, 1971, 1382 (*CA* 98 610).

P. M. Borodin and Yu. I. Mitchenko. Complexing in organophosphorus compounds studied by n.m.r. *Paramagn. Rezonans 1944—1969, Vses. Yubileinaya Konf. 1969*, 1971, 180 (*CA* 98 923).

M. Noshiro and Y. Jitsugiri. Determination of boron in glass by fluorine-19 n.m.r. *Asahi Garasu Kenkyu Hokoku*, 1971, **21**, 47 (*CA* 103 200).

F. J. C. Rossotti. Hydration and structure of copper(II) complexes in solution. *Proc. Conf. Co-ordination Chem., 3rd*, 1971, 283 (*CA* 104 560).

P. M. Borodin, K. Z. Nguyen, and P. P. Andreev. Structure of aqueous solutions of Li_2SiF_6 studied by the fluorine-19 n.m.r. method. *Yad. Magn. Rezonans*, 1971, 103 (*CA* 104 624).

R. F. Snider. Boltzmann equation and gaseous n.m.r. *Kinetic Equations, Papers of 1969 Meeting*, 1971, 129 (*CA* 105 656).

D. Demco and V. Ceausescu. N.m.r. lineshape governed by a time-dependent Hamiltonian. *Rev. Roumaine Phys.*, 1971, **16**, 1093 (*CA* 105 666).

A. P. Potemskaya, G. P. Aleeva, V. Z. Kuprii, and V. A. Lunenok-Brumakina. Structure and decomposition reactions of vanadium peroxide compounds. *Teor. i eksp. Khim.*, 1971, **7**, 757 (*CA* 106 089).

A. S. Tarasevich and Yu. P. Egorov. Determination of P=N bond order in phosphazo-drivatives by a phosphorus-31 n.m.r. method. *Teor. i eksp. Khim.*, 1971, **7**, 828 (*CA* 106 144).

E. I. Angerer and P. M. Borodin. Proton magnetic relaxation in alcohol solutions of antimony trichloride. *Yad. Magn. Rezonans*, 1971, 72 (*CA* 106 168).

Yu. S. Chernyshev and Yu. A. Ignat'ev. Frequency dependence of the spin–lattice relaxation time of benzene on a potassium chloride surface. *Yad. Magn. Rezonans*, 1971, 66 (*CA* 106 169).

P. M. Borodin and G. P. Kondratenkov. Antimony trichloride solutions in isoalipathic alcohols studied by an n.m.r. method. *Yad. Magn. Rezonans*, 1971, 77 (*CA* 106 170).

S. Maricic, M. Cervinka, G. Pifat, and J. Brnjas-Kraljevic. Proton magnetic relaxation studies on haemoproteins in crystals and in concentrated salt solutions. *European Biophys. Congr., Proc., 1st*, 1971, **1**, 77 (*CA* 109 400).

G. Pifat and S. Maricic. Proton magnetic relaxation study of human ferrihaemoglobin and horse myoglobin in dilute salt solutions. *European Biophys. Congr., Proc., 1st*, 1971, **1**, 81 (*CA* 109 401).

Y. Kurimura, E. Tsuchida, and M. Kaneko. Preparations and properties of some water-soluble cobalt(III)–poly-(4-vinylpyridine) complexes. *J. Polymer Sci., Part A-1, Polymer Chem.*, 1971, **9**, 3511 (*CA* 113 592).

S. Tanaka, S. Toda, J. Saito, T. Mitsuishi, C. Nagata, K. Kanohta, S. Hashimoto, Y. Shimizu, and H. Kitazawa. Graphic representation of carbon-13 n.m.r. chemical shifts: master chart. *Bunseki Kagaku*, 1971, **20**, 1573 (*CA* 119 147).

A. A. Tyshchenko, K. H. A. Aslanov, V. B. Leont'ev, and A. S. Sadykov. Correlation between interaction constants in n.m.r. and e.p.r. spectra. *Doklady Akad. Nauk Usb. S.S.S.R.*, 1971, **28**, 37 (*CA* 119 204).

V. A. Atsarkin, M. E. Zhabotinskii, A. F. Mefed, S. K. Morshnev, and M. I. Podak, Role of a spin–spin reservoir in paramagnetic resonance and in the dynamic polarization of nuclei. *Paramagn. Rezonans 1944—1969, Vses. Yubileinaya Konf. 1969*, 1971, 38 (*CA* 119 596).

P. M. Borodin. Kinematic relativism in n.m.r. *Yad. Magn. Rezonans*, 1971, 5 (*CA* 119 636).

L. L. Buishvili. Effect of saturation of the magnetic chromium-53 nuclei resonance on the magnetic aluminium-27 resonance in ruby. *Izvest. V.U.Z., Radiofiz.*, 1971, **14**, 1364 (*CA* 119 637).

H. Lechert and H. J. Hennig. Influence of hydrogen sulphide (adsorption) on the behaviour of proton and sodium-23 resonances in faujasite-type zeolites. *Z. phys. Chem. (Frankfurt)*, 1971, **76**, 319 (*CA* 119 655).

T. G. Pinter. Crystalline electric fields in some thulium–aluminium compounds. *Report*, 1971, 126 (*CA* 131 710).

P. M. Borodin and M. I. Volodicheva. Effect of ions of normal aliphatic alcohols in the $SbCl_3$–ROH system on a chemical shift in an n.m.r. proton signal. *Yad. Magn. Rezonans*, 1971, 38 (*CA* 132 210).

J. W. Neely. Oxygen-17 n.m.r. studies of the first hydration sphere of diamagnetic metal ions in aqueous solution. *Report*, 1971, 72 (*CA* 132 258).

H. Pfeifer. Spin echoes: means for high-resolution n.m.r. in solid bodies. *Wiss. Z. Karl-Marx-Univ. Leipzig., Math.-Naturwiss. Reihe*, 1971, **20**, 549 (*CA* 133 532).

V. D. Doroshev, N. M. Kovtun, E. E. Solov'ev, A. Y. A. Chervonenkis, and A. A. Shemyakov. N.m.r. study of spin reorientation in thulium ferrate single crystals. *Pis'ma Zhur. Eksp. i teor. Fiz.*, 1971, **14**, 501 (*CA* 133 900).

N. S. Biradar and M. A. Punjar. Linkage isomerism in cobalt(III) amine complexes. *Proc. Chem. Symp., 2nd*, 1971, **1**, 209 (*CA* 133 916).

A. Vertes and F. Parak. Relation between the spin relaxation and other properties of paramagnetic iron(III) salts. *Kem. Kozlem.*, 1971, **36**, 429 (*CA* 133 871).

D. F. Gaines and J. Borlin. Internal exchange in new Group III metalloborane derivatives: $Me_2AlB_3H_8$ and $Me_2GaB_3H_8$. *U.S. Nat. Tech. Inform. Serv., AD Rep.*, 1971, 15 (*CA* 127 059).

Z. Michalska and Z. Lasocki. Determination of relative acidities of some substituted phenylsilanetriols by n.m.r. and i.r. spectroscopy. *Bull. Acad. polon. Sci., Sér. Sci. chim.*, 1971, **19**, 757 (*CA* 139 817).

L. I. Zakharkin, B. A. Kvasov, and V. N. Lebedev. Synthesis and study of fluorine-19 n.m.r. spectra of 1-fluoromethyl-*o*-carbaboranes and 2-substituted 1-fluoromethyl-*o*-carbaboranes. *Zhur. obshchei Khim.*, 1971, **41**, 2694 (*CA* 140 936).

P. Salvador Salvador. Applications of n.m.r. in mineralogy. *Bol. Geol. Minero*, 1971, **82**, 543 (*CA* 143 295).

W. Mueller-Warmuth and F. Kraemer. N.m.r. studies on motional processes in glasses. *Internat. Congr. Glass, Sci. Tech. Comm., 9th*, 1971, **1**, 303 (*CA* 144 296).

G. K. N. Reddy and E. G. Leelamani. Carbonyl hydrides of iridium with tertiary arsines. *Proc. Chem. Symp., 2nd*, 1971, **1**, 247 (*CA* 146 924).

N. K. Skvortsov, A. V. Dogadina, G. F. Tereshchenko, N. V. Morkovin, B. I. Ionin, and A. A. Petrov. Protonation of phosphine oxides studied by proton and phosphorus-31 n.m.r. *Zhur. obshchei Khim.*, 1971, **41**, 2807 (*CA* 152 761).

N. N. Magdesieva and V. A. Vdovin. Chemistry of benzo[6-*b*]selenophen: derivatives of selenoindoxyl. *Khim. geterotsikl. Soedinenii*, 1971, 1640 (*CA* 153 554).

C. Rhee and P. J. Bray. N.m.r. studies of the structure of caesium borate glasses and crystalline compounds. *Phys. and Chem. Glasses*, 1971, **12**, 165 (*CA* 157 564).

C. Rhee and P. J. Bray. Effect of ionic motion on the caesium-133 n.m.r. in glassy and crystalline caesium borates. *Phys. and Chem. Glasses*, 1971, **12**, 156 (*CA* 157 565).

R. D. Green and S. P. Sinha. Lanthanide shift reagents: acetylacetone complex of ytterbium(III) ion. *Spectroscopy Letters*, 1971, **4**, 411 (*CA* 160 521).

K. Sugimoto, A. Mizobuchi, K. Matuda, and T. Minamisono. *Hyperfine Interactions of Excited Nuclei, Proc. Conf.*, 1971, **1**, 167 (*CA* 160 569).

D. Spanjaard, R. A. Fox, I. R. Williams, and N. J. Stone. Nuclear spin–lattice relaxation measurements below 0.1 K by nuclear orientation. *Hyperfine Interactions of Excited Nuclei, Prof. Conf.*, 1971, **1**, 345 (*CA* 160 570).

H. S. Gutowsky and D. F. S. Natusch. Influence of paramagnetic species on the internuclear Overhauser effect. *U.S. Nat. Tech. Inform. Serv., AD Rep.*, 1971, 55 (*CA* 160 582).

Chemical Abstracts, 1972, Volume 77.

R. T. Ogata and H. M. McConnell. Binding of a spin-labelled triphosphate to haemaglobin. *Cold Spring Harbor Symp. Quant. Biol.*, 1971, **36**, 325 (*CA* 2158).

A. G. Redfield and R. K. Gupta, Pulsed n.m.r. study of the structure of cytochrome. *Cold Spring Harbor Symp. Quant. Biol.*, 1971, **36**, 405 (*CA* 2159).

J. Reuben. Gadolinium(III) as a paramagnetic probe for magnetic resonance studies of biological macromolecules. *Proc. Rare Earth Res. Cong., 9th*, 1971, **2**, 514 (*CA* 2462).

A. N. Nesmeyanov, O. V. Nogina, V. A. Dubovitskii, B. A. Kvasov, P. V. Petrovskii, and N. A. Lazareva. Transmission of the electronic effects of ligands in mono- and bis-cyclopentadienyl derivatives of titanium in various solvents. *Izvest. Akad. Nauk S.S.S.R., Ser. khim.*, 1971, 2729 (*CA* 4724).

Yu. S. Stark. Nuclear magnetic resonance. *Metody Ispyt., Kontr. Issled. Mashinostroit. Mater.*, 1971, **1**, 512 (*CA* 11 534).

S. P. Sinha and R. D. Green. N.m.r. studies of lanthanide(III) complexes: high-field shifts in complexes of 1,10-phenanthroline. *Spectroscopy Letters*, 1971, **4**, 399 (*CA* 12 077).

B. Matic and A. Brumnic. Time-mode system for measuring nuclear resonance. *Automatika*, 1971, **12**, 381 (*CA* 12 185).

D. G. Davis, S. Charache, and C. Ho. N.m.r. studies of haemoglobins. *Genet., Funct., Phys. Stud. Haemoglobins, Proc. Inter-Amer. 1969*, 1971, 280 (*CA* 15 792).

V. V. Kosovtsev, T. N. Timofeeva, B. I. Ionin, and V. N. Chistokletov. ^1H n.m.r. spectra of alkenylphosphines and derivatives of alkenylphosphonous acid. *Zhur. obschei Khim.*, 1971, **41**, 2638 (*CA* 18 805).

L. I. Zakharkin, V. N. Kalinin, E. G. Rys, and B. A. Kvasov. Electron effect of the 1-(1,8)-dicarba-*closo*-undecaborane(11) group 1-(1,8-$B_9C_2H_{10}$). *Izvest. Akad. Nauk S.S.S.R., Ser. khim.*, 1971, 2570 (*CA* 19 011).

A. M. Bondar and O. P. Revokatov. N.m.r. study of borates. *Eksp. Issled. Mineraloobrazov., Mater. Vses. Soveshch. Eksp. Tekh. 1968*, 1971, 143 (*CA* 22 769).

K. Arnold, G. Klose, and P. Herrmann. Calculation of time-sharing n.m.r. spectra with the help of an analog computer. *Wiss. Z. Karl-Marx-Univ. Leipzig., Math.-Naturwiss. Reihe*, 1971, 20, 573 (*CA* 27 085).

B. Sheard, R. G. Shulman, and T. Yamane. High-resolution n.m.r. studies of sperm whale cyanoferrimyoglobin in water solutions. *Genet., Funct., Phys. Stud. Haemoglobins, Proc. Inter-Amer. 1969*, 1971, 296 (*CA* 30 532).

O. P. Mchedlov-Petrosyan, A. V. Usherov-Marshak, A. G. Brekhunets, and V. V. Mank. Mineral formation during the hydration of tricalcium aluminate and tricalcium silicate studied by proton magnetic resonance methods. *Eksp. Issled. Mineraloobrazov., Mater. Vses. Soveshch. Eksp. Tekh. 1968*, 1971, 137 (*CA* 38 603).

D. Deininger, D. Fruede, G. Geschke, H. Pfeifer, F. Prysyborowski, W. Schirmer, and H. Stach. Results of studying proton magnetic resonance in zeolites. *Adsorbenty, Ikh Poluch., Svoistva Primen., Tr. Vses. Soveshch. 1969*, 1971, 113 (*CA* 39 555).

V. M. Sarnatskii, G. L. Antokol'skii, and V. A. Shutilov. Acoustic saturation of nonequidistant nuclear levels with a mixed mechanism of spin–lattice relaxation. *Yad. Magn. Rezonans*, 1971, 50 (*CA* 41 086).

A. A. Antipin, A. N. Katyshev, and I. N. Kurkin. Spin-lattice relaxation of rare-earth ions in single crystals of scheelite and fluorite structure. *Paramegn. Rezonans*, 1971, 3 (*CA* 40 724).

J. L. Kolopus, D. Kline, A. Chatelain, and R. A. Weeks. Magnetic resonance properties of lunar samples, mostly Apollo 12. *Proc. Lunar Sci. Conf., 2nd*, 1971, 3, 2501 (*CA* 51 250).

J. C. Soyfer, P. Audibert, and N. Giacchero. Spectral (i.r. and n.m.r.) characteristics of methoxyethylmercuric chloride. *Bull. Soc. Pharm. Marseille*, 1971, 20, 77 (*CA* 54 460).

F. Udo. New features in a n.m.r. detection system for measuring polarization of highly polarized substances. *Report*, 1971, 397 (*CA* 54 602).

D. Beckert, D. Michel, and H. Pfeifer. N.m.r. study of sorbed molecules. *Izvest. Fiz. Inst. ANEB (At. Nauchnoeksp. Baza), Bulg. Akad. Nauk.*, 1971, 21, 293 (*CA* 79 835).

G. E. Peterson, A. Carnevale, and H. W. Verleur. Gallium-69 n.m.r. as a sensitive probe of perfection in gallium phosphide crystals. *Mater. Res. Bull.*, 1971, 6, 51 (*CA* 80 669).

A. M. Evdokimov, V. O. Reikhsfel'd, and O. F. Bezrukov. Effect of nonspecific interactions on the chemical shift of hydrogen-1 n.m.r. of organosilanes in solutions. *Kremniiorg, Mater., Tr. Soveshch. Khim. Prakt. Primen. Kremniiorg. 1969*, 1971, 33 (*CA* 81 786).

K. Muthukrishnan and J. Ramakrishna. N.m.r. study of internal motions in zinc fluorosilicate hexahydrate. *J. Indian Inst. Sci.*, 1971, 53, 327 (*CA* 81 859).

D. R. Beuerman. Use of aliphatic α-hydroxy-oximes for the separation of metal ions. *Report*, 1971, 112 (*CA* 83 161).

O. L. Knunyants, V. I. Georgiev, I. V. Galakhov, L. I. Ragulin, and A. A. Neimysheva. The *p–d* bonding in phosphoryl and thiophosphoryl groups of organophosphorus compounds and electron shielding of the phosphorus atom nucleus. *Doklady Akad. Nauk S.S.S.R.*, 1971, 201, 862 (*CA* 87 259).

B. I. Obmoin, N. K. Moroz, D. S. Mirinskii, and I. A. Belitskii. High-pressure bomb for studying n.m.r. *Eksp. Issled. Mineral.*, 1971, 188 (*CA* 95 267).

I. Todo and T. Kado. Apparatus for the measurement of n.m.r. relaxation times by the pulse techniques. *Res. Inst. Appl. Electr.*, 1971, 23, 57 (*CA* 95 274).

B. Gostisa-Mihelcic and Z. Veksli. Proton magnetic resonance study of an allophane found at the Vonji Do Deposit, Macedonia. *Trav. Com. Int. Ftude Bauxites, Oxydes, Hydroxydes Alum.*, 1971, 7, 41 (*CA* 103 895).

A. L. Derzhanski and K. Kotev. Autogenerator apparatus for measuring longitudinal relaxation time during n.m.r. *Izvest. Fiz. Inst. ANEB (At. Nauchnoeksp. Baza), Bulg. Akad. Nauk.*, 1971, 21, 315 (*CA* 107 615).

M. Cohn, J. S. Leigh jun., and G. H. Reed. Mapping active sites of phosphoryl-transferring enzymes by magnetic resonance methods. *Cold Spring Harbor Symp. Quant. Biol.*, 1971, 36, 533 (*CA* 110 898).

S. I. Papko. Determination of trace elements and their comparative characteristics. *Fiz.-Khim. Metody Issled. Anal. Biol. Ob'ektov Nekot. Tekh. Mater.*, 1971, 3 (*CA* 111 130).

G. Kamai, N. A. Chadaeva, I. I. Saidashev, and N. K. Tazeeva. N.m.r. spectra, configuration, and conformation of cyclic esters of arsenic(III) acids. *Vop. Stereokhim.*, 1971, **1**, 127 (*CA* 113 730).

V. P. Khan, R. N. Pletnex, A. P. Stepanov, and I. A. Dmitriev. N.m.r. of boron-11 in beryllium-borosilicate glasses. *Trudy Ural. Politekh. Inst.*, 1971, **193**, 99 (*CA* 117 433).

Z. Luz. Ion solvation in solution studied by n.m.r. spectroscopy. *Israel J. Chem.*, 1971, **9**, 293 (*CA* 118 906).

H. Graenicher. N.m.r. investigations of ice crystals: proton magnetic resonance in ice. *Pulsed Magn. Opt. Resonance, Proc. Ampere Int. Summer Sch. 2, 1972*, 1971, 223 (*CA* 120 068).

N. E. Aimbinder, B. F. Amirkhanov, and A. N. Osipenko. Treatment of experimental data in radiospectroscopy of solids. *Trudy Estestvennonauch. Inst. Perm. Univ.*, 1971, **12**, 67 (*CA* 120 535).

B. Matinenas, K. Eringis, and R. Dagys. Spectra of free ions of transition metals and their complexes. *Teor. Elekron. Obolochek At. Mol., Dokl. Mezhdunar. Simp., 1969*, 1971, 294 (*CA* 145 655).

K. Wuethrich, R. G. Shulman, T. Yamane, and S. Ogawa. Studies of structure–function correlations in haemoglobin by n.m.r. *Genet., Funct., Phys. Stud. Hemoglobins, Proc. Inter-Amer. Symp. Hemoglobins, 1st, 1969*, 1971, 273 (*CA* 161 014).

2
Nuclear Quadrupole Resonance Spectroscopy

BY J. H. CARPENTER

1 Introduction

In this chapter, papers dealing with the nuclear quadrupole resonance (n.q.r.) spectra of solid-state species are reported. These include studies in which a small magnetic field is applied (that is, the quadrupole interaction is much greater than the interaction with the magnetic field), but not n.m.r. studies of quadrupolar nuclei, which are reported in the chapter on n.m.r. In addition, reference should be made to the chapter on microwave spectroscopy for information on quadrupole coupling constants in gaseous molecules.

The number of papers reported on this subject, although still small, is somewhat greater than last year, and although the halogens, especially chlorine, continue to be the main species studied there is a rather more widely distributed selection of nuclei reported this year, in addition to an increase in the number of papers dealing with such nuclei as ^{14}N, ^{75}As, ^{121}Sb, and ^{123}Sb. One reason for this welcome diversification is the increasing use of double-resonance techniques – in which one transition is n.q.r. while the other may be n.q.r., n.m.r., or e.s.r. The use of such techniques is particularly useful for the study of transitions which occur at low frequencies and hence are too weak to be easily observed directly.

Chihara and Nakamura have written one chapter of the volume [1] concerned with magnetic resonance in a new series of reviews covering the whole field of physical chemistry; this chapter deals with n.q.r. in general.

One volume of a periodical is devoted to articles on various n.q.r. topics; these include crystal field effects in n.q.r.,[2] the determination of nuclear quadrupole coupling constants from microwave spectroscopy,[3] a review of nitrogen n.q.r.,[4] which deals with both organic and inorganic compounds, and the theoretical interpretation of n.q.r.[5]

A theoretical study [6] of ^{14}N, ^{17}O, ^{33}S, and ^{73}Ge quadrupole coupling constants in some diatomic radicals and diamagnetic molecules showed

---

[1] 'MTP International Review of Science: Physical Chemistry', Series One, Vol. 4: 'Magnetic Resonance', ed. C. A. McDowell, Butterworths, London, 1972.
[2] A. Weiss, *Fortschr. Chem. Forsch.*, 1972, **30**, 1.
[3] W. Zeil, *Fortschr. Chem. Forsch.*, 1972, **30**, 103.
[4] L. Guibe, *Fortschr. Chem. Forsch.*, 1972, **30**, 77.
[5] E. A. C. Lucken, *Fortschr. Chem. Forsch.*, 1972, **30**, 155.
[6] P. Machmer, *Z. Naturforsch.*, 1972, **27b**, 1123.

that the π-bond order p_π can be expressed in terms of $\xi = |(eQq)_{mol}/(eQq)_{atom}|$ as:

$$p_\pi = a_0 - 4\xi + 4\xi^2$$

Resonances are reported for twenty-six nuclides of eighteen elements; in order of increasing atomic number and mass these are: ^{14}N; ^{17}O; ^{23}Na; ^{27}Al; ^{35}Cl; ^{37}Cl; ^{39}K; ^{40}K; ^{41}K; ^{55}Mn; ^{59}Co; ^{63}Cu; ^{65}Cu; ^{69}Ga; ^{71}Ga; ^{75}As; ^{79}Br; ^{81}Br; ^{85}Rb; ^{87}Rb; ^{93}Nb; ^{121}Sb; ^{123}Sb; ^{127}I; ^{181}Ta; and ^{209}Bi. They are reported according to their vertical group in the Periodic Table, with main group elements first, then transition metals.

2 Instrumentation and Techniques

By using a group-theoretical approach to n.q.r., Boyle[7] was able to discuss the splitting of nuclear spin states without prior wave-mechanical calculations. N.q.r. moments are compared with molecular quadrupole moments and selection rules are deduced.

A new method of determining the elements of the electric field gradient tensor for nuclei with half-integral spin from the Zeeman splitting of the n.q.r. spectrum has been published.[8] The method is applied to the chlorine nuclei in $Ba(ClO_3)_2, H_2O$, in which there are two sets of nuclei with the same values of eQq and η but different orientation of the electric field gradient tensor.

A method whereby the n.q.r. frequency at 0 K and the rotary lattice mode wavenumber are obtained using the Bayer–Kushida theory, by direct numerical analysis of the temperature dependence of the n.q.r. frequency, has been described.[9]

The description of an apparatus in which a digital procedure was used indicated that the system was stable, and the signal accumulation which was possible gave an improvement in sensitivity.[10] This technique was particularly useful for rapid scanning.

The use of a pulsed n.q.r. spectrometer for high-frequency measurements was discussed with particular emphasis on the possibility of determining relaxation times and observing broad lines.[11] Examples of its application to several heavy nuclei were given.

A description of a pulsed double-resonance spectrometer operating in the 10—400 MHz region has also been published[12] and the use of such a spectrometer in the detection of multiplet spectra was discussed[13] and applied to the ^{121}Sb n.q.r. resonances in $2SbCl_3, C_6H_7OH$.

[7] L. L. Boyle, *Internat. J. Quantum Chem.*, 1972, **6**, 313.
[8] S. Sengupta, R. Roy, and A. K. Saha, *J. Phys. Soc. Japan*, 1972, **32**, 1078.
[9] J. A. S. Smith and F. P. Temme, *Mol. Phys.*, 1972, **24**, 441.
[10] N. Watanabe and E. Niki, *Chem. Letters*, 1972, 899.
[11] B. G. Ignatov, A. L. Aleksandrov, I. A. Ekimovskikh, E. V. Bryukhova, B. N. Pavlov, and G. K. Semin, *Zavod. Lab.*, 1972, **38**, 432; B. N. Pavlov, *ibid.*, p. 1007; B. N. Pavlov and V. M. Prokopenko, *Pribory i tekhn. Eksp.*, 1972, 133.
[12] A. D. Gordeev, V. S. Grechishkin, V. A. Shishkin, and E. M. Shishkin, *Pribory i tekhn. Eksp.*, 1972, 234.
[13] V. S. Grechishkin, E. M. Shishkin, V. A. Shishkin, and I. A. Kyuntsel, *Optika i Spektroskopiya*, 1972, **32**, 715 (*Optics and Spectroscopy*, 1972, **32**, 377).

3 Main Group Elements

Group I (Sodium-23, Potassium-39, Potassium-40, Potassium-41, Rubidium-85, and Rubidium-87).—The alkali metals have not been extensively studied by n.q.r. because of their low resonance frequencies and hence weak signals. Two studies have been reported in which a double-resonance method is used to increase the sensitivity. A spin-echo technique is used to observe the n.q.r. of alkali-metal nuclei in some bromates;[14] the application of a pulse at the alkali-metal resonance frequency destroys the spin-echo signal of a ^{79}Br or ^{81}Br nucleus. Another method involving the spin-locked free-decay signal of the bromine nucleus appeared to be less useful. Resonant frequencies are reported for various transitions of ^{23}Na, ^{39}K, ^{85}Rb, and ^{87}Rb at room temperature and 77 K.

In another study a steady-state n.q.r.–n.q.r. double-resonance technique was used.[15] The potassium n.q.r. transition in KClO$_3$ was pumped and the ^{35}Cl transition was used as a monitor to detect the change in absorption. By use of this method transitions have been observed from ^{39}K (93% natural abundance), ^{40}K (0.012%), and ^{41}K (7%) without isotopic enrichment. The steady-state technique appears to have advantages over a pulsed method; the data are gathered in $\sim T_1/T_2$ of the time required for the pulsed method, which not only makes it shorter but also eases the stringent conditions on apparatus stability if T_1 is long.

The n.q.r. frequency of ^{23}Na in NaNO$_3$ has been studied as a function of temperature.[16] From 4.2 to 125 K, a positive $(\partial \nu / \partial T)_p$ is observed. This unusual behaviour is a consequence of the anisotropic thermal expansion of the tetragonal crystal and the particular angle of $\sim 54.7°$ between the c-axis and the Na—O vector. From 125 to 308 K the more usual decrease in resonance frequency occurs, and there is no evidence for a phase transition near 243 K, as had been previously suggested on the basis of dielectric constant and resistivity measurements.

Group III (Aluminium-27, Gallium-69, and Gallium-71).—The ^{27}Al n.q.r. frequency, as well as the magnetic relaxation time, in (La,Gd)Al$_2$ alloys containing 2%, 5%, and 10% Gd has been studied in zero and small magnetic fields.[17] For less than 5% Gd, eQq is unaffected, but relaxation times are affected; at 10% both are affected. The results are consistent with the onset of magnetic order in the Gd random-spin system.

When gallium is allowed to crystallize from the liquid phase in silicone oil dispersions, both the usual orthorhombic α-gallium and the metastable β-gallium are formed. Resonances due to ^{69}Ga and ^{71}Ga in the latter (as well as in the former which has previously been studied) were observed.[18]

---

[14] J. Okada and R. Kado, *J. Phys. Soc. Japan*, 1972, **32**, 1412.
[15] E. P. Jones and S. R. Hartmann, *Phys. Rev. (B)*, 1972, **6**, 757.
[16] G. J. D'Alessio and T. A. Scott, *J. Chem. Phys.*, 1972, **56**, 3724.
[17] D. E. MacLaughlin and M. Daugherty, *Phys. Rev. (B)*, 1972, **6**, 2502.
[18] S. L. Segel, R. D. Heyding, and E. F. W. Seymour, *Phys. Rev. Letters*, 1972, **28**, 970.

These occur at 6.25 MHz for ^{69}Ga and 3.91 MHz for ^{71}Ga. The temperature dependence between 77 and 256.9 K, at which temperature β-gallium melts, gave $\nu^{-1}(\partial\nu/\partial T)_p = 0.002$ K^{-1}, which is about twenty times that observed for the α-phase.

Group V (Nitrogen-14, Arsenic-75, Antimony-121, Antimony-123, and Bismuth-209).—The Stark effect on the ^{14}N n.q.r. in an asymmetric field gradient has been analysed theoretically and applied to various organic compounds.[19]

A review of the advantages and disadvantages of ^{14}N n.q.r. (in conjunction with ^{14}N and ^{15}N n.m.r. and ESCA) for the elucidation of the structure of nitrogen-containing compounds, compared with i.r. or ^{1}H n.m.r. techniques, shows the considerable potentiality of the technique in this field.[20]

The use of an n.m.r.–n.q.r. double-resonance method, in which the sample is transferred periodically from an area of high magnetic flux to one of low magnetic flux where it is irradiated at the n.q.r. frequency, has been described. The power absorbed by the quadrupolar nucleus is transferred to a proton in the magnetic field and can be observed by plotting the amplitude of the sampled free induction decay against the r.f. frequency.[21] This technique is very sensitive, particularly for ^{14}N nuclei directly bound to protons, and its use here is described for several molecules of biochemical interest.

A similar ^{1}H n.m.r.–^{14}N n.q.r. double-resonance technique is used to study the quadrupole coupling of ^{14}N in $(NH_4)_2SO_4$ in its paraelectric phase.[22] Rotation of the crystal enabled eQq and η to be determined. Four non-equivalent nitrogen atoms (corresponding to two chemically different nuclei, each in two crystallographically inequivalent sites) occurred. The two chemically different sets had mean values of eQq of 117 and 91 kHz and η of 0.648 and 0.67 respectively, implying that the NH_4^+ tetrahedra are highly distorted and that the reorientational motion is far from isotropic.

The ^{14}N n.q.r. frequencies of various four-co-ordinate nitrogen compounds containing $\overset{+}{N}$—O, $\overset{+}{N}$—N, $\overset{+}{N}$—N, and $\overset{+}{N}$—C bonds have been analysed using the Townes–Dailey method.[23] A plot of eQq against x_p (the Pauling electronegativity of the substituent) gave a straight line, which was attributed to a polarization of the σ-bond which was linear with x_p. The nuclear relaxation of the ^{14}N n.q.r. of N_2H_4 has been studied in the range 4.2—170 K by the saturation recovery method.[24] At higher temperatures the activation energy for the motion was 2.9 kcal mol^{-1}, which agrees well

[19] J. L. Colot, *J. Mol. Structure*, 1972, **11**, 475; *ibid.*, p. 483; *ibid.*, 1972, **13**, 129.
[20] H. G. Fitsky, D. Wendisch, and R. Holm, *Angew. Chem.*, 1972, **84**, 1037 (*Angew. Chem. Internat. Edn.*, 1972, **11**, 979).
[21] D. T. Edmonds and P. A. Speight, *J. Magn. Resonance*, 1972, **6**, 265.
[22] R. Blinc, M. Mali, R. Osredkr, A. Prelesnik, J. Seliger, and I. Zupancic, *Chem. Phys. Letters*, 1972, **14**, 49.
[23] R. A. Marino and T. Oja, *J. Chem. Phys.*, 1972, **56**, 5453.
[24] Y. Abe, N. Okubo, and S. Kojima, *J. Phys. Soc. Japan*, 1972, **32**, 1444.

with the microwave and i.r. data, but below 100 K the activation energy had the unusually low value of 0.35 kcal mol^{-1}, suggesting that a low-temperature mode of molecular motion was becoming dominant. Resonances from several organic derivatives of hydrazine, $R^1R^2NNH_2$, as well as various hydrazides, $RCONHNH_2$ or $R^1CONHNHCOR^2$, have been reported.[24a] In the hydrazines the results suggested that the hydrogen-bonding ability of the amino nitrogen exceeded that of the substituted nitrogen, and decreased in the order N_2H_4 > alkylhydrazine > arylhydrazine. In the hydrazides, extensive delocalization of the charge in the π-orbital of the substituted nitrogen was inferred.

The compound $Li(N_2H_5)SO_4$ is a protonic semiconductor. Two ^{14}N n.q.r. lines due to the NH_2 group were observed [25] between 4.2 and 360 K; eQq decreased with increasing temperature from 5580 kHz at 4.2 K, whereas η increased from 0.899 at 4.2 K to a maximum at 250 K and then decreased again. This behaviour was interpreted in terms of various motions of the $N_2H_5^+$ ion: below 150 K, oscillation about the N—N axis occurred, with an average angle of 18° between the lone pair and the z-axis of the field gradient. Above 150 K, exchange of the NH_2 protons across the bisector of the HNH angle occurred in addition, and above 320 K large-amplitude motions about the N—N axis predominate.

The temperature dependence of the ^{14}N n.q.r. relaxation time T_1 in $NaNO_2$ showed three minima.[26] The sharp minimum at 35 K is due to tunnelling, the broad one at 77 K is due to an internal molecular motion over a barrier of about 0.6 kcal mol^{-1}, and a flat minimum from about 180 K upwards corresponds to a change in crystal structure previously reported at 178 K.

Six lines were observed in the ^{14}N n.q.r. spectrum of $P(NMe_2)_3$, whereas for $PO(NMe_2)_3$ only four were seen.[27] The line separation decreased with increasing temperature but the lines did not coalesce below the m.p. (200 K), implying that there were three crystallographically inequivalent nitrogen atoms in the P^{III} compound and two in the P^V compound. There was no suggestion of any d_π–p_π bonding.

When crystals of ferroelectric CsH_2AsO_4 and RbH_2AsO_4 were partially deuteriated, new lines appeared in the ^{75}As n.q.r. spectrum which were much broader than those of the undeuteriated compounds.[28] The shift in frequency of the central component was approximately proportional to the deuterium content of the sample in both cases. In a 1 : 1 solid solution of CsH_2AsO_4 and RbH_2AsO_4 with no deuteriation, a weakly structured line of considerable width appeared at a frequency intermediate between those of the pure compounds. The results were explained in terms of inter-molecular hydrogen bonds.

[24a] E. G. Sauer and P. J. Bray, *J. Chem. Phys.*, 1972, **56**, 820; *ibid.*, p. 2788.
[25] R. N. Hastings and T. Oja, *J. Chem. Phys.*, 1972, **57**, 2139.
[26] Y. Abe, Y. Ohneda, S. Abe, and S. Kojima, *J. Phys. Soc. Japan*, 1972, **33**, 864.
[27] L. Krause and M. A. Whitehead, *Mol. Phys.*, 1972, **23**, 547.
[28] G. K. Semin, L. S. Golovchenko, and A. P. Zhukov, *Zhur. strukt. Khim.*, 1972, **13**, 153 (*J. Struct. Chem.*, 1972, **13**, 139).

The [75]As n.q.r. spectra of 1-arsacarba-*closo*-dodecaborane(11), with the carbon atom in the 2, 7, or 12 position, have been reported;[29] the structure is shown in (1). The resonance frequencies and linewidths are given in

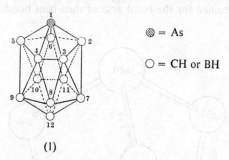

(1)

Table 1; the high value of the frequency for the *ortho*-compound indicates a considerable change in the electron distribution at the arsenic nucleus when a carbon atom is substituted for boron in its first co-ordination sphere, whereas the other two compounds have resonances which are close to each other. The much larger linewidth in the *ortho*- and *meta*-compounds is a consequence of the statistical disordering of the molecules in the solid which is possible; this cannot occur for the *para*-isomer.

Table 1   [75]As n.q.r. frequencies and linewidths in 1-*arsacarba*-closo-dodecaborane(11)

Carbon position[a]	Frequency/MHz	Linewidth/MHz
2 (*ortho*)	97.3	2
7 (*meta*)	58.2	2
12 (*para*)	61.36	0.2

[a] For structure see text.

Several arsenic sulphides have been investigated using n.q.r. The first report of a n.q.r. resonance in a vitreous material has been published;[30] a line at 71.6 MHz, with a linewidth of 7 MHz, was observed in vitreous As_2S_3. This frequency is midway between those of the two lines observed in crystalline orpiment. The broadening is attributed to distortion of the pyramidal angle of the sulphur atoms around the arsenic atom, the observed linewidth being consistent with a distribution of apex angles of 2° about the mean position.

Both the α- and β-forms of synthetic crystals of As_4S_3 gave three resonances; at 77 K these were 64.87, 65.94, and 79.56 MHz for the α-form and 65.42, 67.16, and 79.65 MHz for the β-form.[31] In both forms the molecule has the C_{3v} symmetry shown in Figure 1; the difference is in

---

[29] E. V. Bryukhova, V. I. Kyskin, and L. I. Zakharkin, *Izvest. Akad. Nauk S.S.S.R., Ser. khim.*, 1972, 532.

[30] M. Rubinstein and P. C. Taylor, *Phys. Rev. Letters*, 1972, **29**, 119.

[31] T. J. Bastow, I. D. Campbell, and H. J. Whitfield, *Austral. J. Chem.*, 1972, **25**, 2291.

the packing in the crystal, which also causes the differences between corresponding frequencies in the two forms. A Townes–Dailey treatment showed an occupancy of ~1 for the sp^3 hybrid orbital of the apical arsenic; if this is also assumed for the basal arsenic then bent bonds are formed in the basal plane.

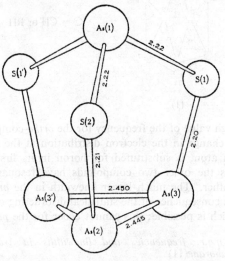

Figure 1 *The molecular structure of* As_4S_3. *Interatomic distances are in* Å (Reproduced by permission from *Austral. J. Chem.*, 1972, **25**, 2292)

A similar study of synthetic α- and β-As_4S_4 has been performed.[32] The α-form shows four close lines, implying a symmetry in the crystal of C_1, compared with D_{2d} in the gas phase. This spectrum is very similar to that of the naturally occurring α-form, realgar. The metastable β-form was made by heating α-As_4S_4 to 540 K *in vacuo* for 24 hours, then quenching to room temperature. Two lines were observed for this form, at 89.84 MHz and 90.54 MHz; this implied a higher symmetry (C_2) in the crystal than in the α-form. The far-i.r. spectra were also measured.

N.q.r. data at 77 K for ^{75}As, ^{79}Br, ^{81}Br, ^{121}Sb, ^{123}Sb, ^{127}I, and ^{209}Bi nuclei, including resonance frequencies, eQq, and, where applicable, η, have been tabulated[33] for the following compounds: As_2S_3, AsI_3, AsSI, AsSeI, Sb_2S_3, SbI_3, SbSI, SbSBr, Sb_2Se_3, $SbBr_3$, Bi_2S_3, BiI_3, and BiSI. The value of eQq for ^{127}I was less in the MSI compounds than in MI_3, with As < Sb < Bi, whereas η increased, implying an increase in intermolecular co-ordinate interaction on going from arsenic to bismuth. In the trisulphide and triselenide compounds, the two values of eQq imply two non-

---

[32] T. J. Bastow and H. J. Whitfield, *Solid State Comm.*, 1972, **11**, 1015.
[33] G. K. Semin, *Phys. Status Solidi (A)*, 1972, **11**, K61.

equivalent positions for the Group V element. The value of eQq was always higher for a sulphur compound than in the corresponding selenium compound. A more detailed study [34] of both ^{75}As and ^{127}I n.q.r. spectra of AsSI and AsSeI shows that the ^{75}As frequencies are close to those in As_2S_3 and As_2Se_3, and the ^{127}I frequencies are close to those in AsI_3. These imply that bridging bonds of the type $As-I\cdots As$ are much less strong than in the corresponding antimony and bismuth compounds.

The substituent transmission effect has been studied in some substituted triphenylarsine and triphenylstibine compounds;[35] the substituents Me, Cl, Br, F, MeO, or CF_3 were in the o-, m- or p-position. Resonance frequencies are reported for ^{35}Cl, ^{75}As, ^{81}Br, ^{121}Sb, and ^{123}Sb. In the arsenic compounds, frequency increases of up to 4% compared with Ph_3As were observed. No correlation was found between the ^{75}As n.q.r. frequency and the Hammett or Taft σ_i parameters, but for the *para*-compounds the frequency correlated well with Taft's σ_r^0 resonance parameter. In the antimony compounds, η for ^{121}Sb and ^{123}Sb was small, as expected, and the value of eQq correlated better with Hammett's σ function than with Taft's σ_i or σ_r^0 functions. This difference from arsenic was attributed to the larger size of the antimony atom which prevents it from interacting with the aromatic π-electron system as much as arsenic, and thus making $\sigma-\pi$ effects more important.

The effect of unsymmetrical substitution on ^{121}Sb and ^{123}Sb n.q.r. spectra was investigated [36] using Ph_2SbCl, $Ph_2Sb-C\equiv C-SbPh_2$, (*trans*-$ClCH=CH)_3Sb$, and Me_3Sb, as well as data on Ph_3Sb and $SbCl_3$. In the unsymmetrical compounds both eQq and η increased; the asymmetry parameter is higher ($\eta = 0.82$) in Ph_2SbCl than in $Ph_2Sb-C\equiv C-SbPh_2$, in which there are three $Sb-C$ bonds and hence less difference in orbital populations on the antimony atom. In these compounds the z-axis is probably not in the same direction as the lone pair.

Different molar proportions of Na_2S and Sb_2S_3 have been fused together and the ^{121}Sb and ^{123}Sb n.q.r. spectra of the resulting solids investigated using a pulsed system.[37] Between 0 and 33% Sb_2S_3, lines were observed due to Na_3SbS_3. No lines were seen between 40 and 42% Sb_2S_3, where X-ray studies indicate that $Na_6Sb_4S_9$ occurs with fast statistical interchange of sodium and antimony atoms. With 45—80% Sb_2S_3, lines due to $NaSbS_2$ (and, above 58%, due to Sb_2S_3 itself) were seen. Just below 40% Sb_2S_3, lines due to other species fell off sharply in intensity, implying another species in which there is also rapid interchange. The spectra observed are similar to those of the minerals miargyrite ($AgSbS_2$) and pirargyrite

---

[34] E. V. Bryukhova, A. P. Chernov, S. A. Demborskii, and G. K. Semin, *Zhur. strukt. Khim.*, 1972, **13**, 528.

[35] T. B. Brill and G. G. Long, *Inorg. Chem.*, 1972, **11**, 225.

[36] G. K. Semin, E. V. Bryukhova, T. A. Babushkina, V. I. Svergun, A. E. Borisov, and N. V. Novikova, *Izvest. Akad. Nauk S.S.S.R., Ser. khim.*, 1972, 1183.

[37] U. A. Buslaev, E. A. Kravchenko, I. A. Kuz'min, V. B. Lazarev, and A. V. Salov, *Zhur. neorg. Khim.*, 1972, **16**, 3367.

(Ag_3SbS_3), implying that the sodium compounds also contain pyramidal SbS_3 groups.

By studying the Zeeman effect on the ^{35}Cl and ^{123}Sb resonances in $SbCl_3$,PhEt, Okuda et al.[38] were able to measure eQq and η for each nucleus. Assuming that the z-axis of the electric field gradient at the chlorine nucleus was parallel to the Sb—Cl bond, they found Cl—Sb—Cl bond angles of 90.4, 93.1, and 96.0°. These are very similar to those in $SbCl_3$ itself and in $2SbCl_3,C_{10}H_8$, showing that there is little distortion of the $SbCl_3$ pyramid on complexing with PhEt.

The ^{35}Cl, ^{37}Cl, ^{121}Sb, and ^{123}Sb n.q.r. resonances have been measured[39] in $ClCN,SbCl_5$, and compared with those in $MeCN,SbCl_5$. The lowering of the resonant frequencies of both chlorine and antimony in the MeCN complex shows the different donor characteristics of the two ligands which is a consequence of the increase of the electronic charge on the chlorine atom and a lowering of eQq for antimony. The two ^{35}Cl frequencies (intensities 4 : 1) in the MeCN complex, which has C_{4v} symmetry, are replaced by three (intensities 2 : 2 : 1) in the ClCN complex, implying a lowering of the molecular symmetry.

In the compounds (p-XC_6H_4)$_3$Bi, where X is Cl or Br, the ^{35}Cl, ^{79}Br, and ^{209}Bi n.q.r. spectra were observed.[40] From the ^{209}Bi resonances were derived: $eQq = 686.6$ MHz, $\eta = 0.04$ for the chlorine compound and $eQq = 673.0$ MHz, $\eta = 0.06$ for the bromine compound. A single line for each transition implied that in both compounds only one type of bismuth site was present, whereas multiplets in the halogen resonance regions showed that the halogens on each ring were in different environments, in agreement with X-ray studies. The mean C—Bi—C angle of 95.5(5)° is the same as in Ph_3Bi.

The temperature and pressure dependence of eQq and η for ^{209}Bi in $BiCl_3$ was investigated near room temperature.[41] The results were analysed using the Bayer–Kushida theory but agreement between observed and calculated results was poor.

Group VI (Oxygen-17).—By using a n.m.r.-n.q.r. double-resonance technique, Edmonds and Zussman[42] were able to observe ^{17}O resonances in a 17% ^{17}O-enriched sample of ice. Values of $eQq = 6.66(10)$ MHz and $\eta = 0.935(10)$ were obtained. If these are compared with values obtained for free molecules of $HD^{17}O$ by microwave spectroscopy [$eQq = 10.17(7)$ MHz, $\eta = 0.75(1)$], the important role of hydrogen-bonding in ice can readily be seen.

Group VII (Chlorine-35, Chlorine-37, Bromine-79, Bromine-81, and Iodine-127).—An e.s.r.-n.q.r. double-resonance study on a single crystal of

[38] T. Okuda, Y. Furukawa, and H. Negita, *Bull. Chem. Soc. Japan*, 1972, **45**, 2940.
[39] M. Burgard and E. A. C. Lucken, *J. Mol. Structure*, 1972, **14**, 397.
[40] I. A. Kyuntsel and V. A. Shishkin, *Zhur. strukt. Khim.*, 1972, **13**, 530.
[41] G. C. Gillies and R. J. C. Brown, *Canad. J. Chem.*, 1972, **50**, 2586.
[42] D. T. Edmonds and A. Zussman, *Phys. Letters (A)*, 1972, **41**, 167.

irradiated $NaClO_3$ showed the magnetic dipole interaction between the electron spin and the ^{35}Cl quadrupolar nucleus.[43]

The ^{35}Cl n.q.r. frequencies of $PCl_4^+BCl_4^-$, $SCl_3^+SbCl_6^-$, and $SeOCl_2$, together with those of eleven organic compounds, have been published.[44] In $SeOCl_2$, eleven resonances were observed between 29.1 and 31.1 MHz, with a mean frequency of 30.41 MHz, which compares with 31.986 MHz in $SOCl_2$. In $SCl_3^+SbCl_6^-$, the resonances at 22.3—25.8 MHz were attributed to the $SbCl_6^-$ ion by comparison with other $SbCl_6^-$ compounds, whereas two at 42.185 and 42.932 MHz are due to SCl_3^+. This is the first reported ^{35}Cl resonance for this ion. Lines at 31.65—32.19 MHz in $PCl_4^+BCl_4^-$ were assigned to PCl_4^+ whereas those between 20.38 and 21.54 MHz were assigned to the BCl_4^- ion. There is some evidence for a phase transition in this compound.

A comparison of the mean ^{35}Cl n.q.r. frequencies in $Cl_3SiNHSiCl_3$ (19.66 MHz) with those in $Cl_3SiOSiCl_3$ (19.89 MHz), Si_2Cl_6 (19.29 MHz), and $SiCl_4$ (20.39 MHz) showed a correlation between the increasing frequency and decreasing ionic character of the Si—Cl bond.[45] However, $SiCl_3F$ has a resonance frequency of 19.753 MHz; this low value may be due to contributions from structures such as (2); the Si—Cl bond length is also shorter in $SiCl_3F$ than in $SiCl_4$, in support of this suggestion.

$$\overset{\underset{|}{Cl}}{\underset{\underset{|}{Cl}}{F-Si=\overset{+}{Cl}}}$$

(2)

The ^{35}Cl n.q.r. spectra of a number of alkyl- and aryl-substituted derivatives of $SnCl_4$ have been reported.[46] The frequency of the ^{35}Cl resonance decreased with increasing alkyl substitution; a plot of this frequency against $\Delta E_Q(^{119}Sn)$ from Mössbauer studies gave a straight line for the di- and tri-chlorides, but $SnCl_4$ and the monohalides were not correlated in this way. The crystallographic splitting of the resonance lines implied some intermolecular co-ordination in the solid phase. From the spectrum of $Cl_2Sn(MeCOO)_2$ it was deduced that the structure was (3) or, more probably, (4).

A single ^{35}Cl resonance was observed in the cubic phase of K_2SnCl_6 above 262 K, and three occurred below this temperature.[47] Below 140 K the same spin–lattice relaxation rate ($\propto T^2$) occurs for each line, implying a Raman spin–phonon interaction. Above 200 K, the rate is given by $\exp(-E_a/kT)$, suggesting a rate-activated mechanism resulting from

---

[43] C. Dimitripoulos, A. Salvi, and Y. Ayant, *Compt. rend.*, 1972, **275**, B, 153.
[44] R. M. Hart, M. A. Whitehead, and H. L. Krause, *J. Chem. Phys.*, 1972, **56**, 3038.
[45] K. Hamada and E. A. Robinson, *Bull. Chem. Soc. Japan*, 1972, **45**, 2219.
[46] Yu. K. Maksyutin, V. V. Khrapov, L. S. Mel'nichenko, G. E. Semin, N. N. Zemlyanski, and K. A. Kocheskov, *Izvest. Akad. Nauk S.S.S.R., Ser. khim.*, 1972, 602.
[47] K. R. Jeffrey, *J. Magn. Resonance*, 1972, **7**, 184.

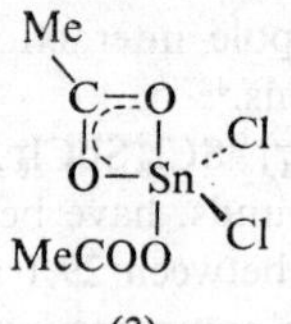

```
       Me
        \
         C=O                          Cl
         ‖  |           Cl            |   O
         O—Sn⟨    ⟩     |         ⟋   |  ⟋
         |    Cl    Me—C⟨   O  ⟩Sn⟨   ⟩C—Me
        MeCOO            ⟍  O ⟋  |  ⟍ O
           (3)                   Cl
                                 (4)
```

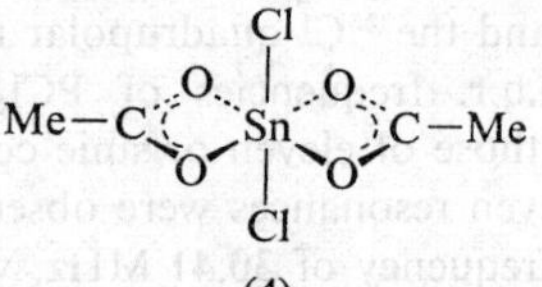

hindered rotation of the $SnCl_6^{2-}$ octahedra with an activation energy of 11.6 kcal mol^{-1}. There is some evidence for a softening of the rotary lattice mode near the phase transition.

The close similarity of the ^{35}Cl n.q.r. frequencies in the compounds p-ClC$_6$H$_4$XMe$_3$, where X is C, Si, Ge, or Sn, suggested that the C—Cl bond is little affected by the XMe$_3$ group.[48]

The P—Cl bond lengths have been correlated with the ^{35}Cl n.q.r. frequencies in a number of chlorocyclophosphazenes (NPCl$_2$)$_n$, and the ^{35}Cl n.q.r. frequencies used as an aid for structural assignment in some cyclodiphosphazanes.[49] In N$_3$P$_3$Cl$_6$, four signals implied a chair conformation, and their closeness implied that the P—Cl bond lengths were equal. In N$_3$P$_3$Cl$_4$Ph$_2$, on the other hand, significantly different P—Cl bond lengths were inferred from the greater spread of frequencies. In N$_3$P$_3$Cl$_2$Ph$_4$ and N$_3$P$_3$Cl$_3$(NMe$_2$)$_3$ the observed lines were attributed to two crystallographically inequivalent molecules. Three different P—Cl bond lengths were deduced for N$_3$P$_3$Cl$_3$(NMe$_2$)$_3$. On combining these data with those for other molecules the relation between ^{35}Cl n.q.r. frequency and P—Cl bond length shown in Figure 2 was obtained. The cyclodiphosphazanes (Cl$_3$PNR)$_2$ (R = Me, Et, or Ph), whose structure is shown in (5), gave axial chlorine resonances of 25.8—26.3 MHz and equatorial

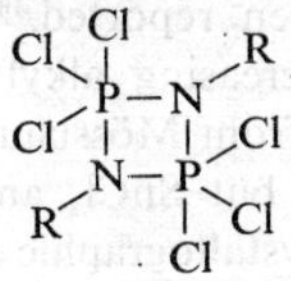

```
              Cl   R
           Cl  \   |
             \  P—N
           Cl⟋ |    |⟍Cl
                N—P
                |   | ⟍Cl
                R   Cl
                (5)
```

chlorine resonances of 30.0—30.8 MHz. Since these compounds have five- rather than four-co-ordinate phosphorus, the plot of $\nu(^{35}Cl)$ against P—Cl bond length did not fall on the same straight line as for the compounds above.

The ^{35}Cl n.q.r. frequencies and, where not previously reported, the X-ray structures, of compounds (6), (7), and (8) have been published.[50] The

---

[48] E. A. C. Lucken, S. Ardjomand, Y. Limouzin, and J. C. Maire, *J. Organometallic Chem.*, 1972, **37**, 247.

[49] R. Keat, A. L. Porte, D. A. Tong, and R. A. Shaw, *J.C.S. Dalton*, 1972, 1648.

[50] T. S. Cameron, C. Y. Cheng, T. Demir, K. D. Howlett, R. Keat, A. L. Porte, C. K. Prout, and R. A. Shaw, *Angew. Chem.*, 1972, **84**, 530 (*Angew. Chem. Internat. Edn.*, 1972, **11**, 510).

P—Cl bond lengths and P—Cl n.q.r. frequencies are shown in Figure 2, and are also included in Figure 2. It can be seen that the frequency of the chlorine found to the phosphoryl group is higher than that of the two chlorines of the groupings of the same frequencies, in fact, in accord with that calculated with the previously-corrected ab initio calculations.

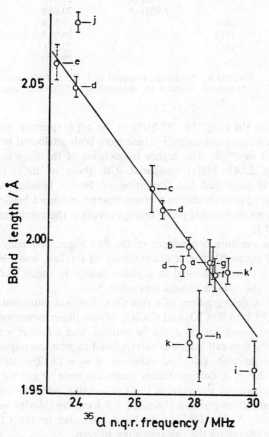

(6) a; X = O
    b; X = S

(7)

(8) a; X¹ = X² = O
    b; X¹ = X² = S
    c; X¹ = O, X² = S

Figure 2 *Plot of P—Cl bond lengths against* ^{35}Cl *n.q.r. frequencies at 77 K for cyclophosphazenes and other phosphorus compounds:* a, $N_3P_3Cl_6$; b, $N_3P_3Cl_4Ph_2$; c, $N_3P_3Cl_2Ph_4$; d, e, $N_3P_3Cl_3(NMe_2)_3$; f, $N_4P_4Cl_8$ (*K-form*); g, $N_4P_4Cl_8$ (*T-form*); h, $N_5P_5Cl_{10}$; i, $NPCl_2(NSOCl)_2$; j, *compound* (6a); k, k', *compound* (7)
(Adapted from *J.C.S. Dalton*, 1972, 1649)

P—Cl bond lengths and ^{35}Cl n.q.r. frequencies are shown in Table 2 and are also included in Figure 2. It can be seen that the frequency of the chlorine bound to the phosphoryl group is lower than that of the thiophosphoryl group, and the lower frequency in (6a) compared with (7) correlates with the P—Cl length as discussed above.[49]

Table 2 P—Cl *bond lengths and* ^{35}Cl *n.q.r. frequencies at 77 K of some phosphorus compounds*

Compound[a]	r(P—Cl)/Å	Mean ν(^{35}Cl)/MHz
(6a)	2.069(3)	23.977[b]
(6b)	2.057(5)	c
(7)	1.968(4)[d]	27.813[d]
	1.991(4)[e]	29.084[e]
(8a)		27.938
(8b)		28.913
(8c)		27.571[d]
		29.310[e]

[a] See text. [b] At 293 K. No signals observed at 77 K. [c] No signals observed at 77 or 293 K. [d] Chlorine attached to phosphoryl. [e] Chlorine attached to thiophosphoryl.

A search in the range 15—55 MHz in the n.q.r. spectrum of BiCl$_3$,H$_2$O revealed only two resonances.[51] These were both attributed to ^{35}Cl rather than to ^{37}Cl or ^{209}Bi. The higher frequencies of the lines in BiCl$_3$,H$_2$O (22.264 and 23.476 MHz) compared with those of BiCl$_3$ (15.955 and 19.155 MHz) suggested the formation of Bi—O bonds. A reversible discontinuity, probably due to a phase change, occurred between 219 and 245 K, and an irreversible phase change occurs if the compound is cooled to below 77 K.

When the pressure dependence of the ^{35}Cl n.q.r. frequency in NaClO$_3$ was studied up to 20 kbar, a discontinuity in $(\partial \nu/\partial p)_T$ was observed near 11 kbar. This was attributed to a phase change in which, because ν was continuous, the volume change was small.[52]

The electric field gradient of a free ClO$_3^-$ ion was estimated[53] from the values of ν(^{35}Cl) in RbClO$_3$ and CsClO$_3$. From these, assuming a value for Q_{Cl} and that η was zero, q could be derived, and a plot of q against a^{-3}, where a is the unit cell side, was extrapolated to zero, corresponding to an infinitely large unit cell. A value of $q = -19.12 \times 10^{24}$ cm^{-3} was obtained. Use of a Townes–Dailey treatment gave an orbital population of 1.3 electrons in the three Cl—O σ-bonding orbitals (assuming no double bonding), implying a charge of $+1.1$ on the chlorine and -0.7 on each oxygen. A 30—40% double-bond character in the Cl—O bonds reduced the charge on the chlorine atom to zero.

[51] Dinesh and M. T. Rogers, *J. Inorg. Nuclear Chem.*, 1972, **34**, 3941.
[52] M. S. Vijaya, J. Ramakrishna, A. S. Reshamwala, and A. Jayaraman, *Materials Res. Bull.*, 1972, **7**, 615.
[53] P. U. Sakellardis and C. A. Kagarakis, *J. Mol. Structure*, 1972, **14**, 127.

In another study, several chlorates (some hydrated) of Group I and Group II elements were investigated.[54] Two signals were observed for the Mg and Ca compounds, but one for the others. The measured frequencies for $Mg(ClO_3)_2,2H_2O$, $Mg(ClO_3)_2,6H_2O$, $Ba(ClO_3)_2$, and $Ba(ClO_3)_2,2H_2O$ differed from previously reported values. With the exception of $KClO_3$, both Mg compounds, and $Ba(ClO_3)_2,2H_2O$, the frequency decreased with Z, which was explained in terms of solid-state effects related to the size of the cation. The discrepancies in the Mg and Ba compounds were due to the effect of the water on the electric field gradient and the low electronic density of K^+ explained that anomaly.

The mean ^{35}Cl frequency in various compounds containing sulphur and chlorine were reported[55] and are shown in Table 3. In SCl_2 and S_2Cl_2, slow

Table 3 Mean ^{35}Cl n.q.r. frequencies in some sulphur compounds

Compound	ν/MHz
SCl_2[a]	39.51
SCl_2[b]	39.2
S_2Cl_2	35.78
Cl_3CSCl	39.91[c]
	39.54[d]
SO_2Cl_2	37.70
SO_2ClF	39.36
$SOCl_2$	31.99

[a] Cooled slowly. [b] Cooled rapidly. [c] CCl_3 resonances. [d] SCl resonance.

cooling to 77 K gave a multiplet spectrum with linewidths of 10—20 kHz, whereas a disordered state of SCl_2, obtained by rapid cooling, gave linewidths of 1—2 MHz.

A study[56] of the ^{35}Cl n.q.r. spectrum of $HICl_4,4H_2O$ showed four lines between 21 and 23 MHz, whose mean value of 22.2 MHz is close to that of other compounds containing the tetrachloroiodate ion. The solid-state structure, obtained by an X-ray study reported in the same paper, is shown in Figure 3. The space group of this centrosymmetric structure is $P2_1/b$, and there are two inequivalent chlorine sites. The n.q.r. data, however, imply four different chlorine sites, and it is deduced that the proton between O(1) and O(1') is in a double-minimum potential; this correlates with the long O(1)—O(1') distance of 2.64 Å. A disordered structure in which the proton stays in one potential well but there is statistical disorder over the lattice, and an ordered model in which the proton jumps from one minimum to the other at a rate slow compared with the 5 kHz linewidth, were suggested as equally good alternatives for explaining the observed spectra. No phase change was observed in the range 130—280 K. In another study,[57] on $NaICl_4,2H_2O$, three of the four ^{35}Cl n.q.r. lines observed showed a

---

[54] P. U. Sakellardis and C. A. Kagarakis, J. Mol. Structure, 1972, 12, 99.
[55] I. P. Biryukov and A. Y. Deich, Zhur. fiz. Khim., 1972, 46, 2385.
[56] R. J. Bateman and L. R. Bateman, J. Amer. Chem. Soc., 1972, 94, 1130.
[57] A. Sasane, D. Nakamura, and M. Kubo, J. Magn. Resonance, 1972, 8, 179.

negative $(\partial \nu/\partial T)_p$, whereas the fourth exhibited a positive gradient between 77 and 365 K, with no observable phase transitions. This behaviour paralleled that of $NaAuCl_4,2H_2O$, suggesting that they are isomorphous, and the X-ray powder photos of the two species indicate that they are isostructural. Thus the ICl_4^- ion is probably square planar in this compound,

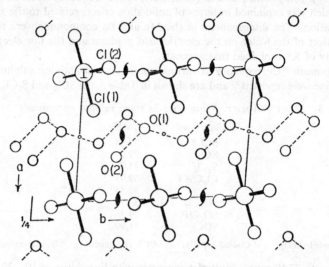

Figure 3 *The unit cell of* $HICl_4,4H_2O$ *viewed parallel to the crystallographic c-axis* (Reproduced by permission from *J. Amer. Chem. Soc.*, 1972, **94**, 1133)

with one chlorine atom involved in weak hydrogen-bonding as in the gold compound.

At 77 K, $Ti(\pi-C_5H_5)Cl_3$ gave two ^{35}Cl signals at 8.042 and 8.145 MHz (intensities 2 : 1) and lines at 11.764, 11.884, 12.028, and 12.208 MHz were observed in the $Ti(\pi-C_5H_5)_2Cl_2$ spectrum.[58] The increase over the mean value in $TiCl_4$ (6.05 MHz) paralleled a similar trend in the bromine compounds and was attributed to participation of the p_π orbital of the halogen in the Ti—Hal bond.

The two ^{37}Cl resonances observed in Cs_2VOCl_4 between 170 and 341 K, which correspond to two sets of equivalent chlorine atoms as found in the crystal structure, were split into doublets at 143 K, suggesting that between 143 and 170 K a phase transition occurs.[59] The frequency data were interpreted to give U_p (the number of unbalanced p-electrons at the chlorine) = 0.177 at 306 K, which agreed well with U_p = 0.167 estimated from e.s.r. data. No resonances could be found between 2 and 15 MHz for Cs_2MoOCl_5 or $CsCrOCl_5$ at 300 or 133 K.

[58] A. N. Nesmeyanov, A. N. Bryukhova, E. V. Bryukhova, I. M. Alymov, O. V. Nogina, and V. A. Dubovitskii, *Izvest. Akad. Nauk S.S.S.R., Ser. khim.*, 1972, 1671.
[59] R. D. Bereman, *Inorg. Chem.*, 1972, **11**, 642.

The ^{35}Cl and ^{37}Cl spectra of both NbCl$_5$,ClCN and TaCl$_5$,ClCN showed a single resonance for each nucleus;[60] for ^{35}Cl these were at 41.930 MHz for the niobium complex and 41.945 MHz for the tantalum complex at 77 K, both having a negative temperature gradient. Since free ClCN gives a resonance at 41.7 MHz, these were attributed to the ClCN chlorine atom. The change in frequency on complex formation is similar to that observed [39] in SbCl$_5$,ClCN.

A linear relation between the ^{35}Cl n.q.r. frequency in salts of aryl-substituted PhFeCl(π-C$_5$H$_5$)$^+$ with BF$_4^-$ or PF$_6^-$ and that for the corresponding PhCl compound was observed.[61] This implied a similar coefficient of σ_i for the substituent in the iron compound as in the chlorobenzene. A random orientation of the phenyl and cyclopentadienyl rings, deduced from diffraction photographs, was manifested as a broadening of the n.q.r. resonances.

The ^{35}Cl n.q.r. frequencies of several compounds R$_3$MCl$_6$ (M = Ir or Rh) and R$_2$MCl$_6$ (M = Rh, Ru, Tc, or Mn) have been studied [62] and are shown in Table 4. The low site-symmetry suggested by the multiplets in

Table 4 ^{35}Cl *n.q.r. frequencies in some hexachlorometallate ions (at 243 K unless otherwise stated)*

Compound	$\nu(^{35}$Cl$)$/MHz					
K$_3$IrCl$_6$,H$_2$O	16.773	16.968	17.533	18.144	18.289	18.863
K$_3$IrCl$_6$	16.648^a	17.378^b	17.508^b	17.733	17.874	18.502
(NH$_4$)$_3$IrCl$_6$,H$_2$O	17.539	17.672	17.780	17.902	18.084	
K$_3$RhCl$_6$,H$_2$O	18.331	18.525	19.536			
Na$_3$RhCl$_6$,2H$_2$O	17.665					
Na$_3$RhCl$_6$,10H$_2$O	17.738					
Cs$_2$RhCl$_6$	21.869^a					
K$_2$RuCl$_6$	17.349^c					
K$_2$MnCl$_6$	18.816^d					
Cs$_2$MnCl$_6$	19.516^d					
K$_2$TcCl$_6$	14.192^d					

^{a} At 273 K. ^{b} At 222 K. ^{c} At 232 K. ^{d} At 298 K.

the Ir compounds and in K$_3$RhCl$_6$,H$_2$O was verified by the far-i.r. spectra, and an X-ray analysis of the Rh compound indicated six different Rh—Cl bond lengths in a RhCl$_6^{3-}$ octahedron in the range 2.302—2.366 Å. In this complex a phase change occurred at about 240 K, below which a single resonance only was observed. In K$_3$IrCl$_6$ three phase changes, one between 201 and 222 K, another between 222 and 243 K, and the third between 243 and 273 K, occurred. A comparison of the mean resonant frequencies (see Table 5) indicated that the frequency increased (i) with

[60] M. Burgard, J. MacCordick, and E. A. C. Lucken, *Inorg. Nuclear Chem. Letters*, 1972, **8**, 185.
[61] A. N. Nesmeyanov, G. K. Semin, T. L. Khopyanova, E. V. Bryukhova, N. A. Vol'kenau, and E. I. Sorotkin, *Doklady Akad. Nauk S.S.S.R.*, 1972, **202**, 854 (*Doklady Chem.*, 1972, **202**, 122).
[62] P. J. Cresswell, J. E. Fergusson, B. R. Penfold, and D. E. Scaife, *J.C.S. Dalton*, 1972, 254.

Table 5 Mean ^{35}Cl n.q.r. frequencies in $K_n^+MCl_6^{n-}$ compounds/MHz

d-electron configuration	Transition-metal series 1st	2nd	3rd		
d^3	Mn^{IV} 18.82	Tc^{IV}	14.19	Re^{IV}	13.89
d^4		Ru^{IV}	17.35	Os^{IV}	16.84
d^5		Rh^{IV}	21.1	Ir^{IV}	20.74
d^6		Pd^{IV}	26.55	Pt^{IV}	25.82
		Rh^{III}	18.76	Ir^{III}	17.65

increasing oxidation state, (ii) on going from third- to second- to first-row transition metals, and (iii) with increasing number of d-electrons. These trends are related to optical electronegativity of the metal ion and to decreasing π-bond formation as the number of d-electrons increases.

In K_2PtCl_6, K_2PdCl_6, and K_2IrCl_6 the temperature dependence of the ^{35}Cl n.q.r. frequency and the spin–lattice relaxation time have been investigated and the results extrapolated to absolute zero.[63] This gave values of ν_0 (the n.q.r. frequency at 0 K) and $\bar{\omega}$ (the rotary lattice frequency, which is temperature dependent). In the related compound K_2PtBr_6 (see below, ref. 68), $\bar{\omega}$ decreased with T as the phase transition at 169 K was approached from above. A similar behaviour was observed here as the temperature approached 0 K, implying that, although there was no actual phase change, an 'incipient' phase transition occurred at 0 K.

In complex chlorides of zinc and copper, the ^{35}Cl n.q.r. frequencies were grouped around 9 MHz, whereas for the bromides the corresponding ^{81}Br frequencies were in the 60 MHz region.[64] The n.q.r. data for the compounds studied are shown in Table 6. The presence of water in the crystal

Table 6 ^{35}Cl, ^{81}Br, and ^{127}I n.q.r. frequencies in some complex halides of copper and zinc at 298 K

Compound	$\nu(Hal)$/MHz	Compound	$\nu(Hal)$/MHz
Rb_3ZnCl_5	8.964	Tl_2ZnBr_4	62.525
Cs_3ZnCl_5	9.065	Tl_2ZnI_4	78.650
Cs_3ZnBr_5	60.520		80.863
Cs_3CuCl_5	9.966	$(Et_4N)ZnBr_3$,py	63.380
Rb_3ZnBr_5	56.064		65.165
	61.475		65.528
	67.200	β-$KZnBr_3,2H_2O$	63.243
$(NH_4)_3ZnBr_5$	55.117		66.640
	60.508		66.673
	61.196	$KZnI_3,2H_2O$	80.334
Cs_3ZnI_5	76.151		85.198
$ZnBr_2,2NH_3$	59.491		86.245
$ZnI_2,2NH_3$	77.542	$ZnBr_2,2H_2O$	62.337
Rb_2ZnI_4	70.416		70.215
	77.399		
	83.727		

[63] R. L. Armstrong and H. M. Van Driel, *Canad. J. Phys.*, **50**, 2048.
[64] W. J. Asker, D. E. Scaife, and J. A. Watts, *Austral. J. Chem.*, 1972, **25**, 2301.

increased the mean frequency and/or the spread in frequencies, probably by hydrogen-bonding. The spectra are discussed in terms of the crystal structures of the compounds. The single line due to Cs_3CuCl_5 was observed from a sample obtained by quenching the melt. When this was annealed above 370 K, disproportionation occurred into Cs_2CuCl_4 and CsCl.

A comparison of the values of eQq in the solid and gas phases for MeHgX compounds gave $(eQq)_{solid}/(eQq)_{gas}$ = 0.70, 0.69, and 0.51 for ^{35}Cl, ^{79}Br, and ^{127}I respectively.[65] This value in MeHgI is the lowest known to date; the large change in eQq is attributed to intermolecular co-ordinate interaction between mercury and the halogen atom, which correlated with the 0.2 Å increase in bond length in MeHgCl when going from gas to solid.

The 'soft sphere' model for n.q.r. has been tested[66] on the trichlorides MCl_3, where M is Ce, Pr, Nd, Sm, or Gd. The ^{35}Cl n.q.r. frequency increased smoothly from 4.3772(3) MHz in $CeCl_3$ to 5.3076(3) MHz in $GdCl_3$, and its pressure dependence, $v_0^{-1}(\partial v/\partial p)_T/10^{-7}$ m^2 N^{-1}, changed regularly from $-5.700(20)$ in $CeCl_3$ to $-3.934(16)$ in $GdCl_3$.

A Zeeman study of the ^{81}Br resonances in the complex Al_2Br_6,C_6H_6 enabled Okuda et al.[67] to estimate eQq and η for the three lines observed. That at 79.99 MHz, with η = 0.289(11), was assigned to the bridging bromine in the Al_2Br_6 dimer, whereas those at 90.10 and 91.18 MHz, with η = 0.086(10) and 0.060(9) respectively, were assigned to terminal bromine atoms. The terminal Br—Al—Br angles of 117.3(5)° agreed reasonably well with the X-ray value of 122(2)°, but the Al—Br—Al bridging angle of 95.5(2)° was much greater than the X-ray value of 87(2)°, suggesting a bent bond. The amount of charge on each bromine atom, estimated by the Townes–Dailey method, was virtually identical to that of the uncomplexed molecule.

The compound K_2PtBr_6 showed phase transitions at 78, 105, 137, 143, and 169 K, and the spin–lattice relaxation time of the ^{79}Br resonance was very sensitive to the last of these, when the lattice changed from cubic to tetragonal. A model involving the softening of the rotary lattice mode and a change in the unit cell without distortion of the $PtBr_6^{2-}$ octahedra was used to explain the observed behaviour.[68]

The resonance frequencies, eQq, and η for ^{127}I in $BaI_2,2H_2O$, which is probably isostructural with $BaBr_2,2H_2O$, have been reported.[69]

A number of mono- and di-iodo- B-substituted derivatives of ortho-, meta-, and para-carbaborane(12), as well as C-substituted p-$B_{10}C_2H_{10}I_2$, have been studied[70] and their ^{127}I n.q.r. frequencies observed. For the 2-, 9-, and 10-iodocarbaborane(12), as well as the C-substituted di-iodo-compound,

[65] E. V. Bryukhova and G. K. Semin, J.C.S. Chem. Comm., 1972, 1216.
[66] D. H. Current, C. L. Foiles, and E. H. Carlson, Phys. Rev. (B), 1972, 6, 737.
[67] T. Okuda, Y. Furukawa, and H. Negita, Bull. Chem. Soc. Japan, 1972, 45, 2245.
[68] H. M. Van Driel, A. Wiszniewska, B. M. Moores, and R. L. Armstrong, Phys. Rev. (B), 1972, 6, 1596.
[69] A. F. Volkov, Phys. Status Solidi (B), 1972, 50, K43.
[70] E. V. Bryukhova, U. V. Gol'tyapin, V. I. Stanko, and G. K. Semin, Zhur. strukt. Khim., 1971, 12, 1095.

eQq and η were determined, but for the B-substituted di-iodo-compounds only the $\pm\frac{1}{2} \leftrightarrow \pm\frac{3}{2}$ transition was observed. The asymmetry parameter η was less than 0.1 for all the mono-iodo-compounds. A linear relation exists between eQq for ^{127}I and the ^{35}Cl resonance frequencies of corresponding compounds:

$$eQq(^{127}\text{I})/\text{MHz} = -638.7 - 83.3\nu(^{35}\text{Cl})/\text{MHz}$$

with eQq increasing $ortho < meta < para$.

Two studies of the IO_3^- ion – a variable-temperature study [71] of KIO_3 and another of the Li, Na, NH_4, Rb, Cs, Mg, Zn, Cd, and Bi compounds [72] – indicated that the crystals had not got the perovskite structure but contained discrete covalently bound IO_3^- pyramidal ions which are slightly distorted. In the chromoiodates $MCrIO_6$ (M = K, Na, NH_4, Cs, or Mg) the asymmetry parameter η was higher,[72] implying structure (9) with one

$$\text{CrO}_3 \overset{O}{\diagdown} \text{IO}_2$$

(9)

I—O bond different from the others. By comparing the ^{35}Cl spectra of I_nAlCl_m compounds with those known to contain the $AlCl_4^-$ ion, Merryman et al.[73] were able to characterize $I_2Cl^+AlCl_4^-$, $I_3^+AlCl_4^-$, and $I_5^+AlCl_4^-$. Both ^{35}Cl and ^{127}I resonances were observed; the central iodine atom showed a regular decrease in resonant frequency in ICl_2^+ (457 MHz), I_2Cl^+ (420 MHz), and I_3^+ (308.6 MHz).

For additional resonances of Group VII elements, see the sections on Group V elements, copper-63 and copper-65, and niobium-93.

4 Transition Metals

Manganese-55.—A study of π-$C_5H_5Mn(CO)_2L$ compounds, where L was one of several ligands, showed that the ($\pm\frac{5}{2} \leftrightarrow \pm\frac{3}{2}$) frequency of ^{55}Mn, which occurred between 17.18 and 20.92 MHz, was closely related to the usually accepted trend in σ-donor ability of the ligand.[74] For some compounds where L was an olefin, the ($\pm\frac{3}{2} \leftrightarrow \pm\frac{1}{2}$) transition could also be measured; this gave η in the range 0.35—0.49, its value being related to the ring strain of the olefin.

Cobalt-59.—From the ^{59}Co transition frequencies, Watanabe [75] derived $eQq = 36.05$ MHz, $\eta = 0.268$ for cis-(Co en$_2$Cl$_2$)Cl and $eQq = 33.7$ MHz,

---

[71] D. F. Baisa, A. I. Barabash, V. P. Demyanko, G. A. Puchkovskaya, and I. S. Rez, *Ukrain. fiz. Zhur.*, 1972, **17**, 1346.

[72] T. G. Balicheva, V. S. Grechishkin, G. A. Petrova, and V. A. Shishkin, *Zhur. neorg. Khim.*, 1972, **17**, 605 (*Russ. J. Inorg. Chem.*, 1972, **17**, 318).

[73] D. J. Merryman, P. A. Edwards, J. D. Corbett, and R. E. McCarley, *J.C.S. Chem. Comm.*, 1972, 779.

[74] W. P. Anderson, T. B. Brill, A. R. Schoenberg, and C. W. Stanger, *J. Organometallic Chem.*, 1972, **44**, 161.

[75] I. Watanabe, *J. Chem. Phys.*, 1972, **57**, 3014.

$\eta = 0.178$ in cis-(Co en$_2$Cl$_2$)NO$_3$. The values of eQq are smaller than for the corresponding trans-compounds and are close to those found in Co(NH$_3$)$_5$Cl^{2+}. The large values of η are attributed to crystal field effects.

A theoretical study of the octahedral complexes of low-spin ^{59}Co^{III} showed that η was sensitive to perturbations which lowered the symmetry whereas eQq was not.[76] This was applied to the species Co(NH$_3$)$_5$CN^{2+} and Co(NH$_3$)$_4$Cl$_2^+$. Thus eQq reflects the grosser features of the structure and bonding whereas η is significantly affected by the symmetry-breaking perturbation.

Copper-63 and Copper-65.—Few resonances due to these nuclei have previously been reported. The ^{63}Cu and ^{65}Cu n.q.r. spectra of eight copper complexes of thiourea (tu), ethylenethiourea (etu), and NN'-dimethylthiourea (dmtu) have been reported,[77] as well as the ^{79}Br and ^{81}Br resonances in Cu(etu)$_2$Br. The values of the frequencies, between 16 MHz and 39 MHz, implied that the Cu—S bond was essentially covalent. Two crystallographically inequivalent sites were found in the tu complexes. From the frequencies of the tu and etu complexes with CuBr and CuCl, it was inferred that the Cu—Br and Cu—Cl bonds were predominantly ionic. The copper resonant frequencies are given in Table 7.

Table 7 ^{63}Cu n.q.r. resonances in some copper compounds

Compound	$\nu(^{63}$Cu)/MHz
Cu$_4$(tu)$_9$(NO$_3$)$_4$[a]	25.088
Cu(tu)$_2$Cl	22.115, 19.296
Cu(tu)$_2$Br	16.443, 16.181
Cu(etu)$_2$ClO$_4$[b]	22.881
{Cu(etu)$_4$}$_2$SO$_4$	31.562
Cu(etu)$_2$Cl	27.860
Cu(etu)$_2$Br	32.010
Cu(dmtu)$_3$Cl[c]	38.804
Cu$_4$(PMe$_3$)$_4$Cl$_4$[d]	e
Cu$_2$(PMe$_3$)$_4$Cl$_2$[f]	30.313, 30.267, 30.197
Cu(PMe$_3$)$_3$Cl[g]	30.280

[a] tu = thiourea.   [b] etu = ethylenethiourea.   [c] dmtu = NN'-dimethylthiourea.
[d] Structure (10).   [e] No resonance observed.   [f] Structure (11).   [g] Structure (12).

In addition to i.r. and ^{1}H n.m.r. spectra, Schmidbauer et al.[78] investigated the ^{63}Cu and ^{35}Cl n.q.r. spectra of compounds (10), (11), and (12). The ^{63}Cu resonances are given in Table 7.

Niobium-93.—Four lines were observed[79] in the ^{93}Nb n.q.r. spectrum of (NMe$_4$)$_2^+$(Nb$_6$Cl$_{12}$)$^{4+}$Cl$_6^-$, as expected for a nucleus with $I = 9/2$, implying that there is only one niobium site. The ^{35}Cl resonance at 12.556 MHz and

[76] M. G. Clark, Chem. Phys. Letters, 1972, **13**, 316.
[77] J. D. Graybeal and S. D. Ing, Inorg. Chem., 1972, **11**, 3104.
[78] H. Schmidbauer, J. Adlkofer, and K. Schwirten, Chem. Ber., 1972, **105**, 3382.
[79] P. A. Edwards, R. E. McCarley, and D. R. Torgeson, Inorg. Chem., 1972, **11**, 1185.

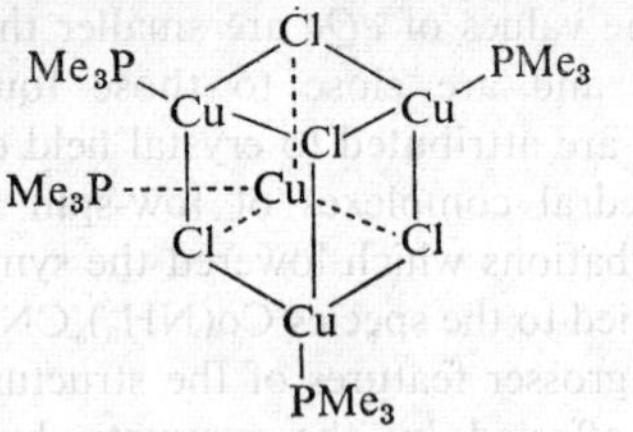

Me_3P—Cu—Cl—Cu—PMe_3 (with bridging Cl's and third Cu-PMe_3)

(10)

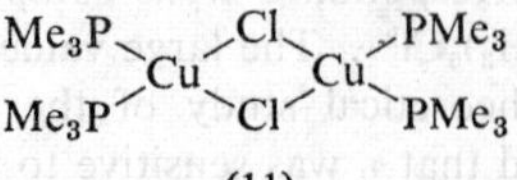

Me_3P\\Cu/Cl\\Cu/PMe_3
Me_3P/ \\Cl/ \\PMe_3

(11)

       Cl
       |
Me_3P—Cu—PMe_3
       |
      PMe_3

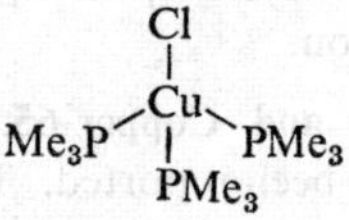

(12)

a corresponding ^{37}Cl resonance at 9.897 MHz were assigned to bridging chlorine atoms; the resonances for the terminal chlorines, which would be highly ionic, were probably too low to be observed. The values of $eQq = 133.48$ MHz, $\eta = 0.008$ for ^{93}Nb implied that 14 electrons were involved in metal–metal bonding.

Tantalum-181.—The first reported resonances for ^{181}Ta ($I = 7/2$) have been observed[80] in Ta_2Cl_{10} and Ta_2Br_{10}, and are shown in Table 8. The assignments are tentative until the ($\pm\frac{5}{2} \leftrightarrow \pm\frac{7}{2}$) transitions are observed,

Table 8 ^{181}Ta n.q.r. resonances in Ta_2Cl_{10} and Ta_2Br_{10}

Compound	Assignment	$\nu(^{181}Ta)$/MHz	eQq/MHz	η
Ta_2Cl_{10}	$\nu_1(\pm\frac{1}{2} \leftrightarrow \pm\frac{3}{2})$	177.6	1786	0.36
	$\nu_1(\pm\frac{1}{2} \leftrightarrow \pm\frac{3}{2})$	183.0	1795	0.38
	$\nu_2(\pm\frac{3}{2} \leftrightarrow \pm\frac{5}{2})$	241.4		
Ta_2Br_{10}	$\nu_1(\pm\frac{1}{2} \leftrightarrow \pm\frac{3}{2})$	156.7	1421	0.45
	$\nu_1(\pm\frac{1}{2} \leftrightarrow \pm\frac{3}{2})$	161.1	1436	0.46
	$\nu_2(\pm\frac{3}{2} \leftrightarrow \pm\frac{5}{2})$	190.0		

and were based on the corresponding resonances for the niobium compounds.

We wish to thank Mr J. D. Cooper for translating several of the Russian publications.

[80] P. A. Edwards and R. E. McCarley, *J.C.S. Chem. Comm.*, 1972, 845.

3
Microwave Spectroscopy

BY J. H. CARPENTER

1 Introduction

The papers reported in this chapter deal with high-resolution gas-phase spectra obtained by the use of radiation in the frequency range from 1 MHz to 1000 GHz. Papers which involve the use of strong magnetic fields for the study of free radicals are excluded, although the Zeeman microwave spectra of diamagnetic species are reported.

The use of the m.b.e.r. technique to obtain very high resolution continues to increase, and various double-resonance techniques, in which one or both of the radiation frequencies falls within the range above, are used as an aid to assignment, to improve sensitivity, or to observe microwave transitions in states unobservable at thermal equilibrium.

The number of papers reported is similar to that of last year. The main text is subdivided as before according to the number of atoms in the molecule, and within each section the ordering is according to the Group in the Periodic Table of the main atom or atoms of interest. Organometallic compounds are included, and some other carbon-containing compounds which were felt to be most suitably described as inorganic. Errors in parameters derived from the spectra are enclosed in parentheses and qualify the least-significant digits; thus 123.4(56) implies 123.4 ± 5.6.

One volume of a recent series of volumes in physical chemistry is devoted to spectroscopy;[1] chapters relevant to microwave spectroscopy include 'The Stark effect' by A. D. Buckingham, 'Phosphorescence-microwave multiple-resonance spectroscopy' by M. A. El-Sayad, 'Millimetre-wave spectroscopy' by G., M., and B. P. Winnewisser, and 'Molecules in Space' by L. E. Snyder.

A monograph[2] entitled 'Nonlinear Spectroscopy of Molecules' deals with double resonance in three- and four-level systems, two photon transitions and Lamb-dip (saturation) spectroscopy.

An introductory review, with particular reference to the analytical applications and potentialities of microwave spectroscopy, has been published.[3]

---

[1] 'MTP International Review of Science: Physical Chemistry', Series One, Vol. 3: 'Spectroscopy', ed. D. A. Ramsay, Butterworths, London, 1972.
[2] K. Shimoda and T. Shimizu, 'Nonlinear Spectroscopy of Molecules', Pergamon, Oxford, 1972.
[3] L. H. Scharpen and V. W. Laurie, *Analyt. Chem.*, 1972, **44**, 378R.

In another review,[3a] the application of microwave spectroscopy to the study of the conformation of small molecules is described and illustrated. While primarily dealing with organic compounds, this should be of interest to all chemists.

2 Instrumentation and Technique

A paper by Kirchhoff[4] describes the calculation and interpretation of centrifugal distortion constants in asymmetric tops, with particular reference to the statistical basis of the method.

A detailed theoretical discussion of radiofrequency–microwave double resonance includes the application of the method to the l-type doublets of carbonyl sulphide,[5] and another use is described by Schwarz and Dreizler,[6] who apply a static Stark field and cause transitions with $\Delta m = \pm 1$ between the components of one level while $\Delta m = 0$ transitions are observed using a swept microwave frequency.

The use of the inverse Lamb-dip technique in two photon transitions in NH_3 using infrared–microwave double resonance is reported;[7] this method promises to be useful for determining accurately the difference between the i.r. laser frequency and the molecular transition frequency.

By measuring the relative intensity of microwave absorption as a function of temperature, the barrier to internal rotation can be determined. This method is more generally applicable than those which measure the frequency of splitting of the lines. A Stark cell designed for such use, with particular emphasis on the minimization of reflection losses, has been described.[8]

A spectrometer is reported which has been designed to measure the shape of a microwave absorption line.[9] This can be used to observe the first to fifth harmonic components of the lineshape; by modulating at a frequency of f/n and detecting at a frequency of f, the nth harmonic is observed.

A microwave spectrometer has been constructed to study the microwave spectra of free radicals and other unstable species.[10] This works in the 20—160 GHz region, and is applied to the study of the NS, SO, and OH radicals.

The design and use of an absorption cell which consists of two Fabry–Pérot type interferometers at right angles are described by Lee and White.[11] This is for use in double-resonance experiments, and it is claimed to have better vacuum-pumping and adsorption characteristics than conventional waveguide cells.

[3a] E. B. Wilson, *Chem. Soc. Rev.*, 1972, **1**, 293.
[4] W. H. Kirchhoff, *J. Mol. Spectroscopy*, 1972, **41**, 333.
[5] P. Hoyng, H. A. Dijkermann, and G. Ruitenberg, *Physica*, 1972, **57**, 57.
[6] R. Schwarz and H. Dreizler, *Z. Naturforsch.*, 1972, **27a**, 708.
[7] S. M. Freund and T. Oka, *Appl. Phys. Letters*, 1972, **21**, 60.
[8] G. Ruitenberg, *J. Mol. Spectroscopy*, 1972, **42**, 161.
[9] R. P. Netterfield, R. W. Parsons, and J. A. Roberts, *J. Phys. (B)*, 1972, **5**, 146.
[10] C. Marlière, J. Burie, and J. L. Destombes, *Compt. rend.*, 1972, **275**, *B*, 315.
[11] M. C. Lee and W. F. White, *Rev. Sci. Instr.*, 1972, **43**, 638.

A commercial spectrometer which is primarily designed for analytical work has been described.[12] Particular attention has been paid to sample-handling facilities, the absorption cell, and the data-processing system.

A backward diode has been used as a sensitive detector.[13] A better signal-to-noise ratio was obtained at 10 GHz than for a conventional diode. The importance of low bias resistance to prevent self-biasing is stressed.

3 Diatomic Molecules

Selected molecular parameters determined from the microwave spectra of diatomic molecules are collected together in Table 1.

The m.b.e.r. spectra of the heteronuclear alkali-metal dimers ^{23}Na^{7}Li, ^{39}K^{7}Li, Rb^{7}Li, ^{39}K^{23}Na, Rb^{23}Na, and ^{133}Cs^{23}Na have been reported.[14] Analysis of the spectra gave values of μ_0^2/B_0 and eQq. The rotational constant B_0 was obtained from other studies or by using Badger's rule, and hence the dipole moments were estimated. From these the ionic character of the bond was deduced, giving values up to 27% for CsNa and 24% for RbLi. The quadrupole coupling constants were much smaller than, and of opposite sign to, those found in the alkali halides, and were related to the value of μ_0/er_0.

A number of studies of alkali halides have been reported. The hyperfine structures of LiCl[15] and LiBr[16] have been measured, using the m.b.e.r. technique, to higher precision than previous measurements. The chloride was studied in the $v = 0$ to $v = 3$ states, and eQq for ^{7}Li, ^{35}Cl, and ^{37}Cl were obtained, as well as the spin–spin and spin–rotation constants. In the bromide, the equivalent parameters were measured for the ground state; recent calculations of $\partial^2 V/\partial z^2$ were used to estimate the nuclear quadrupole moments Q, which were within 10% of tabulated nuclear moments.

In an m.b.e.r. study on ^{39}KF the dependence of the electric dipole moment on the vibrational anharmonicity, centrifugal distortion, vibration–rotation interaction, and electronic polarizability was investigated.[17] The Dunham coefficients were obtained with greater precision than previously published values. A similar study on ^{205}TlF was reported in the same paper.

The electric and magnetic properties of the molecule ^{87}RbF in the $J = 1$, $v = 0$ or $v = 1$ state were studied by m.b.e.r.[18] Parameters

---

[12] J. Cuthbert, E. J. Denney, C. Silk, R. Stratford, J. Farren, T. L. Jones, D. Pooley, R. K. Webster, and F. H. Wells, *J. Phys. (E)*, 1972, **5**, 698.
[13] P. Christen, A. Bauder, and H. H. Günthard, *Rev. Sci. Instr.*, 1972, **43**, 349.
[14] J. Graff, P. J. Dagdigian, and L. Wharton, *J. Chem. Phys.*, 1972, **57**, 710; P. J. Dagdigian and L. Wharton, *ibid.*, p. 1487.
[15] T. F. Gallagher, R. C. Hilborn, and N. F. Ramsey, *J. Chem. Phys.*, 1972, **56**, 5972.
[16] R. C. Hilborn, T. F. Gallagher, and N. F. Ramsey, *J. Chem. Phys.*, 1972, **56**, 855.
[17] H. Dijkermann, W. Flegel, G. Gräff, and B. Mönter, *Z. Naturforsch.*, 1972, **27a**, 100.
[18] J. Heitbaum and R. Schönwasser, *Z. Naturforsch.*, 1972, **27a**, 92.

Table 1 Structural parameters for diatomic molecules

Molecule	B_0/MHz	r_e/Å	μ_0/D	Nucleus	$(eQq)_0$/MHz	Ref.
^{7}Li^{23}Na			0.463(2)	^{7}Li	0.028(4)^a	14
				^{23}Na	−0.749(4)	
^{7}Li^{39}K			3.45(10)	^{7}Li	⩽0.009^a	14
				^{39}K	−1.029(2)	
^{7}Li^{85}Rb			4.00(10)	^{7}Li	0.030(13)^a	14
				^{85}Rb	−9.116(14)	
^{23}Na^{39}K			2.76(10)	^{23}Na	0.171(3)^a	14
				^{39}K	−0.718(2)	
^{23}Na^{85}Rb			3.1(3)	^{23}Na	<13^a	14
				^{85}Rb	<13^a	
^{23}Na^{133}Cs			4.75(20)	^{23}Na	<0.4^a	14
				^{133}Cs	<0.4^a	
^{7}LiCl				^{7}Li	0.250(10)	15
				^{35}Cl	−3.05951(15)	
				^{37}Cl	−2.4109(6)	
^{7}LiBr				^{7}Li	0.21104(10)	16
				^{79}Br	38.368104(36)	
				^{81}Br	32.050860(46)	
^{39}KF	8392.3125(8)^b		8.5926(8)	^{39}K	−7.93387(80)	17
^{87}RbF			8.5453(60)	^{87}Rb	−34.0352(10)	18
^{133}Cs^{35}Cl				^{35}Cl	1.83(15)^b	19
				^{133}Cs	⩽1.1^a	
^{133}Cs^{79}Br				^{79}Br	−6.79(15)^b	20
^{133}Cs^{127}I				^{127}I	−14.28(35)^b	21
				^{133}Cs	<1.0^a	
^{7}Li^{16}O^c			6.84(3)	^{7}Li	0.444(8)	22
Al^{35}Cl	7313.206(4)^b	2.13011		^{35}Cl	−8.8(15)	24
Al^{79}Br	4772.825(2)^b	2.29480		^{79}Br	<100^a	24
Al^{127}I	3528.5533(8)^b	2.53709		^{127}I	−334(10)	24
^{69}Ga^{35}Cl		2.201681(25)		^{35}Cl	−13.20(15)^b	25
				^{69}Ga	−92.40(14)^b	
^{115}InF				^{115}In	−723.7996(2)	26
InCl			3.79(10)			27
^{205}TlF	6689.8736(2)^b		4.2282(8)^b			17
^{205}Tl^{35}Cl			4.515294(1)^b	^{35}Cl	−15.7520(3)^b	28
CO^d			1.06(20)			29
SiS	1.929254(3)	1.73(6)				30
PN	23578.34(8)^b	1.49086(2)	2.7513(9)^b			31, 32
ClF			0.888040(13)	^{35}Cl	−145.87201(43)	35
BrO^e		1.7171(13)	1.77(3)			34a

^{a} Absolute value; ^{b} equilibrium value; ^{c} $X^2\Pi$ state; ^{d} $a'^3\Sigma^+$ state; ^{e} $^2\Pi_{3/2}$ state.

determined include the electric dipole moment, the nuclear quadrupole moment of ^{87}Rb, and the rotational magnetic dipole moment [$\mu_J/J = -0.04170(15)$ kHz G^{-1}].

The $J = 1 \rightarrow 2$ transition at 8.6 GHz of ^{133}Cs^{35}Cl has been observed in the ground state and lowest five vibrationally excited states.[19] For low v, it was found that $eQq(^{35}\mathrm{Cl})/\mathrm{MHz} = 1.830 - 0.118(v + \tfrac{1}{2}) \pm 0.15$, while

[19] J. Hoeft, E. Tiemann, and T. Törring, *Z. Naturforsch.*, 1972, **27a**, 1516.

the absolute value of the ^{133}Cs quadrupole coupling constant was less than 1.1 MHz. For higher v it appeared that $|eQq(^{133}\text{Cs})| > |eQq(^{35}\text{Cl})|$.

From the $J = 1 \rightarrow 2$ transition of ^{133}Cs^{79}Br at 580 °C, the quadrupole coupling constant of ^{79}Br in the $v = 0$ to $v = 5$ states have been determined;[20] these obey the relation: $eQq/\text{MHz} = -6.79 + 0.73(v + \frac{1}{2})$ to within ± 0.15 MHz. In a similar study[21] of ^{133}Cs^{127}I the $J = 2 \rightarrow 3$ transition in the $v = 0$ to $v = 3$ states were measured; here $eQq(^{127}\text{I})/\text{MHz} = -14.28 - 2.10(v + \frac{1}{2})$ to within ± 0.35 MHz. In addition the magnitude of eQq for ^{133}Cs was found to be less than 1 MHz.

The first studies of the free radical LiO in the gas phase were reported.[22] This was formed by heating Li$_2$O to 1800 K; the radiofrequency and X-band spectra of both components $^2\Pi_{3/2}$ and $^2\Pi_{1/2}$ of the ground electronic state of ^{7}Li^{16}O were studied, and several fine-structure and magnetic hyperfine constants were determined. The electric dipole moment, $\mu = 6.84(3)$ D, and the quadrupole coupling constant (diagonal in Λ), $eQq_0(^7\text{Li}) = 0.444(8)$ MHz, were also determined. The high dipole moment, together with the values of the magnetic hyperfine constants, support an ionic structure Li$^+$O$^-$, but the negative Fermi contact constant and the ratio of the measured quadrupole coupling constants imply some covalency in the bond.

In some microwave–optical double-resonance studies of BaO, the $J'' = 1 \rightarrow 2$ microwave transition in the $v = 0$ state of the ground $(X^1\Sigma)$ electronic state and the $J' = 2 \rightarrow 3$ and $2 \rightarrow 1$ transitions in the $v = 7$ state of the first excited $(A^1\Sigma)$ state were observed at 37 403.9, 44 891.4(2) and 29927.6(1) MHz, respectively.[23] The $R(2)$ line of the A–$X(7,0)$ transition is coincident with the 496.5 nm line of an Ar$^+$ laser; a 200 mW beam of the laser radiation was used to excite the transition in BaO, which was formed by reacting barium metal vapour with O$_2$ in an argon carrier gas stream. The intense fluorescence from the A–$X(7,1)$ band was monitored; when the microwave frequency was swept through one of the transitions involved, a change in the luminescence was detected. The $(A^1\Sigma)$, $v = 7$ state is 57.5 kcal mol^{-1} above the ground state and has a radiative lifetime of 326 ns; this is the first observation of a microwave rotational transition in a short-lived electronic state of a diatomic molecule, and the method promises an improvement in sensitivity of 10^2—10^4 over conventional microwave spectroscopy.

Several workers have published studies of Group III monohalides; one is reported above.[17] The millimetre-wave spectra of AlCl and AlBr gave rotational constants to much better accuracy than previous results, and the first reported values of the rotational constants of AlI were also measured.[24]

---

[20] J. Hoeft, E. Tiemann, and T. Törring, *Z. Naturforsch.*, 1972, **27a**, 702.
[21] J. Hoeft, E. Tiemann, and T. Törring, *Z. Naturforsch.*, 1972, **27a**, 1017.
[22] S. M. Freund, E. Herbst, R. P. Mariella, and W. Klemperer, *J. Chem. Phys.*, 1972, **56**, 1467.
[23] R. W. Field, R. S. Bradford, D. O. Harris, and H. P. Broida, *J. Chem. Phys.*, 1972, **56**, 4712; R. W. Field, R. S. Bradford, H. P. Broida, and D. O. Harris, *ibid.*, 1972, **57**, 2209.
[24] F. C. Wyse and W. Gordy, *J. Chem. Phys.*, 1972, **56**, 2130.

Many vibrational states were measured, in addition to the ground states, and four Dunham coefficients were fitted; in addition to the values of B_e (and hence r_e), D_e and hence ω_e were obtained. The nuclear quadrupole coupling of ^{127}I in AlI was $eQq = -334(10)$ MHz; the coupling constants of the other nuclei (including ^{27}Al) could not be determined from the high-J transitions observed, although it was estimated that $|\,eQq(^{79}\text{Br})\,| <$ 100 MHz. By using $eQq(^{127}\text{I})$ and the quadrupole coupling constants for ^{27}Al and ^{35}Cl previously reported, the ionic characters of the Al—X bond were determined to have the high values of 92% for AlCl and 85% for AlI; these imply an electronegativity for univalent Al of $x = 1.3$, compared with $x = 1.5$ for Al^{III}.

The hyperfine structure in the $J = 0 \rightarrow 1$ transition of GaCl at 8.7 GHz was measured, giving improved values of the rotational and hyperfine structure constants.[25] The quadrupole coupling constants of both nuclei were measured:

$$eQq(^{35}\text{Cl})/\text{MHz} = -13.20 - 0.20\,(v + \tfrac{1}{2}) \pm 0.15$$
$$eQq(^{69}\text{Ga})/\text{MHz} = -92.40 + 0.68\,(v + \tfrac{1}{2}) \pm 0.14$$

From an m.b.e.r. study on ^{115}In^{19}F in the $J = 1$ and $J = 2$ states, the variation of the ^{115}In nuclear quadrupole coupling constant with vibrational and rotational states was determined;[26] very little variation with J occurred. The spin–rotation and spin–spin constants were also obtained.

By choosing a vapour pressure at 200—250 °C such that the ^{35}Cl hyperfine structure was not resolved, Tiemann et al.[27] were able to determine the dipole moment of InCl from the $J = 1 \rightarrow 2$ transition.

Another m.b.e.r. study on TlCl, including both Stark and Zeeman effects, gave values of various magnetic and electrical properties, including their variation with vibrational state.[28] As the usual relations between certain of these constants were not obeyed, it was concluded that the molecule obeys Hund's coupling case (c).

The $a'^3\Sigma^+$ state of CO is relatively inaccessible. However, by an analysis of the radiofrequency spectrum of CO in the metastable $a^3\Pi$ state, which is perturbed by the $a'^3\Sigma^+$ state, Wicke et al.[29] were able to determine, from the variation of the electric dipole moment with J, v, and Ω, the transition moment $\langle a^3\Pi\,|\,\mu\,|\,a'^3\Sigma^+\rangle = 0.504(10)$ D and also the permanent dipole in the $a'^3\Sigma^+$ state, $\mu = 1.06(20)$ D (in the sense $^-$C—O$^+$), as well as the matrix elements of the interaction Hamiltonian.

A detailed analysis of isotopic effect in the rotational spectrum of SiS has been published.[30] By considering the isotopic variation of the Dunham constant Y_{01}, corrections resulting from the breakdown of the Born–Oppenheimer approximation were obtained. The electric dipole moment

[25] E. Tiemann, M. Grasshof, and J. Hoeft, Z. Naturforsch., 1972, 27a, 753.
[26] R. H. Hammerle, R. Van Ausdal, and J. C. Zorn, J. Chem. Phys., 1972, 57, 4068.
[27] E. Tiemann, J. Hoeft, and T. Törring, Z. Naturforsch., 1972, 27a, 869.
[28] R. Ley and W. Schauer, Z. Naturforsch., 1972, 27a, 77.
[29] B. G. Wicke, R. W. Field, and W. Klemperer, J. Chem. Phys., 1972, 56, 5758.
[30] E. Tiemann, E. Renwanz, J. Hoeft, and T. Törring, Z. Naturforsch., 1972, 27a, 1566.

was determined to be 1.73(6) D for the ground state of ^{28}Si^{32}S, in the sense $^{+}$Si—S^{-}. The first four potential constants were also determined.

A microwave study of the $J = 0 \rightarrow 1$ transition,[31] and a millimetre-wave study of higher J transitions,[32] of ^{31}P^{14}N gave values of r_e and μ_v which are in excellent agreement. The millimetre-wave work gave estimates of ω_e and $\omega_e x_e$ which were rather worse than those obtained from optical spectroscopy.

The hyperfine Λ-doubling spectra of ^{14}N^{16}O and ^{15}N^{16}O were observed by m.b.e.r. in the $^2\Pi_{1/2}$ and $^2\Pi_{3/2}$ states,[33] and the results analysed using degenerate perturbation theory up to third order.

Laser oscillation of pure rotational transitions in OH and OD radicals has been observed in the 20 μm (1500 GHz) region.[34]

The radical BrO has been investigated in the $^2\Pi_{3/2}$ state.[34a] The $J = 5/2 \leftarrow 3/2$ transition of both the ground and first vibrational state gave an equilibrium bond length of 1.7171(13) Å, while the mean dipole moment of the ^{79}Br and ^{81}Br species, $\mu = 1.77(3)$ D, is significantly greater than the value of 1.61(4) D obtained from gas-phase e.s.r.

Two studies of ClF provided complementary information on its electrical and magnetic properties. An m.b.e.r. study[35] gave more accurate values for $eQq(^{35}\text{Cl})$ and μ than previous values, in addition to the spin–spin and spin–rotation interaction constants. The spin–spin coupling constant was determined to be $J_{\text{ClF}} = 1074(77)$ Hz, its sign being positive. A molecular Zeeman study[36] gave values of the molecular g values, shielded nuclear g values, magnetic susceptibility anisotropy, and molecular quadrupole moment. In addition the sense of the dipole moment was determined, by two independent methods, to be almost certainly $^{-}$Cl—F^{+}, this result being rather surprising. This study also reported a similar analysis of BrF, ClCN, BrCN, and ICN. The sense of the dipole moment could be determined only for $^{-}$ClF^{+} and $^{+}$ClCN^{-}. The molecular quadrupole moments of the fluorides are small and positive, while those for the cyanides are negative, their magnitude increasing from ClCN to ICN.

4 Triatomic Molecules

Several isotopic species of GeF$_2$ had previously been observed, and now the first report on the microwave spectrum of ^{73}GeF$_2$ has been published.[37] The non-zero elements of the quadrupole coupling constant for ^{73}Ge are: $\chi_{aa} = 17.2(10)$ MHz, $\chi_{bb} = 121.7(10)$ MHz, and $\chi_{cc} = -138.9(10)$ MHz.

---

[31] J. Hoeft, E. Tiemann, and T. Törring, *Z. Naturforsch.*, 1972, **27a**, 703.
[32] F. C. Wyse, E. L. Manson, and W. Gordy, *J. Chem. Phys.*, 1972, **57**, 1106.
[33] W. L. Meerts and A. Dymanus, *J. Mol. Spectroscopy*, 1972, **44**, 320.
[34] T. W. Ducas, L. D. Geoffrion, R. M. Osgood, and A. Javan, *Appl. Phys. Letters*, 1972, **21**, 42.
[34a] T. Amano, A. Yoshinaga, and E. Hirota, *J. Mol. Spectroscopy*, 1972, **44**, 594.
[35] R. E. Davis and J. S. Muenter, *J. Chem. Phys.*, 1972, **57**, 2836.
[36] J. J. Ewing, H. L. Tigelaar, and W. H. Flygare, *J. Chem. Phys.*, 1972, **56**, 1957.
[37] H. Takeo and R. F. Curl, *J. Mol. Spectroscopy*, 1972, **43**, 21.

An MO treatment, including back-donation of π-electrons from F to Ge, gave poor agreement with the observed values. An analysis of the spectrum using centrifugal distortion constants calculated from the force field gave slight changes to the previous equilibrium structure: this is now reported as $r_e(\text{Ge}-\text{F}) = 1.73209(2)$ Å, $\alpha_e(\text{F}-\text{Ge}-\text{F}) = 97.1479(4)°$.

Several vibrational states of IC^{14}N and IC^{15}N have been studied in the 30—78 GHz region.[37a] Values of B_v, D_J, q_e, eQq_I (and η for the bending states) were obtained, and an analysis of the resonance between the (100) and (02^{0}0) states gave the anharmonic force constant $k_{122} = 50.2(7)$ cm^{-1}, which fits in with the trend of the other cyanogen halides. The variation of eQq_I with vibrational state was discussed.

The microwave spectra of the nitroxyl radicals HNO and DNO have been observed.[38, 39] The radical was formed from the reaction of H$_2$ or D$_2$ with NO in a flow system; its half-life varied from 1 to 40 s. The rotational constants, calculated from a rigid rotor model, were reported for both species; in addition, for DNO, the electric dipole moment components, $\mu_a = 1.18(4)$, $\mu_b = 1.22(4)$ D, and the nuclear quadrupole coupling tensor elements for ^{14}N, $\chi_{aa} = 1.03(40)$, $\chi_{bb} = -6.13(26)$, and $\chi_{cc} = 5.10(26)$ MHz, were given;[39] the latter implied a similar field gradient to those of FNO and HONO.

The dipole moment and force constants of NSCl are derived from a study of its microwave spectrum.[40] For ^{15}N^{32}S^{35}Cl, the dipole moment components are $\mu_a = 0.5645(85)$ and $\mu_b = 1.747(35)$ D, giving a total dipole moment of 1.836(36) D. By assuming the planarity conditions, four centrifugal distortion constants were obtained for the ground state of ^{14}N^{32}S^{35}Cl, and these were used together with the vibrational frequencies and inertia defects to improve the harmonic force field. It was still found necessary to constrain two force constants for a convergent fit.

Several new transitions have been observed in H$_2$^{16}O and H$_2$^{18}O in the microwave and millimetre-wave regions,[41] and analyses to give rotational and centrifugal distortion constants from i.r. and microwave transitions were performed. Transitions have also been observed in HTO and T$_2$O, and the rotational constants were obtained.[42] The Stark splitting of some millimetre-wave transitions of water have been reported.[43]

The variation of the dipole moment with rotational state and isotopic substitution has been investigated[44] in the microwave spectra of HDO and D$_2$O. The direction of the dipole moment in HDO is found experimentally to be along the bisector of the HOD angle. Increases in μ are observed on

---

[37a] J. B. Simpson, J. G. Smith, and D. H. Whiffen, *J. Mol. Spectroscopy*, 1972, **44**, 558.
[38] S. Saito and K. Takagi, *Astrophys. J.*, 1972, **175**, L47.
[39] K. Takagi and S. Saito, *J. Mol. Spectroscopy*, 1972, **44**, 81.
[40] S. Mizumoto, J. Izumi, T. Beppu, and E. Hirota, *Bull. Chem. Soc. Japan*, 1972, **45**, 786.
[41] F. C. DeLucia, P. Helminger, R. L. Cook, and W. Gordy, *Phys. Rev. (A)*, 1972, **5**, 487; 1972, **6**, 1324.
[42] J. Bellet, G. Steenbeckeliers, and P. Stouffs, *Compt. rend.*, 1972, **275**, B, 501.
[43] Y. Beers and G. P. Klein, *J. Res. Nat. Bur. Stand., Sect. A*, 1972, **76**, 521.
[44] A. H. Brittain, A. P. Cox, G. Duxbury, T. G. Hersey, and R. G. Jones, *Mol. Phys.*, 1972, **24**, 843.

deuteriation (owing to vibrational anharmonicity) and on increasing J (owing to centrifugal distortion).

By measuring 31 new rotational transitions of $H_2{}^{32}S$ up to 766 GHz, Helminger et al.[45] were able to obtain a centrifugal distortion analysis without the use of i.r. data. The use of submillimetre waves, together with the Hamiltonian due to Watson for asymmetric tops, gives a fit of the observations to within experimental error, such that unobserved transitions can be predicted with similar error, and the rotational constants and lower-order centrifugal distortion constants obtained are insensitive to the exact data set and to higher-order terms. Such an analysis is now possible for all light asymmetric tops except water. A similar analysis is reported for D_2S, and the results compared with those obtained from i.r. observations and force field predictions.[46]

The hyperfine structure of the $4_2 \rightarrow 4_1$ rotational transition of $HD^{80}Se$ has been investigated using a high-resolution beam-maser spectrometer.[47] Various electrical and magnetic properties were determined and, by combining the results with earlier work on the $2_2 \rightarrow 2_1$ transition, the diagonal elements of the ^2H quadrupole coupling tensor, $\chi_{aa} = 123.3(6)$ kHz, $\chi_{bb} = -56.5(60)$ kHz, $\chi_{cc} = -66.8(71)$ kHz, as well as of the ^1H and ^2H magnetic coupling tensor, were determined.

The microwave spectrum of SeO_2 has been re-examined,[48] and the dipole moment was determined as 2.62(5) D, compared with a previously published value of 2.7(4) D. In addition, the force field, including cubic terms, was evaluated by using microwave and i.r. data.

Hypofluorous acid, HOF and DOF, has been studied in the millimetre-wave region.[49] The half-life of the species in the absorption cell at room temperature and a pressure of about 20 mTorr increased from 5 to 30 min as the cell became seasoned. Both a- and b-type transitions were observed, and the rotational and some centrifugal distortion constants were derived. From the rotational constants, the substition co-ordinates of the H atom were obtained, and the other structural parameters derived from the first and second moment. The structure thus obtained [$r(O-H) = 0.964(10)$ Å, $r(O-F) = 1.442(1)$ Å, $\theta(HOF) = 97.2(6)°$] showed a longer O—F bond and smaller angle than in OF_2; these correlate with the low O—F stretching frequency and suggest that the fluorine atom is nearly neutral.

5 Tetra-atomic Molecules

Some microwave and millimetre-wave transitions of four isotopic species of non-linear isocyanic acid, HNCO, were recorded and a rotational analysis was performed.[50]

---

[45] P. Helminger, R. L. Cook, and F. C. DeLucia, *J. Chem. Phys.*, 1972, **56**, 4581.
[46] R. L. Cook, F. C. DeLucia, and P. Helminger, *J. Mol. Spectroscopy*, 1972, **41**, 123.
[47] S. Chandra and A. Dymanus, *Chem. Phys. Letters*, 1972, **13**, 105.
[48] H. Takeo, E. Hirota, and Y. Morino, *J. Mol. Spectroscopy*, 1972, **41**, 420.
[49] H. Kim, E. F. Pearson, and E. H. Appelman, *J. Chem. Phys.*, 1972, **56**, 1; E. F. Pearson and H. Kim, *ibid.*, 1972, **57**, 4230.
[50] W. H. Hocking, M. C. L. Gerry, and G. Winnewisser, *Astrophys. J.*, 1972, **174**, L93.

Vibrationally excited states of linear fulminic acid, HCNO, were studied and the *l*-type doublet components in various states observed.[51] The anharmonic vibrational constants g_{ii} (which are the multipliers of l_i^2) were found to vary considerably with vibrational state; the observed anharmonicity in the low-lying vibrational mode is explained in terms of a low hump in the isotropic two-dimensional oscillator potential.

A full account of the microwave spectrum of ClNCO has given rotational and centrifugal distortion constants, the molecular structure, inertia defects, and nuclear quadrupole coupling constants.[52] The inertia defects are

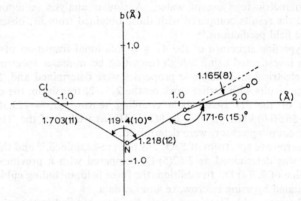

Figure 1 *Molecular structure of chlorine isocyanate*
(Adapted from *J. Mol. Spectroscopy*, 1972, **42**, 563)

small and positive, implying a planar structure; the large change in moments of inertia on substituting at oxygen, together with the value of χ_{cc} for the nitrogen nucleus, imply that the structure of the species observed is ClNCO and not ClOCN. The derived structure is given in Figure 1, in which it can be seen that the N—C—O framework is not linear. The nuclear quadrupole coupling tensor elements were obtained for ^{14}N, ^{35}Cl, and ^{37}Cl; from the latter it was deduced that the main resonance structure is (1), with 4% double-bond character from structures such as (2) and 9% ionic character from structure such as (3) and (4). These contributions agree with the

$$\begin{array}{cccc} \text{Cl}\diagdown & \text{Cl}^+ & \text{Cl}^+ & \text{Cl}^+ \\ \text{N}=\text{C}=\text{O} & \diagdown\text{N}-\text{C}\diagdown_\text{O} & \bar{\text{N}}=\text{C}=\text{O} & \text{N}\equiv\text{C}-\bar{\text{O}} \\ (1) & (2) & (3) & (4) \end{array}$$

near-linearity of the N—C—O chain and the Cl—N bond length, which is close to the sum of the single bond radii.

[51] M. Winnewisser and B. P. Winnewisser, *J. Mol. Spectroscopy*, 1972, **41**, 143.
[52] W. H. Hocking and M. C. L. Gerry, *J. Mol. Spectroscopy*, 1972, **42**, 547.

A detailed centrifugal distortion analysis of *cis-* and *trans-*HONO has enabled Finnigan *et al.*[53] to determine average (r_z) structures and, using a model anharmonic force field, an equilibrium (r_e) structure for the *trans-*isomer. By using the centrifugal distortion constants together with i.r. vibrational frequencies, they were able to determine a harmonic force field in which five of the sixteen potential constants were constrained to zero. The N—O, N=O, and O—H stretching force constants, as well as the N—O/∠NOH interaction constant, differ markedly in the two isomers, correlating with changes in the structure and suggesting the importance of the resonance contributor (5) in the *cis-*isomer.

$$\begin{array}{c} H^+ \quad O^- \\ O=N \end{array}$$
(5)

The molecular structure of NH_2Cl has been determined from the microwave spectra of six isotopic species;[54] the substitution parameters are: r_s(N—H) = 1.017(5) Å, r_s(N—Cl) = 1.7480(1) Å, θ_s(H—N—Cl) = 103.7(4)°, θ_s(H—N—H) = 107(2)°. The quadrupole coupling of both ^{14}N and ^{35}Cl were obtained, and it was found that the z-axis of the coupling tensor coincides, to within experimental error, with the N—Cl bond direction; the elements in the principal axis system of the ^{35}Cl quadrupole coupling tensor are $\chi_z = -99.81(5)$ MHz, χ_y (perpendicular to the symmetry plane) = 47.10(10) MHz, and $\chi_x = 52.71(15)$ MHz. These imply, by use of the Townes–Dailey theory, 2.6% double-bond character in the N—Cl bond.

By using an i.r. and radiofrequency double-resonance technique, Shimizu has measured the dipole moment of PH_3 in the $v_2 = 1$ state;[55] the $P(3,2)$ transition of v_2 is almost coincident with a line from the 10 μm N_2O laser, and by applying a Stark field to produce exact resonance and then tuning in the radiofrequency to a transition between Stark levels, a change in the transmission of the laser was observed. The value of the dipole moment in the $v_2 = 1$ state, $\mu = 0.574(3)$ D is the same, to within experimental accuracy, as in the ground state, in marked contrast to NH_3. Since no broadening of the lines was observed, it was concluded that the inversion splitting in the $v_2 = 1$ state must be less than 400 kHz.

A study of the microwave spectra of PF_2Cl, with ^{35}Cl and ^{37}Cl, gave values of the rotational constants, diagonal quadrupole coupling tensor elements, and dipole moment.[56] The structure was obtained by substitution and first and second moments, giving r(P—F) = 1.571(3) Å, r(P—Cl) =

---

[53] D. J. Finnegan, A. P. Cox, A. H. Brittain, and J. G. Smith, *J.C.S. Faraday II*, 1972, **68**, 548.
[54] G. Cazzoli, D. G. Lister, and P. Favero, *J. Mol. Spectroscopy*, 1972, **42**, 286.
[55] F. Shimizu, *Chem. Phys. Letters*, 1972, **17**, 620.
[56] A. H. Brittain, J. E. Smith, and R. H. Schwendemann, *Inorg. Chem.*, 1972, **11**, 39.

2.030(6) Å, $\theta(F-P-F) = 97.3(2)°$, and $\theta(F-P-Cl) = 99.2(3)°$. A comparison of these values with those in the other molecules of formula PF_nCl_{3-n} and PF_2H showed little change in corresponding bond lengths and angles in the series, but the angles increase HPF < FPF < FPCl < ClPCl. A small asymmetry in the ^{35}Cl quadrupole coupling tensor suggests a minimum π-bond character in the P—Cl bond of 1.5%.

A number of papers [57] have been published on the millimetre-wave spectra of HSSH and its deuteriated analogues. A previous study [57a] had given the molecular structure, but in this series an anomalous K-doubling is discussed. The molecules HSSH and DSSD are very close to being prolate symmetric tops, and in these circumstances a centrifugal term in the Hamiltonian, of the form $(J_+^4 + J_-^4)$, causes splitting, especially of the $K_{-1} = 2$ levels, which is of the same order of magnitude or larger than that due to the more usual asymmetry splitting. It is also concluded that, to fit the structure, an effective bond shortening of 0.003 45(30) Å occurs on replacing S—H by S—D.

Thionyl fluoride has been studied in the ground state [58, 59] and $v_4 = 1$ vibrational state.[59] The quartic centrifugal distortion constants obtained were used by Lucas and Smith [58] to obtain a quadratic force field, which in turn was used to determine a r_z structure (in addition to the r_0 structure) from three isotopic species. The r_z structure is less consistent than the r_0 structure, which is $r_0(O-S) = 1.4127(3)$ Å, $r_0(S-F) = 1.5854(2)$ Å, $\theta_0(F-S-F) = 92.83(2)°$, and $\theta_0(F-S-O) = 106.82(3)°$.

In the microwave spectrum of $SO^{35}Cl^{37}Cl$, there are observed a-, b-, and c-type transitions.[60] The observation of lines up to $J = 68$ gave values of A, B, C, and the quartic and sextic centrifugal distortion terms.

An interesting study [61] describes the radiofrequency and microwave spectra of dimers of hydrogen fluoride, $(HF)_2$, HFDF, and $(DF)_2$, by the m.b.e.r. method. The polymers were formed by use of a supersonic nozzle, and mass spectral data suggested that several polymers of HF were formed. Both $\Delta J = 0$, $\Delta M_J = \pm 1$, and $\Delta J = \pm 1$, $\Delta M_J = 0$ transitions were studied, and the spectrum was analysed in terms of a model in which two rigid HF monomer units are connected by a relatively weak hydrogen bond, as shown in (6). In $(HF)_2$ and $(DF)_2$, the interchange of the two

$$F-H\cdots F\diagup^H$$

(6)

---

[57] G. Winnewisser, *J. Chem. Phys.*, 1972, **56**, 2944; G. Winnewisser and P. Helminger, *ibid.*, p. 2954, 2967; G. Winnewisser, *ibid.*, 1972, **57**, 1803; *J. Mol. Spectroscopy*, 1972, **41**, 534.
[57a] G. Winnewisser, M. Winnewisser, and W. Gordy, *J. Chem. Phys.*, 1968, **49**, 3465.
[58] N. J. D. Lucas and J. G. Smith, *J. Mol. Spectroscopy*, 1972, **43**, 327.
[59] A. Dubrulle and J. L. Destombes, *J. Mol. Structure*, 1972, **14**, 461.
[60] A. Dubrulle and D. Boucher, *Compt. rend.*, 1972, **274**, B, 1426.
[61] T. R. Dyke, B. J. Howard, and W. Klemperer, *J. Chem. Phys.*, 1972, **56**, 2443.

H—F units by breaking the H bond and re-forming another, coupled with rotation by π about the F—F axis, gives an equivalent orientation to the original one. The barrier to such a motion is not high, and 'tunnelling doubling' is observed between the two orientations, the barrier height being calculated to be about 500 cm^{-1}. Such a motion cannot occur in the mixed species HFDF, which has inequivalent hydrogen atoms; of the two possible isomers only the one in which deuterium forms the hydrogen bond is observed. The parameters obtained from the microwave spectrum are shown in Table 2. From the rotational constants, and assuming that the

Table 2 *Parameters obtained from the m.b.e.r. spectrum of hydrogen fluoride dimer*[a]

	(HF)$_2$	HFDF	(DF)$_2$
$\frac{1}{2}(B + C)$/MHz	6504.8(20)	6500.1(1)	6252.194(2)
ν/MHz[b]	19776(12)	—	1579.877(4)
μ_a/D	2.987(3)	3.029(3)	2.9919(6)
$eQq(^2\mathrm{H})$/MHz[c]	—	0.270(30)	0.110(8)[d]

[a] For structural parameters see Table 4; [b] splitting of degenerate vibrational levels by the tunnelling motion; [c] component along the a-axis (F—F axis); [d] measured value is the average for the two sites.

H—F distance of the bound monomers was identical to that of the free monomer (r_e = 0.917 Å), the F—F distance was calculated to be 2.79(5) Å and the F—F—H angle as 108°, this latter being rather sensitive to the motions of both hydrogen atoms. From the deuterium quadrupole coupling constant in HFDF and the 'tunnelling doubling' it can be inferred that the inner hydrogen (or deuterium) oscillates perpendicular to the F—F axis, the H—F motion having an amplitude of about 23°. The quadrupole coupling in (DF)$_2$ then implies that the outer deuterium is at 119(5)° to the F—F axis, in reasonable agreement with the 108° obtained from the rotational constants. The dipole moments are also consistent with the bent model. A study of the microwave spectra of higher polymers of HF is reported in Section 7.

The structure of chloryl fluoride, FClO$_2$, has been determined from the microwave spectrum.[62] The molecule is pyramidal, with the chlorine at the apex, and has r(F—Cl) = 1.664(30) Å, r(Cl—O) = 1.434(15) Å, θ(O—Cl—O) = 113.4(20)°, and θ(F—Cl—O) = 103.2(15)°. The quadrupole coupling constants for ^{35}Cl and ^{37}Cl were also determined. The Cl—F distance is longer than in ClF, which is as expected, but the shorter Cl—O bond and smaller O—Cl—O angle than in ClO$_2$ are somewhat surprising, although they are interpreted in terms of a (p–π^*) σ-bonding formalism.

6 Penta-atomic Molecules

The rotational constants obtained from the microwave spectra of six isotopic species of BF$_2$OH indicate that the molecule is planar.[63] The

[62] C. R. Parent and M. C. L. Gerry, *J.C.S. Chem. Comm.*, 1972, 285.
[63] H. Takeo and R. F. Curl, *J. Chem. Phys.*, 1972, **56**, 4314.

structure shown in Figure 2 was obtained by assuming that the two B—F distances were equal; the hydrogen and oxygen co-ordinates were obtained using the substitution method. The dipole moment is 1.86(2) D.

The microwave spectra of all the monohalogenogermanes, GeH$_3$X, have now been obtained; we here report the first observations on fluorogermane [64] and iodogermane,[64a] together with a re-analysis of bromogermane [64a] to higher accuracy than previous studies. The dipole moment of

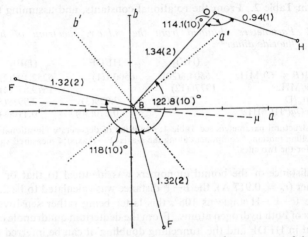

Figure 2 *Molecular structure of* BF$_2$OH. *The principal axis system of the normal species is shown. The dotted lines correspond to the principal axis system of* BF$_2$^{18}OH. *The dashed line gives the orientation of the dipole moment*
(Adapted from *J. Chem. Phys.*, 1972, **56**, 4316)

GeH$_3$F, μ = 2.33(6) D, is slightly higher than that of GeH$_3$Cl, in contrast to the behaviour of the analogous silicon compounds. The Ge—F and Ge—H bond lengths were calculated from the differences between moments of inertia of isotopic species, assuming tetrahedral angles; these gave r(Ge—H) = 1.47(1) and r(Ge—F) = 1.743 Å. On the assumption that the projection of the Ge—H bond length on the *a*-axis was equal to that in GeH$_3$Br, r(Ge—F) = 1.74(1) Å was also obtained, but if the moments of inertia themselves were used, the bond length was estimated as 1.52 Å; this value was considered less reliable. In GeH$_3$Br, the Ge—Br bond length was found by the substitution method, giving r_s(Ge—Br) = 2.2970(2) Å, while in GeH$_3$Br the same projection of the Ge—H bond on to the *a*-axis as for GeH$_3$Br was assumed, giving r(Ge—I) = 2.5075(6) Å. The halogen coupling constants were measured and U_p, the number of unbalanced *p*-electrons, calculated as 0.42, 0.499, and 0.602 for GeH$_3$Cl, GeH$_3$Br, and GeH$_3$I, respectively. The increase in U_p with decreasing

[64] L. C. Krisher, J. A. Morrison, and W. A. Watson, *J. Chem. Phys.*, 1972, **57**, 1357.
[64a] S. N. Wolf and L. C. Krisher, *J. Chem. Phys.*, 1972, **56**, 1040.

electronegativity of the halogen is observed for other Group IV monohalides, and the value for GeH$_3$I lies between that for CH$_3$I ($U_p = 0.844$) and SnH$_3$I ($U_p = 0.555$).

Two studies of cyanogen azide, NCN$_3$, give results in excellent agreement. For the normal species, the dipole moment is determined [65] as $\mu = 2.99(3)$ D. From the rotational constants of three isotopic species the molecular structure was determined,[65a] as shown in Figure 3, on the

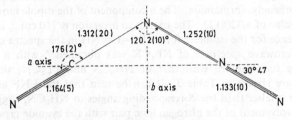

Figure 3 *Molecular structure of cyanogen azide*
(Adapted from *Canad. J. Phys.*, 1972, **50**, 1456)

assumption that the inertia defect was the same for each isotopic species, and that the azide group is linear. The slight deviation of the N—C—N chain from linearity appears to be significant.

Hydroxylamine, NH$_2$OH, and all its deuterated derivatives have been studied in the microwave region at frequencies up to 101 GHz.[66] The r_0 structure is given in Table 3. The quadrupole coupling tensor, in its

Table 3 *Structural parameters for compounds of Group V of formula* MX$_2$YZ

Atom MX$_2$YZ	Bond length/Å			Bond angle/deg			Ref.
	M—X	M—Y	Y—Z	X—M—X	X—M—Y	M—Y—Z	
NH$_2$OH	1.016(8)	1.453(2)	0.962(5)	107.1(5)	103.25(50)	101.37(50)	66
NH$_2$CN	1.001(15)	1.346(5)	1.160(5)	113.5(20)	115.6(11)	180	67
NF$_2$CN	1.398(5)	1.392(9)	1.151(4)	102.9(4)	104.7(6)	169.7(21)[a]	68
PF$_2$CN	1.567(7)	1.811(5)	1.158(3)	99.1(2)	97.1(2)	171.5(8)[a]	69

[a] The nitrogen of the cyanide group is bent away from the fluorine atoms.

principal axis system, is $\chi_x = 5.78(50)$ MHz, $\chi_y = 0.02(50)$ MHz, $\chi_z = -5.76(5)$ MHz, where x is almost along the N—O bond and y is perpendicular to the plane of symmetry. The dipole moment is $\mu = 0.59(5)$ D. Satellite lines were observed for a vibrational state at 342(40) cm^{-1} in NH$_2$OH and 242(50) cm^{-1} in ND$_2$OD, but no inversion splitting was seen,

[65] K. Bolton, R. D. Brown, and F. R. Burden, *Chem. Phys. Letters*, 1972, **15**, 79.
[65a] C. C. Costain and H. W. Kroto, *Canad. J. Phys.*, 1972, **50**, 1453.
[66] S. Tsunekawa, *J. Phys. Soc. Japan*, 1972, **33**, 167.

implying that the barrier to inversion is greater than in ammonia or methylamine.

The analysis of a-type transitions of eight isotopic species of cyanamide, NH_2CN, indicated that it is not a planar molecule.[67] However, the out-of-plane angle of 38°, while greater than the near-zero value in formamide, is less than in nitramide and is close to the value in aniline. The structural parameters are given in Table 3. The $C\equiv N$ bond length is typical for a cyanide group, but the $N-C$ bond length is shorter than in methylamine or, more surprisingly, formamide. The a component of the dipole moment has the high value of 4.32(8) D. The barrier to inversion is 710 cm^{-1}, and there is no evidence for the isomeric form $HN=C=NH$ in the spectra obtained.

The microwave spectrum of NF_2CN was obtained[68] with a view to comparing its structure with that of NH_2CN and PF_2CN; the values obtained are shown in Table 3. It can be seen that the FNF and FNC angles are smaller than the corresponding angles in NH_2CN, suggesting a greater involvement of the nitrogen lone pair with the cyanide group in the latter molecule. This effect is also suggested by the longer $N-C$ bond length in NF_2CN, although it is still about 0.05 Å shorter than in CH_3NF_2. The $N-C\equiv N$ chain is definitely non-linear, in contrast to NH_2CN, and is bent away from the fluorine atoms, suggesting electrostatic repulsion between the fluorine atoms and the nitrogen of the cyanide group. The dipole moment was $\mu = 1.10(2)$ D. Lines belonging to the two lowest vibrational states were observed, their intensities indicating that these states are at 150(30) cm^{-1} and 220(30) cm^{-1}; these may be compared with 190 cm^{-1} predicted for the inversion mode. A similar study[69] was undertaken on PF_2CN. In this the FPF and FPC angles are even smaller than in NF_2CN, as might be expected, but the short $P-C$ distance suggests delocalization of the $C\equiv N$ π-electrons on to the phosphorus atom. The quadrupole coupling constants for ^{14}N, resolved along the $C\equiv N$ bond, were $\chi_z = -4.75(100)$ MHz and $\eta = 1.14(22)$; this latter high value again suggests a perturbation of the triple bond. The bend in the $P-C\equiv N$ angle, away from the fluorine atoms, is similar to that observed in NF_2CN. It is also reported that the $F\cdots F$ distance in PF_2CN is close to the almost constant value found for several other three- and four-co-ordinate phosphorus compounds.

By observing the mixed species $PO^{35}Cl_2{}^{37}Cl$ and $PO^{35}Cl^{37}Cl_2$, Li et al.[70] have been able to determine the molecular structure of $POCl_3$ without making any structural assumptions. The r_0 structure is as follows: $r_0(P-O) = 1.455(5)$ Å, $r_0(P-Cl) = 1.989(2)$ Å, and $\theta_0(Cl-P-Cl) = 103.70(2)°$. A similar study[71] and analysis of the spectra of $VO^{35}Cl_3$,

---

[67] J. K. Tyler, J. Sheridan, and C. C. Costain, *J. Mol. Spectroscopy*, 1972, **43**, 248.
[68] P. L. Lee, K. Cohn, and R. H. Schwendemann, *Inorg. Chem.*, 1972, **11**, 1920.
[69] P. L. Lee, K. Cohn, and R. H. Schwendemann, *Inorg. Chem.*, 1972, **11**, 1917.
[70] Y. S. Li, M. M. Chen, and J. R. Durig, *J. Mol. Structure*, 1972, **14**, 261.
[71] K. Karakida, K. Kuchitsu, and C. Matsumara, *Chem. Letters*, 1972, 293.

$VO^{37}Cl_3$, and $VO^{35}Cl_2{}^{37}Cl$ gave: $r_0(V-O) = 1.595(5)$ Å, $r_0(V-Cl) = 2.131(1)$ Å, and $\theta_0(Cl-V-Cl) = 111.8(2)°$. The values for $VOCl_3$ agree to within experimental error with those obtained from an early electron-diffraction study, but are considerably more precise.

The identification of b-type $\Delta J = +1$ transitions in $SO_2{}^{35}Cl^{37}Cl$ has enabled Dubrulle and Destombes [72] to determine the rotational constants and all the quartic and sextic centrifugal distortion constants for this molecule.

7 Molecules containing Six or More Atoms

The microwave spectrum of π-C_5H_5BeCl showed [73] that it has C_{5v} symmetry, in agreement with the structure obtained by electron diffraction. The spectra of five isotopic species gave substitution co-ordinates for C and Cl, and the structure was obtained by assuming that the hydrogens are coplanar with the C_5 ring: $r(C-H) = 1.090$ Å, $r(C-C) = 1.424$ Å, $r(Be-Cl) = 1.839$ Å, $d(C \cdots Cl) = 3.538$ Å, while the perpendicular distance between Be and the C_5 ring was 1.485 Å. This structure agrees well with that obtained by electron diffraction. The Be—Cl bond is longer than in $BeCl_2$, and the high dipole moment, $\mu = 4.26(16)$ D, and low ^{35}Cl quadrupole coupling constant, $eQq = 22(2)$ MHz, imply that it is fairly ionic, being a polar single bond with little $p_\pi \leftarrow d_\pi$, Be $\leftarrow$ Cl bonding, as might be expected from the participation of the $2p$ orbitals of Be in bonding to the ring.

The microwave spectra of PMe_3,BH_3 and $MePH_2,BH_3$ were studied [74] to compare their structures with that of PF_2H,BH_3. The P—C and P—H bond lengths are shorter, and the bond angles are larger, than in the uncomplexed molecules, as is found in other complexes containing the P—BH_3 linkage. The P—B bond length is longer than in P—F compounds, as expected for $d_\pi \rightarrow p_\pi$ bonding between phosphorus and boron. There is no tilt of the borane group away from the methyl group, as in PF_2Me,BH_3, but the methyl group is tilted 3° away from the PH_2 group, which is explained by interaction with the more negative hydrogen atoms on the BH_3 group. From the dipole moments of 4.99(20) D for PMe_3BH_3 and 4.66(5) D for $MePH_2BH_3$, a bond dipole moment for P—BH_3 of between 3.4 and 4.0 D is estimated.

Several alkyl and halogen derivatives of silane have been investigated. The problem of the interaction between vibrations, internal rotation, and overall rotation in $MeSiH_3$ was analysed, and the coupling between the internal rotation and a degenerate vibration was observed in the microwave spectrum of the ν_6 torsional state.[75] The molecular Zeeman spectra [76]

---

[72] A. Dubrulle and J. L. Destombes, *Compt. rend.*, 1972, **274**, *B*, 181.
[73] A. Bjørseth, D. A. Drew, K. M. Marstokk, and H. Møllendal, *J. Mol. Structure*, 1972, **13**, 233.
[74] P. S. Bryan and R. L. Kuczkowski, *Inorg. Chem.*, 1972, **11**, 553.
[75] E. Hirota, *J. Mol. Spectroscopy*, 1972, **43**, 36.
[76] R. L. Shoemaker and W. H. Flygare, *J. Amer. Chem. Soc.*, 1972, **94**, 684.

of MeSiH$_3$ and MeSiD$_3$ gave the magnitudes and relative signs of the molecular g values and the magnetic susceptibility anisotropy. Two values of the molecular quadrupole moment along the symmetry axis could be derived: $Q_{\parallel}/10^{-26}$ e.s.u. cm^2 = $-(6.31 \pm 0.46)$ or $+(11.74 \pm 0.46)$. Both imply a dipole moment of 0.96 ± 0.40 D, but the negative value of $Q_{\parallel}$ is preferred as this gives the sign of the dipole moment as $^+$CH$_3$SiH$_3^-$.

In the microwave spectrum of EtSiH$_2$Me and EtSiD$_2$Me, lines due to the *trans*-isomers only have been analysed,[77] although two rotamers are present. The transitions observed are b-type transitions, split by internal rotation into doublets. The rotational constants for the 'unperturbed' levels are found and agree with a structure obtained by transference from Me$_2$SiH$_2$ and EtSiH$_3$. The barrier to rotation about the Si—Me bond is V_3 = 1518(30) cal mol^{-1}, which is lower than in SiH$_3$Me (V_3 = 1595 cal mol^{-1}), Me$_2$SiH$_2$ (V_3 = 1640 cal mol^{-1}), or Me$_3$SiH (V_3 = 1830 cal mol^{-1}). The dipole moment, μ = 0.758(5) D, makes an angle of 1°20′ with the bisector of the C—Si—C angle, towards the ethyl group.

In addition to the ground-state spectrum of SiF$_3$·CH:CH$_2$, those of the states ν_{14}, $2\nu_{14}$, ν_{21}, $2\nu_{21}$, and $\nu_{14} + \nu_{21}$ were observed;[78] ν_{14} is the lowest frequency A' vibration, the SiF$_3$ rocking mode, while ν_{21} is the lowest frequency A'' vibration, the SiF$_3$ torsion. From relative intensities the ν_{14} state was estimated to be at about 210 cm^{-1}. The dipole moment of the molecule was found to be 2.38(11) D.

The rotational constants obtained from a recent study[79] of MeSiF$_3$, together with previous values for other isotopic species, enabled Durig *et al.* to determine an r_s structure for the methyl group: r_s(C—H) = 1.081(4) Å, θ_s(H—C—Si) = 111.1(5)°. An r_0 structure for the whole molecule was also obtained: r_0(C—H) = 1.081(4) Å, r_0(C—Si) = 1.812(14) Å, r_0(Si—F) = 1.547(7) Å, θ_0(H—C—Si) = 111.0(5)°, and θ_0(Si—C—F) = 112.3(11)°, showing close agreement for the methyl group with the substitution structure. The dipole moment was 2.33(10) D. The first three excited states in the torsional mode were also studied, and the energy difference between the v = 0 and v = 1 states was 114(6) cm^{-1}, the barrier height being 0.93(9) kcal mol^{-1}. A comparison of the molecule with other molecules MeSiH$_n$F$_{3-n}$ (n = 0, 1, or 2) shows that on increasing fluorine substitution the Si—C and Si—F bond lengths decrease, the dipole moment increases, and the potential barrier to internal rotation V_3 decreases.

The perfluorinated molecule CF$_3$SiF$_3$ has also been investigated[80] in its ground state and several excited torsional (ν_6) states. The frequencies of these torsional states were derived from intensity measurements, and the barrier height was calculated to be 489(50) cm^{-1}, which compares with about 570 cm^{-1} in MeSiH$_3$. The ν_{12} = 1 state and torsionally excited

[77] M. Hayashi and C. Matsumara, *Bull. Chem. Soc. Japan*, 1972, **45**, 732.
[78] H. Jones and R. F. Curl, *J. Mol. Spectroscopy*, 1972, **41**, 226.
[79] J. R. Durig, Y. S. Li, and C. C. Tong, *J. Mol. Structure*, 1972, **14**, 255.
[80] D. R. Lide, D. R. Johnson, K. G. Sharp, and T. D. Coyle, *J. Chem. Phys.*, 1972, **57**, 3699.

states with $v_{12} = 1$, where v_{12} is the SiF$_3$ degenerate rocking vibration, were also observed, and the $v_{12} = 1$ state was estimated to be at 158(12) cm^{-1}. Perturbations were observed for $v_6 \geqslant 13$ in the $v_{12} = 0$ series and $v_6 \geqslant 7$ for the $v_{12} = 1$ series.

The analysis of the microwave spectra of four isotopic species of CD$_3$SiH$_2$Cl gave values of the rotational and centrifugal distortion constants as well as the ^{35}Cl quadrupole coupling constants.[81] Substitution co-ordinates were obtained using two different sets of three isotopic species, giving Si—Cl bond lengths of 2.049(13) Å and 2.052(5) Å, with corresponding angles of 26°53′ and 26°49′ between the Si—Cl bond and the a-axis.

To obtain a structure for SiMe$_3$Cl from the rotational constants of four isotopic species,[82] it was found necessary to assume r(C—H) = 1.095 Å, θ(Si—C—H) = 110.15°; these gave r(Si—Cl) = 2.022(50) Å, r(Si—C) = 1.857(5) Å, and θ(Cl—Si—C) = 110.5(10)°. The quadrupole coupling constant, eQq(^{35}Cl) = −46.9(15) MHz, was compared with those in the molecules SiH$_{3-n}$Cl$_n$ and CH$_{3-n}$Cl$_n$.

A small positive inertia defect and an intensity alternation implied that phosphabenzene, C$_5$H$_5$P, has a planar C_{2v} structure.[83] As none but the normal isotopic species gave an observable spectrum, only two structural parameters could be obtained from its rotational constants. Suitable values for the C—H and C—C bond lengths and the C—C—C and C—C—H bond angles were assumed, these being obtained from the structures of related compounds, and the P—C bond length and CPC angle were allowed to vary. The resulting structure is shown in Figure 4. Although the actual values will vary if the assumed values change, it was considered likely that the P—C bond length would be between 1.70 and 1.73 Å, and the CPC angle between 101 and 104°. This P—C bond length is between that found in PMe$_3$ (1.843 Å) and PPh$_3$ (1.66 Å). Seven vibrational states were also identified, with two fundamentals at 290(40) cm^{-1} and 325(40) cm^{-1}, these being respectively antisymmetric and symmetric with respect to the C_2 axis. The dipole moment of 1.54(2) D is consistent with a π-conjugated system.

The molecular g values and magnetic susceptibility anisotropies of selenophen were obtained from the molecular Zeeman spectra.[84] A comparison of the latter parameters with related molecules showed that selenophen had a similar aromaticity to that of thiophen, which is greater than that of furan.

From the microwave spectra of dimethyl sulphone, Me$_2$SO$_2$, an r_0 structure has been derived.[85] On the assumption that r(C—H) = 1.091 Å

[81] W. Zeil, R. Gegenheimer, S. Pferrer, and M. Dakkouri, *Z. Naturforsch.*, 1972, **27a**, 1150.
[82] J. R. Durig, R. O. Carter, and Y. S. Li, *J. Mol. Spectroscopy*, 1972, **44**, 18.
[83] R. L. Kuczkowski and A. J. Ashe, *J. Mol. Spectroscopy*, 1972, **42**, 457.
[84] W. Czieslik, D. Sutter, H. Dreizler, C. L. Norris, S. L. Rock, and W. H. Flygare, *Z. Naturforsch.*, 1972, **27a**, 1691.
[85] S. Saito and F. Makino, *Bull. Chem. Soc. Japan*, 1972, **45**, 92.

and θ(HCH) = 109.6°, Saito and Makino find r(C—S) = 1.777(6) Å, r(S—O) = 1.431(4) Å, θ(C—S—C) = 103.3(2)°, and θ(O—S—O) = 121.0(3)°. These values agree well with the X-ray structure, except for the O—S—O angle which is 117.9° in the X-ray study. The dipole moment of 4.432(41) D lies along the *b*-axis.

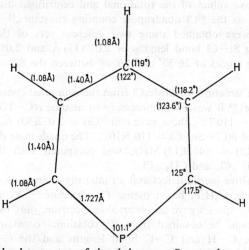

Figure 4 *Molecular structure of phosphabenzene. Structural parameters in parentheses have been transferred from similar molecules*

Transitions have been observed[86] between 9 and 24 GHz in the vapour of hydrogen fluoride at 203 K and 0.01 Torr which have been assigned to hexamer and heptamer molecules. The rotational assignment of two lines were confirmed by the Stark effect, the rest being assigned by comparison with predicted spectra for given structures. The structures assumed were of open chain H—F···H—F polymers, with F···H—F linear. The best fit of the other parameters is shown in Table 4 which includes those for the dimer reported above as well as the monomer. The lower stability of $(HF)_7$ compared with $(HF)_6$ is reflected in its shorter H—F and longer H···F distances.

Table 4 *Structural parameters of polymers of hydrogen fluoride*

Molecule	r(H—F)/Å	r(H···F)/Å	d(F···F)/Å	θ(F—F—F)/deg	Ref.
HF	0.9170	—	—	—	a
$(HF)_2$	0.9170[b]	—	2.79(5)	108[c]	61
$(HF)_6$	0.9997	1.4998	2.4995	104.0	86
$(HF)_7$	0.9640	1.6105	2.5745	103.7	86

[a] D. V. Webb and K. N. Rao, *J. Mol. Spectroscopy*, 1968, **28**, 121; [b] assumed; [c] F—F—H angle.

[86] U. V. Reichart and H. Hartmann, *Z. Naturforsch.*, 1972, **27a**, 983.

The centrifugal distortion K-type doubling in the C_{4v} molecules ^{79}BrF$_5$ and ^{81}BrF$_5$ has been observed for the first time.[87] This effect gives rise to additional terms $\pm R_6 J(J + 1)(J + 2)$ in the energy of the $K = \pm 2$ states; the value of R_6 was calculated to be 0.070(5) kHz. The force field obtained from i.r. data gave reasonable agreement with D_J, D_{JK}, and R_6 obtained here, being fairly insensitive to their values, but the value of D_{JK} enabled one extra interaction constant to be determined.

[87] R. H. Bradley, P. N. Brier, and M. J. Whittle, *J. Mol. Spectroscopy*, 1972, **44**, 536.

4
Vibrational Spectra of Small Symmetric Species and of Single Crystals

BY D. M. ADAMS

1 General Introduction

As in previous years the Reporters on vibrational spectra consider one of their primary responsibilities to be collection of a fairly complete set of references, even though many of these are unlikely to be of much value and make the text scrappy in places. The amount of low-quality spectroscopic work continues to rise and is now of more than nuisance value. This year examples have been found of authors repeating older work (of others) with a few minor variations, sometimes without acknowledgement of the earlier work or with omission of major references of direct relevance, series of papers in which the sole difference from one to the next is change of a central metal atom or a ligand substituent, and lengthy and detailed assignments of complicated spectra on the basis of a few i.r. frequencies and no other evidence, quite apart from correlations claimed with other physical parameters, the basis of the vibrational assignment sometimes being dubious. Papers reporting data for characterization of compounds have generally been listed as references only, as have those for which we consider assignments to be routine or dubious.

The year was marked by the appearance of several books of especial value. Volume I of an extremely comprehensive 'Index of Vibrational Spectra of Inorganic and Organometallic Compounds' by Greenwood, E. Ross, and Straughan [1] will prove invaluable. S. D. Ross's 'Inorganic Infrared and Raman Spectra' [2] ably fills a yawning gap in the review literature, covering the spectroscopy of lattices as well as the simpler and more symmetric complexes, and Sherwood's book [3] provides a painless introduction to aspects of solid-state spectroscopy too often dismissed as of interest to physicists only: it should be read by all chemists dealing with the spectra of solids. We are all indebted to Woodward for what is arguably the clearest account of molecular vibrations yet,[4] and Köningstein's

---

[1] N. N. Greenwood, E. J. F. Ross, and B. P. Straughan, 'Index of Vibrational Spectra of Inorganic and Organometallic Compounds, Vol. 1', Butterworths, London, 1972.

[2] S. D. Ross, 'Inorganic Infrared and Raman Spectra', McGraw-Hill, New York, 1972.

[3] P. M. A. Sherwood, 'Vibrational Spectroscopy of Solids', Cambridge University Press, 1972.

[4] L. A. Woodward, 'Introduction to the Theory of Molecular Vibrations and Vibrational Spectroscopy', Oxford University Press, 1972.

introduction to Raman theory [5] is notable for drawing together a number of more physical and mathematical aspects of the processes. Turrell has published a useful introduction to the vibrational spectroscopy of crystals.[6] The second and much enlarged edition of a well-established tricks-of-the-trade book has appeared.[7] In 1970 Adams and Newton published a set of 'Tables for Factor Group and Point Group Analysis':[8] these are tables of reduced representations and their use makes factor-group analysis of the spectra of solids as simple and reliable as using logarithms. The method is so simple that only four pages of text were needed to explain it. Fately and co-authors [9] have now published a full-length book to explain a highly cumbersome alternative means to the same end, thereby giving the reader the impression that analysis of solid-state spectra is difficult, when it is not. The Adams–Newton method also allows the use of internal co-ordinate sets and carries other tables especially tailored to the needs of polymer spectroscopy.

Other books, not seen by the Reporters, and some of marginal relevance to inorganic spectroscopy, are listed as references 10—22. The new

[5] J. A. Köningstein, 'Introduction to the Theory of the Raman Effect', D. Reidel Publishing Co., Dordrecht, Holland, 1972.
[6] G. Turrell, 'Infrared and Raman Spectra of Crystals', Academic Press, New York, 1972.
[7] 'Laboratory Methods in Infrared Spectroscopy', ed. R. G. J. Miller and B. Stace, 2nd edn., Heyden, London, 1972.
[8] D. M. Adams and D. C. Newton, 'Tables for Factor Group and Point Group Analysis', Beckman–RIIC Ltd., Croydon, 1970.
[9] W. G. Fately, F. R. Dollish, N. T. McDevitt, and F. F. Bentley, 'Infrared and Raman Selection Rules for Molecular and Lattice Vibrations: The Correlation Method', Wiley, New York, 1972.
[10] N. A. Chumaevskii, 'Vibrational Spectra of Group IVB and VB Hetero-organic Compounds', Nauka, Moscow, 1971.
[11] A. N. Lazarev, 'Vibrational Spectra and Structure of Silicates', Russian text, Nauka Press, Leningrad, 1968; English translation, Consultants Bureau, New York, 1972.
[12] A. G. Vlasov, V. A. Florinskaya, A. A. Venediktov, et al., 'Infrared Spectra of Inorganic Glasses and Crystals' (in Russian), Khimiya, Leningrad Otd., Leningrad.
[13] C. Bell, 'Physical Studies of Inorganic Compounds', Pergamon, Oxford, 1972.
[14] R. A. Nyquist and R. O. Kayel, 'Infrared Spectra of Inorganic Compounds', Academic Press, London, 1972.
[14a] 'Vibrational Spectra and Structure', ed. J. R. Durig, Dekker, New York, Vol. 1, 1972.
[15] K. W. F. Kohlrausch, 'Ramanspektren', reprint of the 1943 edition (Akademische Verlagsgesellschaft Becker und Erler Kom. Ges., Leipzig), published by Heyden, London, 1972.
[16] 'Raman Effect, Vol. 1, Principles', ed. A. Anderson, Dekker, New York, 1972.
[17] F. Abeles, 'Optical Properties of Solids', North-Holland Publishing Co., Amsterdam, 1972.
[18] 'Theory of Lattice Dynamics in the Harmonic Approximation', ed. A. A. Maradudin, I. P. Ipatova, E. Montroll, and G. H. Weiss, Academic Press, New York, 1971.
[19] C. J. Bradley and A. P. Cracknell, 'Mathematical Theory of Symmetry in Solids: Representation Theory for Point Groups and Space Groups', Oxford University Press, 1972.
[20] M. D. Harmony, 'Introduction to Molecular Energies and Spectra', Holt, Rinehart, and Winston, New York, 1972.
[21] 'Submillimetre Spectroscopy: Guide to the Theoretical and Experimental Physics of the Far Infrared', ed. G. W. Chantry, Academic Press, London, 1971.
[22] 'Molecular Spectroscopy, 1971' ed. P. W. Hepple, Institute of Petroleum, 1972.

Journal of Raman Spectroscopy [23] is to appear in 1973, and Laser-Raman Spectroscopy Abstracts [24] are already with us. Reviews have appeared on: the metal isotope effect on metal–ligand vibrations;[25] i.r. spectra of coordination compounds;[26, 27] ion-vibration phenomena,[28] dealing with such matters as the identification of sites of ion-pair formation and numbers of sites; vibrational spectroscopy of five-co-ordinated species;[29] i.r. and Raman spectra of solids;[30] neutron scattering and the vibrational spectra of molecular crystals,[31] including a variety of hydrated and ammonium salts; i.r. and Raman spectra of electron donor–acceptor complexes;[32] i.r. and Raman work in relation to the constitution, structure, and bonding in phosphorus compounds;[33] and the application of spectroscopic techniques to the structural analysis of coal and petroleum.[34] Tables of frequencies for 58 polyatomic molecules have been published.[35]

A simple technique has been described for obtaining i.r. emission spectra without instrument modification and illustrated by reference to $CuSO_4,H_2O$, $CuSO_4$, V_2O_3, V_2O_5, VO_2, and $AgNO_3$.[36] Raman spectra of a wide range of organic and inorganic compounds, including graphite, have been obtained using the rotating-sample method.[37] The use of Raman spectroscopy in the study of displacive ferroelectrics has been reviewed.[38] Otherwise it was a quiet year.

2 Spectra of Small Symmetric Species

Diatomic Species.—Various aspects of the spectra of diatomic species continue to attract attention although inevitably most recent work is of a rather physical nature, a situation that is unlikely to be relieved. Thus, nitrogen shows i.r. absorption at 2331 cm^{-1}, $\nu(N{\equiv}N)$, in an electric field of 160 kV cm^{-1}.[39] $\nu(F-F)$ of fluorine is at 891.5 cm^{-1} for the liquid and 0.3 cm^{-1} higher for the gas.[40] The pure rotational Raman spectrum of

---

[23] *Journal of Raman Spectroscopy*, D. Reidel Publishing Co., Dordrecht, Holland.
[24] *Laser-Raman Spectroscopy Abstracts*, ed. I. Sutherland, Science and Technology Agency, London.
[25] K. Nakamoto, *Angew Chem. Internat. Edn.*, 1972, **11**, 666.
[26] K. Nakamoto, *Amer. Chem. Soc. Monogr.*, 1971, No. 168 (Co-ordination Chem. Vol. 1), 134.
[27] M. T. Forel, J. Lascombe, A. Novak, and M. L. Josien, *Colloq. internat. Cent. nat. Rech. sci.*, 1970, No. 191, 167.
[28] W. F. Edgell, *Ions Ion Pairs Org. React.*, 1972, **1**, 153 (*Chem. Abs.*, 1972, **77**, 11 517).
[29] R. R. Holmes, *Accounts Chem. Res.*, 1972, **5**, 296.
[30] H. J. Becher, *Angew. Chem. Internat. Edn.*, 1972, **11**, 26.
[31] T. Kitagawa, *Appl. Spectroscopy Rev.*, 1972, **6**, 79.
[32] C. N. R. Rao, S. N. Bhat, and P. C. Dwivedi, *Co-ordination Chem. Rev.*, 1971, **5**, 1.
[33] E. Steger, *Z. Chem.*, 1972, **12**, 52.
[34] J. G. Speight, *Appl. Spectroscopy Rev.*, 1971, **5**, 211.
[35] T. Shimanouchi, *J. Phys. Chem., Ref. Data*, 1972, **1**, 189.
[36] G. Fabbri and P. Barald, *Appl. Spectroscopy*, 1972, **26**, 593.
[37] H. J. Sloane and R. B. Cook, *Appl. Spectroscopy*, 1972, **26**, 589.
[38] G. Burns, *Materials Res. Bull.*, 1971, **6**, 923.
[39] D. Courtois, C. Thiebeaux, and P. Jouve, *Compt. rend.*, 1972, **274**, *B*, 744.
[40] J.-C. Barral and O. Hartmanshenn, *Compt. rend.*, 1972, **274**, *B*, 981

Vibrational Spectra

chlorine yielded new values for the rotational constants of $^{35}Cl_2$ and $^{37}Cl_2$.[41] Fine structure in the Raman spectra of chlorine,[40] and of chlorine and bromine at 77 K in the ν and 2ν regions has been interpreted in terms of Davydov splittings.[42] Wavenumbers are shown in Table 1.

Table 1 Raman spectra of chlorine and bromine

Vibration	ν/cm^{-1}	2ν/cm^{-1}
$^{35}Cl-^{35}Cl$	539, 537.5	1070
$^{35}Cl-^{37}Cl$	530	1055
$^{37}Cl-^{37}Cl$	522.5	1040
$^{79}Br-^{79}Br$	301, 298	594
$^{79}Br-^{81}Br$	296, 294	590
$^{81}Br-^{81}Br$		587

The depolarization ratio of the fundamental and its first three harmonics has been measured for bromine and iodine and found to correspond roughly with the value (1/3) predicted by Mortensen. There is, superimposed, a small variation of ρ with the order of the harmonic, but this is not understood.[43] Br_2^+, obtained by oxidation of bromine with peroxydisulphuryl difluoride, shows a resonance Raman spectrum with lines at 360 (ν), 710 (2ν), and 1070 (3ν) cm^{-1}.[44] Various spectroscopic constants have been extracted from fine structure of the overtone progression in the resonance Raman spectrum of iodine,[45] which can also be excited in the continuum *above* the convergence limit of the excited molecule.[46] Collisional perturbations of the vibrational spectra of hydrogen and deuterium have been further studied by both Raman [47, 48] and i.r. methods.[49] In CCl_4 solution ν (2204.4), 2ν (4327.4), and 3ν (6371.8) of HI occur at the wavenumbers/cm^{-1} shown in parentheses.[50] The absolute i.r. intensities of the hydrogen halides have been analysed in detail [51] and values of ω_e for AX calculated (A = B, Al, Ga, In, or Tl; X = F, Cl, Br, or I).[52] Depolarization ratios and Raman scattering cross-sections have been deduced from study of the Stokes Q-branch fundamentals of N_2, O_2, CO, NO, CO_2, H_2O, SO_2, and CH_4.[53] Comparison of Raman spectra of HCl and HBr in β-quinol clathrates at room temperature and at 77 K shows that they are in

[41] P. J. Hendra and C. J. Vear, *Spectrochim. Acta*, 1972, **28A**, 1949.
[42] G. G. Dumas, N. Brigot, and J. P. Viennot, *Compt. rend.*, 1972, **275**, B, 379.
[43] M. Berjot, M. Jacon, and L. Bernard, *Compt. rend.*, 1972, **274**, B, 404.
[44] R. J. Gillespie and M. J. Morton, *Inorg. Chem.*, 1972, **11** 586.
[45] W. Kiefer and H. J. Bernstein, *J. Mol. Spectroscopy*, 1972, **43**, 366.
[46] W. Kiefer and H. J. Bernstein, *J. Chem. Phys.*, 1972, **57**, 3017.
[47] J. R. Murray and A. Javan, *J. Mol. Spectroscopy*, 1972, **42**, 1.
[48] Eng-Choon Looi, *Diss. Abs. (B)*, 1972, **32**, 5374.
[49] G. Varghese, S. N. Ghosh, and S. P. Reddi, *J. Mol. Spectroscopy*, 1972, **41**, 291.
[50] N. Van Thank and I. Rossi, *Canad. J. Chem.*, 1972, **50**, 3411.
[51] (a) M. Scrocco, R. Giuliani, and C. Costarelli, *Spectrochim. Acta*, 1972, **28A**, 761; (b) M. Scrocco, *ibid.*, p. 771; (c) M. Scrocco, *ibid.*, p. 777.
[52] V. S. Kushawaha, B. P. Asthana, and C. M. Pathak, *Spectroscopy Letters*, 1972, **5**, 357.
[53] C. M. Penney, L. M. Goldman, and M. Lapp, *Nature Phys. Sci.*, 1972, **235**, 110.

rotational motion. Results with N_2 and O_2 were less conclusive owing to their smaller rotational constants; CO_2, SO_2, H_2S, C_2H_2, HCOOH, MeOH, MeCN, and CD_3CN were also studied in β-quinol.[54] TeS has $\omega_e = 471.18$ cm^{-1},[55] whereas $\nu(Cl-O)$ is at 758 cm^{-1} in $Ca(ClO)_2$.[56]

Triatomic Species.—*Ab initio* calculation of the bending wavenumber of C_3 using the basis sets $(4s\ 3p\ 1d)$ and $(4s\ 2p\ 1d)$ yielded values near the experimental one, 64 cm^{-1}.[57] Various studies of more accessible trinuclear species yielded the assignments of Table 2. Ozone was matrix-isolated,[58]

Table 2 *Vibrational wavenumbers/cm^{-1} for triatomic entities*

Species	ν_1	ν_2	ν_3	$\nu_1 + \nu_3$	Method	Reference
$^{18}O_3$	1038	660	978, 970	1990	I.r.	58
$^{16}O_3$	1100	695	1030, 1025	2110		
O_3^-	—	—	ca. 800	—	I.r.	61, 62
S_3^-	533	232	580		I.r. + Raman	63
Br_3^-	164	53	191	—		
I_3^-	114	52	145	—		
$BrCl_2^-$	278	135	225	—	Raman	64
ICl_2^-	269	127	226	—		
IBr_2^-	162	64	175	—		
TeO_2	831.7	294	848.3			65

but has also been studied in the gas phase (the $\nu_1 + \nu_2 + \nu_3$ combination at *ca.* 2800 cm^{-1}), yielding accurate molecular constants.[59, 60] For O_3^- (Ar matrix, 14 K) ν_3 depends slightly upon the alkali metal used as photoelectron source. A valence angle of $110 \pm 5°$ was deduced; the bonds are considerably weaker than in ozone.[61] The wavenumbers for S_3^- (blue solutions of alkali-metal polysulphides in HMP) are higher than for H_2S_3 (488, 477, and 211 cm^{-1}), where the S_3 unit contains an additional antibonding electron.[63]

The two components of the Fermi doublet $\nu_1, 2\nu_2$ of $CO_2(g)$ at 192 K show complex structure. Two groups of five lines are interpreted in terms of the rotation and vibration of $(CO_2)_2$ dimers held by quadrupole–quadrupole interaction in the locked T position at an intermolecular

---

[54] J. E. D. Davies, *J.C.S. Dalton*, 1972, 1182.
[55] R. F. Barrow, M. A. H. Dudley, M. R. Hitchings, and K. K. Yee, *J. Phys.* (B), 1972, **5**, L.172.
[56] J. L. Arnau, P. A. Giguère, and S. Morissette, *Canad. J. Spectroscopy*, 1972, **17**, 63.
[57] D. H. Liskow, C. F. Bender, and H. F. Schaefer, *J. Chem. Phys.*, 1972, **56**, 5075.
[58] L. Brewer and J. Ling-Fai Wang, *J. Chem. Phys.*, 1972, **56**, 759.
[59] D. E. Snider and J. H. Shaw, *J. Mol. Spectroscopy*, 1972, **44**, 400.
[60] A. Barbe, C. Secroun, and P. Jouve, *Compt. rend.*, 1972, **274**, *C*, 615.
[61] M. E. Jacox and D. E. Milligan, *J. Mol. Spectroscopy*, 1972, **43**, 148.
[62] M. E. Jacox, and D. E. Milligan, *Chem. Phys. Letters*, 1972, **14**, 518.
[63] T. Chivers and I. Drummond, *Inorg. Chem.*, 1972, **11**, 2525.
[64] W. Gabes and H. Gerding, *J. Mol. Structure*, 1972, **14**, 267.
[65] M. Spoliti, S. N. Cesaro, and E. Coffari, *J. Chem. Thermodynamics*, 1972, **4**, 507.

distance of 4.1 Å.[66] The pressure-induced i.r. band of CO_2 at ca. 7.5 nm has been studied at various temperatures.[67]

The ν_2 vibration–rotation band of T_2O [68] has ν_0 (0, 1, 0) = 995.37 cm^{-1}, and the band centres of ν_1 (2299.790 cm^{-1}) and ν_3 (3716.577 cm^{-1}) of HTO [69] were deduced from high-resolution studies. Application of strong magnetic fields to water is said to increase i.r. absorption by 10—15%.[70] Published i.r. absorptivities (10—85 °C) for HDO in liquid D_2O in the ν(OH) region, analysed numerically, show a statistically significant weak band at ca. 3600 cm^{-1}, taken as confirmation of the presence of HDO species (ca. 4.6% at 10 °C) in the liquid state with unbonded OH groups.[71] Earlier conclusions that liquid water at 250 °C consists of freely rotating monomers have been revised on the basis of new work in the 3000—4200 cm^{-1} region at temperatures up to 500 °C and pressures as high as 500 atm; there is evidence of association up to at least the critical temperatures.[72] Further comment on the structure of the liquid state resulted from studies of ν(OD) of HDO both in water [73] and in aqueous water–dioxan, acetone, and THF mixtures.[74] The i.r. spectrum of 'polywater' is said to be similar to those of saturated aqueous solutions of NaOH and HCO_2Na.[75] Previous studies of the structure of aqueous salt solutions have been extended to include HTO in H_2O mixtures.[76] The water-structure-breaking properties of the anions in $NaBF_4$ and the perchlorates of Na, Mg, Zn, and Al are related to the extent of the splitting of the 1.45 μm band of water.[77]

Qualitative and fragmentary investigations of water in inorganic solids continue: compounds studied include Li_2SO_4,H_2O, Li_2SeO_4,H_2O, $LiXO_4,3H_2O$ (X = Cl or Mn), $LiI_3,3H_2O$, $Li_2CrO_4,2H_2O$, $Li_2Cr_2O_7,2H_2O$, and their deuteriates;[78] $LiBF_4,nH_2O$ and $LiBF_4,nD_2O$ (n = 1 or 3);[79] and $CuF_2,2H_2O$ and $M(H_2O)_4F_2$ (M = Fe, Co, Ni, Cd, or Zn).[80] However, more informative studies appear from time to time: rotational and translational modes of the water of crystallization in the certain salts were

[66] L. Mannik, J. C. Stryland, and H. L. Welsh, Canad. J. Phys., 1971, 49, 3056.
[67] A. P. Gal'tsev and M. A. Odishaviya, Soobshch. Akad. Nauk Gruz. S.S.R., 1972, 65, 301.
[68] R. A. Carpenter, N. M. Gailar, H. W. Morgan, and P. A. Staats, J. Mol. Spectroscopy, 1972, 44, 197.
[69] A. Fayt and G. Steenbeckeliers, Compt. rend., 1972, 275, B, 459.
[70] N. G. Klyuchnikov and G. Derygin, Uchen. Zap. Mosk. Gos. Pedagog. Inst., 1971, 345.
[71] E. C. W. Clarke and D. N. Glew, Canad. J. Chem., 1972, 50, 1655.
[72] Yu. E. Gorbatyi, M. B. Epel'baum, and G. V. Bondarenko, Trudy Soveshch. Eksp. Tekh. Mineral Petrogr. 8th, 1968 (publ. 1971), 207 (Chem. Abs., 1972, 76, 92 611).
[73] B. Z. Gorbunov and Yu. I. Naberukhin, J. Struct. Chem., 1972, 13, 16.
[74] B. Z. Gorbunov and Yu. I. Naberukhin, J. Mol. Structure, 1972, 14, 113.
[75] S. Suzuki, Vu Hai, and B. Vodar, J. Chim. phys., 1971, 68, 1385.
[76] G. E. Walrafen and L. A. Blatz, J. Chem. Phys., 1972, 56, 4216.
[77] S. Subramanian and H. F. Fisher, J. Phys. Chem., 1972, 76, 84.
[78] S. A. Shchukarev, T. G. Balicheva, and B. B. Lavrov, Vestnik Leningrad Univ. (Fiz. Khim.), 1971, 83.
[79] M. Manewa and H. P. Fritz, Z. Naturforsch., 1972, 27b, 1127.
[80] K. C. Patil and E. A. Secco, Canad. J. Chem., 1972, 50, 567.

assigned on the basis of H/D shifts.[81] Thus:

	ν_{rot}/cm^{-1}	ν_{trans}/cm^{-1}
Ba(ClO$_3$)$_2$,H$_2$O	451, 390	233, 207
CaSO$_4$,2H$_2$O CuCl$_2$,2H$_2$O	310, 306	271, 243

Similar work with hydrated hexahydroxyantimonates of Li, Na, Mg, Ba, Co, and Ni has shown that vibrational modes are at *ca.* 530 and 680 cm^{-1}, shifting to 405 and 447—520 cm^{-1} respectively upon deuteriation. In addition, ν(Sb—O—H) is sensitive to the nature of the cation, varying in the order Mg (1030 cm^{-1}) < Li < Ni < Co < Ba < Na (1115 cm^{-1}).[81a] Neutron-diffraction work on BaCl$_2$,2H$_2$O has shown that one of the four OH groups is much more weakly hydrogen-bonded than the others. Using the isotope-dilution technique, it has now been confirmed that for HDO in BaCl,2H$_2$O one ν(OX) mode is considerably higher than the other three, *viz.*[82]

ν(OH)	3452	3353	3316	3300 cm^{-1}
ν(OD)	2557	2486	2461	2450 cm^{-1} (all at -175 °C)

Studies of other hydrogen-bonded systems are conveniently included here. A relationship has been found between the angle at hydrogen of a bent hydrogen bond O···H···O and its normal vibrational wavenumbers/cm^{-1}: relative values for minerals such as diaspore, AlO(OH) (1070 and 963 cm^{-1}), and azurite, Cu$_3$(OH)$_2$(CO$_3$)$_2$ (956 and 839 cm^{-1}), are thereby rationalized.[83] In some hydrogen-bonded materials ν(OH) absorption gives rise to a continuum in the range *ca.* 800—2500 cm^{-1}, typical examples being the hydrated proton, strong acid solutions in sulphoxides, phosphine oxides, *etc.*, and a variety of carboxylic acids in amine solvents. An explanation of these bandwidths has now been advanced in terms of the relative lifetimes of the *very* rapidly interconverting hydrogen-bonded structures, AH···B$\rightleftarrows$A$^-$···HB$^+$.[84] The anion in [C$_5$H$_5$NH]$_2$ [MnCl$_4$] has been shown to the hydrogen-bonded to pyridinium.[85]

Raman spectra and depolarization ratios for ^{16}OF$_2$ and ^{18}OF$_2$ in argon matrices at 16 K show wavenumbers within 1 cm^{-1} of those reported earlier from i.r. experiments, thereby confirming the interpretation. Laser-induced detachment of fluorine atoms gave significant yields of OF radicals.[86] Fundamental wavenumbers/cm^{-1} of HOF have been determined from vibration–rotation spectra [87] and for dilute nitrogen matrices at *ca.* 8 K,[88]

---

[81] K. Fukushima and T. Yanagida, *Bull. Chem. Soc. Japan*, 1972, **45**, 2285.
[81a] T. G. Balicheva and N. I. Roi, *J. Struct. Chem.*, 1971, **12**, 384.
[82] G. Brink, *Spectrochim. Acta*, 1972, **28A**, 1151.
[83] F. Zigan, *Ber. Bunsengesellschaft phys. Chem.*, 1972, **76**, 686.
[84] J. Husar and M. M. Kreevory, *J. Amer. Chem. Soc.*, 1972, **94**, 2902.
[85] R. Robert, C. Brassy, and A. Mellier, *Compt. rend.*, 1972, **274**, *B*, 341.
[86] L. Andrews, *J. Chem. Phys.*, 1972, **57**, 51.
[87] E. H. Appelman and H. Kim, *J. Chem. Phys.*, 1972, **57**, 3272.
[88] J. A. Goleb, H. H. Claassen, M. H. Studier, and E. H. Appelman, *Spectrochim. Acta*, 1972, **28A**, 65.

Vibrational Spectra

viz.

	HOF			
	3578.5	1354.8	889.0	Gas
	3537.1	1359.0	886.0	N_2 matrix
DOF	2643.5	1003.9	891.1	Gas

I.r. absorption by OBF in Ne and Ar matrices shows only ν_2 and ν_3, compatible with (but not proving) a linear structure. In Ne wavenumbers/cm^{-1} are:

	ν_3 2151,	(ν_1 966),	ν_2 521
$O^{10}BF$			
$O^{11}BF$	2081,	(964),	502

Values of ν_1 were estimated.[89] The new angular ions FH_2^+, FHD^+, and FD_2^+, formed by reaction of HF with BF_3 or SbF_5, have vibrational wavenumbers/cm^{-1} as shown in Table 3 (BF_4^- salts only).[90] The widths of the

Table 3 Vibrational wavenumbers/cm^{-1} of angular triatomic ions

FH_2^+	FD_2^+	FHD^+	
3200	2400	—	$2\nu_2$
3080	2320	2980	ν_3
2970	2250	2270	ν_1
1680	1240	1495	ν_2

ν_2 (δ) and ν_3 (stretch) i.r. modes of DF_2^- in its Na$^+$ and K$^+$ salts show a striking reduction, together with band frequency shifts, as the percentage of DF_2^- in HF_2^- is lowered towards 'defect' concentrations. The considerable width of these bands in pure $NaHF_2$ and KHF_2 is therefore associated with interionic coupling. The Raman spectrum of $NaHF_2$ shows lines at 630.5 (ν_1, A_{1g} of HF_2^-) and 145 cm^{-1} (E_g, anion libration). Combining vibrational and inelastic neutron-scattering data, the following assignments result (wavenumbers/cm^{-1}):[91]

	$NaHF_2$	KHF_2
$\nu_3 \, A_{2u}$	1500	1450
$\nu_2 \, E_u$	1210	1233
$\nu_1 \, A_{1g}$	630.5	596
$\nu_{rot} \, E_g$	145	143.5
$\nu_{trans} \, E_u, A_{2u}$	228, 180	104
Acoustic, A_{2u}	142, 90	—

I.r. spectra of the products from vacuum-u.v. photolysis of Ar–O_2–HCl and Ar–O_2–H_2O samples (and their deuteriates) at 14 K were identified as due (as appropriate) to DCl_2^-, $DHCl^-$, HOCl, DOCl, D_2O_2, HCl_2^-, DOOH, O_3, DO_2, HO_2, H_2O_2, HOD, and ClO_2. In particular, a detailed study and normal-co-ordinate analysis was made of the spectra of the free radicals HO_2 and DO_2 [δ(OH) 1389, ν(O—O) 1101 cm^{-1}], confirming previous assignments. The best fit for DO_2 was obtained with a D—O—O

---

[89] A. Snelson, *High Temp. Sci.*, 1972, **4**, 141.
[90] M. Couzi, J.-C. Cornut, and P. V. Huong, *J. Chem. Phys.*, 1972, **56**, 426.
[91] J. J. Rush, L. W. Schroeder, and A. J. Melveger, *J. Chem. Phys.*, 1972, **56**, 2793.

bond angle of 105 ± 5°. The relatively small OH stretch force constant and the almost equal values of $k(O-O)$ for HO_2 and O_2^- imply closely related electronic structures.[92] Passing mixtures of I_2, H_2, and Ar through a glow discharge is believed to yield the linear radical IHI. I.r. absorption wavenumbers/cm^{-1} are:[93]

	$\nu_1 + \nu_3$	ν_3	ν_1 (estimated)
IH_2	802.8	682.1	120.7
ID_2	594	470	124

A large quartic term is associated with ν_3. Frequencies for all observed fundamentals, overtones, and combinations of $^{14}N_2^{16}O$, $^{15}N_2^{16}O$, $^{14}N^{15}N^{16}O$, $^{15}N^{14}N^{16}O$, and $^{14}N_2^{18}O$ have been listed and used to determine accurate force constants: $k(N-N)$ 9.1999, $k(N-O)$ 6.0501, $k(\delta)$ 0.3319 mdyn Å^{-1}.[94] When N_2O is produced *in situ* in alkali-metal halide matrices by heating NH_2OH,HCl or NH_4NO_3 it is apparently trapped in sites of two different symmetries. In a KBr matrix at -180 °C the fundamentals have values ν_1 1294, ν_2 589, ν_3 2241 cm^{-1}.[95] The integrated intensities of ν_3 and the overlapping $(\nu_2 + \nu_3 - \nu_2)$ band of N_2O have been measured at 300 K.[96]

The 14 μm region of the HCN spectrum has been studied at high resolution.[97] I.r. spectra of isotopic variants of HCN isolated in argon matrices at 8 K show discrepancies of *ca.* 2 cm^{-1} in the isotope shifts in comparison with analogous gas-phase experiments. Analysis shows that these cannot be attributed to experimental error, changes in anharmonic character, or a change in the geometry of HCN in the matrix, but are caused by the limitations of the linear XYZ vibrational model as a description of the motions of a matrix-isolated molecule.[98,99] The matrix environment has also been shown to affect the i.r. spectra of ClCN and BrCN; these molecules have their degenerate bending modes split, the first overtones of these two bending components forming a Fermi resonance triad with the ν_1 fundamental.[100] The following assignments (wavenumbers/cm^{-1}) are reported for the anions in $Ph_4As^+NCO^-,2H_2O$ [101] and $Me_4N^+TeCN^-$:[102]

	$\nu(CN)$, ν_1, Σ^+	$\nu(NCE)s$, ν_2, Σ^+	Def., ν_3, π
NCO^-	2155	1282, 1202	630
$NCTe^-$	2075	451	366

[92] M. E. Jacox and D. E. Milligan, *J. Mol. Spectroscopy*, 1972, **42**, 495.
[93] P. N. Noble, *J. Chem. Phys.*, 1972, **56**, 2088.
[94] D. F. Smith, J. Overend, R. C. Spiker, and L. Andrews, *Spectrochim. Acta*, 1972, **28A**, 87.
[95] I. C. Hisatsune, *J. Chem. Phys.*, 1972, **57**, 2631.
[96] J. E. Lowder, *J. Quant. Spectroscopy Radiative Transfer*, 1972, **12**, 873.
[97] P. K. L. Yin and N. Rao, *J. Mol. Spectroscopy*, 1972, **42**, 385.
[98] J. Pacansky and G. V. Calder, *J. Mol. Structure*, 1972, **14**, 363.
[99] J. Pacansky and G. V. Calder, *J. Phys. Chem.*, 1972, **76**, 454.
[100] T. B. Freedman and E. R. Nixon, *J. Chem. Phys.*, 1972, **56**, 698.
[101] O. H. Ellestad, P. Klaeboe, E. E. Tucker, and J. Songstad, *Acta Chem. Scand.*, 1972, **26**, 1721.
[102] O. H. Ellestad, P. Klaeboe, and J. Songstad, *Acta Chem. Scand.*, 1972, **26**, 1724.

v_2 of NCO⁻ is complicated by Fermi resonance with $2v_3$ and all of the frequencies of NCTe⁻ are cation-dependent.

In a valuable contribution, data from i.r. (gas-phase and matrix-isolated) work with dihalides of Group IIA and IIB elements have been tabulated and critically reviewed; particular attention is given to determination of molecular geometry.[103] Data for the matrix-isolated entities $SiCl_2$, $SiBr_2$ (i.r.), and MX_2, MXY (Raman), where M = Ge, Sn, or Pb, are summarized in Table 4. Stepwise polymerization of $GeCl_2$ was observed on warming

Table 4 *Vibrational wavenumbers/cm⁻¹ for some dihalide species MX_2*

Species	v_1	v_2	v_3	Reference
$^{28}Si^{35}Cl_2$	512.5	—	501.4	
$^{28}Si^{35}Cl^{37}Cl$	510.0	—	497.9	
$^{28}Si^{37}Cl_2$	505.4	—	496.1	
$^{30}Si^{35}Cl_2$	n.o.	—	491.0	105 [a]
$^{28}SiBr_2$	402.6	—	399.5	
$^{29}SiBr_2$	397.4	—	394.0	
$^{30}SiBr_2$	392.7	—	388.8	
$GeCl_2$	390	163	362	
$SnCl_2$	341	124	320	
$PbCl_2$	305	104	281	
$SnBr_2$	237	84	223	104 [b]
$PbBr_2$	208	n.o.	189	
SnClBr	—	328, 228	—	
PbClBr	—	295, 200	—	

[a] Argon matrices. Data for He and N_2 matrices are in the original paper. [b] N_2 matrices.

up.[104] Using isotopic data the bond angles were estimated as $105 \pm 3°$ for $SiCl_2$ and $109 \pm 3°$ for $SiBr_2$.[105] The v_2 and v_3 fundamentals of the metaborate ion dispersed in alkali-metal halide matrices, as well as various combinations, have been identified for $^nB^mO^lO^-$, where n = 10 or 11 and m and l = 16, 17, or 18.[106] Typically, for $^{10}B^{16}O_2^-$ in KBr, v_2 = 608 and v_3 = 2029 cm⁻¹.

Co-condensation of metal atoms and sundry gaseous species in a matrix has been a fruitful source of new entities for some time. Prominent among these experiments has been work with alkali-metal atoms, and this was further represented in 1972. The new emphasis is to extend such work with transition-metal atoms and to create, among other things, simple binary analogues of the carbonyl and nitrogeno-complexes which are normally stabilized by other ligands. Thus, co-condensation of CO and N_2 with Cr, Ni, or Pt atoms has been investigated by Turner and co-workers,[107] and

---

[103] I. Eliezer and A. Reger, *Co-ordination Chem. Rev.*, 1972, **9**, 189.
[104] G. A. Ozin and A. Vander Voet, *J. Chem. Phys.*, 1972, **56**, 4768.
[105] G. Maass, R. H. Hague, and J. L. Margrave, *Z. anorg. Chem.*, 1972, **392**, 295.
[106] H. W. Morgan and P. A. Staats, *Spectrochim. Acta*, 1972, **28A**, 600.
[107] J. K. Burdett, M. A. Graham, and J. J. Turner, *J.C.S. Dalton*, 1972, 1620.

similar activity continues apace in Toronto.[108, 109] Most groups working on matrix isolation make extensive use of isotopic substitution coupled with normal-co-ordinate analysis in settling assignments. The snag is that it is usual to minimize ν(observed) $-$ ν(calculated) to give the best overall *frequency* fit. However, this does not reproduce exactly the isotopic *shifts* and *patterns* which are known with considerable accuracy. A useful method has now been given for *direct* constraint of isotopic frequency shifts in least-squares determination of force constants, which accurately reproduces the *patterns* of isotopic frequencies; the method was illustrated with reference to di-imide, HNNH, and its deuterates.[110] ν(N≡N) bands of nitrogeno-complexes appear in the characteristic region *ca.* 2100 cm^{-1}; use of ^{15}N labelling helps to identify the number of molecules co-ordinated to each metal atom. For CrN$_2$ and NiN$_2$ there is also some band splitting due to occupation of different sites in the matrix. Of especial interest is the assignment of ν(N≡N) absorption due to nitrogen co-ordinated to clusters of atoms. Wavenumbers/cm^{-1} are:

	Cr	Ni	Pt	Cu
M^{14}N$_2$	2126.8	2169.4	2205.6	—
M(cluster)^{14}N$_2$	2089.8	2206.0	2230	2272.2

The failure to observe dinitrogen complexes M(N$_2$)$_n$ with $n > 1$ is attributed to the large positive charge created on the metal atom when bound to the N$_2$ ligand.[107] The series Pd(CO)$_n$ ($n = 1$—4) has been obtained by Ozin and co-workers by co-condensation of the constituents in argon at 10 K. Their characteristic ν(CO) i.r. wavenumbers/cm^{-1} are: $n = 1$, linear $C_{\infty v}$, 2050; $n = 2$, linear $D_{\infty h}$, 2044; $n = 3$, trigonal planar D_{3h}, 2060; $n = 4$, tetrahedral T_d, 2070. No ν(Pd—C) bands were found to a lower limit of 200 cm^{-1}.[109] Similar i.r. experiments using oxygen led to the formation of Ni(O$_2$) and Ni(O$_2$)$_2$ in which the oxygen molecules were shown to be co-ordinated sideways, in contrast with the end-on attachment of nitrogen. No ν(Ni—O) was found down to 200 cm^{-1}. ν(O—O) wavenumbers/cm^{-1} are: Ni(^{16}O$_2$) 966.2, Ni(^{16}O^{18}O) 940.1, and Ni(^{18}O$_2$) 913.6; bands in the range 1000—1055 cm^{-1} are due to various Ni(O$_2$)$_2$ species.[108] Matrix-isolated (NiF$_2$)$_2$CO and NiF$_2$CO have been identified by the appearance of ν(CO) and changes in the ν(Ni—F) region, and analogous results were obtained for NiF$_2$N$_2$ and NiCl$_2$N$_2$.[111] Andrews' group have also got laser-Raman matrix work off the ground and are continuing their fruitful studies using alkali-metal atoms: computer averaging of the spectra is sometimes necessary.[112] Reaction with oxygen yields (M^{I})$^+$(O$_2$)$^-$,

[108] H. Huber and G. A. Ozin, *Canad. J. Chem.*, 1972, **50**, 3746.
[109] E. P. Kündig, M. Moskovits, and G. A. Ozin, *Canad. J. Chem.*, 1972, **50**, 3587.
[110] J. W. Nibler and D. M. Barnhart, *J. Mol. Spectroscopy*, 1972, **44**, 236.
[111] C. W. DeKock and D. A. Van Leirsburg, *J. Amer. Chem. Soc.*, 1972, **94**, 3235.
[112] D. A. Hatzenbuhler, R. R. Smardzewski, and L. Andrews, *Appl. Spectroscopy*, 1972, **26**, 479.

believed to be an ionically bonded triangular species.[113, 114] In contrast with i.r. experiments, in which $\nu(M^I-O)$ shows up strongly, no Raman evidence was found for it, cf. Ozin's similar experience. However, O_2KO_2 shows a Raman band at 305 cm^{-1} due to $\nu(K-O)s$, implying more covalent bonding than in KO_2.[113] Observed values for $\nu(O-O)$ are shown in Table 5. I.r. absorption at 1800 and 1535 cm^{-1} is due to $\nu(N=N)$ modes in

Table 5 $\nu(O-O)$ *for species* M^IO_2

Species	$\nu(O-O)$/cm^{-1}	Species	$\nu(O-O)$/cm^{-1}
LiO_2	1093	KO_2	1108
$Na^{16}O_2$	1094		
$Na^{16}O^{18}O$	1064	$(Na^+O_2^-)_2$	1079
$Na^{18}O_2$	1034	$(K^+O_2^-)_2$	1098—1104

the products of co-condensation of lithium and nitrogen. The lower feature is attributed to $N_2Li_2N_2$ in which the N_2 *groups* are equivalent but have non-equivalent atoms: the higher absorption is due to LiN_2.[115] Many Li, C, and N isotopic variants of the gaseous molecule LiNC have been studied in matrix-isolated form and shown to be linear. Typically, for $^6Li^{14}N^{12}C$ observed wavenumbers are: $\nu(CN)$ 2080.5, $\nu(Li-N)$ 722.9, and def. 121.7 cm^{-1}.[116] The unobserved ν_3 stretch of matrix-isolated $UO_2(g)$ has been estimated to be at 755 cm^{-1}, and an angle of 105° has been deduced.[117]

Tetra-atomic Species.—A band at *ca.* 1000 cm^{-1} has been attributed to matrix-isolated *trans*-O_4^-.[62] Using in-cavity excitation, the Raman wavenumbers/cm^{-1} of $As_4(g)$ at 610 °C were found to be: $\nu_1(a_1)$ 344, $\nu_2(e)$ 210, and $\nu_3(t_2)$ 255.[118] Salts of the tetrahedral ions M_4^{4-} show i.r. absorption wavenumbers/cm^{-1} as follows: K_4Si_4 359 and 334, with Rb^+ and Cs^+ salts similar, K_4Ge_4 202 and 188. The T_2 stretching mode has been split $(B + E)$ by the S_4 site field.[119] Pb_4O_4 is also thought to be tetrahedral and shows i.r. absorption (N matrix) at 474.4 and 374.4 cm^{-1}, whereas Pb_2O_2 (557.4 and 467.7 cm^{-1}) is probably of D_{2h} symmetry. For comparison, the stretch in PbO is at 718.4 cm^{-1}.[120]

Further details on trigonal and planar AX_3 species have accumulated. Selected Raman data for the boron trihalides are shown in Table 6. Both liquid and solid samples were studied.[121] An independent and direct observation of ν_2 of BBr_3 places it at 370 cm^{-1} (^{11}B) and 385 cm^{-1} (^{10}B)

[113] R. R. Smardzewski and L. Andrews, *J. Chem. Phys.*, 1972, **57**, 1327.
[114] D. A. Hatzenbuhler and L. Andrews, *J. Chem. Phys.*, 1972, **56**, 3398.
[115] R. C. Spiker, L. Andrews, and C. Trindle, *J. Amer. Chem. Soc.*, 1972, **94**, 2401.
[116] Z. K. Ismail, R. H. Hague, and J. L. Margrave, *J. Chem. Phys.*, 1972, **57**, 5137.
[117] S. Abramowitz and N. Acquista, *J. Phys. Chem.*, 1972, **76**, 648.
[118] S. B. Brumbach and G. M. Rosenblatt, *J. Chem. Phys.*, 1972, **56**, 3110.
[119] H. Bürger and R. Eujen, *Z. anorg. Chem.*, 1972, **394**, 19.
[120] J. S. Ogden and M. J. Ricks, *J. Chem. Phys.*, 1972, **56**, 1658.
[121] R. J. H. Clark and P. D. Mitchell, *J. Chem. Phys.*, 1972, **56**, 2225.

Table 6 Raman wavenumbers/cm^{-1} for BX$_3$ in the liquid state [121]

Species	$\nu_1(a'_1)$	$\nu_3(e')$	$\nu_4(e')$	$2\nu_2(A'_1)$
$^{10}BCl_3$	⎫	989.7	—	934.4
$^{11}BCl_3$	⎬ 472.7	950.7	253.7	895.3
$^{10}BBr_3$	⎫	856	—	(788)
$^{11}BBr_3$	⎬ 278.8	816	150.9	754.5
$^{10}BI_3$	⎫	724.8	—	647.0
$^{11}BI_3$	⎬ 192	691.8	101.0	618.5

respectively.[122] The ν_4 region of the i.r. spectra of $^{11}BF_3$ and $^{10}BF_3$ (ca. 480 cm^{-1}) has been studied at high resolution.[123] The Raman spectrum of SbF$_3$(g) is in good agreement with earlier matrix-isolation data.[124]

I.r. absorption wavenumbers/cm^{-1} for pyramidal (C_{3v}) AlCl$_3$ in an argon matrix, viz. $\nu_1(a_1)$ 382.2, $\nu_2(a_1)$ 149.2, $\nu_3(e)$ 594.7, and $\nu_4(e)$ 182.8, may be compared with earlier Raman data for AlCl$_3$(g), viz. 375 (pol.) and 150 cm^{-1}.[125] A normal-co-ordinate analysis was performed. Good agreement with the Raman work was also found for Al$_2$Cl$_6$ in the same matrix for which the wavenumbers/cm^{-1} 620.0, 480.2, and 418.3 are quoted.[126] The i.r. intensities of gaseous and solid NH$_3$ and ND$_3$ have been analysed,[127] and assignments of Table 7 are offered as a result of Raman measurement on liquid mixtures of deuteriated ammonias.[128]

Table 7 Assignments and Raman wavenumbers/cm^{-1} for various ammonias [128]

	NH$_3$	ND$_3$	NH$_2$D	ND$_2$H	
ν_2	1055	812	980	895	ν'_2
ν_4	1645	1204	1610	1258	ν'_4
$\nu_1 + 2\nu_4$	⎧ 3303	2410	1386	1468	ν''_4
	⎩ 3218	2358	2474	3368	ν'_3
ν_3	3386	2529	3380	2490	ν''_3
			3327	2415	ν'_1

The isotopic variants P^{35}Cl$_3$ (260.5), P^{35}Cl$_2$^{37}Cl (258.6) and P^{35}Cl^{37}Cl$_2$ (256.8 cm^{-1}) have the ν_2 wavenumbers shown, determined from i.r. absorption in the gas phase.[129] Normal-co-ordinate analysis of the rather complete i.r. and Raman data for PF$_2$I shows that internal co-ordinates are heavily mixed (ca. 55 : 45%) for ν_3 and ν_4: a', ν_1 851 cm^{-1} ν(PF), ν_2 413 cm^{-1} δ(FPF), ν_3 375, and ν_4 198 cm^{-1} ν(PI) + δ(FPI); a'', ν_5 846 cm^{-1} ν(PF), ν_6 204 cm^{-1} δ(FPI).[130] An i.r. matrix-isolation study of rare-earth trifluorides reported last year showed spectra consistent with planar (D_{3h})

[122] M. C. Pomposiello and A. Brieux de Mandirola, *J. Mol. Structure*, 1972, **11**, 191.
[123] F. N. Masri, *J. Mol. Spectroscopy*, 1972, **43**, 168.
[124] L. E. Alexander and I. R. Beattie, *J.C.S. Dalton*, 1972, 1745.
[125] I. R. Beattie and J. R. Horder, *J. Chem. Soc. (A)*, 1969, 2655.
[126] M. L. Lesiecki and J. S. Shirk, *J. Chem. Phys.*, 1972, **56**, 4171.
[127] M. Uyemura and S. Maeda, *Bull. Chem. Soc. Japan*, 1972, **45**, 2225.
[128] B. deBettignies and F. Wallart, *Compt. rend.*, 1972, **275**, B, 283.
[129] A. Ruoff, H. Bürger, S. Biedermann, and J. Cichon, *Spectrochim. Acta*, 1972, **28A**, 953.
[130] C. R. S. Dean, A. Finch, and P. N. Crates, *J.C.S. Dalton*, 1972, 1384.

structures, except for PrF$_3$ which exhibited *two* bands (542 and 458 cm^{-1}) and was thought to be pyramidal. However, recent electron-diffraction work shows that it is planar. In an attempt to reconcile these pieces of evidence the Raman and i.r. spectra have been further investigated and clearly rule out the C_{3v} suggestion. In terms of D_{3h} rules the assignment of the bands (cm^{-1}) now made is:

Raman	526 (pol.) A'_1	458 (depol.) E'	99 (depol.) E'	
I.r.	542	458	99	86 A''_2

Since the 542 cm^{-1} feature is not present in the Raman spectrum (it should be both strong and polarized), the pyramidal structure is untenable; the band is shown not to be due to either dimers or impurities. The following unusual explanation is advanced. $v_3(E')$ has been perturbed from its expected location *ca*. 493 cm^{-1} by resonance interaction with an E' *electronic* level giving 458 (E') and 542 ($A'_1 + A'_2$). The observed activities can occur only if the ground electronic state is E' (since other possible symmetries would lead to only one i.r. absorption). In support of this explanation it is noted that the position of v_3 is raised in nitrogen matrices and is also sensitive to nitrogen doping of the argon matrix. The effect is found only for PrF$_3$ since it requires an unlikely combination of events, *viz*. a degenerate ground state and resonance with v_3.[131]

The SF$_3^+$ cation has been characterized and its assignment established in two concordant studies.[132, 133] In HF solution Raman wavenumbers/cm^{-1} are 943 (pol.) $v_1(a_1)$, 920 (depol.) $v_3(e)$, 539 $v_2(a_1)$, and 411 (depol.) $v_4(e)$ for the BF$_4^-$ salt.[132] Data for several AXY$_2$ species of C_s symmetry are shown in Table 8. The two independent sets of data for OClF$_2^+$ are in good

Table 8 *Vibrational wavenumbers/cm^{-1} for species* AXY$_2$ [a]

	a'			a''			
Species	v_1 $v(AY_2)s$	v_2 $v(AX)$	v_3 $\delta(AY_2)$	v_4 $\delta(XAY)$	v_5 $v(AY_2)a$	v_6 $\delta(XAY)$	Ref.
OClF$_2^+$	735	1330	512	383	700	403	134, 135
SFO$_2^-$	1100	598	496	378	1170	280	
SClO$_2^-$	1120	215	535	—	1290	—	136
SBrO$_2^-$	1109	201	530	—	1285	—	
SIO$_2^-$	1104, 1085	182	530	—	1275	—	
SeOF$_2$ [b]	667	1049	362	253	637	282	124

[a] Raman and i.r. data are mixed in this Table. Solution data for the SXO$_2^-$ ions are in the original: solid-state data are listed here. [b] Raman gas-phase study.

[131] M. Lesiecki, J. W. Nibler, and C. W. DeKock, *J. Chem. Phys.*, 1972, **57**, 1352.
[132] M. Brownstein and J. Shamir, *Appl. Spectroscopy*, 1972, **26**, 77.
[133] D. D. Gibler, C. J. Adams, M. Fischer, A. Zalkin, and N. Bartlett, *Inorg. Chem.*, 1972, **11**, 2325.
[134] R. Bougon, *Compt. rend.*, 1972, **274**, C, 696.
[135] K. O. Christe, E. C. Curtis, and C. J. Schack, *Inorg. Chem.*, 1972, **11**, 2212.
[136] D. F. Burow, *Inorg. Chem.*, 1972, **11**, 573.

agreement,[134, 135] as are the assignments except for a difference over $\nu_4(a')$ and $\nu_6(a'')$: in HF solution the state of polarization of the two lowest lines could not be determined. Although there is, therefore, no hard evidence for the order $\nu_6 > \nu_4$, comparison with isoelectronic SOF_2 and normal-co-ordinate analyses both suggest that it is correct. The order of SO_2–X^- interaction in the series XSO_2^- (X = F, Cl, Br, or I), *viz*. F > Cl < Br < I, determined by normal-co-ordinate analysis, is supported by the relative intensities of the electronic transitions.[136] The vapours of both SOF_2 and SO_2F_2 have been studied by i.r. absorption at moderate resolution. A notable feature is the extensive satellite absorption associated with upper-state transitions (established by their temperature dependence and by intensity calculations). Hot band sequences were found to originate from: $\nu_4(a_1)$ 385 cm^{-1} torsion, $\nu_5(a_2)$ 389 cm^{-1} deformation for SO_2F_2, and $\nu_4(a')$ 377.8 and $\nu_6(a'')$ 392.5 cm^{-1} deformations for SOF_2.[137]

The assignments below are reported for aqueous solutions of KXO_3 salts (X = Cl, Br, or I). ν_2 has the following wavenumbers/cm^{-1} for the isotopic variants shown: Br^{16}O$_3^-$ 418, Br^{18}O^{16}O$_2^-$ 414, Br^{18}O$_2^{16}$O$^-$ 407, and Br^{18}O$_3^-$ 401 cm^{-1}.[138]

	$\nu_1(a_1)$	$\nu_2(a_1)$	$\nu_3(e)$	$\nu_4(e)$
ClO$_3^-$	933	608	977	477
BrO$_3^-$	805	418	805	358
IO$_3^-$	805	358	775	320

In molten $KClO_3$ and $NaClO_3$ the anion frequencies (Raman) are close to those found in solutions and there is no lifting of degeneracy;[139] in contrast, in a similar Raman experiment with alkali-metal carbonates the 'forbidden' $\nu_2(a_2'')$ mode was observed at *ca*. 880 cm^{-1} and $\nu_3(e')$ split.[140]

The new compound BrNCO shows i.r. absorption (solid) at 3440, 2256, 2164, 2120, 1289, 690, 566, and 473 cm^{-1}.[141]

Penta-atomic Species.—The accumulation of more complete and accurate frequencies for tetrahedral species AB_4 continues, with emphasis on new Raman data. Use of isotopic substitution coupled with force-field calculations is much in evidence, both in regular and in substituted entities.

Raman and i.r. frequencies have been listed for liquid and crystalline samples of SiH_4–SiD_4 mixtures.[142] SiH_4 shows a transition at 63.75 K but no crystallographic data are available on either phase. Spectra of phase I are similar to those of the liquid and presumably arise from orientationally

---

[137] A. J. Sumodi and E. L. Pace, *Spectrochim. Acta*, 1972, **28A**, 1129.
[138] D. J. Gardiner, R. B. Girling, and R. E. Hester, *J. Mol. Structure*, 1972, **13**, 105.
[139] A. B. Lece, A. J. Kale, and K. Sathianandan, *High Temp. Sci.*, 1972, **4**, 231.
[140] J. B. Bates, M. H. Brooker, A. S. Quist, and G. E. Boyd, *J. Phys. Chem.*, 1972, **76**, 1565.
[141] W. Gottardi, *Monatsch.*, 1972, **103**, 1150.
[142] R. P. Fournier, R. Savoie, N. D. The, R. Belzile, and A. Cabana, *Canad. J. Chem.*, 1972, **50**, 35.

disordered molecules, but those of phase II are highly complex and compatible with a structure having one of the following factor groups: S_4, C_4, C_{4h} (tetragonal); C_{3v}, D_3, D_{3d} (trigonal); C_{3h}, C_6, C_{6h} (hexagonal). High-resolution spectra of the ν_3 bands of ^{28}SiH$_4$ [143] and GeH$_4$ [144] have been analysed, as have ν_2 and ν_4 of GeH$_4$.[143a]

Table 9 *Vibrational wavenumbers*/cm^{-1} *for tetrahalogeno-species* [a]

Species	$\nu_1(a_1)$	$\nu_2(e)$	$\nu_3(t_2)$	$\nu_4(t_2)$	Conditions	Ref.
PCl$_4^+$	458	178	662	255	Raman, PCl$_6^-$ salt	148
TiF$_4$(g)	712	185	793	209	Raman (g)	124
BeF$_4^{2-}$	547	255	800	385	Raman, melts	149
GeCl$_4$	—	(147)[b]	463	172		
SnCl$_4$	—	(100)[b]	—	131		
GeBr$_4$	—	—	324	113	I.r., solutions	145
SnBr$_4$	—	—	279	86		
TiCl$_4$	389	114	498	136		
TiBr$_4$	231.5	68.5	393	88		
TiI$_4$	162	51	323	67		
ZrCl$_4$	377	98	418	113		
ZrBr$_4$	225.5	60	315	72	Raman, gas phase	150
ZrI$_4$	158	43	254	55		
HfCl$_4$	382	101.5	390	112		
HfBr$_4$	235.5	63	273	71		
HfI$_4$	158	55	224	63		
CoCl$_4^-$	—	—	304, 290		{NH$_2$(Ph$_2$)P}$_2$N^{2+} salt	151
CoCl$_4^-$	—	—	297			
CrCl$_4^-$	—	—	380		I.r., phenH$^+$ salts	152
ZnCl$_4^{2-}$	—	—	273			
CdBr$_4^{2-}$	164	—	182		I.r. and Raman, R$_4$N$^+$ salts	153
CdI$_4^{2-}$	120	—	146			

[a] Further frequencies (not new) are listed in references 154 and 155 for Group IV tetrahalides. [b] From combinations.

Assorted vibrational data for tetrahalogeno-species of T_d symmetry are gathered in Table 9 and require little comment. Full details have now been published of the vapour-phase Raman spectra of 21 Group IV tetrahalides MX$_4$ (M = C to Sn or Ti to Hf; X = F, Cl, Br, or I).[153a] Fine structure

[143] A. Cabana, L. Lambert, and C. Pepin, *J. Mol. Spectroscopy*, 1972, **43**, 429.
[143a] H. W. Kattenberg, W. Gabes, and A. Oskam, *J. Mol. Spectroscopy*, 1972, **44**, 425.
[144] R. J. Corice jun., K. Fox, and W. H. Fletcher, *J. Mol. Spectroscopy*, 1972, **41**, 95.
[144a] A. V. Pleshkov, *Optika i Spektroskopiya*, 1972, **32**, 819.
[145] T. E. Thomas and W. J. Orville-Thomas, *J. Inorg. Nuclear Chem.*, 1972, **34**, 839.
[146] R. A. Work and M. L. Good, *Spectrochim. Acta*, 1972, **28A**, 1537.
[147] W. Gabes, K. Olie, and H. Gerding, *Rec. Trav. chim.*, 1972, **91**, 1367.
[148] P. Van Huong and B. Desbat, *Bull. Soc. chim. France*, 1972, 2631.
[149] A. S. Quist, J. B. Bates, and G. E. Boyd, *J. Phys. Chem.*, 1972, **76**, 78.
[150] R. J. H. Clark, B. K. Hunter, and D. M. Rippon, *Inorg. Chem.*, 1972, **11**, 56.
[151] R. M. Clipsham and M. A. Whitehead, *Canad. J. Chem.*, 1972, **50**, 75.
[152] S. N. Ghosh, *J. Inorg. Nuclear Chem.*, 1972, **34**, 1456.
[153] S. D. Ross, I. W. Siddiqi, and H. J. V. Tyrrell, *J.C.S. Dalton*, 1972, 1611.
[153a] R. J. H. Clark and D. M. Rippon, *J. Mol. Spectroscopy*, 1972, **44**, 479.

shown in the Raman spectra of MCl_4 (M = C, Si, Ti, or Sn) and MBr_4 (M = C or Sn) at -196 °C has been discussed [144a] but adds little to similar work reported last year. Values of ν_1 and ν_3 for ACl_4 (A = C, Ge, or Sn) and ABr_4 (A = Ge or Sn) have been plotted against various measures of electronegativity and used to support those of Gordy and of Orville-Thomas but not Allred and Rochow's. It is further concluded that Si, Ge, and Sn do not use d-orbitals for π-bonding in these tetrahalides.[145] Some distortion from tetrahedral accompanies ion-pair formation (in benzene solution) between $GaCl_4^-$ or $GaBr_4^-$ and long-chain tertiary or quaternary ammonium cations: the extent of the disturbance revealed by the spectra is dependent upon the particular cation and the concentration. Typically, for [aliquat 336][$GaCl_4$] $\nu_3 = 377$ and $\nu_4 = 151$ cm^{-1} (T_d), but when the cation is $[(C_7H_{15})_3NH]^+$ C_{3v} rules apply: ν_1 362, ν_2 339, ν_3 160, ν_4 383, and ν_6 151 cm^{-1}.[146] The Raman spectrum of solid PBr_4Cl shows that it is to be regarded as $PBr_4^+Cl^-$, isomorphous with PBr_5: the fundamental frequencies of the cation are considerably split by site and correlation effects but are comparable with those reported earlier for PBr_5 and PBr_7.[147]

Although not always providing new vibrational data, several other studies of tetrahalides have deepened our understanding of the spectra of these molecules. It is confirmed that there is no variation of the depolarization ratio for $\nu_1(a_1)$ of the ^{35}Cl and ^{37}Cl variants of MCl_4 (M = C, Si, Ge, or Sn); it further appears that the value (0.003) is independent of the central atom.[154] The depolarization ratios obtained by others [155] for the series of liquids MCl_4 (M = Si, Sn, or Te) are apparently a little larger (ca. 0.05) than Griffiths'; however, the contribution is valuable in listing also the absolute Raman intensities, spectral linewidths, mean-square amplitudes, and force constants. The crucial F_{34} stretch–bend interaction force constant of CCl_4 was estimated from a matrix-isolation investigation of the isotopic splitting pattern in the ν_4 region.[156] An exact force field has been obtained for $^{116}SnCl_4$ and $^{124}SnCl_4$ using new data from an i.r. study of the vapour spectra. From this, frequencies for $^{116}Sn^{35}Cl_3^{37}Cl$, $^{116}Sn^{35}Cl_2^{37}Cl_2$, and $^{116}Sn^{35}Cl^{37}Cl_3$ were estimated and used to account for the observed band contours of ν_3 for $^{116}SnCl_4$ and $^{124}SnCl_4$.[157] A similar exercise has been conducted for a selection of isotopic variants of $SiCl_4$ using previously published matrix-isolation data.[158]

Reaction of bromine and TlCl yields a material formulated as $Tl^I[Tl^{III}Cl_2Br_2]$. Its i.r. spectrum shows several more bands than previously reported for that anion, although this is not necessarily incompatible with the structure claimed. I.r. bands are at 325w, 293s, 273s, 239m, 221s, 212s,

---

[154] J. E. Griffiths, *Spectrochim. Acta*, 1972, **28A**, 1029.
[155] V. S. Dernova, I. F. Kovalev, and M. G. Voronkov, *Doklady Phys. Chem.*, 1972, **202**, 66.
[156] I. W. Levin and W. C. Harris, *J. Chem. Phys.*, 1972, **57**, 2715.
[157] A. Müller, F. Königer, K. Nakamoto, and N. Ohkaku, *Spectrochim. Acta*, 1972, **28A**, 1933.
[158] N. Mohan and A. Müller, *J. Mol. Spectroscopy*, 1972, **42**, 203.

and 198s cm^{-1}.[159] ICl$_4^-$ (NO$^+$ salt) has Raman bands at 284 [$\nu_1(a_{1g})$], 265 [$\nu_2(b_{1g})$], and 127.5 cm^{-1} [$\nu_4(b_{2g})$].[160]

A consideration of the $\nu_1 : \nu_3$ ratios of tetrahedral oxoanions led Baran to estimate ν_1 of XeO$_4$ at 800 cm^{-1},[161] but events appear to have overtaken him.[162] The data are shown in Table 10. The force field for XeO$_4$ differs

Table 10 *Vibrational wavenumbers/cm^{-1} for some MA$_4$ species*

Species	$\nu_1(a_1)$	$\nu_2(e)$	$\nu_3(t_2)$	$\nu_4(t_2)$	Ref.
Ru^{16}O$_4$	885.3	319 ± 2	921	336	164
Ru^{18}O$_4$	834.9	303 ± 2	878.8	320.9	
Xe^{16}O$_4$	775.7	267 ± 5	879.2	305.9	162
Xe^{18}O$_4$	732.9	—	838.0	291.4	
MnO$_4^{2-}$	812	325	820	332	
FeO$_4^{2-}$	832	340	790	322	163 ^{a}
RuO$_4^{2-}$	840	331	804	336	
CrO$_4^{3-}$	834	260	860	324	
[VS$_4$]$^{3-}$	375	—	460	—	
[NbS$_4$]$^{3-}$	408	—	421	—	
[TaS$_4$]$^{3-}$	424	—	399	—	
[MoS$_4$]$^{2-}$	460	—	480	—	
[WS$_4$]$^{2-}$	485	—	465	—	165 ^{a}
[VSe$_4$]$^{3-}$	232	—	365	—	
[NbSe$_4$]$^{3-}$	239	—	316	—	
[TaSe$_4$]$^{3-}$	249	—	277	—	
[MoSe$_4$]$^{2-}$	255	—	340	—	
[WSe$_4$]$^{2-}$	281	—	309	—	

^{a} Raman data for aqueous solutions.

significantly from those of RuO$_4$ and OsO$_4$; in particular, the negative value of F_{rr} emphasizes the differences in bonding (basically sp^3 in XeO$_4$ but d^3s in the others). A considerable collection of Raman and i.r. data for RuO$_4^-$, MO$_4^{2-}$ (M = Mn, Fe, or Ru), MO$_4^{3-}$ (M = Cr, Mn, Re, or Fe), and MO$_4^{4-}$ (M = Ti, V, Cr, Mo, W, Fe, or Co), as Li$^+$, K$^+$, Cs$^+$, Mg^{2+}, or Ba^{2+} salts, has been analysed and MVFF constants have been tabulated. Some of the data shown in Table 10 were from solutions but the bulk came from solid samples (most of the salts are unstable in water or are insoluble): consequently various site- and factor-group split modes were observed. From knowledge of the symmetry correlations, values of the affected fundamentals were estimated and used in calculations. Combining the new observations with earlier data, trends in metal–oxygen force constants (MVFF) were demonstrated: (i) F(M—O) drops with oxidation state for a

[159] R. P. Rastogi, B. L. Dubey, and N. K. Pandey, *J. Inorg. Nuclear Chem.*, 1972, **34**, 831.
[160] J. P. Huvenne and P. Legrand, *Compt. rend.*, 1972, **274**, C, 2073.
[161] E. J. Baran, *Z. Naturforsch.*, 1972, **27a**, 1000.
[162] R. S. McDowell and L. B. Asprey, *J. Chem. Phys.*, 1972, **57**, 3062.
[163] F. Gonzalez-Vilchez and W. P. Griffith, *J.C.S. Dalton*, 1972, 1416.
[164] R. S. McDowell, L. B. Asprey, and L. C. Hoskins, *J. Chem. Phys.*, 1972, **56**, 5712.
[165] A. Müller, N. Weinstock, N. Mohan, C. W. Schläpfer, and K. Nakamoto, *Z. Naturforsch.*, 1972, **27a**, 542.

given metal (e.g. $Ru^{VIII} > Ru^{VII} > Ru^{VI}$) and (ii) a similar progression occurs within an isoelectronic series ($Os^{VIII} > Re^{VII} > W^{VI}$, $Cr^{VI} > V^V > Ti^{IV}$). An explanation is given in terms of varying extents of σ- and π-bonding.[163] Wavenumbers for tetra-thio- and -seleno-anions are also in Table 10; further material appears in Chapter 6, Tables 8 and 18. Sundry data for substituted tetrahedral entities are shown in Tables 11 and 12.

Table 11 Vibrational wavenumbers/cm^{-1} for some substituted tetrahedral entities MAB$_3$ of C_{3v} symmetry

Species	a_1			e			Ref.
	ν_1 $\nu(MB_3)s$	ν_2 $\delta(MB_3)s$	ν_3 $\nu(MA)$	ν_4 $\nu(MB_3)a$	ν_5 $\delta(MB_3)a$	ν_6 Rock	
VOF$_3$(g)	720.5	256.0	1055	801	204.0	309	
VOCl$_3$(g)	409.5	163.0	1042.5	503	124.5	248	169
VOBr$_3$ ^{c}	272.0	118.5	1029.0	401.0	82.0	213.0	
TcO$_3$Cl(l)	948	445	299	932	340	197	166
ReO$_3$Cl(l)	1002	433	303	962	345	197	
MOS$_3^{2-}$ ^{d}	468	—	858	490, 474	—	—	167 ^{a}
WOS$_3^{2-}$	460	—	869	477	—	—	
MoOSe$_3^{2-}$	293	120	858	355	188	120	168 ^{b}
WOSe$_3^{2-}$	292	(120)	878	312	194	(120)	

^{a} Raman, aqueous solutions. ^{b} Average of i.r. and Raman frequencies for Cs$^+$ salts.
^{c} Raman solution values. ^{d} Further data for MOS$_3^{n-}$ species are quoted in Tables 7 and 8 of Chapter 6 and for SPO$_3^{3-}$ and S$_3$PO^{3-} in Table 37 of Chapter 5.

Table 12 Vibrational wavenumbers/cm^{-1} for substituted tetrahedral species MO$_2$X$_2$, of C_{2v} symmetry ^{a}

		VO$_2$F$_2^-$	VO$_2$Cl$_2^-$	^{92}MoO$_2$S$_2^{2-}$	^{100}MoO$_2$S$_2^{2-}$	MoO$_2$Se$_2^{2-}$	WO$_2$Se$_2^{2-}$
a_1	$\nu(MO_2)s$	970	972	819 ^{b}	815 ^{b}	836 ^{c}	863 ^{c}
	$\nu(MX_2)s$	664	ca. 453	473	470	359	335
	$\delta(MO_2)$	330	316	307	305	280	283
	$\delta(MX_2)$	—	—	199.5	199	—	—
a_2	Torsion	—	—	267	267	—	—
b_1	$\nu(MO_2)a$	962	961	801	793	799	808
	$\delta(OMX)$	295	232	246	246	—	—
b_2	$\nu(MX_2)a$	631	435	506	499	342	315
	$\delta(OMX)$	295	232	267	267	114	119
Ref.		170a	170b	170	170	168	168

^{a} Further data on MO$_2$S$_2$ types are in Chapter 6, Table 8 and PO$_2$S$_2^{3-}$ is dealt with in Chapter 5, Table 37. ^{b} Raman spectra of NH$_4^+$ salts. ^{c} I.r. spectra of NH$_4^+$ salts.

[165a] A. Müller, K. H. Schmidt, K. A. Tytko, J. Bouwma, and F. Jellinek, *Spectrochim. Acta*, 1972, **28A**, 381.
[166] A. Guest, H. E. Howard-Lock, and C. J. L. Lock, *J. Mol. Spectroscopy*, 1972, **43**, 273.
[167] A. Müller, N. Weinstock, and H. Schulze, *Spectrochim. Acta*, 1972, **28A**, 1075.
[168] K. H. Schmidt and A. Müller, *Spectrochim. Acta*, 1972, **28A**, 1829.
[169] R. J. H. Clark and P. D. Mitchell, *J.C.S. Dalton*, 1972, 2429.
[170] A. Müller, N. Weinstock, K. H. Schmidt, K. Nakamoto, and C. W. Schläpfer, *Spectrochim. Acta*, 1972, **28A**, 2289.
[170a] E. Ahlborn, E. Diemann, and A. Müller, *J.C.S. Chem. Comm.*, 1972, 378.
[170b] E. Ahlborn, E. Diemann, and A. Müller, *Z. anorg. Chem.*, 1972, **394**, 1.

Table 13 Vibrational wavenumbers/cm⁻¹ for BH_3X^{-174}

		[BH₃F]⁻	[BD₃F]⁻	[¹¹BH₃CN]⁻	[¹¹BH₃NC]⁻	[¹⁰BH₃CN]⁻	[¹¹BD₃CN]⁻	[¹¹BD₃NC]⁻	[¹⁰BD₃CN]⁻
a_1	$\nu(BH_3)s$	2291	1673	2285	2290	2290	1661	1640	1671
	$\delta(BH_3)s$	1125	824	1135	1105	1145	906	940	926
	$\nu(B-X)$	1081	1059	890	760	(910)a	800	665	—
	$\nu(CN)$	—	—	2179	2070	—	2180	2075	1775
e	$\nu(BH_3)a$	2380	1748	2350	2350	2365	1761	1745	—
	$\delta(BH_3)a$	1177	921	1197	1175	(1205)	870	855	(690)
	$\rho(BH_3)$	802	597	872	645	(880)	675	525	—
	$\delta(BCN)$	—	—	335	330	355	330	300	—

a Parentheses indicate a tentative assignment.

Further data for CrO_3X^- species are in Chapter 6. At very low temperatures the spectra of TcO_3Cl and ReO_3Cl become more complex, possibly owing to a phase change.[166] From a Raman vapour-phase study of $VOCl_3$ (and VOF_3) it was possible to resolve the existing contradiction regarding the lowest fundamental, 125 cm^{-1}: it is an e-mode. Mixing $VOCl_3$ and $VOBr_3$ results in immediate redistribution at room temperature. The molecules $VOCl_2Br$ and $VOClBr_2$ were accordingly identified by their Raman spectra and most fundamentals assigned.[169] Oriented polycrystalline films of MeI were studied down to 10 K. Significantly, the first harmonics have the order of their symmetry species inverted with respect to those of the fundamentals, in discord with Davydov's theory as applied to molecular crystals.[171] Wavenumbers/cm^{-1} are:

C_{2v}^{12}	ν_1	ν_2	$2\nu_2$	ν_3	$2\nu_3$
A_1	2934.5	1238.5	2454.2	524.2	1037.9
B_2	2933	1235	2455.2	519.2	1039.1

The ν_4 fundamental (ν_0 = 2296.46 cm^{-1}) of CD_3Br has been studied at high resolution.[172] ν(S—H) is at 2435 and δ(SH) at $ca.$ 1110 cm^{-1} in $[R_4^nN][HSO_3]$ salts.[173] New data for salts of BH_3X^- (Table 13) have been analysed using the hybrid-orbital force field, together with earlier work on BH_4^-, BH_3CO, BH_3PF_3, and BH_3PH_3. The order of magnitude of B—X force constants was found to be: X = F$^-$ > CN$^-$ > NC$^-$ > H$^-$ > CO > PF$_3$ > PH$_3$.[174] The $\nu_4 + \nu_8 - \nu_8$ band of $^{11}BH_3CO$ at $ca.$ 700 cm^{-1} has been analysed.[174a]

Molecules with stereochemically active lone pairs have been further investigated. Assignments for matrix-isolated SeF_4 and TeF_4 (i.r.) (Table 14), as well as vapour- and solid-phase data, have been analysed to show that

Table 14 I.r. absorption wavenumbers/cm^{-1} for SeF_4 and TeF_4 in nitrogen matrices [175]

Mode	$^{78}SeF_4$	TeF_4	Mode	$^{78}SeF_4$	TeF_4
$\nu_1\ a_1$	742.6	695.0	$\nu_6\ b_1$	598.0	588.9, 587.9, 586.9
$\nu_2\ a_1$	588.5	572	$\nu_7\ b_1$	364.0	333.2
$\nu_3\ a_1$	405.5	293.3	$\nu_8\ b_2$	725.2	682.2
$\nu_4\ a_1$	—	—	$\nu_9\ b_2$	254.0	—

the axial bonds are weaker than the equatorial bonds. Oligomers formed upon diffusion, or in more concentrated matrices, appear to be bridged via axial fluorines only.[175] SeF_4 has been studied in all three phases by Raman

---

[171] J. Aubard and G. G. Dumas, *Compt. rend.*, 1972, **275**, B, 419.
[172] R. W. Peterson and T. H. Edwards, *J. Mol. Spectroscopy*, 1972, **41**, 137.
[173] R. Maylor, J. B. Gill, and D. C. Goodall, *J.C.S. Dalton*, 1972, 2001.
[174] J. R. Berschied, jun. and K. F. Purcell, *Inorg. Chem.*, 1972, **11**, 930.
[174a] L. Lambert, C. Pepin, and A. Cabana, *J. Mol. Spectroscopy*, 1972, **44**, 578.
[175] C. J. Adams and A. J. Downs, *Spectrochim. Acta*, 1972, **28A**, 1841.

Vibrational Spectra

Table 15 *Correlation of the fundamental modes of vibration of SeF_4 and SF_4 with those of AsF_5 and BrF_3* [a,124]

BrF₃		AsF₅		SeF₄		SF₄
Mode	Assignment/cm⁻¹	Mode	Assignment/cm⁻¹	Mode	Assignment/cm⁻¹	
a_1	675	a_1' / e_a'	734 / 811	a_1	749	892
b_2 (translation)		e_b'	811	b_2	736	867
b_1	612	a_2''	784	b_1	622	730
a_1	552	a_1'	644	a_1	574	559
b_1	350	a_2'' / e_a''	400 / 386	b_1	—	532
a_2 (rotation)		e_b''	386	a_2	—	414
a_1 (translation)		e_a'	372	a_1	365	464
b_2 (rotation)		e_b'	372	b_2 (rotation)		
a_1	233	e_a'	130	a_1	162	226
b_2	242	e_b'	130	b_2	—	353

[a] Symmetry co-ordinates are used to describe approximately the fundamental modes. Coupling will be serious, notably for the two e' modes of AsF_5.

spectroscopy (using a sapphire cell) and by i.r.(g). The final assignments, correlated with those of BrF_3 and AsF_5, are shown in Table 15.[124] Rather complete assignments of i.r. and Raman data have been made for ClF_3O (1),[176] $ClO_2F_2^+$,[177] and $ClO_2F_2^-$ (2) [178] (Table 16) and for $XNSF_2$ [179]

[176] K. O. Christe and E. C. Curtis, *Inorg. Chem.*, 1972, **11**, 2196.
[177] K. O. Christe, *Inorg. Nuclear Chem. Letters*, 1972, **8**, 453.
[178] K. O. Christe and E. C. Curtis, *Inorg. Chem.*, 1972, **11**, 35.
[179] R. Kebabcioglu, R. Mews, and O. Glemser, *Spectrochim. Acta*, 1972, **28A**, 1593.

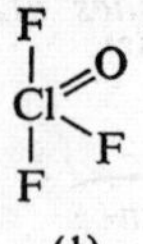

```
    F                 ⎡ F      ⎤⁻
    |   O             | |   O  |
    Cl                | Cl     |
    |   F             | |   O  |
    F                 ⎣ F      ⎦
   (1)                   (2)
```

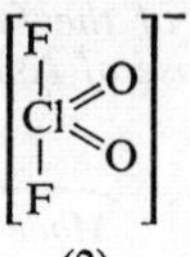

Table 16 Assignments for ClO_nF_{4-n} species ($n = 1$ or 2)

C_{2v}			$[ClO_2F_2]^-$	$[ClO_2F_2]^+$	C_s [a]	$ClOF_3$
a_1	ν_1	$\nu(ClO_2)s$	1065	1249	a'	1223
	ν_2	$\delta(ClO_2)s$	559	—	a'	486
	ν_3	$\nu(ClF_2)s$	363	757	a'	686
	ν_4	$\delta(ClF_2)s$	198	—	a'	478
a_2	ν_5	Torsion	480?	—	a''	414
b_1	ν_6	$\nu(ClF_2)a$	510	829	a''	641
	ν_7	Rock	~335	—	a''	499
b_2	ν_8	$\nu(ClO_2)a$	1220	1484	a'	224
	ν_9	Wag	~335	—	a'	323
Reference			178	177		176

[a] Most of the mode descriptions change from $n = 1$ to $n = 2$.

(Table 17). The quasi-linear C_3S_2 has: $\nu_1(\Sigma_g^+)$ 1665, $\nu_2(\Sigma_g^+)$ 489, $\nu_5(\Pi_g)$ ca. 408, and $\nu_7(\Pi_u)$ 126 cm^{-1}.[180]

Table 17 I.r. and Raman wavenumbers/cm^{-1} and assignments for $XNSF_2$ [179] (i.r. data in parentheses)

	Mode	X = F	X = Cl	X = Br
a'	$\nu(N=S)$	1133 (1150)	1176 (1200)	1208 (1214.7)
	$\nu(SF_2)s$	770 (770)	741 (752)	735 (745)
	$\nu(N-X)$	809 (822)	539 (548.5)	465 (468)
	$\delta(SF_2)$	611 (615)	640 (644.5)	593
	$\rho\omega(SF_2)$	435	409 (409.6)	405
	$\delta(XNS)$	200	165	135
a''	$\nu(SF_2)a$	696 (712)	(694)	660 (699)
	$\rho r(SF_2)$	435	409 (409.6)	405
	$\pi(XNS)$	150	145	135

Hexa-atomic Species.—Five-co-ordinate molecules were relatively unpopular in 1972. An *ab initio* calculation on hypothetical PH_5 was used to show that the lowest-frequency e' mode is effectively a bend of equatorial and axial bonds in the approximate proportions 2 : 1.[181] Such motion is, of course, well adapted to axial-equatorial interchange. In another attempt to bang nails quantitatively into the pseudorotation coffin, Coriolis constants were sought from a high-resolution gas-phase study of PF_5. Band centres of ν_3 and ν_4 are at 946.38 ± 0.01 and 575.511 ± 0.002 cm^{-1} respectively, and ν_7 was estimated as 166 ± 15 cm^{-1} from the combination ($\nu_3 + \nu_7$). However, it proved impossible to resolve the perpendicular bands ν_5 and ν_6.[182]

[180] J. B. Bates and W. H. Smith, *Chem. Phys. Letters*, 1972, **14**, 362.
[181] W. Walker, *J. Mol. Spectroscopy*, 1972, **43**, 411.
[182] J. Schatz and S. Reichman, *J. Chem. Phys.*, 1972, **57**, 4571.

Argument continues over the constitution of pentafluoride vapours and although all of the physical evidence has not been accommodated coherently, a consistent story seems to be emerging from Beattie's work. In the gas phase at 350 °C the Raman spectrum of SbF_5 shows only two lines, *viz.* 683 (pol.) and 272 cm^{-1}, attributed to a monomeric (possibly trigonal-bipyramidal) species. At a lower temperature (140 °C) a more complex pattern is evident, *viz.* 720 (pol.), 672 (pol.), 298, 268, 250, 220, 175, and 150 (pol.) cm^{-1}. It was interpreted as due to a C_{2v} (SbF_4) residue, with octahedral co-ordination at Sb, resulting from the *cis*-fluorine-bridged polymer. Significantly, the i.r. spectra in gas, liquid, and matrix-isolated states are all similar to each other and correspond to that of the low-temperature Raman species; they cannot therefore be due to monomer.[183] NbF_5 and TaF_5 undergo similar reversible dissociation on heating; at higher temperatures (*e.g.* > 300 °C) each vapour shows a single polarized Raman band at 727 (Nb) and 742 (Ta) cm^{-1} respectively. However, at lower temperatures, as with SbF_5, polymeric species are present and an analogous interpretation of the more complex Raman data is offered in terms of octahedrally co-ordinated metal atoms with *cis*-fluorine bridges.[184] However, others have concurrently interpreted their new i.r. (matrix-isolated and vapour) spectra of NbF_5 in terms of a C_{4v} monomer:[185] their spectra clearly correspond to those of Beattie's polymeric species.

The following assignment is given on the basis of Raman measurements on PCl_5 in non-aqueous solvents:[148] A'_1: ν_1 394, ν_2 385 cm^{-1}; A''_2: ν_3 444, ν_4 299 cm^{-1}; E': ν_5 580, ν_6 278, ν_7 98 cm^{-1}; E'': ν_8 261 cm^{-1}. New assignments for SF_5^- and SeF_5^- have appeared, and are shown in Table 18,

Table 18 Wavenumbers/cm^{-1} and assignments for square-pyramidal entities MXY_4

C_{4v}		SF_5^- ^{a}	SeF_5^- ^{a}	$ClOF_4^-$	$MoNCl_4^-$
a_1	$\nu(M-X)$	796	666	1202	1052
	$\nu(MY_4)$	522	515	457	352
	$\pi(MY_4)$	469	332	339	—
b_1	$\nu(MY_4)$	(435)	460	345	—
	$\pi(MY_4)$	269	236	—	—
b_2	$\delta(MY_4)$	342	282	280	—
e	$\nu(MY_4)$	590	480	593	347
	$\delta(MY_4)$	241	202	204	—
	M—X wag	430	399	*ca.* 400	—
Reference		187	187	186	186a

^{a} Cs$^+$ salts.

[183] L. E. Alexander and I. R. Beattie, *J. Chem. Phys.*, 1972, **56**, 5829.
[184] L. E. Alexander, I. R. Beattie, and P. J. Jones, *J.C.S. Dalton*, 1972, 210.
[185] N. Acquista and S. Abramowitz, *J. Chem. Phys.*, 1972, **56**, 5221.
[186] K. O. Christe and E. C. Curtis, *Inorg. Chem.*, 1972, **11**, 2209.
[186a] R. D. Bereman, *Inorg. Chem.*, 1972, **11**, 1148.
[187] K. O. Christe, E. C. Curtis, C. J. Schack, and D. Pilipovich, *Inorg. Chem.*, 1972, **11**, 1679.

together with those for C_{4v} $ClOF_4^-$. The bonding in the latter ion is described as mainly covalent Cl=O with two semi-ionic three-centre four-electron p–p Cl–F bond pairs.[186] An assignment for $SnCl_5^-$ is in Chapter 5, Table 25. I.r. and Raman spectra of HPF_4(g) and H_2PF_3(g) are consistent with a trigonal-bipyramidal structure, and solid-state spectra indicate that this is retained there also. The hydrogen in HPF_4 is equatorial so the symmetry is C_{2v}: data are presented in Table 19. I.r. band contours shown

Table 19 Wavenumbers/cm^{-1} for some five-co-ordinate species of C_{3v} symmetry

C_{2v}		HPF_4	H_2PF_3			ClO_2F_3
a_1	ν(P—H)	2482	2482	a_1	$\nu(ClO_2)s$	1096
	ν(P—F)$eq.$	882	1005		ν(Cl—F)$eq.$	687
	ν(P—F)$ax.$	629	864		$\delta(ClO_2)$	530
	$\delta(PF_2)ax.$	525	614			
	$\delta(eq.)$	200	1233			
a_2	Rock		377			
b_1	$\nu(eq.)$	1024	2549	b_1	ν(Cl—F)$ax.$	699
	$\delta(eq.)$	650	767		ClO_2 wag	601
	PF_2 axial bend	537	472			
b_2	$\pi(eq.)$	1528	1291	b_2	$\nu(ClO_2)a$	1334
	$\nu(PF_2)ax.$	795	825			
	Rock	317	335			
Reference		188	188			189

by the new compound ClF_3O_2 are in excellent agreement with those expected for the C_{2v} model (3).[189]

$$F-Cl(=O)(=O)F_2$$

(3)

The gas-phase i.r. spectra of $ReOCl_4$[189a] and $OsOCl_4$[189b] afford evidence slightly in favour of a square-pyramidal structure (rather than trigonal-bipyramidal). The monomeric nature of solid $ReOCl_4$ is shown by the absence of ν(ReORe) bands, but whereas one group of workers reports a single ν(Re=O) i.r. band (1033 cm^{-1}),[189a] another finds a doublet (1036 and 1016 cm^{-1}), which is attributed to a correlation field splitting.[189c]

Electron diffraction shows B_2Cl_4 to be of D_{2d} symmetry in the vapour phase with a barrier of ca. 1.85 kcal mol^{-1}. The torsional mode (b_1) is Raman-active only and has an estimated position of 20—25 cm^{-1} but has

[188] R. R. Holmes and C. J. Hora, jun., *Inorg. Chem.*, 1972, **11**, 2506.
[189] K. O. Christe, *Inorg. Nuclear Chem. Letters*, 1972, **8**, 457.
[189a] C. G. Barraclough and D. J. Kew, *Austral. J. Chem.*, 1972, **25**, 27.
[189b] C. Calvo, P. W. Frais, and C. J. L. Lock, *Canad. J. Chem.*, 1972, **50**, 3607.
[189c] K. I. Petrov, V. V. Kravchenko, D. V. Drobot, and V. A. Aleksandrova, *Russ. J. Inorg. Chem.*, 1972, **16**, 928.

Vibrational Spectra

not been seen previously owing to the difficulty of working so close to the exciting line. An extremely broad feature has now been found in this region. The $0 \to 1$ transition is calculated to be at 22.8 cm^{-1}, decreasing monotonically to 8 cm^{-1} for $34 \to 35$. The breadth observed is therefore consistent with a barrier of 7.53 ± 0.4 kJ mol^{-1}.[190] Although the symmetry of B$_2$Cl$_4$ apparently changes from D_{2d} (liquid) to D_{2h} upon crystallization (reported last year) it is now said that the bromide retains D_{2d} symmetry throughout.[191] For the crystal, the following assignment (^{11}B modes only) is quoted (values in cm^{-1}):

a_1	ν_1	1077	b_2	ν_5	592	e	ν_7	777
	ν_2	245		ν_6	175		ν_8	289
	ν_3	106					ν_9	65

Both i.r. and Raman spectra of P$_2$Cl$_4$ are rather simple and can be interpreted on the basis of a C_{2h} (*trans* conformation) model, at least so far as the solid is concerned; an i.r. band at 400 cm^{-1} found only for liquid and vapour samples implies the presence of a small proportion of another conformer. There is no correlation splitting, perhaps indicating the presence of a unimolecular cell, as has been shown for P$_2$I$_4$. Raman lattice modes are at 71, 61, and 46 cm^{-1} but none were found in the i.r. Internal modes are assigned as follows: ν(P—P)a_g 397 cm^{-1} (Raman); PCl$_2$ rock a_u 142 cm^{-1} (i.r.); PCl$_2$ wag b_u 203 cm^{-1} (i.r.).[192]

The anions BH$_3$X$^-$ (X = CN$^-$ or NC$^-$) were considered under 'pentaatomic species'. CF$_3$CN and CF$_3$CH$_3$ were included in a study of Raman gas-phase contours for some prolate symmetric tops: it turns out that better Coriolis ζ constants can sometimes be obtained from Raman than from i.r. experiments.[193] Vibrational data for azodicarbonitrile have been assigned on the basis of the C_{2h} *trans* structure (4), *viz.* a_g: 2176, 1422,

$$N{\equiv}C{-}N{=}N{-}C{\equiv}N$$
(4)

1002, (741), and 282 cm^{-1}; b_{2g}: (108 cm^{-1}); b_u: 2204, 982, 904, and 596 cm^{-1}; a_u: 574 and 133 cm^{-1}, where figures in parentheses indicate an assignment deduced from combinations.[194]

Hepta-atomic Species.—Raman and i.r. frequencies for most stable hexahalogeno-anions and molecules must surely have been accumulated by now, although there is a trickle of new numbers, shown in Table 20. The i.r. spectrum of UF$_6$ vapour has been investigated in detail and assignments

---

[190] L. H. Jones and R. R. Ryan, *J. Chem. Phys.*, 1972, **57**, 1012.
[191] J. D. Odom, J. E. Saunders, and J. R. Durig, *J. Chem. Phys.*, 1972, **56**, 1643.
[192] J. D. Odom, J. E. Saunders, and J. R. Durig, *J. Cryst. Mol. Structure*, 1972, **2**, 169.
[193] F. N. Masri, *J. Chem. Phys.*, 1972, **57**, 2472.
[194] B. Bak and P. Jansen, *J. Mol. Structure*, 1972, **11**, 25.

have been made for a variety of second-order bands.[199] A new species,[197] ClF_6^+, formed in the reaction of $FClO_2$ with PtF_6 at $-78\,°C$, is warmly welcomed: see Table 20. Force-field calculations on regular octahedral

Table 20 Vibrational wavenumbers/cm^{-1} for hexahalogeno-ions

Species	Raman			I.r.		Footnote	Ref.
	$\nu_1(a_{1g})$	$\nu_2(e_g)$	$\nu_5(t_{2g})$	$\nu_3(t_{1u})$	$\nu_4(t_{1u})$		
$[OsCl_6]^{2-}$	352	—	177	—	—	a	195
$[OsBr_6]^{2-}$	218	162	—	—	—		
$[ReCl_6]^{2-}$	—	—	—	297	170		
$[OsCl_6]^{2-}$	—	—	—	304	174		
$[IrCl_6]^{2-}$	—	—	—	311	180		
$[PtCl_6]^{2-}$	—	—	—	324	189	b	196
$[ReBr_6]^{2-}$	—	—	—	208	—		
$[OsBr_6]^{2-}$	—	—	—	211	—		
$[IrBr_6]^{2-}$	—	—	—	220	—		
$[PtBr_6]^{2-}$	—	—	—	230	—		
$[ClF_6]^+$	679	580	513	890	582		197
$[SbBr_6]^-$	195	164 ca.	120	241	120	c	198
$[AuF_6]^-$	595	520	224	—	—		198a

^{a} K^+ salts. ^{b} R_4N^+ salts. ^{c} Et_4N^+ salt.

molecules and ions appear every year (*e.g.* refs. 200—205) and are not normally discussed in these Reports. However, the exceptional scope of one paper demands some mention. The authors have made a systematic comparison of the application to hexahalogeno-species of the UBFF, OVFF, MUBFF, MOVFF, and GVFF. Included were: all known hexafluorides MF_6 (other than Xe); MF_6^- (M = P, As, Nb, or Ta); MF_6^{2-} (M = Mn, Ni, Pt, Si, Ge, Sn, or Nb); WCl_6; MCl_6^- (M = Nb, Ta, P, As, or Sb); MCl_6^{2-} (M = Ti, Zr, Hf, Ge, Sn, Pb, Se, Te, Pd, Re, Os, Ir, Pt, Ce, or U); $InCl_6^{3-}$; MBr_6^{2-} (M = Ti, Zr, Hf, Re, Pt, Pd, Sn, Se, or Te); MBr_6^- (M = Nb or Ta); and SnI_6^{2-}. Both observed and calculated frequencies were listed. It was concluded that (*a*) the MOVFF has several advantages, particularly with respect to best overall fit; (*b*) there are trends in relationship between force constants and both mass and oxidation state of the central atom, with the oxidation state the more important; (*c*) *F*, *K*, and

---

[195] G. L. Bottger and C. V. Damsgard, *Spectrochim. Acta*, 1972, **28A**, 1631.
[196] D. A. Kelly and M. L. Good, *Spectrochim. Acta*, 1972, **28A**, 1529.
[197] K. O. Christe, *Inorg. Nuclear Chem. Letters*, 1972, **8**, 741.
[198] C. J. Adams and A. J. Downs, *J. Inorg. Nuclear Chem.*, 1972, **34**, 1829.
[198a] K. Leary and N. Bartlett, *J.C.S. Chem. Comm.*, 1972, 903.
[199] E. Bar-Ziv, M. Freiberg, and S. Weiss, *Spectrochim. Acta*, 1972, **28A**, 2025.
[200] E. Wendling and S. Mahmoudi, *Optika i Spektroskopiya*, 1972, **32**, 492.
[201] M. L. Mehta, *Canad. J. Spectroscopy*, 1972, **17**, 22.
[202] M. N. Avasthi and M. L. Mehta, *Z. Naturforsch.*, 1972, **27a**, 700.
[203] S. P. So and F. T. Chau, *J. Mol. Structure*, 1972, **12**, 113.
[204] M. L. Mehta, *J. Mol. Spectroscopy*, 1972, **42**, 208.
[205] M. N. Avasthi and M. L. Mehta, *Z. Naturforsch.*, 1972, **27a**, 700.

Vibrational Spectra

$H(D)$ from the MUBFF and MOVFF correlate with the number of non-bonding electrons and with crystal-field stabilization energies for the MF_6 and MCl_6^{2-} series.[206] Using the OVFF the non-Jahn–Teller-perturbed frequencies of ReF_6 have been estimated as: ν_2 626 (587) and ν_5 316 (248 cm^{-1}) respectively, where values in parentheses refer to observed band positions.[207]

Another fascinating contribution to the story of XeF_6 comes from the Argonne laboratory. New Raman and i.r. spectra have been interpreted in terms of relative contributions from three electronic isomers (Figure 1a),

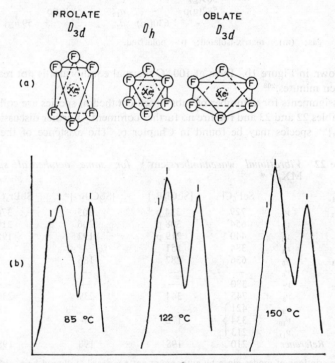

Figure 1 (a) *Three possible electonic isomers of* XeF_6; (b) *Raman spectra, 500—625 cm^{-1}, for three different sample temperatures of* XeF_6 *vapour.* (Reproduced by permission from *J. Chem. Phys.*, 1972, **56**, 5042)

as detailed in Table 21. Convincing evidence in support of this analysis comes in the form of spectra of the vapour and of matrix-isolated XeF_6 for samples with a variety of thermal histories. A typical series of spectra

---

[206] P. Labonville, J. R. Ferraro, M. C. Wall, and L. J. Basile, *Co-ordination Chem. Rev.*, 1972, **7**, 257.

[207] I. W. Levin, S. Abramowitz, and A. Müller, *J. Mol. Spectroscopy*, 1972, **41**, 415.

Table 21 Wavenumbers/cm^{-1} and assignments for XeF$_6$ [208]

O_h			D_{3d}(oblate)			D_{3d}(prolate)		
Raman	I.r.		Raman	I.r.		Raman	I.r.	
69(g)	—	t_{2g}	69(g)	—	e_g			
—	252(m)	t_{1u}	102(g)p	—	a_{1g}			
509(g)	—	e_g	—	302(m)	e_u	—	352(m)	a_{2u}
613(g)p	—	a_{1g}	—	326(m)	a_{2u}	—	365(m)	e_u
—	619(g)	t_{1u}	—	384(m)	e_u	509(g)	—	e_g
			—	507(g,m)	e_u	—	520(m)	e_u
			509(g)	—	e_g			
			582(g)p	—	a_{1g}	582(g)p	—	a_{1g}
			—	630(m)	a_{2u}	—	599(g)	a_{2u}

(g) = gas; (m) = matrix-isolated; p = polarized.

is shown in Figure 1b. Even at 100 °C thermal equilibrium is not reached for ten minutes.[208]

Assignments for a number of substituted octahedral species are collected in Tables 22 and 23 and require no further comment. Further discussion of [IrX$_6$]$^{n-}$ species may be found in Chapter 6. The existence of the ions

Table 22 Vibrational wavenumbers/cm^{-1} for some octahedral species MX$_5$Y [a]

C_{4v}		SeF$_5$Cl	[SbCl$_5$Br]$^-$	[SbClBr$_5$]$^-$ [b]	SbBr$_5$OEt$_2$ [c]
a_1	ν_1	729	334 p [d]	305	376
	ν_2	654	308 p	206	210
	ν_3	440	219 p	192	195
	ν_4	384	151	—	—
b_1	ν_5	636	287	186	—
	ν_6	—	—	—	—
b_2	ν_7	380	—	—	—
e	ν_8	745	344	239	236
	ν_9	421 ⎫			
	ν_{10}	334 ⎬	172	—	—
	ν_{11}	213 ⎭			
Reference		210	198	198	198

[a] Descriptions of modes given by authors are not necessarily followed here, although each is assigned to the correct species. [b] Bands not assigned: 171, 150, 118, 104, and 79 cm^{-1}. [c] Bands not assigned: 140 and 118 cm^{-1}. [d] p = polarized.

N$_2$H$_5^+$ and N$_2$D$_5^+$ in solid hydrazinium fluoride has been confirmed. ν(N—N) is at 968 (^{1}H$_5$) and 959 cm^{-1} (^{2}H$_5$), respectively, and ν(N—N) in the fluoride is lowered with respect to its position in N$_2$H$_5$Cl owing to greater hydrogen-bond strength.[209]

[208] H. H. Claasen, G. L. Goodman, and H. Kim, *J. Chem. Phys.*, 1972, **56**, 5042.
[209] P. Glavic and D. Hadzi, *Spectrochim. Acta*, 1972, **28A**, 1963.
[210] K. O. Christe, C. J. Schack, and E. C. Curtis, *Inorg. Chem.*, 1972, **11**, 583.

Vibrational Spectra

Table 23 Vibrational wavenumbers/cm^{-1} of hexahalogenoantimonate anions in MeCN solution [198]

cis-[SbF$_4$Br$_2$]$^-$			[SbCl$_3$Br$_3$]$^-$
Raman	I.r.		Raman
	637 ⎫		337 p
620 p	620 ⎬ ν(Sb—F)		308 p
601 p	596 ⎭		292
559			256
	290 ⎫ δ(Sb—F)		220 p
	256 ⎭		205
236 p		ν(Sb—Br)	180
220	222 ⎫		
206 p?	205 ⎬ Deformations		
	175 ⎭		

Larger Symmetric Species.—It is said that although the fundamental vibrations of IF$_7$ are accommodated by a D_{5h} model, there are some violations of the selection rules for combinations. This is interpreted to mean that the molecule undergoes minor dynamic distortions from D_{5h} symmetry.[211] Since IF$_5$ shows considerable frequency and intensity changes upon condensation, the spectra of IF$_7$ have also been investigated for gas, liquid, and HF solutions: only rather minor changes were found.[212] A normal-co-ordinate analysis of IF$_7$ led to an assignment of the observed data [213] (Table 24) in considerable discord with those of others. Practitioners in this field should agree on a numbering system for IF$_7$. A similar

Table 24 Vibrational wavenumbers/cm^{-1} and assignments for IF$_7$ and UO$_3$F$_5^{3-}$

D_{5h}			IF$_7$			UO$_2$F$_5^{3-}$	ReF$_7$ [a]
a_1'	ν_1	ν(I—F)ax.	675	676	675 p	805	736
	ν_2	ν(I—F)eq.	629	635	632 p	455	736
a_2''	ν_3	ν(I—F)ax.	672	746	—	859	800
	ν_4	δ(I—F)ax.	257	425	—	⎱ 380,	353
e_1'	ν_5	ν(I—F)eq.	746	670	—	⎰ 287,	703
	ν_6	δ(I—F)eq.	425	257	—	⎧ 225,	217
	ν_7	(I—F)ax. wag	363	365	—	⎩ 195	229
e_1''	ν_8	δ(I—F)	308	310	311	427,	352
e_2'	ν_9	ν(I—F)eq.	509	510	508	⎱275,	597
	ν_{10}	δ(I—F)	342	352	351	⎰ 255	489
e_2''	ν_{11}	δ(I—F)	200 [a]	174 [a]	—	(232 or 138)	241
Reference			211	213	212	215	214

[a] Calculated.

[211] H. H. Eysel and K. Seppelt, *J. Chem. Phys.*, 1972, **56**, 5081.
[212] M. Brownstein and H. Selig, *Inorg. Chem.*, 1972, **11**, 656.
[213] E. Wendling and S. Mahmoudi, *Bull. Soc. chim. France*, 1972, 33.
[214] E. Wendling and S. Mahmoudi, *Rev. Chim. minérale*, 1972, **9**, 291.
[215] E. Wendling and S. Mahmoudi, *Bull. Soc. chim. France*, 1972, 40.

calculation for ReF_7 gave the results shown.[214] I.r. and Raman spectra for $UO_2F_5^{3-}$, D_{5h} symmetry, have been tentatively assigned;[215] see Table 24.

Rather complete i.r. and Raman data are reported for $Fe(CO)_5$, $Fe(^{12}C^{16}O)_5$, $Fe(^{13}C^{16}O)_5$, and $Fe(^{12}C^{18}O)_5$. The entire spectrum is discussed exhaustively and minor changes are made in the current assignment, mainly on the basis of evidence from second-order transitions. Difficulties were experienced in refinement of the normal-co-ordinate analysis but it was concluded that axial M—C bonds are very slightly weaker than equatorial ones, despite the fact that the M—C potential constants are not significantly different.[216] ν(CN) spectra of $M(CN)_5^{3-}$ (M = Co or Ni) in aqueous solution are consistent with a square-based pyramidal arrangement of ligands: the solid $[Cr(en)_3][Ni(CN)_5]$,$1.5H_2O$ contains, it will be recalled, equal proportions of anions in square-pyramidal and trigonal-bipyramidal configurations.[217] Assignments are presented in Table 25. The i.r. and

Table 25 ν(CN) wavenumbers/cm^{-1} for $M(CN)_5^{3-}$ in solution [217]

C_{4v}	M = Ni(aq.)		M = Co(aq. MeOH)	
	Raman	I.r.	Raman	I.r.
a_1	2130 p	2123	2115 p	2105
b_1	2117	—	2110	—
e	2106	2112	2096	2095
a_1	2090 p	2083	2080 p	2085

Raman data for $M(CN)_6^{5-}$ (M = Mn, Tc, or Re), shown in Table 25, have been analysed and reveal a rather low CN stretching force constant, indicative of relatively strong M—C π-bond character, especially for the technetium compound.[218]

In-cavity excitation of the Raman spectrum of $As_4O_6(g)$ at ca. 440 °C gave a number of lines (wavenumbers/cm^{-1}), which are compared in Table 26 with those for P_4O_6.[118] The data are in good agreement with those

Table 26 Raman wavenumbers/cm^{-1} for P_4O_6 and As_4O_6

	P_4O_6	As_4O_6		P_4O_6	As_4O_6
a_1	620	555	t_2	959	782
	562	381		642	496
e	691	409		408	346
	285	185		305	253

---

[216] L. H. Jones, R. S. McDowell, M. Goldblatt, and B. I. Swanson, *J. Chem. Phys.*, 1972, **57**, 2050.
[217] W. P. Griffith and J. R. Lane, *J.C.S. Dalton*, 1972, 158.
[218] W. Krasser, E. W. Bohres, and K. Schwochau, *Z. Naturforsch.*, 1972, **27a**, 1193.

of Beattie and co-workers [220] except that it proved impossible to find their 99 cm^{-1} line, which was also shown to be inconsistent with the thermodynamic data. A line was found at 85 cm^{-1} for the *solid*, but it is a lattice mode.

The $v = 1 \to 2$ transition of the ring-puckering mode of B_2H_6 is at 389.5 ± 0.5 cm^{-1}, identified on the basis of relative intensity measurements of overlapped $0 \to 1$ and $1 \to 2$ bands. For $0 \to 1$ the band centre is at 369.3 cm^{-1} (261.3 cm^{-1} in B_2D_6).[219]

3 Single-crystal and other Solid-state Spectroscopy

The vibrational spectroscopy of single crystals continues to increase in popularity, with some emphasis on the Raman side. In this section we outline the principal results obtained using these methods, including also some non-crystal work which fits in here more logically.

'Simple' Lattice Types.—The i.r.-active TO mode of CaO is at 300 cm^{-1};[220a] its second-order Raman spectrum shows intense emission in the region 400—750 cm^{-1}, with other bands both above and below it.[221] The second-order Raman spectrum of SrO shows a high-frequency cut-off which gives a value of *ca.* 487 cm^{-1} for $LO(\Gamma)$.[222] Using finely *powdered* SrO one-phonon modes activated by the grain surfaces have been observed and interpreted in terms of the critical points of the perfect crystal.[223] Kramers–Krönig-analysed far-i.r. reflectance data for materials with rock-salt, fluorite, and rutile structures are given in Table 27.[224] From these figures,

Table 27 *Lattice wavenumbers*/cm^{-1} *for rock-salt, fluorite, and rutile types*

	I.r.[224]			I.r.[224]				
	ν_{TO}	ν_{LO}		ν_{TO}	ν_{LO}	Raman[225]		Raman[225]
LiF	305	663	CaF$_2$	258	473	330	EuF$_2$	292
NaF	245	427	BaF$_2$	181	334	249	PbF$_2$	255
NiO	390	542	MgF$_2$	448	597	—	SrF$_2$	290
			CdF$_2$	200	381	—	SrCl$_2$	188
							BaCl$_2$	190

and the usual ancillary data, effective ionic charges were calculated. That for NiO is considerably lower than the formal value (2), whereas NaF has 0.8 (*cf.* 1); electron delocalization is therefore much greater in NiO. The

---

[219] W. C. Pringle and A. L. Meinzer, *J. Chem. Phys.*, 1972, **57**, 2920.
[220] I. R. Beattie, K. M. S. Livingstone, G. A. Ozin, and D. J. Reynolds, *J. Chem. Soc. (A)*, 1970, 449.
[220a] S. P. Srivastava and R. D. Singh, *J. Phys. Soc. Japan*, 1971, **31**, 615.
[221] M. Voisin and J. P. Mon, *Phys. Stat. Solidi (B)*, 1971, **48**, K185.
[222] J. P. Mon, *J. Phys. and Chem. Solids*, 1972, **33**, 1257.
[223] L. Novak, *Solid State Comm.*, 1971, **9**, 2129.
[224] I. Nakagawa, *Bull. Chem. Soc. Japan*, 1971, **44**, 3014.
[225] R. Srivastava, H. V. Lauer, L. L. Chase, and W. E. Bron, *Phys. Letters (A)*, 1971, **36**, 333.

i.r. reflectance spectrum of NaF (single crystal) has also been studied as a function of temperature in the range 160—660 cm^{-1}.[226] In a technically demanding study, Bottger and Geddes obtained vacuum-deposited films of anhydrous AgF (rock-salt structure) and determined the phonon wavenumbers/cm^{-1} by both absorption and reflectance techniques. The high-frequency refractive index, obtained by ellipsometry, is 1.73 ± 0.002.[227] At 143 K:

322	$LO(\Gamma)$	221	$TO + TA(L)$
273	$TO + LA(L)$	176	$TO(\Gamma)$

The following assignment is reported for the first- and second-order Raman spectra (wavenumbers/cm^{-1}) of the cuprous halides (zinc blende structure, $B3$). At 190 K:[228]

CuCl	CuBr	CuI	Assignment
75	~ 75	96	$2TA(X)$ or $LA(L)–TA(L)$
131	131	—	$LA(L) + TA(L)$
166	121	128	TO
212	169	147	LO

All of the theoretically predicted Raman modes of β-AgI were observed and assigned in a single-crystal study at 300, 80, and 4.2 K. In agreement with the Lydanne–Sachs–Teller relationship, $LO = 1.19TO$. The two inactive B_1 modes were calculated to be at 97 and 15 cm^{-1}. The assignment of the 80 K spectrum is:[229]

A_1	$\begin{cases} 124 \\ 106 \end{cases}$	LO TO	E_2 E_1	17 $\begin{cases} 124 \\ 106 \end{cases}$	LO TO
E_2	112				

The far-i.r. reflectance spectrum of MnTe has been studied in the range 350—30 K.[229a]

Further data [230] on the EO_1 (PbFCl) lattice were assigned with the aid of some oriented thin-crystal flakes, extending work reported last year (see Table 28).

Raman-active frequencies for several materials with the fluorite structure (Table 27) show variations which apparently cannot be accounted for on the basis of changes of unit-cell volume or mass, in agreement with the known fact that both force constants and polarizabilities vary markedly among the various fluorite lattices.[225]

---

[226] I. F. Chang and S. S. Mitra, *Phys. Rev. (B)*, 1972, **5**, 4094.
[227] G. L. Bottger and A. L. Geddes, *J. Chem. Phys.*, 1972, **56**, 3735.
[228] B. Prevot, C. Carabatos, and M. Leroy, *Compt. rend.*, 1972, **274**, B, 707.
[229] G. L. Bottger and C. V. Damsgard, *J. Chem. Phys.*, 1972, **57**, 1215.
[229a] L. V. Povstyanyi, V. I. Kut'ko, and A. I. Zvyagin, *Fiz. tverd. Tela*, 1972, **14**, 1561.
[230] A. Rulmont, *Spectrochim. Acta*, 1972, **28A**, 1287.

Vibrational Spectra

Table 28 Vibrational wavenumbers/cm^{-1} for the EO$_1$ lattice [230]

D_{4h}	A_{2u}	E_u	A_{2u}	E_u	B_{2g}
BiOF	560—565	350—355	145—155	109	55
BiOCl	515—520	300—325	186	107—115	58.5
BiOBr	525	260	125	65—70	58
BiOI	490—495	250	100	68	60
LaOCl	506	375	192	135	121
LaOBr	515	324	138	100	125.5
NdOCl	535	350	208	126	123
NdOBr	535—540	350	130	98	120
LaOCl	535—540	350—375	208	123	129
LaOBr	545	340—385	131	86	122

The main features of the second-order Raman spectra of single crystals of CaF$_2$, SrF$_2$, BaF$_2$, and PbF$_2$ have been assigned with the aid of the calculated phonon frequencies for critical points of the zone.[231, 232] SrCl$_2$ doped with H$^-$ or D$^-$ (*i.e.* U-centres) shows i.r. absorption (wavenumbers/cm^{-1}) interpreted as follows:[233]

A$^-$ in substitutional positions	571 (H)	400 (D)
A$^-$ in interstitial positions	800 and 970 (H)	580 and 700 (D)

Single-crystal i.r. reflectance and absorption work on the $D0_6$ (LaF$_3$ type) lattices of the trifluorides of Nd, La, Pr, and Ce are reported. Typical values, for NdF$_3$ at 25 K, are 34 (*TO*), 106 (*TO*), 109 (*LO*), and 133 (*TO*) cm^{-1}.[234] Oxides M$_2$O$_3$ with the $D5_3$ (Mn$_2$O$_3$ type) lattice, *viz.* M = Mn, Sc, Y, In, Sm, Eu, Dy, or Yb, showed the number of i.r. bands predicted (16) but only half of the 22 allowed Raman bands were observed: this is not really surprising since $Z = 8$ in the primitive cell. The highest-frequency band varies linearly with M—O distance.[235] No assignment was possible as only powder samples were used.

Structural differences between zinc blende and the 2H and 4H variants of ZnS are reflected in the detailed Raman spectra.[236] Mixed crystals CdS$_{0.6}$Se$_{0.4}$ [237] and Mg$_x$Cd$_{1-x}$Te ($0.2 \leq x \leq 0.6$)[238] show 'two-mode' behaviour in their Raman spectra (*i.e.* frequencies typical of both components). For example, the former crystal shows *LO* and 2*LO* bands of CdSe-like phonons at 196 and 392 cm^{-1}, and CdS-like phonons at 295 and 590 cm^{-1}, as well as the *LO*(CdSe) plus *LO*(CdS) combination at 491 cm^{-1}.

[231] N. Krishnamurthy and V. Soots, *Canad. J. Phys.*, 1972, **50**, 849.
[232] N. Krishnamurthy and V. Soots, *Canad. J. Phys.*, 1972, **50**, 1350.
[233] D. Jumeau and S. Lefrant, *Compt. rend.*, 1972, **275**, *B*, 161.
[234] J. Claudel, A. Hadni, and P. Strimer, *Compt. rend.*, 1972, **274**, *B*, 943.
[235] W. B. White and V. G. Keramidas, *Spectrochim. Acta*, 1972, **28A**, 501.
[236] J. Schneider and R. D. Kirby, *Phys. Rev.* (*B*), 1972, **6**, 1290.
[237] T. Fukumoto, H. Yoshida, S. Nakashima, and A. Mitsuishi, *J. Phys. Soc. Japan*, 1972, **32**, 1674.
[238] S. Nakashima, T. Fukumoto, A. Mitsuishi, and K. Itoh, *J. Phys. Soc. Japan*, 1972, **32**, 1438.

An unambiguous assignment for the Raman-active vibrational modes of β-quartz was deduced from single-crystal measurements,[239] and followed up with a normal-co-ordinate analysis of both it and the isostructural BeF$_2$.[240] Above and below the $\alpha \leftrightarrow \beta$ transition, Raman spectra of polycrystalline crystobalite are strikingly different. Strong bands at 416 and 230 cm^{-1} characterize the α-phase but three (267, 772, and 1065 cm^{-1}) are exhibited by the β-phase. I.r. reflectance data for the α-form, together with the Raman data, were given a qualitative assignment based upon quartz. The sole predicted (T_{2g}) Raman band of the β-phase is at 772 cm^{-1}; the band at 267 cm^{-1} is probably due to a two-phonon process and that at 1065(w) cm^{-1} to an i.r. (T_{1u}) mode made active by disordering.[241]

Factor-group analysis predicts $2A_2 + 4E$ i.r. modes for the distinctive structure of CdAs$_2$, D_4^{10} symmetry. It consists of distorted CdAs$_4$ tetrahedra linked *via* fourfold helical chains of arsenic atoms. Analysis of the reflectance spectrum (wavenumbers/cm^{-1}) gave:[242]

	ν_{TO}	ν_{LO}		ν_{TO}	ν_{LO}
A_2	83	88	E	51	55
	202	213		120	126
				202	208
				245	247

Mixed Oxides and Fluorides.—The perovskite fluoride CsMnF$_3$ has, according to theory, $6A_{2u} + 8E_{1u}$ i.r.-active modes. All but one of these were observed in reflectance and their positions determined by Kramers–Krönig analysis: A_{2u} 73.6, 96.6, 185.8, 208.5, 314.1, and 406.4 cm^{-1}; E_{1u} 108.5, 150.0, 173.4, 207.1, 242.0, 284.3, and 339.8 cm^{-1}.[243] A relation has been developed, analogous to that of Szigetti for diatomic cubic crystals, which relates elastic constants, dielectric permittivity, refractive index, and volume compressibility for crystals of the perovskite type. Using it [244] the following values (cm^{-1}) were obtained:

	ω_{TO} (calc.)	ω_{TO} (obs.)
RbCoF$_3$	128	130
KNiF$_3$	139	147

A particularly precise study of the i.r. reflectance (30—4000 cm^{-1}) of SrTiO$_3$, both as single crystals and polycrystalline powders, showed that the two differ in their reflectivity. In polycrystalline samples damping of the vibrations is greater, probably owing to phonon scattering by defects.[245]

---

[239] J. B. Bates and A. S. Quist, *J. Chem. Phys.*, 1972, **56**, 1528.
[240] J. B. Bates, *J. Chem. Phys.*, 1972, **56**, 1910.
[241] J. B. Bates, *J. Chem. Phys.*, 1972, **57**, 4042.
[242] I. Gregora and J. Petzelt, *Phys. Stat. Solidi (B)*, 1972, **49**, 271.
[243] J. T. R. Dunsmuir, I. W. Forrest, and A. P. Lane, *Materials Res. Bull.*, 1972, **7**, 525.
[244] M. Rousseau and J. Nouet, *Compt. rend.*, 1972, **275**, B, 25.
[245] V. P. Zakharov, A. S. Knyazev, Yu. M. Poplavko, M. V. Rozhdestvenskaya, and V. P. Rubtsov, *Izvest. Akad. Nauk S.S.S.R., Ser. fiz.*, 1971, **35**, 1816.

Vibrational Spectra

New i.r. and Raman data for $KNbO_3$ have been partially assigned by analogy with $BaTiO_3$:[246] however, differences between the two spectra are considerable and confirmation is desirable. The Raman spectrum of $KTa_{0.64}Na_{0.36}O_3$ has been investigated as a function of temperature in both para- and ferro-electric phases.[247] Seven of the thirteen i.r. bands predicted by factor-group analysis have been observed for a series of antimonates $M^{II}Sb_2O_6$ (M^{II} = Ca, Sr, Ba, Cd, or Pb).[248] I.r. spectra are reported for a considerable array of solids with the α-$NaFeO_2$ structure, $A^IB^{III}X_2$ (where X = O or S; A^I = Li, Na, or K; B^{III} = Rh, Cr, In, or various lanthanides).[249] For the sulphides the two predicted Raman bands were found (*e.g.* $KLaS_2$, 197 and 245 cm^{-1}) but only two to four of the six expected i.r. bands could be seen (*e.g.* $KLaS_2$, 168, 218, and 260 cm^{-1}). Variation of these frequencies with change of A and B is discussed.

Sheet and Chain Structures.—A further single-crystal Raman study has been made of Me_4NCdCl_3,[250] reported on last year. I.r. data for α-sodium and magnesium monouranates (powder samples) have been assigned to the chain structure modes of $[(UO_2)O_2^{2-}]_\infty$ with the aid of a normal-co-ordinate calculation.[251] Assignment of the phonon spectra of α-HgS (cinnabar) presents severe problems owing to its exceptionally high optical rotation (*ca.* 480° mm^{-1} at 6000 Å). Broadly, there are two experimental approaches to such a problem. Firstly, circularly rather than linearly polarized light may be used, a remarkably little-known method due to Mathieu. Secondly, as in the present work,[252] the effects of optical rotation can be minimized by low cunning. For the majority of such materials it is sufficient to work with a thin slice of crystal so that the plane of polarization is not significantly rotated in traversing it. With α-HgS a slice 0.3 mm thick was used, in itself too thick to prevent rotation, but this was examined *only* in directions normal to the optic axis (since specific rotation takes maximum and minimum values respectively along and normal to the optic axis). Collection was limited to a 5° cone using 180° scattering. Although the extinctions were not perfect, the assignments are clear-cut and were supported by i.r. reflection measurements. All of the predicted modes were found and are shown in Table 29. A concurrent and independent i.r. reflectance study [253] yielded slightly less complete data, but the two assignments are in concord except for a disagreement over the 40 cm^{-1} band; here Dawson's assignment is clearly preferable.

All three Raman-active modes predicted for the known planar zigzag chain structure present in AuI have been observed, *viz.* 157 (a_{1g} symmetric

---

[246] N. Q. Dao, E. Husson, and Y. Repelin, *Compt. rend.*, 1972, **275**, C, 609.
[247] S. K. Manlief and H. Y Fan, *Phys. Rev.* (*B*), 1972, **5**, 4046.
[248] R. Franck and C. Rocchiccioli-Deltcheff, *Compt. rend.*, 1972, **274**, B, 245.
[249] P. Tarte, M. Tromme, and A. Rulmont, *Spectrochim. Acta*, 1972, **28A**, 1709.
[250] P. S. Peercy and B. Morosin, *Opt. Comm.*, 1971, **4**, 94.
[251] K. Ohwada, *J. Chem. Phys.*, 1972, **56**, 4951.
[252] P. Dawson, *Spectrochim. Acta*, 1972, **28A**, 2305.
[253] J. Barcelo, M. Galtier, and A. Montaner, *Compt. rend.*, 1972, **274**, B, 1410.

Table 29 I.r. and Raman wavenumbers/cm^{-1} and assignment for α-HgS (D_4^3, $Z = 3$)

Raman [252]	I.r.[252]	I.r.[253]	Assignment	
44				⎫
256				⎬ A_1
36	36	34	TO	⎭
	41	40 [a]	LO	⎫
	102	106	TO	⎬ A_2
	139		LO	
	336	340	TO	
	361		LO	⎭
88	83	86	TO	⎫
	91	90	LO	
105	100	107	TO	
147	145	144	LO	
285	280	283	TO	⎬ E
292	289	290	LO	
345	343	344	TO	
353	349	350	LO	
202	n.o.[b]	n.o.	TO, LO	⎭

[a] Erroneously assigned to E species. [b] n.o. = not observed.

chain stretching), 108 (b_{1g} asymmetric chain stretching), and 31 cm^{-1} (chain libration).[253a]

Theory predicts two i.r.-active modes for the para- and eight for the ferro-electric phase of SbSI. Using i.r. reflectance from oriented polycrystalline specimens at various temperatures, the two paraelectric phase modes were located at 4 and 173 cm^{-1}. Five of the ferroelectric phase modes were found, but their frequencies were not listed, although ε' and ε'' curves are shown.[254]

Eighteen of the 21 Raman-active optical modes of D_4^4 TeO$_2$ were found at 85 and 295 K (1A_1 and 2E missing). The measured LO–TO splittings of the six observed E modes account for 63% of the total oscillator strength of the ionic contribution to the ordinary ray dielectric constant.[255]

One of the first fruits of the new generation of triple-monochromator Raman instruments is observation of an 11 cm^{-1} band in yellow HgI$_2$ (above the transition temperature, 126 °C). Published spectra of the yellow phase are notably poorer than those of red HgI$_2$, but the new machines allow a substantial improvement. The new band is attributed to an external mode.[256] The very low-frequency Raman modes of red HgI$_2$ (17 and 29 cm^{-1}) have been attributed to modes internal to the sheets, rather than to the expected lattice modes corresponding to sheet translations (see Volume 5). Only very minor displacement of the atoms would raise the unit-cell symmetry from D_{4h}^5 to O_h^5. This has led Hollebone

[253a] D. Breitinger and K. Köhler, *Inorg. Nuclear Chem. Letters*, 1972, **8**, 957.
[254] F. Sugawara and T. Nakamura, *J. Phys. and Chem. Solids*, 1972, **33**, 1665.
[255] A. S. Pine and G. Dresselhaus, *Phys. Rev. (B)*, 1972, **5**, 4087.
[256] G. Arie, E. DaSilva, H. Rozanska, and C. Sebenne, *Compt. rend.*, 1972, **274**, B, 536.

Vibrational Spectra

and Lever to suggest that a good approximation to the *intensities* expected in a factor-group calculation under D_{4h}^5 can be had by considering a correlation from the more restrictive O_h^5 case.[257] Because of the different dimensionalities involved, it is necessary to follow through separately the individual components of each degenerate representation: the authors introduce their own nomenclature for this, which some readers may find obscure at first. They find that the lattice modes allowed under D_{4h}^5 correlate with components forbidden in O_h^5 and therefore conclude that the published assignment is correct. This method is an important addition to the practical use of group theory and deserves wide attention.

TO frequencies in MoS_2 and $MoSe_2$ are at 380 and 245 cm^{-1} respectively.[257a]

The optical properties of amorphous As_2S_3 and As_2Se_3 have been studied by i.r. absorption and reflection in the range 30—8000 cm^{-1}. Both compounds show a 38 cm^{-1} band due to motion of the arsenic sublattice only; other bands (cm^{-1}) are assigned as follows:[258]

ν(As—E)	301 (E = S)	226 (E = Se)
2 × ν(As—E)	698, 991 (E = S)	487, 732 (E = Se)

ω_{TO} for $MoSe_2$ is at 283 cm^{-1} (Raman).[259]

Me_2SnF_2 and Me_2TlBr have similar sheet structures. A full single-crystal Raman study of them has been made and is especially notable for the very high quality of the experimental technique. Apart from internal modes of methyl, only two Raman bands are expected and their previous assignment from powder spectra is confirmed, *viz.* 144 (E_g) and 531 cm^{-1} (A_{1g}) in D_{4h}^{17} for the tin compound and 96 (E_g) and 488 cm^{-1} (A_{1g}) for Me_2TlBr. The real purpose of the study was the collection of accurate data on derived bond polarizabilities, $\partial\alpha/\partial r$, which are necessary for the discussion of intensities. The paper concludes with an interesting and contentious suggestion. In aqueous solution Me_2Sn^+ shows Raman bands at 529 and 176 cm^{-1} (*cf.* solid-state values), and Me_2Tl^+ has 488 and 96 cm^{-1}. In the crystals the lowest modes are due to rotatory lattice vibrations: it is now suggested that their aqueous-solution counterparts are due to rotations of the ions restricted by hydrogen-bonding, and not to the allowed (at least in the double group $G_{36}^\dagger$ applicable for the case of free methyl rotors) skeletal deformations.[260]

Complex Halides.—$Rb[ICl_2]$ (*Pnma*, $Z = 4$) has been treated to a full single-crystal i.r. and Raman study at 40 K and ambient temperature. Approximately one-quarter of the predicted modes were observed, *i.e.* 24 in the range 42—288 cm^{-1}; consequently most bear multiple symmetry

---

[257] B. R. Hollebone and A. B. P. Lever, *Inorg. Chem.*, 1972, **11**, 1158.
[257a] O. P. Agnihotri, *J. Phys. and Chem. Solids*, 1972, **33**, 1173.
[258] M. Onomichi, T. Arai, and K. Kudo, *J. Non-cryst. Solids*, 1971, **6**, 362.
[259] O. P. Agnihotri and H. K. Sehgal, *Phil. Mag.*, 1972, **26**, 753.
[260] V. B. Ramos and R. S. Tobias, *Inorg. Chem.*, 1972, **11**, 2451.

labels.[261] No attempt was made to fit the data to a higher pseudo-symmetry model.

Single-crystal Raman data for Cs[ICl$_4$] (D_{2h}^{17}, $Z = 4$) and NO[AlCl$_4$] are shown in Table 30. The assignment for the ICl$_4^-$ ion is more useful when viewed in conjunction with the correlation diagram:

$$\begin{array}{ccc} D_{4h} & C_{2h}\text{ (site)} & D_{2h}^{17} \\ \nu_1 \quad a_{1g} & \!\!\!\!\!\!\!\!\diagdown\!\!\!A_g & \!\!\!\!\!\!\!\!\rightarrow A_g + B_{3g} \\ \nu_3 \quad b_{1g} & \!\!\!\!\!\!\!\!\diagup\!\!\!B_g & \!\!\!\!\!\!\!\!\rightarrow B_{1g} + B_{2g} \\ \nu_4 \quad b_{2g} & & \end{array}$$

Table 30 *Single-crystal Raman wavenumbers/cm^{-1} and assignments for some tetrahalogeno-salts*

Cs[ICl$_4$] [262]

$\nu_1 \begin{cases} A_g & 286 \\ B_{3g} & 284 \end{cases}$ $\nu_3 \begin{cases} B_{1g} & 262 \\ B_{2g} & 260 \end{cases}$ $\nu_4 \begin{cases} A_g & 147 \\ B_{3g} & 145 \end{cases}$

$T_x \quad B_{3g} \quad 100$ $\qquad T_y \quad B_{1g} \quad 49$ $\qquad T_z \quad A_g \quad 66$

$R_x \begin{cases} B_{2g} & 115 \\ B_{1g} & 109 \end{cases}$ $R_y \begin{cases} B_{1g} & 86 \\ B_{2g} & 76 \end{cases}$ $R_z \begin{cases} B_{3g} & 43 \\ A_g & 41 \end{cases}$

NO[AlCl$_4$] [263]

$\nu'_1, \nu(\text{NO}^+) \begin{cases} A_g & 2246 \\ B_{2g} & 2243 \end{cases}$ $\nu_1(\text{ex-}A_1) \begin{cases} A_g \\ B_{2g} \end{cases} 356$ $\nu_2(\text{ex-}E) \begin{cases} B_{2g} & 162, & B_{3g} & 145 \\ A_g & 153, & B_{1g} & 140 \end{cases}$

$\nu_3(\text{ex-}T_2) \begin{cases} B_{3g} & 528 \quad B_{2g} \quad 501 \\ B_{2g} & 519 \quad B_{2g} \quad 485 \\ A_g & 504 \quad A_g \quad 480 \end{cases}$ $\nu_4(\text{ex-}T_2) \begin{cases} A_g + B_{2g} & 233 \\ A_g + B_{2g} & 218 \\ B_{1g} + B_{3g} & 210 \end{cases}$

A considerable list of external modes was assigned for NO[AlCl$_4$] in addition to those shown here. On the basis of a normal-co-ordinate analysis, reassignment of the Raman feature at 135 cm^{-1} in Cs$_2$CuCl$_4$ to a B_1 deformation mode is proposed.[264] The lattice dynamics of M$_2$XY$_6$ compounds have been investigated in detail using a MUBFF. Specifically for Cs$_2$UBr$_6$:[265]

Wavenumbers/cm^{-1}	A_{1g}	E_g	T_{1g}	T_{2g}	T_{2u}	T_{1u}		T_{2u}
Obs.	197	—	—	43	87	51	84 195	60
Calc.	196	152	44	43	86.8	51	84 195	60

The choice of appropriate scattering geometries allowed observation of both longitudinal and transverse components of A_1 and E phonons in crystals of Na$_2$ZnCl$_4$,3H$_2$O and its deuteriate, C_{3v}^2.[266] Some splittings were >10 cm^{-1}. Internal modes of H$_2$O having large *TO–LO* splittings exhibit

---

[261] J. P. Coignac and M. Debeau, *Compt. rend.*, 1972, **275**, B, 211.
[262] J. P. Huvenne, P. Legrand, and F. Wallart, *Compt. rend.*, 1972, **275**, C, 83.
[263] P. Barbier, G. Mairesse, F. Wallart, and J. P. Wignacourt, *Compt. rend.*, 1972, **275**, C, 475.
[264] J. A. McGinnety, *J. Amer. Chem. Soc.*, 1972, **94**, 8406.
[265] S. L. Chodos, *J. Chem. Phys.*, 1972, **57**, 2712.
[266] G. L. Cessac, R. K. Khanna, E. R. Lippincott, and A. R. Bandy, *Spectrochim. Acta*, 1972, **28A**, 917.

Vibrational Spectra

small correlation field splittings. The partial results shown in Table 31 are quoted as typical.

Table 31 Values (cm^{-1}) of TO and LO for internal H_2O modes in $Na_2ZnCl_4,3H_2O$ and Na_2ZnCl_4,D_2O

	$Na_2ZnCl_4,3H_2O$		$Na_2ZnCl_4,3D_2O$		
	TO	LO	TO	LO	
A_1	3505	3519	2609	2610	$\nu_3(H_2O)$
E	3503	3516	2604	2618	
A_1	3458	3458	2537	2537	$\nu_1(H_2O)$
E	3436	3442	2525	2526	
A_1	265	271	263	268	$\nu_3(ZnCl_4^{2-})$
E	287	299	287	290	

Raman spectra for eight oriented crystals $M^I_2[FeCl_5(H_2O)]$ and $M^I_2[InCl_5(H_2O)]$ showed a number of bands due to anion internal modes comparable with that predicted by factor-group analysis of the tetramolecular cell, D_{2h}^{16}. Fewer than the predicted number of lattice modes were observed. Assignments are summarized in Table 32. The most unusual feature is that $\pi(MCl_4)$ modes apparently come above $\delta(MCl_4)$ modes. A mechanism involving cancellation of polarizabilities within the unit cell is proposed to account for the exceeding weakness of the B_{2g} modes.[267]

Table 32 Raman wavenumbers/cm^{-1} and assignments for $[MCl_5(H_2O)]^{2-}$ [267]

Mode	C_{4v}	$K_2[FeCl_5(H_2O)]$	$Cs_2[InCl_5(H_2O)]$
$\nu(M-Cl)$	$\nu_2(A_1)$	300	280
	$\nu_3(A_1)$	300	271
	$\nu_5(B_1)$	300	280
	$\nu_8(E)$	276	256
$\pi(MCl_4)$	$\nu_6(B_1)$	226	215
	$\nu_4(A_1)$	190	187
$\delta(MCl_4)$	$\nu_{10}(E)$	180	162
	$\nu_7(B_2)$	176	149
$\delta(MCl)$	$\nu_{11}(E)$	130	124
$\nu(M-O)$	$\nu_1(A_1)$	371	310

Oxoanion-containing Crystals.—Since the ordered perovskites $Ba_2M^{2+}M^{6+}O_6$ (M = Mo, W, or Te) contain one $(M^{6+}O_6)$ group per primitive cell and none of the oxygens are shared with other (MO_6) groups, the i.r. and Raman spectra are both simple and can be assigned on the basis of O_h labels for an isolated octahedron.[268] Typically, for Ba_2MgWO_6,

Raman			I.r.		
ν_1	ν_2	ν_5	ν_3	ν_4	'External'
817	680	444	622	388	319 cm^{-1}

[267] D. M. Adams and D. C. Newton, *J.C.S. Dalton*, 1972, 681.
[268] A. F. Corsmit, H. E. Hoefdraad, and G. Blasse, *J. Inorg. Nuclear Chem.*, 1972, **34**, 3401.

The Raman shifts of Table 33 have been assigned to phonons (as well as some electronic Raman shifts) for the spinels and garnets shown.

Table 33 Raman wavenumbers/cm^{-1} and assignments for YIG garnets [269] and normal spinels [269a]

YIG [a]			Spinels			
			Ag_2MoO_4	Co_2GeO_4	$ZnAl_2O_4$	
130		$420\ E_g + T_{2g}$	88	—	—	T_{2g}
175	T_{2g}	$447\ T_{2g} + A_{1g}$	273	302	242	E_g
193		$507\ A_{1g}$	347	—	—	T_{2g}
237		$593\ T_{2g}$	757	643	655	T_{2g}
274	E_g	$698\ A_{1g} + E_g$	869	757	—	A_{1g}
315		$740\ A_{1g}$				
347	$E_g + A_{1g}$					
380	T_{2g}					

[a] Theory: $3A_{1g} + 8E_g + 14T_{2g}$. Others studied: GdIG, $Gd_{2.4}Tb_{0.66}IG$.

In a recently reported Raman single-crystal study of YVO_4 (D_{4h}^{19}, $Z = 2$, zircon structure), extinctions between the various spectra were so poor as to imply complete breakdown of the selection rules; the authors attributed the poor results to crystal misalignment. Porto and co-workers have now shown that the real reason for the apparent difficulty lies in the exceptionally large birefringence of the material. By stopping down the collection cone successively improved extinctions were observed. Not only has the following assignment now been substantiated, but some previously missing weak phonons have been found.[270]

YVO_4 A_g 968, 1030, 1098, 1138, and 1246 cm^{-1};
E_g 921 and 933 cm^{-1}; B_{1g} 1056 cm^{-1}.

Raman and far-i.r. reflectance data for the isostructural $DyVO_4$ have also been assigned.[271]

Spinels continue to attract attention, as do a variety of other structure types containing oxoanions. Apart from the production of impressive lists of numbers,[272, 273] the following results of more general value appear to emerge. New single-crystal Raman assignments are reported: see Table 33. Far-i.r. and Raman spectra of several Scheelite-type perrhenates, at 77 K, were assigned with the aid of single-crystal Raman work with $NaReO_4$. Of especial importance is the detection of an NH_4^+ librational mode at 264 cm^{-1} in NH_4ReO_4 at 77 K (194 cm^{-1} in the ND_4^+ salt). This

[269] P. A. Grunberg, J. A. Köningstein, and L. G. Van Uitert, *J. Opt. Soc. Amer.*, 1971, **61**, 1613.
[269a] J. A. Köningstein, P. A. Grunberg, J. T. Hoff, and J. M. Preudhomme, *J. Chem. Phys.*, 1972, **56**, 354.
[270] A. Chaves and S. P. S. Porto, *Solid State Comm.*, 1972, **10**, 1075.
[271] F. D'Ambrogio, P. Brueesch, and H. Kalbfleisch, *Phys. Stat. Solidi (B)*, 1972, **49**, 117.
[272] A. Mellier and P. Graverau, *Compt. rend.*, 1972, **274**, B, 1025.
[273] P. P. Cord, P. Courtine, G. Pannetier, and J. Guillermet, *Spectrochim. Acta*, 1972, **28A**, 1601.

disappears on warming and is then not found in either i.r. or Raman spectra.[274] Revised assignments for other Scheelite-type lattices have been deduced by Tarte with the aid of metal-isotope substitution (^{40,44}Ca and ^{92,100}Mo).[275, 276] [On a historical note we observe that Tarte was switching metal isotopes (in solids) years before the better-known, more recent, and very valuable work of Nakamoto and co-workers.] Although the new assignments are suggestive, their confirmation by single-crystal methods is desirable. In particular, it is found in $CaMoO_4$, Na_2WO_4, and Ag_2MoO_4 that $\nu_4 > \nu_2$ for Raman-active modes.[276] There is also considerable interaction between internal anion bending modes and external translatory modes.[275] From computed polariton dispersion curves for $CaWO_4$ it is shown that even modes due to anion internal vibrations have considerable dispersion.[277] Vibrational spectra of powdered Ag_2CrO_4 (orthorhombic) have ν_1 and ν_3 some 20—40 cm^{-1} lower than in K_2CrO_4; this has been interpreted in terms of appreciable Ag–O interaction.[277a]

Complex Cationic Salts.—The NH_4^+ librational mode in NH_4ReO_4 was noted above. The Raman spectrum of NH_4Cl shows a band at *ca.* 93 cm^{-1} which behaves in a curious way on passing through an order/disorder transition at 242.8 K. At the transition temperature it splits into two: this is said to be due to short-range order.[278] NH_4Br has a phase transition at 234.5 K (T_λ). Above T_λ Raman scatter at *ca.* 56 cm^{-1} is anomalous in the sense that it is associated with a zone-boundary *TA* phonon. Below T_λ the tetragonal distortion halves the size of the Brillouin zone, relocating the above phonon to the zone centre, where it becomes Raman-active and shows unusual thermal dependence, increasing in intensity on cooling below T_λ. It is accompanied by an intensity-decreasing mode (not predicted by group theory) attributed, as for NH_4Cl, to short-range ordering. The evolution of long-range order governs the temperature behaviour of these bands.[279] A phase change occurs at 100—110 K in NH_4ClO_4 (detected by i.r.): an assignment of i.r. and Raman data has been made for NH_4ClO_4 and its deuteriate.[280]

An especially interesting single-crystal Raman and i.r. study is reported for [M(en)$_3$]Cl$_3$ (M = Cr, Co, or Rh) using both racemic (*dl*) and optically active (*d*) forms.[281] The *d*-forms have symmetry D_4^8, $Z = 4$, but it is not stated how they were handled experimentally in polarized light; *dl*-forms have trigonal symmetry D_{3d}^4, $Z = 4$, and show much simpler spectra. Assignment of the MN$_6$ A_{1g} breathing mode is clear-cut and shows strong

[274] R. A. Johnson, M. T. Rogers, and G. E. Leroi, *J. Chem. Phys.*, 1972, **56**, 789.
[275] P. Tarte and M. Liegeois-Duyckaerts, *Spectrochim. Acta*, 1972, **28A**, 2029.
[276] M. Liegeois-Duyckaerts and P. Tarte, *Spectrochim. Acta*, 1972, **28A**, 2037.
[277] V. C. Sahni and G. Venkataraman, Proceedings of the 15th Symposium on Nuclear Physics and Solid State Physics, 1970, Vol. 3, p. 447.
[277a] R. L. Carter, *Spectroscopy Letters*, 1972, **5**, 401.
[278] C. H. Wang and R. B. Wright, *J. Chem. Phys.*, 1972, **56**, 2124.
[279] C. H. Wang and R. B. Wright, *J. Chem. Phys.*, 1972, **57**, 4401.
[280] D. J. J. Van Rensburg and C. J. H. Schutte, *J. Mol. Structure*, 1972, **11**, 229.
[281] J. Gouteron-Vaissermann, *Compt. rend.*, 1972, **275**, B, 149.

dependence upon the metal cation in accord with the order of LFSE. Thus, for the dl-forms the values obtained were: 547 (Rh), 527 (Co), and 494 (Cr) cm^{-1}. Although symmetry species have been assigned, some doubt exists as to which bands are due principally to MN$_6$ and which to (en) modes, if, indeed, such distinction can be made. Some results are shown in Table 34. A separate report has appeared dealing with the differences between spectra of the dl- and d-forms of these materials, and of some related tris-oxalates, in the region below 200 cm^{-1}.[282]

Table 34 *Vibrational wavenumbers/cm^{-1} and assignments for dl-[M(en)$_3$]Cl$_3$* [281]

M = Co		M = Rh		M = Cr		Assignment [a]
I.r.	Raman	I.r.	Raman	I.r.	Raman	
578 E_u	582 E_g	574 E_u	576 E_g	550 E_u		} ν(M—N)
544 A_{2u}		542 A_{2u}		517 A_{2u}		} T_{1u}
	527 A_{1g}		547 A_{1g}		494	ν(M—N), A_{1g}
495 E_u	501 E_g	498 E_u	507 E_g	480 E_u		
468 A_{2u}		452 A_{2u}		450 A_{2u}		} δ(ring)
438 E_u	441 E_g	445 ?	447 E_g		442	ν(M—N), E_g
368 E_u	372 E_g	354 E_u	356 E_g			
	338 A_{1g}		329 A_{1g}	~320		} δ(ring)

[a] O_h labels used in this column; all other labels refer to unit-cell symmetry D_{3d}.

Complex Anionic Salts.—The Raman-active modes of Ca(OH)$_2$ and its deuteriate have been assigned by single-crystal methods; all predicted modes were found.[283] The assignment, Table 35, is in complete discord with an earlier one due to Krishmarthi, upon which others have based calculations. Inelastic neutron scattering data for Ca(OH)$_2$ [284] are shown for comparison.

In a Raman study of *TO* phonons in NaNO$_2$ (20—250 °C) it was found that the ferroelectric phase obeys the rules of C_{2v}^{20}. In the paraelectric phase six (not three as expected) extra resonances were observed. This

Table 35 *Vibrational wavenumbers/cm^{-1} and assignments for* Ca(OH)$_2$

	Ca(OH)$_2$ [283]	Ca(OD)$_2$ [283]	Ca(OH)$_2$ [284]	
A_{1g} ν(OH)	3610	2664	530—550	} R of types E_g, E_u
A_{1g} translatory	357	350	ca. 340	
E_g translatory	254	252	310	T + R of types A_{1g}, A_{1u}, E_u
E_g rotatory	680	475	ca. 250	T(acoustic) + T(E_g)
			ca. 100	T(acoustic)

[282] J.-P. Mathieu and J. Gouteron-Vaissermann, *Compt. rend.*, 1972, **274**, B, 880.
[283] P. Dawson, *Solid State Comm.*, 1972, **10**, 41.
[284] A. Bajorek, J. A. Janik, J. M. Janik, I. Natkaniec, T. Stanek, and T. Wasiutynski, *Acta Phys. Polon.* (A), 1971, **40**, 431.

anomaly is accounted for by recognizing the effect of NO_2^- disordering along the b-axis. The nitrogen atom jumps between two equivalent sites by the flipping of NO_2^- about the a-axis: temperature-dependent coupling between normal modes precipitates this motion.[285]

Iqbal continues his courageous studies on solid azides. In a single-crystal study [286] of KN_3 (400—5000 cm^{-1}) the technical quality of the spectra was higher than in an earlier investigation (by Bryant). The very considerable fine structure associated with the main absorption regions was interpreted in terms of internal ± external mode combinations, including those from critical points and symmetry lines in the Brillouin zone. This work forms a good basis for elucidation of the optical phonon frequencies determined by neutron spectroscopy, which must be completed before the lattice dynamics of the solid are fully understood. I.r. and Raman single-crystal methods have been applied to a study of the phase transition in TlN_3.[287] Below the transition temperature the low-frequency spectrum becomes very complex and a non-centrosymmetric structure may be formed. At 90 K internal modes have wavenumbers/cm^{-1}:

$$\nu_2 \begin{cases} A_{2u} & 622 \\ E_u & 628 \end{cases} \quad \nu_1 \begin{cases} B_{1g} & — \\ A_{1g} & 1323 \end{cases} \quad \nu_3 \quad E_u \quad 1995 \text{ cm}^{-1}$$

whereas at 298 K librational lattice modes are at 178 (B_{1g}) and 47 cm^{-1} (E_g), and a translational mode is at 33 cm^{-1} (E_g). About half the number of *internal* modes of KNCS predicted by factor-group analysis were observed in a single-crystal i.r. (to 400 cm^{-1}) and Raman study. However, the main interest lies in the lattice-mode study. Analysis of the $\nu_3 \pm$ lattice mode complex, Figure 2, yielded values of mode energies at points away from $k \sim 0$. This is possible because conservation of momentum in a two-mode process is satisfied if component modes have non-zero equal and opposite wave vectors, k and $-k$. Thus, k values can range throughout the Brillouin zone, although only regions of high density of states will make significant contributions. Typically, for the $\| c$ i.r. experiment:[288]

39	$T(A)$ along Σ, Λ, Δ	109	$T(B_{1u})$ along Σ, Λ, Δ
69	$B_{2g} + B_{3g}$ at Γ	126	$R(B_{2g} + B_{3g})$ at Γ
94	$B_{2g} + B_{3g}$ along Σ, Λ, Δ	142	$T(B_{1u})$ at X, Y, Z

The assignment of Table 36 was deduced from i.r. and Raman work with single-crystal $LiIO_3$, C_6^6. The dielectric constant determined by the LST relation does not reveal any large contribution by lattice dynamics to the low-frequency dielectric response.[289] Raman spectra of $NaClO_3$ and $KClO_3$

---

[285] C. M. Hartwig, E. Wiener-Avnear, and S. P. S. Porto, *Phys. Rev.* (*B*), 1972, **5**, 79.
[286] Z. Iqbal, *J. Chem. Phys.*, 1972, **57**, 2422.
[287] Z. Iqbal and M. L. Malhotra, *J. Chem. Phys.*, 1972, **57**, 2637.
[288] Z. Iqbal, L. H. Sarma, and K. D. Möller, *J. Chem. Phys.*, 1972, **57**, 4728.
[289] W. Otaguro, E. Wiener-Avnear, C. A. Arguello, and S. P. S. Porto, *Phys. Rev.* (*B*), 1971, **4**, 4542.

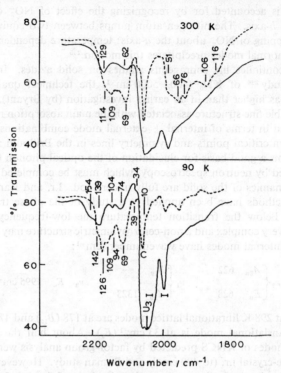

Figure 2 I.r. absorption of single-crystal KCNS in the $\nu_3 \pm$ external mode region: parallel to c, dotted line; perpendicular to c, full line.
(Reproduced by permission from *J. Chem. Phys.*, 1972, **57**, 4728)

Table 36 Raman wavenumbers/cm^{-1} and assignments for LiIO$_3$, C_6^6 [289]

A_1		E_1		E_2
TO	LO	LO	TO	98
	148		180	200
	238	330	340	332
358	468	370	460	347
795	817	769	848	765

Predicted: $4A_1 + 5B_1$ (inactive) $+ 4E_1 + 5E_2$. Frequencies above the lines are due to 'external' modes.

before and after X-ray treatment showed that new lines appeared *ca.* 680 and 980 cm^{-1}: these are attributed to O_3^-, formed by loss of Cl from the anions.[290]

[290] S. Radhakrishna and A. M. Karguppikar, *Proc. Indian Acad. Sci.* (*A*), 1972, **75**, 132.

Single-crystal Raman spectra reported for $CsNO_3$ [291] complete the assignment of all alkali-metal nitrates. In the disordered phase of $CsNO_3$ the discrete lattice spectrum is replaced by a broad band at *ca.* 90 cm^{-1} superimposed on an anisotropic wing. The internal vibrations retain their polarization characteristics in the high-temperature phase, and the melt spectra are very similar to those of the disordered crystal at elevated temperatures. For the room-temperature structure (C_3), (wavenumbers/cm^{-1})

	ν_L	ν_4	ν_1	$\nu_3(TO)$	$\nu_3(LO)$
A_1	29, 50, 113	715	1051	1383	1396
E	39, 116	704	1048	1345	1420

I.r. spectra of NO_3^- isolated in KX and NaX matrices show that the site symmetry is C_{3v}. Harmonic frequencies were shifted to higher values as the lattice constant of the matrix decreased. In KI, bands due to ν(internal) $\pm$ localized lattice modes were found.[292] The vibrational spectrum of Na_2SO_3 has been assigned, from powder data, on the basis of unit-cell symmetry. Anomalous results previously reported for its i.r. spectrum in aqueous solution are shown to be due to a reaction with AgCl cell windows.[293]

$CaCO_3$ III is a phase formed by compression of calcite below 200—300 °C. The Raman spectrum at 77 K is reported but interpretation is difficult in the absence of structural information. The large number of bands indicates low symmetry.[294] ^{13}C substitution has been used to determine coupling constants for ν_2 (*ca.* 850 cm^{-1}) of CO_3^{2-} in some compounds with the aragonite structure (M = Ca, Sr, Ba, Eu, or Pb) and in Li_2CO_3 and Na_2CO_3,H_2O. A relationship found between the value of the coupling constant and the fifth power of the anion distances shows that the coupling is governed by short-range repulsive forces, not by dipole–dipole interaction.[295] B_{2u} and B_{3u} modes in azurite, $2CuCO_3,Cu(OH)_2$, were distinguished with the aid of a normal-co-ordinate analysis.[296]

$BeSO_4,4H_2O$, a uniaxial piezoelectric crystal, has been studied by single-crystal Raman technique in the ν_3 SO_4^{2-} region.[297] The new values (cm^{-1}) are:

$B_2(T)$	1081		
$E(T)$	1124	$E(L)$	1110
$E(T)$	1084	$E(L)$	1162

A full Raman single-crystal assignment for K_2SO_4 is available.[298] I.r. spectra confirm the presence of a continuous solid solution between

---

[291] D. W. James and J. P. Devlin, *J. Chem. Phys.*, 1972, **56**, 4688.
[292] M. Tsuboi and I. C. Hisatsune, *J. Chem. Phys.*, 1972, **57**, 2087.
[293] J. D. Brown and B. P. Straughan, *J.C.S. Dalton*, 1972, 1750.
[294] M. Nicol and W. D. Ellenson, *J. Chem. Phys.*, 1972, **56**, 677.
[295] W. Sterzel and W.-D. Schnee, *Z. anorg. Chem.*, 1972, **392**, 173.
[296] B. Taravel, G. Chauvet, P. Delorme, and V. Lorenzelli, *Compt. rend.*, 1972, **274**, *B*, 882.
[297] B. Unger, *Phys. Stat. Solidi (B)*, 1972, **49**, 107.
[298] M. Debeau, *Rev. Phys. Appl.*, 1972, **7**, 49.

$CaSO_4,2H_2O$ and anhydrite [299] and ν_1 and ν_4 SO_4^{2-} frequencies are listed for a series of metastable crystals in the $BaSO_4$–$CaSO_4$ system.[300] A full assignment of the i.r. single-crystal spectra (30—4000 cm^{-1}) of Li_2SO_4,H_2O and its deuteriate at ambient and liquid-nitrogen temperatures has been supported by a force-constant calculation. The assignments for external modes of the water are compared with inelastic neutron-scattering data in Table 37.[301] Others, also comparing vibrational data with those from i.n.s. for the same compounds, claim that acoustic modes are showing through in the vibrational spectra owing to disorder associated with the water molecules.[302]

Table 37 External water modes (cm^{-1}) in Li_2SO_4,H_2O

	I.r.		I.n.s.
	H_2O	D_2O	H_2O
ρ_r	600	440	769
$\rho_\omega + \rho_\tau$	533	—	556
$\rho_\tau + \rho_\omega$	338	269	343
Translations	196	194	235
	180	176	186
	143	141	164

Full single-crystal Raman assignments are reported for KH_2PO_4,[303, 304] KD_2PO_4, RbH_2PO_4, and $NH_4H_2PO_4$.[303] In addition to the ferroelectric-phase modes (at 90 K) observed by Kaminov and Damen, a sharp line at 515 cm^{-1} was found in that phase for KH_2PO_4 and is related to phosphorus-atom movement.[304] I.r. absorption data have been listed for $FeAsO_4$, $FeAs_{0.995}P_{0.005}O_4$, and $FeAs_{0.995}P_{0.005}O_4,2H_2O$.[305]

Previous assignments of the vibrations of the formate ion have been confirmed by a single-crystal i.r. and Raman study of $Sr(HCOO)_2$. I.r.-active external modes were estimated from combinations but Raman-active ones were observed directly. Below 250 cm^{-1} $9A + 8B_1 + 8B_2 + 8B_3$ modes were found in exact correspondence with predictions for translatory lattice modes. None of the expected $6(A + B_1 + B_2 + B_3)$ rotatory modes were found.[306] No evidence is given to justify such a distinction. A qualitative assignment of i.r. and Raman spectra of $Cu(HCOO)_2,4H_2O$ has been given.[307] Raman spectra of oriented crystals and i.r. spectra of thin films of $CsH(CF_3COO)_2$ and $KH(CF_3COO)_2$ and their deuteriated

[299] M. Soustelle, B. Guilhot, and J.-J. Gardet, *Compt. rend.*, 1972, **274**, C, 853.
[300] O. Vojtech, J. Moravec, and I. Krivy, *J. Inorg. Nuclear Chem.*, 1972, **34**, 3345.
[301] S. Meshitsuka, H. Takahashi, and K. Higasi, *Bull. Chem. Soc. Japan*, 1971, **44**, 3255.
[302] E. Mikuli, J. M. Janik, G. Pytaz, J. A. Janik, J. Sciesinki, E. Sciesinska, and A. Mazurkiewicz, *Inst. Nuclear. Phys.*, Cracow, Rep., 1972, INP-791/PS, 15pp.
[303] E. A. Popova, *Izvest. Akad. Nauk. S.S.S.R., Ser. fiz.*, 1971, **35**, 1812.
[304] J. P. Coignac and H. Poulet, *J. Phys. (Paris)*, 1971, **32**, 679.
[305] F. d'Yvoire, *Compt. rend.*, 1972, **275**, C, 949.
[306] N. R. McQuaker and K. B. Harvey, *Canad. J. Chem.*, 1972, **50**, 1453.
[307] R. S. Krishnan and P. S. Ramanujan, *Spectrochim. Acta*, 1972, **28A**, 2227.

analogues have been fully assigned, the particular interest of this work being modes involving the hydrogen atom of the symmetric hydrogen bond. For the Cs⁺ salt they are:[308]

I.r.	Raman		
1168 cm⁻¹	1200 cm⁻¹	COH	δ bend
870	825	COH	γ bend
—	798	OHO	*sym.* stretch
1830	—	OHO	*asym.* stretch
1020	—	OHO	δ bend
940	—	OHO	γ bend
—	135	OHO	twist

A particularly complete study of oriented crystals (i.r. reflectance and Raman spectra) of Li(CH₃COO),2H₂O has appeared.[308a] Both ⁶Li/⁷Li and ¹H/²H substitution was used. The complete i.r. range 20—3600 cm⁻¹ was covered and a full assignment deduced. In combination with other recent work on hydrates, this forms an important contribution to our knowledge of the lattice dynamics of hydrated salts. I.r. reflectance from beryl in the region 280—1400 cm⁻¹ allowed location of $6A_{2u} + 13E_{1u}$ modes of the $P6/mcc$ structure, leaving only $3E_{1u}$ undetermined.[309] In a single-crystal Raman study of K₃[Fe(CN)₆] most of the factor-group-predicted modes were observed. In particular, the δ(CFeC) bend becomes completely mixed in with the large number of lattice modes present in the range 51—185 cm⁻¹ and loses its identity.[310]

Molecular Crystals.—Work on the Raman spectra of solid hydrogen has been reviewed at length,[311] and the densities of phonon and libron states have been investigated by neutron spectroscopy.[312] Vibrational spectra of the solid halogens have been re-assigned.[313] An assignment of the i.r. and Raman data for crystalline ClO₂ (C_{2v}^{17}, $Z = 4$) was made by analogy with those of SO₂: ν_1, A_1 912—917, A_2 939—944; ν_2, A_1 465, A_2 456—459; ν_3, B_1, B_2 1052—1121 cm⁻¹.[314]

Although the i.r. spectrum of phosphonitrilic chloride trimer, (PNCl₂)₃, has been rather fully studied, the vibrational assignment has been ill-founded for years because of lack of definitive Raman data. Using single-crystal methods most of the Raman-active modes have been assigned unambiguously.[315, 316] In particular, the long-sought $\nu_4(a_1')$ mode was located at 172 cm⁻¹. A full assignment for Ru(π-C₅H₅)₂ crystals has been deduced from i.r. and Raman measurements.[317]

[308] P. J. Miller, R. A. Butler, and E. R. Lippincott, *J. Chem. Phys.*, 1972, **57**, 5451.
[308a] M. Cadene and A. M. Vergnoux, *Spectrochim. Acta*, 1972, **28A**, 1663.
[309] F. Gervais and B. Piriou, *Compt. rend.*, 1972, **274**, B, 252.
[310] D. M. Adams and M. A. Hooper, *J.C.S. Dalton*, 1972, 160.
[311] H. L. Welsh, E. J. Allin, and V. Soots, *Phys. Solid State*, 1969, 343.
[312] H. Stein, H. Stiller, and R. Stockmeyer, *J. Chem. Phys.*, 1972, **57**, 1726.
[313] Y. S. Jain and H. D. Bist, ref. 277, p. 485.
[314] J.-L. Pascal, A. Pavia, and J. Potier, *J. Mol. Structure*, 1972, **13**, 381.
[315] D. M. Adams and W. S. Fernando, *J.C.S. Dalton*, 1972, 2503.
[316] J. Klosowski and E. Steger, *Spectrochim. Acta*, 1972, **28A**, 2189.
[317] D. M. Adams and W. S. Fernando, *J.C.S. Dalton*, 1972, 2507.

Thiourea undergoes a transition at 202 K to a ferroelectric phase characterized by a very low-frequency 'soft mode' which is strongly temperature-dependent (35 cm^{-1} at 178 K).[318] Assignment of the complicated vibrational spectra of $CdCl_2(thiourea)_2$ and $CdCl_2(thiourea)_4$ has been considerably strengthened by single-crystal i.r. and Raman work.[319]

Others.—Two studies have been made of single-crystals of $Bi_{12}GeO_{20}$ by Raman technique,[320, 321] and one of $Bi_{12}SiO_{20}$.[321] They are isomorphous, T^3. The germanium compound shows 36 lines in the range 40—720 cm^{-1} at 15 K and the silicon crystal 43 in the range 40—850 cm^{-1}; all were classified.

The two low-frequency bands of $HgCr_2Se_4$, determined at 300 K by i.r. absorption and reflection, are at 289.9, 281.8 (LO), and 286.8, 268.6 (TO) cm^{-1}.[322] Pyrargyrite, Ag_3AsS_3, should show $3A_1 + 4E$ modes: single-crystal Raman experiments located bands at 362 and 186 cm^{-1} (A_1) and at 332 and 120 cm^{-1} (E). A series of isomorphously substituted samples (Sb replacing As) was also investigated qualitatively.[323]

[318] J. P. Benoit, M. Deniau, and J. P. Chapelle, *Compt. rend.*, 1972, **275**, B, 665.
[319] D. M. Adams and M. A. Hooper, *J.C.S. Dalton*, 1972, 631.
[320] B. K. Bairamov, B. P. Zakharchenya, R. V. Pisarev, and Z. M. Khashkhozhev, *Fiz. tverd. Tela*, 1971, **13**, 3366.
[321] S. Venugopalan and A. K. Ramdas, *Phys. Rev. (B)*, 1972, **5**, 4065.
[322] T. H. Lee, T. Coburn, and R. Gluck, *Solid State Comm.*, 1971, **9**, 1821.
[323] D. K. Arkhipenko, A. A. Godovikov, S. N. Nenasheva, B. G. Nenashev, B. A. Ovekhov, V. S. Pavlynchenko, and M. G. Serbulenko, *Kvantovaya Elektron (Moscow)*, 1971, 69.

5
Characteristic Vibrational Frequencies of Compounds containing Main-group Elements

BY S. R. STOBART

Divisions within this chapter remain unchanged from those employed in previous volumes of the series, vibrational data for Main-group compounds being divided into eight sections, one devoted to each of the Main Groups of elements in the Periodic Table. Further subdivision within each of these sections has been achieved where this is considered to be useful by dealing with each element of the Group in turn (in order of increasing atomic number) or by collecting together references concerning compounds possessing similar groups of atoms.

1 Group I Elements

Monolithium derivatives of acetonitrile and benzonitrile have been isolated and shown by molecular weight measurements to be respectively tetrameric and dimeric,[1] lowering in frequency of $\nu(C{\equiv}N)$ as well as other evidence suggesting association through Li$\cdots$CN interaction. Skeletal modes for the acetonitrile derivative are found (i.r.) at 432 and 575 cm^{-1} for ^{7}LiCH$_2$CN, shifting to 455, 605 cm^{-1} for ^{6}LiCH$_2$CN and to 410, 575 cm^{-1} for ^{7}LiCD$_2$CN. The 432 cm^{-1} band is identified as $\nu(\text{Li}-\text{C})$, that at 575 cm^{-1} being attributed to $\nu(\text{Li}\cdots\text{N})$ between the lithium atom and an adjacent nitrogen atom in the tetramer, on the basis that in the proposed structure a Li$-$C mode should be sensitive to deuteriation whereas a Li$-$N mode should not. Similar observations for PhCH(Li)CN give $\nu(\text{Li}-\text{C})$ at 460 and $\nu(\text{Li}\cdots\text{N})$ at 575 cm^{-1}. The compounds Me$_3$SiOM^1,-OPMe$_3$ (M^1 = Li, Na, or K) have been assigned[2] the cubane-like structures (1) on the basis of physical measurements, including i.r. data. A

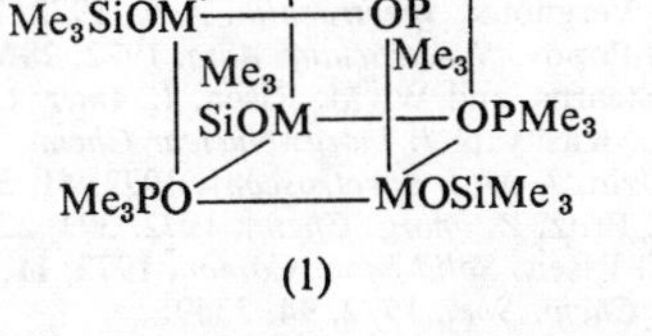

```
              PMe3
               |
               O——————MOSiMe3
              /|     /|
    Me3SiOM—————OP    |
            |  Me3|  |Me3|
            |   SiOM—————OPMe3
            |  /       | /
          Me3PO————————MOSiMe3
```

(1)

[1] R. Das and C. A. Wilkie, *J. Amer. Chem. Soc.*, 1972, **94**, 4555.
[2] H. Schmidbaur and J. Adlkofer, *Chem. Ber.*, 1972, **105**, 1956.

study of the vibrational spectra of ^{7}LiCH$_3$COO,2H$_2$O, ^{6}LiCH$_3$COO,2H$_2$O, and ^{7}LiCH$_3$COO,2D$_2$O is consistent[3] with the known chain structure which includes Li$_2$O$_4$(H$_2$O)$_2$ units. An analysis of fundamentals expected for such units is related to the observed spectra, placing ν_{sym}(Li—O) at 340 cm^{-1} (i.r.) and ν_{asym}(Li—O) modes at 429 (Raman) and 498 (i.r.) cm^{-1}; bands at 210, 241 cm^{-1} are assigned to LiO$_2$ deformation modes.

Far-i.r. spectra of Li$^+$, NH$_4$$^+$, and Na$^+$ salts in 4-Mepy have been measured by Handy and Popov.[4] Frequencies of ion–solvent vibration bands are 390—350 (Li$^+$), 200 (NH$_4$$^+$), and 175 (Na$^+$) cm^{-1}; the position of the Li$^+$ solvent band is strongly affected by 2-substitution of Mepy. Characteristic frequencies observed in the far-i.r. for alkali-metal ions encaged in crown ethers in solution have been reported.[5] With dibenzo-18-crown-6, there is no frequency-dependence of bands attributed to the (Na$^+$-crown) or (K$^+$-crown) motions (respectively at 214 and 168 cm^{-1}) on either anion or solvent, allowing calculation of ion–crown forces based on an M^{1}O$_6$S$_2$ aggregate with D_{6h} symmetry. Raman spectra of concentrated aqueous solutions of NaOH and KOH show low-frequency bands (282—322 cm^{-1}) which have been associated[6] with the stretch of an ion pair M$^+\cdots$OH$^-$.

Matrix-isolation Raman spectroscopy has been used to show[7] that cocondensation of an atomic beam of lithium with molecular oxygen at 4.2—15 K affords LiO$_2$ [ν(O—O) at 1097 cm^{-1}]. In LiCl,nH$_2$O, ν(Li—O) is assigned[8] at 520, 456 ($n = 1$), 482 ($n = 2$), and 476 cm^{-1} ($n = 3$). The vibrational frequencies of alkali-metal cations in their equilibrium positions in various metal–alkali-oxide glasses have been observed[9] as cation-mass-dependent bands in far-i.r. spectra.

2 Group II Elements

The vibrational spectrum of beryllium borohydride has again received attention. Gaseous and matrix-isolation studies by both i.r. and Raman methods on BeB$_2$H$_8$, BeB$_2$D$_8$, and BeB$_2$HD$_7$ are reported,[10] and their interpretation emphasizes the structural complexity of this system. Gas-phase measurements suggest an equilibrium between two coexisting structures, but only one form appears to be present at 20 K, probably with C_{3v} symmetry (2). The vibrational spectrum of (Et$_2$Be)$_2$ has been assigned[11] in terms of D_{2h} pseudosymmetry; similar data for gaseous and liquid Bu$_2^t$Be are consistent[12] with the point group D_{3h}, lower symmetry (D_3) being

---

[3] M. Cadene and A. M. Vergnoux, *Spectrochim. Acta*, 1972, **28A**, 1663.
[4] P. R. Handy and A. I. Popov, *Spectrochim. Acta*, 1972, **28A**, 1545.
[5] A. T. Tsatsas, R. W. Stearns, and W. M. Risen, *J. Amer. Chem. Soc.*, 1972, **94**, 5247.
[6] S. K. Sharma and S. C. Kashyap, *J. Inorg. Nuclear Chem.*, 1972, **34**, 3623.
[7] H. Hüber and G. A. Ozin, *J. Mol. Spectroscopy*, 1972, **41**, 595.
[8] M. Manewa and H. P. Fritz, *Z. anorg. Chem.*, 1972, **392**, 227.
[9] G. Exarhos and W. M. Risen, *Solid State Comm.*, 1972, **11**, 755.
[10] J. W. Nibler, *J. Amer. Chem. Soc.*, 1972, **94**, 3349.
[11] N. Atam, H. Müller, and K. Dehnicke, *J. Organometallic Chem.*, 1972, 37, 15.
[12] J. Mounier, *J. Organometallic Chem.*, 1972, **38**, 7.

$$H-B\overset{H}{\underset{H}{\cdots}}\overset{H}{\underset{H}{Be}}\overset{H}{\underset{H}{\cdots}}B-H$$

(2)

suggested for the solid state. For the first of these compounds $\nu(Be-C)$ frequencies are given as 925 (B_{1u}) and 875 cm^{-1} (A_g). The azido-derivatives [BeCl(N$_3$),OEt$_2$]$_2$ and [Be(N$_3$)$_2$]$_n$ exhibit i.r. bands at 814, 685, and 824, 710 cm^{-1} attributed [13] to $\nu(Be-N)$ modes. Dinitratotriberyllium tetra-t-butoxide shows i.r. bands at 1505, 1515 cm^{-1}, indicating [14] unidentate nitrate co-ordination; i.r. frequencies for other X$_2$Be$_3$(OBu^t)$_4$ are also listed. The adduct [(Et$_2$Be)$_2$SCN]$^-$ has a vibrational spectrum which suggests [12] donation through S: ν(SCN), 2100, 740; ν_{sym}/ν_{asym}(Be$_2$S), 865; and δ(Be$_2$S), 228 cm^{-1}. Raman spectra of vitreous, polycrystalline, and molten BeF$_2$ have been obtained [15] in the range 25—630 °C. These data are compared with the i.r. spectrum of vitreous BeF$_2$, and lead to an assignment based on a model isostructural with β-quartz (Table 1). Reaction of BeCl$_2$ with nitrosyl chloride affords [16] (NO$^+$)$_2$BeCl$_4^-$, with $\nu(Be-Cl)$ assigned at 610, 584, and 553 cm^{-1}.

Table 1 *Raman spectrum of vitreous* BeF$_2$ (D_6^4 *lattice structure*)

Wavenumber/cm^{-1}	Assignment	
282 (R)	a_1	δ(BeFBe)
385 (R)	e_2	δ(FBeF)
410 (i.r.)		
770 (i.r.)	e_1	
805 (R)	e_2	$\nu(Be-F)$
910 (i.r.)	e_1	

Wannagat and his co-workers have reported [17] i.r. frequencies for the compound [(Me$_3$Si)$_2$N]$_2$Mg; a band at 465 cm^{-1} is assigned to ν_{asym}(MgN$_2$). For [Mg(NH$_3$)$_6$]Cl$_2$ and [Mg(ND$_3$)$_6$]Cl$_2$, ν_3 [ν(Mg—N)] and ν_4 [δ(Mg—N)] for the cation are found [18] at 363, 198 (NH$_3$ complex) and 343, 185 cm^{-1} (ND$_3$ complex). The magnesium nitrate–water system has been examined [19] by Raman spectroscopy, covering the entire range from highly dilute solution to an anhydrous molten salt mixture with NaNO$_3$. In near-saturated solution, ν(Mg—OH$_2$) is observed at 354 cm^{-1}.

I.r. absorptions due to vibrations of the hydroxy-groups have been reported for Sr(OH)$_2$ and Sr(OD)$_2$, and a shift of OH modes to lower frequency between M(OH)$_2$ (M = Sr or Ba) is discussed in relation to

---

[13] N. Wiberg, W.-Ch. Joo, and K. H. Schmid, *Z. anorg. Chem.*, 1972, **394**, 197.
[14] R. A. Andersen, N. A. Bell, and G. E. Coates, *J.C.S. Dalton*, 1972, 577.
[15] A. S. Quist, J. B. Bates, and G. E. Boyd, *Spectrochim. Acta*, 1972, **28A**, 1103.
[16] J. MacCordick, *Naturwiss.*, 1972, **59**, 421.
[17] U. Wannagat, H. Autzen, H. Kuckertz, and H. J. Wismar, *Z. anorg. Chem.*, 1972, **394**, 254.
[18] R. Plus, *Compt. rend.*, 1972, **275**, *B*, 345.
[19] M. Peleg, *J. Phys. Chem.*, 1972, **76**, 1019.

hydrogen-bonding.[20] Some related data for Ba(OH)$_2$ hydrates have also been given.[21] In a review [22] the i.r. spectra of Group II dihalides have been critically discussed. Several other papers [23-25] deal with use of i.r. spectroscopy in investigation of oxide surfaces (Be, Mg, and Zn) and other solid-state properties.

3 Group III Elements

Compounds containing B—H Bonds. Fluorination of Me$_3$N,BH$_3$ with HF affords the new fluoroboranes Me$_3$N,BH$_2$F and Me$_3$N,BHF$_2$, for which unassigned i.r. frequencies have been listed;[26] those for Me$_3$N,BF$_3$ are identical with literature values. Other borane adducts for which i.r. data have been reported are PF$_2$(N$_3$),BH$_3$,[27] XPF$_2$,BH$_3$ (X = F, Cl, Br, or I),[28] and diborane dihydrate,[29] which at low temperatures shows four bands attributable to B—H stretching modes in agreement with the formulation BH$_2$(H$_2$O)$_2$$^+BH_4$$^-$ rather than BH$_3$(H$_2$O). In the Raman spectrum of a Et$_2$O adduct of S$_7$NBH$_2$, ν(B—H) modes are found [30] at 2419, 2409 cm^{-1} with ν(S—N) at 766, 750 cm^{-1}. Unassigned i.r. data have been given [31] for (3). I.r. spectra for M(BH$_4$)$_4$ (M = Zr or Hf) are nearly identical, only one ν(B—H)$_{terminal}$ being observed [32] as expected for terdentate BH$_4$ groups, indicating similar structures for the two compounds. Keller has found [33] that reaction of Al(NMe$_2$)$_3$ with excess diborane in ether gives,

```
       Me₂N—BH₂                    NMe₂
      /       \                   /    \
   H₂B        NMe₂            H₂B     Al(BH₄)₂
      \       /                   \    /
       MeS=BH₂                     NMe₂
         (3)                        (4)
```

in addition to known products, two new compounds for which cyclic structures are proposed on the basis of n.m.r. evidence: for one of these, (4), i.r. bands at 2510, 2420, and 2340 cm^{-1} are assigned to ν(B—H) modes whereas a further band at 2120 cm^{-1} is attributed to ν(B—H—Al).

Reactions of substituted lithium amides with diborane yield new μ-aminodiboranes R$_2$NB$_2$H$_5$ (R = Pr^n or Pr^i) and RNHB$_2$H$_5$ (R = Pr^n,

[20] H. D. Lutz, R. Heider, and R. A. Becker, *Spectrochim. Acta*, 1972, **28A**, 871.
[21] G. M. Habashy and G. A. Kolta, *J. Inorg. Nuclear Chem.*, 1972, **34**, 57.
[22] I. Eliezer and A. Reger, *Co-ordination Chem. Rev.*, 1972, **9**, 189.
[23] A. A. Tsyganenko and V. N. Filimonov, *Doklady Phys. Chem.*, 1972, **203**, 257.
[24] L. Genzel and T. P. Martin, *Phys. Status Solidi (B)*, 1972, **51**, 91.
[25] A. V. Sofronova and L. V. Kolobova, *Russ. J. Inorg. Chem.*, 1971, **16**, 794.
[26] J. M. Van Paasschen and R. A. Geanangel, *J. Amer. Chem. Soc.*, 1972, **94**, 2680.
[27] E. L. Lines and L. F. Centofanti, *Inorg. Chem.*, 1972, **11**, 2269.
[28] R. T. Paine and R. W. Parry, *Inorg. Chem.*, 1972, **11**, 1237.
[29] P. Finn and W. L. Jolly, *Inorg. Chem.*, 1972, **11**, 1941.
[30] M. A. Mendelsohn and W. L. Jolly, *Inorg. Chem.*, 1972, **11**, 1944.
[31] A. B. Burg, *Inorg. Chem.*, 1972, **11**, 2283.
[32] T. J. Marks and L. A. Shimp, *J. Amer. Chem. Soc.*, 1972, **94**, 1542.
[33] P. C. Keller, *J. Amer. Chem. Soc.*, 1972, **94**, 4020.

Pr^i, Bu^n Bu^i, Bu^t, or C$_5$H$_{11}$) for which i.r. frequencies have been listed.[34] B—H fundamentals have been assigned [35] for F$_3$PB$_3$H$_7$ and F$_2$ClPB$_3$H$_7$, with ν(B—H)$_{bridge}$ near 2110 and 1570 cm^{-1}. Borlin and Gaines have given [36] unassigned gas-phase i.r. data for Me$_2$AlB$_3$H$_8$ and Me$_2$GaB$_3$H$_8$, the similarity of the spectra indicating closely related structures (5). The i.r. spectra of π-borallyl complexes of transition metals (Ni, Pd, and Pt) show distinctive features,[37] absorptions attributable to the π-B$_3$H$_7^{2-}$ ligand being identified as follows: ν(B—H), 2493—2288 cm^{-1} (complex contour); also 1897, 1642, and 1577 cm^{-1} (the last of these bands is absent for B$_3$H$_8^-$ and its metal complexes).

Partly assigned i.r. frequencies for XPF$_2$,B$_4$H$_8$ (X = F, Cl, Br, or I) have been reported.[28] For B$_4$H$_9$Br, the following assignment of observed i.r. bands has been proposed [38] (cm^{-1}): 2592, 2502, ν(B—H); 2165, ν_{asym}(BHB); 2100, ν_{sym}(BHB); 1145, 1018, $\delta_{skeletal}$; 970, BH$_2$ rocking; 870, BH$_2$ torsion; 693, 683 ? ν(B—Br). Reaction of KSEt with diborane in THF affords K[EtS(BH$_3$)$_2$], with [39] ν(B—H) at 2375, 2330, 2295, and 2202 cm^{-1}. Macrocyclic cyanoboranes (BH$_2$CN)$_n$ have been prepared and partially separated, that with n = 5 (>90% pure) showing [40] i.r. bands at 2469, 2441, and 2429 [ν(B—H)] and 2295 cm^{-1} [ν(C≡N)]. The appearance of only one ν(C≡N) absorption at a relatively high frequency is considered to be consistent with the presence of a single bridging cyano-group as required in a cyclic oligomer (6). In the i.r. spectrum of the salt K$^+$[H$_3$BC(OH)=N(H)CH$_2$COO]$^-$ a band at 2280 cm^{-1} is described as typical [40a] for boranocarboxylates.

Rapid exchange of bridging hydrogen atoms of B$_6$H$_{10}$ with DCl has been demonstrated using i.r. spectroscopy,[41] attenuation of B—H—B

---

[34] L. D. Schwartz and P. C. Keller, *J. Amer. Chem. Soc.*, 1972, **94**, 3015.
[35] R. T. Paine and R. W. Parry, *Inorg. Chem.*, 1972, **11**, 268.
[36] J. Borlin and D. F. Gaines, *J. Amer. Chem. Soc.*, 1972, **94**, 1367.
[37] L. J. Guggenberger, A. R. Kane, and E. L. Muetterties, *J. Amer. Chem. Soc.*, 1972, **94**, 5665.
[38] J. Dazord and H. Maigedt, *Bull. Soc. chim. France*, 1972, 950.
[39] J. J. Mielcarek and P. C. Keller, *J.C.S. Chem. Comm.*, 1972, 1090.
[40] B. F. Spielvogel, R. F. Bratton, and C. G. Moreland, *J. Amer. Chem. Soc.*, 1972, **94**, 8597.
[40a] M. J. Zetlmeisl and L. J. Malone, *Inorg. Chem.*, 1972, **11**, 1245.
[41] H. D. Johnson, V. T. Brice, G. L. Brubaker, and S. G. Shore, *J. Amer. Chem. Soc.*, 1972, **94**, 6711.

bands being accompanied by the appearance of $\nu(B-D-B)_{bridge}$ at 1125 cm^{-1}. Studies on monosubstituted decaboranes have established that the frequencies of $\nu(B-H)_{bridge}$ (1600—1300 cm^{-1}) and $\delta(BH)$ (800—650 cm^{-1}) are characteristic of the site of decaborane skeleton substitution.[42] Controlled hydrolysis of hexadecaborane(20) affords [43] a new stable hydride $B_{14}H_{18}$ with (i.r.) $\nu(B-H)_{terminal}$ at 2562, 2595, $\nu(B-H)_{bridge}$ at 1830—1925 and 1540—1600 cm^{-1}. Reaction of hexaborane(10) with $Fe_2(CO)_9$ gives μ-$Fe(CO)_4$—B_6H_{10} in which the iron centre is co-ordinated to the unique basal B—B bond of the borane;[44] i.r. frequencies for $\nu(B-H)$ modes are 2578, 2555, 2495 (terminal), and 1935, 1850 cm^{-1} (bridge). Bruce and Ostazewski have synthesized [45] a stable copper carbonyl, HB(pyrazolyl)$_3$-Cu(CO), with $\nu(B-H)$ at 2465 and $\nu(CO)$ at 2083 cm^{-1}.

Brown and co-workers have reported [46] identification by i.r. spectroscopy of the first monomeric dialkylboranes. Hydroboration of tetramethylethylene (tme) yields known trialkyl diboranes, but when the tme/borane ratio reaches 16:1 the reaction mixture shows only $\nu(B-H)_{terminal}$ at 2470 cm^{-1}. Similar reaction in a deuteriated system gives a product with $\nu(B-D)_{terminal}$, 1820 cm^{-1}. For 2,7-dimethyl-1,6-diboracyclodecane (7), and bis-borinan (8), $\nu_{sym}\left(B\genfrac{}{}{0pt}{}{H}{H}B\right)$ absorptions are observed [47] at 1595 and 1560 cm^{-1} respectively.

(7)   (8)

Carbaborane chemistry has received considerable attention during 1972 and a number of papers including vibrational data have been published. In the i.r. spectrum of 1,2-dicarba-*nido*-pentaborane(7), obtained from the reaction of acetylene with B_4H_{10}, two bands at unusually high frequency (3150, 3075 cm^{-1}) for $\nu(C-H)$ are consistent [48] with a relatively large C—C bond order in the proposed structure (9). Raman spectra of 4,5-dicarba-*nido*-hexaborane(8) and its C-methyl derivatives $C_2B_4H_7Me$ and $C_2B_4H_6Me_2$ (10) have been measured; assignments made by comparison with remeasured i.r. frequencies are discussed in detail. Tentative separation of

---

[42] F. Hanousek, B. Štíbr, S. Heřmánek, J. Plešek, A. Vítek, and F. Haruda, *Coll. Czech. Chem. Comm.*, 1972, **37**, 3001.
[43] S. Heřmánek, K. Fetter, and J. Plešek, *Chem. and Ind.*, 1972, 606.
[44] A. Davison, D. D. Traficante, and S. S. Wreford, *J.C.S. Chem. Comm.*, 1972, 1155.
[45] M. I. Bruce and A. P. P. Ostazewski, *J.C.S. Chem. Comm.*, 1972, 1124.
[46] E. Negishi, J. J. Katz, and H. C. Brown, *J. Amer. Chem. Soc.*, 1972, **94**, 4025.
[47] E. Negishi, P. L. Burke, and H. C. Brown, *J. Amer. Chem. Soc.*, 1972, **94**, 7431.
[48] D. A. Franz, V. R. Miller, and R. N. Grimes, *J. Amer. Chem. Soc.*, 1972, **94**, 412.

the cage-modes (450—850 cm⁻¹ region) suggests that the framework bonding in these molecules is not as strong as in *closo*-carbaborane analogues.[49] A novel species, μ-1,2-trimethylene-1,2-dicarba-*closo*-dodecaborane(10), with proposed structure (11), has been reported by Hawthorne *et al.*, and i.r. frequencies for (11), the derived anion, and for

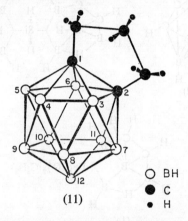

several sandwich-bonded complexes between the anion and Ni or Co, have been listed.[50] Plešek and Heřmánek [51] have reported another new type of carbaborane, tetracarba-di-*nido*-docosaborane(22) with ν(B—H)$_{terminal}$ at 2590, 2600 (sh) and (B—H)$_{bridge}$ at 1830, 1890 cm⁻¹. The corresponding anion $C_4B_{18}H_{22}^-$ has also been obtained, and the second member of the $C_2B_nH_{n+6}$ series, 6,9-dicarba-*nido*-decaborane(14) has been characterized.[52] Lewis-base adducts of 1,2-dicarba-*closo*-dodecaboranes show [53] a shift in the i.r. to 2000—2220 cm⁻¹ for ν(B—H). Solvent effects on ν(C—H) in the i.r. spectra of B-chlorinated *o*-carboranes have been described.[54]

[49] R. W. Jotham, J. S. McAvoy, and D. J. Reynolds, *J.C.S. Dalton*, 1972, 473.
[50] T. E. Paxson, M. K. Kaloustian, G. M. Tom, R. J. Wiersema, and M. F. Hawthorne, *J. Amer. Chem. Soc.*, 1972, **94**, 4882.
[51] J. Plešek and S. Heřmánek, *Chem. and Ind.*, 1972, 890.
[52] B. Štíbr, J. Plešek, and S. Heřmánek, *Chem. and Ind.*, 1972, 649.
[53] J. Plešek, T. Hanslík, F. Hanousek, and S. Heřmánek, *Coll. Czech. Chem. Comm.*, 1972, **37**, 3403.
[54] L. A. Leites, L. E. Vinogradova, N. A. Ogorodnikova, and L. I. Zakharkin, *Zhur. priklad. Spectroskopii*, 1972, **16**, 488.

Novel carbaboranes possessing bridging hetero-atoms have featured in a number of reports. Grimes and co-workers have listed i.r. frequencies (not assigned) for 1-methyl-1-galla-2,4-dicarba-*closo*-heptaborane(7) and its indium analogue. The similarity of these spectra suggest that both share the structure (12), determined crystallographically for the gallium compound.[55] The structure of μ,μ'-silylenebis-(2,3-dicarba-*nido*-hexaboraryl) is thought from n.m.r. studies to be (13); i.r. bands for $\nu(C-H)$, $\nu(B-H)$, and $\nu(Si-H)$ occur[56] at 3020, 2590 and 2530, and 2150 cm^{-1}, respectively. Wavenumbers for i.r. absorptions for 2,3-μ-trimethylsilyl-CC'-dimethyl-4,5-dicarba-*nido*-hexaborane(8) and related compounds with

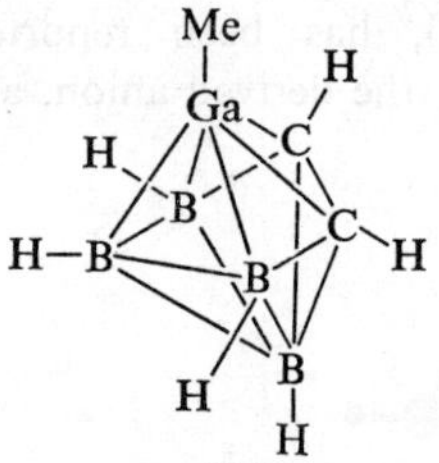

(12) 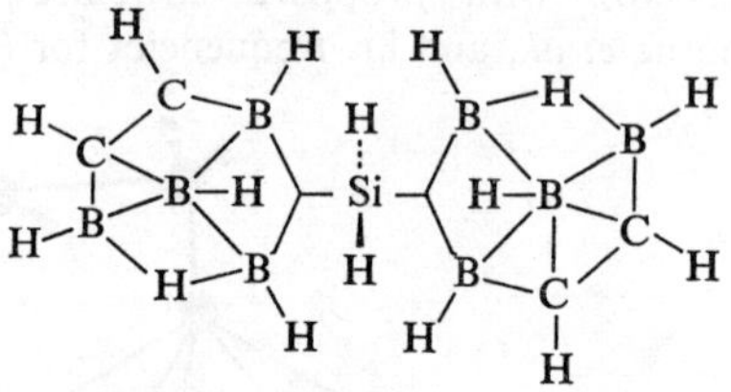   (13)

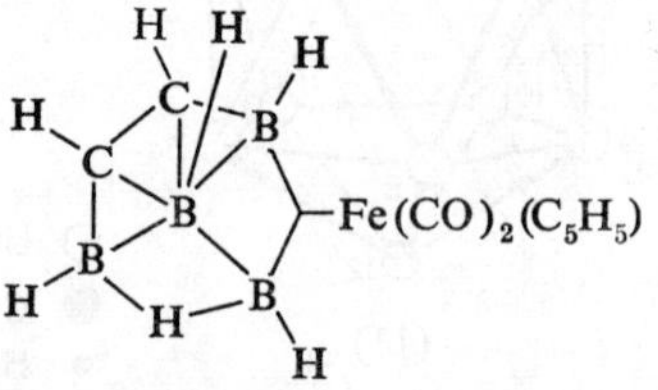

(14)

Me$_3$Ge, Me$_2$ClSi, or Me$_2$B groups in the 2,3-bridging position have been listed.[57] A rather similar iron-bridged structure (14) derived from *nido*-C$_2$B$_4$H$_8$ has[58] $\nu(C-H)_{\text{carbaboranyl}}$ at 3115, $\nu(B-H)$ at 2580, and $\nu(CO)$ at 2010, 1965 cm^{-1}.

Some metallocarbaboranes for which mainly unassigned i.r. data have been given[59-61] are listed in Table 2.

---

[55] R. N. Grimes, W. J. Rademaker, M. L. Denniston, R. F. Bryan, and P. T. Greene, *J. Amer. Chem. Soc.*, 1972, **94**, 1865.
[56] A. Tabereaux and R. N. Grimes, *J. Amer. Chem. Soc.*, 1972, **94**, 4768.
[57] C. G. Savory and M. G. H. Wallbridge, *J.C.S. Dalton*, 1972, 918.
[58] L. G. Sneddon and R. N. Grimes, *J. Amer. Chem. Soc.*, 1972, **94**, 7161.
[59] J. L. Spencer, M. Green, and F. G. A. Stone, *J.C.S. Chem. Comm.*, 1972, 1178.
[60] M. K. Kaloustian, R. J. Wiersema, and M. F. Hawthorne, *J. Amer. Chem. Soc.*, 1972, **94**, 6679.
[61] C. J. Jones, J. N. Francis, and M. F. Hawthorne, *J. Amer. Chem. Soc.*, 1972, **94**, 8391.

Table 2 *Some metallocarbaboranes characterized by i.r. spectroscopy*

Compound	Ref.
$(1,5\text{-}C_8H_{12})\text{Ni}[Me_2C_2B_9H_9]$	59
$(\pi\text{-}C_5H_5)\text{Co}[\pi\text{-}(3)\text{-}1,2\text{-}B_9C_2H_{11}]$	
$(\pi\text{-}C_5H_5)\text{Co}[\pi\text{-}(3)\text{-}1,2\text{-}Me_2\text{-}1,2\text{-}B_9C_2H_9]$ and related isomeric 60
$(\pi\text{-}C_5H_5)\text{Co}[\pi\text{-}(3)\text{-}1,2\text{-}(CH_2)_3\text{-}1,2\text{-}B_9C_2H_9]$ compounds
$(C_5H_5)\text{Co}[\pi\text{-}(2)\text{-}6,7\text{-}B_7C_2H_9]$	
$(C_5H_5)\text{Co}(\pi\text{-}B_7C_2H_9)$	
$(C_5H_5)\text{Co}[\pi\text{-}(1)\text{-}2,4\text{-}B_8C_2H_{10}]$	
$Me_4N[\{\pi\text{-}(3)\text{-}1,2\text{-}B_9C_2H_{11}\}\text{Co}\{\pi\text{-}(1)\text{-}2,4\text{-}B_8C_2H_{10}\}]$	61
$(C_5H_5)\text{Co}(\pi\text{-}B_7C_2H_{11})$	
$Me_4N[\{\pi\text{-}(3)\text{-}1,2\text{-}B_9C_2H_{11}\}\text{Co}\{\pi\text{-}(2)\text{-}1,6\text{-}B_7C_2H_9\}]$	
$Me_4N[\{\pi\text{-}(3)\text{-}1,2\text{-}B_9C_2H_{11}\}\text{Co}\{\pi\text{-}(2)\text{-}1,10\text{-}B_7C_2H_9\}]$	

Compounds containing Al—H or Ga—H Bonds.—Comparison with data for related boranes and alanes has allowed the following i.r. bands to be assigned [62] for $[Bu^n{}_4N]BH_4,AlH_3,NMe_3$: $\nu(B-H)$, 2390—2140 (5 bands); $\nu(Al-H)$, 1720; $\delta(BH)$, 1075; and $\delta(AlH)$, 780 cm^{-1}. The BH_4^- group in the complex is deduced to be bound to Al by a single bridging H. Deuteriation-sensitive bands at 1880—1620 and 800—600 cm^{-1} in the i.r. spectra of complexes $[AlH_3,(Et_2O)_x]_n$ ($x = 0.3$—3.32) have been attributed [63] to $\nu(Al-H)$ and $\delta(AlH)$ modes. I.r. spectroscopy suggests [64] that the final product of the reaction of $AlCl_3$ with $LiBH_4$ in benzene is the complex AlH_2BH_4,C_6H_6. Observation of an i.r. band at 1785 cm^{-1} identified as $\nu(Al-H)$ supports the conclusion that a complex between AlH_3 and (15) is formed by $LiAlH_4$ reduction of the corresponding homocubyl spirophosphonium salt.[65] In a study of the complexes $Al^mH_3,{}^nNMe_3$ ($m = 1$ or 2; $n = 14$ or 15) and also AlH_3,NR_3 and AlD_3,NR_3 (R = Et, Pr^n, or Bu^n),

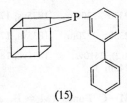

(15)

alane modes have been found [66] in the following ranges (wavenumber/cm^{-1}): $\nu_{sym}(Al-H)$, 1630—1625; $\nu_{asym}(Al-H)$, 1790—1775; $\nu_{sym}(Al-D)$, 1180—1182; $\nu_{asym}(Al-D)$, 1290—1310; $\delta_{sym}(AlH)$, 765—650; $\delta_{asym}(AlH)$, 762—784; $\delta_{sym}(Al-D)$, 555—462; and $\delta_{asym}(Al-D)$, 559—567. Davies and

[62] M. Ehemann, N. Davies, and H. Nöth, *Z. anorg. Chem.*, 1972, **389**, 235.
[63] K. N. Semenenko, Kh. A. Taisumov, A. P. Savchenkova, and V. N. Surov, *Russ. J. Inorg. Chem.*, 1971, **16**, 1104.
[64] V. I. Mikheeva, N. N. Mal'tseva, N. S. Kedrova, and E. T. Makhova, *Russ. J. Inorg. Chem.*, 1971, **16**, 798.
[65] E. W. Turnblom and D. Hellwinkel, *J.C.S. Chem. Comm.*, 1972, 404.
[66] K. N. Semenenko, B. M. Bulychev, and V. B. Polyakova, *Russ. J. Inorg. Chem.*, 1971, **16**, 949.

Wallbridge have listed mainly unassigned i.r. data for THF, diethyl ether, and bis(methylamine) complexes of aluminium hydroborate, $Al(BH_4)_3$,L. Reactions of LiX (X = H, D, BH_4, or Cl) with $Al(BH_4)_3$ to give products with Al—H bonds [ν near 1880 cm^{-1}: probably $HAl(BH_4)_2$,Et_2O and H_2AlBH_4,Et_2O] have also been investigated,[67] and two other papers [67a, 67b] are also concerned with aluminium borohydride. Strong i.r. bands are observed at ~1700 and 760 cm^{-1} in solutions of $LiAlMe_2H_2$, due to $\nu(Al—H)$ and $\delta(AlH)$, respectively.[68] The broadness of an i.r. band centred at 1860 cm^{-1} for the novel compound $Al_4Me_8(NMe_2)_2H_2$ is consistent with bridging Al—H bonds, in agreement with the established eight-membered (Al—H—Al—N—Al—H—Al—N) ring structure; the corresponding deformation mode is identified[69] at 876 cm^{-1}. In the related compound $Al_3Me_6(NMe_2)_2H$, $\nu(Al—H)$ is located [69a] at 1796 cm^{-1}. By contrast, for $2(Bu^i_3Al)—Bu^i_2AlH$, $\nu(Al—H)$ is found at lower frequency (1400 cm^{-1}), attributed to the presence [70] of bridging Al—H—Al units.

Synthesis [71] of a series of novel, volatile, cyclogallata-azonianes $[\overline{CH_2(CH_2)_xNGaH_2}]_n$ (x = 1, 2, 3, or 4; n = 2 or 3), by Storr, Thomas, and Penland provides the only i.r. data for gallanes reported during the year. Corresponding boron and aluminium derivatives were also prepared, and M—H stretching frequencies are compared in Table 3.

Table 3 Wavenumbers/cm^{-1} for $\nu(M—H)$ modes of $[\overline{CH_2(CH_2)_xNMH_2}]_n$

Compound	M = B	M = Al	M = Ga
$\overline{CH_2CH_2NMH_2}$	2430vs, 2390vs, 2320s, 2260vs, 2210m	1835s, 1775vs	1885vs, sh, 1855vs 1800vs, sh
$\overline{CH_2(CH_2)_2NMH_2}$	2395vs, 2330s, 2200m	1835vs, 1775vs	1850vs, br, 1800vs, s
$\overline{CH_2(CH_2)_3NMH_2}$	2400vs, 2330s, 2220m	1835vs, 1780vs	1865vs, br, 1800vs, s
$\overline{CH_2(CH_2)_4NMH_2}$	(2430s, sh?), 2410vs, 2350s, 2240m	1835vs, 1800s, sh	1870vs, br, 1800s

Compounds containing M—C Bonds (M = B, Al, Ga, In, or Tl).—I.r. frequencies for $CF_2=CHBX_2$, and cis- and trans-$CFCl=CHBX_2$ (X = F or Cl) have been listed,[72] and for propynylboron difluoride, $MeC\equiv CBF_2$,

---

[67] N. Davies and M. G. H. Wallbridge, *J.C.S. Dalton*, 1972, 1421.
[67a] K. N. Semenenko, O. V. Kravchenko, and E. B. Lobkovskii, *Zhur. strukt. Khim.*, 1972, **13**, 540.
[67b] K. N. Semenenko, O. V. Kravchenko, S. P. Shilkin, and V. B. Polyakova, *Doklady Chem.*, 1971, **201**, 1063.
[68] E. C. Ashby and J. Watkins, *J.C.S. Chem. Comm.*, 1972, 998.
[69] J. D. Glore, R. E. Hall, and E. P. Schram, *Inorg. Chem.*, 1972, **11**, 550.
[69a] J. D. Glore and E. P. Schram, *Inorg. Chem.*, 1972, **11**, 1532.
[70] J. J. Eisch and S. G. Rhee, *J. Organometallic Chem.*, 1972, **42**, C73.
[71] A. Storr, B. S. Thomas, and A. D. Penland, *J.C.S. Dalton*, 1972, 326.
[72] J. J. Ritter, T. D. Coyle, and J. M. Bellama, *J. Organometallic Chem.*, 1972, **42**, 25.

all twenty fundamentals have been assigned [73] with $\nu(B-C)$ at 782 cm^{-1}. In Cl$_2$B(CH$_2$)$_3$BCl$_2$, ν_{sym}(B—C) is found [74] at 920 cm^{-1}, with ν_{asym}(BCl$_2$) at 1095 cm^{-1}. A detailed vibrational study has been made of the phenylboron halides Ph$_2$BX and PhBX$_2$ (X = F, Cl, Br, or I) leading to assignments for B—Ph and B—X modes [75] collected in Table 4.

Table 4 Boron–phenyl and boron–halogen vibrations for Ph$_2$BX and PhBX$_2$ (wavenumbers/cm^{-1})

	ν(B—Ph)	δ(B—Ph)	ν(B—X)	δ(B—X)
Ph$_2$BF	1375, 1338	671, 582	1348	696
Ph$_2$BCl	1335, 1222	631, 580	910	565
Ph$_2$BBr	1261, 1167	647, 580	837	550
Ph$_2$BI	1260, 1166	633, 571	811	438
PhBF$_2$	1340	646	1378	570
PhBCl$_2$	1240	640	949, 910	551
PhBBr$_2$	1225	607	864, 808	525
PhBI$_2$	1208	585	717, 691	500

Assignments for the vibrational spectra of [(Me$_3$Al)$_2$N$_3$]$^-$ and [Me$_3$AlN$_3$]$^-$ have been proposed and compared with those for isoelectronic Me$_3$Si species. The symmetric and asymmetric Al—C stretching modes give rise to strong Raman bands near 520 and 610 cm^{-1}, respectively.[76] Unassigned i.r. data for Et$_2$IAl,NHMe$_2$ and [EtIAlNMe$_2$]$_2$ have been listed.[77] Haaland and Weidlein have suggested that the i.r. and Raman spectra of solid Et$_2$Al(C$_5$H$_5$) are consistent with the presence of a *pentahapto*-C$_5$H$_5$ ligand. The following assignments (wavenumber/cm^{-1}) are given:[78] ν_{asym}(AlC$_2$), 690; ν_{sym}(AlC$_2$), 580; δ_{sym}(AlC$_2$), 345; and ν(Al—Cp), 210.

Renewed interest in organogallium chemistry has resulted in the measurement of ν(Ga—C) for a number of compounds, including the frequencies listed in Table 5 for alkylgallium halides,[79–81] and in Table 6 for the bridged species (R$_2$M)$_2$O$_4$C$_2$ (M = Ga or In) (obtained [82] by reaction of the trialkylgallium with oxalic acid) and [Me$_2$GaO$_2$PX$_2$]$_2$ (X = F, Cl, or Me).[83]

Bands at 435—446 cm^{-1} in the i.r. spectra of M$^+$(InMe$_4$)$^-$ (M = Li—Cs) are assigned [84] to ν(In—C). The vibrational spectra of alkylindium halides have been examined by two groups of investigators: Poland and Tuck

[73] P. R. Reed and R. W. Lovejoy, *J. Chem. Phys.*, 1972, **56**, 183.
[74] M. Zeldin and A. Rosen, *J. Organometallic Chem.*, 1972, **34**, 259.
[75] F. C. Nahm, E. F. Rothergy, and K. Niedenzu, *J. Organometallic Chem.*, 1972, **35**, 9.
[76] F. Weller and K. Dehnicke, *J. Organometallic Chem.*, 1972, **35**, 237.
[77] K. Gosling and A. L. Bhuiyan, *Inorg. Nuclear Chem. Letters*, 1972, **8**, 329.
[78] A. Haaland and J. Weidlein, *J. Organometallic Chem.*, 1972, **40**, 29.
[79] W. Lind and I. J. Worrall, *J. Organometallic Chem.*, 1972, **40**, 35.
[80] W. Lind and I. J. Worrall, *J. Organometallic Chem.*, 1972, **36**, 35.
[81] M. J. S. Gynane and I. J. Worrall, *J. Organometallic Chem.*, 1972, **40**, C59.
[82] H. U. Schwering, H. D. Hausen, and J. Weidlein, *Z. anorg. Chem.*, 1972, **391**, 97.
[83] B. Schaible and J. Weidlein, *J. Organometallic Chem.*, 1972, **35**, C7.
[84] K. Hoffmann and E. Weiss, *J. Organometallic Chem.*, 1972, **37**, 1.

Table 5 Assignment of Ga—C stretching modes (wavenumber/cm^{-1}) in alkylgallium halides

Compound	ν(Ga—C)	Ref.
MeGa$_2$Cl$_5$	609	
MeGa$_2$Br$_5$	596	
MeGa$_2$I$_5$	577	79
MeGa$_2$Cl$_4$I	609	
MeGa$_2$Br$_4$I	599	
Me$_3$Ga$_2$Br$_3$	621, 587, 545	
Et$_3$Ga$_2$Br$_3$	580, 551, 509	81
Me$_3$Ga$_2$I$_3$	612, 575, 538	
Et$_3$Ga$_2$I$_3$	577, 548, 508	

Table 6 I.r. data (wavenumber/cm^{-1}) for oxo-bridged dialkylgallium derivatives

Compound	ν_{asym}(Ga—C)	ν_{sym}(Ga—C)
(Me$_2$Ga)$_2$(C$_2$O$_4$)	623	559
(Et$_2$Ga)$_2$(C$_2$O$_4$)	615, 583	528
[Me$_2$GaO$_2$PF$_2$]$_2$	593	546
[Me$_2$GaO$_2$PCl$_2$]$_2$	616	553
[Me$_2$GaO$_2$PMe$_2$]$_2$	614	553

report ν(In—C) frequencies as follows [85] (wavenumber/cm^{-1}): [Me$_2$In][InI$_4$], 561 (ν_{asym}: i.r.), 483 (ν_{sym}: R); (EtInI$_2$)$_2$, 489; (Bu^{n}InI$_2$)$_2$, 487; whereas Worrall et al. find [86, 87] two bands for (BuInI$_2$)$_2$ and also for (PrInI$_2$)$_2$, near 485 and 577 cm^{-1}. These workers agree on the ionic formulation [Me$_2$In]$^+$[InI$_4$]$^-$ for the methyl compound, their observations parallelling published data for Me$_2$In$^+$, but for the higher alkyl derivatives the more usual dimeric structure is proposed. In Et$_2$In(OOCMe), ν_{sym} and ν_{asym}(InC$_2$) are found [88] at 468 and 517 cm^{-1}, respectively, and for Me$_2$In(SSCNMe$_2$), MeIn(SSCNMe$_2$)$_2$, and EtIn(SSCNMe$_2$)$_2$, where ν(In—C) modes occur at 518 and 482, 506 and 486 cm^{-1}, it is suggested that these low frequencies are consistent with strongly bonded dithiocarbamate ligands.[89] For R$_2$In(O$_3$SMe), Raman bands at 552, 510 and 520, 474 cm^{-1} are assigned to ν_{sym} and ν_{asym}(In—C) for R = Me or Et, respectively,[90] and similar bands have been distinguished [91] for adducts of MeIn(tdt) (tdt = toluene-3,4-dithiolate). The vibrational spectra of cyclopentadienyl derivatives of indium are consistent with σ-bonded Cp rings;[92] some assignments are shown in Table 7.

[85] J. S. Poland and D. G. Tuck, J. Organometallic Chem., 1972, **42**, 315.
[86] M. J. S. Gynane, L. G. Waterworth, and I. J. Worrall, J. Organometallic Chem., 1972, **43**, 257.
[87] M. J. S. Gynane and I. J. Worrall, Inorg. Nuclear Chem. Letters, 1972, **8**, 547.
[88] H. D. Hausen, J. Organometallic Chem., 1972, **39**, C37.
[89] T. Maeda and R. Okawara, J. Organometallic Chem., 1972, **39**, 87.
[90] H. Olapinski and J. Weidlein, J. Organometallic Chem., 1972, **35**, C53.
[91] A. F. Berniaz and D. G. Tuck, J. Organometallic Chem., 1972, **46**, 243.
[92] J. S. Poland and D. G. Tuck, J. Organometallic Chem., 1972, **42**, 307.

Table 7 Wavenumber/cm^{-1} of ν(In—C) modes in Cp$_3$In and related species

Compound	ν(In—C)
InCp$_3$	339, 316
In(CpMe)$_3$	349, 322
InCp$_3$,PPh$_3$	321, 303
InCp$_3$,bipy	325, 300
InCp$_3$,phen	293
Li[In(indenyl)$_4$]	371, 352, 340
Li[In(indenyl)$_4$],Et$_2$O	371, 351, 340

For dimethylthallium diphosphinite,[93] ν_{sym} and ν_{asym}(TlC$_2$) are at 491 and 532 cm^{-1}, whereas for the corresponding diphosphinate these fundamentals occur at 492 and 541 cm^{-1}. For R(ClCH$_2$)TlX (R = Ph, X = Cl, OCOMe, or OCOPr^i; R = p-MeC$_6$H$_4$, X = Cl, OCOMe, or OCOPr^i; R = Me, X = OCOMe or OCOPr^i; R = OCOMe, X = OCOMe; R = X = OCOPr^i), bands assigned [94] to ν(Tl—C) range from 493 to 537 cm^{-1}, and similarly ν_{asym}(TlC$_2$) are at 531—552 cm^{-1} for [16; R = Me,

$$R_2Tl-N\begin{matrix}C(=O)\\ \\ C(=O)\end{matrix}X$$

(16)

X = (CH$_2$)$_2$ or o-C$_6$H$_4$].[95] Aromatic ring vibrations for a variety of R$_2$TlY compounds (R = Ph, o-MeC$_6$H$_4$, m-MeC$_6$H$_4$ or p-MeC$_6$H$_4$; Y = electronegative group) are sensitive to the nature of Y.[96]

Compounds containing M—N Bonds (M = B, Al, or Ga) or B—P Bonds.—In the i.r. spectra of the t-butylmethyleneaminoboranes, Bu$^t_2$C=NBX$_2$ (X = Cl, Ph, or Bu^n), the high energy of absorptions due to azomethine stretching modes (respectively at 1839, 1820, and 1821 cm^{-1}) is consistent with their assignment to ν_{asym} of a linear C=N⇌B skeleton.[97] An examination of the effect on the frequencies of characteristic ring vibrations of varying X in the B-substituted Δ^2-tetrazaborolines (17; X = Cl, Br, CN, SCN, or SeCN) has revealed that the five stretching modes are remarkably consistent throughout (at 1410, 1362, 1337, 980, and 906 cm^{-1} for X = Cl).[98]

Partial assignments of observed i.r. and Raman spectra have been given [99] for (Me$_3$Si)$_2$NBF$_2$, (Me$_3$Si)$_2$NB(F)NHSiMe$_3$, and the related borazines F$_3$B$_3$N$_3$(SiMe$_3$)$_3$ and F$_3$B$_3$N$_3$H(SiMe$_3$)$_2$, with ν(B—N) modes in

---

[93] B. Walther, *J. Organometallic Chem.*, 1972, **38**, 237.
[94] T. Abe and R. Okawara, *J. Organometallic Chem.*, 1972, **43**, 117.
[95] B. Walther and C. Rockstroh, *J. Organometallic Chem.*, 1972, **42**, 41.
[96] T. N. Srivastava, S. K. Tandon, and K. K. Bajpai, *Spectrochim. Acta*, 1972, **28A**, 455.
[97] M. R. Collier, M. F. Lappert, R. Snaith, and K. Wade, *J.C.S. Dalton*, 1972, 370.
[98] B. Hessett, J. B. Leach, J. H. Morris, and P. G. Perkins, *J.C.S. Dalton*, 1972, 131.
[99] G. Elter, O. Glemser, and W. Herzog, *Chem. Ber.*, 1972, **105**, 115.

MeN−B(X)−NMe with N=N (structure (17))

the expected range (1250—1480 cm^{-1}). The borazines $H_2ClB_3N_3H_2Me$ and $H_2(Me_2N)B_3N_3H_2Me$ have been partially separated into *ortho*- and *para*-isomeric forms, for which the i.r. spectra show some differences in the 950—650 region (BN bending).[100] Other B—N bonded compounds (a number of them borazine derivatives) for which i.r. frequencies have been given are listed in Table 8; in some cases assignments including $\nu(B-N)$ modes have been offered.

Table 8 *References to compounds with B—N bonds characterized by i.r. spectroscopy*

Compound	Ref.
$[Me_3NBH_2OC(R^1)NHR^2]^+$ (R^1 = Me or Ph; R^2 = H or Ph)	
$[Me_3NBH_2O-R]^+$ [R = $CHNMe_2$, $C(Me)NMe_2$, $C(Me)N(H)C_6H_4OEt$, C_4HNPh, C_5H_3N, C_5H_4NMe, or $C(Me)NC_6H_4Cl$]	101
$[Me_3NBCl_2OC(X)NMe_2]^+$ (X = H or I)	
$[Me_3NBH_2CNX]^+$ (X = Me, BH_2NMe_3, or BH_2PMe_3)	
$B(OC_2H_4)_3N$	102
$B(OC_2H_4)_3N,SbCl_5$	
$R_3B_3N_3Me_3$ (R = Ph, Me, or Cl: all ten possible compounds)	
$RMe_2B_3N_3Me_3$ (R = Et or Me_5C_6)	103
$[Me_3N_3B_3Me_2]_2$	
$[Me_3N_3B_3Me_2]_2O$	
$(C_5H_5)_3B_3N_3Me_3$	104
$(MeC_5H_4)_3B_3N_3Me_3$	
(naphthyl-substituted B—N ring compound with Cl, H shown)	105

The vibrational spectra of $[(Me_3Al)_2N_3]^-$ and $[Me_3AlN_3]^-$ contain bands due to Al—N stretching at 424 (ν_{asym}) and 302 (ν_{sym}), and 420 cm^{-1} respectively, and are consistent with C_s symmetry, implying structure (18) for the bistrimethylaluminium compound with no evidence for Al ← N π-bonding.[76] A band at rather higher frequency (500 cm^{-1}) is attributed [106]

[100] O. T. Beachley, *J. Amer. Chem. Soc.*, 1972, **94**, 4223.
[101] D. L. Reznicek and N. E. Miller, *Inorg. Chem.*, 1972, **11**, 858.
[102] D. Fenske and H. J. Becher, *Chem. Ber.*, 1972, **105**, 2085.
[103] L. A. Melcher, J. L. Adcock, and J. J. Lugowski, *Inorg. Chem.*, 1972, **11**, 1247.
[104] B. L. Therrell and E. K. Mellon, *Inorg. Chem.*, 1972, **11**, 1137.
[105] J. Cueilleron and B. Frange, *Bull. Soc. chim. France*, 1972, 107.
[106] C. H. Chan and F. P. Olsen, *Inorg. Chem.*, 1972, **11**, 2836.

$$\text{Me}_3\text{Al} \diagdown \text{N-N-N} \diagup \text{Me}_3\text{Al}$$

(18)

to $\nu(\text{Al}-\text{N})$ for $\text{S}_4\text{N}_4,\text{AlCl}_3$, and similar modes are rather doubtfully assigned[13] near 775 cm^{-1} for $[\text{AlCl}_2(\text{N}_3)]_n$ and $[\text{AlCl}(\text{N}_3)_2]_n$. Trimethylaluminium-trimethylamine has been found to decompose at 50 °C to give the *cis-* and *trans-*isomers of the trimeric species $(\text{Me}_2\text{AlNHMe})_3$ (19);

(*cis-*)   (*trans-*)

(19)

strong i.r./Raman bands at 505 (*trans-*) and 579, 460 cm^{-1} (*cis-*) are thought[107] to arise mainly from $(\text{Al}-\text{N})_3$ ring vibrations. Corresponding ethyl aluminium compounds were obtained similarly. Trimeric cyclopropyl derivatives $[(\text{C}_3\text{H}_5)_2\text{M}(\text{NC}_2\text{H}_4)]_3$ (M = Al or Ga) have also been investigated by i.r. and Raman spectroscopy, the results indicating non-planar ring structures with C_2 symmetry.[108] Schram *et al.* have given partial i.r. assignments[69, 69a] for the cyclic derivatives $\text{Al}_3\text{Me}_6(\text{NMe}_2)_2\text{H}$ and $\text{Al}_4\text{Me}_8(\text{NMe}_2)_2\text{H}_2$, placing $\nu_{\text{asym}}(\text{Al}-\text{N})$ at 553, $\nu_{\text{sym}}(\text{Al}-\text{N})$ at 539 cm^{-1} for the second of these. I.r. spectroscopy has been used to confirm that in $[\text{R}_4\text{N}]\text{Ga}(\text{NCS})_4$ (R = Et or Bu^n) and $[(\text{R}_4\text{N})_3\text{Ga}(\text{NCS})_6]$ (R = Me or Bu^n) the NCS$^-$ groups are N-bonded, and bands near 350 cm^{-1} for the tetraisothiocyanato-complexes are assigned[109] to $\nu(\text{Ga}-\text{N})$.

Reaction of phosphine or phosphonium iodide with BI$_3$ affords H$_3$P,BI$_3$, for which[110] $\nu(^n\text{B}-\text{P})$ is assigned to i.r. absorptions at 487 ($n = 11$) and 500 cm^{-1} ($n = 10$). Fleming and Parry have found that whereas F$_2$PNMe$_2$ reacts to form an N-bonded adduct with BF$_3$, [$\nu_{\text{sym}}(\text{N}-^{10}\text{BF}_3)$ at 654 cm^{-1}] the same base reacts with tetraborane(10) to give the complex H$_7$B$_3$,-F$_2$PNMe$_2$, the i.r. spectrum of which includes features at 592 and 542 cm^{-1} attributable to $\nu(\text{B}-\text{P})$, implying co-ordination through phosphorus.[111] The i.r. spectra of the related species F$_3$PB$_3$H$_7$ and F$_2$ClPB$_3$H$_7$ have also been assigned quite fully,[35] with $\nu(\text{B}-\text{P})$ at 590 cm^{-1}. A review of spectroscopic aspects of metal–phosphorus bonding includes tabulation and

---

[107] K. J. Alford, K. Gosling, and J. D. Smith, *J.C.S. Dalton*, 1972, 2203.
[108] J. Müller, K. Margiolis, and K. Dehnicke, *J. Organometallic Chem.*, 1972, **46**, 219.
[109] L. M. Mikheeva, L. N. Auerman, A. I. Tarasova, and L. N. Komissarova, *Russ. J. Inorg. Chem.*, 1971, **16**, 1126.
[110] M. Schmidt and H. H. J. Schröder, *Z. anorg. Chem.*, 1972, **394**, 290.
[111] S. Fleming and R. W. Parry, *Inorg. Chem.*, 1972, **11**, 1.

discussion of B—P stretching vibrations in complexes between boranes and phosphorus donors.[112]

Compounds containing M—O Bonds (M = B, Al, Ga, In, or Tl).—The alkoxyboranes $(Bu^tO)_3B$, $(Bu^tO)_2BMe$, and $(Bu^tO)BMe_2$ show i.r. absorptions assigned [113] as follows (wavenumber/cm^{-1}): respectively at 1347, 1329, and 1346, $\nu(B-O)$; at 1188, 1183, and 1194, $\nu(C-O)$; and for the methyl compounds, at 1238, and 1306/1118, $\nu(B-C)$. Bands attributed to $\nu(B-O-C)$ have been observed at 1350—1310 cm^{-1} in the i.r. spectra of some alkylborate derivatives of monosaccharides.[114] A number of papers concerning the vibrations of BO_3 units have appeared.[115-119] In particular,[118] the behaviour towards water of B_2O_3 surfaces has been investigated by i.r. techniques, in conjunction with a study of $B(OH)_3$, HBO_2, and B_2O_3. Structures involving the formation of B_3O_9 rings have also been proposed on the basis of vibrational data for a variety of rare-earth borates.[119]

^{18}O-Labelling has been used in an examination of $B(OH)_4^-$. The ν_3 i.r.-active mode was found at 982 cm^{-1} (^{18}O compound), compared with 958 cm^{-1} for the ^{16}O compound, whereas ν_1 was observed in the Raman effect at 744 cm^{-1} for both species. The lack of change in ν_1 leads to the unusual suggestion [119a] that hydrogen-bonding between $B(OH)_4^-$ ions and solvating water molecules is appreciably weaker in the case of ^{18}O-labelled ions than for $B(^{16}OH)_4^-$.

I.r. spectra of films of binary borosilicate glasses ranging in composition from SiO_2 to B_2O_3 have been discussed in terms of features arising from the presence of Si—O—Si, B—O—Si, or B—O—B bonds, a band at 670 cm^{-1} being assigned to $\nu(B-O-Si)$.[120] Halogen, methyl, amino, and phosphine derivatives of $\overline{O(CH_2)_2OB}-$ (1,3,2-dioxaborolan) and $\overline{S(CH_2)_2SB}-$ have been examined by i.r. spectroscopy.[121] References 122—126 contain i.r. data for several other species with B—O bonds.

[112] J. G. Verkade, *Co-ordination Chem. Rev.*, 1972, **9**, 1.
[113] I. Kronawitter and H. Nöth, *Chem. Ber.*, 1972, **105**, 2423.
[114] V. V. Gertsev and L. A. Frolova, *Doklady Chem.*, 1971, **200**, 834.
[115] I. W. Shepherd, *Phys. Rev. (B)*, 1972, **5**, 4524.
[116] P. F. Rza-Zade, G. K. Abdullaev, F. R. Samedov, and Kh. K. Zeinalova, *Russ. J. Inorg. Chem.*, 1971, **16**, 1221.
[117] G. Heller and D. A. Marquard, *Inorg. Nuclear Chem. Letters*, 1972, **8**, 663.
[118] P. Broadhead and G. A. Newman, *Spectrochim. Acta*, 1972, **28A**, 1915.
[119] J. H. Denning and S. D. Ross, *Spectrochim. Acta*, 1972, **28A**, 1775.
[119a] S. Pinchas and J. Shamir, *J. Chem. Phys.*, 1972, **56**, 2017.
[120] A. S. Tenney and J. Wong, *J. Chem. Phys.*, 1972, **56**, 5516.
[121] S. G. Shore, J. L. Crist, B. Lockman, J. R. Long, and A. D. Coon, *J.C.S. Dalton*, 1972, 1123.
[122] H. Häni and J. D. Russell, *Nature Phys. Sci.*, 1972, **235**, 13.
[123] J. Frohnecke and G. Heller, *J. Inorg. Nuclear Chem.*, 1972, **34**, 69.
[123a] H. Gode, I. Zuika, and G. Adijano, *Latv. PSR Zinat. Akad. Vestis, Kim. Ser.*, 1971, 538.
[124] R. Larsson and G. Nunziata, *Acta Chem. Scand.*, 1972, **26**, 1503.
[125] G. J. Barrett and D. T. Haworth, *Inorg. Chim. Acta*, 1972, **6**, 504.
[126] A. N. Maitra and D. Sen, *J. Inorg. Nuclear Chem.*, 1972, **34**, 3643.

Reaction of $AlBr_3$ with a solution of $CsNO_3$ in HNO_3–N_2O_5 affords $Cs[Al(NO_3)_4]$ and $Cs_2[Al(NO_3)_5]$, characterized by X-ray diffraction patterns and i.r. spectroscopy. The spectra are indicative of covalently bonded NO_3 groups and differ in the $\nu(Al-O)$ region (440—500 cm^{-1}) but no firm structural conclusions have been made.[127, 128] Partial i.r. assignments for the isopropoxide species $Al(Pr^iO)_3$, $Ga(Pr^iO)_3$, $In[Al(Pr^iO)_4]_3$, and $In[Ga(Pr^iO)_4]_3$ include [129] $\nu(Al-O)$ at 695 and $\nu(Ga-O)$ near 670 cm^{-1}. The vibrational spectra of the cyclic species $[Me_2MOOEMe_2]_2$ (M = Al, Ga, or In; E = P or As) have been assigned, the dimethylphosphinates containing puckered eight-membered ($M_2O_4P_2$) rings with C_{2h} symmetry.[130] In a detailed study of $[Al(DMSO)_6]^{3+}$ and $[Al([^2H_6]DMSO)_6]^{3+}$ as $AlCl_4^-$, $AlBr_4^-$, Cl^-, and Br^- salts, $\nu(Al-O)$ modes are located at 470, 496 (DMSO complex) and 440, 472 cm^{-1} ($[^2H_6]DMSO$ complex).[131] Reaction of tris(dimethylamino)alane with $Fe(CO)_5$ yields a carbene complex formulated as $[(CO)_4FeC(NMe_2)OAl(NMe_2)_2]_2$; an i.r. band at 1512 cm^{-1} is attributed to $\nu(C-O-Al)$, a relatively high frequency being rationalized [132] in terms of substantial C=O character arising from four-co-ordination at Al. Sato has used i.r. spectroscopy in an investigation of the formation of aluminium hydroxy-species,[133, 134] and the frequencies for ν_1 and ν_4 of $Si(Al)O_4$ units in Linde-A type synthetic zeolites have been correlated with constituent ion radii.[135]

The vibrational spectrum of $Ga(SO_3F)_3$ includes a band at 445 cm^{-1} tentatively assigned [136] to $\nu(Ga-O)$. Similarly, for M_2GaF_5,H_2O (M = NH_4, K, Rb, or Cs), $MGaF_4,2H_2O$ (M = NH_4, Rb, or Cs), and $GaF_3,3H_2O$ and some deuterio-derivatives, $\nu(Ga-O)$ modes give rise to i.r. absorptions in the range [137] 431—483 cm^{-1}. The far-i.r. spectrum of $[Ga_2(OH)_2X_2$-$(Me_2dpma)_2]^{2+}(X^-)_2,H_2O$ [X = Cl or Br; Me_2dpma = methyl-(6-methyl-2-pyridylmethyl)-(2-pyridylmethyl)amine] includes a strong band at 530 (X = Cl) or 532 cm^{-1} (X = Br) not present for $MX_3(Me_2dpma)$ (M = In or Tl; X = Cl or Br) and which is therefore attributed to a $\nu_{asym}(Ga-O)$ vibration.[138] The compounds $(C_6F_5)_3In,(OSMe_2)_2$ and $(C_6F_5)_3In,OPPh_3$ show [139] $\nu(In-O)$ near 414 cm^{-1}. Some vibrational data for nitroparaffin derivatives of Tl^I have been listed.[140]

[127] G. N. Shirokova and V. Ya. Rosolovskii, *Russ. J. Inorg. Chem.*, 1971, **16**, 1106.
[128] G. N. Shirokova and V. Ya. Rosolovskii, *Russ. J. Inorg. Chem.*, 1971, **16**, 808.
[129] A. Mehrotra and R. C. Mehrotra, *Inorg. Chem.*, 1972, **11**, 2170.
[130] H. Olapinski, B. Schaible, and J. Weidlein, *J. Organometallic Chem.*, 1972, **43**, 107.
[131] J. Meurier and M. T. Forel, *Canad. J. Chem.*, 1972, **50**, 1157.
[132] W. Petz and G. Schmid, *Angew. Chem. Internat. Edn.*, 1972, **11**, 934.
[133] T. Sato, *Z. anorg. Chem.*, 1972, **391**, 69.
[134] T. Sato, *Z. anorg. Chem.*, 1972, **391**, 167.
[135] J. J. P. M. de Kanter, I. E. Maxwell, and P. J. Trotter, *J.C.S. Chem. Comm.*, 1972, 733.
[136] A. Storr, P. A. Yeats, and F. Aubke, *Canad. J. Chem.*, 1972, **50**, 452.
[137] K. I. Petrov, I. V. Tananaev, and T. B. Vorotilina, *Russ. J. Inorg. Chem.*, 1971, **16**, 811.
[138] K. Dymock, G. J. Palenik, and A. J. Carty, *J.C.S. Chem. Comm.*, 1972, 1218.
[139] G. B. Deacon and J. C. Parrott, *Austral. J. Chem.*, 1972, **25**, 1169.
[140] A. G. Lee, *Spectrochim. Acta*, 1972, **28A**, 133.

Compounds containing M—S Bonds (M = Al or In) or M—Se Bonds (M = B or Al).—The species [(Me$_3$Al)$_2$SCN]$^-$, [Me$_3$AlSCN]$^-$, [(Me$_3$Al)$_2$SeCN]$^-$, and [(Me$_3$Al)$_2$CN]$^-$ have been examined by i.r. and Raman spectroscopy. The results are consistent with Al—S, and Al—Se bonding (Table 9) in the first three complexes, but with the occurrence of Al—CN-

Table 9 Al—S *and* Al—Se *stretching fundamentals in trimethylaluminium-pseudohalogen complexes*

Compound	ν(Al—E)^a	
[(Me$_3$Al)$_2$SCN]$^-$	178 (ν_{sym})	315 (ν_{asym})
[(Me$_3$Al)SCN]$^-$		336
[(Me$_3$Al)$_2$SeCN]$^-$	154 (ν_{sym})	274 (ν_{asym})

^{a} E = S or Se.

—Al bridging in the fourth.[141] Vibrational bands in the 230—290 cm^{-1} region [142] for R$_2$InX,SSbMe$_3$ (R = Me or Et; X = Cl or Br) and MeInCl$_2$,SSbMe$_3$, and at 379 cm^{-1} for Me$_2$In(SSCNMe$_2$), RIn(SSCNMe$_2$)$_2$ (R = Me or Et) [89] have been assigned to ν(In—S) modes, the higher wavenumber for the dithiocarbamate complexes being consistent with strong bonding to In. Related data for toluene-3,4-dithiolate complexes derived from trimethylindium have also been reported.[91] Unassigned i.r. spectra of amorphous and crystalline B$_2$Se$_3$ have been recorded.[143]

Compounds containing M—Halogen Bonds (M = B, Al, Ga, In, or Tl).— Novel polyboron and silicon–boron derivatives have been synthesized using the reactive species BF, SiCl$_2$, and SiF$_2$. Products obtained in this way for which unassigned i.r. and Raman frequencies have been reported [144, 145] are BF$_2$(SiF$_3$), B$_8$F$_{12}$, and (SiCl$_3$)$_2$,BCl$_2$,BCO. The reaction systems [metal chloride–PCl$_3$–Bu^{t}Cl] (metal = B, Al, or Sn) and [metal chloride–MePCl$_2$–Bu^{t}Cl] (metal = B or Sn) yield solid products characterized as ionic complexes of alkyl chlorophosphonium cations with the species BCl$_4^-$, AlCl$_4^-$, and SnCl$_5^-$. Observed i.r. bands and assignments for the anions [145a] are listed in Table 10. Further similar data have been

Table 10 *I.r. wavenumbers/cm^{-1} for* BCl$_4^-$ *and* AlCl$_4^-$ *species*

Assignment	[Bu^{t}PCl$_3$][AlCl$_4$]	[Bu^{t}PCl$_3$][BCl$_4$]
ν_3	491vs	723s (^{10}B), 696vs (^{11}B)
ν_1	349m	396mw
ν_4	183/175s	275w
ν_2		196vw, sh (?)

[141] F. Weller and K. Dehnicke, *J. Organometallic Chem.*, 1972, **36**, 23.
[142] T. Maeda, G. Yoshida, and R. Okawara, *J. Organometallic Chem.*, 1972, **44**, 237.
[143] R. Hillel and J. Cueilleran, *Bull. Soc. chim., France*, 1972, 98.
[144] D. L. Smith, R. Kirk, and P. L. Timms, *J.C.S. Chem. Comm.*, 1972, 295.
[145] R. W. Kirk, D. L. Smith, W. Airey, and P. L. Timms, *J.C.S. Dalton*, 1972, 1392.
[145a] J. I. Bullock, N. J. Taylor, and F. W. Parrett, *J.C.S. Dalton*, 1972, 1843.

reported for $PCl_4^+BCl_4^-$, where differences in the Raman (but not i.r.) spectra from those given previously have been found,[146] due to overlap of ν_1(cation) with ν_4, ν_5, ν_6(anion), and for several BF_4 and also EF_6 (E = P or Sb) species.[147] Solutions of $Na[BF_3OH]$ in $NaBF_4$ (crystalline or molten) show i.r. absorptions at 3641 (2688 on deuteriation) and 767 cm^{-1}, assigned[148] to ν(OH) and ν_{sym}(B—F) of BF_3OH^-. Reaction of molten $NaCl$–$AlCl_3$ with As_2O_3 at 170 °C affords a homogeneous metastable oxide solution, up to 6% oxygen being present, but the only species detectable[149] by Raman spectroscopy is $AlCl_4^-$.

It has been suggested[150] that melts consisting of gallium trichloride–caesium chloride in various mol. ratios contain $GaCl_4^-$, $Ga_2Cl_7^-$, and Ga_2Cl_6. Frequencies of Raman shifts for $GaCl_4^-$ were close to those reported previously for aqueous solutions (ν_1, 343; ν_2, 120; ν_3, 370; and ν_4, 153 cm^{-1}); for $Ga_2Cl_7^-$, bands were detected as follows (wavenumber/cm^{-1}): 393m, sh(dp), 366vs(p), 316w, 140s(dp), and 90s(dp), these resembling known data for $Al_2Cl_7^-$. The Raman spectrum of solid $InBr_2$ is consistent with the formulation $In^I(In^{III}Br_4)$, the e stretching fundamental (ν_3) for the anion (at 239 cm^{-1} in solution) splitting into three components at 228, 236, and 241 cm^{-1}. By contrast, solid In_2Br_3 appears to exist as $(In^+)_2(In_2Br_6^{2-})$, a band at 139 cm^{-1} being attributed to stretching of a metal–metal bond in the anion.[151] A full assignment for the vibrational spectrum of $Cs_3In_2Cl_9$ has been proposed[152] by analogy with previously reported single-crystal data for $Cs_2Tl_2Cl_9$, and the i.r. spectrum for Cs_3InBr_6 has been illustrated[153] in the range 400—30 cm^{-1}. Alkylindium halides have also received some attention as noted earlier;[85, 86] some conflicting conclusions have been reached regarding ν(In—I) vibrations, as is shown in Table 11.

Table 11 In—I *stretching modes* (cm^{-1}) *for alkylindium iodides*

Compound	ν(In—I)	Ref.
$(Me_2In)(InI_4)$	138(ν_1); 191(ν_3); 62(ν_4)	85
$(EtInI_2)_2$	170($\nu_{terminal}$); 140(ν_{bridge})	
$(EtInI_2)_2$	250($\nu_{terminal}$); 171(ν_{bridge})	86
$(PrInI_2)_2$	196($\nu_{terminal}$); 167(ν_{bridge})	
$(Bu^nInI_2)_2$	169($\nu_{terminal}$); 141(ν_{bridge})	85
$(BuInI_2)_2$	184($\nu_{terminal}$); 168(ν_{bridge})	86

[146] K. Niedenza, I. A. Boenig, and E. B. Bradley, *Z. anorg. Chem.*, 1972, **393**, 88.
[147] V. K. Akimov, A. I. Busev, and D. I. Andzhaparidze, *Russ. J. Inorg. Chem.*, 1971, **16**, 1427.
[148] J. B. Bates, J. P. Young, M. M. Murray, H. W. Kohn, and G. E. Boyd, *J. Inorg. Nuclear Chem.*, 1972, **34**, 2721.
[149] H. Kühnl and U. Geffarth, *Z. anorg. Chem.*, 1972, **391**, 280.
[150] H. A. Øye and W. Bues, *Inorg. Nuclear Chem. Letters*, 1972, **8**, 31.
[151] L. Waterworth and I. J. Worrall, *Inorg. Nuclear Chem. Letters*, 1972, **8**, 123.
[152] F. J. Brinkman, *J. Inorg. Nuclear Chem.*, 1972, **34**, 394.
[153] A. G. Dudareva, Yu. E. Bogatov, B. N. Ivanov-Emin, and P. I. Fedorov, *Russ. J. Inorg. Chem.*, 1971, **16**, 1378.

The distribution of $\nu(M-X)$ bands in the vibrational spectra of MX_3-(terpy) complexes (M = Al, X = Cl or Br; M = Ga or In, X = Cl, Br, or I; M = Tl, X = Cl) has been shown to be of limited value as a criterion for distinguishing between alternative structures for these systems: similar patterns were found for *trans*-octahedral and five-co-ordinate, MX_2-(terpy)$^+$X$^-$, arrangements.[154] A shift to lower wavenumber of *ca.* 70—80 cm^{-1} for ν(CO) in complexes of aromatic aldehydes with BF$_3$ confirms that co-ordination occurs *via* the carbonyl oxygen atom.[155] Other complexes of Group III element trihalides for which vibrational data have been given [156-159] are listed in Table 12.

Table 12 References containing vibrational data for complexes of Group III halides

Compound	Ref.
tmu,BF$_3$ (tmu = tetramethylurea)	156
R^{1}R^{2}O,AlX$_3$ (R^1, R^2 = CH$_3$ or CD$_3$; X = Cl or Br)	157
3(bipy),GaX$_3$ 3(phen),GaX$_3$ (X = Br or ClO$_4$) 2(bipy),GaCl$_3$ bipy,GaBr$_3$ phen,GaX$_3$ (X = Cl or Br)	158
{GaX$_2$[Ni(salen)$_2$]}$^+$GaX$_4^-$ (X = Cl or Br) and related Ga^{III} and In^{III} halide complexes	159

Broadening and splitting of ν(Tl—Cl) fundamentals in the i.r. spectrum of *trans*-(CoCl$_2$en$_2$)(TlCl$_4$) has been discussed in terms of Tl—Cl—H hydrogen bonding.[160] A review of the co-ordination chemistry of thallium(I) includes some reference to vibrational spectroscopy.[161]

4 Group IV Elements

Preparation of thirty-three crystalline adducts of tetra-alkylammonium halides with carbon tetrahalides and with C$_2$Br$_4$ and C$_2$I$_4$ has been reported by Creighton and Thomas. I.r. and Raman observations are discussed in terms of site-symmetry for the halogenocarbon molecule; consideration of results, including bands assigned to lattice vibrations in conjunction with X-ray data, leads to the conclusion that these species should be regarded as

---

[154] G. Beran, K. Dymock, H. A. Patel, A. J. Carty, and P. M. Boorman, *Inorg. Chem.*, 1972, **11**, 896.
[155] M. Rabinovitz and A. Grinvald, *J. Amer. Chem. Soc.*, 1972, **94**, 2724.
[156] J. S. Hartman and G. J. Schrobilgen, *Canad. J. Chem.*, 1972, **50**, 713.
[157] J. Derouault, M. Fouassier, and M. T. Forel, *J. Mol. Structure*, 1972, **11**, 423.
[158] F. Ya. Kul'ba, V. L. Stolyarov, and A. P. Zharkov, *Russ. J. Inorg. Chem.*, 1971, **16**, 1712.
[159] M. D. Hobday and T. D. Smith, *J.C.S. Dalton*, 1972, 2287.
[160] K. Brodersen, J. Rath, and G. Thiele, *Z. anorg. Chem.*, 1972, **394**, 13.
[161] A. G. Lee, *Coord. Chem. Rev.*, 1972, **8**, 289.

donor-acceptor complexes.[162] The same authors have suggested [163] that the Raman spectra of similar systems in solution in non-hydrogen-bonding solvents show evidence for the existence of 1:1 complexes, X^-,CBr_4, present as ion-pairs with the cation. It is further proposed, on the basis of changes in vibrational frequencies and depolarization ratios for CBr_4 on complexation, that in CBr_4Cl^-, the chloride ion is bound to a face rather than an apex or edge of the CBr_4 tetrahedron. I.r. and Raman spectra for CSClBr have been recorded, with stretching modes ν_1, ν_2, and ν_3 at 1130, 764, and 438 cm^{-1}, respectively,[164] and the absolute intensities of i.r. absorptions for the related species CSX_2 (X = F or Cl) have been measured.[165] The first report of methyl hydrogen carbonate, MeOCO(OH), includes [166] the i.r. data listed in Table 13.

Table 13 *I.r. spectrum of methyl hydrogen carbonate,* MeOCO(OH)

Wavenumber/cm^{-1}	Assignment	Wavenumber/cm^{-1}	Assignment
3484s	ν(OH)	1277s	δ_{sym}(Me)
1779m	ν_{asym}(CO)	1255s	δ_{sym}(Me)
1730s	ν_{asym}(CO)	823s	π(OCO$_2$)
		567w	δ_{sym}(OCO$_2$)

Virtually complete vibrational assignments for several cyclic fluorocarbon derivatives have been reported: the tetrafluorocyclopropanes $C_3F_4H_2$, C_3F_4HD, and $C_3F_4D_2$ possess, respectively, C_{2v}, C_s, and C_{2v} symmetry;[167] tetrafluoroethylene oxide [168] and 1-H-trifluorocyclopropene [169] have been the subjects of detailed gas-phase i.r. studies; observed data for perfluorocyclohexane are satisfactorily accounted for [170] on the basis of the point group D_{3d}.

The N-(trichloromethyl)chloroformimidium salt $Cl_3CN(H)=CCl_2^+,-SbCl_6^-$ and its neutral relative $Cl_3CN=CCl_2,SbCl_5$ have been examined by i.r. and Raman techniques.[171] Raman spectra of solutions of MeCOCl in HFSO$_3$, HClSO$_3$, HClO$_4$, and HCF$_3$SO$_3$ are consistent [172] with the equilibria shown in Scheme 1.

Compounds containing M—H Bonds (M = Si, Ge, or Sn).—The vibrational spectra of phenylsilane and phenyl[^{2}H$_3$]silane have been subjects of a detailed study.[173] Vapour-phase i.r. measurements suggest that the silyl group can

---

[162] J. A. Creighton and K. M. Thomas, *J.C.S. Dalton*, 1972, 403.
[163] J. A. Creighton and K. M. Thomas, *J.C.S. Dalton*, 1972, 2254.
[164] J. L. Brema and D. C. Moule, *Spectrochim. Acta*, 1972, **28A**, 809.
[165] M. J. Hopper, J. W. Russell, and J. Overend, *Spectrochim. Acta*, 1972, **28A**, 1215.
[166] G. Gattow and W. Behrendt, *Angew. Chem. Internat. Edn.*, 1972, **11**, 534.
[167] N. C. Craig, G. J. Anderson, E. Cuellar-Ferreira, J. W. Koepke, and P. H. Martyn, *Spectrochim. Acta*, 1972, **28A**, 1175.
[168] N. C. Craig, *Spectrochim. Acta*, 1972, **28A**, 1195.
[169] N. C. Craig and J. W. Koepke, *Spectrochim. Acta*, 1972, **28A**, 180.
[170] F. A. Miller and B. M. Harney, *Spectrochim. Acta*, 1972, **28A**, 1059.
[171] A. Schmidt, *Chem. Ber.*, 1972, **105**, 3050.
[172] R. J. P. Corriu, G. Dabosi, and A. Germain, *Bull. Soc. chim. France*, 1972, 1617.
[173] J. R. Durig, K. L. Hellams, and J. H. Mulligan, *Spectrochim. Acta*, 1972, **28A**, 1039.

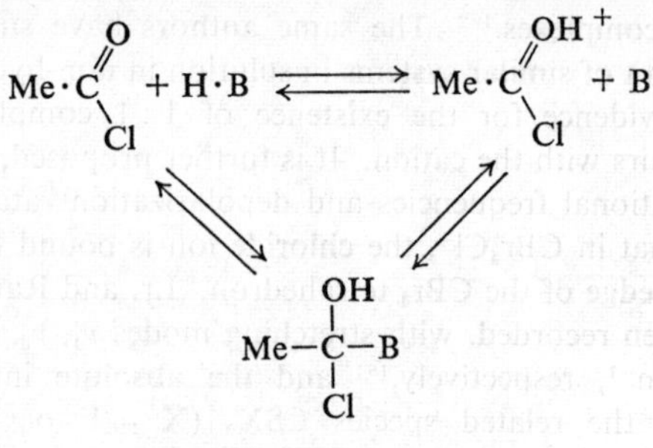

$$\text{Me} \cdot \underset{\underset{\text{Cl}}{|}}{\overset{\overset{\text{O}}{\|}}{\text{C}}} + \text{H} \cdot \text{B} \rightleftarrows \text{Me} \cdot \underset{\underset{\text{Cl}}{|}}{\overset{\overset{\text{OH}^+}{\|}}{\text{C}}} + \text{B}^-$$

$$\text{Me} - \underset{\underset{\text{Cl}}{|}}{\overset{\overset{\text{OH}}{|}}{\text{C}}} - \text{B}$$

(B = FSO$_3$, ClSO$_3$, ClO$_4$, or CF$_3$SO$_3$)

Scheme 1

rotate freely, leading to an assignment of ring-vibrations in terms of local C_{2v} symmetry, and of Si—H(D) fundamentals in the way shown in Table 14. Si—H stretching wavenumbers for Me$_2$EtSiH, MeEt$_2$SiH, and Et$_3$SiH have been measured and compared.[174]

Table 14 Wavenumbers/cm^{-1} of fundamentals of the silyl group in phenylsilane

	PhSiH$_3$ ($n = 1$)	PhSiD$_3$ ($n = 2$)
ν_{asym}(Si—^{n}H) }	2158	1584
ν_{sym}(Si—^{n}H)		1558
δ_{asym}(Si^{n}H$_3$)	952	671
δ_{sym}(Si^{n}H$_3$)	928	662
ρ(Si^{n}H$_3$) [o.p.]	678	540
ρ(Si^{n}H$_3$) [i.p.]	645	507

Co-condensation of SiF$_2$ with diborane affords two thermally very unstable diborane derivatives, tentatively formulated as the novel heterocycles (SiF$_2$)$_n$B$_2$H$_4$ (20) on the basis of, as well as other evidence, their

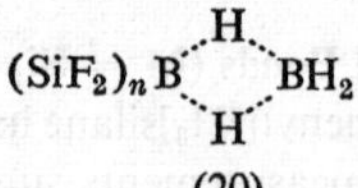

$$(\text{SiF}_2)_n \text{B} \underset{\text{H}}{\overset{\text{H}}{\cdots}} \text{BH}_2$$

(20)

gas-phase i.r. spectra: the latter include bands assigned to ν(B—H)$_{\text{terminal}}$ (near 2600 cm^{-1}) and ν(B—H)$_{\text{bridge}}$ (at 1570 cm^{-1}). Decomposition of these species yields, in addition to known products, the new compounds H$_3$SiSiF$_3$ and H$_2$Si(SiF$_3$)$_2$ for which gas-phase i.r. absorptions have been measured and assigned. Those given for 1,1,1-trifluorodisilane are summarized in Table 15. The authors suggest that the large change in wave-

---

[174] I. V. Shevchenko, I. F. Kovalev, V. S. Dernova, M. G. Voronkov, Yu. I. Khudobin, N. A. Andreeva, and N. P. Kharitonov, *Izvest. Akad. Nauk. S.S.S.R., Ser. khim.*, 1972, 98.

Table 15 Infrared spectrum of 1,1,1-trifluorodisilane (wavenumber/cm^{-1})

Suggested assignment	H$_3$SiSiF$_3$	D$_3$SiSiF$_3$
ν_{asym}(Si—^{n}H)	2191	1601
ν_{sym}(Si—^{n}H)	2180	1568
ν_{asym}(Si—F)	} 961	959
δ_{asym}(Si^{n}H$_3$)		774
ν_{sym}(Si—F)	} 826	866
δ_{sym}(Si^{n}H$_3$)		654
ν(Si—Si)	532	511
δ_{asym}(SiF$_3$)	505	493
δ_{sym}(SiF$_3$)	306	299

number for ν_{sym}(Si—F) observed between H$_3$SiSiF$_3$ and D$_3$SiSiF$_3$ can be accounted for in terms of strong interaction of this vibration with δ_{sym}(SiH$_3$), the corresponding antisymmetric modes not coupling so effectively.[175]

I.r. bands for Si^{n}H$_3$,N^{n}HPF$_2$ (n = 1 or 2) have been recorded and partly assigned. Doubling of N—H stretching and bending modes in the vapour phase at room temperature is taken to indicate the existence of two different conformers, but solid-phase measurements at 77 K suggest that at this temperature one of these is much the more stable.[176] Partial fluorination with PF$_5$ of Si—H bonds in hydrosiloxanes has yielded the following new compounds: SiH$_3$OSiH$_2$F, (SiH$_2$F)$_2$O, MeSiHFOSiH$_2$Me, and (MeSiHF)$_2$O. Gas-phase i.r. absorptions have been reported and partially assigned for each of these.[177] Trisilylamine reacts with H$_2$E (E = S or Se) at room temperature to give, in addition to (SiH$_3$)$_2$E, solid adducts which have been formulated as NH$_4$$^+$(ESiH$_3$)$^-$, and related salts of the MeNH$_3$$^+$ and Me$_2$NH$_2$$^+$ cations have also been prepared.[178] Some i.r. frequencies for the silyl anions are listed in Table 16. Reaction of atomic silicon with silanes affords the polysilanes Me$_2$SiHSiH$_2$SiHMe$_2$, MeH$_2$SiSiH$_2$SiH$_2$Me, and n-Si$_5$H$_{12}$, for which unassigned i.r. data have been given.[179] SiHCl$_3$

Table 16 Assignment of SiH$_3$ fundamentals (wavenumber/cm^{-1}) for salts of silanethiol and silaneselenol

Assignment	R$^+$(SSiH$_3$)$^-$		R$^+$(SeSiH$_3$)$^-$	
	R = NH$_4$	R = ND$_4$	R = NH$_4$	R = Me$_2$NH$_2$
ν(Si—H)	2100	2100	2120	2050
δ(SiH$_3$)	930	935	930	935
ρ(SiH$_3$)	635	635	615	—
ν(Si—E)^a	550	550	430	430

^{a} E = S or Se.

[175] D. Solan and A. B. Burg, *Inorg. Chem.*, 1972, **11**, 1253.
[176] D. E. J. Arnold, E. A. V. Ebsworth, H. F. Jessep, and D. W. H. Rankin, *J.C.S. Dalton*, 1972, 1681.
[177] E. W. Kifer and C. H. Van Dyke, *Inorg. Chem.*, 1972, **11**, 404.
[178] S. Cradock, E. A. V. Ebsworth, and H. F. Jessep, *J.C.S. Dalton*, 1972, 359.
[179] P. S. Skell and P. W. Owen, *J. Amer. Chem. Soc.*, 1972, **94**, 5434.

and MeSiHCl$_2$ and their deuteriated analogues have been identified as products from the reaction of chlorosilylnickel complexes with HCl or DCl in benzene solution through observation of appropriate Si—H (Si—D) stretching fundamentals.[180]

Silylgermylmethane, SiH$_3$CH$_2$GeH$_3$, synthesized in 35% yield by the action of NaGeH$_3$ on SiH$_3$CH$_2$Cl, and its derivatives GeH$_3$CH$_2$SiH$_2$Cl, GeH$_3$CH$_2$SiHCl$_2$, (GeH$_3$CH$_2$SiH$_2$)$_2$O, and (SiH$_3$CH$_2$)$_2$GeH$_2$ have been examined by Van Dyke et al. Partial assignments of i.r. bands presented in this work include those of Table 17 for the parent compound.[181]

Table 17 I.r. data for silylgermylmethane, SiH$_3$CH$_2$GeH$_3$

Assignment	Wavenumber/cm^{-1}
ν(Si—H)	2150—2210
ν(Ge—H)	2056—2095
δ(CH$_2$) 'scissor'	1362—1382
δ(CH$_2$) 'wag'	1049—1055
δ(SiH$_3$)	941
δ(GeH$_3$)	850
ν(Si—C)	750—774
ν(Ge—C)	650 (?)

A complete interpretation of the vibrational spectra of p-FC$_6$H$_4$GeH$_3$ and p-ClC$_6$H$_4$GeH$_3$ indicates a very low barrier to rotation about the C—Ge bond.[182] The i.r. and Raman spectra of the methylhalogenogermanes MeGe^{n}H$_2$X (n = 1 or 2; X = F, Cl, Br, or I) have been the subject of a thorough study by Drake and co-workers, including valence force-constant calculations.[183] The authors were able to propose assignments for all fundamentals except the torsion for each molecule on the basis of C_s symmetry, including those of Table 18 for GeH$_2$ bending modes.

Table 18 Wavenumbers/cm^{-1} for Ge—H bending modes in methylhalogenogermanes, MeGeH$_2$X or MeGeD$_2$X

	$\nu_6(a')$ (GeH$_2$ 'bend')		$\nu_7(a')$ (GeH$_2$ 'wag')		$\nu_{16}(a'')$ (GeH$_2$ 'twist')		$\nu_{17}(a'')$ (GeH$_2$ 'rock')	
X	H	D	H	D	H	D	H	D
F	900	625	721	565	705	565	472	379
Cl	875	624	717	544	717 (?)	544 (?)	463	360
Br	875	622	705	539	705 (?)	531	456	350
I	873	618	694	520	694 (?)	513	442	336

Identification of the latter was facilitated by reference to data obtained through synthesis of additional deuteriated species CD$_3$GeH$_2$X (X = Cl

[180] Y. Kiso, K. Tamao, and M. Kumada, J.C.S. Chem. Comm., 1972, 105.
[181] C. H. Van Dyke, E. W. Kifer, and G. A. Gibbon, Inorg. Chem., 1972, 11, 408.
[182] J. R. Durig and J. B. Turner, J. Phys. Chem., 1972, 76, 1558.
[183] G. K. Barker, J. E. Drake, R. T. Hemmings, and B. Rapp, Spectrochim. Acta, 1972, 28A, 1113.

or Br). Substantial mixing of ν_7 and ν_{16}, indicated by the appearance of the spectra, featured prominently in the calculations. Two strong absorptions of equal intensity attributable to symmetric GeH$_3$ deformation modes occur in the i.r. spectrum of $(GeH_3)_2Fe(CO)_4$, at 809 and 835 cm^{-1}. This situation contrasts with that encountered for other digermyl compounds, which show only one such band, but similar spectra have been observed for certain disilyl derivatives.[184]

References 185—192 give wavenumbers for Si—H or Ge—H vibrations for the compounds collected in Table 19.

Table 19 *References to Si—H and Ge—H compounds characterized by i.r. and/or Raman spectroscopy*

Compound	Ref.	Compound	Ref.
H$_2$Si (cyclopropane)	184a	Me$_3$SiSiH$_2$Ph Me$_3$SiSiH$_2$I Me$_3$SiSiH$_2$(OMe) Me$_3$SiSiH(OMe)$_2$	187
(benzocycloheptene)—SiH$_2$	185	SiH$_3$SiH$_2$SiH$_2$Br SiH$_3$SiHBrSiH$_3$	188
Me—Si (adamantane-like cage with Si—H)	185a	GeH$_3$AsMe$_2$ (GeH$_3$)$_2$AsMe GeH$_3$AsPh$_2$ (GeH$_3$)$_2$AsPh	189
		GeH$_3$Co(CO)$_4$ Ge^{2}H$_3$Co(CO)$_4$	190
		GeH$_3$Mn(CO)$_5$ Ge^{2}H$_3$Mn(CO)$_5$	191
Me(Cl)SiHSiH$_3$ MeSiH$_2$SiH$_2$Cl Me(Cl)SiHSiH$_2$Cl MeSiCl$_2$SiH$_3$ Me(Cl)SiHSiHCl$_2$	186	H$_2$Ge═GeH$_2$	192

Hydrostannylation of styrene, using R$_2$SnH$_2$ (R = Et, Bu^n, or Ph) or RPhSnH$_2$ (R = Et or Bu^n), has been followed by monitoring the decrease in intensity of i.r. absorptions due to ν(Sn—H), wavenumbers for the latter for each stannane also being listed.[193]

[184] S. R. Stobart, *J.C.S. Dalton*, 1972, 2442.
[184a] B. N. Cyvin, S. J. Cyvin, L. V. Vilkov, and V. S. Mastryukov, *Rev. Chim. Minérale*, 1971, **8**, 877.
[185] L. Birkofer and H. Haddad, *Chem. Ber.*, 1972, **105**, 2101.
[185a] C. L. Frye and J. M. Klosowski, *J. Amer. Chem. Soc.*, 1972, **94**, 7186.
[186] A. J. Vanderwielen and M. A. Ring, *Inorg. Chem.*, 1972, **11**, 246.
[187] E. Hengge, G. Bauer, and M. Marketz, *Z. anorg. Chem.*, 1972, **394**, 93.
[188] T. C. Geisler, C. G. Cooper, and A. D. Norman, *Inorg. Chem.*, 1972, **11**, 1710.
[189] J. W. Anderson and J. E. Drake, *J.C.S. Dalton*, 1972, 951.
[190] R. D. George, K. M. Mackay, and S. R. Stobart, *J.C.S. Dalton*, 1972, 974.
[191] R. D. George, K. M. Mackay, and S. R. Stobart, *J.C.S. Dalton*, 1972, 1505.
[192] J. V. Scibelli and M. D. Curtis, *J. Organometallic Chem.*, 1972, **40**, 317.
[193] L. S. Mel'nichenko, A. N. Rodionov, N. N. Zemlyanskii, and K. A. Kocheshkov, *Doklady Chem.*, 1971, **201**, 996.

Compounds containing M—C Bonds (M = Si, Ge, Sn, or Pb).—I.r. spectroscopy has been used (in the range 400—1100 cm^{-1}) in a study of SiC single-crystal films formed by bombardment of crystalline Si with C$^+$ ions.[194] Flash photolysis of 1,1-dimethyl-1-silacyclobutane (21) at 650 °C, followed by deposition of the pyrolysate at -196 °C onto an NaCl plate, gives in addition to features due to (21) a new sharp band in the i.r. at 1407 cm^{-1}. Warming to -120 °C results in non-reversible disappearance of this band,

Me$_2$Si⎯⎤
　　　　⎦
(21)

$\dfrac{\text{Me}}{\text{Me}}$Si=CH$_2$

(22)

which is tentatively attributed to a fundamental of the species (22), presumably ν(Si=CH$_2$).[195] Co-condensation of tin with carbon monoxide at 20 K in an Ar matrix affords a phase with a large number of i.r. bands in the CO stretching region.[196] At low CO concentration in the matrix, however, a very prominent band appears at 1921 cm^{-1}, thought to be due to the metal-rich species SnCO; a band at 1908 cm^{-1} observed after reaction of Ge under similar conditions is attributed to GeCO.

I.r. and Raman spectra of the ethynyls M(C≡CMe)$_4$ (M = Si, Ge, or Pb) have been discussed on the basis of T_d symmetry, ν(M—C) modes being assigned[197] as shown in Table 20. Similar data for tetrabenzyl-germane and -stannane and some halogeno-derivatives have also been reported,[198] and are listed in Table 21. Incidence of more than the predicted

Table 20 *Metal–carbon stretching wavenumber/cm^{-1} for tetra(methylethynyl) derivatives of Si, Ge, and Pb*

M	ν(M—C), a_1	ν(M—C), f_2
Si	384	605
Ge	371	440
Pb	331	—

Table 21 *Raman bands assigned to Ge—C and Sn—C stretching fundamentals for tetrabenzyl and related derivatives*

Compound	Wavenumber/cm^{-1}
(PhCH$_2$)$_4$Ge	615(dp), 588(p)
(PhCH$_2$)$_4$Sn	572(dp), 558(p)
(PhCH$_2$)$_3$SnCl	583(dp), 565(p)
(PhCH$_2$)$_2$SnCl$_2$	589(dp), 570(p)
(PhCH$_2$)$_2$SnBr$_2$	587(dp), 568(p)

[194] E. K. Baranova, K. D. Demakov, K. V. Stavinin, L. I. Strel'tsov, and I. B. Khaibullin, *Doklady Phys. Chem.*, 1971, **200**, 847.
[195] T. J. Barton and C. L. McIntosh, *J.C.S. Chem. Comm.*, 1972, 861.
[196] A. Bos, *J.C.S. Chem. Comm.*, 1972, 26.
[197] R. E. Sacher, B. C. Pant, F. A. Miller, and F. R. Brown, *Spectrochim. Acta*, 1972, **28A**, 1361.
[198] L. Verdonck and Z. Eeckhaut, *Spectrochim. Acta*, 1972, **28A**, 433.

number of features for the latter compounds in the range 600—200 cm^{-1} is thought to arise through hindered rotation about the Ph—C bonds, and for (PhCH$_2$)$_2$Hg, also examined, a single polarized band was found at 564 cm^{-1}. Vibrational spectra for (Me$_3$SiCH$_2$)$_4$M (M = Sn or Pb) and Me$_3$SiCH$_2$Cl have been recorded and discussed as a part of an investigation of the properties of some trimethylsilylmethyl–metal compounds.[199] Consideration of Raman spectra for R$^1{}_n$Sn(C≡CR^2)$_{4-n}$ (R^1, R^2 = alkyl, CH=CH$_2$, or Ph; n = 1—3) and R$_3$SnC≡CSnR$_3$ (R as above) have led to the proposal that in these species, interactions between separate substituent π-systems are completely inhibited by the tin atoms.[200] Shifts and broadening of ν(C—H)$_{ethynyl}$ in dioxan and acetone solutions for a series of related ethynyl–tin derivatives have been interpreted in terms of R$^1{}_3$SnC≡CH···OR^2 hydrogen-bonding.[201] Unassigned i.r. frequencies for (23) and (24) have been presented.[202, 203]

(23) M = Si, Ge, Sn, or Pb;
R = Me or Ph

A comprehensive study below 350 cm^{-1} of trimethylchlorosilane by microwave as well as far-i.r. and Raman techniques has provided a number of new assignments for this molecule. A value of 3.0 kcal mol^{-1} for the barrier to internal rotation has been derived from frequencies of 233, 208 cm^{-1} [176, 148 cm^{-1} for (CD$_3$)$_3$SiCl] for i.r. absorptions attributed to, respectively, e and a_2 torsions.[204] Some new data have also been given for Me$_3$GeF, including normal-co-ordinate calculations.[205] The i.r. and Raman spectra of Pr^{n}SnCl$_3$ are consistent with the presence in the liquid state of both *trans*- and *gauche*-isomers, ν(Sn—C) being markedly different (599 as opposed to 523 cm^{-1}) for the two rotamers.[206] Investigation of the compounds (p-YC$_6$H$_4$CH$_2$)$_n$SnCl$_{4-n}$ (Y = H, F, or Cl; n = 2 or 3) has shown that although *para*-substitution has virtually no effect on ν(Sn—Cl), references are collected in Table 22.

[199] W. Mowat, A. Shortland, G. Yagupsky, N. J. Hill, M. Yagupsky, and G. Wilkinson, *J.C.S. Dalton*, 1972, 533.
[200] O. A. Zasyadko, R. G. Mirskov, N. P. Ivanova, and Yu. L. Frolov, *Zhur. priklad. Spektroskopii*, 1971, **15**, 718.
[201] O. A. Zasyadko, Yu. L. Frolov, and R. G. Mirskov, *Zhur. priklad. Spektroskopii*, 1971, **15**, 939.
[202] J. Y. Corey, M. Dueber, and M. Malaidza, *J. Organometallic Chem.*, 1972, **36**, 49.
[203] G. Fritz and M. Hähnke, *Z. anorg. Chem.*, 1972, **390**, 104.
[204] J. R. Durig, R. O. Carter, and Y. S. Li, *J. Mol. Spectroscopy*, 1972, **44**, 18.
[205] C. Peuker and K. Licht, *Z. phys. Chem. (Leipzig)*, 1971, **248**, 103.
[206] H. Geissler, Chr. Peuker, R. Heess, and H. Kriegsmann, *Z. anorg. Chem.*, 1972, **393**, 230.
[207] L. Verdonck and G. P. van der Kelen, *J. Organometallic Chem.*, 1972, **40**, 135.

a decrease in ν(Sn—C) is evident.[207] Unassigned i.r. wavenumbers have been listed for some perfluoro-alkyl and -aryl silanes and germanes;

Table 22 *References containing i.r. wavenumbers for perfluoro-alkyl and -aryl silanes and germanes*

Compound	Ref.
CF_3SiF_3	208
$C_2F_5SiF_3$	
Ph_FMBr_3	
$(Ph_F)_2MBr_2$	
$(Ph_F)_3MBr$ (M = Si or Ge)	
Ph_FMF_3	
$(Ph_F)_2MF_2$	
$(Ph_F)_3SiF$	
$(Ph_F)_2SiEt_2$	209
Ph_FSiEt_3	
$(Ph_F)_2GeR_2$ (R = Me or Et)	
Ph_FGeMe_3	
$(Ph_F)_2Si(OH)_2$	
$[(Ph_F)_2SiO]_n$	
$[(Ph_F)_2GeO]_3$	

Normal-co-ordinate calculations for the cyclobutanes (25) have been published.[184a] Studies on 1,3,5,7-tetrasila-adamantanes (26) (named carborundanes) include some i.r. data.[185a] Butadiene reacts with SiF_2 to give (27) for which partially assigned i.r. bands have been given.[210] Reactions of tin(II) halides with dimethylacetylenedicarboxylate afford (28), with ν(Sn—C) assigned [211] at 390—415 cm^{-1}. A band at 570 cm^{-1} in the i.r. spectrum of Me_2Sn(salen) is due to ν_{asym}(Sn—C), but no feature attributable to ν_{sym}(Sn—C) could be found, suggesting the structure (29) for this complex, with an approximately linear C—Sn—C grouping.[213] Harrison and co-workers have obtained low-frequency i.r. data for several five- and six-co-ordinate methyltin halides which can be interpreted in terms of different stereochemistries: planar SnC_3 groups (*e.g.* in Me_3SnI_2) give rise to a single ν(Sn—C) absorption near 550 cm^{-1}, but Me_2SnX_3 systems usually show two bands at *ca.* 570 and 515 cm^{-1}, indicating the presence of both axial and equatorial Sn—C bonds.[214]

Various other papers contain vibrational information for compounds with M—C bonds, some including assignments for ν(M—C) fundamentals (M = Si, Ge, or Sn). These refer to $Me_nSi(OR)_{4-n}$ (n = 1—3; R = Me

[208] K. G. Sharp and T. D. Coyle, *Inorg. Chem.*, 1972, **11**, 1259.
[209] M. Weidenbruch and N. Wessal, *Chem. Ber.*, 1972, **105**, 173.
[210] J. C. Thompson and J. L. Margrave, *Inorg. Chem.*, 1972, **11**, 913.
[211] P. G. Harrison, *Inorg. Nuclear Chem. Letters*, 1972, **8**, 555.
[212] H. Schmidbaur and W. Kapp, *Chem. Ber.*, 1972, **105**, 1203.
[213] R. Barbieri and R. H. Herber, *J. Organometallic Chem.*, 1972, **42**, 65.
[214] M. K. Das, J. Buckle, and P. G. Harrison, *Inorg. Chim. Acta*, 1972, **6**, 17.

(25) X = H, Cl, or F

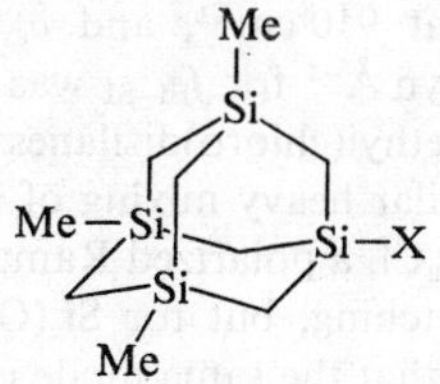

(26) X = H, Cl, NEt$_2$, or Me

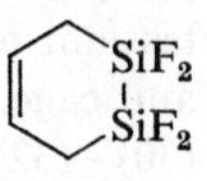

(27)

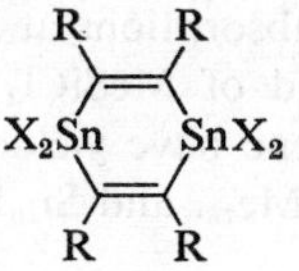

(28) X = Cl, Br, or I; R = MeO·CO

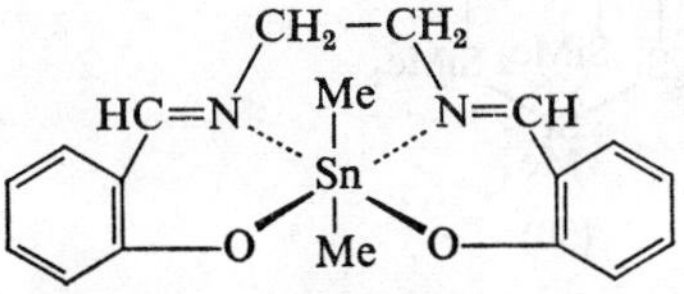

(29)

or Et);[215] Me$_n$Si(ON:CR^{1}R^2)$_{4-n}$ (n = 0—3; R^1, R^2 various);[216] the silylated ylides Me$_n$SiR$_{4-n}$ [n = 0—3; R = CH=S(O)Me$_2$];[212] Et$_3$GeCH-(CN)CH$_2$PEt$_2$;[217] some organotin esters;[218] R$_2$Sn(PO$_2$F$_2$)$_2$ (R = Me, Et, Pr^n, Bu^n, or n-C$_8$H$_{17}$);[219] and the sulphonates [220] Me$_2$Sn(SO$_3$F)$_2$, Me$_2$Sn-(SO$_3$CF$_3$)$_2$, Me$_3$SnSO$_3$F, MeSnCl$_2$SO$_3$F, Me$_2$SnCl(SO$_3$F), MeSnCl(SO$_3$F)$_2$, and MeSnCl(SO$_3$CF$_3$)$_2$ for which ν_{asym}(Sn—C) [or ν(Sn—C)] and ν_{sym}-(Sn—C), respectively, were found within the ranges 576—585 and 523—535 cm^{-1}.

Compounds containing M—M Bonds (M = Si, Ge, or Sn).—Within this category only polysilanes have received significant attention during 1972. I.r. and Raman spectra for hexafluorodisilane have been recorded and assigned by Hoffler, Hengge, and Waldhor: normal-co-ordinate calculations indicated that a band at 541 cm^{-1} with 56% ν(Si—Si) character also involved substantial contributions from both of the other a_{1g} modes

[215] N. V. Kozlova, V. P. Bazov, I. F. Kovalev, and M. G. Voronkov, *Latv. PSR Zinat. Akad. Vestis, Kim. Ser.*, 1971, 604.
[216] A. Singh, A. K. Rai, and R. C. Mehrotra, *J.C.S. Dalton*, 1972, 1911.
[217] J. Satgé, C. Couret, and J. Escudié, *J. Organometallic Chem.*, 1972, **34**, 83.
[218] N. W. G. Debye, D. E. Fenton, and J. J. Zuckerman, *J. Inorg. Nuclear Chem.*, 1972, **34**, 352.
[219] T. H. Tan, J. R. Dalziel, P. A. Yeats, J. R. Sams, R. C. Thompson, and F. Aubke, *Canad. J. Chem.*, 1972, **50**, 1843.
[220] P. A. Yeats, J. R. Sams, and F. Aubke, *Inorg. Chem.*, 1972, **11**, 2634.

[ν_{sym}(Si—F) at 910 cm^{-1}, and δ_{sym}(SiF$_3$) at 220 cm^{-1}], and a value of 2.4 ± 0.2 mdyn Å^{-1} for f_{Si-Si} was derived.[221] Further work by the same authors on methyl(chloro)disilanes [222] and methyl(methoxy)disilanes [223] has suggested similar heavy mixing of ν(Si—Si) with other fundamentals. For Me$_2$ClSiSiMe$_2$Cl, a polarized Raman band at 398 cm^{-1} is attributed mainly to Si—Si stretching, but for Si$_2$(OMe)$_6$ (f_{Si-Si} = 2.25 mdyn Å^{-1}) calculations showed that the same mode contributed substantially to three bands, at 760, 523, and 229 cm^{-1}. 1,1,1-Trifluorosilane has already been mentioned,[175] and for this compound and its deuteriated counterpart ν(Si—Si) were assigned to i.r. absorptions at 532 and 511 cm^{-1}, respectively. Reactions of Me$_2$SiCl$_2$ and of MeSiCl$_3$ with sodium–potassium alloy in the presence of naphthalene have yielded novel bicyclic and cage polysilanes Si$_8$Me$_{14}$, Si$_9$Me$_{16}$, Si$_{10}$Me$_{16}$, and Si$_{10}$Me$_{18}$. Structures (30)—(32) have been

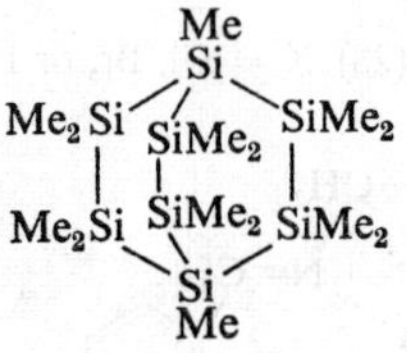

(30)

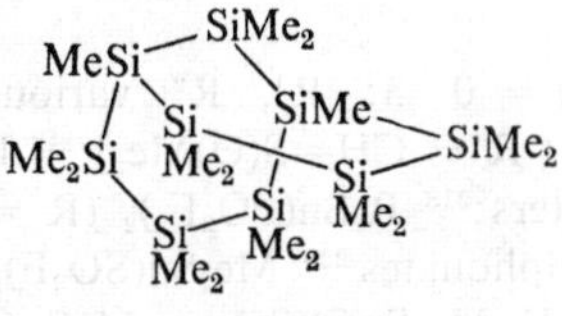

(31) 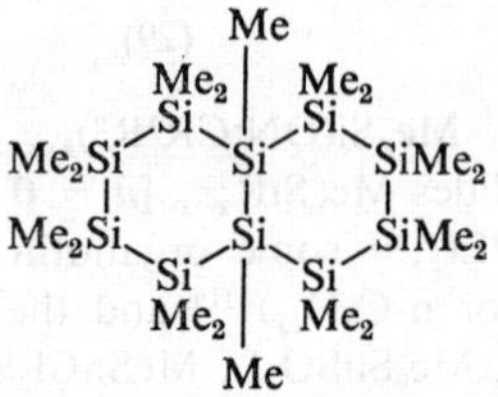   (32)

proposed for three of these on the basis of n.m.r. and u.v. data, and although measured i.r. frequencies have not been assigned it is suggested that bands in the 350—450 cm^{-1} region arise from ν(Si—Si) modes.[224] Further unassigned data have been published, for polysilanes listed earlier in Table 19, and also for Me$_3$SiSiIPh$_2$, Me$_3$SiSi(OMe)$_2$Ph, and Me$_3$SiSi(OMe)$_3$ (ref. 187), and Me$_3$P=N—SiMe$_2$SiMe$_3$ and Me$_3$P=N—SiMe$_2$SiMe$_2$—CH=PMe$_3$ (ref. 224a).

An improved synthesis has been reported for GeBr$_2$ which, on reaction with GeBr$_4$, affords Br$_3$GeGeBr$_3$. Observation for this digermane of six Raman-active fundamentals (315, 196, and 78, species A_{1g}; 339, 111, and

---

[221] F. Höfler, S. Waldhör, and E. Hengge, *Spectrochim. Acta*, 1972, **28A**, 29.
[222] F. Höfler and E. Hengge, *Monatsh.*, 1972, **103**, 1506.
[223] F. Höfler and E. Hengge, *Monatsh.*, 1972, **103**, 1513.
[224] R. West and A. Indriksons, *J. Amer. Chem. Soc.*, 1972, **94**, 6110.
[224a] H. Schmidbaur and W. Vornberger, *Chem. Ber.*, 1972, **105**, 3187.

Characteristic Vibrational Frequencies of Compounds

68 cm^{-1}, species E_g) together with non-coincident i.r. bands at 327 and 247 cm^{-1} establishes a D_{3d} (staggered) conformation in the solid state. Assignment of an individual band to ν(Ge—Ge) was not attempted.[225] I.r. spectra for the permethylcyclopolygermanes (Me$_2$Ge)$_n$ (n = 5, 6, or 7) have been measured, but only one absorption, at 367 cm^{-1} for the first of these, could be assigned to a Ge—Ge stretching mode.[226] Similar data for R$_3$SnSnR$_3$ (R = Bu^n or Ph) have been reported but no assignments were given.[227]

Compounds containing M—N Bonds (M = Si, Ge, Sn, or Pb), M—P Bonds (M = Si or Sn), or Si—As Bonds.—The compounds Me$_3$N^{n}H-[(Cl$_3$Si)$_2$N] have been prepared by treating bis(trichlorosilyl)amine with Me$_3$N. I.r. bands for these species at 1185, 785 (n = 1) and 1192, 792 cm^{-1} (n = 2) have been assigned to ν_{asym} and ν_{sym}(SiNSi), respectively, and it has been proposed [228] that the anion possesses a structure corresponding to [Cl$_3\bar{\text{Si}}$=$\overset{+}{\text{N}}$=$\bar{\text{Si}}$Cl$_3$]$^-$. Bis(trimethylsilyl)acetamide and monotrimethylacetamide have been shown to have structures (33) and (34) using ^{15}N

```
         OSiMe₃                    O
         ∥                         ∥
  Me—C                      Me—C
         ＼                         ＼
          NSiMe₃                    NH(SiMe₃)

          (33)                      (34)
```

substitution, both by n.m.r. and i.r. investigation. The occurrence of small shifts in i.r. bands near 930, 720 cm^{-1} for (33) and 1030, 990 cm^{-1} for (34) between different isotopomers is tentatively ascribed to involvement of ν(Si—N) modes.[229] Some assignments for Me$_3$SiN(R)BF$_2$ (R = Me, Et, Pr, Pr^i, or Bu) [230] and Me$_3$P=NSi$_2$Me$_5$, Me$_3$P=NSi$_2$Me$_4$N=PMe$_3$, Me$_3$P=NSi$_2$Me$_4$Cl, and Me$_3$P=NSi$_2$Me$_4$CH=PMe$_3$ have been suggested, including ν(Si—N) near 570 cm^{-1} in the (disilaryl)phosphineimides.[224a] The azomethine derivatives Bu$^t_2$C=NR (R = SiMe$_3$, SiMe$_2$Cl, SiMeCl$_2$, or SiCl$_3$) show ν(C=N) at ca. 1735 and ν(Si—N) at ca. 960 cm^{-1}, corresponding bands for Ph$_2$C=NR (R = SiMe$_3$ or SiMe$_2$Cl) occurring at 1640 and 905 cm^{-1}, respectively.[231] Wannagat and Meier have characterized the novel cyclic systems (35)—(38), for which [232] either one or two i.r. bands at 895—950 cm^{-1} are assigned to ν_{asym}(SiNSi) with ν_{sym} at 530—550 cm^{-1}.

---

[225] M. D. Curtis and P. Wolber, *Inorg. Chem.*, 1972, **11**, 431.
[226] E. Carberry, B. D. Dombek, and S. C. Cohen, *J. Organometallic Chem.*, 1972, **36**, 61.
[227] H. Prakash and H. H. Sisler, *Inorg. Chem.*, 1972, **11**, 2258.
[228] H. H. Moretto, P. Schmidt, and U. Wannagat, *Z. anorg. Chem.*, 1972, **394**, 125.
[229] C. H. Yoder and D. Bonelli, *Inorg. Nuclear Chem. Letters*, 1972, **8**, 1027.
[230] G. Elter, O. Glemser, and W. Herzog, *J. Organometallic Chem.*, 1972, **36**, 257.
[231] J. B. Farmer, R. Snaith, and K. Wade, *J.C.S. Dalton*, 1972, 1501.
[232] U. Wannagat and S. Meier, *Z. anorg. Chem.*, 1972, **392**, 179.

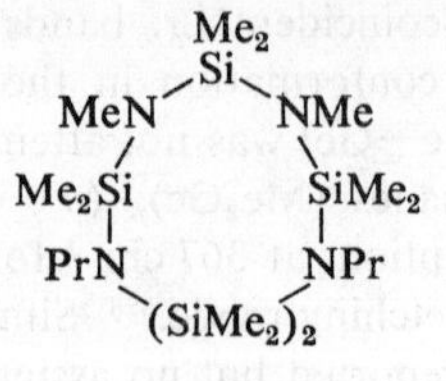

(35)

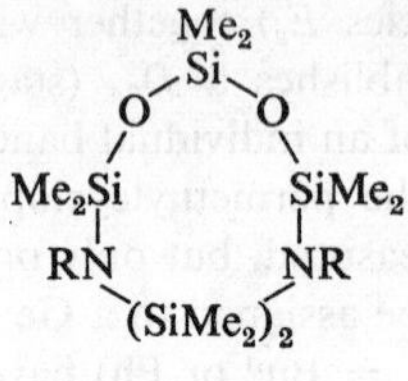

(36) R = Me or Pr^n

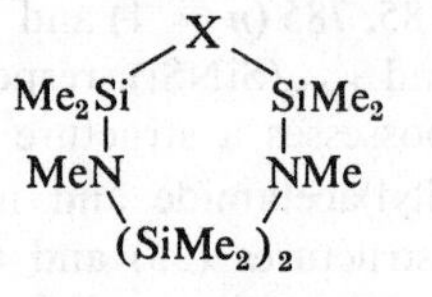

(37) X = NMe or O

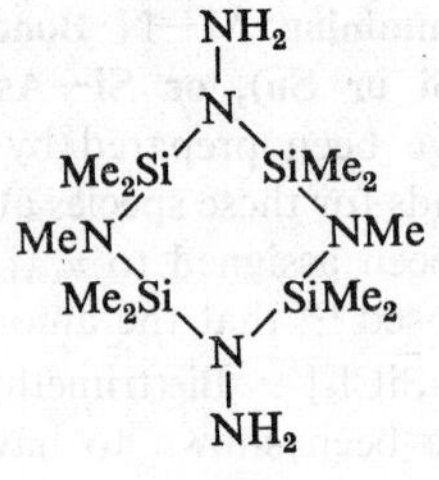

(38)

Decreasing basicity for increased n in the series of compounds $(R_2N)_{4-n}$-GeCl$_n$ (R = Me or Et; n = 1—3) is suggested from measurements of changes in ν(C—D···N) in deuteriochloroform solution.[233] I.r. and Raman spectra have been listed and assigned for a range of aminostannanes, including those of the types (Bu$^n_3$Sn)$_2$NEt, Bu$^n_3$SnNR^{1}R^2, and Bu$^n_2$Sn(NR$_2$)$_2$. For compounds with one tin–nitrogen bond, ν(Sn—N) were distinguished [234] near 590 cm^{-1}, and for the Sn$_2$N species ν_{asym} were at *ca.* 690 with ν_{sym} at 600 cm^{-1}. Keterimino-derivatives of tin (and also Si, Ge, and B) synthesized by 1,4-insertion reactions of the olefins (CF$_3$)$_2$C=C(CN)$_2$ and (CF$_3$)(CF$_2$Cl)C=C(CN)$_2$ have been characterized spectroscopically but ν(M—N) frequencies have not been given.[235] I.r. bands at 490—540 and 410—475 cm^{-1} have been dubiously assigned, to ν(Pb—N) and ν(Pb—O) modes respectively, for several Pb^{IV}–Schiff-base complexes whose structures remain uncertain.[236]

Compounds with Si or Sn bonded to other elements of Group V are referred to in two papers. Measurement of the i.r. spectra of Me$_3$MP(H)Ph (M = Si, Sn, or C) has been reported, but only in the region above 500 cm^{-1} so that ν(M—P) modes were not observed.[237] Fairly complete tabulation and assignments for the vibrations of the silylarsines (Me$_3$Si)$_{3-n}$-As(SiH$_3$)$_n$, (Me$_3$Si)$_{3-n}$AsMe$_n$, (Me$_3$Si)$_{3-n}$AsPh$_n$, and (SiH$_3$)$_{3-n}$AsPh$_n$

---

[233] J. Rejhon, J. Hetflejš, M. Jakoubková, and V. Chvalovský, *Coll. Czech. Chem. Comm.*, 1972, **37**, 3054.
[234] A. Marchand, C. Lemerle, M. T. Forel, and M. H. Soulard, *J. Organometallic Chem.*, 1972, **42**, 353.
[235] E. W. Abel, J. P. Crow, and J. N. Wingfield, *J.C.S. Dalton*, 1972, 787.
[236] N. S. Biradar, V. H. Kulkarni, and N. N. Sirmokadam, *J. Inorg. Nuclear Chem.*, 1972, **34**, 3651.
[237] P. G. Harrison, S. E. Ulrich, and J. J. Zuckerman, *Inorg. Chem.*, 1972, **11**, 25.

(n = 1 or 2), and $Me_3SiAsHMe$ and $Me_3SiAsMe(SiH_3)$ have been presented by Anderson and Drake; observation of Si—As stretching modes was restricted, however, to $Me_3SiAsMe_2$ (354 cm^{-1}, Raman pol.) and to $(Me_3Si)_2AsMe$, where a Raman band at 352 cm^{-1} was judged to contain both symmetric and antisymmetric components.[238]

Compounds containing M—O Bonds (M = Si, Ge, Sn, or Pb).—The cases for and against the involvement of siloxy-species in the formation of so-called 'polywater' have been argued in adjacent notes. Thus while Bascum has assigned i.r. bands found at 1650, 1425 cm^{-1} to ν_{asym} and ν_{sym}(O—C—O), and at 1000—1100 cm^{-1} to ν(Si—O—Si), arising from dissolved silicate-bicarbonate residues of vitreous origin,[239] this interpretation is dismissed by Brummer and co-workers.[240] I.r. intensities and bandwidths for ν(O—H) have been correlated with relative acidities for a series of silanols $(XC_6H_4)_3SiOH$ (X = H, p-Me, m-Me, p-F, p-Cl, m-F, m-Cl, or p-CF$_3$).[241] The vibrational spectra of $C(OR)_4$ (R = Me or Et) and the corresponding silicon tetra-alkoxides have been measured, with ν_{asym} and ν_{sym}(Si—O), respectively, at 843 and 640 cm^{-1}, and δ(SiO$_4$) modes at 310, 237, and 208 cm^{-1}. $C(OMe)_4$ was deduced to exist in the liquid state as two rotamers with S_4 and D_{2d} symmetries.[242]

I.r. spectroscopy has been used to show that Cl_3Si or $(MeO)_3Si$ groups can be introduced on to SiO_2 surfaces by treatment with $SiCl_4$ or MeOH, respectively.[243] Vibrational data for rare-earth pyrosilicates have been discussed in relation to the existence of both linear (D_{3d}) and non-linear (C_{2v}) structures.[244] I.r. dichroism of oriented crystalline films of hexamethyl-cyclotrisiloxane indicates no significant deviation from molecular D_{3h} symmetry in the crystal.[245] Further data for $(Me_2SiO)_3$, $(Me_2SiO)_4$, and $(Me_3Si)_2O$, and for $Si_3O_9^{3-}$ and $Si_2O_7^{2-}$ have also been reported,[246, 247] and some approximate assignments for $Me_nSi(OR)_{4-n}$ [n = 1—3; R = OCH(CH$_2$Cl)CH$_2$OPh] and $Me_nSi(OR)_{4-n}$ [n = 1 or 2; R = CH(CH$_2$Cl)CH$_2$OCH$_2$CH=CH$_2$] have been listed.[248]

MacDiarmid's interesting conclusion that the complex $(Me_3Si)_2Fe(CO)_4$ possesses the novel dimeric structure (39) is supported by the appearance

---

[238] J. W. Anderson and J. I. Drake, *J. Inorg. Nuclear Chem.*, 1972, **34**, 2455.
[239] W. D. Bascom, *J. Phys. Chem.*, 1972, **76**, 456.
[240] S. B. Brummer, J. I. Pradspies, G. Entine, C. Leung, and H. Lingertat, *J. Phys. Chem.*, 1972, **76**, 457.
[241] E. Popowski, H. Kelling, and G. Schott, *Z. anorg. Chem.*, 1972, **391**, 137.
[242] J. W. Ypenburg and H. Gerding, *Rec. Trav. chim.*, 1972, **91**, 1245.
[243] V. A. Tertykh, V. M. Mashchenko, and A. A. Chuiko, *Doklady Chem.*, 1971, **200**, 824.
[244] J. H. Denning, R. F. Hudson, D. R. Laughlin, S. D. Ross, and A. M. Sparasci, *Spectrochim. Acta*, 1972, **28A**, 1787.
[245] S. Dobos, G. Fogarasi, and E. Castellucci, *Spectrochim. Acta*, 1972, **28A**, 877.
[246] R. Konopka and B. Stojczyk, *Acta Phys. Polon. (A)*, 1971, **40**, 537.
[247] A. N. Lazarev, I. Ignat'ev, and A. P. Mirgorodskii, *Vysokotemp. Khim. Silikat. Okisov, Tr. Vsesl Soveshch, 3rd*, 1968 (pub. 1972), 12.
[248] P. Bajaj, R. C. Mehrotra, J. C. Maire, and R. Ouaki, *J. Organometallic Chem.*, 1972, **40**, 301.

$$\begin{array}{c} \text{Me}_3\text{Si} \quad \text{CO} \\ \text{OC}-\text{Fe}-\text{CO} \\ \text{Me}_3\text{SiOC}-|-\text{COSiMe}_3 \\ \text{OC}-\text{Fe}-\text{CO} \\ \text{OC} \quad \text{SiMe}_3 \\ (39) \end{array}$$

in the i.r. spectrum of bands at 1228m, 1098s, and 986s cm^{-1} assigned to stretching modes of an Si—O—C linkage; ν(O—C—C—O) was also tentatively identified.[249]

Three papers [250-252] dealing with germanates and related structures are of rather limited interest, but in one [251] vibrational evidence for the existence of non-planar cyclic $\text{Ge}_3\text{O}_9^{6-}$ anions is discussed.

Davies and his co-workers have reported wavenumbers for ν(O—H) and δ(OH) for a series of triorganotin hydroxides R$_3$SnOH (R = Me, Et, Pr, Bu, or Ph) and also ν_{asym}(SnOSn) for (R$_3$Sn)$_2$O at 740 (R = Me or Et), 760 (R = Pr), or 770 cm^{-1} (R = Bu, octyl, or Ph).[253] Action of 100% HNO$_3$ on bis(methylcyclopentadienyl)tin(II) to give a product with an i.r. band at 410 cm^{-1} assignable to ν(Sn—O) may indicate formation of a tin(II) nitrate.[254] On the basis of ν(CO$_2$) data, both PhCO$_2^-$ groups in Me$_2$Pb(O$_2$CPh)$_2$ are thought to be bidentate.[255] ν(Pb—O) for Me$_2$Pb(acac)$_2$ has been found to shift from 388 in CHCl$_3$ solution to 378 cm^{-1} in HMPA, an effect attributed to polarization of the Pb—O bonds through solvent–Pb co-ordination.[256]

Compounds containing M—S Bonds (M = Si, Ge, or Sn), M—Se Bonds (M = Si, Ge, or Sn), or Sn—Te Bonds.—A lengthy paper concerning the i.r. and Raman spectra of (RS)$_4$C (R = Me or Et) and (MeS)$_4$M (M = Si, Ge, or Sn) provides the assignments shown in Table 23. 'Doubling' of some fundamentals in the liquid phase was noted for all except the stannane, suggesting the presence of more than one rotational isomer in each case.[257] The first compound with two SH groups bonded to Si, (Bu^{t}O)$_2$Si(SH)$_2$, has been obtained from the reaction of SiS$_2$ with Bu^{t}OH, but although the i.r.

[249] M. A. Nasta, A. G. MacDiarmid, and F. E. Saalfeld, *J. Amer. Chem. Soc.*, 1972, **94**, 2449.
[250] J. Preudhomme and P. Tarte, *Spectrochim. Acta*, 1972, **28A**, 69.
[251] A. N. Lazarev, I. S. Ignat'ev, R. G. Grebenshchikov, and A. A. Shirvinskaya, *Izvest. Akad. Nauk. S.S.S.R., Neorg. Materialy*, 1972, **8**, 1101.
[252] G. Odent and F. Arnabi, *Compt. rend.*, 1972, **275**, C, 1275.
[253] J. M. Brown, A. C. Chapman, R. Harper, D. J. Mowthorpe, A. G. Davies, and P. J. Smith, *J.C.S. Dalton*, 1972, 338.
[254] P. G. Harrison, M. I. Khalil, and N. Logan, *Inorg. Nuclear Chem. Letters*, 1972, **8**, 551.
[255] M. Aritomi, Y. Kawasaki, and R. Okawara, *Inorg. Nuclear Chem. Letters*, 1972, **8**, 1053.
[256] M. Aritomi, Y. Kawasaki, and R. Okawara, *Inorg. Nuclear Chem. Letters*, 1972, **8**, 69.
[257] J. W. Ypenburg and H. Gerding, *Rec. Trav. chim.*, 1972, **91**, 1117.

Table 23 M—S *stretching fundamentals* (*wavenumber*/cm^{-1}) *for tetrakis-(methylthio)–Group* IV *derivatives*

M in M(SMe)$_4$	C	Si	Ge	Sn
ν_{asym}(M—S)	780	529	401	356, 368
ν_{sym}(M—S)	{ ca. 430 { ca. 420	389 381	364 356	334 332

spectrum (5000—400 cm^{-1}) has been illustrated only one band was listed, at 2550 cm^{-1} assigned [258] to ν(S—H). For R$_2$P(E)SSiMe$_3$ (R = F or CF$_3$; E = O or S) i.r. absorptions at 525 cm^{-1} were attributed to ν(P—S—Si) modes, corresponding ν(P—O—Si) bands being found near 1040 cm^{-1} for the analogous oxygen-bridged species.[259] The compounds Cd$_4$SiS$_6$ and Cd$_4$SiSe$_6$ have been shown by *X*-ray methods to exist as three-dimensional arrays of distorted SiE$_4$ and CdE$_4$ tetrahedra, consistent with the assignments [260] for observed vibrations of Table 24.

Table 24 Si—E *stretching modes for tetrahedral* SiE$_4$ *units in* Cd$_4$SiE$_6$ (E = S *or* Se)

	E = S	E = Se
ν_{sym}(E—S) (Raman)	413	378
ν_{asym}(E—S) (i.r.)	{ 556 { 532 { 515	{ 460 { 443 { 425

In the anions [M(mnt)Ph$_3$]$^-$ (M = Ge or Sn; mnt = dicyanoethylene-1,2-dithiolate), ν(Ge—S) modes were distinguished [261] at 327, 319, and 300 cm^{-1} with ν(Sn—S) at 305, 290 cm^{-1}. The range 345—377 cm^{-1} is spanned by ν(Sn—S) for N-substituted *N'*-cyano-*S*-(triphenylstannyl)isothioureas R—N=CSSnPh$_3$ (R = *p*-O$_2$NC$_6$H$_4$, Ph, PhCH$_2$, *p*-EtOC$_6$H$_4$, or
      |
   NHCN
Et).[262] Similarly, diethylthiocarbamate complexes (40; X = Y = alkyl or halogen, or X = alkyl, Y = halogen) show [263] a characteristic ν(Sn—S)

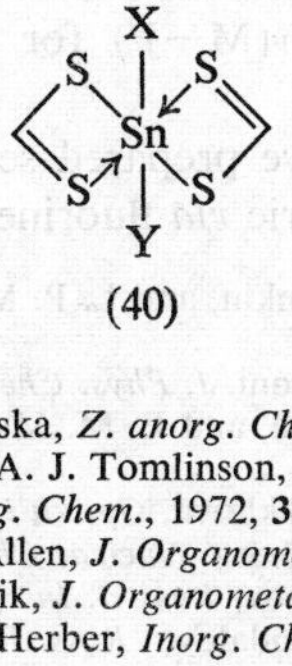

(40)

---

[258] W. Wojnowski and M. Wojnowska, *Z. anorg. Chem.*, 1972, **389**, 302.
[259] R. G. Cavell, R. D. Leary, and A. J. Tomlinson, *Inorg. Chem.*, 1972, **11**, 2573.
[260] B. Krebs and J. Mandt, *Z. anorg. Chem.*, 1972, **388**, 193.
[261] E. S. Bretschneider and C. W. Allen, *J. Organometallic Chem.*, 1972, **38**, 43.
[262] R. A. Cardona and E. J. Kupchik, *J. Organometallic Chem.*, 1972, **43**, 163.
[263] J. L. K. F. De Vries and R. H. Herber, *Inorg. Chem.*, 1972, **11**, 2458.

band near 380 cm^{-1}, although consideration of ν(C=S) bands and also of Mössbauer parameters implies unequal interaction between the different S atoms of the ligand and the central tin atom. Partial i.r. data for [(Ph$_F$)$_3$Sn]$_2$E and [(Ph$_F$)$_3$Sn]E(GeEt$_3$) (E = S or Se) and [(Ph$_F$)$_3$Sn]$_2$Te have been reported.[264]

Compounds containing M—Halogen Bonds (M = Si, Sn, or Pb).—An examination of pure molten tin(II) chloride at 250 °C and of SnCl$_2$–KCl melts at several temperatures by Raman spectroscopy has revealed bands at 275 ± 5 (pol), 225 ± 6 (pol), and 110 ± 6 cm^{-1} (dp). Intensity variations of these bands as a function of composition have been interpreted [265] in terms of a complex equilibrium involving polymeric aggregates of SnCl$_2$ molecules and a monomeric species of the type SnCl$_n{}^{2-n}$, the most likely of the latter being SnCl$_3{}^-$. For Sn^{II}ClX (X = Br or I) ν(Sn—Cl) modes have been found [266] at ca. 230 cm^{-1}. Raman intensity studies of liquid silicon-, germanium-, and tin-tetrachlorides have been reported.[267]

The anion SnCl$_5{}^-$ has been obtained as its t-butyltrichlorophosphonium salt, and characterized by means of the data [145a] of Table 25. Si—F

Table 25 *I.r. and Raman wavenumbers*/cm^{-1} *and assignments for* SnCl$_5{}^-$

Assignment	I.r.	Raman
$\nu_5(E')$	350vs	353m
$\nu_1(A'_1)$	340sh	338vs
$\nu_3(A''_2)$	314vs	312w
$\nu_2(A'_1)$	—	269w
$\nu_8(E'')$ (?)	169vs	171w
$\nu_4(A''_2)$ (?)	160vs	163m
$\nu_6(E')$ (?)	150vs	152m
$\nu_7(E')$	66vw	—

stretching frequencies have been located for a variety of salts formed by dehydrofluorination of primary amine–silicon tetrafluoride adducts, some band-listings also being given for the latter.[268] In particular the following data (wavenumber/cm^{-1}) are of interest: SiF$_5{}^-$, 880, 788; Si$_2$F$_{11}{}^{3-}$, 877, 790, 720; Si$_3$F$_{16}{}^{4-}$, 878, 788, 719. Wavenumbers of i.r. bands for SiF$_6{}^{2-}$ have been reported elsewhere,[269] and in addition i.r. absorptions at 564, 503 cm^{-1} are attributed to ν(M—F) for (NH$_4$)$_2$SnF$_6$ and (NH$_4$)$_2$PbF$_6$, respectively.[270]

Aukbe and co-workers have prepared several new fluorotin(IV) derivatives, all found to be polymeric *via* fluorine or fluorosulphonate bridging;

[264] M. N. Bochkarev, N. S. Vyazankin, and L. P. Maiorova, *Doklady Chem.*, 1971, **200**, 830.
[265] E. J. Hathaway and V. A. Maroni, *J. Phys. Chem.*, 1972, **76**, 2796.
[266] S. S. Batsomov, V. F. Lyakhova, and E. M. Moroz, *Russ. J. Inorg. Chem.*, 1971, **16**, 1233.
[267] N. K. Sidorov and L. S. Stal'nakhova, *Optika i Spektroskopiya*, 1972, **32**, 829.
[268] J. J. Harris and B. Rudner, *J. Inorg. Nuclear Chem.*, 1972, **34**, 75.
[269] V. N. Krylov and E. V. Komarov, *Russ. J. Inorg. Chem.*, 1971, **16**, 827.
[270] R. L. Davidovich and T. A. Kaidalova, *Russ. J. Inorg. Chem.*, 1971, **16**, 1354.

wavenumbers for $\nu(\text{Sn}-\text{F})$ bands are shown in Table 26. Among the methyltin fluorides investigated, $\nu(\text{Sn}-\text{F})_{\text{terminal}}$ could be detected only for MeSnF$_3$, for which six-co-ordination about tin was deduced. The observation that $\nu(\text{Sn}-\text{F})_{\text{bridge}}$ moves to higher wavenumber as methyl groups are

Table 26 Sn—F *stretching modes*[a] *(wavenumbers/cm^{-1}) for methyltin fluorides and related compounds*

	$\nu(\text{Sn}-\text{F})_{\text{terminal}}$	$\nu(\text{Sn}-\text{F})_{\text{bridge}}$
Me$_3$SnF	—	335
Me$_2$SnF$_2$	—	360
MeSnF$_3$	644(R)	425
Me$_2$SnClF	—	365
MeSnCl$_2$F	—	398
SnF$_2$(SO$_3$F)$_2$	$\begin{cases} 612(\text{R}, \nu_{\text{sym}}) \\ 691(\nu_{\text{asym}}) \end{cases}$	

[a] I.r. except for (R), Raman.

replaced by Cl or F is explained by a concomitant decrease in Sn—F$_{\text{bridge}}$ bond polarity, and additionally the mutual absence of $\nu_{\text{asym}}/\nu_{\text{sym}}$ from Raman/i.r. spectra in the case of F$_2$Sn(SO$_3$F)$_2$ provides evidence for a linear or near-linear F—Sn—F configuration in this compound.[271]

Other compounds containing M—C or M—H as well as M—halogen bonds have been referred to in foregoing sections.

As usual a number of adducts of Group IV tetrahalides and related compounds (mainly involving six-co-ordination at the metal) have been examined by vibrational spectroscopy, but discussion of reported data is frequently very superficial. Stereochemistries for SiCl$_3$X(py)$_2$ (X = H, F, or Br) and SiCl$_2$X$_2$(py)$_2$ (X = F or Br) have been deduced,[272] using a combination of spectral data and normal-co-ordinate calculations. On formation of adducts with cyclic amines, $\nu(\text{Sn}-\text{Cl})$ for Ph$_3$SnCl, Ph$_2$SnCl$_2$, and Pr^{n}SnCl$_3$ shifts to lower wavenumber.[273] I.r. frequencies have been published for a profusion of complexes Ph$_2$SnCl$_2$,2L, Me$_2$SnCl$_2$,2L, Ph$_2$SnCl$_2$,L, and Me$_2$SnCl$_2$,L [L = Me$_2$SO, Et$_2$SO, Pr$_2$SO, Bu$_2$SO, Ph$_2$SO, (CH$_2$)$_4$SO, pyO, and Ph$_3$PO][274] and Sn(OR)Cl$_3$,L (R = Me, Et, Pr, or C$_2$H$_4$Cl; L = HMPA, Ph$_3$PO, pyNO, or other phosphoryl or amine-oxide ligand).[275] For twenty-five further similar complexes SnX$_4$,L$_2$ (X = Cl, Br, or I) far-i.r. bands have been observed which establish the incidence of two to four $\nu(\text{Sn}-\text{X})$ absorptions for *cis*-isomers, but only one for *trans*-structures.[276] Mainly unassigned vibrational spectra for SnX$_4$,2PhNH$_2$ (X = Cl or Br) indicate bonding to Sn through the carbonyl oxygen

[271] L. E. Levchuk, J. R. Sams, and F. Aubke, *Inorg. Chem.*, 1972, **11**, 43.
[272] D. H. Boal and G. A. Ozin, *Canad. J. Chem.*, 1972, **50**, 2484.
[273] K. L. Jaura and K. K. Sharma, *J. Indian Chem. Soc.*, 1971, **48**, 965.
[274] B. V. Liengme, R. S. Randall, and J. R. Sams, *Canad. J. Chem.*, 1972, **50**, 3212.
[275] R. C. Paul, V. Nagpal, and S. L. Chadha, *Inorg. Chim. Acta*, 1972, **6**, 335.
[276] P. G. Harrison, B. C. Lane, and J. J. Zuckerman, *Inorg. Chem.*, 1972, **11**, 1537.

atom.[277] Other $SnCl_4$ adducts formed by extraction from aqueous solution into various ethers and ketones have also been studied.[278]

Formulation of a product from the reaction of SiF_4 with $(Ph_3P)_3Pt$ as $(Ph_3P)_2Pt,SiF_4$, an acid–base adduct involving five-co-ordinate silicon, relies partly on the observation of $\nu(Si-F)$ modes at relatively low wavenumber (875, 780 cm^{-1}), with $\delta(SiF)$ at 477, 443 cm^{-1}. In the same work, i.r. frequencies for a new SiF_5^- salt, that with Ph_3MeP^+, were listed.[279] The i.r. spectrum of Cl_2Si(phthalocyanine) includes a band at 180 cm^{-1}, attributed to $\delta(SiCl_2)$.[280]

5 Group V Elements

Compounds containing E—H Bonds (E = N or P).—Three shifts at 3267 (dp), 3218 (pol), and 1550 cm^{-1} (pol) in the Raman spectrum of molten $NaNH_2$ at 220 °C can be assigned to the three fundamentals of NH_2^- having C_{2v} symmetry, and a further broad band in the 300—700 cm^{-1} region consisting of several components is considered to arise from hindered rotations.[281] In $Na_2Ga(NH_2)_5$, $\nu(N-H)$ modes have been found [282] at 3260, 3210 cm^{-1} with $\delta(NH)$ at 1580, 1545 cm^{-1}. From measurement of i.r. spectra of NH_4^+ salts in aqueous solution using a multipass Fourier transform system, it is suggested that ν_3 for the ammonium ion is at 3050 cm^{-1}, not 2300 cm^{-1} as accepted previously.[283] Examination of a number of complex ammonium halides by i.r. spectroscopy in the region of fundamental and lattice vibrations of the NH_4^+ ion leads to the proposal [284] that the following can be used to estimate the extent of hydrogen-bonding in these materials: (i) the presence of combination bands involving a torsional mode of the NH_4^+ ion; (ii) the breadth of lattice modes; (iii) the frequency of ν_4 for NH_4^+. The far-i.r. spectra of several tertiary ammonium salts (octyl$_3$NHX, X = Cl, Br, or ClO_4; Et_3NHX, X = Cl or Br) have been studied in the solid state and in solution. The frequencies of two broad bands in the 50—200 cm^{-1} range exhibit anion-dependency and solvent and concentration effects which indicate that the one at higher wavenumber is due to a cation–anion stretching mode, whereas that at lower wavenumber is the $N^+-H\cdots X^-$ bending mode of a hydrogen-bonded ion pair.[285] Changes in fundamental frequencies for liquid ammonia on adding either chlorate or iodide ions have been found to correspond to the expected effect of these solutes on the degree of $N\cdots H$ hydrogen-bonding.[286] Intramolecular H-bonding has been suggested for

---

[277] K. M. Ali, J. Charalambous, and M. J. Frazer, *J.C.S. Dalton*, 1972, 206.
[278] M. J. Taylor, J. R. Milligan, and D. L. Parnell, *J. Inorg. Nuclear Chem.*, 1972, **34**, 2133.
[279] T. R. Durkin and E. P. Schram, *Inorg. Chem.*, 1972, **11**, 1048.
[280] S. C. Mathur, J. Singh, and A. C. Krupnick, *Indian J. Phys.*, 1971–72, **44**, 657.
[281] P. T. Cunningham and V. A. Moroni, *J. Chem. Phys.*, 1972, **57**, 1415.
[282] P. Molinié, R. Brec, and J. Rouxel, *Compt. rend.*, 1972, **274**, C, 1388.
[283] M. J. D. Low and R. T. Young, *Spectroscopy Letters*, 1972, **5**, 245.
[284] J. T. R. Dunsmuir and A. P. Lane, *Spectrochim. Acta*, 1972, **28A**, 45.
[285] J. R. Kludt, G. Y. W. Kwong, and R. L. McDonald, *J. Phys. Chem.*, 1972, **76**, 339.
[286] J. H. Roberts, A. T. Lemley, and J. J. Lagowski, *Spectroscopy Letters*, 1972, **5**, 271.

aminotetrafluorophosphorane, H_2NPF_4, for which gas-phase i.r. features are assigned [287] as follows (wavenumber/cm^{-1}): ν(N—H), 3465, 3575; ν(P—F)$_{axial}$, 867, 977, 1030, 1039, 1058; ν(P—F)$_{equatorial}$, 840, 850, 858. Unassigned i.r. frequencies have been listed [288] for the series of compounds Cl_2FCSNH_2, ClF_2CSNH_2, $(ClF_2CS)_2NH$, $(Cl_2FCS)_2NH$, $CF_3S(ClF_2CS)NH$, $CF_3S(Cl_2FCS)NH$, and $Cl_2FCS(ClF_2CS)NH$.

Reaction of HF and water with organic dichlorophosphines has yielded the hitherto unknown phosphonous acid fluorides, RPHOF, for which some i.r. frequencies and assignments [289] are given in Table 27. For

Table 27 *I.r. wavenumbers/cm^{-1} and assignments for phosphonous acid fluorides, RP(O)HF*

R	Me	Et	Ph
ν(P—H)	2350	2350	2350
δ(PH)	1030	1050	1120
ν(P—F)	870	835	835

$Me_3CP(H)Ph$ and $Me_3MP(H)Ph$ (M = Si or Sn), ν(P—H) modes give rise to i.r. absorptions at 2288, 2290, and 2295 cm^{-1}, respectively.[237]

Compounds containing E—C Bonds (E = N, P, As, Sb, or Bi).—The reaction of F_2PBr with Ag_2CN_2 has been investigated by Rankin, who has shown by gas-phase electron diffraction that the product is a carbodi-imide (41) rather than a cyanamide. From this, assignments for the observed

$$\begin{array}{c} F \\ | \\ F-P \end{array} N-C-N \begin{array}{c} P-F \\ | \\ F \end{array}$$

(41)

i.r. and Raman frequencies have been deduced,[290] some of which are set out in Table 28. The presence of an absorption at 2250 cm^{-1} attributable to

Table 28 *Wavenumbers/cm^{-1} for selected vibrations of bis(difluorophosphino)carbodi-imide*

Assignment	I.r.	Raman
ν_{asym}(NCN)	2199	
ν_{sym}(NCN)	1498	1495
ν(P—F)	{ 858 830	{ 845 816
ν(P—N)	{ 748 	 644
δ(NCN)	{ 570 	 556

[287] A. H. Cowley and J. R. Schweiger, *J.C.S. Chem. Comm.*, 1972, 560.
[288] A. Haas and R. Lorenz, *Chem. Ber.*, 1972, **105**, 273.
[289] U. Ahrens and H. Falius, *Chem. Ber.*, 1972, **105**, 3317.
[290] D. W. H. Rankin, *J.C.S. Dalton*, 1972, 869.

$\nu(C\equiv N)$ in the i.r. spectrum of poly(cyanodithioformic acid), $[NCCS(SH)]_x$, shows that polymerization occurs *via* a C=S grouping.[291] Table 29 lists some other compounds with N—C bonds for which i.r. wavenumbers have been given.[292, 293]

Table 29 *Compounds with* N—C *bonds characterized by i.r. spectroscopy*

Compound	Ref.	Compound	Ref.
$[(CF_3)_2C=NS]_2$		$Me_3SiN=C(CF_3)_2$	
$(CF_3)_2C=NSCl$		$X[N=C(CF_3)_2]_3$ (X = As, P, PCl_2, PO, or B)	
$(CF_3)_2C=NSNMe_2$		$S[N=C(CF_3)_2]_2$	
$[(CF_3)_2C(Cl)N=]_2S$	292	$CF_3SN=C(CF_3)_2$	293
$(CF_3)_2C=NSNH_2$		$CF_3S(O)-N=C(CF_3)_2$	
$(CF_3)_2C=NSCN$		$(CF_3)_2C=NC(CF_3)_2NSO$	
$(CF_3)_2C=NSSMe$		$(CF_3)_2C(F)NCO$	
		$(CF_3)_2NC(CF_3)_2NCO$	

Gas-phase i.r. and liquid Raman spectra for $E(CF_3)_3$ (E = P, As, or Sb) have been measured and assigned. Normal-co-ordinate calculations indicated substantial mixing among fundamentals, but intense Raman shifts in the range 470—270 cm^{-1} were shown to arise mainly from $\nu(E-C)$ modes. Values arrived at for f_{E-C} were 2.86 (E = P), 2.55 (E = As), and 2.01 mdyn Å^{-1} (E = Sb), rather smaller than those for the corresponding trimethyl compounds.[294] I.r. spectra for MePClF and MePBrF have been compared [295] with information already available for $MePX_2$ (X = F, Cl, or Br), leading to the following conclusions (assignment/wavenumber/cm^{-1}): for MePClF, $\nu(P-F)$, 797; $\nu(P-C)$, 701; and $\nu(P-Cl)$, 512 cm^{-1}; for MePBrF, $\nu(P-F)$, 796; $\nu(P-C)$, 700; and $\nu(P-Br)$, 413 cm^{-1}. The action of dimethylzinc on bis(trifluoromethyl)phosphine at ambient temperature affords $[F_3CPCF_2]_2$, evidently the first example of a P—C—P—C heterocycle. I.r. bands observed in the vapour-phase included those at 502, 434, and 371 cm^{-1}, tentatively attributed to $\nu(P-C)_{ring}$ modes.[296] The ylide $Me_3P=CH_2$ reacts with HF to give tetramethylfluorophosphorane, Me_4PF, crystalline samples of which (25 °C) showed no characteristic $\nu(P-F)$ band in either the i.r. or Raman spectrum, suggesting it to be largely ionic under these conditions. A very strong i.r. band with only a weak Raman counterpart at 783 cm^{-1}, and a Raman shift with no coincident i.r. absorption at 660 cm^{-1}, could be assigned to ν_{asym} and $\nu_{sym}(PC_4)$, respectively, supporting this conclusion.[297] Likewise, the spectrum of liquid Me_4POMe is complex [298] but does not rule out an ionic formulation $Me_4P^+OMe^-$.

[291] R. Engler and G. Gattow, *Z. anorg. Chem.*, 1972, **390**, 73.
[292] S. G. Metcalf and J. M. Shreeve, *Inorg. Chem.*, 1972, **11**, 1631.
[293] R. F. Swindell, D. P. Babb, T. J. Ouellette, and J. M. Shreeve, *Inorg. Chem.*, 1972, **11**, 242.
[294] H. Bürger, J. Cichon, J. Grobe, and F. Höfler, *Spectrochim. Acta*, 1972, **28A**, 1275.
[295] H. W. Schiller and R. W. Rudolph, *Inorg. Chem.*, 1972, **11**, 187.
[296] Dae-Ki Kang and A. B. Burg, *J.C.S. Chem. Comm.*, 1972, 763.
[297] H. Schmidbaur, K.-H. Mitschke, and J. Weidlein, *Angew. Chem. Internat. Edn.*, 1972, **11**, 144.
[298] H. Schmidbaur and H. Stühler, *Angew. Chem. Internat. Edn.*, 1972, **11**, 145.

Simon and Schumann have studied a variety of organoarsenic oxide derivatives.[299, 300] For mono-methyl and -ethyl compounds, $\nu(As-C)$ were found in the range 594—629 cm^{-1}, and $R^1AsO(OR^2)_2$ (R^1, R^2 = Me, Et) were deduced to have $CAsO(OC)_2$ skeletons possessing C_s symmetry. Following a detailed vibrational study of pentaphenylarsenic and pentaphenylantimony, it has been proposed that in solution these compounds adopt respectively trigonal-bipyramidal and square-pyramidal configurations.[301] The organoantimony(v) derivatives $Me_3Sb(OCOR)_2$ (R = CF_3, CF_2H, CFH_2, CCl_3, CCl_2H, CBr_2H, $CBrH_2$, Me, CD_3, or CH_2CN) show [302] $\nu_{asym}(SbC_3)$ in the range 578—589 cm^{-1}. Interpretation of the vibrational spectrum of triacetatodimethylantimony(v) as shown in Table 30 leads to

Table 30 Assignment of bands (600—200 cm^{-1})^a for triacetatodimethylantimony (wavenumber/cm^{-1})

Assignment	I.r.	Raman
$\nu_{asym}(Sb-C)$	550	540
$\nu_{sym}(Sb-C)$		526
	511	
$\rho(COO)$	491	
		480
$\nu_{asym}(Sb-O)_{unidentate}$	296	299
$\nu_{sym}(Sb-O)_{unidentate}$		261
$\nu_{asym}(Sb-O)_{bidentate}$	245	
$\nu_{sym}(Sb-O)_{bidentate}$	220	221

^{a} Bands characteristic of $\nu_{asym}(COO)$ for both uni- and bi-dentate CH_3COO also observed at higher wavenumber.

the conclusion that the structure of this compound involves both uni- and bi-dentate acetate groups (42).[303] Unassigned i.r. data have been reported for $(Ph_F)_3Bi$,[304] and for $PhBiX_2,L$ (X = Cl, Br, or I; L = phen or bipy).[305]

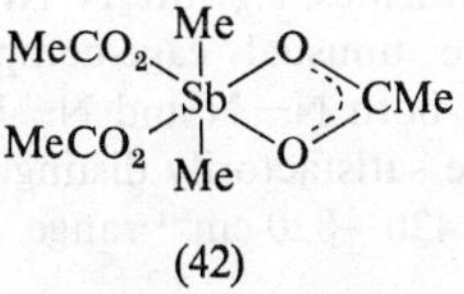

(42)

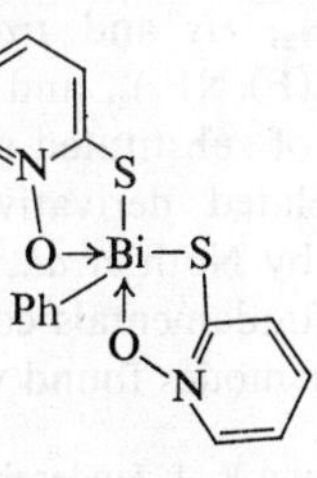

(43)

[299] A. Simon and H.-D. Schumann, Z. anorg. Chem., 1972, **393**, 39.
[300] A. Simon and H.-D. Schumann, Z. anorg. Chem., 1972, **393**, 23.
[301] I. R. Beattie, K. M. S. Livingston, G. A. Ozin, and R. Sabine, J.C.S. Dalton, 1972, 784.
[302] R. G. Goel and D. R. Ridley, J. Organometallic Chem., 1972, **38**, 83.
[303] H. A. Meinema and J. G. Noltes, J. Organometallic Chem., 1972, **36**, 313.
[304] G. B. Deacon and I. K. Johnson, Inorg. Nuclear Chem. Letters, 1972, **8**, 271.
[305] S. Faleschini, P. Zanella, L. Doretti, and G. Faraglia, J. Organometallic Chem., 1972, **44**, 317.

The i.r. spectrum of bis-(1-oxopyridine-2-thiolato)phenylbismuth (43) contains[306] characteristic ν(Bi—C) and ν(Bi—O) bands, respectively, at 442 and 330, 350 cm^{-1}.

Compounds containing N—N or N—P Bonds.—In the i.r. spectrum of $N_2H_5N_3$, ν(N—N) for the hydrazinium ion was observed[307] at 950 cm^{-1}, with azide ν_2 and ν_3 at 635 and 2035 cm^{-1}. Torsional modes have been identified (Table 31) in the far-i.r. spectra of five unsymmetrical disubstituted hydrazines.[308] An assignment has been proposed for the vibrations

Table 31 Wavenumber/cm^{-1} for far-i.r. bands assigned to N—N torsional modes

Compound	τ(N—NH$_2$)	τ(N—ND$_2$)
Me$_2$NNH$_2$	278	
Me$_2$NND$_2$		208
(CH$_2$)$_5$NNH$_2$	282	
(CH$_2$)$_5$NND$_2$		205
O(CH$_2$)$_4$NNH$_2$	286	
(CH$_2$)$_6$NNH$_2$	277	
(CH$_2$)$_6$NND$_2$		209
(CH$_2$)$_3$(CHMe)$_2$NNH$_2$	292	

of the adduct 1,1-dimethylhydrazine-monoborane. Without evidence from a series of isotopic variants this work can only be regarded as highly speculative; however, it is concluded that intense bands at 865 cm^{-1} both in the i.r. and the Raman effect arise from ν(N—N), and that the structure of the adduct is represented by Me$_2$NNH$_2$,BH$_3$, with the borane group attached to the 'NH$_2$' nitrogen atom.[309] Characteristic frequencies for (Me$_3$Si)$_2$N—N=N—NH(SiMe$_3$) have been reported but no assignment was attempted.[310] Other unassigned i.r. data have been offered for[311] [FN=C(F)]$_2$N$_2$, cis and trans-[FN=C(NF$_2$)]N=N[C(F)=NF], (NF$_2$)$_3$-C—N=N—C(F)(NF$_2$)$_2$, and (NF$_2$)$_2$C(F)—N=NC(NH$_2$)=NF, and for a large number of substituted germyl hydrazines, e.g. Me$_2$N·N(R)GeMe$_3$,[312] and other related derivatives.[313] The unusual cage-compounds (44), characterized by Noth et al.,[314] contain both N—N and N—P bonds, but ν(N—N—P) fundamentals could not be satisfactorily distinguished in the i.r. from other modes found within the 420—820 cm^{-1} range.

[306] J. D. Curry and R. J. Jandacek, J.C.S. Dalton, 1972, 1120.
[307] G. Pannetier, F. Margineau, A. Dereign, and R. Bonnaire, Bull. Soc. chim. France, 1972, 2623.
[308] S. M. Craven, F. F. Bentley, and D. F. Pensenstadler, Appl. Spectroscopy, 1972, 26, 646.
[309] J. R. Durig, S. Chatterjee, J. M. Casper, and J. D. Odom, J. Inorg. Nuclear Chem., 1972, 34, 1805.
[310] N. Wiberg and W. Uhlenbrock, Chem. Ber., 1972, 105, 63.
[311] J. B. Hynes, T. E. Austin, and L. A. Bigelow, Inorg. Chem., 1972, 11, 418.
[312] L. K. Peterson and K. I. Thé, Canad. J. Chem., 1972, 50, 553.
[313] L. K. Peterson and K. I. Thé, Canad. J. Chem., 1972, 50, 562.
[314] R. Goetze, H. Nöth, and D. S. Payne, Chem. Ber., 1972, 105, 2637.

Examination of Me_2NPF_2 in the far-i.r. region has established that only one molecular conformation exists in all phases in the temperature range -190 to $+150$ °C. Several low-frequency fundamentals were also located,[315] including a methyl torsion at 189 cm^{-1}. P—N stretching modes for

(44a)

(44b) X = O, S, Se, or BH_3

$(Me_2N)_nPCl_{4-n}{}^+SbCl_6{}^-$ have been observed in the Raman effect at 747 ($n = 1$), 765 and 731 ($n = 2$), and 772 and 678 cm^{-1} ($n = 3$).[316] In an elegant demonstration of the value of simultaneous investigation by electron diffraction and vibrational techniques, Rankin and Cyvin have made assignments for all the fundamentals of F_2PNCO and F_2PNCS, using i.r. and Raman data in association with normal-co-ordinate analysis. Extensive mixing revealed by the latter, particularly for $\nu(P-N)$, was held to account for the large shift from 622 (F_2PNCO) to 714 cm^{-1} (F_2PNCS) of the band nominally ascribed to this vibration, leading on to the conclusion that the P—N stretch cannot be adequately represented by any single normal mode.[317] For the corresponding azide F_2PN_3 and some similar phosphorus(v) species $F_2P(E)N_3$ and $FP(E)(N_3)_2$ (E = O or S), $\nu(P-N)$ were assigned [318] in the range 740—825 cm^{-1}. Low-temperature i.r. and Raman spectroscopy, in conjunction with valence force-field calculations, have been used in a determination of the shapes of the thermally unstable complexes PX_3,NMe_3 (X = Cl or Br) and their [$^2H_9$]trimethylamine isotopomers. Comparison of the results with those for $AsCl_3,NMe_3$ suggests that, like the latter, the P compounds are pseudo-trigonal-bipyramidal with a stereochemically active lone pair of electrons in the equatorial plane. A force-constant of 1.25 mdyn Å^{-1} was estimated for the phosphorus-ligand interaction by analogy with other main-group trimethylamine complexes.[319] Keat has found, by reference to the series of compounds $[Cl_2P(S)]_2NMe$, $Cl_2P(S)NMeP(O)Cl_2$, $Cl_2PNEtP(S)Cl_2$, $[Cl_2P(S)]_2NEt$, $[Cl_2P(S)]_2NPh$, and $Cl_2P(S)NPhP(O)Cl_2$, that as for related dichlorophosphinyl derivatives $\nu_{asym}(P-N-P)$ (in the range 828—918 cm^{-1}) increases in wavenumber on going from $P^{III}-N-P^{III}$ to $P^{III}-N-P^V$ to P^V-N-P^V. The corresponding ν_{sym} modes [with $\nu(P-S)$]

---

[315] J. R. Durig and J. M. Casper, *J. Cryst. Mol. Struct.*, 1972, **2**, 1.
[316] K. Pressl and A. Schmidt, *Chem. Ber.*, 1972, **105**, 3518.
[317] D. W. H. Rankin and S. J. Cyvin, *J.C.S. Dalton*, 1972, 1277.
[318] S. R. O'Neill and J. M. Shreeve, *Inorg. Chem.*, 1972, **11**, 1629.
[319] D. H. Boal and G. A. Ozin, *J.C.S. Dalton*, 1972, 1824.

were detected in the 730—770 cm^{-1} region.[320] The same author has also reported that replacement of Cl by NMe$_2$ on P results in a reduction in strength of bridging P—N bonds, as evidenced by the measurements of Table 32: ν_{asym}(P—N—P) and ν(P=O) move to lower wavenumber,

Table 32 Assignments for selected bands (wavenumber/cm^{-1}) in the i.r. spectra of some compounds with P—N—P linkages

Compound	ν(P=O)	ν(C—N)	ν_{asym}(P—N—P)
[Cl$_2$P(O)]$_2$NMe	1310, 1290	1040	912
Cl$_2$P(O)NMeP(O)ClNMe$_2$	1294, 1253	1055	901
($\pm$ meso)-Cl(Me$_2$N)P(O)NMeP(O)ClNMe$_2$	1250	1060	890
Cl$_2$P(O)NMeP(O)(NMe$_2$)$_2$	1310, 1200 (?)	1060	918
Cl(Me$_2$N)P(O)NMeP(O)(NMe$_2$)$_2$	1250	1063	888
[(Me$_2$N)$_2$P(O)]$_2$NMe	1236	1063	887

while ν(C—N) increases.[321] P—N and P=N stretching modes (respectively at 1418—1430 and 757—959 cm^{-1}) have been given for a series of phosphoric acids substituted with ureas and related ligands.[322] Careful reaction of 'phospham' [phosphornitridimide, (PN$_2$H)$_n$] with KNH$_2$ gives the salt (KPN$_2$)$_n$. Some i.r. data for reactant and product are shown in Table 33.

Table 33 I.r. wavenumbers/cm^{-1} and assignments for (PN$_2$H)$_n$ and derived anion

Assignment	(PN$_2$H)$_n$	(PN$_2$K)$_n$
ν(N—H)	3110, 2700	
ν_{asym}(P=N—P)	1250	
ν_{asym}(PN$_2^-$)$_n$		1175, 1085
ν_{asym}(P—NH—P)	950	
ν_{sym}(PN$_2^-$)$_n$		860, 802
δ(PN$_2^-$)$_n$		512, 470
δ(P=N—P)	480	

The vibrational frequencies for modes of the (PN$_2^-$)$_n$ ion were also compared [323] with those for isoelectronic (SiO$_2$)$_n$ [for example ν(PN$_2^-$)$_n$ at 1085, ν(SiO$_2$)$_n$ at 1078 cm^{-1}]. References to other acyclic P—N derivatives giving mostly unassigned i.r. frequencies are collected in Table 34.

As in previous years, much journal space has been occupied by vibrational data for cyclic compounds with P—N bonds (especially phosphonitrilic derivatives) but the majority of the papers involved simply catalogue i.r. frequencies and include few worthwhile assignments. Appropriate references are contained in Table 35. Elsewhere, i.r. spectra for P$_3$N$_3$Cl$_6$,

---

[320] R. Keat, J.C.S. Dalton, 1972, 2189.
[321] I. Irvine and R. Keat, J.C.S. Dalton, 1972, 17.
[322] F. Markalous, J. Zerman, J. Beránek, M. Černík, and J. Toužín, Coll. Czech. Chem. Comm., 1972, **37**, 725.
[323] J. Goubeau and R. Pantzer, Z. anorg. Chem., 1972, **390**, 25.

Table 34 Compounds with P—N bonds characterized by i.r. spectroscopy

Compound	Ref.
$RCON=PF_2Ph$ (R = CF_3 or C_3F_7) $CF_3CON(NMe_2)PF_3Ph$	324
$RSP(X)_2=NP(X_2)=O$ (R = Me or Et; X = F and/or Cl) $SPXNHP(O)X_2$ $SPX_2NHC(O)Me$ (X = F and/or Cl)	325
$O=PF_2N=PCl_2N=PX_3$ (X_3 = Cl_3, Cl_2NMe_2, or Cl_2NEt_2) $O=PF_2N=PCl_2NR_2$ (R = Me or Et, or R_2 = $MeSiMe_3$) $O=PF_2N=PCl_2NCS$ $O=PFClN=PCl_2NR_2$ (R = Me or Et) $O=PFClN=PClN(NEt_2)_2$ $O=PPhFN=PCl_3$	326
$S=PX_2NMeSCCl_3$ $O=PX_2NMeSCCl_3$ (X = F or Cl, or X_2 = ClF) $FSO_2NMeSCCl_3$ $CF_3SO_2NMeSCCl_3$	327
$[PhCH_2P(NH_2)(C_2H_4)_2P(NH_2)CH_2Ph]Cl$ $[PhCH_2P(C_2H_4)_3PNH_2]Cl_2$ $[PhP(NH_2)(PhCCH)_2]Cl$	328
$S=P(Cl_2)N=P(Cl_2)N=P(Cl_2)NRSiMe_3$ (R = H or Me) $S=P(Cl_2)N=P(Cl_2)N=P(Cl)_2X$ (X = Cl or $NHSiMe_3$)	329

Table 35 References to published vibrational data for cyclic P—N compounds

Compound	Ref.
Cl_2P–N=PCl_2 / N–$P(S)(Cl)$–NMe (cyclic)	329
cis-$P_3N_3Ph_4(H)NEt_2$	330
$P_3N_3(OPhNO_2)_6$	331
cis-$P_3N_3F_4(NMe_2)_2$ trans-$P_3N_3F_4(NMe_2)_2$	332
R^1_2P–N=PR^2_2 / N–$P(R^3)(H)$–N (cyclic) (R^1, R^2, R^3 = Me, Ph, OPh, or NMe_2: seven compounds)	333

[324] G. Czieslik and O. Glemser, *Z. anorg. Chem.*, 1972, **394**, 26.
[325] H. W. Roesky, B. H. Kuhtz, and L. F. Grimm, *Z. anorg. Chem.*, 1972, **389**, 167.
[326] H. W. Roesky and W. Kloker, *Z. Naturforsch.*, 1972, **27b**, 486.
[327] H. W. Roesky, W. Schaper, and S. Tutkunkardes, *Z. Naturforsch.*, 1972, **27b**, 620.
[328] S. E. Frazier and H. H. Sisler, *Inorg. Chem.*, 1972, **11**, 1431.
[329] H. W. Roesky, *Chem. Ber.*, 1972, **105**, 1439.
[330] M. Bermann and J. R. Van Wazer, *Inorg. Chem.*, 1972, **11**, 209.
[331] H. R. Allcock and E. J. Walsh, *J. Amer. Chem. Soc.*, 1972, **94**, 4538.
[332] E. Niecke, H. Thamm, and D. Böhler, *Inorg. Nuclear Chem. Letters*, 1972, **8**, 261.
[333] A. Schmidpeter, J. Ebeling, H. Stary, and C. Weingand, *Z. anorg. Chem.*, 1972, **394**, 171.

Table 35 (cont.)

Compound	Ref.
$P_nN_nX_{2n-2x}\begin{bmatrix}-NMe\\ \diagdown\\ (CH_2)_2\\ \diagup\\ -NMe\end{bmatrix}_x$ ($n = 3$ or 4; $X =$ Cl or F; $x = 1$ or 2)	334
$P_4N_4Cl_n(NCS)_{8-n}$ (mixture of n)	335
$(MeNPF_3)_4$	
$[(MeN)_4P_3F_6]^+PF_6^-$	336
$(MeN)_4P_3F_7$	
$[P_3N_3F_5NH]_3B$ [also $(S{=}PF_2NH)_3B$]	337
$Cl_2P\overset{N}{\underset{N\diagdown S\diagup N}{\diagup}}PCl_2$ $\;$ Cl	338
$Cl_2P\overset{N}{\underset{N\diagdown S\diagup N}{\diagup}}PCl_2$ $\;\overset{\|}{\underset{O}{S}}$—Cl	339
$O_2S[NPCl_2NHR]_2$ ($R =$ Et, Pr, or Bu)	
$O_2S[NPCl_2]_2NR$ ($R =$ Pr or Bu)	340
$O_2S[NPClNEt_2]_2NMe$	

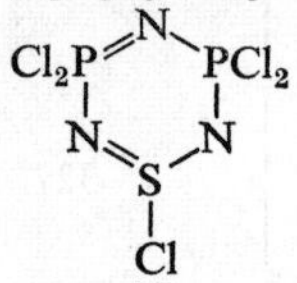

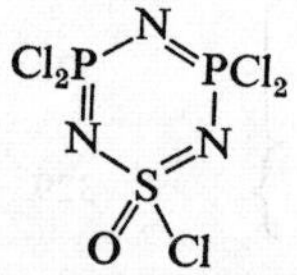

α-(NSOCl)$_3$ (45), and the related mixed-ring systems PNCl$_2$(NSOCl)$_2$ and (PNCl$_2$)$_2$NSOCl have been correlated.[341] For (46), ν(P=N) and ν(P—N) were assigned[342] respectively at 1230/1120 and 855 cm^{-1}. Isomeric phos-

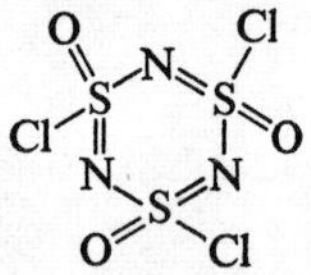

(45) 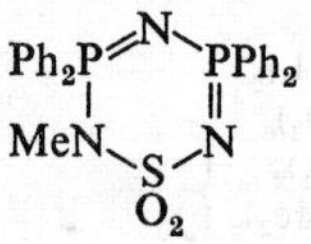   (46)

phonitriles P$_4$N$_4$Cl$_n$(NMe$_2$)$_{8-n}$ have been found to exhibit little change in ν(NC$_2$) with isomer-type (unlike corresponding trimeric species) but the presence of *gem*- or non-*gem*-groups can be distinguished by examination

[334] T. Chivers and R. Hedgeland, *Canad. J. Chem.*, 1972, **50**, 1017.
[335] R. L. Dieck and T. Moeller, *Inorg. Nuclear Chem. Letters*, 1972, **8**, 763.
[336] K. Utrory and W. Czysch, *Monatsh.*, 1972, **103**, 1048.
[337] H. W. Roesky, *Chem. Ber.*, 1972, **105**, 1726.
[338] H. W. Roesky, *Angew. Chem. Internat. Edn.*, 1972, **11**, 642.
[339] U. Klingebiel and O. Glemser, *Z. Naturforsch.*, 1972, **27b**, 467.
[340] U. Klingebiel and O. Glemser, *Chem. Ber.*, 1972, **105**, 1510.
[341] H. H. Baalmann, H. P. Velvis, and J. C. van de Grampel, *Rec. Trav. chim.*, 1972, **91**, 935.
[342] M. Bermann and J. R. Van Wazer, *Inorg. Chem.*, 1972, **11**, 2515.

of $\nu(P-N)$ modes. Thus *gem*-amine groups give two widely separated i.r. bands, whereas when only non-*gem*-groups are present smaller splittings or single bands are observed.[343] Reaction of $P_4N_4Me_8$ or $P_5N_5Me_{10}$ with $M(CO)_6$ (M = Mo or W) yields (ring)-tricarbonyl-metal derivatives, but only $\nu(CO)$ wavenumbers have been reported.[344] Polymeric bis(amino)-phosphazenes $[NP(NHR)_2]_n$ (R = Me, Pr, Bu, or CH_2CF_3) have been prepared[345] and in each case $\nu(P=N)$ lies between 1320 and 1100 cm^{-1}.

Compounds containing As—N, Sb—N, or P—P Bonds.—The vibrational spectra of a number of organoarsenic azides $R^1{}_2AsN_3$, $R^2R^3AsN_3$, or $R(X)AsN_3$ (R^1, R^2, R^3 = Me, Et, Ph; X = Cl or Br) and of $(R_2N)PhAsN_3$ (R = Me or Et) have been reported by Revitt and Sowerby;[346] $\nu(As-N_3)$ modes are assigned in the 400—450 cm^{-1} region, and in the amino-derivatives $\nu(As-N_{amino})$ are at 580—600 cm^{-1}. Rotational isomerism in several of these compounds was deduced from the distribution of bands attributable to $\nu(As-C)$ fundamentals. For the related amino-derivatives R_2AsNEt_2 (R = alkyl) and Cl_2AsNEt_2, $\nu(As-N)$ occurs[347] at 545—610 cm^{-1}. Unassigned i.r. wavenumbers have been listed for the following As—N bonded compounds:[348] $[R_3AsNH_2]Cl$ (R = Et or Pr^n); $[R_3AsNMe_2]Cl$ (R = Me, Et, or Pr^n); and

$$\left[HN \begin{array}{c} C_6H_4 \\ \diagdown \\ C_6H_4 \end{array} As \begin{array}{c} Y \\ \diagup \\ NH_2 \end{array} \right]^+ Cl^-,$$

Y = Me or Cl. Wavenumbers for $\nu(Sb-N)$ fundamentals for several tris(dialkylamino)antimony(III) species have been reported[349] and are shown in Table 36.

Table 36 Sb—N *stretching wavenumbers*/cm^{-1} *for tris(dialkylamino)-antimony compounds*

R in Sb(NR$_2$)$_3$	Me	Et	Pr	Bu
ν_{asym}(Sb—N)	520	570	548	598
ν_{sym}(Sb—N)	—	460	495	535

The four-co-ordinate phosphorus anions $Na_3(P_2O_5F),12H_2O$ and $K_2(P_2O_4F_2)$ have been obtained from red phosphorus and H_2O_2 in the presence of fluoride ions. Assignments for vibrations of the second of these are based on the point group C_{2v}, with $\nu(P-P)$ at 303 cm^{-1} in the Raman effect.[349a] A similar shift at 340 cm^{-1} for $[Cl_2P(O)PPh_2(CH_2)_2PPh_2P(O)$-

---

[343] D. Millington and D. B. Sowerby, *J.C.S. Dalton*, 1972, 2035.
[344] N. L. Paddock, T. N. Ranganathan, and J. N. Wingfield, *J.C.S. Dalton*, 1972, 1578.
[345] H. R. Allcock, W. J. Cook, and D. P. Mack, *Inorg. Chem.*, 1972, **11**, 2584.
[346] D. M. Revitt and D. B. Sowerby, *J.C.S. Dalton*, 1972, 847.
[347] L. S. Sagam, R. A. Zurgaro, and K. J. Irgdlic, *J. Organometallic Chem.*, 1972, **39**, 301.
[348] L. K. Kranuich and H. H. Sisler, *Inorg. Chem.*, 1972, **11**, 1226.
[349] A. Kiennemann, G. Levy, F. Schué, and C. Taniélian, *J. Organometallic Chem.*, 1972, **35**, 143.
[349a] H. Falius, *Z. anorg. Chem.*, 1972, **394**, 217.

$Cl_2]Cl_2$ has been assigned to $\nu(P-P)$, and in the same paper [350] various $\nu(P=O)$, $\nu(P=S)$, and $\nu(As=S)$ frequencies were also given.

Compounds containing E—O Bonds (E = N, P, As, Sb, or Bi).—Reaction of $AlBr_3$ with excess N_2O_5 has been reported to give $NO_2[Al(NO_3)_4]$, with $\nu(NO_2^+)$ at 2380 and $\delta(NO_2^+)$ at 575 cm^{-1} in the i.r. The anion apparently possesses D_{2d} symmetry.[351] Two papers concerning the Raman spectra of polycrystalline [352] and molten [353] Ag^I and Tl^I nitrates have been published, together with some results for molten MNO_3–KNO_3 eutectics (M = Li or Na).[353a]

I.r. and Raman spectra for crystalline $POBr_3$ suggest [354] that its crystal structure belongs to the space group D_{2h}^{16}. The i.r. spectra of $[Co(en)_3]$-SPO_3, $[Co(en)_3]S_2PO_2$, and $[Co(en)_3]S_3PO$ have been compared with those for corresponding sodium salts.[355] This has allowed assignment of the fundamentals for the thiophosphate anions in the way shown in Table 37.

Table 37 Wavenumbers/cm^{-1} for some fundamental vibrations of thiophosphate anions

SPO_3^{3-}		$S_2PO_2^{3-}$		S_3PO^{3-}	
$\nu_{asym}(P-O)$	1005	$\nu_{asym}(P-O)$	1040	$\nu(P-O)$	1020
$\nu_{sym}(P-O)$	935	$\nu_{sym}(P-O)$	965	$\nu_{asym}(P-S)$	560
$\delta_{asym}(PO_3)$	605	$\delta(PO_2)$	585	$\nu_{sym}(P-S)$	465
$\delta_{sym}(PO_3)$	470	$\nu_{asym}(P-S)$	565	$\delta(OPS)$	360
$\nu(P-S)$	425	$\nu_{sym}(P-S)$	420	$\delta_{asym}(PS_3)$	310
$\delta(OPS)$	320	$\rho(PO_2)$	400	$\delta_{sym}(PS_3)$	230
		$\delta(PS_2)$	266		

Some new i.r. data have been obtained [356] for $(CF_3)_2P(O)Br$, for which $\nu(P=O)$ and $\delta(P=O)$ are at 1320 and 502 cm^{-1}. Goubeau et al. have reported i.r. and Raman wavenumbers and simplified valence force field calculations for each of $P(O)F_2Me$, $P(O)FMe_2$, $P(S)F_2Me$, $P(S)FMe_2$, and $P(S)F_2Et$. All force constants were found to decrease on substitution of F by Me, one suggestion being that this is due to Me—P inductive effects; $\nu(P=O)$ were found at 1334 and 1255 cm^{-1}, respectively, for the di- and mono-fluorophosphine oxides, and $\nu(P=S)$ at 640 and 602 cm^{-1} in the related thio-derivatives.[357] Oxo-2-dimethyl-5,5-dioxaphosphorinanes have been shown to exist as mixtures of conformational isomers (47a and b)

---

[350] E. Lindner and H. Beer, Chem. Ber., 1972, **105**, 3261.
[351] G. N. Shirokova and V. Ya. Rosolovskii, Russ. J. Inorg. Chem., 1971, **16**, 1699.
[352] K. Balasubrahmanyam and G. J. Janz, J. Chem. Phys., 1972, **57**, 4084.
[353] K. Balasubrahmanyam and G. J. Janz, J. Chem. Phys., 1972, **57**, 4089.
[353a] S. V. Vokov, N. P. Evtushenko, and S. P. Baranov, Teor. i eksp. Khim., 1972, **8**, 124.
[354] E. Huler, A. Burgos, E. Silberman, and H. W. Morgan, J. Mol. Structure, 1972, **12**, 121.
[355] D. B. Powell and J. G. V. Scott, Spectrochim. Acta, 1972, **28A**, 1067.
[356] R. G. Cavell, R. D. Leary, and A. J. Tomlinson, Inorg. Chem., 1972, **11**, 2578.
[357] D. Köttgen, H. Stole, R. Pantzer, A. Lentz, and J. Goubeau, Z. anorg. Chem., 1972, **389**, 269.

(47a) ⇌ (47b)

with two bands observed [358] due to $\nu(P \to O)$, at 1280—1320 (νPO_{eq}) and 1260—1300 cm^{-1} (νPO_{ax}). I.r. bands characteristic of the P—O—C group of (48), identified [359] at 1040, 1015, and 920 cm^{-1}, are shifted to 1070, 1030,

$$\begin{array}{c} H_2CO \\ | \quad PNHNMe_2 \\ H_2CO \end{array}$$

(48)

and 950 cm^{-1} in the chloramine adduct ($-OCH_2-CH_2O-$)PNHNMe$_2$,-NH$_2$Cl. I.r. spectral data for the following have also been published:[360] $R_nP(OCOCF_3)_{3-n}$ (n = 1 or 2), PhP(O)(COCF$_3$)$_2$, R(CF$_3$)P(O)OP(O)-(CF$_3$)R, R$_2$P(O)OPR$_2$(CF$_3$)$_2$ (R = Et or Ph throughout), Et$_2$P(O)OPEt$_2$-(COCF$_3$)$_2$, and Et(COCF$_3$)P(O)OP(O)(COCF$_3$)Et; and also [360a] Cl$_2$P(O)-NHCOCl$_3$ and Cl$_2$P(O)NHCO$_2$Me.

The thermal decomposition [361] of PbHPO$_4$ and Pb(H$_2$PO$_4$)$_2$, and the formation of complexes between TiCl$_4$ and trialkylphosphates,[362] have each been studied using vibrational spectroscopy. A number of phosphates and related anions have been investigated similarly,[363-367] indicating symmetries of C_{2v} (eclipsed) and C_{2h} (staggered), respectively,[363] for Na and K salts of H$_2$P$_2$O$_6^{2-}$, of C_{2v} for P$_2$O$_7$ units in α-pyrophosphates,[364] and of C_{3v} for certain *ortho*-phosphates and related oxo-anions.[366]

Determination of the crystal structures of E$_2$F$_{10}$O^{2-} (E = As or Sb) As$_2$F$_8$O$_2^{2-}$, and Sb$_3$F$_{12}$O$_3^{3-}$ has allowed some spectroscopic interpretations to be made which lead to the following values for stretching force constants, using a simplified model:[368] f_{As-O} = 5.1 (As—O—As bridge), f_{Sb-O} = 4.55 (Sb—O—Sb bridge or Sb$_3$O$_3$ ring), and f_{As-O} = 3.8 mdyn Å^{-1} (As$_2$O$_2$ ring). I.r. spectra for the arsonic acids RAsO(OH)$_2$ (R = Bu,

---

[358] J. P. Majoral, R. Pujol, and J. Navech, *Bull. Soc. chim. France*, 1972, 606.
[359] S. E. Frazier and H. H. Sisler, *Inorg. Chem.*, 1972, **11**, 1223.
[360] P. Sartori and M. Thomzik, *Z. anorg. Chem.*, 1972, **394**, 157.
[360a] Yu. A. Nuzhdina and Yu. P. Egorov, *Zhur. strukt. Khim.*, 1972, **13**, 72.
[361] M. V. Goloshchapov, Yu. T. Tovgashin, and B. V. Martgrienko, *Russ. J. Inorg. Chem.*, 1971, **16**, 1485.
[362] G. Roland, B. Gilbert, and G. Duyckaerts, *Spectrochim. Acta*, 1972, **28A**, 835.
[363] P. Klíma, J. Stejskal, and B. Hájek, *Spectrochim. Acta*, 1972, **28A**, 1909.
[364] J. Hanuza, B. Jeżowska-Trzebiatowska, and K. Kukaszewicz, *J. Mol. Structure*, 1972, **13**, 391.
[365] P. Rémy, J. Fraissard, and A. L. Boullé, *Bull. Soc. chim. France*, 1972, 2222.
[366] P. Tarte and J. Thelen, *Spectrochim. Acta*, 1972, **28A**, 5.
[367] G. Brun and G. Jourdan, *Compt. rend.*, 1972, **275**, C, 821.
[368] W. Haase, *Ber. Bunsengesellschaft phys. Chem.*, 1972, **76**, 1000.

hexyl, heptyl, octyl, decyl, Ph, o-MeC$_6$H$_4$, p-MeC$_6$H$_4$, p-EtC$_6$H$_4$, or p-biPh) have been recorded.[369] The internal modes of AsO$_4$$^{3-}$ in KH$_2$AsO$_4$ have been assigned in terms of S_4 site symmetry.[370]

Factor-group splitting has been found to be slight for solid SbOCl, explaining observation of a fewer-than-predicted number of vibrational bands, but the spectra are in any case quite complicated owing to the complexity of the Sb$_6$O$_6$Cl$_4$ repeat unit.[371] Bridging Sb—O—Sb stretches span the range 727—360 cm^{-1}. Poly(triphenylstibine oxide) has been compared spectroscopically with Ph$_3$Sb, leading to the assignment of ν_{asym} and ν_{sym}(Sb—O) at 744 and 669 cm^{-1}, respectively.[372] A study of the hydrolysis of the SbF$_6$$^-$ ion in aqueous HF confirms [372a] the primary product to be SbF$_5$(OH)$^-$, with ν(Sb—O) at 692 cm^{-1}. Some other antimonates have also been studied.[372b] By contrast, ν(Sb—O) has been located [373] between 330 and 422 cm^{-1} for a series of oxybis(triorganoantimony)-diperchlorates, all of which contain the five-co-ordinate cations [(R$_3$SbL)$_2$-O]$^{2+}$ (L = DMSO, DPSO, py N-oxide, Ph$_3$PO, or Ph$_3$AsO). In [(Ph$_3$Bi)$_2$O]X$_2$ (X = ClO$_4$, NO$_3$, Cl, Br, NCO, or CF$_3$COO), ν_{asym}(Bi—O—Bi) occurs at 620—632 cm^{-1}, with the symmetric stretch between 382 and 338 cm^{-1}; i.r. spectroscopy indicates covalent binding of the anions.[374] Approximate assignments for Bi(PO$_2$Cl$_2$)$_3$,POCl$_3$ include that of i.r. and Raman bands at 430, 410, 400, and 380 cm^{-1} to ν(Bi—O) modes.[375] I.r. spectroscopy has been used in a study of extraction of Bi^{III} from bromide solutions using di-2-ethylhexylphosphoric acid.[376]

Compounds containing N—S or P—S Bonds.—Formation of the novel thiocyanatidate linkage has been achieved, in (Me$_3$CCH$_2$O)$_2$P(O)SC≡N which is isolable at −5 °C, when ν(SC≡N) at 2170 cm^{-1} is observed;[377] at room temperature, rapid isomerization to (Me$_3$CCH$_2$O)$_2$P(O)N=C=S occurs, with (N=C=S) at 2010 cm^{-1}. Wavenumbers for ν(S≡N), in the region of 1480 cm^{-1}, have been reported for N≡SF$_2$—NMe$_2$ and related species.[378] Bis(sulphinylamino)sulphide, S(NSO)$_2$, exhibits i.r. absorptions at 1184 and 988 cm^{-1} due to ν_{asym} and ν_{sym}, respectively, of the NSO group.[379] For the similar but structurally distinct sulphur(VI) species

---

[369] U. Dietze, *J. prakt. Chem.*, 1971, **313**, 889.
[370] H. Ratajczak, *J. Mol. Structure*, 1972, **11**, 267
[371] K. I. Petrov, V. V. Fomichev, G. V. Zimina, and V. E. Plyushchev, *Russ. J. Inorg. Chem.*, 1971, **16**, 1006.
[372] D. L. Venezky, C. W. Sink, B. A. Nevett, and W. F. Fortescue, *J. Organometallic Chem.*, 1972, **35**, 131.
[372a] J. E. Griffiths and G. E. Walrafn, *Inorg. Chem.*, 1972, **11**, 427.
[372b] I. N. Lisichkin, A. V. Kerimbekov, N. A. Kerimbekova, and O. Ya. Manashirov, *Russ. J. Inorg. Chem.*, 1971, **16**, 1138.
[373] R. G. Goel and H. S. Prasad, *Inorg. Chem.*, 1972, **11**, 2141.
[374] R. G. Goel and H. S. Prasad, *J. Organometallic Chem.*, 1972, **36**, 323.
[375] A. Klopsch and K. Dehnicke, *Z. Naturforsch.*, 1972, **27b**, 1304.
[376] I. S. Levin, Yu. M. Yukhin, and I. A. Vorsina, *Russ. J. Inorg. Chem.*, 1971, **16**, 1191.
[377] A. Lopusínski and J. Michalski, *Angew. Chem. Internat. Edn.*, 1972, **11**, 838.
[378] O. Glemser, J. Wegener, and R. Höfer, *Chem. Ber.*, 1972, **105**, 474.
[379] D. A. Armitage and A. W. Sinden, *Inorg. Chem.*, 1972, **11**, 1151.

$Me_3SiN=S(O)=NSiMe_3$ and $Me_3SiN=S(F_2)O$, corresponding $\nu(NSO)$ modes have been assigned at 1230 (asym) and 1132 (sym), and 1486 (asym) and 1261 cm^{-1} (sym), respectively.[380] $S=N$ stretching has been attributed to *three* i.r. bands (at 1340, 1309, and 1281 cm^{-1}) observed [381] for $F_2S=NC-(O)NCO$, and $\nu(N=S=N)$ is at 1130 cm^{-1} for $Ph_FSN=S=NSPh_F$.[382]

Two publications concern derivatives of tetrasulphur tetranitride. In one, the preparation of the first oxide derivative, (49), is described giving $\nu_{sym}(S=O)$ at 1420 and 1400 cm^{-1} as well as some other (unassigned) i.r. frequencies.[383] The second [384] reports the formation of adducts S_4N_4-(olefin)$_2$ [olefin = norbornene, nbd, $(C_5H_6)_2$, 5-norborneol, or 5-methylene-norbornene] where a shift in $\nu(S-N)$ from 705 to 690 cm^{-1} on adduction is consistent with a weakening of S—N bonds; similarly, $\delta(SSN)$

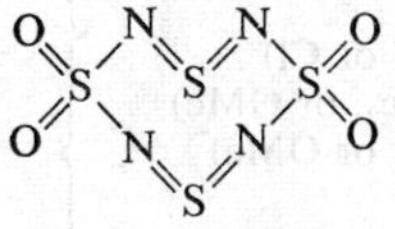

(49) 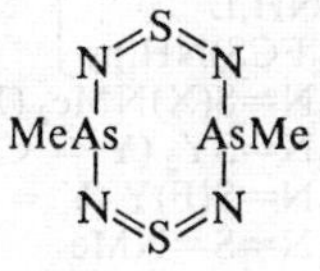   (50)

moves from 341 to 327 cm^{-1}, and this is rationalized by assuming some loss of the rigid cage character of S_4N_4. Ring $\nu(N=S=N)$ modes occur [385] at 1050 and 930 cm^{-1} in the i.r. spectrum of (50); and $\nu(N-S-N)$ are at 797 (asym) and 634 cm^{-1} (sym) for sulphinylbis-(1-aziridine).[386] References to other compounds with S—N bonds, most of them containing only unassigned i.r. listings, are collected in Table 38.

Table 38 *References to compounds with* S—N *Bonds*

Compound	Ref.
NF_2SO_3F	387
$(CF_3)_2S(O)NH$	
$(CF_3)(C_2F_5)S(O)NH$	388
$(C_2F_5)_2S(O)NH$	
$RSSN=CCl_2$ (R = CF_3, CF_2Cl, $CFCl_2$, or CCl_3)	
$CF_3SN=CCIY$ (Y = SCl, F, or $SSCF_3$)	
$CF_2ClSN=C(Cl)SSCF_2Cl$	389
$F_3CN(SCF_3)_2$	
$F_3CSN=SOF_2$	

[380] O. Glemser, M. F. Feser, S. P. von Halasz, and H. Saran, *Inorg. Nuclear Chem. Letters*, 1972, **8**, 321.
[381] A. Clifford, J. S. Harman, and C. A. McAuliffe, *Inorg. Nuclear Chem. Letters*, 1972, **8**, 567.
[382] A. Golloch and M. Kuss, *Z. Naturforsch.*, 1972, **27b**, 1280.
[383] H. W. Roesky and O. Petersen, *Angew. Chem. Internat. Edn.*, 1972, **11**, 918.
[384] M. R. Brinkman and C. W. Allen, *J. Amer. Chem. Soc.*, 1972, **94**, 1550.
[385] O. J. Scherer and R. Wies, *Angew. Chem. Internat. Edn.*, 1972, **11**, 529.
[386] H. L. Spell and J. Laane, *Appl. Spectroscopy*, 1972, **26**, 86.
[387] A. M. Qureshi, *Pakistan J. Sci. Res.*, 1970, **22**, 164.
[388] D. T. Sauer and J. M. Shreeve, *Inorg. Chem.*, 1972, **11**, 238.
[389] P. Gielow and A. Haas, *Z. anorg. Chem.*, 1972, **394**, 53.

Table 38 (cont.)

Compound	Ref.
$(CF_3)_2CFN=S=NC(CF_3)_2N=C(CF_3)_2$	390, 391
$(CF_3)_2CFN=S=NCF(CF_3)_2$	
$[(CF_3)_2C=NC(CF_3)_2N=]_2S$	
$(CF_3)_2C=NS-N=C(CF_3)_2$ $\overset{\|}{\underset{}{NH}}$	391
$CF_3S(F)=NCF(CF_3)_2$	
$(CF_3)_2S=NCF(CF_3)_2$	
$(CF_3)_2S=NC(CF_3)_2N=C(CF_3)_2$	
Cl_3CSNH_2	
$(F_3CS)_2NH,L$	
$(F_3CS)(ClF_2CS)NH,L$ $\quad$ (L = NMe_3 or py)	392
$(ClF_2CS)_2NH,L$	
$(F_3CS)(Cl_2FCS)NH,py$	
$CF_2ClCF_2N=S(X)NMe_2$ (X = F or Cl)	
$CF_2ClCF_2N=SY_2$ (Y = Cl, NMe_2, or OMe)	
$CFCl_2CF_2N=S(F)Y$ (Y = NMe_2 or OMe)	393
$CF_2ClCF_2N=S=NMe$	
$CF_2ClCF_2N=S$	
$S_3N_2Cl^+MCl_4^-$	
$S_4N_3^+MCl_4^-$ $\quad$ (M = Fe or Al)	394
$S_5N_5^+MCl_4^-$	
cyclo-$[(NSOCl)_2(NSOF)]$ $\quad$ (isomer mixtures)	395
cyclo-$[(NSOCl)(NSOF)_2]$	
![structure] N=S-N / S-N-S six-membered ring with $PF_2=O$ substituent	396

For $(CF_3)_2P(S)OH$, $\nu(P=S)$ has been observed in the i.r. at 773 cm^{-1}. The absence of a corresponding band in the spectrum of the O-deuteriated molecule was explained in terms of a Fermi resonance involving $\delta(POD)$, giving rise with $\nu(P=S)$ to a strong absorption at 810 cm^{-1}. In the related oxide $[(CF_3)_2P(S)]_2O$, $\nu(P=S)$ occurred at 803 cm^{-1}, and $\delta(P=S)$ was identified at 414 cm^{-1} for all three compounds.[397] Splitting of several fundamentals, notably $\nu(P=S)$, at low temperature in the Raman spectra of liquid C^{n}H$_3$OPSClF (n = 1 or 2) has been ascribed to the incidence of two

[390] R. F. Swindell and J. M. Shreeve, *Inorg. Nuclear Chem. Letters*, 1972, **8**, 759.
[391] R. F. Swindell and J. M. Shreeve, *J. Amer. Chem. Soc.*, 1972, **94**, 5713.
[392] A. Haas and R. Lorenz, *Chem. Ber.*, 1972, **105**, 3161.
[393] R. Mews and O. Glemser, *Inorg. Chem.*, 1972, **11**, 2521.
[394] A. J. Banister and P. J. Dainty, *J.C.S. Dalton*, 1972, 2658.
[395] T. P. Lin, U. Klingebiel, and O. Glemser, *Angew. Chem. Internat. Edn.*, 1972, **11**, 1095.
[396] H. W. Roesky and L. F. Grimm, *Angew. Chem. Internat. Edn.*, 1972, **11**, 642.
[397] A. A. Pinkerton and R. G. Cavell, *J. Amer. Chem. Soc.*, 1972, **94**, 1870.

isomeric configurations.[398] A fairly detailed spectroscopic study of the anions $OSPF_2^-$, $S_2PF_2^-$, S_2PFMe^-, and $S_2P(CN)_2^-$ resulted [399] in assignments for $\nu(PS_2)$ modes in the ranges 630—665 (asym) and 529—573 cm^{-1} (sym). I.r. and Raman wavenumbers have also been given [400] for Me_2PSMe and the cationic species $[Me_3PSMe]X$ (X = Cl or I).

Other Compounds containing a Group V Element Bonded to a Group VI Element.—The Raman spectrum as well as crystal data for $\alpha\text{-}As_4S_4$ suggests that it is identical with the mineral realgar, whereas a distinguishable Raman spectrum has been found for $\beta\text{-}As_4S_4$. Spectral changes for both α- and β-forms during prolonged laser illumination suggest formation of polymeric products.[401]

Gas-phase laser Raman spectra of As–S mixed vapours at temperatures up to 800 °C have been studied. In sulphur-rich mixtures the spectra of various S_n species predominate, but for As-rich mixtures As_4, As_4S_3, As_4S_4, and As_2S_2 could be detected.[401a]

Wavenumbers for the ligand $\nu(C{=}C)$, $\nu(C{-}S)$, and $\nu(R{-}C{\langle}^C_S)$ vibrations are consistent [402] with some π-delocalization over ES_2C_2 rings in the following dicyanoethylene-1,2-dithiolate (mnt) and toluene-3,4-dithiolate (tdt) complexes of Sb^{III} and Bi^{III}: $[Sb(mnt)_2]^-$, $E_2(tdt)_3$, (E = Sb or Bi) $[Bi_2(mnt)_2X_4]^{2-}$ and $[Bi_2(mnt)_3X_2]^{2-}$ (X = Cl, Br, or I), and $[Bi_2(mnt)_5]^{4-}$. As part of a study of some metal complexes of ethanedithiol and dimercaptodiethyl sulphide, i.r. data for which are consistent with a chelated ligand in the *gauche* configuration, $\nu(Bi{-}S)$ modes for $Bi(S_2C_2H_4)X$ (X = Cl or Br) have been assigned at *ca.* 320 and 270 cm^{-1} in the Raman effect.[403]

For the three compounds $Et_2NP(Me)C{\equiv}CP(Me)NEt_2$, $(R_2N)_2P(Se)$-
$$\overset{\|}{Se} \qquad \overset{\|}{Se}$$
$C{\equiv}CP(Se)(NR_2)_2$ (R = Me or Et), i.r. absorptions, respectively at 545, 538, and 528 cm^{-1} have been assigned to $\nu(P{=}Se)$ fundamentals.[404] An analysis of the vibrational spectra of vitreous As_2E_3 (E = S or Se) suggests [405] that ν_{asym} and $\nu_{sym}(As{-}Se{-}As)$ occur at 237 and 156 cm^{-1}, with $\nu_{asym}(As{-}S{-}As)$ at 340 cm^{-1}.

Tellurium tetrachloride has been found to react with Me_3SiN_3 to give Cl_3TeN_3 and $Cl_2Te(N_3)_2$; assignments for some i.r. bands have been given [406] for these derivatives, including $\nu(Te{-}N)$ at 412 cm^{-1}.

Compounds containing Group V–Halogen Bonds.—Jander and his coworkers have devoted considerable attention to the structural chemistry of

[398] J. R. Durig and J. W. Clark, *J. Cryst. Mol. Structure*, 1971, **1**, 43.
[399] H. W. Roesky, R. Pantzer, and J. Goubeau, *Z. anorg. Chem.*, 1972, **392**, 42.
[400] F. Seel and K. D. Velleman, *Chem. Ber.*, 1972, **105**, 406.
[401] E. J. Porter and G. M. Sheldrick, *J.C.S. Dalton*, 1972, 1347.
[401a] A. Rogstad, *J. Mol. Structure*, 1972, **14**, 421.
[402] G. Hunter, *J.C.S. Dalton*, 1972, 1496.
[403] M. Ikram and D. B. Powell, *Spectrochim. Acta*, 1972, **28A**, 59.
[404] W. Kuchen and K. Koch, *Z. anorg. Chem.*, 1972, **394**, 74.
[405] V. M. Bermudez, *J. Chem. Phys.*, 1972, **57**, 2793.
[406] N. Wiberg, G. Schwenk, and K. H. Schmid, *Chem. Ber.*, 1972, **105**, 1209.

derivatives and adducts of nitrogen tri-iodide, mainly using i.r. spectroscopy. In the first of three papers [407-409] it is concluded that, whereas Me_2NI is probably a monomer, $MeNI_2$ and its adducts with $MeNH_2$, Me_3N, or py are polymeric, involving tetrahedral nitrogen with one free and two bridging iodine atoms. Further results for $^nNI_3, ^nNH_3$ (n = 14 or 15), $^{14}NI_3$,py, and $^{14}NI_3, ^{14}ND_3$ have been correlated to the known structure of NI_3, NH_3 (shown crystallographically to be polymeric, with chains of units linked by three bridging I atoms per nitrogen),[408] and a related structure has been deduced for $NNN'N'$-tetraiodo-ethylenediamine from similar observations.[409] Wavenumbers and assignments for $\nu(N-I)_{terminal}$ and $\nu(N-I)_{bridge}$ presented for all of these compounds can be set out as shown in Table 39, and provide the evidence for the proposed structures. A covalent 1 : 2 complex of hexamethylenetetramine with bromine has been obtained,[410] for which $\nu(N \cdots Br)$ is at 130 cm^{-1}.

Table 39 N—I *stretching wavenumbers*/cm^{-1} *for nitrogen tri-iodide derivatives*

Compound	$\nu(N-I)_{terminal}{}^a$	$\nu(N-I)_{bridge}{}^a$
Me_2NI	467	n.o.
$MeNI_2$	545	415(a) / 346(s)
$MeNI_2, MeNH_2$	535	414(a) / 345(s)
$MeNI_2, Me_3N$	518	382(a) / 320(s)
$MeNI_2$,py	544	404(a) / 349(s)
$^{14}NI_3, ^{14}NH_3$	558(a) / 488(s)	382(a) / 175(s)
$^{14}NI_3, ^{14}ND_3$	556(a) / 491(s)	375(a) / 173(s)
$^{15}NI_3, ^{15}NH_3$	540(a) / 470(s)	377(a) / 174(s)
$^{14}NI_3$,py	567(a) / 486(s)	372(a) / 164(s)
$H_2NCI_2CI_2NH_2$	520	405, 303

^{a} (a) = ν_{asym}, (s) = ν_{sym} mode.

The gas-phase redistribution of PF_2Cl has been shown [411] to take place only to a limited extent after 24 h at 300 °C, by monitoring i.r. bands due to PF_3. The stoicheiometry and stability of complexes formed between $AsCl_3$ and aromatic hydrocarbons have been studied in cyclopentane solution, by observing [412] changes in wavenumber for Raman shifts of ν_1

[407] J. Jander, K. Knuth, and W. Renz, *Z. anorg. Chem.*, 1972, **392**, 143.
[408] K. Knuth, J. Jander, and U. Engelhardt, *Z. anorg. Chem.*, 1972, **392**, 279.
[409] J. Jander, K. Knuth, and K. U. Trommsdorff, *Z. anorg. Chem.*, 1972, **394**, 225.
[410] G. A. Bowmaker and S. F. Hannan, *Austral. J. Chem.*, 1972, **25**, 1151.
[411] A. D. Jordan and R. G. Cavell, *Inorg. Chem.*, 1972, **11**, 564.
[412] B. Gilbert and G. Duyckaerts, *Spectrochim. Acta*, 1972, **28A**, 825.

and ν_3 of $AsCl_3$. Variable-temperature Raman spectroscopy has been used in an investigation [413] of transition points in solid $SbCl_3$.

Three publications have shed some light on the structures of complex halogenoantimonates. Donaldson et al. have measured far-i.r. frequencies for MSb_2X_9 salts (M = Rb or Cs; X = Cl, Br, or I) and examined their assignment in terms of either O_h or C_{3v} symmetry; in view of accompanying Mössbauer data, the latter is preferred.[414] Determination of the crystal structure of the compound $SbCl_2F_3$ has shown that it consists of $SbCl_4^+$ tetrahedra and $[F_4Sb(Cl)-F-Sb(Cl)F_4]^-$ anions; this has allowed a re-interpretation of the vibrational spectrum,[415] as shown in Table 40.

Table 40 Observed vibrational fundamentals for $SbCl_2F_3$

Wavenumber/cm^{-1}	Assignment
655 ⎫	
628 ⎬	$\nu(Sb-F)$ of $Sb_2Cl_2F_9^-$
609 ⎭	
444	$\nu_{asym}(Sb-Cl)$ of $SbCl_4^+$
393	$\begin{cases} \nu_{sym}(Sb-Cl) \text{ of } SbCl_4^+ \\ \nu(Sb-Cl) \text{ of } Sb_2Cl_2F_9^- \end{cases}$
294	$\delta(Sb-F)$ of $Sb_2Cl_2F_9^-$
145	$\delta_{asym}(Sb-Cl)$ of $SbCl_4^+$

The existence of genuine chlorobromoantimonates cat[$SbCl_5Br$] (cat = PCl_4 or Et_4N) has been postulated by Bentley et al. on the basis of Raman data, bands near 330, 310, 290, 175, and 157 cm^{-1} being attributed to the anion.[416] Some band-frequencies and assignments given for butyrolactam complexes of SbX_3 (X = Cl or Br) and $BiCl_3$, in which the ligand is $>C=O\cdots E$ bonded, include $\nu(E-O)$ at 480—500 cm^{-1} and $\nu(E-X)$ modes in the usual ranges.[417]

6 Group VI Elements

Compounds containing O—H, S—H, or Te—H Bonds.—The presence of the $H_5O_2^+$ ion in $HCl,2H_2O$, reported by other workers during 1971, has been confirmed by Rozière and Potier.[418] I.r. and Raman spectra of polycrystalline rubidium hydrogen di-trichloroacetate and its deuterio-derivative can be interpreted [419] assuming a symmetrical hydrogen-bonded anion $Cl_3COO-H-OOCCl_3$, $\nu(O-H-O)$ being assigned to a strong, broad feature near 800 cm^{-1}. Two papers containing normal-co-ordinate calculations on liquid water have appeared.[420, 421]

[413] W. M. O. Julien and H. Gerding, Rec. Trav. chim., 1972, **91**, 743.
[414] J. D. Donaldson, M. J. Tricker, and B. W. Dale, J.C.S. Dalton, 1972, 893.
[415] H. Preiss, Z. anorg. Chem., 1972, **389**, 254.
[416] F. F. Bentley, A. Fingh, P. N. Gates, and F. J. Ryan, Inorg. Chem., 1972, **11**, 413.
[417] S. T. Yuan and S. K. Madan, Inorg. Chim. Acta, 1972, **6**, 463.
[418] J. Rozière and J. Potier, J. Mol. Structure, 1972, **13**, 91.
[419] D. Hadži, M. Obradovič, B. Orel, and T. Šolmajer, J. Mol. Structure, 1972, **14**, 439.
[420] B. Curnette and J. Bandekar, J. Mol. Spectroscopy, 1972, **41**, 500.
[421] J. B. Bryan and B. Curnette, J. Mol. Spectroscopy, 1972, **41**, 512.

Engler and Gattow have reported $v(S-H)$ at 2500 cm^{-1} in the i.r. for [HCS(SH)]$_3$ and at 2540 cm^{-1} for HCOSH, together with some other assignments [422, 423] and also analogous data for [HCS(SH)]$_x$, HCO(SMe), cat$^+$[HCOS]$^-$ (cat = Na, Ph$_4$P, or Ph$_4$As) and [HCS(OMe)]$_x$. The results of a systematic investigation of the vibrational spectrum of methanetellurol have been published.[424] Liquid Raman and matrix i.r. measurements for the four isotopic variants C^{n}H$_3$Te^{m}H (n or m = 1 or 2) are included, together with a normal-co-ordinate analysis based on C_s molecular symmetry; Table 41 contains wavenumbers for Raman bands assigned to the three A' fundamentals v_3, v_7, and v_8.

Table 41 Raman wavenumbers/cm^{-1} for v_3, v_7, and v_8 fundamentals for methanetellurol

		CH$_3$TeH	CH$_3$TeD	CD$_3$TeH	CD$_3$TeD
$v_3(A')$	v(Te—H)	2016		2017	
	v(Te—D)		1449		1449
$v_7(A')$	δ(CTeH)	608		541	
	δ(CTeD)		447		425
$v_8(A')$	v(C—Te)	516	525	479	487

Compounds containing E—C Bonds (E = S, Se, or Te).—Force constants have been calculated [425] for (CH$_3$)$_2$S, CH$_3$SCD$_3$, and (CD$_3$)$_2$S using published frequency data. An assignment has been proposed for the fundamental vibrations of ethylene sulphide.[426] In dithioformates M^I[HCS$_2$] (M^I = K$^+$, Rb$^+$, Cs$^+$, or Tl$^+$; or Ph$_4$P$^+$, Ph$_4$As$^+$, Et$_4$N$^+$, or Ph$_3$Sn$^+$), M^{II}[HCS$_2$]$_2$ (M^{II} = Pb^{2+}, Zn^{2+}, Cd^{2+}, or Hg^{2+}), and In(HCS$_2$)$_3$, i.r. bands at ca. 1250 and 988 cm^{-1} have been assigned [427] to v_{asym}(SCS) modes, with the corresponding v_{sym} at 848 cm^{-1}. Similar results for HC(SMe)$_3$ include those given as follows [428] (wavenumber/cm^{-1}): v(C—S—Me), 1177/1145; ρ(SMe), 961/872; v(Me—S), 708; δ_{sym}(CS$_3$), 642; and δ_{asym}(CS$_3$), 627. Along with the latter, unassigned i.r. frequencies for related compounds such as [HCS(SR)]$_3$ (R = Me or Et), HC(SEt)$_3$, and HC(SCH$_2$Ph)$_3$ were also listed.[428] Partially assigned vibrational data have been published [429] for (CF$_3$S)$_2$PCF$_3$. Ellestad et al. have measured i.r. and Raman spectra for (51)—(53);[430, 431] force-constant calculations provided evidence for very extensive mixing of skeletal modes in all four compounds. Unassigned i.r. wavenumbers have been reported [432] for tellurophen (54). In several

[422] R. Engler and G. Gattow, Z. anorg. Chem., 1972, **389**, 145.
[423] R. Engler and G. Gattow, Z. anorg. Chem., 1972, **388**, 78.
[424] C. W. Sink and A. B. Harvey, J. Chem. Phys., 1972, **57**, 4434.
[425] J. W. Ypenburg, Rec. Trav. chim., 1972, **91**, 671.
[426] V. T. Aleksanyan and G. M. Kuzyants, J. Struct. Chem. U.S.S R., 1971, **12**, 243.
[427] R. Engler, G. Gattow, and M. Dräger, Z. anorg. Chem., 1972, **388**, 229.
[428] R. Engler, G. Gattow, and M. Dräger, Z. anorg. Chem., 1972, **390**, 64.
[429] I. B. Mishra and A. B. Burg, Inorg. Chem., 1972, **11**, 664.
[430] O. H. Ellestad, P. Klaboe, and G. Hagan, Spectrochim. Acta, 1972, **28A**, 1855.
[431] O. H. Ellestad, P. Klaboe, G. Hagan, and T. Stroyer-Hansen, Spectrochim. Acta, 1972, **28A**, 149.
[432] T. J. Barten and R. W. Roth, J. Organometallic Chem., 1972, **39**, C66.

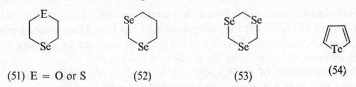

(51) E = O or S   (52)   (53)   (54)

interesting papers concerning phenyl-selenium and -tellurium halides, most emphasis has been placed on vibrations associated with the E–halogen bonds; their discussion is therefore deferred to a later section.

Compounds containing O—O, S—O, Se—O, Te—O, S—S, S—Se, or Te—Te Bonds.—In a paper emphasizing the value of laser Raman spectroscopy in detecting O—O bonds in fluoroperoxides, Melveger et al. have quoted [433] the following O—O stretching wavenumbers/cm^{-1}: 870 (CF_3-OOH); 943 (CF_3OOCl); 1300 (FOOF); and 880 (HOOH). The i.r. spectrum of CF_3OOF has been recorded by DesMarteau, with absorptions at 950, 870, and 755 cm^{-1} tentatively attributed to CO, OO, and OF stretching modes.[434] The same author reports that hydrolysis of $F_5SOOC(O)F$ gives pentafluorosulphur hydroperoxide, F_5SOOH, in nearly quantitative yield, for which a very intense Raman shift at 735 cm^{-1} is assigned to ν(O—O), while a band in the i.r. at 1385 cm^{-1} is attributed to δ(OOH).[435] An independent study of CF_3OOF by other workers has provided i.r. data which were not assigned.[436] The vibrational spectrum of dimethyl peroxide has also been thoroughly investigated, with the aid of force-constant calculations. In the two isotopic varients CH_3OOCH_3 and CD_3OOCD_3, ν(O—O) was assigned at 779 and 720 cm^{-1}, respectively, but the O—O motion was found to be highly coupled, leading the authors to suggest that the stretching wavenumbers cannot be considered to be an accurate reflection of bond strength.[437] I.r. spectra for some alkali-metal peroxoborates have been found to be complex, but all show a band near 850 cm^{-1} attributed [438] to ν(O—O).

Four publications have appeared containing rather unexciting information collected *via* vibrational spectroscopy for a number of sulphur oxyhalide species.[439—442] I.r. data (500—1500 cm^{-1}) have been tabulated for 159 sulphides, sulphoxides, and sulphones.[443]

[433] A. J. Melveger, L. R. Anderson, C. T. Ratcliffe, and W. B. Fox, *Appl. Spectroscopy*, 1972, **26**, 381.
[434] D. D. DesMarteau, *Inorg. Chem.*, 1972, **11**, 193.
[435] D. D. DesMarteau, *J. Amer. Chem. Soc.*, 1972, **94**, 8933.
[436] I. J. Solomon, A. J. Kacmarek, W. K. Sumida, and J. K. Raney, *Inorg. Chem.*, 1972, **11**, 195.
[437] M. E. B. Bell and J. Laane, *Spectrochim. Acta*, 1972, **28A**, 2239.
[438] U. Sommer and G. Heller, *J. Inorg. Nuclear Chem.*, 1972, **34**, 2713.
[439] R. A. Suthers and T. Henshall, *Z. anorg. Chem.*, 1972, **388**, 257.
[440] R. A. Suthers and T. Henshall, *Z. anorg. Chem.*, 1972, **388**, 269.
[441] V. A. Tertykh, V. V. Pavlov, V. M. Mashchinko, and A. A. Chuiko, *Doklady Phys. Chem.*, 1971, **201**, 1038.
[442] P. Nanni and F. Viani, *J. Mol. Structure*, 1972, **14**, 413.
[443] J. Brunn and K. Doerffel, *Wiss. Z. Tech. Hochsch. Chem. 'Carl Schlorlemmer' Leuna-Merseburg*, 1971, **13**, 101.

Ring-closure reactions between polysulphanes and $SOCl_2$ afford cyclopolysulphur oxides S_nO; the compound S_8O has thus been isolated pure, and in its i.r. spectrum $\nu(SO)$ is at 1134 cm^{-1}, as expected by analogy with other compounds containing the S—SO—S linkage.[444] Characterization of $(NH_4)(SO_3CF_3)$ has led to the assignment of a number of i.r. bands, including those following [445] (wavenumber/cm^{-1}): 1270, $\nu_{asym}(SO_3)$; 1035, $\nu_{sym}(SO_3)$; 763, $\nu(C-S)$; and 650, $\delta_{asym}(SO_3)$. Chlorine fluoride is reported to react with SbF_5-3SO_3F to yield $ClOSO_2F$, part of the vibrational spectrum for which is reproduced in Table 42; no

Table 42 Sulphur–oxygen vibrations for $ClOSO_2F$

Raman (wavenumber/cm^{-1})	I.r. (wavenumber/cm^{-1})	Assignment
1477w	1458vs	$\nu_{asym}(SO_2)$
1242m	1238vs	$\nu_{sym}(SO_2)$
	837ms	$\nu(S-O)$
	532m	$\rho(SO_2)$
487w	487mw	$\delta(SO)$
	390w	$\tau(SO_2)$

evidence for the species Cl_2^+ or ClF^+ was observed.[446] In the i.r. spectra of ISO_3F and $I(SO_3F)_3$, ν_{asym} and $\nu_{sym}(SO_3)$ have been distinguished, respectively at [447] 1390 and 1395, and at 1200 and 1195 cm^{-1}. Two compounds containing skeletons bonded by O—SO$_2$—O units have been studied by i.r. and Raman methods: they are dimethylsulphate,[448] and $F_5SeOSO_2O-SO_2F$, the latter prepared through action of SO_3 on pentafluoro-orthoselenic acid.[449] Selected assignments for each of these are included in Table 43.

Table 43 Raman wavenumbers/cm^{-1} for sulphur–oxygen stretching modes in compounds with OSO_2O linkages

	$CH_3OSO_2OCH_3$	$CD_3OSO_2OCD_3$	$F_5SeOSO_2OSO_2F$
$\nu_{asym}(SO_2)(O)$	1389	1394	1439
$\nu_{sym}(SO_2)(O)$	1196	1204	1241
$\nu_{asym}(SO_2)(F)$			1486
$\nu_{sym}(SO_2)(F)$			1251
$\nu_{asym}(O-S-O)$	828	790	
$\nu_{sym}(O-S-O)$	757	727	

I.r. spectra of the adducts $SO_2Cl_2,(py)_2$ and SO_2Cl_2,PPh_3 indicate N → S and P → S bonding, the main evidence for this being the minimal effect on $\nu(S=O)$ of complex formation. A broad absorption centred at 2500 cm^{-1} has been tentatively ascribed to a lowered $\nu(C-H)$, as in (55).[450] Some

[444] R. Stendel and M. Rebsch, *Angew. Chem. Internat. Edn.*, 1972, **11**, 302.
[445] F. A. Schröder, B. Gänswein, and G. Brauer, *Z. anorg. Chem.*, 1972, **391**, 295.
[446] R. J. Gillespie and M. J. Morton, *Inorg. Chem.*, 1972, **11**, 591.
[447] C. Chung and G. H. Cady, *Inorg. Chem.*, 1972, **11**, 2528.
[448] K. O. Christe and E. C. Curtis, *Spectrochim. Acta*, 1972, **28A**, 1889.
[449] K. Seppelt, *Chem. Ber.*, 1972, **105**, 3131.
[450] A. J. Banister, B. Bell, and L. F. Moore, *J. Inorg. Nuclear Chem.*, 1972, **34**, 1161.

(55)

solid-state vibrational data for salts containing the EO_4 unit (E = S or Se) have been published during the year.[451–453] Two other papers[388, 454] list $\nu(S=O)$ wavenumbers for a large number of compounds, as shown in Table 44.

Table 44 *Compounds with S—O bonds characterized by i.r. spectroscopy* [$\nu(S=O)$, *wavenumbers*/cm^{-1}]

Compound	$\nu(S=O)$	Compound	$\nu(S=O)$
$(CF_3)_2S(O)NH$	1353	$PhOS(O)F_3$	1323
$(CF_3)(C_2F_5)S(O)NH$	1351	$p\text{-MeC}_6H_4OS(O)F_3$	1324
$(C_2F_5)_2S(O)NH$	1350	$m\text{-MeC}_6H_4OS(O)F_3$	1323
$(CF_3)_2S(O)NMe$	1243	$p\text{-ClC}_6H_4OS(O)F_3$	1327
$(CF_3)_2S(O)NSCF_3$	1331	$m\text{-ClC}_6H_4OS(O)F_3$	1327
$(CF_3)_2S(O)NCN$	1369	$p\text{-FC}_6H_4OS(O)F_3$	1331
$(CF_3)_2S(O)NS(O)CF_3$	1339	$m\text{-FC}_6H_4OS(O)F_3$	1334
$(CF_3)_2S(O)NC(O)CF_3$	1312	$(PhO)_2S(O)F_2$	1288
$(CF_3)_2S(O)NCl$	1328	$(p\text{-MeC}_6H_4O)_2S(O)F_2$	1284
$(CF_3)_2S(O)F_2$	1325	$(p\text{-ClC}_6H_4O)_2S(O)F_2$	1294, 1287
$(CF_3)_2S(O)NSiMe_3$	1266	$(p\text{-NO}_2C_6H_4O)_2S(O)F_2$	1283

Pentafluoro-orthoselenic acid, $HOSeF_5$, has been synthesized by several routes from oxyfluorides of selenium. Its gas-phase i.r. spectrum has been found to be very simple, inviting interpretation in terms of a slightly perturbed octahedral structure,[455] with absorptions at the following wavenumbers/cm^{-1}: 3609s, ν(O—H); 1171s, δ(OH); 750vs, ν(Se—F); and 436s, δ(SeF). Conversion into the salts M^ISeOF_5 (M^I = Li$^+$, Na$^+$, K$^+$, Rb$^+$, Cs$^+$, or NH$_4^+$) provides i.r. data which include[456] assignments for ν(Se—O), at 898, 921, 922, and 923 cm^{-1}, respectively, for M^I = NH$_4^+$, K$^+$, Rb$^+$, and Cs$^+$. By observation of vibrations arising from the SeO$_4$ unit, i.r. spectroscopy has been used to study[457] the thermal decomposition of $(NH_4)_3Sc(SeO_4)_3$. Unassigned i.r. spectra have been listed[458] for

[451] S. Peytavin, G. Brun, L. Cot, and M. Maurin, *Spectrochim. Acta*, 1972, **28A**, 1995.
[452] S. Peytavin, G. Brun, J. Guillermet, L. Cot, and M. Maurin, *Spectrochim. Acta*, 1972, **28A**, 2005.
[453] A. A. Belyaeva, M. I. Dvorkin, and L. D. Shcherba, *Optika i Spektroskopiya*, 1971, **31**, 585.
[454] D. S. Ross and D. W. A. Sharp, *J.C.S. Dalton*, 1972, 34.
[455] K. Seppelt, *Angew. Chem. Internat. Edn.*, 1972, **11**, 630.
[456] K. Seppelt, *Chem. Ber.*, 1972, **105**, 2431.
[457] K. Tetsu, B. N. Ivanov-Emin, and L. G. Korotaeva, *Russ. J. Inorg. Chem.*, 1971, **16**, 1416.
[458] F. Sladky and H. Kropshofer, *Inorg. Nuclear Chem. Letters*, 1972, **8**, 195.

ROTeF$_5$ (R = Me, Et, or EtOCOCH$_2$). Reasonably detailed assignments for the observed fundamentals of the TeO$_6^{6-}$ ion have been offered by Lentz,[459] including calculation of Te—O force constants, and the i.r. spectrum of Na$_2$Te$_4$O$_9$,4.5H$_2$O has been illustrated.[460]

Several features near 670 cm^{-1} in low-temperature matrices hitherto ascribed to the S$_2$ molecule have been re-assigned to vibrations of the S=S$\langle^S_S$ unit of a branched sulphur chain.[461] Unassigned vibrational data for (56),[462] and also for the compounds [463] CF$_3$SC(ClF)SSR$_F$ (R$_F$ = CF$_3$,

(56) x = 2, 3, or 5

CF$_2$Cl, or CFCl$_2$), CF$_3$SC(Cl$_2$)SSR$_F$ (R$_F$ = CF$_3$ or CF$_2$Cl), and (CF$_3$)$_2$-CClSSCF$_2$Cl have been published. Dibromo- and dichloro-(tetramethylthiourea)selenium(II) have been prepared,[464] and show ν(Se—S) respectively at 240 and 233 cm^{-1}. The vibrational spectrum of Ph$_2$Te$_2$ (and those of some substituted phenyl derivatives) below 400 cm^{-1} have been reported.[465] An i.r. absorption at 169 cm^{-1}, with a very strong Raman counterpart, was attributed to Te—Te stretching, with ν(Ph—Te) tentatively placed at 255 and 202 cm^{-1}.

Compounds containing Group VI–Halogen Bonds.—A short paper by Züchner and Glemser contains a variety of information concerning chlorine oxide fluorides:[466] for ClOF$_3$, ν(^{35}Cl—O) is at 1226 with ν(^{37}Cl—O) at 1216 cm^{-1}; in addition, Raman spectra of salts of [ClOF$_2$]$^+$ were compared with literature data, and Raman wavenumbers for the hitherto unknown [ClOF$_4$]$^-$ ion were assigned as follows: 1202w, ν_1; 455vs, ν_2; 410s, ν_3; 349vs, ν_4; 285m, ν_6; and 204w, ν_9. The I—O bonds in H$_5$IO$_6$ have been shown by vibrational spectroscopy to be non-equivalent, but in strongly acidic solution the Raman spectrum of the regular O_h ion I(OH)$_6^+$ is observed,[467] with ν(I—O) at 642 (A_{1g}) and 662 cm^{-1} (E_g). I.r. and Raman data for M^{I}IO$_2$F$_2$ (M^I = Na or K) have been published,[468] and some unassigned i.r. frequencies for IF$_4$OMe have been listed.[469] Synthesis of two novel iodine perchlorates, I(OClO$_3$)$_3$ and Cs$^+$[I(OClO$_3$)$_4$]$^-$, has been

[459] A. Lentz, *Z. anorg. Chem.*, 1972, **392**, 218.
[460] S. A. Malyutin, K. K. Samplavskaya, and M. Kh. Karapet'yants, *Russ. J. Inorg. Chem.*, 1971, **16**, 1559.
[461] R. Steudel, *Z. Naturforsch.*, 1972, **27b**, 469.
[462] F. Fehér, M. Langer, and R. Volkert, *Z. Naturforsch.*, 1972, **27b**, 1006.
[463] A. Haas, W. Klug, and H. Marsmann, *Chem. Ber.*, 1972, **105**, 820.
[464] K. J. Wynne, P. S. Pearson, M. G. Newton, and J. Golen, *Inorg. Chem.*, 1972, **11**, 1192.
[465] W. R. McWhinnie and P. Thavornyutikarn, *J. Organometallic Chem.*, 1972, **35**, 149.
[466] K. Züchner and O. Glemser, *Angew. Chem. Internat. Edn.*, 1972, **11**, 1094.
[467] H. Siebert and G. Wieghardt, *Z. Naturforsch.*, 1972, **27b**, 1299.
[468] A. Finch, P. N. Gates, and M. A. Jenkinson, *J. Fluorine Chem.*, 1972, **2**, 111.
[469] G. Oates and J. M. Winfield, *Inorg. Nuclear Chem. Letters*, 1972, **8**, 1093.

achieved.[470] The first compound appears from its Raman spectrum to be polymeric, but the vibrational spectrum of the second can be interpreted in a manner consistent with a simple square-planar IO_4 skeleton for the anion (Table 45).

Table 45 *The vibrational spectrum of solid* $Cs^+[I(OClO_3)_4]^-$

I.r.[a] (wavenumber/cm^{-1})	Raman[b] (wavenumber/cm^{-1})	Approximate description
1230vs	1243w,sh ⎫ 1207mw ⎭	$\nu_{asym}(ClO_3)$
1015vs	1038s ⎫ 1016vw ⎭	$\nu_{sym}(ClO_3)$
630vs	630s	$\nu(O-Cl)$
570—650vs	607mw	$\delta_{asym}(ClO_3), \delta(ClO_3)$ ('scissor')
485s	489s ⎫	
430vw	430mw ⎭	$\delta(ClO_3)$ ('umbrella')
	261vs	$\nu_{sym}(IO_4)$ (in-phase)
	240s	$\nu_{sym}(IO_4)$ (out-of-phase)
	131ms ⎫ 106ms ⎭	$\delta(IOCl)$

[a] Ambient temperature; [b] $-60\ °C$ to avoid explosive decomposition.

Vibrational assignments published previously for SF_3^+ have been confirmed[471] by Raman measurements on BF_4^-, PF_6^-, and AsF_6^- salts of this cation in anhydrous HF. Confirmation by i.r. spectroscopy of the formation of $ECl_3^+SO_3F^-$ on dissolution of ECl_4 (E = Se or Te) in HSO_3F has been claimed.[472] SeF_5Cl has been prepared and some unassigned i.r. data have been given.[473] The vibrational spectra of some mixed halogenotellurium(IV) anions have been correlated and discussed.[474]

Remaining reports concern aryl-selenium and -tellurium halides, the spectroscopic properties of which have been investigated by Wynne and McWhinnie and their respective co-workers. The most interesting of these papers[475] gives data for salts containing the $[PhSeBr_2]^-$ ion. The latter is formulated with a T-shaped configuration, with $\nu_{asym}(Se-Br)$ at 180 and $\nu_{sym}(Se-Br)$ near 150 cm^{-1}. For $RSeCl_3$ (R = Ph or p-MeC$_6$H$_4$) and $RTeCl_3$ (R = Ph, p-MeC$_6$H$_4$, or p-MeOC$_6$H$_4$) some tentative assignments have been made, placing $\nu(E-Cl)$ (E = Se or Te) in the 300—400 cm^{-1} range.[476] Fairly full assignment for $\nu(Te-X)$ fundamentals has been attempted[477] for some diaryl-tellurium dihalides, as shown in Table 46; and consideration of vibrational as well as other physical data for $RTeX_3$ (R = Ph, p-MeC$_6$H$_4$, p-MeOC$_6$H$_4$, p-EtOC$_6$H$_4$, or p-PhOC$_6$H$_4$;

---

[470] K. O. Christe and C. J. Schack, *Inorg. Chem.*, 1972, **11**, 1682.
[471] M. Brownstein and J. Shamir, *Appl. Spectroscopy*, 1972, **26**, 77.
[472] R. C. Paul, K. K. Paul, and K. C. Malhotra, *J. Inorg. Nuclear Chem.*, 1972, **34**, 2523.
[473] C. J. Schack, R. D. Wilson, and J. F. Hon, *Inorg. Chem.*, 1972, **11**, 208.
[474] G. A. Ozin and A. Vander Voet, *J. Mol. Structure*, 1972, **13**, 435.
[475] K. J. Wynne and P. S. Pearson, *Inorg. Chem.*, 1972, **11**, 1196.
[476] K. J. Wynne, A. J. Clark, and M. Berg, *J.C.S. Dalton*, 1972, 2370.
[477] W. R. McWhinnie and M. G. Patel, *J.C.S. Dalton*, 1972, 199.

Table 46 *Tellurium–halogen stretching bands[a] (wavenumbers/cm^{-1}) in diaryl-tellurium dihalides, R_2TeX_2*

R	X = Cl	X = Br	X = I
Ph	287, 262	159, 186	116, 129
p-MeC$_6$H$_4$	264, 248	167/157, 173	110, 149
o-MeC$_6$H$_4$	275, 255	166, 172	109, 148/145
p-MeOC$_6$H$_4$	271, 247	164, 193	
C$_6$F$_5$	269, 264	168, 184	

[a] Listed as ν_{sym}, ν_{asym} for each R under each X, as assigned in ref. 477.

X = Cl, Br, or I) suggests association in these compounds, possibly to give dimeric structures involving five-co-ordinate Te in the case of the chlorides and iodides.[478]

7 Group VII Elements

Carbon disulphide reacts with [PtF(PPh$_3$)$_3$][HF$_2$] to give an insertion product whose crystal structure has been determined, and for which a broad i.r. band near 1700 cm^{-1} is possibly due to the (HF$_2$)$^-$ anion.[479] Direct chlorination of I$_2$ at -78 °C in the presence of SbCl$_5$ (1 mol equiv.) affords [I$_2$Cl$^+$][SbCl$_6^-$]; in addition to anion bands, this shows Raman shifts at 359 and 196 cm^{-1}, assigned to ν(I—Cl) and ν(I—I) of an (IICl)$^+$ cation.[480] Seven bands in the Raman spectrum of [BrF$_4^+$][Sb$_2$F$_{11}^-$] have been assigned to vibrations of a C_{2v} cation as follows [481] (wavenumbers/cm^{-1}): 726, $\nu_1(A_1)$; 702, $\nu_6(B_1)$; 606, $\nu_2(A_1)$; 424, $\nu_3(A_1)$; 385, $\nu_7(B_1)$; 365, $\nu_9(B_2)$; and 216, $\nu_4(A_1)$. A compound of composition [ClF$_6$PtF$_6$,ClF$_4$PtF$_6$] has been isolated from reaction between PtF$_6$ and excess ClF$_5$ at 23 °C. I.r. bands at 890, 540 cm^{-1} not assignable to [ClF$_4^+$][PtF$_6^-$] were taken to provide evidence for the ClF$_6^+$ ion.[482]

The existence of at least three new chlorine–fluorine species has been postulated through i.r. spectroscopy of a photolysed Cl–F–(Ar or N$_2$) matrix system. Observed bands (wavenumbers/cm^{-1}) were assigned [483] as follows: (a) 630, ν_{asym}(ClF$_2$); 462, ν(Cl—Cl); and 270, ν_{sym} for (57); (b) 557, ν_{asym}(ClF$_2$) for (58); and (c) 536, ν(Cl—F) and 302, ν(Cl—Cl) for Cl$_2$F. The i.r. and Raman spectra of solid M^I(IF$_6$) (M^I = K, Rb, or Cs) show a large number of bands, indicating a relatively low symmetry (possibly C_{2v}) for the structure of the anion,[484] contrasting with that of BrF$_6^-$.

Resonance-enhanced Raman spectra of I$_3^-$, I$_2$Br$^-$, and IBr$_2^-$ and of their complexes with amylose have been obtained. For the free anions,

[478] W. R. McWhinnie and P. Thavornyutikarn, *J.C.S. Dalton*, 1972, 551.
[479] J. A. Evans, M. J. Hacker, R. D. W. Kemmitt, D. R. Russell, and J. Stocks, *J.C.S. Chem. Comm.*, 1972, 72.
[480] J. Shamir and M. Lustig, *Inorg. Nuclear Chem. Letters*, 1972, **8**, 985.
[481] T. Surles, A. Perkins, L. A. Quarterman, H. H. Hyman, and A. I. Popov, *J. Inorg. Nuclear Chem.*, 1972, **34**, 3561.
[482] F. Q. Roberto, *Inorg. Nuclear Chem. Letters*, 1972, **8**, 737.
[483] M. R. Clarke, W. H. Fletcher, G. Mamantov, E. J. Vasini, and D. G. Vickroy, *Inorg. Nuclear Chem. Letters*, 1972, **8**, 611.
[484] K. O. Christe, *Inorg. Chem.*, 1972, **11**, 1215.

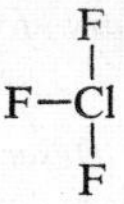

```
       F                    F
       |                    |
     F—Cl              F—Cl—Cl
       |                    |
       F                    F
      (57)                (58)
```

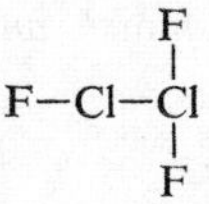

assuming an asymmetric, linear structure in each case, v_1 and v_3 have been assigned [485] respectively at 113, 158 cm^{-1} for KI$_3$ and at 112, 180 cm^{-1} for KI$_2$Br. Only one band, at 167 cm^{-1}, was found for KIBr$_2$. Far-i.r. spectra for a number of charge-transfer complexes between IX (X = I, Br, or Cl) and organic sulphides have been measured in polyethylene matrices. In a number of instances bands could be attributed to v(S—I) and in all cases shifts in v(I—X) were observed from the values for the free molecules.[486] Force constants f_{SI} in the range 0.79—0.96 and 1.31—1.32 mdyn Å^{-1} were computed for I$_2$ and ICl complexes, respectively, from the data of Table 47.

Table 47 Raman bands (wavenumbers/cm^{-1}) assigned to v(I—S) modes for complexes of I$_2$ or ICl with organic sulphides

Compound	X = I	X = Cl
Ph$_2$S,IX	95	105
MeSPh,IX	130	160
Me$_2$S,IX	160	n.o.
MeSEt,IX	157	205
Et$_2$S,IX	149	195
Bu$^t_2$S,IX	142	n.o.

The origin of far-i.r. absorptions at 78—140 cm^{-1} in the spectra of dioxan adducts of I$_2$, Br$_2$Cl$_2$, ICl, IBr, and ICN has been discussed in a further paper.[487]

8 Group VIII Elements

1972 has been an exceptional year for xenon chemistry so far as spectroscopic measurements are concerned, with seven publications reporting worthwhile and sometimes highly novel results.

The reaction between Xe and O$_2$BF$_4$ has been found to give a white solid, which decomposes at 243 K: vibrational spectroscopy, with observed data assigned as shown in Table 48, suggests that this product is FXe—BF$_2$, possessing a xenon–boron bond.[488] Raman spectra for XeF$_6$,UF$_5$ in the solid state and in HF solution cannot be clearly interpreted in terms of the presence of ClF$_6^-$ and in a similar way no obvious evidence for AsF$_6^-$ in [XeF$_5$,AsF$_6$] has been observed.[489]

[485] M. E. Heyde, L. Rimai, R. G. Kilponen, and D. Gill, *J. Amer. Chem. Soc.*, 1972, **94**, 5222.
[486] M. Yamada, H. Saruyama, and K. Aida, *Spectrochim. Acta*, 1972, **28A**, 439.
[487] G. W. Brownson and J. Yarwood, *Spectroscopy Letters*, 1972, **5**, 185.
[488] C. T. Goetschel and K. R. Loos, *J. Amer. Chem. Soc.*, 1972, **94**, 3018.
[489] B. Frlec, M. Bohinc, P. Charpin, and M. Drifford, *J. Inorg. Nuclear Chem.*, 1972, **34**, 2938.

Table 48 Wavenumbers/cm^{-1} and assignments for vibrational data for FXeBF$_2$

I.r.	Raman	Description
1250w		
1180 ⎱ s		ν_{asym}(B—F)
1149 ⎰		
885 ⎱ m		ν_{sym}(B—F)
857 ⎰		
694 ⎱ s		δ_{asym}(BF$_2$)
672 ⎰		
588s	590m	ν(Xe—F)
524 ⎱		
504 ⎰ m		δ_{sym}(BF$_2$)
492 ⎰		
379w	370w	ν(Xe—B)

Xenon difluoride reacts with HOSeF$_5$ to yield Xe(OSeF$_5$)$_2$ as a stable crystalline solid, which can be equilibrated with excess XeF$_2$ giving FXe(OSeF$_5$). For the latter, strong polarized Raman shifts at 501 and 173 cm^{-1} were assigned to [ν(Xe—F) + ν(Xe—O)] and δ(OXeF) modes, respectively; further unassigned i.r. and Raman wavenumbers were reported for the bis(pentafluoro-orthoselenate).[490] Eisenberg and DesMarteau have measured the vibrational spectra of F$_5$Xe(OSO$_2$F),[491] and of FXe(OPOF$_2$) and Xe(OPOF$_2$)$_2$.[492] For the first of these, ν(Xe—F$_{ax}$) and ν(Xe—F$_{eq}$) of XeF$_5^+$ are assigned at 650 and 595 cm^{-1}; for the second, ν(Xe—O) could not be clearly distinguished from PO$_2$F$_2$ deformations in the 560—670 cm^{-1} region; and for the third, features at 675, 560, and 525 cm^{-1} were tentatively attributed to ν(Xe—F) modes, possibly involving bidentate difluorophosphate. Raman spectra for the solid adducts XeOF$_4$,2SbF$_5$ and XeO$_2$F$_2$,2SbF$_5$ show these to be Sb$_2$F$_{11}^-$ species involving [493] respectively the cations (59) and (60); for the former, ν(Xe=O) is located at 942 cm^{-1}

$$
\begin{array}{c}
\text{F} \\
| \\
^+\text{Xe} {\diagup \text{O}} \\
| \quad \diagdown \text{F} \\
\text{F}
\end{array}
\qquad
\begin{array}{c}
\quad\;\; + \\
\text{O} \diagup \overset{\text{Xe}}{\underset{\text{O}}{\|}} \diagdown \text{F} \\
\text{F}
\end{array}
$$

(59) (60)

with ν(Xe—F) at 634 (eq), 589 (sym. ax.) and 554 cm^{-1} (asym. ax.), whereas for (60), ν_{asym}(Xe=O) occurs at 924 with ν_{sym} at 865 cm^{-1}, and ν(Xe—F) is assigned at 595 cm^{-1}. Finally, a broad i.r. band at 640—720 cm^{-1} for Li$_4$XeO$_6$,nH$_2$O (0.6 ⩽ n ⩽ 2) has been attributed to the octahedral perxenate anion.[494]

[490] K. Seppelt, *Angew. Chem. Internat. Edn.*, 1972, **11**, 723.
[491] D. D. DesMarteau and M. Eisenberg, *Inorg. Chem.*, 1972, **11**, 2641.
[492] M. Eisenberg and D. D. DesMarteau, *Inorg. Chem.*, 1972, **11**, 1901.
[493] R. J. Gillespie, B. Landa, and G. J. Schrobilgen, *J.C.S. Chem. Comm.*, 1972, 607.
[494] I. S. Kirin, Yu. K. Gusev, V. K. Isupov, L. I. Molkanov, V. Ya. Mishin, and A. V Krupinskaya, *Russ. J. Inorg. Chem.*, 1971, **16**, 1549.

6
Vibrational Spectra of Transition-element Compounds

BY M. GOLDSTEIN

1 Introduction

There are two main differences in the treatment of material in this chapter compared with previous years. Firstly, as a logical development, less prominence is given to correlations which are now well established. Secondly, in view of the rapid increase in the volume of data collected, more emphasis is placed on papers concerned with the spectroscopy of transition-element compounds and its use in structural analysis at the expense of reporting details of data used mainly for characterizing purposes. However, no papers collected in the normal way have been omitted from the Report.

The material has been arranged in the same way as in previous years. Information on vibrations of M—X bonds is presented in the sequence of vertical groups of the transition elements M in the Periodic Table, although lanthanides and actinides are considered at the end of the chapter. Within each group the elements X are generally also arranged according to vertical groups in the Periodic Table. A number of studies reported span more than one transition-metal group and are conveniently dealt with at this stage unless the emphasis is on internal vibrations of co-ordinated ligands, in which case the data are discussed in the next chapter.

2 General

Vibrational spectroscopy allows a reasonably clear distinction to be made between different types of co-ordination of the $[BH_4]^-$ ion to metal atoms; in $(\pi-C_5H_5)_2Ti(BH_4)$, $(\pi-C_5H_5)_2Zr(BH_4)$, and $(\pi-C_5H_5)_2ZrH(BH_4)$ the $[BH_4]^-$ ligand is bidentate, whereas in $(\pi-C_5H_5)_3U(BH_4)$ and $(\pi-C_5H_5)_3Th(BH_4)$ it is terdentate.[1]

When MeHgI, $(Me_2AuI)_2$, or $(Me_3PtI)_4$ is dissolved in liquid NH_3, the cations $[MeHg(NH_3)]^+$, $[Me_2Au(NH_3)_2]^+$, or $[Me_3Pt(NH_3)_3]^+$ are formed. These have similar Raman spectra to aquo-analogues but $\nu(MC)$ modes are 12—19 cm^{-1} lower because of the greater strength of the M—NH$_3$ compared with the M—OH$_2$ bond.[2]

---

[1] J. T. Marks, W. J. Kennelly, J. R. Kolb, and L. A. Shimp, *Inorg. Chem.*, 1972, **11**, 2540.
[2] H. Hagnauer, G. C. Stocco, and R. S. Tobias, *J. Organometallic Chem.*, 1972, **46**, 179.

Two studies of the i.r. spectra of the species $[M(CN)_5(NO)]^{n-}$ have been reported, where M = Co, Ni, Cu, or Zn (ref. 3) and M = V, Cr, Mo, Mn, Fe, or Ru (ref. 4). In the latter case [4] the assignments were assisted by studies of the isotopic species $Na_2[Fe(CN)_5(^{15}NO)],2H_2O$, $Na_2[Fe(CN)_5(NO)],2D_2O$, and $Na_2[Fe(CN)_5(NO)],2HDO$.

A review of transition-metal nitrido-complexes (compounds of the N^{3-} ligand which involve discrete molecular units) includes a section on vibrational spectra, with particular reference to $\nu(M \equiv N)$; a summary of data is given in Table 1.[5]

Table 1 *Vibrational data on metal–nitrido-species*/cm^{-1}

Species	$\nu_{asym}(M_xN)$	$\nu_{sym}(M_xN)$	$\delta(M_xN)$
$[MO_3N]^{n-}$	ca. 1020	—	ca. 380
$[MNY_y]^{n-}$	1000—1120	—	300—370
$[M_2NY_y]^{n-}$	1040—1130	240—370	ca. 120
$[M_3NY_y]^{n-}$	ca. 770	ca. 220	ca. 130
$[M_4NY_y]^{n-}$	ca. 600	ca. 140	—

A thorough and careful attempt has been made to provide firm evidence for assignment of $\nu(ML)$ in series of twenty-eight N-aryl- and nine N-alkyl-salicylaldimine complexes.[6,7] The principal means used were noting the effects of (a) ^{15}N-labelling, (b) electron-withdrawing substituents in either the N-aryl- or salicylaldimine rings, and (c) change in M and the accompanying ligand-field stabilization energy. Despite the complexity of the spectra the arguments are well presented and there is good internal consistency. Ranges deduced are $\nu(MN)$ at 408—559 cm^{-1} (N-aryl series) or 392—513 cm^{-1} (N-alkyl series), with (tentatively) $\nu(MO)$ [coupled to $\nu(MN)$] at 370—575 cm^{-1} or 508—595 cm^{-1}, respectively.[6,7]

Convincing assignments of $\nu(MN)$ have also been presented for several other series of transition-metal complexes.[8–14] In chelate complexes of ligand (1), ring-strain reduces $\nu(MN)$ [M = Mn, Co, ^{62}Ni, or ^{64}Zn] by about 15—20% compared with the values for corresponding bipy compounds.[8] UBFF Calculations indicate substantial mixing of $\nu(MN)$ and $\nu(MO)$ with internal ligand modes in Co^{II}, Ni^{II}, and Pb^{II} complexes of the NN'-ethylenebis(acetylacetoneiminato) group; the M—N and M—O stretching force constants (Co < Ni < Pd) vary in the same manner as the

---

[3] L. Tosi, *Compt. rend.*, 1972, **275**, C, 439.
[4] J. Ziolkowski, B. Jezowska-Trzebiatowska, and B. B. Kedzia, *Bull. Acad. polon. Sci., Sér. Sci. chim.*, 1972, **20**, 231.
[5] W. P. Griffith, *Co-ordination Chem. Rev.*, 1972, **44**, 369.
[6] G. C. Percy and D. A. Thornton, *J. Inorg. Nuclear Chem.*, 1972, **34**, 3357.
[7] G. C. Percy and D. A. Thornton, *J. Inorg. Nuclear Chem.*, 1972, **34**, 3369.
[8] B. Hutchinson and A. Sunderland, *Inorg. Chem.*, 1972, **11**, 1948.
[9] A. Bigotto, V. Galasso, and G. de Alti, *Spectrochim. Acta*, 1972, **28A**, 1581.
[10] H. Ogoshi, Y. Saito, and K. Nakamoto, *J. Chem. Phys.*, 1972, **57**, 4194.
[11] Y. Saito, J. Takemoto, B. Hutchinson, and K. Nakamoto, *Inorg. Chem.*, 1972, **11**, 2003.
[12] M. Goldstein and W. D. Unsworth, *Spectrochim. Acta*, 1972, **28A**, 1107.
[13] M. Goldstein and W. D. Unsworth, *Spectrochim. Acta*, 1972, **28A**, 1297.
[14] M. Goldstein, F. B. Taylor, and W. D. Unsworth, *J.C.S. Dalton*, 1972, 418.

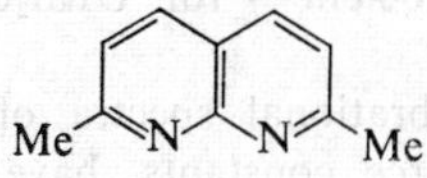

Me ⟨structure⟩ Me

(1)

calculated total bond-overlap populations.[9] I.r. spectra (to 100 cm^{-1}) of porphin complexes of ^{64}Zn, ^{68}Zn, Cu, and Ni have been assigned using a normal-co-ordinate analysis of the eighteen in-plane vibrations; the largest contribution to ν(MN) is in bands found at ca. 203 (Zn), 246 (Cu), and 295 cm^{-1} (Ni).[10] The metal isotope technique (Cr, Fe, and Cu isotopes) has also been used to aid assignment of ν(MN) in tris(bipy) and tris(phen) complexes of Cr, Fe, V, Ti, Cu, Co, and Mn.[11] For adducts of chromium(III, II, I, or 0), iron(III or II), vanadium(II or 0), titanium(0 or $-$I), and cobalt(III), in which there are no valence e_g* electrons, ν(MN) is in the range 300—390 cm^{-1}. However, for the complexes of copper(II), nickel(II), zinc(II), cobalt(II, I, or 0), and manganese(II, 0, or $-$I), the effect of e_g* electrons is to decrease ν(MN) to 180—290 cm^{-1}.

Refs. 12—14 are discussed below. Values for metal–nitrogen stretching modes have also been suggested for complexes of 2-(2'-pyridyl)benzimidazole[15] and 8-amino-7-hydroxy-4-methylcoumarin,[16] and partial i.r. data (900—1300 cm^{-1}) listed for [M(phen)$_3$]$^{2+}$ xanthates (M = Mn, Fe, Co, Ni, or Zn).[17]

A review of spectroscopic studies of metal–phosphorus bonding in co-ordination complexes (including phosphoryl adducts) includes a comprehensive tabulation and discussion on ν(MP) assignments for more than 150 complexes.[18]

The i.r. spectra of bis(tropolonates), [MT$_2$] (M = Ni or Zn), have been interpreted in terms of octahedral polymeric structures involving oxygen bridges, largely on the basis of values of modes described as ν(MO) in comparison with octahedral species [MT$_2$L$_2$] (L = H$_2$O or py). Thiotropolonates [M(ST)$_2$] are said to be planar (Ni or Cu), octahedral polymeric (Zn), or tetrahedral (Co).[19]

Neutral acetylacetone (acacH) forms complexes with Cr, Mn, Co, Ni, or Zn which show ν(MO) values at 184—264 cm^{-1}, lower than those associated with related acetylacetonato-chelates (400—450 cm^{-1}).[20]

Suggestions for ν(MO) values have been made for adducts of 2,6-lutidine N-oxide[21] and of 8-amino-7-hydroxy-4-methylcoumarin.[22] Un-

---

[15] D. M. L. Goodgame and A. A. S. C. Machado, *Inorg. Chim. Acta*, 1972, **6**, 317.
[16] D. K. Rastogi, *J. Inorg. Nuclear Chem.*, 1972, **34**, 619.
[17] D. G. Holah and C. N. Murphy, *Inorg. Nuclear Chem. Letters*, 1972, **8**, 1069.
[18] J. G. Verkade, *Co-ordination Chem. Rev.*, 1972, **9**, 1.
[19] L. G. Hulett and D. A. Thornton, *Spectroscopy Letters*, 1972, **5**, 323.
[20] Y. Nakamura, K. Isope, H. Morita, S. Yamazaki, and S. Kawaguchi, *Inorg. Chem.*, 1972, **11**, 1573.
[21] N. M. Karayannis, C. M. Mikulski, L. L. Pytlewski, and M. M. Labes, *J. Inorg. Nuclear Chem.*, 1972, **34**, 3139.
[22] D. K. Rastogi, A. K. Srivastava, P. C. Jain, and B. R. Agarwal, *J. Inorg. Nuclear Chem.*, 1972, **34**, 1449.

assigned i.r. data (400—700 cm^{-1}) for URh_2O_6, UCr_2O_6, and $ZnSb_2O_6$ have been published.[23]

Assignments of the vibrational spectra of $[VS_4]^{3-}$, $[MoSe_4]^{2-}$, and $[WSe_4]^{2-}$, and derived force constants, have been given based on T_d models.[24] Data for v_1 and v_3 for these and related ions are included in Chapter 4 (v_2 could not be located).[24] For NbS_2X_2 (X = Cl, Br, or I) and the polysulphides VS_4 and MS_3 (M = Ti, Zr, Nb, or U), v(M—S—M) bridge-stretching modes give complex i.r. and Raman patterns in the 200—400 cm^{-1} region.[25] Specific v(MS) assignments have been proposed for v(MS) in four-co-ordinate complexes containing bidentate chelating thiocarbonato-ligands $[M(CS_3)_2]^{2-}$, $[L_2M(CS_3)]$, and $[L_2M(COS_2)]$. For example, in $[Ph_4As]_2[Pd(CS_3)_2]$, v(MS) is at 375 (A_{1g}), 339 (B_{3u}), 321 (B_{1g}), and 290 cm^{-1} (B_{2u}).[26] Other assignments given for v(MS) in series of compounds are 370—380 cm^{-1} in quinazoline-[^{1}H,^{3}H]-2,4-dithione complexes of rhodium(I, II, or III) and ruthenium(II or III),[27] and ca. 310 cm^{-1} in $M(S_2PPh_2)_2$ and $M(SeSPPh_2)_2$ complexes (M = Co, Ni, or Zn).[28]

The i.r. spectra and geometry of dihalides of elements of Groups IIA and IIB (Zn, Cd, and Hg) have been critically reviewed, and data included on first-transition series (Sc to Cu) analogues. Data obtained from the gas phase and from matrix isolation are tabulated and discussed, with particular emphasis on the reliability of the geometry deduced from i.r. data.[29]

Low-frequency i.r. measurements have been used to augment n.q.r. studies on alkali-metal salts of $[MCl_6]^{3-}$ (M = Rh or Ir) and $[MCl_6]^{2-}$ (M = Mn, Tc, Ru, or Rh) anions, by confirming that the anions are in low-symmetry sites. Conclusions regarding the extent of π-bonding in these systems have been reached by comparing the variation of n.q.r. frequency and of v_3(MCl) data with the d-electron configuration of M.[30]

An analysis has been presented of the Raman (10—300 cm^{-1}) and extended far-i.r. (20—100 cm^{-1}) spectra of complexes MX_2L_2 {M = Mn, Fe, or Ni; X = Cl or Br; L = py, [^{2}H$_5$]py, or $\frac{1}{2}$(pyrazine)} in terms of the octahedral halogen-bridged structures present.[12] These data supplement existing i.r. results (to 100 cm^{-1}). Data on similar complexes containing terminal M—X bonds, MX_2L_4 {M = Mn, Co, or Ni; X = Cl, Br, or I; L = py, [^{2}H$_5$]py, pyrazole, or $\frac{1}{2}$(hydrazine)} have also been discussed, and clarify previous discrepancies in assignments of v(MX) and v(MN) in such

---

[23] J. Omaly and J. P. Badand, *Compt. rend.*, 1972, **275**, *C*, 371.
[24] A. Müller, K. H. Schmidt, K. H. Tytko, J. Bouwma, and F. Jellinek, *Spectrochim. Acta*, 1972, **28A**, 381.
[25] C. Perrin, A. Perrin, and J. Prigent, *Bull. Soc. chim. France*, 1972, 3086.
[26] J. M. Burke and J. P. Fackler, jun., *Inorg. Chem.*, 1972, **11**, 2744.
[27] U. Agarwala and L. Agarwala, *J. Inorg. Nuclear Chem.*, 1972, **34**, 241.
[28] A. Müller, V. V. K. Rao, and P. Christophliemk, *J. Inorg. Nuclear Chem.*, 1972, **34**, 345.
[29] I. Eliezer and A. Reger, *Co-ordination Chem. Rev.*, 1972, **9**, 189.
[30] P. J. Cresswell, J. E. Fergusson, B. R. Penfold, and D. E. Scaife, *J.C.S. Dalton*, 1972, 254.

Table 2 Metal–metal vibrations/cm⁻¹ of dinuclear complexes

Complex	$\nu(M^1-M^2)^a$	Ref.
Mo₂(O₂CMe)₄(py)₂	363[b]	37
Mo₂(O₂CCF₃)₄(py)₂	367[b]	37
H₃GeMn(CO)₅	219	38[c]
D₃GeMn(CO)₅	220	38[c]
Me₃SnMn(CO)₅	178	39
Me₂ClSnMn(CO)₅	194	39
Me₂SiCH₂CH₂CH₂Fe(CO)₄	302[d]	40
H₃Ge(H)Fe(CO)₄	226	41
H₃GeCo(CO)₄	228	42
D₃GeCo(CO)₄	224	42
trans-[(py)₂ClPtMo(CO)₃(π-C₅H₅)]	142[e]	43
Hg₂(O₂CMe)₂	166	44
Hg₂(O₂CCD₃)₂	161	44

[a] Raman data unless otherwise stated. [b] These values, being ca. 30 cm⁻¹ lower than in the parent compounds Mo₂(O₂CCX₃)₄, suggest weakening of the Mo–Mo bond on axial substitution, in agreement with X-ray data. [c] Assignments given for other modes in terms of local C_{3v} and C_{4v} symmetry at Ge and Mn, respectively. [d] Tentative i.r. assignment. [e] I.r. value.

Table 3 Metal–metal vibrations/cm⁻¹ of trinuclear and polynuclear complexes

Complex	$\nu_{sym}(M^1M^2_2)$ I.r.	$\nu_{sym}(M^1M^2_2)$ Raman	$\nu_{asym}(M^1M^2_2)$ I.r.	$\nu_{asym}(M^1M^2_2)$ Raman	Ref.
trans-[L₂Pd{Mn(CO)₅}₂]	—	—	148—149	—	
trans-[L₂Pt{Mn(CO)₅}₂]	—	—	153—155	—	
trans-[py₂Pd{Co(CO)₄}₂]	—	—	168	—	
trans-[py₂Pt{Co(CO)₄}₂]	—	—	177	—	43[a]
trans-[py₂Pd{Mo(CO)₃(π-C₅H₅)}₂]	—	—	134	—	
trans-[L₂Pt{Mo(CO)₃(π-C₅H₅)}₂]	—	—	140—144	—	
(H₃Ge)₂Fe(CO)₄	—	229	—	216	41[b]
H₂Ge[Mn(CO)₅]₂	—	200	—	226	
[Hg{Ru(π-C₅H₅)₂}₂]²⁺	—	110—113	—	—	45[c]
[XHg{Co(CO)₄}₂]⁻	151	—	158	—	46[d]
(Ph₃Ge)₂Zn(bipy)	—	—	212	—	
(Ph₃Ge)₂Zn(tmed)	—	—	221	—	
(Ph₃Sn)₂Zn(bipy)	—	—	195	—	
(Ph₃Sn)₂Zn(tmed)	—	—	211	—	47
(Ph₃Ge)₂Cd(bipy)	—	—	182	—	
(Ph₃Ge)₂Cd(tmed)	—	—	192	—	
(Ph₃Sn)₂Cd(bipy)	—	—	168	—	
(Ph₃Sn)₂Cd(tmed)	—	—	178	—	
In[Mn(CO)₅]₃	—	—	168, 162, 155 [e]	—	48[h]
(OC)₅MnGaGa[Mn(CO)₅]₃	265[f]	265[f]	205, 201 [g]	—	

[a] L = py, 3-methylpyridine, or 4-methylpyridine; simplified force-constant calculations carried out for a linear triatomic M²—M¹—M² model. [b] Raman polarization data confirm the unusual order $\nu_{sym}(FeGe_2) > \nu_{asym}(FeGe_2)$. [c] ClO₄⁻, PF₆⁻, and BF₄⁻ salts. [d] Identical values for the series X = Cl, Br, or I. [e] Undistinguished i.r. values for ν(InMn) of a trigonal-planar InMn₃ skeleton. [f] ν(GaGa). [g] Undistinguished ν(GaMn). [h] These and other vibrational data used to identify the compounds as products of reaction of Mn₂(CO)₁₀ with In or Ga.

compounds.[13] Full details have now appeared of the vibrational spectroscopic evidence for assigning a pyrazine-bridged sheet structure, with terminal halogen atoms, to complexes $MX_2(pyrazine)_2$ [M = Co or Ni; X = Cl, Br, or I].[14]

$\nu(MX)$ Modes have also been assigned in the following systems: $MCl_2(Ph_2PO \cdot NMe_2)_2$ [M = Mg, Ca, Mn, Fe, Co, Ni, Cu, Zn, or Cd];[31] $MCl_2(thiophen-2-aldoxime)_2$ [M = Zn, Ni, Cu, or Pd];[32] $[MX_3L]^+$ salts (M = Co, Ni, Cu, or Zn; X = Cl or Br; L = 1,1,4-trimethylpiperazinium cation);[33] a large number of complexes of pyridine-2-thiol, pyridine-4-thiol, 2-methylpyridine-6-thiol, or some oxygen analogues with Sn^{IV}, Bi^{III}, or M^{II} (M = Co, Ni, Zn, Cd, Hg, or Pt);[34] $[MCl_4]^{2-}$ salts (M = Mn, Co, Cu, or Zn) of the 2,4-dimethyl-1H-1,5-benzodiazepinium cation;[35] and in dimethoxyethane adducts of MCl_4 (M = Ti or V) or $CrCl_3$.[36]

Metal–metal vibrational assignments, of which fewer were noted this year, are collected together in Tables 2 and 3.[37-48] Further measurements have also been made on the spectra of $M[Co(CO)_4]_2$ species (M = Zn, Cd, or Hg); metal–metal force constants vary in the order:

$$k(\text{Zn—Co}) \gtrsim k(\text{Cd—Co}) \gtrsim k(\text{Hg—Co})$$

and significant coupling across the trinuclear $M^1M^2{}_2$ unit was found.[49]

3 Scandium

I.r. spectroscopy has been used to show that various $[Sc(NCS)_4]^-$ salts (generally hydrated) contained N-bonded NCS^- ligands; the values proposed for $\nu(M-OH_2)$ and $\nu(M-NCS)$ are remarkably high.[50]

[31] M. W. G. De Bolster and W. L. Groeneveld, *Z. Naturforsch.*, 1972, **27b**, 759.
[32] M. P. Coakley and M. E. Casey, *J. Inorg. Nuclear Chem.*, 1972, **34**, 1937.
[33] A. S. N. Murthy, J. V. Quagliano, and L. M. Vallarino, *Inorg. Chim. Acta*, 1972, **6**, 49.
[34] B. P. Kennedy and A. B. P. Lever, *Canad. J. Chem.*, 1972, **50**, 3488.
[35] P. W. W. Hunter and G. A. Webb, *J. Inorg. Nuclear Chem.*, 1972, **34**, 1511.
[36] E. Hengge and H. Zimmermann, *Monatsh.*, 1972, **103**, 418.
[37] F. A. Cotton and J. G. Norman, *J. Amer. Chem. Soc.*, 1972, **94**, 5697.
[38] R. D. George, K. M. Mackay, and S. R. Stobart, *J.C.S. Dalton*, 1972, 1505.
[39] R. A. Burnham, F. Glockling, and S. R. Stobart, *J.C.S. Dalton*, 1972, 1991.
[40] C. S. Cundy and M. F. Lappert, *J.C.S. Chem. Comm.*, 1972, 445.
[41] S. R. Stobart, *J.C.S. Dalton*, 1972, 2442.
[42] R. D. George, K. M. Mackay, and S. R. Stobart, *J.C.S. Dalton*, 1972, 974.
[43] P. Braunstein and J. Dehand, *J.C.S. Chem. Comm.*, 1972, 164; *Compt. rend.*, 1972, **274**, C, 175.
[44] R. P. J. Cooney and J. R. Hall, *J. Inorg. Nuclear Chem.*, 1972, **34**, 1519.
[45] D. N. Hendrickson, Y. S. Sohn, W. H. Morrison, jun., and H. B. Gray, *Inorg. Chem.*, 1972, **11**, 808.
[46] H. L. Conder and W. R. Robinson, *Inorg. Chem.*, 1972, **11**, 1527.
[47] F. J. A. Des Tombe, G. J. M. Van Der Kerk, H. M. J. C. Creemers, N. A. D. Carey, and J. G. Noltes, *J. Organometallic Chem.*, 1972, **44**, 247.
[48] H.-J. Haupt and F. Neumann, *Z. anorg. Chem.*, 1972, **394**, 67.
[49] R. J. Ziegler, J. M. Burlitch, S. E. Hayes, and W. M. Risen, jun., *Inorg. Chem.*, 1972, **11**, 702.
[50] L. N. Komissarova, T. M. Sas, and V. G. Gulia, *Russ. J. Inorg. Chem.*, 1971, **16**, 1115.

The i.r. and Raman spectra of $[Sc(CO_3)_4]^{5-}$ in aqueous solution have been compared with those of aqueous Na_2CO_3, and the presence of bidentate CO_3^{2-} groups has been deduced; $\nu(ScO)$ is given as 360 cm^{-1} (i.r.).[51] The thermal decompositions of the following Sc compounds have been studied using i.r. and other techniques: $K_3Sc(SeO_4)_3$ and $MSc(SeO_4)_2$ [M = Rb or Cs];[52] $Sc_2(SeO_4)_3(H_2O)_5$;[53] and $ScI_3(H_2O)_6$ and $Sc(OH)I_2(H_2O)_5$.[54] Assignments of i.r. data for $ScPO_4$ pre-heated to 1000 °C, $ScPO_4(H_2O)_2$, and $Sc(H_2PO_4)_3$ have been given in terms of T_d symmetry of the PO_4 groups, whereas for $ScPO_4$ pre-heated to 800 °C and $[Sc(PO_3)_3]_n$ the symmetry of the PO_4 groups is said to be C_{3v}.[55]

I.r. data have been listed for $ScX_3(phen)_2$ and $ScX_3(bipy)_2$ [X = NCS, NO_3, or Cl], all of which are believed to contain seven-co-ordinate scandium;[56, 57] some of the assignments given for $\nu(ML)$ modes [*e.g.* $\nu(ScCl)$ at *ca.* 470 cm^{-1}] are in spectacular contrast to normally accepted ranges. I.r.-active $\nu(ScX)$ values have been given for anhydrous $ScCl_3$ (310 cm^{-1}) and $ScBr_3$ (280 cm^{-1}).[58]

No vibrational data on yttrium compounds were collected during last year. Compounds of lanthanides and actinides are considered in Sections 13 and 14.

4 Titanium, Zirconium, and Hafnium

The polymeric structure of $[(\pi-C_5H_5)_2TiH]_x$ gives rise to the very low value for $\nu(Ti-H-Ti)$ of 1140 cm^{-1} (800—850 cm^{-1} in the deuteriated compound), compared with the value of 1450 cm^{-1} for $[(\pi-C_5H_5)_2TiH]_2$.[59] In contrast, $[(\pi-C_5H_5)(C_5H_4)TiH]_x$ [one of several products from the action of Na on $(\pi-C_5H_5)_2TiCl_2$] shows $\nu(TiH)$ at 1960 and 1815 cm^{-1} (1355 and 1305 cm^{-1} in the corresponding deuteriate).[60] I.r. bands at 1350 and 1780 cm^{-1} have been assigned to $\nu(ZrH_2Zr)$ and $\nu(ZrHAl)$, respectively, in the tetranuclear compound (2); these bands shift to 980 and 1290 cm^{-1}, respectively, on deuteration.[61]

[51] B. Taravel, F. Fromage, P. Delorme, and V. Lorenzelli, *Compt. rend.*, 1972, **275**, B, 589.
[52] H. Tetsu, L. G. Korotaeva, and B. N. Ivanov-Emin, *Russ. J. Inorg. Chem.*, 1971, **16**, 957.
[53] H. Tetsu, L. G. Korotaeva, and B. N. Ivanov-Emin, *Russ. J. Inorg. Chem.*, 1971, **16**, 1552.
[54] N. P. Shepelev, I. V. Arkangel'skii, L. N. Komissarova, and V. M. Shatskii, *Russ. J. Inorg. Chem.*, 1971, **16**, 1706.
[55] L. N. Komissarova, P. P. Mel'nikov, E. G. Teterin, and V. F. Chuvaev, *Russ. J. Inorg. Chem.*, 1971, **16**, 1414.
[56] L. N. Komissarova, Yu. G. Eremin, V. S. Katochkina, and T. M. Sas, *Russ. J. Inorg. Chem.*, 1971, **16**, 1570.
[57] L. N. Komissarova, Yu. G. Eremin, V. S. Katochkina, and T. M. Sas, *Russ. J. Inorg. Chem.*, 1971, **16**, 1708.
[58] R. W. Stotz and G. A. Melson, *Inorg. Chem.*, 1972, **11**, 1720.
[59] J. E. Bercaw, R. H. Marvich, L. G. Bell, and H. H. Brintzinger, *J. Amer. Chem. Soc.*, 1972, **94**, 1219.
[60] E. E. van Tamelen, W. Cretney, N. Klaentschi, and J. S. Miller, *J.C.S. Chem. Comm.*, 1972, 481.
[61] P. C. Waites, H. Weigold, and A. P. Bell, *J. Organometallic Chem.*, 1972, **43**, C29.

$$(\pi\text{-}C_5H_5)_2Zr\underset{H}{\overset{H}{\diagup\diagdown}}Zr(\pi\text{-}C_5H_5)_2$$
$$\underset{Me_3Al}{\overset{H}{\diagup}} \quad (2) \quad \underset{AlMe_3}{\overset{H}{\diagdown}}$$

Further i.r. data have been reported for Me_2TiCl_2, CD_3TiCl_3, and $(CD_3)_2TiCl_2$, but definitive assignments were not possible.[62] In 1:1 adducts of $MeTiCl_3$ with $MeCOCH_2CH_2NMe_2$, $MeOCH_2CH_2SMe$, Me_2NCH_2-CH_2SMe, and $o\text{-}Me_2NC_6H_4CH_2NMe_2$, $\nu(TiC)$ is in the range 459—484 cm^{-1}; two geometrical isomers were identified for $MeTiCl_3(Me_2\text{-}NCH_2CH_2SMe)$, corresponding to the Me—Ti bond being *trans* to the SMe or NMe_2 groups of the ligand.[63] Further details and some additional examples have now been given [64] of the spectra of methyltitanium halogenoanions included in last year's report, *e.g.* $[Me_2Ti_2Cl_7]^-$, $[Me_2Ti_2ClBr_6]^-$, $[Me_2Ti_2Br_8]^{2-}$, $[Me_2Ti_2Cl_6Br_2]^{2-}$, and $[MeTiCl_2Br_3]^{2-}$. Assignments for the low-frequency vibrations of $(\pi\text{-}C_5H_5)TiX_3$ and $(\pi\text{-}C_5Me_5)TiX_3$ [X = Cl or Br] are listed in Table 4.[65] Similar values for $\nu(Ti\text{—ring})$ (430—440 cm^{-1}) and $\nu(TiCl)$ (310—320 cm^{-1}) have been proposed for complexes $(\pi\text{-}C_5H_5)TiCl_2(L)$ [L = bipy, phen, py, or α-picolylamine].[66] Spectra of $(\pi\text{-}C_5Me_5)Ti(CO)_2$ and $[(\pi\text{-}C_5H_5)_2Ti]_2$ have also been reported.[59] In $(Et_2N)_3Ti$—CH=CMe_2, $\nu(TiC)$ and $\nu(TiN)$ are at 495 and 612 cm^{-1}, respectively;[67] corresponding values for $(Et_2N)_3Ti$—$C(Me)$=CH_2 are 533 and 620 cm^{-1}.

Isothiocyanato-complexes of titanium have been formulated as in Table 5 on the basis of $\nu(CS)$ and $\delta(NCS)$ frequencies.[68] In $K_2[M(NCSe)_6]$,

Table 4 Low-frequency vibrations/cm^{-1} of $(\pi\text{-}C_5H_5)TiX_3$ and $(\pi\text{-}C_5Me_5)TiX$ compounds

Assignment	$(\pi\text{-}C_5H_5)TiCl_3$	$(\pi\text{-}C_5Me_5)TiCl_3$	$(\pi\text{-}C_5H_5)TiBr_3$	$(\pi\text{-}C_5Me_5)TiBr_3$
$\nu(Ti\text{—ring})$	453	460	425	438
$\nu_{asym}(TiX_3)$	404	408	$\begin{cases}345\\355\end{cases}$	330
$\nu_{sym}(TiX_3)$	327	338	$\begin{cases}260\\273\end{cases}$	$\begin{cases}255\\261\\264\end{cases}$
$\delta_{sym}(TiX_3)$	$\begin{cases}125\\137\end{cases}$	160	—	—
$\delta_{asym}(TiX_3)$	—	115	$\begin{cases}81\\87\end{cases}$	110
$\delta(X\text{—Ti—ring})$	$\begin{cases}158\\167\end{cases}$	210	130	174

[62] J. F. Hanlan and J. D. McCowan, *Canad. J. Chem.*, 1972, **50**, 747.
[63] R. J. H. Clark and A. J. McAlees, *Inorg. Chem.*, 1972, **11**, 342.
[64] R. J. H. Clark and M. A. Coles, *J.C.S. Dalton*, 1972, 2454.
[65] O. S. Roshchupkina, V. A. Dubovitskii, and Yu. G. Borod'ko, *J. Struct. Chem.*, 1972, **12**, 928.
[66] R. S. P. Coutts, R. L. Martin, and P. C. Wailes, *Austral. J. Chem.*, 1972, **25**, 1401.
[67] H. Bürger and H.-J. Neese, *J. Organometallic Chem.*, 1972, **36**, 101.
[68] A. M. Sych and V. P. Dem'yanenko, *Russ. J. Inorg. Chem.*, 1971, **16**, 1593.

Vibrational Spectra of Transition-element Compounds 317

Table 5 *Low-frequency modes/cm^{-1} of some isothiocyanato-complexes of titanium*

Compound	ν(Ti—NCS)	ν(Ti—amine)	δ(Ti—NCS)	Symmetry^a
[Bu$^n_4$N]$_2$[Ti(NCS)$_6$]	336	—	146	O_h
[Me$_4$N][TiO(NCS)$_4$]^b	$\begin{cases}354\\284\end{cases}$	—	158	D_{4h}
[Ti(bipy)$_2$(NCS)$_2$](NCS)$_2$	$\begin{cases}364\\310\end{cases}$	232	$\begin{cases}160\\146\end{cases}$	D_{2h}
[Ti(phen)$_2$(NCS)$_2$](NCS)$_2$	$\begin{cases}368\\315\end{cases}$	212	158	D_{2h}

^{a} Skeletal symmetry deduced from the vibrational data. ^{b} A polymeric structure is proposed on the basis of i.r. bands at 800 and 730 cm^{-1} assigned to ν(TiOTi); δ(TiOTi) given as 236 cm^{-1}.

ν(MN) is at 249 (M = Zr) or 228 cm^{-1} (M = Hf).[69] The extraction of Zr and Hf thiocyanato-complexes by solvents such as (Bu^{n}O)$_3$PO has been studied using i.r. spectroscopy.[70]

In alkylphosphine adducts (1:1) of TiCl$_3$, ν(TiCl) is in the expected 298—375 cm^{-1} range but ν(TiP) is significantly dependent on the nature of the phosphine; values are 410 (MePH$_2$), 375 (Me$_2$PH), 370 (Me$_3$P), and 320 cm^{-1} (Et$_3$P).[71]

An accurate and rapid Raman spectroscopic method has been developed for determining low concentrations of anatase in rutile TiO$_2$ pigments (0.03—10%).[72] In Li$_4$TiO$_4$, i.r. values for ν(TiO) (720—740 cm^{-1}) support crystallographic evidence for the presence of TiO$_4$ tetrahedra rather than TiO$_6$ octahedra (similar results were obtained for Li$_4$GeO$_4$).[73] Complexes MeOTiCl$_3$(L) [L = MeOCH$_2$CH$_2$NMe$_2$, MeOCH$_2$CH$_2$SMe, or Me$_2$NCH$_2$CH$_2$SMe][74] show ν(TiO) in the 605—615 cm^{-1} range with ν(TiCl) between 270 and 400 cm^{-1}. Other systems for which ν(M—O) or ν(M=O) data have been given are: co-precipitated hydroxides of Bi^{III} and Ti^{IV};[75] oxytitanium(IV) complexes with various Schiff bases;[76] [TiO(SO$_4$)$_2$]$^{2-}$, [TiO(SO$_4$)$_5$]$^{4-}$, [Ti(SO$_4$)$_3$]$^{2-}$, [Ti$_2$O(SO$_4$)$_4$]$^{2-}$, and [Ti$_4$O(SO$_4$)$_9$]$^{4-}$ salts and their thermal decomposition products;[77] the complexes TiOCl$_2$L$_2$ (L = benzidine, *o*-phenetidine, *etc.*);[78] tetrakis-(*dl*-mandelato)zirconium;[79] and volatile double alkoxides of hafnium(IV) such as NaHf$_2$(OEt)$_9$, K$_2$Hf$_3$(OPr^i)$_{14}$, KHf(OBu^t)$_5$, HfAl(OPr^i)$_7$, and HfGa$_2$(OPr^i)$_{10}$.[80] Vibrations of the biden-

[69] A. Galliart and T. M. Brown, *J. Inorg. Nuclear Chem.*, 1972, **34**, 3568.
[70] O. A. Sinegribova and G. A. Yagodin, *Russ. J. Inorg. Chem.*, 1971, **16**, 1194.
[71] C. D. Schmulbach, C. H. Kolich, and C. C. Hinckley, *Inorg. Chem.*, 1972, **11**, 2841.
[72] R. J. Capwell, F. Spagnolo, and M. A. DeSesa, *Appl. Spectroscopy*, 1972, **26**, 537.
[73] B. L. Dubey and A. R. West, *Nature Phys. Sci.*, 1972, **235**, 155.
[74] R. J. H. Clark and A. J. McAlees, *J.C.S. Dalton*, 1972, 640.
[75] C. Gh. Macarovici and Gh. Morar, *Z. anorg. Chem.*, 1972, **393**, 275.
[76] N. S. Biradar, V. B. Mahale, and V. H. Kulkarni, *Inorg. Nuclear Chem. Letters*, 1972, **8**, 997.
[77] S. A. Filatova, Ya. G. Goroshchenko, E. K. Khandros, and G. S. Semenova, *Russ. J. Inorg. Chem.*, 1971, **16**, 832.
[78] M. M. Khan, *J. Inorg. Nuclear Chem.*, 1972, **34**, 3589.
[79] E. M. Larsen and E. H. Homeier, *Inorg. Chem.*, 1972, **11**, 2687.
[80] R. C. Mehrotra and A. Mehrotra, *J.C.S. Dalton*, 1972, 1203.

tate AcO⁻ ligands in $L_4M(OAc)_2$ $[L_4H_2$ = octaethylporphin; M = Zr or Hf] have also been assigned.[81]

$TiCl_3(acac)$ has been shown by X-ray crystallography to be a centrosymmetric dimer (3); comparison with $TiCl_2(acac)_2$ and $[Ti_2Cl_9]^-$ enabled

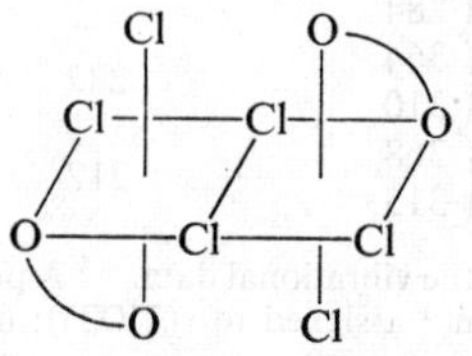

(3)

i.r. bands to be assigned to $\nu(TiCl)$ terminal (399 cm⁻¹) and $\nu(TiCl)$ bridging (276 and 246 cm⁻¹).[82] A partial assignment has been given of i.r. and Raman bands of $[PCl_4]_2[Ti_2Cl_{10}]$ in terms of modes predicted for a bridged M_2Cl_{10} system with D_{2h} symmetry.[83] Complexes of MCl_4 (M = Ti, Zr, or Hf) with chlorinated alkyl cyanides such as $CH_nCl_{3-n}CN$ (n = 1 or 2) have also been studied by i.r. spectroscopy.[84] Spectra of ZrX_3 (X = Cl or Br) and ZrX_3L_2 (X = Cl, Br, or I; L = various) are included in a review of certain d^n complexes.[85]

5 Vanadium, Niobium, and Tantalum

When $(\pi\text{-}C_5H_5)_2NbBH_4$ $[\nu(BH)$ = 2450 cm⁻¹] is treated with Ph_3P or $PhPMe_2$, the complexes $(\pi\text{-}C_5H_5)_2NbH(Ph_3P)$ $[\nu(NbH)$ = 1625 cm⁻¹] or $(\pi\text{-}C_5H_5)_2NbH(PhPMe_2)$ $[\nu(NbH)$ = 1630 cm⁻¹] are formed.[86] In $[(\pi\text{-}C_5H_5)_2NbH_2(PhPMe_2)][PF_6]$,[86] $\nu(NbH)$ is at 1740 cm⁻¹.

Partial assignments [e.g. $\nu(VC)$ = 368—369 cm⁻¹] have been presented[87] for $K_4[V(CN)_7],2H_2O$ and $K_4[V(CN)_6(NO)],H_2O$. In a range of dimethylniobium halides and their adducts, $\nu(NbC)$ is at ca. 490 cm⁻¹ in the i.r.[88]

Unassigned i.r. and Raman data are available for $(Et_2N)MF_4$ and $(Et_2N)_2MF_3$ (M = Nb or Ta).[89] Internal modes (i.r.) of the NCS⁻ groups have assigned for $[TaO(NCS)_3(MeCN)_2]$, $NH_4[TaO(NCS)_4],2MeCN$, $[TaS(NCS)_3(PhNCCl_2)]$, and $NH_4[TaS(NCS)_4(PhNCCl_2)]$,[90] while $\nu(MN)$ i.r. values (overall, 228—328 cm⁻¹; generally, 300—320 cm⁻¹) have been

---

[81] J. W. Buchler and K. Rohbock, *Inorg. Nuclear Chem. Letters*, 1972, **8**, 1073.
[82] N. Serpone, P. H. Bird, D. G. Bickley, and D. W. Thompson, *J.C.S. Chem. Comm.*, 1972, 217.
[83] D. Nicholls and K. R. Seddon, *Spectrochim. Acta*, 1972, **28A**, 2399.
[84] G. W. A. Fowles, K. C. Moss, D. A. Rice, and N. Rolfe, *J.C.S. Dalton*, 1972, 915.
[85] D. A. Miller and R. D. Bereman, *Co-ordination Chem. Rev.*, 1972, **9**, 107.
[86] C. R. Lucas and M. L. H. Green, *J.C.S. Chem. Comm.*, 1972, 1005.
[87] A. Müller, P. Werle, E. Diemann, and P. J. Aymonino, *Chem. Ber.*, 1972, **105**, 2419.
[88] G. W. A. Fowles, D. A. Rice, and J. D. Wilkins, *J.C.S. Dalton*, 1972, 2313.
[89] J. C. Fuggle, D. W. A. Sharp, and J. M. Winfield, *J.C.S. Dalton*, 1972, 1766.
[90] H. Böhland and F. M. Schneider, *Z. anorg. Chem.*, 1972, **390**, 53.

proposed for complexes of the types $Nb(NCE)_4L_2$, $Ta(NCS)_5(py)$, and $Ta(NCE)_5(bipy)$ [L = py or ½bipy; E = S or Se].[91]

New i.r. data [92] (300—1100 cm^{-1}) on *pure* V_2O_3, V_2O_4, and V_2O_5 are quite different from those of previous workers. The absence of coincidences between Raman and i.r. bands shows that from the possible structures proposed for Nb_2O_3 from crystallographic studies, the D_{3d}^3 ($P\bar{3}m1$) space group is correct; assignments of the spectra are given based on quadratic central force-field calculations.[93]

A re-assignment has been put forward for the spectra of the $[V_2O_7]^{4-}$ ion, the principal difference being in ν_{asym}(VOV) (680—780 cm^{-1}) and ν_{sym}(VOV) (500—550 cm^{-1}); the actual values for these modes depend on the cation.[94] The i.r. spectra (300—4000 cm^{-1}) of orthovanadates having the apatite structure, $M_5(VO_4)_3X$ [M = Ca, Sr, or Ba; X = F, Cl, or Br], have been assigned in terms of the known C_{6h} unit cell group as follows (symbols refer to T_d modes of VO_4^{3-}):[95] $\nu_1(E_u)$ = 826—850 cm^{-1}; $\nu_3(A_u + 2E_u)$ = 846—882 and 786—820 cm^{-1}; $\nu_4(A_u + 2E_u)$ = 2—3 bands, 356—417 cm^{-1}. Variations in the VO_4^{3-} internal modes with cation M^{2+} and halide X^- are also discussed. Site-group and unit-cell-group analyses have also been carried out for $M_3(VO_4)_2$ [M = Sr or Ba].[96] Raman powder spectroscopy has been shown to be a sensitive method for determination of stoicheiometric variations in lithium niobates and tantalates.[97] Several other vanadates or heteropolyanions containing vanadium have been studied.[98-102]

Two papers on $VO(acac)_2$ have appeared,[103, 104] and further work is clearly needed to reconcile the differing conclusions drawn. A study of the intensity and frequency variations in the ν(V=O) mode has led to the conclusion that in THF or EtOH no great disruption of the double-bond character of the V=O linkage occurs, whereas in py or $CHCl_3$ there is a significant lowering in VO bond order.[103] According to other workers,[104] *two* series of adducts $VO(acac)_2(L)$ [L = alkylpyridine] are formed depending on the nature of L, differentiated by having ν(V=O) lowered by 42 ± 4 or 29 ± 4 cm^{-1}. *X*-Ray data show that the former series are

[91] J. N. Smith and T. M. Brown, *Inorg. Chem.*, 1972, **11**, 2697.
[92] G. Fabbri and P. Baraldi, *Analyt. Chem.*, 1972, **44**, 1325.
[93] J. H. Denning and S. D. Ross, *J. Phys.* (*C*), 1972, **5**, 1123.
[94] R. G. Brown and S. D. Ross, *Spectrochim. Acta*, 1972, **28A**, 1263.
[95] E. J. Baran and P. J. Aymonino, *Z. anorg. Chem.*, 1972, **390**, 77.
[96] E. J. Baran, P. J. Aymonino, and A. Müller, *J. Mol. Structure*, 1972, **11**, 453.
[97] B. A. Scott and G. Burns, *J. Amer. Ceram. Soc.*, 1972, **55**, 225.
[98] A. S. Povarennykh and S. V. Gevork'yan, *Mineral. Sbornik* (*Lvov*), 1970, **24**, 254.
[99] B. Reuter and G. Colsmann, *Z. anorg. Chem.*, 1972, **394**, 138.
[100] A. I. Ivakin and A. P. Yatsenko, *Russ. J. Inorg. Chem.*, 1971, **16**, 893.
[101] D. U. Begalieva, A. B. Bekturov, and A. K. Il'yasova, *Russ. J. Inorg. Chem.*, 1971, **16**, 1464.
[102] A. A. Fotiev, A. G. Rustamov, A. A. Mambetov, *Russ. J. Inorg. Chem.*, 1971, **16**, 1604.
[103] R. Larsson, *Acta Chem. Scand.*, 1972, **26**, 549.
[104] M. R. Caira, J. M. Haigh, and L. R. Nassimbeni, *J. Inorg. Nuclear Chem.*, 1972, **34**, 3171.

cis-isomers (4) whereas the others have the trans-configuration; the cis-compounds have more complicated spectra in the 300—600 cm^{-1} range because of the lower symmetry and the presence of two different V—O bond lengths.[104]

$$\begin{array}{c} \text{structure (4): L--V with multiple V=O and V--O bonds} \end{array}$$

(4)

Assignments of ν(M=O) have also been given for many other complexes containing vanadyl(IV or V) groups,[105-115] and for some niobyl(V) species.[81, 105, 116] Oxohalogenovanadium species have received particular attention. Assignments for the four [Ph$_4$E][VO$_2$X$_2$] salts (E = P or As; X = F or Cl) have been given in terms of C_{2v} symmetry for the anions (Chapter 4), and simple valence force constants calculated.[112, 113] The ammonium compound NH$_4$[VO$_2$F$_2$], on the other hand, is considered to contain polymeric [—VF$_2$(O)—O—VF$_2$(O)—]$_n$ chains, and the following assignments have been made on this basis: ν(V=O) (968, 975, and 982 cm^{-1}), ν_{asym}(—V—O—V—) (738 and 806 cm^{-1}), ν_{sym}(—V—O—V—) (486 and 520 cm^{-1}), ν_{sym}(VF$_2$) (578 cm^{-1}), and ν_{asym}(VF$_2$) (448 cm^{-1}).[114] The [VOCl$_5$]$^{2-}$ ion has also been characterized [ν(V=O) at 917 cm^{-1}].[115] Absence of a band attributable to ν(V=O) indicates that VOCl and M$_3$[VOCl$_4$] (M = K, Rb, or Cs) are [—V—O—V—]$_n$ polymers.[117]

The product C$_{10}$H$_{25}$P$_4$O$_{12}$V, from reaction of VCl$_3$ with tetraethyl methylenediphosphonate, has been formulated as a polynuclear six-co-ordinate species, partly on the basis of i.r. data.[118]

[105] A. T. Casey, D. J. Mackey, R. L. Martin, and A. H. White, *Austral. J. Chem.*, 1972, **25**, 477.
[106] E. Higginbotham and P. Hambright, *Inorg. Nuclear Chem. Letters*, 1972, **8**, 747.
[107] A. Hodge, K. Nordquest, and E. L. Blinn, *Inorg. Chim. Acta*, 1972, **6**, 491.
[108] G. O. Carlisle and D. A. Crutchfield, *Inorg. Nuclear Chem. Letters*, 1972, **8**, 443.
[109] L. V. Kobets, N. I. Vorob'ev, V. V. Pechkovskii, and A. I. Komyak, *Zhur. priklad. Spektroskopii*, 1971, **15**, 682.
[110] R. G. Cavell, E. D. Day, W. Byers, and P. M. Watkins, *Inorg. Chem.*, 1972, **11**, 1591.
[111] K.-H. Thiele, W. Schumann, S. Wagner, and W. Brüser, *Z. anorg. Chem.*, 1972, **390**, 280.
[112] E. Ahlborn, E. Diemann, and A. Müller, *J.C.S. Chem. Comm.*, 1972, 378.
[113] E. Ahlborn, E. Diemann, and A. Müller, *Z. anorg. Chem.*, 1972, **394**, 1.
[114] R. Mattes and H. Rieskamp, *Z. Naturforsch.*, 1972, **27b**, 1424.
[115] J. Selbin, C. J. Ballhausen, and D. G. Durret, *Inorg. Chem.*, 1972, **11**, 510.
[116] K. A. Uvarova, Yu. I. Usatenko, N. V. Mel'nikova, and Zh. G. Klopova, *Russ. J. Inorg. Chem.*, 1971, **16**, 1141.
[117] V. T. Kalinnikov, A. I. Morozov, V. G. Lebedev, and O. D. Ubozhenko, *Russ. J. Inorg. Chem.*, 1971, **16**, 1088.
[118] C. M. Mikulski, N. M. Karayannis, L. L. Pytlewski, R. O. Hutchins, and B. E. Maryanoff, *Inorg. Nuclear Chem. Letters*, 1972, **8**, 225.

In the i.r. spectrum of $Nb[S_2P(OEt)_2]_4$, bands at 403, 356, 274, and 215 cm^{-1} are all assigned as ν(NbS) because they are absent from spectra of the sodium salt of the ligand.[119] Data on $[MS_4]^{3-}$ and $[MSe_4]^{3-}$ (M = V, Nb, or Ta)[24] have been included in Chapter 4.

A review of the chemistry of certain d^n complexes contains references to vibrational spectroscopic studies of e.g. MX_4L_2 (M = Nb or Ta; X = F, Cl, Br, or I; L = unidentate ligand), $Nb(NR_2)_4$, $[Nb(NCS)_6]^{2-}$, $[Nb(OR)Cl_5]^{2-}$, and $Nb(S_2CNR_2)_4$ [R = various], with particular reference to metal–ligand modes.[85]

In the far-i.r. spectra of MVBr$_3$ compounds (M = NH$_4$, Rb, or Cs), a doublet in the 210—240 cm^{-1} region has been assigned to ν_{asym}(VBr); in CsVBr$_3$(H$_2$O)$_3$, ν(VBr) is at 240 cm^{-1}, while for MVBr$_3$(H$_2$O)$_6$ [M = NH$_4$, K, Rb, or Cs] broad bands approximately at 470, 620, and 720 cm^{-1} are attributed to ν(V—OH$_2$).[120] New vanadium(II) halide complexes VX$_2$-(4-picoline)$_4$ have been characterized, showing ν(VX) in the i.r. at 306, 286 [superimposed on ν(VN)], or 220 cm^{-1} for X = Cl, Br, or I, respectively.[121] Possible structures have been proposed for $[M_6X_{12}]Y_2,nH_2O$ (M = Nb or Ta; X = Cl or Br; Y = Cl, Br, or I) partly on the basis of i.r. data.[122] In the complexes L$_4$MF$_3$ and L$_4$NbOF (L$_4$H$_2$ = octaethylporphin; M = Nb or Ta), ν(MF) is in the 540—600 cm^{-1} range.[81]

6 Chromium, Molybdenum, and Tungsten

Two groups of workers have reported ν(WH) in $(R_2PhP)_3WH_6$ [R = Me [123] or Et [124]]. Data are also available for ν(MH) in $(\pi$-$C_5H_5)_2WH_2$ and $(\pi$-$C_5H_5)_2W(H)CHCl_2$,[125] and in $(\pi$-$C_5H_5)_2M(H)\{-C(CO_2Me)=CHCO_2$-Me\} (M = Mo or W), $(\pi$-$C_5H_5)_2Mo(H)\{-C(CF_3)=CHCF_3\}$, and $(\pi$-$C_5H_5)_2Mo(H)\{-CH(CN)CH_2X\}$ (X = H or CN).[126] Complexes [WH$_4$-(PR$_3$)$_4$] \{PR$_3$ = PMe$_2$Ph, PMePh$_2$, or $\frac{1}{2}$(diphos)\},[127] shown to be stereochemically rigid in C_6D_6 at room temperature (n.m.r.), all show complex i.r. ν(WH) patterns in the region 1700—1850 cm^{-1}.

Skeletal modes of Bu^t_4Cr and $(neo$-$C_5H_{11})_4Cr$ have been located in the range 300—600 cm^{-1}, with ν(CrC) probably at 385 and 375 cm^{-1}, respectively.[128] I.r. spectra in the ν(CO) and 300—700 cm^{-1} regions have been discussed for solutions of 37 compounds LM(CO)$_5$ [L = various; M = Cr, Mo, or W]; the E (in C_{4v}) ν(M—C) mode is taken to be in the range 360—390 (M = Mo or W) or 440—470 cm^{-1} (M = Cr), and shows a

---

[119] R. N. McGinnis and J. B. Hamilton, *Inorg. Nuclear Chem. Letters*, 1972, **8**, 245.
[120] H. J. Seifert and A. Wüsteneck, *Inorg. Nuclear Chem. Letters*, 1972, **8**, 949.
[121] M. M. Khamar and L. F. Larkworthy, *Chem. and Ind.*, 1972, 807.
[122] H. Schäfer, B. Plautz, and H. Plautz, *Z. anorg. Chem.*, 1972, **392**, 10.
[123] J. R. Moss and B. L. Shaw, *J.C.S. Dalton*, 1972, 1910.
[124] B. Bell, J. Chatt, and G. J. Leigh, *J.C.S. Dalton*, 1972, 2492.
[125] Kon Swee Chen, J. Kleinberg, and J. A. Landgrebe, *J.C.S. Chem. Comm.*, 1972, 295.
[126] A. Nakamura and S. Otsuka, *J. Amer. Chem. Soc.*, 1972, **94**, 1886.
[127] B. Bell, J. Chatt, G. J. Leigh, and T. Ito, *J.C.S. Chem. Comm.*, 1972, 34.
[128] W. Kruse, *J. Organometallic Chem.*, 1972, **42**, C39.

linear correlation with the electronegativity of the donor atom in L.[129] The same authors have studied 26 complexes cis-[M(CO)$_4$(chel)] (M = Cr, Mo, or W; chel = bidentate chelate having N, P, As, or S donor atoms); bands in the region 365—563 cm^{-1} are (questionably) attributed to ν_{asym}(M—C) of two trans M—C bonds.[130]

X-Ray studies have shown that in {N$_4$P$_4$(NMe$_2$)$_8$}W(CO)$_4$ the phosphonitrile acts as a bidentate σ-ligand through one ring-N atom and one exocyclic group; in agreement with this it is found that its four ν(CO) values are close to those of (en)W(CO)$_4$.[131] Factor-group splitting is believed to be the reason why (N$_3$P$_3$Cl$_6$)Cr(CO)$_3$ shows twice as many i.r. ν(CO) bands (numerical data not given)[132] as (B$_3$N$_3$Me$_6$)Cr(CO)$_3$.

A vibrational analysis has been carried out on K$_3$[Cr(CN)$_5$(NO)], assuming the free-ion symmetry C_{4v} since crystal-field effects on the spectra were not observed. I.r. bands (to 100 cm^{-1}) were assigned by analogy with existing data on the Fe and Mn analogues, and by using ^{15}NO substitution. Dichroism observed in crystal polarized spectra could not be explained in terms of either of the two (previously proposed) alternative crystal structures, but no firmer conclusions could be drawn.[133] The i.r. spectrum of the hydrated compound K$_3$[Cr(CN)$_5$(NO)],H$_2$O has been found[134] to differ from that previously given. Assignments for this and the related compounds K$_4$[M(CN)$_5$(NO)] (M = Cr [134, 135] or Mo [134]) have been proposed.

In isothiocyanato-complexes, ν(M—NCS) has been assigned as follows: K$_2$[M(NCS)$_6$] (315 cm^{-1}, M = Mo; 281 cm^{-1}, M = W),[136] K[W(NCS)$_6$] (305 cm^{-1}),[136] [Mo(NCS)$_4$(bipy)] (305 cm^{-1}),[137] and [W(NCS)$_4$L$_2$] (275 cm^{-1}, L = py; 282 cm^{-1}, L$_2$ = bipy).[137] In the latter two series, ν(W—py) and ν(M—bipy) are in the 275—305 cm^{-1} range.[137] Assignments suggested[138] for [Ph$_4$As]$_3$[Cr(NCO)$_6$] include ν(CrN) at 345 cm^{-1}. In [MoNCl$_4$]$^-$, ν(MN) is at 1052 cm^{-1}, with ν(MCl$_4$) at 352 (sym) and 347 cm^{-1} (asym).[139]

I.r. absorption spectra of vapours over MoO$_3$ have been measured in an argon matrix, and bands attributed to ring vibrations of (MoO$_3$)$_3$, (MoO$_3$)$_4$, and (MoO$_3$)$_5$ species at 837, 856, and 865 cm^{-1}, respectively.[140] In complexes of the type MOX$_3$(chel) and MO$_2$X$_2$(chel) [M = Mo or W; chel = phen or bipy], the M—O force constants decrease as X is successively

---

[129] R. A. Brown and G. R. Dobson, *Inorg. Chim. Acta*, 1972, **6**, 65.
[130] G. R. Dobson and R. A. Brown, *J. Inorg. Nuclear Chem.*, 1972, **34**, 2785.
[131] H. P. Calhoun, N. L. Paddock, J. Trotter, and J. N. Wingfield, *J.C.S. Chem. Comm.*, 1972, 875.
[132] N. K. Hota and R. O. Harris, *J.C.S. Chem. Comm.*, 1972, 407.
[133] G. Paliani and A. Poletti, *Spectroscopy Letters*, 1972, **5**, 105.
[134] L. Tosi, *J. Chim. phys.*, 1972, **69**, 1052.
[135] L. Tosi, *Compt. rend.*, 1972, **274**, B, 249.
[136] C. J. Horn and T. M. Brown, *Inorg. Chem.*, 1972, **11**, 1970.
[137] T. M. Brown and C. J. Horn, *Inorg. Nuclear Chem. Letters*, 1972, **8**, 377.
[138] R. A. Bailey and T. W. Michelsen, *J. Inorg. Nuclear Chem.*, 1972, **34**, 2935.
[139] R. D. Bereman, *Inorg. Chem.*, 1972, **11**, 1148.
[140] P. A. Perov, V. N. Novikov, and A. A. Mal'tsev, *Vestnik Moskov. Univ., Khim.*, 1972, **13**, 89.

varied in the sequence F, Cl, Br; values (cm^{-1}) for ν(MoO) in the (phen) series are as follows:[141] MoOX$_3$(phen) [980, X = F; 975, X = Cl; 967, X = Br]; MoO$_2$X$_2$(phen) [945 and 925, X = F; 943 and 905, X = Cl; 933 and 900, X = Br; these values refer to ν_{sym} and ν_{asym} respectively].

As in previous years, ν(M=O) modes have been apparently identified in a wide range of compounds, generally with the usual confidence and despite the fearsome complexity of many of the spectra in the appropriate regions, as follows: [L^{1}H$_2$][MoOX$_5$], MoOX$_3$L^1, Mo$_2$O$_3$Br$_4$L$^2{}_2$, and Mo$_2$O$_4$-X$_2$L$^1{}_2$ (L^1 = phen, X = Cl [142] or Br;[143] L^1 = bipy, X = Br;[144] L^2 = bipy [144] or phen [143]); {MoO(salen)}$_2$O [ν(MoOMo) at 750 and 430 cm^{-1}];[145] some oxomolybdenum(v) complexes of quinolin-8-ol;[146] L$_4$MoO-(OMe) and L$_4$WO(OPh) [L$_4$H$_2$ = octaethylporphin; ν(MOM) of {L$_4$M-(O)}$_2$O (M = Mo or W) also given];[81] MoO$_2$(dtc)$_2$ and Mo$_2$O$_3$(dtc)$_4$ [dtc = diethyldithiocarbamato; ν(MoOMo) at 752 (asym) and 428 cm^{-1} (sym)];[105] WOF$_n$(OEt)$_{4-n}$ [n = 0, 1, or 2; ν(W—OEt) at 600 cm^{-1}];[147] WOX$_2$(PR$_3$)$_2$ [X = Cl or NCS; PR$_3$ = various; δ(WO) at ca. 250 cm^{-1}];[148] WO(O$_3$SF)$_4$ [ν(W—O) at 280, 273, and 214 cm^{-1}; δ(WO) at 184, 154, and 141 cm^{-1}];[149] MoO$_2$Cl$_2$(H$_2$O) [ν(MoOMo) at 789 and 867 cm^{-1}; structure determined by X-ray diffraction];[150] and (NH$_4$)$_2$[MoOBr$_5$].[151] A further opinion has been expressed on the position of bands characteristic of M$\langle{}^O_O\rangle$M bridges (M = Mo or W) (cf. previous years' reports),[152] and modes of the Cr$_3$O group in Cr$_3$O(O$_2$CMe)$_4$ have been tentatively assigned [153] to i.r. bands at ca. 650 cm^{-1}.

More detailed assignments have been made for oxyhalogeno-anions. For [Ph$_4$E][CrO$_3$X] (E = P or As; X = F or Cl),[154] ν_{sym}(CrO$_3$) = 905—910 cm^{-1}, ν(CrX) = ca. 638 (X = F) or ca. 437 cm^{-1} (X = Cl), and ν_{asym}(CrO$_3$) = ca. 950 and ca. 930 cm^{-1}. In the case of Rb[Mo$_2$O$_2$F$_9$], i.r. and Raman spectra (Table 6) are said to be consistent with bridging F atoms but terminal O atoms.[155] A similar situation (bridging F, terminal O) was deduced [156] for M[MoO$_2$F$_3$] (M = NH$_4$, K, Rb, Cs, or Tl) and M[WO$_2$F$_3$] (M = Rb or Cs), and the structure (5) proposed on the

---

[141] R. Kergoat and J. E. Guerchais, *Bull. Soc. chim. France*, 1972, 1746.
[142] H. K. Saha and M. C. Halder, *J. Inorg. Nuclear Chem.*, 1972, **34**, 3097.
[143] H. K. Saha and A. K. Banerjee, *J. Inorg. Nuclear Chem.*, 1972, **34**, 1861.
[144] H. K. Saha and A. K. Banerjee, *J. Inorg. Nuclear Chem.*, 1972, **34**, 697.
[145] A. Van Den Bergen, K. S. Murray, and B. O. West, *Austral. J. Chem.*, 1972, **25**, 705.
[146] W. Andruchow and R. D. Archer, *J. Inorg. Nuclear Chem.*, 1972, **34**, 3185.
[147] Yu. A. Buslaev, Yu. V. Kokunov, and V. A. Bochkaveva, *Russ. J. Inorg. Chem.*, 1971, **16**, 1393.
[148] A. V. Butcher, J. Chatt, G. J. Leigh, and P. L. Richards, *J.C.S. Dalton*, 1972, 1064.
[149] R. Dev and G. H. Cady, *Inorg. Chem.*, 1972, **11**, 1134.
[150] F. A. Schröder and A. N. Christensen, *Z. anorg. Chem.*, 1972, **392**, 107.
[151] G. Y.-S. Lo and C. H. Brubaker, jun., *J. Co-ordination Chem.*, 1972, **2**, 5.
[152] P. C. H. Mitchell and R. D. Scarle, *J.C.S. Dalton*, 1972, 1809.
[153] R. Grecu and D. Lupu, *Rev. Roumaine Chim.*, 1971, **16**, 1811.
[154] E. Diemann, E. Ahlborn, and A. Müller, *Z. anorg. Chem.*, 1972, **390**, 217.
[155] A. Benter and W. Sawodny, *Angew. Chem. Internat. Edn.*, 1972, **11**, 1020.
[156] R. Mattes, G. Müller, and H. J. Becher, *Z. anorg. Chem.*, 1972, **389**, 177.

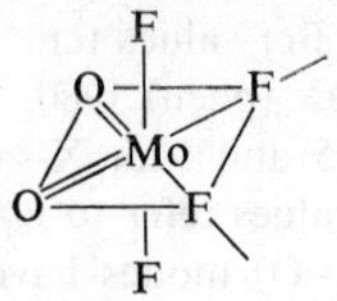

```
    F
O—|—F
 \\ Mo /
 // | \
O   | F
    F
```
(5)

basis of the vibrational data (*e.g.* Table 6) was confirmed by a single-crystal X-ray study of $Cs[MoO_2F_3]$.[156]

In the i.r. spectra of $[Cr(Me_2SO)_6]X_3$ (X = ClO_4 or I), splitting of the t_{1u} (in O_h) $\nu_{asym}(CrO_6)$ and $\delta_{asym}(CrO_6)$ modes, each into two components, has been observed.[157]

Table 6 Assignments/cm^{-1} of the vibrational spectra of $Rb[Mo_2O_2F_9]$ and $Cs[MoO_2F_3]$

Assignment	$Rb[Mo_2O_2F_9]$[155]		$Cs[MoO_2F_3]$[156]	
	I.r.	Raman	I.r.	Raman
$\nu_{sym}(MoO)$	1033	1033	970	974
$\nu_{asym}(MoO)$	1022	—	919	912
$\nu_{sym}(MoF)_t$	689	685	—	—
$\nu_{asym}(MoF)_t$	660	—	581^a	580^a
$\nu(MoF)_t$	635	576	—	—
$\nu_{asym}(MoFMo)_b$	422	—	{449, 411}	—
Bending modes	{310, 284, 270}	{321, 310}	{393, 308}	403

^{a} Assigned to an *axial* MoF_2 grouping by analogy with $K_2MoO_2F_4(H_2O)$.

Optical and i.r. spectroscopy have been used to study the symmetry of CrO_4^{2-} ions in alkali halides.[158] I.r. spectra (400—1300 cm^{-1}) of tungstates of the scheelite ($CaWO_4$) type (Sr, Ba, and Pb cations) and of the wolframite ($FeWO_4$) type (Mg, Cu, Ni, Co, Zn, Mn, and Cd cations) have been discussed.[159] Polymolybdates,[160] polytungstates,[161, 162, 163] and their thermal decomposition products[161, 162] have been studied by i.r. and other techniques. The formation of molybdates[164, 165] and tungstates[165, 166] by interaction of component oxides[164, 166] or by hydrolysis of $MO_3(dien)$ complexes (M = Mo or W)[165] has also been followed using vibrational

[157] E. Koglin and W. Krasser, *Ber. Bunsengesellschaft phys. Chem.*, 1972, **76**, 401.
[158] S. C. Jain, A. V. R. Warrier, and S. K. Agrawal, *Chem. Phys. Letters*, 1972, **14**, 211.
[159] M. L. Zorina and L. F. Syritso, *Zhur. priklad. Spektroskopii*, 1972, **16**, 1043.
[160] A. B. Kiss, S. Holly, and E. Hild, *Acta Chim. Acad. Sci. Hung.*, 1972, **72**, 147.
[161] R. Ripan, D. Stănescu, and M. Puşcaşin, *Z. anorg. Chem.*, 1972, **391**, 187.
[162] R. Ripan, M. Puşcaşin, D. Stănescu, and P. Boian, *Z. anorg. Chem.*, 1972, **391**, 183.
[163] M. V. Mokhosoev, N. A. Taranets, and M. N. Zayats, *Russ. J. Inorg. Chem.*, 1971, **16**, 1012.
[164] T. G. Alkhazov, V. M. Khiteeva, Sh. A. Feizullaeva, and M. S. Belen'kii, *Russ. J. Inorg. Chem.*, 1971, **16**, 902.
[165] R. S. Taylor, P. Gans, P. F. Knowles, and A. G. Sykes, *J.C.S. Dalton*, 1972, 24.
[166] R. Albrecht and R. Möbius, *Z. anorg. Chem.*, 1972, **392**, 62.

spectroscopy. Other miscellaneous applications to compounds in this section include a study of adsorption of py and H_2O onto α-chromia,[167] and an investigation of short-time polymerization of C_2H_4 with Cr^{II} and Cr^{VI} surface species on oxide carriers.[168]

Assignments of the Raman spectra of $K_3[MOS_3]X$ species (M = Mo or W; X = Cl or Br) are listed in Table 7.[169] Co-ordination of the oxytri-

Table 7 Assignments of the Raman spectra/cm^{-1} of $K_3[MOS_3]X$ species

Mode	Description	$K_3[MoOS_3]Cl$	$K_3[MoOS_3]Br$	$K_3[WOS_3]Cl$	$K_3[WOS_3]Br$
$\nu_1 (A_1)$	$\nu(MO)$	858	874	875	872
$\nu_2 (A_1)$	$\nu_{sym}(MS_3)$	459	458	469	452
$\nu_4 (E)$	$\nu_{asym}(MS_3)$	{481, 469}	{486, 479, 468}	453	{471, 468}
$\nu_5 (E)$	$\rho(MS_3)$	{277, 263}	{277, 269}	{277, 269}	280
$\nu_3 (A_1)$ $\nu_6 (E)$	$\delta_{sym}(MS_3)$ $\delta_{asym}(MS_3)$	{185, 178}	181	184	182

thiotungstate(VI) anion to some metal(II) ions in $[M(WOS_3)_2]^{2-}$ salts (M = Co, Ni, or Zn) occurs via sulphur atoms only [$\nu(WS)_t$ at ca. 490, $\nu(WS)_b$ at ca. 440, $\nu(WO)_t$ at ca. 920 cm^{-1}].[170] Some assignments for other thio-anions of Mo and W are given in Table 8;[171, 172] the polarizing

Table 8 Raman data/cm^{-1} on thiomolybdates and thiotungstates

Compound	$\nu(MO)$	$\nu(MS)$	Ref.
$TlMoO_2S_2$	845^a, 829^b	461^a, 468$^{b, c}$	171
$TlWO_2S_2$	875^a, 824^b	465^a, 455^b	171
$TlMoOS_3$	831	444^a, 459^b	171
$TlWOS_3$	860, 851	465^a, 450^b	171
$CuWOS_3{}^d$	910^c	445^c	172
$TlMoS_4$	—	445^a, 465^b	171
$CuMoS_4{}^d$	—	495^c, 485^c, 446^c	172
$TlWS_4$	—	475^a, 457^b, 446^b	171
$Cu_2WS_4{}^e$	—	474, 452, 413	172
$Cu(NH_4)WS_4{}^f$	—	452^c, 466, 433	172
$PbWS_4{}^d$	—	ca. 458^c	172
$ZnWS_4{}^d$	—	ca. 435^c	172
$CuWSe_4{}^d$	—	ca. 284^g	172

^{a} Symmetric mode. ^{b} Asymmetric mode. ^{c} I.r. value (cm^{-1}). ^{d} Impure compound. ^{e} $\nu(CuS)$ also assigned at 282 (and 217 ?) cm^{-1} in the Raman spectrum. ^{f} $\nu(CuS)$ also assigned at 248 cm^{-1} in the i.r. and 257 (and 233 ?) cm^{-1} in the Raman spectrum. ^{g} I.r. value (cm^{-1}) for $\nu(WSe)$.

[167] A. Zecchina, E. Cuglielminotti, L. Cerruti, and S. Coluccia, *J. Phys. Chem.*, 1972, **76**, 571.
[168] H. L. Krauss and H. Schmidt, *Z. anorg. Chem.*, 1972, **392**, 258.
[169] A. Müller, W. Sievert, and H. Schulze, *Z. Naturforsch.*, 1972, **27b**, 720.
[170] A. Müller and H. H. Heinsen, *Chem. Ber.*, 1972, **105**, 1730.
[171] A. Müller, Ch. K. Jørgensen, and E. Diemann, *Z. anorg. Chem.*, 1972, **391**, 38.
[172] A. Müller and R. Menge, *Z. anorg. Chem.*, 1972, **393**, 259.

effects of the Tl^+ cation on the anions is discussed in ref. 171. Other data on $[MX_4]^{2-}$ (M = Mo or W; X = S or Se)[24] have already been mentioned.

The transverse optical phonon frequencies of MoS_2 and $MoSe_2$ occur in the i.r. at 380 and 245 cm^{-1}, respectively.[173] The i.r. spectra below 500 cm^{-1} of Cr(ethylxanthato)$_3$ and Cr(dimethyldithiocarbamato)$_3$ have been listed and compared with data obtained from the single-crystal absorption and emission electronic spectra of these compounds.[174] The formation of Cr_2Te_2O from the action of heat on CrTe in air has been followed by i.r. and X-ray methods.[175]

The Raman spectrum of $[CrCl_6]^{3-}$ has now been obtained and the i.r. spectrum re-investigated for $[M(NH_3)_6]^{3+}$ salts (M = Co, Rh, Cr, or Ir); the nearly regular octahedral symmetry of the anion has been thereby confirmed.[176] The compounds were shown to crystallize in space group T_h^6 (Z = 4), and a unit cell group analysis was used in making the following assignments (cm^{-1}; M = Rh):[176]

$\nu_1 = 286$ (Raman) $\nu_4 = 199$ (i.r.)

$\nu_2 = 237$ (Raman) $\nu_5 = 162$ (Raman)

$\nu_3 = 315$ (i.r.) $\nu_6 = 182$ (i.r., Rb$^+$ salt)

Complete assignments of the allowed vibrations of $[M_2Cl_9]^{3-}$ ions (M = Cr or W) have been proposed on the basis of D_{3h} symmetry as shown in Table 9;[177] a full normal-co-ordinate analysis was carried out, and the force constant for direct W—W interaction in $[W_2Cl_9]^{3-}$ estimated as 1.15 ± 0.1 mdyn Å^{-1}. Some comments have been made on the i.r. spectra [internal H_2O and $\nu(M-OH_2)$ modes] of $[M_6X_8]X_4(H_2O)_2$ compounds (M = Mo or W; X = Cl, Br, or I).[178] For $[Mo_6Cl_8]Cl_2(MeCO_2)_2$, a polymeric structure with bridging acetato-groups has been proposed on the basis of i.r. data; $\nu(MoCl)$ is said to be at 333 cm^{-1} ('inner' Mo—Cl bonds) and 255 cm^{-1} ('outer' Mo···Cl interactions).[179]

The observation of $\nu(Mo-Cl)$ bands in the i.r. region (243—341 cm^{-1}) expected for octahedral co-ordination of Mo^{IV} has led to the proposal that the ligand L is unidentate in $MoCl_4L_2$ (L = pyrazine, quinoxaline, or 4,4'-bipyridyl) but bidentate in $MoCl_4L$ (L = bipy).[180] In the seven-co-ordinate complexes $[M(CO)_3(T)I]^+$ [M = Mo or W; T = bis-(2-pyridylmethyl)amine, bis-(2-pyridylmethyl)methylamine, or bis-(2-pyridylethyl)-amine],[181] $\nu(MI)$ is in the range 139—142 cm^{-1}. Other halogeno-compounds

---

[173] O. P. Agnihotri, *J. Phys. and Chem. Solids*, 1972, **33**, 1173.
[174] W. J. Mitchell and M. K. DeArmond, *J. Mol. Spectroscopy*, 1972, **41**, 33.
[175] S. S. Batsanov, L. M. Doronina, T. A. Volkova, and V. E. Borodaevskii, *Russ. J. Inorg. Chem.*, 1971, **16**, 1545.
[176] H. H. Eysel, *Z. anorg. Chem.*, 1972, **390**, 210.
[177] R. J. Ziegler and W. M. Risen, jun., *Inorg. Chem.*, 1972, **11**, 2796.
[178] H. Schäfer and H. Plautz, *Z. anorg. Chem.*, 1972, **389**, 57.
[179] G. Holste and H. Schäfer, *Z. anorg. Chem.*, 1972, **391**, 263.
[180] W. M. Carmichael and D. A. Edwards, *J. Inorg. Nuclear Chem.*, 1972, **34**, 1181.
[181] J. G. Dunn and D. A. Edwards, *J. Organometallic Chem.*, 1972, **36**, 153.

Table 9 Assignments for the fundamental modes of vibration/cm^{-1} in [M$_2$Cl$_9$]$^{3-}$ ions

Assignment[a]		Cs$_3$Cr$_2$Cl$_9$[b]		K$_3$W$_2$Cl$_9$[c]	
		I.r.	Raman	I.r.	Raman
A_1'	ν_1	—	375	—	332
	ν_2	—	280	—	257
	ν_3	—	161	—	139
	ν_4	—	121	—	115
A_2''	ν_7	360	—	313	—
	ν_8	261	—	232[d]	—
	ν_9	184	—	158	—
E'	ν_{10}	342	335	285	281
	ν_{11}	234	233	215	209
	ν_{12}	196	196	181	178
	ν_{13}	138	{143, 138}	96	91
	ν_{14}	—	ca. 78	76	ca. 76
E''	ν_{15}	—	320	—	294
	ν_{16}	—	222	—	226
	ν_{17}	—	131	—	123
	ν_{18}	—	113	—	107

[a] Based on D_{3h} symmetry; A_1'' (ν_5) and A_2' (ν_6) species are forbidden. [b] Assignments also given for Et$_4$N$^+$ and Bu$^n_4$N$^+$ salts. [c] Assignments also given for Cs$^+$ and Bu$^n_4$N$^+$ salts. [d] Value for Cs$^+$ salt.

of Mo or W studied by vibrational spectroscopy include Mo(chel)Cl$_2$ [chel = NN-ethylenebis(salicylaldimine) or N-substituted salicylaldimine] [145] and various W^{IV} halide complexes of N- and S-donor ligands.[182]

7 Manganese, Technetium, and Rhenium

Crystallographic data on the new carbonyl hydride H$_2$Re$_2$(CO)$_8$ are consistent with structure (6); Raman bands at 1382 and 1272 cm^{-1} (shifting to 974 and 924 cm^{-1} in the deuteriate) are consistent with the presence of bridging H atoms.[183]

In K$_3$[Re(CN)$_6$],3H$_2$O, ν(ReC) is given as 907 cm^{-1} in the i.r.[184] For the carbonyl derivatives (7), ν(Mn—CO) modes are at 475 and 450 cm^{-1}

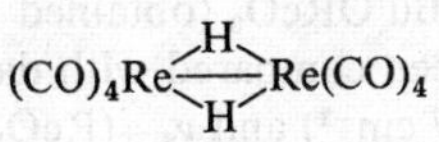

(CO)$_4$Re⟨H,H⟩Re(CO)$_4$

(6)

Me$_3$M   Mn(CO)$_5$
 F$_3$C――CF$_3$
        |
 F$_3$C――CF$_3$

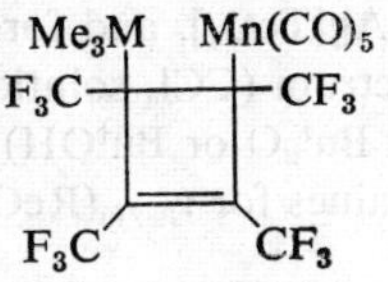

(7)

[182] M. A. Schaefer, *Diss. Abs. Internat. (B)*, 1972, **32**, 5678.
[183] M. J. Bennett, W. A. G. Graham, J. K. Hoyano, and W. L. Hutcheon, *J. Amer. Chem. Soc.*, 1972, **94**, 6232.
[184] O. E. Skolozdra, A. N. Sergeeva, and K. N. Mikhalevich, *Russ. J. Inorg. Chem.*, 1971, **16**, 861.

(M = Ge) or 450 and 445 cm^{-1} (M = Si).[185] On the basis of an i.r. study it has been concluded that [Mn(CO)$_4$(NO)] has a C_{3v} trigonal-bipyramidal structure (axial NO group) in the vapour phase and in solution (C_6H_6 and CCl_4), rather than the C_{2v} form suggested by X-ray work on the crystalline form at $-110\ °C$; assignments made {largely by comparison with [Fe-(CO)$_5$]} are given in Table 10.[186]

Table 10 Vibrational assignments for [Mn(CO)$_4$(NO)]

Mode		Description	ν/cm^{-1}	Mode	Description	ν/cm^{-1}
A_1	ν_1	$\nu(CO)_{eq}$	2111	E ν_{10}	$\nu(CO)_{eq}$	2020
	ν_2	$\nu(CO)_{ax}$	1996	ν_{11}	$\delta(MnNO)$	657
	ν_3	$\nu(NO)$	1781	ν_{12}	$\delta(MnCO)$	641
	ν_4	$\delta(Mn-CO)$	615	ν_{13}	$\nu(MnC)_{eq}$	456
	ν_5	$\nu(MnN)$	524	ν_{14}	$\delta(MnCO)$	417
	ν_6	$\nu(MnC)_{eq}$	398	ν_{15}	$\delta(CMnC)$	102
	ν_7	$\nu(MnC)_{ax}$	357	ν_{16}	$\delta(CMnN)$	102
	ν_8	$\delta(CMnC)$	66	ν_{17}	$\delta(CMnC)$	52
A_2	ν_9	$\delta(MnCO)$	—	ν_{18}	$\delta(CMnN)$	52

I.r. bands (to 650 cm^{-1}) have been listed for species [ReCl$_{3+n}$(py)$_m$(NO)]$^{n-}$ ($n = 0$, $m = 2$; $n = 1$, $m = 1$; or $n = 2$, $m = 0$), and $\nu(NO)$ has been assigned in the 1725—1770 cm^{-1} range.[187] N-Co-ordination has been suggested for (Bu$^n_4$N)$_2$[Re$_2$(NCSe)$_8$] on the basis of i.r. bands at 273 and 300 cm^{-1} arising from $\nu(ReN)$,[188] while in complexes Mn(NCS)$_2$(L) [L = (8)], $\nu(MnN)$ is in the 250—260 cm^{-1} region.[189]

(8) R^1 = H or Me, R^2 = Me or Et

A general review of the chemistry of Mn in higher oxidation states includes references to a range of spectroscopic results, *e.g.* on [MnO$_4$]$^{n-}$ ions ($n = 1$, 2, or 3).[190] Assignments of the i.r. spectra have been given for K[TcO$_4$] and Ag[TcO$_4$], and force constants calculated.[191]

The i.r. spectrum (CCl$_4$ solution) of Bu^{t}OReO$_3$ (obtained from reaction of Re$_2$O$_7$ with Bu$^t_2$O or Bu^{t}OH) has been compared with data for Me$_3$Si-OReO$_3$, and values for ν_{asym}(ReO$_3$) (967 cm^{-1}) and ν_{sym}(ReO$_3$) (1006 cm^{-1})

[185] H. C. Clark and T. L. Hauw, *J. Organometallic Chem.*, 1972, **42**, 429.
[186] G. Barna and I. S. Butler, *Canad. J. Spectroscopy*, 1972, **17**, 2.
[187] D. K. Hait, B. K. Sen, and P. Bandyopadhyay, *Z. anorg. Chem.*, 1972, **388**, 184.
[188] R. R. Hendriksma, *Inorg. Nuclear Chem. Letters*, 1972, **8**, 1035.
[189] B. Chiswell and K. W. Lee, *Inorg. Chim. Acta*, 1972, **6**, 567.
[190] W. Levason and C. A. McAuliffe, *Co-ordination Chem. Rev.*, 1972, **7**, 353.
[191] J. Hanuza and B. Jezowska-Trzebiatowska, *Bull. Acad. polon. Sci., Sér. Sci. chim.*, 1972, **20**, 271.

have been discussed in terms of the high formal charge on Re^{VII}.[192] I.r. and Raman data on $ReO_2Cl(py)_4(H_2O)_2$, which was recently assigned the structure $[Re(OH)_4(py)_4]Cl$, have been shown to be consistent only with the octahedral *trans*-dioxo-structure $[ReO_2(py)_4]^+$; $K_3[ReO_2(CN)_4]$ and $[ReO_2(en)_2]Cl,2H_2O$ are also believed to have this geometry.[193]

Suggestions have been made [194] for $\nu(ReO)$ in $Re(acac)_3$ and $Re(hfac)_3$. I.r. and Raman data have been reported for *trans*-$[ReCl_2(acac)_2]$, shown to be a monomer and not a dimer as previously believed; the new *cis*-$[ReX_2(acac)_2]$ compounds (X = Cl, Br, or I) were also studied.[195]

Values for $\nu(ReORe)$ [81, 105, 196, 197] or $\nu(Re=O)$ [81, 105, 197] have been given for $L_4ReO(OPh)$ and $(L_4ReO)_2O$ [L_4H_2 = octaethylporphin],[81] [(dtc)$_4$-ReO]$_2$O (dtc = diethyldithiocarbamato),[105] $[ReO(NH_2)_4]_n$,[196] and some dithiocarbamato-complexes of Re^{III} and Re^V.[197]

More detailed studies have been carried out on oxohalogeno-compounds of Re, and three reports [198-200] have been concerned with $ReOCl_4$ (see Chapter 4). Data are also available for ReO_3Cl [199] and for $Re_2O_4Cl_5$,[199, 200] for which the structure $O_3Re-O-Re(Cl)_4=O$ is proposed.[200] The compound previously described as 'β-$ReOCl_3$' has now been shown by X-ray diffraction to be $(ReO_3Cl)_2Re_2O_3Cl_6$,[199] having $\nu(Re=O)$ at 1039, 1003, 970, and 930 cm^{-1}, $\nu(ReORe)$ at 830 cm^{-1}, and $\nu(ReCl)$ at 350 cm^{-1}. The compounds $ReOCl_4(MeCN)$,[196] $[Ph_4P][ReOBr_5]$,[196] and *trans*-$[ReOCl_4(OH_2)]$ {$\nu(ReCl)$ at *ca.* 350 cm^{-1}} [201] also show $\nu(Re=O)$ in the expected region.

Oxohalogeno- and halogeno-anions of technetium have been characterized by i.r. spectroscopy, *viz.* TcO_3Cl, $TcOCl_3$, and $TcOCl_4$;[202] K_2TcOCl_5;[191, 203] $(NH_4)_2[TcOCl_5]$, $Cs_2[TcOCl_5]$, $[TcO_2(en)_2]Cl$, $K_2[TcCl_6]$, and $Cs_2[TcCl_6]$;[191] and $K_2[TcO(OH)Cl_4]$.[203] Force constants have been computed for several of these species.[191]

The far-i.r. reflection spectrum of MnTe over the temperature range 30—350 K has been discussed.[204]

Electronic reflectance and i.r. spectroscopy [$\nu(MnBr)$ at 230 cm^{-1}] show that Cs_2MnBr_4 and Rb_2MnBr_4 contain tetrahedrally co-ordinated anions.[205] The presence of two $\nu(MnCl)$ bands (360, 335sh cm^{-1}) in the i.r.

[192] C. Ringel and G. Boden, *Z. anorg. Chem.*, 1972, **393**, 65.
[193] N. P. Johnson, *J. Inorg. Nuclear Chem.*, 1972, **34**, 2875.
[194] W. D. Courrier, W. Forster, C. J. L. Lock, and G. Turner, *Canad. J. Chem.*, 1972, **50**, 8.
[195] W. D. Courrier, C. J. L. Lock, and G. Turner, *Canad. J. Chem.*, 1972, **50**, 1797.
[196] D. A. Edwards and R. T. Ward, *J.C.S. Dalton*, 1972, 89.
[197] J. F. Rowbottom and G. Wilkinson, *J.C.S. Dalton*, 1972, 826.
[198] C. G. Barraclough and D. J. Kew, *Austral. J. Chem.*, 1972, **25**, 27.
[199] C. Calvo, P. W. Frais, and C. J. L. Lock, *Canad. J. Chem.*, 1972, **50**, 3607.
[200] K. I. Petrov, V. V. Kravchenko, D. V. Drobot, and V. A. Aleksandrova, *Russ. J. Inorg. Chem.*, 1971, **16**, 928.
[201] P. W. Frais and C. J. L. Lock, *Canad. J. Chem.*, 1972, **50**, 1811.
[202] A. Guest and C. J. L. Lock, *Canad. J. Chem.*, 1972, **50**, 1807.
[203] V. I. Spitsyn, M. I. Glinkina, and A. F. Kuzina, *Doklady Chem.*, 1971, **200**, 875.
[204] L. V. Povstyanyi, V. I. Kut'ko, and A. I. Zvyagin, *Fiz. tverd. Tela*, 1972, **14**, 1561.
[205] H.-J. Seifert and E. Dau, *Z. anorg. Chem.*, 1972, **391**, 302.

spectrum of $[Co(en)_3][MnCl_6],2H_2O$ has been attributed to a Jahn–Teller effect, but possible crystal symmetry effects were ignored.[206] I.r. data have been given for $[Re_3Cl_{12}]^{3-}$ salts [207, 208] and a normal-co-ordinate calculation has been performed.[208] The computed normal modes show that none of the vibrations is localized in any bond or angle; the Re—Re stretching force constant calculated (1.35 mdyn Å^{-1}) indicates multiple bonding.[208] I.r. data are also available for Re_3Cl_9, Re_3Br_9, and the new compound $Re_3Br_6Cl_3$ (9).[207]

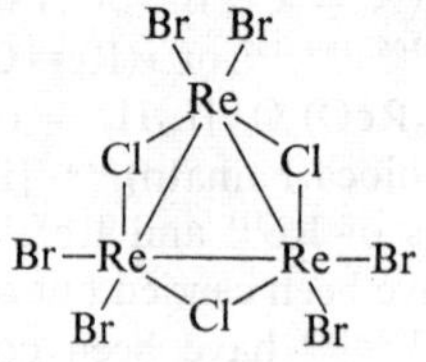

(9)

The observation of two i.r.-active ν(ReF) bands for $ReF_3(CO)_3$ (650 and 580 cm^{-1}) is consistent with *fac*-geometry (A and E modes under C_{3v} symmetry).[209] Other complexes for which ν(MX) values (cm^{-1}) have been given include the following: $MnCl_3L_3$ [L = pyO (270, 310), Ph_3PO (340), or Ph_3AsO (280, 330); ν(MnO) at 410—455 cm^{-1}],[210] $Mn(CO)_2\{PhP$-$(CH_2CH_2PPh_2)_2\}X$ [X = Br (209) or I (190)],[211] $[Re(CO)_3X(Ph_2PCH_2$-$CH_2CN)]_2$ {X = Cl (288) or Br (198)},[212] and $M(CO)_3LX$ [L = a π-bonded dinitrile, $NCCH_2CN$ or $NCCH_2CH_2CN$; M = Mn, X = Cl (287) or Br (220—228); M = Re, X = Cl (288) or Br (195—205)].[213] General ranges for ν(MnX) in $[Mn^{III}(porphyrin)X(OH_2)]$ (X = halide or pseudo-halide) have been given in a review of manganese porphyrin complexes.[214]

8 Iron, Ruthenium, and Osmium

Assignments (often supported by deuteriation studies) have been given for ν(MH) or ν(MD) in several compounds of these elements as follows: H_2FeCO_4,[41] H_2FeL_4 [L = $P(OEt)_3$, $PhP(OEt)_2$, $\frac{1}{2}$(diphos), Ph_2PMe, $PhPMe_2$, $PhP(OPr^i)_2$, *etc.*],[215] $D_2Fe(PF_3)_4$,[216] $H_2Fe(PPh_2H)_4$,[217] H_3Ge-

---

[206] W. Levason, C. A. McAuliffe, and S. G. Murray, *Inorg. Nuclear Chem. Letters*, 1972, **8**, 97.
[207] M. A. Bush, P. M. Druce, and M. F. Lappert, *J.C.S. Dalton*, 1972, 500.
[208] K. I. Petrov and V. V. Kravchenko, *Russ. J. Inorg. Chem.*, 1971, **16**, 930.
[209] T. A. O'Donnell and K. A. Phillips, *Inorg. Chem.*, 1972, **11**, 2562.
[210] E. Contreras, V. Riera, and R. Usón, *Inorg. Nuclear Chem. Letters*, 1972, **8**, 287.
[211] I. S. Butler, N. J. Coville, and H. K. Spendjian, *J. Organometallic Chem.*, 1972, **43**, 185.
[212] B. N. Storhoff, *J. Organometallic Chem.*, 1972, **43**, 197.
[213] M. F. Farona and K. F. Kraus, *J.C.S. Chem. Comm.*, 1972, 513.
[214] L. J. Boucher, *Co-ordination Chem. Rev.*, 1972, **7**, 289.
[215] D. H. Gerlach, W. G. Peet, and E. L. Muetterties, *J. Amer. Chem. Soc.*, 1972, **94**, 4545.
[216] T. Kruck and R. Kobelt, *Chem. Ber.*, 1972, **105**, 3765.
[217] J. R. Sanders, *J.C.S. Dalton*, 1972, 1333.

FeH(CO)$_4$,[41] K[FeH(PF$_3$)$_4$],[216] trans-[FeHX(PPh$_2$H)$_4$] (X = Cl, Br, I, NCS, or SnCl$_3$),[217] EtFeH(Ph$_2$PEt)$_3$,[218] [FeH(L)(diphos)$_2$]X (L = N$_2$, Me$_2$CO, MeCN, PhCN, or NH$_3$; X = BPh$_4$ or ClO$_4$) and [FeH(diphos)$_2$]-BPh$_4$,[219] H$_2$RuL$_4$ [L = P(OMe)$_3$, P(OEt)$_3$, P(OPr^i)$_3$, PhP(OEt)$_2$, or PhPMe$_2$],[215] H$_2$Ru(PPh$_2$H)$_4$ and trans-[RuHX(PPh$_2$H)$_4$] (X = Cl, Br, I, NCS, or SnCl$_3$),[217] RuHCl(CO)(PR$_3$)$_2$ and RuHCl(CO)(PR$_3$)$_2$(py) (R = C$_6$H$_{11}$),[220] (HCO$_2$)RuH(PPh$_3$)$_3$(PhMe),[221] [Ru(H){P(OMe)$_2$Ph}$_5$]BPh$_4$, [(C$_8$H$_{12}$)RuHL$_3$]BPh$_4$ (L = N$_2$H$_4$, py, or 4-picoline), and two isomers of [(C$_8$H$_{12}$)RuH(Me$_2$NNH$_2$)$_3$]BPh$_4$,[222] and OsHCl(CO)(PR$_3$) and OsHCl-(CO)(PR$_3$)(py) (R = C$_6$H$_{11}$).[220]

Assignment of the Fe—H bending modes in H$_2$Fe(CO)$_4$ (gas-phase) has been attempted in terms of C_{2v} symmetry.[41]

In (C$_2$H$_4$)Fe(CO)$_4$,[223] ν(Fe—C$_2$H$_4$) is at 356 cm^{-1}. In related compounds corresponding modes are as follows: (C$_4$H$_4$)Fe(CO)$_3$ (398 cm^{-1}),[224] {C(CH$_2$)$_3$}Fe(CO)$_3$ (372 cm^{-1}),[225] and (C$_4$H$_6$)$_2$Fe(CO) [394 (ν_{asym}) and 297 cm^{-1} (ν_{sym})].[226]

Partial i.r. assignments [e.g. ν(CO), δ(MCO), and/or ν(M—CO)] have been given for {C(CH$_2$)$_3$}Fe(CO)$_3$,[225] (C$_4$H$_6$)$_2$Fe(CO),[226] (OC)$_4$Fe{Pt(py)$_2$-Cl}$_2$,[227] (C$_3$F$_7$)Fe(CO)$_2$(NH$_3$)$_2$I and [(C$_3$F$_7$)Fe(CO)$_2$(NH$_3$)$_3$]$^+$,[228] (π-C$_5$H$_5$)-(OC)$_2$Fe—C=CMeC(O)N(SO$_2$Cl)CH$_2$ and (π-C$_5$H$_5$)(OC)$_2$Fe—CHCMe$_2$C-(O)N(SO$_2$Cl)CH$_2$,[229] and [Os(NH$_3$)$_5$CO]X$_2$ and cis-[Os(NH$_3$)$_4$CO(N$_2$)]X$_2$ (X = Cl, Br, or I).[230]

An examination of the ν(CO) region in the spectra of Os$_4$O$_4$(CO)$_{12}$ suggests that the molecule has T_d symmetry in solution; although a partial normal-co-ordinate analysis was carried out, ν(OsO$_4$) could not be clearly distinguished from ν(OsC) and δ(OsCO).[231]

An i.r. and Mössbauer spectroscopic study has led to the suggestion that there is little interaction between ions in [Co(NH$_3$)$_6$]$_4$[Fe(CN)$_6$]$_3$,-10H$_2$O, Li$_4$[Co(NH$_3$)$_6$]$_8$[Fe(CN)$_6$]$_7$,H$_2$O, and Na[Co(NH$_3$)$_6$][Fe(CN)$_6$],-4H$_2$O, e.g. because no ν(CN)$_b$ frequencies were found.[232]

---

[218] V. D. Bianco, S. Doronzo, and M. Aresta, *J. Organometallic Chem.*, 1972, **42**, C63.
[219] P. Giannoccaro, M. Rossi, and A. Sacco, *Co-ordination Chem. Rev.*, 1972, **8**, 77.
[220] F. G. Moers and J. P. Langhout, *Rec. Trav. chim.*, 1972, **91**, 591.
[221] S. Komiya and A. Yamamoto, *J. Organometallic Chem.*, 1972, **46**, C58.
[222] J. J. Hough and E. Singleton, *J.C.S. Chem. Comm.*, 1972, 371.
[223] D. C. Andrews and G. Davidson, *J. Organometallic Chem.*, 1972, **35**, 161.
[224] D. C. Andrews and G. Davidson, *J. Organometallic Chem.*, 1972, **36**, 349.
[225] D. C. Andrews and G. Davidson, *J. Organometallic Chem.*, 1972, **43**, 393.
[226] G. Davidson and D. A. Duce, *J. Organometallic Chem.*, 1972, **44**, 365.
[227] B. Munchenbach and J. Dehand, *Naturwiss.*, 1972, **59**, 647.
[228] H. Krohberger, J. Ellermann, and H. Behrens, *Z. Naturforsch.*, 1972, **27b**, 890.
[229] Y. Yamamoto and A. Wojcicki, *Inorg. Nuclear Chem. Letters*, 1972, **8**, 833.
[230] A. D. Allen and J. R. Stevens, *Canad. J. Chem.*, 1972, **50**, 3093.
[231] W. Van Bronswyk and R. J. H. Clark, *Spectrochim. Acta*, 1972, **28A**, 1429.
[232] N. A. Verendyakina, G. B. Siefer, Yu. Ya. Karitonov, and B. V. Borshagovskii, *Russ. J. Inorg. Chem.*, 1971, **16**, 1447.

A MUBFF calculation, using a full 13-atom model, has been carried out on the [Fe(CN)$_5$NO]$^{2-}$ ion.[233] The frequencies, forms of the modes, and force constants of [Ru(NO)X$_5$]$^{2-}$ salts (X = Cl, Br, or I) have been calculated and significant mixing of ν(NO) and ν(RuN) stretching vibrations has been found; the effect of varying X or the cation (K$^+$, Rb$^+$, or Cs$^+$) on various frequencies and force constants was discussed.[234] Partial assignments [*e.g.* ν(NO) or ν(FeN)] have been given for several mono- and bis-nitrosyliron(I) complexes with such ligands as diethyldithiocarbamato, butylxanthato, or dithiophosphato.[235]

The compound (π-C$_5$H$_5$)(CO)$_2$Fe—P(O)(CF$_3$)$_2$ has been so formulated on the basis of i.r. and n.m.r. data.[236]

The new nitrido-complexes Cs$_2$[RuNX$_5$] (X = Cl or Br) show ν(Ru≡N) at ca. 1047 cm^{-1}.[237]

Deuteriation of [Ru(NH$_3$)$_5$(N$_2$)]X$_2$ (X = Br or I) has shown[238] that of the four i.r. bands in the 390—500 cm^{-1} region the highest three are ν(Ru—NH$_3$) and the other band [424 (X = Br) or 415 cm^{-1} (X = I) in the NH$_3$ series] is ν(Ru—N$_2$), thus invalidating previous assignments. However, the existing assignment of ν(Os—N$_2$) at 518 cm^{-1} in [Os(NH$_3$)$_5$-(N$_2$)]Cl$_2$ is confirmed.[238]

The metal isotope effect (^{54}Fe and ^{57}Fe) has been very effectively used to identify metal–ligand vibrations in the two forms of [Fe(NCS)$_2$(phen)$_2$].[239] At 298 K the compound is in the high-spin state whereas at 105 K it adopts the low-spin configuration. In addition to the effect of isotopic substitution, the assignments in Table 11 were made by analogy with the spectra

Table 11 *Assignments of metal–ligand stretching modes/cm^{-1} in* [Fe(NCS)$_2$-(phen)$_2$] *isomers*[a]

	High-spin form (298 K)		Low-spin form (105 K)	
	$\nu(^{54}$Fe)	$\nu(^{54}$Fe)$-\nu(^{57}$Fe)	$\nu(^{54}$Fe)	$\nu(^{54}$Fe)$-\nu(^{57}$Fe)
ν(Fe—NCS)	252.0	4.0	532.6 / 528.5	1.6 / 1.7
ν(Fe—phen)	220.0	4.5	379.0 / 371.0	5.0 / 6.0

[a] All other isotopic shifts observed were ⩽ ±0.5 cm^{-1}.

of *trans*-[Fe(NCS)$_2$(py)$_4$] (high-spin) and [Fe(phen)$_3$]$^{2+}$ (low-spin). The results show that the Fe—(NCS) and Fe—(phen) bonds are stronger in the

---

[233] B. B. Kedzia, B. Jezowska-Trzebiatowska, and J. Ziolkowski, *Bull. Acad. polon. Sci.*, *Sér. Sci. chim.*, 1972, **20**, 237.
[234] B. P. Khalepp, *Sbornik Aspir. Rab. Kazan. Gos. Univ. Tochnye. Nauki: Mekh., Fiz.*, 1970, **2**, 150.
[235] B. P. Khalepp and S. A. Luchkina, *Sbornik Aspir. Rab. Kazan. Gos. Univ., Mat., Mekh., Fiz.*, 1970, 91.
[236] R. C. Dobbie, P. R. Mason, and R. J. Porter, *J.C.S. Chem. Comm.*, 1972, 612.
[237] D. Pawson and W. P. Griffith, *Chem. and Ind.*, 1972, 609.
[238] M. W. Bee, S. F. A. Kettle, and D. B. Powell, *J.C.S. Chem. Comm.*, 1972, 767.
[239] J. H. Takemoto and B. Hutchinson, *Inorg. Nuclear Chem. Letters*, 1972, **8**, 769.

low-spin complexes,[239] in agreement with bond-length data previously obtained crystallographically for the two spin types of [Fe(NCS)$_2$(bipy)$_2$].

I.r. assignments are available for the ammines [Os(NH$_3$)$_5$(N$_2$)]X$_2$, [Os(NH$_3$)$_5$CO]X$_2$, cis-[Os(NH$_3$)$_4$(N$_2$)$_2$]Br$_2$, and cis-[Os(NH$_3$)$_4$(CO)(N$_2$)]X$_2$ (X = Cl, Br, or I),[230] and [Ru$_2$(Cl)$_3$(NH$_3$)$_6$Cl$_2$],nH$_2$O [n = 2 or 0].[240]

Assignments have been given for ν(M=O) in the new complexes trans-Cs$_2$[RuO$_2$X$_4$] (X = Cl, Br, CN, or $\frac{1}{2}$C$_2$O$_4$), trans-[RuO$_2$(NH$_3$)$_4$]Cl$_2$, and trans-[RuO$_2$(py)$_2$Cl$_2$] (790—872 cm^{-1}),[237] and in OsO$_4$(py), Os$_2$O$_6$(py)$_4$, RuO$_2$(OH)$_2$(py), OsO$_2$(py)$_2$(C$_2$H$_4$O$_2$), and some related anionic species [ν(RuO$_2$) at ca. 800 cm^{-1}; ν(OsO$_2$) at 850—900 (asym) and 790—850 cm^{-1} (sym)].[241] OsOCl$_4$ is probably square pyramidal.[198]

I.r. spectroscopy (60—5000 cm^{-1}) has been used to show that the thermal conversion of lepidocrocite into hematite takes place via maggernite (γ-Fe$_2$O$_3$).[242] In some oxo-bridged dimeric iron complexes,[243] ν_{asym}(FeOFe) is near 840 cm^{-1}, whereas in basic trinuclear acetates of Fe^{III}, bands at ca. 530 cm^{-1} are assigned to vibrations of the (Fe$_3$O) group.[153]

ν(FeS) is found at 325—330 and 370—376 cm^{-1} in [Fe(NN-di-isopropyl-dithiocarbamato)$_3$]$^{n+}$ (n = 0 or 1).[244]

A review [245] of some aspects of the co-ordination chemistry of iron^{III} includes references to previously reported i.r. and Raman data [with emphasis on ν(Fe—ligand)] on such complexes as FeCl$_3$(H$_2$O)$_6$, FeCl$_3$(dioxan), FeCl$_3$(DMSO)$_2$, FeCl$_3$(py)$_4$, FeCl$_3$(pyrazole)$_3$, FeX$_3$(Ph$_3$M) [X = Cl or Br; M = As or P], [FeL$_6$](ClO$_4$)$_3$ (L = pyO or phenazine), [FeX$_4$]$^-$ (X = Cl or Br), [FeX$_6$]$^{3-}$ (X = F, Cl, or CN), and FeX(dtc)$_2$.

A Raman spectral study has been made of the species present in aqueous FeCl$_3$ solutions.[246] In near-saturated solutions, bands due to [FeCl$_4$]$^-$ are found. At lower concentrations there is evidence for formation of FeCl$_3$(H$_2$O)$_2$ [Raman bands at 318(pol), 165, ca. 390, and ca. 120 cm^{-1}] having two axial H$_2$O molecules and three equatorial Cl groups. Under conditions of high Fe^{3+} concentration but low Cl$^-$ concentration, [FeCl$_2$(H$_2$O)$_n$]$^+$ is thought to be present.[246] The species believed to be [R$_4$NCl,R$_4$N][FeCl$_4$] (R = n-octyl) shows a far-i.r. band at 189 cm^{-1} (in addition to [FeCl$_4$]$^-$ modes) in benzene solution, attributed to an anion–cation mode.[247]

Proposals have been made for the position of ν(FeX) modes in iron(III) fluoro-complexes (e.g. [FeF$_4$]$^-$, [FeF$_5$]$^{2-}$, and [FeF$_6$]$^{3-}$)[248] and some FeX$_2$ (X = Cl or Br) complexes of amides.[249]

[240] E. E. Mercer and L. W. Gray, *J. Amer. Chem. Soc.*, 1972, **94**, 6426.
[241] W. P. Griffith and R. Rossetti, *J.C.S. Dalton*, 1972, 1449.
[242] G. S. Sakash and L. S. Solntseva, *Zhur. priklad. Spektroskopii*, 1972, **16**, 741.
[243] H. J. Schugar, G. R. Rossman, C. G. Barraclough, and H. B. Gray, *J. Amer. Chem. Soc.*, 1972, **94**, 2683.
[244] E. A. Pasek and D. K. Straub, *Inorg. Chem.*, **11**, 259.
[245] S. A. Cotton, *Co-ordination Chem. Rev.*, 1972, **8**, 185.
[246] A. L. Marston and S. F. Bush, *Appl. Spectroscopy*, 1972, **26**, 579.
[247] R. A. Work and R. L. McDonald, *J. Inorg. Nuclear Chem.*, 1972, **34**, 3123.
[248] E. N. Deichman, Yu. Ya. Kharitonov, and A. A. Shakhnazaryan, *Russ. J. Inorg. Chem.*, 1971, **16**, 1731.
[249] T. Birchall and M. F. Morris, *Canad. J. Chem.*, 1972, **50**, 201.

The far-i.r. spectra of $RuX_2(py)_4$ complexes (X = Cl, Br, or I) have been assigned in terms of *cis* configurations with ν(MX) at 313 and 325 cm^{-1} (X = Cl) or 168 and 175 cm^{-1} (X = Br); ν(MN) appeared as two bands in the 281—308 cm^{-1} range.[250] Assignments for ν(RuX) in the expected regions have also been given for *cis*-[RuCl$_2$(CO)$_n$(py)$_{4-n}$] (n = 2 or 3),[251] *cis*-[RuX$_2$(CO)$_3$L] (X = Cl or Br; L = EtCN, PhCN, or H$_2$C=CHCN),[251] (π-C$_6$H$_6$)RuCl$_2$(L) [L = PPh$_3$, PMePh$_2$, *etc.*],[252] *cis*- and *trans*-[RuX$_2$-(EtCN)$_2$(MPh$_3$)$_2$] (X = Cl or Br; M = P, As, or Sb),[253] RuCl$_2$(HDMA) and Ru$_2$Cl$_3$(HDMA) [DMA = *NN*-dimethylacetamide],[254] (π-C$_6$H$_6$)RuCl-(PPh$_3$)(R) [R = Me or Ph] and (π-C$_6$H$_6$)RuCl(π-C$_3$H$_5$),[255] and twenty-seven compounds of types such as *trans*-[RuCl$_2$L$_2$], *cis*-[Ru(CO)$_2$Cl$_2$L$_2$], *mer*-[Ru(CO)Cl$_2$L$_3$], *cis*-[Ru(CO)$_2$Cl$_2$L], and Ru(NO)Cl$_3$L$_2$ [L = Ph$_2$MCH$_2$MPh$_2$, *cis*-Ph$_2$MCH=CHMPh$_2$, or *cis*- or *trans*-Ph$_2$MCH=CHPh; M = P or As].[256] (π-C$_6$H$_6$)RuCl$_2$ is believed to be dimeric, partly on the basis of ν(RuCl) data.[252, 257] Bridging [260 and 290 (X = Cl) or 195 and 211 cm^{-1} (X = Br)] and terminal [331 (X = Cl) or 234 cm^{-1} (X = Br)] ν(RuX) modes have been identified in the i.r. spectra of the dimers [Ru(CO)$_3$X$_2$]$_2$ (X = Cl or Br).[251] I.r. spectral studies of ν(RuX) have been used in the determination of configuration of two isomers (*cis* and *trans*) of RuCl$_2$(CO)$_2$(PPh$_3$)$_2$,[258] and of RuX$_3$(MPh$_3$)L$_2$ complexes (10) [X = Cl or Br; M = P or As; L = Me$_2$S, py, $\frac{1}{2}$(phen), or $\frac{1}{2}$(bipy)].[259]

The new arylamido-complexes [OsCl$_2$(NC$_6$H$_4$X)(PPh$_3$)$_2$] (X = H, Cl, or OMe)[260] show ν(OsCl) at 310—320 cm^{-1}.

$$\begin{array}{c} X \\ X - | - MPh_3 \\ / Ru / \\ L - | - X \\ L \end{array}$$

(10)

---

[250] D. W. Raichart and H. Taube, *Inorg. Chem.*, 1972, **11**, 999.
[251] E. Benedetti, G. Braca, G. Sbrana, F. Salvetti, and B. Grassi, *J. Organometallic Chem.*, 1972, **37**, 361.
[252] R. A. Zelonka and M. C. Baird, *J. Organometallic Chem.*, 1972, **35**, C43.
[253] B. E. Prater, *J. Organometallic Chem.*, 1972, **34**, 379.
[254] B. R. James, R. S. McMillan, and E. Ochiai, *Inorg. Nuclear Chem. Letters*, 1972, **8**, 239.
[255] R. A. Zelonka and M. C. Baird, *J. Organometallic Chem.*, 1972, **44**, 383.
[256] J. T. Mague and J. P. Mitchener, *Inorg. Chem.*, 1972, **11**, 2714.
[257] R. A. Zelonka and M. C. Baird, *Canad. J. Chem.*, 1972, **50**, 3063.
[258] S. Cenini, A. Fusi, and G. Capparella, *Inorg. Nuclear Chem. Letters*, 1972, **8**, 127.
[259] E. S. Switkes, L. Ruiz-Ramirez, T. A. Stephenson, and J. Sinclair, *Inorg. Nuclear Chem. Letters*, 1972, **8**, 593.
[260] B. Bell, J. Chatt, J. R. Dilworth, and G. J. Leigh, *Inorg. Chim. Acta*, 1972, **6**, 635.

9 Cobalt, Rhodium, and Iridium

Compounds of these elements for which $\nu(MH)$ assignments have been given are listed in Table 12.[261-277]

I.r. spectroscopy has been used to show that in the solid state [(π-dienyl)FeCo(CO)$_4$(π-diene)] complexes (diene = norbornadiene, cyclohexa-1,3-diene, or 2,3-dimethylbuta-1,3-diene) exists as either *cis* or *trans* carbonyl-bridged tautomers, an equilibrium being established in solution.[278] Reaction between AlCl$_3$ and NaCo(CO)$_4$ (in excess) gives a compound AlCo$_3$(CO)$_9$ whose ν(CO) bands closely resemble those of Co$_4$(CO)$_{12}$ and which is tentatively formulated as (11).[279] Values of ν(CO)

(11)

have been listed for uncharacterized addition products from reaction of SPF$_2$Br, SPF$_2$Cl, or OPF$_2$Br with such complexes as Ir(CO)X(PPh$_3$)$_2$ or Ir(CO)Cl(PMePh$_2$)$_2$ [X = Cl or Br].[280]

In complexes IrIMeX(CO)(PBu^{t}R$_2$) [X = Cl or Br; R = Et, Pr^n, or Bu^n], ν(IrMe) is said to be in the 1220—1232 cm^{-1} region.[276]

The following variation of ν(Co—allyl) in (XC$_3$H$_4$)Co(CO)$_3$ has been discussed:[281]

---

[261] G. M. Intille, *Inorg. Chem.*, 1972, **11**, 695.
[262] I. Ojima, M. Nihonyanagi, and Y. Nagai, *J.C.S. Chem. Comm.*, 1972, 938.
[263] P.-C. Kong and D. M. Roundhill, *Inorg. Chem.*, 1972, **11**, 1437.
[264] J. V. Kingston, F. T. Mahmond, and G. R. Scollary, *J. Inorg. Nuclear Chem.*, 1972, **34**, 3197.
[265] J. T. Mague, *Inorg. Chem.*, 1972, **11**, 2558.
[266] E. K. Barefield and G. W. Parshall, *Inorg. Chem.*, 1972, **11**, 964.
[267] C. Cocevar, G. Mestroni, and A. Camus, *J. Organometallic Chem.*, 1972, **35**, 389.
[268] E. R. Birnbaum, *J. Inorg. Nuclear Chem.*, 1972, **34**, 3499.
[269] G. P. Khare and R. Eisenberg, *Inorg. Chem.*, 1972, **11**, 1385.
[270] M. S. Fraser and W. H. Baddley, *J. Organometallic Chem.*, 1972, **36**, 377.
[271] C. Masters, B. L. Shaw, and R. E. Stainbank, *J.C.S. Dalton*, 1972, 664.
[272] B. L. Shaw and R. E. Stainbank, *J.C.S. Dalton*, 1972, 2109.
[273] S. A. Smith, D. M. Blake, and M. Kubota, *Inorg. Chem.*, 1972, **11**, 660.
[274] S. Doronzo and V. D. Bianco, *Inorg. Chem.*, 1972, **11**, 466.
[275] A. Aràneo and T. Napoletano, *Inorg. Chim. Acta*, 1972, **6**, 363.
[276] B. L. Shaw and R. E. Stainbank, *J.C.S. Dalton*, 1972, 223.
[277] A. Aràneo, T. Napoletano, and P. Fantucci, *J. Organometallic Chem.*, 1972, **42**, 471.
[278] A. R. Manning, *J.C.S. Dalton*, 1972, 821.
[279] K. E. Schwarzhans and H. Steiger, *Angew. Chem. Internat. Edn.*, 1972, **11**, 535.
[280] C. B. Colburn, W. E. Hill, and D. W. A. Sharp, *Inorg. Nuclear Chem. Letters*, 1972, **8**, 625.
[281] A. L. Clarke and N. J. Fitzpatrick, *J. Organometallic Chem.*, 1972, **43**, 405.

X =	2-Me	1-Me	H	2-Cl	1-Cl
ν/cm^{-1} =	360	364	365	369	370

In $[MCl(C_2H_4)_2]_2$, i.r. bands at 398 and 502 cm^{-1} (M = Rh) or 448 and 537 cm^{-1} (M = Ir) have been assigned to $\nu(M-C_2H_4)$.[282] I.r. spectra of $[(\pi-C_5H_5)_2Co][CoI_3L]$ (L = PPh$_3$ or OPPh$_3$) have been given.[283]

Table 12 Compounds for which ν(MH) assignments have been given (M = Rh or Ir)

Compound	Ref.
RhHCl$_2$(PR$_3$)$_2$	261[a]
RhH(PPh$_3$)$_2$(SiEt$_3$)Cl	262[b]
[RhHCl(MeOPPh$_2$)$_4$]$^+$	263
RhH(CO)X$_2$(chel)	264[c]
[RhHCl(chel)$_2$]X	
[RhH$_2$(diars)$_2$]BPh$_4$	} 265[d]
[RhH$_2$\{P(OPh)$_3$\}$_4$]$^+$	
RhH$_2$[P(OPh)$_3$]$_2$(chel)	} 266[e]
[RhH$_2$(PPh$_3$)$_2$(bipy)]$^+$	267
[IrH(piperidine)$_4$X]X	268[f]
IrHCl(CO)(PPh$_3$)$_2$L	269[g]
IrH(CO)(PPh$_3$)$_2$L	
IrH(CO)(AsPh$_3$)$_2$(fumaronitrile)	} 270[h]
IrHCl$_2$(PBu$^t_2$R)$_2$	
[PBu$^t_2$MeH][Ir$_2$Cl$_2$H(PBu$^t_2$Me)$_2$]	} 271[i]
IrHCl$_2$(PBu$^t_2$R)$_2$L	272[j]
IrHCl(PPh$_3$)$_2$(O$_2$CR^1)	
IrHCl(PPh$_3$)$_2$(L)(O$_2$CR^2)	} 273[k]
[IrHX(chel)$_2$]Y	
[IrH$_2$(chel)$_2$]Y	} 274[l]
IrH$_2$(MPh$_3$)$_2$(chel)	275[m]
[IrH$_2$(CO)(PPh$_3$)$_3$][(CF$_3$CO$_2$)$_2$H]	263
IrH$_2$X(CO)(PBu^{t}Me$_2$)	
IrH$_2$Y(CO)(PBu^{t}R$_2$)$_2$	} 276[n]
IrH$_2$X(AsPh$_3$)$_2$(CNC$_6$H$_4$-p)	277[o]

[a] PR$_3$ = PMe$_3$, PEt$_3$, PMe$_2$Ph, PEt$_2$H, PEtPh$_2$, or PPh$_3$; isomeric forms with H *trans* to Cl or to PR$_3$ distinguished by their ν(RhH) values. [b] This substance may be a mixture of stereoisomers having a different configuration to the compound of the same formula previously obtained. [c] X = Cl or Br; chel = phen or bipy. [d] chel = PhMePCH$_2$CH$_2$-PMePh, X = [Rh(CO)$_2$Cl$_2$] or Cl; chel = diars, X = PF$_6$. [e] chel = the *ortho*-carbon-bonded ligand o-(PhO)$_2$POC$_6$H$_4$. [f] L = 2-mercapto-5-methylbenzenethio; configuration given. [h] L = fumaronitrile, cinnamonitrile, benzylidenemalonitrile, fumaric acid, or dimethyl fumarate. [i] R = Me, Et, or Pr^n. [j] R = Me; L = CO, MeNC, py, 4-methylpyridine, or P(OMe)$_3$. R = Et; L = CO, MeNC, or MeCN. R = Pr^n; L = CO or MeNC. Abnormally high values were noted for the compounds with L = py or 4-methylpyridine. [k] R^1 = R^2 or p-O$_2$NC$_6$H$_4$. L = CO; R^2 = Me, Et, Pr^n, Ph, H, CF$_3$, or MeCHCl. L = py; R^2 = Me, Et, or Ph. L = PhCN; R^2 = Me or Pr^n. L = PMe$_2$Ph; R^2 = Ph. L = p-MeC$_6$H$_4$NC; R^2 = Me. [l] X = Cl, Br, or I; Y = X, ClO$_4$, or BPh$_4$; chel = vinylenebis(diphenylphosphine). [m] M = P or As; chel = S$_2$CNEt$_2$ or S$_2$COEt; structure having *trans* MPh$_3$ groups, with *cis* H atoms given. [n] X = Cl or Br. Y = Cl; R = Et, Pr^n, or Bu^n. Y = Br; R = Pr^n. [o] X = H, F, Cl, Br, I, or N$_3$.

[282] A. L. Onderdelinden and A. van der Ent, *Inorg. Chim. Acta*, 1972, **6**, 420.
[283] M. Van den Akker, R. Olthof, F. Van Bothuis, and F. Jellinek, *Rec. Trav. chim.*, 1972, **91**, 75.

In [Co(CN)$_5$L]$^{3-}$ complexes there is a linear relationship between ν(CN) and polar substituent constant of L:[284]

L =	none or H$_2$O	ν(CN)/cm^{-1} =	2082
	alkyl		2087—2090
	alkenyl		2090—2094
	benzyl		2094
	H		2097
	thioalkyl		2110
	I		2116
	Br		2122
	CN		2127

Assignments of the i.r. spectra of *cis*-[Co(NH$_3$)$_3$X$_3$] (X = Cl or Br) and *trans*-[Co(NH$_3$)$_3$Cl$_3$] include assignments for ν(CoN) (472—500 cm^{-1}) and a mode described as ν(CoX) and δ(NCoN) (273—358 cm^{-1}).[285] I.r. data (375—5000 cm^{-1}) are also available [286] for [Co(NH$_3$)$_6$]$^{3+}$, [Co(NH$_3$)$_5$(NO$_2$)]$^{2+}$, *cis*- and *trans*-[Co(NH$_3$)$_4$(NO$_2$)$_2$]$^+$, 1,2,3- and 1,2,4-[Co(NH$_3$)$_3$(NO$_2$)$_3$], *trans*-[Co(NH$_3$)$_2$(NO$_2$)$_4$]$^-$, and [Co(NO$_2$)$_6$]$^{3-}$. In [Co(CO)$_2$(NH$_3$)(PPh$_3$)(CONH$_2$)],[287] ν(Co−NH$_3$) has been assigned to a medium intense i.r. band at 329 cm^{-1}. Very approximate normal-co-ordinate calculations, using five-atom M−NH$_3$ models, have been used to aid assignments of rhodium [288] and iridium [289] ammines and their deuteriates; proposed values (i.r., cm^{-1}) for ν(MN) are:

[Rh(NH$_3$)$_5$Cl]Cl$_2$	471, 478, *ca.* 489, 506
[Rh(ND$_3$)$_5$Cl]Cl$_2$	434, 442, 453, 470
[Rh(NH$_3$)$_5$Br]Br$_2$	470, 500
[Rh(ND$_3$)$_5$Br]Br$_2$	433, 466
[Ir(NH$_3$)$_5$Cl]Cl$_2$	477, 486, 500, 516, 527
[Ir(ND$_3$)$_5$Cl]Cl$_2$	442, 450, *ca.* 460, 477, 491

Tentative assignments for ν(CoN) (*ca.* 230—240 cm^{-1}) have been offered for complexes CoX$_2$(2,2′-dithiodipyridine) and CoX$_2$(4,4′-dithiodipyridine) [X = Cl or Br].[290] The internal vibrations of the cations in [Co(en)$_3$]$^{3+}$ salts (and the *N*-deuteriated species) are essentially the same in the solid state and in solution in H$_2$O or D$_2$O, indicating little cation–anion interaction.[291]

X-Ray data show that the trinuclear basic acetates [Rh$_3$O(CX$_3$CO$_2$)$_6$(H$_2$O)$_3$]ClO$_4$,H$_2$O (X = H or D) are isostructural with the known Cr^{III}

---

[284] T. Funabiki and K. Tarama, *Bull. Chem. Soc. Japan*, 1972, **45**, 2945.
[285] M. Linhard, H. Siebert, B. Breitenstein, and G. Tremmel, *Z. anorg. Chem.*, 1972, **389**, 11.
[286] J. Csaszar, *Magyar Kém. Folyóirat*, 1972, **78**, 154.
[287] H. Krohberger, H. Behrens, and J. Ellermann, *J. Organometallic Chem.*, 1972, **46**, 139.
[288] Yu. Ya. Kharitonov, N. A. Knyazeva, G. Ya. Mazo, I. B. Baranovskii, and N. B. Generalova, *Russ. J. Inorg. Chem.*, 1971, **16**, 1050.
[289] Yu. Ya. Kharitonov, N. A. Knyazeva, G. Ya. Mazo, I. B. Baranovskii, and N. B. Generalova, *Russ. J. Inorg. Chem.*, 1971, **16**, 1172.
[290] J. R. Ferraro, B. B. Murray, and N. J. Wieckowicz, *J. Inorg. Nuclear Chem.*, 1972, **34**, 231.
[291] R. W. Berg and K. Rasmussen, *Spectroscopy Letters*, 1972, **5**, 349.

analogue; group-frequency assignments given include $\nu_{sym}(Rh_3O)$ at 302 (X = H) or 298 cm^{-1} (X = D) and $\nu_{asym}(Rh_3O)$ at 382 and 397 (X = H) or 380 cm^{-1} (X = D).[292] I.r. data are also available for (MeCO)Rh(tetraphenylporphyrin),[293] NH$_4$CoPO$_4$,H$_2$O and its thermal decomposition products,[294] and two geometrical isomers of each of the complexes [CoX(PPh$_3$)(dimethylglyoximato)$_2$] (X = Cl, Br, or I).[295]

In dialkyl sulphide complexes MX$_3$L$_3$ [M = Rh, X = Cl, Br, or I; M = Ir, X = Cl or Br; L = Me$_2$S, Et$_2$S, (CH$_2$)$_4$S, or (CH$_2$)$_5$S], ν(MS) is in the 270—325 cm^{-1} i.r. range; i.r. [ν(MS) and ν(MX)], u.v., ^{1}H n.m.r., and dipole moment measurements all indicate *mer* geometry.[296]

Splitting into three components of ν_3 (in T_d) of the anion in [R$_3$NH][CoCl$_4$] in benzene solution has been attributed to lowering of symmetry to C_{2v} as a result of R$_3$NH$\cdots$ClCoCl$_3$ interactions since the [R$_4$N]$^+$ salt shows a single ν(CoCl) i.r. band at 297 cm^{-1}; [R$_4$N][CoCl$_3$] is said to be a pseudo-tetrahedral polymer even in solution (R = n-octyl).[297] Raman and i.r. spectra of alkali-metal and silver salts of some hexahalogenoiridates have been assigned. The $\nu_1(a_{1g})$ fundamental shows some variation with the nature of the cation (M$^+$):[298]

	$\Delta\nu$/cm^{-1} (M$^+$)
[IrCl$_6$]$^{2-}$	352 (K$^+$), 341 (Cs$^+$)
[IrCl$_6$]$^{3-}$	323 (K$^+$), 331 (Ag$^+$)
[IrBr$_6$]$^{2-}$	216 (K$^+$), 207 (Cs$^+$)
[IrBr$_6$]$^{3-}$	200 (K$^+$), 198 (Ag$^+$)
[IrI$_6$]$^{2-}$	156 (K$^+$)

Dramatic differences between relative Raman intensities of ν_1 and ν_2 for [IrX$_6$]$^{2-}$ species (X = Cl, Br, or I) suggest that the supposed occurrence of dynamic Jahn–Teller effects may in some cases be quenched by spin-orbit interaction and/or by charge transfer.[298] A variation of ν_3(IrCl) (308—323 cm^{-1}) in the isomorphous series M$_2$[IrCl$_6$] (M = K, Rb, Cs, or NH$_4$) is recorded.[299] See also ref. 30.

Two geometrical isomers of [CoCl(NH$_3$)(tren)]Cl$_2$ have been isolated, showing ν(CoCl) at 342 (purpureo) or 366 cm^{-1} (red isomer).[300] Observations on the i.r. spectra [285] of *cis*-[Co(NH$_3$)$_3$X$_3$] (X = Cl or Br) and *trans*-[Co(NH$_3$)$_3$Cl$_3$] have been mentioned above; another compound of formula 'Co(NH$_3$)$_3$Cl$_3$', described by Werner, has been found to have ν(CoCl)

---

[292] I. B. Baranovskii, G. Ya. Mazo, and L. M. Dikareva, *Russ. J. Inorg. Chem.*, 1971, **16**, 1388.
[293] B. R. James and D. V. Stynes, *J.C.S. Chem. Comm.*, 1972, 1261.
[294] L. N. Shchegrov, V. V. Pechkovskii, A. G. Ryadchenko, and R. Ya. Mel'nikova, *Russ. J. Inorg. Chem.*, 1971, **16**, 1622.
[295] A. V. Ablov, A. M. Gol'dman, O. A. Bologna, Yu. A. Simonov, and M. M. Botoshanskii, *Russ. J. Inorg. Chem.*, 1971, **16**, 1167.
[296] E. A. Allen and W. Wilkinson, *J.C.S. Dalton*, 1972, 613.
[297] M. G. Kuzina, A. A. Lipovskii, and S. A. Nikitina, *Russ. J. Inorg. Chem.*, 1971, **16**, 1313.
[298] G. L. Bottger and A. E. Salwin, *Spectrochim. Acta*, 1972, **28A**, 925.
[299] G. Pannetier and D. Macarovici, *J. Thermal Anal.*, 1972, **4**, 187.
[300] C.-H. L. Yang and M. W. Grieb, *J.C.S. Chem. Comm.*, 1972, 656.

values of 270 and 282 cm^{-1} and is therefore formulated [301] as the chloro-bridged cis-[{$Co_2Cl_3(NH_3)_6$}Cl_3]$_x$. Fairly complete i.r. and Raman spectral assignments have been given for [$Co(NH_3)_4X_2$]$^+$ and [$Co(en)_2X_2$]$^+$ salts; ν_{asym}(CoX) is at wavenumbers (cm^{-1}) of 280 (X = Cl) or 170 (X = Br) with ν_{sym}(CoX) at 530 (X = F), 360 (X = Cl), or 230 cm^{-1} (X = Br).[302] Values of ν(CoX) in the expected regions have been noted for the tetrahedral complexes CoX_2L_2 (X = Cl or Br; L = 2-bromothiazole).[303] Other references listing ν(CoX) modes have already been mentioned.[12-15, 31, 33-35, 290]

Assignments of ν(RhX) and ν(IrX) modes, generally made in structural characterization, have been given in a range of complexes as listed in Table 13.[263-265, 273, 276, 282, 288, 296, 304-316]

Table 13 *Compounds for which ν(RhX) or ν(IrX) assignments have been made*

Compound	Ref.
L(Ph$_3$P)RhClC{N(R)CH$_2$}$_2$	304^a
[(OC)Cl$_3$Rh—C(Ph)N(R^1)C(Ph)=NMe]$_2$	
(Me$_2$PhP)$_2$Cl$_3$Rh—C(Ph)N(Me)C(Ph)=NMe	
[(OC)Cl$_3$Rh—C(Ph)NHR^2]$_n$	305^b
[(OC)Cl$_3$Rh—C(Me)NHC$_6$H$_4$Me-o]$_n$	
(Me$_2$PhP)$_2$Cl$_3$Rh—C(Me)NHC$_6$H$_4$Me-o	
(Ph$_3$P)(OC)Cl$_3$Rh—C(Ph)NHR^1	
(chel)$_2$RhCl	
(chel)RhCl$_2$Rh(cod)	306^c
(chel)RhCl(cod)	
(chel)$_2$RhCl(CO)	
RhClL$_3$	
RhCl(CO)$_2$L	307^d
Rh$_2$Cl$_2$L$_4$	
RhCl(TPPO)L	
RhCl(TPPO)(CO)(Ph$_3$P)	308^e
[RhCl(TPPO)]$_2$	

[301] K. Wieghardt and H. Siebert, *Z. Naturforsch.*, 1972, **27b**, 349.
[302] G. Chottard, *Compt. rend.*, 1972, **274**, C, 1116.
[303] E. J. Duff, M. N. Hughes, and K. J. Rutt, *Inorg. Chim. Acta*, 1972, **6**, 408.
[304] D. J. Cardin, M. J. Doyle, and M. F. Lappert, *J.C.S. Chem. Comm.*, 1972, 927.
[305] M. F. Lappert and A. J. Oliver, *J.C.S. Chem. Comm.*, 1972, 274.
[306] P. R. Brookes, *J. Organometallic Chem.*, 1972, **42**, 459.
[307] D. G. Holah, A. N. Hughes, and B. C. Hui, *Canad. J. Chem.*, 1972, **50**, 3714.
[308] D. G. Holah, A. N. Hughes, and B. C. Hui, *Canad. J. Chem.*, 1972, **50**, 2442.
[309] P. Hong, N. Nishii, K. Sonogashira, and N. Hagihara, *J.C.S. Chem. Comm.*, 1972, 993.
[310] F. Faraone, F. Cusmano, P. Piraino, and R. Pietropaolo, *J. Organometallic Chem.*, 1972, **44**, 391.
[311] B. Corain and M. Martelli, *Inorg. Nuclear Chem. Letters*, 1972, **8**, 39.
[312] H. Onoue and I. Moritani, *J. Organometallic Chem.*, 1972, **44**, 189.
[313] B. L. Booth, R. N. Haszeldine, and G. R. H. Neuss, *J.C.S. Chem. Comm.*, 1972, 1074.
[314] G. Pannetier, P. Fougeroux, and R. Bonnaire, *J. Organometallic Chem.*, 1972, **38**, 421.
[315] R. Ugo, A. Pasini, A. Fusi, and S. Cenini, *J. Amer. Chem. Soc.*, 1972, **94**, 7364.
[316] B. E. Mann, C. Masters, and B. L. Shaw, *J.C.S. Dalton*, 1972, 48.

Table 13 (*cont.*)

Compound	Ref.
[RhCl(CO)(CPh$_2$)]$_n$	309^f
[RhCl(CO)(CPh$_2$)py]$_2$,2CH$_2$Cl$_2$	
[RhCl(H)(MeOPPh$_3$)$_4$]$^{2+}$	263
(π-C$_5$H$_5$)RhCl(PPh$_3$)(CH$_2$CN)	310
[RhCl$_2$(PPh$_3$)$_2${−C(S)NMe$_2$}]$_2$	311^g
[RhCl(ND$_3$)$_5$]Cl$_2$	288^h
HRhX$_2$(CO)L	264^i
RhX(CO)$_2$(opd)	
[HRhCl(PhMePCH$_2$CH$_2$PMePh)$_2$]X	265^j
[(chel)$_2$RhCl]$_2$	312^k
RhX$_3$(sulphide)$_3$	296^l
[RhX(ol)$_2$]$_2$	
RhX(alk)$_2$	282^m
[RhX(cod)]$_2$	
IrX$_3$(sulphide)$_3$	296^n
[IrX(ol)$_2$]$_2$	
IrX(alk)$_2$	282^m
[IrX(cod)]$_2$	
IrCl(OOBu^t)$_2$(CO)(MPh$_3$)$_2$	313°
IrCl(cod)(CH$_2$=CHCH$_2$Cl)	314^p
trans-[IrCl(CO)L$_2$]	315^q
IrCl$_2$X(PMe$_2$Ph)(diphos)	
[IrCl$_2$L^1(PMe$_2$Ph)(diphos)]ClO$_4$	
IrCl$_3$(AsMe$_2$Ph)L^2	316^r
IrCl$_3$(PMe$_2$Ph)(Ph$_2$AsCH$_2$CH$_2$AsPh$_2$)	
IrCl$_3$(PMePh$_2$)(diphos)	
IrCl$_3$(CO)L^1$_2$	
IrCl$_2$R(CO)(PBu^{t}Me$_2$)	
IrClH$_2$(CO)L^2$_2$	276^s
IrCl(O$_2$)(CO)L^2$_2$	
IrCl(Me)IL^2$_2$	
HIrCl(PPh$_3$)$_2$(O$_2$CR^1)	273^t
HIrCl(PPh$_3$)$_2$(L)(O$_2$CR^2)	

^{a} L = Ph$_3$P, R = Ph or *p*-tolyl; L = CO, R = Ph. ^{b} R^1 = Me or Et; R^2 = Me or Pr^i. ^{c} chel = *o*-Ph$_2$PC$_6$H$_4$CH=CH$_2$; cod = cyclo-octa-1,5-diene. ^{d} L = phosphole (12) or (13). ^{e} TPPO = 1,2,5-triphenylphosphole 1-oxide; L = Ph$_3$P, Me$_2$S, phen, 2(py), or 2(CO) (all bonded to Rh), or L = FeCl$_3$ (bonded to the phosphoryl group); the TPPO ligand is bonded to Rh *via* its π-system. ^{f} The solvate is believed to contain CO as the only bridging group, whereas for the polymeric compound both CO and Cl bridges are suggested. ^{g} Suggested to be a chloro-bridged dimer. ^{h} Isotopic (^{35,37}Cl) splitting observed. ^{i} X = Cl or Br; L = phen or bipy; (opd) = *o*-phenylenediamine. ^{j} X = Cl$^-$ or [Rh(CO)$_2$Cl$_2$]$^-$. ^{k} Chloro-bridged complexes, where chel = the *o*-carbon-bonded ligands —C$_6$H$_4$C(R)=NOH (R = Me, Et, Pr^n, or Ph). ^{l} X = Cl, Br, or I; (sulphide) = Me$_2$S, Et$_2$S, (CH$_2$)$_4$S, or (CH$_2$)$_5$S; *mer* geometry proposed. ^{m} (ol) = C$_2$H$_4$, C$_3$H$_6$, or C$_8$H$_{14}$; (alk) = C$_4$H$_6$ or isoprene; (cod) = cyclo-octa-1,5-diene. ^{n} X = Cl or Br; (sulphide) = Me$_2$S, Et$_2$S, (CH$_2$)$_4$S, or (CH$_2$)$_5$S. ^{o} M = P or As. ^{p} (cod) = cyclo-octa-1,5-diene. ^{q} L = Ph$_3$P, EtPh$_2$P, Et$_2$PhP, or (*p*-XC$_6$H$_4$)$_3$P; X = F, Cl, Me, or MeO. ^{r} X = Cl, Br, I, NO$_3$, NO$_2$, N$_3$, NCO, or SCN; ν(IrX) or internal models of X given in some cases; L^1 = PMePh$_2$, CO, or py; L^2 = diphos or Ph$_2$AsCH$_2$CH$_2$AsPh$_2$. ^{s} L^1 = PBu^{t}Et$_2$, PBu^{t}Pr^n$_2$, PBu^{t}Bu^n$_2$, or PBu^t$_2$Me; R = Cl, CCl$_3$, CH$_2$CH=CH$_2$, PhCH=CHCH$_2$, MeCO, Me$_2$C=CHCO, MeOCO, PhN$_2$, or PhSO$_2$; L^2 = L^1 or PBu^{t}Me$_2$; ν(Ir—O$_2$) at 824—850 cm^{-1}. ^{t} R^1 = R^2 or *p*-O$_2$NC$_6$H$_4$. L = CO; R^2 = Me, Et, Pr^n, Ph, H, CF$_3$, or MeCHCl. L = py; R^2 = Me, Et, or Ph. L = PhCN; R^2 = Me or Pr^n. L = PMe$_2$Ph; R^2 = Ph. L = *p*-MeC$_6$H$_4$NC; R^2 = Me.

(12) X = H or Ph   (13)

10 Nickel, Palladium, and Platinum

A linear plot of ν(PtH) in trans-[(p-YC$_6$H$_4$S)PtH(PPh$_3$)$_2$] with the Hammett σ_p parameter of Y shows that changes in ν(PtH) simply reflect changes in electron density at Pt as Y is varied.[317] Assignments have been given for ν(PtH) in complexes of the type trans-[HPtXL$_2$] (X = Cl, Br, I, NO$_3$, CN, NCO, NCS, SCN, or NCSe; L = Ph$_3$P, Ph$_2$MeP, Ph$_2$EtP, (p-tolyl)$_3$P, Et$_3$P, Bu$^n_3$P, Et$_3$As, etc.);[318-320] noteworthy is the existence of Pt—NCS and Pt—SCN isomers. Similar values for ν(PtH) are available for [HPtX-(1H,3H-quinazoline-2,4-dithione)] (X = NO$_3$ or I),[321] [HPt(PPh$_3$)$_3$]$^+$ salts,[322] and [HPtL(PPh$_3$)$_2$] (L = SiPh$_3$, SiPh$_2$Me, SiPh$_2$H, SiEt$_3$, Si(OEt)$_3$, SiMe$_2$OSiMe$_2$H, or SiMe{OSiMe$_3$}$_2$).[323]

Methylnickel complexes have been characterized. In the series MeNiX-(PMe$_3$)$_2$ [X = Cl, Br, or I], ν(NiC) is near 520 cm^{-1} with δ(NiCH$_3$) at ca. 1150 cm^{-1}; the lower value of 487 cm^{-1} for ν(NiC) is obtained in Me$_2$Ni(PMe$_3$)$_3$.[324] For the complexes [Me$_2$Pt(PMe$_2$Ph)$_2$L$^1_2$]$^{2+}$, [Me$_2$Pt-(PMe$_2$Ph)$_2$IL^2]$^+$, and [Me$_2$Pt(PMe$_2$Ph)$_2$(tetrapyrazolylborato)]$^+$ {L^1 = $\frac{1}{2}$(phen), $\frac{1}{2}$(bipy), $\frac{1}{3}$(terpy), $\frac{1}{2}$(diphos), $\frac{1}{2}$(diars), MeCN, py, P(OMe)$_3$, S$_2$CNEt$_2$, HN=COMe(C$_6$F$_5$)$_2$, p-MeOC$_6$H$_4$CN, p-MeOC$_6$H$_4$NC, p-MeC$_6$-H$_4$NC, or EtNC; L^2 = py, PPhMe$_2$, EtNC, p-MeOC$_6$H$_4$NC, or p-MeC$_6$-H$_4$NC}, two ν(PtC) i.r. bands have been observed in most cases (503—586 cm^{-1}), in agreement with structures involving cis methyl groups.[325] The values for ν(Pt—CH$_3$) arising from methyl groups trans to lutidine (578—585 cm^{-1}) are significantly different from those due to the trans Me—Pt—X system (552—580 cm^{-1}) in complexes Me$_3$PtX(3,5-lutidine) [X = Cl, Br, I, NCO, NCS, NO$_3$, NO$_2$, or MeCO$_2$]; for the Pt—CH$_3$ bonds trans to X, the following order of ν(PtC) modes is observed:[326]

$$NO_3 \sim MeCO_2 \sim NCO > Cl > Br \sim NO_2 > I \;(\gg Me)$$

Data on [Me$_3$Pt(NH$_3$)$_3$]$^+$ have been mentioned in an earlier section.[2]

---

[317] A. E. Keskinen and C. V. Senoff, J. Organometallic Chem., 1972, **37**, 201.
[318] H. C. Clark and A. Kurosawa, J. Organometallic Chem., 1972, **36**, 399.
[319] M. W. Adlard and G. Socrates, J. Inorg. Nuclear Chem., 1972, **34**, 2339.
[320] M. W. Adlard and G. Socrates, J.C.S. Dalton, 1972, 797.
[321] U. Agarwala and Lakshmi Agarwala, J. Inorg. Nuclear Chem., 1972, **34**, 251.
[322] K. Thomas, J. T. Dumler, B. W. Renoe, C. J. Nyman, and D. M. Roundhill, Inorg. Chem., 1972, **11**, 1795.
[323] C. Eaborn, A. Pidcock, and B. Ratcliff, J. Organometallic Chem., 1972, **43**, C5.
[324] H. F. Klein and H. H. Karsch, Chem. Ber., 1972, **105**, 2628.
[325] H. C. Clark and L. E. Manzer, Inorg. Chem., 1972, **11**, 2749.
[326] D. F. Clegg, J. R. Hall, and G. A. Swile, J. Organometallic Chem., 1972, **38**, 403.

A novel mode of bonding of acetoacetate esters, through the terminal aceto-carbon atom, has been reported; the complexes ClPd(CH$_2$COCH$_2$-CO$_2$CH$_2$R)L$_2$ [R = Me, L = py or ½(bipy); R = Ph, L = py, 4-methyl-pyridine, 2,6-lutidine, or ½(bipy)] have been characterized, showing ν(PdC) in the 498—543 cm^{-1} region.[327]

Vibrational data on Zeise's salt (i.r. and Raman) and some related complexes (i.r.) support previous assignments (*Spectrochim. Acta*, 1969, **25A**, 749), for example:[328]

	ν(PtC)/cm^{-1}	ν(PtX)/cm^{-1}
K[PtCl$_3$(C$_2$H$_4$)],H$_2$O	480, 390	336, 330
K[PtCl$_3$(C$_2$H$_4$)]	490, 400	336, 328
K[PtCl$_3$(C$_2$D$_4$)]	450, 384	336, 326
K[PtBr$_3$(C$_2$H$_4$)]	484, 393	241, 225

Unassigned i.r. data are available[329] for PdC(CF$_3$)$_2$OL$_2$ [L = P(OPh)$_3$, PMePh$_2$, or ½(diphos)], PdC(CF$_3$)$_2$OC(CF$_3$)$_2$OL$_2$ [L = P(OMe)$_3$, P(OMe)$_2$Ph, or Me$_2$AsCH$_2$Ph], and PdC(CF$_3$)$_2$OC(CF$_3$)$_2$O(PMePh$_2$)$_2$.

Species M(CO)$_x$ [M = Ni, Pd, or Pt; x = 1—4] have been identified by i.r. spectroscopy in the ν(CO) region on matrices produced from co-condensation of (M + CO) or (M + CO + Ar); although wavenumbers are given there are no detailed assignments to specific species given in this preliminary note;[330] see also Chapter 4. In NiI$_2$(CO)(PMe$_3$)$_2$, ν(NiC) is at 483 cm^{-1}.[331] Reactions of M(PPh$_3$)$_x$ [M = Pd, x = 4; M = Pt, x = 3 or 4] with CO at high pressures have been monitored by i.r. spectroscopy, and ν(CO) data used to identify Pt(CO)$_2$(PPh$_3$)$_2$, Pt(CO)$_3$(PPh$_3$), and Pd(CO)$_2$(PPh$_3$)$_2$, but in no cases were M(CO)$_4$ species formed.[332]

Reaction of Ni(CNBu^t)$_4$ with ClCO$_2$Et gives[333] 'NiCl(CNBu^t)$_2$' having ν(NC) = 2180 cm^{-1}. Vibrational data are available for some cyano-complexes of nickel(II)[334] and platinum(IV).[335]

The metal isotope effect has been used in making i.r. assignments for chelate complexes M{S(CH$_2$)$_n$NH$_2$}$_2$ and M{S(CH$_2$)$_2$NMe$_2$}$_2$ (M = Ni or Pd; n = 2 or 3),[336] and *trans*-[Ni(chel)$_2$X$_2$] {chel = 2,5-dithiahexane,

---

[327] S. Baba, T. Sobata, T. Ogura, and S. Kawaguchi, *Inorg. Nuclear Chem. Letters*, 1972, **8**, 605.

[328] J. Hubert, P. C. Kong, F. D. Rochon, and T. Theophanides, *Canad. J. Chem.*, 1972, **50**, 1596.

[329] H. D. Empsall, M. Green, and F. G. A. Stone, *J.C.S. Dalton*, 1972, 96.

[330] H. Huber, P. Kündig, M. Moskovits, and G. A. Ozin, *Nature Phys. Sci.*, 1972, **235**, 98.

[331] M. Pankowski and M. Bigorgne, *J. Organometallic Chem.*, 1972, **35**, 397.

[332] T. Inglis and M. Kilner, *Nature Phys. Sci.*, 1972, **239**, 13.

[333] S. Otsuka, M. Naruto, T. Yoshida, and A. Nakamura, *J.C.S. Chem. Comm.*, 1972, 396.

[334] C. A. McAuliffe, M. O. Workman, and D. W. Meek, *J. Co-ordination Chem.*, 1972, **2**, 137.

[335] M. N. Memering, *Diss. Abs. Internat.* (*B*), 1972, **32**, 5674.

[336] C. W. Schläpfer and K. Nakamoto, *Inorg. Chim. Acta*, 1972, **6**, 177.

2-(ethylthio)ethylamine, 2-(methylthio)ethylamine, or NN-dimethylethyl-enediamine}.[337] Typical results for metal–ligand stretching modes are shown in Table 14. Other assignments of ν(MN) relevant to this section

Table 14 *Metal isotope effects on the far-i.r. spectra of some complexes of Ni and Pd*

Complex	ν(MN)/cm^{-1}	ν(MS)/cm^{-1}
^{58}Ni(SCH$_2$CH$_2$NMe$_2$)$_2$	371.2	397.3
^{62}Ni(SCH$_2$CH$_2$NMe$_2$)$_2$	367.5	394.1
Pd(SCH$_2$CH$_2$NMe$_2$)$_2$ [a]	394	367
trans-[^{58}Ni(2,5-dithiahexane)$_2$Cl$_2$]	266.7[b]	232,[c] 209.6
trans-[^{62}Ni(2,5-dithiahexane)$_2$Cl$_2$]	261.7[b]	232,[c] 205.6

[a] Assignments for Pd in natural abundance made on intensity grounds; note the reverse order of ν(MS) and ν(MN) compared with the Ni analogue. [b] ν(NiCl) modes.
[c] Symmetric mode involving no motion of metal atom.

are for the five-co-ordinate complexes [Ni(tscR$_2$)X]X (tscR$_2$ = cyclo-pentanone-, cyclohexanone-, or cycloheptanone-thiosemicarbazone; X = Cl or Br; ca. 290 and ca. 240 cm^{-1}),[338] adducts of *trans*-[Ni(β-ketoenolato)$_2$] with piperidine, piperazine, methylpiperazine, or morpholine [300—385 cm^{-1}; ν(NiO) in acetylacetonato-adducts at 570—568 and 412—416 cm^{-1}],[339] *trans*-[PdCl$_2$(morpholine)$_2$] (supposedly at 500 cm^{-1}),[340] and *trans*-[PtL$_2$I$_2$] (L = py, 3- or 4-methylpyridine, 4-ethylpyridine, or 3,5-dimethylpyridine; tentatively at 245—294 cm^{-1}).[341] The variation in ν(PtN) noted in ref. 341 has been examined in more detail (with some disagreement in assignments) in a study of the i.r. spectra of *trans*-[MX$_2$L$_2$] (M = Pd or Pt; X = Cl or I; L = py, 4-methylpyridine, or 4-ethylpyridine) [bromides were excluded because of the proximity of ν(PtBr) to ν(PtN)].[342] The variations of ν(MX) with L, and of ν(MN) with X (Table 15) are considered to be too large to be due to differing basicity in L but may possibly arise from different molecular packing in the crystalline state.[342]

The mode of co-ordination of pseudohalide ligands to elements of this section has continued to be an active field.[318–320, 334, 343–346] Thus on the basis of i.r. (or with other) data [generally just ν(CN) values] structural

[337] C. W. Schläpfer, Y. Saito, and K. Nakamoto, *Inorg. Chim. Acta*, 1972, **6**, 284.
[338] B. Beecroft, M. J. M. Campbell, and R. Grzeskowiak, *Inorg. Nuclear Chem. Letters*, 1972, **8**, 1097.
[339] G. Marcotrigiano, R. Battistuzzi, and G. C. Pellacani, *Canad. J. Chem.*, 1972, **50**, 2557.
[340] R. A. Singh and E. B. Singh, *J. Inorg. Nuclear Chem.*, 1972, **34**, 769.
[341] G. W. Watt, L. K. Thompson, and A. J. Pappas, *Inorg. Chem.*, 1972, **11**, 747.
[342] M. Pfeffer, P. Braunstein, and J. Dehand, *Inorg. Nuclear Chem. Letters*, 1972, **8**, 497.
[343] L. Sacconi and D. Gatteschi, *J. Co-ordination Chem.*, 1972, **2**, 107.
[344] M. V. Artemenko, E. A. Chistyakova, P. A. Suprunenko, and G. I. Kal'naya, *Russ. J. Inorg. Chem.*, 1971, **16**, 1026.
[345] K. K. Chow and C. A. McAuliffe, *Inorg. Nuclear Chem. Letters*, 1972, **8**, 1031.
[346] J. L. Lauer, M. E. Peterkin, J. L. Burmeister, K. A. Johnson, and J. C. Lim, *Inorg. Chem.*, 1972, **11**, 907.

Table 15 $v(MX)$ and $v(MN)$ assignments in Pd and Pt halide complexes of pyridine and substituted pyridines[a]

Complex[b]	$v(MX)$/cm^{-1}	$v(MN)$/cm^{-1}
trans-[PdCl$_2$(py)$_2$]	358	278
trans-[PdI$_2$(py)$_2$]	177	278
trans-[PdCl$_2$(4-Mepy)$_2$]	356	302
trans-[PdI$_2$(4-Mepy)$_2$]	208	320
trans-[PdCl$_2$(4-Etpy)$_2$]	373	306
trans-[PdI$_2$(4-Etpy)$_2$]	195	278
trans-[PtCl$_2$(py)$_2$]	342	282
trans-[PtI$_2$(py)$_2$]	183	293
trans-[PtCl$_2$(4-Mepy)$_2$]	350	317
trans-[PtI$_2$(4-Mepy)$_2$]	199	327[c]
trans-[PtCl$_2$(4-Etpy)$_2$]	354	328
trans-[PtI$_2$(4-Etpy)$_2$]	193	264

[a] Taken from ref. 342. [b] 4-Mepy = 4-methylpyridine; 4-Etpy = 4-ethylpyridine.
[c] A substantially different value (256 cm^{-1}) was tentatively proposed in ref. 341.

conclusions have been drawn regarding [HPt(NCS)(Bu$^n_3$P)$_2$] (N- and S-bonded isomers),[319] [Ni(NCS)$_2$(Ph$_2$AsCH$_2$CH$_2$NMeCH$_2$−)$_2$] (N-bonded and bridged [N- plus S-bonded] isomers),[343] the 1:2 complex of Ni(NCS)$_2$ with 2-benzylbenzimidazole {as [L$_2$Ni(SCN)$_2$NiL$_2$][SCN]$_2$},[344] and [Pd(SCN)$_2$(cis-Ph$_2$PCH=CHPPh$_2$)] (S-bonded on the basis of δ(SCN) at 410 cm^{-1}).[345] Values for v(PdX) in N-bonded [Pd(Et$_4$dien)X]$^+$ complexes [Et$_4$dien = $NNN'N'$-tetraethyldiethylenetriamine; X = NCS (365), NCSe (360), NCO (365), N$_3$ (380), NO$_2$ (320 cm^{-1})] have been compared with those in the S-(or Se-)bonded [Pd(dien)X]$^+$ salts [X = SCN (320) or SeCN (318 cm^{-1})].[346]

Some i.r. data are also available for Cs$_2$[PtI$_3$(NO$_2$)(NH$_3$)$_2$] and its reaction products with Br$_2$ or I$_2$,[347] and for clathrates involving [Ni(CN)$_4$]$^{2-}$.[348, 349]

The following assignments of metal–phosphorus modes have been made:

	$v(MP)$/cm^{-1}	Ref.
NiI$_2$(CO)(PMe$_3$)$_2$	228	331
Ni(PH$_3$)$_4$	296	350
[Ni(PMe$_3$)$_4$]$^{2+}$	215	351
Ni(PMe$_3$)$_2$X$_2$	215—216	351
(X = Cl, Br, CN, or NO$_2$)		
Ni(PMe$_3$)$_2$(NCS)$_2$	218, 190	351
trans-[ClPd(CHCl$_2$)(PPh$_3$)$_2$]	ca. 445	352

[347] G. S. Muraveiskaya and I. I. Antokol'skaya, *Russ. J. Inorg. Chem.*, 1971, **16**, 868.
[348] S. Akyüz, A. B. Dempster, R. L. Morehouse, and N. Zengin, *J.C.S. Chem. Comm.*, 1972, 307.
[349] A. Sopkova, E. Matejcikova, J. Chomic, and J. Skorsepa, 'Proceedings of the 3rd Conference on Co-ordination Chemistry', 1971, p. 331.
[350] M. Trabelsi, A. Loutelier, and M. Bigorgne, *J. Organometallic Chem.*, 1972, **40**, C45.
[351] A. Merle, M. Dartiguenave, and Y. Dartiguenave, *J. Mol. Structure*, 1972, **13**, 413.
[352] Ya. M. Kimel'fel'd, E. M. Smirnova, N. I. Pershikova, O. L. Kaliya, O. M. Temkin, and R. M. Flid, *J. Struct. Chem.*, 1971, **12**, 1014.

A preliminary report of the formation of binary CO_2 complexes of Ni in Ar matrices [353] was quickly shown by the same authors to be erroneous.[354] The bands originally attributed to $Ni(CO_2)_n$ species were shown in fact to be due to $Ni(N_2)_n$ complexes, the N_2 being an impurity in the gas mixture; this clearly illustrates the caution needed in interpreting results from matrix-isolation studies.

Partial i.r. data have been given for $Pd(hmcc)_2$ [(hmcc)H = ligand (14); $\nu(PdO) = 570$ cm^{-1}],[355] [(Ph$_3$P)Pd—OC(O)CH$_2$]$_n$ [ν(C=O) at 1545

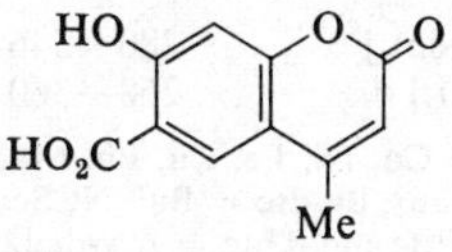

(14)

cm^{-1}],[356] the bis(hydroxy-bridged) complexes $[L_2M(OH)_2ML_2](BF_4)_2$ [M = Pd; L = Ph$_3$P. M = Pt; L = py, Ph$_3$P, or Et$_3$P. ν(PtO) at ca. 480 cm^{-1}],[357] and nickel–silica gels.[358]

Metal–sulphur vibrational assignments made are listed in Tables 14 [336, 337] and 16.[346, 359–363]

Two new ν(Ni—halogen) assignments reported are worthy of note. In [Ni(NN-di-n-butyldithiocarbamato)$_2$I], ν(NiI) is at the acceptably high value of 275 cm^{-1} for this Ni^{III} complex.[361] In the five-co-ordinate Ni^{II} species [Ni(tscR$_2$)X]X (X = Cl or Br; tscR$_2$ = cyclopentanone-, cyclohexanone-, or cycloheptanone-thiosemicarbazone) the ν(NiCl) (256—270 cm^{-1}) and ν(NiBr) (219—224 cm^{-1}) values are intermediate between the ranges normally found for four- and six-co-ordination;[338] similarly, ν(NiI) is at 153 and 194 cm^{-1} in NiI$_2$(CO)(PMe$_3$)$_2$.[331]

A single-crystal X-ray study has shown PtCl$_4$ to be isostructural with PtBr$_4$ and α-PtI$_4$. Fifteen of the 21 i.r. bands predicted for the octahedral chain were found (377s, 364sh, 345wm, 337s, 324sh, 310w, 290m, 274s, 235wbr, 200w, 182s, 167sh, 140w, 130w, and 110wm), and the data were said to be consistent with the existence of terminal and bridging Cl

---

[353] H. Huber, M. Moskovits, and G. A. Ozin, *Nature Phys. Sci.*, 1972, **236**, 127.
[354] H. Huber, M. Moskovits, and G. A. Ozin, *Nature Phys. Sci.*, 1972, **239**, 48.
[355] D. K. Rastogi, *Austral. J. Chem.*, 1972, **25**, 729.
[356] S. Baba, T. Ogura, S. Kawaguchi, H. Tokunan, Y. Kai, and N. Kasai, *J.C.S. Chem. Comm.*, 1972, 910.
[357] G. W. Bushnell, K. R. Dixon, R. G. Hunter, and J. J. McFarland, *Canad. J. Chem.*, 1972, **50**, 3694.
[358] V. V. Sviridov, G. A. Popkovich, and S. A. Serova, *Russ. J. Inorg. Chem.*, 1971, **16**, 925.
[359] R. A. Bailey and T. W. Michelsen, *J. Inorg. Nuclear Chem.*, 1972, **34**, 2671.
[360] J. Willemse and J. A. Cras, *Rec. Trav. chim.*, 1972, **91**, 1309.
[361] J. Willemse, P. H. F. M. Feuwette, and J. A. Cras, *Inorg. Nuclear Chem. Letters*, 1972, **8**, 389.
[362] B. J. McCormick and B. P. Stormer, *Inorg. Chem.*, 1972, **11**, 729.
[363] E. A. Allen and W. Wilkinson, *Spectrochim. Acta*, 1972, **28A**, 725.

Table 16 Assignments of ν(MS) modes[a]

Compound	ν(MS)/cm^{-1}	Ref.
M[Pt(SCN)$_6$]	267—293	359[b]
[Pd(dien)(SCN)]	320	346
[Ni(Bu$_2$dtc)$_3$]$^+$	370, 387, 410	
[Ni(Bu$_2$dsc)$_3$]$^+$	327, 317, 284[c]	
[Pd(Bu$_2$dtc)$_3$]$^+$	333, 360, 383	360[d]
[Pt(Bu$_2$dtc)$_3$]$^+$	341, 365	
[Ni(Bu$_2$dtc)$_2$]I	377	361[d, e]
[Ni(R$^1_2$tc)$_2$]		
[Ni(R$^2_2$tc)$_2$(py)$_2$]	377—394	362[f]
trans-[MX$_2$(sulphide)$_2$]	280—346	
cis-[PtY$_2$(sulphide)$_2$]	268—360	363[g]

[a] See also Table 14. [b] M^{II} = Co, Ni, Fe, Cu, Pb, Mn, Zn, or Cd. [c] ν(NiSe). [d] (Bu$_2$dtc) = NN-dibutyldithiocarbamato; Bu$_2$dsc = Bu$^n_2$NCSe$_2$. [e] Square-pyramidal geometry suggested by an e.s.r. study. [f] R$^1_2$tc and R$^2_2$tc = NN-dialkylthiocarbamato; R^1 = Me, Et, or Pr^n, or R$^1_2$ = (CH$_2$)$_4$; R^2 = Pr^n, or R$^2_2$ = (CH$_2$)$_4$; ν(NiO) at 520—539 cm^{-1}. [g] M = Pd or Pt; X = Cl, Br, or I; Y = Cl or Br; (sulphide) = Me$_2$S, Et$_2$S, (CH$_2$)$_4$S, or (CH$_2$)$_5$S; the slightly wider range of ν(MS) in the cis-compounds was noted.

atoms.[364] Mixed Pt^{II} halides have been shown (using X-ray measurements) to be individual compounds and have the following i.r. bands:[365]

	ν(PtCl)/cm^{-1}	ν(PtBr)/cm^{-1}
PtICl	275, 340	—
PtCl$_2$	319	—
PtBrCl	307, 360	240
PtIBr	—	238
PtBr$_2$	—	240

Assignments of bridging ν(PdX) or ν(PtX) modes have been proposed in [(Bu$^n_3$P)$_2$PtX$_2$Pt(PBu$^n_3$)$_2$][BX$_4$]$_2$ (X = Cl, 303 and 279 cm^{-1}; X = Br, 209 and 200 cm^{-1}),[366] [(PhMe$_2$P)$_2$PtX$_2$HgY$_2$] (X = Y = Cl, 295 and 270 cm^{-1}; related compounds also studied),[367] [(Ph$_2$PCH$_2$PPh$_2$)PtCl$_2$Pt(Ph$_2$PCH$_2$-PPh$_2$)] (249 cm^{-1}),[368] [(PhMe$_2$P)RPdCl$_2$PdR(PMe$_2$Ph)] {R = H$_2$C=C-(Me)CH$_2$C(CF$_3$)=C(CF$_3$), 261 cm^{-1}},[369] and the o-carbon-bonded chelates [o-R$^1_2$N=C(R^2)C$_6$H$_4$PdCl]$_2$ (R^1 or R^2 = H, Me, Ph, etc.; 254—320 cm^{-1}).[370] The bis-arene complexes cis-[Pd{C(NHR)Y}$_2$Cl$_2$] (Y = MeO, PhNH, Me$_2$N, p-MeC$_6$H$_4$NH, etc.; R = various) appear to have loosely bound Cl atoms,[371] ν(PdCl) having the low values of 263—305 cm^{-1}.

ν(PtCl) values have been used to suggest the magnitude of the trans influence of the appropriate groups in trans-[PtL$_2$(ArSO$_2$)Cl] (296—304 cm^{-1}; L = Et$_3$P, Et$_3$As, Et$_2$Se, or Et$_2$Te; Ar = Ph, Me, or p-ClC$_6$H$_4$;

---

[364] M. F. Pilbrow, J.C.S. Chem. Comm., 1972, 270.
[365] S. S. Batsanov and L. A. Vostrikova, Russ. J. Inorg. Chem., 1971, **16**, 1792.
[366] P. M. Druce, M. F. Lappert, and P. N. K. Riley, J.C.S. Dalton, 1972, 438.
[367] R. W. Baker, M. J. Braithwaite, and R. S. Nyholm, J.C.S. Dalton, 1972, 1924.
[368] F. Glockling and R. J. I. Pollock, J.C.S. Chem. Comm., 1972, 467.
[369] T. G. Appleton, H. C. Clark, R. C. Poller, and R. J. Puddephatt, J. Organometallic Chem., 1972, **39**, C13.
[370] H. Onoue and I. Moritani, J. Organometallic Chem., 1972, **43**, 431.
[371] B. Crociani, C. Boschi, C. G. Troilo, and U. Croatto, Inorg. Chim. Acta, 1972, **6**, 655.

the $ArSO_2$ groups are S-bonded to Pt),[372, 373] trans-[Pt(PPh$_3$)$_2$(COR)Cl] (254—272 cm^{-1}; R in the acyl ligand = Ph, p-MeOC$_6$H$_4$, CH$_2$=CH, or CH$_2$=CMe),[374] and trans-[Pt(PMe$_2$Ph)$_2$(R)Cl] (272 cm^{-1}, R = Me$_3$SiCH$_2$; 340 cm^{-1}, R = Cl).[375] In trans-[Pd(PEt$_3$)$_2$ClL]$^+$, the variation of ν(PdCl) (330—275 cm^{-1}) with the nature of the trans ligand L [= CO, p-MeC$_6$-H$_4$NC, P(OPh)$_3$, P(OEt)$_3$, PPh$_3$, PEt$_3$, py, or 2,4,6-trimethylpyridine] parallels the known σ-donor strengths of these ligands.[376]

Other ν(PdX) and ν(PtX) assignments have already been mentioned [328, 342] or are listed in Table 17.[327, 341, 346, 352, 363, 377—383]

Table 17 *Further references to compounds for which ν(PdX) or ν(PtX) assignments have been proposed*

Compound[a]	Ref.
[Pd(Et$_4$dien)X]$^+$	346
[H$_2$C=CHCH$_2$S(CH$_2$)$_3$SPdBr]$_2$	377[b]
[MeS(CH$_2$)$_3$SPdI]$_2$	
[ClPd(CH$_2$COCH$_2$CO$_2$CH$_2$R)L^1$_2$]	327
trans-[PdCl(CHCl$_2$)(PPh$_3$)$_2$]	352
trans-[PdCl$_2$L^2$_2$]	
trans-[PdCl$_2$L^3$_2$]	
[Pd$_2$Cl$_4$L^2$_2$]	
[Pd$_2$Cl$_2$(P^2—C)$_2$]	378
[Pd$_2$Cl$_2$(P^1—C)$_2$]	
trans-[PdCl(P^1—C)(PPh$_3$)]	
trans-[PdCl(P^2—C)(PPh$_3$)]	
[PdCl(P^1—C)py][c]	
trans-[PdX$_2$(sulphide)$_2$]	
trans-[PtX$_2$(sulphide)$_2$]	363
cis-[PtY$_2$(sulphide)$_2$]	
trans-[PtI$_2$(amine)$_2$]	341
[(cis-Ph$_2$PCH=CHPPh$_2$)PtCl$_2$]	379
cis-[{H$_2$C=CH(CH$_2$)$_2$C(O)Me}PtY$_2$]	380
[(o-NCC$_6$H$_4$PPh$_2$)$_2$PtCl$_2$]	381[d]
[(o-NCC$_6$H$_4$PPh$_2$)PtCl$_2$]$_3$	
(Ph$_3$P)$_2$PtCl(CCl=CCl$_2$)	382[e]

[372] F. Faraone, L. Silvestro, S. Sergi, and R. Pietropaolo, *J. Organometallic Chem.*, 1972, **34**, C55.
[373] F. Faraone, L. Silvestro, S. Sergi, and R. Pietropaolo, *J. Organometallic Chem.*, 1972, **46**, 379.
[374] S. P. Dent, C. Eaborn, A. Pidcock, and B. Ratcliff, *J. Organometallic Chem.*, 1972, **46**, C68.
[375] M. R. Collier, C. Eaborn, B. Jovanović, M. F. Lappert, L. Manojlović-Muir, K. W Muir, and M. M. Truelock, *J.C.S. Chem. Comm.*, 1972, 613.
[376] W. J. Cherwinski, H. C. Clark, and L. E. Manzer, *Inorg. Chem.*, 1972, **11**, 1511.
[377] L. Cattalini, J. S. Coe, S. Degetto, A. Dendoni, and A. Vigato, *Inorg. Chem.*, 1972, **11**, 1519.
[378] A. J. Cheney and B. L. Shaw, *J.C.S. Dalton*, 1972, 860.
[379] R. B. King and P. N. Kapoor, *Inorg. Chem.*, 1972, **11**, 1524.
[380] B. T. Heaton and D. J. A. McCaffrey, *J. Organometallic Chem.*, 1972, **43**, 437.
[381] D. H. Payne and H. Frye, *Inorg. Nuclear Chem. Letters*, 1972, **8**, 73.
[382] D. T. Clark and D. Briggs, *Nature Phys. Sci.*, 1972, **237**, 16.
[383] A. J. Cheney and B. L. Shaw, *J.C.S. Dalton*, 1972, 754.

348 Spectroscopic Properties of Inorganic and Organometallic Compounds

Table 17 (cont.)

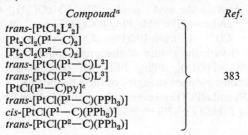

Compound[a]	Ref.
trans-[PtCl$_2$L$^2{}_2$]	
[Pt$_2$Cl$_2$(P^1—C)$_2$]	
[Pt$_2$Cl$_2$(P^2—C)$_2$]	
trans-[PtCl(P^1—C)L^2]	
trans-[PtCl(P^2—C)L^3]	383
[PtCl(P^1—C)py][c]	
trans-[PtCl(P^1—C)(PPh$_3$)]	
cis-[PtCl(P^1—C)(PPh$_3$)]	
trans-[PtCl(P^2—C)(PPh$_3$)]	

[a] X = Cl, Br, or I; Y = Cl or Br; Et$_4$dien = $NNN'N'$-tetraethyldiethylenetriamine; R = Me or Ph; L^1 = py, ½(bipy), etc.; L^2 = PBu^t(o-tolyl)$_2$; L^3 = PBu$^t{}_2$(o-tolyl); (P^1—C) = o-CH$_2$C$_6$H$_4$PBu^t(o-tolyl); (P^2—C) = o-CH$_2$C$_6$H$_4$PBu$^t{}_2$; (sulphide) = Me$_2$S, Et$_2$S, (CH$_2$)$_4$S, or (CH$_2$)$_5$S; (amine) = py or 3-methyl-, 4-methyl-, 4-ethyl-, or 3,5-dimethyl-pyridine. [b] Some closely related complexes also studied. [c] Mixture of cis- and trans-isomers. [d] Structures proposed on the basis of ν(PtCl) and ν(CN) data. [e] Compound formed by surface isomerism of (Ph$_3$P)$_2$Pt(C$_2$Cl$_4$) during X-ray photoelectron spectroscopic study.

11 Copper, Silver, and Gold

Assignments have been given for ν(AuC$_2$) (530—570 cm^{-1}) and some other modes in a range of dimethylgold complexes Me$_2$AuCl(EMe$_2$) [E = S or Se], Me$_2$AuX{RS(CH$_2$)$_n$RS}AuMe$_2$X [X = Cl, n = 2 or 3, R = Me or Et; X = Br, I, or SCN, n = 2, R = Me or Et; X = SCN, n = 3, R = Me], and [Me$_2$Au(MeSCH$_2$CH$_2$SMe)](NO$_3$).[384] Reaction of MeAu(PMe$_2$Ph) with F$_3$C—C≡C—CF$_3$ gives a stable 2:1 adduct formulated as in (15) on the basis of i.r. and n.m.r. data;[385] ν(AuMe) is at 502 cm^{-1} with ν(CC) of the bridging acetylene at 1565 and 1586 cm^{-1}.

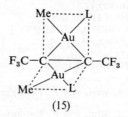

(15)

Unassigned i.r. data have been listed for fluorocarbon complexes derived from tertiary-phosphine–gold methyls.[386] Data on [Me$_2$Au(NH$_3$)$_2$]$^+$ have been mentioned in an earlier section.[2]

Metal isotopic substitution has been used in making assignments of the i.r. spectra of en and alkyl-substituted en complexes of copper, for example:[387]

[384] H. Schmidbaur and K. C. Dash, *Chem. Ber.*, 1972, **105**, 3662.
[385] A. Johnson, R. J. Puddephatt, and J. L. Quirk, *J.C.S. Chem. Comm.*, 1972, 938.
[386] C. M. Mitchell and F. G. A. Stone, *J.C.S. Dalton*, 1972, 102.
[387] G. W. Rayner-Canham and A. B. P. Lever, *Canad. J. Chem.*, 1972, **50**, 3866.

	ν(CuN)/cm^{-1}	ν(CuCl)/cm^{-1}
^{63}Cu(en)Cl$_2$	375, 317.5	265
^{65}Cu(en)Cl$_2$	373, 315	264

In N-alkyl derivatives, ν(CuN) varies in the range 269—400 cm^{-1}. For bis[pyridine-2-(N-cyanocarboxamidato)]aquocopper(II), the crystal structure of which was determined,[388] ν(CuN) has been given as 302 cm^{-1}. In the Cu^I complexes L$_3$Cu(ClO$_4$), wavenumbers/cm^{-1} of ν(CuN) are appreciably *higher* than in related Cu^{II} systems: 425 and 375 (L = 2-methylpyridine), 400 and 306 (L = 2,5-dimethylpyridine), 277 and 259 (L = 2-ethylpyridine), and 260 and 248 (L = 2-isopropylpyridine).[389]

I.r. (NO$_3^-$ modes), electronic, and (especially) e.s.r. spectral studies indicate distorted *cis*-octahedral structures for [Cu(chel)$_2$(NO$_3$)]NO$_3$ [chel = 2,6-dimethyl-o-phenanthroline or 3,3'-dimethylene-4,4'-dimethyl-2,2'-biquinoline].[390] Partial i.r. data are also available for some new dinuclear Cu^{II} chelates such as of the [NN'-bis-(3-aminopropyl)oxamidato]copper(II) ligand.[391]

I.r. bands in the region *ca.* 330—390 cm^{-1} are assigned as ν(AuP) in (Me$_3$P)AuX, [(Me$_3$P)$_4$Au]Y, and [(Me$_3$P)$_2$Au]Y (X = NO$_3$ or BF$_4$; Y = Cl, Br, I, or X).[392]

I.r. bands at 798 and 810 cm^{-1} have been assigned to ν_{asym}(CuOCu) in [Cu(NH$_3$)(C$_2$O$_4$)]$_n$.[393] For (DPPH)Cu(chel) [DPPH = 2,2-diphenyl-1-picrylhydrazyl; chel = acac, trifluoroacetylacetonato, or N-methylsalicylaldi-iminato],[394] values for ν(CuO) are said to be 449—458 cm^{-1}. Reaction of an Ag$_2$O surface with O$_3$, CO$_2$, and H$_2$O vapour has been studied using i.r. techniques.[395] When [Cu(arginine)$_2$(ClO$_4$)$_2$] (shown by i.r. to contain bidentate ClO$_4^-$ groups) is slowly recrystallized from water, the product contains unco-ordinated ClO$_4^-$, *viz.* [Cu(arginine)$_2$(H$_2$O)](ClO$_4$)$_2$, and is converted back into the anhydrous material when pressure is applied during KBr disc preparation.[396]

Assignments for ν(MS) and ν(MSe) modes are listed in Table 18.[384, 397—402]

[388] A. C. Bonamartini, A. Montenero, M. Nardelli, C. Palmieri, and C. Pelizzi, *J. Cryst. Mol. Structure*, 1971, **1**, 389.
[389] A. H. Lewin, R. J. Michl, P. Ganis, and U. Lepore, *J.C.S. Chem. Comm.*, 1972, 661.
[390] Ph. Thomas, D. Rehorek, H. Spindler, R. Kirmse, and H. Hennig, *Z. anorg. Chem.*, 1972, **392**, 241.
[391] H. Ojima and K. Nonoyama, *Z. anorg. Chem.*, 1972, **389**, 75.
[392] H. Schmidbaur and R. Franke, *Chem. Ber.*, 1972, **105**, 2985.
[393] L. Cavalca, A. C. Villa, A. G. Manfredotti, A. Mangia, and A. A. G. Tomlinson, *J.C.S. Dalton*, 1972, 391.
[394] F. Leh and J. K. S. Wan, *Canad. J. Chem.*, 1972, **50**, 999.
[395] T. L. Slager, B. J. Lindgren, A. J. Mallmann, and R. G. Greenler, *J. Phys. Chem.*, 1972, **76**, 940.
[396] S. T. Chow and C. A. McAuliffe, *Inorg. Nuclear Chem. Letters*, 1972, **8**, 913.
[397] K. H. Schmidt, A. Müller, J. Bouwma, and F. Jellinek, *J. Mol. Structure*, 1972, **11**, 275.
[398] J. G. M. van der Linden and P. J. M. Geurts, *Inorg. Nuclear Chem. Letters*, 1972, **8**, 903.
[399] J. G. M. van der Linden and W. P. M. Nijssen, *Z. anorg. Chem.*, 1972, **392**, 93.
[400] G. Marcotrigiano, R. Battistuzzi, and G. Peyronel, *Inorg. Nuclear Chem. Letters*, 1972, **8**, 399.
[401] R. A. Potts, *J. Inorg. Nuclear Chem.*, 1972, **34**, 1749.
[402] E. A. Allen and W. Wilkinson, *Spectrochim. Acta*, 1972, **28A**, 2257.

Table 18 Assignments for ν(MS) and ν(MSe) vibrations/cm^{-1} (M = Cu or Au)

Compound[a]	ν(MS) or ν(MSe)	Ref.
Cu$_3$VS$_4$	285	
Cu$_3$NbS$_4$	267, 244	
Cu$_3$TaS$_4$	267, 244	397[b]
Cu$_3$NbSe$_4$	169, ca. 144	
Cu$_3$TaSe$_4$	170, 141	
Cu(S$_2$CNEt$_2$)$_2$	358	
[Cu(S$_2$CNEt$_2$)$_2$][I$_3$]	398	
Cu(Se$_2$CNEt$_2$)$_2$	240	398[c]
[Cu(Se$_2$CNEt$_2$)$_2$][I$_3$]	310	
[Au(S$_2$CNEt$_2$)$_2$]Br	378	
[Au(S$_2$CNBu$^n_2$)$_2$][AuBr$_2$]	380	
Br$_2$Au(S$_2$CNEt$_2$)	378	399[c]
[Au(Se$_2$CNEt$_2$)$_2$]Br	253	
[Au(Se$_2$CNBu$^n_2$)$_2$][AuBr$_2$]	250	
Br$_2$Au(Se$_2$CNEt$_2$)	259	
Au[SC(NH$_2$)$_2$]$_2$X	262—284	400[d]
Au[SC(NH$_2$)$_2$]$_2$Y,H$_2$O		
AuCl$_3$(Me$_2$SO)	430	401[e]
AuCl$_3$[(CD$_3$)$_2$SO]	399	
AuCl(sulphide)	270—286	402[f]
AuBr(sulphide)		
Me$_2$AuCl(Me$_2$S)	263, 295[g]	
Me$_2$AuCl(Me$_2$Se)	278[g]	384
Me$_2$AuZ[RS(CH$_2$)$_n$SR]AuZMe$_2$	258—315[g]	
[Me$_2$Au{MeS(CH$_2$)$_2$SMe}]NO$_3$	270, 306	

[a] X = Cl, Br, or I; Y = BF$_4$, CF$_3$CO$_2$, or ClO$_4$; (sulphide) = Me$_2$S, (CH$_2$)$_4$S, or (CH$_2$)$_5$S; Z = Cl (n = 2 or 3, R = Me or Et), Br (n = 2, R = Me or Et), I (n = 2, R = Me or Et), or SCN (n = 2, R = Me or Et; n = 3, R = Me). [b] Considerable covalent character in the Cu—S or Cu—Se bonds is indicated. [c] ν(CN) Values given for these and some related compounds. [d] Conductance data show compounds to contain ionic X or Y groups; internal modes of SC(NH$_2$)$_2$ ligands assigned, showing S-co-ordination to Au. [e] ν(SO) at 1198 cm^{-1} shows S-co-ordination to Au. [f] ν(AuS) could not be observed for analogous AuX$_3$(sulphide) complexes (X = Cl or Br; expected region ca. 300 cm^{-1}). [g] Modes described as ν(AuSQ) or ν(AuSeQ), where Q = Cl or Z.

Complexes [CuL$_3$]X (L = Ph$_3$PS, Ph$_3$AsS, etc.; X = ClO$_4$ or BF$_4$) have been so formulated on the basis of i.r. spectral (internal modes of X) and conductivity data.[403]

I.r. bands of matrix-isolated CuCl have been attributed to trimeric (393.5, 383, 285, and 101 cm^{-1}) and tetrameric species (324, 248, 234, and 218 cm^{-1}), since mass spectral data indicated these to be the main components of CuCl vapour; simple normal-co-ordinate calculations were performed.[404]

The observation of three ν(CuCl) i.r. bands in benzene solutions of [R$_4$N]CuCl$_3$ (387, 316, 215 cm^{-1}; R = n-octyl) is said to indicate that the

---

[403] J. A. Tiethof, A. H. Hetey, P. E. Nicpon, and D. W. Meek, *Inorg. Nuclear Chem. Letters*, 1972, **8**, 841.

[404] S. N. Cesaro, E. Coffari, and M. Spoliti, *Inorg. Chim. Acta*, 1972, **6**, 513.

anion is polymeric; under similar conditions the [R$_3$NH]$^+$ salt shows two ν(CuCl) modes (313, 263 cm^{-1}), apparently as a result of R$_3$N—H···Cl interactions.[297]

Fluorination of AuF$_3$ in the presence of XeF$_6$ in excess affords [Xe$_2$F$_{11}$]$^+$[AuF$_6$]$^-$, which has been characterized partly by Raman data.[405] In addition to modes due to the (known) cation, Raman bands at 586 and 223 cm^{-1} are found, which are attributed to the [AuF$_6$]$^-$ ion; the corresponding caesium salt gives values of 595 (ν_1, a_{1g}), 520 (ν_2, e_g), and 224 cm^{-1} (ν_5, t_{2g}). Also reported are Raman data for Cs[AuF$_4$]: 588 (ν_1, a_{1g}), 561 (ν_4, b_{2g}), and 237 and 230 cm^{-1} (ν_3, b_{1g}).[405] AuI has been studied by Raman spectroscopy;[406] see Chapter 4.

Copper(I) halide complexes have received some attention. ν(CuX) values [232 and 162 (X = Cl), 202 and 145 (X = Br), or 175 and 128 (X = I)] and other data show that CuX(o-Me$_2$NC$_6$H$_4$AsMe$_2$) complexes are dimers, (chel)CuX$_2$Cu(chel), with X-bridges (for X = Cl this has been shown previously by X-ray diffraction).[407] On the other hand, the high value (375 cm^{-1}) of ν(CuCl) in Cu$_2$Cl$_2$(cis,trans-cyclo-octa-1,3-diene), together with the observation of a single ν(C=C) band at 1505 cm^{-1}, indicates the presence of terminal Cu—Cl bonds, viz. ClCu(diene)CuCl.[408] In CuCl complexes (2:1, 3:2, 4:3, 1:1, or 2:3) with Ph$_2$P(CH$_2$)$_n$PPh$_2$ (n = 1 or 2), ν(CuCl) is in the 236—287 cm^{-1} range.[409]

Other references giving ν(CuX) data have already been mentioned.[303, 387] Far-i.r. bands have been listed for Cu$_4$Cl$_6$O(Ph$_3$PO)$_4$ but not assigned.[410] A new Cu^{II} iodide complex [Cu(py)I$_2$,H$_2$O] shows ν(OH$_2$) at ca. 3400 cm^{-1} in the i.r., showing the H$_2$O molecule to be co-ordinated.[411]

Notable developments in ν(Au—halogen) correlations have taken place. In a large series of complexes LAuX (L = neutral unidentate ligand), ranges of ν(AuX) are 310—342 (X = Cl), 205—235 (X = Br), or 155—190 cm^{-1} (X = I).[402, 412] The order of decreasing trans influence of L in these compounds, as indicated by the ν(AuCl) values, is (in part):[412]

$$Ph_3AsS > Ph_3PS > Ph_3As > Ph_3Sb \sim Ph_3P > Et_3P.$$

For related compounds LAuX$_3$, ν(AuX) falls in the range 300—370 (X = Cl) or 202—270 cm^{-1} (X = Br).[402, 412] The most definitive assignments of ν(AuX) in Au^{III} complexes have been made using both i.r. and Raman spectral measurements for the complexes (sulphide)AuX$_3$ [X = Cl or Br; (sulphide) = Me$_2$S, Et$_2$S, (CH$_2$)$_4$S, or (CH$_2$)$_5$S]:[402]

---

[405] K. Leary and N. Bartlett, J.C.S. Chem. Comm., 1972, 903.
[406] D. Breitinger and K. Köhler, Inorg. Nuclear Chem. Letters, 1972, **8**, 957.
[407] L. Volponi, B. Zarli, G. G. De Paoli, and E. Celon, Inorg. Nuclear Chem. Letters, 1972, **8**, 309.
[408] H. A. Tayim and A. Vassilian, Inorg. Nuclear Chem. Letters, 1972, **8**, 215.
[409] N. Marsich, A. Camus, and E. Cebulec, J. Inorg. Nuclear Chem., 1972, **34**, 933.
[410] B. Carr and J. F. Harrod, Canad. J. Chem., 1972, **50**, 2792.
[411] B. K. Mohapatra, Chem. and Ind., 1972, 383.
[412] D. R. Williamson and M. C. Baird, J. Inorg. Nuclear Chem., 1972, **34**, 3393.

		X = Cl	X = Br
(b_1)	$\nu_{asym}(trans\text{-}AuX_2)$:	360—366	258—270
(a_1)	$\nu_{sym}(trans\text{-}AuX_2)$:	335—340	232—260
(a_1)	$\nu(AuX)$ trans to L:	312—324	202—235

Some mixed halogeno-complexes have also been studied,[412] e.g. $(Ph_3P)AuClBr_2$.

The new air-stable adducts $(p\text{-}XC_6H_4)AuCl_2L$ (X = H, Me, or Cl; L = $Pr^n{}_2S$, Ph_3P, or Me_3P) are apparently cis,[413] $\nu(AuCl)$ for the Au—Cl bonds trans to the aryl group falling in the range 276—294 cm^{-1} whereas the other $\nu(AuCl)$ modes are at 310—365 cm^{-1}. However, when L = py there are no bands in the 276—294 cm^{-1} region, suggesting a trans formulation. As measured by $\nu(AuCl)$, the trans influence of the groups in these Au^{III} complexes is (cf. ref. 412 cited above):[413]

$$Ar > PR_3 > SPr^n{}_2 > Cl > py$$

Other assignments of $\nu(AuX)$ are for certain $[AuBr_2]^-$ salts (250—255 cm^{-1}), $Br_2Au(S_2CNEt_2)$ (216 and 243 cm^{-1}), and $Br_2Au(Se_2CNEt_2)$ (222 and 247 cm^{-1}).[399] (See also footnote g to Table 18.)

12 Zinc, Cadmium, and Mercury

Chemisorption of hydrogen on to ZnO gives rise to an i.r. band at 1705 cm^{-1}, attributed to $\nu(ZnH)$ [$\nu(ZnD)$ at 1225 cm^{-1}].[414]

The new complex mercury(II) cations $[MeHgL]^+$ (L = PMe_3, $AsMe_3$, SMe_2, or py) show $\nu(HgC)$ in the 536—563 cm^{-1} range.[415] In MeHgSMe (537 cm^{-1})[416] and $(XCH_2)_2Hg$ [510 and 530 (X = Cl), 488 and 510 (X = Br), or 475 and 488 cm^{-1} (X = I)],[417] somewhat lower values are found for $\nu(HgC)$. In the latter series,[417] trans-conformations dominate in the solid state but the melts contain appreciable amounts of gauche-conformers. The difference between $\nu(HgC)$ in $[MeHg(NH_3)]^+$ (547 cm^{-1}) and $[MeHg(OH_2)]^+$ (566 cm^{-1}) has already been mentioned,[2] and reference to the spectrum of $(PhCH_2)_2Hg$[418] is made in the preceding chapter in relation to the spectra of related germanium and tin compounds.

A careful comparison of the i.r. spectra of $Hg(C_5H_5)_2$ and $Hg(C_5D_5)_2$ in the $\nu(CH)$ region has confirmed the presence of five fundamentals, and that the σ-bonded structure $Hg(h^1\text{-}C_5H_5)_2$ is correct.[419] The vibrational spectra of Ph_2Hg, $(C_6D_5)_2Hg$, and $PhHgX$ (X = Cl, Br, or I) have been

---

[413] K. S. Liddle and C. Parkin, J.C.S. Chem. Comm., 1972, 26.
[414] R. J. Kokes, A. L. Dent, C. C. Chang, and L. T. Dixon, J. Amer. Chem. Soc., 1972, 94, 4429.
[415] P. L. Goggin, R. J. Goodfellow, S. R. Haddock, and J. G. Eary, J.C.S. Dalton, 1972, 647.
[416] R. A. Nyquist and J. R. Mann, Spectrochim. Acta, 1972, 28A, 511.
[417] Y. Imai and K. Aida, Spectrochim. Acta, 1972, 28A, 517.
[418] M. Dräger and G. Gattow, Spectrochim. Acta, 1972, 28A, 425.
[419] J. Mink, L. Bursics, and G. Végh, J. Organometallic Chem., 1972, 34, C4.

studied and a normal-co-ordinate calculation has been performed for the in-plane vibrations of the PhHgX system.[420]

I.r. data are available for $MeOCH_2CH_2HgCl$ [421] and $PhCH_2HgCH_2I$.[422] The pyrolysis of $XHg(CCl_3)$ [X = CCl_3, Cl, or Ph] has been studied by i.r. spectroscopy using matrix-isolation methods.[423]

Complexes $Zn(CN)_2,2RNH_2$ (R = Pr^n or Bu^n) are deduced as being octahedral chain polymers on the basis of $\nu(CN)$ and other i.r. data.[424] In $Hg(CN)_2L^1$ and $Hg(CN)_2L^2{}_2$ [L^2 = py or methyl-substituted pyridine; $L^1 = L^2$ or methyl-substituted pyridine N-oxide], $\nu(HgC)$ is in the 424—486 cm^{-1} range.[425, 426]

Metal-isotopic substitution (^{64,68}Zn) has been used in making assignments for $ZnX_2(py)_2$,[427] $ZnX_2(2,2'$-dithiodipyridine),[428] and $ZnX_2(4,4'$-dithiodipyridine) [428] [X = Cl or Br]. Substitution of [$^2H_5$]py for py enabled $\nu(ZnN)$ to be distinguished from $\nu(ZnX)$,[427] and the application of high pressures was used to discriminate between $\nu_{sym}(ZnX)$ and $\nu_{asym}(ZnX)$.[428] Some assignments are given in Table 19.

Table 19 *Some vibrational assignments*/cm^{-1} *for* ^{64}Zn *and* ^{68}Zn *halide complexes*

Complex^a	$\nu_{asym}(ZnX)$	$\nu_{sym}(ZnX)$	$\nu(ZnN)^b$
$^{64}ZnCl_2(py)_2$	329.2	296.5	222.4, 203.9
$^{68}ZnCl_2(py)_2$	324.4	294.1	218.8, 201.5
$^{64}ZnCl_2([^2H_5]py)_2$	328.4	296.3	218.6, 199.8
$^{64}ZnBr_2(py)_2$	257.0	223.0^c	obsc.,^{d} 184.5^c
$^{68}ZnBr_2(py)_2$	251.7	219.3^c	obsc., 183.8^c
$^{64}ZnBr_2([^2H_5]py)_2$	255.3	221.7^c	obsc., 181.3^c
$^{64}ZnI_2(py)_2$	227.4	obsc.^{c}	217.0, 168.8^e
$^{68}ZnI_2(py)_2$	223.0	obsc.^{c}	214.5, 168.1^e
$^{64}ZnI_2([^2H_5]py)_2$	225.5	obsc.^{c}	213.9, 161.6^e
$^{64}ZnCl_2(2,2'$-dtdp)	322^e	295	224
$^{68}ZnCl_2(2,2'$-dtdp)	321	291	220
$^{64}ZnBr_2(2,2'$-dtdp)		267	212
$^{68}ZnBr_2(2,2'$-dtdp)		261	210
$^{64}ZnCl_2(4,4'$-dtdp)	340	299	213
$^{68}ZnCl_2(4,4'$-dtdp)	335	296	211
$^{64}ZnBr_2(4,4'$-dtdp)	248	201	226
$^{68}ZnBr_2(4,4'$-dtdp)	244	197	221

^{a} dtdp = dithiodipyridine. ^{b} Where two values are given, these are $\nu_{asym}(ZnN)$ and $\nu_{sym}(ZnN)$, respectively. ^{c} Coupled modes. ^{d} obsc. = obscured. ^{e} Includes contribution from ligand mode.

[420] J. Mink, G. Végh, and Yu. A. Pentin, *J. Organometallic Chem.*, 1972, **35**, 225.
[421] J. C. Soyfer, P. Audibert, and N. Giacchero, *Bull. Soc. pharm. Marseille*, 1971, **20**, 77.
[422] R. Scheffold and U. Michel, *Angew. Chem. Internat. Edn.*, 1972, **11**, 231.
[423] A. K. Mal'tsev, R. G. Mikaélyan, and O. M. Nefedov, *Doklady Phys. Chem.*, 1971, **201**, 1027.
[424] K. Möckel and W. Müller-Litz, *Z. anorg. Chem.*, 1972, **393**, 81.
[425] I. S. Ahuja and A. Garg, *J. Inorg. Nuclear Chem.*, 1972, **34**, 2074.
[426] I. S. Ahuja and A. Garg, *J. Inorg. Nuclear Chem.*, 1972, **34**, 2681.
[427] Y. Saito, M. Cordes, and K. Nakamoto, *Spectrochim. Acta*, 1972, **28A**, 1459.
[428] J. R. Ferraro, B. Murray, A. Quattrochi, and C. A. Luchetti, *Spectrochim. Acta*, 1972, **28A**, 817.

ν(MN) modes have been located in generally accepted regions for the following complexes: M(SCN)L$_2$ (M = Zn or Cd; L = py, PhNH$_2$, or derivatives of these);[429] CdX$_2$(pyrazine), CdX$_2$(py)$_2$, and CdI$_2$(PhNH$_2$)$_2$ [X = Cl, Br, or I];[430] M(chel)$_2$,nH$_2$O [M = Zn, n = 4; M = Cd, n = 8; M = Hg, n = 1; chel = 3-(4-pyridyl)triazoline-5-thione];[431] MX$_2$(triam), Zn$_2$X$_4$(triam), and Hg$_2$I$_4$(triam) [X = Cl, Br, or I; triam = NNN'-$N''N'''N'''$-hexamethyl-3,6-diazaoctane-1,8-diamine];[432] and [MeHg(py)]-(NO$_3$).[415]

In the novel compound Hg(PBu$^t_2$)$_2$, a product from reaction of secondary phosphines with HgBu$^t_2$, a Raman band (absent from the i.r. spectrum) at 370 cm^{-1} is attributed to ν_{sym}(HgP$_2$); the derived force constant of 2.16 mdyn Å^{-1} is said to be consistent with considerable Hg—P double-bond character.[433]

In NO$_3^-$ or BF$_4^-$ salts of the cations [XHg(PMe$_3$)]$^+$ and [Hg(PMe$_3$)$_2$]$^{2+}$ (X = Cl, Br, I, CN, or Me), ν(HgP) is in the 347—375 cm^{-1} range, while some related AsMe$_3$ complexes give ν(HgAs) at *ca.* 250 cm^{-1}.[415]

Raman spectra of the crystalline polymeric compounds [HgOH]NO$_3$, [Hg$_3$O$_2$](NO$_3$)$_2$, and [Hg$_3$O$_2$](NO$_3$)$_2$,H$_2$O [from hydrolysis of aqueous Hg(NO$_3$)$_2$] show that the frequency of the most intense ν(HgO) mode depends on the number of Hg atoms bonded to a common oxygen atom.[434]

From a Raman spectral study of concentrated aqueous zinc nitrate solutions at 50, 70, and 88 °C, contact-ion-paired nitrate, solvent-separated nitrate, and 'free' NO$_3^-$ have been shown to be present. Apart from various NO$_3^-$ internal modes, bands were identified as due to ν(Zn—OH$_2$) (390—399 cm^{-1}) and ν(Zn—ONO$_2$) (260 cm^{-1}).[435] A comparison of the i.r. spectra of (MeCO$_2$)$_2$Zn(NH$_3$)$_2$ and (HCO$_2$)$_2$Zn(NH$_3$)$_2$ has led to the suggestion that the former compound has octahedral co-ordination at Zn (bidentate acetato-ligands giving a *trans*-[ZnO$_4$N$_2$] skeleton), whereas the formate is tetrahedrally co-ordinated (unidentate formato-ligands giving a C_{2v} [ZnO$_2$N$_2$] structure).[436]

Fairly complete i.r. and Raman data have been presented for (MeCO$_2$)$_2$-Hg$_2$, (MeCO$_2$)$_2$Hg, and their fully deuteriated derivatives. Bands assigned as ν(HgO) are as follows:[44]

	I.r./cm^{-1}	Raman/cm^{-1}
(CH$_3$CO$_2$)$_2$Hg$_2$	268	295
(CD$_3$CO$_2$)$_2$Hg$_2$	250	283
(CH$_3$CO$_2$)$_2$Hg	313	279
(CD$_3$CO$_2$)$_2$Hg	—	255

---

[429] I. S. Ahuja and A. Garg, *J. Inorg. Nuclear Chem.*, 1972, **34**, 1929.
[430] M. Goldstein and W. D. Unsworth, *J. Mol. Structure*, 1972, **14**, 451.
[431] B. Singh and R. Singh, *Indian J. Chem.*, 1971, **9**, 1013.
[432] A. Cristini and G. Ponticelli, *J.C.S. Dalton*, 1972, 2602.
[433] M. Bandler and A. Zarkadas, *Chem. Ber.*, 1972, **105**, 3844.
[434] R. P. J. Cooney and J. R. Hall, *Austral. J. Chem.*, 1972, **25**, 1159.
[435] A. T. G. Lemley and R. A. Plane, *J. Chem. Phys.*, 1972, **57**, 1648.
[436] A. I. Grigor'ev and E. G. Pogodilova, *J. Struct. Chem.*, 1971, **12**, 240.

The series $[XHg(O_2CMe)]_n$ (X = Cl, Br, I, or CN) appear to be true 'mixed compounds', whereas for X = SCN the probable formulation is $Hg(SCN)_2,Hg(O_2CMe)_2$ [ν(HgO) at 313 cm^{-1}, ν(HgS) at 276 and 300 cm^{-1}].[44]

In complexes of various sulphoxides (R^1R^2SO) with mercuric chloride, ν(HgO) is said to be in the 392—423 cm^{-1} range.[437] For the bis(monothiocarbamato) chelates $M(OSCNR)_2$ [M = Zn or Cd; R = piperidyl or pyrrolidinyl], ν(MO) and ν(MS) are given as 471—491 and 375—388 cm^{-1}, respectively; the mercury complex $Hg(OSCNC_4H_8)_2$ shows an i.r. band at 1571 cm^{-1} assigned as ν(C=O), indicating that the oxygen atom is not co-ordinated.[438]

ν(MS) assignments (cm^{-1}) have been given for: MeHgSMe (329),[416] [MeHg(SMe$_2$)](NO$_3$) (304),[415] M[Cd(SCN)$_4$] (M = Co or Ni) (227),[359] $M(SCN)_2L_2$ [L = py, PhNH$_2$, or derivatives of these; M = Cd (229—270) or Hg (248—290)],[429] $Hg(SCN)_2(LO)$ [290—305; (LO) = methyl-substituted pyridine N-oxide],[425] $HgX_2\{S=C(SR^2)(NR^1_2)\}$ (215—225; X = Cl, Br, or I; R^1 = Me, R^2 = Me, Et, PhCH$_2$, or PhCOCH$_2$; R^1 = Et, R^2 = PhCOCH$_2$),[439] $HgX_2\{SC(NH_2)_2\}$ (232—254; X = Cl, Br, or I),[440] and $M\{MeC(S)CHC(E)COEt\}_2$ [332—390; E = S (M = Zn, Cd, or Hg) or O (M = Zn or Cd)].[441]

I.r. data for the gas-phase and matrix-isolated MX_2 species (M = Zn, Cd, or Hg; X = F, Cl, Br, or I) have been reviewed with emphasis on the reliability of deduced geometry.[29] The i.r. and Raman spectra of HgX_2 (X = Cl, Br, or I) have been measured in dioxan or benzene solution.[442]

The Raman spectra of molten or glassy $ZnCl_2$–$AlCl_3$ mixtures indicate that $[ZnCl_4]^{2-}$ ions are not present, although gradual changes in band positions and intensities occur when $AlCl_3$ is progressively added to molten $ZnCl_2$, suggesting some breakdown of the polymeric $[ZnCl_2]_n$ structure.[443] The Raman spectrum of molten Cs_2ZnCl_4 at 620 °C has been illustrated in a paper describing a simple furnace for obtaining high-temperature Raman spectra;[444] this spectrum has also been discussed.[443] Raman spectra of $Cd(NO_3)_2$ with either KCl or KBr in equimolar $NaNO_3$–KNO_3 melts indicate that the predominant complex ions present are $[CdCl_3]^-$ or $[CdBr_4]^{2-}$, respectively.[445]

Assignments in terms of the appropriate line group or sheet group symmetry have been given for the i.r. and Raman spectra [ν(CdN) and $\nu(CdX_2)_n$ modes] of $CdX_2(py)_2$, CdX_2(dioxan), and CdY_2(pyrazine)

---

[437] F. Biscarini, L. Fusina, and G. D. Nivellini, *J.C.S. Dalton*, 1972, 1003.
[438] B. J. McCormick and D. L. Greene, *Inorg. Nuclear Chem. Letters*, 1972, **8**, 599.
[439] H. C. Brinkhoff and J. M. A. Dautzenberg, *Rec. Trav. chim.*, 1972, **91**, 117.
[440] G. Marcotrigiano and R. Battistuzzi, *Inorg. Nuclear Chem. Letters*, 1972, **8**, 969.
[441] A. R. Hendrickson and R. L. Martin, *Austral. J. Chem.*, 1972, **25**, 257.
[442] T. B. Brill, *J. Chem. Phys.*, 1972, **57**, 1534.
[443] C. M. Begun, J. Brynestad, K. W. Fung, and G. Mamantov, *Inorg. Nuclear Chem. Letters*, 1972, **8**, 79.
[444] G. M. Begun, *Appl. Spectroscopy*, 1972, **26**, 400.
[445] J. H. R. Clarke, P. J. Hartley, and Y. Kuroda, *Inorg. Chem.*, 1972, **11**, 29.

[X = Cl or Br; Y = Cl, Br, or I], which are deduced to have octahedrally co-ordinated halogen-bridged structures. Tetrahedral co-ordination was determined for CdI_2L_2 [L = py, $PhNH_2$, or $\frac{1}{2}$(dioxan)].[430]

Addition of iodide ions to solutions in 40% aqueous HF containing HgO and KMF_6 salts (M = Ti or Sn) affords the species $(HgI)_2MF_6$, which have been shown[406] to belong to space group D_{2h}^{17}. The Raman spectra of these species contain the $[MF_6]^{2-}$ fundamentals and bands attributed to planar zig-zag $[HgI^+]_n$ chains (the HgI^+ ion is isoelectronic with AuI):[406]

	$(HgI)_2TiF_6$	$(HgI)_2SnF_6$
a_{1g} [ν(sym)]	140 ⎫	136
b_{1g} [ν(asym)]	135 ⎭	
MF_6^{2-} rotation	74	64
Chain libration	38, 26	39, 23

In addition to $\nu(HgX)_t$ modes, i.r. bands for the complexes $(PhMe_2P)_2$-PtY_2HgX_2 (Y = Cl or Br, X = Y or I) have been assigned to $\nu(PtClHg)_b$ [294—298 and 268—278 cm^{-1}].[367] Values for $\nu(HgXHg)_b$ in the presumed dimers $HgX_2\{SC(NH_2)_2\}$ have been given as 192 (X = Cl), 144 (X = Br), or 106 cm^{-1} (X = I), but no account was taken of the possible existence of geometrical isomers.[440]

Other compounds for which $\nu(MX)$ data are available are as follows (see also Table 19 [427, 428]): $[XHgO_2CMe]_n$ (X = Cl, Br, or I);[44] [XHg-$\{Co(CO)_4\}_2]^-$ (X = Cl, Br, or I);[46] $[XHg(PMe_3)]^+$ (X = Cl, Br, or I) and $[ClHg(AsMe_3)]^+$;[415] $(R^1R^2SO)_n HgCl_2$ (n = 1, 1.5, or 2; R^1 and R^2 = various);[437] $HgX_2[S=C(SR^2)(NR^1_2)]$ (X = Cl, Br, or I; R^2 = Et, R^1 = $PhCOCH_2$; R^2 = Me, R^1 = Me, Et, $PhCH_2$, or $PhCOCH_2$);[439] HgX_2-(dioxan) [X = Cl, Br, or I];[442] MX_2L_2 (M = Zn, Cd, or Hg; X = Cl, Br, or I; L = quinoline or isoquinoline);[446] $HgCl_2[S\text{-}\beta\text{-}(2\text{-pyridylethyl})\text{-L-}$ cysteine];[447] and CdX_2(dien) [X = Cl, Br, or I].[448] Data are also available for $HgBr_2$-$AgNO_3$ melts[449] and $ZnCl_2,2RNH_2$ (R = Pr^n or Bu^n).[424]

13 Lanthanides

Anhydrous cyanides $M^1(CN)_2$ and $M^2(CN)_3$ [M^1 = Eu or Yb; M^2 = Ce, Pr, Sm, Eu, Ho, or Yb] have been prepared and $\nu(CN)$ values given.[450, 451] In some cases a weak i.r. band at 450—510 cm^{-1} has been identified, probably $\nu(MC)$ or $\nu(MN)$.[450]

The absence of coincidences between Raman and i.r. bands of La_2O_3 shows that of the possible structures previously proposed, that having the

---

[446] I. S. Ahuja and A. Garg, *Inorg. Chim. Acta*, 1972, **6**, 453.
[447] R. H. Fish and M. Friedman, *J.C.S. Chem. Comm.*, 1972, 812.
[448] G. Cova, D. Galizzioli, D. Giusto, and F. Morazzoni, *Inorg. Chim. Acta*, 1972, **6**, 343.
[449] V. D. Prisyazhnyi and S. P. Baranov, *Ukrain. khim. Zhur.*, 1972, **38**, 385.
[450] I. J. McColm and S. Thompson, *J. Inorg. Nuclear Chem.*, 1972, **34**, 3801.
[451] I. Colquhoun, N. N. Greenwood, I. J. McColm, and G. E. Turner, *J.C.S. Dalton*, 1972, 1337.

D_{3d}^3 space group is probably correct. Assignments given on this basis were made using quadratic central force-field calculations.[93] The far-i.r. spectra of CeO_2 and the non-stoicheiometric Pr and Tb oxides have been studied over the range 10—250 cm^{-1} at liquid-helium temperatures.[452]

I.r. data have been given for several series of oxyacid salts of lanthanides: simple and double sulphates of Eu, Gd, and Tb;[453] $Tm_2(SO_4)_3,8H_2O$;[454] twenty-three phosphato-complexes (including hydrates, basic salts, and double phosphates) of Pr, Sm, and Yb;[455] $Ln_2(CO_3)_3,nH_2O$ [Ln = La, Ce, Pr, or Nd (n = 8); Ln = Sm, Eu, Gd, Tb, Dy, Ho, Er, or Tm (n = 2—3)] and some basic carbonates;[456] and $CsSm(C_2O_4)_2,H_2O$ and its thermal decomposition products.[457]

A study of the bending modes of the ClO_4^- groups in complexes $Nd(ClO_4)_3,4[Ph_n(PhCH_2)_{3-n}PO]$ (n = 0, 1, 2, or 3), together with conductance data, has led to the formulation $[NdL_4(ClO_4)_2]ClO_4$.[458] When $Nd(O_2PCl_2)_3$ [$\nu(PO_2)$ at 1090 and 1230 cm^{-1}, $\nu(PCl)$ at 550 and 585 cm^{-1}, $\nu(NdO)$ at 390 and 410 cm^{-1}] is treated with $ZrCl_4$ in $POCl_3$, a liquid laser solution is formed, but the nature of the interaction is not certain.[459] In a study of the i.r. and Raman spectra of $Ln(NO_3)_3$(hexamethylphosphoramido)$_3$ [Ln = Dy, Er, or Yb] and some related compounds, suggestions were made for $\nu(LnO)$ values.[460, 461] Assignments for $\nu(LnO)$ and $\nu(LnN)$ have been proposed for some lanthanide complexes of (16; R^1 or R^2 = H or Me).[462]

$$R^1 \diagup\!\!\!\diagdown_{N} \diagdown\!\!\!\diagup P(O)(OEt)_2$$
with R^2 at the 4-position

(16)

The i.r. spectra of the tropolonates $[LnT_3]_n$[463] and $[LnT_4]^-$ salts[463, 464] have been studied. Bands attributed to $\nu(LnO)$ vary with the 4f orbital

---

[452] D. Bloor and J. R. Dean, *J. Phys. (C)*, 1972, **5**, 1237.
[453] L. V. Lipis, V. S. Il'yashenko, V. N. Egorov, G. V. Yukhnevich, and V. I. Volk, *Zhur. priklad. Spektroskopii*, 1971, **14**, 1044.
[454] B. W. Berringer, J. B. Gruber, and E. A. Karlow, *J. Inorg. Nuclear Chem.*, 1972, **34**, 2084.
[455] K. I. Petrov, Yu. B. Kirillov, and S. M. Petushkova, *Russ. J. Inorg. Chem.*, 1971, **16**, 970.
[456] P. E. Caro, J. O. Sawyer, and L. Eyring, *Spectrochim. Acta*, 1972, **28A**, 1167.
[457] Zh. Sh. Kublashvili and E. G. Davitashvii, *Russ. J. Inorg. Chem.*, 1971, **16**, 807.
[458] O. A. Serra, M. L. Ribeiro Gibran, and A. M. B. Galindo, *Inorg. Nuclear Chem. Letters*, 1972, **8**, 673.
[459] C. Y. Liang, E. J. Schimitschek, and J. A. Trias, *J. Inorg. Nuclear Chem.*, 1972, **34**, 1099.
[460] J. A. Sylvanovich, jun. and S. K. Madan, *J. Inorg. Nuclear Chem.*, 1972, **34**, 1675.
[461] J. A. Sylvanovich, jun. and S. K. Madan, *J. Inorg. Nuclear Chem.*, 1972, **34**, 2569.
[462] A. N. Speca, N. M. Karayannis, and L. L. Pytlewski, *Inorg. Chim. Acta*, 1972, **6**, 639.
[463] L. G. Hulett and D. A. Thornton, *J. Mol. Structure*, 1972, **13**, 115.
[464] L. G. Hulett and D. A. Thornton, *Chimia (Switz.)*, 1972, **26**, 72.

population in the lanthanide ion Ln^{3+}, apparently as a result of small but finite crystal-field stabilization (except $4f^0$, $4f^7$, and $4f^{14}$ ions).

The vibrational spectra of europium dihalides (gas-phase) are mentioned in a review.[29] The i.r. spectra of the solid compounds EuX_2 and LnX_3 (Ln = La, Nd, Sm, Eu, Gd, Er, or Yb; X = halide) have been measured to 200 cm^{-1}.[465] The complex i.r. absorption patterns (20—400 cm^{-1}) observed for $TmCl_3,6H_2O$ and $HoCl_3,6H_2O$ agree well with frequency values previously deduced from vibronic spectra.[466]

14 Actinides

The i.r. spectrum of $U(BH_4)_4$ in the vapour phase is consistent with the presence of triply bridging [UH_3BH] units with an overall symmetry of T_d; assignments made are analogous to those for $Zr(BH_4)_4$ (cf. last year's report).[467]

I.r. assignments have been given for $\nu(UO_2)$ modes in a wide range of compounds: $[Et_4N]_2[UO_2(NCO)_4(H_2O)]$;[138] $UO_2(SO_4),nH_2O$ and their deuteriates ($n = 1$ or 4);[468] $UO_2(SO_4),(PhNHNH_2)$ and $UO_2(SO_4)$,-(benzidine),(THF);[469] $UO_2\{NN'$-ethylenebis(salicylideneiminato)$\}$(MeOH) (crystal structure determined);[470] $UO_2(S_2CNEt_2)_2(Me_3NO)$ (crystal structure determined);[471] $UO_2(HL)_2X$, $UO_2(L)_2$, and $UO_2(H_2O)_2(L)_2,PdCl_2$ [X = Cl or NO_3; HL = (17); R^1 or R^2 = H, Me, Et, or Ph];[472] [$UO_2\{CO(NH_2)_2\}_5]X(NO_3)$ [X = Cl or I] and $[UO_2\{CO(NH_2)_2\}_4(H_2O)]X_n$-$(NO_3)_{2-n}$ [X = Cl, $n = 0.5$; X = Br, $n = 1$; X = I, $n = 1.1$ or 1.7];[473]

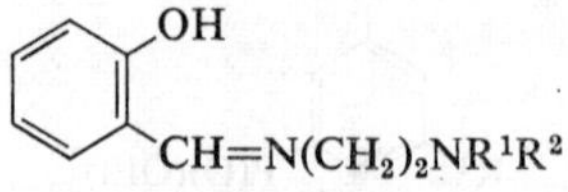

(17)

$UO_2\{NN'$-diaminebis(salicylideneimine)$\}$(L) [L = EtOH, py, Me_2SO, or $HCONMe_2$];[474] and $(UO_2)_2MO_4$ [$\nu(O_2U-OMO_3) = 420$ (M = Si) or 440 cm^{-1} (M = Ge)].[475]

[465] M. D. Taylor, T. T. Cheung, and M. A. Hussein, J. Inorg. Nuclear Chem., 1972, **34**, 3073.
[466] B. W. Berringer, J. B. Gruber, D. N. Olsen, and J. Stöhr, J. Inorg. Nuclear Chem., 1972, **34**, 373.
[467] B. D. James, B. E. Smith, and M. G. H. Wallbridge, J. Mol. Structure, 1972, **14**, 327.
[468] R. Delobel and J.-M. Leroy, Compt. rend., 1972, **274**, C, 1286.
[469] S. M. F. Rahman, J. Ahmad, and M. M. Haq, Z. anorg. Chem., 1972, **392**, 316.
[470] G. Bandoli, D. A. Clemente, U. Croatto, M. Vidali, and P. A. Vigato, Inorg. Nuclear Chem. Letters, 1972, **8**, 961.
[471] E. Forsellini, G. Bombieri, R. Graziani, and B. Zarli, Inorg. Nuclear Chem. Letters, 1972, **8**, 461.
[472] M. Vidali, P. A. Vigato, G. Bandoli, D. A. Clemente, and U. Casellato, Inorg. Chim. Acta, 1972, **6**, 671.
[473] G. V. Ellert, I. V. Tsapkina, O. M. Evstaf'eva, V. F. Zolin, and P. S. Fisher, Russ. J. Inorg. Chem., 1971, **16**, 1640.
[474] A. Pasini, M. Gullotti, and E. Cesarotti, J. Inorg. Nuclear Chem., 1972, **34**, 3821.
[475] J.-P. Legros, R. Legros, and É. Masdupuy, Bull. Soc. chim. France, 1972, 3051.

A Raman and i.r. study of aqueous solutions of $Th(NO_3)_4$ has revealed a complex equilibrium between several species, including NO_3^-, $[Th(NO_3)]^{3+}$, and $[Th(NO_3)_2]^{2+}$; a polarized Raman band at 230 cm^{-1} is assigned as $\nu(Th-ONO_2)$.[476] The i.r. spectrum (NO_3^- modes) of $K_3[Th(NO_3)_7]$ suggests that all the NO_3^- groups are co-ordinated and have C_{2v} symmetry.[477] The extraction of $UO_2(NO_3)_2$ with $(Bu^nO)_3PO$ and $(C_8H_{17})_3PO$ has been studied using i.r. spectroscopy.[478] Partial i.r. data are available for $(Cl_2CHCO_2)_4U$ [$\nu(U-O)$ at 460 cm^{-1}][479] and about 20 neptunium compounds such as $NpO_2(OH)_2,xH_2O$, $Cs_2Np(NO_3)_6$, Cs_nNpO_2-$(NO_3)_3$ [$n = 1$ or 2], $(NH_4)NpO_2SO_3,H_2O$, $NpO_2(CO_3),xH_2O$, Na_4Np-$(CO_3)_4,xH_2O$, $K_6Np(CO_3)_5,xH_2O$, $Np(C_2O_4)_2,6H_2O$, $K_4Np(C_2O_4)_4,4H_2O$, and $CsNpO_2(C_2O_4),3H_2O$.[480]

Far-i.r. bands of UO_2F_2 have been assigned on the basis of a previously published normal-co-ordinate analysis (*J. Inorg. Nuclear Chem.*, 1971, **33**, 1615) as follows: 260 (ν_7, E_u), 234 (ν_8, E_u), and 146 cm^{-1} (ν_4, A_{2u}).[481] Unassigned i.r. bands (460, 550, and 655 cm^{-1}) have been given for UOF_4 (prepared by reaction of UF_6 with H_2O in an HF slurry).[482] The splitting of $\nu_{asym}(UO_2)$ into two components (930 and 960 cm^{-1}) in the i.r. spectra of $NH_4[(UO_2)_2F_4],nH_2O$ ($n = 3$ or 4) has been commented upon.[483]

Force-constant calculations have been carried out on the $[UO_2F_5]^{3-}$ ion based on data from the luminescent spectrum[484] or vibrational spectrum[485] of the potassium salt. Assignments given are: $\nu(UO)$, 860; $\nu(UF)$, 376; $\delta(UO)$, 289; $\delta(UF)$, 220 (in-plane) and 195 cm^{-1} (out-of-plane).[485] Suggestions for $\nu(UF)$ values have also been presented for $[enH_2][UF_6]$ and $[pnH_2][UF_6]$ (420—425 cm^{-1}),[486] $[RNH_3][UF_5]$ (460—470 cm^{-1}; R = Me, Et, Bu^n, or PhCH$_2$),[486] K_nUF_{4+n} (290—380 cm^{-1}),[487] and Na_nUF_{4+n} (360—404 cm^{-1}) ($n = 3, 2, \frac{7}{6}$, or $\frac{1}{2}$).[488]

A normal-co-ordinate analysis has been carried out on the $[NpCl_6]^{2-}$ ion using a modified seven-parameter UBFF; comparison with other data shows that the M—Cl stretching force constants are in the sequence:[489]

$$[UCl_6]^{2-} < [NpCl_6]^{2-} < [PuCl_6]^{2-}$$

[476] B. G. Oliver and A. R. Davis, *J. Inorg. Nuclear Chem.*, 1972, **34**, 2851.
[477] A. K. Molodkin, Z. V. Belyakova, and O. M. Ivanova, *Russ. J. Inorg. Chem.*, 1971, **16**, 835.
[478] A. M. Rozen, D. A. Denisov, and Z. I. Nikolotova, *Zhur. fiz. Khim.*, 1972, **46**, 566.
[479] T. S. Lobanova, A. V. Ivanova, and K. M. Dunaeva, *Russ. J. Inorg. Chem.*, 1971, **16**, 1087.
[480] Yu. Ya. Kharitonov and A. I. Moskvin, *Doklady Chem.*, 1971, **200**, 613.
[481] K. Ohwada, *J. Inorg. Nuclear Chem.*, 1972, **34**, 2357.
[482] P. W. Wilson, *J.C.S. Chem. Comm.*, 1972, 1241.
[483] V. P. Seleznev, A. A. Tsvetkov, B. N. Sudarikov, and B. V. Gromov, *Russ. J. Inorg. Chem.*, 1971, **16**, 1174.
[484] Phan Dinh Kien and A. I. Komyak, *Zhur. priklad. Spetroskopii*, 1972, **16**, 1052.
[485] K. Ohwada, T. Soga, and M. Iwasaki, *Spectrochim. Acta*, 1972, **28A**, 933.
[486] K. C. Satapathy and B. Sahoo, 'Proceedings of the 2nd Chemistry Symposium, Bombay, India', 1970, vol. 1, p. 289.
[487] T. Soga, K. Ohwada, and M. Iwasaki, *Appl. Spectroscopy*, 1972, **26**, 482.
[488] K. Ohwada, T. Soga, and M. Iwasaki, *J. Inorg. Nuclear Chem.*, 1972, **34**, 363.
[489] N. K. Sanyal and L. Dixit, *Current Sci.*, 1972, **41**, 562.

The far-i.r. and Raman spectra of $UCl_4,2(CNCl)$ [$\nu(UCl)$ at 170—220 cm^{-1}] and $ThCl_4,2(CNCl)$ [$\nu(ThCl)$ at 270—315 cm^{-1}] are said to be consistent with structures containing only bridging Cl atoms, although it was necessary to postulate the existence of 'long' (*ca.* 290 pm) and 'short' (*ca.* 270 pm) M—Cl bonds.[490] The i.r. spectra of $Cs_2PuCl_6,5NH_3$, $PuCl_3,5NH_3$, and $PuI_3,8NH_3$ have been shown; weak features at *ca.* 500 cm^{-1} may be $\nu(PuN)$.[491]

[490] J. MacCordick and G. Kaufmann, *Bull. Soc. chim. France*, 1972, 23.
[491] J. M. Cleveland, G. H. Bryan, and R. J. Sironen, *Inorg. Chim. Acta*, 1972, **6**, 54.

7
Vibrational Spectra of some Co-ordinated Ligands

BY G. DAVIDSON

The ligands have been subdivided according to the position of the donor atom in the Periodic Table, except for some which contain more than one possible donor atom (*e.g.* —NCS, —SCN); these have been treated separately. Each paper is referred to only once in this chapter, and it will be necessary for a reader interested in a complex containing several different ligands to check through all of the possible sections in which it might be mentioned.

1 Carbon Donors

Bands due to $\nu(C=C)$ are found at 1594 and 1574 cm^{-1} in the i.r. spectrum of 3-neopentylallyl-lithium. It is suggested [1] that these are due to the presence of *trans*- and *cis*-isomers (1a) and (1b).

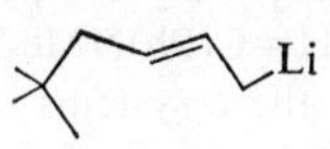

(1a) 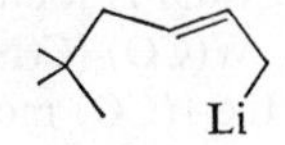   (1b)

An i.r. band at 1500 cm^{-1}, characteristic of the π-allyl group, is observed in (π-1-methylallyl)(cyclo-octatetraene)titanium, (h^3-C$_3$H$_4$Me)Ti(h^8-C$_8$H$_8$), together with absorptions due to the h^8-C$_8$H$_8$ ring.[2]

$\nu(C=O)$ modes have been assigned to i.r. absorptions for Cp$_2$Ti(COR)Cl (1620 cm^{-1}, R = Me; 1595 cm^{-1}, R = Ph), and for Cp$_2$Ti(COR)I (1610 cm^{-1}, R = Me; 1605 cm^{-1}, R = Et).[3]

Unassigned i.r. data have been listed [4,5] for (C$_5$H$_5$)$_2$TiCl$_2$, (C$_5$D$_5$)$_2$TiCl$_2$, (C$_5$H$_4$Me)TiCl, (Me-⟨C$_5$H$_2$D$_2$⟩)$_2$TiCl$_2$, and (C$_5$H$_5$)$_2$TiR (R = Ph; *o*-, *m*-, or *p*-MeC$_6$H$_4$; 2,6-Me$_2$C$_6$H$_3$; 2,4,6- Me$_3$C$_6$H$_2$; C$_6$F$_5$; or CH$_2$Ph).

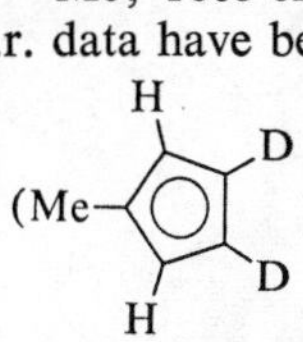

[1] W. H. Glaze, J. E. Hanicak, M. L. Moore, and J. Chaudhuri, *J. Organometallic Chem.*, 1972, **44**, 39.
[2] H. K. Hofstee, H. O. van Oven, and H. J. de Liefde Meijer, *J. Organometallic Chem.*, 1972, **42**, 205.
[3] C. Floriani and G. Fachinetti, *J.C.S. Chem. Comm.*, 1972, 790.
[4] H. A. Martin, M. van Gorkom, and R. O. de Jongh, *J. Organometallic Chem.*, 1972, **36**, 93.
[5] J. H. Teuben and H. J. de Liefde Meijer, *J. Organometallic Chem.*, 1972, **46**, 313.

(h^5-C_5H_5)TiX$_2$ (X = Cl, Br, or I) all give the characteristic h^5-C_5H_5 bands. The THF adducts give a similar spectrum, with the addition of bands due to O-co-ordinated THF, e.g. ν_{as}(COC) has shifted from 1070 to 1035 ($\pm$5) cm^{-1}.[6]

I.r. spectra have been drawn (with no numbers or assignments) for (h^5-C_5H_5)$_2$ZrMe$_2$, (h^5-C_5H_5)Zr(Me)(ONNO), and (h^5-C_5H_5)$_2$Zr(Me)(ONNO).[7]

The Raman and i.r. spectra of (C$_5$H$_5$)$_4$Zr and (C$_5$H$_5$)$_4$Hf are rather complex, and cannot be assigned in detail, but it is clear that both h^1- and h^5-cyclopentadienyl rings are present in each one.[8]

Partially assigned i.r. and Raman spectra for the trimethylsilylmethyl complexes Cr(CH$_2$SiMe$_3$)$_4$, Mo$_2$(CH$_2$SiMe$_3$)$_6$, W$_2$(CH$_2$SiMe$_3$)$_6$, V(CH$_2$SiMe$_3$)$_4$, and VO(CH$_2$SiMe$_3$)$_3$ support certain structural conclusions made regarding these compounds that were arrived at by use of other spectroscopic techniques.[9]

I.r. absorptions have been listed, but not assigned, for bis(cycloheptatriene)vanadium, V(C$_7$H$_8$)$_2$, while bands characteristic of the tropyliumring ligand were found at 3030, 2940, 1438, 1302, 860, and 810 cm^{-1} in V(C$_7$H$_7$)$^{2+}$.[10]

In the acetylenic carbene complexes M(CO)$_5$[C(OEt)C≡CPh], where M = Cr or W, ν(C≡C) is found at 2153, 2154 cm^{-1}, respectively.[11] The ethylenic complex W(CO)$_5${C(NMe$_2$)CH=C(Ph)NMe$_2$} has ν(C=C) at 1548 cm^{-1}. All of the ν(CO) modes for these systems may be assigned in terms of C_{4v} symmetry.

The following ν(CN) values have been quoted: Cr(CHNMe$_2$)(CO)$_5$ 1545; Fe(CHNMe$_2$)(CO)$_4$ 1555; RhCl$_3$(CHNMe$_2$)(PEt$_3$)$_2$ 1587; RhCl$_3$(CO)[C(NHPh)$_2$](PPh$_3$) 1531 [with ν(C≡O) at 2090, 2100 cm^{-1}]; PtCl$_4$(CHNMe$_2$)(PEt$_3$) 1630 cm^{-1}.[12]

The carbonyl stretching modes of (butadiene)tetracarbonylchromium (2) are found at 2040, 1977, 1946, and 1932 cm^{-1}.[13] This is the first example of a 1,3-diene-tetracarbonylchromium complex.

(2) — Cr(CO)$_4$

---

[6] R. S. P. Coutts, R. L. Martin, and P. C. Wailes, *Austral. J. Chem.*, 1971, **24**, 2533.
[7] P. C. Wailes, H. Weigold, and A. P. Bell, *J. Organometallic Chem.*, 1972, **34**, 155.
[8] B. V. Lokshin and E. M. Brainina, *J. Struct. Chem.*, 1971, **12**, 923.
[9] W. Mowat, A. Shortland, G. Yagupsky, N. J. Hill, M. Yagupsky, and G. Wilkinson, *J.C.S. Dalton*, 1972, 533.
[10] J. Müller and B. Mertschenk, *J. Organometallic Chem.*, 1972, **34**, C41.
[11] E. O. Fischer and F. R. Kreissl, *J. Organometallic Chem.*, 1972, **35**, C47.
[12] B. Cetinkaya, M. F. Lappert, and K. Turner, *J.C.S. Chem. Comm.*, 1972, 851.
[13] E. Koerner von Gustorf, O. Jaenicke, and O. E. Polansky, *Angew. Chem. Internat. Edn.*, 1972, **11**, 532.

The i.r. spectrum has been listed and some assignments have been proposed, especially the cyclopentadienyl characteristic vibrations, for $[Li(C_5H_5)CrCl_3,2THF]_2$(dioxan).[14]

Three carbonyl stretching bands are observed for (tellurophen)$Cr(CO)_3$, at 1967 (A_1); 1895, 1872 (E) cm^{-1}. Splitting of the E mode shows that the local symmetry approximation is inadequate. The dinuclear complex (3) was also prepared [ν(CO) at 2064, 2033, and 1995 cm^{-1}].[15]

(3)

Kjellstrup et al. have published [16a] a complete set of symmetry co-ordinates for (h^6-C_6H_6)$Cr(CO)_3$. A normal co-ordinate analysis for this complex suggests [16b] that many of the observed frequency shifts for C_6H_6 modes from the free-ligand values may be explained by kinematic coupling effects.

The Raman spectrum of (benzene)chromium dicarbonyl has been reported,[16c] as have the Raman and low-temperature i.r. spectra of ([$^2H_6$]benzene)$Cr(CO)_3$.[17] Some changes in earlier assignments were suggested.

Unassigned i.r. bands have been reported for (phenylacetylene)- and (styrene)-tricarbonylchromium.[18]

ν(CH-phenyl), ν(CH-aliphatic), ν(C≡O), and some other characteristic absorptions have been assigned [19] in ArCr(CO)$_3$, where Ar = 2-phenyl-ethanol [also ν(OH)], 2-phenylethyl toluene-p-sulphonate [also ν(OSO$_2$)], or 2-phenylethyl bromide.

I.r. spectra in the ν(CO) region have been used to study the protonation of arenechromium tricarbonyl and arenechromium dicarbonyl triphenyl-phosphine complexes.[20] It was shown that the former are not protonated in CF$_3$CO$_2$H [no shift to higher frequency in ν(CO)], but only in BF$_3$,H$_2$O. The PPh$_3$ complexes, however, showed evidence for protonation in CF$_3$CO$_2$H, and in CF$_3$CO$_2$H–CH$_2$Cl$_2$ mixtures. From the proportions of non-protonated and protonated species in these mixtures, it was shown that protonation ability in ArCr(CO)$_2$(PPh$_3$) follows the sequence: Ar = Me$_3$C$_6$H$_3$ > MeOC$_6$H$_5$ > MeC$_6$H$_5$ > C$_6$H$_6$ > MeO$_2$CC$_6$H$_5$.

[14] B. Müller and J. Krausse, *J. Organometallic Chem.*, 1972, **44**, 141.
[15] K. Öfele and E. Dotzauer, *J. Organometallic Chem.*, 1972, **42**, C87.
[16] (a) S. Kjellstrup, S. J. Cyvin, J. Brunvoll, and L. Schäfer, *J. Organometallic Chem.*, 1972, **36**, 137; (b) J. Brunvoll, S. J. Cyvin, and L. Schäfer, *ibid.*, p. 143; (c) L. Schäfer, G. M. Begun, and S. J. Cyvin, *Spectrochim. Acta*, 1972, **28A**, 803.
[17] I. J. Hyams and E. R. Lippincott, *Spectrochim. Acta*, 1972, **28A**, 1741.
[18] G. R. Knox, D. G. Leppard, P. L. Pauson, and W. E. Watts, *J. Organometallic Chem.*, 1972, **34**, 347.
[19] A. Ceccon and G. S. Biserni, *J. Organometallic Chem.*, 1972, **39**, 313.
[20] B. V. Lokshin, V. I. Zdanovich, N. K. Baranetskaya, V. N. Setkina, and D. N. Kursanov, *J. Organometallic Chem.*, 1972, **37**, 331.

Two novel cyclopentadienyl-molybdenum complexes have been prepared by Treichel and Dean.[21] Compound (4) gives ν(CO) at 1934, 1841 cm^{-1}, with ν(CN) at 1573 and ν(CS) at 1167 cm^{-1}; (5) possesses analogous bands at 2006, 1935; 1616, 1155 cm^{-1}.

A number of new species containing the di-π-cyclopentadienyl-molybdenum or -tungsten grouping have been prepared.[22] Thus, (π-Cp)$_2$Mo(CO) gives ν(CO) at 1905 cm^{-1}, while ν(C=C) modes are found at 1465, 1430, 1450 cm^{-1} respectively, for (π-Cp)$_2$M(C$_2$H$_3$R), where M = Mo, R = H; M = W, R = H; and M = W, R = Me.

(π-Cp)Mo(CO)$_3$(COPh) possesses two ν(C≡O) bands in its i.r. spectrum (2022, 1934 cm^{-1}), while ν(C=O) is found at 1640 cm^{-1}. The tungsten analogue gives absorptions due to these modes at 2126, 1935; 1606 cm^{-1}.[23]

Solid-state i.r. spectra indicate the presence of two forms of aza-allyl-allene complexes of Mo and W {e.g. (π-C$_5$H$_5$)Mo(CO)$_2$[(p-tol)$_2$CNC-(p-tol)$_2$]}; the precise nature of and relationship between these two forms is being elucidated by crystallographic methods.[24]

Unassigned i.r. data have been listed for LiW(C$_6$F$_5$)$_5$,2Et$_2$O.[25]

An assignment of the vibrational spectrum of the h^1-allyl complex (σ-C$_3$H$_5$)Mn(CO)$_5$ has been made.[26] The Mn(CO)$_5$ fragment possesses effective C_{4v} symmetry, and frequencies very close to those observed for (CH$_3$)Mn(CO)$_5$.

The local-symmetry concept could not, however, be applied to the Mn(CO)$_4$ vibrations in the closely related h^3-allyl complex (π-C$_3$H$_5$)Mn(CO)$_4$ (6).[27] The ν(CO), ν(MnC), and δ(MnCO) modes all gave more bands than predicted for C_{4v} symmetry, and the effective, overall symmetry of C_s must

[21] P. M. Treichel and W. K. Dean, *J.C.S. Chem. Comm.*, 1972, 804.
[22] F. W. S. Benfield, B. R. Francis, and M. L. H. Green, *J. Organometallic Chem.*, 1972, 44, C13.
[23] A. N. Nesmeyanov, L. G. Makarova, N. A. Ustynyuk, and L. U. Bogatyreva, *J. Organometallic Chem.*, 1972, 46, 105.
[24] H. R. Keable and M. Kilner, *J.C.S. Dalton*, 1972, 153.
[25] E. Kinsella, V. B. Smith, and A. G. Massey, *J. Organometallic Chem.*, 1972, 34, 181.
[26] H. L. Clarke and N. J. Fitzpatrick, *J. Organometallic Chem.*, 1972, 40, 379.
[27] G. Davidson and D. C. Andrews, *J.C.S. Dalton*, 1972, 126.

be used in their assignment. In addition, a complete and almost unambiguous assignment was made of the internal vibrations of the π-C_3H_5 unit [see Table 1, which includes also data from $(\pi$-$C_3H_5)Co(CO)_3$; see below].

Table 1 *Assignment of vibrations of the π-allyl group in $(\pi$-$C_3H_5)Mn(CO)_4$ and $(\pi$-$C_3H_5)Co(CO)_3$ (all figures are wavenumber/cm^{-1})*

Vibration		$(\pi$-$C_3H_5)Mn(CO)_4$	$(\pi$-$C_3H_5)Co(CO)_3$
$\nu(CH_2)$	A''	3078	3087
$\nu(CH)$	A'	3025	3023
$\nu(CH_2)$	A'	2973	2971
$\nu(CH_2)$	A''	2964	—
$\nu(CH_2)$	A'	2948	2940
$\delta(CH_2)_{as}$	A''	1503	1487
$\delta(CH_2)_s$	A'	1462	1473
$\nu(CCC)_{as}$	A''	1397	1389
$\pi(CH)$	A'	1214	1228
$\delta(CH)$	A''	1155	1189
$\nu(CCC)_s$	A'	1017	1020
$\rho_t(CH_2)_s$	A'	1007	1020
$\rho_w(CH_2)_s$	A'	920	951
$\rho_w(CH_2)_{as}$	A''	883	934
$\rho_t(CH_2)_{as}$	A''	980	927
$\rho_r(CH_2)_{as}$	A''	788	805
$\rho_r(CH_2)_s$	A'	774	775
$\delta(CCC)$	A'	521	525

$\nu(C=C)$ has been listed for a number of transition-metal perfluoro-1-methylpropenyl complexes, *i.e.* containing the $M-C(CF_3)=CF(CF_3)$ grouping. Examples were $C_4F_7Mn(CO)_5$, $C_4F_7Re(CO)_5$, $C_4F_7Fe(CO)_2Cp$, $C_4F_7Cr(NO)_2Cp$, *etc.*, and $\nu(C=C)$ in all cases was found to be in the range 1614—1636 cm^{-1}.[28]

I.r. spectra in the $\nu(CO)$ region have been used to study protonation [giving higher-frequency $\nu(CO)$ bands] of phosphine derivatives of CpMn-$(CO)_3$ in solution in CF_3CO_2H and CF_3CO_2H–CH_2Cl_2.[29] Protonation of the metal atom is believed to occur, and the ease of protonation increases with increasing electron-releasing properties of both π-ring substituents and the phosphine ligands.

I.r. investigations of $\nu(C=O)_{ketone}$ and $\nu(C=O)_{carbonyl}$ in a number of complexes of the types (7) and (8) have been made.[30] Empirical rules for distinguishing between possible isomers were drawn up.

In $(\pi$-$C_5H_5)Mn(CO)(CS)_2$, $\nu(C=O)$ is found at 1991 cm^{-1}, with $\nu(C\equiv S)$ at 1305, 1235 cm^{-1}, and in $(\pi$-$C_5H_5)Mn(CS)_3$, $\nu(C\equiv S)$ is at 1338, 1240 cm^{-1}.[31]

[28] R. B. King and W. R. Zipperer, *Inorg. Chem.*, 1972, **11**, 2119.
[29] B. V. Lokshin, A. G. Ginzberg, V. N. Setkina, D. C. Kursanov, and I. B. Nemirovskaya, *J. Organometallic Chem.*, 1972, **37**, 347.
[30] M. Le Plouzennec and R. Dabard, *Bull. Soc. chim. France*, 1972, 3600.
[31] A. E. Fenster and I. S. Butler, *Canad. J. Chem.*, 1972, **50**, 598.

Structures (7) and (8): cyclopentadienyl Mn(CO)₃ complexes with Me and CO(CH₂)₂CO₂H substituents.

Some approximate assignments have been listed for (9; M = Si, Ge, or Sn), while more detailed assignments were given for $Me_3M(CHCCF_3)_3$-$Mn(CO)_5$ (M = Ge or Si): e.g. ν(Mn—CO) at 475, 450 cm^{-1} (Ge), 450—445 cm^{-1} (Si).[32]

Structures (9) and (10).

The i.r. band of medium intensity at ca. 1700 cm^{-1}, due to the amido ν(C=O) in compounds of the type (10) (e.g. $R^1 = R^2$ = Ph), is approximately 100 cm^{-1} higher than in cis-[M(CO)$_4$(NH$_2$R)(CONHR)] (M = Mn or Re) owing to conjugation and ring-size effects.[33a]

I.r. and Raman spectra of solids and solution samples of $(h^5$-C$_5$H$_5)$-Re(CO)$_3$ have been assigned [33b] in terms of 'local' symmetry, and compared with data for the related Mn complex. Wavenumbers and force constants for the M—CO bands demonstrate the enhanced strength of Re—CO compared with Mn—CO bands.

The complex FeEt(acac)(PPh$_3$) gives characteristic i.r. bands of acac and PPh$_3$, together with ν(CH) of the ethyl ligand at 2830 cm^{-1}, and CH deformations at 1450, 1350 cm^{-1}. FeMe$_2$(PPh$_3$)$_2$ gives, in addition to PPh$_3$ bands, ν(CH) of the methyl group at 2850, 2770 cm^{-1}, and CH deformations at 1510, 1270, 1180 cm^{-1}.[34]

I.r. spectra have been listed and approximately assigned for C$_3$F$_7$Fe(CO)$_2$-(NH$_3$)$_2$I and [C$_3$F$_7$Fe(CO)$_2$(NH$_3$)]X, X = I, and Cr(NCS)$_4$(NH$_3$)$_2$.[35] ν(Fe—C) is found in the range 500—425 cm^{-1}.

Ph$_2$MC≡CCF$_3$ (where M = P or As) react with Fe$_3$(CO)$_{12}$ to give the complex (11), showing no ν(C≡C) band, and ν(CO) at 1969w, 1993w, 2006s, 2036m, 2062s, and 2081m (all cm^{-1}).[36]

---

[32] H. C. Clark and T. L. Hauw, J. Organometallic Chem., 1972, **42**, 429.
[33] (a) T. Inglis, M. Kilner, and T. Reynoldson, J.C.S. Chem. Comm., 1972, 774; (b) B. V. Lokshin, Z. S. Klemenkova, and Ya. V. Makarov, Spectrochim. Acta, 1972, **28A**, 2209.
[34] Y. Kubo, A. Yamamoto, and S. Ikeda, J. Organometallic Chem., 1972, **46**, C50.
[35] H. Krohberger, J. Ellermann, and H. Behrens, Z. Naturforsch., 1972, **27b**, 890.
[36] T. O'Connor, A. J. Carty, M. Mathew, and G. J. Palenik, J. Organometallic Chem., 1972, **38**, C15.

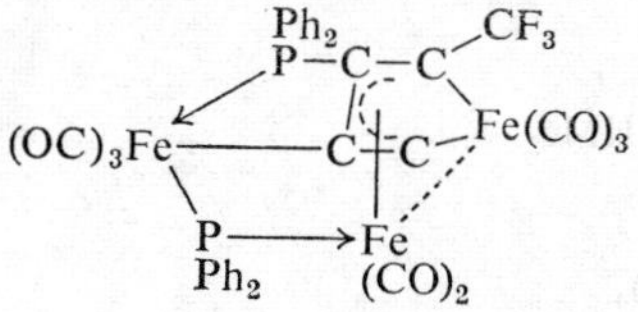

Vibrational Spectra of some Co-ordinated Ligands

(11) 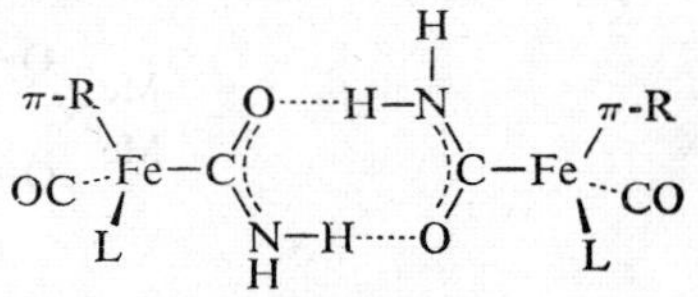   (12)

The i.r. spectra of a series of iron carbonyl carbamoyl complexes are consistent with dimerization *via* hydrogen-bridges in the solid state, *i.e.* (12), where R = C_5H_5, $C_5H_4CHPh_2$, or $C_7H_8CPh_3$ and L = CO, PPh_3, or PEt_3.[37]

The ease of formation of carboxamido-complexes from corresponding carbonyls (of Fe, Ru, Mn, Re, Pd, Pt, Mo, and W):

$$L_mM(CO)_n + 2RNH_2 \rightleftharpoons L_mM(CO)_{n-1}(CONHR)^- + RNH_3^+$$

has been related to the electron density at the carbonyl carbon, and hence to the C≡O stretching force constant. Thus, if $f > 17$ mdyn Å^{-1}, carboxamido formation occurs readily, and if $f < 16$ mdyn Å^{-1} it does not occur at all.[38]

Bands in the i.r. spectrum of bis(methylamino)carbene-tetracarbonyliron, $(OC)_4FeC(NHMe)_2$, have been assigned[39] by comparison with O=C-$(NHMe)_2$.

In $L(OC)_2Fe(C=O)_2Fe(CO)_3$, where R = Me or Ph and L = CO, H_2NNHPh, py, NC_5H_{11}, or NH_3, $\nu(C=O)_{keto}$ is found at 1534—1540 (R = Me) and 1492—1506 (R = Ph) cm^{-1}.[40]

A study has been made of the i.r. and Raman spectra (the latter for the solid phase only) of $(C_2H_4)Fe(CO)_4$.[41] The following assignments of ethylene modes were proposed: (A_1) $\nu(CH_2)$ 2920; $\delta(CH_2)$ 1508, $\nu(C=C)$ 1193 (these two strongly coupled): $\rho_w(CH_2)$ 939; $\nu(Fe-C_2H_4)$ 356; (A_2) $\nu(CH_2)$?; $\rho_r(CH_2)$ 1082; $\rho_t(CH_2)$ 781; $\tau(C_2H_4)$?; (B_1) $\nu(CH_2)$ 2980; $\delta(CH_2)$ 1447; $\rho_w(CH_2)$ 1023; C_2H_4 tilt 305; (B_2) $\nu(CH_2)$ 3078; $\rho_r(CH_2)$ 710; C_2H_4 tilt 400 (all in cm^{-1}).

In (13), the $\nu(C=C)$ band in the free ligand (1638 cm^{-1}) disappears on complex formation.[42] The complex shows two $\nu(C=O)$ absorptions at 1744, 1715 cm^{-1}, and four $\nu(C\equiv O)$ bands (2109, 2049, 2047, 2004 cm^{-1}, together with a number of weaker features in the same region, whose origins are unclear).

---

[37] J. Ellermann, H. Behrens, and H. Krohberger, *J. Organometallic Chem.*, 1972, **46**, 119.

[38] R. J. Angelici and L. J. Blacik, *Inorg. Chem.*, 1972, **11**, 1754.

[39] K. Öfele and C. G. Kreiter, *Chem. Ber.*, 1972, **105**, 529.

[40] V. Kiener and E. O. Fischer, *J. Organometallic Chem.*, 1972, **42**, 447.

[41] D. C. Andrews and G. Davidson, *J. Organometallic Chem.*, 1972, **35**, 161.

[42] E. Koerner von Gustorf, O. Jaenicke, and O. E. Polansky, *Z. Naturforsch.*, 1972, **27b**, 575.

(13) [structure: Me₂C(O-)₂ dimedone-like ring with =CHCHMe₂ and Fe(CO)₄]

In $Fe(bipy)_2(tcne)_2$, $\nu(C\equiv N)$ drops from 2220 cm⁻¹ (free tcne) to 2110 cm⁻¹ on co-ordination.[43]

The complex (14) gives an i.r. band at 1598 cm⁻¹, which is assigned to $\nu(C=C)$ [$\nu(CO)$ 2021, 1952; $\nu(NO)$ 1750 cm⁻¹ also observed]. Compound

(14) ON, CO / OC—Fe—CH₂CCl=CH₂ / PPh₃

(15) H₂C=CCl / CH₂ / Fe(ON)(CO)(PPh₃)

(15), however, which is known to be an unsymmetrical π-allyl complex from n.m.r. data, gives no $\nu(C=C)$ absorption at ca. 1600 cm⁻¹, in agreement with this formulation [$\nu(CO)$ 1950; $\nu(NO)$ 1708 cm⁻¹].[44]

The i.r. and Raman spectra of (cyclobutadiene)iron tricarbonyl, $(C_4H_4)Fe(CO)_3$, have been reported and almost completely assigned. The vibrations of both the C_4H_4—Fe and the $Fe(CO)_3$ moieties appeared to obey the selection rules appropriate for 'local symmetry' of C_{4v}, C_{3v}, respectively. Most of the internal modes of cyclobutadiene were at similar frequencies to those of other π-bonded C_nH_n systems, except that the ring-breathing mode was somewhat higher (1234 cm⁻¹); $\nu(Fe-C_4H_4)$ was at 398 cm⁻¹, and the Fe—C_4H_4 tilt at 471 cm⁻¹.[45]

A vibrational assignment has been proposed for the trimethylene-methane molecule, in $[C(CH_2)_3]Fe(CO)_3$, for the first time.[46] CC_3 stretches were found at 1348 (*E*) and 918 (A_1) cm⁻¹, with skeletal deformations at 802 (A_1, out-of-plane) and 471 (*E*, in-plane) cm⁻¹. The $Fe(CO)_3$ vibrations were closely similar to those in analogous complexes.

(*trans,trans*-2,4-Hexadiene)iron tricarbonyl shows $\nu(C=O)$ bands at 1687 and 1699 cm⁻¹ in CS_2 solution, due to the presence of the isomers (16a) and (16b).[47]

$\nu(C\equiv O)$ has been plotted against σ_p of the *para*-substituent for a series of (*para*-substituted 1-phenyl-1,3-butadiene)iron tricarbonyl complexes (17; R = NH_2, OMe, H, NHCOMe, Br, COMe, or CN). A straight-line

---

[43] T. Yamamoto, A. Yamamoto, and S. Ikeda, *Bull. Chem. Soc. Japan*, 1972, **45**, 1104.
[44] G. Cardacci, S. M. Murgia, and A. Foffani, *J. Organometallic Chem.*, 1972, **37**, C11.
[45] D. C. Andrews and G. Davidson, *J. Organometallic Chem.*, 1972, **36**, 349.
[46] D. C. Andrews and G. Davidson, *J. Organometallic Chem.*, 1972, **43**, 393.
[47] M. Brookhart and D. L. Harris, *J. Organometallic Chem.*, 1972, **42**, 441.

(16a) Me—CH=CH—CH=CH—C(=O)H Fe(CO)₃

(16b) Me—CH=CH—CH=CH—C(H)=O Fe(CO)₃

(17) R-C₆H₄-CH=CH-CH=CH₂ · Fe(CO)₃

plot was observed, clearly showing the transmission of electronic effects, but the exact mechanism of this transmission could not be ascertained.[48]

The reaction between the azabullvalene (18) and $Fe_2(CO)_9$ affords the complexes (19), $\nu(CO)$ at 2016, 2000, 1942 cm^{-1}, $\nu[C(OMe)=N]$ at 1625 cm^{-1}, and (20), $\nu(CO)$ at 2056, 2000, 1974 cm^{-1}, $\nu(C=O)$ at 1720 cm^{-1}.[49]

(18) · (19) · (20)

A vibrational assignment has been proposed for $(C_4H_6)_2Fe(CO)$. The internal modes of the two C_4H_6 ligands show that there is very little interaction between them, and that electronically they are very similar to C_4H_6 in $(C_4H_6)Fe(CO)_3$. $\nu(CO)$ occurs at 1982 cm^{-1} (in C_6H_{12} solution), with the two $\delta(Fe-C-O)$ modes at 532 and 576 cm^{-1}.[50]

I.r. bands have been listed for di-isoprene-, di-1,3-pentadiene-, isoprenecyclo-octatetraene-, and 1,3-pentadienecyclo-octatetraene-carbonyl-iron.[51]

I.r. intensity studies on $(h^5\text{-}C_5H_5)Fe(CO)_2X$, where X = Cl, I, CN, $SnCl_3$, C(O)Me, or $(C_6H_5CH_2C_5H_5)Fe(CO)_2PPh_3$, led Darensbourg to the conclusion that the cyclopentadienyl ligand is acting primarily as a donor.[52]

[48] J. M. Landesberg and L. Katz, *J. Organometallic Chem.*, 1972, **35**, 327.
[49] Y. Becker, A. Eisenstadt, and Y. Shvo, *J.C.S. Chem. Comm.*, 1972, 1156.
[50] G. Davidson and D. A. Duce, *J. Organometallic Chem.*, 1972, **44**, 365.
[51] A. Carbonaro and F. Cambisi, *J. Organometallic Chem.*, 1972, **44**, 171.
[52] D. J. Darensbourg, *Inorg. Chem.*, 1972, **11**, 1606.

The i.r. spectra of $[(\pi\text{-Cp})M(CO)_2]_2$ (M = Fe or Ru), have been investigated as functions of solvent and temperature.[53] The data for the iron compound are consistent with those obtained previously, while for the Ru complex they are assigned on the basis of four isomers being present, *cis*-bridged, *trans*-bridged, *trans*-nonbridged, and a polar non-bridged structure, probably (21).

A number of σ-carbaborane complexes of iron have been prepared,[54] *e.g.* $CpFe(CO)_2\text{-}\sigma\text{-}(B_{10}C_2H_{11})$ and its 1,2-dimethyl derivative, together

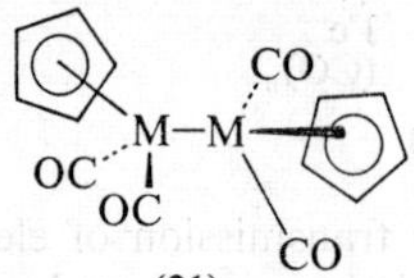

(21) 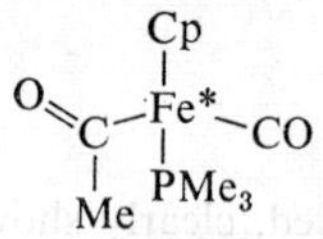   (22)

with some derivatives containing the π-CpFe unit, and the carbaborane σ-bonded to the cyclopentadienyl ring. Numbers of characteristic i.r. absorptions were listed in each case.

Two optical isomers of (22) have been isolated. They show $\nu(C\equiv O)$ at 1920 cm^{-1}, and $\nu(C=O)$ (of the acetyl group) at 1595 cm^{-1}.[55]

Using previously published data, Schäfer *et al.* have published a normal co-ordinate analysis on $Fe(C_5H_5)_2$ and $Fe(C_5D_5)_2$.[56]

Stereoisomers of the ferrocenic alcohols $(h^5\text{-}C_5H_5)Fe\{h^5\text{-}1,2\text{-}MeC_5H_3CH\text{-}(OH)Me\}$ have been isolated.[57] They give rise to distinctive spectra in the $\nu(OH)$ region.

The ferrocenic alcohols (23; R^1, R^2 = Me or Ph) all give a band in the $\nu(OH)$ region that is characteristic of an Fe···H—O interaction (*ca.* 3570 cm^{-1}), *i.e.* in all cases the OH is *cis* with respect to the Fe atom.[58]

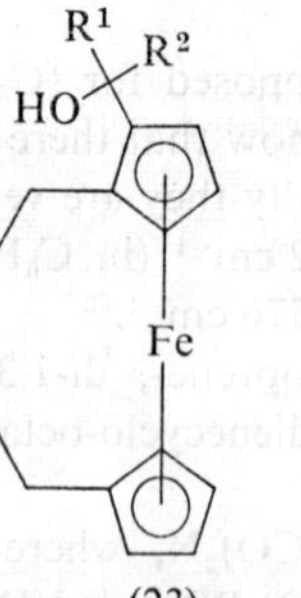

(23) 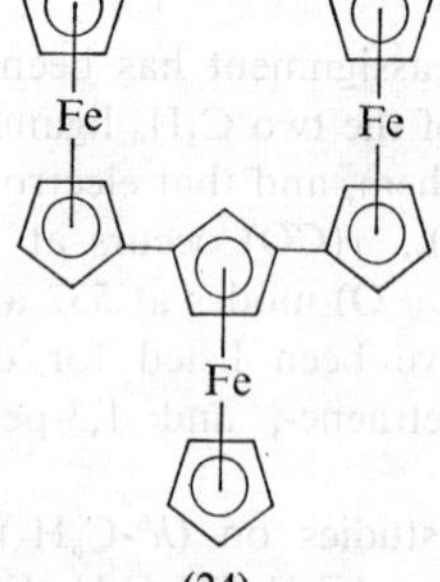   (24)

[53] J. G. Bullitt, F. A. Cotton, and T. J. Marks, *Inorg. Chem.*, 1972, **11**, 672.
[54] L. I. Zakharkin, L. V. Orlova, B. V. Lokshin, and L. A. Fedorov, *J. Organometallic Chem.*, 1972, **40**, 15.
[55] H. Brunner and E. Schmidt, *J. Organometallic Chem.*, 1972, **36**, C18.
[56] L. Schäfer, J. Brunvoll, and S. J. Cyvin, *J. Mol. Structure*, 1972, **11**, 459.
[57] C. Moise, D. Sautrey, and J. Tirouflet, *Bull. Soc. chim. France*, 1971, 4562.
[58] C. Moise, J.-P. Monin, and J. Tirouflet, *Bull. Soc. chim. France*, 1972, 2048.

$\nu(C=O)$ has been reported for a number of α-carbonyl ferrocenes, e.g. formylferrocene (1663), acetylferrocene (1663), ferrocenoyl chloride (1766), ferrocenoic acid (1692), and methyl ferrocenoate (1712 cm^{-1}).[59] Their products from photodecomposition in H_2O-Me_2SO or H_2O-py were all believed to contain carboxylate groups.

Some characteristic i.r. bands have been listed for 1,3-terferrocenyl (i.e. 1,3-diferrocenylferrocene) (24).[60]

$(\pi\text{-}C_6H_6)RuCl(\pi\text{-}C_3H_5)$ gives i.r. bands at 2910, 2870, 1201, 992, 969, 944, 910, 784, and 582 cm^{-1}, characteristic of the π-allyl group.[61a]

A full assignment for ruthenocene has been arrived at through i.r. and Raman studies (including single-crystal data) at ambient and liquid-nitrogen temperatures.[61b] Inactive modes were located as weak i.r. bands at -196 °C. It was shown that, contrary to previous assignments, the i.r.-active ring 'tilt' mode of $(h^5\text{-}C_5H_5)_2Ru$ is of lower energy than the Ru–Ring bond stretch.

Electrochemical oxidation of Cp_2Ru at a mercury anode forms $[Cp_2Ru-Hg-RuCp_2]^{2+}(ClO_4)_2^{2-}$.[62] I.r. and Raman spectra of this and the PF_6^- amd BF_4^- salts have been reported, with a Raman band at 110 cm^{-1} assigned to $\nu(Ru-Hg)$. Other assignments included: 338 cm^{-1}, symmetric Ring–Ru–Ring stretch; 190 cm^{-1}, Ring–Ru–Ring bend.

The Cp—Ru—Cp deformation was assigned similarly (194 cm^{-1}) in the adducts Cp_2Ru,HgX_2 (X = Cl or Br).[63] These assignments should be contrasted with that of Bodenheimer (Chem. Phys. Letters, 1970, **6**, 519) for ruthenocene itself, to a band at 111.5 cm^{-1}.

Characteristic i.r. frequencies have been listed and partly assigned for the carbamoyl complexes $Co(CO)_3(PPh_3)(CONMe_2)$ and $Co(CO)_2(PPh_3)$-$(NH_3)(CONH_2)$.[64]

The vibrational spectrum of $(h^3\text{-}C_3H_5)Co(CO)_3$ has been assigned.[65] The internal π-allyl modes were very similar to those of $(h^3\text{-}C_3H_5)Mn$-$(CO)_4$ – see above, Table 1. Local symmetry of C_{3v} sufficed in the assignment of the $Co(CO)_3$ modes, in contrast to the situation for the Mn complex, where overall symmetry had to be used. Thus, no splitting of the $\nu(CO)$ mode (E) at 2025 cm^{-1} (vapour phase) could be detected.

Clarke and Fitzpatrick have obtained i.r. spectra for a series of π-allyl cobalt complexes $(XC_3H_4)Co(CO)_3$, where X = H, 1-Me, 2-Me, 1-Cl, or 2-Cl.[66] Some splitting of the E mode $\nu(CO)$ band was detectable for X = 2-Me or 1-Cl. Assignments were proposed, and trends in CO force

---

[59] L. H. Ali, A. Cox, and T. J. Kemp, J.C.S. Chem. Comm., 1972, 265.
[60] E. W. Neuse and R. K. Crossland, J. Organometallic Chem., 1972, **43**, 385.
[61] (a) R. A. Zelonka and M. C. Baird, J. Organometallic Chem., 1972, **44**, 383; (b) D. M. Adams and W. S. Fernando, J.C.S. Dalton, 1972, 2507.
[62] D. N. Hendrickson, Y. S. Sohn, W. H. Morrison, jun., and H. B. Gray, Inorg. Chem., 1972, **11**, 808.
[63] W. H. Morrison jun. and D. N. Hendrickson, Inorg. Chem., 1972, **11**, 2912.
[64] H. Krohberger, H. Behrens, and J. Ellermann, J. Organometallic Chem., 1972, **46**, 139.
[65] D. C. Andrews and G. Davidson, J.C.S. Dalton, 1972, 1381.
[66] H. L. Clarke and N. J. Fitzpatrick, J. Organometallic Chem., 1972, **43**, 405.

constants compared with the results of Self-Consistent Charge and Configuration MO (SCCCMO) calculations.

2-Acetyl-π-allylcobalt tricarbonyl can be protonated in concentrated H_2SO_4 solution, giving the trimethylenemethane-cobalt derivative (25),

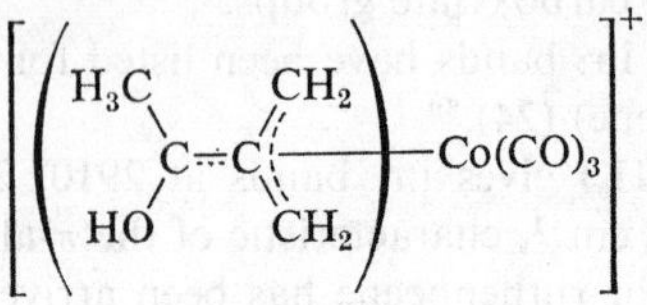

$$\left[\left(\begin{array}{c}H_3C\\ \\HO\end{array}\!\!\!\!>\!\!C\!=\!C\!\!\!<\!\!\!\begin{array}{c}CH_2\\ \\CH_2\end{array}\right)\!\!-\!\!Co(CO)_3\right]^+$$
(25)

which gives CO stretching bands at 2125, 2084 cm^{-1} (the shift to higher frequency compared to the initial complex is consistent with cation formation).[67]

I.r. maxima have been listed for (h^4-cyclobutadiene)(h^5-cyclopentadienyl)cobalt.[68] Those associated with the four-membered-ring were similar to those in $(C_4H_4)Fe(CO)_3$.

Unassigned i.r. spectra have been listed for (26a) and (26b), where L = $SiMe_3$ or Si_2Me_5.[69]

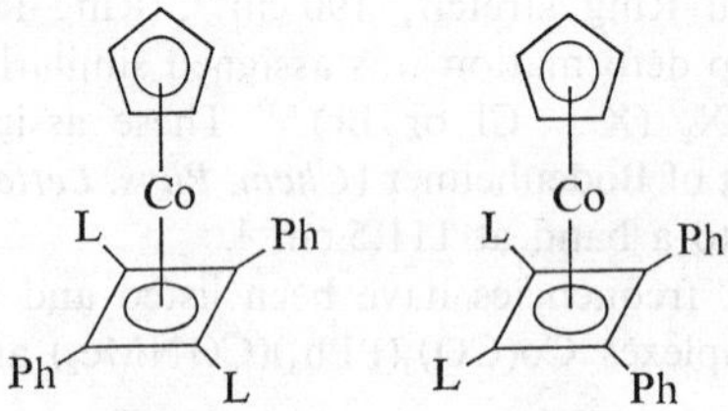

(26a) (26b)

$\nu(C=O)_{acyl}$ is assigned to a weak i.r. band at 1655—1670 cm^{-1} in $[NEt_4][Rh_6(CO)_5(COR)]$, R = Et or Pr.[70] Wavenumbers of $\nu(C\equiv O)$ were also listed (2080—1724 cm^{-1}).

$\nu(C\equiv O)$, $\nu(C=O)$, and $\nu(C=C)$ have been assigned[71] in Rh(CO)-$(PPh_3)_2L$, where L = $C(CO_2Me)=CH(CO_2Me)$, $C(CO_2H)=CH(CO_2H)$, or CPh=CHPh. No co-ordination of C=C or C=O to the metal was indicated, hence these are four-co-ordinate complexes (27).

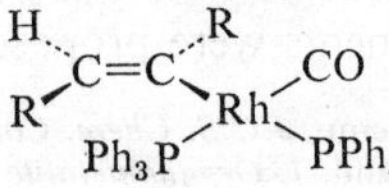

(27)

[67] S. Otsuka and A. Nakamura, *Inorg. Chem.*, 1972, **11**, 644.
[68] M. Rosenblum, B. North, D. Wells, and W. P. Giering, *J. Amer. Chem. Soc.*, 1972, **94**, 1239.
[69] H. Sakurai and J. Hayashi, *J. Organometallic Chem.*, 1972, **39**, 365.
[70] P. Chini, S. Martinengo, and G. Garlaschelli, *J.C.S. Chem. Comm.*, 1972, 709.
[71] B. L. Booth and A. D. Lloyd, *J. Organometallic Chem.*, 1972, **35**, 195.

$\nu(C{\equiv}C)$ was found in the region 2128—2092 cm^{-1} for all of the following: $(Ph_3P)_2Rh(C{\equiv}CPh)_2L$ (L = CO or SnMe$_3$), $(Ph_3P)_2Ir(C{\equiv}CPh)_2CO$, $(Ph_3P)_2Ir(C{\equiv}CPh)_2(SnMe_3)CO$, $(Ph_3P)_2Pt(C{\equiv}CPh)SnR_3$ (R = Me or Et), and $(Ph_2MeP)_2Pt(C{\equiv}CPh)SnMe_3$.[72a]

The new carbene complexes L(Ph$_3$P)RhCl{C[N(R)CH$_2$]$_2$} show $\nu(C-N)$ at 1498 cm^{-1} (L = PPh$_3$, R = Ph), 1495 cm^{-1} (L = CO, R = Ph), and 1513 cm^{-1} (L = PPh$_3$, R = p-tolyl).[72b]

In [Rh(C$_2$H$_4$)$_3$(MeCN)$_2$]$^+$ (28), the i.r. spectrum of the bound C$_2$H$_4$ is very similar to that in Zeise's salt, except that no strong bond is seen at ca. 400 cm^{-1}.[73]

(28)

A number of new complexes have been prepared from the cation [Rh(vp)$_2$]$^+$ [vp = (o-vinylphenyl)diphenylphosphine].[74] Thus: [Rh(CO)(vp)$_2$]$^+$, which is five-co-ordinate, gives $\nu(CO)$ at 2035 cm^{-1}; [Rh(O$_2$)(vp)$_2$]$^+$, $\nu(Rh\langle{}^O_O|)$ at 849 cm^{-1}; [Rh(CO$_2$)(vp)$_2$]$^+$, $\nu(CO_2)$ at 1358, 1480 cm^{-1} (typical of co-ordinated CO$_2$); and [Rh(SO$_2$)(vp)$_2$]$^+$, $\nu(SO_2)$ at 1060 cm^{-1} (symmetric), 1140, 1161 (antisymmetric), i.e. the SO$_2$ is S-bonded.

For tris-o-vinylphenyl-phosphine (tvpp) and -arsine (tvpa) and phenyl-(bis-o-vinylphenyl)phosphine (dvpp) complexes of Rh^I, $\nu(C{=}C)$ frequencies (ca. 1250 cm^{-1}) have been listed.[75a] The structure of the ligand in the complexes [RhX(tvpa)(AsPh$_3$)], [RhX(tvpa)(py)], and [RhX(tvpa)CO] is different from that in RhX(tvpa): $\nu(C{=}C)$, at 1270 cm^{-1} (vs) in RhCl(tvpa), is of only medium intensity (at 1251 cm^{-1}) in RhCl(tvpa)(AsPh$_3$). The change in $\nu(C{=}C)$ frequency is attributable to the use by the rhodium atom of different orbitals in forming the two different bonding arrangements.

Powell and Leedham have investigated the vibrational spectra of a series of isoelectronic d^8 ions complexed to 1,5-cyclo-octadiene.[75b] A lowering in frequency of two ligand bands [to which $\nu(C{=}C)$ character is attributed] is found, and the order is that of the expected metal–olefin

---

[72] (a) B. Cetinkaya, M. F. Lappert, J. McMeeking, and D. Palmer, *J. Organometallic Chem.*, 1972, **34**, C37; (b) D. J. Cardin, M. J. Doyle, and M. F. Lappert, *J.C.S. Chem. Comm.*, 1972, 927.
[73] F. Maspero, E. Perrotti, and F. Simonetti, *J. Organometallic Chem.*, 1972, **38**, C43.
[74] P. R. Brookes, *J. Organometallic Chem.*, 1972, **43**, 415.
[75] (a) D. I. Hall and (the late) R. S. Nyholm, *J.C.S. Dalton*, 1972, 804; (b) D. B. Powell and T. J. Leedham, *Spectrochim. Acta*, 1972, **28A**, 337.

bond strength, viz: [(cod)RhCl]$_2$ (1476, 1241 cm^{-1}) > (cod)PtCl$_2$ (1500, 1267 cm^{-1}) > (cod)PdCl$_2$ (1522, 1271 cm^{-1}). The values in free cod are 1644, 1280 cm^{-1}.

ν(C=O) is in the region 1651—1701 cm^{-1} for several cyclo-dienone complexes of Rh, viz. [(dp)RhCl]$_2$, [(dp)Rh(π-C$_5$H$_5$)], [(cq)RhCl]$_2$, [(cq)Rh(acac)], and [(cq)Rh(π-C$_5$H$_5$)] (where dp = tricyclo[5,2,1,0^{2,6}]deca-4,8-dien-3-one; cq = tricyclo[6,2,1,0^{2,7}]undeca-4,9-diene-3,6-dione).[76] In [(dp)Rh(π-C$_5$H$_5$)], there is no band attributable to ν(C=C) of an uncomplexed C=C bond, and the structure is believed to be (29).

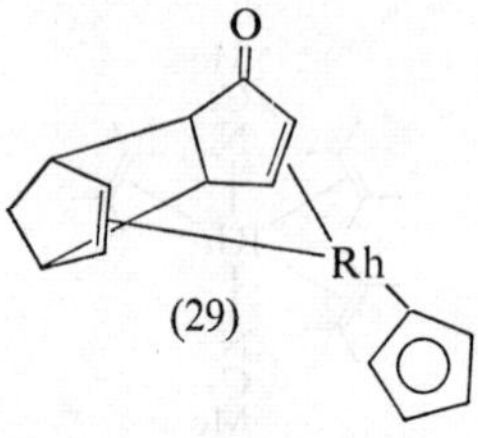

(29)

ν(C≡C) has been assigned as follows in a series of alkynyliridium(I) complexes:[77] Ir(C≡CPh)(CO)(PPh$_3$)$_2$ 2091 cm^{-1}; Ir(C≡CPh)(O$_2$)(CO)(PPh$_3$)$_2$ 2133 cm^{-1}; Ir(C≡CPh)(PPh$_3$)$_3$ 1988 cm^{-1}.

The structure (30) is consistent with the absence of bands due to non-complexed C=C, and with the presence of ν(Ir—Cl) at ca. 250 cm^{-1} in the i.r. spectrum of this complex.[78]

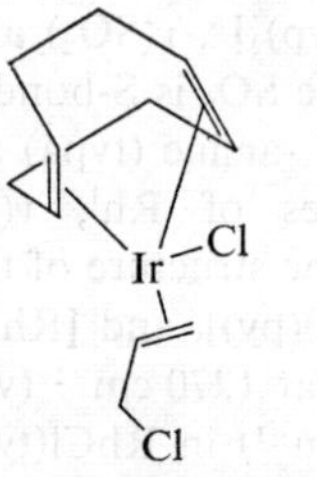

(30) 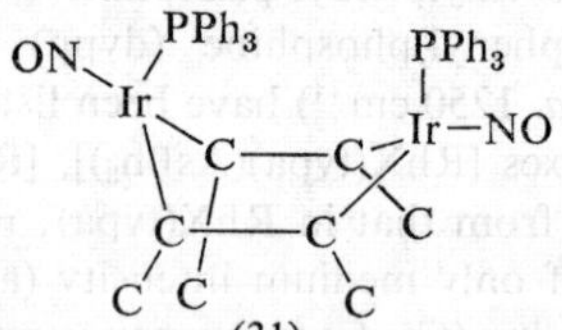   (31)

The reaction of hexafluorobut-2-yne with Ir(NO)(PPh$_3$)$_3$ affords a complex whose identity has been established by X-ray crystallography as (31), i.e. di-μ-hexafluorobut-2-enyl bis[cis-triphenylphosphinenitrosyl-iridium(I)]. This shows ν(NO) at 1780 cm^{-1}.[79]

The following assignments have been made for the tcne complexes (32): ν(C≡N) 2233, 2220; ν_{as}(N=C=C) 2168; ν(C≡O) 2080; ν_s(N=C=C)

---

[76] B. F. G. Johnson, H. V. P. Jones, and J. Lewis, *J.C.S. Dalton*, 1972, 463.
[77] R. Nast and L. Dahlenburg, *Chem. Ber.*, 1972, **105**, 1456.
[78] G. Pannetier, P. Fougeroux, and R. Bonnaire, *J. Organometallic Chem.*, 1972, **38**, 421.
[79] J. Clemens, M. Green, M.-C. Kuo, C. J. Fritchie, jun., J. T. Mague, and F. G. A. Stone, *J.C.S. Chem. Comm.*, 1972, 53.

1355 cm^{-1}, and (33): ν(CN) of tcne 2230; ν(C≡O) 2080; ν(CN) of Ir—CN not observed.[80]

ν(C≡O) and ν(C≡N) have been listed for IrH(CO)(X)L$_2$, where X = fumaronitrile, cinnamonitrile, benzylidenemalononitrile, fumaric acid, or dimethyl fumarate and L = PPh$_3$, e.g. (34).[81]

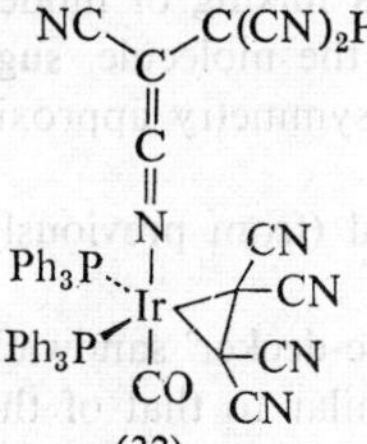

(32) 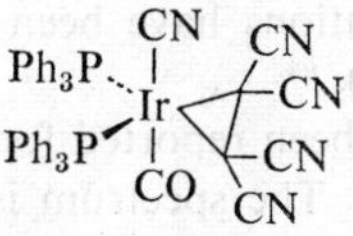 (33) 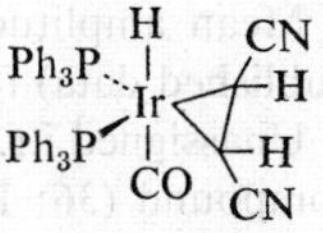 (34)

Cu$_4$Ir$_2$(PPh$_3$)$_2$(C≡CPh)$_8$ has been obtained from the reaction of Ir(CO)Cl(PPh$_3$)$_2$ and [Cu(C≡CPh)]$_n$.[82] A full X-ray structure determination shows that the structure is (35; L = C≡CPh). ν(C≡C) occurs at

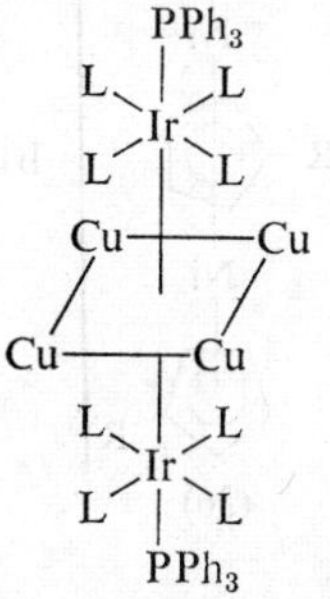

(35)

2017, 2001, and 1975 cm^{-1}, and there is an interaction between two C≡C bonds from each Ir and each of the four copper atoms.

M(CNMe)$_4^{2+}$ (where M = Ni, Pd, or Pt) all give ν(C≡N) at ca. 2300 cm^{-1} (2290, Ni; 2305, Pd; 2308, Pt), whereas

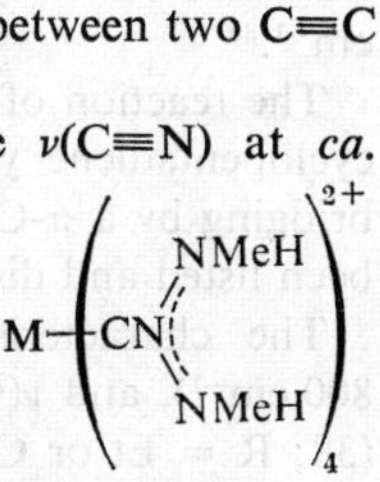

$$\left(M\!-\!\!\!-\!\!\!-\!CN\!\!\begin{array}{c}NMeH\\ \\NMeH\end{array}\right)_4^{2+}$$

(M = Pd or Pt) gives ν(C⋯N) at 1580 cm^{-1} (shifting to 1550 cm^{-1} on deuteriation), and no bands due to co-ordinated isocyanide.[83]

---

[80] M. S. Fraser, G. F. Everitt, and W. H. Baddley, *J. Organometallic Chem.*, 1972, **35**, 403.
[81] M. S. Fraser and W. H. Baddley, *J. Organometallic Chem.*, 1972, **36**, 377.
[82] O. M. Abu Salah, M. I. Bruce, M. R. Churchill, and S. A. Bezman, *J.C.S. Chem. Comm.*, 1972, 858.
[83] J. S. Miller and A. L. Balch, *Inorg. Chem.*, 1972, **11**, 2069.

(RNC)$_2$Ni(PhC≡CPh), where R = 2,6-dimethyl-4-bromophenyl, has ν(N≡C) at 2160 and 2100 cm^{-1}, and ν(C≡C) at 1830 cm^{-1}.[84]

The i.r. and Raman spectra of C$_5$H$_5$NiNO have been studied, and, with the aid of C$_5$D$_5$NiNO and C$_5$H$_5$Ni^{15}NO species, it has been possible to reassign some fundamentals. While these modifications are relatively slight, an analysis of combination bands clearly shows mixing of modes localized on both the organic and inorganic parts of the molecule, suggesting that caution must be exercised in using local-symmetry approximations.[85]

Mean amplitudes of vibrations have been calculated (from previously published data) for Ni(C$_5$H$_5$)$_2$.[86]

Unassigned i.r. data have been reported for the 'triple-decker' sandwich compound (36; R = Bu^t).[87] The spectrum is very similar to that of the corresponding nickelocene.

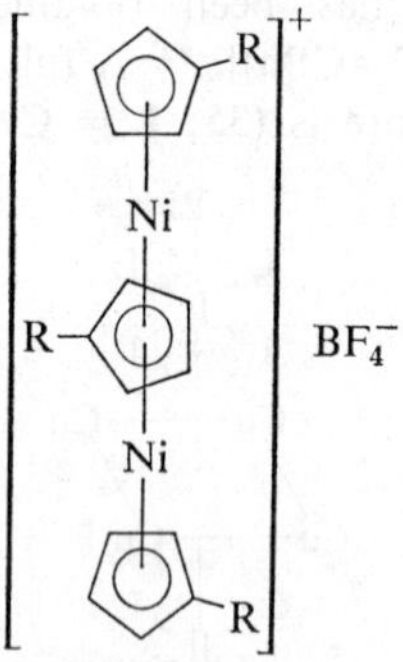

(36)

Cp(PBu$_3$)NiS(CH$_2$)$_n$SNi(PBu$_3$)Cp, where n = 2, 4, or 6, all show the characteristic π-cyclopentadienyl band (o.o.p. deformation) at $ca.$ 790 cm^{-1}.[88]

The reaction of PdBr$_2$ or NiI$_2$ with alkali-metal cyclopentadienides and cyclopentadiene yields products whose spectroscopic properties support bridging by a π-C$_5$H$_6$ ring, $e.g.$ (37; R = H or C$_5$H$_5$).[89] I.r. spectra have been listed and discussed.

The characteristic out-of-plane C—H deformation of π-C$_5$H$_5$ ($ca.$ 800 cm^{-1}), and ν(C$\cdots$S) (995—1015 cm^{-1}) are found in the i.r. spectra of (38; R = Et or CH$_2$Ph) and (39; n = 2, 4, or 6).[90]

---

[84] Y. Suzuki and T. Takizawa, $J.C.S. Chem. Comm.$, 1972, 837.
[85] G. Paliani, R. Cataliotti, A. Poletti, and A. Foffani, $J.C.S. Dalton$, 1972, 1741.
[86] J. Brunvoll, S. J. Cyvin, J. D. Ewbank, and L. Schäfer, $Acta Chem. Scand.$, 1972, **26**, 2160.
[87] A. Salzer and H. Werner, $Angew. Chem. Internat. Edn.$, 1972, **11**, 930.
[88] F. Sato, T. Yoshida, and M. Sato, $J. Organometallic Chem.$, 1972, **37**, 381.
[89] E. O. Fischer, P. Meyer, C. G. Kreiter, and J. Müller, $Chem. Ber.$, 1972, **105**, 3014.
[90] F. Sato, K. Iida, and M. Sato, $J. Organometallic Chem.$, 1972, **39**, 197.

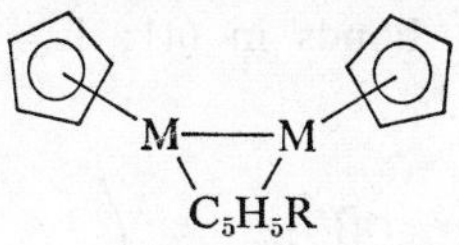

Vibrational Spectra of some Co-ordinated Ligands

(37)

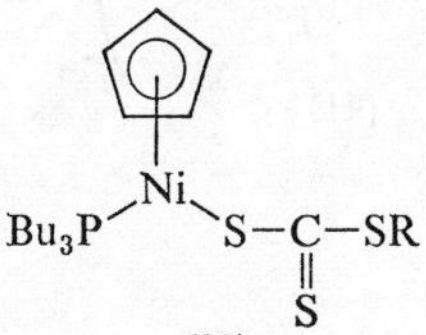

(38)

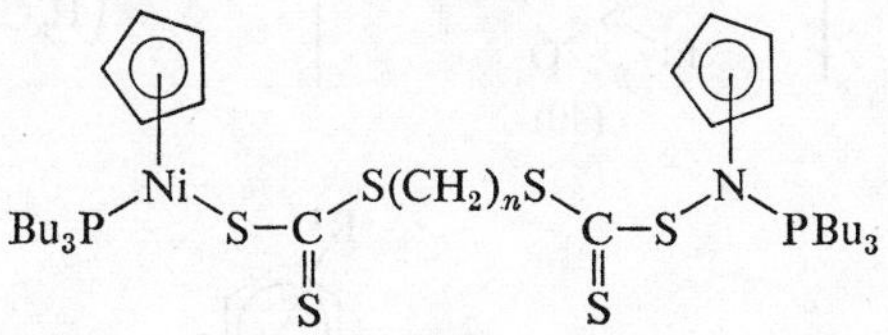

(39)

Some assignments of i.r. bands have been proposed for a series of new Pd–carbene complexes;[91] see Table 2.

Characteristic ligand-frequencies [ν(NH), ν(N≡C), ν(N⋯C), and ν(M—X)] have been listed for a large number of cationic carbene complexes of Pd^{II} and Pt^{II}, trans-[MX{C(NHR^1)NR^{2}R^3}L$_2$]$^+$.[92]

Table 2 Vibrational assignments for some Pd carbene complexes (all figures are wavenumber/cm^{-1})

Complex	ν(N≡C)	ν(N⋯C)	ν(N—H)
(Bu^{t}NC)PdCl$_2$(C$\underset{\text{NMe}_2}{\overset{\text{NHBu}^t}{<}}$)	2230	1565	3310
(Bu^{t}NC)PdBr$_2$(C$\underset{\text{NMe}_2}{\overset{\text{NHBu}^t}{<}}$)	2225	1565	3310
(Bu^{t}NC)PdI$_2$(C$\underset{\text{NMe}_2}{\overset{\text{NHBu}^t}{<}}$)	2210	1555	3350

The dialkylcarboxamido complexes L$_2$M(X)CONR^{1}R^2 (L = PPh$_3$, AsPh$_3$, or PMePh$_2$; M = Pd or Pt; X = Cl, I, or NCO; R^1 = H or Me; R^2 = Me or Pr^i) show characteristic ν(C=O) bands in the 1565—1615 cm^{-1} range.[93]

The complex (40) gives ν(CH), of OMe, at 2820 cm^{-1}, together with a band at 1107 cm^{-1} assignable to S-co-ordinated sulphoxide group.[94]

---

[91] G. A. Larkin, P. P. Scott, and M. G. H. Wallbridge, *J. Organometallic Chem.*, 1972, **37**, C21.
[92] L. Busetto, A. Palazzi, B. Crociani, U. Belluco, E. M. Badley, B. J. L. Kilby, and R. L. Richards, *J.C.S. Dalton*, 1972, 1800.
[93] C. R. Green and R. J. Angelici, *Inorg. Chem.*, 1972, **11**, 2095.
[94] Y. Takahashi, A. Tokuda, S. Sakai, and Y. Ishii, *J. Organometallic Chem.*, 1972, **35**, 415.

Two different $\rangle$C=C$\langle$ bonds in (41; L = PMe$_2$Ph) give stretching modes at 1650, 1610 cm^{-1}.[95]

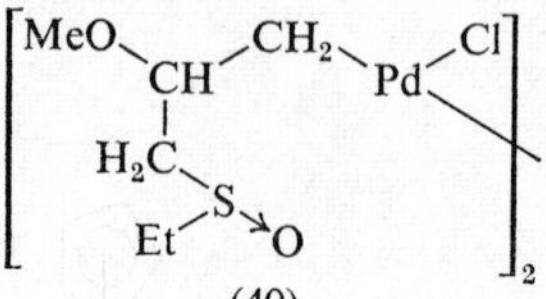

(40)

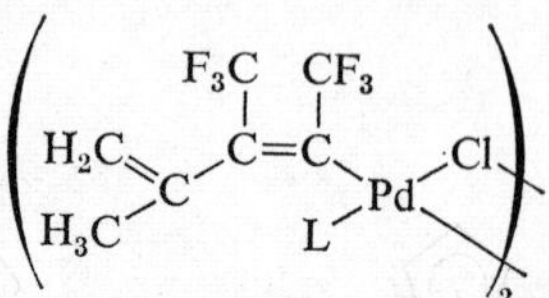

(41)

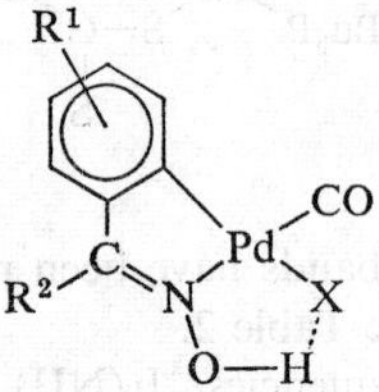

(42)

A series of complexes (42) has been prepared, having ν(OH) in the range 3170—3286 cm^{-1}, indicative of hydrogen-bonding. Trends in the OH stretching frequency, X = I > Br > Cl, and ν(CO), X = Cl > Br > I, are consistent with known inductive effects.[96]

Complexes in which allylammonium groups are co-ordinated to Pd include Na[Cl$_2$PdCl$_2$PdCl(CH$_2$=CHCH$_2$NH$_3$Cl)] and Pd(CH$_2$=CH—CH$_2$-NH$_3$)Cl$_3$. The i.r. and Raman spectra of the latter indicate considerable mixing of the ν(C=C) stretch and δ(CH$_2$) modes.[97] The extent of interaction can be examined qualitatively by reference to the free CH$_2$=CH-CH$_2$NH$_3^+$ ion.

ν(C≡C) in (43) is found at 1845 cm^{-1} (i.r., KBr disc).[98] A characteristic allyl vibration is found at 1485—1490 cm^{-1} in the following complexes

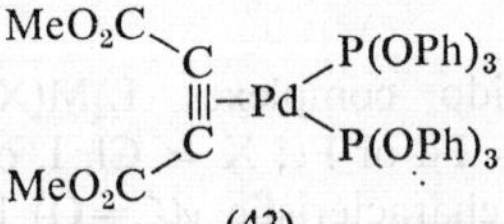

(43)

[mdp = methylenebis(diphenylphosphine), Ph$_2$PCH$_2$PPh$_2$]: [(π-C$_3$H$_5$)-PdCl,(mdp)], [(π-C$_3$H$_5$)PdCl,(mdp)$_2$], [(π-C$_3$H$_5$)Pd(mdp)$_2$][AlClBr$_3$], [(π-C$_3$H$_5$)PdCH(PPh$_2$)$_2$]$_2$, [(π-C$_3$H$_5$)PdCH(PPh$_2$)$_2$,(mdp)].[99]

---

[95] T. G. Appleton, H. C. Clark, R. C. Poller, and R. J. Puddephatt, *J. Organometallic Chem.*, 1972, **39**, C13.
[96] H. Onoue, K. Nakagawa, and I. Moritani, *J. Organometallic Chem.*, 1972, **35**, 217.
[97] F. R. Hartley and J. L. Wagner, *J.C.S. Dalton*, 1972, 2282.
[98] T. Ito, S. Hasegawa, Y. Takahashi, and Y. Ishii, *J.C.S. Chem. Comm.*, 1972, 629.
[99] K. Issleib, H. P. Abicht, and H. Winkelmann, *Z. anorg. Chem.*, 1972, **388**, 89.

No bands due to free C=C can be detected in (44), but weak features near 1500 cm^{-1} can be assigned to ν(C=C) of the complexed double bond.[100] Complex (45), on the other hand, gives ν(−CCl=CH$_2$) at 1640 cm^{-1} in the i.r. spectrum.[101]

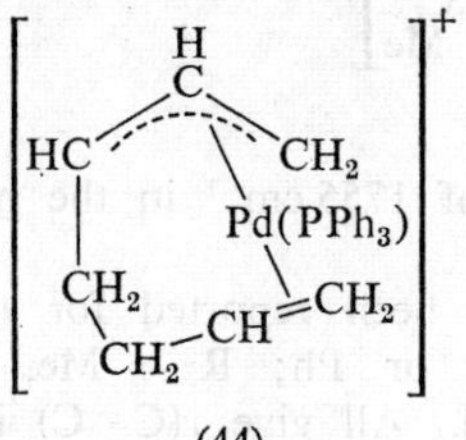

(44) 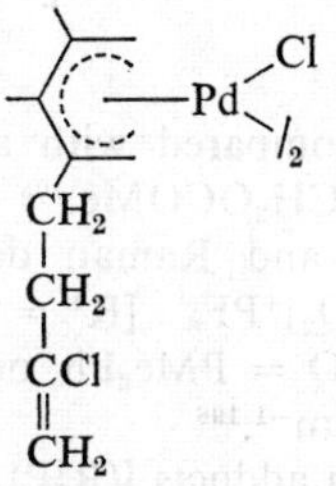 (45)

Some new π-allylic complexes of Pd have been prepared; thus the 'free' C=C bond gives stretching frequencies at 1661 cm^{-1} (46) and 1662 cm^{-1} (47).[102]

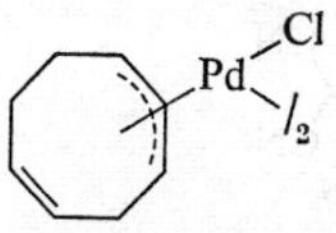

(46) 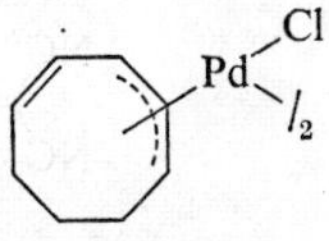 (47)

Solid PdCl$_2$(cod) (cod = *cis,trans*-1,3-cyclo-octadiene) gives an i.r. band due to co-ordinated >C=C< bonds only (1475 cm^{-1}), but in CHCl$_3$ solution a further band is seen, at 1640 cm^{-1}, suggesting that the equilibrium

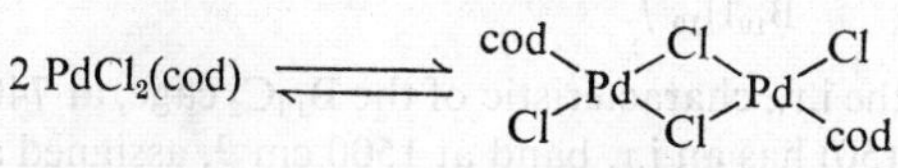

$$2\ PdCl_2(cod) \rightleftharpoons \underset{Cl}{\overset{cod}{Pd}}\underset{Cl}{\overset{Cl}{Pd}}\overset{Cl}{\underset{cod}{}}$$

is established. Solid PtCl$_2$(cod) shows bands due to both free and co-ordinated double bonds (1640, 1510 cm^{-1}, respectively).[103]

ν(CF) modes have been assigned in a wide variety of complexes containing the Pt−CF$_3$ unit. ν_s occurs at *ca.* 1080—1100 cm^{-1}, with the degenerate stretch (for isolated CF$_3$) split, at *ca.* 980—1000, *ca.* 1010—1050 cm^{-1}.[104]

In [48; (a) R$_3^1$ = Ph$_2$Me, R^2 = H, X = BF$_4$; (b) R$_3^1$ = Ph$_2$Me, R^2 = Me, X = BF$_4$; (c) R$_3^1$ = Ph$_3$, R^2 = H, X = ClO$_4$], ν(C=O) is at

---

[100] Y. Takahashi, S. Sakai, and Y. Ishii, *Inorg. Chem.*, 1972, **11**, 1516.
[101] D. J. S. Guthrie and S. M. Nelson, *Co-ordination Chem. Rev.*, 1972, **8**, 139.
[102] C. Agami, J. Levisalles, and F. Rose-Munch, *J. Organometallic Chem.*, 1972, **35**, C59.
[103] H. A. Tayim and A. Vassilian, *Inorg. Nuclear Chem. Letters*, 1972, **8**, 659.
[104] T. G. Appleton, M. H. Chisholm, H. C. Clark, and L. E. Manzer, *Inorg. Chem.*, 1972, **11**, 1786.

$$\left[\begin{array}{c} \text{MeCHR}^2 \\ R^1_3P\diagdown\!\!\diagup\text{CH} \\ \text{Pt} \diagdown\!\!\diagup\text{O} \\ R^1_3P\diagup\!\!\uparrow \\ \quad\quad O{=}C \\ \quad\quad\quad\quad\text{Me} \end{array}\right]^+ X^-$$

(48)

1575 cm^{-1}, compared with a value of 1755 cm^{-1} in the parent ligand CH$_2$=CHR^{2}CH$_2$OCOMe.[105]

Some i.r. and Raman data have been reported for *trans*-[PtCH$_3$-(R^{1}C≡CR^2)Q$_2$]$^+$PF$_6^-$ [R^1 = Me, Et, or Ph; R^2 = Me, Et, Ph, or CPh$_2$(OH); Q = PMe$_2$Ph or AsMe$_3$]. All give ν(C≡C) in the range 2024—2116 cm^{-1}.[106]

In the tcnq adducts [(R$^1_3$P)$_2$Pt(C≡CR^2)$_2$]tcnq (R^1 = Me or Et; R^2 = H or Me), the i.r. spectrum due to the tcnq is rather similar to that of the free ligand, except for the presence of a strong band at 1600 cm^{-1} due to ν(C=C). This is i.r.-forbidden in free tcnq (D_{2h}), and its presence here suggests a lowering of symmetry to C_{2v}, for example. The postulated structure is (49).[107]

$$\begin{array}{c} \text{NC}\diagdown\quad\quad\quad\diagup\text{CN} \\ \quad\quad\text{C}\!=\!\bigcirc\!=\!\text{C} \\ \text{NC}\diagup\quad\quad\quad\diagdown\text{CN} \\ \quad\quad\quad\;\; | \\ \quad\quad\quad R^1_3P \\ R^2C≡C-M-C≡CR^2 \\ \quad\quad\quad R^1_3P \end{array}$$

(49)

$$\textit{cis}\text{-Pt(PPh}_2)_2\left(\begin{array}{c} -C\!-\!\!-\!CMe \\ \diagdown\!\!\diagup \\ B_{10}H_{10} \end{array}\right)_2 \; (\textit{i.e.}\text{ containing }\sigma\text{-bonded carborane units})$$

shows bands in the i.r., characteristic of the B$_{10}$C$_2$ cage, at 740, 2550 cm^{-1}.[108]

The complex (50) has an i.r. band at 1500 cm^{-1}, assigned as ν(N⋯C⋯N) [with a contribution from δ(NH)]. Compound (51) has a similar band at 1495 cm^{-1}.[109]

$$\begin{array}{c} \quad\quad\text{H}\;\;\text{H} \\ \quad\quad\text{N}\!-\!\text{N} \\ \text{HMeNC}\diagdown\quad\diagup\text{CNMeH} \\ \quad\quad\text{Pt} \\ \quad\quad\diagup\;\;\diagdown \\ \quad\quad\text{I}\quad\text{I} \end{array} \quad\quad \begin{array}{c} \quad\quad\text{N}\!-\!\text{N} \\ \text{HMeNC}\diagdown\;|\;\diagup\text{CNMeH} \\ \quad\quad\text{Pt} \\ \quad\;\;\diagup\,|\,\diagdown \\ \quad\;\text{I}\;\;\text{I}\;\;\text{I} \end{array}$$

(50)   (51)

---

[105] H. C. Clark and H. Kurosawa, *J.C.S. Chem. Comm.*, 1972, 150.
[106] M. H. Chisholm and H. C. Clark, *Inorg. Chem.*, 1971, **10**, 2557.
[107] H. Masai, K. Sonogashira, and N. Hagihara, *J. Organometallic Chem.*, 1972, **34**, 397.
[108] R. Rogarski and K. Cohn, *Inorg. Chem.*, 1972, **11**, 1429.
[109] A. L. Balch, *J. Organometallic Chem.*, 1972, **37**, C19.

ν(NH) and ν(N≡C) have been recorded for the Pt^{II} carbene complexes *trans*-[Pt(CNC$_2$H$_5$){P(CH$_3$)$_2$C$_6$H$_5$}$_2$(carbene)]$^+$PF$_6^-$, where carbene is C(OEt)NHEt, C(NHPh)NHEt, C(NHC$_6$H$_4$Me)NHEt, or C(SCH$_2$Ph)-NHEt, and for a number of other Pt^{II} cyano-complexes.[110]

Barriers to rotation of ethylene in complexes PtXYL(C$_2$H$_4$) (X = Cl, Br, or CF$_3$CO$_2$, *trans* to C$_2$H$_4$; Y = Cl or Br, *cis* to C$_2$H$_4$; L = phosphine, arsine, phosphite, or primary amine) have been measured by n.m.r. techniques.[111] Comparison of the derived $\Delta G^{\neq}$ values with ν(CO) for analogous PtXYL(CO) demonstrates a significant steric contribution to the rotational barrier.

The Pt–cyclopropenone complex (52; L = PPh$_3$), which gives no band assignable to ν(C=C), and ν(C=O) at 1750 cm^{-1}, isomerizes at -30 °C to give the Pt insertion product (53). This complex gives ν(C=C) at 1675

cm^{-1}, ν(C=O) at 1640 cm^{-1}. The analogous reactions with cyclopropenones (54; R^1, R^2 = Me or Ph) gave only the latter type (the π-complexes could not be isolated).[112]

Complexes of thiiren 1,1-dioxide (55) with L$_2$PtX [L = PPh$_3$; X = C$_2$H$_4$, CS$_2$, or (PPh$_3$)$_2$] and *trans*-Ir(CO)Cl(PPh$_3$)$_2$ show that the >C=C<

bond is co-ordinated [no ν(C=C) in complexes; 1614 cm^{-1} in the free ligand].[113]

Compound (56) gives ν(C=O) at 1615 cm^{-1} (X = Cl), 1620 cm^{-1} (X = Br), with ν(C=C) at 1510 cm^{-1} (X = Cl or Br).[114] The hex-1-en-5-one ligand can also give rise to π-allyl complexes from the form (57).

---

[110] H. C. Clark and L. E. Manzer, *Inorg. Chem.*, 1972, 11, 503.
[111] J. Ashley-Smith, I. Douek, B. F. G. Johnson, and J. Lewis, *J.C.S. Dalton*, 1972, 1776.
[112] J. P. Visser and J. E. Ramakers-Blom, *J. Organometallic Chem.*, 1972, 44, C63.
[113] J. P. Visser, C. G. Lehveld, and D. N. Reinhoudt, *J.C.S. Chem. Comm.*, 1972, 178.
[114] B. T. Heaton and D. J. A. McCaffrey, *J. Organometallic Chem.*, 1972, 43, 437.

Dimeric complexes of this ligand give no identifiable $\nu(C=C)$, and a $\nu(C=O)$ at ca. 1680 cm^{-1} (i.e. the $>C=O$ group is not complexed).

$\nu(C\equiv C)$ is in the range 1880—1904 cm^{-1} for a series of complexes Pt(ac)$_2$, where (ac) is one of six disubstituted acetylenes with highly complex substituents, e.g. $(CH_3)(C_5H_{11})C(OH)-C\equiv C-C(OH)(C_5H_{11})(CH_3)$.[115]

The 5-methylenecycloheptene complex (58), in which two double bonds perpendicular to one another are both co-ordinated to Pt, gives $\nu(C=C)$ ca. 100 cm^{-1} lower than in the free ligand.[116]

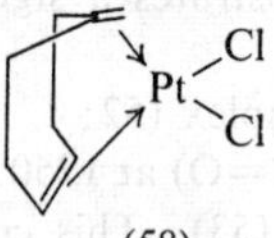

(58)

PtCl$_2$(1,4-cod)$_2$ has $\nu(C=C)$ at 1500 and 1480 cm^{-1} in the i.r., suggesting that both olefinic bonds are co-ordinated.[117]

$\nu(C=C)$ in complexes of a series of allylic alcohols with Cu^I perchlorate is decreased by ca. 110 cm^{-1} compared to the value for the free alcohol. This indicates a stronger co-ordination than to CuCl (where the drop is ca. 95 cm^{-1}).[118]

An i.r. study (1300—3500 cm^{-1}) has been made [119a] of the absorption of butenes on to a Cu$_2$O catalyst. Evidence was presented for the initial formation of, for example, a but-1-ene π-complex with Cu$^+$ ions.

CuCl$_2$(cis,trans-1,3-cod) has $\nu(C=C)$ at 1505 cm^{-1} (i.r.); both double bonds are, therefore, co-ordinated.[119b] $\nu(C=C)$ wavenumbers were also reported for [RhCl(1,5-cod)]$_2$ (1480 cm^{-1}), AuCl$_3$(cis,trans-1,3-cod) (1625, 1530 cm^{-1}), and AgNO$_3$(cis,trans-1,3-cod) (1580 cm^{-1}).

$\nu(C=C)$ has been quoted for CuL$_n$(OTf), where OTf = trifluoromethanesulphonate, and $n = 1$, L = cyclo-octa-1,3,5,9-tetraene or $n = 2$, L = endo-dicyclopentadiene.[120]

$\nu(C\equiv C)$ in two isomers of (PhC≡CAg)$_2$,AgNO$_3$ is 40 cm^{-1} lower than in PhC≡CAg.[121a]

Raman spectra of aqueous solutions of some silver–olefin complexes AgR$^+$ (where R = C$_2$H$_4$, cis-2-butene, trans-2-butene, or cyclohexene) and the i.r. spectrum of the new complex [PtCl$_2$(Me$_2$C=CMe$_2$)]$_2$ have been

---

[115] F. D. Rochon and T. Theophanides, Canad. J. Chem., 1972, **50**, 1325.
[116] C. B. Anderson and J. T. Michalowski, J.C.S. Chem. Comm., 1972, 459.
[117] H. A. Tayim, A. Bouldoukian, and M. Kharboush, Inorg. Nuclear Chem. Letters, 1972, **8**, 231.
[118] Y. Ishino, T. Ogura, K. Noda, T. Hirashima, and O. Manabe, Bull. Chem. Soc. Japan, 1972, **45**, 150.
[119] (a) S. V. Gerei, E. V. Rozhkova, and Ya. B. Gorokhvatskii, Doklady. Phys. Chem., 1971, **201**, 968; (b) H. A. Tayim and A. Vassilian, Inorg. Nuclear Chem. Letters, 1972, **8**, 215.
[120] R. G. Solomon and J. K. Kochi, J.C.S. Chem. Comm., 1972, 559.
[121] (a) T. G. Sukhova, O. L. Kaliya, O. N. Temkin, and R. M. Flid, Russ. J. Inorg. Chem., 1971, **16**, 816; (b) D. B. Powell, J. G. V. Scott, and N. Sheppard, Spectrochim. Acta, 1972, **28A**, 327.

examined in detail.[121b] It was concluded that while a band at 1240 cm^{-1} in the Pt–ethylene complex can be assigned mainly to C=C stretching (as has been suggested previously), a further band at 1500 cm^{-1} should be assigned mainly to ν(C=C) for substituted olefins. The degree of interaction between fundamentals with wavenumbers *ca.* 1240 cm^{-1} and *ca.* 1500 cm^{-1} for the complexed olefins was discussed.

ν(C≡N) in R^{1}R^{2}C(Br)—C≡N is shifted about 30 cm^{-1} to lower wavenumbers on forming compounds of the type RRC(ZnBr)—C≡N.[122] This shift is inconsistent with the alternative formulation R^{1}R^{2}C=C=N—ZnBr.

A detailed vibrational assignment for diallylmercury (CH$_2$=CH—CH$_2$)$_2$-Hg suggests a symmetry of C_i for the molecule in the liquid state.[123] A similar detailed assignment has been proposed for the vibrations of the allylmercuric halides CH$_2$=CH—CH$_2$HgX (X = Cl, Br, or I).[124] In solution, the only rotational isomer present possesses C_1 symmetry.

Unassigned listings of i.r. bands have been made for fluorenylmercuric chloride and bis(fluorenyl)mercury,[125] and for (59).[126]

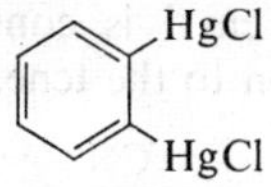

(59)

I.r. and Raman spectral data have been listed for bis(cyclo-octatetraenyl)thorium, Th(C$_8$H$_8$)$_2$, and some tentative assignments given.[127] These data, together with electronic spectra and *X*-ray powder patterns, were compared with those for U(C$_8$H$_8$)$_2$, and it was concluded that the thorium compound has a sandwich structure, with rings having aromatic character.

Unassigned i.r. bands have been listed for Ce(C$_8$H$_8$)$_2$ [128] and tetrafluorenylcerium(IV), (C$_{13}$H$_9$)$_4$Ce.[129]

Assignments of δ_s(CH$_3$) (1045—1106 cm^{-1}) and ρ(CH$_3$) (694—750 cm^{-1}) have been proposed for the Li$^+$, Na$^+$, K$^+$, Rb$^+$, and Cs$^+$ salts of InMe$_4^-$.[130]

Assignments of characteristic π-cyclopentadienyl bands have been made for In(C$_5$H$_5$), In(C$_5$H$_5$)$_3$, and In(C$_5$H$_4$Me)$_3$.[131]

I.r. spectra have been assigned (without discussion) for CH$_3$(C$_5$H$_5$)TlX (X = OCOPr^i, OCOMe, OCOEt, tropolonate, or 4-isopropyl tropolo-

---

[122] N. Goasdoué and M. Gaudemar, *J. Organometallic Chem.*, 1972, **39**, 17.
[123] C. Sourisseau and B. Pasquier, *J. Organometallic Chem.*, 1972, **39**, 65.
[124] C. Sourisseau and B. Pasquier, *J. Organometallic Chem.*, 1972, **39**, 51.
[125] E. Samuel and M. D. Rausch, *J. Organometallic Chem.*, 1972, **36**, 29.
[126] D. Seyferth and D. L. White, *J. Organometallic Chem.*, 1972, **34**, 119.
[127] J. Goffart, J. Fuger, B. Gilbert, B. Kanellakopulos, and G. Duyckaerts, *Inorg. Nuclear Chem. Letters*, 1972, **8**, 403.
[128] B. L. Kalsotra, R. K. Multani, and B. D. Jain, *Chem. and Ind.*, 1972, 389.
[129] B. L. Kalsotra, R. K. Multani, and B. D. Jain, *J. Inorg. Nuclear Chem.*, 1972, **34**, 2679.
[130] K. Hoffmann and E. Weiss, *J. Organometallic Chem.*, 1972, **37**, 1.
[131] J. S. Poland and D. G. Tuck, *J. Organometallic Chem.*, 1972, **42**, 307.

nate).[132] For X = OCOMe, the following were proposed: $\nu(Tl-C_5H_5)$ 318; $\nu(Tl-CH_3)$ 513; $\delta(CH)$ 648; $\pi(CH)$ 752; $\rho(Tl-Me)$ 785; ring def. 820; $\delta(CH)_{C_5H_5}$ 990; $\delta(CH)_{C_5H_5}$ 1021; $\nu_s(COO)$ 1420; $\nu_{as}(COO)$ 1530 cm^{-1}, and several $\nu(CH)$ bands.

A normal co-ordinate analysis has been performed on the hypothetical system (60). Using force fields of free ethylene and H_2O, a predicted set of

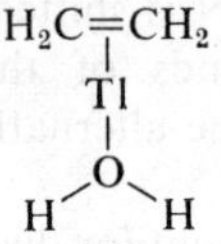

$$H_2C=CH_2$$
$$|$$
$$Tl$$
$$|$$
$$O$$
$$H \quad H$$
(60)

frequencies was calculated. Significant shifts from free-ligand values were found in a few modes (e.g. CH_2 rocking), and this was ascribed to kinematic coupling effects. It should be pointed out, however, that very variable frequencies have been calculated for these modes in free ethylene by numerous workers in the past.[133]

An adduct of Ph_3P with tcne has been prepared.[134] The presence of a single $\nu(C\equiv N)$ band at 2190 cm^{-1} is consistent with (61), appreciable negative charge being shifted on to the tcne.

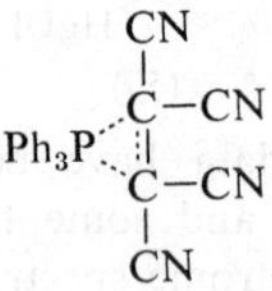

$$\begin{array}{c} CN \\ | \\ C-CN \\ Ph_3P\cdots \| \\ C-CN \\ | \\ CN \end{array}$$
(61)

2 Carbonyls

A simplified model has been described for dealing with interactions between carbonyl groups, especially those bonded to different metal atoms.[135a] A consistent interpretation was given of solution and solid-state spectra for a series of binuclear metal carbonyl complexes.

Kettle and co-workers have described the application of Wolkenstein's bond polarizability approach to Raman intensities of the terminal $\nu(CO)$ vibrations for some metal carbonyl species. General intensity formulae were given and discussed in detail for $M(CO)_6$ (M = Cr, Mo, or W), $RM(CO)_3$ [RM = (C_5H_5)Mn or (arene)Cr], and $RM(CO)_5$ (RM = BrMn or Ph_3SnRe).[135b] It was concluded that not only do totally symmetric

---

[132] T. Abe and R. Okawara, J. Organometallic Chem., 1972, 35, 27.
[133] L. Schäfer, J. D. Ewbank, S. J. Cyvin, and J. Brunvoll, J. Mol. Structure, 1972, 14, 185.
[134] J. E. Douglas, Inorg. Chem., 1972, 11, 654.
[135] (a) J. G. Bullitt and F. A. Cotton, Inorg. Chim. Acta, 1971, 5, 637; (b) S. F. A. Kettle, I. Paul, and P. J. Stamper, J.C.S. Dalton, 1972, 2413; (c) J. H. Darling and J. S. Ogden, ibid., p. 2496; (d) M. J. Cleare, H. P. Fritz, and W. P. Griffith, Spectrochim. Acta, 1972, 28A, 2019; (e) S. Cenini, B. Ratcliff, A. Fusi, and A. Pasini, Gazzetta, 1972, 102, 141.

$\nu(CO)$ modes frequently give rise to *weak* Raman bands, but that non-totally symmetric $\nu(CO)$ modes can be of comparable band intensity to totally symmetric non-carbonyl vibrations.

$\nu(CO)$ band patterns to be expected for binary metal carbonyls produced from $C^{16}O$–$C^{18}O$ mixtures in matrix-isolation experiments have been calculated.[135c]

In the complexes $Cs_2M(CO)X_5$ (M = Os, Ru, Ir, or Rh, X = Cl or Br; M = Ir or Rh, X = I), the values of $\nu(CO)$ and $\nu(M-C)$ may be rationalized on the grounds of more effective M—C π-overlap from third-row transition elements. A close comparison of these data with those from related nitrosyls is consistent with more effective π-accepting properties for NO than for CO.[135d]

Solvent (CH_2Cl_2, MeCN, CS_2, or C_6H_{12}) effects on $\nu(CO)$ in the i.r. spectra of $[Mn(CO)_5]_nSnR_{4-n}$, $[(\pi\text{-}Cp)Fe(CO)_2]_nSnR_{4-n}$, $[Co(CO)_3L]_n\text{-}SnR_{4-n}$, and $[(\pi\text{-}Cp)Mo(CO)_3]_nSnR_{4-n}$ (where n = 1 or 2; L = CO, phosphines, or arsines; R = Cl, Br, I, Ph, Et, or Me) are interpreted[135e] in terms of direct dipole–dipole interactions with the solvent. These are dependent upon R and L. Coupling of CO groups across the M—M'—M system (M = Mo, Mn, Fe, or Co; M' = Sn or also Hg) was confirmed and discussed.

Transition-metal carbonyls in which the oxygen is also co-ordinated to an acidic centre (*e.g.* AlR_3) have been reviewed, and the use of $\nu(CO)$ data in this field has been illustrated.[136]

$\nu(CO)$, and some lower-frequency bands, have been assigned for the following:[137] $[V(CO)_5(NH_3)]^-$ $[C_{4v}, \nu(CO)$ 1979 (A_1), 1785 (E) cm^{-1}; $\delta(VCO)$, 649 cm^{-1}; $\nu(VC)$ 460 cm^{-1}]; $[V(CO)_5(CN)]^{2-}$ $[C_{4v}, \nu(CO)$ 1852 (A_1), 1793 (E), 1744 (A_1) cm^{-1}]; $[V(CO)_4(CN)_2]^{4-}$ $[D_{2h}, \nu(CO)$ 1794 (B_{1u}), 1748 (B_{2u}) cm^{-1}].

In matrix-isolation experiments, bands have been observed at 1855, 1852, and 1838 cm^{-1}, assigned to $\nu(CO)$ of $Cr(CO)_5^-$ [C_{4v}; produced by co-condensation of Na and $Cr(CO)_6$].[138]

Using an approximate MO calculation, the CO force constants of $M(CO)_{6-x}L_x$ (M = Cr, Mn, or Fe; L = Cl or Br; x = 1 or 2), have been correlated with calculated occupancies of the carbonyl 5σ and 2π orbitals.[139] Back-bonding to the 2π *and* σ-donation to 5σ affect the force constant. Also, direct donation of electron density from a halogen σ-orbital to the CO 2π orbital is the most important mechanism by which a change in halogen effects a change in the carbonyl force constant.

$\nu(CO)$ and $\nu(NO)$ have been listed for $(C_5H_5)Cr(CO)(NO)L$, where L = cyclo-octene, ethylene, acetylene, $C_2(CO_2Me)_2$, acenaphthylene, norbornene, or maleic anhydride.[140a]

---

[136] D. F. Shriver and A. Alichi, *Co-ordination Chem. Rev.*, 1972, **8**, 15.
[137] D. Rehder, *J. Organometallic Chem.*, 1972, **37**, 303.
[138] P. A. Breeze and J. J. Turner, *J. Organometallic Chem.*, 1972, **44**, C7.
[139] M. B. Hall and R. F. Fenske, *Inorg. Chem.*, 1972, **11**, 1619.
[140] (*a*) M. Herberhold and H. Alt, *J. Organometallic Chem.*, 1972, **42**, 407; (*b*) R. Pince and M. Poilblanc, *Spectrochim. Acta*, 1972, **28A**, 907; (*c*) R. T. Jernigan, R. A. Brown, and G. R. Dobson, *J. Co-ordination Chem.*, 1972, **2**, 47.

Raman spectra of liquid $M(CO)_6$ (M = Cr, Mo, or W) have been recorded and assigned in relation to earlier data. An automatic computational method for the assignment of harmonics and combinations was described, and valence force constants were calculated.[140b]

Two C—O stretching force constants and three CO–CO stretch–stretch interaction constants have been determined from the four $\nu(CO)$ wavenumbers of 26 complexes cis-$L_2M(CO)_4$ (L_2 = bidentate chelating ligand with N, P, As, or S donor atoms, M = Cr, Mo, or W).[140c] No relationships among the interaction constants were assumed. The results were in good agreement with force-constant data previously derived from isotope-enrichment studies.

$\nu(CO)$ has been listed for a series of pentacarbonyl complexes $M(CO)_5L$ (M = Cr, Mo, or W; L = o-, m-, or p-tritolylphosphine). All show a medium-intensity i.r. band at ca. 2070 cm^{-1}. In addition, the complexes containing m- or p-tolyl groups also gave a sharp singlet at ca. 1950 cm^{-1}, while the o-tolyl complexes gave a doublet in that region, due to the steric effects of the *ortho*-methyl groups.[141] $\nu(CO)$ values were also listed for a variety of cis- and trans-$M(CO)_4L_2$ complexes of similar type.

In $[C_5H_5(CO)_3Mo—SnCl_2(S_2CNMe_2)]$, $\nu(CN)$ occurs at 1525 cm^{-1}, indicative of partial C—N double-bonding. In addition, five $\nu(CO)$ bands are seen in CS_2 solution, suggesting that isomers (62a) and (62b) are present

(62a) (62b)

(R = S_2CNMe_2, the third CO on the Mo atom is staggered), *i.e.* there is hindered rotation about the Mo—Sn bond.[142]

Knox et al. have shown [143] that earlier reports (A. N. Nesmeyanov, C. G. Dvoryantseva, Yu. N. Sheinker, N. E. Kolobova, and K. N. Anisimov, *Doklady Akad. Nauk S.S.S.R., Ser. khim.*, 1966, **169**, 843) of $\nu(CO)$ modes in $MnRe(CO)_{10}$ and $(h^5-C_5H_5)W(CO)_3Mn(CO)_5$ included a number of bands due to decomposition products of $Mn_2(CO)_{10}$ with the CCl_4 solvent.

Mono-^{13}CO-substituted $Mn_2(CO)_{10}$ and $Re_2(CO)_{10}$ have been studied by i.r. and Raman spectroscopy in the $\nu(CO)$ region.[144] The spectra and assignments of the all-^{12}CO-species agree with those of earlier workers. When one ^{13}CO is placed in an axial position, one A_1 mode (in C_{4v}) drops from 1983 to 1950 cm^{-1} (Mn), from 1978 to 1943 cm^{-1} (Re), whereas the 'equatorial' spectrum is unaffected. Substitution in the equatorial position

---

[141] J. A. Bowden and R. Colton, *Austral. J. Chem.*, 1971, **24**, 2471.
[142] W. K. Glass and T. Shiels, *J. Organometallic Chem.*, 1972, **35**, C64.
[143] S. A. R. Knox, R. J. Hoxmeier, and H. D. Kaesz, *Inorg. Chem.*, 1971, **10**, 2636.
[144] W. T. Wozniak and R. K. Sheline, *J. Inorg. Chem.*, 1972, **34**, 3765.

affects the spectrum more profoundly and the symmetry is now only C_s. Using the frequencies so assigned, sets of wavenumbers were computed for all-^{12}CO- and mono-^{13}CO-species (^{13}CO axial or equatorial).

The i.r. spectrum of crystalline Mn(CO)$_4$NO has been obtained using polarized radiation.[145] In conjunction with solution, liquid-, and vapour-phase data this gives additional support to the view that the molecule belongs to the point group C_{2v}. A probable space group for the crystal of C_s^4 (Cc) was also indicated.

Exchange of ^{13}CO with Mn(CO)$_5$Br has been studied by following intensity changes in the ν(CO) region in hexane solution. The results were only consistent with a CO dissociative mechanism, with a slight preference for loss of radial CO.[146]

(h^5-C$_5$H$_5$)Mn(CO)$_2$ is produced by photolysis of (h^5-C$_5$H$_5$)Mn(CO)$_3$.[147] The C≡O stretching frequencies are decreased on the removal of one carbonyl group: CpMn(CO)$_3$: $\nu(A_1)$ 2026, $\nu(E)$ 1938; CpMn(CO)$_2$: ν_s 1955, ν_{as} 1886 (all cm^{-1}). Very similar figures were found for the (MeC$_5$H$_4$) derivatives.

I.r. intensities of ν(CO) bands for a series of olefin and Group V donor-atom complexes of the type (π-Cp)Mn(CO)$_2$L have been measured.[148] The ratio of the i.r. intensities for ν_{sym}, $\nu_{antisym}$ is rather insensitive to changes in the donor ligand, which suggests that the π-bonded ring acts as a buffer to the vibronic contributions which the ligand L can make to the CO groups (^{55}Mn n.q.r. measurements support this observation).

The interaction between (π-MeCp)Mn(CO)$_3$, [(π-Cp)Fe(CO)$_2$]$_2$, or (π-Cp)Cr(NO)$_2$Cl with organolanthanides (C$_5$H$_5$)$_3$Ln (Ln = Sm, Er, or Yb), or their C$_5$H$_4$Me analogues, in CH$_2$Cl$_2$ solution has been studied by i.r. spectroscopy in the ν(CO) and ν(NO) regions. Decreases in ν(NO) and ν(CO) indicate co-ordination of the Ln compound to the oxygen lone pairs. No, or very little, adduct formation was found with (C$_5$H$_5$)YbCl or (C$_5$H$_4$Me)YbCl, supporting the supposed dimeric structures of these species.[149]

(h^5-C$_5$H$_5$)Mn(CO)(NO)$_2$ gives i.r. bands (in C$_6$H$_{12}$ solution) at 1993, 1970 cm^{-1} [ν(CO)$_{terminal}$], 1813 cm^{-1} [ν(CO)$_{bridge}$], 1732 cm^{-1} [ν(NO)$_{terminal}$], and 1534 cm^{-1} [ν(NO)$_{bridge}$]. This is explained by the presence of cis- and trans-isomers (63a) and (63b), with overlapping of the bridging (CO) and terminal and bridging ν(NO) bands.[150a]

Solid-state i.r. and Raman studies on Ph$_3$SnMn(CO)$_5$, Ph$_3$SnMn(CO)$_4$-PPh$_3$, and Ph$_3$SnFe(h^5-C$_5$H$_5$)(CO)$_2$ in the carbonyl-stretching region have

[145] A. Poletti, G. Paliani, R. Cataliotti, A. Foffani, and A. Santucci, *J. Organometallic Chem.*, 1972, **43**, 377.
[146] A. Berry and T. L. Brown, *Inorg. Chem.*, 1972, **11**, 1165.
[147] P. S. Braterman and J. D. Black, *J. Organometallic Chem.*, 1972, **39**, C3.
[148] W. P. Anderson, T. B. Brill, A. R. Schoenberg, and C. W. Stanger, *J. Organometallic Chem.*, 1972, **44**, 161.
[149] A. E. Crease and P. Legzdins, *J.C.S. Chem. Comm.*, 1972, 268.
[150] (a) T. J. Marks and J. S. Kristoff, *J. Organometallic Chem.*, 1972, **42**, C90; (b) H. J. Buttery, S. F. A. Kettle, G. Keeling, I. Paul, and P. J. Stamper, *J.C.S. Dalton*, 1972, 2487.

```
        O                                  O
        ‖                                  ‖
π-Cp    C      π-Cp         π-Cp           C           NO
     \ / \   /                  \        /   \        /
     Mn   Mn                     Mn              Mn
    / \ / \                     /   \        /       \
   OC   N   NO                OC     N           π-Cp
        ‖                            ‖
        O                            O
       (63a)                         (63b)
```

been interpreted using a new type of symmetry group, called the *situs* group. This represents a modification of the crystallographic unit cell for vibrational purposes, and the relationship between the symmetry properties of the two was discussed.[150b]

Shifts in ν(CO) bands have been used to detect formation of M–(CO)–AlR$_3$ adducts.[151] Stepwise interactions of tri-isobutylaluminium with both bridging carbonyls of [(π-Cp)Fe(CO)$_2$]$_2$, all four bridging carbonyls of [(π-Cp)Fe(CO)]$_4$, but with only one of the two carbonyls of (π-Cp)Ni$_3$(CO)$_2$ were observed. The following orders of relative affinities for AlR$_3$ were established: [(π-Cp)Fe(CO)$_2$]$_2$ > [(π-Cp)NiCO]$_2$;   (π-Cp)Ni$_3$(CO)$_2$ > [(π-Cp)NiCO]$_2$; [(π-Cp)Fe(CO)]$_4$ > [(π-Cp)Fe(CO)$_2$]$_2$.

A study of the temperature dependence of the relative intensities of the two ν(CO) bands at *ca.* 1980 cm^{-1} for (π-Cp)Fe(CO)$_2$SiCl$_2$Me has shown that the entropy difference between the symmetric and unsymmetric isomers is 1.4 ± 0.5 e.u., with the sign favouring the unsymmetrical isomers. The enthalpy difference suggests a lower limit to the barrier to rotation about the Fe—Si bond of *ca.* 0.8 kcal mol^{-1}.[152]

The new pentanuclear carbidocarbonyl of iron [Fe$_5$C(CO)$_{14}$]$^{2-}$ gives ν(CO) bands at 2021 (vw), 1966 (vs), 1930 (mw), 1897 (w), and 1773 (mw) cm^{-1}. These are consistent with the structure (64), having terminal and edge-bridging CO groups.[153]

```
                    (CO)₂
                     Fe
            (OC)₂   /│\    CO
               \\  / │ \  /
                Fe───┼──Fe(CO)₂
            CO /    │
                \   C
         (OC)₂Fe \  │  ─── CO
                  \ │    /
                   Fe
                   │ (CO)₂
                   C
                   ‖
                   O
                  (64)
```

Phenylpentacenedi-iron pentacarbonyl gives ν(CO) bands at 2040, 2010, 1975 (terminal), and 1785 (bridging) cm^{-1}. It was not possible to distinguish between the structures (65a) and (65b).[154a]

---

[151] A. Alichi, N. J. Nelson, D. Strope, and D. F. Shriver, *Inorg. Chem.*, 1972, **11**, 2976.
[152] J. Dalton, *Inorg. Chem.*, 1972, **11**, 915.
[153] A. T. T. Hsieh and M. J. Mays, *J. Organometallic Chem.*, 1972, **37**, C53.
[154] (a) D. F. Hunt and J. W. Russell, *J. Organometallic Chem.*, 1972, **46**, C22; (b) D. A. Duddell, S. F. A. Kettle, and B. T. Kontnik-Matecka, *Spectrochim. Acta*, 1972, **28A**, 1571; (c) O. A. Gansow, D. A. Schexnayder, and B. Y. Kimura, *J. Amer. Chem. Soc.*, 1972, **94**, 3406.

$(OC)_2Fe\text{———}Fe(CO)_2$ with CO bridge

(65a)

$(OC)_2Fe\text{———}Fe(CO)_2$ with CO bridge

(65b)

The vibrational spectrum of (butadiene)tricarbonyliron has been investigated in the 2000 cm^{-1} region.[154b] A simple factor-group analysis, which gives an inadequate description of the solid-state i.r. and Raman spectra, was modified by a detailed study based on the crystal structure. Thus, the absence of one A_1-derived ν(CO) band can be related to the symmetry properties of the crystal structure.

^{13}C n.m.r. spectra of 18 monomeric derivatives of the $(h^5\text{-}C_5H_5)Fe(CO)_2$ unit have been measured.[154c] Interdependence of ^{13}C chemical shifts and ν(CO) frequencies indicates that δ(CO) values are determined by changes in the paramagnetic screening term.

Benedetti *et al.* have shown that earlier workers' reports on the i.r. spectrum in the ν(CO) region for $[Ru(CO)_3X_2]_2$ (X = Cl or Br) in CHCl$_3$ solution, which showed a time-dependence for the spectrum, were invalidated by the reaction of the bridged complex with stabilizing agents in the chloroform.[155]

cis-Ru(CO)$_4$(SiCl$_3$)$_2$ has been shown to exchange with ^{13}CO in a manner which is completely stereospecific.[156] Both equatorial CO groups exchange, while the axial CO groups are unaffected. The persistence of stereospecificity as the second ^{13}CO group was introduced established that the initially axial and equatorial CO groups are at all times differentiated in the proposed five-co-ordinate intermediate (see Figure 1). ^{13}CO enrichment is not observed in the i.r. when (Me$_3$Si)$_2$Os(CO)$_4$ is treated with ^{13}CO at 55 °C for 165 minutes.[157]

Pyrolysis of Os$_3$(CO)$_{12}$ gives Os$_4$(CO)$_{13}$, Os$_5$(CO)$_{16}$, Os$_6$(CO)$_{18}$, Os$_7$(CO)$_{12}$, Os$_8$(CO)$_{23}$, and Os$_5$(CO)$_{15}$C$_4$.[158] All show relatively simple ν(CO) spectra, with no evidence for bridging CO groups.

Vibrational spectra have been reported and assigned for M[Co(CO)$_4$]$_2$ (M = Zn, Cd, or Hg).[159] Symmetries of D_{3d} were adequate to explain the spectra, although some near-coincidences between i.r.- and Raman-active frequencies were observed, because of the very weak interactions across the Co—M—Co bridge. The metal–metal force constants were all almost

---

[155] E. Benedetti, G. Braca, G. Sbrana, F. Salvetti, and B. Grassi, *J. Organometallic Chem.*, 1972, **37**, 361.
[156] R. K. Pomeroy, R. S. Gay, G. O. Evans, and W. A. G. Graham, *J. Amer. Chem. Soc.*, 1972, **94**, 272.
[157] R. K. Pomeroy and W. A. G. Graham, *J. Amer. Chem. Soc.*, 1972, **94**, 274.
[158] C. R. Eady, B. F. G. Johnson, and J. Lewis, *J. Organometallic Chem.*, 1972, **37**, C39.
[159] R. J. Ziegler, J. M. Burlitch, S. E. Hayes, and W. M. Risen, *J. Inorg. Chem.*, 1972, **11**, 702.

equal (1.30—1.26 mdyn Å^{-1}), with a slight but definite trend: k(Zn—Co) $\gtrsim$ k(Cd—Co) $\gtrsim$ k(Hg—Co).

The proportion of bridged-CO isomers in [(Et$_3$M)Co(CO)$_3$]$_2$ increases in the sequence M = P < As < Sb [160] [monitored by i.r. observations in the ν(CO) region].

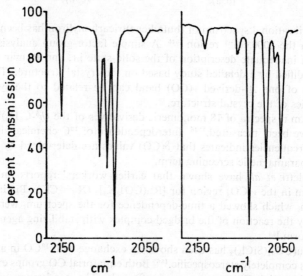

Figure 1 *I.r. spectra (in n-heptane solution) of cis-*Ru(CO)$_4$(SiCl$_3$)$_2$. *At left, before exchange, showing major bands assigned to the all-*^{12}CO *species at* 2150, 2103, 2084, *and* 2094 cm^{-1}; *weaker bands at* 2140, 2054, *and* 2057 cm^{-1} *assigned to mono-*^{13}CO *species in natural abundance. At right, after exchange with* 95.5% ^{13}CO *for 18 h at room temperature, bands at* 2140, 2084, 2064, *and* 2046 cm^{-1} *assigned to equatorially di-*^{13}CO*-substituted molecule; bands at* 2145, 2100, *and* 2054 cm^{-1} *assigned to equatorially mono-*^{13}CO*-substituted molecule*
(Reproduced by permission from *J. Amer. Chem. Soc.*, 1972, **94**, 272)

The i.r. spectrum (together with mass spectral, magnetic, and ^{11}B n.m.r. measurements) of Co$_3$(CO)$_6$(COBCl$_2$NEt$_3$)$_2$ is consistent with the structure (66; L = BCl$_2$NEt$_3$). ν(CO) bands are seen at 2092, 2054, 2040, and 2022 cm^{-1}, with ν(B—O) at 1224 cm^{-1}.[161]

The complex (67; Y = Cl or Br) gives ν(CO)$_{terminal}$ at 2045, 2002, and 1985 cm^{-1}, with ν(CO)$_{bridging}$ at 1799 cm^{-1}.[162]

ν(CO) bands have been assigned for MCo(CO)$_9$ (M = Mn or Re) and MnCo(CO)$_8$(PPh$_3$),[163] and (arene)Co$_4$(CO)$_9$ (68) (of C_{3v} symmetry; arene = toluene, tetrahydronaphthalene, or mesitylene).[164]

[160] D. J. Thornhill and A. R. Manning, *J. Organometallic Chem.*, 1972, **37**, C41.
[161] G. Schmid and B. Stutte, *J. Organometallic Chem.*, 1972, **37**, 375.
[162] P. A. Elder and B. H. Robinson, *J. Organometallic Chem.*, 1972, **36**, C45.
[163] G. Sbrignadello, G. Bor, and L. Maresca, *J. Organometallic Chem.*, 1972, **46**, 345.
[164] G. Bor, G. Sbrignadello, and F. Marcati, *J. Organometallic Chem.*, 1972, **46**, 357.

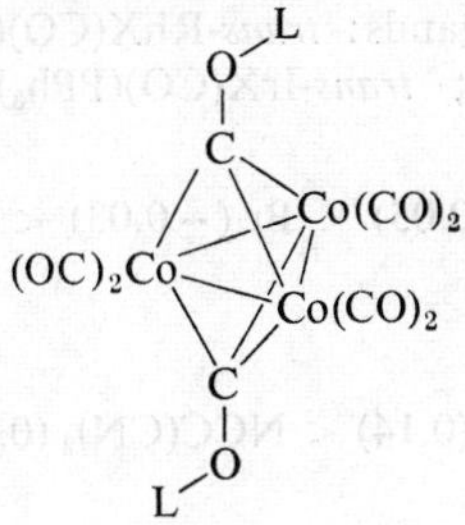

Vibrational Spectra of some Co-ordinated Ligands

(66) 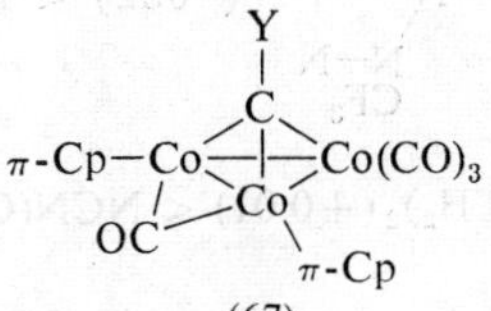 (67)

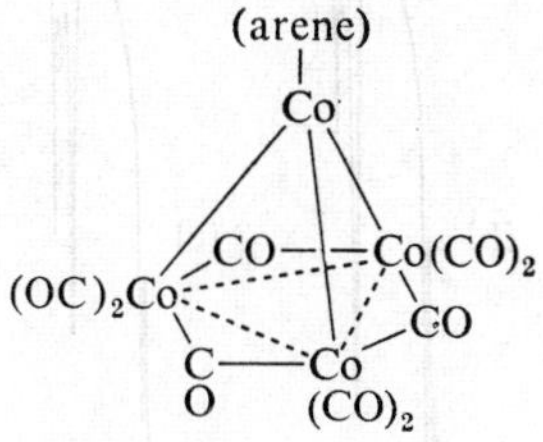

(89)

The reaction

$$[Rh(CO)_2Cl]_2 + CO + H_2O + NaHCO_3 \longrightarrow Rh_4(CO)_{12}$$

proceeds *via* a new (possibly bridging) carbonyl species, having $\nu(CO)$ at 1886 cm^{-1}.[165a]

On cooling a liquid paraffin–heptane solution of $Rh_4(CO)_{12}$ under 490 atm pressure of CO, the i.r. spectrum of a new Rh carbonyl species is observed.[165b] It is proposed that this is the bridged isomer of the previously unsubstantiated parent carbonyl $Rh_2(CO)_8$ (see Figure 2). On decreasing the pressure of CO (400, 300, and 200 atm) the new i.r. frequencies are observed but are weaker; at temperatures above $-7\,°C$, reversible disappearance of these bands occurs, consistent with the stability of the $Rh_2(CO)_8$ species only at low temperature and high CO pressure.

A careful examination of changes which occur in the $\nu(CO)$ region of the i.r. spectrum of a solution of $[Rh(CO)_2Cl]_2$ upon addition of PPh$_3$ (1 : 2 molar ratio) provides evidence that the product should be reformulated as the dimeric complex $[Rh(CO)Cl(PPh_3)]_2$.[166] This is formed *via* an initial reaction to give *cis*-$[RhCl(CO)_2(PPh_3)]_n$, where n is probably 2, followed by CO loss.

I.r. intensities of CO and CN stretching modes in a number of planar Rh^I, Ir^I, and Pt^{II} complexes have been used [167] to calculate the following

---

[165] (*a*) P. E. Cattermole and A. G. Osborne, *J. Organometallic Chem.*, 1972, **37**, C17; (*b*) R. Whyman, *J.C.S. Dalton*, 1972, 1375.
[166] D. F. Steele and T. A. Stephenson, *J.C.S. Dalton*, 1972, 2161.
[167] R. Schlodder, S. Vogler, and W. Beck, *Z. Naturforsch.*, 1972, **27b**, 462.

scales of donor ability for anionic ligands: *trans*-RhX(CO)(PPh$_3$) X = Cl (0.000), NCO (0.15), or N$_3$ (0.33); *trans*-IrX(CO)(PPh$_3$)$_2$ X = CN

$(-0.15) < -\text{N}\begin{smallmatrix}\text{N=N}\\ |\\ \text{N—N}\\ \text{CF}_3\end{smallmatrix}$ $(-0.12) < \text{I}(-0.05) < \text{Br}(-0.03) < \text{Cl}(0.00) <$

$-\text{N}\begin{smallmatrix}\text{CO}\\ \text{CO}\end{smallmatrix}(\text{CH}_2)_2(+0.01) < \text{NCN(CN)}(0.14) < \text{NCC(CN)}_2(0.15) < \text{NCO}$

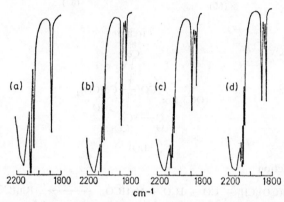

Figure 2 *The reaction of* Rh$_4$(CO)$_{12}$ *with carbon monoxide in liquid paraffin–heptane:* (a) 490 atm *pressure and* 20 °C, (b) 460 atm *and* 4 °C, (c) 445 atm *and* −7 °C, (d) 430 atm *and* −19 °C
(Reproduced from *J.C.S. Dalton*, 1972, 1375)

(0.25) < NCS (0.34) < N$_3$ (0.37); *trans*-PtX(CN)(PPh$_3$)$_2$ X = CN (−0.30) < I (−0.17) < Br (−0.05) < Cl (0.00) < N$_3$ (0.10) (negative values correspond to π-acceptor behaviour).

The reactions of Ir$_4$(CO)$_{12}$ under pressures of CO and H$_2$ have been investigated by i.r. spectroscopy, using a high-pressure cell, in order to obtain evidence for the species HIr(CO)$_4$ and Ir$_2$(CO)$_8$. No reaction is observed with CO alone but with added H$_2$ a new species could be detected at 430 atm pressure (1:1 CO–H$_2$ mixture) and 200 °C. This gave ν(CO) bands at 2054m, 2031s, 1999w cm^{-1}, very similar to HCo(CO)$_4$, and it was identified as HIr(CO)$_4$.[168]

The ions (69a) and (69b) give ν(CO) bands at 2126, 2079 cm^{-1} and 2120, 2070 cm^{-1}, respectively.[169]

Both of the complexes (70) and (71), where L = PPh$_3$, give ν(CO) absorptions in the region typical of Ir^{III} species (2030 cm^{-1}), *cf.* Ir^I analogues with ν(CO) *ca.* 1965 cm^{-1}.[170a]

[168] R. Whyman, *J.C.S. Dalton*, 1972, 2294.
[169] D. Forster, *Inorg. Chem.*, 1972, **11**, 473.
[170] (*a*) G. P. Khare and R. Eisenberg, *Inorg. Chem.*, 1972, **11**, 1385; (*b*) M. F. Koenig and M. Bigorgne, *Spectrochim. Acta*, 1972, **28A**, 1693.

Vibrational Spectra of some Co-ordinated Ligands

[structures (69a), (69b), (70), (71)]

Koenig and Bigorgne have investigated the Raman intensities of $\nu(CO)$ bands in $Ni(CO)_{4-n}L_n$ [$n = 0$; $n = 1$ or 2 for $L = PMe_3$; $n = 1, 2$, or 3 for $L = P(OMe)_3$].[170b]

$\nu(CO)$ wavenumbers have been listed for the following alkyl(fluorocarbon)phosphine nickel carbonyls:[171] $LNi(CO)_3$ $L = (CF_3)_3P$ (2116, 2062) > $CH_3P(CF_3)_2$ (2105, 2046) ≈ $EtP(CF_3)_2$ (2103, 2058) > $Me_2P(CF_3)$ (2081, 2004) > Me_3P (2064, 1982); and for $L_2Ni(CO)_2$ $L = (CF_3)_3P$ (2100, 2065) > $MeP(CF_3)_2$ (2076, 2034) ≈ $EtP(CF_3)_2$ (2074, 2031).

Two groups of workers have studied the co-condensation of Pd and/or Pt atoms with CO in inert matrices.[172, 173] Darling and Ogden[172] observed the following i.r.-active $\nu(CO)$ bands of isotopic species of $Pd(CO)_4$: T_2, $Pd(C^{16}O)_4$ 2070.3, A_1, $Pd(C^{16}O)(C^{18}O)_3$ 2047.5; A_1, $Pd(C^{16}O)_2(C^{18}O)_2$ 2037.0; A_1, $Pd(C^{16}O)_3(C^{18}O)$ 2029.0; T_2, $Pd(C^{18}O)_4$ 2022.0 cm^{-1}. Kundig et al.[173] obtained $Ni(CO)_4$, $Pd(CO)_4$, and $Pt(CO)_4$ by similar means, and investigated i.r. and Raman spectra of the matrices. The CO wavenumbers and Cotton–Kraihanzel force constants derived from them are listed in Table 3.

Table 3 *Vibrational assignments for* $M(CO)_4$ (M = Ni, Pd, or Pt)

Compound	A_1, wavenumber/ cm^{-1}	T_2, wavenumber/ cm^{-1}	k_{CO}/ mdyn Å^{-1}	k_{CO-CO}/ mdyn Å^{-1}
$Ni(CO)_4$	2130	2043	17.23	0.37
$Pd(CO)_4$	2122	2066	17.48	0.24
$Pt(CO)_4$	2119	2049	17.25	0.30

[171] D.-K. Kang and A. B. Burg, *Inorg. Chem.*, 1972, **11**, 902.
[172] J. H. Darling and J. S. Ogden, *Inorg. Chem.*, 1972, **11**, 666.
[173] P. Kündig, M. Moskovits, and G. A. Ozin, *J. Mol. Structure*, 1972, **14**, 1371.

3 Nitrogen Donors

Molecular Nitrogen, Azido-, and Related Complexes.—For a number of M—N_2 complexes, Darensbourg has obtained a straight-line relationship between $\nu(N_2)$ and the absolute integrated i.r. intensity of the absorption.[174] Thus, the i.r. intensity is largely determined by the extent of π-electron charge transfer from M to N_2 during the N_2 stretching motion.

The dinuclear complex $(\pi\text{-Cp})_2\text{TiN}_2\text{Ti}(\pi\text{-Cp})_2$, isolated from the nitrogen-fixing system $(\pi\text{-Cp})_2\text{TiCl} + \text{MeMgI} + N_2$, shows $\nu(N_2)$ at 1280 cm^{-1} in the i.r. (which shifts to 1240 cm^{-1} in the $^{15}N_2$ complex).[175] This is by far the lowest value so far reported for a dinitrogen complex, and a structure of the type (72) is suggested to account for the i.r. activity of $\nu(N_2)$.

$$Cp_2Ti\diagdown\overset{N=N}{\underset{(72)}{}}\diagup TiCp_2$$

(Benzene)chromium dicarbonyl dinitrogen, $(C_6H_6)Cr(CO)_2N_2$, is prepared by the peroxide oxidation of $(C_6H_6)Cr(CO)_2N_2H_4$.[176] $\nu(N_2)$ is found at 2145 cm^{-1}, with $\nu(CO)$ at 1941 (A_1), 1898 (B_1) cm^{-1}. [Note the decrease compared to $(C_6H_6)Cr(CO)_3$]. An analogous hexamethylbenzene complex, together with the dinuclear (hmb)$Cr(CO)_2$—N=N—(CO)-Cr(hmb) [the latter shows no $\nu(N_2)$ in the i.r.], were also reported. A more complete tabulation of similar data has been published by the same workers[177] (benzene, mesitylene, and hexamethylbenzene complexes – mononuclear and dinuclear for the last two).

$\nu(N_2)$ frequencies have been listed for $Mo(N_2)_2(L-L)_2$ [L—L = $(Ph_2PCH_2)_2$, $(Ph_2AsCH_2)_2$, or $Ph_2AsCH_2CH_2PPh_2$].[178] A weak band is found at *ca.* 2040 cm^{-1}, and a strong absorption at *ca.* 1970 cm^{-1} in the i.r. The diphos compound is known to possess a *trans*-structure, and the others are believed to be the same, the weak, higher-frequency band being the Raman-active mode. Although the $\nu(N_2)$ values are all rather similar within this series, the chemical stabilities differ markedly.

A rather similar series of complexes has been studied by Hidai *et al.*, viz. $Mo(N_2)_2[Ph_2P(CH_2)_nPPh_2]_2$, where $n = 1$, 2, or 3.[179a] $\nu(N_2)$ moves to lower wavenumber as the chain length of the chelating ligand increases (1995, 1970, 1925 cm^{-1} for $n = 1$, 2, and 3 respectively), possibly suggesting increased electron-donating power with increased n.

Dinitrogen complexes of Mo^I have been prepared by the oxidation of $Mo(N_2)_2(diphos)_2$, *e.g.* in $[Mo(N_2)_2(diphos)_2]^+$ I_3^-, $\nu(N\equiv N)$ is found at

---

[174] D. J. Darensbourg, *Inorg. Chem.*, 1972, **11**, 1436.
[175] Yu. G. Borodko, I. N. Ivleva, L. M. Kachapina, S. I. Salienko, A. K. Shilova, and A. E. Shilov, *J.C.S. Chem. Comm.*, 1972, 1178.
[176] D. Sellmann and G. Maisel, *Z. Naturforsch.*, 1972, **27b**, 465.
[177] D. Sellmann and G. Maisel, *Z. Naturforsch.*, 1972, **27b**, 718.
[178] T. A. George and C. D. Seibold, *Inorg. Nuclear Chem. Letters*, 1972, **8**, 465.
[179] (*a*) M. Hidai, K. Tominari, and Y. Uchida, *J. Amer. Chem. Soc.*, 1972, **94**, 110; (*b*) T. A. George and C. D. Seibold, *ibid.*, p. 6859.

2043 cm^{-1}, cf. 1976 cm^{-1} in the neutral precursor. This increase is consistent with an increased formal oxidation state of the Mo.[179b]

[B(pz)$_4$](π-Cp)Mo(CO)$_2$, where B(pz)$_4$ = tetrakis(pyrazolyl) borate, gives four ν(CO) bands in solution (at 1950, 1935, 1865, 1845 cm^{-1} in C$_6$D$_5$CD$_3$ solution).[180] This is explained on the basis of the presence of two conformers (73a) and (73b).

(73a)   (73b)

The complex WCl(diphos)$_2$(N$_2$COR), prepared from trans-[W(N$_2$)$_2$-(diphos)$_2$] by successive treatment with the RCOCl and Me$_3$N, gives ν(N$_2$) at 1338 cm^{-1} (1298 cm^{-1} in the ^{15}N$_2$ derivative).[181]

Series of compounds [WX$_2$(N$_2$H$_2$)(diphos)$_2$], X = Cl or Br, and [WX(N$_2$H$_2$)(diphos)$_2$]$^+$Y$^-$, X = Cl or Br, Y = ClO$_4$ or BPh$_4$, have been prepared, with N-deuteriated, ^{15}N, Et$_2$PCH$_2$CH$_2$PEt$_2$, and Mo analogues.[182] ν(NH), and ν(ND) were listed. The cations [WH(N$_2$)$_2$(diphos)$_2$]$^+$ and [WH(^{15}N$_2$)(diphos)$_2$]$^+$ gave ν(N$_2$) at 1995, 1935 cm^{-1}, respectively.

The di-imine complex (74) is assigned the trans-structure on the basis of the non-observance of ν(N=N) in the i.r. spectrum.[183] ν(NH) (3250 cm^{-1}) and δ(NH) (1338 cm^{-1}) were seen, however, with ν(CO) at 1880 and 1915 cm^{-1} (in benzene solution).

(74)

(π-Cyclopentadienyl)rhenium dicarbonyl dinitrogen, (h^5-C$_5$H$_5$)Re(CO)$_2$-(N$_2$), gives ν(N$_2$) at 2141 and ν(CO) at 1970(A_1) and 1915(B_1) cm^{-1}.[184a] Interaction of a dinitrogen complex of Re^I, Re^I(PhMe$_2$P)$_4$(N$_2$)Cl, with Re^V(O)Cl$_2$(OMe)(PPh$_3$)$_2$ in solution yields an equilibrium mixture containing the bridged species Re^I—N=N—Re^V.[184b] In the initial complex ν(N$_2$) is at 1921 cm^{-1}, and this drops to 1837 cm^{-1} in the adduct.

AlMe$_3$ forms adducts with transition-metal carbonyls via the oxygen atom – Chatt et al. have shown that a similar process can occur with dinitrogen complexes, leading to a lowering of ν(N$_2$).[185] Thus in trans-

[180] J. L. Calderon, F. A. Cotton, and A. Shaver, J. Organometallic Chem., 1972, 37, 127.
[181] J. Chatt, G. A. Heath, and G. J. Leigh, J.C.S. Chem. Comm., 1972, 444.
[182] J. Chatt, G. A. Heath, and R. L. Richards, J.C.S. Chem. Comm., 1972, 1010.
[183] D. Sellman, J. Organometallic Chem., 1972, 44, C47.
[184] (a) D. Sellmann, J. Organometallic Chem., 1972, 36, C27; (b) D. J. Darensbourg, Inorg. Chim. Acta, 1972, 6, 527.
[185] J. Chatt, R. H. Crabtree, and R. L. Richards, J.C.S. Chem. Comm., 1972, 534.

[ReCl(N$_2$)(PMe$_2$Ph)$_4$], ν(N$_2$) falls from 1923 to 1894 cm^{-1} on the formation of an AlMe$_3$ adduct.

[ReCl$_3$(NCOPh)(PPh$_3$)$_2$] has ν(C=O) at 1520 cm^{-1}.[186]

Assignments for ν(N≡N) have been proposed as follows:[187a] FeH$_2$-(^{14}N≡^{14}N)(PPh$_3$)$_3$ 2074; FeH$_2$(^{14}N≡^{15}N)(PPh$_3$)$_3$ 2042; FeH$_2$(^{15}N≡^{15}N)-(PPh$_3$)$_3$ 2008; FeH$_2$(^{14}N≡^{14}N)(PPh$_2$Et)$_3$ 2047; FeH$_2$(^{14}N≡^{14}N)(PPhEt$_2$)$_3$ 2020 cm^{-1}.

In [FeH(N$_2$)(diphos)$_2$]$^+$, as BPh$_4^-$ or ClO$_4^-$ salts, ν(N$_2$) occurs in the 2120—2130 cm^{-1} region.[187b]

The product (75) of the first reported insertion of chlorosulphanyl isocyanate into a transition-metal–carbon bond has ν(C≡O) at 2075, 2028 cm^{-1}; ν(C=O) at 1670 cm^{-1}; ν(SO) at 1353, 1133 cm^{-1}.[188]

$$(\pi\text{-Cp})-\text{Fe}\begin{matrix}\text{OC}\\ \\ \text{OC}\end{matrix}\!\!-\!\!\text{N}\!\!-\!\!\overset{\overset{\text{O}}{\|}}{\text{C}}\!\!-\!\!\text{CH}_2\!\!-\!\!\text{C}\begin{matrix}\text{Me}\\ \\ \text{CH}_2\end{matrix}$$
$$\qquad\qquad\ \ \text{SO}_2\text{Cl}$$
(75)

(Ph$_3$P)$_3$Ru(N$_2$)H$_2$ gives ν(N≡N) at 2147 cm^{-1}.[189]

The i.r. spectra of *trans*-[RuCl(NO)(das)$_2$]$^{2+}$, *trans*-[RuI(NO)(das)$_2$]$^{2+}$, *trans*-[RuCl(N$_2$)(das)$_2$]$^+$, and *trans*-[RuCl(CO)(das)$_2$]$^+$, das = *o*-phenylenebis(dimethylarsine), have been obtained (250—4000 cm^{-1}).[190a] ^{15}N substitution of the NO,N$_2$ ligands assisted in the assignment of RuXY modes. Force constants (using a three-body model) for Ru—X—Y were calculated, and on the basis of these the band at *ca.* 490 cm^{-1} in the N$_2$ complex was reassigned as a Ru—N—N bend (not a Ru—N stretch). All of the results are consistent with Ru—(XY) $d_\pi p_\pi$-bonding.

The osmium dinitrogen complex *cis*-[Cl(NH$_3$)$_4$OsNNOs(NH$_3$)$_5$]$^{4+}$ gives ν(N≡N) at 2000 cm^{-1} (i.r.), 1999 cm^{-1} (Raman), while *cis*-[{Cl(NH$_3$)$_4$Os}$_2$-N$_2$]$^{3+}$ has no i.r. band in this region (1995 cm^{-1} in the Raman), showing the equivalence of the two Os atoms on the vibrational time-scale.[190b]

Pentacyanonitrosylcobalt(III), once considered as containing NO$^-$, was shown by J. B. Raynor [*J. Chem. Soc.* (*A*), 1966, 997] to be [(NC)$_5$Co-(N$_2$O$_2$)Co(CN)$_5$]$^{6-}$, with a bridging hyponitrite group. This work has now been repeated,[191] with no acknowledgement, and only minor additions.

K$_2$[Rh$_2$(OH$_2$)(NO$_2$)$_6$(NH$_3$)$_2$(N$_2$)] shows a Raman line at 2070 cm^{-1} [which shifts to 2045 cm^{-1} when (^{15}NH$_4$)$_2$SO$_4$ is used in the preparation],

---

[186] J. Chatt and J. R. Dilworth, *J.C.S. Chem. Comm.*, 1972, 549.
[187] (*a*) Yu. G. Borodko, M. O. Broitman, L. M. Kachapina, A. K. Shilova, and A. E. Shilov, *J. Struct. Chem.*, 1971, **12**, 498; (*b*) P. Giannocaro, M. Rossi, and A. Sacco, *Co-ordination Chem. Rev.*, 1972, **8**, 77.
[188] Y. Yamamoto and A. Wojcicki, *J.C.S. Chem. Comm.*, 1972, 1088.
[189] W. H. Knoth, *J. Amer. Chem. Soc.*, 1972, **94**, 104.
[190] (*a*) M. S. Quinby and R. D. Feltham, *Inorg. Chem.*, 1972, **11**, 2468; (*b*) R. N. Magnuson and H. Taube, *J. Amer. Chem. Soc.*, 1972, **94**, 7213.
[191] B. Jezowska-Trzebiatowska, J. Hanuza, M. Ostern, and J. Ziólkowski, *Inorg. Chim. Acta*, 1972, **6**, 141.

while there is no significant i.r. absorption at this position.[192] Hence a linear Rh—N≡N—Rh group is present.

v(N≡N) in IrX(N$_2$)(PPh$_3$)$_2$ is a 2095 cm^{-1} (X = Cl, Br, or I), and v(NO) in [IrX(NO)(PPh$_3$)$_2$]$^+$ is at 1902 (X = Cl or Br) or at 1895 (X = I) cm^{-1}.[193]

The complex (76) shows v(N$_2$) at *ca*. 1415 cm^{-1}, substantially lower than the value (*ca*. 2300 cm^{-1}) in the corresponding diazonium salt.[194] A wider

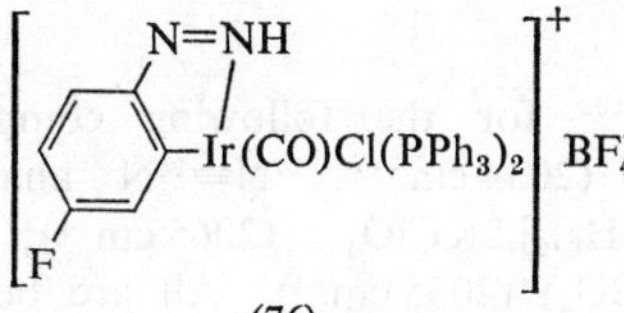

[structure (76): benzene ring with F substituent, N=NH group, and —Ir(CO)Cl(PPh$_3$)$_2$, with BF$_4^-$ counterion, overall cationic]

[structure (77): benzene ring with X substituent, fused with N=N—N linkage to IrCl(CO)(PPh$_3$)$_2$]

(76) 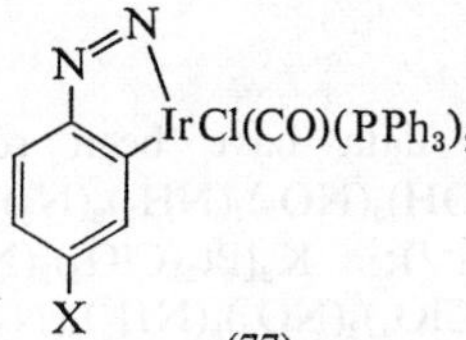   (77)

range of this type of complex[195] gives very similar results, while (77; X = F or Br) possesses v(N=N) at 1450 cm^{-1}.

Ni(CO)$_3$(N$_2$), produced by photolysis of Ni(CO)$_4$ with N$_2$ at 20 K in a N$_2$ matrix, gives the following i.r. wavenumbers: Ni(CO)$_3$(^{14}N$_2$), v(CO) 2027/2031 (*E*), 2098 (*A*$_1$) cm^{-1}, v(N$_2$), 2266 (*A*$_1$) cm^{-1}; Ni(CO)$_3$(^{15}N$_2$), v(CO) 2027/2031 (*E*), 2096 (*A*$_1$) cm^{-1}, v(N$_2$) 2193 (*A*$_1$) cm^{-1}.[196] The intensity of the v(N$_2$) band is consistent with the structure (78), although the decrease from free N$_2$ is rather small, *i.e.* the Ni—N$_2$ bond is weak.

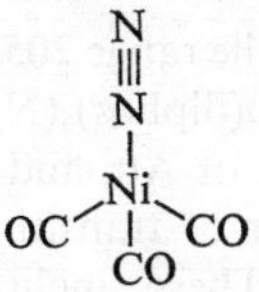

N
|||
N
|
Ni
OC / | \ CO
CO

(78)

Diazoalkane-nickel(0) complexes (R$_2$CN$_2$)NiL$_2$, where R$_2$C = fluorenylidene, Ph$_2$C, or (NC)$_2$C and L = Bu^{t}NC; R$_2$C = fluorenylidene, L$_2$ = cod, show a strong i.r. band at 1520—1540 cm^{-1} due to v(C=N=N). This excludes linear end-on co-ordination involving a lone pair on the terminal nitrogen (which would give a band at *ca*. 2000—2200 cm^{-1}).[197]

Co-condensation of Pd atoms with ^{14}N$_2$ at 4.2—10 K leads to formation of a binary complex, believed to be Pd(N$_2$)$_3$. Using isotopic variants of

[192] L. S. Volkova, V. M. Volkov, and S. S. Chernikov, *Russ. J. Inorg. Chem.*, 1971, **16**, 1383.
[193] R. J. Fitzgerald and H. M. W. Lin, *Inorg. Chem.*, 1972, **11**, 2270.
[194] A. B. Gilchrist, G. W. Rayner-Canham, and D. Sutton, *Nature*, 1972, **235**, 42.
[195] F. W. B. Einstein, A. B. Gilchrist, G. W. Rayner-Canham, and D. Sutton, *J. Amer. Chem. Soc.*, 1972, **94**, 645.
[196] A. J. Rest, *J. Organometallic Chem.*, 1972, **40**, C76.
[197] S. Otsuka, A. Nakamura, T. Koyama, and Y. Tatsumo, *J.C.S. Chem. Comm.*, 1972, 1105.

N_2, shifts in $\nu(N{\equiv}N)$ were observed, and shown to agree closely with those calculated for $Pd(N_2)_3$.[198]

Unassigned i.r. data have been listed for several 1,2,3-benzotriazole (79) complexes of Pd^{II}.[199]

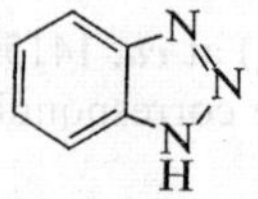

(79)

$\nu(N_2)$ data have been reported[200] for the following complexes: $K_2[Pt_2(OH)_4(NO_2)_4(NH_3)_2(N_2)],2H_2O$ (2034 cm^{-1}; $^{15}N{\equiv}^{14}N$ analogue, 2009 cm^{-1}); $K_2[Pt_2(ClO_4)_2(NO_2)_6(NH_3)_2],2KClO_4$ (2065 cm^{-1}); and $K_2[Pt_2(ClO_4)_2(NO_2)_6(NH_3)_2(N_2)],\frac{1}{2}(K_2SO_4)$ (2055 cm^{-1}). All are believed to contain a linear Pt—N≡N—Pt grouping.

The reaction of $Pt(Ph_3P)_n$ ($n = 3$ or 4) with NO gives a monomeric hyponitrite complex (80), $Pt(N_2O_2)(Ph_3P)_2$. It shows i.r. bands at 1285, 1240, and 1062 cm^{-1}.[201]

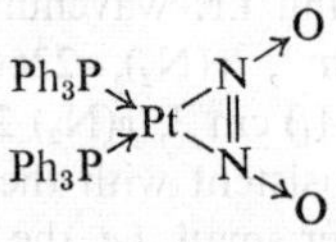

(80)

$\nu_{as}(N_3)$ has been reported (in the range 2053—2090 cm^{-1}) for $M(PPh_3)_2N_3$, M = Cu, Ag, or Au, and $Cu(diphos)_3(N_3)_2$, together with $\nu_{as}(NCS)$ in $M(PPh_3)_2(NCS)$, M = Cu, Ag, or Au, and $Ag[P(OEt)Ph_2]_2NCS$.[202]

The i.r. spectra of a number of transition-metal azide complexes have been studied by Agrell.[203] These include $Cu(N_3)_2$, $[Cu(N_3)_2(NH_3)_2]$, $[Cu(N_3)_2(py)_2]$, $[Zn(N_3)_2(NH_3)_2]$, $[Zn(N_3)_2(py)_2]$, and $[Cd(N_3)_2(py)_2]$. Assignments were proposed, where appropriate, for the NH_3 and py vibrations, and a correlation was found between the value of ν_3 of the N_3 group and the degree of asymmetry of that group – increased asymmetry (determined crystallographically) gave rise to a higher value of ν_3.

Wavenumbers for the azido-group vibrations in four new diarylthallium(III) azides are listed:[204] $\nu_{as}(N_3)$ 2000—2200 cm^{-1}; $\nu_s(N_3)$, 1325—1330 cm^{-1}; $\delta(N_3)$ 648—652 cm^{-1}.

[198] G. A. Ozin, M. Moskovits, P. Kündig, and H. Huber, *Canad. J. Chem.*, 1972, **50**, 2385.
[199] Y. Watanabe, I. Mitsudo, M. Tanaka, K. Yamamoto, and Y. Takegami, *Bull. Chem. Soc. Japan*, 1972, **45**, 925.
[200] V. M. Volkov and L. S. Volkova, *Russ. J. Inorg. Chem.*, 1971, **16**, 1382.
[201] S. Cenini, R. Ugo, G. LaMonica, and S. D. Robinson, *Inorg. Chim. Acta*, 1972, **6**, 182.
[202] R. F. Ziolo, J. A. Thich, and Z. Dori, *Inorg. Chem.*, 1972, **11**, 626.
[203] I. Agrell, *Acta Chem. Scand.*, 1971, **25**, 2965.
[204] T. N. Srivastava and K. K. Bajpai, *J. Inorg Nuclear Chem.*, 1972, **34**, 1458.

The i.r. spectrum of bis(benzalhydrazono)tin(IV) chloride is consistent with the structure (81).[205]

$$\begin{array}{c} H_2N-N=CH-C_6H_5 \\ | \\ Cl\diagdown_{Sn}\diagup Cl \\ Cl\diagup \diagdown Cl \\ | \\ C_6H_5-HC=N-NH_2 \\ (81) \end{array}$$

Amines and Related Ligands.—ν(NH) has been reported as follows: LiClO$_4$(cyclam) 3295; (LiBr)$_2$(cyclam) 3250; (LiI)$_2$(cyclam) 3250; (LiClO$_4$)$_2$-(cyclam) 3275, 3155 (the latter being due to hydrogen-bonded species); free cyclam 3260, 3185 (all wavenumbers/cm^{-1})[206a] (cyclam = 1,4,8,11-tetra-azacyclo-tetradecane).

The i.r. and Raman spectra of LiNO$_3$,2NH$_3$ and LiNO$_3$,4NH$_3$ at -180 °C can be interpreted in terms of [Li(NH$_3$)$_n$]$^+$ complex ions with different Li$\cdots$N distances in the crystalline structure.[206b] ν(Li$^+\cdots$N) are in the range 460—570 cm^{-1} (NH$_3$), 400—500 cm^{-1} (ND$_3$). Similar data were also reported on the closely similar systems NaX,nNH$_3$ (X = Br or I).[206c]

The complexes MX$_4$L, MX$_4$L$_2$ (M = Ti or Sn; X = Cl, Br, or I; L = N-allylthiourea) involve co-ordination via N to the metals. Wavenumbers were listed for ν(NH), ν(C=C), and several skeletal bands of the ligand.[207]

I.r. wavenumbers have been listed (down to 400 cm^{-1}) but not assigned for CrCl$_3$L$_3$, where L = 2-, 3-, or 4-cyanopyridine.[208a]

The i.r. spectra of tris-(2,2'-bipyridyl) complexes of Cr, V, and Ti in low oxidation states have been reported.[208b] Trends in ligand fundamentals with changes in metal formal oxidation states were discussed.

In the complexes fac-M(CO)$_3$T and [M(CO)$_3$(T)I]$^+$, where M = Mo or W; T = bis-(2-pyridylmethyl)amine, bis-(2-pyridylmethyl)methylamine, or bis-(2-pyridylethyl)amine, the presence of just two bands in the 1550—1650 cm^{-1} region (ν_{8a}, ν_{8b} of the pyridine ring) indicates that both pyridine rings are co-ordinated to the metal.[209] ν(CO) values were also listed.

Approximate assignments have been proposed by Behrens et al. for the following complexes: Mn(CO)$_4$(NH$_3$)(CONH$_2$), Mn(CO)$_3$(PPh$_3$)(NH$_3$)-(CONH$_2$), and Mn(CO)$_3$(NH$_3$)$_2$(CN).[210]

---

[205] C. H. Stapfer, R. W. D'Andrea, and R. H. Herber, *Inorg. Chem.*, 1972, **11**, 204.
[206] (a) D. E. Fenton, C. Nave, and M. R. Truter, *J.C.S. Chem. Comm.*, 1972, 1303; (b) A. Regis and J. Corset, *J. Chim. phys.*, 1972, **69**, 707; (c) A. Regis, J. Limouzi, and J. Corset, *ibid.*, p. 696.
[207] R. P. Singh and I. M. Pande, *J. Inorg. Nuclear Chem.*, 1972, **34**, 1131.
[208] (a) J. C. Chang, M. A. Haile, and G. R. Keith, *J. Inorg. Nuclear Chem.*, 1972, **34**, 360; (b) E. König and E. Lindner, *Spectrochim. Acta*, 1972, **28A**, 1393.
[209] J. G. Dunn and D. A. Edwards, *J. Organometallic Chem.*, 1972, **36**, 153.
[210] H. Behrens, E. Lindner, D. Martens, P. Wild, and R. J. Lampe, *J. Organometallic Chem.*, 1972, **34**, 367.

Some amine ligand vibrations have been assigned in Mn(acac)$_2$L$_2$, where L$_2$ = en or L = propylamine, allylamine, methylallylamine, or *trans*-crotylamine.[211]

Mn^{II}, Fe^{II}, and Ni^{II} complexes of the onium ion of (82) show i.r. spectra which contain, in addition to free ligand bands, a number of intense

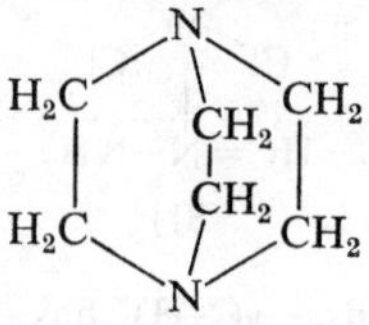

<chemical structure of (82): a bicyclic diamine with two N atoms bridged by CH$_2$ groups>

(82)

absorptions due to ν(N$^+$—H) (3000—2250 cm^{-1}) and to N$^+$—H deformations (probably coupled with CH$_2$ deformations *etc.*) (1600—700 cm^{-1}).[212]

A 1:1 adduct of Mn(acac)$_2$ and allylamine is shown to be a dimer in Et$_2$O solution.[213] I.r. bands associated with the NH$_2$ group are at 3340, 3240, 3160, and 1553 cm^{-1}. These represent shifts to lower wavenumbers of *ca.* 40 cm^{-1} compared to the free-ligand values, showing that the NH$_2$ group is co-ordinated. ν(C=C), however, is unshifted, and a single-crystal *X*-ray study confirms the presence of bridging acac ligands and unidentate allylamines.

The i.r. spectra of Re$_3$Cl$_9$(bipy)$_{1.5 \text{ or } 2}$ and Re$_3$Cl$_8$(bipy)$_2$ are all very similar.[214] In addition to characteristic bands of co-ordinated 2,2'-bipyridyl, bands due to the 2,2-bipyridinium cation were seen at 1528, 990, and 887 cm^{-1}.

trans-ReCl$_3$(NMe)(PPh$_2$R)$_2$, where R = Me, Et, or Ph, have an i.r. band at *ca.* 1310 cm^{-1} which appears to be associated with the NMe group.[215]

Unassigned i.r. spectra have been listed for [(MeCN)$_4$Fe(*trans*-tetramine)]$^{2+}$, [Fe(*trans*-tetramine)]$^{2+}$, [(MeCN)$_2$Fe(*trans*-tetramine)]$^{4+}$, and [HFe(*trans*-tetramine)]$^+$, where *trans*-tetramine = (83).[216]

The complex (84) gives ν(C≡O) at 2082, 2046, 2002, 1991 cm^{-1}, ν(C=O) at 1729 cm^{-1} (R = n-C$_3$H$_7$); ν(C≡O) at 2082, 2045, 1998 cm^{-1}, ν(C=O) at 1716 cm^{-1} (R = CMe$_3$).[217]

I.r. bands have been listed for the purpose of the characterization of Fe(mephen)$_2$X$_2$, where X = NCS, N$_3$, or CN and mephen = 2-methyl-1,10-phenanthroline.[218]

---

[211] Y. Nishikawa, Y. Nakamura, and S. Kawaguchi, *Bull. Chem. Soc. Japan*, 1972, **45**, 155.
[212] L. M. Vallorino, V. L. Goedken, and J. V. Quagliano, *Inorg. Chem.*, 1972, **11**, 1466.
[213] S. Koda, S. Ooi, H. Kuroya, Y. Nishikawa, Y. Nakamura, and S. Kawaguchi, *Inorg. Nuclear Chem. Letters*, 1972, **8**, 89.
[214] D. G. Tisley and R. A. Walton, *Inorg. Chem.*, 1972, **11**, 179.
[215] J. Chatt, R. J. Dosser, and G. J. Leigh, *J.C.S. Chem. Comm.*, 1972, 1243.
[216] D. C. Olson and J. Vasilevskis, *Inorg. Chem.*, 1972, **11**, 980.
[217] H. Alper, *Inorg. Chem.*, 1972, **11**, 976.
[218] E. König, G. Ritter, K. Madeja, and A. Rosenkranz, *J. Inorg. Nuclear Chem.*, 1972, **34**, 2877.

Vibrational Spectra of some Co-ordinated Ligands

```
        CH₂
MeHC⁄      ⁄CMe₂
    |       |
    ⁄NH  HN⁄
H₂C         CH₂
|           |
H₂C         CH₂
    ⁄NH  HN⁄
    |       |                            O
Me₂C⁄      ⁄CHMe                         ‖
        CH₂                    R—N⁄   ⁄N—R
        (83)                        ⁄C⁄
                              (OC)₃Fe----Fe(CO)₃
                                      (84)
```

Unassigned i.r. bands have been listed for [M(pccbf)]⁺BF₄⁻, where M = Fe, Ni, or Zn and pccbf = fluoroborotris-(2-aldoximo-6-pyridyl)-phosphine. The cations are believed to possess trigonal-prismatic co-ordination about the metal.[219]

cis-Geometry is deduced for the complexes [Co(en)$_2$(RNH$_2$)Cl]Cl$_2$, where R = Bu^n, Bu^i, or Bu^s, from the observation of two CH$_2$ rocking modes in their i.r. spectra (at 879—871, 899—890 cm^{-1}).[220a]

Bands in the 600—400 cm^{-1} region of the i.r. spectra for trans-[CoX$_2$-(en)$_2$]X (X = Cl, Br, or I) have been attributed to ν(Co—N) and chelate deformation modes.[220b]

I.r. data were recorded for o-aminophenylarsenic acid complexes of Co, Ni, Cu, Zn, and Cd.[220c]

It has been shown that i.r. spectra in the 800—950 cm^{-1} and 2800—3000 cm^{-1} regions are useful for differentiating fac- and mer-co-ordination of dien in octahedral Co(dien)$_2^{3+}$.[221]

ν(NH) is at ca. 3220 cm^{-1} (cf. 3272 cm^{-1} in the free ligand) and ν(CN) is unshifted (ca. 2260 cm^{-1}) in CoL$_4^{2+}$, NiL$_4^{2+}$, and NiL$_6^{4+}$, where L = hydrogen cyanamide, H$_2$NCN.[222]

Selected i.r. wavenumbers have been listed for the characterization of the amino-complexes Co(oxalate)L$_2$, where L = py, aniline, isoquinoline, or ½(phen).[223]

Co^{II}, Ni^{II}, and Zn^{II} complexes of NN'N"-tris-(2-picolyl)-cis,cis-1,3,5-triaminocyclohexane [= (pcc)$_3$tach] are basically octahedral. They give ν(NH) in the range 3205—3275 cm^{-1}, with the acyclic CH$_2$ bending mode between 1455 and 1465 cm^{-1}.[224]

I.r. spectra have been listed for 7 isomeric forms of [Co(tmd)(dien)Cl]-ZnCl$_4$,xH$_2$O and 6 of [Co(tmd)(dpt)Cl]ZnCl$_4$,xH$_2$O, where tmd = 1,3-

---

[219] J. E. Parks, R. Wagner, and R. H. Holm, Inorg. Chem., 1971, 10, 2472.
[220] (a) S. C. Chan and K. M. Chan, Z. anorg. Chem., 1972, 389, 205; (b) M. Nakahara and M. Mitsuya, Bull. Chem. Soc. Japan, 1972, 45, 2209; (c) I. S. Maslennikova and V. N. Shemyakin, Zhur. fiz. Khim., 1972, 46, 1004.
[221] F. R. Keene and G. H. Searle, Inorg. Chem., 1972, 11, 148.
[222] W. C. Wolsey, W. H. Huestis, and T. W. Theyson, J. Inorg. Nuclear Chem., 1972, 34, 2358.
[223] G. P. Singh, P. R. Shukla, and R. N. Srivastava, J. Inorg. Nuclear Chem., 1972, 34, 3251.
[224] R. A. D. Wentworth, Inorg. Chem., 1971, 10, 2615.

diaminopropane; dpt = dipropylenetriamine; dien = diethylenetriamine.[225]

The i.r. spectra of [Co(dh)$_2$(sam)$_2$]NO$_3$, where dh = dimethylglyoximato and sam = p-NH$_2$·C$_6$H$_4$·SO$_2$NHR (various R), are said to indicate that the sulphanilamide ligands are co-ordinated *via* the p-NH$_2$ group.[226]

The shifts in ν(NH) on co-ordination of allylamine with MSO$_4$ (M = Co^{II}, Ni^{II}, Cu^{II}, or Zn^{II}) have been correlated with the heats of formation of the crystalline complexes, and related to the Irving–Williams series.[227]

A listing has been made of free- and co-ordinated-ligand i.r. wavenumbers for Co^{II}, Ni^{II}, and Cu^{II} halide derivatives of 6-methyl-2,3-di-(6-methyl-2-pyridyl)quinoxaline (85; R^1 = Me, R^2 = H) and 6,7-dimethyl-2,3-di-(6-methyl-2-pyridyl)quinoxaline (85; R^1 = R^2 = Me).[228]

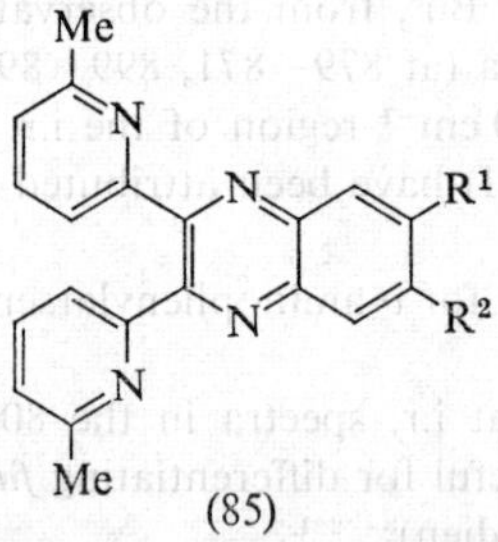

(85)

I.r. wavenumbers (to 250 cm^{-1}) have been reported for MCl$_2$L$_2$ (M = Co, Ni, Cu, Zn, or Cd) and M(SO$_4$)L$_2$ (M = Cu or Cd), where L = p-aminoazobenzene.[229] No attempt at assignment was made.

Some ligand vibrations have been assigned for RhCl$_3$L$_3$, where L = 4-CN-, 3-CN-, 4-Et-, or 3-Et-py, and for [RhCl$_2$L$_4$]$^+$, where L = 4-NH$_2$- or 3-NH$_2$-py.[230] These are all consistent with unidentate co-ordination of the pyridine ligand *via* the ring nitrogen.

ν(NH) wavenumbers for about 20 complexes of Rh^{III} with *C-rac*- and *C-meso*-5,5,7,12,12,14-hexamethyl-1,4,8,11-tetra-azacyclotetradecane have been reported.[231]

The Ir(en)$_2$(NH$_3$)$_2^{3+}$ ion, produced by the reduction of the chloramine complex Ir(en)$_2$(NH$_2$Cl)$_2^{3+}$, is the *trans*-isomer, since only one i.r. band due

---

[225] A. R. Gainsford and D. R. House, *Inorg. Chim. Acta*, 1972, **6**, 227.
[226] V. N. Shafromskii and J. L. Fusii, *Russ. J. Inorg. Chem.*, 1971, **16**, 1171.
[227] M. S. Barvinok, Yu. B. Kalugin, and L. A. Obozova, *Russ. J. Inorg. Chem.*, 1971, **16**, 1617.
[228] D. F. Colton and W. J. Geary, *J.C.S. Dalton*, 1972, 547.
[229] L. V. Kinovalov, I. S. Maslennikova, and V. N. Shemyakin, *Russ. J. Inorg. Chem.*, 1971, **16**, 1528.
[230] C. McRobbie and H. Frye, *Austral. J. Chem.*, 1972, **25**, 893.
[231] N. F. Curtis and D. F. Cook, *J.C.S. Dalton*, 1972, 691.

to the NH_2 symmetric deformation, and one due to CH_2 rocking, were seen.[232a]

Some i.r. data have been published relating to $[Ni(en)_3]X_2$ (X = Cl⁻, Br⁻, I⁻, or $\frac{1}{2}S_2O_3^{2-}$),[232b] to γ-picoline complexes obtained from $Ni(ClO_4)_2$,[232c] and to $Ni(diethanolamine)_2X_2$ and $[Ni(diethanolamine)(H_2O)X]$, (X = Cl, Br, or NO_3).[232d]

I.r. spectra have been shown for $[Ni(acac)(tetramen)]^+$, $[Ni(acac)(tetramen)(OH_2)]^+$, and $[Ni(acac)(tetramen)NO_3]$, where tetramen = $NNN'N'$-tetramethylethylenediamine.[233]

Negative shifts in the wavenumbers of $\nu(NH_2)$ and $\nu(NH)$ of triethylenetetramine upon formation of Ni^{II} and Cu^{II} complexes show that all of the amino-groups of this ligand are co-ordinated to the metal.[234]

A partial assignment has been proposed[235] for ligand vibrations in $[Ni(pyDPT)X]X$ and $[Ni(pyDPT)Y]PF_6$ {X = Cl, Br, I, NO_3, or SCN; Y = Cl, Br, NO_3, or SCN; pyDPT = $(2\text{-}C_5H_4N)CH=N(CH_2)_3NH(CH_2)_3\text{-}N=CH(2\text{-}C_5H_4N)$}.

The trimethylhydrazinium cation $H_2\overset{+}{N}NMe_3$ co-ordinates to Ni^{II}, forming yellow, paramagnetic complexes $NiCl\{H_2NNMe_3\}_2$ with tetragonally distorted octahedral geometry. On heating to 145 °C, the co-ordination number changes reversibly, with formation of $[H_2NNMe_3]_2^{2+}\text{-}[NiCl_4]^{2-}$. The NH_2 vibrations change markedly on co-ordination to the metal, e.g. $\nu(NH_2)$ is lowered by ca. 100 cm⁻¹, $\delta(NH_2)$ is lowered by 10—20 cm⁻¹, $\rho_t(NH_2)$ is raised by 20 cm⁻¹, and $\rho_w(NH_2)$ is raised by ca. 100 cm⁻¹. In addition, $\rho_r(NH_2)$ appears as a new, medium-strong i.r. band (ca. 600 cm⁻¹), although $\nu(Ni-N)$ could not be detected.[236]

Analysis of i.r. bands attributable to vibrations of co-ordinated primary and secondary amino-groups has proved useful in studying complexes of Ni^{II} and Cu^{II} with polydentate nitrogen-containing ligands produced by the reaction of acetone with $bis(diaminoethane)M^{II}$ (M = Ni or Cu).[237]

Determination of the crystal structure of $[Ni(en)_2(H_2O)(BF_4)]BF_4$ confirms the presence of a unidentate co-ordinated BF_4^- unit. I.r. evidence was ambiguous, however, although features not attributable to free BF_4^- are ν_1 at 765 cm⁻¹, ν_4 (split) at 516, 521 cm⁻¹, and ν_3 at 1050 cm⁻¹ (signs of splitting). ν_2 could not be detected.[238]

Unassigned i.r. data have been given for two isomeric forms of $Pd(2\text{-chloropropane-1,3-diamine})Cl_2$.[239]

---

[232] (a) T. R. Weaver, B. C. Lane, and F. Basolo, *Inorg. Chem.*, 1972, **11**, 2277; (b) J. Csàszar, *Magyar Kém. Folyóirat.*, 1972, **78**, 219; (c) V. M. Bhatnagar, *Rev. Roumine Chim.*, 1972, **17**, 477; (d) M. N. Hughes, B. Waldron, and K. J. Rutt, *Inorg. Chim. Acta*, 1972, **6**, 619.
[233] Y. Fukuda and K. Soue, *J. Inorg. Nuclear Chem.*, 1972, **34**, 2315.
[234] E. Cara, A. Cristini, A. Diaz, and G. Ponticelli, *J.C.S. Dalton*, 1972, 527.
[235] C. T. Spencer, *Inorg. Chem.*, 1971, **10**, 2407.
[236] V. L. Goedken, L. M. Vallarino, and J. V. Quagliano, *Inorg. Chem.*, 1971, **10**, 2682.
[237] N. F. Curtis, *J.C.S. Dalton*, 1972, 1357.
[238] A. A. G. Tomlinson, M. Bonamico, G. Dessy, V. Fares, and L. Scaramuzza, *J.C.S. Dalton*, 1972, 1671.
[239] T. G. Appleton and J. R. Hall, *Inorg. Chem.*, 1972, **11**, 112.

The i.r. and Raman spectra of a number of Pt^{II} and Pd^{II} complexes with trimethylamine {trans-MX$_2$(NMe$_3$)$_2$ and [Pr$_4^n$N][MX$_3$NMe$_3$], where M = Pd, X = Cl; M = Pt, X = Cl or Br} have been listed and discussed.[240] All of the wavenumbers identified with ν(M—N) and ν_s(NC$_3$) modes are substantially lower for the Pd complexes than for the Pt analogues, despite the smaller mass of Pd. This is similar to the behaviour of M—P and M—As modes, and suggests that the weakness of NMe$_3$ as a ligand, by comparison with PMe$_3$ or AsMe$_3$, is not obviously attributable to differences in the orbitals available for bonding to the metal.

Some i.r. data on Pt and Pd complexes of 1,3-diaminopropan-2-ol, NH$_2$CH$_2$CH(OH)CH$_2$NH$_2$ (tnOH), and of 2-chloropropane-1,3-diamine, NH$_2$CH$_2$CHClCH$_2$NH$_2$ (tnCl), have been obtained.[241a]

For the complexes [M(en)$_2$]X$_2$ (M = Pd or Pt; X = Cl, Br, or I), the preparation and vibrational examination of: (a) normal, C-, N-perdeuteriated chlorides, (b) normal bromides and iodides, (c) N-deuteriated Pd(en)$_2$Br$_2$ and Pd(en)$_2$I$_2$, and (d) C-deuteriated Pt(en)$_2$I$_2$, has made possible quite a complete analysis of the vibrational fundamentals of these complexes.[241b]

Identification of only two i.r. bands due to ν(Cu—N) has been taken to imply a cis-geometry for solid bis(glycinamidato)copper(II) (85a).[241c]

(85a)

Partial i.r. assignments have been given for CuL$_2$Cl$_2$, CuL$_2$(OH)Cl, CuL$_4$Cl$_2$, Cu$_4$OBr$_6$L$_5$, CuL$_2$(OH)Br, CuL$_4$Br$_2$, CuL$_2$(OH)NO$_3$, CuL$_4$(NO$_3$)$_2$, CuL$_2$(OH)ClO$_4$, and CuL$_3$(OH)ClO$_4$, where L = cyclohexylamine.[241d]

I.r. spectra (550—200 cm^{-1}) were presented for fifteen Cu^{II} and Cu^{I} complexes of bipy and phen. Empirical spectral correlations could be made for distinguishing octahedral, tetragonal, trigonal-bipyramidal, tetrahedral, pseudo-tetrahedral, and square-planar co-ordination.[242]

Six new mixed Cu^{II} chelates with NNN'N'-tetramethylenediamine (tmen) have been prepared.[243] They are [Cu(tmen)(en)](ClO$_4$)$_2$, [Cu(tmen)(en)]SO$_4$,4H$_2$O, [Cu(tmen)(en)](NO$_3$)$_2$,H$_2$O, [Cu(tmen)(gly)]ClO$_4$, [Cu(tmen)(ox)],4H$_2$O, and [Cu(tmen)(aca)]ClO$_4$. The i.r. spectra show that there is no significant interaction of the anions with the complexes.

---

[240] P. L. Goggin, R. J. Goodfellow, and F. J. S. Reed, *J.C.S. Dalton*, 1972, 1298.
[241] (a) T. G. Appleton and J. R. Hall, *Inorg. Chem.*, 1972, **11**, 117; (b) R. W. Berg and K. Rasmussen, *Spectrochim. Acta*, 1972, **28A**, 2319; (c) G. R. Dukes and D. W. Margerum, *J. Amer. Chem. Soc.*, 1972, **94**, 8414; (d) G. Ondrejovič, L. Macáškova, and J. Gažo, *Z. anorg. Chem.*, 1972, **393**, 173.
[242] G. C. Percy and D. A. Thornton, *J. Mol. Structure*, 1972, **14**, 313.
[243] Y. Fukuda and K. Soue, *Bull. Chem. Soc. Japan*, 1972, **45**, 465.

A number of complexes of dibenzylethylenediamine, dben, have been examined.[244] In [Cu(dben)$_2$Br]ClO$_4$ there is no apparent interaction of the perchlorate ion with the complex, but in [Cu(dben)$_2$(SeCN)]ClO$_4$ a slight splitting of the ClO$_4^-$ band at ca. 1050 cm^{-1} leads the authors to suggest that it is 'semi-co-ordinated'.

The i.r. spectra of some bipy and 4,4'-bipy complexes of Cd have been discussed in terms of some rather unlikely structures.[245, 246]

Raman and i.r. spectra have been obtained [247] for M(en)X$_2$ (M = Zn, Cd, or Hg; X = Cl, Br, or SCN). The ligand vibrations (M = Cd or Hg) conform to the expectations for a bridging ethylenediamine group (C_{2h} symmetry). The Zn complexes were less straightforward, and several different solid phases were obtained, some with bridging and non-bridging en ligands. A full assignment of the en vibrations under C_{2h} symmetry was proposed. A listing (with partial assignment) of vibrational frequencies for Cd(en)M(CN)$_4$,2C$_6$H$_6$ (M = Cd, Hg, or Ni) has been published.[248]

Lanthanide(III) nitrate complexes of $\beta\beta'\beta''$-triaminotriethylamine (tren) have been prepared.[249] The mono-tren complexes [Ln(tren)(NO$_3$)$_3$] appear to contain only C_{2v} co-ordinated nitrate groups. The di-tren complexes are either [Ln(tren)$_2$](NO$_3$)$_3$ (Ln = Sm to Yb inclusive), containing only D_{3h}, unco-ordinated NO$_3^-$, or [Ln(tren)$_2$NO$_3$](NO$_3$)$_2$ (Ln = La to Nd). Assignments of characteristic bands for NO$_3^-$ (D_{3h} and C_{2v}) and amino-groups were proposed.

I.r. wavenumbers have been listed for lanthanide complexes of malondihydrazide [250] and phenylhydrazine.[251]

ν(C=O) is lowered in dimethylacetamide, N-methylpyrrolidone, and dimethylpropionamide by 52, 56, and 56 cm^{-1}, respectively, on co-ordination to AlEt$_3$.[252]

Fairly complete assignments for the i.r. spectra of the three-co-ordinate complexes M[N(SiMe$_3$)$_2$]$_3$ (M = Ga, Si, Ti, V, Cr, or Fe) have been proposed.[253] Thus, ν_{as}(MNSi$_2$) is at ca. 900 cm^{-1}, ν_s(MNSi$_2$) at ca. 800 cm^{-1}, ν_s(MN$_3$) at ca. 420 cm^{-1}, and ν_{as}(MN$_3$) at ca. 380 cm^{-1}.[253]

In [86; X = $-$(CH$_2$)$_2-$ or o-C$_6$H$_4$], two ν(C=O) bands are observed, the weaker (> 1600 cm^{-1}) being due to ν_s(CO)$_2$, the stronger (< 1600 cm^{-1}) due to ν_{as}(CO)$_2$.[254]

---

[244] K. C. Patel and D. E. Goldberg, Inorg. Chem., 1972, 11, 759.
[245] M. K. Alyaviya and A. N. Zueva, Russ. J. Inorg. Chem., 1971, 16, 948.
[246] M. K. Alyaviya and A. N. Zueva, Russ. J. Inorg. Chem., 1971, 16, 1079.
[247] T. Iwamoto and D. F. Shriver, Inorg. Chem., 1971, 10, 2428.
[248] T. Iwamoto and D. F. Shriver, Inorg. Chem., 1972, 11, 2570.
[249] J. H. Forsberg, T. M. Kubik, and R. G. Gucwa, Inorg. Chem., 1971, 10, 2656.
[250] N. K. Dutt and A. Sengupta, J. Inorg. Nuclear Chem., 1971, 33, 4185.
[251] S. M. F. Rahman, J. Ahmad, and M. M. Haq, J. Inorg. Nuclear Chem., 1971, 33, 4351.
[252] E. Herbeuval, J. Jozefowicz, G. Roques, and J. Néel, Compt. rend., 1972, 275, C, 351.
[253] E. C. Alyea, D. C. Bradley, and R. G. Copperthwaite, J.C.S. Dalton, 1972, 1580.
[254] B. Walther and C. Rockstroh, J. Organometallic Chem., 1972, 42, 41.

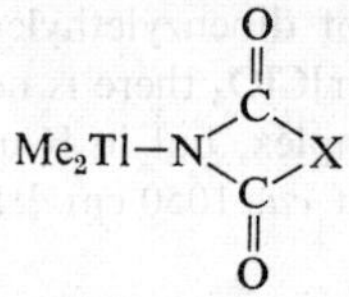

(86)

Structure (86): Me₂Tl—N with C=O groups and X, forming a chelate ring.

Bu$^n_3$Sn(NSO) gives i.r. bands at 1260, 1090 cm^{-1}, characteristic of the sulphinylamino-group.[255]

Oximes.—New complexes of phenyl-2-pyridylketoxime [syn-PhC(=NOH)-C$_5$H$_4$N] have been reported.[256]

ν(C=N) is assigned to a strong broad band at ca. 1650 cm^{-1} in several complexes of Fe^{II} with the tetraoximes MeC(=NOH)C(=NOH)—R—C(=NOH)C(=NOH)Me (R = ·C$_6$H$_4$·O·C$_6$H$_4$·, ·C$_6$H$_4$·C$_6$H$_4$·, or pyridine-2,6-diyl).[257]

Several polymeric Fe^{II} chelates with tetraoxime ligands MeC(=NOH)-C(=NOH)·CH$_2$·CH$_2$·(CH$_2$)$_n$·CH$_2$·CH$_2$·C(=NOH)C(=NOH)Me (n = 2 or 6) have been characterized, having 5 or 6 Fe—N$_4$ units within the polymer chain.[258] ν(C=N) (within the chelating group) is found at ca. 1620 cm^{-1}, and ν(OH) (of the free end-group oximes) at 3200 cm^{-1}.

Some ligand assignments [ν(NH), ν(C=N), in particular] have been made for Cu(LH)$^+$, Cu(LH$_2$)$^{2+}$, where LH$_2$ = 4,4,9,9-tetramethyl-5,8-diazadodecane-2,11-dione dioxime (87).[259]

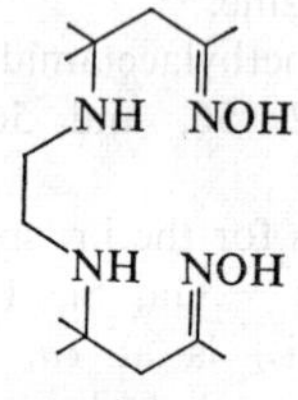

(87)

Ag[CoL$_2$(CN)$_2$] (where L = anion derived from the α-dioximes, dimethylglyoxime, 1,2-cyclohexanedione dioxime, or 1,2-cyclopentanedione dioxime) give i.r. bands which can be assigned as follows: ν(C≡N) (cyano-group) ca. 2130 cm^{-1}; ν(C=N) (oximato-group) ca. 1580 cm^{-1}; ν(N—OH) ca. 1230 cm^{-1}; ν(N—O) ca. 1090 cm^{-1}; ν(Co—C) ca. 865 cm^{-1}; ν(Co—N) ca. 510 cm^{-1}.[260]

---

[255] D. A. Armitage and A. W. Sinden, *J. Organometallic Chem.*, 1972, **44**, C45.
[256] B. Sen and D. Malone, *J. Inorg. Nuclear Chem.*, 1972, **34**, 3509.
[257] J. Backes, I. Masuda, and K. Shinra, *Bull. Chem. Soc. Japan*, 1972, **45**, 1061.
[258] J. Backes, I. Masuda, and K. Shinra, *Bull. Chem. Soc. Japan*, 1972, **45**, 1724.
[259] J. W. Fraser, G. R. Hedwig, H. K. J. Powell, and W. T. Robinson, *Austral. J. Chem.*, 1972, **25**, 747.
[260] C. Várhelyi, I. Gănescu, and L. Szobyori, *Z. anorg. Chem.*, 1971, **386**, 232.

I.r. wavenumbers have been listed but not assigned for MA_2 and $MA_2(py)_2$ (M = Zn or Ni; A = salicylaldoxime).[261]

Ligands containing $>C=N<$ Groups.—Unassigned i.r. data have been listed for a number of 2-methylimidazole complexes,[262] and also for Li_3CrL_6,THF (L = the imidazolato ligand).[263]

In the 'double-bond' region (1700—1200 cm^{-1}) the spectra of the Mo–Schiff-base complexes $Mo(sal-NR)_2X_2$ and $Mo(salen)Cl_2$ were indicative of chelation of the ligand in the salicylaldimine form, thus the highest-wavenumber band (ca. 1610 cm^{-1} in the free ligands) drops on co-ordination [sal-NR = N-substituted salicylaldimines; salen = NN-ethylenebis-(salicylaldimine)].[264]

Structurally diagnostic i.r. bands [ν(OH), ν(NH), ν(CO), and ν(C=N)] have been listed for some complexes between $Mo(CO)_6$ and Schiff bases.[265]

The ligand {$(py)_2$tame} [= (88)] forms complexes with Mn^{II} and Cu^{II}. For $Mn[\{(py)_2tame\}]^{2+}$, it is believed that no co-ordination of NH_2 occurs

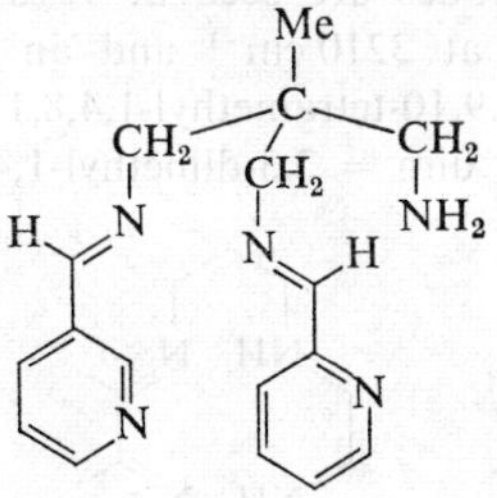

(88)

[$\nu(NH_2)$ = 3358 cm^{-1}], whereas for $Cu[\{(py)_2tame\}]Cl^+$ [having $\nu(NH_2)$ at 3330 cm^{-1}] the NH_2 group is co-ordinated.[266]

ν(NH) (3230 cm^{-1}) has been reported [267] for $Na_2[Fe(NN'$-dimethyl-ethylenediamine)(CN)_4],4H_2O$. Oxidation of this complex affords an α-di-imine species with ν(CN) at 1520 and 1620 cm^{-1}.

I.r. wavenumbers have been listed for some imide complexes of Fe^{II}, Ni^{II}, and Cu^{II},[268] and for 2-methylimidazole (L) complexes ML_2X_2 (M = Co, Ni, Cu, or Zn; X = Cl, Br, I, or NO_3).[269]

On complexing to form (89), the ligand N-[1-methoxy-(6-methyl-2-pyridyl)methyl]benzothiazolin-2-ylideneimine (formed by the reaction of

---

[261] A. P. Rao and S. P. Dubey, *J. Inorg. Nuclear Chem.*, 1972, **34**, 2041.
[262] J. Reedijk, *Rec. Trav. chim.*, 1972, **91**, 507.
[263] D. Tille, *Z. anorg. Chem.*, 1972, **390**, 234.
[264] A. van den Bergen, K. S. Murray, and B. O. West, *Austral. J. Chem.*, 1972, **25**, 705.
[265] P. C. H. Mitchell and D. A. Parker, *J.C.S. Dalton*, 1972, 1828.
[266] S. O. Wandiga, J. E. Sarneski, and F. L. Urbach, *Inorg. Chem.*, 1972, **11**, 1349.
[267] V. L. Goedken, *J.C.S. Chem. Comm.*, 1972, 207.
[268] W. V. Malik, C. L. Sharma, M. C. Jain, and Y. Ashraf, *J. Inorg. Nuclear Chem.*, 1971, **33**, 4333.
[269] D. M. L. Goodgame, M. Goodgame, and G. W. Rayner-Canham, *Inorg. Chim. Acta*, 1972, **6**, 245.

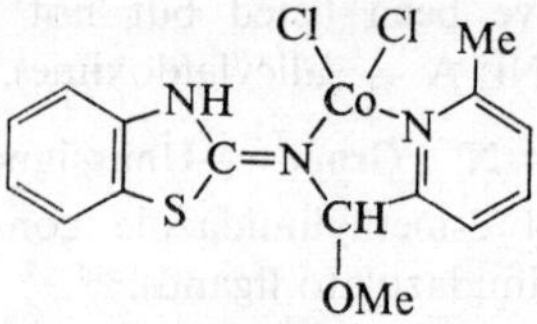

(89)

2-aminobenzothiazole and 6-methylpyridine-2-aldehyde) shows changes in its i.r. spectrum, including the disappearance of the band at 1528 cm^{-1} attributed to ν(C=N) in the thiazole ring. A new band appears at 1560 cm^{-1} in the complex.[270]

The complex Co(tim)Cl$_2^+$ gives ν_s(C=N) at 1550—1600 cm^{-1}, and ν_{as}(C=N) at 1640 cm^{-1}, together with a further characteristic, but unassigned, band at ca. 1210 cm^{-1}. Co(dim)X$_2^+$ (X = Cl, Br, or NO$_2$) and Co(dmc)X$_2$ (X = Cl, Br, or NO$_2$) show no bands due to ν(C=O) or ν(NH$_2$), but ν(C=N) modes are seen at 1570—1590 cm^{-1} and 1630—1640 cm^{-1}, with ν(NH) at 3210 cm^{-1} and an unassigned band at ca. 1195 cm^{-1} [tim = 2,3,9,10-tetramethyl-1,4,8,11-tetra-azacyclotetradeca-1,3,8,10-tetraene (90); dim = 2,3-dimethyl-1,4,8,11-tetra-azacyclotetra-

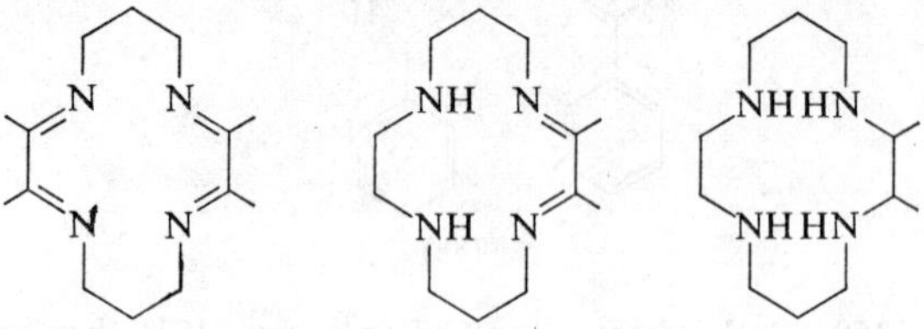

(90)   (91)   (92)

deca-1,3-diene (91); dmc = 2,3-dimethyl-1,4,8,11-tetra-azacyclotetradecane (92)].[271]

Characteristic ligand-band i.r. wavenumbers of 2-aminobenzimidazole complexes of Co and Ni have been listed.[272]

Unassigned i.r. spectra were listed for [Rh(salen)(Cl)(py)], (pyH)$^+$-[Rh(salen)Cl$_2$]$^-$, and the Rh—Rh-bonded [Rh(salen)(py)]$_2$ [salen is the dianion of NN'-ethylenebis(salicylaldimine)].[273] Slight differences were observed in the 1300—1600 cm^{-1} region of the i.r. spectra of M(3,3'-dinitro-salen) and M(5,5'-dinitro-salen) (M = Ni^{II} or Cu^{II}).[274]

Selected i.r. data for reaction products resulting from complexing of Ni^{II} or Cu^{II} N-allylsalicylaldimine complexes with Ni^{II} or Cu^{II} nitrates

[270] A. Mangia, M. Nardelli, C. Pelizzi, and G. Pelizzi, *J.C.S. Dalton*, 1972, 996.
[271] S. C. Jackels, K. Farmery, E. K. Barefield, N. J. Rose, and D. H. Busch, *Inorg. Chem.*, 1972, **11**, 2893.
[272] M. J. M. Campbell, D. W. Card, R. Grzeskowiak, and M. Goldstein, *J.C.S. Dalton*, 1972, 1687.
[273] R. J. Cozens, K. S. Murray, and B. O. West, *J. Organometallic Chem.*, 1972, **38**, 391.
[274] M. Tamaki, I. Masuda, and K. Shinra, *Bull. Chem. Soc. Japan*, 1972, **45**, 1400.

include bands near 800, 1020, 1280, and 1500 cm^{-1}, attributed to $-ONO_2$ modes.[275]

The complex cation (93) gives ν(NH) at 3160 cm^{-1}, and ν(C=N) at 1629 cm^{-1}.[276]

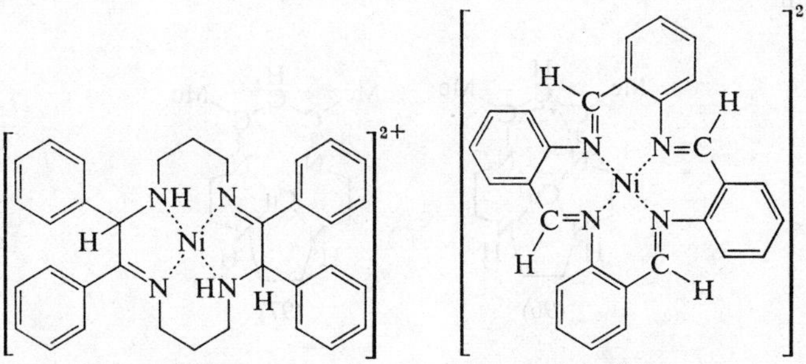

(93)   (94)

I.r. spectra of the nickel complex of the Schiff base of o-aminobenzaldehyde, Ni(taab)$^{2+}$ (94), and of its reduced analogue Ni(H$_8$taab)$^{2+}$, show that the ν(C=N) mode at 1568 cm^{-1} in the former is missing in the latter, indicating hydrogenation of all four Schiff-base linkages.[277]

ν(C=N) is found in the region 1580—1619 cm^{-1} in the complexes (95), where R^1 = H, R^2 = Ph; R^1 = Me, R^2 = Ph; R^1 = R^2 = Ph; R^1 = H, R^2 = p-Me·C$_6$H$_4$; R^1 = H, R^2 = Me. If Cl$^-$ is replaced by MeCOO$^-$, two bands characteristic of bridging acetato-groups (ca. 1580, ca. 1410 cm^{-1}) are seen.[278]

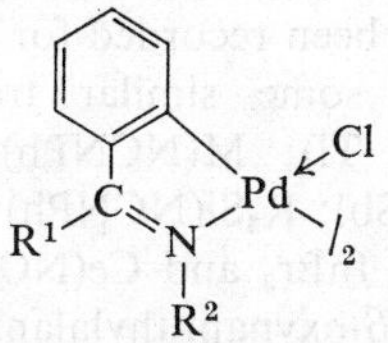

(95)

Raman data have been obtained for a series of nitrile and imino-ether complexes containing (CH$_3$)Pt^{II},[279] e.g. [Pt(CH$_3$)Q$_2${NH=C(OMe)C$_6$F$_5$}]$^+$, [Pt(CH$_3$)Q$_2${NH=C(OC$_3$H$_7$)C$_6$F$_4$CN}Pt(CH$_3$)Q$_2$]$^{2+}$, [Pt(CH$_3$)Q$_2$(NCC$_6$F$_4$-CN)Pt(CH$_3$)Q$_2$]$^{2+}$, and [Pt(CH$_3$)Q$_2$(NC$_6$F$_5$)]$^+$. ν(C=N) was found between 1662 and 1671 cm^{-1}, ν(Pt—Me) between 551 and 576 cm^{-1}, and ν(C≡N) between 2230 and 2292 cm^{-1}. ν(C≡N) has, therefore, increased on co-ordination, suggesting the presence of Pt—N σ-bonding. The

[275] J. O. Miners, E. Sinn, R. B. Coles, and C. M. Harris, J.C.S. Dalton, 1972, 1149.
[276] D. St. C. Black and H. Greenland, Austral. J. Chem., 1972, 25, 1315.
[277] V. Katović, L. T. Taylor, F. L. Urbach, W. H. White, and D. H. Busch, Inorg. Chem., 1972, 11, 479.
[278] H. Onoue and I. Moritani, J. Organometallic Chem., 1972, 43, 431.
[279] H. C. Clark and L. E. Manzer, Inorg. Chem., 1971, 10, 2699.

variations observed in the value of $\nu(C≡N)$ were related to electronic properties of the nitrile ligand.

Cu(at)$^+$ (96) gives bands due to $\nu(C\text{---}N)$ and $\nu(C\text{---}C)$ at 1550, 1520 cm^{-1}, respectively, while Cu(ath)$^{2+}$ (97) gives a band assigned to $\nu(C=N)$ at 1695 cm^{-1}.[280]

(96) (97)

I.r. spectroscopy in the $\nu(NH)$ region indicates that in both CuCl$_2$(mbi)$_2$ and CuCl$_2$(dmbi), the Cu atoms are co-ordinated to the ligand molecules via the tertiary N atom in the imidazole ring (mbi = 2-methylbenzimidazole, dmbi = 1,2-dimethylbenzimidazole).[281]

Wavenumbers due to the phenolic $\nu(C-O)$ band in a series of Cu^{II} N-hydroxyalkylsalicylideneimines have been listed (1539 ± 14 cm^{-1}).[282] I.r. wavenumbers have also been listed [283a] for some thiocarbazide derivatives of Cu^I and Ag^I.

I.r. wavenumbers have been listed, and assigned for characteristic bands, for dinuclear Cu^{II} complexes of Schiff bases derived from 2,6-diformyl-4-methylphenol and glycine or alanine.[283b]

Values for $\nu(C=N)$ have been recorded for a number of organometallic phenylcarbodi-imides and some similar transition-metal compounds: M(NCNPh) (M = Ag or Tl); M(NCNPh)$_2$ (M = Cu, Cd, or Hg); M(NCNPh)$_3$ (M = As or Sb); R$_3$Si(NCNPh); and Ph$_2$Si(NCNPh)$_2$.[284]

$\nu(CN)$ and $\nu(Ph-O)$ for InBr$_3$ and Ce(NO$_3$)$_3$ of aromatic Schiff bases (such as salicylalanine and β-oxynaphthylalanine) are at 1612—1660 cm^{-1} and 1260—1290 cm^{-1}, respectively.[285]

A qualitative discussion has been made of the i.r. spectrum of the uranium bis(phthalocyanine) complex U(C$_{32}$H$_{16}$N$_8$)$_2$.[286]

ν_{as}(UO$_2$) and $\nu(C=N)$ (of the ligand) have been listed for a series of

---

[280] J. G. Martin, R. M. C. Wei, and S. C. Cummings, *Inorg. Chem.*, 1972, **11**, 475.
[281] M. V. Artemenko and K. F. Slynsarenko, *Russ. J. Inorg. Chem.*, 1971, **16**, 1154.
[282] T. Tokii, Y. Muto, K. Imai, and H. B. Jonassen, *J. Inorg. Nuclear Chem.*, 1972, **34**, 3377.
[283] (*a*) N. K. Dutt and N. C. Chakdar, *Inorg. Chim. Acta*, 1971, **5**, 536; (*b*) H. Okawa, S. Kida, Y. Muto, and T. Tokii, *Bull. Chem. Soc. Japan*, 1972, **45**, 2480.
[284] H. Köhler and H. V. Döhler, *Z. anorg. Chem.*, 1971, **386**, 197.
[285] E. P. Trailina, A. V. Leshchenko, S. S. Lyapina, G. A. Petrova, I. I. Domilideva, and V. I. Spitsyn, *Doklady Phys. Chem.*, 1971, **201**, 1055.
[286] I. S. Kirin, A. B. Kolyadin, and P. N. Moskalev, *Russ. J. Inorg. Chem.*, 1971, **16**, 1455.

complexes UO_2LA, where A = unidentate neutral ligand and L = polydentate Schiff base.[287]

I.r. data, including proposed assignments of $\nu(NH)$, $\nu(CH)$, $\nu(C=N)$ (1655—1626 cm^{-1}), $\nu(C-O)$ (1280—1270 cm^{-1}), $\nu(C-N)$ etc., have been reported[288] for some 1 : 1 and 1 : 2 complexes of SnX_4 with some Schiff bases derived from substituted benzaldehydes and the amines

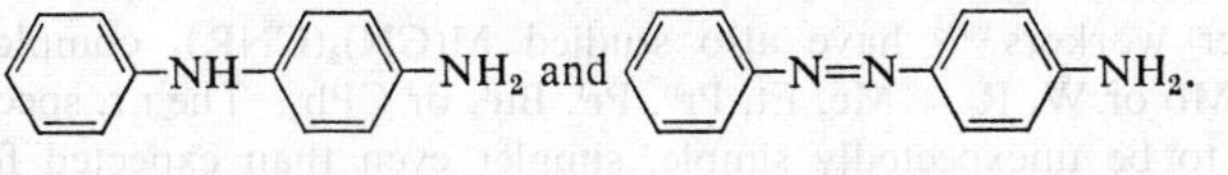

⟨⟩—NH—⟨⟩—NH$_2$ and ⟨⟩—N=N—⟨⟩—NH$_2$.

Similar data were also reported[289] for $Sn^{IV}(bah)_2$ and $Sn^{IV}(sah)_2$, where H_2bah = 3-(o-hydroxyphenylamino)crotonophenone and H_2sah = N-(2-hydroxyphenyl)salicylalanine.

Cyanides and Isocyanides.—The i.r. spectrum of the complex $[TaCl_3(MeCN)_2]_2$ shows five bands in the 2300 cm^{-1} region, indicating the presence of co-ordinated MeCN in significantly different environments, as in the postulated structure (98).[290]

The highest-wavenumber $\nu(NC)$ band (ca. 2065 cm^{-1}) in the i.r. spectrum of $[(ArCN)_6Cr]^0$ (Ar = Ph, p-Tol, or p-Cl·C_6H_4) has been shown[291] to

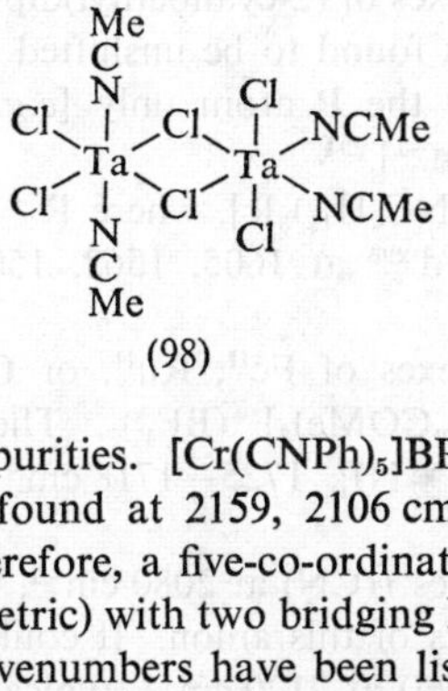

```
              Me
              C
              N
      Cl   |   Cl   |   NCMe
        \  |  /  \  |  /
         Ta     Ta
        /  |  \  /  |  \
      Cl   |   Cl    NCMe
              N    Cl
              C
              Me
             (98)
```

arise from (oxidized) impurities. $[Cr(CNPh)_5]BPh_4$ has been synthesized, and $\nu(NC)$ bands were found at 2159, 2106 cm^{-1} (Raman), 2060, 1980 cm^{-1} (i.r.). It is not, therefore, a five-co-ordinate monomer, but a dimer (presumably centrosymmetric) with two bridging PhNC groups.

$\nu(CN)$ [and $\nu(CO)$] wavenumbers have been listed for the Cr^0 and Mo^0 isonitrile complexes $(RNC)M(CO)_5$, cis-$(RNC)_2M(CO)_4$, and fac-$(RNC)_3M(CO)_3$ (where R = Me, Et, Pr^i, Bu^t, p-Me·C_6H_4, or p-Cl·C_6H_4).[292] For the $(RNC)Cr(CO)_5$ systems, a small shift to higher wavenumber for $\nu(C≡N)$ by comparison with the free ligand is consistent with the behaviour of RNC largely as a σ-ligand in these complexes.

[287] L. Cattalini, S. Degetto, M. Vidali, and P. A. Vigato, Inorg. Chim. Acta, 1972, 6, 173.
[288] N. S. Biradar and V. B. Mahale, Z. anorg. Chem., 1972, 388, 277.
[289] R. Barbieri, Inorg. Nuclear Chem. Letters, 1972, 8, 451.
[290] D. G. Blight, R. L. Deutscher, and D. L. Kepert, J.C.S. Dalton, 1972, 87.
[291] P. Gans and S. M. E. Haque, Chem. and Ind., 1972, 978.
[292] J. A. Connor, E. M. Jones, G. K. McEwen, M. K. Lloyd, and J. A. McCleverty, J.C.S. Dalton, 1972, 1246.

$\nu(C\equiv N)$ also increases by 10—35 cm^{-1} upon formation of MoX$_3$(NCR)$_3$ (R = Me, Ph, CH$_2$Ph, Et, Pr^n, Pr^i, or Bu^n; X = Cl or Br) and MoX$_3$-(NCMe)$_4$ (X = Cl or Br).[293a]

$\nu(C\equiv N)$ and $\nu(RN\equiv C)$ wavenumbers have been listed for (RNC)$_4$Mo-(CN)$_4$ (R = Me, Pr^n, Bu^t, CH$_2$=CHCH$_2$, or Ph$_2$CH).[293b] $\nu(C\equiv N)$ is at a higher energy than that for Mo(CN)$_8^{4-}$, a result consistent with the decrease in negative charge on the complex.

Other workers[293c] have also studied M(CN)$_4$(CNR)$_4$ complexes, for M = Mo or W, R = Me, Et, Pr^n, Pr^i, Bu^t, or CPh$_3$. The i.r. spectra were found to be unexpectedly simple, simpler even than expected for cubic co-ordination, only one ν(CN) and one ν(RN—C) (at higher wavenumbers) being observed. A dodecahedral MA$_4$B$_4$ structure was confirmed by X-ray data.

The complexes (99; X = Cl or Br) give a ν(CN) band at 2295 cm^{-1} (*i.e.* increased on co-ordination), consistent with M ← NC co-ordination.

$$X(CO)_3Re\begin{array}{c}PPh_2(CH_2)_2CN\\ \\NC(CH_2)_2PPh_2\end{array}Re(CO)_3X$$

(99)

Some monomeric complexes of (2-cyanoethyl)diphenylphosphine were also prepared, and ν(CN) was found to be unshifted on co-ordination, *i.e.* the ligand is unidentate, *via* the P atom only [*e.g.* in Re(CO)$_4$LBr, ν(CN) occurred at 2235, 2255 cm^{-1}].[294]

In (π-Cp)Fe(CO)[(C=NC$_6$H$_{11}$)$_3$R], where R = CH$_2$Ph or CH$_2$C$_6$H$_4$Cl, ν(C=N) bands are found[295] at 1605, 1562, 1503 cm^{-1} and 1605, 1574, 1506 cm^{-1}, respectively.

γ-Oxo-isocyano-complexes of Fe^{II}, Ru^{II}, or Os^{II} have been prepared, *i.e.* [M(C≡N—CMe$_2$CH$_2$COMe)$_6$]$^{2+}$(BF$_4^-$)$_2$. The i.r. spectra gave bands at 2215—2206 cm^{-1} [ν(C≡N)], 1725—1718 cm^{-1} [ν(C=O)], and 1058—1053 cm^{-1} [ν_3(BF$_4$)].[296]

[R$_4$N]$^+$[Co(CN)$_5$]$^{3-}$ gives ν(CN) at 2080 cm^{-1}, in agreement with earlier work on alkali-metal salts of this anion. It could be oxidized to (R$_4$N)$_3$-[Co(CN)$_5$O$_2$], having ν(CN) at 2120 cm^{-1}, typical for Co^{III}(CN)$_5$X, and a band at 1138 cm^{-1} assigned to the O$_2$ ligand.[297]

The new complex K$_3$[Co(CN)$_5$(NCO)] gives ν(CN) absorptions at 2194 cm^{-1} (due to —NCO) and 2129 cm^{-1} (due to —CN).[298]

Since the π-acceptor strengths of t-butyl isocyanide and tervalent P or As ligands are similar, it would be expected that successive replacement of

---

[293] (*a*) A. D. Westland and N. Muriithi, *Inorg. Chem.*, 1972, **11**, 2971; (*b*) M. Novotny, D. F. Lewis, and S. J. Lippard, *J. Amer. Chem. Soc.*, 1972, **94**, 6961; (*c*) R. V. Parish and P. G. Simms, *J.C.S. Dalton*, 1972, 2389.
[294] B. N. Storhoff, *J. Organometallic Chem.*, 1972, **43**, 197.
[295] Y. Yamamoto and H. Yamazaki, *Inorg. Chem.*, 1972, **11**, 211.
[296] M. Schaal and W. Beck, *Angew. Chem. Internat. Edn.*, 1972, **11**, 527.
[297] D. A. White, A. J. Solodar, and M. M. Baizer, *Inorg. Chem.*, 1972, **11**, 2160.
[298] M. A. Cohen, J. P. Melpolder, and J. L. Burmeister, *Inorg. Chim. Acta*, 1972, **6**, 188.

the former by the latter should have relatively little effect on the ν(NC) vibration of the Bu^{t}NC ligand. This is indeed observed to be so, since in [(Bu^{t}NC)$_5$Co]$^+$PF$_6^-$, ν(NC) bands are found at 2151, 2120 cm^{-1}, quite close to the single ν(NC) band in {(Bu^{t}NC)Co[P(Pf)$_3$]}$^+$PF$_6^-$ (2113 cm^{-1}) {P(Pf)$_3$ is tris-(2-diphenylphosphinoethyl)phosphine, [Ph$_2$PCH$_2$CH$_2$]$_3$P}.[299]

[Co(CN)$_2$(diphos)$_2$]$^+$ gives rise to only one i.r. band from ν(CN) (2112 cm^{-1}), indicating a *trans*-structure.[300]

ν(NC) bands have been listed for a number of (isocyanide) rhodium cationic complexes.[301, 302]

ν(CN) in [Rh(NH$_3$)$_5$(RCN)]$^{3+}$ (R = CH$_3$, CD$_3$, CH$_3$CH$_2$, CH$_2$=CH, CH$_2$=CMe, Ph, o-F·C$_6$H$_4$, m-F·C$_6$H$_4$, or p-F·C$_6$H$_4$) shows an increase of 50—60 cm^{-1} on complexing. This is characteristic of M—N co-ordination. Characteristic NH$_3$ bands and ν(Rh—NH$_3$) were also observed.[303]

In a series of cationic Ni and Pd complexes [MClLQ$_2$]$^+$ (where L = CNR; Q = phosphine or arsine ligand) it is found [304] that

$$\Delta(N\equiv C) = \nu(N\equiv C)_{free} - \nu(N\equiv C)_{complexed}$$

is greater for Pd (80—100 cm^{-1}) than for Ni (12—70 cm^{-1}), and that it is greater for R = alkyl than for R = aryl. There is no correlation, however, with the nature of the substituents on an aryl group.

Assignments of ν(Pd—C), δ(PdCN), and ν(CN) have been proposed for a number of M^{II}[Pd(CN)$_4$] species (M = Mn, Fe, Co, Ni, Cu, or Zn).[305] Higher ν(CN) values are found than in the related K$^+$ salt, owing to the higher ionic potential of the M^{2+} ions. The Cu^{2+} salt gives two ν(Pd—C) modes, indicative of a lowering in symmetry.

The presence of 2 or 3 strong bands due to ν(NC) in the i.r. spectra of [PtCl$_2$(RNC)$_2$] (R = substituted phenyl, Bu^t, or cyclohexyl) argues against the previously suggested Magnus' salt-type structure for these systems [this would give rise to only one ν(NC) band].[306]

ν(NC) has been assigned in a number of [Me$_2$PtQ$_2$(NCR)$_2$]$^{2+}$ and [Me$_2$PtQ$_2$I(NCR)]$^+$ complexes (Q = PMe$_2$Ph). It has previously been established (*J. Organometallic Chem.*, 1971, **30**, C89) that $\Delta(N\equiv C)$ [= $\nu(N\equiv C)_{complex} - \nu(N\equiv C)_{free}$] is an inverse measure of electron density at the metal, and it is suggested [307] that these Me$_2$Pt^{IV} complexes have a similar electron density at Pt as a number of Pt^{II}–isocyanide complexes.

ν(CN) in [PtX(CNMe)(PPh$_3$)$_2$]$^+$ is in the range 2273—2236 cm^{-1}, and in the sequence X = CN$^-$ > SCN$^-$ > NO$_2^-$ > Cl$^-$ > N$_3^-$ > I$^-$ >

---

[299] R. B. King and M. S. Saran, *Inorg. Chem.*, 1972, **11**, 2112.
[300] P. Rigo and A. Turco, *Co-ordination Chem. Rev.*, 1972, **8**, 175.
[301] P. R. Branson and M. Green, *J.C.S. Dalton*, 1972, 1303.
[302] R. V. Parrish and P. G. Simms, *J.C.S. Dalton*, 1972, 809.
[303] R. D. Foust jun. and P. C. Ford, *Inorg. Chem.*, 1972, **11**, 899.
[304] W. J. Cherwinski, H. C. Clark, and L. E. Manzer, *Inorg. Chem.*, 1972, **11**, 1511.
[305] H. Siebert and W. Weise, *Z. Naturforsch.*, 1972, **27b**, 865.
[306] H. J. Keller, H. Lorentz, H. H. Rupp, and J. Weiss, *Z. Naturforsch.*, 1972, **27b**, 631.
[307] H. C. Clark and L. E. Manzer, *Inorg. Chem.*, 1972, **11**, 2749.

$CN_4Me \gg C_6F_5^- > OH^- > C(OMe)NMe^- \approx C(OCH_2CH=CH_2)NMe^- \approx C(NHC_6H_4NO_2)NMe^- > C(NHC_6H_4Me)NMe^- > Ph^- > Me^-$. For the complexes $[Pt(L)(CNMe)(PPh_3)_2]^{2+}$ it lies between 2280 and 2263 cm^{-1}, with the order L = $PPh_3 \gtrsim MeCN \gtrsim P(OMe)_3 \gtrsim Me_2S \approx py > MeNC > C(OCH_2CH=CH_2)NHMe > C(NHC_6H_4Me)NHMe > NMe_3$.[308]

A number of NCBH$_3^-$ complexes have been prepared, e.g. $(R_3M)_3$Cu-(NCBH$_3$), M = P, As, or Sb, and $(R_3P)_3$Ag(NCBH$_3$).[309] These show an increase of ν(NC) by comparison with the free ion (1—24 cm^{-1}), indicating M—N bonding. The BH$_3$ frequencies are all very similar to those of the free ion.

I.r. spectra (350—400 cm^{-1}) have been listed for 2CuX,L [X = Cl or Br; L = Me$_2$C=C(CN)$_2$ or (CD$_3$)$_2$C=C(CN)$_2$].[310] The increase in ν(CN) and decreases in ν(C=C) on co-ordination are apparently consistent with the formulation (100), although a dimeric structure involving bonding of the C=C bond to Cu cannot be completely excluded.

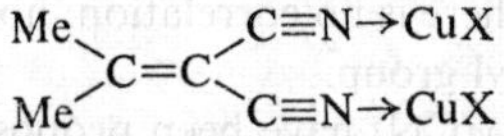

The complex (101) has ν(N≡C) of the isocyanide ligand at 2146 cm^{-1}.[311] ν(CN) has been listed for all the lanthanide complexes (π-Cp)$_3$Ln-(CNC$_6$H$_{11}$), and the variation in this with 4f-orbital occupancy interpreted in terms of $f \to \pi^*$ back-bonding.[312]

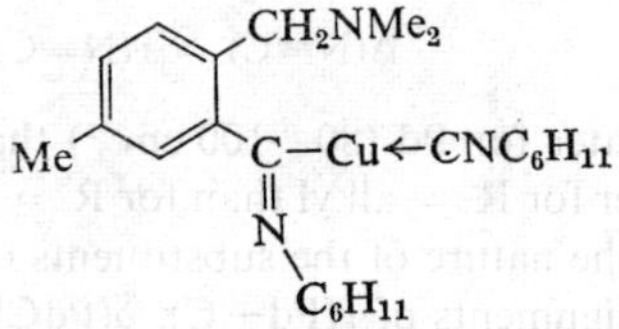

In ThCl$_4$,2CNCl and UCl$_4$,2CNCl, ν(C≡N) is found at 2242 cm^{-1} (2219 cm^{-1} in free CNCl). Thus, bonding of CNCl occurs via the nitrogen atom.[313]

Papers reporting ν(CN) data, with little or no comment, are listed in Tables 4 and 5.[314—329]

[308] P. M. Treichel and W. J. Kuebel, *Inorg. Chem.*, 1972, **11**, 1289.
[309] S. J. Lippard and P. S. Welcker, *Inorg. Chem.*, 1972, **11**, 6.
[310] S. K. Smirnov, O. G. Strukov, S. S. Dubov, A. M. Gribov, and E. L. Gal'perin, *Russ. J. Inorg. Chem.*, 1971, **16**, 1159.
[311] G. van Koten and J. G. Noltes, *J.C.S. Chem. Comm.*, 1972, 59.
[312] R. von Ammon and B. Kanellakopulos, *Ber. Bunsengesellschaft. phys. Chem.*, 1972, **76**, 995.
[313] J. MacCordick and G. Kaufmann, *Bull. Soc. chim. France*, 1972, 23.
[314] P. M. Treichel, G. E. Dirreen, and H. J. Mueh, *J. Organometallic Chem.*, 1972, **44**, 339.
[315] R. A. Bailey and E. N. Balko, *J. Inorg. Nuclear Chem.*, 1972, **34**, 2668.
[316] S. Papp, S. Kovács, and I. Liszi, *J. Inorg. Nuclear Chem.*, 1972, **34**, 3111.
[317] H. Alper and R. A. Partis, *J. Organometallic Chem.*, 1972, **35**, C41.
[318] H. Brunner and M. Vogel, *J. Organometallic Chem.*, 1972, **35**, 169.
[319] J. A. Ferguson and T. J. Mayer, *Inorg. Chem.*, 1972, **11**, 631.
[320] A. L. Balch and J. Miller, *J. Amer. Chem. Soc.*, 1972, **94**, 417.
[321] B. E. Prater, *J. Organometallic Chem.*, 1972, **34**, 379.

References continued on facing page.

Table 4 Cyanide and isocyanide complexes of Mn, Fe, and Ru for which $\nu(CN)$ data have been published

Compound	Ref.
$[Mn(CO)_n(CNMe)_{6-n}]^+$ ($n = 0\text{—}5$)	314
$[MnBr(CO)_n(CNMe)_{5-n}]$ ($n = 1\text{—}4$)	
$K_2[Mn(CN)_5(H_2O)]$, KCN	315
$[Fe(CN)_6]^{3-}$, $[Fe(CN)_6]^{4-}$, $[Fe(CN)_5(NO)]^{2-}$ (as phosphonium salts)	316
$[Fe(CO)_4(CNAr)]$	317
$[Fe(CO)_3(CNAr)_2]$ (Ar = o-, m-, or p-Me·C$_6$H$_4$)	
$(\pi\text{-Cp})Fe(CO)(CNC_6H_{11})I$	318
$(\pi\text{-Cp})Fe(CO)[CNCH(Me)Ph]I$	
$(\pi\text{-Cp})Fe[CNC_6H_{11}]_2I$	
$(\pi\text{-Cp})Fe[CNCH(Me)Ph]_2I$	
$\{[(\pi\text{-Cp})Fe(CO)(NCMe)]_2(Ph_2PCH_2CH_2PPh_2)\}^+$	319
$(MeNC)_4Fe(C_4H_9RN_4)^{2+}$ (R = H, Me, or Ph)	320
cis- and trans-Ru(CNEt)$_2$(MPh$_3$)$_2$X$_2$ (M = P, As, or Sb; X = Cl or Br)	321

Table 5 Cyanide and iso-cyanide complexes of Co, Rh, Ir, Ni, Pd and Pt for which $\nu(CN)$ data have been published

Compound	Ref.
$Et_4N[Co(CN)_2(CO)_2\{P(C_6H_{11})_3\}], H_2O$	322
$Et_4N[Co(CN)_2(CO)(PPh_3)_2], 2H_2O$	
$K[Co(CN)_2(CO)_3L_2], 3H_2O$ (L = MePPh$_2$ or Me$_2$PPh)	
$K[Co(CN)_2(CO)L_2], (Me_2CO\ or\ H_2O)$ (L = PEt$_3$)	
$K[Co(CN)_2(CO)(diphos)], H_2O$	
$[Co(dpe)_2(CN)_2]^+ClO_4^-$	
$Co(dpe)_2(CN)_2MX_3$ (M = Co; X = Cl, Br, or NCS; M = Mn, Fe, Ni, or Zn; X = Cl)	323
$[(ArNC)_4Rh(tcne)]^+$ (Ar = p-Me·C$_6$H$_4$, p-MeO·C$_6$H$_4$, or o-Me·C$_6$H$_4$)	324
$IrL(CO)(MPh_3)_2$ (M = P or As)	
$IrL(CO)(PPh_3)_2L'$ (L' = tcne, fumn, or SO$_2$)	
$RhL(CO)(PPh_3)_3$	325
$RhL(CO)(PPh_3)_2(tcne)$	
$PtEt(L)(PPh_3)_2$	
$PtH(L)(PPh_3)_2$	
$PtH(L)(PEt_3)_2$	
$[Pd(L)(Me_5dien)]^+$ (L = dicyanoketeniminato, C$_4$N$_3$)	
$[Ir(CO)(MeNC)_4]^+$	326
$[Ir(diphos)_2(MeNC)]^+$	
$[IrY(diphos)_2(MeNC)]^{2+}$ (Y = Cl, I, H, or HgCl)	
$[Ir(MeNC)_4]^{2+}$	

[322] J. Halpern, G. Cuastalla, and J. Bercaw, Co-ordination Chem. Rev., 1972, **8**, 167.
[323] P. Rigo, B. Longato, and G. Favero, Inorg. Chem., 1972, **11**, 300.
[324] K. Kawakami, T. Komeshima, and T. Tanaka, J. Organometallic Chem., 1972, **34**, C21.
[325] M. Lenarda and W. H. Baddley, J. Organometallic Chem., 1972, **39**, 217.
[326] W. M. Bedford and G. Rouschias, J.C.S. Chem. Comm., 1972, 1224.
[327] B. Corain, Co-ordination Chem. Rev., 1972, **8**, 159.
[328] B. Crociani, T. Boschi, M. Nicolini, and U. Belluco, Inorg. Chem., 1972, **11**, 1292.
[329] P. M. Treichel and W. J. Kuebel, Inorg. Chem., 1972, **11**, 1285.

Table 5 (cont.)

Compound	Ref.
[Ni(CN)$_2$L]$_2$ Ni(CN)$_2$L$_{1.5}$ Ni(CN)$_2$(PPr$_3$)$_2$ [L = Ph$_2$P(CH$_2$)$_4$PPh$_2$]	327
PdLL'Cl$_2$ (L = PPh$_3$, L' = p-Me·C$_6$H$_4$NC, p-MeO·C$_6$H$_4$NC, C(NHPh)NHC$_6$H$_4$·OMe, C(NHPh)$_2$, C(NHPh)NHC$_6$H$_4$·Cl, $etc.$; L = AsPh$_3$, L' = C(NHPh)NHC$_6$H$_4$·Me)	328
[Pt(OH)(CNMe)$_3$]$^+$ [Pt(CSNHMe)(CNMe)(PPh$_3$)$_2$]$^+$ {Pt[C(NC$_6$H$_4$·Me)NHMe](CNMe)(PPh$_3$)$_2$}$^+$ {Pt[C(NPh)NHMe](CNMe)(PPh$_3$)$_2$}$^+$ {Pt(NCMe)(CNMe)(PPh$_3$)$_2$}$^+$	329

Nitrosyls.—ν(CO) and ν(NO) have been listed for a number of transition-metal complexes, containing CO and NO groups, which are bonded to a triphenylphosphine substituted on to a resin substrate, (polymer)—C$_6$H$_4$—PPh$_2$, where (polymer) is a cross-linked polystyrene.[330a]

The previously unknown mononuclear nitrosyl Cr(NO)$_4$ has been prepared [330b] by slowly streaming NO through irradiated Cr(CO)$_6$ in hydrocarbon solution. Only three bands were found in the i.r. spectrum of the black-brown solid product, at 1721 cm^{-1} [ν_{as}(NO)], 650 cm^{-1} [ν_{as}(Cr—N)], and 496 cm^{-1} [δ_{as}(NO—Cr—NO)].

ν(NO) has been listed [330c] for a large number of π-cyclopentadienyl-molybdenum nitrosyl complexes.

The following ν(NO) wavenumbers were reported for molybdenum- and tungsten-nitrosyl halides:[331] [Mo(NO)Cl$_3$]$_n$ 1590 cm^{-1}; [W(NO)Cl$_3$]$_n$ 1590 cm^{-1} (both bridging); Mo(NO)Cl$_3$(Ph$_3$PO)$_2$ 1710 cm^{-1} [with ν(PO) at 1180 cm^{-1} and 1130—1140 cm^{-1}]; Mo(NO)Cl$_3$(bipy) 1705 cm^{-1}; W(NO)Cl$_3$(Ph$_3$PO)$_2$ 1650 cm^{-1} [ν(PO) at 1170, 1140 cm^{-1}].

In (π-Cp)ReX(CO)(NO), ν(CO) is at 1972 cm^{-1} (X = Me) and 1979 cm^{-1} (X = H); ν(NO) is at 1715 cm^{-1}, 1722 cm^{-1}, respectively.[332]

The variations of ν(NO) in neutral radical anion, dianion, and radical cation tetrahedral derivatives related to Co(CO)$_3$(NO) and Fe(CO)$_2$(NO)$_2$ have been discussed in terms of the electric charge distribution in these complexes, and have been related to data obtained by other techniques (especially e.s.r. and n.m.r.).[333]

ν(CO), ν(NO), and ν(CN) have been listed for a variety of isocyanide complexes and carbonyl–nitrosyl derivatives related to {Fe(CO)$_3$[S$_2$C$_2$(CF$_3$)$_2$]}$_n$.[334]

---

[330] (a) J. P. Collman, L. S. Hegedus, M. P. Cooke, J. R. Norton, G. Dolcetti, and D. N. Marquardt, *J. Amer. Chem. Soc.*, 1972, **94**, 1789; (b) M. Herberhold and A. Razavi, *Angew. Chem. Internat. Edn.*, 1972, **11**, 1092; (c) J. A. McCleverty and D. Seddon, *J.C.S. Dalton*, 1972, 2526, 2588.

[331] R. Davis, B. F. G. Johnson, and K. H. Al-Obaichi, *J.C.S. Dalton*, 1972, 508.

[332] R. P. Stewart, N. Okamoto, and W. A. G. Graham, *J. Organometallic Chem.*, 1972, **42**, C32.

[333] R. E. Dessy, J. C. Charkoudian, and A. L. Rheingold, *J. Amer. Chem. Soc.*, 1972, **94**, 738.

[334] C. J. Jones, J. A. McCleverty, and D. G. Orchard, *J.C.S. Dalton*, 1972, 1109.

Table 6 shows the assignments proposed by Schreiner et al. for $\nu(Ru-N)/\delta(Ru-N-O)$ and $\nu(NO)$ in a series of complexes trans-$[Ru(NH_3)_4(NO)L]^{n+}$.[335] These figures suggest the existence of a vibrational trans-effect series: $NH_3 < NCO^- < N_3^- < MeCO_2^- < Cl^- < Br^- < OH^-$.

Table 6 *Some vibrational assignments in trans-$[Ru(NH_3)_4(NO)L]^{n+}$ of wavenumbers/cm^{-1}*

Complex	$\nu(Ru-N)$ or $\delta(Ru-N-O)$	$\nu(NO)$
trans-$[Ru(NH_3)_4(NO)(NH_3)]Cl_3$	602	1903
trans-$[Ru(NH_3)_4(NO)(NCO)]I_2$	615	1890
trans-$[Ru(NH_3)_4(NO)(N_3)]I_2$	595	1884
trans-$[Ru(NH_3)_4(NO)(OAc)]I_2$	595	1882
trans-$[Ru(NH_3)_4(NO)Cl]Cl_2$	608	1880
trans-$[Ru(NH_3)_4(NO)Br]Br_2$	591	1870
trans-$[Ru(NH_3)_4(NO)(OH)]Cl_2$	628	1834

$Ru_3(CO)_{10}(NO)_2$ and $Os_3(CO)_{10}(NO)_2$ give $\nu(CO)$ absorptions in solution characteristic of terminal carbonyls in trimetallic carbonyls of C_{2v} symmetry. Their similarity suggests that they are isostructural. $\nu(NO)$ is found in the solid-phase spectra at 1517, 1500 cm^{-1} (Ru), 1503, 1484 cm^{-1} (Os), together with peaks at 723 (Ru), 739 (Os) cm^{-1}, believed to be associated with the double nitrosyl bridge (102).[336]

$$\begin{array}{c} (CO)_4 \\ M \\ / \quad \backslash \\ (OC)_3M \underset{NO}{\overset{NO}{\diagup\diagdown}} M(CO)_3 \end{array}$$
(102)

$\nu(NO)$ in a variety of tertiary phosphine- and arsine-nitrosyl halide derivatives of Ru and Os, mostly of the form $MX_3(NO)(ER_3)_2$, spans the range 1829—1876 cm^{-1} (M = Ru), 1811—1853 cm^{-1} (M = Os).[337a]

Vibrational spectra have been assigned for $M^I{}_n[M'(NO)X_5]$ (M^I = K or Cs; M' = Os^{II}, Ru^{II}, or Ir^{III}; X = Cl, Br, or I) with some assistance from ^{15}N-substituted species.[337b] The NO stretch increases in wavenumber in the order $Os^{II} < Ru^{II} < Ir^{III}$, while $\nu(M-N)$ decrease in the same sequence (for constant X). This is explicable on the basis that the $M'-N$ bond has more π-character in the Os complex, and that the higher oxidation state in Ir^{III} decreases the $M'-N$ bonding.

In Table 7 are listed $\nu(NO)$ for some neutral and cationic osmium nitrosyl carbonyls.[338]

---

[335] A. F. Schreiner, S. W. Lin, P. J. Hauser, E. A. Hopous, D. J. Hamm, and J. D. Gunter, *Inorg. Chem.*, 1972, **11**, 880.
[336] J. R. Norton, J. P. Collman, G. Dolcetti, and W. T. Robinson, *Inorg. Chem.*, 1972, **11**, 382.
[337] (a) S. D. Robinson and M. F. Uttley, *J.C.S. Dalton*, 1972, 1; (b) M. J. Cleare, H. P. Fritz, and W. P. Griffith, *Spectrochim. Acta*, 1972, **28A**, 2013.
[338] G. R. Clark, K. R. Grundy, W. R. Roper, J. M. Waters, and K. R. Whittle, *J.C.S. Chem. Comm.*, 1972, 119.

$\nu(NO)$ in $Co^{II}(dmgH)_2(NO)$ is at 1641 cm^{-1} (shifted to 1615 cm^{-1} on ^{15}N substitution).[339] $Co^{III}(dmgH)_2(N\dot{O}_2),OH_2$ gives the following wavenumbers associated with the nitro-ligand: $\nu_{as}(NO_2)$ 1453; $\nu_s(N\dot{O}_2)$ 1321;

Table 7 Nitrosyl stretching wavenumbers/cm^{-1} for some osmium nitrosyl carbonyls

Complex	$\nu(NO)$
$OsH(CO)(NO)(PPh_3)_2$	1620
$[Os(CO)_2(NO)(PPh_3)_2]^+$	1750
$[Os(CO)(NO)(PPh_3)_3]^+$	1705
$[Os(CO)(NO)(PPh_2Me)_3]^+$	1700
$[Os(CO)(NO)(Ph_2PCH_2CH_2PPh_2)(PPh_3)]^+$	1700
$[Os(CO)(NO)(RNC)(PPh_3)_2]^+$	1700 (NC 2150)

$\delta(ONO)$ 819; $\rho_w(NO_2)$ 615 cm^{-1} (at 1416, 1300, 812, 604 cm^{-1} in the $^{15}NO_2$ complex).

In $[(\pi-Cp)Co(NO)(PPh_3)]^+PF_6^-$, $\nu(NO)$ is at 1848 cm^{-1}, in $[(\pi-Cp)Co\{P(C_6H_{11})_3\}]^+PF_6^-$ 1840 cm^{-1}, and in $[(\pi-Cp)Rh(NO)(PPh_3)]^+PF_6^-$ at 1831 cm^{-1}.[340a]

$\nu(NO)$ wavenumbers have been listed for a range of oxidative–addition products obtained from $M(NO)L_3$ (M = Co, Rh, or Ir; L = PR$_3$), typically $CoI_2(NO)(PPh_3)_3$.[340b]

The nitrosyl stretching frequency in $RhCl(NO)(PPh_3)_2$, $RhCl(NO)(AsPh_3)_2$, $RhCl_3(NO)(PPh_3)_2$, and $RhCl_3(NO)(AsPh_3)_2$ is at 1629 ± 1 cm^{-1}, *i.e.* it is independent of the EPh$_3$ ligand and of the oxidation state of the Rh.[341]

The cationic nitrosyls $[IrCl_3(NO)L_2]^+$ (L = PPh$_3$ or AsPh$_3$) react with alcohols to give neutral Ir^{III} complexes containing bound alkyl nitrates, with characteristic absorptions at 1550, 1400, 1100, 970, 880—835 cm^{-1} $[IrCl_3(RONO)L_2]$.[342]

$[Ni(tep)NO]^+BF_4^-$ [tep = (103)] gives a nitrosyl stretch at 1760 cm^{-1}, together with characteristic tep and BF_4^- bands.[343]

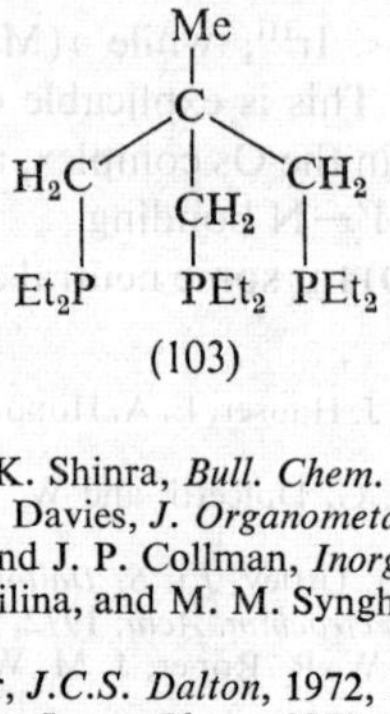

```
            Me
            |
            C
           /|\
       H₂C | CH₂
           |
          CH₂
       /   |   \
     Et₂P PEt₂ PEt₂
```
(103)

---

[339] M. Tamaki, I. Masuda, and K. Shinra, *Bull. Chem. Soc. Japan*, 1972, **45**, 171.
[340] (a) N. G. Connelly and J. D. Davies, *J. Organometallic Chem.*, 1972, **38**, 385; (b) G. Dolcetti, N. W. Hoffmann, and J. P. Collman, *Inorg. Chim. Acta*, 1972, **6**, 531.
[341] Yu. N. Kukushkin, L. I. Danilina, and M. M. Syngh, *Russ. J. Inorg. Chem.*, 1971, **16**, 1449.
[342] C. A. Reed and W. R. Roper, *J.C.S. Dalton*, 1972, 1243.
[343] D. Berglund and D. W. Meek, *Inorg. Chem.*, 1972, **11**, 1493.

A study has been made of ν(NO) in the complexes trans-[PtX(NO)A$_4$]X$_2$ (X = Cl, Br, HSO$_4$, or NO$_3$; A = NH$_3$, MeNH$_2$, Pr^{n}NH$_2$, or $\frac{1}{2}$en). ν(NO) increases when Cl$^-$ or Br$^-$ are replaced by HSO$_4^-$ or NO$_3^-$, as does the ease of hydrolysis of the complex. The NO was said to be acting as a one-electron ligand.[344]

A number of other papers reported ν(NO) data (see Table 8).[345-349]

Table 8 *Complexes for which nitrosyl stretching wavenumbers have been assigned*

Complex	Ref.
[Ni(NO)(PPh$_3$)$_3$]$^+$	
[Ni(NO)(CO)(PPh$_3$)$_2$]$^+$	
{Ni(NO)(PPh$_3$)$_2$[P(OPh)$_3$]}$^+$	
Co(NO)(PPh$_3$)$_3$	
Co(NO)(CO)(PPh$_3$)$_2$	
[Co(NO)$_2$(PPh$_3$)$_2$]$^+$	345
[Co(NO)$_2$(diphos)]$^+$	
Co(NO)(diphos)(THF)	
Fe(NO)$_2$(PPh$_3$)$_2$	
Fe(NO)$_2$(diphos)	
[Mo(NO)(CO)$_3$(diphos)]$^+$	
[W(NO)(CO)$_3$(diphos)]$^+$	
[Fe(CO)$_2$(NO)L$_2$]$^+$PF$_6^-$ [L = P(OPh)$_3$, P(OMe)$_3$, PPh$_3$, PPh$_2$Me, PMe$_2$Ph, PEt$_3$, AsPh$_3$, etc.]	
[Fe(CO)(NO)L$_3$]$^+$PF$_6^-$ [L = P(OMe)$_3$, PMePh$_2$, or PMe$_2$Ph]	346
{[Fe(CO)(NO)(dppe)]$_2$(dppe)}(PF$_6$)$_2$	
[Fe(NO)(dppe)]$_2$BPh$_4$	
Fe(CO)(NO)(PPh$_3$)$_2$(CO$_2$Me)	
Fe(CO)(NO)(dppe)(CO$_2$Me)	
cis-[FeL(S$_2$CNEt$_2$)$_2$(NO)] (L = NO$_2$, Cl, Br, or I)	347
[Ru(NO)Cl$_3$(L$_2$)] (L$_2$ = Ph$_2$PCH$_2$PPh$_2$, Ph$_2$AsCH$_2$AsPh$_2$, cis-Ph$_2$AsCH=CHAsPh$_2$, etc.)	348
[Mesoporphyrin IX dimethylesterato]-dinitrosylruthenium(II)	349

4 Phosphorus and Arsenic Donors

ν(PF) [and ν(CO) where appropriate] have been listed for (π-Cp)V(CO)$_2$-(PF$_2$NC$_6$H$_{10}$)$_2$, (π-Cp)Mo(CO)$_2$(COMe)(PF$_2$NMe$_2$), (π-Cp)Mo(CO)(C$_3$H$_5$)-(PF$_2$NEt$_2$)$_2$, (π-Cp)W(CO)$_2$(PF$_2$NC$_5$H$_{10}$)$_2$, (π-Cp)Mn(CO)(PF$_2$NC$_5$H$_{10}$)$_2$, (π-Cp)$_2$Fe$_2$(CO)$_3$(PF$_2$NEt$_2$), (π-Cp)Fe(PF$_2$NEt$_2$)I, and related complexes.[350]

[344] A. I. Stetsenko, N. V. Ivannikova, and V. M. Kiseleva, *Russ. J. Inorg. Chem.*, 1972, **16**, 865.
[345] B. F. G. Johnson, S. Bhaduri, and N. G. Connelly, *J. Organometallic Chem.*, 1972, **40**, C36.
[346] B. F. G. Johnson and J. A. Segal, *J.C.S. Dalton*, 1972, 1268.
[347] H. Büttner and R. D. Feltham, *Inorg. Chem.*, 1972, **11**, 971.
[348] J. T. Mague and J. P. Mitchener, *Inorg. Chem.*, 1972, **11**, 2714.
[349] T. S. Srivastava, L. Hoffmann, and M. Tsutsui, *J. Amer. Chem. Soc.*, 1972, **94**, 1385.
[350] R. B. King, W. C. Zipperer, and M. Ishaq, *Inorg. Chem.*, 1972, **11**, 1361.

P—H stretching wavenumbers are found between 2295 and 2420 cm^{-1} in $(\pi\text{-Cp})\text{Fe(CO)}_2\text{L}$, $[(\pi\text{-Cp})\text{Mo(CO)}_3\text{L}']^+$, $[(\pi\text{-Cp})\text{Mo(CO)}_2\text{L}]^+$, $\text{Mn(CO)}_3\text{L}'(\text{Br})$, $\text{PhMn(CO)}_4\text{L}'$, $\text{Cr(CO)}_4\text{L}_2$, and $\text{Mo(CO)}_3\text{L}_3$ (L = PPhH$_2$; L' = PPh$_2$H).[351]

Some assignments have been made of i.r. bands of Mo(NO)$_2$(dpm)X$_2$ (dpm = bisdiphenylphosphinomethane; X = Cl, Br, or I), M(NO)$_2$(dam)$_2$X$_2$, and M(NO)$_2$(dam)X$_2$ (M = Mo or W; X = Cl, Br, or I; dam = bisdiphenylarsinomethane).[352]

The complex Mn$_2$(CO)$_9$(PH$_3$) has been prepared and the position of the PH$_3$ ligand, determined by i.r. spectroscopy in the ν(CO) region, is that of an equatorial substituent on a dimanganese decacarbonyl structure.[353]

The following i.r. data have been reported[354] for $(\pi\text{-Cp})\text{Mn(CO)}_n\text{-}(\text{PF}_3)_{3-n}$ (n = 2, 1, or 0): n = 2, ν(CO) 1993, 1930 cm^{-1}, ν(PF) 846 cm^{-1}; n = 1, ν(CO) 1953 cm^{-1}, ν(PF) 846 cm^{-1}; n = 0, ν(PF) 910, 836 cm^{-1}, δ(PF$_3$) 560, 545 cm^{-1}.

ν(PF) wavenumbers are found to increase as shown in the series of iron–PF$_3$ complexes: Fe(PF$_3$)$_4^{2-}$ < Fe(PF$_3$)$_5$ < Fe(PF$_3$)$_4$X$_2$ (X = Cl, Br, or I).[355] This can be explained in terms of changes in Fe basicity, leading to less Fe → PF$_3$ back-bonding, with a consequent increase in the P—F bond order.

The complex (104) gives ν(PF) at 867 and 853 cm^{-1}, with ν(CO) at 2104, 2065, 2042, 2031, and 2018 cm^{-1} (C_{2v} symmetry requires the presence of five CO stretches).[356a]

$$(OC)_3Fe\overset{\overset{F_2}{P}\diagup\overset{F_2}{P}}{\cdots\cdots\cdots\cdots}Fe(CO)_3$$
(104)

ν(PF) wavenumbers have been listed[356b] for $(h^2\text{-C}_5\text{H}_6)\text{Fe(PF}_3)_3$, $(h^5\text{-C}_5\text{H}_5)\text{Fe(PF}_3)_2\text{H}$, and $\text{K}^+[(h^5\text{-C}_5\text{H}_5)\text{Fe(PF}_3)_2]^-$.

ν(CO) and ν(PF) bands in the series of complexes Ru(PF$_3$)$_x$(CO)$_{5-x}$, where x = 1—5, have been reported by Udovich and Clark.[357] In each case there are more bands than could be explained by the presence of only one isomer.

Yellow [cis; 2ν(CO) at 2001 cm^{-1}] isomers have been isolated for the complex Ru(CO)$_2$(PPh$_3$)$_2$I$_2$.[358]

---

[351] P. M. Treichel, W. K. Dean, and W. M. Douglas, *J. Organometallic Chem.*, 1972, **42**, 145.
[352] J. A. Bowden, R. Colton, and C. J. Commons, *Austral. J. Chem.*, 1972, **25**, 1393.
[353] E. O. Fischer and W. A. Herrmann, *Chem. Ber.*, 1972, **105**, 286.
[354] T. Kruck and V. Krause, *Z. Naturforsch.*, 1972, **27b**, 302.
[355] T. Kruck, R. Kubelt, and A. Prasch, *Z. Naturforsch.*, 1972, **27b**, 344.
[356] (a) W. M. Douglas and J. K. Ruff, *Inorg. Chem.*, 1972, **11**, 900; (b) T. Kruck and L. Knoll, *Chem. Ber.*, 1972, **105**, 3783.
[357] C. A. Udovich and R. J. Clark, *J. Organometallic Chem.*, 1972, **36**, 353.
[358] J. Jeffery and R. J. Mawby, *J. Organometallic Chem.*, 1972, **40**, C42.

In $CoX_2[Me_2NPF_2]_3$ (X = Br or I) the i.r. spectra show [359] that the absorptions in the region 3000—2700 cm^{-1} are almost identical to those in the free ligand. This is believed to indicate the presence of unco-ordinated N, and therefore co-ordination to the Co *via* the P atom. This conclusion is supported by the increase in wavenumbers of $\nu_s(PF_2)$ on complexing (from 770 cm^{-1} to 812, 805 cm^{-1} for X = I or Br, respectively), which is known to be associated with the presence of M—P bonds. An unassigned i.r. spectrum has also been reported [360] for the closely related complex $Co\{[Me_2N]_2PF\}_3I_2$.

The presence of an intense i.r. band at *ca.* 522 cm^{-1} in the i.r. spectra of $[CoX(hdf)_2(PPh_3)]$ complexes (where X = Cl, Br, I, or NO_3; hdf = benzildioximato) is said to be characteristic of the co-ordinated PPh_3 group.[361] An analogous band at 455 cm^{-1} is found in $[CoX(hdf)(SbPh_3)]$.

Ligand-exchange reactions occur in solution between the dimeric complexes $[RhCl(L)_2]_2$ (L = PF_3 or CO), to give $Rh_2Cl_2(PF_3)_x(CO)_{4-x}$ (x = 1, 2, or 3) as shown by changes in the i.r. spectrum in the $\nu(CO)$ region.[362]

The series of complexes (105; Y = I, CF_3, C_2F_5, n-C_3F_7, or n-C_7F_{15}) all show bands characteristic of complexed PF_3 at 891 ± 8 cm^{-1} and

(105)

873 ± 3 cm^{-1}.[363] The same authors have also reported $\nu(PF)$ data on the following:[364] $(h^5$-$C_5Me_5)Rh(PF_3)_2$, $(h^5$-$C_5Me_5)IrF(PF_2)(PF_3)$, $Re(CO)_3$-$(PF_3)_2Br$, $(h^5$-$C_5H_5)_2Mo_2(CO)_5(PF_3)$, $(h^5$-$C_5H_5)Mo(CO)_2(PF_3)_2$, $(h^5$-$C_5H_5)$-$Fe(CO)(PF_3)I$, $MeMo(CO)_2(PF_3)(h^5$-$C_5H_5)$, and $MeW(CO)_2(PF_3)(h^5$-$C_5H_5)$.

In the series $[RhX(PF_3)_2]_2$, where X = Cl, Br, or I, $\nu(PF)$ falls in the order Cl > Br > I, a trend paralleling that generally found in transition-metal carbonyl halides.[365] Similar data were also reported for $IrCl(PF_3)_2$, $IrCl(PF_3)_4$, $Rh(acac)(PF_3)_2$, $Ir(acac)(PF_3)_2$, $HRh(PF_3)_4$, $HIr(PF_3)_4$, Rh_2-$(PF_3)_8$, $Ir_2(PF_3)_8$, $Rh_3(PF_3)_9$, *etc.*

---

[359] T. Nowlin and K. Cohn, *Inorg. Chem.*, 1971, **10**, 2387.
[360] T. Nowlin and K. Cohn, *Inorg. Chem.*, 1972, **11**, 560.
[361] A. V. Ablov, A. M. Gol'dman, and O. A. Bologa, *Russ. J. Inorg. Chem.*, 1971, **16**, 937.
[362] J. F. Nixon and J. R. Swain, *J.C.S. Dalton*, 1972, 1044.
[363] R. B. King and A. Efraty, *J. Organometallic Chem.*, 1972, **36**, 371.
[364] R. B. King and A. Efraty, *J. Amer. Chem. Soc.*, 1972, **94**, 3768.
[365] M. A. Bennett and D. J. Patmore, *Inorg. Chem.*, 1971, **10**, 2387.

The olefinic CH deformation has been assigned to an i.r. band in the range 971—915 cm⁻¹ in the following:[366] RhCl(bdps),CH₂Cl₂; RhBr-(bdps); IrCl(bdps),CH₂Cl₂; Ir(CO)(bdps),CH₂Cl₂; and IrCl(bdps)(PPh₃), where bdps = 2,2′-bis(diphenylphosphino)stilbene (106).

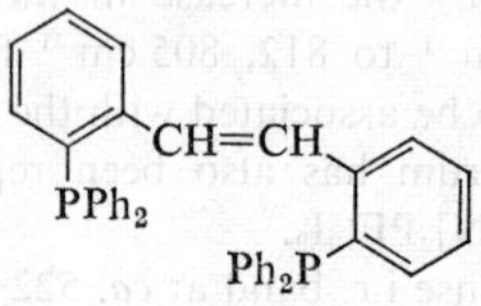

$$\underset{PPh_2}{\bigcirc}CH=CH\underset{Ph_2P}{\bigcirc}$$

(106)

Ni(PH₃)₄ has been prepared and a low-temperature solid-phase Raman spectrum was obtained.[367] This shows ν(PH) at 2299 cm⁻¹, δ(PH) at 1057 cm⁻¹, and ν(Ni—P) at 296 cm⁻¹, characteristic of co-ordinated PH₃.

Internal vibrations of the PMe₃ ligand have been assigned in Ni(PMe₃)₄²⁺ and Ni(PMe₃)₂X₂.[368]

The following assignments have been proposed for NiI₂(CO)(PMe₃)₂: ν(CH) 2975, 2912, 2905 cm⁻¹; ν(CO) 2020, 1974 cm⁻¹; δ(CH₂) 1416, 1408, 1403, 1386, 1297, 1285, and 1279 cm⁻¹; ρ(CH₃) 941, 856, 849 cm⁻¹; ν(P—C) 738, 679 cm⁻¹; δ(CPC) 374, 277, and 266 cm⁻¹.[369]

Some approximate assignments have been given for the PPh₃ bands in (Ph₃P)₂Pt,SiF₄,[370] and (107).[371]

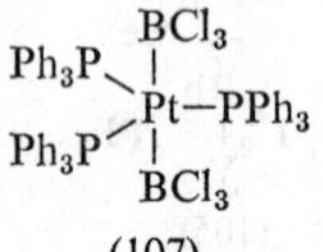

$$\begin{array}{c} \text{BCl}_3 \\ \text{Ph}_3\text{P} \quad | \\ \phantom{\text{Ph}_3\text{P}}\!\text{Pt}\!-\!\text{PPh}_3 \\ \text{Ph}_3\text{P} \quad | \\ \text{BCl}_3 \end{array}$$

(107)

I.r. bands in [PtCl₂{P(OEt)₃}₂] at 800, 980, 1050, and 1165 cm⁻¹ are assigned to vibrations of the P—O—C group.[372]

Unassigned i.r. spectra have been listed for a number of Pt^{II} and Pt^{IV} complexes of cis- and trans-Ph₂PCH=CHPPh₂.[373]

Assignments have been made to the so-called ν(As—Ph) and ν(P—Ph) modes, together with ν(CN), ν(CS), and ν(CO) as appropriate, for [Ag(AsPh₃)₄]X (X = NO₃, ClO₄, or BrO₃) and [Ag(EPh₃)₃X] (X = SCN or NCO; E = P or As).[374]

---

[366] M. A. Bennett, P. W. Clark, G. B. Robertson, and P. O. Whimp, *J.C.S. Chem. Comm.*, 1972, 1011.
[367] M. Trabelsi, A. Louterlier, and M. Bigorgne, *J. Organometallic Chem.*, 1972, **40**, C45.
[368] A. Merle, M. Dartiguenave, and Y. Dartiguenave, *J. Mol. Structure*, 1972, **13**, 413.
[369] M. Pańkowski and M. Bigorgne, *J. Organometallic Chem.*, 1972, **35**, 397.
[370] T. R. Durkin and E. P. Schram, *Inorg. Chem.*, 1972, **11**, 1048.
[371] T. R. Durkin and E. P. Schram, *Inorg. Chem.*, 1972, **11**, 1054.
[372] A. D. Troitskaya and Z. L. Shmakova, *Russ. J. Inorg. Chem.*, 1971, **16**, 872.
[373] R. B. King and P. N. Kapoor, *Inorg. Chem.*, 1972, **11**, 1524.
[374] R. N. Dash and D. V. Ramana Rao, *Z. anorg. Chem.*, 1972, **393**, 309.

5 Oxygen Donors

Molecular Oxygen, Peroxo-, and Hydroxy-complexes.—The potentially terdentate ligand thiodiethanol forms 1 : 1 complexes with the chlorides of Cr^{III}, Mn^{II}, Ni^{II}, Co^{II}, and Cu^{II}, for which i.r. data show that all the hydroxy-groups are co-ordinated to the metal. 1 : 2 Complexes are also formed, in which the presence of i.r. bands as low as 2680 cm^{-1} shows that two of the hydroxy-groups are not co-ordinated to the metal.[375]

ν(OH) wavenumbers and several bands in the 840—1000 cm^{-1} region have been listed for FeL_3^{2+} and RuL_3^{2+} (where L = phenanthroline or bathophenanthroline) and for $Cu(neocuproin)_2^+$. It is claimed that water is co-ordinated to the metal cations, in addition to the three bidentate ligands, although this conclusion appears to be open to question.[376]

The ClO_4^- ion is believed to be co-ordinated to the Fe in $Fe(pq)_2(ClO_4)_2$, in a unidentate fashion [ν_1 = 1040, ν_2 = 925 cm^{-1}; pq = (108)].[377]

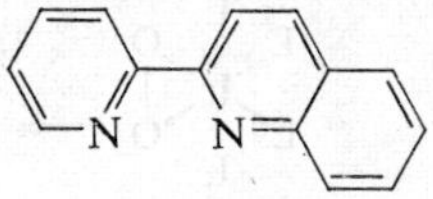

(108)

The main i.r. absorptions of $[RuO_4(py)_2]$, $[Ru(OH)_2(py)_2(bipy)]$, and $[Ru(OH)_2(py)_2(phen)]$ have been listed.[378] The characteristic bands of py were assigned, and for the OH complexes ν(OH) at 3380, 3390 cm^{-1}, respectively.

The dioxygenyl complexes $Ru(O_2)(CO)_2(PPh_3)_2$ and $Os(O_2)(CO)_2(PPh_3)_2$ give bands assignable as $\nu(M-O_2)$ at 849 cm^{-1} (M = Ru), 820 cm^{-1} (M = Os).[379] The closely related complex $Ru(PhC\equiv CPh)(CO)_2(PPh_3)_2$ possesses an i.r. band assigned to $\nu(C\equiv C)$ at 1776 cm^{-1}.

Alkyldioxycobaloximes (109; L = py or H_2O), prepared by the irradiation of some alkyl(pyridinato)cobaloximes in the presence of oxygen, give

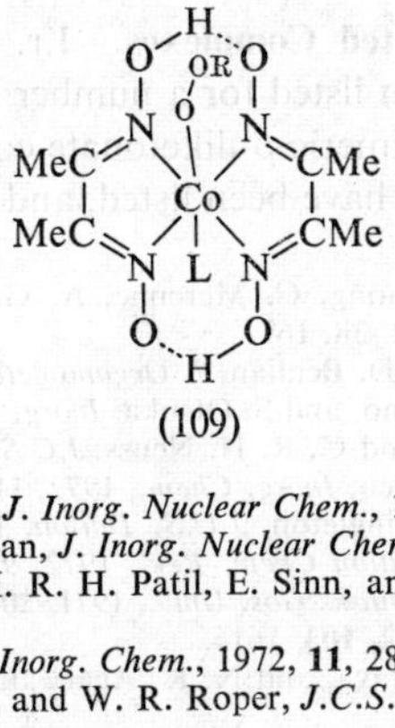

(109)

[375] B. Sen and D. A. Johnson, *J. Inorg. Nuclear Chem.*, 1972, **34**, 609.
[376] S. Burchett and C. E. Meloan, *J. Inorg. Nuclear Chem.*, 1972, **34**, 1207.
[377] C. M. Harris, S. Kokot, H. R. H. Patil, E. Sinn, and H. Wang, *Austral. J. Chem.*, 1972, **25**, 1631.
[378] T. Ishiyama and Y. Koda, *Inorg. Chem.*, 1972, **11**, 2837.
[379] B. E. Cavit, K. R. Grundy, and W. R. Roper, *J.C.S. Chem. Comm.*, 1972, 60.

a characteristic i.r. band due to the peroxy-grouping, in the 800—900 cm^{-1} region.[380, 381]

In the dioxygen complexes $RhX(O_2)(PPh_3)_2L$, where X = Cl, Br, or I and L = an isocyanide, $\nu(O-O)$ and $\nu(Rh-O)$ (presumably strongly coupled) are found at ca. 890 cm^{-1}, ca. 580 cm^{-1}, respectively. These assignments are based upon the use of ^{18}O-substitution, and the spread in the values over all the complexes studied was very small.[382a]

$\nu(O-O)$ in $IrX(OOBu^t)_2(CO)L_2$, where X = Cl or Br, L = PPh$_3$ or AsPh$_3$, is between 880 and 890 cm^{-1}.[382b]

$[Ir(O_2)(dp)_2]Cl$, where dp = cis-vinylenebis(diphenylphosphine), gives an i.r. band assigned as $\nu(Ir-O_2)$ at 843 cm^{-1}.[383]

A band described as $\nu(O-O)$ is seen at 838 cm^{-1} in the solid-state i.r. spectrum of $[Ir(O_2)(AsMe_2Ph)_4]^+BPh_4^-$. The proposed structure of the cation is (110).[384a]

$$\begin{array}{c} L \\ L \diagdown | \diagup O \\ Ir \\ L \diagup | \diagdown O \\ L \end{array}$$

(110)

A review of compounds of the Pt metals containing chelating dioxygen includes data for and discussion of the 800—900 cm^{-1} i.r. band assigned to $\nu(O-O)$.[384b]

Partial i.r. data are given for the complexes $MCl_3,7H_2O$ (M = La, Ce, or Pr) and $M'Cl_3,6H_2O$ (M' = Pr, Nd, Sm, Eu, Gd, Tb, Dy, Ho, Er, Tm, Yb, Lu, and Y) and discussed in terms of the extent of hydration of the M^{3+} ions.[384c]

I.r. wavenumbers have been listed for $La(IO_3)_3,3H_2O$, $La(IO_3)_3$, and $LaIO_5,nH_2O$ (where n = 1, 2, or 3). The only suggested assignments were to $\nu(I-O)$ (ca. 800 cm^{-1}), $\nu(I-O-I)$ (ca. 550 cm^{-1}), and to $\delta(I-O)$ (ca. 450 cm^{-1}).[385]

Acetylacetonates and Related Complexes.—I.r. data, for the purposes of characterization, have been listed for a number of Cr^{III} β-diketonates.[386]

The i.r. spectra of the dimeric β-diketonate complexes $[(\beta\text{-dik})_2Fe(OR)]_2$ (111; R = Me, Et, or Pr^i) have been listed, and bands characteristic of the

---

[380] C. Fontaine, K. N. V. Duong, C. Merenne, A. Gaudemer, and C. Giannotti, *J. Organometallic Chem.*, 1972, **38**, 167.
[381] C. Giannoti, B. Septe, and D. Benlian, *J. Organometallic Chem.*, 1972, **39**, C5.
[382] (a) A. Nakamura, Y. Tatsuno, and S. Otsuka, *Inorg. Chem.*, 1972, **11**, 2058; (b) B. L. Booth, R. N. Haszeldine, and G. R. H. Neuss, *J.C.S. Chem. Comm.*, 1972, 1074.
[383] S. Doronzo and V. D. Bianco, *Inorg. Chem.*, 1972, **11**, 466.
[384] (a) L. M. Haines and E. Singleton, *J.C.S. Dalton*, 1972, 1891; (b) V. J. Choy and C. H. O'Connor, *Co-ordination Chem. Rev.*, 1972, **9**, 145; (c) S. E. Kharzeeva and V. V. Serebrennikov, *Tr. Tomsk. Gos. Univ.*, 1971, **204**, 350.
[385] M. Odehnal, *Monatsh.*, 1972, **103**, 1615.
[386] A. D. Taneja, K. P. Srivastava, and N. K. Agarwal, *J. Inorg. Nuclear Chem.*, 1972, **34**, 3573.

```
    O   R   O
   ╱ ╲ ╱ ╲ ╱ ╲
  O   O   O
  │   Fe  │   Fe  │
  O   O   O
   ╲ ╱ ╲ ╱ ╲ ╱
    O   R   O
```
(111)

various —OR groupings identified [β-dik = dpm (2,2,6,6-tetramethylheptane-2-5-dione) or acac].[387]

The following complexes have been prepared which contain acetylacetone as a neutral ligand:[388, 389] $CoCl_2(acacH)$, $CoBr_2(acacH)$, $ZnCl_2(acacH)$, $NiBr_2(acacH)_2$, $CrCl_2(acac)(acacH)$, $CrBr_2(acac)(acacH)$, and $MnBr_2$-$(acacH)_2$. The Co, Zn, and Ni complexes gave ketonic $\nu(C=O)$ bands at ca. 1720 cm^{-1} and ca. 1700 cm^{-1}; the Cr species gave these and others due to the acac$^-$ chelate, while the Mn complex gave bands only at 1627 and 1564 cm^{-1}, assigned to $\nu(C=O)$ and $\nu(C=C)$ of the enolic form of acacH.

The preparations of $CoCl_2(L)$, $CoBr_2(L)$, and $ZnCl_2(L)$, where L = ethyl acetoacetate (etacH) or ethyl malonate (etmalH), and of $MnBr_2$-(etmalH) have been reported. Small shifts in the keto $\nu(C=O)$ were observed on co-ordination, and $\nu(C=O)$ of the enolic tautomer of etmalH was also seen (1618—1658 cm^{-1}), although this is not present in the free ligand. For the etacH complexes, the keto/enol ratios were similar to those in the free ligand.[390]

The i.r. spectra of (L)Co(salen), where L = β-diketonates such as acac, bzac (benzylacetonate), or tfac (trifluoroacetylacetonate) and salen = NN-ethylenebis(salicylaldimine), are all consistent with the presence of oxygen-chelating β-diketonate ligands.[391]

Some i.r. data have been listed for a number of Co^{II} and Co^{III} complexes of 1,5-dialkylpentane-2,4-diones,[392] and for a closely related series of Ni^{II} complexes [$NiL_2(H_2O)_2]_2$. (L = heptane-3,5-dionato, nonane-4,6-dionato, 2,6-dimethylheptane-3,5-dionato, 2,8-dimethylnonane-4,6-dionato, and tridecane-6,8-dionato).[393] In the latter, co-ordinated water was detected by the presence of a sharp, strong i.r. band in the range 3350—3400 cm^{-1}; the electronic spectra were consistent with six-co-ordination at the Ni, hence the dimeric formulation (two bridging oxygens).

A number of hexafluoroacetylacetonato complexes of silver(I), of the general type Ag(hfacac)(olefin), have been prepared.[394] In all cases the β-diketonate ligand was acting as an OO'-chelate [giving bands due to $\nu(C\doteq O)$ in the region 1630—1670 cm^{-1}]. This behaviour is in contrast to

[387] C.-H. Wu, G. R. Rossman, H. B. Gray, G. S. Hammond, and H. J. Schugar, *Inorg. Chem.*, 1972, **11**, 990.
[388] Y. Nakamura, M. Gotani, and S. Kawaguchi, *Bull. Chem. Soc. Japan*, 1972, **45**, 457.
[389] Y. Nakamura, K. Isobe, H. Morita, S. Yamazaki, and S. Kawaguchi, *Inorg. Chem.*, 1972, **11**, 1573.
[390] H. Morita, Y. Nakamura, and S. Kawaguchi, *Bull. Chem. Soc. Japan*, 1972, **45**, 2468.
[391] R. J. Cozens and K. S. Murray, *Austral. J. Chem.*, 1972, **25**, 911.
[392] I. Yoshida, H. Kobayashi, and K. Ueno, *Bull. Chem. Soc. Japan*, 1972, **45**, 2768.
[393] I. Yoshida, H. Kobayashi, and K. Ueno, *Bull. Chem. Soc. Japan*, 1972, **45**, 1411.
[394] W. Partenheimer and E. H. Johnson, *Inorg. Chem.*, 1972, **11**, 2840.

that of Pd(hfacac)$_2$PPh$_3$ for example, in which the β-diketonate is bonded *via* one oxygen and the γ-carbon atoms, giving bands at 1768, 1723 cm^{-1} [ν(C=O)] and 1656, 1629 cm^{-1} [ν(C$\cdots$O)].

In the complexes (112; R^1 = R^2 = H), (112; R^1 = Me, R^2 = H), and (112; R^1 = R^2 = Me), one band is found at *ca.* 1720 cm^{-1} that is due to

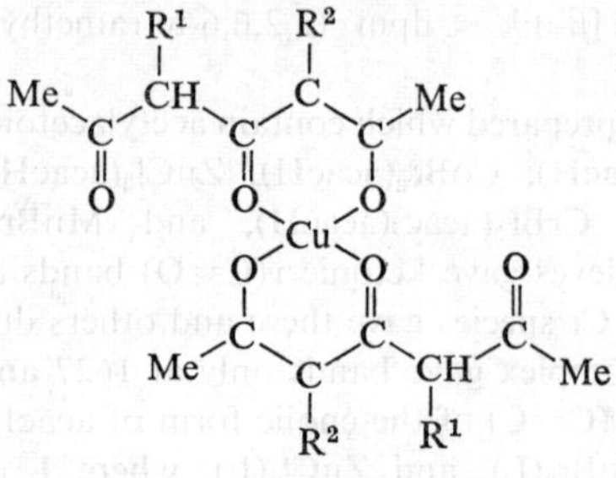

(112)

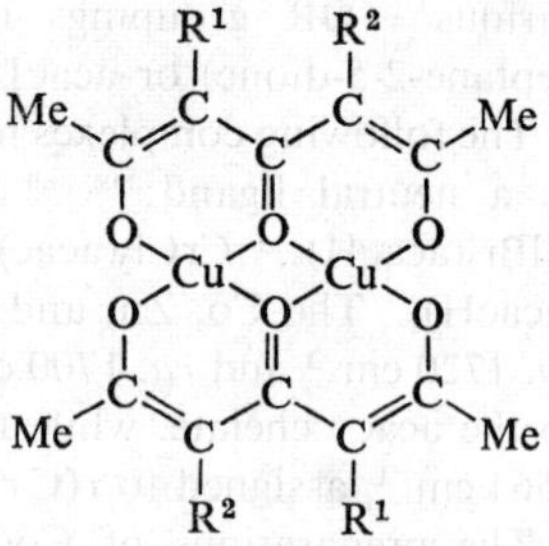

(113)

unco-ordinated ν(C=O), with a complex pattern between 1600 and 1500 cm^{-1} due to chelate ring stretches.[395] The dinuclear series (113; R^1 = R^2 = H), and (113; R^1 = H, R^2 = Me) lack the former feature.

I.r. data have been listed for a number of Hg^{II} β-diketonate complexes.[396]

Raman (M = La, Pr, or Gd) and i.r. (M = La, Ce, Pr, Nd, or Gd) spectra have been obtained and assigned for the complexes M(dpm)$_3$, where dpm = (114).[397] Detailed assignments were proposed for internal

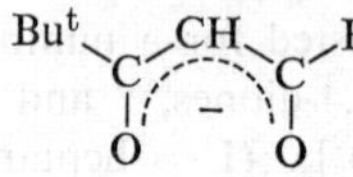

(114)

ligand vibrations (> 400 cm^{-1}), with 'ν(M—O)' at 392—403 cm^{-1} and δ(o.o.p.)(OMO) at 245—248 cm^{-1}.

I.r. spectra were published (with no lists of assignments, or even frequencies) for Nd(thd)$_3$, Nd(thd)$_3$,DMF, Pr(thd)$_6$, Er(thd)$_3$,DMF, and Er(thd)$_3$ (thd = the anion derived from 2,2,6,6-tetramethylheptane-3,5-dione, *i.e.* the 'dpm' of the previous reference).[398]

ν(C=O) wavenumbers for (C$_6$F$_5$)$_2$TlX, (C$_6$F$_5$)$_2$(Ph$_3$PO)TlX, and (C$_6$F$_5$)$_2$-(Ph$_3$AsO)TlX [X = CH$_3$COCHCOCH$_3$ (acac), CF$_3$COCHCOCH$_3$ (tfac), CF$_3$COCHCOCF$_3$ (hfac), PhCOCHCOCH$_3$ (bzac), PhCOCHCOPh (dbm), or quinolin-8-olate] are in the region expected for chelating

[395] F. Sagara, H. Kobayashi, and K. Ueno, *Bull. Chem. Soc. Japan*, 1972, **45**, 794.
[396] A. D. Taneja, K. R. Srivastava, and N. K. Agarwal, *J. Inorg. Nuclear Chem.*, 1972, **34**, 2980.
[397] H.-Y. Lee, F. F. Cleveland, J. S. Ziomek, and F. Jarke, *Appl. Spectroscopy*, 1972, **26**, 251.
[398] V. A. Mode and D. H. Sisson, *Inorg. Nuclear Chem. Letters*, 1972, **8**, 357.

β-diketonate groups. In the second and third series, $\nu(P=O)$ or $\nu(As=O)$ is lowered, indicating co-ordination *via* the O atom.[399]

The anionic 2,4-pentanedionato complexes $[X_4Sn(C_5H_7O_2)]^-$ (X = Cl, Br, or I) all show bands at *ca.* 1560 cm^{-1} and *ca.* 1580 cm^{-1} due to the OO'-chelated ligands.[400]

$\nu(C=O)$, $\nu(C=C)$, and $\nu(Sb-O)$ were listed for the monomeric dihalogenodiaryl(acetylacetonato)Sb^V compounds $(p\text{-}Y\cdot C_6H_4)_2SbX_2(acac)$, where X = F, Cl, or Br and Y = NO$_2$, Cl, H, Me, or MeO.[401]

I.r. data for a series of compounds $R_nSbCl_{4-n}(acac)$ [$\nu(Sb-O)$, $\nu(C=O)$, $\nu(C=C)$] show that increased substitution of Cl by R leads to a weakening of the Sb–(acac) interaction [as shown by decreased $\nu(Sb-O)$ and increased $\nu(C=O)$].[402]

Carboxylates.—A normal co-ordinate analysis has been performed to investigate the influence of ligand polarization and the strength of the M—O bonds on the frequencies of the in-plane vibrations of the group (115; M = C or N).[403a]

$$\begin{array}{c} O \\ \| \\ X \\ / \ \backslash \\ O \quad\quad O \\ | \quad\quad | \\ M \quad\quad M \end{array}$$
(115)

I.r. spectra have been reported for $C_{17}H_{35}CO_2H$ and CCl_3CO_2H, their Co, Fe, Ni, Mn, Mg, and Cu salts, and complexes of these salts with the acids.[403b]

Assignments have been proposed [404] for the $\nu_{as}(CO_2)$ and $\nu_s(CO_2)$ vibrations in a number of eight-co-ordinate lactates, mandelates, and isopropylmandelates of Zr^{IV} and Hf^{IV}.

Some i.r. wavenumbers were listed in a paper [405] on Cr, Mn, Fe, and Al complexes of $C(OH)(CF_3)_2CO_2H$.

The i.r. spectra of the bis(phenylglycollates) $(PhCHOHCO_2)_2M$, where M = Mn^{II}, Co^{II}, Ni^{II}, Cu^{II}, or Zn^{II}, show bands due to $\nu(OH)$ at 3130—3270 cm^{-1}, $\nu(CO_2)$ at 1545—1585 cm^{-1}, and $\delta(OH)$ at 1003—1025 cm^{-1}. The relevance of these data to possible modes of co-ordination was discussed.[406]

[399] G. B. Deacon and V. N. Garg, *Austral. J. Chem.*, 1971, **24**, 2519.
[400] D. W. Thompson, J. F. Lefelhocz, and K. S. Wong, *Inorg. Chem.*, 1972, **11**, 1139.
[401] N. Nishii and R. Okawara, *J. Organometallic Chem.*, 1972, **38**, 335.
[402] H. A. Meinema, A. Mackor, and J. G. Noltes, *J. Organometallic Chem.*, 1972, **37**, 285.
[403] (*a*) B. Taravel, G. Chauvet, P. Delorme, and V. Lorenzelli, *J. Mol. Structure*, 1972, **13**, 283; (*b*) O. E. Lavenevskii and E. G. Yarkova, *Izvest. Akad. Nauk. Kirg. S.S.R.*, 1972, 56.
[404] E. M. Larsen and E. H. Homeier, *Inorg. Chem.*, 1972, **11**, 2687.
[405] J. T. Price, A. J. Tomlinson, and C. J. Willis, *Canad. J. Chem.*, 1972, **50**, 939.
[406] K. N. Kovalenko, D. V. Kazachenko, V. P. Kurbatov, and L. G. Kovaleva, *Russ. J. Inorg. Chem.*, 1971, **16**, 1303.

Fe(O$_2$CH)L$_2$, where L = PEtPh$_2$, has been prepared by the insertion of CO$_2$ into the Fe—H bond of FeH$_4$L$_3$ or FeH$_2$(N$_2$)L$_3$.[407] The presence of the formate group is indicated by the i.r. absorption at 1590 cm^{-1} [ν_{as}(CO$_2$)] and 1370 cm^{-1} [ν_s(CO$_2$)].

The Ru carboxylato-complexes [Ru$_2$(HCO$_2$)$_4$Cl], [Ru$_2$(MeCO$_2$)$_4$Cl], [Ru$_2$(CH$_2$ClCO$_2$)$_4$Cl], and [Ru(OH)(CCl$_3$CO$_2$)$_2$(H$_2$O)] all give rise to characteristic ν_{as}(CO$_2$), ν_s(CO$_2$) bands of co-ordinated carboxylate.[408]

In (HCOO)Ru(PPh$_3$)$_3$(H)(toluene), ν_{as}(CO$_2$) is seen at 1553 cm^{-1}, with ν_s(CO$_2$) at 1310 cm^{-1} and ν(CH) (of the formate ligand) at 2895 and 2805 cm^{-1}. In Rh$_2$H$_2$(CO$_2$)(PPh$_3$)$_6$(toluene), the co-ordinated carbon dioxide gives ν_{as}(CO$_2$), ν_s(CO$_2$) bands at 1460, 1300 cm^{-1}, respectively.[409]

The distinction, using i.r. spectroscopy, between unco-ordinated, unidentate, and bidentate carbonate ion has been examined during a study of carbonato-cobaltate complexes, e.g. in [Co(NH$_3$)$_6$]$^{3+}$(Cl,CO$_3$)$^{3-}$, ν_3 of free CO$_3^{2-}$ is at ca. 1380 cm^{-1}, whereas in [Co(NH$_3$)$_5$(CO$_3$)]Br (unidentate carbonate) this feature is split, giving absorptions at 1370 and 1450 cm^{-1}. In [Co(NH$_3$)$_4$(CO$_3$)]Cl (with bidentate carbonate) this splitting is even more pronounced (1255, 1592 cm^{-1}).[410]

As part of a study of the kinetics of reactions involving μ-amido-μ-carboxylato-cobalt(III) complexes, Scott and Sykes have listed the values of ν_{as}(CO$_2$) and ν_s(CO$_2$) for acetate and formate ligands, and the separations of these bands with different degrees of co-ordination[411] (see Table 9).

Table 9 Carboxylate stretching frequencies (wavenumbers/cm^{-1}) in some acetato- and formato-complexes of cobalt(III)

Complex	ν_{as}(CO)	ν_s(CO$_2$)	$\Delta\nu$
Na$^+$MeCO$_2^-$	1578	1425	153
[Co(NH$_3$)$_5$(MeCO$_2$)]$^{2+}$	1603	1380	223
[(NH$_3$)$_4$Co-μ-(NH$_2$, MeCO$_2$)-Co(NH$_3$)$_4$]$^{4+}$	1530	1410	120
[(NH$_3$)$_3$Co-μ-(OH, OH, MeCO$_2$)-Co(NH$_3$)$_3$]$^{3+}$	1535	1440	95
Na$^+$HCO$_2^-$	1590	1355	235
[Co(NH$_3$)$_5$(HCO$_2$)]$^{2+}$	1640	1345	295
[(NH$_3$)$_4$Co-μ-(NH$_2$, HCO$_2$)-Co(NH$_3$)$_4$]$^{4+}$	1570	1365	205
[(NH$_3$)$_3$Co-μ-(OH, OH, HCO$_2$)-Co(NH$_3$)$_3$]$^{3+}$	1550	1355	195

I.r. spectroscopy has been used to confirm that the acetato C=O group is co-ordinated to Co in the ethylenediamine-NN'-diacetato-complexes of Co^{III}.[412]

ν(C—O) wavenumbers of co-ordinated carbonate ion in complexes [Co(N)$_4$CO$_3$]$^+$, where (N)$_4$ = 4NH$_3$, [2(en)], etc., show a linear relationship with the C—O bond length.[413]

[407] V. D. Bianco, S. Doronzo, and M. Rossi, J. Organometallic Chem., 1972, 35, 337.
[408] M. Mukaida, T. Nomura, and T. Ishimori, Bull. Chem. Soc. Japan, 1972, 45, 2143.
[409] S. Komiya and A. Yamamoto, J. Organometallic Chem., 1972, 46, C58.
[410] R. D. Gillard, P. R. Mitchell, and M. G. Price, J.C.S. Dalton, 1972, 1211.
[411] K. L. Scott and A. G. Sykes, J.C.S. Dalton, 1972, 2364.
[412] K. Kuroda, Bull. Chem. Soc. Japan, 1972, 45, 2176.
[413] V. S. Sastri, Inorg. Chim. Acta, 1972, 6, 264.

I.r. (4000—200 cm^{-1}) spectra have been shown for the Co^{II} and Cu^{II} salts of d-, l-, and dl-forms of mandelic acid, M(PhCHOHCO$_2$)$_2$, and bands have been listed with some proposed assignments.[414]

Two different forms of each of the ClO$_4^-$ and Br$^-$ salts of (116) have been obtained, showing different i.r. spectra. The spectra of the unprotonated compounds, of the new complexes (117), and of various deuterio-derivatives in the ν(CO) ranges (1100—1800 cm^{-1}) were discussed in detail,

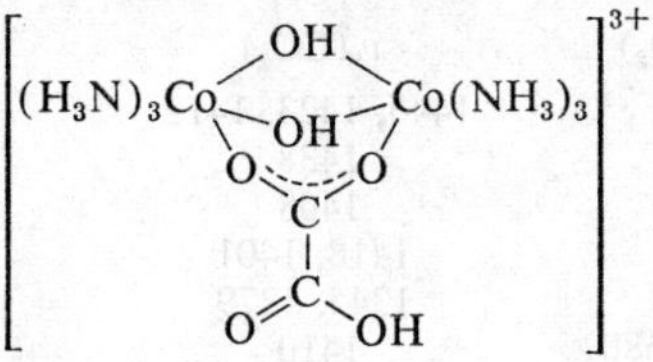

(116)

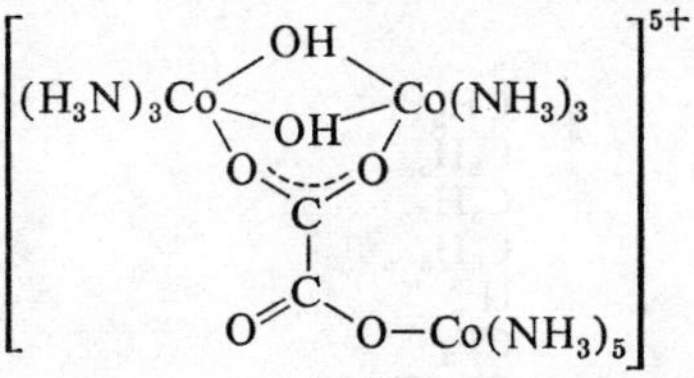

(117)

including the possibility of distinguishing between uni-, bi-, and ter-dentate oxalato-cobalt(III) amines.[415]

ν_{as}(CO$_2$) of the carboxylato-groups in the complexes [Co(RCO$_2$)(NH$_3$)$_5$]$^{2+}$, [Co$_2${RCO$_2$,(OH)$_2$}(NH$_3$)$_6$]$^{3+}$, and [Co{(RCO$_2$)$_2$(OH)}(NH$_3$)$_3$]$^{3+}$, where R = H, Me, CH$_2$Cl, CHCl$_2$, CCl$_3$, CH$_2$F, CHF$_2$, CF$_3$, or CO$_2^-$, have been listed.[416]

In the i.r. spectrum of [CoL(H$_2$O)],2H$_2$O, where H$_3$L = trimethylenediaminetriacetic acid, the presence of a strong sharp absorption at 1620 cm^{-1} is believed to indicate that all the three carboxylato-groups of L are co-ordinated.[417]

Unassigned i.r. spectra have been listed for potassium *cis*- and *trans*-bis(oxalato)diaquorhodate(III).[418]

The complex (118) gives bands due to ν(C≡O) at 2075, 2005, and 1988 cm^{-1}, and ν(C=O) of the bridging carboxylate group at 1580, 1440 cm^{-1} [$\nu_{as,s}$(CO$_2$), respectively]. The monomeric compound (119) gives ν(C≡O) at 1980 cm^{-1}, with ν(C=O) at 1610, 1470 cm^{-1}. These are

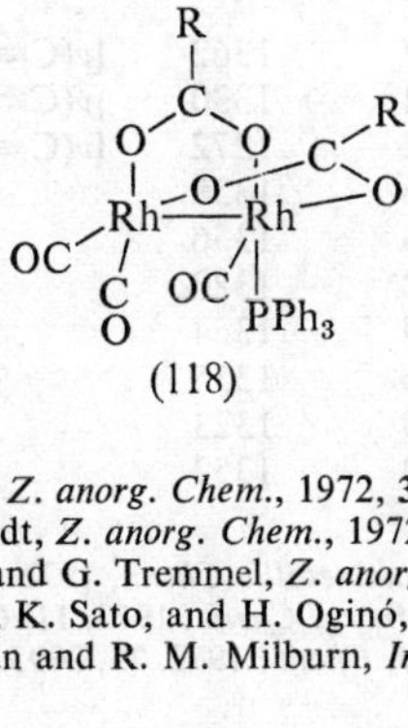

(118) 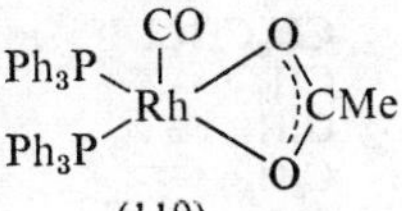 (119)

[414] A. Ranade, *Z. anorg. Chem.*, 1972, **388**, 105.
[415] K. Wieghardt, *Z. anorg. Chem.*, 1972, **391**, 142.
[416] H. Siebert and G. Tremmel, *Z. anorg. Chem.*, 1972, **390**, 292.
[417] M. Tanaka, K. Sato, and H. Oginó, *Inorg. Nuclear Chem. Letters*, 1972, **8**, 93.
[418] N. S. Rowan and R. M. Milburn, *Inorg. Chem.*, 1972, **11**, 639.

in agreement with the structure shown, *i.e.* the acetate is bidentate and the rhodium is five-co-ordinate.[419]

$\nu_{as}(CO_2)$ and $\nu_s(CO_2)$ have been listed for $IrClH(PPh_3)_2(RCO_2)$, where R = Me, Et, Pr, Ph, H, CF_3, CH_3CHCl, or p-$O_2N \cdot C_6H_4$, and $IrClO$-$(PPh_3)_2(CH_3CO_2)$ (see Table 10) and for a number of complexes of the

Table 10 Carboxyl stretching vibrations (wavenumbers/cm^{-1}) for $IrClH(PPh_3)_2(RCO_2)$

R	$\nu_{as}(CO_2)$	$\nu_s(CO_2)$
CH_3	1535	1445, 1423, 1412
C_2H_5	1525	1438
C_3H_7	1530	1408
C_6H_5	1520	1418, 1401
H	1550	1345, 1279
CF_3	1710, 1680	1410
CH_3CHCl	1547	1410
p-$O_2N \cdot C_6H_4$	1534	1420
$IrClD(PPh_3)_2(CH_3CO_2)$	1535	1338

general formula $IrClHL_2L'(RCO_2)$, where L = PPh_3, L′ = CO, py, PhCN, PMe_2Ph, or p-Me·C_6H_4CN, and R = various allyl groups, *etc.* (see Table 11).[420]

Cookson and Deacon have reported similar data for a number of nickel carboxylates[421] (see Table 12). The separations are rather large,

Table 11 Carboxyl stretching vibrations (wavenumbers/cm^{-1}) for $IrClH(PPh_3)_2L'(RCO_2)$

L′	R	$\nu_{as}(CO_2)$	$\nu_s(CO_2)$	
CO	CH_3	1631 1613	1310	[$\nu(C\equiv O)$ 2026]
CO	CH_3 (D complex)	1634 1618	1311 1351	[$\nu(C\equiv O)$ 2060]
CO	C_2H_5	1626	1325	[$\nu(C\equiv O)$ 2010, 2018]
CO	C_3H_7	1622	1335	[$\nu(C\equiv O)$ 2035]
CO	C_6H_5	1621 1612	1338	[$\nu(C\equiv O)$ 2025]
CO	H	1687	1362	[$\nu(C\equiv O)$ 2036]
CO	CF_3	1654	1330	[$\nu(C\equiv O)$ 2005]
CO	CH_3CHCl	1632	1272	[$\nu(C\equiv O)$ 2025]
C_5H_5N	CH_3	1636	1355	
C_5H_5N	C_2H_5	1636	1356	
C_5H_5N	C_6H_5	1632	1320	
C_6H_5CN	CH_3	1639	1354	
C_6H_5CN	C_3H_7	1626	1355	
PMe_2Ph	C_6H_5	1630	1323	
p-Me·C_6H_4CN	CH_3	1630	1352	

[419] G. Csontos, B. Heil, and L. Markó, *J. Organometallic Chem.*, 1972, **37**, 183.
[420] S. A. Smith, D. M. Blake, and M. Kubota, *Inorg. Chem.*, 1972, **11**, 660.
[421] P. G. Cookson and G. B. Deacon, *Austral. J. Chem.*, 1972, **25**, 2095.

Table 12 *Carboxyl stretching vibrations (wavenumbers/cm⁻¹) for some Ni carboxylate complexes*

Compound	$\nu_{as}(CO_2)$	$\nu_s(CO_2)$	$\Delta\nu$
$(C_6F_5CO_2)_2Ni(bipy),2H_2O$	1610	1365	245
$(C_6F_5CO_2)_2Ni(bipy),H_2O$	1620	1373	247
$(p\text{-MeO}\cdot C_6F_4\cdot CO_2)_2Ni(bipy),2H_2O$	1640	1370	270
$(p\text{-MeO}\cdot C_6F_4\cdot CO_2)_2Ni(bipy)$	1644	1381	263
$(p\text{-MeO}\cdot C_6F_4\cdot CO_2)_2Ni(phen),H_2O$	1627	1378	249
$(p\text{-MeO}\cdot C_6F_4\cdot CO_2)_2Ni(phen)$	1620	1375	245
$(p\text{-EtO}\cdot C_6F_4\cdot CO_2)_2Ni(bipy),2H_2O$	1639	1370	269
$(p\text{-EtO}\cdot C_6F_4\cdot CO_2)_2Ni(bipy)$	1644	1385	259
$(p\text{-EtO}\cdot C_6F_4\cdot CO_2)_2Ni(phen),H_2O$	1620	1376	244

but the complexes appear to involve bidentate co-ordination of the carboxylates, and so it is suggested that they must be unsymmetrically bonded to the metal.

$\nu(CO)$ wavenumbers have been reported [422] for (chel)(bipy)Pd⁰ (at 1696, 1650 cm⁻¹), where chel = dimethylfumarato anion.

$[Pt(PPh_2Me)_2CO_3]$ gives bands due to the co-ordinated carbonate ion at 1670, 1630, 1290, 985, and 820 cm⁻¹; $[Pt(PPh_2Me)_2(C_2O_4)]$,EtOH gives features characteristic of the oxalate at 1705, 1680, 1660, 1370, and 790 cm⁻¹, and $\{Pt(PPh_3)_2[C(O)OEt]\}_2$ gives bands due to C(O)OEt at 1630 and 1015 cm⁻¹.[423]

Low-temperature, solid-phase i.r. spectra have been obtained for copper(II) acetate monohydrate, and also for the anhydrous salt.[424] Nearly all of the acetate fundamentals were observed to be split, in similar manner to those in the acetic acid dimer. Temperature-dependence studies of the i.r. bands confirm the antiferromagnetic transitions occurring in these compounds.

The only differences in the i.r. spectra of copper formate tetrahydrate in its antiferro- and para-electric phases lie in the librational and translational modes of the crystal H_2O molecules.[425]

In $(C_6Cl_5)Hg(OCOCF_3)$, $\nu_{as}(CO_2)$ is at 1684, 1675 cm⁻¹, $\nu_s(CO_2)$ is at 1420 cm⁻¹, and $\delta(OCO)$ is at 740 cm⁻¹. For $(C_6Cl_5)Hg(OCOCHF_2)$, the corresponding modes give bands at 1647, 1635 cm⁻¹, 1426 cm⁻¹, and 724 cm⁻¹, respectively.[426]

I.r. spectra of neodymium monoethylamine diacetate nitrate, $Nd(L)NO_3$,·1.33H_2O, suggest that both unidentate [$\nu(C=O)$ 1578 cm⁻¹] and bidentate [$\nu(C=O)$ 1609 cm⁻¹] carboxylate groups are present.[427] Shifts in $\nu(NH)$ on complexing are also diagnostic of Nd—N bonding.

---

[422] T. Ito, Y. Takahashi, and Y. Ishii, *J.C.S. Chem. Comm.*, 1972, 629.
[423] D. M. Blake and L. M. Leung, *Inorg. Chem.*, 1972, **11**, 287.
[424] A. M. Heyns, *J. Mol. Structure*, 1972, **11**, 93.
[425] J. Hiraishi, *Bull. Chem. Soc. Japan*, 1972, **45**, 128.
[426] R. J. Bertino, G. B. Deacon, and F. B. Taylor, *Austral. J. Chem.*, 1972, **25**, 1645.
[427] N. I. Sevost'yanova, K. F. Belaeva, and L. I. Martynenko, *J. Struct. Chem.*, 1971, **12**, 162.

A few assignments of $\nu(CO_2)$ modes have been made, from the i.r. spectra of lanthanum iminodiacetates, $LnHZ_2$.[428] I.r. wavenumbers of the CO_3^{2-} ion in $La_3Cl(CO_3)_4$ have been reported.[429]

The following assignments have been made for $Nd_2(C_2O_4)_3,10H_2O$, $Nd_2(C_2O_4)_3,10D_2O$, and $Nd_2(C_2O_4)_3,6D_2O$: $\nu_s(CO_2)$ 1312—1310 cm^{-1} and 1355—1340 cm^{-1}; $\delta(H_2O) + \nu_{as}(CO)_2$ 1603—1590 cm^{-1}; $\delta(OCO)$ 800—796 cm^{-1} and 480—470 cm^{-1}; $\nu(C-C)$ 900 cm^{-1}.[430a]

I.r. spectra of hydrated and anhydrous pimelates, azelates, and sebacates of La, Sm, Gd, and Lu have been obtained.[430b]

Plutonium formate, $Pu(O_2CH)_3$, has i.r. absorptions at 2910 cm^{-1} [$\nu(CH)$], 1585 cm^{-1} [$\nu_{as}(CO_2)$], 1423, 1401 cm^{-1} [$\delta(CH)$], 1350 cm^{-1} [$\nu_s(CO_2)$], and 779 cm^{-1} [$\delta(OCO)$]. No $\nu(Pu-O)$ band was found (above 200 cm^{-1}), which is in agreement with the highly ionic bonding.[431]

In (120), the Me_2NCO_2 group gives i.r. absorptions at 1605 and 1510 cm^{-1}.[432]

(120)

A number of assignments (summarized in Table 13) have been proposed for the i.r. and Raman spectra of diethylindium acetate.[433]

$\nu(C=O)$, $\nu(C-O)$, and $\delta(O=C-O)$ are found in the usual spectral ranges for $(R_4N)_2[MAr_2(ox)_2]$, where ox = oxalato; R = Bu^n, M = Si,

Table 13 Vibrational assignments for $Et_2In(OOCMe)$ (all figures are wavenumber/cm^{-1})

I.r.	Raman	Assignment
1525 vs, br	1520 vw, br	$\nu_{as}(CO_2)$
1465 vs	1465 ⎱ s	$\nu_s(CO_2)$
	1462 ⎰	
1176 m	1179 s	$\omega_s(In-CH_2)$
644 vs	640 vvw, br	$\rho(In-CH_2) + \delta(CO_2)$
515 s	518 w	$\nu_{as}(InC_2)$
466 ms	469 vs	$\nu_s(InC_2)$

[428] R. I. Badalova, G. N. Kupriyanova, N. D. Mitrofanova, L. I. Martynenko, and V. I. Spitsyn, Russ. J. Inorg. Chem., 1971, **16**, 1556.
[429] R. Aumont, F. Gouet, M. Passaret, and M. P. Bottorel, Compt. rend., 1972, **275**, C, 491.
[430] (a) G. V. Bezdenezhnykh, E. I. Krylov, and V. A. Sharov, Russ. J. Inorg. Chem., 1971, **16**, 1563; (b) B. S. Azikov, S. E. Kharzeeva, and V. V. Serebrennikov, Tr. Tomsk Gos. Univ., 1971, **204**, 289, 294, 299.
[431] L. R. Crisler, J. Inorg. Nuclear Chem., 1972, **34**, 3263.
[432] W. Petz and G. Schmid, J. Organometallic Chem., 1972, **35**, 321.
[433] H.-D. Hansen, J. Organometallic Chem., 1972, **39**, C37.

Ar = ½(ox), Ph, p-F·C_6H_4, p-Cl·C_6H_4, or p-Me·C_6H_4; R = Bu^n, M = Ge, Ar = Ph or Cl; R = Et, M = Ge, Ar = ½(ox).[434]

(−)-Ethyl-(1-naphthyl)phenylgermanecarboxylic acid gives an intense carbonyl band at 1650 cm^{-1}, with a strong, unassigned band at 1210 cm^{-1}.[435]

I.r. spectra have been reported [436] for the allyl (= R) tin carboxylates $R_3Sn(OOCMe)$, $R_3Sn(OOCCH_2Cl)$, $R_2Sn(OOCCH_2Cl)_2$, $R_2Sn(OOCCHCl_2)_2$, $[R_2Sn(OOCCH_2Cl)]_2O$, $[R_2Sn(OOCCHCl_2)]_2O$, and $[R_2Sn(OOCCCl_3)]_2O$. The solid diallyldicarboxylates are believed to be polymeric, with five-co-ordinate tin, and bridging carboxylate groups, whereas the solution data are consistent with the presence of dimers. In the distannoxane series, bands due to non-bridging (ca. 1630 cm^{-1}, ca. 1360 cm^{-1}) and bridging (ca. 1560, ca. 1420 cm^{-1}) carboxylates were present both in the solid and in solution.

The trimethyltin complexes $Me_3SnO_2CCH_2OH$ and $Me_3SnSCH_2CH_2CO_2SnMe_3$ both show (solid-state) i.r. bands due to $\nu_{as}(CO_2)$ at ca. 1580 cm^{-1}, i.e. due to bridging carboxylates. A number of similar dimethyltin carboxylates also gave i.r. evidence for bridging carboxylate groups.[437]

In a series of organotin acrylates $R_3SnO_2C(CN)C=CPh_2$ (where R = Me, Et, Pr^n, Bu^n, or Ph), $\nu(CN)$ was found between 2215 and 2220 cm^{-1}, with $\nu(C=O)$ 1640—1650 cm^{-1}.[438]

A reasonably detailed assignment has been proposed [439] for the vibrations of $Me_2Sn(Cl)OCOMe$. In addition, $\nu_{as}(CO_2)$, $\nu_{as}(SnC_2)$, $\nu_s(SnC_2)$, and $\nu(Sn-Cl)$ were listed for $Me_2SnCl(O_2CR)$, where R = Me, CH_nCl_{3-n} (n = 1—3), CH_2Br, CH_2I, CF_3, C_2F_5, C_3F_7, or CF_2Cl.

A number of characteristic bands have been assigned in $[Me_2Sn(O_2CR)]_2O$, where R = CH_nCl_{3-n} (n = 0—3), CH_2Br, CH_2I, CF_3, C_2F_5, C_3F_7, or CF_2Cl.[440] The carboxylate CO_2 stretches are consistent with the presence of bidentate, bridging O_2CR groups.

Morris and Rockett have assigned $\nu_{as}(CO_2)$ and $\nu_s(CO_2)$ for a number of organometallic carboxylates (see Table 14) [Fc = —(h^5-C_5H_4)Fe(h^5-C_5H_5)].[441]

Table 14 *Carboxyl stretching vibrations (wavenumbers/cm^{-1}) in some organo-tin and -titanium carboxylates*

Compound	$\nu_{as}(CO_2)$	$\nu_s(CO_2)$
$Bu_2Sn(SCH_2CO_2CH_2Fc)_2$	1601	1365
$Bu_2Sn(OCOFc)_2$	1581	1315
$Cp_2Ti(OCOPh)_2$	1630	1370

[434] G. Schott and D. Lange, *Z. anorg. Chem.*, 1972, **391**, 27.
[435] C. Eaborn, R. E. E. Hill, and P. Simpson, *J. Organometallic Chem.*, 1972, **37**, 267.
[436] V. Peruzzo, G. Plazzogna, and G. Tagliavini, *J. Organometallic Chem.*, 1972, **39**, 121.
[437] M. Wada, S.-I. Sato, M. Aritomi, M. Harakawa, and R. Okawara, *J. Organometallic Chem.*, 1972, **39**, 99.
[438] R. A. Cummins, P. Dunn, and D. Oldfield, *Austral. J. Chem.*, 1971, **24**, 2257.
[439] C. S.-C. Wang and J. M. Shreeve, *J. Organometallic Chem.*, 1972, **38**, 287.
[440] C. S.-C. Wang and J. M. Shreeve, *J. Organometallic Chem.*, 1972, **46**, 271.
[441] D. R. Morris and B. W. Rockett, *J. Organometallic Chem.*, 1972, **35**, 179.

'$\nu(C=O)$' and '$\nu(C-O)$' were listed for the five-co-ordinate species $R^1_3Sb(OCOR^2)_2$, where R^1 = Me or Ph; R^2 = CH_nF_{3-n} (n = 1—3), CH_nCl_{3-n} (n = 1 or 2), CH_nBr_{3-n} (n = 1 or 2), CD_3, or CH_2CN.[442]

I.r. data have been listed for the new trifluoroacetates $NaBiL_4$, $NaAsOL_2$, Na_2TeL_6, NH_4VOL_3, BiL_3, IO_2L, and VO_2L, where L = CF_3CO_2. The data for $Bi(CF_3CO_2)_3$ are typical: $\nu_{as}(CO_2)$ 1632 cm^{-1}; $\nu_s(CO_2)$ 1444 cm^{-1}; $\nu(C-C)$ 844 cm^{-1}; $\delta(CO_2)$ 733 cm^{-1}; $\nu_{as}(CF_3)$ 1207, 1120 cm^{-1}; $\nu_s(CF_3)$ 806 cm^{-1}.[443]

Keto-, Alkoxy-, Phenoxy-, and Ether Ligands.—Characteristic i.r. bands have been listed for a number of transition-metal complexes of 1,4-dioxan, 1,4-thioxan, and 1,2-dimethoxyethane.[444]

A quantity of i.r. data has been listed for methanol complexes $M(MeOH)_6^{2+}$, where M = Mg, Mn, Co, Ni, Zn, or Cd.[445]

By comparison with 2-naphthol, some of the stronger i.r. bands of 1-nitroso-2-naphthol have been assigned,[446] both for the free ligand and its complexes with divalent metals. Some of the assignments conflict with previous work.

A number of alkylchromium dichloride complexes containing co-ordinated tetrahydrofuran, $RCrCl_2(THF)$, have been prepared.[447] $\nu_s(COC)$ and $\nu_{as}(COC)$ shift to lower wavenumbers on complexing, with the shifts in the sequence: R = $CH_3 > C_2H_5 >$ n-$C_3H_7 >$ i-C_4H_9, consistent with the trend in the inductive effects of R. $\nu(CH)$ of the alkyl group is found at ca. 2800 cm^{-1}.

Tris-(2-acylpyrrolato)chromium(III) complexes $Cr(RCOC_4H_3N)_3$ (R = H, Me, or Ph), have $\nu(C=O)$ at ca. 1550 cm^{-1}, compared to 1652 cm^{-1} in the free ligand (and R = H).[448]

$\nu_s(COC)$ and $\nu_{as}(COC)$ of tetrahydrofuran shift to lower wavenumber on formation of $FeCl_3,THF$, consistent with Fe—O co-ordination.[449]

For 9,10-phenanthrenequinone (quon) complexes of Fe, Co, and Ni $[M(quon)_n]$ (n = 2 for Co or Ni; n = 3 for Fe), co-ordination through the oxygen is clear from the fact that large shifts (ca. 215 cm^{-1}) of $\nu(C=O)$ occur from that of the free quinone on complex formation.[450]

The chelates $Co(chel)_2$ and $Ni(chel)_2$ derived from benzoin (chelH, PhCHONCOPh) show i.r. bands due to $\nu(C=O)$ at 1618 cm^{-1} (Ni), 1625 cm^{-1} (Co), due to $\nu(C-O)$ at 1040 cm^{-1} (Ni), 1050 cm^{-1} (Co) (both

---

[442] R. G. Goel and D. R. Ridley, *J. Organometallic Chem.*, 1972, **38**, 83.
[443] P. V. Radheshwar, R. Dev, and G. H. Cady, *J. Inorg. Nuclear Chem.*, 1972, **34**, 3913.
[444] N. M. Karayannis, C. M. Mikulski, A. N. Speca, J. T. Cronin, and L. L. Pytlewski, *Inorg. Chem.*, 1972, **11**, 2330.
[445] A. D. van Ingen Schenau, W. L. Groeneveld, and J. Reedijk, *Rec. Trav. chim.*, 1972, **91**, 88.
[446] S. Gurrieri and G. Siracusa, *Inorg. Chim. Acta*, 1971, **5**, 650.
[447] K. Nishimura, H. Kuribayashi, A. Yamamoto, and S. Ikeda, *J. Organometallic Chem.*, 1972, **37**, 317.
[448] C. S. Davies and N. J. Gogan, *J. Inorg. Nuclear Chem.*, 1972, **34**, 2791.
[449] L. S. Benner and C. A. Root, *Inorg. Chem.*, 1972, **11**, 652.
[450] C. Floriani, R. Henzi, and F. Calderazzo, *J.C.S. Dalton*, 1972, 2640.

lower than in free benzoin), and probably due to $\nu(M-O)$ at 514 cm^{-1} (Ni), 490 cm^{-1} (Co).[451]

Salient i.r. wavenumbers have been listed [452] for 1-amino-3,3,3-trifluoro-2-propanol and its complexes with Co^{III}, Ni^{II}, and Cu^{II}. All of the shifts which occurred on complex formation were small, but $\nu(NH)$ remained in the complexes, whereas $\nu(OH)$ was absent, hence the bonding is *via* the O atom.

Oxidative addition reactions of Rh^I and Ir^I complexes have yielded products with *o*-quinones as ligands.[453] I.r. bands characteristic of the *o*-diolato ligands have been listed for such compounds, and correlated with bands in the spectra of the free quinones.

Some assignments of ligand vibrations have been proposed for Rh(ahmc)Cl$_4^{2-}$, Ir(ahmc)(OH)(H$_2$O), Ce(ahmc)$_3$(H$_2$O)$_2$, Pr(ahmc)$_3$(H$_2$O)$_2$, Pd(hmcc)$_2$, Ce(hmcc)$_3$(H$_2$O)$_2$, and Pr(hmcc)$_3$(H$_2$O)$_2$, where ahmcH = 8-amino-7-hydroxy-4-methylcoumarin (121) and hmcc = 7-hydroxy-4-methylcoumarin-6-carboxylic acid (122).[454]

(121) (122)

A study of the 3-(carboxyalkyl)salicylaldehyde chelate complexes of Cu^{II} includes some i.r. data for the co-ordinated aldehydes.[455] Frequencies for ν_s and ν_{as} of the carboxylate CO$_2$ and the (shifted) $\nu(C=O)$ of the aldehyde (near 1611 cm^{-1}) were listed.

I.r. spectra of the glycerates Ln(C$_3$H$_7$O$_3$)$_3$,nH$_2$O (n = 0 or 1; Ln = 1 of 12 lanthanides) have been assigned using the group-frequency approach, *e.g.* C—C—C skeletal stretches, 820—865 cm^{-1} and 1200—1265 cm^{-1}; ν(CO) (primary alcohol) 1010—1060 cm^{-1}; ν(CO) (secondary alcohol) 1110—1105 cm^{-1}.[456]

Unassigned i.r. data have been listed for some rare-earth mixed-ligand complexes of salicylaldehyde, M(sal)$_3$L$_2$, where M = La, Pr, Nd, Sm, Eu, or Tb and L = py, quinoline, $\frac{1}{2}$(bipy), or $\frac{1}{2}$(*o*-phen).[457a]

ν_{as}(COC) (960—1115 cm^{-1}) and ν_s(COC) (850—995 cm^{-1}) have been assigned in the i.r. spectra of ether complexes (R$_2$O)UCl$_5$ (R = Me, Et, Pr^n, Bu^n, or n-C$_5$H$_{11}$; or R$_2$O = C$_4$H$_8$O or C$_4$H$_8$O$_2$).[457b] The dioxan adduct contains unidentate C$_4$H$_8$O$_2$.

[451] K. C. Malhotra and S. C. Chaudhry, *Chem. and Ind.*, 1972, 606.
[452] I. Yoshida and H. Kobayashi, *Bull. Chem. Soc. Japan*, 1972, **45**, 2448.
[453] Y. S. Sohn and A. L. Balch, *J. Amer. Chem. Soc.*, 1972, **94**, 1144.
[454] D. E. Rastogi, *Austral. J. Chem.*, 1972, **25**, 729.
[455] T. Tanaka, *Bull. Chem. Soc. Japan*, 1972, **45**, 2113.
[456] D. V. Pakhomova, V. N. Kuniok, and V. V. Serebrennikov, *Russ. J. Inorg. Chem.*, 1971, **16**, 1588.
[457] (a) K. K. Rohatgi and S. K. Sen Gupta, *J. Inorg. Nuclear Chem.*, 1972, **34**, 3061; (b) J. D. Ortega and W. P. Tew, *J. Co-ordination Chem.*, 1972, **3**, 13.

Oertel has studied the Raman spectra of aqueous solutions containing $B(OH)_4^-$ and various polyols, *e.g.* 1,2-ethanediol, 1,3-propanediol, and 1,2,3-propanetriol.[458] The spectra can only be explained in terms of chelate formation, involving species such as (123), and similar six-membered

$$\left[\begin{array}{c} H_2C-O\quad O-CH_2 \\ \quad\ \ \diagdown B \diagup \\ H_2C-O\quad\ O-CH_2 \end{array}\right]^-$$

(123)

chelates. Assignments were proposed for the five- and six-membered rings. For example, in the monoborate–1,2-ethanediol chelate, ring stretches were found at 1134, 1065, 942, 885, and 764 cm^{-1} (the last being the A_1 mode, primarily B—O stretching in character). In the six-membered ring system, the band analogous to this last was at *ca.* 710 cm^{-1}.

In (aldehyde)BF$_3$ complexes, ν(CH) of the aldehyde is shifted by *ca.* 150 cm^{-1} to higher wavenumbers, and its intensity decreases upon co-ordination. ν(C=O) is reduced by *ca.* 70 cm^{-1}. The aldehydes studied were CH$_3$CHO, CD$_3$CHO, C$_2$H$_5$CHO, C$_3$H$_7$CHO, C$_6$H$_5$CHO, and C$_6$D$_5$CHO.[459]

O-Bonded Amides and Ureas.—The vibrations of a number of salts and complexes of *N*-hydroxyurea, H$_2$N—C(=O)NH(OH), have been listed and assigned.[460]

I.r. spectra of the dimethylformamide complexes NbX$_4$(DMF)$_2$ (X = Cl or Br) and NbI$_4$(DMF)$_2$,6DMF show downward shifts (*ca.* 35 cm^{-1}) for ν(C=O) and upward shifts (*ca.* 40 cm^{-1}) of δ(NCO) on complexing. Thus, co-ordination *via* the carbonyl oxygen is occurring.[461a]

I.r. spectra of the MnSO$_4$ complex with urea show that the ligand is O-bonded.[461b]

I.r. spectra of a number of complexes of the terdentate chelating agent 1,10-phenanthroline-2-carboxamide (with Fe^{II}, Co^{II}, Ni^{II}, and Cu^{II}) have been obtained.[462] In each case ν(C=O) is in the range 1660—1690 cm^{-1}, with ν(CN) 1415—1425 cm^{-1}. These data indicate co-ordination of the amide group to the metal *via* the oxygen atom, not the nitrogen.[462]

The i.r. wavenumbers of [Co(urea)$_4$](NO$_3$)$_2$ have been assigned, but without any real evidence. Both bidentate and O-bonded unidentate urea are believed to be present.[463a]

I.r. spectra have been given for CuCl$_2$, CuBr$_2$, and Cu(NO$_3$)$_2$ complexes with MeCONH$_2$, MeCONHMe, and MeCONHMe$_2$.[463b]

---

[458] R. P. Oertel, *Inorg. Chem.*, 1972, **11**, 544.
[459] E. Taillandier, J. Liquier, and M. Taillandier, *J. Mol. Structure*, 1971, **10**, 463.
[460] R. Berger and H. P. Fritz, *Z. Naturforsch.*, 1972, **27b**, 608.
[461] (*a*) K. Kirksey and J. B. Hamilton, *Inorg. Chem.*, 1972, **11**, 1945; (*b*) N. N. Rumov, *Uch. Zap. Yaroslav. Gos. Pedagog. Inst.*, 1970, **79**, 142.
[462] H. A. Goodwin and F. E. Smith, *Austral. J. Chem.*, 1972, **25**, 37.
[463] (*a*) P. S. Gentile, P. Carfagno, and S. Haddad, *Inorg. Chim. Acta*, 1972, **6**, 296; (*b*) M. A. A. Beg and M. A. Hashmi, *Pakistan J. Sci. Ind. Res.*, 1971, **14**, 458.

I.r. data have been listed for the characterization of $Ln(ClO_4)_3,4DA$, where DA = diacetamide.[464a]

Group-frequency assignments have been proposed for urea modes in the i.r. spectra of $[UO_2\{CO(NH_2)_2\}_4(H_2O)]X_n(NO_3)_{2-n}$ ($X = Cl$, $n = 0.5$; $X = Br$, $n = 1$; $X = I$, $n = 1.1$ or 1.7) and of $[UO_2\{CO(NH_2)_2\}_5]X(NO_3)$ ($X = I$ or Cl).[464b]

Nitrates and Nitrato-complexes.—Nitrate bands are found at 1345 and 830 cm^{-1} in a number of $M(nipa)_3(NO_3)_{2\text{ or }3}$ complexes, where nipa = nonamethylimidodiphosphoramide and M = Mg, Ca, Sr, Ba, Al, Cr, Fe, or In.[465]

NO_3^- spectra in $M(tripa)_2(NO_3)_2$, where M = Mg, Co, or Ni, and $M'(tripa)_2(NO_3)_3$, where M' = Al or Fe, are consistent with the presence of unco-ordinated, ionic nitrate. In $M(tripa)(NO_3)_2$, where M = Ca, Mn, Cu, Zn, or Cd, however, co-ordination of NO_3^- occurs, and this is reflected in the i.r. spectra [tripa = $(Me_2N)_2P(O)N(Me)P(O)NMe_2N(Me)P(O)$- $(NMe_2)_2$, bis(methylimido)triphosphoric acid pentakis-dimethylamide].[466]

The i.r. spectra of $Mn(NO_3)_3$ and $(NO_2)^+[Mn(NO_3)_4]^-$, and the i.r. and Raman spectra of $Na_2[Mn(NO_3)_4]$ and $K_2[Mn(NO_3)_4]$, have been obtained.[467] In each case, the nitrate vibrations could be assigned in terms of C_{2v} symmetry [bridging for $Mn(NO_3)_3$, bidentate for the rest]. Thus, in $Mn(NO_3)_3$, $\nu_1[A_1, (NO)] = 1540$ cm^{-1}; $\nu_2[A_1, (NO_2)] = 977$ cm^{-1}, $\nu_4[B_1, (NO_2)] = 1270, 1255$ cm^{-1}, and $\nu_3(A_1$, ring def.) and $\nu_6[B_2, (TlO_2NO)] = 794, 768, 743$ cm^{-1}. In $Na_2[Mn(NO_3)_4]$, $\nu_1 = 1490$ cm^{-1}, $\nu_2 = 1041, 1036$ cm^{-1}, $\nu_3 = 765, 750$ cm^{-1}, $\nu_4 = 1280$ cm^{-1}, $\nu_5[B_1, (ONO)] = 719$ cm^{-1}, and $\nu_6 = 820, 812, 807$ cm^{-1}.

cis- and *trans*-Isomers of $Re(CO)_4(PPh_3)(NO_3)$ have been characterized.[468] *cis*-Isomer: $\nu(CO)$ 2070, 2010, 1990, and 1968 cm^{-1}; $\nu_{as}(NO_2)$ 1520 cm^{-1}; $\nu_s(NO_2)$ 1265 cm^{-1}; $\nu(N=O)$ 995 cm^{-1}; and $\delta(ONO_2)$ 790 cm^{-1}; *trans*-isomer: $\nu(CO)$ 2100, 2012 cm^{-1}, $\nu_{as}(NO_2)$ 1505 cm^{-1}; $\nu_s(NO_2)$ 1270 cm^{-1}; $\nu(N=O)$ 990 cm^{-1}; and $\delta(ONO_2)$ 790 cm^{-1}.

From various pieces of physical evidence, including i.r. spectra, the presence of unidentate NO_3^- in $Ni(etu)_4(NO_3)_2$ and asymmetric bidentate nitrate in $Co(etu)_2(NO_3)_2$ and $Co(tmtu)_2(NO_3)_2$ is deduced.[469] The deductions were made on rather slight evidence (etu = ethylenethiourea: tmtu = tetramethylthiourea).

Vibrational spectroscopy has been used to investigate the nature of the metal–ligand environment of $Cu(py)_n(NO_3)_2$ (where $n = 2, 3$, or 4) and

---

[464] (a) C. Airoldi and Y. Gushikem, *J. Inorg. Nuclear Chem.*, 1972, **34**, 3921; (b) G. V. Ellert, I. V. Tsapkina, O. M. Ewstaf'eva, V. F. Zolin, and P. F. Fisher, *Russ. J. Inorg. Chem.*, 1971, **16**, 1640.
[465] M. W. G. de Bolster and W. L. Groeneveld, *Rec. Trav. chim.*, 1972, **91**, 95.
[466] M. W. G. de Bolster, J. den Heijer, and W. L. Groeneveld, *Z. Naturforsch.*, 1972, **27b**, 1324.
[467] D. W. Johnson and D. Sutton, *Canad. J. Chem.*, 1972, **50**, 3326.
[468] R. Davis, *J. Organometallic Chem.*, 1972, **40**, 183.
[469] E. C. Devore and S. L. Holt, *J. Inorg. Nuclear Chem.*, 1972, **34**, 2303.

$M(py)_3(NO_3)_2$, where M = Co, Ni, or Zn.[470] Extensive use of isotopic substitution (^{62}Ni, ^{58}Ni, ^{63}Cu, and ^{65}Cu) and related techniques was made, to assign low-frequency i.r. spectra. It has been observed for the $Cu(py)_n$-$(NO_3)_2$ system that as the covalency of the nitrate bond increases, the Cu—O stretch shifts to higher energy, while the Cu—N stretch moves to lower energy.

The highest-wavenumber (1641 cm^{-1}) i.r. band assigned to a ν(N—O) fundamental in $Ir_3(NO_3)_{10}$ indicates the presence of bridging or bidentate nitrate groups, as well as unidentate NO_3^- [$\nu_{as}(NO_2)$ 1549 cm^{-1}] in this complex.[471]

The complex $(bipy)_2Pd(NO_3)_2, H_2O$ has i.r. bands due to the nitrate group as follows: ν_3 1310—1372 cm^{-1}, ν_1 1025 cm^{-1}; ν_2 822 cm^{-1}, indicating, it is said, either a low site symmetry for the nitrate or interaction with the Pd or the H_2O.[472]

The i.r. spectrum of $K_2[Pt(NO_2)_3(NO_3)_3]$ shows bands due to coordinated NO_3^- (1520, 1290, 980, and 825 cm^{-1}) and NO_2^- (ν_{as} 1510 cm^{-1}, ν_s 1390 cm^{-1}, δ 845 cm^{-1}, ρ 600 cm^{-1}).[473]

Raman intensity measurements on the nitrate bands at 722 and 1052 cm^{-1} in aqueous copper(II) nitrate solutions have been used, after corrections due to the strong absorption of the Raman-scattered light (an Ar^+ laser was used as excitation source), to calculate the concentration quotient of $CuNO_3^+$ (K_{assoc} = 0.07 ± 0.02 at 25 °C).[474]

Fairly full assignments for the i.r. and Raman spectra of tetranitratoaurates(III) have been given [i.e. for $M^I Au(NO_3)_4$, where M^I = H, Na, K, Rb, Cr, NO^+, or NO_2^+].[475] An apparently unequivocal means of distinguishing between uni- and bi-dentate nitrate co-ordination was discussed; of the three ν(NO) fundamentals derived from the NO_3^- group, in the bidentate case that at ca. 1300 cm^{-1} [$\nu_s(NO_2)$] gives rise to strong bands, while for unidentate NO_3^-, $\nu_{as}(NO_2)$ is found near 1300 cm^{-1} as a fundamental which is only weakly observable.

I.r. data (including the combination band region ca. 1750 cm^{-1}) indicate the presence of co-ordinated nitrate groups in the rare-earth complexes $Ln(napy)_3(NO_3)_3$, where napy = 1,8-naphthyridine, and in $Ln(napy)_2$-$(NO_3)_3$.[476] They are believed to be twelve- and ten-co-ordinate, respectively.

The presence of co-ordinated NO_3^-, both uni- and bi-dentate, in M_2Sc-$(NO_3)_5$ (M = K, Rb, or Cs) has been inferred from i.r. measurements.[477]

[470] M. Chorea, J. R. Ferraro, and K. Nakamoto, J.C.S. Dalton, 1972, 2297.
[471] B. Harrison and N. Logan, J.C.S. Dalton, 1972, 1587.
[472] A. J. Carty and P. C. Chieh, J.C.S. Chem. Comm., 1972, 158.
[473] L. K. Shubochkin, E. F. Shubochkina, M. A. Golubnichaya, and L. D. Sorokina, Russ. J. Inorg. Chem., 1972, 16, 877.
[474] A. R. Davis and C. Chong, Inorg. Chem., 1972, 11, 1891.
[475] C. C. Addison, G. S. Brownlee, and N. Logan, J.C.S. Dalton, 1972, 1440.
[476] J. Foster and D. G. Hendricker, Inorg. Chim. Acta, 1972, 6, 371.
[477] L. N. Komissarova, G. Ya. Pushkina, and V. I. Spitsyn, Russ. J. Inorg. Chem., 1971, 16, 1262.

Vibrational Spectra of some Co-ordinated Ligands 439

The nitrato-groups in $M(NO_3)_3L_3$ (M = Y or La—Yb, L = 2,7-dimethyl-1,8-naphthyridine) are co-ordinated to M, but uni- or bi-dentate modes of bonding could not be distinguished. Characteristic nitrato-group wavenumbers were listed for each complex.[478]

I.r. bands due to the NO_3^- (of C_{2v} symmetry) and H_2O have been assigned[479] for $Eu(NO_3)_2,xH_2O$ (x = 0—6, inclusive). Approximate assignments of NO_3^- or ClO_4^- bands have been made[480] for $Ce(NO_3)_4$,-$2(Ph_3PO)$, $Ce(NO_3)_4$,7DMSO, $Ce(NO_3)_4$,6(1,10-phen), and $Ce(ClO_4)_4$,-7DMSO.

The nitrato-group in $U(NO_3)(dbp)$ (solid) appears to be covalently bonded to the uranium (bands at 1520, 1275 cm^{-1}).[481] Unassigned i.r. spectra were given for $U(NO_3)(dbp)_3$ and $U(dbp)_4$ (dbp = di-n-butyl phosphate).

Molecular weight and i.r. spectroscopic data have been used[482] to show that in Ph_4BiX (X = NO_3^- or $CCl_3CO_2^-$) the X is bonded (unidentate) to the Bi, giving a five-co-ordinate complex. Thus, in Ph_4BiONO_2, $\nu_{as}(NO_2)$ 1442 cm^{-1}; $\nu_s(NO_2)$ 1295 cm^{-1}; $\nu(BiO-NO_2)$ 1032 cm^{-1}; o.o.p. deformation 825 cm^{-1}; and for $Ph_4BiOCOCl_3$, $\nu_{as}(CO_2)$ 1680 cm^{-1}; $\nu_s(CO_2)$ 1300 cm^{-1}.

Ligands containing O—N, O—P, or O—As Bonds.—$\nu(NO)$ bands (1210—1225 cm^{-1}) have been listed for a number of complexes of pyridine N-oxide.[483]

Empirical assignments of the i.r. spectra of the cupferronates of Cu^{II}, Hg^{II}, Al^{III}, Fe^{III}, Ga^{III}, Bi^{III}, Ti^{IV}, V^{IV}, Zr^{IV}, Th^{IV}, U^{IV}, V^V, and Nb^V (124) have been made.[484]

$$\left[Ph-N\underset{N-O}{\overset{O}{\diagdown}}M \right]_n$$
(124)

An amount of unassigned i.r. data on complexes of hexamethylphosphoramide and nonamethylimidodiphosphoramide has been published.[485]

$\nu(P=O)$ and $\nu(As=O)$ bands have been assigned (in the expected ranges) for complexes of ditertiary phosphine or arsine oxides, $Ph_2E(O)$-$(CH_2)_nE(O)Ph_2$ (E = As or P).[486]

---

[478] D. G. Hendricker and R. J. Foster, *J. Inorg. Nuclear Chem.*, 1972, **34**, 1949.
[479] K. E. Mironov, A. P. Popov, V. Ya. Vorob'eva, and Z. A. Grankina, *Russ. J. Inorg. Chem.*, 1972, **16**, 1476.
[480] F. Březina, *Coll. Czech. Chem. Comm.*, 1972, **37**, 3174.
[481] E. R. Schmid and V. Satrawaka, *Monatsh.*, 1972, **103**, 442.
[482] R. E. Beaumont and R. G. Goel, *Inorg. Nuclear Chem. Letters*, 1972, **8**, 989.
[483] C. P. Prabhakaran and C. C. Patel, *J. Inorg. Nuclear Chem.*, 1972, **34**, 3485.
[484] A. T. Pilipenko, L. L. Shevchenko, and V. N. Strokan, *Russ. J. Inorg. Chem.*, 1971, **16**, 1279.
[485] M. W. G. de Bolster and W. L. Groeneveld, *Rec. Trav. chim.*, 1972, **91**, 171.
[486] S. S. Sandhu and R. S. Sandhu, *J. Inorg. Nuclear Chem.*, 1972, **34**, 2295.

I.r. spectra of several phosphites of Group IA and IIA metals have been reported.[487] Similar data have been given [488] for di-n-octyl phenylphosphonate, mono- and di-n-decylphosphoric acids, and some of their Ca^{2+} salts.

In $Ca(bapo)_4(ClO_4)_2$, where bapo = bisphenyldimethylaminophosphine oxide, the ν_4 mode of ClO_4^- is split, possibly suggesting that the anion is participating in the co-ordination around the Ca^{2+} ion.[489a]

I.r. and Raman spectra of the complexes of MoO_2Cl_2 and $SbCl_5$ with Me_3XO (X = N, P, or As) have been studied in an attempt to establish their molecular symmetries.[489b]

I.r. wavenumbers have been partially assigned for some co-ordination complexes of $PH_2O_2^-$ with Mn and V.[490]

$\nu(NO)$ is in the range 1205—1276 cm^{-1} in the i.r. spectra of Fe^{II}, Fe^{III}, and Cu^{II} complexes of 2-, 3-, or 4-cyanopyridine N-oxide.[491]

$HRu_2(CO)_3[P(OC_6H_4)(OC_6H_5)_2]_2[OP(OC_6H_5)_2]$ has been shown by single-crystal X-ray studies to be (125).[492] The bridging $(C_6H_5O)_2PO$ group gives $\nu(PO)$ at 1075 cm^{-1}.

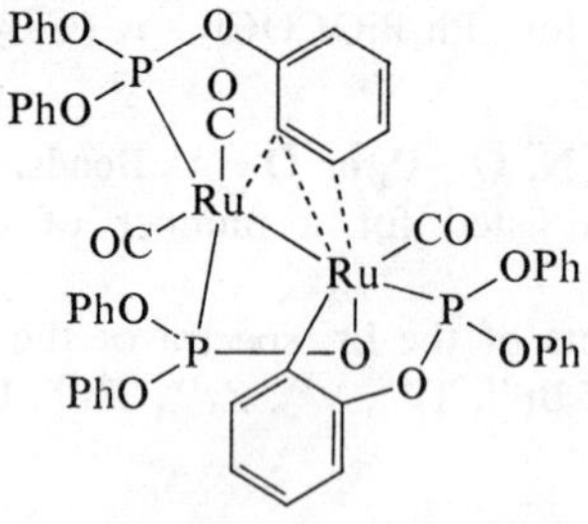

(125)

$\nu(P=O)$ and $\nu(M-hal)$ have been listed [493] for the Co and Ni complexes $[ML_3][MX_4]$, where X = Cl, Br, or I and L = $Ph_2P(O)(CH_2)_nP(O)Ph_2$ (n = 1 or 2).

A listing of $\nu(NO)$ wavenumbers has been made [494] (all ca. 1200 cm^{-1}) for CoL_2X_2, where X = Cl, Br, I, NCS, or ClO_4 and L = 2,6-lutidine N-oxide.

$\nu(As-O)$ bands are at ca. 760 cm^{-1} and 855 cm^{-1} in M^{II} salts of phenylarsonic acid and of o-arsanilic acid (M = Co, Ni, Cd, or Zn).[495]

---

[487] M. Ebert and J. Eysseltová, *Monatsh.*, 1972, **103**, 188.
[488] G. H. Griffiths, G. J. Moody, and J. D. R. Thomas, *J. Inorg. Nuclear Chem.*, 1972, **34**, 3043.
[489] (a) M. W. G. de Bolster and W. L. Groeneveld, *Z. Naturforsch.*, 1972, **27b**, 759; (b) F. Choplin, M. Burgard, J. Hildbrand, and G. Kaufmann, *Colloq. Int. Cent. Nat. Rech. Sci.*, 1970, No. 191, p. 213.
[490] J. Sala-Pala, R. Kergoat, and J. E. Guerchais, *Compt. rend.*, 1972, **274**, C, 595.
[491] G. W. Watt and W. R. Strait, *J. Inorg. Nuclear Chem.*, 1972, **34**, 947.
[492] M. I. Bruce, J. Howard, I. W. Newell, G. Shaw, and P. Woodward, *J.C.S. Chem. Comm.*, 1972, 1041.
[493] F. Mani and M. Bacci, *Inorg. Chim. Acta*, 1972, **6**, 487.
[494] D. W. Herlocker, *Inorg. Chim. Acta*, 1972, **6**, 211.
[495] S. S. Sandhu and G. K. Sandhu, *J. Inorg. Nuclear Chem.*, 1972, **34**, 3249.

Bands in the region 1140—1205 cm^{-1} have been assigned to ν(PO) in ethylenebis(diphenylphosphine oxide) complexes of Co^{II}, Ni^{II}, and Cu^{II}.[496]

I.r. data for M^{II} complexes of (126) and (127), where M = Co, Ni, Cu, or Mg, show that ν(P=O) decreases by 10—40 cm^{-1} upon co-ordination, *i.e.* co-ordination occurs *via* the $\overset{\diagdown}{\diagup}$P—O groups.[497]

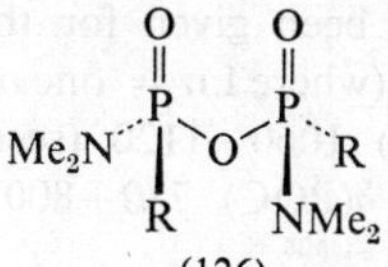

(126) 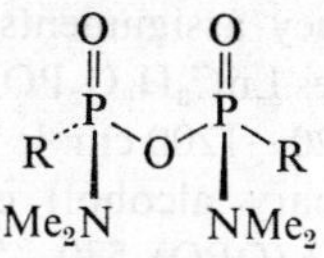   (127)

Bands due to ν(P=O) and ν(As=O) and internal vibrations of the ligand (not assigned in detail) have been reported [498] for NiX$_2$L, NiBr$_2$L$_{1.5}$, and NiL$_2$(ClO$_4$)$_2$,2H$_2$O [X = Cl or NO$_3$; L = Ph$_2$E(O)(CH$_2$)$_n$E(O)Ph$_2$, E = P for n = 2 or 4, E = As for n = 4].

The complexes CuL$_2$X$_2$ [X = ClO$_4$, NO$_3$, or BF$_4$; L = diethyl 4-methyl-pyridine-2-phosphonate or diethyl pyridine-2-phosphonate (128; R = Me or H)] have been prepared.[499] Wavenumbers purporting to be the ν(P=O), pyridine modes, and anion modes have been listed.

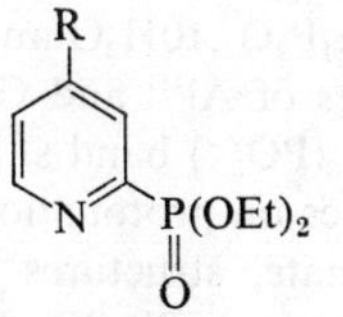

(128)

The i.r. spectra of the organic phases in the extraction of Cu^{2+} from aqueous HCl (containing LiCl) using Oct$^n{}_3$PO in benzene or kerosene have been investigated.[500] A band at 1110 cm^{-1} is assigned to ν(P=O) of the co-ordinated phosphine oxide.

Approximate assignments of ligand vibrations have been proposed for NH$_4$L, Cu(L$_2$)$_2$, Cu(L)$_2$(py), and Cu(L)$_2$(4Mepy), where L = *N*-nitroso-*N*-phenylhydroxylamine, cupferron, (129).[501]

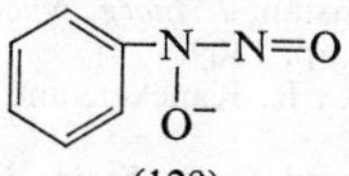

(129)

[496] B. J. Brisdon, *J.C.S. Dalton*, 1972, 2247.
[497] M. D. Joesten and Y. T. Chen, *Inorg. Chem.*, 1972, **11**, 429.
[498] S. S. Sandhu and R. S. Sandhu, *Inorg. Chim. Acta*, 1972, **6**, 383.
[499] A. N. Speca, L. L. Pytlewski, and N. M. Karayannis, *J. Inorg. Nuclear Chem.*, 1972, **34**, 3671.
[500] T. Sato and M. Yamatake, *Z. anorg. Chem.*, 1972, **391**, 174.
[501] D. P. Graddon and C. Y. Hsu, *Austral. J. Chem.*, 1971, **24**, 2267.

Lists of unassigned i.r. absorptions have been presented for complexes of the general formula $Ln(ClO_4)_3,4DDPA$ (where $Ln = Y$ or $La—Lu$ inclusive; $DDPA = NN$-dimethyldiphenylphosphinamide) and $LnCl_3$,-$5DPPA$ = diphenylphosphinamide).[502a, b] Similar data have been given for the cupferronates of La, Ce, Gd, and Yb [cupferron = (129)],[503] and for LnP_3O_9,xH_2O and $Ln_4(P_4O_{12})_3,xH_2O$.[504]

Group-frequency assignments have been given for the i.r. spectra of glycerophosphates $LnC_3H_6O_2PO$,H_2O (where Ln = one of 12 lanthanides), e.g. $\nu(POC)$ 1120—1200 cm^{-1}; $\nu(CO)$ 1060—1120 (secondary alcohol), 970—1030 (primary alcohol) cm^{-1}; $\delta(POC)$ 780—800 cm^{-1}; $\delta_{as}(OPO)$ 590—660 cm^{-1}; $\delta_s(OPO)$ 530—540 cm^{-1}.[505]

$\nu(PO)$ is lowered by 145 cm^{-1} (compared with the free ligand) in $U(ClO_4)_4,5HMPA$, and the ClO_4^- ions are unco-ordinated, according to i.r. data; the complex is therefore considered to be five-co-ordinate (HMPA = hexamethylphosphoramide).[506]

I.r. and Raman spectra of $(hmpt)_2ZnCl_2$, of $(hmpt)^{10}BF_3$, and of $(hmpt)^{11}BF_3$ have been recorded and assigned with the aid of isotopic shifts (hmpt = hexamethylphosphotriamide): $\nu(O—^{10}BF_3)$ 921 cm^{-1}; $\nu(O—^{11}BF_3)$ 890 cm^{-1}. Small shifts only were found in the hmpt fundamentals.[507]

I.r. spectra have been studied, over the restricted range 1300—1000 cm^{-1}, for aqueous solutions of $Na_4P_2O_7,10H_2O$ and $Na_5P_3O_{10},6H_2O$ at different pH values, and of complexes of Al^{III} and Ga^{III} with these phosphates.[508] It was concluded that the $\nu_{as}(PO_3^{2-})$ band shifts on co-ordination to Al or Ga in the same way as it does on protonation.

Monomeric, four-co-ordinate, structures have been proposed for the complexes $(C_6F_5)_3InL$, where $L = Ph_3PO, Ph_3AsO, Ph_3P, Ph_3As$, or py, from their i.r. spectra.[509] $(C_6F_5)InL_2$, $L = OSMe_2$ or THF, are also monomeric, and five-co-ordinate; the complexes $[(C_6F_5)_3In]_2L$, L = bipy or $Ph_2PCH_2CH_2PPh_2$, and $[(C_6F_5)_3In]_2(tmed)$ have bridging ligands, with four-, five-co-ordinate indium, respectively.[509] The $\nu(P=O)$, $\nu(As=O)$ bands of the Ph_3PO or Ph_3AsO are lowered (to 1159, 874 cm^{-1}, respectively) from free-ligand values, i.e. there is In—O co-ordination. O-Bonding is also postulated for the $OSMe_2$ complex [$\nu(S=O)$ being found at ca. 1200 cm^{-1}].

---

[502] (a) G. Vicentini and P. O. Dunstan, J. Inorg. Nuclear Chem., 1972, 34, 1303; (b) G. Vicentini and J. C. Prado, ibid., p. 1309.
[503] N. V. Thakur, V. B. Kartha, C. R. Kanekar, and V. R. Maratke, J. Inorg. Nuclear Chem., 1972, 34, 2831.
[504] Y. Gushiken, E. Giesbrecht, and O. A. Serra, J. Inorg. Nuclear Chem., 1972, 34, 2179.
[505] D. V. Pakhomova, V. N. Kumok, and V. V. Serebrennikov, Russ. J. Inorg. Chem., 1971, 16, 1586.
[506] J. G. H. du Preez and H. E. Rohmer, Inorg. Nuclear Chem. Letters, 1972, 8, 921.
[507] M. T. Forel, S. Volf, and M. Fouassier, Spectrochim. Acta, 1972, 28A, 1321.
[508] I. A. Sheka, L. P. Barchuk, and G. S. Semenova, Russ. J. Inorg. Chem., 1971, 16, 1701.
[509] G. B. Deacon and J. C. Parrott, Austral. J. Chem., 1972, 25, 1169.

The gas-phase value for $\nu(P=O)$ in $SnCl_4,2POCl_3$ is at 1207 cm^{-1} (i.r.), almost the same value as reported previously for the solid.[510]

A number of assignments have been proposed for $Ph_3Sn[ON=C(CN)-CN]$, i.e. Ph_3SnX and related complexes [511] (see Table 15).

Table 15 *Some vibrational assignments for $Ph_3Sn[O-N=C(CN)_2]$ and related complexes (all figures are wavenumber/cm^{-1})*

Vibration	Ph_3SnX	Ph_2SnX_2	Bu^n_3SnX	$Bu^n_2SnX_2$
$\nu(CC)$	1248	1242	1245	1245
$\nu_s(CNO)$	1120	1123	1150	1140
$\nu_{as}(CNO)$	1382	1440	1450	1435
$\nu(CN)$	2245	2245	2245	2238

The i.r. spectra of the nitrosoalkane complexes $[R_2Pb(RNO)_2]^{2+}(NO_3)_2^{2-}$ are consistent only with the presence of ionic, i.e. non-co-ordinated, NO_3^- ions.[512]

Complexes containing the pentaco-ordinated cations $[(R_3SbL)_2O]^{2+}$ have been prepared, where L = pyO, Ph_3PO, Ph_3AsO, diphenyl sulphoxide, dimethylacetamide, or DMSO.[513] Co-ordination of L to Sb takes place *via* the ligand O-atom in each case, as shown by a significant drop in $\nu(X-O)$ from the free ligand to the complex, *e.g.* for Ph_3AsO, $\nu(AsO)$ is at 890 cm^{-1} (free), 845 cm^{-1} (complex).

Ligands containing O-S Bonds. Wavenumbers of SO_4^{2-} have been listed for the series $M^I_3M^{III}(SO_4)_3$, where M^I = Na or Ag; M^{III} = Ga, V, Cr, Fe, or Rh.[514]

By i.r. spectroscopy, 1,4-dithian monosulphoxide has been shown to co-ordinate to metals (*e.g.* Mn, Fe, Co, Ni, Cu, Zn, Cd, Pt, or Pd) *via* the oxygen atom of the sulphoxide group.[515]

The complexes (130) and (131) give bands assigned to $\nu_{as}(SO_2)$, $\nu_s(SO_2)$ at 1267, 1153 cm^{-1} and 1249, 1128 cm^{-1}, respectively.[516]

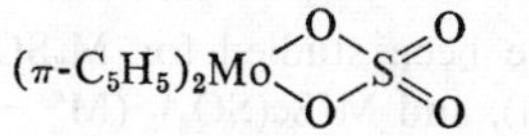

$(\pi\text{-}C_5H_5)_2Mo\begin{smallmatrix}O\\O\end{smallmatrix}S\begin{smallmatrix}O\\O\end{smallmatrix}$ $(\pi\text{-}C_5H_5)_2W\begin{smallmatrix}O\\O\end{smallmatrix}S\begin{smallmatrix}O\\O\end{smallmatrix}$

(130) 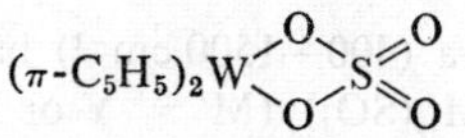   (131)

$\left(O\begin{smallmatrix}O\\\|\\S\\|\\F\end{smallmatrix}O\begin{smallmatrix}O\\\|\\W\\|\\X_3\end{smallmatrix}O\begin{smallmatrix}O\\\|\\S\\|\\F\end{smallmatrix}O\begin{smallmatrix}O\\\|\\W\\|\\X_3\end{smallmatrix}\right)_n$

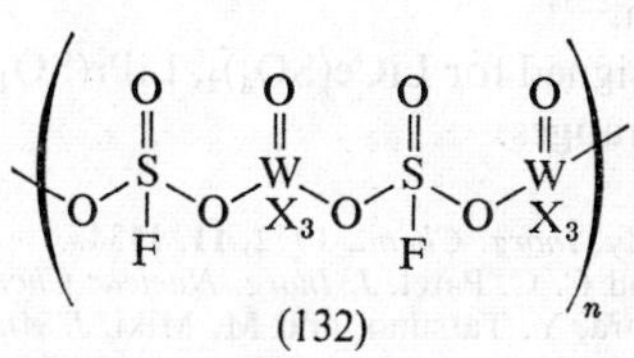

(132)

[510] E. K. Krzhizhanovskaya and A. V. Suvorov, *Russ. J. Inorg. Chem.*, 1971, **16**, 1355.
[511] H. Köhler, V. Lange, and B. Eichler, *J. Organometallic Chem.*, 1972, **35**, C17.
[512] K. C. Williams and D. W. Imhoff, *J. Organometallic Chem.*, 1972, **42**, 107.
[513] R. G. Goel and H. S. Prasad, *Inorg. Chem.*, 1972, **11**, 2141.
[514] R. Perret and P. Couchot, *Compt. rend.*, 1972, **274**, C, 1735.
[515] A. H. M. Fleur and W. L. Groeneveld, *Rec. Trav. chim.*, 1972, **91**, 317.
[516] M. L. H. Green, A. H. Lynch, and M. G. Swanwick, *J.C.S. Dalton*, 1972, 1445.

A number of assignments have been made to fluorosulphate vibrations in exotetrakis(fluorosulphate)W^{VI} (132), *i.e.* $\nu_{as}(SO_2)$ 1464, 1416 cm^{-1}; $\nu_s(SO_2)$ 1240 cm^{-1}; $\nu_s(SO_2$, bridging) 1163, 1040 cm^{-1}; $\nu(S-F)$ 873, 853, 826, 801 cm^{-1}; $\delta_{as}(SO_3$, bridging) 707, 700 cm^{-1}; $\nu_s(SO_3)$ 644 cm^{-1}; $\delta(SO_2$, bridging) 552 cm^{-1}; S—F wag 455 cm^{-1}; S—F deformation 421 cm^{-1}.[517]

Due to Jahn–Teller distortions, the Mn^{III} complexes MnL_6, where L = DMSO, DMF, pyO, or antipyrine, have tetragonally distorted configurations.[518] This is indicated, *inter alia*, by the presence of two $\nu(E=O)$ absorptions – thus when L = DMSO, $\nu(S=O)$ at 960, 915 cm^{-1}; L = antipyrine, $\nu(C=O)$ at 1630, 1610 cm^{-1}, although when L = pyO or DMF only one analogous band is found in each case.

Reactions of dioxygen complexes of Ni and Pd afford products containing, for example, co-ordinated SO_4, NO_3, NO, CO, or CO_2.[519] Thus, for the sulphato-complexes $MSO_4(Bu^tNC)_2$, where M = Ni^{II} or Pd^{II}, $\nu(SO)$ is at 1170—1010 cm^{-1} (ν_3) and 680—580 cm^{-1} (ν_4), in accord with C_{2v} local symmetry of *cis*-chelated SO_4. Analogous results were found for the nitrato- and carbonato-complexes.

A large number of Pd^{II} and Pt^{II} cationic complexes of the type $[PdL_4]$-$(BF_4)_2$ (L = DMSO; tetramethylene sulphoxide TMSO; diethyl sulphoxide DESO; di-n-propyl sulphoxide NPSO; di-n-butyl sulphoxide NBSO; or di-iso-amyl sulphoxide IASO) have been prepared,[520] some containing mixed *S*- and *O*-co-ordination sites. Cationic Pd^{II} and Pt^{II} complexes of 2,5-dithiahexane 2,5-dioxide, (dthO$_2$) have also been prepared and polymeric structures have been proposed. Assignments of $\nu(SO)$ and $\nu(M-S)$, $\nu(M-O)$ were made.

Gold(III) fluorosulphate gives a very strong complex Raman spectrum, and it is believed to be covalent and polymeric.[521]

I.r. data have been given for the characterization of $La(TMSO)_4(NO_3)_3$, $Ln_2(TMSO)_7(NO_3)_6$, and $Ln_2(TMSO)_6(NO_3)_6$ (TMSO = tetramethylene sulphoxide).[522a]

I.r. spectra (400—1500 cm^{-1}) have been studied for M_2SO_4 (M = Li, K, or Cs), $M'_2(SO_4)_3$ (M' = Y or La), and $M''Sc(SO_4)_2$ (M'' = Rb or Cs), and some integrated absorption intensities have been discussed in terms of sulphate co-ordination.[522b]

SO_4^{2-} bands were assigned for $LiCe(SO_4)_2$, $LiPr(SO_4)_2$, and $LiSm(SO_4)_2$.[523] All were in the usual ranges.

---

[517] R. Dev and G. H. Cady, *Inorg. Chem.*, 1972, **11**, 1134.
[518] C. P. Prabhackaran and C. C. Patel, *J. Inorg. Nuclear Chem.*, 1972, **34**, 2371.
[519] S. Otsuka, A. Nakamura, Y. Tatsuno, and M. Miki, *J. Amer. Chem. Soc.*, 1972, **94**, 3761.
[520] J. H. Price, A. N. Williamson, R. F. Schramm, and B. B. Wayland, *Inorg. Chem.*, 1972, **11**, 1280.
[521] W. M. Johnson, R. Dev, and G. N. Cady, *Inorg. Chem.*, 1972, **11**, 226.
[522] (*a*) P. B. Bertan and S. F. Madan, *J. Inorg. Nuclear Chem.*, 1972, **34**, 3081; (*b*) B. E. Zaitsev, V. G. Remizov, M. A. Galiullin, L. G. Korotaeva, and B. N. Ivanov-Emin, *Izvest. V.U.Z. Khim. i khim. Tekhnol.*, 1972, **15**, 473.
[523] V. I. Volk and L. L. Zaitseva, *Russ. J. Inorg. Chem.*, 1971, **16**, 1513.

The i.r. spectra of trifluoro(fluorosulphato)Ce^{IV}, CeO(SO$_3$F)$_2$, have been reported.[524] That of the former is very similar to that of K$^+$SO$_3$F$^-$, for example, indicating a largely ionic system. The oxo-compound gives a much more complex spectrum, and greater covalency was suggested.

Several uranium(III) sulphates and double sulphates have been examined by i.r. spectroscopy, with some discussion of characteristic sulphate vibrations.[525]

I.r. spectra of a number of monoalkyltin orthosulphites show characteristic absorptions at 1110, 880, and 685 cm^{-1}.[526]

The fluorosulphate vibrations in SnF$_2$(SO$_3$F)$_2$ were consistent with bridging SO$_3$F units of C_s symmetry (see Table 16).[527] Similar data have

Table 16 *Assignment of fluorosulphate vibrations in* SnF$_2$(SO$_3$F)$_2$

Assignment		Wavenumber/cm^{-1}
ν(SO$_3$)	(A'')	ca. 1420 (doublet in i.r. and Raman)
ν(SO$_3$)	(A')	ca. 1110 (doublet in i.r. and Raman)
ν(SO$_3$)	(A')	1069
ν(SF)	(A')	859
δ(SO$_3$)	(A')	629
δ(SO$_3$)	(A'')	590
δ(SO$_3$)	(A')	550
ρ(SO$_3$)	(A'')	433
τ(SO$_3$F)	(A')	279

been listed for MeSnCl(SO$_3$F)$_2$, Me$_2$SnCl(SO$_3$F), MeSnCl$_2$(SO$_3$F), Me$_2$Sn(SO$_3$F)$_2$, Me$_3$Sn(SO$_3$F), SnCl$_2$(SO$_3$F)$_2$, and SnF$_2$(SO$_3$F)$_2$ (see Table 17).[528] All contain only bridging, bidentate SO$_3$F groups, resulting in polymeric chain- or sheet-like structures. It was suggested that a greater splitting of the E mode of ν(SO$_3$) of free SO$_3$F$^-$ (*i.e.* the two highest bands in co-ordinated SO$_3$F) corresponded to greater covalent character of the M—OSO$_2$F bond.

In PhPb(O$_2$CMe)$_3$,H$_2$O,DMSO, the ν(SO) of DMSO is found at 936 cm^{-1} [$\Delta\nu$(SO) = -119 cm^{-1} on complex formation].[529] A similar position is found for PhPb(O$_2$CMe)$_3$,2DMSO.

6 Sulphur and Selenium Donors

Metal carbonyl derivatives containing the trifluoromethylthio ligand show differences in the ν(C—F) region between bridging and terminal SCF$_3$ groups.[530]

---

[524] R. Dev, W. M. Johnson, and G. H. Cady, *Inorg. Chem.*, 1972, **11**, 2259.
[525] R. Barnard, J. I. Bullock, and L. F. Larkworthy, *J.C.S. Dalton*, 1972, 964.
[526] C. H. Stapfer and R. H. Herber, *J. Organometallic Chem.*, 1972, **35**, 111.
[527] L. E. Levchuk, J. R. Sams, and F. Aubke, *Inorg. Chem.*, 1972, **11**, 43.
[528] P. A. Yeats, J. R. Sams, and F. Aubke, *Inorg. Chem.*, 1972, **11**, 2634.
[529] H.-J. Haupt and F. Huber, *Z. Naturforsch.*, 1972, **27b**, 724.
[530] J. L. Davidson and D. W. A. Sharp, *J.C.S. Dalton*, 1972, 107.

Table 17 Vibrational modes of the SO$_3$F group in various tin and methyltin fluorosulphates (all figures are wavenumber/cm^{-1})

Compound	ν(SO$_3$)A''	ν(SO$_3$)A'	ν(SO$_3$)A'	ν(SF)	δ(SO$_3$F)A'	δ(SO$_3$F)A'	δ(SO$_3$F)A''	ρ(SO$_3$)	τ(SO$_3$F)
Me$_3$Sn(SO$_3$F)	1350	1180	1076	827	620	590	554	417	304
Me$_3$Sn(SO$_3$F)	1355	1207	1068	820	630	596	555	410	298
Me$_2$SnCl(SO$_3$F)	1344	1190	1072	820	607	590	555	409	306
MeSnCl(SO$_3$F)$_2$	1361	1165	1072	830	620	590	555	420	305
MeSnCl$_2$(SO$_3$F)	1350	1250	1080	825	605	588	555	405	300
SnCl$_2$(SO$_3$F)$_2$	1385	1130	1087	864	628	586	555	446	312
SnF$_2$(SO$_3$F)$_2$	1420	1101	1068	855	630	590	548	430	280

Vibrational Spectra of some Co-ordinated Ligands

Metal–sulphur co-ordination in complexes of the diphosphinothioyl ligands (133; $R^1 = R^2 = $ Me; X = O or S), (133; $R^1 = R^2 = $ Ph; X = CH_2, O, or S), and (133; $R^1 = $ Me, $R^2 = $ Ph; X = CH_2, O, or S) has been detected by decreases in $\nu(P=S)$ observed on complex formation.

$$R_2^2\overset{S}{\overset{\|}{P}}-X-\overset{S}{\overset{\|}{P}}R_2^1$$
(133)

Thus, when $R^1 = $ Me, $R^2 = $ Ph, X = CH_2 [*i.e.* (diphenylphosphinothioyl)(dimethylphosphinothioyl)methane, pmm], $\nu(P=S)_{methyl}$ decreases from 579 to 543 cm^{-1}, and $\nu(P=S)_{phenyl}$ decreases from 601 to 581 cm^{-1}. The smaller effect on the P=S bond associated with the phenyl groups is expected, since the methyl-substituted thiophosphoryl group should be the better donor to a metal.[531]

I.r. data have been listed, and approximately assigned, for $M(S_2PX_2)_n$, as follows: X = CF_3, CH_3, or Ph; M = Fe^{3+}, Co^{3+}, Mn^{2+}, Co^{2+}, Zn^{2+}, Cd^{2+}, or Hg^{2+}: X = OEt or F; M = Co^{3+}, Co^{2+}, or Zn^{2+}; X = F, M = Hg^{2+}; n = valence of the metal.[532] All of the $n = 2$ complexes show pseudo-tetrahedral co-ordination of M by the four sulphur atoms of the two dithiophosphinate ligands.

Vibrational studies of octahedral [M—(S—S)$_2$—M] cages (134) in the compounds NbS_2X_2 (X = Cl, Br, or I) and VS_4, and of the trigonal-bipyramidal [M—(S—S)SM] (135) in MS_3 (M = Ti, Zr, Nb, or U) show

(134)

(135)

that $\nu(S-S)$ is i.r.- and Raman-active, and occurs in the region 560—600 cm^{-1} [*cf.* pyrites, where $\nu(S-S)$ is below 500 cm^{-1}].[533]

The complexes PbL_3, CoL_3, MnL_3, FeL_3, ZnL_2, CuL_2, NiL_2, PdL_2, and PtL_2 (L = $Et_2NCS_2^-$) have been prepared and examined by i.r. spectroscopy.[534a] $\nu(S_2)C-N$ increases from 1490 to 1538 cm^{-1} in the order listed, corresponding to increasing double-bond character.

$\nu(CN)$ is found between 1495 and 1530 cm^{-1} in the i.r. spectra of $M(chel)_3$ (M = Fe, Co, Cr, or Mn), $M(chel)_2$ (M = Zn, Cd, Pd, Ni, Cu, or Pb),

---

[531] D. A. Wheatland, C. H. Clapp, and R. W. Waldron, *Inorg. Chem.*, 1972, **11**, 2340.
[532] R. G. Cowell, E. D. Day, W. Byers, and P. M. Watkins, *Inorg. Chem.*, 1972, **11**, 1759.
[533] C. Perrin, A. Perrin, and J. P. Prigent, *Bull. Soc. chim. France*, 1972, 3086.
[534] (a) D. V. Sokol'skii, L. M. Kurashvili, and I. A. Zavorokhina, *Izvest. Akad. Nauk Kazakh. S.S.R., Ser. khim.*, 1971, **21**, 10; (b) R. Heber, R. Kirmse, and E. Hoyer, *Z. anorg. Chem.*, 1972, **393**, 159.

and M(chel) (M = Tl or Ag), where chel = NN-diethylthioselenocarbamato, $Et_2NC(S)Se^-$.[534b]

I.r. spectra of dimethylaminoethanethiol complexes of Ni^{II}, Co^{II}, Rh^{III}, Pd^{II}, Os^{IV}, and Pt^{IV} confirm co-ordination through S and N atoms.[535]

I.r. spectra of a series of dithiobenzoato-complexes of bivalent (Ni, Pd, and Pt) and tervalent (Cr, Fe, Rh, and In) metals, and some deuterio-analogues, have been reported.[536] The two stretching wavenumbers of the dithiocarboxylic acid group have been located in the range 900—1000 cm^{-1}, with ν(M—S) between 300 and 400 cm^{-1}.[536] The regular decrease of the highest wavenumber substituent-sensitive band of the phenyl group with the optical electronegativity of the chelated metal ion is discussed.

Vanadium dithiocarboxylates VL_4, where L = $PhCS_2^-$, p-Me·C_6H_4·CS_2^-, $MeCS_2^-$, or $PhCH_2CS_2^-$, give i.r. spectra consistent with symmetrical bidentate bonding of the ligand, *i.e.* (136), *e.g.* in the case where R = Ph, $\nu_s(CS_2) = 945$ cm^{-1}, $\nu_{as}(CS_2) = 1020$ cm^{-1}.[537]

$$\left(R-C \begin{array}{c} S \\ S \end{array} V \right)_4 \qquad \left[Cp_2V \begin{array}{c} S \\ S \end{array} C-OR \right]^+ BF_4^-$$

(136)   (137)

The first xanthates of V^{IV} have been prepared from $(h^5\text{-}C_5H_5)_2VCl_2$ + Na_2SCOR (R = Me, Et, Pr^i, Bu^n, or C_6H_{11}).[538] They are formulated as (137). I.r. spectra were listed and an assignment was proposed, although ν(C—O), ν(C=S), and ν(CO—R) are believed to be strongly coupled (all are in the range 1000—1250 cm^{-1}).

Characteristic i.r. bands of the tetramethylenedithiocarbamates $M[S_2CN\text{-}(CH_2)_4]_4$ of Mo, W, Nb, and Ta are found[539] as follows: ν(C—N) *ca.* 1500 cm^{-1}; $\nu(NC_2)$ *ca.* 1170 cm^{-1}; ν(C—S) *ca.* 1000 cm^{-1}; ν(M—S) *ca.* 330 cm^{-1}.

The i.r. spectra of the complexes VOL_3, $NbOL_3$, $Mo_2O_3L_4$, and MoO_2L_2 (L = NN'-diethyldithiocarbamate, S_2CNEt_2) are consistent with S,S-chelation of the ligand.[540]

I.r. spectra have been listed and partially assigned for the following dithiophosphinate complexes: $NbCl_3L_3$, $TaCl_3L_3$, CrL_3, MoL_3, $MoOL_2$, $WOClL_2$, and WCl_2L_3 (L = $F_2PS_2^-$).[541, 542]

---

[535] P. C. Jain, D. K. Rastogi, and H. L. Nigam, *Indian J. Chem.*, 1971, **9**, 1368.
[536] M. Maltese, *J.C.S. Dalton*, 1972, 2664.
[537] O. Piovesana and G. Cappuccilli, *Inorg. Chem.*, 1972, **11**, 1543.
[538] A. T. Casey and J. R. Thackeray, *Austral. J. Chem.*, 1972, **25**, 2085.
[539] T. M. Brown and J. N. Smith, *J.C.S. Dalton*, 1972, 1614.
[540] A. T. Casey, D. J. Mackey, R. L. Martin, and A. H. White, *Austral. J. Chem.*, 1972, **25**, 477.
[541] R. G. Cowell and A. R. Sanger, *Inorg. Chem.*, 1972, **11**, 2011.
[542] R. G. Cowell and A. R. Sanger, *Inorg. Chem.*, 1972, **11**, 2016.

In $WCl_6(Me_2S)$, the $\nu_{sym}(CSC)$ is shifted from 691 cm^{-1} (free ligand) to 667 cm^{-1}, while $\nu_{asym}(CSC)$ shifts from 741 to 730 cm^{-1}.[543]

The presence of four $\nu(PS)$ bands in the 480—640 cm^{-1} range of the spectrum of $[R_2PSSM(CO)_3]_2$ (R = Et or Ph; M = Mn or Re) suggests the structure (138) for these complexes. In the derivatives $(R_2PS_2)M(CO)_3L$

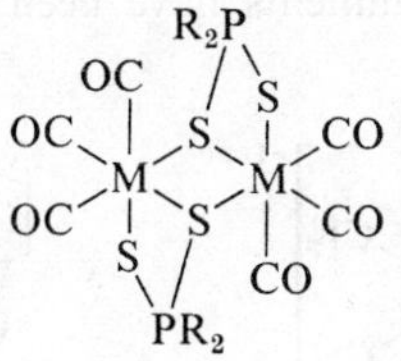

(138)

(L = py or PPh_3), one $\nu_s(PS_2)$ and one $\nu_{as}(PS_2)$ band can be seen (ca. 635, ca. 490 cm^{-1}, respectively).[544]

The i.r. spectrum of tris-(NN-diethyldithiocarbamato)manganese(III) has been compared with that of the Co^{III} analogue in the light of structural differences revealed by X-ray crystallographic determinations.[545] Some of the expected broadening of C—S and M—S stretching bands is observed for the Mn (distorted co-ordination) over that for Co (regular, D_3, co-ordination).

I.r. data have been listed [546] for Mn, Fe, Co, Ni, and Cu salts of ethylidene tetrathiotetra-acetic acid, $(HO_2CCH_2S)_2CH-CH(SCH_2CO_2H)_2$.

I.r. evidence has been used to show that hydrometallation of a C=S bond occurs when CS_2 reacts with cis-$[MH(CO)_3(Ph_2PC_2H_4PPh_2)]$ (M = Mn or Re).[547] The structure M—S—C(=S)H is suggested for the product, this group giving bands at 1245 cm^{-1} [δ(CH)] and at 1008, 780 cm^{-1} (i.r.) and 999, 781 cm^{-1} (Raman) [for ν(C—S)] for the Re complex.

Wavenumbers assigned to $\nu(CO)$ and $\nu(PS_2)$ have been listed for the complexes (139; L = CO, py, PPh_3, $AsPh_3$, or $SbPh_3$), (140), (141; L—L = bipy), and $Et_2PS_2Re(CO)_4(NH_3)$.[548]

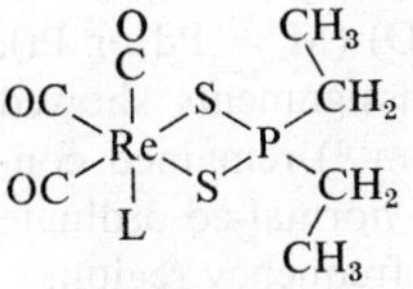

(139) 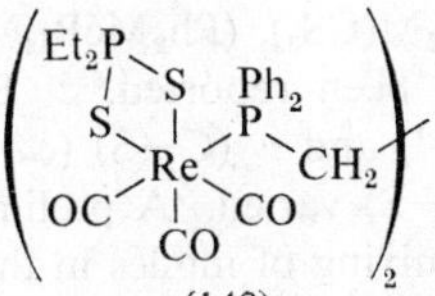   (140) 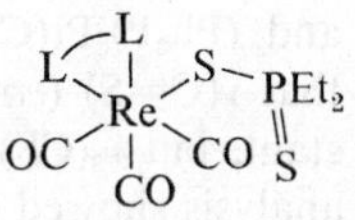   (141)

[543] P. M. Boorman, M. Islip, M. M. Reimer, and K. J. Reimer, J.C.S. Dalton, 1972, 890.
[544] E. Lindner and K.-M. Matejcek, J. Organometallic Chem., 1972, **34**, 195.
[545] P. C. Healy and A. H. White, J.C.S. Dalton, 1972, 1883.
[546] P. Petráš and J. Podlaha, Inorg. Chim. Acta, 1972, **6**, 253.
[547] F. W. Einstein, E. Enwall, N. Flitcroft, and J. M. Leach, J. Inorg. Nuclear Chem., 1972, **34**, 885.
[548] E. Lindner and H. Berke, J. Organometallic Chem., 1972, **39**, 145.

$\nu(C\!\!\cdots\!\!N)$ bands in the dialkyldithiocarbamate complexes [Fe^{III}(R$_2$dtc)$_3$] and [Fe^{IV}(R$_2$dtc)$_3$]$^+$ are noticeably dependent upon the oxidation state of the iron.[549] Thus when R = Me, Et, Pr^i, or C$_6$H$_{11}$, the $\nu(C\!\!\cdots\!\!N)$ band is found at 1560, 1520, 1500, and 1490 cm^{-1} for the Fe^{IV} and at 1520, 1480, 1470, 1490, and 1470 cm^{-1} for the Fe^{III} complexes.

Some approximate assignments have been proposed for ligand bands in the complexes (142).[550]

(142)   (143)

The reaction of SO$_2$ with (h^5-C$_5$H$_5$)Fe(CO)$_2$Na gives [(h^5-C$_5$H$_5$)Fe-(CO)$_2$]$_2$SO$_2$, shown by single-crystal X-ray studies to be the first known molecule in which SO$_2$ alone bridges two transition-metal atoms (143). In the i.r. spectrum, ν(SO) bands are found at 1135, 993 cm^{-1}, and ν(CO) at 2027, 2015, 1965, and 1953 cm^{-1}.[551]

For the chelate complexes of diphenylselenothiophosphinate [Ph$_2$P-(Se)S]$^-$ with Cr^{III}, Co^{II}, Ni^{II}, Zn^{II}, Cd^{II}, Pb^{II}, and Sb^{III}, stretching vibrations of the four-membered chelate ring MS(Se)P have been assigned by analogy with MS$_2$P and MSe$_2$P ring systems. The strongly coupled PS and PSe stretching modes are found at 576—548 cm^{-1} and 525—510 cm^{-1}, respectively.[552]

Selected i.r. bands for Co^{III} thioxanthate complexes have been given, including $\nu(C\!\!\cdots\!\!S)$ at ca. 980 and 950 cm^{-1}.[553]

The complexes formulated as MH(SO$_2$)(CO)(PPh$_3$)$_2$ (M = Rh or Ir) give ν_s and ν_{as}(SO$_2$) at 1038, 1183 cm^{-1} (Rh), 1037, 1175 cm^{-1} (Ir). In M(SO$_2$)(PPh$_3$)$_3$ (M = Pd or Pt), the observed values are 1056, 1215 cm^{-1} (Pd), 1053, 1201 cm^{-1} (Pt).[554]

I.r. and Raman spectra of M'[M(CS$_3$)$_2$] (M = Ni, Pd, or Pt; M' = various cations), (Ph$_2$MeP)$_2$M(CS$_3$), (Ph$_2$MeP)$_2$M(CS$_2$O) (M = Pd or Pt), and (Ph$_3$P)$_2$Pt(CS$_2$O) have been reported.[555] Band assignments showed that ν(C=S) (ca. 1030 cm^{-1}) and ν_{as}(C—S) (ca. 850 cm^{-1}) remained constant, but ν_s(CS) and ν(M—S) varied. A preliminary normal-co-ordinate analysis showed extensive mixing of modes in the low-frequency region.

[549] E. A. Pasek and D. K. Straub, *Inorg. Chem.*, 1972, **11**, 259.
[550] L. H. Pignolet, R. A. Lewis, and R. H. Holm, *Inorg. Chem.*, 1972, **11**, 99.
[551] M. R. Churchill, B. G. DeBoer, K. L. Kalra, P. Reich-Rohrwig, and A. Wojcicki, *J.C.S. Chem. Comm.*, 1972, 981.
[552] P. Christophliemk, V. V. K. Rao, I. Tossidis, and A. Müller, *Chem. Ber.*, 1972, **105**, 1736.
[553] D. F. Lewis, S. J. Lippard, and J. A. Zubieta, *J. Amer. Chem. Soc.*, 1972, **94**, 1563.
[554] J. J. Levison and S. D. Robinson, *J.C.S. Dalton*, 1972, 2013.
[555] J. M. Burke and J. P. Fackler jun., *Inorg. Chem.*, 1972, **11**, 2744.

A number of ligand vibrations have been assigned in the complexes $M[S_2PX_2]_2$ (M = Ni, Pd, or Pt; X = Me, Ph, F, or CF_3).[556]

The Ni^{II} complexes of the cyclic dithioethers 1,4-dithiacycloheptane (dtch) and 1,5-dithiacyclo-octane (dtco), $Ni(dtch)_2^{2+}X_2^{2-}$ and $Ni(dtco)_2^{2+}X_2^{2-}$ (X = ClO_4^- or BF_4^-), gave spectra indicating only a very weak perturbation of the tetrahedral anions.[557] These were entirely consistent with lattice force effects, suggesting a basically four-co-ordinate, square-planar structure for the complexes.

The complex (144) was shown to be an S-dithiocarbamate by the absence of ν(C=S), but the presence of ν(C=N) at 1530 cm^{-1}.[558]

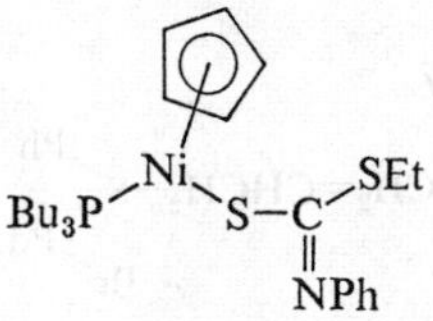

Bu_3P\\Ni/SEt
  S—C
  ||
  NPh
(144)

ν(PO) and ν(PS) in $Ni(dtp)_2(PPh_3)$ [where dtp = $(EtO)_2PS_2^-$] increase slightly compared with those in the square-planar complex $Ni(dtp)_2$, as would be expected if back π-bonding from Ni is decreased by pushing the plane of the sulphur atoms below the Ni.[559]

Dithiocumate complexes [*i.e.* containing (145)] of Ni, Pd, Pt, or Zn give ν(C—C) (of S_2C—Ar) in the range 1245—1289 cm^{-1}; $\nu(S_2C)$ lies between 900 and 1100 cm^{-1}; and in complexes containing (146), ν(SS) is at *ca.* 480 cm^{-1}.[560]

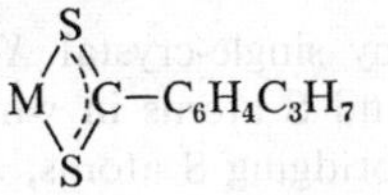

(145) 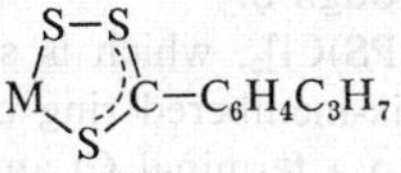   (146)

The use of $(CD_3)_2S$ and $(CD_3)_2SO$, as well as the hydrogen analogues, has enabled Tranquille and Forel to present[561] a very thorough investigation of the vibrational spectra of *trans*-PdL_2X_2 (L = sulphide or sulphoxide; X = Cl or Br), with particular emphasis on the positions of the ligand fundamentals. Small shifts in ν(Pd—S) modes for Me_2S complexes were also noted, whereas similar fundamentals are virtually identical in frequency for DMSO and [$^2H_6$]DMSO.

[556] R. G. Cowell, W. Byers, E. D. Day, and P. M. Watkins, *Inorg. Chem.*, 1972, **11**, 1598.
[557] W. K. Musker and N. L. Hill, *Inorg. Chem.*, 1972, **11**, 710.
[558] F. Sato and M. Sato, *J. Organometallic Chem.*, 1972, **46**, C63.
[559] N. Yoon, M. J. Incorvia, and J. I. Zink, *J.C.S. Chem. Comm.*, 1972, 499.
[560] J. M. Burke and J. P. Fackler, jun., *Inorg. Chem.*, 1972, **11**, 3000.
[561] M. Tranquille and M. T. Forel, *Spectrochim. Acta*, 1972, **28A**, 1305.

The complexes (147), (148), and (149) all give $\nu(C=C)$ at 1628 cm^{-1}, from the free olefinic bond.[562]

Vibrational spectra of dimethyl sulphide complexes of Pd^{II}, Pt^{II}, and Au^{I} have been examined and assigned.[563]

$$\left(\begin{array}{c} CH_2=CH \\ \quad \diagdown \\ CH_2 \end{array} S \diagup\!\!\!\diagup\!\!\!\diagdown\!\!\!\diagdown\!\!\! \begin{array}{c} (CH_2)_3 \\ \diagup\!\!\!\diagdown S \\ \diagup\!\!\!Pd \\ Br \end{array} \right)_2$$

(147)

$$\left[\begin{array}{c} CH_2=CHCH_2 \\ \diagdown S \end{array} \diagup\!\!\!\diagdown\!\!\! \begin{array}{c} (CH_2)_3 \\ \diagup\!\!\!\diagdown S \\ \diagup Pd \\ Br \end{array} \right]_3$$

(148)

$$\left(CH_2=CHCH_2-S \diagup\!\!\!\diagdown\!\!\! \begin{array}{c} Ph \\ \diagup Pd \\ Br \end{array} \right)_2$$

(149)

In the following complexes, i.r. data indicate that the thioacetamide ligand (taa) is unidentate, and S-co-ordinated: Pt(taa)$_4$Cl$_2$, Pt(taa)$_4$I$_2$, Pd(taa)$_4$Cl$_2$, Pd(taa)$_4$Br$_2$, Pd(taa)$_4$(ClO$_4$)$_2$, Rh(taa)$_3$Cl$_3$, and Ir(taa)$_3$Cl$_3$.[564]

The reaction of Pt(OCOMe)$_2$(PPh$_3$)$_2$ with SO$_2$ at room temperatures produces a complex Pt(SO$_3$Me)$_2$(PPh$_3$)$_2$, whose i.r. spectrum contains bands at 1270 and 1100 cm^{-1}, assigned to ν_{as}(SO$_2$) and ν_s(SO$_2$) respectively. In addition, there is another band at 990 cm^{-1} attributed to ν(S—O) and the complex is thus formulated as Pt(SO$_2$OMe)$_2$(PPh$_3$)$_2$, with co-ordination through S.[565]

In [Cu(Me$_3$PS)Cl]$_3$, which is shown by single-crystal X-ray studies to consist of a six-membered ring of Cu and S atoms in which each Cu is co-ordinated to a terminal Cl and two bridging S atoms, ν(PS) is shifted *ca.* 50 cm^{-1} to lower frequency compared with the free ligand (compared with a shift of 15—35 cm^{-1} when non-bridging co-ordination to Cu^I occurs).[566a]

Some assignments for characteristic ligand vibrations have been made[566b] for Cu^{II} complexes of 1,5-disubstituted 2,4-dithiobiuret ligands (square-planar structures).

---

[562] L. Cattalini, J. S. Coe, S. Degetto, A. Dondoni, and A. Vigato, *Inorg. Chem.*, 1972, **11**, 1519.
[563] P. L. Goggin, R. J. Goodfellow, S. R. Haddock, F. J. S. Reed, J. G. Smith, and K. M. Thomas, *J.C.S. Dalton*, 1972, 1904.
[564] Yu. N. Kukushkin, S. A. Simanova, N. N. Knyazeva, V. P. Alashkevich, S. I. Bakhireva, and E. P. Leonenko, *Russ. J. Inorg. Chem.*, 1971, **16**, 1327.
[565] D. M. Barlex and R. D. W. Kemmitt, *J.C.S. Dalton*, 1972, 1436.
[566] (*a*) J. A. Tiethof, J. K. Stalick, P. W. R. Corfield, and D. W. Meek, *J.C.S. Chem. Comm.*, 1972, 1141; (*b*) K. P. Srivastava and N. K. Agarwal, *Z. anorg. Chem.*, 1972, **393**, 168.

Comparison of the i.r. spectra of 2-mercaptobenzothiazole and its complex with Cu^I confirms co-ordination through S.[567]

Assignments of ν(CN) (1420—1560 cm^{-1}) and ν(CS) (808—882 cm^{-1}) have been made for Ag(dithio-oxamide)$_2^+$ and analogous complexes of substituted dithio-oxamides.[568]

The i.r. spectra of Zn, Cd, Hg, and Co complexes of selenourea have been discussed in relation to earlier normal-co-ordinate calculations on urea and thiourea.[569] Considerable mixing of ligand vibrations is indicated, and co-ordination is through Se rather than N [ν(M—Se) is in the range 245—167 cm^{-1}, and the ratio of ν(M—Se) to ν(M—S) in related seleno- and thio-urea complexes is near to 0.8].

ν(C$\cdots$C) and ν(C—O) have been assigned[570] to 1571—1499 cm^{-1}, 1202—1184 cm^{-1}, respectively, in the complexes M(OEtsacsac)$_2$ (M = Zn, Cd, or Hg) and M(OEtacsac)$_2$ (M = Zn or Cd), where OEtsacsac = (150a), OEtacsac = (150b).

<pre>
       H   H
   Me   OEt   Me   OEt

    S   S   S   O
    (150a)   (150b)
</pre>

Group-frequency i.r. assignments have been made[571] for {(O$_2$NO)Zn-[S=C(NH$_2$)$_2$]$_2$} and {Zn[S=C(NH$_2$)$_2$]$_4$}$^{2+}$ (NO$_3$)$_2^{2-}$.

I.r. spectra (4000—400 cm^{-1}) have been studied[572] for M(CN)$_2$(tu)$_2$ (M = Zn, Cd, or Hg), and the ν(C≡N) and ν(CS) wavenumbers have been discussed. S-co-ordination was proposed, and the data were said to be consistent with thermogravimetric results for the relative strengths of M—S bonds: Zn > Cd > Hg.

Unassigned i.r. data were listed for Zn, Cd, Hg, Pb, and Cu^{II} complexes of dithiolenephthalic acid (151).[573]

Raman spectra have been reported for bis(thiourea)dichlorocadmium and tetrakis(thiourea)dichlorocadmium.[574] The bis-complex is discussed

<pre>
            O   O
            ‖   ‖
   M—S—C—⟨benzene⟩—C—S—M—S
                                    n
            (151)
</pre>

[567] M. M. Khan and A. U. Malik, *J. Inorg. Nuclear Chem.*, 1972, **34**, 1847.
[568] G. C. Pellacani and T. Feltri, *Inorg. Nuclear Chem. Letters*, 1972, **8**, 325.
[569] G. B. Aitken, J. L. Duncan, and G. P. McQuillan, *J.C.S. Dalton*, 1972, 2103.
[570] A. R. Hendrickson and R. L. Martin, *Austral. J. Chem.*, 1972, **25**, 257.
[571] P. I. Protsenko, A. G. Glinina, and G. P. Protsenko, *Russ. J. Inorg. Chem.*, 1971, **16**, 1748.
[572] A. N. Sergeeva, L. A. Kisleva, and S. M. Galitskaya, *Russ. J. Inorg. Chem.*, 1971, **16**, 945.
[573] A. V. Pandey and M. E. Mittal, *Inorg. Chim. Acta*, 1972, **6**, 135.
[574] D. M. Adams and M. A. Hooper, *J.C.S. Dalton*, 1972, 631.

on the basis of a factor-group analysis, but for the tetrakis-complex a molecular model is adequate.

Some unassigned i.r. data were given for $HgCl_2L_2$ complexes, where L = 1,5-disubstituted-2,4-dithiobiurets.[575]

$Me_2In(dtc)$ and $RIn(dtc)_2$ (dtc = dithiocarbamate, $-S_2CNMe_2$) all give $\nu(CN)$ at ca. 1500 cm^{-1}, with *one* band associated with $\nu(CS)$ (1000 ± 10 cm^{-1}), *i.e.* the dtc ligands are bidentate, giving four- or five-co-ordinate indium, respectively. $R_2In(ox)$ (ox = oxinate) is known to be dimeric, and it shows bands characteristic of bidentate oxinate (five-co-ordinate In).[576]

The complexes R_2InX,L, $MeInCl_2,L$ and $InCl_3,L$ (L = Me_3SbS; R = Me or Et; X = Cl, Br, or I) all contain In—S bonding of the trimethylstibine sulphide ligand.[577] Thus $\nu(Sb-S)$ drops from 431 cm^{-1} (free ligand) to the range 393—407 cm^{-1}.

In N-substituted N'-cyano-S-(triphenylstannyl)isothioureas, RN=C-(NHCN)SSnPh$_3$ (R = p-$O_2N \cdot C_6H_4$, Ph, PhCH$_2$, p-EtO$\cdot C_6H_4$, or Et), $\nu(C\equiv N)$ lies in the range 2203—2183 cm^{-1}, $\nu(C=N)$ 1534—1515 cm^{-1}, with $\nu(NH)$ 3401—3356 cm^{-1}.[578]

7 Potentially Ambident Ligands

Cyanate, Thiocyanate Complexes, *etc.*, and Iso-analogues.—The complexes NbX_4L_2, TaX_5(bipy) (X = NCS$^-$ or NCSe$^-$, L = bipy or 4,4'-dimethylbipy), $Nb(NCS)_4(py)_2$, and $Ta(NCS)_5(py)$ all contain N-bonded $-NCS$ or $-NCSe$.[579] Shifts in the CN, CS, and CSe stretches all point to that conclusion.

For $Nb(NCS)_2(OR)_3$(bipy) and $Ta(NCS)_2(OR)_3$(bipy) (R = Me or Et), some characteristic vibrations of NCS, OR, and bipy have been listed and discussed briefly.[580] CN stretching modes below 2100 cm^{-1} exclude the possibility of $-NCS-$ bridging, and are consistent with the presence of N-bonded $-NCS$. Splitting of these bands is said to indicate that the M—NCS system is bent.

N-bonded thiocyanate is indicated for $K[Cr(NCS)_2(acen)]$ by i.r. bands at 2085, 785 cm^{-1} [$\nu(CN)$, $\nu(CS)$, respectively; acen = NN'-ethylenebis-(acetylacetone iminate)].[581]

$\nu(CN)$ in $[(OC)_3Mo(SCN)_3Mo(CO)_3]^{3-}$ has been assigned to bands at 2132 and 2096 cm^{-1}. The former is in agreement with the presence of bridging SCN, while the latter is ambiguous (not being in the region usually associated with M—NCS—M systems).[582]

---

[575] K. P. Srivastava and N. K. Agarwal, *J. Inorg. Nuclear Chem.*, 1972, **34**, 3926.
[576] T. Maeda and R. Okawara, *J. Organometallic Chem.*, 1972, **39**, 87.
[577] T. Maeda, G. Yoshida, and R. Okawara, *J. Organometallic Chem.*, 1972, **44**, 237.
[578] R. A. Cordona and E. J. Kupchik, *J. Organometallic Chem.*, 1972, **43**, 163.
[579] J. N. Smith and T. M. Brown, *Inorg. Chem.*, 1972, **11**, 2697.
[580] N. Vuletić and C. Djordjević, *J.C.S. Dalton*, 1972, 2322.
[581] K. Yamanouchi and S. Yamada, *Bull. Chem. Soc. Japan*, 1972, **45**, 2140.
[582] J. F. White and M. F. Farona, *J. Organometallic Chem.*, 1972, **37**, 119.

ν(CN) and δ(NCS) bands have been assigned in M(NCS)$_6^{2-}$ (M = Mo or W) and W(NCS)$_6^-$: all were consistent with M—N bonding.[583]

The complex Re$_2$(CO)$_8$(NCO)$_2$ gives ν_{as}(NCO) at 2197 cm^{-1}, suggesting the presence of bridging NCO groups [no ν(CO) was found below 1950 cm^{-1}]. The structure (152a) was therefore proposed.[584]

$$\begin{array}{c}
\text{O} \\
\| \\
\text{C} \\
\text{OC} \quad | \quad \text{CO} \\
\text{OC} \diagdown | \diagup \text{N} \diagdown | \diagup \text{CO} \\
\quad \text{Re} \quad \quad \text{Re} \\
\text{OC} \diagup | \diagdown \text{N} \diagup | \diagdown \text{CO} \\
\text{OC} \quad | \quad \text{CO} \\
\text{C} \\
\| \\
\text{O}
\end{array}$$

(152a)

ν(CN) is at 2050 cm^{-1} in [(C$_4$H$_9$)$_4$N]$_2$[Re$_2$(SeCN)$_8$], but it is not known whether the bonding is via the Se or the N.[585]

The i.r. and Raman spectra of [(Bu$^n_4$N)]$_3$[Fe(CN)$_5$(NCX)] (X = S or Se) are consistent with Fe—N co-ordination in both cases.[586]

The complexes [Co{PhP(OEt)$_2$}$_4$(NCS)]$^+$, [Co{PhP(OEt)$_2$}$_3$(NCS)$_2$], and [Co{Ph$_2$P(CH$_2$)$_2$PPh$_2$}$_2$(NCS)]$^+$ all give one ν(CN) band (2083—2060 cm^{-1}). The position of this is characteristic of an N-bonded, non-bridging NCS group – therefore these complexes contain five-co-ordinate Co. The presence of only one band in the second complex indicates that the two NCS groups are trans.[587a]

N- and S-bonded isomers of (4-Bu^{t}py)Co(dmg)$_2$(thiocyanato) have been characterized (dmgH = dimethylglyoxime).[587b] The N-bonded form gives ν(CN) at 2110 cm^{-1}, with the S-form having ν(CN) at 2055 cm^{-1}, the latter band being much weaker than the former.

Two isomeric bridging thiocyanato-complexes of Co have been prepared, showing the following absorptions in the 2000—2200 cm^{-1} region (N.B. the identities of the isomers are known from their preparations; i.r. spectra cannot be used to distinguish them). (NH$_3$)$_5$Co(NCS)Co(CN)$_5$: 2175 cm^{-1} [ν(Co—NCS—Co)]; 2141, 2131, 2114 cm^{-1} [ν(CN)]; and (NH$_3$)$_5$Co(SCN)-Co(CN)$_5$: 2170 cm^{-1} [ν(Co—SCN—Co)]; 2120 [ν(CN)].[588]

The δ(NCS) or δ(NCSe) region (ca. 400—450 cm^{-1}) has been used to distinguish N- from S- or Se-bonded systems, in some NCS and NCSe complexes of Co^{II}, Ni, and Cu with N-benzylethylenediamine and NN'-dibenzylethylenediamine.[589]

[583] C. J. Horn and T. M. Brown, Inorg. Chem., 1972, 11, 1970.
[584] R. B. Saillant, J. Organometallic Chem., 1972, 39, C71.
[585] R. R. Hendriksma, J. Inorg. Nuclear Chem., 1972, 34, 1581.
[586] D. F. Gutterman and H. B. Gray, Inorg. Chem., 1972, 11, 1727.
[587] (a) A. Bertacco, U. Mazzi, and A. A. Orio, Inorg. Chem., 1972, 11, 2547; (b) L. A. Epps and L. G. Marzilli, J.C.S. Chem. Comm., 1972, 109.
[588] R. C. Buckley and J. G. Wardeska, Inorg. Chem., 1972, 11, 1723.
[589] K. C. Patel and D. E. Goldberg, J. Inorg. Nuclear Chem., 1972, 34, 637.

Ni(NCS)$_2$L$_2$ has two isomeric forms; one is red, with square-planar nickel and Ni—NCS bonding; the other is green, and polymeric, with octahedral Ni, and bridging NCS groups. They may be distinguished by their i.r. spectra (L = 4-methylquinoline).[590a]

ν(CN) and δ(NCS) of the NCS ligand, and ν(C=N) of L (ca. 1600 cm^{-1}) have been listed for Ni(NCS)$_2$L compounds where L = (152b) or (152c).[590b]

(152b)

(152c)

Unassigned i.r. data have been listed for [Ni(dpt)L]X$_2$ and [Ni(dpt)L-(NCS)]$^+$(NCS)$^-$, where L = ethylenediamine or propane-1,3-diamine; dpt = 1,5,9-triazanonane.[591]

A number of four- and five-co-ordinated complexes of tep [1,2-bis-(diethylphosphino)ethane] with Ni^{II} and four-co-ordinate complexes with Pd^{II} have been prepared. In [Ni(tep)(NCS)$_2$], ν(CN) is at 2095 cm^{-1}, i.e. the NCS is N-bonded; in [Pd(tep)(CN)$_2$], ν(C≡N) is at 2130 and 2135 cm^{-1}, i.e. the cyano-groups are cis; in [Pd(tep)(NCS)$_2$], ν(CN) of NCS occurs at 2110, 2083 cm^{-1}; this splitting is too great for cis-(NCS)$_2$, so the presence of one N-, and one S-bonded group is suggested; and, finally, for [Ni(tep)$_2$-(NCS)], ν(CN) is at 2070 cm^{-1} (N-bonded).[592]

The i.r. spectra of the complexes [Pd(sbtas)(SCN)]$^+$X$^-$ [where X = BPh$_4$ or NCS and sbtas = tris-(o-dimethylarsinophenyl)stibine] have been compared.[593] Both show bands characteristic of S-bonded thiocyanate ligands [2105, 715, 420 cm^{-1} for ν(CN), ν(CS), δ(NCS)], while the second also shows vibrations of unco-ordinated NCS$^-$ at 2052 cm^{-1} [ν(CN)], and 735 cm^{-1} [ν(CS)].

In the compounds Cu(en)$_n$X$_2$,nH$_2$O and Cu(bipy)X$_2$,nH$_2$O (X = NCS, SO$_4$, NO$_3$, HCO$_2$, OAc, or oxalate), i.r. data are used to deduce the mode of co-ordination of X.[594a]

The crystal structure of the 1:1 complex of Cu(SCN)$_2$ with 1-(2-aminoethyl)biguanide, NH$_2$CH$_2$CH$_2$N=C(NH$_2$)NHC(NH)NH$_2$ (= L), shows that it is [CuL(NCS)](SCN).[594b] The splitting of the NCS$^-$ i.r. bands is consistent with this formulation: co-ordinated NCS (ν_1 763, ν_3 2075 cm^{-1}); ionic SCN$^-$ (ν_1 719, ν_3 2050 cm^{-1}).

---

[590] E. Jóna, M. Jannický, T. Šramko, and J. Gažo, Coll. Czech. Chem. Comm., 1972, **37**, 3679; (b) B. Chiswell and K. W. Lee, Inorg. Chim. Acta, 1972, **6**, 583.
[591] G. Ponticelli and C. Preti, J.C.S. Dalton, 1972, 708.
[592] E. C. Alyea and D. W. Meek, Inorg. Chem., 1972, **11**, 1029.
[593] L. Baracco and C. A. McAuliffe, J.C.S. Dalton, 1972, 948.
[594] (a) D. M. Proctor, B. J. Hathaway, and P. G. Hodgson, J. Inorg. Nuclear Chem., 1972, **34**, 3689; (b) G. D. Andreetti, L. Coghi, M. Nardelli, and P. Sgarabotto, J. Cryst. Mol. Structure, 1971, **1**, 147.

The three i.r.-active modes of the cyanato ligand have been listed for the complexes $Cu(NCO)_2L$ and $Cu(NCO)_2L_2$, where L = a nitrogen-donor, e.g. py, lutidines.[595]

The dimethylgold pseudohalides $[Me_2Au(NCO)]_2$ and $[Me_2Au(NCSe)]_2$ have been prepared and their i.r. and Raman spectra investigated.[596] These were found to be very similar to those of $[Me_2Au(NCS)]_2$ and $[Me_2Au(N_3)]_2$, respectively, and therefore the new complexes were believed to possess analogous structures, of C_{2h}, D_{2h} symmetries.

The complex trans-$[Me_4N][Au(CN)_2(SCN)_2]$ gives an i.r. spectrum indicative of S-bonding only,[597] with no evidence for the presence of an N-bonded isomer (cf. D. Negoiu and L. M. Băloiu, Z. anorg. Chem., 1971, 382, 92). The selenium analogue, while also Se-bonded in the solid state, contains a certain amount of the N-isomer in $MeNO_2$ solution [a new $\nu(SeCN)$ band appears at 2117 cm^{-1}].

Mixed cyanothiocyanato complexes $K_2[M(CN)_2(SCN)_2],nH_2O$ (M = Zn, n = 2; M = Cd, n = 4; M = Hg, n = 0) have been studied [598a] by t.g.a. and i.r. spectroscopy. $\nu(CS)$ and $\nu(CN)$ modes were listed, the former being taken to indicate that the Zn complex is N-bonded, whereas the Cd and Hg complexes are S-bonded.

I.r. spectra of $K_2M(NCS)_2X_2,nH_2O$ (M = Zn or Cd; n = 1—3; X = Cl, Br, or I) have been discussed from the point of view of the nature of the NCS to M bonding.[598b]

$\nu_s(NCS)$, $\nu_{as}(NCS)$, and $\delta(NCS)$ in $Cd(NCS)_2L$ are consistent with N-co-ordination when L = o-phen, nitrophen, dimephen, or bipy. When L = 8-aminoquinoline (8-amq), however, $\nu_{as}(NCS)$ is at 2110, 2085 cm^{-1}, and $\delta(NCS)$ at 465, 415 cm^{-1}, and these are consistent with the formulation $Cd(NCS)(SCN)(8\text{-}amq)$, i.e. one N- and one S-bonded NCS ligand.[599]

I.r. and Raman studies on aqueous solutions containing MeHgSCN and alkali-metal thiocyanates show that very little perturbation of the MeHg-SCN molecule occurs [evidence from $\nu(HgS)$, ν_{as}, and $\nu_s(SCN)$]. Hence, any complex formation which occurs must be of ion–dipole type.[600]

The previously ill-characterized compound $Ni[Hg(SCN)_3]_2,2H_2O$ shows i.r. absorptions at 2182, 2138, 1595, 740, 470, and 438 cm^{-1}.[601a]

I.r. spectra of $MHg(XCN)_2,Y_2,nA$ (M = Mn, Co, Ni, Zn, or Cu; X = S or Se; Y = Cl, Br, or SCN; A = Me_2CO or EtOH; n = 0—4) have been

---

[595] J. Kohout, M. Quastlerová-Hrastijova, and J. Gažo, Coll. Czech. Chem. Comm., 1971, 36, 4026.
[596] F. Stocco, G. C. Stocco, W. M. Scovell, and R. S. Tobias, Inorg. Chem., 1971, 10, 2639.
[597] J. B. Melpolder and J. L. Burmeister, Inorg. Chem., 1972, 11, 911.
[598] (a) A. N. Sergeeva, L. A. Kiseleva, and S. M. Galitskaya, Russ. J. Inorg. Chem., 1971, 16, 1098; (b) G. V. Tsintsadze, Tr. Gruz. Politekh. Inst., 1970, 3, 81.
[599] A. Syamal, Z. Naturforsch., 1972, 27b, 1002.
[600] J. Relf, R. P. Cooney, and H. F. Henneike, J. Organometallic Chem., 1972, 39, 75.
[601] (a) R. D. Gillard and M. V. Twigg, Inorg. Chim. Acta, 1972, 6, 150; (b) G. V. Tsintsadze, A. Yu. Tsivadze, and A. S. Managadze, Soobsch. Akad. Nauk. Gruz. S.S.R., 1972, 65, 329.

studied.[601b] Assignments were given for ν(CN), ν(CS), ν(CSe), δ(NCS), or δ(NCSe).

I.r. wavenumbers have been listed for $(h^5\text{-}C_5H_5)_3\text{CeX}$ and $(h^5\text{-indenyl})\text{-}CeX_2$, where X = NCO, NCS, N_3, or CN.[602]

Octathiocyanato-actinide(IV) complexes $(NEt_4)_4M(NCS)_8$ (M = Th, Pa, U, Np, or Pu), $Cs_4M(NCS)_8$ (M = U or Pu), and also $(NEt_4)_4M\text{-}(NCSe)_8$ (M = Pa or U) all show a single strong i.r. absorption due to ν(CN), as expected for O_h symmetry.[603a]

I.r. spectra of microcrystalline $Eu(NCS)_3,4C_4H_8O_2$ and $M^IEu(NCS)_4,\text{-}4EtOH$ (M^I = K, Rb, or Cs) have been obtained, and are consistent with co-ordination via N.[603b]

Characteristic NCS bands were assigned for $Ga(NCS)_3,2H_2O$, $MGa(NCS)_4$ (M = Na, K, Rb, Cs, or NH_2), and $M_3Ga(NCS)_6$.[604]

Assignments have been proposed for ν_{as}, ν_s(NCS), δ(NCS), and 2δ(NCS) in $Me_3Pb(NCS)$, $Ph_3Pb(NCS)$, $Ph_2Pb(NCS)_2$, $[Ph_3Pb(NCS)_2]^-$, and $[Ph_2Pb(NCS)_4]^{2-}$. The three neutral complexes contain N-bonded NCS, monomeric in solution but polymeric (trigonal-bipyramidal co-ordination at Pb) in the solid, the anions also containing N-bonded NCS, with trigonal-bipyramidal and *trans*-octahedral structures.[605]

Ligands containing N and O Donor Atoms.—I.r. data have been given (unassigned) for Cr, Fe, Co^{III}, Ni, and Cu^{II} complexes of 8-amino-7-hydroxy-4-methylcoumarin (153).[606] O-/N-chelation was indicated in all cases.

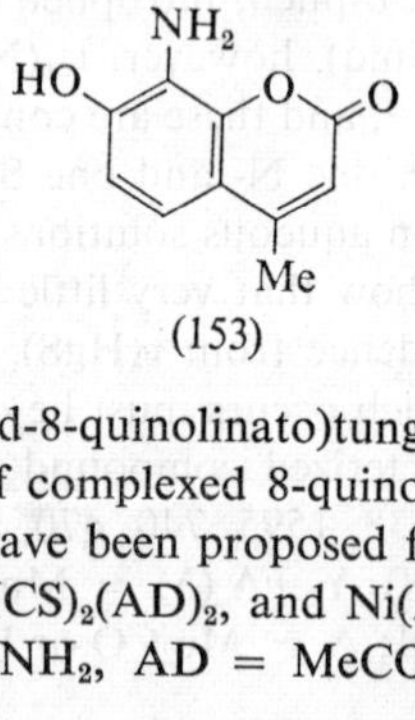

(153)

Tetrakis-(5,7-disubstituted-8-quinolinato)tungsten(V) salts give an i.r. absorption characteristic of complexed 8-quinolinates at 1115 cm^{-1}.[607]

Fairly full assignments have been proposed for MnA_3^{2+}, FeA_3^{3+}, CoA_3^{2+}, $Co(NCS)_2A_2$, NiA_3^{2+}, $Co(NCS)_2(AD)_2$, and $Ni(AD)_3^{2+}$ in the range 4000—400 cm^{-1} (A = $MeCONHNH_2$, AD = $MeCONDND_2$).[608]

---

[602] B. L. Kalsotra, R. K. Multani, and B. D. Jain, *J. Inorg. Nuclear Chem.*, 1972, **34**, 2265.

[603] (a) Z. M. S. Al-Kazzaz, K. W. Bagnall, D. Brown, and B. Whittaker, *J.C.S. Dalton*, 1972, 2273; (b) G. V. Tsintsadze, A. N. Borshch, and E. A. Krezerli, *Soobsch. Akad. Nauk. Gruz. S.S.R.*, 1971, **64**, 69.

[604] L. M. Mikheeva, G. I. Elfimeva, and L. N. Komissarova, *Russ. J. Inorg. Chem.*, 1971, **16**, 1787.

[605] N. Bertazzi, G. Alonzo, A. Silvestri, and G. Consiglio, *J. Organometallic Chem.*, 1972, **37**, 281.

[606] D. K. Rastogi, A. K. Srivastava, P. C. Jain, and B. R. Agarwal, *Inorg. Chim. Acta*, 1972, **6**, 145.

[607] R. D. Archer, W. D. Bonds jun., and R. A. Pribush, *Inorg. Chem.*, 1972, **11**, 1550.

[608] Yu. Ya. Kharitonov and R. I. Machkhoshvili, *Russ. J. Inorg. Chem.*, 1971, **16**, 1438.

The two iron(III) complexes of the amino-acid DL-methionine (= met) Fe(met)(OH)Cl,2MeOH and Fe(met)(OH)NO$_3$,2MeOH have been isolated.[609] Both give ν_{as}(CO$_2$) at 1610 cm^{-1}, with ν_s(CO$_2$) at 1420—1435 cm^{-1}.

ν(NO) bands have been listed (see Table 18) for the nitro- and nitrito-isomers of [Ru(bipy)$_2$L(NO$_2$)]$^+$, where L = OH$_2$, NCMe, MeOH, OC(CH$_3$)$_2$, or py, and Ru(bipy)$_2$(NO$_2$)X, where X = Cl or I.[610]

Table 18 (N—O) *Bands for some ruthenium nitro- and nitrito-complexes (all figures are wavenumber/cm^{-1})*

Complex	ν(N—O)	
	Nitro	Nitrito
[Ru(bipy)$_2$(NCMe)(NO$_2$)]$^+$	1337, 1292	—
[Ru(bipy)$_2$(OH$_2$)(NO$_2$)]$^+$	1338, 1294	1314, 1134
[Ru(bipy)$_2$(HOMe)(NO$_2$)]$^+$	1337, 1291	1397, 1133
[Ru(bipy)$_2$(OCMe$_2$)(NO$_2$)]$^+$	1340, 1297	1394, 1131
[Ru(bipy)(py)(NO$_2$)]$^+$	1340, 1299	1394, 1130
Ru(bipy)$_2$(NO$_2$)Cl,H$_2$O	1338, 1298	1396, 1133
Ru(bipy)$_2$(NO$_2$)I,2H$_2$O	1339, 1297	1395, 1134

I.r. spectra were reported for [RuNH$_2$(NH$_3$)$_4$NO]X$_2$ (X = Br or I) and [Ru(NH$_3$)$_5$NO$_2$]X,H$_2$O (X = Cl, Br, or I; X = I, N is ^{15}N; X = I, H is ^{2}H). ν(NO$_2$) is assigned at 1203 cm^{-1} (1169 cm^{-1} for ^{15}N analogue) but δ(NO$_2$) vibrations could not be seen.[611]

The complex μ-superoxo-bis[bis-(L-histidinato)cobalt(III)]$^{3+}$ gives a band due to an uncomplexed carboxy-group at 1740 cm^{-1}.[612] Thus, all of the histidine ligands cannot be terdentate.

I.r. spectra of Co^{III}, Ni^{II}, and Cu^{II} complexes of btat (154) and of the Ni^{II} complex of bhat (155) have been obtained.[613] NH stretching (3380—3160 cm^{-1}), C⋯O and/or C⋯C stretching (ca. 1620, 1550 cm^{-1}), and CF$_3$ stretching modes (1195—1080 cm^{-1}) were assigned.

(154)   (155)

[609] E. J. Halbert and M. J. Rogerson, *Austral. J. Chem.*, 1972, **25**, 421.
[610] S. A. Adeyemi, F. J. Miller, and T. J. Meyer, *Inorg. Chem.*, 1972, **11**, 994.
[611] F. Bottomley and J. R. Crawford, *J.C.S. Dalton*, 1972, 2145.
[612] M. Woods, J. A. Weil, and J. K. Kinnaird, *Inorg. Chem.*, 1972, **11**, 1713.
[613] S. C. Cummings and R. E. Sievers, *Inorg. Chem.*, 1972, **11**, 1483.

The complexes of *N*-hydroxyethylethylenediamine with Co^{II}, Ni^{II}, Cu^{II}, and Zn^{II} show no band due to ν(OH), hence the hydroxy-group is co-ordinated to the metal.[614]

The new quadridentate ligand 1,5-diazacyclo-octane-*NN'*-diacetate (dacoda) possesses an unusual steric arrangement which allows the co-ordination of only one further ligand. A series of five-co-ordinate complexes (156) has therefore been prepared, [M(dacoda)(H$_2$O)],2H$_2$O (M =

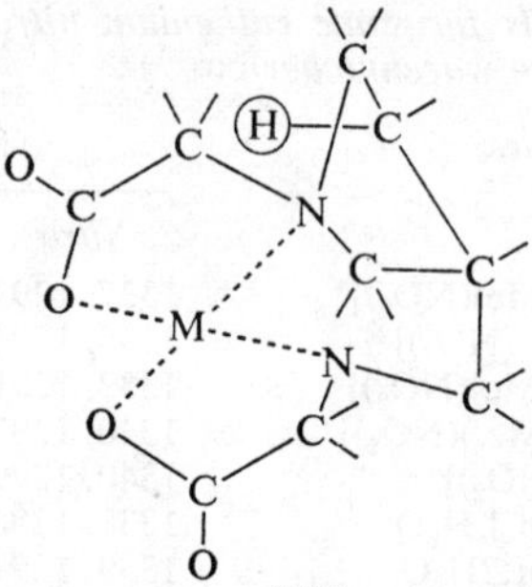

(156)

Co, Ni, Cu, or Zn), in which the values of the ν_{as}(CO$_2$) wavenumbers (1600—1630 cm^{-1}) indicate that both carboxylates are indeed co-ordinated to the metal.[615]

Ni^{II} and Cu^{II} complexes of 2-acetamidothiazole and 2-acetamidobenzothiazole have been separated into three types through a study of their i.r. spectra.[616] These are (*a*) carbonyl-oxygen and thiazole-nitrogen donor atoms, (*b*) amide-nitrogen and thiazole-nitrogen donor atoms, and (*c*) carbonyl-oxygen, amide-nitrogen, and thiazole-nitrogen donor atoms. In (*a*) and (*c*), ν(C=O) is shifted down by *ca.* 30 cm^{-1}, whereas for (*b*) an upward shift (*ca.* 5 cm^{-1}) was noted on complex formation. Confirmation of these conclusions arose from assignments to ν(M—O) (350—410 cm^{-1}) and ν(M—N) (230—280 cm^{-1}) as appropriate.

Characteristic ν(C=C) and ν(N=C) i.r. absorptions were reported for (157; R^1 = R^2 = Me or R^{1}R^2 = —CH$_2$CH$_2$CH$_2$CH$_2$—; X = Cl, H, or Bu^t) and [158; M = Ni, Cu, or Co; R^1, R^2, and X as for (157)].[617]

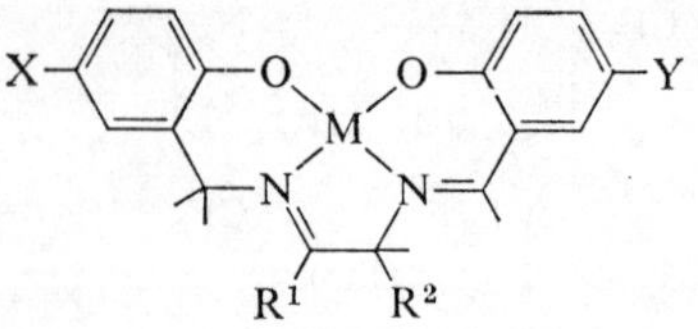

(157) 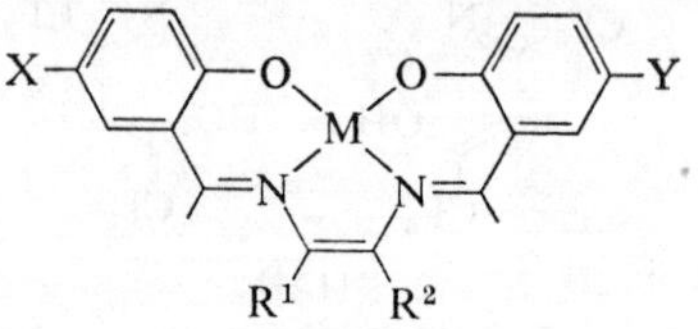   (158)

[614] M. N. Hughes, M. Underhill, and K. J. Rutt, *J.C.S. Dalton*, 1972, 1219.
[615] D. F. Averill, J. I. Legg, and D. L. Smith, *Inorg. Chem.*, 1972, **11**, 2344.
[616] M. N. Hughes and K. J. Rutt, *J.C.S. Dalton*, 1972, 1311.
[617] H. Kanatomi and I. Murase, *Inorg. Chem.*, 1972, **11**, 1356.

In the complexes (159; R^1 = Et; R^2 = Me, Et, Pr^n, Bu^n, or Ph) and (159; R^1 = Me; R^2 = Pr^n, Bu^n, or Ph), prepared from benzilmonohydrazone and R^1R^2CO in the presence of Ni^{2+}, $\nu(NH_2)$, $\delta(NH_2)$, and $\nu(C=O)$ of the monohydrazone have disappeared. Observed instead are bands due to $\nu(C-N) + \nu(C-O)$ (*ca.* 1280 cm^{-1}) and to $\nu(C=C)$ (*ca.* 1530 cm^{-1}).[618]

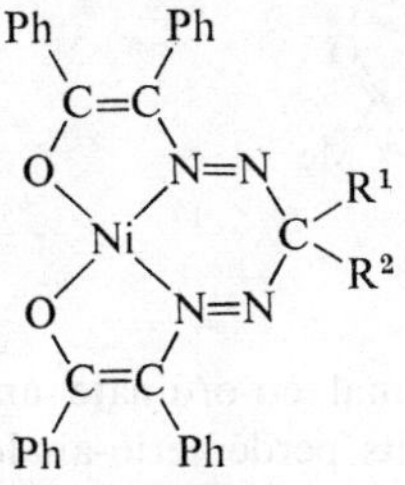

(159)

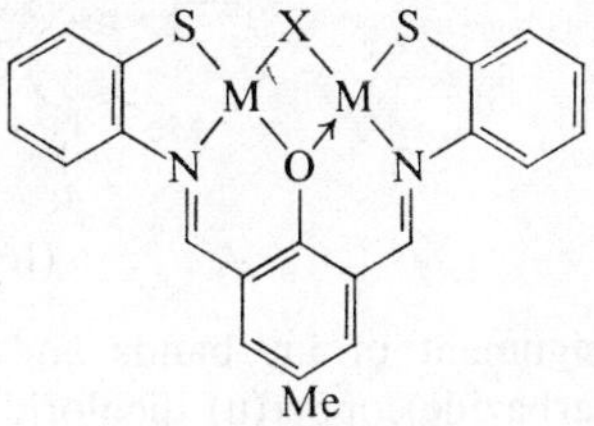

(160)

The complexes (160; M = Ni or Cu; X = N_3) show a pair of i.r. bands at 1535, 1575 cm^{-1}, both associated primarily with C=N stretching.[619]

Ni and Cu complexes of the types (161)—(163) have been prepared.[620] (161) give $\nu(C=O)$ at 1670 cm^{-1} and $\nu(C=N)$ at 1635 cm^{-1}; (162) give two bands in the 1645—1625 cm^{-1} region due to unco-ordinated and co-ordinated azomethine groups, and (163) give only one band due to $\nu(C=N)$.

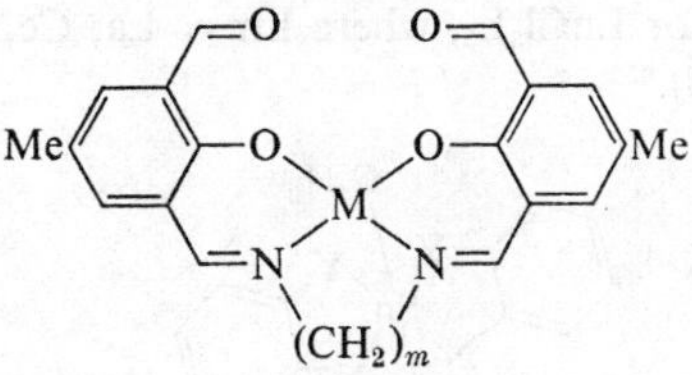

(161)

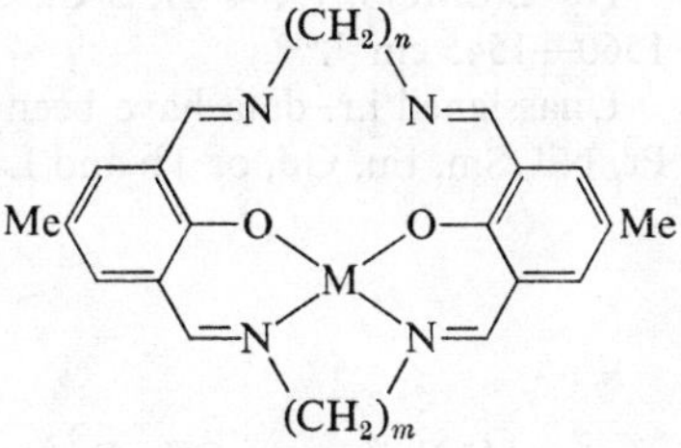

(162)

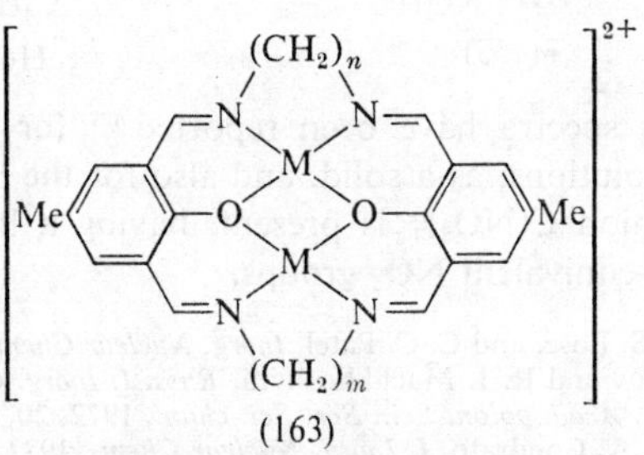

(163)

[618] C. M. Kerwin and G. A. Melson, *Inorg. Chem.*, 1972, **11**, 726.
[619] W. D. McFadyen, R. Robson, and H. Schaap, *Inorg. Chem.*, 1972, **11**, 1777.
[620] H. Okawa and S. Kida, *Bull. Chem. Soc. Japan*, 1972, **45**, 1759.

I.r. and n.m.r. spectra of some Pd^{II} complexes are believed[621] to be consistent with the structure (164; R = Pr^i, C_6H_{11}, Ph, or p-tolyl).

Fairly complete assignments have been made for $Pt(A)_4^{2+}$, $Pt(AD)_4^{2+}$, $Cu(A)_2^{2+}$, and $Cd(A)_2^{2+}$, where A = $MeCONHNH_2$ and AD = $MeCONDND_2$.[622]

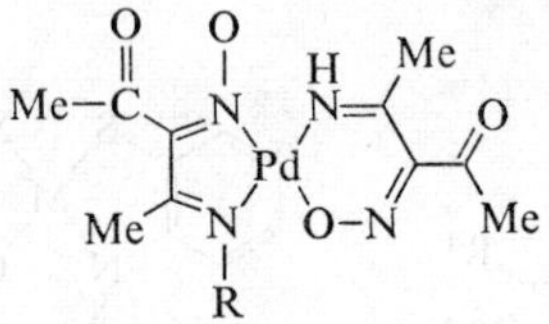

(164)

An assignment of i.r. bands and a normal co-ordinate analysis for bis(semicarbazide)copper(II) dichloride and its perdeuterio-analogue have been reported.[623] The semicarbazide forms O- and N-bonded chelate rings.

Changes observed on heating the glycine–Cu–montmorillonite complex to 190 °C can be understood in terms of a change from unidentate to bidentate Cu–glycine bonding together with elimination of H_2O that was originally co-ordinated to Cu.[624]

$\nu(CO)$ of the Cu complexes of the ONS-Schiff bases of substituted salicylaldehydes and S-methyl dithiocarbazate, $Cu\{X \cdot C_6H_3(O)CH=N-N=C(SMe)S\}$ (X = H, 5-Cl, 5-Br, 3-MeO, or 5-NO_2), is in the range 1560—1545 cm^{-1}.[625]

Unassigned i.r. data have been given for $LnCl_3L_3$, where Ln = La, Ce, Pr, Nd, Sm, Eu, Gd, or Tb and L = (165).[626]

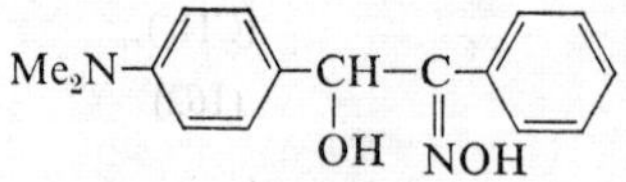

(165) 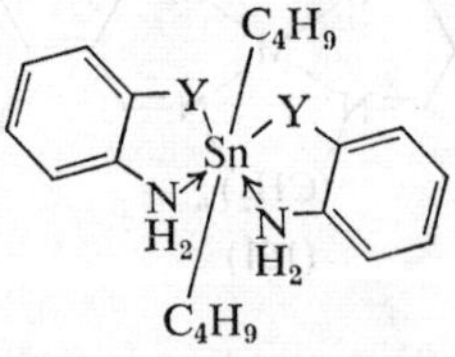 (166)

I.r. and Raman spectra have been reported[627] for $Na^+[C(NO_2)_3]^-$ in H_2O and MeCN solutions, as a solid, and also for the ^{15}N analogue. The trinitromethane anion $C(NO_2)_3^-$ is present, having a low symmetry (less than C_3) with non-equivalent NO_2 groups.

---

[621] B. C. Sharma, K. S. Bose, and C. C. Patel, *Inorg. Nuclear Chem. Letters*, 1972, **8**, 805.
[622] Yu. Ya. Kharitonov and R. I. Machkhoshvili, *Russ. J. Inorg. Chem.*, 1971, **16**, 847.
[623] B. B. Kedzia, *Bull. Acad. polon. Sci., Sér. Sci. chim.*, 1972, **20**, 557.
[624] S. D. Jang and R. A. Condrate, *J. Inorg. Nuclear Chem.*, 1972, **34**, 3282.
[625] M. A. Ali, S. E. Livingstone, and D. J. Phillips, *J.C.S. Chem. Comm.*, 1972, 909.
[626] R. Pastorek, *Monatsh.*, 1972, **103**, 1542.
[627] V. A. Shlyapochnikov, G. I. Oleneva, and S. S. Novikov, *Izvest. Akad. Nauk S.S.S.R., Ser. khim.*, 1971, 2603.

Unassigned i.r. listings have been published for the o-aminophenol, o-phenylenediamine, and o-aminothiophenol derivatives of dibutyltin(IV)[628] (166; Y = O, NH, or S).

Ligands containing either N and As or N and S Donor Atoms.—A few isolated i.r. wavenumbers have been listed for ligand vibrations in the Pd and Pt complexes MLX_2 and ML_2X_2, where X = halide, SCN, or ClO_4 and L is a hybrid bidentate ligand containing both N and As donor atoms.[629]

Unassigned i.r. data have been given for Co, Ni, and Cu complexes (167) of 3-(4-pyridyl)-triazoline-5-thione.[630]

(167)

The compounds $NiL_2Cl_2,2HOAc$, NiL_2Br_2, NiL_2I_2, and $NiL_2(ClO_4)_2$, where L = NN-dicyclohexyldithio-oxamide or NN'-dibenzyldithio-oxamide, have ν(CN) in the region 1600—1430 cm^{-1} and ν(CS) in the region 886—755 cm^{-1}. The ligands were believed to be NS-chelated.[631]

Thiosemicarbazole forms N- and S-bonded chelate rings in bis(thiosemicarbazide)nickel(II) dichloride.[632] The i.r. spectrum was assigned on the basis of C_{2h} symmetry.

Band assignments have been proposed for the i.r. spectra of trans-bis(thiosemicarbazidato)nickel(II) and its perdeuterio-analogue.[633a]

A normal co-ordinate analysis of the 1 : 1 metal–ligand models has been carried out in order to assist in the assignment of the vibrational spectra of $[Cu(NH_2CONHNH_2)_2]Cl_2$, $[Ni(NH_2CSNHNH_2)_2]Cl_2$, $[Ni(NH_2CSNNH_2)_2]$, and their fully deuteriated analogues.[633b]

Some ligand vibrations have been assigned in the Zn, Cd, and Hg complexes of thiosemicarbazide (168),[634] and in Pr, Nd, Ce, and Sm complexes of 1-substituted-tetrazoline-5-thiones $Ln(L)_3$.[635]

Ligands containing O and S Donor Atoms.—The potentially ambidentate ligand phenylsulphoxy-acetic acid $PhS(O)CH_2CO_2H$ forms simple salts with Mg, Ca, Ba, Mn, Co, Ni, Cu, Zn, Ag^I, and Hg^I. With Pt^{II}, however,

---

[628] R. C. Mehrotra and B. P. Dachlas, *J. Organometallic Chem.*, 1972, **40**, 129.
[629] B. Chiswell, R. A. Plowman, and K. Verrall, *Inorg. Chim. Acta*, 1972, **6**, 275.
[630] B. Singh and R. Singh, *J. Inorg. Nuclear Chem.*, 1972, **34**, 3449.
[631] G. C. Pellancani and G. Peyronel, *Inorg. Nuclear Chem. Letters*, 1972, **8**, 299.
[632] B. B. Kedzia, *Bull. Acad. polon. Sci.*, *Sér. Sci. chim.*, 1972, **20**, 565.
[633] (a) B. B. Kedzia, *Bull. Acad. polon. Sci.*, *Sér. Sci. chim.*, 1972, **20**, 579; (b) B. B. Kedzia, *Pr. Nauk. Inst. Chem. Nieorg. Met. Pierwiastkow Rzadkich. Politech. Wroclaw*, 1972, No. 11.
[634] D. S. Mahadevappa and A. S. Ananda Murthy, *Austral. J. Chem.*, 1972, **25**, 1565.
[635] Lakshmi and U. Agarwala, *J. Inorg. Nuclear Chem.*, 1972, **34**, 2255.

$$\begin{array}{c}\text{H}_2\text{N}-\text{C}=\text{S} \quad \overset{\text{H}_2}{\text{N}}-\text{NH} \\ \quad | \quad\quad\quad \text{M}^{2+} \quad | \\ \text{HN}-\text{N} \quad\quad \text{S}=\text{C}-\text{NH}_2 \\ \quad\quad \text{H}_2 \end{array}$$

(168)

it complexes *via* the S-atom.[636] ν(S=O) and ν(C=O) wavenumbers were listed for all of the compounds studied.

Linkage isomers of the sulphinato-complexes [M(bipy)$_2$(L)$_2$], where L = toluene-*p*-sulphonyl and M = Fe, Co, or Ni, show slightly different ν(SO) wavenumbers.[637] Thus, for example, when M = Fe, the S-bonded complex gives ν_{as}(SO) at 1219, 1199 cm^{-1}, ν_s(SO) at 1034, 1012 cm^{-1}, while the O-bonded species have ν_{as}(SO) at 1054 cm^{-1} and ν_s(SO) at 918 cm^{-1}. Similar results were quoted by the same authors elsewhere.[638a]

Complexes of *meso*- and *rac*-PhS(O)CH$_2$CH$_2$S(O)Ph and of *meso*- and *rac*-PhS(O)CH$_2$S(O)Ph with Co, Ni, Pt, and Cu have been studied.[638b] $\Delta\nu$(SO) values indicate O-co-ordination (shifts to lower wavenumber) in all but the Pt compounds, for which shifts to higher wavenumber indicate S-co-ordination.

The tris(sulphinato-*S*-) complexes (RS)$_3$Rh(OH$_2$)$_3$, where R = Me,

$$\begin{array}{c}\text{O}\\ \|\\ \\ \|\\ \text{O}\end{array}$$

Ph, or *p*-tolyl, show ν_{as}(SO$_2$) at 1230—1212 and 1188—1202 cm^{-1}, ν_s(SO$_2$) at 1101—1090 and 1069—1049 cm^{-1}, and δ(SO$_2$) at 579—576 and 538—520 cm^{-1}.[639] *cis*-Octahedral (C_{3v}) co-ordination was proposed.

Values of ν_s(SO$_2$), ν_{as}(SO$_2$), and δ(SO$_2$) for a number of *S*-sulphinates of Ir^{III}, namely (169) and (170), where L = PPh$_3$ and R = Me, Et, Pr^n,

$$\begin{array}{cc}\text{RSO}_2 & \text{RSO}_2 \\ | & \text{Cl}\diagdown | \diagup \text{L} \\ \text{Cl}-\text{Ir}-\text{L} & \text{Ir} \\ \diagup\diagdown & \text{L}\diagup | \diagdown \text{Cl} \\ \text{L} \quad \text{Cl} & \text{CO} \\ (169) & (170)\end{array}$$

p-MeO·C$_6$H$_4$, *p*-Me·C$_6$H$_4$, Ph, *p*-Cl·C$_6$H$_4$, or *p*-O$_2$N·C$_6$H$_4$, have been tabulated by Kubota and Loeffler.[640]

(Di-n-propylmonothiocarbamato)nickel(II) gives i.r. bands at 1545 cm^{-1} [ν(CO)], 1525 cm^{-1} [ν(C=N)], 668 cm^{-1} [ν(CS)], 450 cm^{-1} [ν(MO)], and 385 cm^{-1} [ν(MS)], indicating co-ordination through the O and S atoms.[641]

[636] N. J. Gogan, M. J. Newlands, and B.-Y. Tan, *Canad. J. Chem.*, 1972, **50**, 3202.
[637] E. König, E. Lindner, I. P. Lorenz, and G. Ritter, *Inorg. Chim. Acta*, 1972, **6**, 123.
[638] (*a*) E. Lindner and I. P. Lorenz, *Chem. Ber.*, 1972, **105**, 1032; (*b*) T. R. Musgrave and G. D. Kent, *J. Co-ordination Chem.*, 1972, **2**, 23.
[639] E. Lindner and I. P. Lorenz, *Inorg. Nuclear Chem. Letters*, 1972, **8**, 979.
[640] M. Kubota and B. M. Loeffler, *Inorg. Chem.*, 1972, **11**, 469.
[641] J. Willemse, *Inorg. Nuclear Chem. Letters*, 1972, **8**, 45.

(h^5-C_5H_5)Ni(PBu$_3$)(SO$_2$Ph) gives i.r. bands in the region 1000—1200 cm^{-1} due to S-bonded sulphinate.[642]

In the Ni^{II} complexes of the NN-dialkylthiocarbamates (171), bands assigned to ν(C⋯N) and ν(C⋯O) were in the region 1500—1566 cm^{-1}. No evidence for bands due to unco-ordinated >C=O was found, and hence the ligands appear to be co-ordinating via S and O.[643]

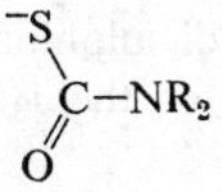

(171) 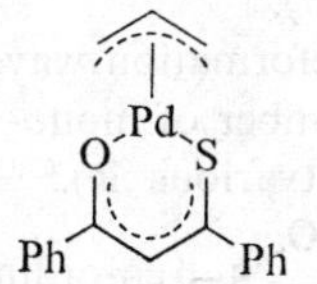 (172)

A band at 1260 cm^{-1} in (π-allyl)Pd(sdbm) (172), sdbm = monothiodibenzoylmethane, has been assigned to ν(C⋯S).[644]

The series of complexes trans-PtL$_2$(RSO$_2$)Cl, where L = PEt$_3$, AsEt$_3$, SeEt$_2$, or TlEt$_2$ and R = Ph, Me, or p-Cl·C$_6$H$_4$, all give ν_{as}(SO$_2$) in the range 1185—1204 cm^{-1}, and ν_s(SO$_2$) between 1095 and 1089 cm^{-1}, i.e. these are all S-sulphinates.[645]

Some i.r. and Raman data have been reported for two different crystal modifications of p-Cl·C$_6$H$_4$S(O$_2$)HgCl.[646] In both, the S-bonded formulation was confirmed.

The triorganogermanium sulphinates R$_3^1$Ge(O$_2$SR^2) (R^1 = Et or Ph; R^2 = Ph or p-Me·C$_6$H$_4$) have been prepared.[647] They give two bands characteristic of SO stretching. These are at ca. 1125 cm^{-1} and ca. 800—840 cm^{-1}, assigned as ν(SO) and ν_{as}(SOGe) respectively – indicating an O-bonded sulphonate structure (173).

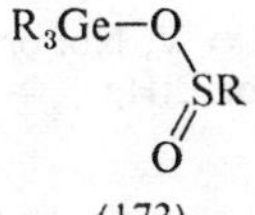

(173)

The i.r. spectrum of Me$_2$Sn(O$_2$SMe)$_2$ is in accord with six-co-ordinate tin, having a linear Me$_2$Sn unit and weak co-ordination of the methanesulphinate: 3018, 2930 cm^{-1} [ν(CH$_3$)]; 1415, 1405 cm^{-1} [δ_{as}(Me—Sn) + δ_{as}(Me—S)]; 1298 cm^{-1} [δ_s(Me—S)]; 1195 cm^{-1} [δ_s(Me—Sn)]; 972 cm^{-1}

---

[642] M. Sato, F. Sato, N. Takemoto, and K. Iida, J. Organometallic Chem., 1972, 34, 205.
[643] B. J. McCormick and B. P. Stormer, Inorg. Chem., 1972, 11, 729.
[644] S. J. Lippard and S. M. Morehouse, J. Amer. Chem. Soc., 1972, 94, 6949.
[645] F. Faraone, L. Silvestro, S. Sergi, and R. Pietropaolo, J. Organometallic Chem., 1972, 34, C55; 1972, 46, 379.
[646] J. R. Brush, P. G. Cookson, and G. B. Deacon, J. Organometallic Chem., 1972, 34, C1.
[647] E. Lindner and K. Schardt, J. Organometallic Chem., 1972, 44, 111.

[$\nu_{as}(SO_2)$]; 938 cm^{-1} [$\nu_s(SO_2)$]; 787 cm^{-1} [$\rho(CH_3-Sn)$]; 700 cm^{-1} [$\nu(CS)$]; 582 cm^{-1} [$\nu_{as}(SnC_2)$]; 541 cm^{-1} [$\delta(SO_2)$]; 424, 395 cm^{-1} [$\nu_{as}(SnO_2)$ + $\nu_s(SnO_2)$].[648]

A number of characteristic i.r. and Raman bands have been listed for the following series of alkyltin derivatives:[649] R_3SnO_2SR (R = Me, Et, Pr^n, or Bu^n), $R_2Sn(O_2SR)_2$ (R = Et, Pr^n, Pr^i, or Bu^n), $(R_3Sn)_2SO_4$ (R = Me, Et, Pr^n, or Pr^i), R_2SnSO_4 (R = Et, Pr^n, Pr^i, or Bu^n), and R_2SnSO_3 (R = Et, Pr^n, or Bu^n).

Stretching and deformation wavenumbers due to the SO_2 groups have been given for a number of mono- and di-sulphinates of tin $Bz_3Sn(O_2SR)$ and $Bz_2Sn(O_2SR)_2$ (various R).[650] All of the $\nu(SO_2)$ wavenumbers are consistent with Sn$\underset{O}{\overset{O}{\diagup\diagdown}}$S— co-ordination.

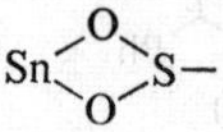

$\nu_s(SO_2)$ and $\nu_{as}(SO_2)$ have been listed for a number of bismuth trisarenesulphinates $Bi(O_2SAr)_3$ [Ar = Ph, p-Me·C_6H_4, p-Cl·C_6H_4, p-MeCONH·-C_6H_4, or 2,4,6-$(Me_2CH)_3C_6H_2$].[651] All lie in the region expected for O-bonding, and the small separations indicate bidentate or bridging bidentate bonding.

X-Ray analysis of $Pt(PPh_3)_2MeI(SO_2)$ shows not the expected five-co-ordinate structure at Pt but co-ordination of the SO_2 molecule to iodine. ν_1 and ν_3 for the SO_2 unit were found at 1130 and 1305 cm^{-1}, quite different values from the species $PtCl(SO_2Me)(PPh_3)_2$ (1070 and 1215 cm^{-1}).[652]

8 Appendix: Additional References to Metal Carbonyl Complexes

Vanadium, Niobium, and Tantalum Carbonyl Complexes

		Ref.
$M(CO)_6L$	M = V, Nb, or Ta; L = Ph$_3$Sn, Ph$_3$PAu, or EtHg	⎫
$[(Ph_3P)_3Au]^+[Ta(CO)_6]^-$		
$M(CO)_5(PPh_3)L$	M = V, Nb, or Ta; L = Ph$_3$Sn, Ph$_3$PAu, or EtHg	⎬ 653
$V(CO)_5(PBu_3)(Ph_3Sn)$		
$Ta(CO)_5[P(OPh)_3](Ph_3Sn)$		
$M(CO)_4(Ph_2PCH_2)_2(Ph_3Sn)$	M = V, Nb, or Ta	⎭
$XV(CO)_4(Ph_2PCH_2)_2$	X = H or I	⎫ 654
$V(CO)_4(Ph_2PCH_2)_2$		⎭
$V(CO)_3(C_7H_7X)$	X = H, 7-Me, 7-Ph, 7-CN, 7-CH$_2$CO$_2$Me, 3-OMe, 3-OEt, or 3-OC$_3$H$_7$	655

[648] E. Lindner and D. Frembs, *J. Organometallic Chem.*, 1972, **34**, C12.
[649] U. Kunze, E. Lindner, and J. Koola, *J. Organometallic Chem.*, 1972, **38**, 51.
[650] U. Kunze, E. Lindner, and J. Koola, *J. Organometallic Chem.*, 1972, **40**, 327.
[651] G. B. Deacon and G. D. Fallon, *Austral. J. Chem.*, 1972, **25**, 2107.
[652] M. R. Snow, J. McDonald, F. Basolo, and J. A. Ibers, *J. Amer. Chem. Soc.*, 1972, **94**, 2526.
[653] A. Davison and J. E. Ellis, *J. Organometallic Chem.*, 1972, **36**, 113.
[654] A. Davison and J. E. Ellis, *J. Organometallic Chem.*, 1972, **36**, 131.
[655] J. Muller and B. Mertschenk, *J. Organometallic Chem.*, 1972, **34**, 165.

Chromium Carbonyl Complexes

		Ref.
$[Fe\{1,7-B_9H_9CHAs-Cr(CO)_5\}_2]^{2+}$		656
$Cr(CO)_5(dam)$	dam = bis(diphenylarsino)methane	657
$(CO)_5CrC(Fc)X$	$X = O^-NMe_4^+$, OMe, OEt, NH_2, NMe_2, or NC_4H_8; Fc = ferrocenyl	658
$Cr(CO)_5(amine)$	amine = piperidine, pyrrolidine, diethylamine, n-butylamine, cyclohexylamine, morpholine, 3-picoline, pyridine, aniline, or pyrrole	659
$Cr(CO)_5(SMeL)$	$L = Me_3Sn$ or Ph_3PAu	660
$(CO)_5CrC(Me)OMe$		
$(CO)_5CrC(Me)OTi(\pi\text{-}Cp)_2Cl$		661
$(CO)_5CrC(Me)OTi(\pi\text{-}Cp)_2O(Me)CCr(CO)_5$		
$(CO)_5CrC(R)Me$	R = OMe, $HNC_6H_4 \cdot C_6H_4NH_2$, or $HNC_6H_3(Me) \cdot C_6H_3(Me)NH_2$	
$(CO)_5CrC(Me)HNR \cdot NH(Me)CCr(CO)_5$	$R = C_6H_{10}CH_2C_6H_{10}$ or $(CH_2)_n$, $n = 10, 6, 5, 4,$ or 3	662
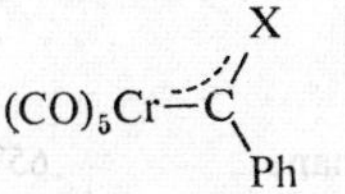 $(CO)_5Cr=C\overset{X}{\underset{Ph}{}}$	X = OEt, SPh, or OPh	663
$(CO)_5CrC(SR)Me$ $(CO)_5CrC(SR)Ph$	R = Me, Et, or Ph	664
$(CO)_5CrC(NR^1R^2)Ph$	$R^1 = R^2 = H$ or Me; or $R^1 = H$, $R^2 = $ Me, Et, Ph, or $MeOC_6H_4$; or $NR^1R^2 = \overline{NCH_2CH_2CH_2CH_2}$	665
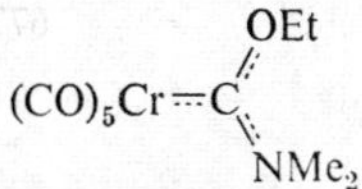 $(CO)_5Cr=C\overset{OEt}{\underset{NMe_2}{}}$		666
$Cr(CO)_5(dtsm)$ $Cr(CO)_5(dpsm)$	dtsm = bis(di-p-tolylstibino)methane dpsm = bis(diphenylstibino)methane	667
$Cr(CO)_5 \cdot PPh_2L$	$L = SnMe_3, SnMe_2Cl, SnMe_2Br,$ or $SnMeBr_2$	668

---

[656] D. C. Beer and L. J. Todd, *J. Organometallic Chem.*, 1972, **36**, 77.
[657] R. Colton and C. J. Rix, *Austral. J. Chem.*, 1972, **24**, 2461.
[658] J. A. Connor and J. P. Lloyd, *J.C.S. Dalton*, 1972, 1470.
[659] R. J. Dennenberg and D. J. Darensbourg, *Inorg. Chem.*, 1972, **11**, 72.
[660] W. Ehrl and H. Vahrenkamp, *Chem. Ber.*, 1972, **105**, 1471.
[661] E. O. Fischer and S. Fontana, *J. Organometallic Chem.*, 1972, **40**, 159.
[662] E. O. Fischer and S. Fontana, *J. Organometallic Chem.*, 1972, **40**, 367.
[663] E. O. Fischer and W. Kalbfus, *J. Organometallic Chem.*, 1972, **46**, C15.
[664] E. O. Fischer, M. Leupold, C. G. Kreiter, and J. Müller, *Chem. Ber.*, 1972, **105**, 150.
[665] E. O. Fischer and M. Leupold, *Chem. Ber.*, 1972, **105**, 599.
[666] E. O. Fischer, E. Winkler, C. G. Kreiter, G. Huttner, and B. Krieg, *Angew. Chem. Internat. Edn.*, 1971, **10**, 922.
[667] T. Fukumoto, Y. Matsumura, and R. Okawara, *J. Organometallic Chem.*, 1972, **37**, 113.
[668] H. Nöth and S. N. Sze, *J. Organometallic Chem.*, 1972, **43**, 249.

Chromium Carbonyl Complexes (*cont.*)

		Ref.
$Cr(CO)_5(tfp)$	tfp = triferrocenylphosphine, $P(C_5H_4FeC_5H_5)_3$	669
$Cr(CO)_5(napy)$	napy = 1,8-naphthyridine	
$Cr(CO)_5(\text{2-mnapy})$	2-mnapy = 2-methyl-1,8-naphthyridine	670
$Cr(CO)_5(dhnapy)$	dhnapy = *trans*-decahydro-1,8-naphthyridine	
$[Cr(CO)_5SCF_3]^-$		671
$Cr(CO)_5[(Ph_2P)_3P]$		672
$Cr(CO)_5(Ph_3P=CHR)$	R = Ph or CH=CHMe	673
$Cr(CO)_5L$	L = isothiazole, 1,2-benzisoxazole, 2-methylbenzoxazoles, 2-methylbenzothiazoles, or 2-methylbenzoselenazoles	674

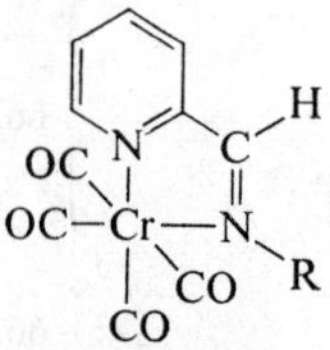 [Cr complex structure with pyridine and N–R ligand]	R = Me, Pr^i, cyclohexyl, or Ph	675
$Cr(CO)_4(dam)$	} dam = bis(diphenylarsino)methane	657
$Cr(CO)_4(dam)_2$		
$Cr(CO)_4L$	L = f$_4$fars = $Me_2As-\overline{C=C-CF_2CF_2}$ with AsMe$_2$	676
$(CO)_4Cr{\scriptstyle\diagdown}{\atop C-Me}{\scriptstyle\diagup}{PEt_3 \atop OMe}$		677
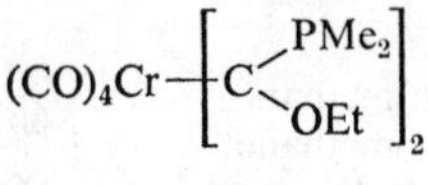 $(CO)_4Cr\!\left[\!C{\scriptstyle\diagup PMe_2 \atop \diagdown OEt}\!\right]_2$		678
$Cr(CO)_4(L-L)$	$L-L = PhP{\scriptstyle\diagup C_2H_4AsPh_2 \atop \diagdown C_2H_4AsPh_2}$	679
cis-$(CF_3PH_2)Cr(CO)_4$		680

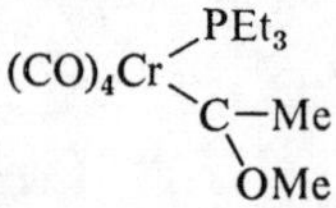

---

[669] C. V. Pittman jun. and G. O. Evans, *J. Organometallic Chem.*, 1972, **43**, 361.
[670] T. E. Reed and D. G. Hendricker, *J. Co-ordination Chem.*, 1972, **2**, 83.
[671] W. J. Schlientz and J. K. Ruff, *Inorg. Chem.*, 1972, **11**, 2265.
[672] H. Schumann and E. von Deuster, *J. Organometallic Chem.*, 1972, **40**, C27.
[673] K. A. O. Starzewski, H. T. Dieck, K. D. Franz, and F. Hohmann, *J. Organometallic Chem.*, 1972, **42**, C35.
[674] J. C. Weis and W. Beck, *J. Organometallic Chem.*, 1972, **44**, 325.
[675] H. Brunner and W. A. Herrmann, *Chem. Ber.*, 1972, **105**, 770.
[676] J. P. Crow, W. R. Cullen, and F. L. Hou, *Inorg. Chem.*, 1972, **11**, 2125.
[677] E. O. Fischer, H. Fischer, and H. Werner, *Angew. Chem. Internat. Edn.*, 1972, **11**, 644.
[678] E. O. Fischer, F. R. Kreissl, C. G. Kreiter, and E. W. Meineke, *Chem. Ber.*, 1972, **105**, 2558.
[679] R. B. King and P. N. Kapoor, *Inorg. Chim. Acta*, 1972, **6**, 391.
[680] J. F. Nixon and J. R. Swain, *J.C.S. Dalton*, 1972, 1038.

Chromium Carbonyl Complexes (cont.)

		Ref.
$Cr(CO)_4(napy)$	napy = 1,8-naphthyridine	
$Cr(CO)_4(2\text{-mnapy})$	2-mnapy = 2-methyl-1,8-naphthyridine	
$Cr(CO)_4(2,7\text{-dmnapy})$	2,7-dmnapy = 2,7-dimethyl-1,8-naphthyridine	670
$Cr(CO)_4(dhnapy)$	dhnapy = *trans*-decahydro-1,8-naphthyridine	
$[Ph_4As]^+[(CO)_4CrL]^-$	L = CF_3CN_4 or $MeSCN_4$	681
$Cr(CO)_3(dam)_3$	dam = bis(diphenylarsino)methane	657

X—⌬—Cr(CO)₃   X = CH₂—(pyrrole) or (pyrrole)   682

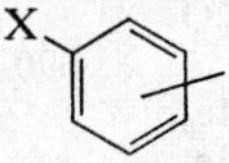

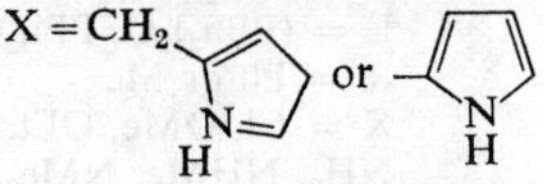

$M(tpp)Cr(CO)_3$ M = Cr^{II}, $Mn^{III}Cl$, Co^{II}, Ni^{II}, Zn^{II}, or H_2; tpp = $\alpha\beta\gamma\delta$-tetraphenylporphin   683

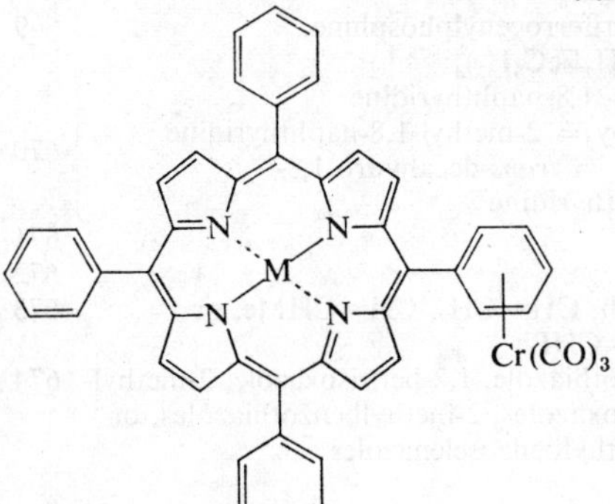

$Cr(CO)_3(\pi\text{-}Cp)Hg(S_2CNEt_2)$		684
$Cr(CO)_3(\pi\text{-}Cp)SnX_3$	X = Cl, Br, or I	685
$[Cr(CO)_3(\pi\text{-}Cp)]_2SnX_2$	X = F, Cl, Br, or I	
$Cr(CO)_3[PhP(C_2H_4AsPh_2)_2]$		679
$Cr(CO)_3(napy)$	napy = 1,8-naphthyridine	670
$Cr(CO)_2(C_7H_8)L$	L = PPh_3 or $P(OPh)_3$	686
$Cr(CO)_2(dam)$	dam = bis(diphenylarsino)methane	657
$[Cr(CO)_2(\pi\text{-}Me_5C_5)]_2$		687
$Cr(CO)L_5$	L = $P(OMe)_3$, $MeP(OMe)_2$, or $Me_2P(OMe)$	688

Molybdenum Carbonyl Complexes

$[Fe\{1,7\text{-}B_9H_9CHP\text{—}Mo(CO)_5\}_2]^{2+}$		656
$Mo(CO)_5(dam)$	dam = bis(diphenylarsino)methane	657

[681] J. C. Weis and W. Beck, *Chem. Ber.*, 1972, **105**, 3203.
[682] N. J. Gogan and C. S. Davies, *J. Organometallic Chem.*, 1972, **39**, 129.
[683] N. J. Gogan and Z. U. Siddiqui, *Canad. J. Chem.*, 1972, **50**, 720.
[684] W. K. Glass and T. Shiels, *Inorg. Nuclear Chem. Letters*, 1972, **8**, 257.
[685] P. Hackett and A. R. Manning, *J.C.S. Dalton*, 1972, 2434.
[686] W. P. Anderson, W. G. Blanderman, and K. A. Drews, *J. Organometallic Chem.*, 1972, **42**, 139.
[687] R. B. King and A. Efraty, *J. Amer. Chem. Soc.*, 1972, **94**, 3773.
[688] R. Mathieu and R. Poilblanc, *Inorg. Chem.*, 1972, **11**, 1858.

Molybdenum Carbonyl Complexes (*cont.*)

		Ref.
$Mo(CO)_5$(amine)	amine = piperidine, pyrrolidine, diethylamine, n-butylamine, cyclohexylamine, morpholine, 3-picoline, pyridine, aniline, or pyrrole	659
$Mo(CO)_5$(dtsm)	dtsm = bis(di-*p*-tolylstibino)methane	⎫
$Mo(CO)_5$(dpsm)	dpsm = bis(diphenylstibino)methane	⎬ 667
$[Mo(CO)_5]_2$(dmsm)	dmsm = bis(dimethylstibino)methane	⎭
$Mo(CO)_5L$	L = PPh_2Cl, PMe_2Cl, PPh_2OH, PMe_2OH, or PPh_2OMe	⎫ ⎬ 689
$[Mo(CO)_5]_2L$	L = $(PPh_2)_2O$, PPh_2OPMe_2, or $(PMe_2)_2O$	⎭
$Mo(CO)_5(R_2PX)$	R = Ph or Me X = Cl, OMe, OEt, OPr^n, OPr^i, $OSiMe_3$, NH_2, NHMe, NMe_2, NHPh, SEt, or Me	690
$Mo(CO)_5 \cdot PPh_2L$	L = $SnMe_3$, $SnMe_2Cl$, $SnMe_2Br$, or $SnMeBr_2$	668
$Mo(CO)_5$(tfp)	tfp = triferrocenylphosphine, $P(C_5H_4FeC_5H_5)_3$	669
$Mo(CO)_5$(napy)	napy = 1,8-naphthyridine	⎫
$Mo(CO)_5$(2-mnapy)	2-mnapy = 2-methyl-1,8-naphthyridine	⎬ 670
$Mo(CO)_5$(dhnapy)	dhnapy = *trans*-decahydro-1,8-naphthyridine	⎭
$[Mo(CO)_5SCF_3]^-$		671
$Mo(CO)_5[(Ph_2P)_3P]$		672
$Mo(CO)_5(Ph_3P=CHR)$	R = Ph, $CH=CH_2$, CH=CHMe, or CH=CHPh	673
$Mo(CO)_5L$	L = isothiazole, 1,2-benzisoxazole, 2-methylbenzoxazoles, 2-methylbenzothiazoles, or 2-methylbenzoselenazoles	674

(structure: pyridyl-imine chelate with Mo(CO)_4 core)	R = Me, Pr^i, Ph, or OH	675
$Mo(CO)_4$(dam)	⎱ dam = bis(diphenylarsino)methane	⎰ 657
$Mo(CO)_4$(dam)$_2$		
$Mo(CO)_4$(5-Me-phen)		⎫
$Mo(CO)_4$(5-Cl-phen)		⎬ 691
$Mo(CO)_4$(5-NH_2-phen)		⎭
$Mo(CO)_4[PhP(C_2H_4AsPh_2)_2]$		679
cis-$[(CF_3)_2PX]_2Mo(CO)_4$	X = H, Cl, Br, I, NMe_2, or NCS	⎱ 680
cis-$(CF_3PX_2)Mo(CO)_4$	X = H, Cl, or Br	⎰
$Mo(CO)_4$(napy)	napy = 1,8-naphthyridine	⎫
$Mo(CO)_4$(2-mnapy)	2-mnapy = 2-methyl-1,8-naphthyridine	⎬ 670
$Mo(CO)_4$(2,7-dmnapy)	2,7-dmnapy = 2,7-dimethyl-1,8-naphthyridine	
$Mo(CO)_4$(dhnapy)	dhnapy = *trans*-decahydro-1,8-naphthyridine	⎭

[689] C. S. Kraihanzel and C. M. Bartish, *J. Amer. Chem. Soc.*, 1972, **94**, 3572.
[690] C. S. Kraihanzel and C. M. Bartish, *J. Organometallic Chem.*, 1972, **43**, 343.
[691] R. T. Jernigan and G. R. Dobson, *Inorg. Chem.*, 1972, **11**, 81.

Molybdenum Carbonyl Complexes (cont.)

		Ref.
$Mo(CO)_4(PPh_3)NHC_5H_{10}$		⎫
$Mo(CO)_4(PPh_3)NC_5H_5$		⎬ 692
$Mo(CO)_4(Ph_3P=CHPh)_2$		⎫
$Mo(CO)_4(Ph_3P=CH-CR=CH_2)$		⎬ 673
	R = H or Me	⎭
$Mo(CO)_4(PPh_2Me)_2$		693
$[Ph_4As]^+[(CO)_4MoL]^-$	L = CF_3CN_4 or $MeSCN_4$	⎫ 681
$[Ph_4As]_2^+[Mo_2(CO)_8(C_3H_3N_2)_2]^{2-}$		⎭
$Mo(CO)_3(dae)I_2$	dae = 1,2-bis(diphenylarsino)ethane	694
$Mo(CO)_3L_2X_2$	X = Cl or Br;	695
	L = tri-(m-tolyl)phosphine or tri-(p-tolyl)phosphine	
$[ROZnMo(CO)_3(\pi\text{-}Cp)]_4$	R = Me or Et	696
$Mo(CO)_3(dam)_3$	dam = bis(diphenylarsino)methane	657
$Mo(CO)_3L(P\text{-en})$	(P-en) = $H_2P \cdot C_2H_4 \cdot PH_2$;	697
	L = $P(OPh)_3$ or $P(OEt)_3$	
$Mo(CO)_3(\pi\text{-}Cp)SMe$		660
$Mo(CO)_3(\pi\text{-}Cp)Hg(S_2CNR_2)$	R = Me or Et	684
$Mo(CO)_3(\pi\text{-}Cp)I$		⎫
$Mo(CO)_3(\pi\text{-}Cp)SnX_3$	X = Cl, Br, or I	⎬ 685
$Mo(CO)_3(\pi\text{-}Cp)SnX_2$	X = F, Cl, Br, or I	⎭
$[Mo(CO)_3(\pi\text{-}Cp)]_nInCl_{3-n}$	n = 1, 2, or 3	698
$Mo(CO)_3[PhP(C_2H_4AsPh_2)_2]$		679
$Mo(CO)_3(\pi\text{-}Cp)CX=C(CN)_2$	X = H, Cl, or CN	699
$Mo(CO)_3(\pi\text{-}Cp)CONR_2$	R_2 = Me_2 or $(CH_2)_5$	⎫ 700
$Mo(CO)_3(\pi\text{-}Cp)CONHMe$		⎭
$Mo(CO)_3(Cl)L(HgCl)$	L = bipy, phen, 5-Me-phen, 5-Cl-phen, 5-NO_2-phen, or 5-NH_2-phen	⎫ 691
$Mo(CO)_3(Cl)(bipy)SnMeCl_2$		⎭
$Mo(CO)_3(\pi\text{-}Cp)SiMe_2SiMe_2X$		
	X = Me, Cl, or Br	701
$Mo(CO)_3(\pi\text{-}Cp)(CH_2-CH_2-CH=CMe_2)$		⎫ 702
$Mo(CO)_3(Br)(MeCN)_2(CH_2-CH_2-CH=CH_2)$		⎭
$Mo(CO)_3(napy)$	napy = 1,8-naphthyridine	⎫
$Mo(CO)_3(2\text{-mnapy})$	2-mnapy = 2-methyl-1,8-naphthyridine	⎬ 670
$Mo(CO)_3(napy)L$	napy = 1,8-naphthyridine; L = 2,7-dimethyl-1,8-naphthyridine, phen, bipy, or 2,9-dimethyl-1,10-phenanthroline	⎭
$(Mo(CO)_3(\pi\text{-}Cp))TlMe_2$		703

[692] G. Schwenzer, M. Y. Darensbourg, and D. J. Darensbourg, *Inorg. Chem.*, 1972, **11**, 1967.
[693] P. M. Treichel, W. M. Douglas, and W. K. Dean, *Inorg. Chem.*, 1972, **11**, 1615.
[694] M. W. Anxer and R. Colton, *Austral. J. Chem.*, 1971, **24**, 2223.
[695] J. A. Bowden and R. Colton, *Austral. J. Chem.*, 1972, **25**, 17.
[696] J. M. Burlitch and S. E. Hayes, *J. Organometallic Chem.*, 1972, **42**, C13.
[697] G. R. Dobson and A. J. Rettenmaier, *Inorg. Chim. Acta*, 1972, **6**, 507.
[698] A. T. T. Hsieh and M. J. Mays, *J. Organometallic Chem.*, 1972, **37**, 9.
[699] R. B. King and M. S. Saran, *J. Amer. Chem. Soc.*, 1972, **94**, 1784.
[700] W. Jetz and R. J. Angelici, *J. Amer. Chem. Soc.*, 1972, **94**, 3799.
[701] W. Malisch, *J. Organometallic Chem.*, 1972, **39**, C28.
[702] J. Y. Mérour, C. Charrier, J. Benaim, J. L. Roustan, and D. Commereuc, *J. Organometallic Chem.*, 1972, **39**, 321.
[703] B. Walther and C. Rockstroh, *J. Organometallic Chem.*, 1972, **44**, C4.

Molybdenum Carbonyl Complexes (cont.)

Compound	Description	Ref.
$[Mo(CO)_2(dae)_{1.5}X_2]_2$	$X = Cl$, Br, or I;	694
$Mo(CO)_2(dae)_2X_2$	dae = 1,2-bis(diphenylarsino)ethane	
$Mo(CO)_2L_2X_2$	$X = Cl$ or Br; L = tri-(m-tolyl)phosphine or tri-(p-tolyl)phosphine	695
$Mo(CO)_2(\pi$-Cp$)(PPh_3)L$	L = Cl, I, or $OC(CF_3)_2$	704

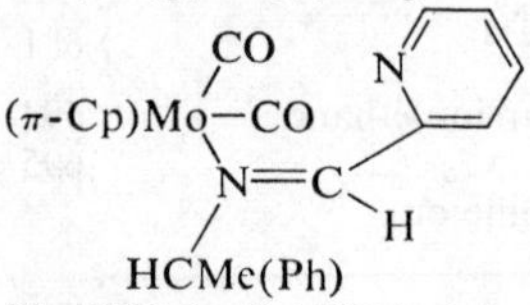

(π-Cp)Mo(CO)(CO)(N=C(H)(HCMe(Ph)))—pyridyl (diastereoisomers) 705

$Mo(CO)_2(\pi$-Cp$)[HB(pz)_3]$	$HB(pz)_3$ = hydrido-tris(pyrazolyl)borate	706
$Mo(CO)_2(\pi$-Cp$)[Et_2B(pz)_2]$	$Et_2B(pz)_2$ = diethyl-bis(pyrazolyl)borate	
$Mo(CO)_2(dam)$	dam = bis(diphenylarsino)methane	657
$Mo(CO)_2(\pi$-Cp$)(PhPMe_2)R$	R = MeCO or EtCO	707
cis-(P-en)$L_2Mo(CO)_2$	(P-en) = $H_2P \cdot C_2H_4 \cdot PH_2$; L = $P(OEt)_3$	697
$Mo(CO)_2(\pi$-Cp$)L(SnMe_3)$	L = PPh_3, $P(OPh)_3$, $PPhMe_2$, $AsPh_3$, or $SbPh_3$	708
$Mo(CO)_2(\pi$-Cp$)[P(OCH_2)_3CMe]I$		
$[Mo(CO)_2(\pi$-Cp$)L]_2SnMe_2$	L = PPh_3 or $P(OCH_2)_3CMe$	
$Mo(CO)_2(\pi$-Cp$)R$	R = $[(p$-tolyl$)_2C]_2N$, $[(p$-$CF_3 \cdot C_6H_4)_2C]_2N$, $[(p$-tolyl$)(p$-$MeO \cdot C_6H_4)C]_2N$, or $[(Ph)(p$-tolyl$)C]_2N$	709
$Mo(CO)_2(COMe)(\pi$-Cp$)(L-L)$	$L-L = PhP(C_2H_4AsPh_2)_2$	679
$[Mo(CO)_2(\pi$-Cp$)(L-L)]^+[PF_6]^-$		
$Mo(CO)_2(\pi$-Cp$)(COMe)(L'-L')$	$L'-L' = Ph_2AsC_2H_4PPh_2$	
$R^+[Mo(CO)_2(\pi$-Cp$)(COMe)CN]^-$	R = K or Ph_4As	710
$Mo(CO)_2(\pi$-Cp$)(COMe)(CNMe)$		
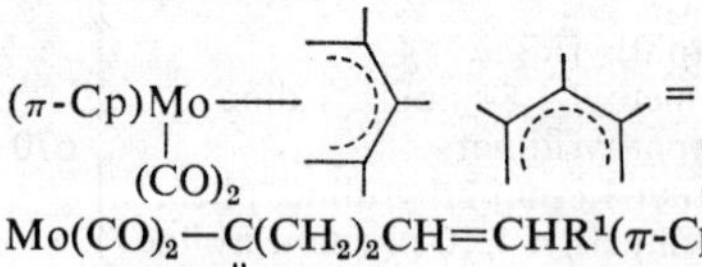 (π-Cp)Mo(CO)$_2$—allyl	= exo-1-methyl-, endo-1-methyl-, endo-1-CHD_2-, or exo,endo-1,3-dimethylallyl	702
$Mo(CO)_2-C(CH_2)_2CH=CHR^1(\pi$-Cp$)(PPh_3)$ with $\|O$	R^1 = H or Me	
$Mo(CO)_2(NO)(L)Cl$	L = diphos [1,2-bis(diphenylphosphino)ethane], bipy, phen, or dpm [bis(diphenylphosphino)methane]	711

---

[704] M. I. Bruce, B. L. Goodall, D. N. Sharrocks, and F. G. A. Stone, *J. Organometallic Chem.*, 1972, **39**, 139.
[705] H. Brunner and W. A. Herrmann, *Chem. Ber.*, 1972, **105**, 3600.
[706] J. L. Calderon, F. A. Cotton, and A. Shauer, *J. Organometallic Chem.*, 1972, **38**, 105.
[707] P. J. Craig and J. Edwards, *J. Organometallic Chem.*, 1972, **46**, 335.
[708] T. A. George, *Inorg. Chem.*, 1972, **11**, 77.
[709] H. R. Keable and M. Kilner, *J.C.S. Dalton*, 1972, 1535.
[710] T. Kruck, M. Höfler, and L. Liebig, *Chem. Ber.*, 1972, **105**, 1174.
[711] W. R. Robinson and M. E. Swanson, *J. Organometallic Chem.*, 1972, **35**, 315.

Molybdenum Carbonyl Complexes (cont.)

		Ref.
(π-Cp)(CO)$_2$Mo ← As(Me)$_2$CH$_2$CH$_2$CH$_2$ (cyclic)		⎫
(π-Cp)(CO)$_2$(I)Mo ← As(Me)$_2$CH$_2$CH$_2$CH$_2$Cl		⎬ 712
(π-Cp)(CO)$_2$(CN)Mo ← As(Me)$_2$CH$_2$—CH=CH$_2$		⎥
[(π-Cp)(CO)$_2$Mo ← As(Me)$_2$CH$_2$—CH=CH$_2$]$^+$[PF$_6$]$^-$ (cyclic)		⎭
Mo(CO)(C$_7$H$_7$)LI	L = P(OPh)$_3$, P(OMe)$_3$, P(OEt)$_3$, P(OPr^i)$_3$, PPh$_3$, or PBu$^n_3$	713
Mo(CO)L$_5$	L = P(OMe)$_3$, MeP(OMe)$_2$, Me$_2$P(OMe), or PMe$_3$	688
Mo(CO)(NO)(MeCN)(PPh$_3$)$_2$Cl		⎫ 711
Mo(CO)(NO)(py)$_2$(PPh$_3$)Cl		⎭
Mo(CO)(π-Cp)$_2$		714

Tungsten Carbonyl Complexes

[Fe{1,7-B$_9$H$_9$CHP—W(CO)$_5$}$_2$]$^{2+}$		656
W(CO)$_5$(dam)	dam = bis(diphenylarsino)methane	657
(CO)$_5$WC(Fc)X	Fc = ferrocenyl; X = O$^-$NMe$_4^+$, OMe, OEt, NH$_2$, NMe$_2$, or NC$_4$H$_8$	658
W(CO)$_5$(amine)	amine = piperidine, pyrrolidine, diethylamine, n-butylamine, cyclohexylamine, morpholine, 3-picoline, pyridine, aniline, or pyrrole	659
W(CO)$_5$(Me$_3$SnSMe)		⎫
W(CO)$_5$-SMe-W(CO)$_3$(π-Cp)		⎬ 660
W(CO)$_5$-SMe-AuPPh$_3$		⎭
(CO)$_5$WC(SMe)Me		664
W(CO)$_5$(dtsm)	dtsm = bis(di-p-tolylstibino)methane	⎫
W(CO)$_5$(dpsm)	dpsm = bis(diphenylstibino)methane	⎬ 667
W(CO)$_5$(dmsm)	dmsm = bis(dimethylstibino)methane	⎭
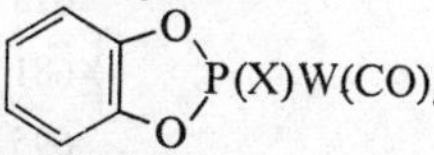 (catecholato)P(X)W(CO)$_5$	X = Cl or NEt$_2$	715
W(CO)$_5$L	L = PPh$_2$CH$_2$CH$_2$PPh$_2$, PPh$_2$PPh$_2$, or PPh$_3$	⎫
(CO)$_5$W[PPh$_2$CH$_2$CH$_2$PPh$_2$]W(CO)$_5$		⎬ 716
[(CO)$_5$W{PPh$_2$CH$_2$CH$_2$PPh$_2$(CH$_2$Ph)}]$^+$[PF$_6$]$^-$		⎥
[(CO)$_5$W(PPh$_2$CH$_2$PPh$_2$Me)]$^+$[I]$^-$		⎭
W(CO)$_5$·PPh$_2$L	L = SnMe$_3$, SnMe$_2$Cl, SnMe$_2$Br, or SnMeBr$_2$	668
W(CO)$_5$(tfp)	tfp = triferrocenylphosphine, P(C$_5$H$_4$FeC$_5$H$_5$)$_3$	669
W(CO)$_5$(napy)	napy = 1,8-naphthyridine	⎫
W(CO)$_5$(2-mnapy)	2-mnapy = 2-methyl-1,8-naphthyridine	⎬ 670
W(CO)$_5$(dhnapy)	dhnapy = trans-decahydro-1,8-naphthyridine	⎭

[712] K. P. Wainwright and S. B. Wild, *J.C.S. Chem. Comm.*, 1972, 571.
[713] T. W. Beall and L. W. Houk, *Inorg. Chem.*, 1972, **11**, 915.
[714] J. L. Thomas and H. H. Brintzinger, *J. Amer. Chem. Soc.*, 1972, **94**, 1386.
[715] A. D. George and T. A. George, *Inorg. Chem.*, 1972, **11**, 892.
[716] R. L. Keiter and D. P. Shah, *Inorg. Chem.*, 1972, **11**, 191.

Tungsten Carbonyl Complexes (cont.)

		Ref.
$[W(CO)_5L]^-$	L = SCF_3, SCH_3, or SPh	}671
$W(CO)_5S(Me)SnMe_3$		
$[W(CO)_5SiI_2]_2$		717
$W(CO)_5[(Ph_2P)_3P]$		672
$W(CO)_5(Ph_3P=CHPh)$		673
$W(CO)_5L$	L = isothiazole, 1,2-benzisoxazole, 2-methylbenzoxazoles, 2-methylbenzothiazole, or 2-methylbenzoselenazole	674

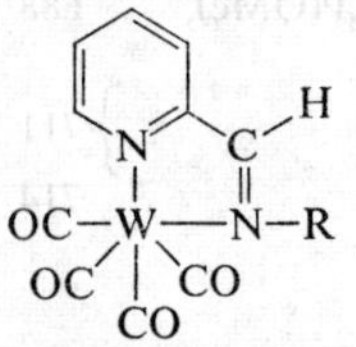

[Structure: pyridyl-CH=N-R coordinated to W with 3 CO groups; R = Me, Pr^i, or Ph]   675

$W(CO)_4(dam)$	dam = bis(diphenylarsino)methane	}657
$W(CO)_4(dam)_2$		
$(CO)_4W-[C(PMe_2)(OEt)]_2$		678
$W(CO)_4(phen)$		}691
$W(CO)_4(phen),2HgCl_2$		
$W(CO)_4(2\text{-mnapy})$	2-mnapy = 2-methyl-1,8-naphthyridine	
$W(CO)_4(2,7\text{-dmnapy})$	2,7-dmnapy = 2,7-dimethyl-1,8-naphthyridine	}670
$W(CO)_4(\text{dhnapy})$	dhnapy = trans-decahydro-1,8-naphthyridine	
$[W(CO)_4(PPh_3)L]^-$	L = SCH_3 or SCF_3	671
$W(CO)_4(PPh_3)NHC_5H_{10}$		}692
$W(CO)_4(PPh_3)NC_5H_5$		
$W(CO)_4(Ph_3P=CHPh)_2$		673
$[Ph_4As]^+[W(CO)_4L]^-$	L = CF_3CN_4 or $MeSCN_4$	}681
$[Ph_4As]_2^+[W_2(CO)_8(C_3H_3N_2)_2]^{2-}$		
$W(CO)_3L_2X_2$	X = Cl, Br, or I; L = tri-(m-tolyl)phosphine or tri-(p-tolyl)phosphine	695
$W(CO)_3(dam)_3$	dam = bis(diphenylarsino)methane	657
$W(CO)_3(\pi\text{-Cp})SMe$		660
$W(CO)_3(\pi\text{-Cp})Hg(S_2CNR_2)$	R = Me or Et	684
$W(CO)_3(\pi\text{-Cp})I$		
$W(CO)_3(\pi\text{-Cp})SnX_3$	X = Cl, Br, or I	}685
$[W(CO)_3(\pi\text{-Cp})]SnX_2$	X = F, Cl, Br, or I	
$[W(CO)_3(\pi\text{-Cp})]_3In$		698
$W(CO)_3(\pi\text{-Cp})CONHR$	R = Me or Bu^t	}700
$W(CO)_3(\pi\text{-Cp})CONR_2$	R_2 = Me_2 or $(CH_2)_5$	
$W(CO)_3(C_5Me_5)Me$		687
$W(CO)_3(\pi\text{-Cp})CX=C(CN)_2$	X = H, Cl, or CN	699
$W(CO)_3(\pi\text{-Cp})SiMe_2SiMe_2X$	X = Me, Cl, or Br	701

[717] A. Schmid and R. Boese, *Chem. Ber.*, 1972, **105**, 3306.

Tungsten Carbonyl Complexes (cont.)

		Ref.
W(CO)$_3$(napy)	napy = 1,8-naphthyridine	} 670
W(CO)$_3$(2-mnapy)	2-mnapy = 2-methyl-1,8-naphthyridine	
W(CO)$_2$L$_2$X$_2$	X = Cl, Br, or I; L = tri-(m-tolyl)phosphine or tri-(p-tolyl)phosphine	695

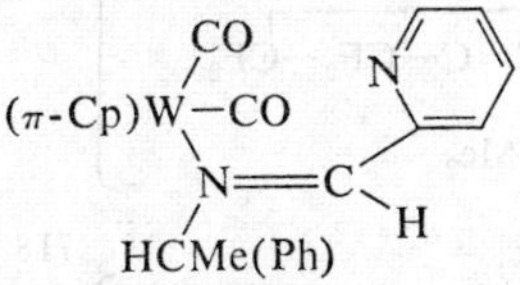

(π-Cp)W(CO)$_2$(N=CH-py) with HCMe(Ph) group (diastereoisomers) 705

W(CO)$_2$(π-Cp)R	R = (p-tolyl)$_2$CNC(p-tolyl)$_2$ or (p-CF$_3$·C$_6$H$_4$)$_2$CNC(p-CF$_3$·C$_6$H$_4$)$_2$	709
R$^+$[W(CO)$_2$(π-Cp)(COMe)CN]$^-$	R = K or Ph$_4$As	710
W(CO)$_2$(NO)(L)Cl	L = diphos [1,2-bis(diphenylphosphino)ethane], bipy, or phen	711
W(CO)(C$_7$H$_7$)LI	L = P(OEt)$_3$ or P(OPr^i)$_3$	713
W(CO)L$_5$	L = P(OMe)$_3$, MeP(OMe)$_2$, Me$_2$P(OMe), or PMe$_3$	} 688

Manganese Carbonyl Complexes

		Ref.
[ROZnMn(CO)$_5$]$_4$	R = Me or Et	696
Mn(CO)$_5$(ClCH$_2$CH$_2$OCO)		
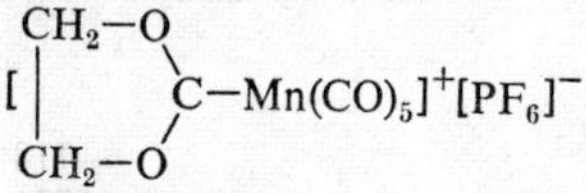 [CH$_2$-O, CH$_2$-O / C—Mn(CO)$_5$]$^+$[PF$_6$]$^-$		} 718
[Mn(CO)$_5$]$_3$In		
[Mn(CO)$_5$]$_2$InX	} X = Cl or Br	
[Mn(CO)$_5$]InX$_2$		
[(MeCN)$_2$InMn$_2$(CO)$_{10}$][ClO$_4$]		
[R$_4$N][XInMn$_3$(CO)$_{15}$]		} 719
[R$_4$N][X$_2$InMn$_2$(CO)$_{10}$]	} R = Me or Et; X = Cl or Br	
[R$_4$N][X$_3$InMn(CO)$_5$]		
(oxine)In[Mn(CO)$_5$]$_2$		
Mn(CO)$_5$(π-Cp)CX=C(CN)$_2$	X = H, Cl, or CN	699
Mn(CO)$_5$X	X = Cl, Cl,FeCl$_3$, Br, Br,AlBr$_3$, I, or I,AlI$_3$	720
o-Ph$_2$P·C$_6$H$_4$·CH$_2$SiMe$_2$Mn(CO)$_4$		721
Mn(CO)$_4$L	L = bza (benzylidenedianilinyl) or bzm (benzylidenemethylaminyl)	722

---

[718] M. Green, J. R. Moss, I. W. Nowell, and F. G. A. Stone, *J.C.S. Chem. Comm.*, 1972, 1339.

[719] A. T. T. Hsieh and M. J. Mays, *J.C.S. Dalton*, 1972, 516.

[720] M. Pankowski, B. Demerseman, G. Bouquet, and M. Bigorgne, *J. Organometallic Chem.*, 1972, **35**, 155.

[721] H. G. Ang and P. T. Lau, *J. Organometallic Chem.*, 1972, **37**, C4.

[722] R. L. Bennett, M. I. Bruce, B. L. Goodall, M. Z. Iqbal, and F. G. A. Stone, *J.C.S. Dalton*, 1972, 1787.

Manganese Carbonyl Complexes (cont.)

		Ref.
$Mn_2(CO)_8L$	$L = f_4fars = Me_2As-\overset{\frown}{C=C}-CF_2CF_2$ with $AsMe_2$ branch	
$Mn_2(CO)_8LI_2$		676
$Mn_2(CO)_8L'$	$L' = f_4fos = Me_2P-\overset{\frown}{C=C}-CF_2-CF_2$ with PMe_2 branch	
cis-[$Mn(CO)_4Cl(\overset{\frown}{COCH_2CH_2O})$]		718
$Mn_2(CO)_8(dpm)$	$dpm = Ph_2PCH_2PPh_2$	723
$Mn_2(CO)_8(dpm)Br_2$		
$Mn(CO)_3L_2X$	$X = Me$ or Br; $L = PhCH_2PMe_2$ or $PhCH_2AsMe_2$	724
[$Mn(CO)_3(PhCH_2PMe_2)_2$]$_2$		
$Mn(CO)_3[P(OR)_3]Br$	$R = Et$, CH_2CH_2Cl, or $CH_2CH=CH_2$	725
$Mn(CO)_3LX$	$X = Cl$ or I; $L = f_4fars = Me_2As\overset{\frown}{C=C}CF_2CF_2$ with $AsMe_2$	676

2-Ar-pyridine · $Mn(CO)_3$: $Ar = CH_2Ph$ or Ph — 682

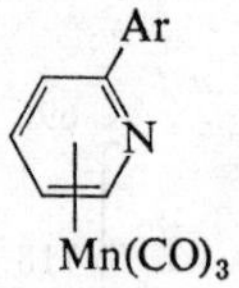

fac-[$Mn(CO)_3Cl(PPh_3)(\overset{\frown}{COCH_2CH_2O})$]		718
$Mn(CO)_3(\pi\text{-}2\text{-}CH_3C_3B_8H_5)$		726
$Mn(CO)_3(Me_5C_5)$		687
[$Mn(CO)_3(R_2PS_2)$]$_2$	$R = Et$ or Ph	727
$Mn(CO)_3(R_2PS_2)(py)$		
[$Mn(CO)_3(dpm)$]$_2$		
[$Mn(CO)_3(dpe)$]$_2$	$dpe = Ph_2P(CH_2)_2PPh_2$	723
fac-$Mn(CO)_3(dpm)Br$	$dpm = Ph_2PCH_2PPh_2$	
fac-$Mn(CO)_3(dpe)Br$		
$Mn(CO)_3(dpe)$		
fac-[$Mn(CO)_3(PMe_3)_2Br$]		728
mer-[$Mn(CO)_3(PMe_3)_2Br$]		
$Mn(CO)_2(\pi\text{-Cp})[E(OR)_3]$	$E = P$, As, or Sb; $R = Me$ (not Sb), Et, Bu^n, or Ph	729

[723] R. H. Reimann and E. Singleton, *J. Organometallic Chem.*, 1972, **38**, 113.
[724] R. L. Bennett, M. I. Bruce, and F. G. A. Stone, *J. Organometallic Chem.*, 1972, **38**, 325.
[725] I. S. Butler, N. J. Coville, and H. K. Spendjian, *J. Organometallic Chem.*, 1972, **43** 185.
[726] J. W. Howard and R. N. Grimes, *Inorg. Chem.*, 1972, **11**, 263.
[727] E. Lindner and K. M. Matejcek, *J. Organometallic Chem.*, 1972, **34**, 195.
[728] R. H. Reimann and E. Singleton, *J. Organometallic Chem.*, 1972, **44**, C19.
[729] T. B. Brill, *J. Organometallic Chem.*, 1972, **40**, 373.

Manganese Carbonyl Complexes (*cont.*)

		Ref.
$Mn(CO)_2[P(OMe)_3]X$	$X = Cl$ or Br	
$Mn(CO)_2(triphos)X$	$X = Cl, Br,$ or I	
$Mn(CO)_2[P(OMe)_3](diphos)X$	$\}X = Cl$ or Br	
$Mn(CO)_2[P(OMe)_3](triphos)X$		725
	diphos = 1,2-bis(diphenylphosphino)-ethane	
	triphos = bis-[2-(diphenylphosphino)-ethyl]phenylphosphine	
$(\pi\text{-}C_5H_4Me)(CO)_2MnC(Fc)OMe$	Fc = ferrocenyl	658
$Mn(CO)_2(\pi\text{-}Cp)(Me_3SnSMe)$		660
$Mn(CO)_2(\pi\text{-}Cp)S(Me)AuPPh_3$		
$Mn(CO)_2(\pi\text{-}Cp)L$	$L = PPh(NEt_2)_2, PPhF_2, PPhFCl, PPhCl_2,$ $PPhBr_2, PPhI_2, PPhFNEt_2, PPhClNEt_2,$ $PPhBrNEt_2$, or $PPhINEt_2$	730
$Mn(CO)_2Br[PhP(C_2H_4AsPh_2)_2]$		679
$mer\text{-}[Mn(CO)_2L_3Br]$	$L = PMe_3, PMe_2Ph, AsMe_2Ph,$ or $P(OEt)_3$	728
$trans\text{-}\{Mn(CO)_2[P(OR)_3]_3Br\}$	$R = Me$ or Et	
$Mn(CO)_2LY$	$L = \pi\text{-}Cp, pzB(pz)_3$ [tetrakis-(1-pyrazolyl)-borate], or $HB(CH_3pzCH_3)_3$ [hydrotris-(3,5-dimethyl-1-pyrazolyl)borate], $Y = CO, PF_3, PCl_3, P(OPh)_3, P(OMe)_3,$ $PPh_3, PMe_3, etc.$	731
$Mn(CO)_2(\pi\text{-}C_5H_4Me)(PPh_2Me)$		693
$Mn(CO)(dpe)_2$	dpe = $Ph_2P(CH_2)_2PPh_2$	
$Mn(CO)(dpm)_2Br$	dpm = $Ph_2PCH_2PPh_2$	723
$Mn(CO)(dpe)_2Br$		

Technetium and Rhenium Carbonyl Complexes

$o\text{-}\overline{Ph_2PC_6H_4CH_2SiMe_2}Re(CO)_5$		721
$Re(CO)_5X$	$X = Cl, Br,$ or I	732
$[Re(CO)_5]_n InCl_{3-n}$	$n = 1, 2,$ or 3	698
$(CO)_9M_2C(OMe)R$	$M = Tc$ or Re; $R = Me$ or Ph	733
$Re_2(CO)_9L$	$L = Cl_3SiH, Cl_3SiD, PhCl_2SiH,$ or $MeCl_2SiH$	734
$Re(CO)_4(bza)$	bza = benzylidene-dianilinyl	722
$[Re(CO)_4X]_2$	$\}X = Cl, Br,$ or I	732
$[Re(CO)_4X_2]^-$		
$Re_2(CO)_8L$	$\}L = f_4fars = Me_2As\overline{-C=C-CF_2CF_2}$ with $AsMe_2$	
$Re_2(CO)_8I_2L$		676
$Re_2(CO)_8L'$	$L' = f_4fos = Me_2P\overline{-C=C-CF_2CF_2}$ with PMe_2	

[730] M. Höfler and M. Schnitzler, *Chem. Ber.*, 1972, **105**, 1133.
[731] A. R. Schoenberg and W. P. Anderson, *Inorg. Chem.*, 1972, **11**, 85.
[732] R. Colton and J. E. Knapp, *Austral. J. Chem.*, 1972, **25**, 9.
[733] E. O. Fischer, E. Offhaus, J. Müller, and D. Nöthe, *Chem. Ber.*, 1972, **105**, 3027.
[734] J. K. Hoyano and W. A. G. Graham, *Inorg. Chem.*, 1972, **11**, 1265.

Technetium and Rhenium Carbonyl Complexes (cont.)

		Ref.
$[Re(CO)_3X_3]^{2-}$ $Re(CO)_3(H_2O)_2X$	$X = Cl$ or Br	732
$Re(CO)_3IL$	$L = f_4fars = Me_2As-C=C-CF_2CF_2$ $\qquad\qquad\qquad\qquad\quad\ \|$ $\qquad\qquad\qquad\qquad\ AsMe_2$	676
$[Re(CO)_3(R_2PS_2)]_2$ $Re(CO)_3(R_2PS_2)(py)$ $Re(CO)_3(R_2PS_2)(PPh_3)$	$R = Et$ or Ph	727
$Re(CO)_3(LH)$ $(OC)_3Re(L)Re(CO)_3$	$LH_2 = $ mesoporphyrin IX dimethyl ester	735
$[Re(CO)_2I_4]^+$		736

Iron Carbonyl Complexes

		Ref.
$Fe(CO)_4L$	$L = PhCH_2PMe_2$ or $PhCH_2AsMe_2$	724
$Fe(CO)_4L$	$L = $ chalcone or $2'$-hydroxychalcone	737
$Fe(CO)_4(f_4AsP)$ $Fe_2(CO)_8(f_4AsP)$	$f_4AsP = CF_2-CF_2-C=C$ $\qquad\qquad\qquad\qquad \| \quad\ \|$ $\qquad\qquad\qquad\ Ph_2P\ \ AsMe_2$	738
$Fe(CO)_4(f_6AsP)$	$f_6AsP = CF_2-(CF_2)_2-C=C$ $\qquad\qquad\qquad\qquad\qquad \|\quad\ \ \|$ $\qquad\qquad\qquad\qquad\ Ph_2P\ \ AsMe_2$	
$Fe(CO)_4(PEtPh_2)$		739
$Fe(CO)_4[P(CF_3)_2H]$		740
$trans$-$Fe(CO)_4(SnBr_3)_2$		741
$(CO)_4FeC(OMe)Ph$		
$(CO)_4FeC(OEt)R$	$R = C_6F_5, C_6Cl_5,$ or Ph	
$(CO)_4FeC(NMe_2)R$	$R = Ph$ or C_6F_5	742
$(CO)_4FeC(NH_2)Ph$		
$(CO)_4FeC(OEt)NR_2$	$R = Me$ or Et	
$Fe(CO)_4(dpsm)$	$dpsm = $ bis(diphenylstibino)methane	
$[FeCO)_4]_2(dmsm)$	$dmsm = $ bis(dimethylstibino)methane	667
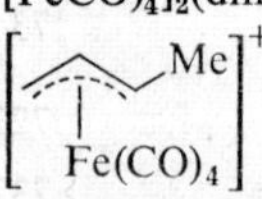 $\left[\begin{array}{c}\diagup\!\!\diagdown\!\!\diagup^{Me}\\ \| \\ Fe(CO)_4\end{array}\right]^+ BF_4^-$		743
$Fe(CO)_4PMe_2CH_2CH_2SiR_3$	$R = Me$ or F	744

---

[735] D. Ostfeld, M. Tsutsui, C. P. Hrung, and D. C. Conway, *J. Co-ordination Chem.*, 1972, **2**, 101.

[736] M. Freni, P. Romiti, V. Valenti, and P. Fantucci, *J. Inorg. Nuclear Chem.*, 1972, **34**, 1195.

[737] A. M. Brodie, B. F. G. Johnson, P. L. Josty, and J. Lewis, *J.C.S. Dalton*, 1972, 2031.

[738] L. S. Chia, W. R. Cullen, and D. A. Harbourne, *Canad. J. Chem.*, 1972, **50**, 2182.

[739] D. J. Darensbourg, *Inorg. Nuclear Chem. Letters*, 1972, **8**, 529.

[740] R. C. Dobbie, M. J. Hopkinson, and D. Wittaker, *J.C.S. Dalton*, 1972, 1030.

[741] N. Dominelli, E. Wood, P. Vasuder, and C. H. W. Jones, *Inorg. Nuclear Chem. Letters*, 1972, **8**, 1077.

[742] E. O. Fischer, H. J. Beck, C. G. Kreiter, J. Lynch, J. Müller, and E. Winkler, *Chem. Ber.*, 1972, **105**, 162.

[743] D. H. Gibson and R. L. Vonnahme, *J. Amer. Chem. Soc.*, 1972, **94**, 5090.

[744] J. Grobe and U. Möller, *J. Organometallic Chem.*, 1972, **36**, 335.

Iron Carbonyl Complexes (cont.)

		Ref.
$Fe(CO)_4$(chalcone)		745
$Fe(CO)_4(BrC_2F_4)Br$		
$Fe(CO)_4(C_2F_2Br_2)$		746
$Fe(CO)_4(C_4F_4Br_2)$		
$[Fe(CO)_4C(=O)R](Ph_3P)_2N$	R = Me, Bu^n, Et, CH_2Ph, Ph, or $CH_2=CH$	
$[Fe(CO)_4R(Ph_3P)_2N]$	R = $(CH_2)_2Ph$, CH_2Ph, Et, $NCCH_2$, $EtOC(O)CH_2$, or $MeOCH_2$	747
$Fe(CO)_4L$	L = $MePH_2$, $PhPH_2$, Me_2PH, Et_2PH, $PhMePH$, or $(p\text{-}Me \cdot C_6H_4)_2PH$	748
$Fe(CO)_4L$	L = PPh_2Me or $PPhMe_2$	693
$Fe_3(CO)_{10}(f_4AsP)$	$f_4AsP = \overset{\displaystyle\ }{CF_2-CF_2-C=C}$ 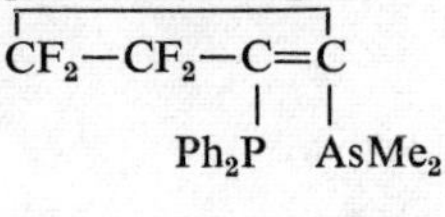 $\ |\ \ \ \ \ \ |$ $\ Ph_2P\ \ \ AsMe_2$	738

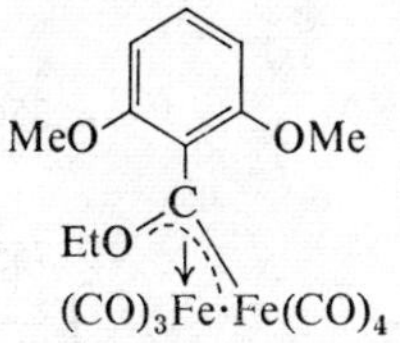

MeO   OMe
    C
EtO ⇃
$(CO)_3Fe \cdot Fe(CO)_4$ 749

$(\mu_3\text{-}Me_3SiN)\text{-}[Fe_3(CO)_{10}]$ (structure with $Fe(CO)_3$ groups bridged by N–$SiMe_3$ and CO)   750

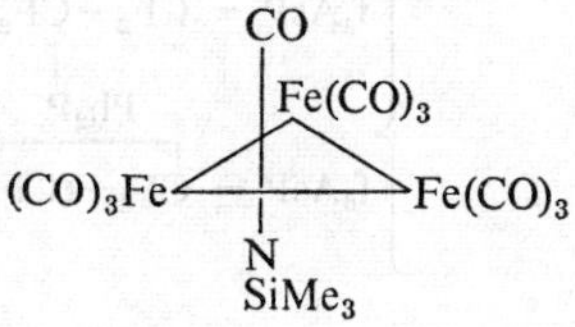

$(CO)_3Fe(CF_2)_2(CO)Fe(CO)_3$ 746

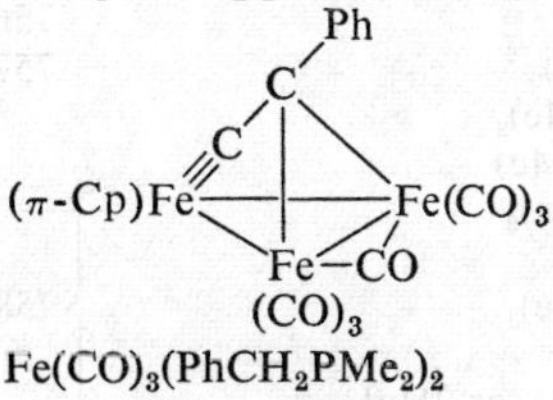

$(\pi\text{-}Cp)Fe\text{---}Fe(CO)_3$ with Ph–C≡C, Fe–CO, $(CO)_3$ 751

$Fe(CO)_3(PhCH_2PMe_2)_2$ 724

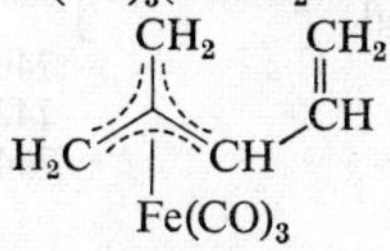

$H_2C\overset{CH_2}{=}\overset{CH_2}{\text{---}}\overset{}{CH}\text{---}CH$ with $Fe(CO)_3$ 752

[745] J. A. S. Howell, B. F. G. Johnson, P. L. Josty, and J. Lewis, *J. Organometallic Chem.*, 1972, **39**, 329.
[746] F. Seel and G. V. Röschenthaler, *Z. anorg. Chem.*, 1971, **386**, 297.
[747] W. O. Siegl and J. P. Collman, *J. Amer. Chem. Soc.*, 1972, **94**, 2516.
[748] P. M. Treichel, W. K. Dean, and W. M. Douglas, *Inorg. Chem.*, 1972, **11**, 1609.
[749] E. O. Fischer, E. Winkler, G. Huttner, and D. Regler, *Angew. Chem. Internat. Edn.*, 1972, **11**, 238.
[750] E. Koerner von Gustorf and R. Wagner, *Angew. Chem. Internat. Edn.*, 1971, **10**, 910.
[751] K. Yasufuku and H. Yamazaki, *Bull. Chem. Soc. Japan*, 1972, **45**, 2664.
[752] W. E. Billups, L. P. Lin, and O. A. Gansow, *Angew. Chem. Internat. Edn.*, 1972, **11**, 637.

Iron Carbonyl Complexes (cont.)

		Ref.
Fe(CO)$_3$L	L = benzylideneacetone, chalcone, 2′-hydroxychalcone, 2,6-dibenzylidene-cyclohexanone or p-substituted cinnamylideneaniline	737

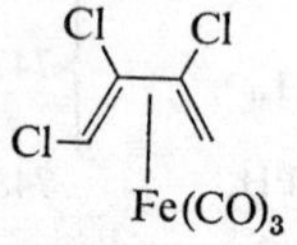

Cl₂C=C(Cl)−C(Cl)=CCl₂·Fe(CO)$_3$ (two isomers shown) }753

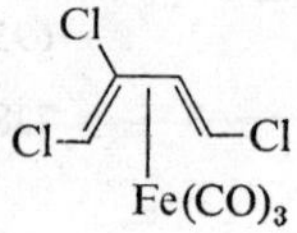

(cyclopentadienyl-Br)Fe(CO)$_3$ 754

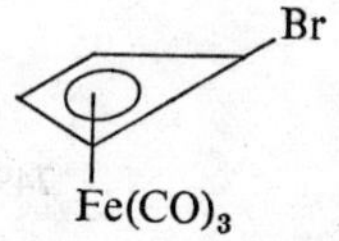

anti- and syn-[Fe(CO)$_3$(SMe)]$_2$ 755

Fe(CO)$_3$(f$_4$AsP)
Fe(CO)$_3$(f$_4$AsP)$_2$
Fe(CO)$_3$(f$_6$AsP)
Fe$_2$(CO)$_6$(f$_4$AsP)
Fe$_2$(CO)$_6$(f$_6$AsP)
Fe$_3$(CO)$_9$(f$_4$AsP)

f_4AsP = CF$_2$−CF$_2$−C=C with Ph$_2$P and AsMe$_2$ substituents
f_6AsP = CF$_2$−(CF$_2$)$_2$−C=C with Ph$_2$P and AsMe$_2$ substituents }738

trans-Fe(CO)$_3$(PEtPh$_2$)$_2$		739
Fe(CO)$_3$(PhCH$_2$C$_5$H$_5$)		756
Fe(CO)$_3$L$_2$,mBF$_3$	m = 0, L = P(OMe)$_3$	757
	m = 2, L = PF(OMe)$_2$	
	m = 4, L = PF$_2$(OMe)	
Fe(CO)$_3$L$_2$	L = PMe$_3$ or P(OMe)$_3$	
Fe(CO)$_3$L$_2$,HgX$_2$	X = Cl, Br, or I	
[Fe(CO)$_3$L$_2$X]$^+$[FeCl$_4$]$^-$	L = PMe$_3$ or P(OMe)$_3$	758
	X = Cl or HgCl	
Fe(CO)$_3$[P(OMe)$_3$]$_2$Y	Y = 4HgCl$_2$, 2HgBr$_2$, or 2HgI$_2$	
R$_2$Fe$_2$(CO)$_6$[P(CF$_3$)$_2$]$_2$	R = H or D	740
Ph$_3$P(CO)$_3$FeC(OEt)C$_6$F$_5$		742
Fe(CO)$_3$(PMe$_2$CH$_2$CH$_2$SiMe$_3$)$_2$		744

---

[753] H. A. Brune, G. Horlbeck, and P. Müller, *Z. Naturforsch.*, 1972, **27b**, 911.
[754] H. A. Brune, G. Horlbeck, and H. Röttelle, *Z. Naturforsch.*, 1972, **27b**, 505.
[755] H. Büttner and R. D. Feltham, *Inorg. Chem.*, 1972, **11**, 971.
[756] M. Y. Darensbourg, *J. Organometallic Chem.*, 1972, **38**, 133.
[757] B. Demerseman, G. Bouquet, and M. Bigorgne, *J. Organometallic Chem.*, 1972, **35**, 125.
[758] B. Demerseman, G. Bouquet, and M. Bigorgne, *J. Organometallic Chem.*, 1972, **35**, 341.

Iron Carbonyl Complexes (cont.)

		Ref.
$Fe_2(CO)_6(\pi\text{-}Cp)L$	L = PPh_2, SMe, or Ph	759
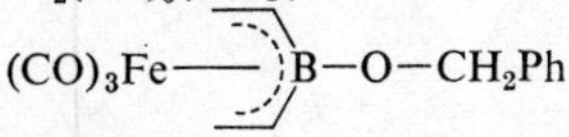 $(CO)_3Fe\text{———}B\text{—}O\text{—}CH_2Ph$		760
$Fe(CO)_3L$	L = Benzylideneacetone, chalcone, dypnone, or 2,6-dibenzylidenecyclohexanone	

[structure: cycloheptatrienyl with $R^1R^2C=$ exocyclic and $Fe(CO)_3$] R = Me or Ph

[structure: with $Fe(CO)_3$]

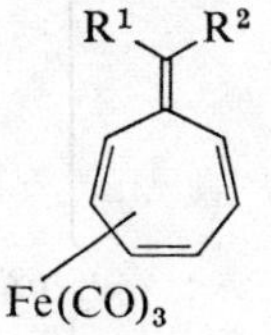

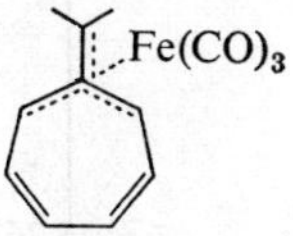

⎱ 745

[structure: cationic cycloheptatrienyl with R^1R^2 and $Fe(CO)_3$]
R^1 = Me, R^2 = Et; R^1 = H, R^2 = p-tolyl
R^1 = Me, R^2 = p-tolyl

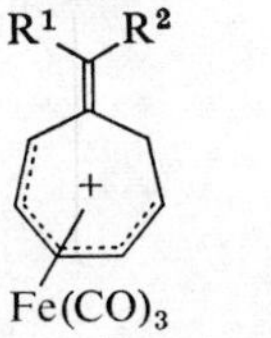

[structure: with R, R and $Fe(CO)_3$]
R^1 = H, R^2 = Me; R^1 = Me, R^2 = Me
R^1 = Me, R^2 = Et; R^1 = H, R^2 = p-tolyl
R^1 = H, R^2 = Ph; R^1 = Me, R^2 = p-tolyl

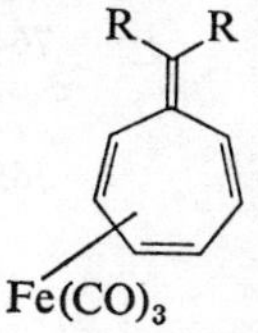

[structures: Me, $(CN)_2$, $(CN)_2$, MeO, $Fe(CO)_3$; and acetyl cycloheptadiene with Et and $Fe(CO)_3$] (Also some other related compounds)

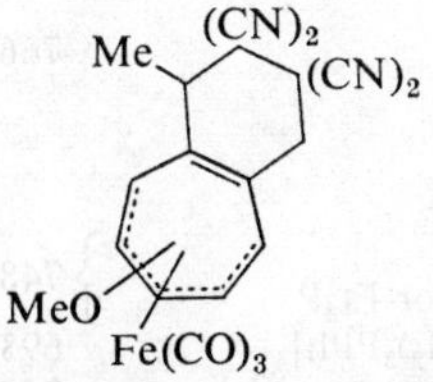

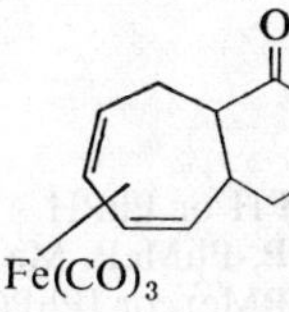

[structure: $(OC)_3Fe$ ... $Fe(CO)_3$ bicyclic]

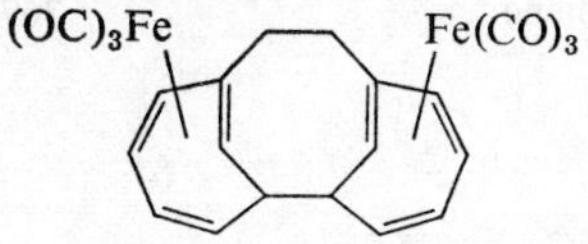

⎱ 761

---

[759] R. J. Haines and C. R. Nolte, *J. Organometallic Chem.*, 1972, **36**, 163.
[760] G. E. Herberich and H. Müller, *Angew. Chem. Internat. Edn.*, 1971, **10**, 937.
[761] B. F. G. Johnson, J. Lewis, P. McArdle, and G. L. P. Randall, *J.C.S. Dalton*, 1972, 2076.

Iron Carbonyl Complexes (cont.)

Ref.

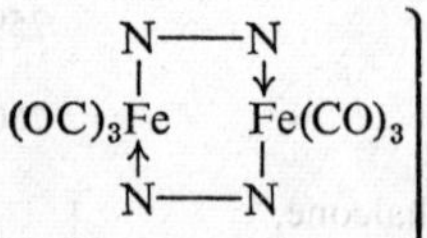

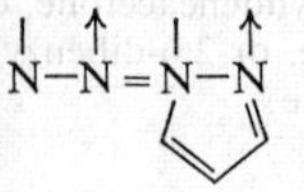

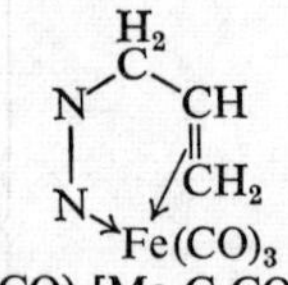

$(OC)_3Fe\underset{N—N}{\overset{N—N}{\square}}Fe(CO)_3$ $N-N=N-N$

$\underset{Fe(CO)_3}{\underset{N}{\overset{N}{\square}}\overset{H_2}{\underset{CH_2}{\overset{CH}{\square}}}}$ } 762

$Fe(CO)_3[Me_5C_5COMe]$ 687

$Fe_2(CO)_6(C_{23}H_{18}N_2)$ $C_{23}H_{18}N_2 = $ [diazepine with Ph groups]

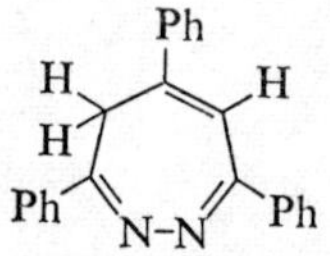

$Fe_2(CO)_6(C_{23}H_{20}N_2)$ $C_{23}H_{20}N_2 = $ [diazepine with Ph groups] } 763

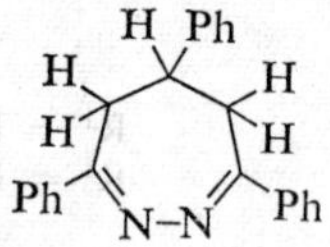

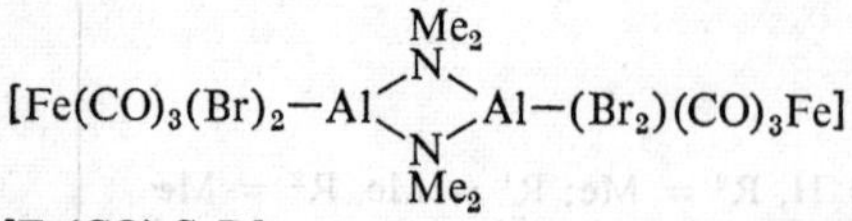

$[Fe(CO)_3(Br)_2-Al\underset{N}{\overset{N}{\underset{Me_2}{\overset{Me_2}{\square}}}}Al-(Br_2)(CO)_3Fe]$ 764

$[Fe(CO)_3SeR]_2$ $R = Me, Et, Pr^i, CF_3,$ or C_2F_5 765

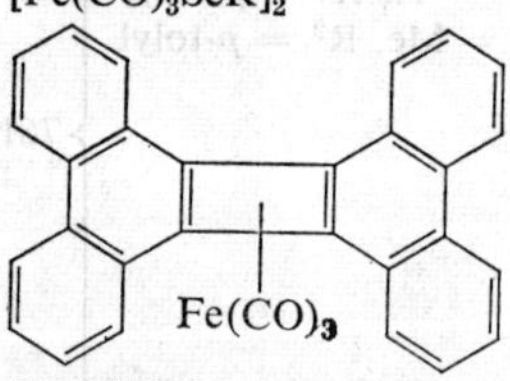

[structure with Fe(CO)₃] 766

$Fe_2(CO)_6L_2$ $L = MePH$ or $PhPH$
$Fe_2(CO)_6LH$ $L = Ph_2P, PhMeP, Me_2P,$ or Et_2P } 748
$Fe_2(CO)_6(L)$ $L = (PhPMe)_2$ or $[PhP(CH_2)_3PPh]$ 693
$Fe(CO)_3(C_6H_8)$ $C_6H_8 = $ cyclohexadiene 767

---

[762] R. B. King and A. Bond, *J. Organometallic Chem.*, 1972, **46**, C53.
[763] D. P. Madden, A. J. Carty, and T. Birchall, *Inorg. Chem.*, 1972, **11**, 1453.
[764] W. Petz and G. Schmid, *J. Organometallic Chem.*, 1972, **35**, 321.
[765] P. Rosenbuch and N. Welcman, *J.C.S. Dalton*, 1972, 1963.
[766] H. A. Staab, E. Wehinger, and W. Thorwart, *Chem. Ber.*, 1972, **105**, 2290.
[767] J. D. Warren, M. A. Busch, and R. J. Clark, *Inorg. Chem.*, 1972, **11**, 452.

Iron Carbonyl Complexes (cont.)

Complex	Substituents	Ref.
$(\pi\text{-Cp})\text{Ni}\overset{R^1\diagdown C=C\diagup R^2}{\underset{Ph_2}{\diagdown P \diagup}}\text{Fe(CO)}_3$	$R^1 = Ph, R^2 = Ph, Me, H,$ or CO_2Me $R^1 = Me, R^2 = CO_2Me$ $R^1 = H, R^2 = CO_2Me$; or $R^1 = R^2 = CO_2Me$	768
$Fe_2(CO)_5(PY_3)(SPh)_2$	$Y = Et, Ph, OMe, OEt, OPr^i$, or OPh	769
$Fe_2(CO)_5PX_3(SBu^t)_2$	$X = Et, Ph, OMe,$ or OPh	⎫
$Fe(CO)_3(SBu^t)_2Fe(CO)(Ph_2PC_2H_2PPh_2)$		⎬ 770
$Fe_2(CO)_5(\pi\text{-Cp})L$	$L = PPh_2$ or Bu^tS	⎫
$Fe_2(CO)_5(\pi\text{-Cp})PPh_2 \cdot X$	$X = PEt_3, P(OPr^i)_3,$ or $P(OPh)_3$	⎬ 759
(CO)$_2$Fe—Fe(CO)$_2$ with bridging CO and indacene-NMe$_2$ ligand		771
$Fe_3(CO)_8LXY$	$X, Y = S, Se,$ or Te but $X \neq Y$,	⎫
$Fe_3(CO)_7L_2XY$	$L = Ph_3As$ or $(PhO)_3P$	⎬ 772
$(\pi\text{-Cp})Fe(CO)_2$–C(Ph)=C(Ph)–Cu–Cl–Cu–Cl–C(Ph)≡C–Fe(CO)$_2$(π-Cp) (bridged structure)		773
$[ROZnFe(CO)_2(\pi\text{-Cp})]_4$	$R = Me$ or Et	696
$[Fe(CO)_2(S_2CNEt_2)(SMe)]_2$		⎫
cis-$[Fe(CO)_2(CNR_2)]$	$R = Me$ or Et	⎬ 755
$Fe_2(CO)_4(f_4AsP)_2$	$f_4AsP = \overset{\overline{}}{CF_2CF_2-C=C}$ $Ph_2P \;\; AsMe_2$	738
$\triangleright\!-Fe(CO)_2(\pi\text{-Cp})$		774
$Fe(CO)_2PPh_3(ArC_5H_5)$	$Ar = Ph$ or $PhCH_2$	756
$[Fe(CO)_2(PY_3)(SPh)]_2$	$Y = Et, Ph, OMe, OEt, OPr^i,$ or OPh	⎫
$[Fe(CO)_2(SbPh_3)(SPh)]_2$		⎬ 769

---

[768] K. Yasufuku and H. Yamazaki, *J. Organometallic Chem.*, 1972, **35**, 367.
[769] J. A. De Beer and R. J. Haines, *J. Organometallic Chem.*, 1972, **36**, 297.
[770] J. A. De Beer and R. J. Haines, *J. Organometallic Chem.*, 1972, **37**, 173.
[771] D. F. Hunt and J. W. Russell, *J. Amer. Chem. Soc.*, 1972, **94**, 7198.
[772] R. Rossetti, P. L. Stanghellini, O. Gambino, and G. Cetini, *Inorg. Chim. Acta*, 1972, **6**, 205.
[773] M. I. Bruce, R. Clark, J. Howard, and P. Woodward, *J. Organometallic Chem.*, 1972, **42**, C107.
[774] A. Cutler, R. W. Fish, W. P. Giering, and M. Rosenblum, *J. Amer. Chem. Soc.*, 1972, **94**, 4354.

Iron Carbonyl Complexes (cont.)

Compound		Ref.
$[Fe(CO)_2PX_3(SBu^t)]_2$	X = Et, Ph, OMe, or OPh	
$[Fe(CO)_2(SBu^t)]_2(Ph_2PYPPh_2)$		
	Y = CH_2, C_2H_4, or NEt	770
$[Fe(CO)_2\{P(OMe)_3\}(SZ)]_2$	Z = Me, Pr^i, or Bz	
$Fe_2(CO)_4(\pi\text{-}Cp)PPh_2(X)$	X = PEt_3, PPh_3, $P(OMe)_3$, $P(OEt)_3$,	
	$P(OPr^i)_3$, $P(OPh)_3$, or $Ph_2PCH_2PPh_2$	
$Fe_2(CO)_4(\pi\text{-}Cp)PPh_2X_2$	X = PEt_3, $P(OMe)_3$, $P(OPr^i)_3$, or $P(OPh)_3$	759
$Fe_2(CO)_4(\pi\text{-}Cp)PPh_2(Ph_2PCH_2PPh_2)$		
$Fe(CO)_2(PMe_3)_2X$	X = Cl_2, $Cl_2,HgCl_2$, I_2, or $I_2,HgCl_2$	758
$Fe(CO)_2(\pi\text{-}Cp)[(CF_3)_2P]$		740
$Fe(CO)_2(\pi\text{-}Cp)GeX_3$	X = Cl, Br, or I	775
$Fe(CO)_2(\pi\text{-}Cp)(SMe)$		660
$[Fe(CO)_2(\pi\text{-}Cp)(Ph_2PCH_2CH_2PPh_2)]^+$		776
$[Fe(CO)_2(\pi\text{-}MeC_5H_4)]_2SnBr_2$		
$[Fe(CO)_2(\pi\text{-}Cp)]_2SnX_2$	X = F, Br, or Cl	
$[Fe(CO)_2(\pi\text{-}Cp)]_3SnX$	X = F or Cl	777
$[Fe(CO)_2(\pi\text{-}MeC_5H_4)][Fe(CO)_2(\pi\text{-}Cp)]_2SnCl$		
$Fe(CO)_2(\pi\text{-}Ar)SnX_3$	Ar = Cp, MeC_5H_4, or C_9H_7	
$[Fe(CO)_2(\pi\text{-}Ar)]_2SnX_2$	X = Cl, Br, or I	
$[Fe(CO)_2(\pi\text{-}Cp)][Fe(CO)_2(\pi\text{-}Ar)]SnX_2$		
	Ar = MeC_5H_4 or C_9H_7	
	X = Cl, Br, or I	778
$[Fe(CO)_2(\pi\text{-}Cp)]_3SnX$	X = F or Cl	
$[Fe(CO)_2(\pi\text{-}C_9H_7)]_3SnX$	X = F or Br	
$[Fe(CO)_2(\pi\text{-}Cp)][Fe(CO)_2(\pi\text{-}Ar)]SnCl$		
	Ar = MeC_5H_4 or C_9H_7	
$[Rh\{Fe(CO)_2(\pi\text{-}RC_5H_4)(PPh_2)\}_2]^+[X]^-$		
	R = H or Me	
	X = BPh_4, PF_6, or SbF_6	779
$[Rh\{Fe(CO)_2(\pi\text{-}Cp)(PPh_2)\}_2\{P(OMe)_3\}]^+[BPh_4]^-$		
$[Fe(CO)_2(\pi\text{-}Cp)]_nInCl_{3-n}$	n = 1 or 2	698
$Fe(CO)_2(\pi\text{-}Cp)(CONHCMe_3)$		780
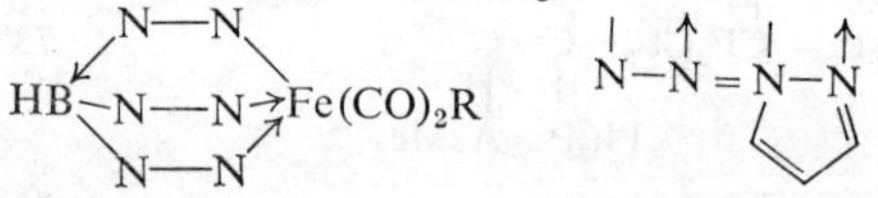 $HB\begin{pmatrix}N\text{---}N\\N\text{---}N\\N\text{---}N\end{pmatrix}Fe(CO)_2R$ $N\text{---}N = \begin{pmatrix}\uparrow\\N\text{---}N\\ \end{pmatrix}$		762
	R = $COCH=CHMe$ or $CH=CHMe$	
$Fe(CO)_2(\pi\text{-}Me_5C_5)(MeCO)$		687
$Fe(CO)_2(\pi\text{-}Cp)C(CN)=C(CN)_2$		699
$Fe(CO)_2(\pi\text{-}Cp)SiMe_2SiMe_2X$	X = Cl or Br	701

---

[775] R. C. Edmonson, E. Eisner, M. J. Newlands, and L. K. Thompson, *J. Organometallic Chem.*, 1972, **35**, 119.
[776] J. A. Ferguson and T. J. Meyer, *Inorg. Chem.*, 1972, **11**, 631.
[777] P. Hackett and A. R. Manning, *J. Organometallic Chem.*, 1972, **34**, C15.
[778] P. Hackett and A. R. Manning, *J.C.S. Dalton*, 1972, 1487.
[779] R. J. Haines, R. Mason, J. A. Zubieta, and C. R. Nolte, *J.C.S. Chem. Comm.*, 1972, 990.
[780] W. Jetz and R. J. Angelici, *J. Organometallic Chem.*, 1972, **35**, C37.

Vibrational Spectra of some Co-ordinated Ligands

Iron Carbonyl Complexes (cont.)

Ref.

benzotriazole–$Fe(CO)_2(\pi\text{-}Cp)$ (N-bonded)

benzotriazole–$Fe(CO)_2(\pi\text{-}Cp)$ (N-bonded, isomer) } 781

Compound	Substituents	Ref.
$Fe(CO)_2(\pi\text{-}Cp)X$	X = Cl, Cl,AlCl$_3$, Cl,FeCl$_3$, Cl,SbCl$_5$, Br, Br,AlBr$_3$, I, I,FeCl$_3$, or I,AlI$_3$	720
$Fe(CO)_2(\pi\text{-}Cp)SeR_f$	R_f = CF$_3$, C$_2$F$_5$, or C$_3$F$_7$	765
$Fe(CO)_2(PF_3)(C_6H_8)$	C_6H_8 = cyclohexadiene	767
$Fe_2(CO)_3(\pi\text{-}Cp)_2(CNBu^t)$		780
$[Fe(CO)(\pi\text{-}Cp)]_2(Ph_2PCH_2CH_2PPh_2)$		
$\{[Fe(CO)(\pi\text{-}Cp)]_2(Ph_2PCH_2CH_2PPh_2)\}^+$		776
$\{[Fe(CO)(\pi\text{-}Cp)(NCMe)]_2(Ph_2PCH_2CH_2PPh_2)\}^{2+}$		
$Fe(CO)(NO)_2[C(OEt)NR^1R^2]$	$R^1 = R^2$ = Me or Et; or R^1 = H, R^2 = Me	782a
$[Fe(CO)H(diphos)_2]^+[BPh_4]^-$		782b
$[\{Fe(CO)(\pi\text{-}Cp)\}_2L]^+[A]^-$	L = Ph$_2$PCH$_2$PPh$_2$, Ph$_2$PC$_2$H$_4$PPh$_2$, or (I)(Ph$_2$PCH$_2$PPh$_2$); A = BPh$_4$ or SbF$_6$	
$[\{Fe(CO)(\pi\text{-}Cp)\}_2L]^+[SbF_6]^-$	L = Ph$_2$PC$_2$H$_2$PPh$_2$ or Ph$_2$PN(Et)PPh$_2$	783
$[\{Fe(CO)(\pi\text{-}Cp)\}_2Ph_2PC_3H_6PPh_2]^+[BPh_4]^-$		
$Fe(CO)(\pi\text{-}Cp)(CNBu^t)(CONHBu^t)$		780
$Fe(CO)(\pi\text{-}Cp)LH$	L = PMe$_3$, P(OMe)$_3$, PMe$_2$Ph, PPh$_3$, or P(NMe$_2$)$_3$	784
$Fe(CO)(\pi\text{-}Cp)LCl$		
$Fe(CO)(COMe)(\pi\text{-}Cp)[PhP(C_2H_4AsPh_2)_2]$		679
$R^{1+}[Fe(CO)(\pi\text{-}Cp)(COR^2)CN]^-$	R^1 = K or Ph$_4$As	
$Fe(CO)(\pi\text{-}Cp)(COR^3)(CNMe)$	R^2 = Me or Et; R^3 = Me or Et	710
$[Fe(CO)(\pi\text{-}Cp)SeR]_2$	R = Et or Pr^n	765
$Fe(CO)(PF_3)_2(C_6H_8)$	C_6H_8 = cyclohexadiene	767
$Fe(CO)(\pi\text{-}Cp)L(COMe)$	L = C$_6$H$_{11}$NC or Bu^{t}NC	
$Fe(CO)(\pi\text{-}Cp)[(C=NC_6H_{11})_3R]$	R = CH$_2$Ph or CH$_2$C$_6$H$_4$Cl	785

Ruthenium Carbonyl Complexes

Compound		Ref.
$Ru_3(CO)_{12-n}L_n$	L = tertiary phosphine or tertiary arsine; n = 1, 2, 3, or 4	786

---

[781] A. N. Nesmeyanov, V. N. Babin, N. S. Kochetkova, E. I. Mysov, Yu. A. Belousov, and L. A. Fedorov, *Doklady Chem.*, 1971, **200**, 838.
[782] (a) E. O. Fischer, F. R. Kreissl, E. Winkler, and C. G. Kreiter, *Chem. Ber.*, 1972, **105**, 588; (b) P. Giannoccaro, M. Rossi, and A. Sacco, *Co-ordination Chem. Rev.*, 1972, **8**, 77.
[783] R. J. Haines and A. L. du Preez, *Inorg. Chem.*, 1972, **11**, 330.
[784] P. Kalck and R. Poilblanc, *Compt. rend.*, 1972, **274**, C, 66.
[785] Y. Yamamoto and H. Yamazaki, *Inorg. Chem.*, 1972, **11**, 211.
[786] M. I. Bruce, G. Shaw, and F. G. A. Stone, *J.C.S. Dalton*, 1972, 2094.

Ruthenium Carbonyl Complexes (cont.)

Compound	Notes	Ref.
$[Et_4N]^+[HRu_3(CO)_{11}]^-$		789
$Ru_3(CO)_9X_3$	$X = PhCH_2PMe_2$ or $PhCH_2AsMe_2$	⎫
$Ru(CO)_3(PhCH_2PMe_2)_2Cl_2$		⎬ 724
$Ru(CO)_3(PhCH_2AsMe_2)_2Cl_2$		⎭
$HRu_3(CO)_9(C_{12}H_{15})$	Hydrocarbons are cyclododeca-1,5,9-triene	⎫ 787
$HRu_3(CO)_9(C_{12}H_{17})$	derivatives	⎭
$H_2Ru_3(CO)_9(C_2Ph_2)$		788
$X_2Ru_6(CO)_{18}$	$X = H$ or D	789
$(C_{23}H_{20}N_2)Ru_3(CO)_9$	$C_{23}H_{20}N_2 = $ [structure: diazacycloheptadiene with two Ph groups and CH-Ph bridge]	763
$H_4Ru_4(CO)_{12}$		790
$H_2Ru_3(CO)_9L$	$L = S$, Se, or Te	791
$Ru_3(CO)_9X$	$X = Bu^tC\equiv CH$ or $PhC\equiv CH$	792
$Ru_3(CO)_9$(1,3-hexadiene)		793

[Structure diagram of a diruthenium complex with cyclopentadienyl, GeMe_2 bridges, and CO/GeMe$_3$ ligands]   794

Compound	Notes	Ref.
$HRu_3(CO)_8(C_{12}H_{15})(PMe_2Ph)$	Hydrocarbons are cyclododeca-1,5,9-triene derivatives	⎫
$Ru_4(CO)_{10}(C_{12}H_{16})$		
$HRu_3(CO)_7(C_{24}H_{34})$		⎬ 787
$HRu_3(CO)_7(C_{12}H_{15})L_2$	$L = P(OMe)_3$, $P(OCH_2)_3CEt$, or PMe_2Ph	
$HRu_3(CO)_6(C_{12}H_{15})L'_3$	$L' = P(OMe)_3$ or $P(OCH_2)_3CEt$	⎭
$H_4Ru_4(CO)_{11}[P(C_4H_9)_3]$		⎫
$H_4Ru_4(CO)_9[P(C_4H_9)_3]_3$		⎬ 790
$H_4Ru_4(CO)_8[P(C_4H_9)_3]_4$		⎭
$Ru(CO)_2(bza)_2$	bza = benzylidenedianilinyl;	⎫
$[Ru(CO)_2(bzm)Cl]_2$	bzm = benzylidenemethylaminyl	⎬ 722
$Ru(CO)_2(PhCH_2PMe_2)Cl_2$		724
$[Ru(CO)_2H(PPh_3)_3]^+$		⎫
$Ru(CO)_2(PPh_3)_3$		⎬ 795
$Ru(CO)_2(PPh_3)_2L$	$L = O_2$, C_2H_4, or $PhC\equiv CPh$	⎭

---

[787] M. I. Bruce, M. A. Cairns, and M. Green, *J.C.S. Dalton*, 1972, 1293.
[788] O. Gambino, E. Sappa, and G. Cetini, *J. Organometallic Chem.*, 1972, **44**, 185.
[789] J. Knight and M. J. Mays, *J.C.S. Dalton*, 1972, 1022.
[790] F. Piacenti, M. Bianchi, P. Frediani, and E. Benedetti, *Inorg. Chem.*, 1971, **10**, 2710.
[791] E. Sappa, O. Gambino, and G. Cetini, *J. Organometallic Chem.*, 1972, **35**, 375.
[792] E. Sappa, O. Gambino, L. Milone, and G. Cetini, *J. Organometallic Chem.*, 1972, **39**, 169.
[793] M. Valle, O. Gambino, L. Milone, G. A. Vaglio, and G. Cetini, *J. Organometallic Chem.*, 1972, **38**, C46.
[794] S. A. R. Knox, R. P. Phillips, and F. G. A. Stone, *J.C.S. Chem. Comm.*, 1972, 1227.
[795] B. E. Cavit, K. R. Grundy, and W. R. Roper, *J.C.S. Chem. Comm.*, 1972, 60.

Ruthenium Carbonyl Complexes (cont.)

		Ref.
[Ru(CO)$_2$(tetraphenylporphinato)]		796
Ru(CO)$_2$(π-Cp)X	X = Cl, Br, I, CN, SCN, Ph, or NCBPh$_3$	⎫
[{Ru(CO)$_2$(π-Cp)}$_2$Cl]$^+$[SbF$_6$]$^-$		⎬ 797
[{Ru(CO)$_2$(π-Cp)}$_2$X]$^+$[Y]$^-$	X = Br or I; Y = PF$_6$ or BPh$_4$	⎭
Ru(CO)$_2$I$_2$L$_2$	L = py, Ph$_3$P, or Ph$_3$As	⎫ 798
[NEt$_4$]$_2^+$[Ru$_2$(CO)$_4$I$_6$]$^{2-}$		⎭

(indenyl-Ru(CO)(GeMe$_3$) structure) 794

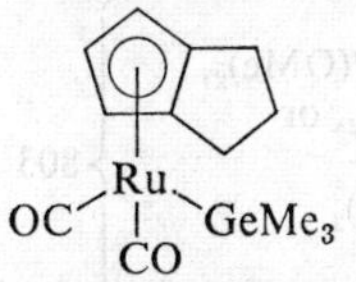

	Ref.
Ru(CO)(PhCH$_2$PMe$_2$)$_3$Cl$_2$	724
[Ru(CO)H(MeCN)$_2$(PPh$_3$)$_2$]$^+$	795
Ru(CO)HI(CNAr)(PPh$_3$)$_2$	

(Ru complex diagram with Ar = p-tolyl; X = O, Y = Me; X = S, Y = NEt$_2$) 799

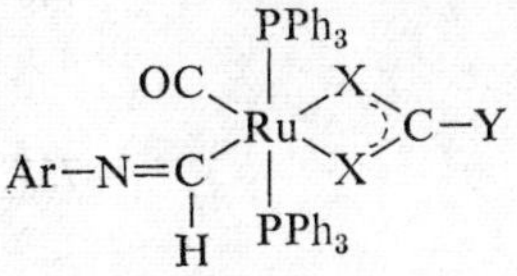

Ru(CO)I$_2$(SbPh$_3$)$_3$ 798

Osmium Carbonyl Complexes

		Ref.
Os$_3$(CO)$_9$(R$_2$C$_2$)$_2$		⎫
Os$_2$(CO)$_6$(R$_2$C$_2$)$_2$	R = p-Cl·C$_6$H$_4$ or p-Me·C$_6$H$_4$	⎬
Os$_3$(CO)$_8$(R$_2$C$_2$)$_2$		⎬ 800
Os$_3$(CO)$_8$(R$_2$C$_2$)$_2$AsPh$_3$		⎬
Os$_3$(CO)$_8$(Ph$_2$C$_2$)L	L = PPh$_3$, AsPh$_3$, SbPh$_3$, or PF$_3$	⎭
[Os(CO)$_2$(NO)(PPh$_3$)$_2$]$^+$		801
Os(CO)$_2$(PPh$_3$)$_3$		⎫
Os(CO)$_3$(PPh$_3$)$_2$(O$_2$)		⎬ 795
Os(CO)H(NO)(PPh$_3$)$_2$		⎭
[Os(CO)(NO)X$_3$]$^+$	X = PPh$_3$ or PPh$_2$Me	⎫
[Os(CO)(NO)(Ph$_2$PCH$_2$CH$_2$PPh$_2$)(PPh$_3$)]$^+$		⎬ 801
[Os(CO)(NO)(RNC)(PPh$_3$)$_2$]$^+$	R = p-tolyl	⎬
[Os(CO)(NO$_2$)(RNC)$_2$(PPh$_3$)$_2$]$^+$		⎭

---

[796] D. Cullen, E. Meyer, jun., T. S. Srivastava, and M. Tsutsui, *J.C.S. Chem. Comm.*, 1972, 584.
[797] R. J. Haines and A. L. du Preez, *J.C.S. Dalton*, 1972, 944.
[798] J. V. Kingston and G. R. Scollary, *J. Inorg. Nuclear Chem.*, 1972, **34**, 227.
[799] D. F. Christian, G. R. Clark, W. R. Roper, J. M. Waters, and K. R. Whittle, *J.C.S. Chem. Comm.*, 1972, 458.
[800] R. P. Ferrari, C. A. Vaglio, O. Gambino, M. Valle, and G. Cetini, *J.C.S. Dalton*, 1972, 1998.
[801] G. R. Clark, K. R. Grundy, W. R. Roper, J. M. Waters, and K. R. Whittle, *J.C.S. Chem. Comm.*, 1972, 119.

Cobalt Carbonyl Complexes

		Ref.
$[MeOZnCo(CO)_4]_4$		696
$Hg[Co(CO)_4]_3^-$		
$XHg[Co(CO)_4]_2^-$	$X = Cl, Br, or I$	802
$Co(CO)_4SnCl_3$		803
$Co(CO)_4(BrC_2F_4)$		746
$[(CO)_9Co_3CCO]^+[BF_4]^-$		804
$Co_4(CO)_{12}$		805
$[Co(CO)_3L]_2Hg$	$L = (PhO)_3P, PhOP(OCH_2)_2,$ $(2-ClC_2H_4O)_3P, (MeO)_3P, PhP(OMe)_2,$ $Ph_2P(OMe), (Et_2N)_3P, PhPPr^i_2,$ or Ph_2MeAs	803
$[Co_2(CO)_6L']Hg$	$L' = (Ph_2PCH_2)_2$ or $(Ph_2AsCH_2)_2$	
$[Co(CO)_3HgBr]_2(Ph_2AsCH_2)_2$		
$Co(CO)_3(Ph_2MeAs)X$	$X = HgBr$ or $SnCl_3$	
$Co_3(CO)_9Y$	$Y = OBCl_2NEt_3, OBBr_2NEt_3,$ or $OAlBr_2NEt_3$	806
$(CO)_3Co(CF_2)_2Co(CO)_3$		746

$(\pi\text{-Cp})Fe \overset{\displaystyle C\text{---}Ph}{\underset{\displaystyle Co(CO)_3}{\diagdown\!\!\!\diagup}} Co$ 751

$Co_4(CO)_{12-n}L_n$	L = tertiary phosphine or tertiary phosphite: $n = 1, 2, 3,$ or 4	805
$Co(CO)_2L_3^+$	$L = P(OMe)_3, PMe_3,$ or PEt_3	807
$Co(CO)_2L'_2(OCMe)$	$L' = P(OMe)_3$	

$$F_3C-C \cdots N=N \cdots Ph$$
$$F_3C-C \cdots Co \cdots CO$$
$$\underset{O}{C} \cdots CO$$
808

$Co_2(CO)_4[C_6(CF_3)_x(CH_3)_yH_{6-x-y}]$		809
	$x = 6, y = 0; x = 5, y = 0$	
	$x = 5, y = 1; x = 4, y = 0$	
	$x = 3, y = 0; x = 3, y = 1$	
	$x = 3, y = 3; x = 2, y = 2$	
$Co(CO)_2(NO)[C(OEt)NR^1R^2]$	$R^1 = R^2 = Me$ or Et; or $R^1 = H, R^2 = Me$	782a

[802] H. L. Conder and W. R. Robinson, *Inorg. Chem.*, 1972, **11**, 1527.
[803] J. Newman and A. R. Manning, *J.C.S. Dalton*, 1972, 241.
[804] J. E. Hallgren, C. S. Eschbach, and D. Seyferth, *J. Amer. Chem. Soc.*, 1972, **94**, 2547.
[805] D. Labroue and R. Poilblanc, *Inorg. Chim. Acta*, 1972, **6**, 387.
[806] G. Schmid and V. Bätzel, *J. Organometallic Chem.*, 1972, **46**, 149.
[807] S. Attali and R. Poilblanc, *Inorg. Chim. Acta*, 1972, **6**, 475.
[808] M. I. Bruce, B. L. Goodall, A. D. Redhouse, and F. G. A. Stone, *J.C.S. Chem. Comm.*, 1972, 1228.
[809] R. S. Dickson and P. J. Fraser, *Austral. J. Chem.*, 1972, **25**, 1179.

Vibrational Spectra of some Co-ordinated Ligands

Cobalt Carbonyl Complexes (cont.)

		Ref.
$[Et_4N]^+[Co(CO)_2(CN)_2\{P(C_6H_{11})_3\}]^-, H_2O$		⎫
$Co(CO)_2(CN)L_2$	$L = PPh_3$, PEt_3, or $P(C_6H_{11})_3$	⎬ 810
$[Co(CO)_2L_2]_2Hg$	$L = (PhO)_3P$, $PhOP(OCH_2)_2$, $(2\text{-}ClC_2H_4O)_3P$, $(MeO)_3P$, $PhP(OMe)_2$, Ph_2POMe, Ph_2MeP, or Et_3P	⎬ 803
$[Co(CO)_2L']_2Hg$	$L' = (Ph_2PCH_2)_2$ or $(Ph_2AsCH_2)_2$	
$Co(CO)_2(Ph_2PCH_2)_2X$	$X = HgCl$ or $SnCl_3$	
$Co(CO)[P(OMe)_3]_4$		807
$[Et_4N]^+[Co(CO)(CN)_2(PPh_3)_2]^-, 2H_2O$		⎫
$K^+[Co(CO)(CN)_2L_2]^-, 3H_2O$	$L = MePPh_2$ or Me_2PPh	⎬ 810
$K^+[Co(CO)(CN)_2(PEt_3)_2]^-, Me_2CO, H_2O$		
$K^+[Co(CO)(CN)_2(diphos)]^-, H_2O$		⎭

Rhodium Carbonyl Complexes

		Ref.
$Rh(CO)_2(bzaH)Cl$	bza = benzylidenedianilinyl	⎫ 722
$Rh(CO)_2(bzmH)Cl$	bzm = benzylidenemethylaminyl	⎭
$Rh(CO)_2Cl(PhNH_2)$		
$Rh(CO)_2(azb)$	⎱ azb = azobenzene anion	⎬ 811
$[Rh(CO)_2(azb)Cl]_n$	⎰	
$[Rh(CO)_2L]^+$	$L = $ bipy, acac, or salicylaldehyde phenylimine (sal=NPh)	812
$Rh(CO)_{3-n}ClL_n$	⎱ $L = PMe_3$, PMe_2Ph, or PPh_3	⎫ 813a
$Rh_2(CO)_{4-m}Cl_2L_m$	⎰ $n = 1, 2,$ or 3; $m = 0, 1, 2, 3,$ or 4	⎭
$Rh(CO)_2X(opd)$	$X = Cl$ or Br; opd = o-phenylenediamine	813b
$Rh(CO)_2(C_{23}H_{18}N_2)X$	$C_{23}H_{18}N_2 = $ 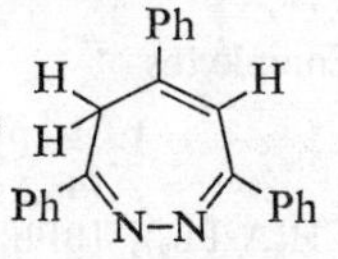 (structure with Ph, H, N–N, Ph groups)	⎬ 763
$[Rh(CO)_2X]_2(C_{23}H_{20}N_2)$	$C_{23}H_{20}N_2 = $ 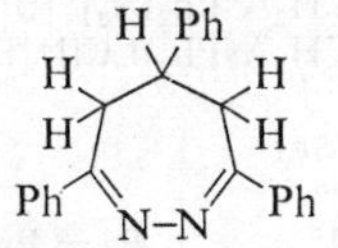 (structure with H, Ph, N–N, Ph groups) $X = Cl$ or Br	
cis-$[Rh(CO)_2Cl(R_3P)]$	$R_3 = Ph_3$, $MePh_2$, or Et_2Ph; or $R_3P = Ph_3PO$	⎬ 814
trans-$[Rh(CO)_2Cl(PhBu_2P)]$		
$Rh_2(CO)_4Cl_2(bipy)_x$	$x = 1, 2,$ or 3	815

---

[810] J. Halpern, G. Guastalla, and J. Bercaw, *Co-ordination Chem. Rev.*, 1972, **8**, 167.
[811] M. I. Bruce, M. Z. Iqbal, and F. G. A. Stone, *J. Organometallic Chem.*, 1972, **40**, 393.
[812] C. Cocevar, G. Mestroni, and A. Camus, *J. Organometallic Chem.*, 1972, **35**, 389.
[813] (a) J. Gallay, D. de Montauzon, and R. Poilblanc, *J. Organometallic Chem.*, 1972, **38**, 179; (b) J. V. Kingston, F. T. Mahmoud, and G. R. Scollary, *J. Inorg. Nuclear Chem.*, 1972, **34**, 3197.
[814] L. D. Rollmann, *Inorg. Chim. Acta*, 1972, **6**, 137.
[815] Yu. S. Varshavskii, N. V. Kiseleva, and N. A. Buzina, *Russ. J. Inorg. Chem.*, 1971, **16**, 862.

Rhodium Carbonyl Complexes (cont.)

		Ref.
$Rh(CO)_2(PPh_3)_2(CO_2Me)$		816
$Rh_2(CO)_3Cl(PhP{<}^{C_2H_4AsPh_2}_{C_2H_4AsPh_2})$		679
$Rh(CO)X(LPh_3)_2$	X = Cl or Br	} 817a
$[Rh(CO)X(LPh_3)]_2$	L = P, As, or Sb	
$[Rh(CO)(bza)Cl]_2$	bza = benzylidenedianilinyl	722
$Rh(CO)L_2Cl$	L = $PhCH_2PMe_2$ or $PhCH_2AsMe_2$	724
$[Rh(CO)(azb)Cl]_2$	azb = azobenzene anion	811
$[Rh(CO)(Ph_3P)ClC[NPhCH_2]_2$		817b
$[Rh(CO)(PPh_3)L]^+$	L = bipy, acac, or salicylaldehyde phenylimine (sal=NPh)	812
$Rh(CO)PR_3Cl_x$	x = 1 or 3; R_3 = Me_3, Et_3, Pr_3^n, Bu_3^n, $Pent_3^n$, Oct_3^n, Pr_3^i, cyclo-Hex_3, Bz_3, Me_2Ph, Et_2Ph, or $EtPh_2$	} 813a
$Rh(CO)HX_2L$	X = Cl or Br; L = phen or bipy	813b
$[(OC)Cl_3RhC(Ph)N(Et)C(Ph)N(Me)]_2$		
$[(OC)X_3RhC(Ph)N(Me)C(Ph)N(Me)]_2$	X = Cl, Br, or I	
$[(OC)Cl_3RhC(Ph)NHR]_n$	R = Me or Pr^i	} 818b
$[(OC)Cl_3RhC(Me)NH(C_6H_4 \cdot Me{-}o)]_n$		
$(Ph_3P)(OC)Cl_3RhC(Ph)NHR$	R = Me or Et	
$[Rh(CO)(Ph_3P)(tnb)(oxq)]$	tnb = 1,3,5-trinitrobenzene oxq = 8-quinolinolato ligand	819
$[Rh(CO)(PPh_3)_2\{OC(NH_2)OEt\}]^+[BF_4]^-$		} 816
$Rh(CO)(PPh_3)_2(NHCO_2Et)$		

Iridium Carbonyl Complexes

$Ir(CO)_2L_2Cl$	L_2 = $Ph_2PCH_2CH_2PPh_2$ or $Ph_2AsCH_2CH_2AsPh_2$	
$[Ir(CO)_2(Ph_2AsCH_2CH_2AsPh_2)]^+[BPh_4]^-$		
$[Ir(CO)_2(Ph_2AsCH_2CH_2AsPh_2)X_2]^+[BPh_4]^-$	X = Cl or I	} 820
$[Ir(CO)_2(Ph_2AsCH_2CH_2AsPh_2)LCl]^+[BPh_4]^-$	L = MeCO or $PhSO_2$	
$[Ir(CO)_2(YEt_2)Cl]_n$	Y = S, Se, or Te	
$Ir(CO)_2(PPh_3)_2(CO_2Me)$		816
$Ir(CO)(Cl)_2(PPh_3)_2Pr^n$		
$[Ir(CO)(Cl)_2R(\mu\text{-}Cl)]_2$	R = Pr^i or Pr^n	} 821a
$Ir(CO)XL_2(OOBu^t)_2$	X = Cl or Br; L = PPh_3 or $AsPh_3$	
$Ir(CO)I_2(PPh_3)_2(OOBu^t)$		} 821b
$Ir(CO)Cl_2(CH_2CHXCH_3)(PMe_2Ph)_2$	X = OMe, OEt, OAc, or OH	} 822
$Ir(CO)Cl_2(CH_2CHYCH_3)(PEt_2Ph)_2$	Y = Cl, OMe, or OAllyl	

---

[816] K. von Werner and W. Beck, *Chem. Ber.*, 1972, **105**, 3947.
[817] (a) D. M. Barlex, M. J. Hacker, and R. D. W. Kemmitt, *J. Organometallic Chem.*, 1972, **43**, 425; (b) D. J. Cardin, M. J. Doyle, and M. F. Lappert, *J.C.S. Chem. Comm.*, 1972, 927.
[818] (a) G. M. Intille, *Inorg. Chem.*, 1972, **11**, 695; (b) M. F. Lappert and A. J. Oliver, *J.C.S. Chem. Comm.*, 1972, 274.
[819] Yu. S. Varshavskii, T. G. Cherkasova, and N. A. Buzina, *Russ. J. Inorg. Chem.*, 1971, **16**, 915.
[820] P. Piraino, F. Faraone, and R. Pietropaolo, *J.C.S. Dalton*, 1972, 2319.
[821] (a) M. A. Bennett and R. Charles, *J. Amer. Chem. Soc.*, 1972, **94**, 666; (b) B. L. Booth, R. N. Haszeldine, and G. R. H. Neuss, *J.C.S. Chem. Comm.*, 1972, 1074.
[822] J. M. Duff, B. E. Mann, E. M. Miller, and B. L. Shaw, *J.C.S. Dalton*, 1972, 2337.

Iridium Carbonyl Complexes (cont.)

		Ref.
$Ir(CO)H_2(Ph_3P)_2L$	L = Me_3Si, Me_3Ge, Me_3Sn, Cl_3Si, or Cl_3Ge	823
$Ir(CO)(Ph_3P)_2(NHCO_2Et)$		⎫ 816
$[Ir(CO)(Ph_3P)_2\{OC(NH_2)OEt\}]^+[BF_4]^-$		⎭

Nickel, Palladium, and Platinum Carbonyl Complexes

		Ref.
$Ni(CO)_3L, mBF_3$	$m = 0$, L = $P(OMe)_3$	757
	$m = 1$, L = $PF(OMe)_2$	
	$m = 2$, L = $PF_2(OMe)$	
	$m = 0$, L = PF_3	
$Ni(CO)_3[C(OEt)NR^1R^2]$	$R^1 = R^2$ = Me or Et	782a
	R^1 = H, R^2 = Me	
$Pd(CO)Cl_2L$	L = DMSO, Et_2S, or tetrahydrothiophen	824
$[Pt(CO)Cl(bipy)]^+[Pt(CO)Cl_3]^-$		⎫
$[Pt(CO)Cl_2(bipy)(CO)PtCl_2], x(DMF)$	$x = 1$—1.5	⎬ 825
$[Bu_4N]^+[Pt(CO)Cl_3]^-$		⎭
$Ni(CO)(\pi\text{-Cp})X$	X = $GeCl_3$, $GeBr_3$, GeI_3, $SnCl_3$, or $SnBr_3$	⎫ 775
$[Ni(CO)(\pi\text{-Cp})]_2Y$	Y = $GeCl_2$ or $SnBr_2$	⎭
$[Ni(CO)RBr]_2$	R = $h^3\text{-}C_3Ph_3$, $h^3\text{-}C_3Bu^t_3$, or $h^3\text{-}C_3Bu^t_2Me$	826
$[Pd(CO)PPh_3]_3$		⎫ 816
$[Pd(CO)(PPh_3)_2CO_2Me]^+[BF_4]^-$		⎭

$$(\pi\text{-Cp})_2Ti\begin{matrix}C\equiv CPh\\ \diagdown\\ \diagup\\ C\equiv CPh\end{matrix}Ni(CO) \qquad 751$$

Mixed Transition-metal Carbonyls

		Ref.
$[(OC)_5Cr\text{-NCS-Mo}(CO)_5]^-$		827
$(OC)_5Cr\text{-SMe-Mo}(CO)_3(\pi\text{-Cp})$		660
$[Cr(CO)_3(\pi\text{-Cp})][Mo(CO)_3(\pi\text{-Cp})]SnX_2$	X = Br or I	685
$[(OC)_5Cr\text{-NCS-W}(CO)_5]^-$		827
$[CrW(CO)_{10}(SMe)]^-$		671
$(OC)_5Cr\text{-SMe-W}(CO)_3(\pi\text{-Cp})$		660
$[Cr(CO)_3(\pi\text{-Cp})][W(CO)_3(\pi\text{-Cp})]SnX_2$	X = F or Cl	685

[structure: 2-picolyl-type ligand with CH_2 linker to a benzene ring bearing $Cr(CO)_3$; pyridine N coordinated to $Mn(CO)_3$]   682

[823] F. Glockling and J. G. Irwin, *Inorg. Chim. Acta*, 1972, **6**, 355.
[824] E. A. Andronov, Yu. N. Kukushkin, and V. G. Churakov, *Russ. J. Inorg. Chem.*, 1971, **16**, 1235.
[825] I. B. Bondarenko, N. A. Buzina, Yu. S. Varshavskii, M. I. Gel'fman, V. V. Razumovskii, and T. G. Cherkasova, *Russ. J. Inorg. Chem.*, 1971, **16**, 1629.
[826] W. K. Olander and T. L. Brown, *J. Amer. Chem. Soc.*, 1972, **94**, 2139.
[827] H. Behrens, D. Uhlig, and E. Lindner, *Z. anorg. Chem.*, 1972, **394**, 8.

Mixed Transition-metal Carbonyls (*cont.*)

	Ref.
$Cr(CO)_5(triphos)Mn(CO)_3Br$	828
$(OC)_5Cr$-SMe-$Fe(CO)_2(\pi$-$Cp)$	660
$[Cr(CO)_3(\pi$-$Cp)][Fe(CO)_2(\pi$-$Cp)]SnCl_2$	685
$[(OC)_5Mo$-SCN-$W(CO)_5]^-$	827
$Mo(CO)_3(\pi$-$Cp)$-SMe-$W(CO)_5$	660
$[Mo(CO)_3(\pi$-$Cp)][W(CO)_3(\pi$-$Cp)]SnX_2$ $\quad$ X = F, Cl, Br, or I	685
$Mo(CO)_3(\pi$-$Cp)$-SMe-$Mn(CO)_2(\pi$-$Cp)$	660
$[Mo(CO)_3(\pi$-$Cp)][Fe(CO)_2(\pi$-$Cp)]SnCl_2$	685, 777

$$(\pi\text{-Cp})Mo\underset{(CO)_2}{\overset{(\pi\text{-Cp})\diagdown Mo(CO)_2}{\diagup}}\underset{Fe(CO)_3}{\overset{\diagup}{\diagdown}}Fe(CO)_3 \qquad 829$$

$W(CO)_3(\pi$-$Cp)$-SMe-$Mn(CO)_2(\pi$-$Cp)$	⎫
$W(CO)_5$-SMe-$Fe(CO)_2(\pi$-$Cp)$	⎬ 660
$[W(CO)_3(\pi$-$Cp)][Fe(CO)_2(\pi$-$Cp)]SnCl_2$	685

$$(\pi\text{-Cp})W\underset{(CO)_2}{\overset{(\pi\text{-Cp})\diagdown W(CO)_2}{\diagup}}\underset{Fe(CO)_3}{\overset{\diagup}{\diagdown}}Fe(CO)_3 \qquad 829$$

$Mn(CO)_2(\pi$-$Cp)$-SMe-$Fe(CO)_2(\pi$-$Cp)$	660
$[Me_4N]^+[MnOs_2(CO)_{12}]^-$	⎫
$HMnOs_3(CO)_{16}$	
$H_3MnOs_3(CO)_{13}$	
$[Me_4N]^+[ReOs_2(CO)_{12}]^-$	
$HReOs_2(CO)_{12}$	⎬ 787
$HReOs_3(CO)_{15}$	
$H_3ReOs_3(CO)_{13}$	
$[Me_4N]^+[ReRu_3(CO)_{16}]^-$	
$H_2Re_2Ru(CO)_{12}$	⎭

$$(\pi\text{-Cp})\text{-Fe}\underset{CO}{\overset{CO}{\diagdown}}\text{-C}\overset{Me}{\underset{C}{\diagdown}}\underset{(CO)_3}{\overset{Co(CO)_3}{\diagup}} \qquad 751$$

$$(\pi\text{-Cp})Ni\underset{\overset{\|}{C}\!\!\atop O}{\overset{\overset{O}{\|}\atop C}{\diagdown\!\!\!\diagup}}Fe\underset{L}{\overset{(\pi\text{-Cp})}{}} \qquad L = CO, PPh_3, Ph_2PMe_2, PhPMe_2, PMe_3, \text{ or } P(OPh)_3 \qquad 830$$

---

[828] M. L. Schneider, N. J. Coville, and I. S. Butler, *J.C.S. Chem. Comm.*, 1972, 799.
[829] A. T. T. Hsieh and M. J. Mays, *J. Organometallic Chem.*, 1972, **39**, 157.
[830] K. Yasufuku and H. Yamazaki, *J. Organometallic Chem.*, 1972, **38**, 367.

Mixed Transition-metal Carbonyls *(cont.)*

		Ref.
$Fe_2Pt(CO)_9L$	$L = AsPh_3$, PPh_3, PPh_2Me, $PPhMe_2$, or PMe_3	831
$Fe_2Pt(CO)_8L'_2$	$L' = PPh_3$, PPh_2Me, $PPhMe_2$, PMe_3, $PPh(OMe)_2$, $P(OPh)_3$, diphos, or diars	
$FePt_2(CO)_5[P(OPh)_3]_3$		
$Ru_2Pt(CO)_8(diphos)$		
$Ru_2Pt(CO)_7(PPhMe_2)_3$		
$RuPt_2(CO)_5L_3$	$L = AsPh_3$, PPh_3, PPh_2Me, $PPhMe_2$, or $PPh(OMe)_2$	832
$RuPt_2(CO)_4L'_4$	$L' = PPh(OMe)_2$ or $P(OPh)_3$	
$Os_2Pt_2(CO)_8H_2(PPh_3)_2$		
$Os_2Pt(CO)_7(PPh_2Me)_3$		
$OsPt_2(CO)_5L_3$	$L = PPh_3$ or PPh_2Me	
$[Ni_2Co_4(CO)_{14}]^{2-}$		833
$[NiCo_3(CO)_{11}]^-$		
$(\pi\text{-Cp})NiCo(CO)_4L$	$L = PPh(C_6H_{11})_2$, PPh_3, or $P(p\text{-}F\cdot C_6H_4)_3$	834

[831] M. I. Bruce, G. Shaw, and F. G. A. Stone, *J.C.S. Dalton*, 1972, 1082.
[832] M. I. Bruce, G. Shaw, and F. G. A. Stone, *J.C.S. Dalton*, 1972, 1781.
[833] P. Chini, A. Cavalieri, and S. Martinengo, *Co-ordination Chem. Rev.*, 1972, **8**, 3.
[834] A. R. Manning, *J. Organometallic Chem.*, 1972, **40**, C73.

8
Mössbauer Spectroscopy

BY R. GREATREX

1 Introduction

Following the pattern adopted in last year's Report, the material in this chapter is organized into seven main sections. After this opening section, which lists the resonances studied and covers books and review articles, there is a brief survey of theoretical aspects followed by developments in instrumentation and methodology. The main body of the work is reported in Sections 4—6, which deal in turn with iron-57, tin-119, and other elements. Section 7 takes the form of a Bibliography and contains for the most part papers on alloys of iron and tin not discussed in the text.

Four resonances have been reported for the first time during the year, namely ^{73}Ge (13.3 keV), ^{190}Os (187 keV), ^{236}U (45.3 keV), and ^{239}Pu (57.3 keV). In addition the following forty resonances have received attention: ^{57}Fe (14.4 keV), ^{61}Ni (67.4 keV), ^{67}Zn (93.3 keV), ^{83}Kr (9.3 keV), ^{99}Ru (90 keV), ^{119}Sn (23.9 keV), ^{121}Sb (37.1 keV), ^{125}Te (35.5 keV), ^{127}I (57.6 keV), ^{129}I (27.8 keV), ^{133}Cs (81.0 keV), ^{149}Sm (22.5 keV), ^{151}Eu (21.7 keV), ^{153}Eu (83.4, 97.4, and 103.2 keV), ^{155}Gd (86.5 keV), ^{157}Gd (64.0 keV), ^{161}Dy (25.6 keV), ^{165}Ho (94.7 keV), ^{166}Er (80.6 keV), ^{170}Yb (84.3 keV), ^{171}Yb (66.7 keV), ^{178}Hf (93.2 keV), ^{180}Hf (93.3 keV), ^{180}W (104 keV), ^{181}Ta (6.2 keV), ^{182}W (100.1 keV), ^{183}W (46.5 and 99.1 keV), ^{184}W (111.1 keV), ^{186}W (122.5 keV), ^{186}Os (137.2 keV), ^{188}Os (155.0 keV), ^{189}Os (69.6 keV), ^{191}Ir (129.5 keV), ^{195}Pt (130 keV), ^{197}Au (77.3 keV), ^{237}Np (59.5 keV), and ^{238}U (44.5 keV).

Books and Reviews.—It is convenient to group together at this point all books and review articles, and it should be noted that these references may not be mentioned again in subsequent sections of the report. When searching for references on a specific topic it may therefore be wise to scan this section as well as the later pages.

A book dealing with chemical applications of Mössbauer spectroscopy has been published during the year [1] and a uniform theoretical treatment of magnetic resonance has appeared, covering e.s.r., n.m.r., and quadrupole resonance, in addition to the Mössbauer effect.[2] The Mössbauer-effect data index covering the year 1970 is now available,[3] and its authors have

---

[1] H. Sano, 'Mössbauer Spectroscopy. Its Chemical Applications', Kodansha, Tokyo, 1972.
[2] C. P. Poole and H. A. Farach, 'Theory of Magnetic Resonance', Wiley, New York, 1972.
[3] J. G. Stevens and V. E. Stevens, 'Mössbauer Effect Data Index, covering the 1970 Literature', Plenum, New York, 1972.

also produced an extensive compilation of nuclear parameters for observed Mössbauer transitions, together with a survey of numerous applications of the technique.[4] Standard substances for Mössbauer calibration have been catalogued by the I.U.P.A.C. Commission.[5] A number of reviews covering general principles of Mössbauer spectroscopy and various applications of the technique in chemistry and physics have also appeared.[6-11]

A panel of the International Atomic Energy Agency met in Vienna in 1972 to review the present status of the Mössbauer technique and its applications in different fields of science and technology. The proceedings of this conference have now appeared [12] and contain papers on instrumentation,[13] phase analysis, site population, and lattice defects,[14] atomic motion in metals,[15] implantation studies,[16] coherence effects,[17] the actinide elements,[18] transferred hyperfine interactions in $5s-p$ elements,[19] biological systems,[20,21] organometallics,[22] co-ordination compounds,[23] chemical effects of nuclear transformations,[24] mineralogical applications,[25] and a number of topics of a more general nature.[26-28]

[4] J. G. Stevens, J. C. Travis, and J. R. Devoe, *Analyt. Chem.*, 1972, **44**, 384R.
[5] 'Catalogue of physiochemical standard substances' (IUPAC Comm. on Physiochemical Measurements and Standards), *Pure Appl. Chem.*, 1972, **29**, 599.
[6] J. R. Sams, 'Some applications of Mössbauer spectroscopy in chemistry and chemical physics', ed. C. A. McDowell, in 'MTP Review of Chemistry: Physical Chemistry', Series One, Vol. 4, Chap. 3, Butterworths, London, 1972.
[7] R. L. Cohen, 'Mössbauer spectroscopy recent developments', *Science*, 1972, **178**, 828.
[8] U. Gonser, 'Mössbauer spectroscopy', *Radex Rundsch*, 1972, 171.
[9] D. Barb, 'Mössbauer effect and its applications', *Rev. Fiz. Chim.* (*A*), 1972, **9**, 151.
[10] A. Andreeff and M. Schenk, 'Nuclear physical methods in crystallography and solid state physics', *Krist. Tech.*, 1972, **7**, 317.
[11] A. Szabo, 'Mössbauer spectrometry study of molybdenum trioxide–iron(III) oxide catalysts', *Rev. Roumaine Chim.*, 1972, **23**, 550.
[12] 'Mössbauer Spectroscopy and its Applications', International Atomic Energy Agency, Vienna, 1972.
[13] G. M. Kalvius and E. Kankeleit, 'Recent improvements in instrumentation and methods of Mössbauer spectroscopy', ref. 12, p. 9.
[14] U. Gonser, 'Phase analysis, site population and lattice defects', ref. 12, p. 89.
[15] C. Janot, 'Mossbauer effect and atomic motion in metals', ref. 12, p. 109.
[16] H. de Waard, 'Lattice location of implanted impurities derived from Mössbauer effect measurements', ref. 12, p. 123.
[17] Yu. Kagan and A. M. Afanas'ev, 'Coherence effects during nuclear resonant interaction of gamma quanta in perfect crystals', ref. 12, p. 143.
[18] G. M. Kalvius, 'Mössbauer spectroscopy in the actinides', ref. 12, p. 169.
[19] M. Pasternak, 'Transferred hyperfine interactions in $5s-p$ elements', ref. 12, p. 197.
[20] G. Lang, 'Interpretation of paramagnetic Mössbauer spectra of biological molecules', ref. 12, p. 213.
[21] H. Frauenfelder, I. C. Gunsalus, and E. Münck, 'Iron–sulphur proteins and Mössbauer spectroscopy', ref. 12, p. 231.
[22] R. H. Herber, 'Mössbauer spectroscopy of organometallic compounds', ref. 12, p. 257.
[23] J. Danon, 'Mössbauer effect applications to co-ordination chemistry of transition elements', ref. 12, p. 281.
[24] J. P. Adloff and J. M. Friedt, 'Mössbauer studies of chemical effects of nuclear transformations', ref. 12, p. 301.
[25] A. G. Maddock, 'Mössbauer spectroscopy in mineralogy', ref. 12, p. 329.
[26] S. Hüfner and E. Matthias, 'Advantages and limitations of Mössbauer spectroscopy in comparison to other methods', ref. 12, p. 349.
[27] A. Simopoulos and A. Kostikas, 'Mössbauer spectroscopy in Greece', ref. 12, p. 381.
[28] I. Dézsi, 'Mössbauer studies in developing countries', ref. 12, p. 389.

An extensive review has appeared, entitled "Mössbauer spectra of inorganic compounds: bonding and structure". It deals critically with recent studies chosen to illustrate the use of the technique. The concepts of partial isomer shift and partial quadrupole splitting are examined in detail, but the authors have chosen not to discuss magnetic hyperfine data.[29]

Experimental aspects of Mössbauer spectroscopy have been covered [29a] and the interpretation of tin-119 spectra has been discussed at length.[30] A substantial survey of applications in the study of the chemical effects of nuclear reactions in solids has appeared.[31] Applications in co-ordination chemistry,[32, 32a, 33] organometallic chemistry,[34] and biological systems,[35, 36] and studies of frozen solutions,[37] spin-crossover systems,[38] and intermolecular interactions [39] have also been reviewed.

Other areas surveyed include lattice-defect studies [40] and applications in surface science [41] and ore mining.[42] Mössbauer studies have also been mentioned briefly in several general reviews.[43−46]

---

[29] R. H. Herber and Y. Hazony, 'Experimental aspects of Mössbauer spectroscopy', in 'Techniques of Chemistry', ed. A. Weissberger and B. W. Rossiter, Wiley, New York, 1972, Vol. 1.
[29a] G. M. Bancroft and R. H. Platt, *Progr. Inorg. Chem.*, 1972, **15**, 59.
[30] R. V. Parish, *Progr. Inorg. Chem.*, 1972, **15**, 101.
[31] A. G. Maddock, 'Mössbauer spectroscopy in the study of the chemical effects of nuclear reactions in solids', ed. A. G. Maddock, in 'MTP International Review of Science: Inorganic Chemistry', Series One, Butterworths, London, 1972, Vol. 8, p. 253.
[32] K. Burger, 'Co-ordination Chemistry', Butterworths, London, 1972.
[32a] K. Burger, *Inorg. Chim. Acta, Rev.*, 1972, **6**, 31.
[33] M. L. Good and C. A. Clausen, 'Applications of Mössbauer spectroscopy in the study of coordination compounds', in 'Co-ordination Chemistry', ed. A. E. Martell, Van Nostrand, 1972, Vol. 1.
[34] J. C. Maire, *Afinidad*, 1972, **29**, 177.
[35] A. J. Bearden and W. R. Dunham, 'Iron electronic configurations in proteins: studies by Mössbauer spectroscopy', in *Structure and Bonding*, 1972, Vol. 8.
[36] J. L. Holtzman, *Methods Pharmacol.*, 1972, **2**, 157.
[37] I. Dézsi, *Kozp. Fiz. Kut. Intez.*, 1972, 72.
[38] E. König, *Ber. Bunsengesellschaft phys. Chem.*, 1972, **76**, 975.
[39] H. Sano, *Kagaku No Ryoiki*, 1972, **26**, 58.
[40] H. Ino, *Oyo Butsuri*, 1972, **41**, 735.
[41] M. C. Hobson, *Progr. Surface Membrane Sci.*, 1972, **5**, 1.
[42] L. Simon, *Rudy*, 1972, **20**, 46.
[43] J. J. Zuckerman, 'Synthesis and properties of the tin–halogen and tin–halogenoid bond', in 'The Bond to Halogens and Halogenoids – Organometallic Compounds of the Group IV Elements', ed. A. G. MacDiarmid, Marcel Dekker, New York, 1972, Vol. 2.
[44] G. G. Libowitz, 'Solid state properties of metallic and saline hydrides', ed. L. E. J. Roberts, in 'MTP International Review of Science: Inorganic Chemistry', Series One, Butterworths, London, 1972, Vol. 10, Chap. 3.
[45] J. J. Turner, 'Physical and spectroscopic properties of the halogens', ed. V. Gutmann, in 'MTP International Review of Science: Inorganic Chemistry', Series One, Butterworths, London, 1972, Vol. 3, Chap. 9.
[46] J. D. M. McConnell, 'Binary and complex oxides', ed. D. W. A. Sharp, in 'MTP International Review of Science: Inorganic Chemistry', Series One, Butterworths, London, 1972, Vol. 5, Chap. 2.

2 Theoretical

This section includes only papers of a general theoretical nature. Theoretical studies relevant to a particular isotope are discussed where appropriate in later sections of the Report.

The possibility of using Mössbauer radiation to create a γ-laser has been discussed. The expressions derived for the cross-section of the stimulated emission of γ-quanta and for the amplification coefficients show that a laser may be made using crystals containing long-lived ($\tau \gtrsim 10^6$ s) nuclear isotopes.[47]

Approximations used in the literature in the calculation of Mössbauer isomer shifts have been discussed and compared with the results obtained using mathematically exact expressions incorporating Dirac–Slater orbitals and a three-parameter Fermi-type nuclear charge distribution. It was shown that the non-uniformity of the electronic density over the nuclear volume affects the results and that the quantities $\Delta \mid \psi(0) \mid^2$ and $\Delta \langle r^2 \rangle$ are not entirely independent variables. Furthermore, $\Delta \langle r^2 \rangle$ is not always an adequate description for the change of the nuclear radius; higher terms, $\Delta \langle r^n \rangle$, should be included.[48]

A simple relationship connecting the temperature shift, the recoil-free fraction, and the force constant has been derived. The force constant experienced by an impurity atom in a host lattice is different from that experienced by the host atoms themselves.[49]

The exponent of the Debye–Waller factor and the energy shift for γ-rays resonantly absorbed by a deformed crystal have been calculated using the technique of double-time Green's function. Deformation introduces an observable broadening of the Mössbauer line and an additional energy shift also appears.[50] Theoretical considerations of Mössbauer fraction experiments indicate that in the neighbourhood of a structural phase transition the Debye–Waller factor exhibits a cusp-shaped anomaly, rather than a divergence, for a displacive-type transition and no anomaly for an order–disorder transformation.[51] General analytical expressions have been obtained for the Debye frequency tensor in terms of the elastic constants of single crystals for all crystallographic systems. For tetragonal tin the ratio of the probabilities of the Mössbauer effect in the direction of the c-axis and normal to the c-axis at $T = 0$ K was estimated to be 1.03, in good agreement with experiment. For ^{67}Zn a much larger anisotropy of the Mössbauer effect is predicted, the corresponding ratio being 2.8.[52]

For most Mössbauer nuclides the excited nuclear level is populated by a γ-de-excitation which, being non-recoil-free, leads to the formation of a

[47] R. V. Khokhlov, *Pis'ma Zhur. eksp. i teor. Fiz.*, 1972, **15**, 580.
[48] B. Fricke and J. T. Waber, *Phys. Rev.* (*B*), 1972, **5**, 3445.
[49] G. P. Gupta and K. C. Lal, *Phys. Status Solidi* (*B*), 1972, **51**, 233.
[50] B. P. Srivastava, H. N. K. Sharma, and D. L. Bhattacharya, *Phys. Status Solidi* (*B*), 1972, **53**, 683.
[51] F. Borsa and A. Rigamonti, *Phys. Letters* (*A*), 1972, **40**, 399.
[52] Ya. A. Losilevskii, *Phys. Status Solidi* (*B*), 1972, **53**, 405.

non-stationary vibrational state. In most cases the lifetimes (10^{-13}—10^{-14} s) of these excited vibrations are much shorter than the Mössbauer nuclear lifetime (e.g. 10^{-7} s for ^{57}Fe) and the recoil-free emission is unaffected. However, there are many nuclides whose Mössbauer levels have lifetimes of ca. 10^{-10} s (e.g. ^{153}Eu, ^{155}Gd, ^{191}Ir, ^{188}Os, ^{195}Pt, ^{187}Re, ^{183}W, ^{232}Th, and ^{238}U) and if the Mössbauer atom is an impurity in a suitably chosen host lattice, there could occur in its phonon spectrum sharp lines whose typical lifetimes are ca. 10^{-10} s also. In such cases the emission recoil-free fraction of the Mössbauer active impurity atom is likely to be affected and it is suggested that time-delayed coincidence techniques may then be used to measure the lifetimes of the non-stationary vibrational states.[53]

Polarization effects in Mössbauer experiments have been discussed on the basis of an s-matrix and d-matrix formalism. Equations were derived for both Mössbauer transmission and scattering and the generality of the formalism was demonstrated by extending the discussion to include interference between Mössbauer and electronic scattering.[54] This effect has been shown elsewhere to be dependent on the absorber thickness,[55] and experimental observations of the phenomenon are mentioned later (see refs. 114 and 115). The interference of hyperfine components in the scattering of γ-irradiation by Mössbauer nuclei has been discussed elsewhere.[56]

Mössbauer diffraction by magnetic crystals has received theoretical treatment. The proposed theory takes into account Rayleigh scattering and includes expressions for the amplitudes of coherent scattering and for the differential cross-section of coherent scattering of polarized radiation. The Mössbauer diffraction method is similar to neutron diffraction but is simpler and more highly selective.[57, 58] Nuclear diffraction of 14.4 keV γ-rays from an iron–nickel single crystal[59] and a sodium nitroprusside single crystal[60] have been observed experimentally; in contrast to X-ray diffraction the intensity of the γ-ray diffracted beam was shown to be angle-dependent. Mössbauer dispersion in paramagnetic crystals has been treated theoretically.[61]

The preliminary account of the production of Mössbauer sidebands by radio-frequency excitation of magnetic materials, noted in last year's Report, has now been elaborated[62] (see also refs. 109—112). In addition,

---

[53] A. Szczepański, *Solid State Comm.*, 1972, **10**, 447.
[54] B. L. Chrisman, *Nuclear Phys. (A)*, 1972, **186**, 264.
[55] P. West, *Nuclear Instr. and Methods.*, 1972, **101**, 243.
[56] A. S. Ivanov and A. V. Kolpakov, *Soviet Phys. Solid State*, 1972, **14**, 1537.
[57] V. A. Belyakov and R. Ch. Bokun, *Izvest. Akad. Nauk S.S.S.R., Ser. fiz.*, 1972, **36**, 1476.
[58] V. A. Belyakov, *Soviet Phys. Solid State*, 1972, **13**, 1824.
[59] V. S. Zasimov, R. N. Kuz'min, and A. I. Firov, *Kristallografiya*, 1972, **17**, 864.
[60] R. M. Mirzababaev, V. V. Sklyarevskii, and G. V. Smirnov, *Phys. Letters (A)*, 1972, **41**, 349.
[61] A. V. Mitin, *Phys. Status Solidi (B)*, 1972, **53**, 93.
[62] L. Pfeiffer, N. D. Heiman, and J. C. Walker, *Phys. Rev. (B)*, 1972, **6**, 74.

an independent theoretical treatment of this effect has appeared [63] and it has been demonstrated experimentally that the sidebands can be quenched by the addition of a suitable damping medium.[64] The effect of radiofrequency fields on the Mössbauer spectra of paramagnetic materials has also been discussed. It is proposed that the field acts as an additional relaxation mechanism by inducing e.s.r. transitions in the electronic sub-system. It should therefore be possible to match the spectra obtained at a given temperature in the absence of the r.f. field with those obtained at a lower temperature in the presence of the field. An interesting situation arises when $S > \frac{1}{2}$; the hyperfine pattern is now made up of components from several electronic levels and it is suggested that the r.f. field can be used selectively to quench these individual components.[65]

The effect of vacancy diffusion on the Mössbauer linewidth of impurity atoms in body-centred and simple cubic lattices has been considered.[66]

Calculations have been made of the possible effect on the Mössbauer spectrum of the distribution of the electric-field gradient which is possible in very small crystallites.[67]

Estimates have been made of the optimum absorber thickness for a single-line Mössbauer effect measurement in the presence of strong non-resonant absorption. Numerical results for different magnitudes of the background noise were presented.[68] The dependence of resonance linewidth, resonance area, and magnitude of the Mössbauer effect on film thickness has been solved analytically for Mössbauer back-scattering spectra obtained *via* the conversion electrons.[69]

It has been pointed out that because the transmitted intensity has a Lorentzian shape only for an absorber of zero effective thickness and vanishing self-absorption, then systematic errors must be expected if attempts are made to fit a superposition of Lorentzians to spectra in which the quadrupole splitting [70] or the magnetic hyperfine structure [71-73] is only partially resolved. In each case procedures were discussed for determining the actual line positions and widths. Experiments with ^{133}Cs, ^{165}Ho, and ^{178}Hf,[73] designed to test the calculations, are discussed in later sections of this chapter. A linear combination of Lorentzian and Gaussian profiles is

[63] Yu. V. Baldokhin, S. A. Borshch, L. M. Klinger, and V. A. Povitskii, *Zhur. eksp. i teor. Fiz.*, 1972, **63**, 708.
[64] G. Albanese, G. Asti, and S. Rinaldi, *Lett. Nuovo Cimento Soc. Ital. Fis.*, 1972, **4**, 220.
[65] I. B. Bersucker and S. A. Borshch, *Phys. Status Solidi (B)*, 1972, **49**, K71.
[66] S. P. Repetskii, *Ukrain. fiz. Zhur.*, 1972, **17**, 1112.
[67] A. Z. Hrynkiewicz, A. J. Pustowka, B. Sawicka, and J. A. Sawicki, *Phys. Status Solidi (A)*, 1972, **9**, 607.
[68] J. Pelzl, *Nuclear Instr. and Methods*, 1972, **102**, 349.
[69] R. A. Krakowski and R. B. Miller, *Nuclear Instr. and Methods.*, 1972, **100**, 93.
[70] R. Meads, B. M. Place, F. W. D. Woodhams, and R. C. Clark, *Nuclear Instr. and Methods*, 1972, **98**, 29.
[71] S. A. Wender and N. Hershkowitz, *Nuclear Instr. and Methods*, 1972, **98**, 105.
[72] N. Hershkowitz, R. D. Ruth, S. A. Wender, and A. B. Carpenter, *Nuclear Instr. and Methods*, 1972, **102**, 205.
[73] E. Gerdau, W. Rath, and H. Winkler, *Z. Phys.*, 1972, **257**, 29.

considered to be a convenient approximation of a Voigt profile and thus suitable for fitting Mössbauer spectra.[74]

An optimal method of extracting hyperfine parameters from complex Mössbauer spectra has been presented for the specific case of $^{129}I_2$ at 77 K, although the method can be adapted for other isotopes. Rather than fit the data to twelve line-positions and then fit these line-positions to theory to derive the hyperfine parameters, the theoretical spectrum is calculated directly from an initial guess of the parameters which are then optimized simultaneously with a least-squares fit to the data, the required derivatives with respect to each parameter being determined numerically rather than analytically. The actual results are given on p. 596.[75] Computer processing of complex Mössbauer spectra has been discussed elsewhere,[76] and an algorithm has been presented for estimating local field parameters from Mössbauer spectra in the general case of a combination of magnetic dipole and electric quadrupole interactions for Mössbauer nuclei with arbitrary ground- and excited-state spins.[77] The analytical determination of hyperfine coupling parameters for ^{57}Fe has also been discussed.[78]

Expressions have been derived for estimating the order of magnitude of so-called 'magnetic anomalies' in the temperature dependence of certain Mössbauer effect characteristics such as the recoil-free fraction and the spectrum centroid for ferromagnetic and antiferromagnetic materials.[79]

3 Instrumentation and Methodology

The application of computerization to Mössbauer spectroscopy has been discussed, with reference to the recent widespread availability and reasonable cost of computers for elementary data handling and reduction, and of various sensors which can be equipped with digital readout.[80]

Two constant-velocity Mössbauer spectrometers [81, 82] and a constant-acceleration instrument [83] which allows one to observe the first derivative, with respect to velocity, of the Mössbauer spectrum, have been described.

An in-beam spectrometer has been modified to allow Coulomb excitation Mössbauer measurements in a back-scattering geometry, rather than in the usual transmission arrangement. Data obtained with the 122.6 keV γ-rays of ^{186}W were presented and compared with the results from conventional

---

[74] J. J. Peyre and G. Principi, *Nuclear Instr. and Methods*, 1972, **101**, 605.
[75] R. Robinette, J. G. Cosgrove, and R. L. Collins, *Nuclear Instr. and Methods*, 1972, **105**, 509.
[76] V. N. Belogurov, V. A. Bylinkin, and V. V. Surikov, *Latv. P.S.R. Zinat. Akad. Vestis, Fiz. Teh. Zinat. Ser.*, 1972, 27.
[77] G. N. Belozerskii, V. N. Gittsovich, A. N. Murin, and Yu. P. Khimich, *Soviet Phys. Solid State*, 1972, **13**, 2249.
[78] L. Dabrowski, J. Piekoszewski, and J. Suwalski, *Nuclear Instr. and Methods*, 1972, **103**, 545.
[79] V. M. Belova and V. I. Nikolaev, *Soviet Phys. Solid State*, 1972, **14**, 111.
[80] R. C. Axtmann, *Chem. Eng. Comput.*, 1972, **1**, 11.
[81] B. G. Egiazarov and A. I. Shamov, *Pribory Tekhn. Eksp.*, 1972, 55.
[82] S. Reiman and E. Realo, *Eesti N.S.V. Tead. Akad. Toim., Fuus., Mat.*, 1972, **21**, 41.
[83] A. M. Voronin, *Izvest. Akad. Nauk. Kaz. S.S.R., Ser. Fiz.-Mat.*, 1972, **10**, 61.

techniques.[84] A detector for efficient back-scatter Mössbauer spectroscopy, specifically designed for examination of works of art containing iron-bearing pigments, has been described in detail. The detector is simple to construct, durable, and gives results superior to those so far reported.[85] Two other spectrometers suitable for reflection Mössbauer spectroscopy have been described.[86, 87] The results of an application in structural vibration measurement using one of these spectrometers were presented and the performance of the detector was assessed.[87]

Details have been given of a resonance detector for tin-119 studies; it incorporates a windowless electron mutliplier and can be used for studying thin layers by registering the conversion electrons emitted from the sample after γ-irradiation.[88] The advantages of using a high-resolution silicon–lithium drifted detector to detect the 23.9 keV γ-rays of ^{119m}Sn and to resolve them from the 25.8 keV X-rays have been demonstrated. This detector enables a correct background subtraction to be made and it has been possible to obtain accurate values for the maximum resonance cross-section of ^{119m}Sn and for the f-factors of $BaSnO_3$, SnO_2, and white tin (see Section 5).[89]

It is claimed that the accuracy of ^{83}Kr resonance absorption measurements can be nearly doubled by filling the gas proportional counter with a mixture of argon, neon, and methane.[90] It is claimed elsewhere that the use of multiwire proportional counters can possibly reduce running times by a factor of hundreds, especially when the nuclear levels which undergo the Mössbauer γ-transition are populated through a nuclear reaction rather than a radiative decay.[91]

The effect of experimental geometry on the lineshape in Mössbauer spectroscopy has been considered.[92]

A special cryostat has been designed for Mössbauer effect studies of lattice defects formed after high-dose neutron irradiation at 4.2 K. The measurements can be made without warming the samples up to temperatures higher than 10 K between irradiation and measurement so that the defects are retained in their original state.[93] A relatively simple low-temperature apparatus for high-pressure Mössbauer studies has been

[84] J. A. Hicks, W. R. Owens, and R. N. Wilenzick, *Rev. Sci. Instr.*, 1972, **43**, 1086.
[85] B. Keisch, *Nuclear Instr. and Methods*, 1972, **104**, 237.
[86] A. N. Artem'ev, K. P. Aleshin, V. V. Skylarevskii, and E. P. Stepanov, *Pribory Tekhn. Eksp.*, 1972, 48.
[87] J. J. Singh, *Nuclear Instr. and Methods*, 1972, **101**, 199.
[88] P. Kamenov, R. Iordanova, E. Vapirev, and V. F. Gurev, *Rev. Roumaine Phys.*, 1972, **17**, 309.
[89] A. Biran, A. Yarom, P. A. Montano, H. Schechter, and U. Shimony, *Nuclear Instr. and Methods*, 1972, **98**, 41.
[90] S. P. Ekimov, I. A. Yutlandov, L. M. Krizhanskii, and N. K. Cherezov, *Pribory Tekhn. Eksp.*, 1972, 51.
[91] C. Barbero, G. C. Bonazzola, T. Bressani, E. Chiavassa, and A. Musso, *Lett. Nuovo Cimento Soc. Ital. Fis.*, 1972, **4**, 19.
[92] N. N. Delyagin, K. P. Mitrofanov, and V. I. Nesterov, *Nuclear Instr. and Methods*, 1972, **100**, 315.
[93] P. Rosner, W. Vogl, and G. Vogl, *Nuclear Instr. and Methods*, 1972, **105**, 473.

described. The pressure system is of the supported anvil type and can produce pressures in excess of 200 kbar. Either a liquid-nitrogen flow system or a liquid-helium cryostat is used to cool the pressure cell. The pressure-transmitting column is made of a special low-conductive grade of fibre glass. Sample temperatures of 14 K at pressures of up to 170 kbar can be obtained. The apparatus is suitable for source experiments only.[94]

A vacuum furnace for Mössbauer experiments, which fits into the 5 cm room-temperature bore of a superconducting solenoid, has been described. The temperature homogeneity of the absorber, a disc 2.0 cm in diameter, is better than 0.25 °C at 360 °C. As an illustration of its use, the complete temperature dependence of the hyperfine fields at the five iron sites in $SrFe_{15}O_{19}$ were determined.[95]

It has been shown that the Mössbauer effect of ^{119}Sn nuclei in an iron matrix is a suitable method for measuring temperatures down to 6×10^{-3} K. The ^{57}Fe resonance in the same sample can only be used to measure temperatures down to 0.015 K; below this temperature the ^{57}Fe nuclear spins are not in thermal equilibrium with the lattice and cannot act as a temperature probe.[96] The isomer shift of pure iron foil containing 2.19 and 89% ^{57}Fe does not change between 0.015 K and 0.060 K and cannot therefore be used to measure ultra-low temperatures.[97] A ^{161}Dy thermometer may be useful in calibrating other thermometers below 0.6 K.[98]

Calculations have been made of six combinations of source and absorber temperature necessary to eliminate the second-order Doppler shift in spectra obtained with a $^{57}Co/Pd$ source and an $^{57}Fe/Fe$ absorber.[99]

4 Iron-57

General Topics.—*Nuclear Parameters, Hyperfine Interactions, and New Effects.* There have been three new determinations of the fractional increase in the ^{57}Fe nuclear charge radius upon excitation, $\Delta R/R$. The numerical values obtained are $-(3.1 \pm 0.6) \times 10^{-4}$,[100] -4.1×10^{-4},[101] and $-(8.7 \pm 0.3) \times 10^{-4}$,[102] and these can be compared with previous estimates listed in Table 1. The method used for the first determination does not embody any assumptions about the electronic structure, thereby avoiding the main source of error in previous determinations, and is based

---

[94] D. L. Williamson, S. Bukshpan, R. Ingalls, and H. Shechter, *Rev. Sci. Instr.*, 1972, **43**, 194.
[95] J. M. D. Coey, D. C. Price, and A. H. Morrish, *Rev. Sci. Instr.*, 1972, **43**, 54.
[96] D. V. Pavlov, A. Ya. Parshin, V. P. Peshkov, B. G. Egiazarov, A. M. Shamov, and V. P. Romashko, *Pis'ma Zhur. eksp. i teor. Fiz.*, 1972, **16**, 231.
[97] A. Ya. Parshin, V. P. Peshkov, B. G. Egiazarov, A. I. Shamov, and V. P. Pomashko, *Pis'ma Zhur. eksp. i teor. Fiz.*, 1972, **15**, 44.
[98] J. Hess, *Rev. Sci. Instr.*, 1972, **43**, 688.
[99] H. N. K. Sharma, B. P. Srivastava, and D. L. Bhattacharya, *Current Sci.*, 1972, **41**, 327.
[100] R. Rüegsegger and W. Kündig, *Phys. Letters (B)*, 1972, **39**, 620.
[101] R. R. Sharma and A. K. Sharma, *Phys. Rev. Letters*, 1972, **29**, 122 (Erratum *op. cit.*, *ibid.*, 1972, **29**, 1428).
[102] H. Micklitz and P. H. Barrett, *Phys. Rev. Letters*, 1972, **28**, 1547.

Table 1

Author	$-\Delta R/R$ ($\times 10^4$)	Author	$-\Delta R/R$ ($\times 10^4$)
Walker (1961)	18	McNab (1971)	10.6
Danon (1966)	5	Rüegsegger and Kündig	3.1 ± 0.6
Goldanskii (1966)	9	Sharma and Sharma	4.1
Simánek (1967)	4	Micklitz and Barrett	8.7 ± 0.3
Pleiter (1971)	4.5		

on the fact that the internal conversion coefficient, α, is roughly proportional to $|\psi(0)|^2$. The influence of the chemical environment on the radioactive decay constant, λ, is therefore given by the equation

$$\frac{\Delta \lambda}{\lambda} = \frac{\alpha}{\alpha + 1} \frac{\Delta |\psi(0)|^2}{|\psi(0)|^2}$$

and together with the well-known equation for the isomer shift, δ, one obtains

$$\frac{\Delta R}{R} = \frac{5\alpha \lambda \delta}{4\pi Z e^2 R^2 (\alpha + 1) \Delta \lambda |\psi(0)|^2}$$

$\Delta R/R$ can therefore be determined by measurement, on two chemically different sources, of the isomer shift, δ, and the quantity $\Delta\lambda/\lambda$ (by a coincidence technique). In the actual experiment four different sources were used, ^{57}Co in thin foils of copper, cobalt, and gold, and ^{57}Co in the form of CoO powder.[100] The second determination [101] included relativistic effects and was based on an analysis of the isomer shifts for equivalent iron sites in Fe_2O_3 and two inequivalent iron sites in several rare-earth iron garnets. The third determination [102] was based on an analysis of the isomer shifts of the two species, stable $Fe(3d^64s^2)$ and unstable $Fe^+(3d^7)$, produced in the electron-capture decay of ^{57}Co in solid xenon.

The fact that the probability for internal conversion is proportional to the electron density at the nucleus has also been exploited in a completely different and very elegant experiment designed to measure the $2s$ and $3s$ electron-spin densities at the nuclear site of ^{57}Fe in iron metal. The nuclear transitions $\Delta m_I = +1$ and -1 correspond to total angular momentum changes of the emitted internal conversion electron of $\Delta m_e = -1$ and $+1$, respectively. Provided that the electrons are emitted into a final s-state, and this occurs in greater than 90% of all emissions, then $\Delta m_e = \Delta m_s$, where m_s is the z component of the spin angular momentum, and the following relations hold:

$$\alpha_{ns}(-\tfrac{3}{2} \to -\tfrac{1}{2}) \propto |\psi_{ns}\uparrow(0)|^2$$
$$\alpha_{ns}(+\tfrac{3}{2} \to +\tfrac{1}{2}) \propto |\psi_{ns}\downarrow(0)|^2$$

The new technique combines Mössbauer spectroscopy and electron spectroscopy to measure the ratio of internal conversion electrons emitted in the decay of the $m_I = \pm\tfrac{3}{2}$ states and yielded values for the ratio $R = |\psi_{ns}\uparrow(0)|^2/|\psi_{ns}\downarrow(0)|^2$ of 0.9937 ± 0.0015 and 1.012 ± 0.006 for the

2s and 3s shells, respectively. These results are in qualitative agreement with theoretical calculations of the core contributions to the magnetic hyperfine interaction.[103]

Analysis of the magnetic hyperfine interaction of the octahedral iron(III) ion in $K_3Co_{1-x}Fe_x(CN)_6$ has yielded a value of $H_c/2S = +255 \pm 5$ kG (1 T = 10 kG) for the magnetic contact term, and a value of $\langle r^{-3} \rangle = 1.0$ a.u. (1 a.u. = 1.48×10^{-28} m^{-3}) for the $3d(t_{2g})$ wavefunction. Comparison of these results with the corresponding theoretical values for the free Fe^{3+} ion, -126 kG and 5.7 a.u., respectively, indicates that there is an overall expansion of the $3d(t_{2g})$ wavefunction in covalent transition-metal complexes coupled with a significant 4s participation in the bonding of the central transition-metal ion, which is responsible for the reversal of the sign of H_c. Similar results were obtained from an analysis of e.p.r. data for $K_4Fe_{1-x}Mn_x(CN)_6,3H_2O$ and supporting evidence is derived from a consideration of the quadrupole splitting, isomer shift, and molecular vibrations in $K_3Fe(CN)_6$.[104]

Dale has pointed out that a recent paper by Cosgrove and Daniels, which showed the ^{57}Fe quadrupole splitting due to a single 3d electron to be constant for any real wavefunction with real coefficients, ignored the spin parts of the wavefunctions; the treatment is therefore considered to be invalid on technical grounds and the result only applicable to the very limited field where the symmetry is lower than rhombic and where spin–orbit coupling can be ignored.[105] Cosgrove and Collins, while granting that the assumption of real wavefunctions implies the omission of spin–orbit coupling, contest the viewpoint that their original treatment is invalid and suggest that departure from rhombic symmetry is quite common (e.g. trigonal distortion from octahedral symmetry). They reiterate the main point of their original paper, that the electric-field gradient parameters change remarkably as one spans the spin-free T_{2g} sub-space and that the quadrupole splitting in no way indicates this fact.[106]

A time-dependent recoilless fraction has been directly observed with delayed-coincidence Mössbauer spectroscopy. The effect is observed at the Fe^{3+} site but not at the Fe^{2+} site in a $CoSO_4,7H_2O$ source and is thought to be due to the existence of a 'heated localized mode' relaxing at a rate dependent on the rate at which the energy of the localized mode can be transferred into the normal modes. The relaxation time of the recoilless fraction was estimated to be between 10 and 100 ns. The effect is not observed in sources of $CoCl_2,6H_2O$ and $Co(NH_4)_2(SO_4)_2,6H_2O$. The reason for this and for the preferential appearance of the effect at the Fe^{3+} site in $CoSO_4,7H_2O$ is not fully understood.[107]

---

[103] Cheng-jyisong, J. Trooster, and N. Benczer-Koller, *Phys. Rev. Letters*, 1972, **29**, 1165.
[104] Y. Hazony, *J. Phys. (C)*, 1972, **5**, 2267.
[105] B. W. Dale, *J. Chem. Phys.*, 1972, **56**, 4721.
[106] J. G. Cosgrove and R. L. Collins, *J. Chem. Phys.*, 1972, **56**, 4721.
[107] G. R. Hoy and P. P. Wintersteiner, *Phys. Rev. Letters*, 1972, **28**, 877.

For Mössbauer-effect spectra with unresolved or just barely resolved structure it is not possible to establish the broadening mechanism (*i.e.* magnetic, quadrupole, isomer shift, or some combination of these) on the basis of the Mössbauer technique alone. However, time-differential perturbed angular correlation measurements in conjunction with Mössbauer data can often determine unambiguously the origin of the hyperfine interactions. In particular it was demonstrated that the quadrupole doublet present in the spectrum of copper–nickel alloys containing iron impurities is primarily due to a distribution of electric-field gradients acting at the ^{57}Fe nuclei.[108]

Isaak and Preikschat have described an experiment in which a fine gold mesh is moved transversely with a high velocity in a beam of 14.4 keV Mössbauer γ-rays from a ^{57}Co source. When the beam is analysed into energy components by a Mössbauer spectrometer they find that there is an undisturbed main line and two sidebands, displaced in frequency by ν, where ν is the rate at which cells of the mesh cut a line in the direction of the γ-ray beam. They identify these sidebands with the well-known sidebands that arise in amplitude modulation.[109] Cranshaw has raised certain objections to this interpretation and has suggested that the experiment is more satisfactorily explained in terms of the diffraction of radiation by a grating.[110] A number of papers dealing with the production of Mössbauer sidebands by radiofrequency excitation of magnetic materials are mentioned in the Theoretical section,[62–65] and the dependence of this effect on particle size and driving frequency has been studied in Fe_2O_3.[111] The effect of a high-frequency magnetic field on the Mössbauer spectra of iron foils has also been investigated.[112]

Nuclear magnetic resonance has been observed in ^{57}Co dissolved in iron, by measurement of a change in the Mössbauer spectrum at 0.07 K. The spectrum shows strong intensity asymmetry about zero velocity at $\tau \sim gH\beta_N/k = 0.097$ K, which is removed at the n.m.r. resonant frequency for dilute ^{57}Co in iron.[113] Asymmetries observed in the six-line hyperfine spectrum of iron, obtained in 90° scattering of the 14.4 keV γ-rays, have been attributed to interference between electronic (Rayleigh) scattering and nuclear resonant scattering,[114] and similar effects have been observed in Fe_2O_3.[115] Theoretical treatments of this interference phenomenon are mentioned above.[54–56]

Pressure-dependence Studies. In a definitive paper, Slichter and Drickamer have reviewed experiments in which high pressures have been shown to

[108] R. C. Reno and L. J. Swartzendruber, *Phys. Rev. Letters*, 1972, **29**, 712.
[109] G. R. Isaak and E. Preikschat, *Phys. Letters* (*A*), 1972, **38**, 257.
[110] T. E. Cranshaw, *Phys. Letters* (*A*), 1972, **39**, 296.
[111] L. Pfeiffer, *Amer. Inst. Phys. Conf. Proc.*, 1972, No. 5 (Pt. 2), 796.
[112] I. A. Dubovtsev and A. P. Stepanov, *Fiz. Metal. Metalloved*, 1972, **33**, 1108.
[113] G. Cain, *Phys. Letters* (*A*), 1972, **38**, 279.
[114] L. Y. Lee and C. D. Goodman, *Phys. Rev.* (*C*), 1972, **6**, 836.
[115] A. N. Artem'ev, V. V. Sklyarevskii, G. V. Smirnov, and E. P. Stepanov, *Pis'ma Zhur. eksp. i teor. Fiz.*, 1972, **15**, 320.

induce changes in the charge- or spin-state of iron ions. Commonly iron(III) is reduced to iron(II) and iron(II) undergoes changes in spin involving either a decrease or an increase in the multiplicity. A general analysis of these effects was attempted in terms of the changes in Coulomb energy, closed-shell repulsions, and both covalent bonding energy and crystal-field energy accompanying the change in electronic state.[116]

The effect of pressures up to 180 kbar (1 atm = 1.013 bar; 1 bar = 10^5 N m^{-2} = 10^5 kg m^{-1} s^{-2}) on the electronic structure of twelve iron(III) β-diketone complexes has been measured by means of Mössbauer resonance and optical absorption studies. The systematic variations of the electron-donor and -acceptor properties of the ligands permit a more detailed interpretation of the results than in previous studies. The degree of conversion of iron(III) into iron(II) correlates well with the change of isomer shift with pressure (and thus with changes of the electronic properties of the ligands) and correlates also with the decrease in area under the charge-transfer peaks of the optical absorption spectra.[117] Two pressure-dependence studies published last year have been amended; the first was concerned with iron phthalocyanine derivatives[118] and the second with protoporphyrin IX, hemiporphyrins, and related compounds.[119]

There is considerable evidence that h.c.p. iron should order antiferromagnetically at low temperatures and high pressures. A search has now been made for such effects at 20 K and 48 K and at pressures of 148 and 176 kbar, respectively, but no evidence was found for magnetic ordering. Values of the volume coefficients of the isomer shift and magnetic hyperfine splitting of b.c.c. iron were obtained at 82 K and the effects of pressure on the second-order Doppler shift were clearly observed. H.c.p. iron has a Debye temperature of 545 ± 25 K at *ca.* 145 kbar, whereas for b.c.c. iron the value is 410 ± 25 K at 1 atm. Evidence was presented for a hysteresis associated with the b.c.c. ⇌ h.c.p. transformation.[120] The pressure-dependence of the recoil-free fraction and second-order Doppler shift of iron in various host lattices has been studied elsewhere.[121] A new phase, most probably antiferromagnetic, has been revealed in invar at high pressures. In $Fe_{0.7}Ni_{0.3}$ the Néel temperature has a slope of +1.9 ± 0.3 K kbar^{-1} with a zero-pressure intercept of −41 ± 21 K.[122]

Lattice Dynamics. The recently developed Mössbauer scattering technique has been used to study the molecular dynamics of a number of organic substances in the glassy and supercooled liquid states. The technique enables direct information to be obtained on molecules which do not

---

[116] C. P. Slichter and H. G. Drickamer, *J. Chem. Phys.*, 1972, **56**, 2142.
[117] C. W. Frank and H. G. Drickamer, *J. Chem. Phys.*, 1972, **56**, 3551.
[118] D. C. Grenoble and H. G. Drickamer, *J. Chem. Phys.*, 1972, **56**, 1017.
[119] D. C. Grenoble, C. W. Frank, C. B. Bargeron, and H. G. Drickamer, *J. Chem. Phys.*, 1972, **56**, 1017.
[120] D. L. Williamson, S. Bukshpan, and R. Ingalls, *Phys. Rev. (B)*, 1972, **6**, 4194.
[121] B. N. Srivastava and R. N. Tyagi, *Indian J. Pure Appl. Phys.*, 1972, **10**, 89.
[122] D. R. Rhiger and R. Ingalls, *Phys. Rev. Letters*, 1972, **28**, 749.

contain a resonant isotope. The fraction of elastic scattering (f) is obtained from the ratio of the fractional depth of the Mössbauer dip, measured with a 'black' absorber between sample and detector, to the fractional depth with the absorber between source and sample. For the materials n-propylbenzene, n-butylbenzene, diethyl phthalate, di-n-butyl phthalate, di-2-ethylhexyl phthalate, ethylcyclohexane, propane-1,2-diol, and squalane the temperature dependence of the recoilless fraction is different in the glassy and supercooled liquid regions, and this is explained in terms of the extra molecular motion, such as libration, which arises in the supercooled liquid owing to free volume effects. In the glassy state the recoilless fraction varies with temperature in a fashion consistent with an effective Debye temperature of approximately 45 K. In the supercooled liquid state the recoilless fraction falls rapidly with increase in temperature and becomes increasingly small near the melting point.[123] A preliminary study has been made of three organic polymers of differing physical properties, using the same technique. Amorphous polyisobutene shows similar behaviour to that exhibited by the supercooled liquids, *i.e.* a linear region below the glass transition temperature (T_g 200 K) and a sharply decreasing f above T_g, indicative of rapidly increasing molecular freedom. Polytetrafluoroethylene (PTFE) shows behaviour representative of the molecular dynamics of a mixture of crystalline and amorphous components. There is no evidence of line-broadening at 295 K; hence any relaxation process associated with molecular rotation or diffusion must have a characteristic time greater than 6×10^{-7} s at this temperature. Poly(ethylene glycol) gives an unbroadened linewidth with large f, typical of an organic crystal. Water-containing samples of the latter show a decreased f which is interpreted as being due to increased molecular freedom caused by lattice swelling.[124]

The polymer dynamics of polyacrylonitrile in the glass transition region have been studied by conventional absorption experiments on ^{57}Fe incorporated as an impurity into the lattice. Values of the glass transition temperature, T_g, obtained from the recoil-free fraction *vs.* temperature curve, were found to be in very good agreement with those obtained by dilatometric methods. The Mössbauer effect was observable both in the glassy state (when the polymer is characterized by an equivalent Debye θ temperature) and in the 'rubbery' state just above T_g. Typical results for mixtures of polyacrylonitrile and Fe(ClO$_4$)$_2$,6H$_2$O containing (i) 6.2% and (ii) 8.1% by weight of Fe^{2+} were (i) $T_g = 270$ K, $\theta = 97$ K and (ii) $T_g = 265$ K, $\theta = 85$ K.[125]

Alloy-type Systems. Metals and alloys are not discussed explicitly but the relevant papers are listed in the Bibliography at the end of the review. This section deals with carbides (including steels), nitrides, silicides, and antimonides.

---

[123] D. C. Champeney and D. F. Sedgwick, *J. Phys.* (*C*), 1972, **5**, 1903.
[124] D. C. Champeney and D. F. Sedgwick, *Chem. Phys. Letters*, 1972, **15**, 377.
[125] S. Reich and I. Michaeli, *J. Chem. Phys.*, 1972, **56**, 2350.

In agreement with thermodynamic results it has been found that high-carbon iron–carbon austenite is an ideal solid solution of the carbon atoms distributed in the octahedral interstitial sites, whereas in unaged martensite half the carbon atoms occupy tetrahedral rather than octahedral interstitial sites.[126]

By correlating X-ray textures with Mössbauer line intensities it has been shown that the spin in Fe_3C, cementite, is oriented within a solid angle of 20° along the crystallographic c-axis.[127] The ductile–brittle transition in carbon steels has been investigated; in addition to the six-line hyperfine pattern characteristic of low-carbon steels, peaks due to Fe_3C were also observed and it is thought that this stabilization of covalent Fe—C bonds may be one of the reasons for the brittle fracture of steels at low temperatures.[128] It has been shown that quadrupole splitting can be induced in a stainless-steel foil (type 310) simply by folding the foil along a line and then straightening it; the effect is attributed to the formation of a line defect.[129] Radiation damage in iron and steel has been studied.[130]

Both hexagonal and orthorhombic iron nitrides FeN_x show ferromagnetic behaviour at low temperatures. The internal field at a given iron nucleus is thought to be determined predominantly by the configuration of nearest neighbour nitrogen atoms. For compositions with x between 0.5 and 0.33, iron atoms with two and three nitrogen neighbours predominate and the spectra consist of two overlapping six-line patterns.[131]

Magnetic and Mössbauer studies have been performed on the solid solutions $Mn_{5-x}Fe_xSi_3$ ($x = 1, 2, 3, 3.5$, or 4) with D_{8_8} structure. Throughout the series the spectra above the magnetic ordering temperatures are made up of two overlapping quadrupole doublets consistent with the presence of iron(II) ions in $6(g)$ and iron(III) ions in $4(d)$ sites. As the total iron content increases, the iron(II) doublet grows in intensity relative to the iron(III) doublet, indicating that the iron atoms occupy preferentially the $6(g)$ sites, with a small amount of disorder. Below the magnetic ordering temperature two overlapping six-line hyperfine patterns are observed, with hyperfine fields which increase with increasing iron content for both $4(d)$ and $6(g)$ sites, and relative intensities which confirm the site preferences suggested above. The compositional dependence of the hexagonal lattice parameters and their c/a ratio also correlate with this ordering scheme. A previous suggestion of an inversion of site preferences at intermediate compositions was not confirmed.[132]

---

[126] M. Lesoille and P. M. Gielen, *Met. Trans.*, 1972, **3**, 2681.
[127] U. Gonser, M. Ron, H. Ruppersberg, W. Keune, and A. Trautwein, *Phys. Status Solidi (A)*, 1972, **10**, 493.
[128] R. N. Kuz'min, S. V. Nikitina, A. N. Ovchinnikov, N. D. Tyutev, and V. A. Golovnin, *Metalloved. Term. Obrab. Metal.*, 1972, **38**, 43.
[129] B. P. Srivastava, H. N. K. Sharma, and D. L. Bhattacharya, *Phys. Status Solidi (A)*, 1972, **10**, K117.
[130] J. Christiansen, *Atomkernenergie*, 1972, **19**, 161.
[131] M. Mekata, H. Yoshimura, and H. Takaki, *J. Phys. Soc. Japan*, 1972, **33**, 62.
[132] V. Johnson, J. F. Weiher, C. G. Frederick, and D. B. Rogers, *J. Solid State Chem.*, 1972, **4**, 311.

$Fe_{1+\delta}Sb$ (0.08 ⩽ δ ⩽ 0.38), which crystallizes in the NiAs structure, with the excess of iron atoms occupying interstitial positions, has been shown to be antiferromagnetic and not paramagnetic as previously thought. A range of iron environments is present, giving rise to a broadened hyperfine spectrum with an internal field at 4.2 K of 106 kG (98 kG at 77 K). The interstitial iron atoms do not produce a hyperfine pattern; the peaks are probably both too broad and too weak to be observed.[133] The quadrupole splitting of $FeSb_2$ shows an unusual temperature dependence between 6.4 and 540 K, owing to promotion of electrons across an energy gap. The latter was estimated to be 0.033(1) eV, in good agreement with the value derived from electrical conductivity measurements. An analysis of the isomer shift in terms of the Debye approximation yields a value of 380 K for the Debye temperature.[134]

^{57}Fe Impurity Studies. This section covers experiments which involve the doping of non-iron compounds, including inert-gas matrices, with ^{57}Fe. It will be seen that the technique is particularly useful for monitoring a variety of phase transitions in solids.

The suitability of β-rhombohedral boron as a host material for ^{57}Co for use as a Mössbauer source has been investigated. The material has a low atomic number and high Debye temperature but unfortunately does not give a single-line resonance. Samples prepared by diffusing ^{57}Fe at temperatures greater than 1000 °C were shown to give a reproducible quadrupole interaction and a strongly anisotropic f-factor. The spectra also contain a third line near zero velocity, the intensity of which is dependent on the method of sample preparation and reduces slightly with increasing period and temperature of diffusion. In contrast, reproducible doublets can only be obtained for sources of ^{57}Co diffused into boron by raising the temperature to 1500 °C. The possible locations of iron and cobalt within the boron lattice were discussed.[135] Other authors have suggested that the spectrum of 1 at. % iron in boron indicates that the iron atoms populate three different crystallographic positions.[136]

The chemical isomer shifts for ^{57}Fe (and ^{119}Sn) atoms isolated in solid nitrogen are identical with those of rare-gas isolated atoms. Despite an observed quadrupole splitting of the resonance, due to the distortion of the cubic nitrogen lattice by the impurity iron atoms, the relatively small photoelectric cross-section of nitrogen atoms makes this matrix useful for studies of nuclei that require relatively thick Mössbauer absorbers. The nitrogen matrix has a minimal effect on the atomic configurations of the trapped atoms.[137]

[133] K. Yamaguchi, H. Yamamoto, Y. Yamaguchi, and H. Watanabe, *J. Phys. Soc. Japan*, 1972, **33**, 1292.
[134] J. Stetger and E. Kostiner, *J. Solid State Chem.*, 1972, **5**, 131.
[135] F. Stanke and F. Parak, *Phys. Status Solidi (B)*, 1972, **52**, 69.
[136] R. Wappling, L. Haggstrom, and S. Devanarayanan, *Phys. Scripta*, 1972, **5**, 97.
[137] H. Micklitz and P. H. Barrett, *Appl. Phys. Letters*, 1972, **20**, 387.

NaCl doped with $FeCl_3$ has been studied in an attempt to elucidate the mechanism of aggregation of the impurity. The iron, although doped originally as Fe^{3+}, was shown to exist as Fe^{2+} in a single environment after being heated at 650 °C under argon.[138]

Ammonium chloride undergoes a lambda transition of the order–disorder type at 242 K. In the low-temperature ordered state all of the

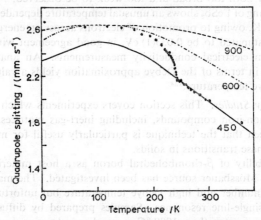

Figure 1 *Theoretical analysis of the quadrupole splitting in terms of an axial splitting parameter. Curves for a t_{2g} splitting of 450, 600, and 900 cm^{-1} are shown*

NH_4^+ tetrahedra have the same orientation with respect to the crystallographic axes, whereas in the high-temperature disordered states they are randomized between the two possible equivalent orientations. This transition has now been investigated using $^{57}Fe^{2+}$ impurity atoms, which are thought to reside on interstitial sites of tetragonal symmetry in an effective $[FeCl_4(H_2O)_2]^{2-}$ complex. The comparatively large value for the room-temperature quadrupole splitting (1.97 mm s^{-1}) was thought to favour a $|d_{xy}\rangle$ singlet ground state and the unexpected angular-independence of the relative intensities of the two components of the doublet for a single crystal was taken to indicate that the impurity sites are all equivalent and that they feature a principal axis pointing in one of three mutually orthogonal directions such as the $\langle 100 \rangle$, $\langle 010 \rangle$, and $\langle 001 \rangle$ axes with equal probability. In a temperature-dependence study on the polycrystalline material there was no discontinuity in the regular decrease in both the spectrum areas and the chemical isomer shift at the lambda point, and the linewidth remained constant at 0.26 mm s^{-1}. However, the quadrupole splitting decreased markedly from about 190 K up to the known lambda point at 242 K, as can be seen in Figure 1. In this respect it is noteworthy that recent thermal-expansion and specific-heat studies on NH_4Cl only

[138] A. J. Bannaghan, D. R. Hayman, and P. L. Pratt, *Proceedings of the Seventh International Symposium on the Reactivity of Solids*, 1972, p. 68.

reveal effects of the phase transition in the range 234—244 K. The quadrupole splitting was analysed theoretically and the ligand-field splitting was found to be approximately 900 cm^{-1} in the low-temperature ordered state compared with 450 cm^{-1} in the disordered state.[139]

No anomaly is observed in either the isomer shift or the normalized resonance intensity at the antiferroelectric transition ($T_c = 148$ K) for $^{57}Fe^{3+}$ doped into $NH_4H_2PO_4$. However, there is a sudden increase in the linewidth around T_c, consistent with the expected increase in the electric-field gradient created by the onset of antiferroelectric ordering, the effect being more pronounced in single-crystal than in powder absorbers. The experiments suggest that the iron atoms do not take part in the coupled ferroelectric mode, in contrast to the observation of Brunstein for $^{57}Co : KH_2PO_4$.[140]

The electronic properties of Fe^{2+} in cubic $KMgF_3$ have been studied. At room temperature the single-crystal spectra contain singlets from Fe^{2+} and Fe^{3+}. At temperatures below 12 K, the Fe^{2+} line splits into a quadrupole doublet and comparison of the data with a random-strain model due to Ham yields a value of 120 cm^{-1} for the position of the first excited spin–orbit level. Experiments in an applied magnetic field at low temperatures yield [141] a value of -495 ± 30 kG for the core-polarization hyperfine field in Fe^{2+} and a value of 4.1 a.u. for $\langle r^{-3} \rangle$. $^{57}Fe^{2+}$ in place of Ca^{2+} in $RbCaF_3$ has been shown to have a large isomer shift relative to FeF_2, owing to the decrease of the overlap-induced electron density at the iron nucleus accompanying the increase in the Fe—F distance (206 pm in FeF_2, $\delta = 1.41$ mm s^{-1}; 236 pm in $RbCaF_3$, $\delta = 1.55$ mm s^{-1}).[142] The magnetic hyperfine interaction of $^{57}Fe^{2+}$ in FeF_2, MnF_2, and ZnF_2 is discussed on p. 516.

Experiments on the trinuclear chromium complex $[Cr_3(MeCO_2)_6(H_2O)_3]$-$Cl,6H_2O$ with $^{57}Fe^{3+}$ substituting for Cr^{3+} up to 2% are described on p. 529, and an analysis of the Mössbauer hyperfine interaction data for the octahedral $3d(t_{2g}^5)$ iron(III) ion in $K_3Co_{1-x}Fe_x(CN)_6$ are discussed on p. 504. Mössbauer spectra taken during the u.v. excitation of rutile, containing adsorbed Fe^{3+}, suggest that the adsorbed Fe^{3+} ions form acceptor surface states which trap holes to become Fe^{4+}. The isomer shift of the induced Fe^{4+} is claimed to be more negative than any value previously reported, indicating a more ionic species.[143]

Preliminary data have been reported of studies of the metal–non-metal transition in the V_nO_{2n-1} system doped with 1 at.% ^{57}Fe. The study was undertaken to elucidate the origin of anomalies in the temperature dependence of the magnetic susceptibility for members of this system. The oxides

[139] T. C. Gibb, N. N. Greenwood, and M. D. Sastry, *J.C.S. Dalton*, 1972, 1896.
[140] M. D. Sastry, *Solid State Comm.*, 1972, **11**, 1671.
[141] R. B. Frankel, J. Chappert, J. R. Regnard, A. Misetich, and C. R. Abeledo, *Phys. Rev. (B)*, 1972, **5**, 2469.
[142] A. Cruset, *Chem. Phys. Letters*, 1972, **16**, 326.
[143] A. J. Nozik, *J. Phys. (C)*, 1972, **5**, 3147.

V_4O_7, V_5O_9, V_6O_{11}, and V_8O_{15} all have two kinks in the χ vs. T curve at temperatures T_A and T_B ($T_A < T_B$), whereas V_7O_{13}, which is metallic, shows only the kink at the lower temperature, T_A = 43 K. V_3O_5, which is a semiconductor, shows neither kink but shows instead a broad maximum at T = 133 K. Mössbauer spectra were obtained for V_3O_5, V_4O_7, and V_7O_{13} which were chosen as representative of the three types of magnetic behaviour. They were all antiferromagnetic at 4.2 K with effective hyperfine fields H_{eff} of 390, 400, and 440 kG, respectively. For V_4O_7 and V_7O_{13}, H_{eff} follows a Brillouin function as the temperature increases, and disappears near T_A, which would therefore appear to be equivalent to the Néel point. For V_3O_5 H_{eff} disappears at about 69 K, which is much lower than the temperature of the broad peak in the χ vs. T curve.[144]

Spectra have been obtained for [57]Fe in ferroelectric $PbTiO_3$ in the temperature range 300—1100 K. The quadrupole interaction in the ferroelectric phase remains nearly constant up to 320 °C, beyond which it decreases gradually, to disappear at the transition temperature T_c = 480 ± 2 °C. The ratio of the quadrupole interaction of [57]Fe in $PbTiO_3$ to that in $BaTiO_3$ at room temperature is 1.41 ± 0.04, in good agreement with the results of perturbed angular correlation studies. Both the centre shift and the area under the resonance show anomalous temperature-variation in the vicinity of the transition temperature and it is suggested that the Debye temperature of the lattice decreases considerably on crossing the transition temperature. The essential validity of the suggested temperature variation of the soft mode was demonstrated.[145] The ferroelectric phase change in $BaTiO_3$ has been discussed elsewhere.[146]

Spectra have also been obtained for samples of ZnS containing 2—14.2% iron.[147]

[57]Co *Source Experiments and Decay After-effect Phenomena.* Experiments to test the suitability of β-rhombohedral boron as a host matrix for a [57]Co Mössbauer source are referred to on p. 509.[135] The Mössbauer effect in impurity atoms of [57m]Fe in silicon has also been studied.[148]

Diffusion of [57]Co in InSb has been studied by means of [57]Fe emission spectroscopy. Two types of iron atoms are present, one of which decreases in relative concentration with increasing depth below the surface. Three different diffusion mechanisms were operative, the corresponding diffusion coefficients being 7×10^{-4}, 3×10^{-12}, and 7×10^{-10} cm² s⁻¹ at 420 °C for $x <$ 10 μ, 10 $< x <$ 50 μ, and x = 50 μ, respectively.[149]

[144] H. Okinaka, K. Kosuge, S. Kachi, M. Takano, and T. Takada, *J. Phys. Soc. Japan*, 1972, **32**, 1148.
[145] V. G. Bhide and M. S. Hegde, *Phys. Rev. (B)*, 1972, **5**, 3488.
[146] H. G. Maguire and L. V. C. Rees, *J. Phys. (Paris), Colloq.*, 1972, (2), 173.
[147] T. M. Aivazyan, L. A. Kocharyan, A. R. Mkrtchyan, and M. Pulatov, *Izvest. Akad. Nauk Uzbek. S.S.R., Ser. fiz.-mat. Nauk*, 1972, **16**, 64.
[148] B. I. Boltaks, M. K. Bakhadyrkhanov, and P. P. Seregin, *Soviet Phys. Solid State*, 1972, **13**, 2358.
[149] D. N. Nasledov, Yu. S. Smetannikova, K. I. Vinogradov, and V. K. Yarmarkin, *Phys. Letters (A)*, 1972, **40**, 224.

Mössbauer studies of ^{57}Co-doped LaCoO$_3$, combined with magnetic susceptibility data in the temperature range 4.2—1200 K, have shown that cobalt ions exist predominantly in the low-spin cobalt(III) state at low temperatures and transform partially to high-spin Co^{3+} ions up to 200 K. Above 200 K the Co^{3+}/Co^{III} ion-pairs transform to Co^{II}/Co^{4+} ion-pairs. At higher temperatures the proportion of Co^{3+} decreases progressively and disappears completely above the first-order localized-electron–collective-electron transition temperature, 1210 K. Isomer shift data coupled with the f-factor changes which occur at the transition temperature indicate that the transition is caused essentially by the change in entropy of the d electrons. The studies are consistent with Goodenough's hypothesis that the crystal field and band limits are distinct thermodynamic states.[150, 151]

The charge states of ^{57}Fe atoms produced in the electron-capture decay of ^{57}Co atoms in solid xenon are discussed on p. 503.

The chemical after-effects of electron-capture decay in ^{57}Co and of the isomeric transition in ^{119}Sn have been shown to be similar. Thus, the emission spectrum of Co$_3$(PO$_4$)$_2$,xH$_2$O doped with ^{57}Co contains a pair of doublets due to Fe^{2+} and Fe^{3+} (the proportion of the latter increasing with increasing x) and the emission spectrum of SnCl$_2$,2H$_2$O labelled with ^{119}Sn shows a small shoulder indicative of Sn^{4+}. By contrast the higher charge states are not present in the emission spectra of Sn$_3$(PO$_4$)$_2$ doped with ^{57}Co and of Sn$_3$(PO$_4$)$_2$ doped with ^{119}Sn, indicating that Fe^{3+} and Sn^{4+} are produced by oxidation of the decay products by OH radicals created during autoradiolysis of the H$_2$O ligands in the hydrated complexes.[152]

The appearance of resonances corresponding to iron(II) species in the Mössbauer emission spectra of ^{57}Fe formed by electron-capture decay of ^{57}Co in cobalt(III) compounds is difficult to explain. It was once thought that the internal pressure experienced by the newly formed ^{57}Fe^{3+} in sites originally occupied by the smaller ^{57}Co^{3+} parent might be the cause, but ^{57}Fe^{2+} resonances have since been observed in ^{57}Co-doped iron(III) complexes, rendering this explanation unlikely. It seems more probable that the emitted Auger electrons cause radiolysis of the parent compound to produce the aliovalent species. The full paper dealing with the Mössbauer study of the electron-irradiation of Fe(acac)$_3$, a preliminary account of which was published in 1970, has now appeared. The paper also describes similar studies on iron(III) citrate, iron(III) edta, and iron(III) trisdipiperidyl perchlorate. In all cases the spectra resemble the emission spectra observed after electron-capture decay in the corresponding ^{57}Co^{3+}-labelled compounds, which suggests that the stabilization of the anomalous iron charge states in these compounds is due to an autoradiolysis mechanism.[153]

---

[150] V. G. Bhide, D. S. Rajoria, Y. S. Reddy, G. R. Rao, G. V. S. Rao, and C. N. R. Rao, *Phys. Rev. Letters*, 1972, **28**, 1133.
[151] V. G. Bhide, D. S. Rajoria, G. R. Rao, and C. N. R. Rao, *Phys. Rev. (B)*, 1972, **6**, 1021.
[152] J. M. Friedt and Y. Llabador, *Radiochem. Radioanalyt. Letters*, 1972, **9**, 237.
[153] E. Baggio-Saitovitch, J. M. Friedt, and J. Danon, *J. Chem. Phys.*, 1972, **56**, 1269.

Emission spectra of the $^{57}Co^{2+}$-doped acetylacetonates of aluminium(III), chromium(III), and cobalt(III) consist of simple quadrupole doublets, whereas the doped acetylacetonates of manganese(III) and iron(III) show three-peak spectra which correspond with that of $^{57}Co(acac)_3$. These

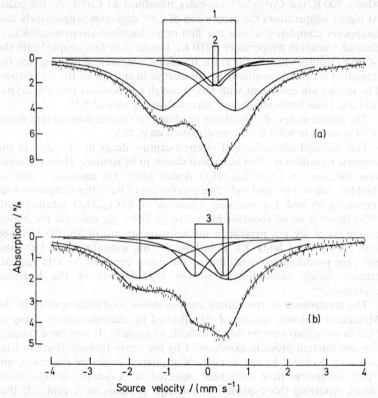

Figure 2 *Mössbauer spectra of* (a) $K_3{}^{57}Co(CN)_6$ *and* (b) $Co(bipy)_3{}^{57}Co(CN)_6,2H_2O$·
The component doublets are ascribed to: 1, *a ferripentacyanide*; 2, $Fe(CN)_6{}^{4-}$ *or* $Fe(CN)_6{}^{3-}$; 3, $Fe(CN)_6{}^{3-}$

results are taken as evidence that cobalt(II) exchanges readily in the solid state with the manganese and iron compounds but not with the aluminium, chromium, and cobalt compounds. Possible reasons for the differences in behaviour were discussed.[154]

The spectra of $[Fe(^{57}Co)(bipy)_3](ClO_4)_3$ (cobalt-doped) and $[^{57}Co(bipy)_3][ClO_4]_3$ (cobalt-labelled) both indicate the presence of Fe^{2+}, $Fe(bipy)_3{}^{2+}$, and $Fe(bipy)_3{}^{3+}$ species, which demonstrates further the unimportance of pressure effects within the lattice. The hexacyanides $K_3Fe(^{57}Co)(CN)_6$, $K_3{}^{57}Co(CN)_6$ [see Figure 2(a)], and $K_4Fe(^{57}Co)(CN)_6$,-

[154] V. Ramshesh, K. S. Venkateswarlu, and J. Shankar, *J. Inorg. Nuclear Chem.*, 1972, **34**, 2121.

3H$_2$O (Prussian Blue) give very similar spectra containing two doublets with intensities in the ratio 80 : 20. The first is assigned to a pentacyanide, the formation of which argues against displacement of ^{57}Fe from the ^{57}Co site, and the second to the species ^{57}Fe(CN)$_6^{3-}$. In a further attempt to check whether an exchange of central atoms can occur between two complexes, measurements were also performed on two specifically ^{57}Co-labelled double complexes, [^{57}Co(bipy)$_3$][Co(CN)$_6$],2H$_2$O and [Co(bipy)$_3$]-[^{57}Co(CN)$_6$],2H$_2$O. The spectrum of the first compound contains only resonances due to ^{57}Fe(bipy)$_3^{2+}$ and ^{57}Fe(bipy)$^{3+}$ and is distinctly different from the spectrum of the second compound [see Figure 2(b)], which contains only two doublets, one from a ferripentacyanide and the other from the ^{57}Fe(CN)$_6^{3-}$ species mentioned earlier. There is therefore no evidence of an exchange of central atoms as a consequence of the electron-capture process.[155]

The decay of ^{57}Co^{2+}[Fe^{III}(CN)$_6$] has been studied by Mössbauer emission spectroscopy and $\gamma-\gamma$ coincidence techniques. In the time interval $\tau = 0$—60 ns both Fe^{2+} and Fe^{3+} were detected, in the proportions 30 : 70, but for $\tau = 0$—160 ns only Fe^{3+} was detected. The iron therefore appears initially in the same valence state as the parent cobalt atom, but decays rapidly to the stable Fe^{3+} state.[156]

The emission spectrum of [Co(phen)$_3$](ClO$_4$)$_2$,2H$_2$O has been reinterpreted. At 4.2 K the spectrum consists of three overlapping doublets which were originally assigned to $S = 0$, $S = 1$, and $S = 2$ states of the iron(II) ion. Both theoretical and experimental evidence has now been presented to demonstrate, firstly, that the existence of the spin triplet ($S = 1$) ground-state is most unlikely and that the doublet originally assigned to this state arises instead from the $S = 2$ ground-state of the intact [^{57}Fe(phen)$_3$(ClO$_4$)$_2$] complex and, secondly, that the doublet originally assigned to the $S = 2$ state arises from a complex involving one defect phenanthroline ligand.[157]

Electron-capture decay of ^{57}Co in the complex Co^{II}L$_2$Cl$_2$ [L = o-(Me$_2$As)$_2$C$_6$H$_4$] has been shown to produce Fe^{II}L$_2$Cl$_2$ complexes containing both high-spin ($t_{2g}^4 e_g^2$) and low-spin ($t_{2g}^6 e_g^0$) iron(II), whereas [Co^{III}L$_2$Cl$_2$]$^+$ gives the corresponding complexes with low-spin iron(III) ($t_{2g}^5 e_g^0$) and low-spin iron(IV) ($t_{2g}^4 e_g^0$).[158] The ^{57}Fe absorption spectra of these complexes are discussed on p. 538 (see ref. 241).

The Mössbauer parameters for an oxygenated haeme complex, produced in a frozen solution by nuclear decay from the isomorphous ^{57}Co-labelled compound, have been shown to agree well with those obtained by Mössbauer absorption spectroscopy of oxyhaemoglobin.[159]

[155] K. E. Siekierska, J. Fenger, and J. Olsen, *J.C.S. Dalton*, 1972, 2020.
[156] V. P. Alekseev, V. I. Goldanskii, V. E. Prusakov, A. V. Nefed'ev, and R. A. Stukan, *Pis'ma Zhur. eksp. i teor. Fiz.*, 1972, **16**, 65.
[157] E. König, P. Gütlich, and R. Link, *Chem. Phys. Letters*, 1972, **15**, 302.
[158] A. Cruset and J. M. Friedt, *Radiochem. Radioanalyt. Letters*, 1972, **10**, 353.
[159] L. Marchant, M. Sharrock, B. M. Hoffman, and E. Münck, *Proc. Nat. Acad. Sci. U.S.A.*, 1972, **69**, 2396.

Compounds of Iron.—*High-spin Iron*(II) *Compounds.* The configuration interaction method has been used to investigate the effects of weak covalency on the crystal-field splittings, the *g*-factors, the spin Hamiltonian, the spin–orbit factors, and the nuclear quadrupole splitting in the salts FeF_2 and $KFeF_3$. For FeF_2 the method is ideal for the calculation of the energies associated with the t_{2g} triplet, as demonstrated by the excellent agreement between the observed and calculated pressure dependence of the quadrupole splitting, which is very sensitive to these low-lying levels.[160] Independent studies on FeF_2 have indicated that the splitting of the ferrous T_{2g} state, due to the axial component of the crystal field, decreases from $E_{axial} = 1300$ K at 300 K to $E_{axial} = 1000$ K at 965 K. Thermal shift and thermodynamic data for FeF_2, $KFeF_3$, and $FeCl_2$ show that the electronic charge density at the iron nucleus is essentially independent of temperature, indicating that the expected increase in this density due to isothermal expansion must be approximately cancelled by an increase due to thermal effects at constant volume. In contrast, FeF_3 displays a significant decrease of electron density at the iron nucleus with increasing temperature. A model which fits quantitatively the high-pressure FeF_2 quadrupole-splitting data of Champion *et al.* below 60 kbar was also given, the new feature of the model being a method for estimating the effect of pressure on the 3*d* radial wavefunction.[161] Analysis of the magnetic hyperfine interaction of Fe^{2+} in FeF_2, $Fe^{2+}:MnF_2$, and $Fe^{2+}:ZnF_2$ has yielded a value for the core polarization hyperfine field of $H_c = -514 \pm 30$ kG and a value of $\langle r^{-3} \rangle_{eff} = 3.9 \pm 0.4$ a.u.[162]

$FeCl_2$ has been studied as the isolated monomer in a solid argon matrix at 4.2 K. The spectrum consists of a doublet with a quadrupole splitting of 0.62 ± 0.05 mm s^{-1} and an isomer shift of 0.88 ± 0.05 mm s^{-1} relative to iron at 300 K. A broad line from small amounts of impurities, possibly hydrated forms of $FeCl_2$, is also present. When the matrix is changed to xenon there is, in addition to these features, a sharp line at $+0.89$ mm s^{-1}, which is thought to correspond to monomeric $FeCl_2$ in a different environment. This line disappears when the sample is annealed.[163]

Three-dimensional magnetic ordering has been shown to occur in the linear chain compound $RbFeCl_3$ at $T_N = 2.55 \pm 0.05$ K.[164]

The value of Mössbauer spectroscopy in conjunction with d.t.a. in studies of crystallization phenomena in aqueous solutions has been demonstrated by measurements on frozen solutions of $FeCl_2$. When the transparent glass, formed by quick cooling of a 30 wt.% solution, is warmed slowly from 77 K, four temperatures are distinguished which correspond to

---

[160] D. M. Silva and R. Ingalls, *Phys. Rev.* (B), 1972, **5**, 3725.
[161] H. K. Perkins and Y. Hazony, *Phys. Rev.* (B), 1972, **5**, 7.
[162] C. R. Abeledo, R. B. Frankel, and A. Misetich, *Chem. Phys. Letters*, 1972, **14**, 561.
[163] T. K. McNab, D. H. W. Carstens, D. M. Gruen, and R. L. McBeth, *Chem. Phys. Letters*, 1972, **13**, 600.
[164] G. R. Davidson, M. Eibschütz, D. E. Cox, and V. J. Minkiewicz, *Amer. Inst. Phys. Conf. Proc.*, 1972, No. 5 (Pt. 1), 436.

(i) melting from glass to undercooled liquid, (ii) crystallization of salt-free ice, (iii) crystallization of $FeCl_2,9H_2O$, (iv) melting of both crystals. The third temperature increases by 22 K as the fraction of D_2O in the water increases from zero to 100%, whereas the other three temperatures remain almost constant. The cause of this large isotope effect is not known.[165, 166] Mössbauer studies on frozen aqueous solutions of FeX_2 (X = Cl, Br, or I)[37] and propane-1,3-diol solutions of $FeSO_4,7H_2O$ have also been described.[167]

It has been claimed that the Mössbauer spectra of siderite, $FeCO_3$, show pronounced anisotropy effects in the recoil-free fraction. At 6 K the ratio of the recoil-free fraction parallel to the axis of symmetry to that in a perpendicular direction is $f_\parallel/f_\perp = 7$. This leads to a ratio for the mean-square displacements in these two directions which is unexpectedly large, especially in view of the fact that the 85 K spectrum is symmetric. The effect is too large to be explained on the basis of lattice vibrations alone and is thought to be associated with the interactions within the magnetic layers.[168] Contributions to the internal field in $FeCO_3$ are discussed on p. 557.

The Mössbauer spectrum of $Fe(ClO_4)_2,6H_2O$ has been studied as a function of temperature and applied magnetic field. On cooling through 243 K the quadrupole interaction changes from $+1.4$ to -3.1 mm s^{-1}, the negative sign in the low-temperature value indicating a $|z^2\rangle$ ground state. When the sample is reheated the transition does not occur until 258 K and is spread over less than 5 K. This hysteresis, coupled with the fact that the f-factor is 20% greater in the low-temperature phase, suggests that the transition is probably first order. The dependence of the effective hyperfine field (H_{eff}) on the applied field (H_{app}) was also studied at various temperatures and the results are shown in Figure 3. At 295 K the induced field (H_i), which is of opposite sign to H_{app}, is negligibly small, with the result that H_{eff} is only slightly less than H_{app}. However, at very low temperatures, H_i predominates over H_{app} and causes H_{eff} to change sign. It was pointed out that determinations of the sign of the electric-field gradient in polycrystalline ferrous salts must be made at a temperature where either H_{app} or H_i predominates, otherwise only broadened, unsplit lines will be obtained. For $Fe(ClO_4)_2,6H_2O$ the two should cancel exactly at 30 K.[169]

New single-crystal measurements have been made on $FeSiF_6,6H_2O$ and are consistent with a d_{z^2} ground state for the Fe^{2+} ion. The observed line intensities are in excellent agreement with values calculated on the basis of V_{zz} being negative in sign, axially symmetric, and directed along the

---

[165] S. L. Ruby, A. Bernabei, and B. J. Zabransky, *Chem. Phys. Letters*, 1972, **13**, 382.
[166] S. L. Ruby, *J. Non-Cryst. Solids*, 1972, **8**—10, 78.
[167] J. J. Höjgaard, *Phys. Kondens. Mater.*, 1972, **13**, 273.
[168] W. Kündig, A. B. Denison, and P. Rüegsegger, *Phys. Letters (A)*, 1972, **42**, 199.
[169] J. M. D. Coey, I. Dézsi, P. M. Thomas, and P. J. Ouseph, *Phys. Letters (A)*, 1972, **41**, 125.

trigonal symmetry axis of the distorted $Fe(H_2O)_6$ octahedron. The powder spectrum is symmetric, indicating that there is no significant Goldanskii–Karyagin effect.[170]

$Fe(HCO_2)_2,2H_2O$ has been studied at room temperature, at ten different temperatures between 4.2 and 1.85 K, and at 0.027 K.[171] In the 4.2 K spectrum [see Figure 4(a)] the two sharp quadrupole doublets arise from

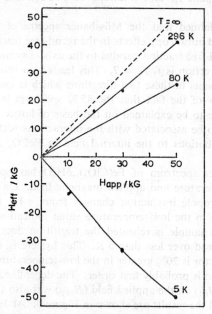

Figure 3 *The effective hyperfine field in* $Fe(ClO_4)_2,6H_2O$
[Reproduced by permission from *Phys. Letters (A)*, 1972, **41**, 125]

two each of two inequivalent Fe^{2+} ions. The inner doublet is assigned to Fe^{2+} ions in *A*-sites surrounded by six oxygen atoms from different carboxy-groups, and the outer doublet to Fe^{2+} ions at *B*-sites, surrounded by two oxygen atoms from different carboxy-groups and four oxygen atoms of water molecules. From the temperature dependence of the quadrupole splitting the *A*- and *B*-site ions are thought to have a spin–orbit doublet ground state and a singlet ground state, respectively. At 1.6 K [Figure 4(b)] and 0.027 K the inner pair of lines become magnetically broadened, whereas the outer lines are essentially unchanged and suffer only a slight broadening as a result of overlap with magnetic components from the *A*-site resonance. Approximate hyperfine parameters for the *B*-site ion, estimated by subtracting the outer doublet from this spectrum

[170] V. K. Garg and K. Chandra, *Phys. Status Solidi (B)*, 1972, **50**, K49.
[171] M. Shinohara, A. Ito, M. Suenaga, and K. Ono, *J. Phys. Soc. Japan*, 1972, **33**, 77.

and comparing the residual pattern [Figure 4(c)] with computer-simulated spectra, were as follows: $H_{eff} = 88$ kG, $\frac{1}{2}e^2qQ = +1.00$ mm s^{-1}, $\eta = 1$, $\theta = 80°$, and $\phi = 0°$, where θ and ϕ specify the directions of the internal magnetic field relative to that of the principal axis of the electric-field

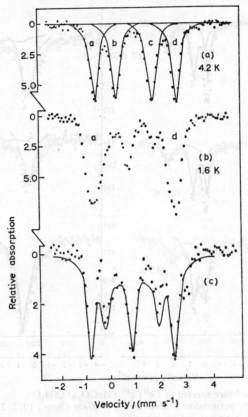

Figure 4 *Mössbauer absorption spectra of powdered* $Fe(HCO_2)_2,2H_2O$ *obtained at* (a) 4.2 K *and* (b) 1.6 K. *The spectrum in* (a) *is decomposed into four lines labelled as* a, b, c, *and* d *and the inner two lines are attributed to A sites and the outer two to B sites. Spectrum* (c) *is obtained by subtracting the absorption lines* a *and* d *in spectrum* (a) *from spectrum* (b). *The curve is the best fit, with* $H_{int} = 88$ kG, $\frac{1}{2}e^2qQ = +1.0$ mm s^{-1}, $\eta = +1.0$, $\theta = 80°$, *and* $\phi = 0°$
(Reproduced by permission from *J. Phys. Soc. Japan*, 1972, **33**, 77)

gradient.[171] Solid-solution formation in the iron(II) formate–magnesium formate system, which is difficult to detect by X-ray diffraction, has been proved with the aid of Mössbauer spectroscopy. The A-site is preferentially occupied by Fe^{2+} ions at low iron concentrations.[172]

[172] K. Nagorny and J. F. March, *Z. Phys. Chem.*, 1972, **78**, 311.

The probability of thermal electron transfer between the iron(II) and iron(III) ions in the mixed valence compounds $Fe^{II}Fe^{III}_2(MeCO_2)_6O,5H_2O$ (A), $[Fe^{II}Fe^{III}_2(MeCO_2)_6O]Cl,5H_2O$, and $[Fe^{II}Fe^{III}_2(MeCO_2)_6O]py_{3.5}$ has been shown to increase with temperature. This is illustrated in Figure 5

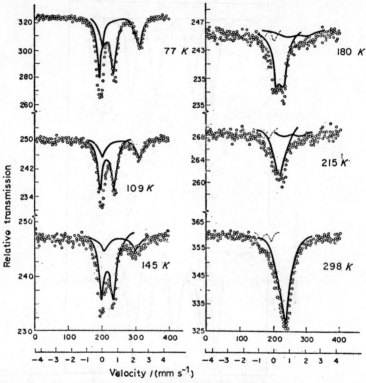

Figure 5 *Mössbauer spectra of* $Fe^{II}Fe^{III}_2(MeCO_2)_6O,5H_2O$
(Reproduced by permission from *J. Inorg. Nuclear Chem.*, 1972, **34**, 2803)

for compound A. At room temperature the relaxation time is shorter than the lifetime of the excited Mössbauer level so that the two oxidation states are no longer distinguishable. The increased shielding of the *s* electrons by the 3*d* electrons with increasing temperature opposes the second-order Doppler shift and causes the slope of the isomer shift *vs.* temperature curve for the iron(III) resonance to be positive.[173]

Iron(II) species formed in the u.v. photolyses of $Fe_2(C_2O_4)_3,5H_2O$, $K_3Fe(C_2O_4)_3,3H_2O$, and $(NH_4)_3Fe(C_2O_4)_3,3H_2O$ have been studied.[174]

---

[173] D. Lupu, D. Barb, G. Filoti, M. Morariu, and D. Tarina, *J. Inorg. Nuclear Chem.*, 1972, **34**, 2803.
[174] G. N. Belozerskii, V. V. Boldyrev, T. K. Lutskina, A. A. Medvinskii, A. N. Murin, Yu. T. Pavlyukhin, and V. V. Sviridov, *Kinetika i Katalitz*, 1972, **13**, 73.

Anisotropy of the probability of the Mössbauer effect has been observed in single crystals of pyrite (FeS_2) but not in vivianite [$Fe_3(PO_4)_2,8H_2O$] and is independent of temperature between 90 and 600 K, both for polycrystalline and single-crystal samples. The temperature dependence of the recoil-free fraction is weaker for pyrite than for vivianite, indicating a larger role of optical branches in the vibrations of iron atoms in pyrite and therefore a larger influence of the local surroundings, which in turn have symmetry lower than cubic.[175]

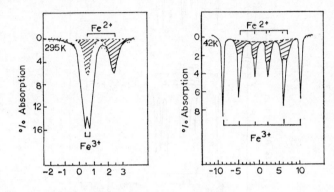

Figure 6 *Spectra of* $K_{0.40}FeF_3$ *at 295 and* 4.2 K
(Reproduced by permission from *J. Solid State Chem.*, 1972, **5**, 402)

The hexagonal, tetragonal, and pyrochlore-type non-stoicheiometric iron fluorides M_xFeF_3 (M = K, Rb, Cs, or NH_4) have been studied over the temperature range 4.2—295 K. In all cases the iron(II) and iron(III) ions remain in discrete oxidation states, indicating the absence of charge hopping at least on the microsecond timescale. The spectra of the hexagonal and tetragonal phases, an example of which is shown in Figure 6, exhibit broadened lines, consistent with the disordering of Fe^{2+} and Fe^{3+} in the structure. The magnetic ordering temperatures could be determined to an accuracy of ±1 K and were found to be in the range 116—135 K. By contrast, the pyrochlore-type phases give much sharper spectra, as illustrated in Figure 7, characteristic of structural ordering between the iron(II) and iron(III) ions, and have magnetic ordering temperatures between 17 and 21 K.[176]

A re-examination of the temperature dependence of the Mössbauer parameters of [NMe_4]$_2$[$FeCl_4$] has revealed the presence of a phase transition at 239 K. This is illustrated in Figure 8, which also shows that

---

[175] I. P. Suzdalev, I. A. Vinogravod, and V. K. Imshennik, *Soviet Phys. Solid State*, 1972, **14**, 1136.
[176] N. N. Greenwood, F. Ménil, and A. Tressaud, *J. Solid State Chem.*, 1972, **5**, 402.

the temperature dependence of the quadrupole splitting deviates markedly from theoretical prediction based on a tetragonal distortion. Possible causes for this discrepency were discussed, including in particular the effects of vibronic admixture of the $|x^2 - y^2\rangle$ and $|3z^2 - r^2\rangle$ levels.[177]

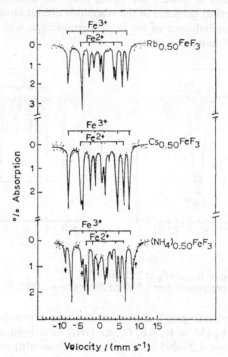

Figure 7 Spectra of $Rb_{0.50}FeF_3$, $Cs_{0.50}FeF_3$, and $(NH_4)_{0.50}FeF_3$ at 4.2 K (Reproduced by permission from *J. Solid State Chem.*, 1972, **5**, 402)

Spectra have been obtained for a number of six-co-ordinate iron(II) complexes of $FeCl_2$ and $FeBr_2$ with amides, ureas, aniline, and benzothiazole, having stoicheiometries FeX_2L, FeX_2L_2, FeX_2L_3, FeX_2L_4, and FeX_2L_6. Complexes of the first three types are polymeric with bridging halogens and have room-temperature quadrupole splittings in the range 2.5—3.0 mm s^{-1}, suggesting that the d_{xy} orbital is stabilized as a result of the distortion from pure octahedral symmetry. The FeX_2L_4-type complexes are monomeric; the formamide derivative has an unusually small quadrupole splitting with a large temperature dependence which points to a nearly regular octahedral structure with only small distortional splittings in the t_{2g} levels.[178] A number of related tetrahedral complexes of the type

[177] T. C. Gibb, N. N. Greenwood, and M. D. Sastry, *J.C.S. Dalton*, 1972, 1947.
[178] T. Birchall and M. F. Morris, *Canad. J. Chem.*, 1972, **50**, 201.

FeX_2L_2 (X = Cl or Br; L = benzothiazole, thioacetamide, thiourea, N-methylthiourea, or NN-dimethylthiourea) have also been studied. They have large (> 3.0 mm s^{-1}) quadrupole splittings with only small temperature dependences, indicative of considerable distortion from purely tetrahedral geometry.[179] It has been shown by X-ray diffraction that $Fe(py)_4Cl_2$,

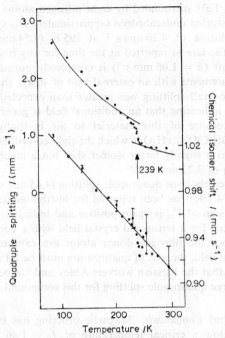

Figure 8 *The temperature dependence of the quadrupole splitting and chemical isomer shift relative to iron metal in* $[NMe_4]_2[FeCl_4]$. *At temperatures below 239 K the solid line represents the theoretical prediction for a tetragonal distortion of 132 cm^{-1}, while above 239 K the distortion is 84 cm^{-1}*

$Co(py)_4Cl_2$, and $Ni(py)_4Cl_2$ are in fact isomorphous, so that $Fe(py)_4Cl_2$ must be considered to have a *trans*-octahedral structure and not a *cis*-octahedral structure as suggested recently (see last year's Report) on the basis of a theoretical and experimental Mössbauer study.[180]

The Mössbauer spectrum of the iron(II) O-bonded $[Fe(bipy)_2(MeC_6H_4-OSO)_2]$ at 295 K has a quadrupole splitting of 2.65 mm s^{-1}, whereas the S-bonded $[Fe(bipy)_2(MeC_6H_4SO_2)],2H_2O$ has the much lower value of 0.31 mm s^{-1}; in each case the chemical isomer shift is 0.31 mm s^{-1} relative to iron. These values are characteristic of $^5T_2$ and $^1A_1$ iron(II) ground

[179] T. Birchall and M. F. Morris, *Canad. J. Chem.*, 1972, **50**, 211.
[180] D. Forster and D. J. Dahm, *Inorg. Chem.*, 1972, **11**, 918.

states and are consistent with the magnetic moments at 292 K of $\mu_{\text{eff}} =$ 5.27 and 0.95 BM, respectively. The temperature dependence of the quadrupole splitting for the O-bonded isomer can be rationalized in terms of an axial ligand field which splits the $^5T_2$ term ($^5T_2 \rightarrow {}^5B_2 + {}^5E$) such that $\delta = -1450 \pm 70$ cm^{-1}.[181]

Tetrakis-(1,8-naphthyridine)iron(II) perchlorate, [Fe(napy)$_4$](ClO$_4$)$_2$, which contains Fe^{2+} surrounded by eight nitrogen atoms arranged at the vertices of a distorted dodecahedron (approximately D_{2d} symmetry), has a quadrupole splitting of 4.26 mm s^{-1} at 295 K (4.54 mm s^{-1} at 4.2 K), thought to be the largest reported at the time for any iron(II) compound. The isomer shift ($\delta = 1.06$ mm s^{-1}) is completely normal, however, for iron(II). Measurements with an external field of 40 kG showed that V_{zz} is positive. The overall splitting was greater than expected for an applied field of 40 kG, indicating that an additional field is generated at the iron nucleus.[182] Exposure of this material to air is thought to produce [Fe(napy)$_4$(H$_2$O)](ClO$_4$)$_2$,xH$_2$O in which the eight co-ordination is retained; this product has a slightly larger isomer shift but a much reduced quadrupole splitting of 3.27 mm s^{-1}.[183]

Another extremely large quadrupole splitting (4.65 mm s^{-1} at 300 K and 4.36 mm s^{-1} at 4.2 K) has been reported for bis(thiosemicarbazide)iron(II) sulphate. The sign of V_{zz} is again positive and indicates a $|d_{xy}\rangle$ ground state, as expected for a tetragonal crystal field with a compression along the z-axis. Since the nitrogen donor atoms are expected to exert the strongest crystal field, the axis of quantization must be NFeN.[184] It should be pointed out that the Russian workers Ablov and Gerbeleu had already reported the large quadrupole splitting for this compound in 1971 (see last year's Report).

High-spin Iron(III) Compounds. Magnetic ordering has been detected in FeCl$_3$,6H$_2$O below a critical temperature of $T_c = 1.46 \pm 0.01$ K. The magnetic hyperfine field is 396 ± 2 kG at 1.16 K, and is oriented at an angle of 73° to the major axis of the electric-field gradient.[185]

There is a continuing interest in the application of Mössbauer spectroscopy to the study of frozen solutions of iron salts, and certain aspects of this work are discussed on p. 516. Extensive studies of paramagnetic iron(III) salt solutions have shown that the nature of the chemical bonding between iron and its ligand sphere influences the magnitude of the internal magnetic field pertaining to the $m_I = \pm\frac{3}{2} \rightarrow \pm\frac{1}{2}$ transition of the Kramers' doublet $S_z = \pm\frac{5}{2}$, and also influences the relaxation time. The method provides information about the number of solvated and complex species

[181] E. König, E. Lindner, I. P. Lorenz, and G. Ritter, *Inorg. Chim. Acta*, 1972, **6**, 123.
[182] E. König, G. Ritter, E. Lindner, and I. P. Lorenz, *Chem. Phys. Letters*, 1972, **13**, 70.
[183] E. Dittmar, C. J. Alexander, and M. L. Good, *J. Co-ordination Chem.*, 1972, **2**, 69.
[184] M. J. M. Campbell, *Chem. Phys. Letters*, 1972, **15**, 53.
[185] T. X. Carroll and M. Kaplan, *Phys. Letters (A)*, 1972, **41**, 145.

of iron present as well as on the relative amounts of these components.[186-188] Independent measurements have been carried out on frozen aqueous solutions of iron(III) perchlorate,[189] and the effects of weak magnetic fields on the paramagnetic hyperfine structure in the Mössbauer spectrum of this material have been studied.[190]

It was established some time ago that, on rewarming an aqueous solution of $FeCl_2$ which had been cooled rapidly to 77 K, the resonance disappeared abruptly at a temperature of 180—200 K but reappeared some time later if the solution was kept at the same temperature. A possible explanation of this behaviour is that the glass formed during the initial cooling transforms at 180—200 K to a supercooled liquid, which then slowly crystallizes. Experiments on frozen aqueous solutions of $FeCl_3$ have now established that this explanation is in fact correct. It was argued that slow crystallization would be accompanied by the formation of an iron-enriched phase, the iron concentration of which could be observed by means of the effect of changes in the spin–spin relaxation on the magnetic hyperfine splitting in the Mössbauer spectrum. An iron-enriched phase would be expected to show fast spin–spin relaxation and therefore no magnetic hyperfine splitting. Accordingly, magnetic lines, which were present in the 90 K spectrum of rapidly cooled aqueous $FeCl_3$ solutions, were no longer present in spectra obtained at 90 K after the solution had been allowed to warm up to 200 K. As a further check, the process was repeated with Fe^{3+}-exchanged resin and anion-exchange resin swelled in $FeCl_3$ solution. As expected, the magnetic hyperfine structure did not disappear in these experiments because aggregation of the Fe^{3+} ions was precluded by the action of the exchange resin.[191]

Experiments on a single crystal of $FeBO_3$ have shown that it is possible to recreate antiferromagnetism in this material above its Néel temperature ($T_N = 348.35$ K) by the application of an external magnetic field. This effect can be observed up to temperatures of more than 15 K above T_N for an applied field of 29.3 kG.[192]

Neutron diffraction and Mössbauer studies have been reported for the planar antiferromagnet $KFeF_4$ in the critical region. There is a discontinuity in the isomer shift, but not in the quadrupole splitting, at the Néel point. In the temperature range $0.96 < T/T_N < 0.998$ the internal magnetic field follows the power law $H_{int}(T) = H(0)D(1 - T/T_N)^\beta$, where $H(0) = H(4.2 \text{ K}) = 540 \pm 4 \text{ kG}$, $T_N = 141.51 \pm 0.05$ K, $D = 1.12 \pm$

---

[186] A. Vértes and F. Parak, *J.C.S. Dalton*, 1972, 2062.
[187] A. Vértes and F. Parak, *Acta Chim. Acad. Sci. Hung.*, 1972, **74**, 293.
[188] A. S. Plachinda, M. Ranogajec-Komor, and A. Vértes, *J. Radioanalyt. Chem.*, 1972, **10**, 89.
[189] T. Ohya and K. Ono, *J. Chem. Phys.*, 1972, **57**, 3240.
[190] A. M. Afanas'ev, V. D. Gorobchenko, I. Dézsi, I. I. Lukashevich, and N. I. Filippov, *Zhur. eksp. i teor. Fiz.*, 1972, **62**, 673.
[191] A. S. Plachinda and E. F. Makarov, *Chem. Phys. Letters*, 1972, **15**, 627.
[192] S. S. Yakimov, V. I. Ozhogin, V. Ya. Gamlitskii, V. M. Cherepanov, and S. D. Pudkov, *Phys. Letters (A)*, 1972, **39**, 421.

0.03, and $\beta = 0.209 \pm 0.008$. The main axis of the electric-field gradient tensor is canted at an angle $\theta = 11 \pm 2°$ with respect to the hyperfine field direction.[193] Mössbauer and magnetic susceptibility data for the new planar fluoride CsFeF$_4$ show that long-range order in three dimensions exists only at temperatures below $T_N = 160 \pm 2$ K. Good fits to the 4.2 K spectrum were obtained with $H_{int} = 540 \pm 3$ kG, $\frac{1}{2}e^2qQ(1 + \frac{1}{3}\eta^2)^{\frac{1}{2}} = -1.54 \pm 0.05$ mm s^{-1}, and $\eta = 0.3 \pm 0.3$. The polar angle of H_{int} in

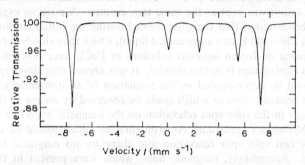

Figure 9 *Mössbauer spectrum of* RbFeF$_4$ *at 80 K. The statistical spread of the data along the vertical axis is comparable with the width of the solid line; the individual points shown in the original spectrum have been omitted for clarity*
(Reproduced by permission from *Phys. Letters*, 1972, **42A**, 199)

the electric-field gradient principal axis system is $\theta = 5 \pm 4°$.[194, 195] Similar anisotropy in the recoil-free fraction to that described for FeCO$_3$ on p. 517 has also been observed for the magnetic layer structure RbFeF$_4$. As can be seen in Figure 9, the ratio of the areas of lines 1 and 2 is much greater than the expected value $\frac{3}{2}$. The observed value of 2.28 ± 0.05 yields a value of $f_{\parallel}/f_{\perp} = 6.5$ in this case. The spectrum in Figure 9 can be fitted with the following parameters: $H(0) = 457 \pm 2$ kG, $\frac{1}{2}e^2qQ = 1.71 \pm 0.02$ mm s^{-1}, and $\theta = 16.7 \pm 0.4°$.[168]

Spectra have been obtained for polycrystalline $(NH_4)_3FeF_6$ in the temperature range 78—348 K and indicate that the tetragonal to cubic phase transition occurs at 263 K, with a hysteresis of 0.5 K.[196] Cs$_3$FeCl$_6$ has been shown to exist in the solid state in two forms with different isomer shifts and quadrupole splittings; one is orange and the other is yellow and metastable.[197]

Magnetically perturbed Mössbauer spectra have been used to establish that five-co-ordinate monomeric products are formed when the Schiff base

---

[193] G. Heger and R. Geller, *Phys. Status Solidi (B)*, 1972, **53**, 227.
[194] M. Eibschütz, H. J. Guggenheim, L. Holmes, and J. L. Bernstein, *Solid State Comm.*, 1972, **11**, 457.
[195] M. Eibschütz, G. R. Davidson, H. J. Guggenheim, and D. E. Cox, *Amer. Inst. Phys. Conf. Proc.*, 1972, No. 5 (Pt. 1), 670.
[196] S. Morup and N. Thrane, *Solid State Comm.*, 1972, **11**, 1319.
[197] E. Frank and D. St. P. Bunbury, *J. Inorg. Nuclear Chem.*, 1972, **34**, 535.

dimer $[Fe(salen)Cl]_2$ is crystallized rapidly from nitromethane and pyridine. The dimer has a negative principal component of the electric-field gradient tensor (V_{zz}), a non-zero asymmetry parameter, and an internal magnetic field equal to the applied field. By contrast, the products obtained by rapid crystallization show positive V_{zz}, an asymmetry parameter of zero which indicates axial symmetry, and large internal fields, in accord with their formulation as monomers. Furthermore, their perturbed spectra are broad, which confirms the presence of a high degree of magnetic anisotropy in these materials.[198]

The Mössbauer spectra of the compounds TPPFeX (TPP = $\alpha\beta\gamma\delta$-tetraphenylporphin; X = Cl, Br, I, or NCS) show a broad, asymmetric peak at 298 K and 78 K, which becomes a resolved doublet at 4.2 K. The high-temperature behaviour is interpreted in terms of slow relaxation within the $\pm\frac{1}{2}$, $\pm\frac{3}{2}$, and $\pm\frac{5}{2}$ Kramers' doublets, which arise from the $^6S_{\frac{5}{2}}$ ground state of the Fe^{3+} ion and are equally populated. At 4.2 K the low-lying $\pm\frac{1}{2}$ substate becomes preferentially occupied and because there is a slight splitting of the $+\frac{1}{2}$ and $-\frac{1}{2}$ levels, owing to magnetic dipoles in the crystal lattice, relaxation between these levels is so rapid compared with the Mössbauer lifetime of 10^{-7} s that there is no magnetic broadening and a symmetric, sharp Mössbauer doublet results. The probable ordering of the zero-field splitting parameter, D, deduced from the spectra was [TPPFeCl] < [TPPFeNCS] < [TPPFeBr] < [TPPFeI]. At 298 and 78 K the less intense, broadened peak, which corresponds to the $\pm\frac{1}{2} \rightarrow \pm\frac{3}{2}$ nuclear transition if the magnetic axis is parallel to the crystal-field axis, occurs at positive velocity, indicating that V_{zz} is positive. At 6 K there is a reversal in the asymmetry for the thiocyanate and bromide, which probably reflects a reorientation of the magnetic axis to a direction perpendicular to the axis of the electric-field gradient, rather than a change in sign of V_{zz}.[199] By contrast, the spectra of eight new μ-oxo-bis[$\alpha\beta\gamma\delta$-tetrakis(aryl, pyridyl, or thienyl)porphinatoiron(III)] complexes all show well-resolved, reasonably symmetric quadrupole doublets at the three temperatures 298, 78, and 4.2 K. Furthermore, there is little dependence of the quadrupole splitting on the meso substituents, in contrast to the haemin cases. The absence of magnetic broadening is due to strong spin–spin coupling between the two $S = \frac{5}{2}$ iron(III) ions via the oxygen bridge, which leads to rapid relaxation.[200] The spectra of the compounds $\alpha\beta\gamma\delta$-tetrakis(p-chlorophenyl)porphinatoiron(III) chloride and iodide, $\alpha\beta\gamma\delta$-tetrakis(p-methoxyphenyl)porphinatoiron(III) chloride, bromide, iodide, azide, thiocyanate, acetate, and trifluoroacetate, and $\alpha\beta\gamma\delta$-tetrakis(pentafluorophenyl)porphinatoiron(III) chloride and bromide show a broad asymmetric peak with a shoulder on the high-energy side at 298 and 78 K; at 4.2 K

---

[198] W. M. Reiff, *Inorg. Chim. Acta*, 1972, **6**, 267.
[199] C. Maricondi, D. K. Straub, and L. M. Epstein, *J. Amer. Chem. Soc.*, 1972, **94**, 4157
[200] M. A. Torréns, D. K. Straub, and L. M. Epstein, *J. Amer. Chem. Soc.*, 1972, **94** 4160.

this asymmetric broadening reverses for several of the haemins, particularly the p-methoxyphenyl derivatives.[201]

Spectra have been obtained at 295 and 77 K for $enH_2[(FeHedta)_2O]$,$6H_2O$, $Na_4[\{Fe(edta)\}_2O], 12H_2O$, $NaFe(edta), 3H_2O$, and $Fe(Hedta), 1.5H_2O$ and compared with established data on two other pairs of oxo-bridged iron(III) dimers and corresponding high-spin monomers $[\{Fe(terpy)\}_2O](NO_3)_4$,$H_2O$, $Fe(terpy)Cl_3$, $[Fe(salen)]_2O,2py$, and $Fe(salen)Cl,(MeNO_2)_x$. However, no definite conclusions were drawn regarding the spin-state of the iron(III) ions in these molecules. The isomer shifts are all similar (ca. 0.78 mm s^{-1} at 77 K) and although the quadrupole splittings show differences they do not follow a set pattern.[202]

A Mössbauer study of the distorted octahedral iron complexes of glycine, leucine, lysine, tryptophan, and methionine at 82 and 298 K has shown that the high-spin Fe^{3+} ion is bound to the carboxylic and amino-acid groups and not to the side-chains of the amino-acids.[203]

The intramolecular antiferromagnet $[Fe\{NN'$-ethylenebis(salicylaldimato)$\}Cl]_2$, which contains two coupled Fe^{3+} ions bridged by oxygen, has been studied in the temperature range 4.2—22 K and in applied magnetic fields of 0—80 kG. At 4.2 K only the ground state, with effective spin $S = 0$, is populated appreciably. An analysis of the magnetic hyperfine field and its temperature dependence yields an exchange constant $J = -6.7$ cm^{-1} and a hyperfine field of -192 kG per unit spin at each iron nucleus.[204]

Comparison of the chemical isomer shifts for $FeX_2(ox)$ and $FeX(ox)_2$ (X = Cl or Br; oxH = quinolin-8-ol) with those of related species has suggested that the compounds contain five- and six-co-ordinate iron, respectively, and are therefore probably dimeric. The two most likely structures for dimeric $FeX(ox)_2$ and $FeX_2(ox)$ involve either halogen bridges [(1) and (2)] or oxygen bridges [(3) and (4)]. Further evidence in

[201] M. A. Torréns, D. K. Straub, and L. M. Epstein, *J. Amer. Chem. Soc.*, 1972, **94**, 4162.
[202] H. J. Schugar, G. R. Rossman, C. G. Barraclough, and H. B. Gray, *J. Amer. Chem. Soc.*, 1972, **94**, 2683.
[203] R. Raudsepp and I. Arro, *Eesti N.S.V. Tead. Akad. Toim., Fuus., Mat.*, 1972, **21**, 187.
[204] R. Lechan, C. R. Abeledo, and R. B. Frankel, *Amer. Inst. Phys. Conf. Proc.*, 1972, No. 5 (Pt. 1), 659.

favour of these suggestions comes from the sign of V_{zz}, which is positive for FeX$_2$(ox) and negative for FeX(ox)$_2$; the five-co-ordinate complexes FeCl(acac)$_2$ and Fe(salen)$_2$O are known to have a positive V_{zz}, whereas the six-co-ordinate [Fe(salen)Cl] has a negative V_{zz}. The complexes show relaxation effects due to the population of states other than the non-magnetic $S = 0$ ground state at temperatures above 4.2 K, and a

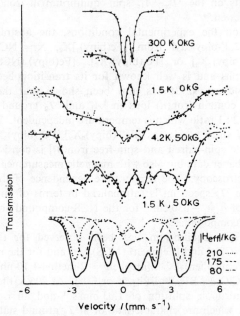

Figure 10 *Experimental and calculated Mössbauer spectra of the compound* [Fe$_3$(MeCO$_2$)$_6$(H$_2$O)$_3$]Cl,6H$_2$O
(Reproduced by permission from *J. Phys. Soc. Japan*, 1972, 33, 1312)

Goldanskii–Karyagin effect may be present in the spectra of FeCl(ox)$_2$.[205]

The magnetism of the trinuclear complex salt [Fe$_3$(MeCO$_2$)$_6$(H$_2$O)$_3$]-Cl,6H$_2$O, and the isomorphous chromium salt with ^{57}Fe^{3+} substituting for Cr^{3+} ions up to 2%, has been studied. In the absence of an external field, both materials show only quadrupole-split spectra (Figure 10). When an external field of 50 kG is applied at 1.5 K the spectrum of the iron salt consists of three sets of magnetically split patterns with effective fields of 210, 175, and 80 kG, but the (Cr$_2$^{57}Fe) clusters in the substituted material show only a single pattern with an effective field of 235 kG. These results were discussed in terms of the antiferromagnetic intracluster interactions, which cause the Fe^{3+} and Cr^{3+} ions to have small positive or negative spin

---

[205] D. Cunningham, M. J. Frazer, A. H. Qureshi, F. B. Taylor, and B. W. Dale, *J.C.S. Dalton*, 1972, 1090.

components along the large polarizing field. In an iron cluster the three interactions are not equivalent and in a (Cr_2Fe) cluster the Fe^{3+}–Cr^{3+} interactions are stronger than the Cr^{3+}–Cr^{3+} one.[206]

A number of other compounds containing high-spin iron(III) which have been studied are discussed elsewhere, see refs. 117, 173, and 176.

Spin-crossover Systems, Unusual Electronic States, and Biological Compounds. Results on the $^5T_{2g}$–$^1A_{1g}$ spin-equilibrium in iron(II) complexes have been reviewed.[38]

Depending on the experimental conditions, the abstraction of one molecule of 2,2'-bipyridyl from [Fe(bipy)$_3$]X$_2$, X = NCS or NCSe, produces [Fe(bipy)$_2$X$_2$] or [Fe$_3$(bipy)$_7$X$_6$]. [Fe(bipy)$_2$(NCS)$_2$] exists in three polymorphs and is well known for its transition between the $^5T_2$ and $^1A_1$ ground states. It has now been shown that the compounds [Fe$_3$(bipy)$_7$X$_6$] contain iron(II) ions in $^1A_1$ and $^5T_2$ ground states simultaneously in a 2 : 1 ratio which is temperature independent. The structural formulation of these species as [Fe^{sp}(bipy)$_2$X$_2$]$_2$[Fe^{sf}(bipy)$_2$X$_2$],bipy [Fe^{sp} and Fe^{sf} denote spin-paired and spin-free iron(II)] is consistent not only with the Mössbauer data but also with magnetic measurements and i.r. and electronic spectroscopy. The temperature dependence of the quadrupole splitting in the $^5T_2$ spectrum is rationalized in terms of an axial ligand-field splitting of $\delta = -630$ cm^{-1} for the NCS compound and -600 cm^{-1} for the NCSe compound.[207]

A $^5T_2$–$^1A_1$ spin equilibrium has been observed for the complexes [Fe(py)$_2$phen(NCS)$_2$] and [Fe(phen)$_2$(NCS)$_2$],[208] and for the novel series of complexes [Fe(mephen)$_3$]X$_2$ [mephen = 2-methyl-1,10-phenanthrolinoiron(II); X = ClO$_4$,[209] BF$_4$,[209] BPh$_4$,[210] or I [210]]. At 295 K the chlorate salt exhibits a quadrupole splitting of 1.03 mm s^{-1} and an isomer shift of $+0.95$ mm s^{-1}, which are characteristic of a $^5T_2$ ground state. At 244 K, line-broadening is observed as a result of slow electronic relaxation between the $^5T_2$ and $^1A_1$ states. At even lower temperature, e.g. 196 K, a second doublet appears with parameters ($\Delta = 0.57$, $\delta = +0.38$ mm s^{-1}), typical of the $^1A_1$ state. At 4.2 K, 72.7% of the resonance is due to the $^1A_1$ state. The other salts show essentially similar behaviour. From the temperature dependence of the quadrupole splitting for the $^5T_2$ state, the axial ligand-field splittings for the four salts were found to be $\delta = -640$, -580, -440, and -440 cm^{-1}, respectively. With the aid of magnetic susceptibility data, the $^5T_2$–$^1A_1$ energy separation ε was shown to vary with temperature and to have the values 300 and 990 cm^{-1} at 294 and 98 K,

[206] M. Takano, J. Phys. Soc. Japan, 1972, 33, 1312.
[207] E. König, G. Ritter, K. Madeja, and W. H. Böhmer, J. Phys. and Chem. Solids, 1972, 33, 327.
[208] P. Spacu, M. Teodorescu, G. Filotti, and P. Telnic, Z. anorg. Chem., 1972, 392, 88.
[209] E. König, G. Ritter, H. Spiering, S. Kremer, K. Madeja, and A. Rosenkranz, J. Chem. Phys., 1972, 56, 3139.
[210] E. König, G. Ritter, B. Braunecker, K. Madeja, H. A. Goodwin, and F. E. Smith, Ber. Bunsengesellshaft phys. Chem., 1972, 76, 400.

respectively. From measurements with an applied magnetic field of 50 kG, the sign of V_{zz} was shown to be positive for the $^1A_1$ component. The signal for the $^5T_2$ component was smeared over a wide velocity range and did not interfere with the $^1A_1$ component. None of the complexes [Fe(mephen)$_3$]X$_2$ (X = CN, Cl, Br, NCS, N$_3$, or malonato) [211] exhibits a spin equilibrium; the cyanide is exclusively in the low-spin $^1A_1$ state, whereas the others adopt the high-spin $^5T_2$ configuration. Measurements of the magnetic susceptibility and Mössbauer spectrum of [Fe(phen)$_2$ox],- 5H$_2$O have now been extended down to 1.2 and 4.2 K, respectively, and are consistent with a $^3A_2$ ground state, characterized by a zero-field splitting of $D = 4.6$ cm^{-1} into an upper $M_S = |\pm 1\rangle$ and a lower $|0\rangle$ level. The high value of $g = 2.80$ indicates significant mixing-in of quintet levels of higher energies, and this is apparently the source of the considerable increase in the high-temperature moment of about 3.96 BM. A complex Mössbauer pattern is obtained on application of an applied magnetic field of 20 and 40 kG, indicating that in at least a fraction of the iron atoms an internal magnetic field is generated.[212]

Spectra have been obtained for six different iron(III) dithiocarbamates of general formula Fe(S$_2$CNR$_2$)$_3$ between 100 and 300 K. In each case only a single quadrupole doublet was observed at all temperatures, and the temperature dependence of the quadrupole splitting of this doublet was considered to be inconsistent with the presence of a $^2T_{2g}$–$^6A_{1g}$ equilibrium.[213] In a closely related study a linear dependence was found between the isomer shift and magnetic moment for a series of 19 compounds of the same general type. To explain this dependence it was suggested that the $^6A_{1g}$ and $^2T_{2g}$ states are both populated to varying degrees from compound to compound and that there is fast relaxation between these two states such that the resulting spectra reflect a weighted average of the two electronic configurations.[214] A study on the chelate bis-(NN-diethyldithiocarbamato)- iron(III) chloride is discussed on p. 539 and some iron(IV) dithiocarbamates are mentioned later in this section.

Full details have now appeared of the $^6A_1$–$^2T_2$ spin equilibrium in four tris(monothio-β-diketonato)iron(III) complexes (5a—d), preliminary studies of which were communicated in 1970. In favourable cases, superimposed spectra are observed in which both spin states can be recognized. Because the total degeneracies of the $^6A_1$ and $^2T_2$ states are the same, a 1 : 1 mixture is expected to be present at sufficiently high temperatures; however, the thermal equilibrium was found to be far from ideal in that the theoretical maximum ratio of isomers is exceeded in some cases. It was suggested that the failure to detect individual spin states in the dithiocarbamates discussed

---

[211] E. König, G. Ritter, K. Madeja, and A. Rosenkranz, *J. Inorg. Nuclear Chem.*, 1972, **34**, 2877.
[212] E. König and B. Kanellakopulos, *Chem. Phys. Letters*, 1972, **12**, 485.
[213] P. B. Merrithew and P. G. Rasmussen, *Inorg. Chem.*, 1972, **11**, 325.
[214] R. R. Eley, N. V. Duffy, and D. L. Uhrich, *J. Inorg. Nuclear Chem.*, 1972, **34**, 3681.

$$\begin{bmatrix} \text{H}-\text{C} \begin{array}{c} \text{R}^1 \\ | \\ \text{C}=\text{S} \\ \diagdown \\ \diagup \\ \text{C}=\text{O} \\ | \\ \text{R}^2 \end{array} \text{Fe} \end{bmatrix}_3$$

(5) a; $R^1 = R^2 = $ Ph
 b; $R^1 = R^2 = $ Me
 c; $R^1 = $ Me, $R^2 = $ Ph
 d; $R^1 = $ Ph, $R^2 = $ Me

earlier may arise simply from an unfavourable similarity in the quadrupole splittings of the two states, rather than from relaxation effects.[215]

The $^6A_1-{}^2T_2$ spin equilibrium has been confirmed in Na[Fe(thsa)$_2$],3H$_2$O and NH$_4$[Fe(sesa)$_2$] (H$_2$thsa = thiosemicarbazone of salicylaldehyde; H$_2$sesa = selenosemicarbazone of salicylaldehyde), whereas the compounds Li[Fe(thsa)$_2$],2H$_2$O, NH$_4$[Fe(sespu)$_2$], NH$_4$[Fe(thpu)$_2$], NH$_4$-[Fe(phthsa)$_2$],0.5H$_2$O, and Li[Fe(thpu)$_2$],3H$_2$O (H$_2$sespu = selenosemicarbazone of pyruvic acid; H$_2$thpu = thiosemicarbazone of pyruvic acid,

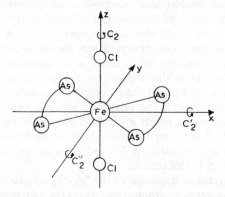

Figure 11 *A simplified diagram of the local environment for the iron ion showing the x- and y-axes as chosen to bisect the Fe—As axes. With these axes, the t_{2g} basis states are $|x^2 - y^2\rangle$, $|xz\rangle$, and $|yz\rangle$*
(Reproduced by permission from *J. Chem. Phys.*, 1972, **57**, 3709)

H$_2$phthsa = phenylthiosemicarbazone of salicylaldehyde) are exclusively low spin, and the compounds [FeCl(thdac)] and [FeBr(thdac)],H$_2$O [H$_2$thdac = H$_2$NC(S)NHN(CH$_2$CO$_2$H)$_2$] are both high spin.[216]

The new complexes [Fe(R$_2$dtc)$_3$]BF$_4$ (R = Me, Et, Pr^i, or cyclohexyl), [Fe(pyr)(dtc)$_3$]BF$_4$, and Fe(Pr^{i}dtc)$_2$BF$_3$ (dtc = dithiocarbamate) have all

---

[215] M. Cox, J. Darken, B. W. Fitzsimmons, A. W. Smith, L. F. Larkworthy, and K. A. Rogers, *J.C.S. Dalton*, 1972, 1192.
[216] V. V. Zelentsov, A. Ablov, K. I. Turta, R. A. Stukan, N. V. Gerbeleu, E. V. Ivanov, A. P. Bogdanov, N. A. Barba, and V. G. Bodyu, *Zhur. neorg. Khim.*, 1972, **17**, 1929.

been found to have isomer shifts which are smaller than expected for either iron(II) or iron(III) and are therefore thought to contain iron(IV). The nature of the last complex mentioned is uncertain; the BF_3 may be present as the ion F_3BO^{2-} derived from F_3BOH_2.[217]

The iron(IV) complex $[Fe(diars)_2X_2](BF_4)_2$ (X = Cl or Br) has been studied in detail in the presence of magnetic fields up to 53 kG and at temperatures ranging from 4.2 to 348 K. The results were interpreted in terms of spin Hamiltonian parameters with an effective spin $S = 1$ and were consistent with a $^3A_2$ ground state, corresponding to a strong compression, or strengthening of the ligand field, along the Cl—Fe—Cl axis. A simplified diagram of the local environment for the iron ion is shown in Figure 11. The strong axial distortion can be understood in terms of the bonding scheme shown in Figure 12. A value for the covalency parameter of $N^2 \sim 0.90$ was estimated from the quadrupole splitting and magnetic hyperfine interaction, indicating that there is very little π-back-bonding to the empty d-orbitals of the diarsine ligands. Because of this the in-plane $d_{x^2-y^2}$ orbital remains essentially non-bonding. However, the high formal charge on the metal is reduced by forward σ-donation from the filled π-orbitals on the chlorine ligands, with a concomitant increase in the energy of the d_{xz} and d_{yz} orbitals relative to $d_{x^2-y^2}$. As indicated in Figure 12, the ligand-field splitting was found to be $\Delta \sim 2800$ cm^{-1}. Spin–orbit coupling in this complex leads to a zero-field splitting of $D = 23$ cm^{-1}, with a one-electron spin–orbit coupling parameter of $\xi = 467$ cm^{-1}.[218]

Paramagnetic hyperfine structure has been observed in the 4.2 K Mössbauer spectra of FeO_4^{2-} diluted in concentrated NaOH solutions in the presence of applied magnetic fields ranging from 0 to 2 kG. The spectra, which are shown in Figure 13, can be interpreted as the superposition of components from each of the three members of a spin triplet. Because of the dilution and low temperatures, the relaxation time among the electronic states is long enough for the ^{57}Fe nucleus to couple to the magnetic moment which is induced in each state by the applied magnetic field. It should be remembered that in the present case the electronic states themselves have no moments in the absence of an external field. This is in contrast to the half-integral spin systems which have Kramers' degeneracy, where each member of a Kramers' doublet does have a non-zero magnetic moment which can couple to the nuclear magnetic moment if the relaxation rate is slow enough. The data yield an isotropic hyperfine interaction $A = -1.17$ mm s^{-1} in the nuclear excited state, $g = 2.0$ (isotropic) and zero electric-field gradient, all of which are consistent with an e_g^2 configuration for the Fe^{VI} ion. The saturation value of the hyperfine field is somewhat smaller than expected (175 kG).[219]

[217] E. A. Pasek and D. K. Straub, *Inorg. Chem.*, 1972, **11**, 259.
[218] E. A. Paez, D. L. Weaver, and W. T. Oosterhuis, *J. Chem. Phys.*, 1972, **57**, 3709.
[219] W. T. Oosterhuis and F. de S. Barros, *J. Chem. Phys.*, 1972, **57**, 4304.

The fluoride, chloride, azide, imidazole, cyanate, and methane-thiol derivatives of iron(III) haemoglobin have been studied in the temperature range 4—195 K. In favourable cases measurements at the higher temperatures distinguish unambiguously between the high- and low-spin forms of

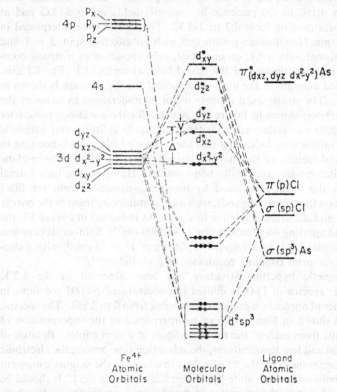

Figure 12 *A MO diagram for the complex* [Fe(diars)$_2$X$_2$](BF$_4$)$_2$ *representing the one-electron energies of the orbitals involved. There is only a small delocalization of the unpaired electrons according to the Mössbauer data.* Δ *is found to be* 2760 cm^{-1}, *and V is less than* 700 cm^{-1}
(Reproduced by permission from *J. Chem. Phys.*, 1972, **57**, 3709)

the compounds. Changes in isomer shift from compound to compound and from spin-state to spin-state are relatively small, but variations in the quadrupole splitting are more appreciable and depend greatly on the nature of the axial ligand. Measurement in an applied field at 4.2 K provides a far more direct method of spin-state determination.[220] An interaction between the electronic magnetic moment of the iron atom and the nuclear

---

[220] M. R. C. Winter, C. E. Johnson, G. Lang, and R. J. P. Williams, *Biochim. Biophys. Acta*, 1972, **263**, 515.

magnetic moments of the ligands has been clearly observed in the paramagnetic hyperfine structure of the zero-field Mössbauer spectrum of acid metmyoglobin, pH 6. Calculated spectra give good agreement with experiment if the nuclear moments are approximated by an effective field of

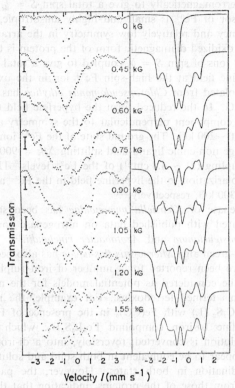

Figure 13 *Mössbauer spectra of FeO_4^{2-} diluted in concentrated NaOH at 4.2 K in a variable magnetic field applied parallel to the gamma beam. The vertical bars at the left indicate 0.5% of the background and the bars at the top indicate the line positions of an Fe impurity which could not be eliminated. The solid curves at the right are calculated from the parameters $D = 0.11$ cm^{-1}, $E = 0.02$ cm^{-1}, $A = -1.17$ mm s^{-1}, $P = 0$, $g = 2.00$, and the value of the applied field is indicated for each case*
(Reproduced by permission from *J. Chem. Phys.*, 1972, **57**, 4304)

10 Gauss.[221] An emission experiment with an oxygenated haeme complex is mentioned on p. 515.

Iron chelated by deferoxamine has been studied by Mössbauer spectroscopy.[222]

[221] W. T. Oosterhuis and P. J. Viccaro, *Biochim. Biophys. Acta*, 1972, **264**, 11.
[222] J. L. Bock and G. Lang, *Biochim. Biophys. Acta*, 1972, **264**, 245.

The iron atoms in the reduced paramagnetic form of putidaredoxin, a 2Fe—S protein from *Pseudomonas putida*, have been shown to give spectra typical of high-spin Fe^{3+} ($S = \frac{5}{2}$) and Fe^{2+} ($S = 2$), respectively. From applied field measurements it was shown conclusively that the spins are coupled antiferromagnetically to give a total spin $S = \frac{1}{2}$. The magnetic hyperfine tensor of Fe^{3+} is small and highly anisotropic, which implies strong covalency and relatively low symmetry in the arrangement of the ligands. The oxidized diamagnetic form of the protein is thought to contain two Fe^{3+} ions of spin $S = \frac{5}{2}$, coupled to give a total spin of zero.[223]

The hyperfine field at the high-spin Fe^{3+} ion in the oxidized form of rubredoxin isolated from *Chloropseudomonas ethylica* has been found to be 370 ± 3 kG. In the reduced form the hyperfine field tensor is anisotropic with a component perpendicular to the symmetry axis of the iron atom of about −200 kG. The ground state of the Fe^{2+} ion is a d_{z^2} orbital. There is a large non-cubic ligand-field splitting $\Delta/k = 900$ K and a small spin-orbit coupling ($D \sim 4.4$ cm^{-1}) of the Fe^{2+} levels. The contribution of the core polarization to the hyperfine field in the Fe^{3+} and Fe^{2+} ions is −370 and −300 kG, respectively.[224]

The nitrogenase of *Klebsiella pneumoniae* has been studied and the results compared with published data on nitrogenase components from *Clostridium pasteurianum* and *Azotobacter vinelandii*.[225] The ferredoxin from the blue-green alga *Microcystis flos-aquae* has also been studied.[226]

Spectra have been reported for a number of iron–sulphur compounds which might be considered as potential models for the non-haeme iron protein systems termed ferredoxins. For example, the reaction of the ligand $(CF_3)_2C_2S_2$ (L) with $Fe(CO)_5$ in the presence of H_2S produces a black crystalline tri-iron compound $Fe_3L_4S_6H_2$, which in disulphide-containing solution is converted, reversibly, into a di-iron species. The iron atoms appear to be equivalent in both solid and solution, with octahedral co-ordination in both states. However, the parameters differ significantly from those of the proteins, indicating that the value of the compound as a model is limited.[227] The behaviour of the isomer shift for the series of compounds $[Fe_4S_4L_4]^{n-}[AsPh_4]_n^+$ [$L = S_2C_2(CF_3)_2$] is similar to that exhibited by the HPI chromatium. In both systems a small increase in isomer shift accompanies the reduction, consistent with a slight decrease in the s-electron density at the iron nuclei, which remain equivalent.[228] Related results on some dinuclear bridged sulphido-derivatives of iron are discussed in the following section (see p. 540).

---

[223] E. Münck, P. G. Debrunner, J. C. Tsibris, and I. C. Gunsalus, *Biochemistry*, 1972, **11**, 855.
[224] K. K. Rao, M. C. W. Evans, R. Cammack, D. O. Hall, C. L. Thompson, P. J. Jackson, and C. E. Johnson, *Biochem. J.*, 1972, **129**, 1063.
[225] R. R. Eady, B. E. Smith, K. A. Cook, and J. R. Postgate, *Biochem. J.*, 1972, **128**, 655.
[226] K. K. Rao, R. V. Smith, R. Cammack, M. C. W. Evans, D. O. Hall, and C. E. Johnson, *Biochem. J.*, 1972, **129**, 1159.
[227] K. A. Rubinson and G. Palmer, *J. Amer. Chem. Soc.*, 1972, **94**, 8375.
[228] I. Bernal, B. R. Davis, M. L. Good, and S. Chandra, *J. Coord. Chem.*, 1972, **2**, 61.

Low-spin and Covalent Complexes. The quadrupole splitting of $Li_3Fe(CN)_6$,$4H_2O$ decreases from 0.80 to 0.25 mm s^{-1} after hydration, presumably because of a redistribution of the Li$^+$ cations around the ferricyanide ion. The spectrum of a frozen aqueous solution of $Na_3Fe(CN)_6$,H_2O shows only a broad singlet, whereas the solid gives a quadrupole splitting of 0.96 mm s^{-1}. The zero quadrupole splitting is attributed to a cancellation of the electric-field gradient arising from the low-spin iron(III) ions by that arising from the lattice.[229] The effect on the Mössbauer parameters of varying the cation in a series of iron cyano complex anions had been studied.[230] Spectra have been obtained for a number of Prussian Blues produced from the thermal decomposition of $H_4Fe(CN)_6$ and $H_3Fe(CN)_6$.[231] Three iron species have been detected in the decomposition products of benzene and anisole solutions of $[Bu_3PCH_2Ph]^+[Fe^{II}(CN)_3]^-$.[232]

It has been shown that hydrated sodium and lithium ferricyanides are reduced by electron irradiation to the corresponding ferrocyanide, but the anhydrous complexes are unaffected. The reduction is therefore attributed to hydrogen radicals produced from the radiolysis of water. Silver ferricyanide shows different behaviour and it is thought that a species of the type $Ag_3Fe^{III}(CN)_5(NC)$, with an iron–isocyanide bond, is produced.[233] Proton irradiation effects in potassium iron cyanides have also been studied. $K_4[Fe(CN)_6]$,$3H_2O$ gives metallic iron, which is in the superparamagnetic state for low beam currents (0.05—0.30 μA) and in the ferromagnetic state for higher beam currents (0.6 μA). It is suggested that a reaction of the type

$$K_4[Fe(CN)_6] \longrightarrow Fe + 4KCN + (CN)_2$$

may occur. The disintegration of $K_3[Fe(CN)_6]$ is more complex. Fe_3C is detected in the Mössbauer spectrum of the products and is believed to be formed in the initial conversion of $K_3[Fe(CN)_6]$ into $K_4[Fe(CN)_6]$, which then decomposes as already described.[234]

Data have been given for addition compounds of $[Fe(CN)_5NO]^{2-}$ with $CS(NH_2)_2$, $MeCSNH_2$, $PhCOMe$, S^{2-}, and SO_3^{2-} and for derivatives in which the nitrosyl group of this anion is substituted by Me_2SO, morpholine, $PhCH_2NH_2$, $HOCH_2CH_2NH_2$, p- and m-$H_2NC_6H_4OMe$, pyrollidine, p-$(H_2N)_2C_6H_4$, 2,4-$(H_2N)_2C_6H_3OMe$, p-$(PhNH)C_6H_4OH$, and pyridine oxide.[235]

The predicted correlation between the ^{57}Fe quadrupole splitting and the $^1T_1$ splitting in the electronic spectra for low-spin iron(II) compounds

---

[229] P. H. Domingues and J. Danon, *Chem. Phys. Letters*, 1972, **13**, 365.
[230] L. Korecz, P. Mag, S. Papp, B. Mohai, and K. Burger, *Magyar Kém. Folyóirat*, 1972, **78**, 479.
[231] J. C. Fanning, C. D. Elrod, B. S. Franke, and J. D. Melnik, *J. Inorg. Nuclear Chem.*, 1972, **34**, 139.
[232] S. Papp and A. Vértes, *Radiochem. Radioanalyt. Letters*, 1972, **9**, 231.
[233] E. Baggio Saitovitch, D. Raj, and J. Danon, *Chem. Phys. Letters*, 1972, **17**, 74.
[234] K. Kisyńska, M. Kopcewicz, and A. Kotlicki, *Phys. Status Solidi (B)*, 1972, **49**, 85.
[235] U. Weihofen, *Z. Naturforsch.*, 1972, **27a**, 565.

has now been observed for the series of compounds cis- and trans-[FeX$_2$(ArNC)$_4$] (X = Cl or SnCl$_3$; ArNC = p-methoxyphenyl isocyanide), cis-[Fe(SnCl$_3$)$_2$(ArNC)$_4$], and [Fe(ArNC)$_5$SnCl$_3$]ClO$_4$. The ^{119}Sn spectra for the compounds containing the SnCl$_3$ ligand are discussed on p. 574.[236]

The sign of V_{zz} has been shown to be negative (Δ = -1.14 mm s^{-1}) in trans-[FeH(L)(depe)$_2$]$^+$BPh$_4^-$ (L = p-MeOC$_6$H$_4$NC; depe = 1,2-bisdiethylphosphinoethane) and on the basis of partial quadrupole splitting considerations the sign of V_{zz} is thought to be negative also in the compounds trans-[FeH(L)(depe)$_2$]$^+$BPh$_4^-$ [L = CO, Me$_3$CNC, P(OPh)$_3$, P(OMe)$_3$, N$_2$, PhCN, or MeCN]. The very small quadrupole splitting for the dinitrogen complex (Δ = 0.33 mm s^{-1}) indicates that N$_2$ is the best ($\pi - \sigma$) ligand, but its centre shift indicates that N$_2$ is one of the poorest ($\sigma + \pi$) ligands. The instability of dinitrogen compounds is therefore associated with the weak σ-donor properties of the N$_2$ ligand.[237]

The hexanitroferrates K$_2$MFe(NO$_2$)$_6$ (M = Ca, Sr, Ba, or Pb) all give well-resolved quadrupole split spectra (Δ = 0.62—0.68 mm s^{-1}) despite their cubic symmetry and the fact that the iron atom is in the low-spin d^6 configuration. The presence of an electric-field gradient at the iron atom is thought to reflect a lower symmetry arising out of the hindered rotations of the nitro-groups. The large isomer shifts of these compounds indicate that π-back-donation between the metal atom and the nitro-group is very weak.[238]

[Fe(bipy)$_2$(NO$_3$)$_2$](ClO$_4$), isolated from blue nitric acid solutions of [Fe(bipy)$_3$](ClO$_4$)$_2$, has been shown to contain low-spin iron(III). It was thought previously that iron(II) species were responsible for the blue colour.[239] The complexes trans-[FeX$_2$(das)$_2$](ClO$_4$) (X = Cl or Br) and trans-[FeBr(NO)(das)$_2$](ClO$_4$) [das = o-phenylene bis(dimethylarsine)] are also thought to contain iron(III), which implies that the nitrosyl group is negatively charged in the latter. Magnetic susceptibility data indicate that there is a tetragonal splitting of the $^2T_{2g}$ ground state in the dihalides and a rhombic splitting in the nitrosyl complex. The quadrupole splittings reflect these structural differences, the values for the dihalides (Δ = 2.2—2.5 mm s^{-1}) being more than double that for the nitrosyl complex (Δ = 1.0 mm s^{-1}).[240]

The diarsine complexes [Fe^{II}L$_2$I$_2$], [Fe^{III}L$_2$Cl$_2$]$^+$, and [Fe^{IV}L$_2$Cl$_2$]$^{2+}$ [L = o-(Me$_2$As)$_2$C$_6$H$_4$] have all been shown to contain iron in low-spin configurations. The effects of electron capture decay in the corresponding cobalt complexes are discussed on p. 515.[241]

[236] G. M. Bancroft and K. D. Butler, *J.C.S. Dalton*, 1972, 1209.
[237] G. M. Bancroft, R. E. B. Garrod, A. G. Maddock, M. J. Mays, and B. E. Prater, *J. Amer. Chem. Soc.*, 1972, **94**, 647.
[238] P. S. Manoharan, S. S. Kaliraman, V. G. Jadhao, and R. M. Singru, *Chem. Phys. Letters*, 1972, **13**, 585.
[239] G. B. Briscoe, M. E. Fernandopulle, and W. R. McWhinnie, *Inorg. Chim. Acta*, 1972, **6**, 598.
[240] R. D. Feltham, W. Silverthorn, H. Wickman, and W. Wesolowski, *Inorg. Chem.*, 1972, **11**, 676.
[241] A. Cruset and J. M. Friedt, *Radiochem. Radioanalyt. Letters*, 1972, **10**, 345.

Experiments with applied magnetic fields have shown that the large temperature-independent quadrupole splitting ($\Delta = 2$ mm s^{-1}) of low-spin tris(dithioacetylacetonato)iron(III) is negative in sign and therefore corresponds to a d_{xy} hole and a ^2A ground term. These results support a recent study of the temperature dependence of the quadrupole splitting and suggest that the origin of the quadrupole splitting is primarily in the distribution of the non-bonding electrons.[242]

A single-crystal study has been carried out on the chelate bis-(NN-diethyldithiocarbamate)iron(III) chloride. This compound is of interest because it is known to exhibit ferromagnetism below 2.43 K, even though the lattice is composed of discrete molecules each containing only one iron atom. The magnetic exchange must be primarily intermolecular in origin because the iron–iron separations (*ca.* 700 pm) rule out significant dipole–dipole contributions. The iron atoms are situated roughly at the centroid of a rectangular pyramid, formed by a base of four sulphur atoms and an apical chlorine atom. The data showed that V_{zz} is parallel to the Fe—Cl direction and that the magnetic axis is parallel to the major axis of the base of the rectangular pyramid. Data taken at 4.2 K showed that the exchange fields are greatest in a direction perpendicular to the Fe—Cl axis. The path of intermolecular exchange is therefore of the form Fe—S$\cdots$S—Fe.[243] N.m.r., magnetic susceptibility, and Mössbauer data have been discussed for some manganic and ferric dithiocarbamates:[244] spin-crossover situations in compounds of this type were discussed in the previous section.

Data have been given for the following mono-iron carbonyl derivatives: Ph$_3$MFe(CO)$_4$ and (Ph$_3$M)$_2$Fe(CO)$_3$ (M = P, As, or Sb);[245] (C$_4$H$_6$)Fe(CO), (C$_4$H$_6$)(C$_8$H$_8$)Fe(CO), and (C$_8$H$_8$)$_2$Fe;[246] *cis*-(OC)$_4$XFeSnX$_3$ (X = Cl, Br, or I) and *cis*- and *trans*-(OC)$_4$Fe(SnX$_3$)$_2$ (X = Cl or Br).[247] The quadrupole splittings of the latter were examined in terms of the partial quadrupole splittings for low-spin iron(II) compounds reported by Bancroft.

The electric-charge distribution in the tetrahedral iron compounds FeL$_2$(NO)$_2$ (L$_2$ = 2CO, 2,2′-bipyridyl, *o*-phenanthroline, or di-2-pyridyl ketone) has been examined by Mössbauer, e.s.r., i.r., and n.m.r. spectroscopic methods. Large-scale one-electron reductions were performed on the L$_2$ = 2,2′-bipyridyl and di-2-pyridyl ketone derivatives, and the spectra of the products were analysed in terms of overlapping contributions from the neutral material, the radical cation, and the radical anion. The isomer shift of the radical cation is more positive than that of the neutral parent,

---

[242] W. M. Reiff and D. Szymanski, *Chem. Phys. Letters*, 1972, **17**, 288.
[243] H. H. Wickmann, *J. Chem. Phys.*, 1972, **56**, 976.
[244] R. M. Golding, *Pure Appl. Chem.*, 1972, **32**, 123.
[245] L. H. Bowen, P. E. Garrou, and G. G. Long, *Inorg. Chem.*, 1972, **11**, 182.
[246] P. Mag, L. Korecz, A. Carbonaro, and K. Burger, *Radiochem. Radioanalyt. Letters*, 1972, **9**, 137.
[247] N. Dominelli, E. Wood, P. Vasudev, and C. H. W. Jones, *Inorg. Nuclear Chem. Letters*, 1972, **8**, 1077.

indicating that the electron is removed from a MO having a substantial contribution from a metal s-type orbital. One-electron reduction has the opposite effect. The e.s.r. data show that the unpaired electron introduced by reduction resides in a MO which is highly ligand π^* in character. Since a negative isomer shift signals an increase in the s-electron density at the iron nucleus, the Mössbauer results indicate a σ-transmission of electric charge from the ligand to the iron atom.[248] Similar results were obtained with $[Fe(CO)_3PMe_2]_2$, the radical anion and dianion having increasingly more negative isomer shifts than the neutral parent.[249]

By contrast, chemical oxidation of the bridged sulphido-derivatives cis-$[Fe(\pi-C_5H_5)(CO)SR]_2$ (R = Me, Et, or Pr^i) to the mono- and dicationic derivatives has been shown to result in a small decrease in chemical isomer shift at the equivalent iron atoms, consistent with removal of electrons from a delocalized MO having predominantly metal d-character.[250] Attention has been focused on these species because of the possible relevance of their chemistry to that of the iron–protein systems termed ferrodoxins. However, the Mössbauer results on the latter indicate that the redox processes occur preferentially at one of the iron centres.

Spectra have been reported for a number of novel fluorocarbon mixed ligand complexes and are consistent with the structures (6)—(9). Nonequivalent iron environments were detected in (7) and (9). Derivatives of

```
         CO                        CO       CO
         |   Ph2  Me2              |   Ph2  Me2  |
    OC—Fe—P    As            OC—Fe—P    As—Fe—CO
        /  \     \                /  \    /  \
       OC  OC    \              OC   CO  OC  CO
               (CF2)n/2            F2 F2
       (6) a; n = 4                  (7)
           b; n = 6
```

```
                                          CO
                                     OC\  |  /CO
    Me2   Ph2  |  Ph2   Me2              Fe
    As\   P—Fe—P    /As            CO   /   \
       \  /    \  /                     Me2
        OC      CO                OC—Fe——As    (CF2)n/2
       F2 F2   F2 F2                   |
              (8)                      CO
                                        \
                                         L
                                   (9) a; n = 4
                                       b; n = 6
```

stoicheiometry $(L^1-L^2)Fe(CO)_3$ and $(L^1-L^2)_2Fe_2(CO)_4$ were also studied but their structures were less precisely defined.[251]

---

[248] R. E. Dessy, J. C. Charkoudian, and A. L. Rheingold, *J. Amer. Chem. Soc.*, 1972, **94**, 738.
[249] R. E. Dessy, A. L. Rheingold, and G. D. Howard, *J. Amer. Chem. Soc.*, 1972, **94**, 746.
[250] J. A. de Beer, R. J. Haines, and R. Greatrex, *J.C.S. Chem. Comm.*, 1972, 1094.
[251] L. S. Chia, W. R. Cullen, and D. A. Harbourne, *Canad. J. Chem.*, 1972, **50**, 2182.

Reaction of 3,6-diphenylpyridazine with $Fe_2(CO)_9$ in benzene has been shown to give $C_{22}H_{12}N_2O_6Fe_2$ (10) in which the iron atoms are equivalent. Reaction of this material with maleic anhydride gives $C_{26}H_{14}N_2O_9Fe_2$,

$$Ph-\underset{N-N}{\diagup\!\!\!\diagdown}-Ph$$
$$(OC)_3Fe\text{------}Fe(CO)_3$$
(10)

which has two equivalent iron atoms and has been shown by X-ray diffraction to contain two $Fe(CO)_3$ groups linked by an Fe—Fe bond, a bridging maleic anhydride, and a bridging pyridazine ligand. Reaction with dimethylacetylenedicarboxylate and diethyl acetylenedicarboxylate gives 1 : 1 adducts which have non-equivalent iron atoms.[252]

The Mössbauer spectra of $(C_{23}H_{18}N_2)Fe_2(CO)_6$, the product of the reaction between $Fe_2(CO)_9$ and 3,5,7-triphenyl-4H-1,2-diazepine, and $(C_{23}H_{20}N_2)Fe_2(CO)_6$, the product of the reaction between $Fe_2(CO)_9$ and 3,5,7-triphenyl-4,5,6-trihydro-1,2-diazepine, indicate non-equivalent iron environments in both complexes. The n.m.r., Mössbauer, and mass spectra were discussed in the light of the X-ray-determined molecular structure.[253]

Preliminary Mössbauer data have been given for $Fe_3(CO)_8(Ph_2PC_2CF_3)_2$,-$0.5C_6H_6$, which was shown by X-ray diffraction to contain three non-equivalent iron environments. The spectrum consists of four lines, with intensities in the ratio 1 : 2 : 1 : 2, which could be fitted with six Lorentzians as required by the known structure. The arsenic analogue gives a similar spectrum.[254]

The correlation between the Mössbauer quadrupole splitting and linewidth of the Fe-$2p_{\frac{3}{2}}$ bands in the photoelectron spectra of ferrocene, 1,1'-dibenzoyl ferrocene, diferrocenylmethylium tetrafluoroborate, and ferrocinium tetrafluoroborate, which was first observed by Cowan et al. in 1971, has been confirmed. The correlation arises because the unpaired electrons, which are the cause of the electric-field gradient, also give rise to line-broadening of the ionization bands. The results show that the iron atoms in (11) have approximately the same charge and that this does not

$$\left[\underset{Fe}{\bigcirc}\text{-CH-}\underset{Fe}{\bigcirc} \right]^+ BF_4^-$$
(11)

---

[252] H. A. Patel, A. J. Carty, M. Mathew, and G. J. Palenik, *J.C.S. Chem. Comm.*, 1972, 810.
[253] D. P. Maddern, A. J. Carty, and T. Birchall, *Inorg. Chem.*, 1972, **11**, 1453.
[254] T. O'Connor, A. J. Carty, M. Mathew, and G. J. Palenik, *J. Organometallic Chem.*, 1972, **38**, C15.

differ appreciably from that in ferrocene. The charge on the iron atom is positive in all of the compounds studied.[255] Data have been given for the 2-substituted ferrocenyltin compound [Fe(π-C$_5$H$_5$){π-C$_5$H$_3$(CH$_2$Me)(SnBuR$_2$)}].[256]

Oxide and Chalcogenide Systems containing Iron.—*Binary Oxides.* A comprehensive Mössbauer investigation of the various forms of non-stoicheiometric Fe$_{1-x}$O, which exist at various temperatures and compositions, has been carried out.[257-259] The spectra of samples covering

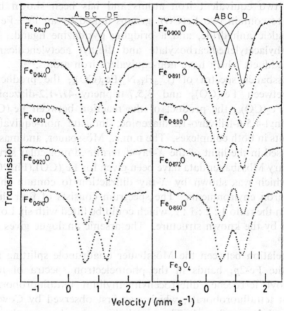

Figure 14 *Room-temperature Mössbauer spectra of Fe$_{1-x}$O quenched from 1520 K into water, showing the progressive change of the resonance profile as a function of composition. For interpretation of peaks A—E see text. Both velocity scales are with respect to Fe metal at 295 K*

the complete range of composition from Fe$_{0.95}$O to Fe$_{0.89}$O and quenched under a variety of conditions were analysed, and variations correlated with the known phase diagram of the system.[257] Typical room-temperature spectra of rapidly quenched samples (Figure 14) show an envelope of overlapping resonances which, for small values of x, can be resolved into five

[255] R. Gleiter, R. Seeger, H. Binder, E. Fluck, and M. Cais, *Angew. Chem.*, 1972, **11**, 1028.
[256] D. R. Morris and B. W. Rockett, *J. Organometallic Chem.*, 1972, **35**, 179.
[257] N. N. Greenwood and A. T. Howe, *J.C.S. Dalton*, 1972, 110.
[258] N. N. Greenwood and A. T. Howe, *J.C.S. Dalton*, 1972, 116.
[259] N. N. Greenwood and A. T. Howe, *J.C.S. Dalton*, 1972, 122.

broad Lorentzian peaks, assignable to a series of Fe^{2+} quadrupole doublets and an Fe^{3+} doublet at a lower velocity. Peak A and its counterpart under peak C are assigned to Fe^{3+} in octahedral oxygen co-ordination, with a small quadrupole splitting (0.40 mm s^{-1}) arising from neighbouring defects. This assignment has the attraction that the Fe^{2+} component of peaks B and C has the same area as the combined area of peaks D and E.

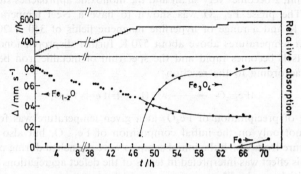

Figure 15 *Changes in the Mössbauer spectrum of $Fe_{0.940}O$ prior to and during decomposition. The increase in intensity of the Fe_3O_4 peaks (right-hand scale) is accompanied by a simultaneous decrease in the quadrupole splitting, Δ, of the residual $Fe_{1-x}O$ (left-hand scale) which finally decomposed to Fe_3O_4 and Fe. The temperatures at which the spectra were recorded are shown on the top scale. Spectra were accumulated for 0.25, 1, or 5 h*

On this basis *ca.* 13% of the total resonance area arises from Fe^{3+} and *ca.* 87% from Fe^{2+}, compared with the proportions of 11.2 and 88.8% calculated from the known composition of the sample. As the phase deviates further from stoicheiometry the range of site symmetries produces a more complicated envelope which cannot be resolved into the sum of a small number of Lorentzian peaks; components from Fe_3O_4 are also evident. The spectra of samples of $Fe_{0.947}O$ which have been most rapidly quenched from 1520 K into water indicate that negligible disproportionation has occurred during quenching. These spectra, when compared with those predicted for various possible defect structures, suggest the presence of clusters of four vacant cation sites around a tetrahedral Fe^{3+}. Changes in the spectra as the Fe^{3+} content is increased reflect the increase in the defect cluster size towards that previously proposed for $Fe_{0.90}O$ (thirteen vacant cation sites with four tetrahedral Fe^{3+}). There is no evidence in the room-temperature spectra for electron hopping between Fe^{2+} and Fe^{3+} and spectra of the magnetically ordered oxide phases at 77 K are consistent with this interpretation.[257]

The second paper describes a Mössbauer investigation of the disproportionation of quenched samples of $Fe_{1-x}O$ between 300 and 700 K.[258] Changes in the Mössbauer spectrum as a sample of $Fe_{0.940}O$ was heated *in vacuo* are summarized in Figure 15. At temperatures above 500 K the prior rearrangement of $Fe_{1-x}O$ into $Fe_{1-x-y}O$ and $Fe_{1-x+y}O$, which had previously been proposed, was too rapid to be recorded on the Mössbauer spectra, but the observed reduction in the quadrupole splitting as a function of time at constant temperature were discussed in terms of the reaction:

$$(1 - 4z)Fe_{1-x}O \longrightarrow (1 - 4x)Fe_{1-z}O + (x - z)Fe_3O_4$$

In the limit, z becomes very small and the monoxide approaches stoicheiometry. The phase $Fe_{0.99}O$ was shown to have a Néel temperature of 196 ± 3 K and a range of hyperfine magnetic fields of 340 ± 20 kG at 77 K. At temperatures above about 570 K further disproportionation of $Fe_{1-z}O$ itself becomes rapid and the spectrum of metallic iron begins to appear, according to the reaction:

$$4Fe_{1-z}O \longrightarrow (1 - 4z)Fe + Fe_3O_4$$

The rate of precipitation of Fe_3O_4 at a given temperature was found to depend not only on the initial composition of $Fe_{1-x}O$, but also on the temperature from which the initial samples were quenched during preparation. This effect was interpreted in terms of the defect aggregations present in the initial samples.[258]

The third paper describes diffusion line-broadening studies on $Fe_{1-x}O$ at 1074 and 1173 K.[259] Activation energies E for the jump process were calculated from the expression $\Delta\Gamma = (\Delta\Gamma_0) \exp(-E/RT)$ and found to be 140 ± 20 and 135 ± 20 kJ mol^{-1} for $Fe_{0.940}O$ and $Fe_{0.910}O$, respectively. The results are in reasonable agreement with those obtained from radioactive tracer studies. The observed reduction in the average cation-jump frequency from 42 to 30 MHz over the range $Fe_{0.940}O$ to $Fe_{0.910}O$ at 1074 K was compared with the number of atoms statistically able to jump in random and clustered defect configurations, but the observed behaviour could only be simulated by the incorporation of an activation energy of jumping which increases with increasing cluster size.[259] Possible mechanisms of the diffusion process have been considered in detail.[260]

Wustite has also been studied independently by three other groups. The first paper describes results on twelve quenched samples of $Fe_{1-x}O$ with compositions in the range $0.002 \leq x \leq 0.250$, but gives no indication of the effect of quench rate.[261] The second presents data for $Fe_{0.92}O$ and for a sample claimed to be stoicheiometric $Fe_{1.0}O$; the latter was prepared by the reduction of haematite below 570 °C and could therefore not be separated from magnetite which accompanies its formation.[262] The final study deals with the compositions FeO_{1+x} ($0.05 \leq x \leq 0.13$) and magnesium wustite $Fe_{1.1}$,aMgO ($0 \leq a \leq 1$); the spectra are believed to show

---

[260] N. N. Greenwood and A. T. Howe, Proceedings of the Seventh International Symposium on the Reactivity of Solids, 1972, p. 240.

[261] H. U. Hrynkiewicz, D. S. Kulgawczuk, E. S. Mazanek, A. M. Pustowka, K. Tomala, and M. E. Wyderko, *Phys. Status Solidi (A)*, 1972, **9**, 611.

[262] V. P. Romanov and L. F. Checherskaya, *Phys. Status Solidi (B)*, 1972, **49**, K183.

evidence of electron exchange between Fe^{2+} and Fe^{3+}.[263] The non-stoicheiometric solid solutions $Fe^{3+}_{2x}Fe^{2+}{}_{(1-2x)(1-y)}M^{2+}_{[(1-2x)y]}O_{(1+x)}$ (M $=Ca^{2+}$ or Mg^{2+}; x = excess oxygen; y = degree of substitution of Fe^{2+} by Ca^{2+} or Mg^{2+}) have also been studied.[264]

Vibrational anisotropy has been detected in haematite (α-Fe_2O_3) and ilmenite ($FeTiO_3$) and shown to decrease as the temperature increases.[265] It has been shown that the magnetic structure of α-Fe_2O_3 is affected by the mechanical processes of cutting and polishing, during the preparation of thin single-crystal slices for use as Mössbauer absorbers. The effect is reflected in the relative intensities of lines 2 and 3 of the six-line hyperfine pattern and is attributed to deformation twinning which results from stresses imposed during sample preparation. The anisotropy is removed by annealing.[266]

The quadrupole splitting of α-Fe_2O_3 microcrystals supported on silica gel has been shown to decrease upon addition of water, methanol, and ammonia. Previous experiments have indicated that the quadrupole splitting is dependent on the oxide particle size and that this is a result either of a lattice expansion throughout the crystallite or of lattice expansion or chemical modification of a surface shell. Furthermore, these models have been used to determine the particle size. However, the observation that the quadrupole splitting is strongly dependent on the chemical environment at the surface now suggests that the lattice expansion model and the shell model are inadequate for this purpose.[267]

The effect on the Morin temperature of diluting haematite with non-magnetic ions such as Al^{3+} [268] and Sn^{4+} [269] has been studied and is discussed in greater detail on p. 555.

Static magnetization and Mössbauer measurements on ultrafine crystallites (~ 6.5 nm) of γ-Fe_2O_3 have shown them to be superparamagnetic at 296 K with an average relaxation time of 2×10^{-9} s. Spectra at 5 K in an applied magnetic field of $\leqslant 50$ kG indicate a cation distribution of $1 : (1.71 \pm 0.05)$ for A and B sites and a non-collinear spin arrangement. The low value of the magnetization, 59 e.m.u. g^{-1} at 4.2 K for $1/H = 0$, is explained by assuming that the spins of cations which lie in the surface layer make random angles between 0° and 90° with the direction of the nett moment. When a polarizing field is applied to the absorber at 296 K the spectrum changes from a broad singlet into two hyperfine patterns

---

[263] B. S. Bokshtein, A. A. Zhukhovitskii, V. A. Kozheurov, A. A. Lykasov, and G. S. Nikol'skii, *Zhur. fiz. Khim.*, 1972, **46**, 878.
[264] V. A. Kozheurov, G. G. Mikhailov, A. A. Zhukhovitskii, B. S. Bokshtein, A. A. Lykasov, Yu. S. Kuznetsov, S. Yu. Gurevich, and G. S. Nikol'skii, *Zhur. fiz. Khim.*, 1972, **46**, 2246.
[265] E. F. Makarov, I. P. Suzdalev, and G. Groll, *Zhur. eksp. teor. Fiz.*, 1972, **62**, 1834.
[266] R. O. Keeling, *J. Appl. Phys.*, 1972, **43**, 4736.
[267] H. M. Gager, M. C. Hobson, and J. F. Lefelhocz, *Chem. Phys. Letters*, 1972, **15**, 124.
[268] J. K. Srivastava and R. P. Sharma, *Phys. Status Solidi (B)*, 1972, **49**, 135.
[269] B. P. Fabrichnyi, E. V. Lamykin, A. M. Babeshkin, and A. N. Nesmeyanov, *Solid State Comm.*, 1972, **11**, 343.

from A and B sites; the lines are broadened because of the range of ordering temperature of the particle distribution.[270]

The multiple ordering theory of the low-temperature transition in magnetite (Fe_3O_4) has been established by the observation of two peaks in the heat capacity and by changes in the shapes of the Mössbauer lines, both occurring between 113 and 123 K.[271] The effect of vacancies on the Mössbauer spectrum of magnetite has been studied.[272] It has been claimed that the B-site pattern in the 300 K spectrum can be resolved into a doublet and that a localized hopping model is therefore not uniquely confirmed by the Mössbauer spectrum.[273]

The oxidation of Fe_3O_4 to α-Fe_2O_3 in the absence of iron has been shown by Mössbauer spectroscopy to have an activation energy of 49 kcal mol^{-1}. Vacuum annealing of Fe_3O_4/α-Fe_2O_3 layers on iron gave an activation energy of 31 kcal mol^{-1} for the transport of Fe through Fe_3O_4.[274] Amorphous ferric hydroxide has been characterized and its conversion into α-FeOOH studied kinetically.[275] Mössbauer spectroscopy has been used to analyse quantitatively the various components such as α-FeOOH, β-FeOOH, γ-FeOOH, γ-Fe_2O_3, and Fe_3O_4, which are present in rust formed on iron after exposure to air containing sulphur dioxide.[276] Analysis of the corrosion products α-FeOOH and γ-Fe_2O_3 by detection of the internal conversion electrons has also been described.[277] The conversions of paramagnetic β-FeOOH into antiferromagnetic α-Fe_2O_3 and of paramagnetic $(Cr_{0.8}Fe_{0.2})(OH)_3$ into $(Cr_{0.8}Fe_{0.2})_2O_3$ have been monitored by Mössbauer spectroscopy.[278] The spectra of haematite, magnetite, limonite, chamosite, and siderite ores have been discussed with regard to the usefulness of the Mössbauer technique for studying slag and furnace clinker.[279] The distribution of Fe^{2+} and Fe^{3+} ions in silicate slags has been studied. In alkaline silicate glasses, Na_2O–SiO_2–Fe_2O_3, all the Fe^{3+} are tetrahedrally co-ordinated, whereas in alkaline-earth silicate glasses, CaO–SiO_2–Fe_2O_3, a large fraction of the Fe^{3+} are octahedrally co-ordinated.[280] The catalytic mechanism of Fe_2O_3 in the chlorosilane–alkoxysilane condensation has been studied.[281]

[270] J. M. D. Coey and D. Khalafalla, *Phys. Status Solidi (A)*, 1972, **11**, 229.
[271] B. J. Evans and E. F. Westrum, *Phys. Rev. (B)*, 1972, **5**, 3791.
[272] V. P. Ramonov, V. Checherskii, and V. V. Eremenko, *Phys. Status Solidi (A)*, 1972, **9**, 713.
[273] B. J. Evans, *Amer. Inst. Phys. Conf. Proc.*, 1972, No. 4 (Pt. 1), 296.
[274] D. A. Channing and M. J. Graham, *Corrosion Sci.*, 1972, **12**, 271.
[275] S. Okamoto, H. Sekizawa, and S. I. Okamoto, Proceedings of the Seventh International Symposium on the Reactivity of Solids, 1972, p. 341.
[276] W. Meisel and G. Kreysa, *Z. Chem.*, 1972, **12**, 301.
[277] H. Onodera, H. Yamamoto, H. Watanabe, and H. Ebiko, *Jap. J. Appl. Phys.*, 1972, **11**, 1380.
[278] V. G. Jadhao, R. M. Singru, and C. N. R. Rao, *Phys. Status Solidi (A)*, 1972, **12**, 605.
[279] A. Pustowka, J. Sawicki, E. Mazanek, and M. Wyderko, *Hutnik*, 1972, **39**, 109.
[280] L. Pargamin, C. H. P. Lupis, and P. A. Flinn, *Met. Trans.*, 1972, **3**, 2093.
[281] A. Vértes, B. Csakvari, P. Gomory, and M. Komor, *Radiochem. Radioanalyt. Letters*, 1972, **9**, 303.

Spinel Oxides and Garnets. Calculations have been made of the effect on the Mössbauer spectrum of electric-field gradient distributions which occur in very small particles of cubic crystals with the normal spinel lattice.[67]

The high- and low-temperature forms of Li_5FeO_4 have both been shown to give large quadrupole splittings and small isomer shifts, consistent with a high degree of covalency in the iron–neighbour bonds. Solid solutions of the type $Li_5Fe_xAl_{1-x}O_4$ were obtained for $0 < x < 1$ in the low-temperature phase but only for $0 < x < 0.33$ in the high-temperature phase. A new compound was detected at $x = 0.43$.[282] The spectra of $LiFe_{0.005}Al_{4.995}O_8$ in applied magnetic fields of <1 kG have been compared with theoretical spectra calculated using e.p.r. parameters. The comparison revealed that the observed resonances which were sensitive to the applied fields are associated with the $\pm\frac{1}{2}$ spin Kramers' doublet of the ionic ground term and that, for zero applied field, the field-sensitive resonances are affected drastically by dipolar fields originating from neighbouring nuclei. The sign of the zero-field splitting parameter D was found to be negative, the value of the ground-state hyperfine coupling constant $A_g = -2.45 \pm 0.03$ mm s^{-1}, and the quadrupole coupling constant $2P = +0.67 \pm 0.03$ mm s^{-1}.[283]

The substituted ferrites $Li_{0.5}Fe_{2.5-x}Al_xO_4$ ($0 \leq x \leq 1.0$) and $Li_{0.5}Fe_{2.5-x}Ga_xO_4$ ($0 \leq x \leq 1.63$) have been studied in the paramagnetic region. The electric-field gradient, V_{zz}, for the octahedral Fe^{3+} ion is thought to be negative in sign and its magnitude reaches a maximum at $x = 0.8$ for the aluminium derivatives and at $x = 0.6$ for the gallium derivatives. The results can be rationalized in terms of a point-charge model.[284]

It has been shown that Mg^{2+} ions substitute for Fe^{3+} ions in the magnetite lattice to yield the solid solutions $Fe_{3-x}Mg_xO_4$. As a result of this the electronic exchange $Fe^{2+} \rightleftharpoons Fe^{3+}$ is apparently characterized by two relaxation times, namely the relaxation time of the localized electronic exchange for a given pair of ions, and the relaxation time for electronic exchange between all pairs of ions in the octahedral sublattice. Partial ordering of the cations in the octahedral sublattice occurs and it is possible to distinguish four distinct magnetic hyperfine components in the Mössbauer spectra.[285] Mössbauer spectroscopy has been used to determine the phase composition of magnesium–manganese ferrite powders prepared by the thermal decomposition of the coprecipitated oxalates.[286]

The cation distributions and magnetic structures of solid solutions of the ferrimagnetic spinels $Fe_{3-x}Cr_xO_4$ ($x = 0$—1.75) have been studied at

[282] G. Le Corre, A. Malve, C. Gleitzer, and J. Foct, *Compt. rend.*, 1972, **274**, *C*, 466.
[283] P. J. Viccaro, F. de S. Barros, and W. T. Oosterhuis, *Phys. Rev. (B)*, 1972, **5**, 4257.
[284] O. A. Bayukov, A. I. Drokin, V. P. Ikonnikov, M. I. Petrov, and V. N. Seleznev, *Soviet Phys. Solid State*, 1972, **13**, 2803.
[285] A. K. Gapeev, T. S. Gendler, R. N. Kuz'min, A. A. Novakova, and B. I. Pokrovskii, *Kristallografiya*, 1972, **17**, 141.
[286] P. L. Gruzin, L. M. Isakov, Yu. A. Sokolov, M. N. Uspenskii, and G. I. Patoka, *Zavod. Lab.*, 1972, **38**, 1097.

77 K and room temperature. The Cr^{3+} ions were found to enter the octahedral sites, displacing the Fe^{2+} ions into the tetrahedral sublattice, the final cation distribution being $Fe^{2+}[Fe^{3+}_{0.25}Cr^{3+}_{1.75}]O_4$. For small values of x a magnetic structure of the Néel type exists; for $x > 0.5$ this structure disorders, and for $x = 1.75$ a partially ordered angular spin structure of the Yafet–Kittel type appears.[287] This system has previously been studied in great detail by Wertheim and co-workers (see Vol. 5, p. 537).

Mössbauer effect studies on cubic and tetragonal $Mn_{1.9}Fe_{1.1}O_4$ have failed to detect the presence of Fe^{2+} ions. This rules out the possibility of an equilibrium of the type $Mn^{2+} + Fe^{3+} \rightleftharpoons Mn^{3+} + Fe^{2+}$, and indicates that the cubic to tetragonal phase transition is not associated with a significant change in the concentration of Fe^{2+} ions in octahedral sites.[288] The manganites $M^{2+}Mn_{1.9}Fe_{0.1}O_4$ (M = Cd, Mg, Zn, or Co) have also been studied both above and below the temperature of the cubic-tetragonal distortion.[289]

The spectra of the nickel ferrites $Fe_{3-x}Ni_xO_4$ ($x = 0, 0.2, 0.4, 0.6$, or 0.8) have been analysed in detail. For $x = 0$ (Fe_3O_4) there is a magnetic hyperfine pattern from Fe^{3+} ions in tetrahedral sites and a second pattern from the octahedral Fe^{3+} and Fe^{2+} ions which are rendered indistinguishable by rapid electron hopping. As x increases, the Ni^{2+} ions are progressively substituted for Fe^{2+} ions in the octahedral sites and the spectra distinguish between four types of iron ions in the octahedral sites: (i) for $0 < x < 1$ a fraction of the Fe^{3+} ions are not involved in electron hopping and give a pattern which is superimposed on that due to the tetrahedral Fe^{3+} ions; (ii) for $0 < x < 0.4$ there is a pattern from an equal number of Fe^{2+} and Fe^{3+} ions which are rendered indistinguishable by electron exchange; (iii) for $0.2 < x < 0.8$ there is a group of Fe^{3+} and Fe^{2+} ions in the proportion 1:2 which are also coupled by electronic exchange; (iv) for $x > 0.6$ some of the octahedral Fe^{3+} ions are not involved in electron exchange and are distinguishable from the first type by a relatively higher isomer shift and internal field. For $x = 1$ (Fe_2NiO_4) only two hyperfine patterns are observed, the one with the lower isomer shift and H_{eff} being assigned to tetrahedral Fe^{3+} ions, and the other to octahedral Fe^{3+} ions. It is proposed that, at low values of x, the itinerant electrons from the Fe^{2+} ions are delocalized in the sublattice leading to metallic conductivity, whereas at high values of x they are in localized groups of exchange-coupled iron ions.[290]

Angular ordering of the Fe^{3+} spins has been detected in the ferrites $NiFe_{2-x}Al_xO_4$. For $x = 0.2$ and 0.5, two six-line patterns of equal intensity are observed, indicating that the Fe^{3+} ions are evenly distributed over the

---

[287] A. S. Bukin, A. K. Gapeev, R. N. Kuz'min, and A. A. Novakova, *Kristallografiya*, 1972, **17**, 799.
[288] M. K. Hucl, F. Van der Woude, and G. A. Sawatzky, *Phys. Letters*, 1972, **42A**, 99.
[289] G. Filoti, A. Gelberg, V. Gomolea, and M. Rosenberg, *Internat. J. Magn.*, 1972, **2**, 65.
[290] J. W. Linnett and M. M. Rahman, *J. Phys. and Chem. Solids*, 1972, **33**, 1465.

tetrahedral and octahedral sites and that all of the Ni^{2+} and Al^{3+} ions are in octahedral sites. For $x > 1$ several distinct hyperfine components are present, three from Fe^{3+} in tetrahedral sites and one from Fe^{3+} in octahedral sites, and these are attributed to an angular ordering of the spins, rather than to the existence of a range of different environments. In nickel ferrite itself the axis of easy magnetization is directed along the [111] axis at an angle of 55° to the direction of V_{zz}; the development of angular ordering is accompanied by a deviation of the angle θ from 55° and the appearance of a quadrupole splitting because the expression $(3 \cos^2\theta - 1)$ now becomes non-zero. For large concentrations of Al^{3+} ($x > 1$) a paramagnetic component also appears in the spectrum with a relative intensity which decreases as the temperature is lowered.[291]

Fe^{2+} has been shown to exist in both the tetrahedral and octahedral sites in the ferrites $Ni_{1-x-y}Fe_x^{2+}Zn_yFe_2^{3+}O_4$.[292]

Possible explanations for the discrepancy between the measured magnetic moments of $NiCr_{2-x}Fe_xO_4$ and the values calculated on the basis of the Néel model have been discussed.[293-295] The isomer shift of the Fe^{3+} ions in tetrahedral sites in $NiCr_{1.2}Fe_{0.8}O_4$ increases at the magnetic ordering temperature, indicating a decrease in the spin density in the region of the iron nuclei.[296]

The processes which occur in nickel–cobalt ferrites during thermomagnetic treatment have been investigated.[297-299]

Relaxation effects have been observed in the spectra of $Co_xZn_{1-x}Fe_2O_4$ ($x = 0.3, 0.5,$ or 0.75) in the temperature range 80—700 K. The theoretical spectra calculated on the basis of the stochastic model (in which the hyperfine interaction is replaced by an interaction between the nuclear spin and a randomly varying time-dependent external magnetic field) show good agreement with the experimental data in all cases. For each composition the relaxation time decreases very slowly with increase in temperature, but in the region of the Néel temperature it decreases rapidly and leads to the total collapse of the magnetic hyperfine splitting. The initial introduction

---

[291] V. F. Belov, M. N. Shipko, T. A. Khimich, V. V. Korovushkin, and L. N. Korablin, *Soviet Phys. Solid State*, 1972, **13**, 1692.
[292] V. A. Potakova, N. D. Zverev, and V. P. Romanov, *Phys. Status Solidi (A)*, 1972, **12**, 623.
[293] V. I. Nikolaev, S. S. Yakimov, F. I. Popov, and V. N. Zarubin, *Soviet Phys. Solid State*, 1972, **14**, 521.
[294] P. P. Kirichok, V. F. Belov, V. A. Trukhtanov, G. S. Pdval'nykh, M. N. Shipko, V. V. Voitkiv, and V. V. Korovushkin, *Ukrain. fiz. Zhur.*, 1972, **17**, 459.
[295] P. P. Kirikchok, G. S. Podval'nykh, and V. F. Belov, *Izvest. Akad. Nauk S.S.S.R., Ser. fiz.*, 1972, **36**, 397.
[296] V. A. Gordienko, V. V. Zubenko, V. I. Nikolaev, and S. S. Yakimov, *Soviet Phys. Solid State*, 1972, **14**, 530.
[297] P. P. Kirichok, G. S. Podval'nykh, N. V. Kobrya, and O. R. Borovskaya, *Izvest. Akad. Nauk S.S.S.R., Ser. Fiz.*, 1972, **36**, 402.
[298] P. P. Kirichok, G. S. Podval'nykh, N. V. Kobrya, O. R. Borovskaya, and L. M. Letyuk, *Izvest. Vyssh. Ucheb. Zaved., Fiz.*, 1972, **15**, 133.
[299] P. P. Kirichok, G. S. Podval'nykh, N. V. Kobrya, O. R. Borovskaya, and L. M. Letyuk, *Zhur. fiz. Khim.*, 1972, **46**, 1550.

of Zn^{2+} ions into the system ($x = 0.7$) increases the relaxation time at low T ($T < T_N$), but further increase in the Zn^{2+} concentration from $x = 0.5$ to $x = 0.3$ has little effect, because of the role of Zn^{2+} ions in increasing the magnetic disorder in the ferrite.[300]

The quadrupole coupling constants associated with the tetrahedral and octahedral sites in the rare-earth iron garnets $R_3Fe_5O_{12}$ (R = Lu, Yb, Dy, Sm, or Gd) have been estimated and compared with experimental values. The good agreement indicates that covalency effects are not important as regards the quadrupole coupling in these oxides. The apparent contradiction of this conclusion with the interpretation of the isomer-shift data was rationalized in terms of cancellation effects of the distant contributions from the various s-orbitals to the electric-field gradient.[301]

A single crystal of yttrium iron garnet magnetized in the [111] direction with a field of 55 kG has been used in a determination of the quadrupole coupling at the octahedral Fe^{3+} site. The result, $\frac{1}{2}e^2qQ = -0.46$ mm s^{-1}, disagreed with that estimated recently on the basis of an LCAO MO calculation. It was suggested that neglect of the electrostatically produced dipoles on the oxide ions might be the cause of the disagreement and a procedure for constructing MO's which takes such dipoles into account was suggested.[302] The change in the direction of easy magnetization in oriented slices of yttrium iron garnet single crystals, which occurs at about 500 °C, has been shown to be due to non-axial magnetic anisotropy induced by tensile stresses generated in the plane of the sample by polishing. The effect was absent in samples which had been annealed at 850 °C for 4 h. In this case the direction of magnetization (along the [111] axis) is determined by the intrinsic magnetic anisotropy energy.[303]

The Mössbauer spectra of diamagnetically substituted yttrium iron garnet have been considered in detail. In favourable circumstances it is possible to use the Mössbauer effect to measure not only the cation distribution among different sites, but also to determine whether the cation distribution in a given sublattice is random. This was found to be the case in $\{Y\}_3(Fe_{1-x}Sc_x)_2[Fe]_3O_{12}$ with $x = 0.10$ and 0.25 (the different brackets refer respectively to dodecahedral, octahedral, and tetrahedral sites). In^{3+} and Rh^{3+} were also found to prefer the octahedral sites, whereas Al^{3+} and Ga^{3+} prefer the tetrahedral sites. At higher temperatures, central peaks appear in the spectra of the mixed oxides, as illustrated in Figure 16 for $\{Y\}_3(Fe)_2[Fe_{0.75}Ga_{0.25}]_3O_{12}$. These peaks grow in intensity at the expense of the magnetically split spectrum which finally disappears at T_c. An applied magnetic field was found to increase the proportion of hyperfine splitting in the spectrum. The mixed spectra could not be explained by a range of ordering temperatures in different parts of the crystal, or by

---

[300] S. C. Bhargava and P. K. Iyengar, *Phys. Status Solidi* (*B*), 1972, **53**, 359.
[301] R. R. Sharma, *Phys. Rev.* (*B*), 1972, **6**, 4310.
[302] R. M. Housley and R. W. Grant, *Phys. Rev. Letters*, 1972, **29**, 203.
[303] G. N. Belozerskii, Yu. P. Khimich, and Yu. M. Yakovlev, *Soviet Phys. Solid State*, 1972, **14**, 993.

independent relaxation of the ionic spins. Instead, the central peaks were thought to arise from iron in clusters of short-range order. The size of the clusters was estimated to be 10^3—10^5 ions just below the temperature at which the last trace of magnetic hyperfine splitting disappears. A study of the geometry of the random lattice, with nearest-neighbour interactions, showed no tendency for localized regions of this magnitude to form and it was suggested that they are evanescent.[304]

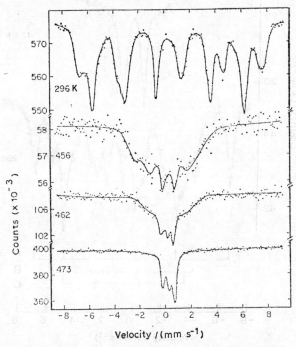

Figure 16 Mössbauer spectra of $\{Y\}_3(Fe)_2[Fe_{0.75}Ga_{0.25}]_3O_{12}$ at different temperatures
[Reproduced by permission from Phys. Rev. (B), 1972, **6**, 3240]

The Fe^{3+} ions in $Ca_3Zr_2TiFe_2O_{12}$ have been shown to occupy tetrahedral positions, whereas in $Ca_3Zr_{1.5}Ti_{1.5}Fe_2O_{12}$ they are partially replaced by Ti^{4+} ions and shifted into octahedral sites.[305] The Fe^{3+} ions in $Ca_3Fe_{2-x}M_xGe_3O_{12}$ and $Cd_3Fe_{2-x}M_xGe_3O_{12}$ (M = Al^{3+}, Ga^{3+}, Cr^{3+}, Sc^{3+}, or In^{3+}) all occupy octahedral sites.[306] Cation distributions in the garnets $Gd_3Fe_{5-x}Al_xO_{12}$, $Gd_3Fe_5O_{12}$, and $Er_3Fe_5O_{12}$ have also been discussed.[307]

[304] J. M. D. Coey, Phys. Rev. (B), 1972, **6**, 3240.
[305] R. Hrichova and J. Lipka, Coll. Czech. Chem. Comm., 1972, **37**, 3352.
[306] I. S. Lyubutin, L. M. Belyaev, R. Grizhikhova, and I. Lipka, Kristallografiya, 1972, **17**, 146.
[307] V. N. Belogurov, B. M. Lebed, V. I. Mosel, and P. E. Senkov, Phys. Status Solidi (A), 1972, **11**, K93.

The Mössbauer effect in antiferromagnetic substances with garnet structures has been considered elsewhere.[308]

Other Oxide Systems. The solid solutions $CaAl_{12-x}Fe_xO_{19}$ ($x \leqslant 4.8$) have been shown to give three resolvable quadrupole doublets from Fe^{3+} ions in tetrahedral, trigonal-bipyramidal, and octahedral sites. Analysis of the

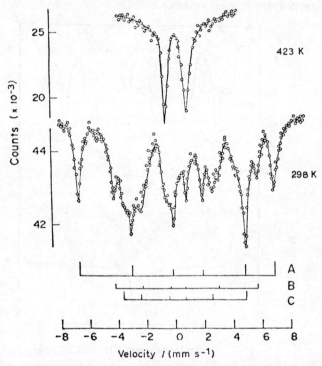

Figure 17 *Mössbauer spectra of $Sr_4Fe_6O_{13}$. The relative intensities are* A : B : C = 3 : 1 : 2
(Reproduced by permission from *J. Phys. and Chem. Solids*, 1972, **33**, 1169)

relative areas indicates a marked preference for iron to occupy the tetrahedral sites. Magnetic ordering is observed for $x = 4.8$ and 6.[309]

The new strontium ferrate $Sr_4Fe_6O_{13}$ gives a complex spectrum at room temperature (see Figure 17), which can be interpreted in terms of three overlapping six-line magnetic hyperfine patterns from iron atoms in non-equivalent environments, with relative intensities in the ratio 3 : 1 : 2. The material is not attracted to a magnet at room temperature and is

---

[308] A. P. Dodokin, I. S. Lyubutin, B. V. Mill, and V. P. Peshkov, *Zhur. eksp. teor. Fiz.*, 1972, **63**, 1002.
[309] F. P. Glasser, F. W. D. Woodhams, R. E. Meads, and W. G. Parker, *J. Solid State Chem.*, 1972, **5**, 255.

therefore antiferromagnetic. The isomer shifts are characteristic of Fe^{3+}. Above the Néel temperature (145 ± 5 °C) the spectrum consists of two absorption bands from three overlapping quadrupole doublets. The quadrupole splitting (ca. 1.3 mm s^{-1}) is approximately double that observed below T_N and is consistent with the magnetization being directed along an axis perpendicular to the major axis of the electric-field gradient.[310]

Spectra have been obtained for the solid-solution series $Sr_{1-x}La_x$-$FeO_{2.5+x/2}$. The intensity of the characteristic six-line absorption pattern of the tetrahedral iron site in the brownmillerite-type phase (0 ⩽ x ⩽ 0.2) decreases as the La content increases. This is attributed to a change in the co-ordination state of Fe^{3+} from tetrahedral to octahedral as a result of the increased oxygen content. The spectrum for the composition $x = 0.15$ contains a new six-line pattern with parameters intermediate between those for the tetrahedral and octahedral sites and is thought to arise from the presence of Fe^{3+} ions in a five-co-ordinate trigonal-bipyramidal oxygen environment. The brownmillerite phase disappears completely at $x = 0.3$ and is replaced by a cubic perovskite structure; however, absorption peaks corresponding to a small proportion of tetrahedrally co-ordinated Fe^{3+} ions still remain. At $x ⩾ 0.5$ only six-co-ordinate Fe^{3+} ions can be detected. The values of the isomer shift, quadrupole splitting, and internal magnetic field for the sample with $x = 0.15$ at room temperature are, respectively, +0.380 mm s^{-1}, +0.109 mm s^{-1}, and 515 kG for the octahedral site, +0.136 mm s^{-1}, −0.150 mm s^{-1}, and 411 kG for the tetrahedral site, and +0.326 mm s^{-1}, +0.326 mm s^{-1}, and 488 kG for the proposed trigonal-bipyramidal site.[311]

The influence of the thermal history on the Mössbauer spectra of samples of the type $SrO_{6-x}Fe_2O_3,xAl_2O_3$ has been studied.[312]

The ordered perovskite Ba_2FeReO_6 has been shown to give a single line with an isomer shift of +0.65 ± 0.03 mm s^{-1}, above the Curie temperature. At 77 K a six-line pattern is obtained, indicating that the material is magnetically ordered with an internal field of 456 ± 4 kG. At room temperature the intermediate behaviour of partial magnetic order is observed. Although the lack of any quadrupole splitting suggests that the iron is tervalent, the isomer shift and hyperfine field are intermediate between the values expected for high-spin Fe^{3+} and Fe^{2+}. Furthermore, this material exhibits metallic conductivity and a high Curie temperature, which may indicate that the Fe^{3+}–Re^{5+} combination is degenerate with the Fe^{2+}–Re^{6+} combination. Ordinary superexchange rules, which assume σ-type interactions to be stronger than π-type interactions, predict that this material should be ferromagnetic. Instead, it is found to be ferrimagnetic, which suggests that π-type interactions may in fact be dominant.[313]

[310] F. Kanamaru, M. Shimada, and M. Koizumi, *J. Phys. and Chem. Solids*, 1972, **33**, 1169.
[311] H. Yamamura and R. Kiriyama, *Bull. Chem. Soc. Japan*, 1972, **45**, 2702.
[312] V. Fluorescu, I. Bunget, D. Barb, M. Morariu, and D. Tarina, *Rev. Roumaine Phys.*, 1972, **17**, 261.
[313] A. W. Sleight and J. F. Weiher, *J. Phys. and Chem. Solids*, 1972, **33**, 679.

Data have been given for nine ferrites in the series $BaFe_{12-2x}Zn_xTi_xO_{19}$. For $x = 0$, four Fe^{3+} six-line patterns are observed, whereas for $x > 0$ there are five patterns from non-equivalent iron environments.[314]

Effects similar to those described earlier for $\{Y\}_3(Fe)_2[Fe_{0.75}Ga_{0.25}]_3O_{12}$ have also been observed for the disordered solid solution $Ba(Fe_{0.35}Al_{0.65})_2O_4$.

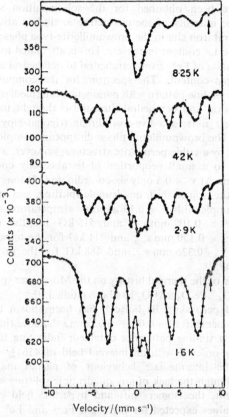

Figure 18 *Mössbauer spectra of* $Ba(Fe_{0.35}Al_{0.65})O_4$. *For explanation of arrowed peaks, see original paper*
(Reproduced by permission from *J. Phys. and Chem. Solids*, 1972, 33, 1631)

Above 20 K the spectra consist entirely of a quadrupole doublet, but below this temperature magnetic lines also appear and their intensity grows as the temperature is lowered further, as can be seen in Figure 18. These results indicate that the solid solutions contain antiferromagnetic clusters of various sizes, the smaller ones being above their blocking

[314] J. P. Mahoney, A. Tauber, and R. O. Savage, *Amer. Inst. Phys. Conf. Proc.*, 1972, No. 5 (Pt. 2), 816.

temperatures and having rapidly fluctuating nett moments, and the larger ones being stable and producing well-defined magnetic splitting. The variation with temperature of the ratio of the intensity of the sextet to that of the doublet is shown in Figure 19 and is seen to be discontinuous,

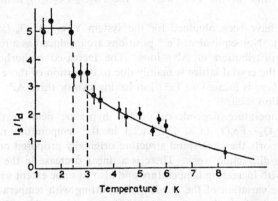

Figure 19 *Temperature variation of the ratio of the intensity of the sextet to that of the doublet, I_s/I_d*
(Reproduced by permission from *J. Phys. and Chem. Solids*, 1972, **33**, 1631)

which suggests successive blocking according to the Néel theory. The theory of random clusters in disordered solid solutions, due to de Gennes *et al.*, was used to determine the blocking temperatures, and the anisotropy constant K values were deduced on the basis of the Néel theory of superparamagnetism.[315]

Mixed oxides prepared by heating the hydroxides coprecipitated from solutions of Fe^{3+} and Al^{3+} have been studied.[316] The system $(1 - x)Fe_2O_3$–xAl_2O_3 ($x = 0$, 0.56, 1.98, 5.27, or 10.17 wt.%) has been studied in detail in the temperature range 10—1000 K. There is a rapid and non-linear decrease in the Morin temperature T_M of the α-Fe_2O_3 lattice. The phase transition is not sharp and becomes more and more diffuse as the magnetic dilution increases. The Fe^{3+} spins are canted at an angle of about 30° away from the *c*-axis at low temperatures for the 0.56 wt.% diluted sample, and the canting angle appears to remain constant with increased dilution. The ratio T_M/T_N (T_N = Néel temperature) and the anisotropy energy, K, deduced from the temperature and concentration dependence of the hyperfine field, decrease with increased dilution, which supports the existing anisotropy theory. Three magnetic phases are thought to be present in the 5.27 wt.% diluted sample.[268] The Morin transition temperature decreases from 260 to 153 K when Sn^{4+} (0.34 at.%) is doped into α-Fe_2O_3.

---

[315] R. Chevalier and C. Do-Dinh, *J. Phys. and Chem. Solids*, 1972, **33**, 1631.
[316] L. Korecz, I. Kurucz, G. Menczel, E. Papp-Molnar, E. Pungor, and K. Burger, *Magyar Kém. Folyóirat*, 1972, **78**, 508.

Spectra taken at room temperature in a magnetic field indicate that the hyperfine field at the ^{119}Sn nucleus is oriented in the (111) plane of the rhombohedral lattice of α-Fe$_2$O$_3$.[269]

The effect of the impurity ions Fe^{2+}, Cr^{3+}, Pr^{3+}, and Gd^{3+} on the spin relaxation time of the Fe^{3+} ions in the Al_2O_3–Fe_2O_3 system has been studied.[317]

Spectra have been obtained for the system $YFe_{1-x}Al_xO_3$ (x = 0, 0.1, 0.2, or 0.3). Non-equivalent Fe^{3+} positions are produced as a result of the statistical distribution of Al^{3+} ions. The increased orthorhombic distortion of the crystal lattice is mainly due to distortion of those octahedra in which there is located an Fe^{3+} ion having two or three Al^{3+} ions in its nearest-cation shell.[318]

The temperature dependence of the hyperfine field in the system $(1 - x)Cr_2O_3$–xFe_2O_3 ($x \leqslant 4.95$ wt.%) in the temperature range 80—300 K supports the cone spiral structure originally proposed on the basis of neutron diffraction work. There is a linear decrease in the cone half-angle, θ, with increasing temperature and also to some extent with increasing x. The variation of the quadrupole splitting with temperature and x follows qualitatively the temperature and composition dependence of the trigonality, c/a, of the rhombohedral lattice. It was suggested that the anomalous spectra observed earlier in the $0.965Cr_2O_3$–$0.035Fe_2O_3$ system were probably due to superparamagnetic effects.[319]

A study of ferrimagnetic cobalt-doped γ-Fe$_2$O$_3$ has failed to detect the presence of either Fe^{2+} or Fe^{4+}, and indicates that cobalt replaces iron on B sites and fills B-site vacancies.[320]

Hyperfine interactions at the Fe^{2+} sites in ilmenite, $FeTiO_3$, have been determined by Mössbauer spectroscopy. The spectrum of a synthetic polycrystalline sample at 5 K is shown in Figure 20. The small internal magnetic field (H_{int} = -43 ± 3 kG) and the large quadrupole coupling ($\frac{1}{2}e^2qQ$ = 1.44 ± 0.01 mm s^{-1}) shifts the $|-\frac{3}{2}\rangle \rightarrow |-\frac{1}{2}\rangle$ transition to higher energy than the $|+\frac{1}{2}\rangle \rightarrow |+\frac{1}{2}\rangle$ transition and leads to relative intensities 2 : 1 : 1 : 2 : 3 : 3 in order of ascending energy. The negative sign of H_{int} was determined by application of an external magnetic field of 55 kG parallel to the c-axis of a single-crystal mineral sample of ilmenite at 87 K. Under these conditions the magnetization is far from saturated and only a small reduction in the internal field (H_{int} = 41 ± 3 kG) was observed. This was sufficient, however, to establish the negative sign. The orbital, dipolar, and core polarization contributions to the internal field in both $FeTiO_3$ and $FeCO_3$ were recalculated to be H_{orb} = $+420$ and $+579$, H_{dip} = $+59$ and $+85$, H_F = -522 and -479 kG, respectively. The

---

[317] J. K. Srivastava and K. G. Prasad, *Phys. Letters*, 972, **40A**, 37.
[318] V. F. Belov, T. A. Khimich, M. N. Shipko, and M. I. Dakhis, *Soviet Phys. Solid State*, 1972, **14**, 437.
[319] J. K. Srivastava and K. G. Prasad, *Phys. Status Solidi (B)*, 1972, **54**, 755.
[320] D. Khalafalla and A. H. Morrish, *J. Appl. Phys.*, 1972, **43**, 624.

theoretical interpretation of the internal fields in terms of these contributions was shown to depend critically upon the lattice contribution to $\frac{1}{2}e^2qQ$, which is still unknown.[321]

The room temperature spectrum of $FeVO_4$ has been shown to contain six resonance lines (see Figure 21), which are assigned to three non-

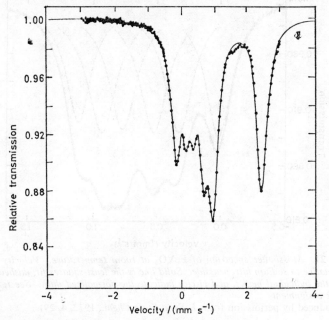

Figure 20 *Mössbauer spectrum of synthetic polycrystalline ilmenite* ($FeTiO_3$) *at 5 K*
[Reproduced by permission from *Phys. Rev.* (*B*), 1972, **5**, 1700]

equivalent Fe^{3+} ions. The outer doublet arises from Fe^{3+} in a distorted trigonal-bipyramidal environment and the inner doublets to Fe^{3+} in distorted octahedral sites.[322]

Eleven different compositions in the system $LiCr_{1-x}Fe_xO_2$, which has the ordered rocksalt structure, have been studied from 4.2 K to 300 K. All compositions order antiferromagnetically at low temperatures. There is a sharp discontinuity in the composition dependence of all the Mössbauer parameters at the cubic-to-rhombohedral phase transition at $x = 0.7$.[323]

Fe_2TeO_6, which crystallizes in a trirutile structure and is magnetoelectric, has been shown to contain Fe^{3+} ions. It is magnetically ordered at 77 K but not at 300 K.[324]

[321] R. W. Grant, R. M. Housley, and S. Geller, *Phys. Rev.* (*B*), 1972, **5**, 1700.
[322] B. Robertson and E. Kostiner, *J. Solid State Chem.*, 1972, **4**, 29.
[323] A. Tauber, W. M. Moller, and E. Banks, *J. Solid State Chem.*, 1972, **4**, 138.
[324] S. Bukshpan, E. Fischer, and R. M. Hornreich, *Solid State Comm.*, 1972, **10**, 657.

A Mössbauer study of the oxides $Fe_2Mo_3O_8$ and $FeZnMo_3O_8$ has confirmed the local trigonal symmetry of the tetrahedral and octahedral sites and has shown that the octahedral Fe^{2+} ions have a strong axial magnetic anisotropy. $FeZnMo_3O_8$ is ferromagnetic below 20 K, whereas

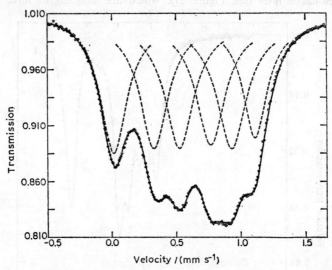

Figure 21 *Mössbauer spectrum of* $FeVO_4$ *at room temperature. Velocity scale is relative to sodium nitroprusside. Solid line is the least-squares fit, dashed lines are the individual peaks, and circles indicate the normalized data. See text for peak assignment*
(Reproduced by permission from *J. Solid State Chem.*, 1972, **4**, 29)

$Fe_2Mo_3O_8$ is antiferromagnetic with $T_N = 60$ K. In both cases the magnetic moments lie along the trigonal axis of the crystal.[325] Spectra have also been reported for six catalysts prepared from MoO_3 and Fe_2O_3.[11]

Minerals. For the third year in succession, this section is dominated by a discussion of Mössbauer studies on lunar samples. Most of the work described here was published very recently in the Proceedings of the Third Lunar Science Conference which met in Houston, Texas, in January 1972 to discuss results on samples from five lunar landings—Apollo 11, 12, 14, and 15 and Luna 16.

The iron-bearing minerals in the four Apollo 14 fines (14003,20, 14162,48, 14163,50, and 14259,17) have been examined. The results parallel earlier data on the Apollo 11 and 12 samples but show significant differences in mineral content. A typical soil spectrum at 78 K is shown at the top of Figure 22 and can be interpreted in terms of three overlapping doublets from Fe^{2+} in (i) olivine and $M1$ sites of pyroxenes, (ii) $M2$ sites

[325] A. Czeskleba, P. Imbert, and F. Varret, *Amer. Inst. Phys. Conf. Proc.*, 1972, No. 5 (Pt. 2), 811.

of pyroxenes, and (iii) ilmenite, with quadrupole splittings which decrease in that order, together with an ill-defined absorption from glassy material. Spectra at higher velocities exhibit a weak magnetic pattern, with broad lines, attributable to iron–nickel alloys (see below). The proportion of the iron as ilmenite is much lower than that of Apollo 11 fines (23%) but is similar to that found for Apollo 12 (5—8%). There is no evidence for

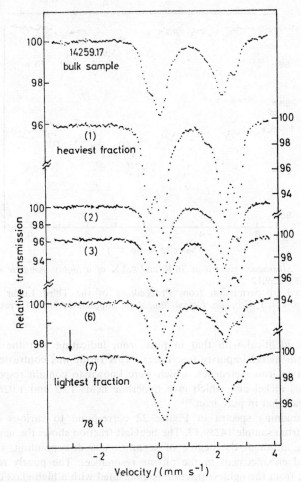

Figure 22 *Mössbauer spectra at 78 K of separations by density from sample 14259,17. The arrow indicates one of the iron–nickel resonance lines which is strongest in the lightest fraction*
(Reproduced by permission from Proceedings of the Third Lunar Science Conference, Vol. 1, *Geochim. Cosmochim. Acta*, Supplement 3, MIT Press, 1972, p. 2479)

the presence of Fe^{3+} or troilite (FeS), although the latter has been reported in previous Apollo samples.[326]

The fines returned by the automatic station Luna 16 give essentially similar spectra to those discussed above. The results indicate that these samples differ from those returned by Apollo 11 in having a lower ilmenite and greater olivine content. The olivine content is also greater than that in Apollo 12 samples. The hyperfine field of the small magnetic component

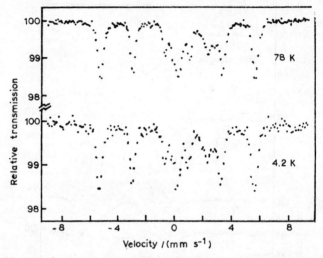

Figure 23 *Mössbauer spectra at 78 K and 4.2 K of a highly magnetic separate from soil 14259,17*
(Reproduced by permission from Proceedings of the Third Lunar Science Conference, Vol. 1, *Geochim. Cosmochim. Acta*, Supplement 3, MIT Press, 1972, p. 2479)

present is identical with that in pure iron, indicating that the sample contains no nickel impurity (see later). This result was contrasted with data for two iron meteorites, which were known to contain respectively 6 and 16% nickel and which give hyperfine fields 1.03 and 1.026 times greater than that in pure iron.[327]

The remaining spectra in Figure 22 correspond to various density fractions from sample 14259,17. The heaviest fraction shows the narrowest lines and, although the degree of separation is disappointing, there is substantial enhancement in the olivine resonance. The poorly resolved resonance from the lightest fraction is associated with a high glass content

[326] T. C. Gibb, R. Greatrex, N. N. Greenwood, and M. H. Battey, Proc. 3rd Lunar Science Conference, Vol. 1, *Geochim. Cosmochim. Acta*, Supplement 3, MIT Press, 1972, p. 2479.
[327] T. V. Malysheva, Proc. 3rd Lunar Science Conference, Vol. 1, *Geochim. Cosmochim. Acta*, Supplement 3, MIT Press, 1972, p. 105.

Mössbauer Spectroscopy

and is also noteworthy in containing a discernible residual quantity of metallic iron. The greater part of the latter is therefore associated with the glassy phases. Fractionation with a hand magnet was more successful and yielded 7 mg of a highly magnetic fraction (from an initial sample weighing 4.5 g), the spectrum of which is shown in Figure 23. The sample contains a comparatively small silicate residue and the majority of the iron is ferromagnetic with a magnetic hyperfine field of 333 ± 1 kG at 295 K, 345 ± 1 kG at 77 K, and 346 ± 1 at 4.2 K. These values compare with fields of 330, 337, and 338 kG for pure iron. The majority of metals which alloy with iron reduce the hyperfine field, the exceptions being cobalt and nickel. The data are consistent with an average nickel content of 3 at.% and this figure was confirmed by X-ray fluorescence and electron microprobe analysis.[326] The spectrum of the soil 14259,69 has also been analysed.[328]

Spectra of partial magnetic separates from Apollo 11 fines 10084, taken at various temperatures and applied magnetic fields, have yielded a value for the magnetic hyperfine field at the iron nucleus at 5 K of H_{hf} = 340.6 ± 1.0 kG, compared with a redetermined value of 339.7 kG for pure iron metal. This determination places an upper limit of 1.5% on any nickel admixture into the iron grains, which is taken as strong evidence that the metal originates from a reduction process rather than by direct meteoritic addition. The iron was believed to be present in the superparamagnetic, as well as in the ferromagnetic, state and it was estimated that approximately 20% of the iron grains were between 13.4 and 8.5 nm in diameter and that there were considerably fewer grains in the size range 8.5—2.0 nm. An upper limit of 0.04% was placed on the amount of magnetite or similar Fe^{3+}-containing magnetic spinel present in the 10084,85 fines and this was estimated to be about an order of magnitude lower than that required to account for the characteristic ferromagnetic resonance observed in the fines.[329]

Spectra for the rock chips 14301,15, 14303,36, 14310,66, 14311,32, 14318,35, and 14321,179 [326] and the rock samples 10048, 12053, 14047,47, 14053,48, 14063,47, 14301,65, and 14303,35 [328] have been analysed in detail. The spectra have significantly narrower resonance lines than those of the soils, attributable to the much lower glass content, and show greater differences from sample to sample. Rock 14310,66 appears to have a site occupancy of $(Mg_{0.90}Fe_{0.10})_{M1}[Mg_{0.02}Fe_{0.46}Ca_{0.52}]_{M2}Si_2O_6$, the high degree of cation order being consistent with equilibration at a temperature of about 900 K.[326] Rock 14053 is exceptional in having a very high content of nearly pure metallic iron (see Figure 24) and it is claimed that in the spectrum at

---

[328] F. C. Shchwerer, G. P. Huffman, R. M. Fisher, and T. Nagata, Proc. 3rd Lunar Science Conference, Vol. 1, *Geochim. Cosmochim. Acta*, Supplement 3, MIT Press, 1972, p. 3173.

[329] R. M. Housley, R. W. Grant, and M. Abdel-Gawad, Proc. 3rd Lunar Science Conference, Vol. 1, *Geochim. Cosmochim. Acta*, Supplement 3, MIT Press, 1972, p. 1065.

86 K there is evidence of an additional magnetic component, with a hyperfine field of about 400 kG, due possibly to a spinel. Rock 14063 has an exceptionally high olivine content, whereas 14047 contains a large amount of glass and gives a broad magnetic component suggestive of a high nickel content. In 14303 most of the iron in the pyroxene phase occupies the $M2$ sites. For samples 10048 and 12053 it was shown that increases which

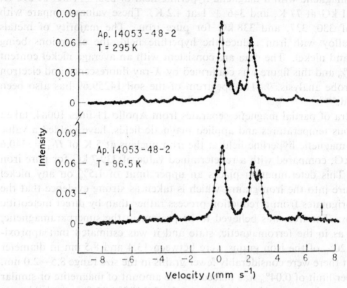

Figure 24 *Mössbauer spectra of* 14053. *The arrows denote the positions of the outermost peaks for the weak magnetic phase discussed in the text*
(Reproduced by permission from Proceedings of the Third Lunar Science Conference, Vol. 1, *Geochim. Cosmochim. Acta*, Supplement 3, MIT Press, 1972, p. 3173)

occur in the electrical conductivity on heating the sample are paralleled by changes in the Mössbauer spectrum, corresponding to a redistribution of about 2% of the $M2$ iron into the $M1$ sites and, for 10048 only, to a dramatic increase of about 30% in the ilmenite resonance. For samples 12053 and 14303 interesting relaxation phenomena were observed at 2 K.[328]

Cation distributions derived from Mössbauer spectra and exsolution relationships from single-crystal X-ray studies have been compared for three pyroxene fractions from Apollo 12 rock 12021, and it is claimed that the Mössbauer measurements indicate an anomalous excess of cations in $M2$ sites. It is suggested that the anomaly is a result of a bias in the Mössbauer measurements owing, firstly, to the presence of large amounts of calcium and other minor impurities, which produce differing local environments for the iron atoms in $M1$ sites and split the $M1$ absorption so that part of it lies beneath the $M2$ doublet and, secondly, to non-superposition of the $M1$ doublets for the exsolved pigeonite and augite in each fraction.[330]

It was pointed out that this effect would be more pronounced in augite than in pigeonite, because samples of the latter contain much less calcium and smaller amounts of exsolved augite lamellae. Furthermore, the cation distributions in augite and pigeonite phases, as determined by Mössbauer spectroscopy, are statistical averages because they both show a wide range in chemical composition as well as extensive exsolution. Despite these difficulties, the technique is still considered to be extremely useful for determining the relative cooling histories of lunar rocks. The pigeonites from rocks 12021, 12053, 12038, and 14310 all show a high degree of cation order, consistent with a final equilibration at temperatures of about 970 K. The presence of secondary generation exsolution lamellae in these samples indicates that the rocks were reheated to these temperatures some time after crystallization from the melt at a much higher temperature (~ 1470 K). Rock 14053 shows the highest degree of cation disorder and probably equilibrated at a temperature of about 1100 K. However, second-generation exsolution lamellae are absent so the duration of reheating must have been very short. The augites from rocks 12021, 12053, and 14053 all show high degrees of cation order which correspond to equilibration temperatures similar to those deduced for the pigeonites from the same rocks.[331]

The distribution of Mg^{2+} and Fe^{2+} ions over the $M1$ and $M2$ positions in pyroxene separated from the basaltic rocks 14053 and 14310 has been studied in order to analyse their subsolidus cooling history. The results suggest that the pigeonite from 14053,47 equilibrated at approximately 1100 K and must have been quenched extremely rapidly by impact at Fra Mauro as a fragment of a larger body from the Imbrian ejecta. By contrast, the orthopyroxene from 14310,116 probably equilibrated at about 870 K at a depth of several metres in a coherent body of appreciable size.[332] This result is in excellent agreement with that discussed earlier for rock 14310,66.[326]

The centre of gravity of the total resonance area of ^{57}Fe in plagioclases from rocks 14053, 14310, and 15415 and from some terrestrial anorthosites and basalts has been tentatively interpreted in terms of the Fe^{3+}/Fe_{total} ratio. The values for the lunar samples range from 0.04 to 0.12 and those for the terrestrial materials from 0.18 to 0.57. The values can be correlated with the oxygen partial pressure conditions which prevailed during crystallization.[333]

---

[330] E. Dowty, M. Ross, and F. Cuttitta, Proc. 3rd Lunar Science Conference, Vol. 1, *Geochim. Cosmochim. Acta*, Supplement 3, MIT Press, 1972, p. 481.

[331] S. Ghose, G. Ng, and L. S. Walter, Proc. 3rd Lunar Science Conference, Vol. 1, *Geochim. Cosmochim. Acta*, Supplement 3, MIT Press, 1972, p. 507.

[332] K. Schürmann and S. S. Hafner, Proc. 3rd Lunar Science Conference, Vol. 1, *Geochim. Cosmochim. Acta*, Supplement 3, MIT Press, 1972, p. 493.

[333] K. Schürman and S. S. Hafner, Proc. 3rd Lunar Science Conference, Vol. 1, *Geochim. Cosmochim. Acta*, Supplement 3, MIT Press, 1972, p. 615

The temperature dependence of the Mg, Fe distribution in a lunar olivine has been discussed.[334]

Previous applications of Mössbauer spectroscopy in mineralogy have been reviewed [25] but, compared with last year, relatively few papers dealing with new work on terrestrial minerals have appeared. Hyperfine interactions of Fe^{2+} in ilmenite, $FeTiO_3$, are discussed in the previous section.[321]

The technique has been used to determine the temperatures of formation of samples of olivine. The degree of ordering was the same in olivine samples heated at 1000 °C for 2 days or at 1100 °C for 5 days but disordering occurred in samples heated at 1150 °C for 5 h. It was therefore concluded that the olivine was formed originally at a temperature of 1100—1150 °C. Thermometric studies indicated that the sample had equilibrated at 1100—1130 °C.[335]

The Fe^{2+} and Fe^{3+} ions in tourmalines have both been shown to occupy six-co-ordinate sites, with substantial amounts of iron in the smaller octahedra generally assumed to be filled with Al^{3+} ions. In titaniferous garnets the Fe^{3+} ions not only occupy the octahedral sites, but also substitute for silicon in the very small tetrahedral sites in amounts proportional to the titanium content of the garnets. Discrepancies between the Fe^{2+} content of garnets as determined by Mössbauer spectroscopy and by chemical analysis are attributed to the presence of Ti^{3+} in the garnets. The geochemical significance of these results lies in the observation that many minerals are in fact chemically resistant or contain significant amounts of titanium, with the result that errors in chemically determined Fe^{2+}/Fe^{3+} ratios and unusual cation site-occupancies may be prevalent in certain rock-forming silicates.[336]

The phosphate minerals triplite, zwieselite, triploidite, and wolfeite have been shown to contain non-equivalent Fe^{2+} ions. Line-broadening which is present in the spectra is thought to indicate microscopic disorder due to half occupied F sites in the fluoride minerals and to sets of closely related metal sites in the hydroxy minerals.[337]

Other minerals studied include biotite,[338] wollastonite,[339] iron–manganese nodules from the Pacific Ocean,[340, 341] and arfvedsonite and aegirine–augite from the Joan Lake agpaitic complex, Labrador.[342] Mössbauer spectroscopy has also been used in the backscattering mode to monitor pyritic oxidation. It is possible to distinguish between unreacted and

[334] D. Virgo and S. S. Hafner, *Earth and Planetry Sci. Letters*, 1972, **14**, 305.
[335] T. V. Malysheva, B. P. Romanchev, and V. D. Shvagerov, *Geokhimiya*, 1972, 496.
[336] R. G. Burns, *Canad. J. Spectroscopy*, 1972, **17**, 51.
[337] E. S. Kostiner, *Amer. Mineral.*, 1972, **57**, 1109.
[338] E. V. Pol'shin, I. V. Matyash, V. E. Tepikin, and V. P. Ivanitskii, *Kristallografiya*, 1972, **17**, 328.
[339] I. Shinno, *Kyushu Daigaku Kyoyobu Chigaku Kenkyu Hokoku*, 1972, **17**, 51.
[340] A. Z. Hrynkiewicz, A. J. Pustowka, B. D. Sawicka, and J. A. Sawicki, *Phys. Status Solidi (A)*, 1972, **9**, K159.
[341] A. Z. Hrynkiewicz, A. J. Pustowka, B. D. Sawicka, and J. A. Sawicki, *Phys. Status Solidi (A)*, 1972, **10**, 281.
[342] S. K. Singh and M. Bonardi, *Lithos*, 1972, **5**, 217.

oxidized pyrite minerals from spectra of the mineral surface obtained through 2 mm of water.[343]

Chalcogenides. Tetragonal FeS has been studied to resolve discrepancies in the literature. The observed spectrum contains only an unresolved quadrupole doublet with an isomer shift of 0.62 mm s^{-1}. It was suggested that magnetic components seen by other workers probably arose from impurities of other sulphides. The results were discussed in terms of a tentative bonding scheme for the compound; the iron is considered to be low-spin.[344] Solid solutions in the wurtzite (ZnS)–troilite (FeS) system have been studied.[345] The anisotropy of the Mössbauer effect in pyrites (FeS$_2$) [175] and pyritic oxidation studies [343] have already been referred to.

The thiospinel Fe[Cr$_2$]S$_4$ has been studied in detail over the temperature range 2—500 K and in zero and externally applied magnetic fields. The magnetically split spectra were fitted with a zero value of the asymmetry parameter over the temperature range 7—170 K, but for $T < 4.2$ K a non-zero value was required. Experiments in applied magnetic fields at 7 and 81 K showed that the quadrupole interaction, which appears below the magnetic ordering temperature, is not produced by a crystallographic distortion. Asymmetric line-broadening, observed in some of the spectra, was attributed to the presence of relaxation effects in the compound. These observations, coupled with the temperature dependence of the isomer shift which shows deviations from the Debye model behaviour, were explained in terms of a model for the compound. It was suggested that the sixth d-electron is localized at both very high and very low temperatures, whereas at intermediate temperatures it occupies a very narrow band formed by electron transfer between octahedral and tetrahedral sites. Below about 179 K most aspects of the spectra can be explained in terms of localized behaviour and a dynamic Jahn–Teller effect which at very low temperatures resolves itself into a static distortion of the tetrahedral sites.[346] More detailed experiments at low temperature have substantiated this suggestion that the electric-field gradient arises from the Jahn–Teller stabilization of the $^5E_g(Fe^{2+})$ ground state. The stabilization energy, $\Delta/k \sim 28$ K ($0 \leqslant T \leqslant 8$ K), is at least an order of magnitude smaller than in the corresponding oxide FeCr$_2$O$_4$.[347] The quadrupole splitting in Cd$_{0.98}$Fe$_{0.02}$Cr$_2$S$_4$ below the Curie temperature (96 K) is thought to be magnetically induced. The observed temperature dependences of the quadrupole and magnetic hyperfine splittings were rationalized in terms of the combined action of a crystal field, an exchange field, and a spin–orbit interaction.[348]

---

[343] R. A. Baker, *Water Res.*, 1972, **6**, 9.
[344] A. Kjekshus, *Acta Chem. Scand.*, 1972, **26**, 1105.
[345] V. V. Kurash, T. V. Malysheva, V. D. Shvagerev, V. I. Goldanskii, and A. Ya. Volkova, *Geokhimiya*, 1972, **5**, 568.
[346] M. R. Spender and A. H. Morrish, *Canad. J. Phys.*, 1972, **50**, 1125.
[347] M. R. Spender and A. H. Morrish, *Solid State Comm.*, 1972, **11**, 1417.
[348] A. M. van Diepen and R. P. van Stapele, *Phys. Rev. (B)*, 1972, **5**, 2462.

The magnetic ordering in cubanite, Cu_2FeSnS_4, has been studied by Mössbauer spectroscopy. The spectrum at 4.2 K is shown in Figure 25 and reveals that the internal magnetic field, $H_{int} = 205$ kG, is perpendicular to the tetragonal electric-field gradient axis and that $\frac{1}{2}e^2qQ = -2.7$ mm s^{-1}. The negative sign of the quadrupole interaction was corroborated by

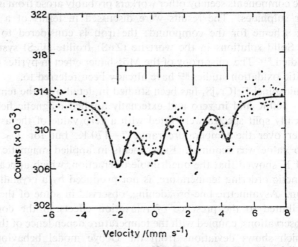

Figure 25 *Mössbauer spectrum of* $Cu_2{}^{57}FeSnS_4$ *at* 4.2 K. *The full line is a computer fit to the case of* $H_{int} \perp z$ *and negative* $e^2qQ/2$
(Reproduced by permission from *J. Phys. and Chem. Solids*, 1972, **33**, 1873)

the spectrum obtained at room temperature in an applied magnetic field of 28 kG and indicates that the orbital ground state for the tetrahedral Fe^{2+} ion is $|3z^2 - r^2\rangle$. This is separated from the first excited state $|x^2 - y^2\rangle$ by 1700 K. The negative sign disagrees with a point-charge calculation, which neglects important dipolar terms, but calculations of the single-ion anisotropy based on the observed negative sign correctly predict the relative orientation of the magnetic- and electric-field gradient axes. The ^{119}Sn spectrum of cubanite is discussed on p. 590.[349] The sulphides $Cu_4Fe_4S_8$, $Cu_{18}Fe_{16}S_{32}$, and $Cu_9Fe_9S_{16}$ have also been studied.[350]

Preliminary measurements have been reported on the ternary selenides listed in Table 2. These six selenides are of the type $M_3\square X_4$ with ordered vacancies. Half of the iron atoms are located in (001) vacancy planes and other iron and M atoms are statistically distributed in full planes. The compounds are all ferrimagnetic and show a metallic type conductivity. In the paramagnetic region the Mössbauer spectra exhibit resonances which correspond to neither ionic Fe^{2+} nor Fe^{3+}, and in the magnetically

[349] U. Ganiel, E. Hermon, and S. Shtrikman, *J. Phys. and Chem. Solids*, 1972, **33**, 1873.
[350] M. G. Townsend, J. L. Horwood, S. R. Hall, and J. L. Cabri, *Amer. Inst. Phys. Conf. Proc.*, 1972, No. 5 (Pt. 2), 887.

ordered region the magnetic hyperfine fields recorded for the two iron sites (see Table 2) are much less than those normally found in ionic oxides. These results indicate that extensive electron delocalization occurs in these compounds and that an ionic model is inappropriate.[351]

Table 2

Compound	Internal fields $(H \pm 5)/\text{kG}$
$TiFe_2Se_4$	221, 130
$CrFe_2Se_4$	248, 141
Fe_3Se_4	230, 110
$CoFe_2Se_4$	115, 75
$NiFe_2Se_4$	70, 35

Spectra have been obtained for the iron tellurides $FeTe_x$ ($x = 0.95$, 1.50, or 1.97) at room temperature and have been fitted with two doublets from two non-equivalent iron atoms. The isomer shifts for each iron atom are similar and are compatible with $3d^2 4s 4p^3$ hybridization.[352]

5 Tin-119

A number of papers containing information relevant to this section are mentioned above.[52, 88, 96]

General Topics.—The magnetic moment of the 89 keV excited state of ^{119}Sn in a cubic Co–Fe matrix has been detected by Mössbauer measurements at 30—75 mK to be $-1.40 \pm 0.08 \mu_N$.[353] The quadrupole moment of this $\frac{11}{2}-$ state has also been determined with sources of ^{119}Sn(OH)$_2$ at temperatures between 14 mK and 4.2 K. The experiment is based on the fact that nuclear orientation in the $\frac{11}{2}-$ state at low temperatures leads to an alignment in the $\frac{3}{2}+$ state which results in an asymmetry in the intensities of the two quadrupole lines. The value obtained was $Q_{11/2} = -0.13 \pm 0.04$ b.[354]

A value has also been calculated for the quadrupole moment of the 23.9 keV $\frac{3}{2}+$ state of ^{119}Sn by comparing ^{119}Sn and ^{121}Sb quadrupole splittings for isoelectronic and isostructural compounds (see Figure 26). All of the antimony compounds and two of the tin compounds, $[Me_3SnCl_2]^-$ and $[Ph_3SnCl_2]^-$, are known to have negative e^2qQ values and it is assumed that the other R_3SnX_2 compounds have negative values also. From the slope of the graph the quadrupole moment was calculated to be $-0.062 \pm 0.02 \times 10^{-28}$ m^2. The quadrupole splitting of the $SnCl_5^-$ ion was deduced to be negative.[355]

---

[351] B. Lambert-Andron, G. Berodias, and D. Babot, *J. Phys. and Chem. Solids*, 1972, **33**, 87.
[352] V. Fano and I. Ortalli, *Phys. Status Solidi (A)*, 1972, **10**, K121.
[353] D. F. Gumprecht, T. E. Katila, L. C. Moberg, and P. O. Lipas, *Phys. Letters (A)*, 1972, **40**, 297.
[354] G. N. Beloserski, D. M. Gumprecht, and P. Steiner, *Phys. Letters*, 1972, **42B**, 349.
[355] G. M. Bancroft, K. D. Butler, and E. T. Libbey, *J.C.S. Dalton*, 1972, 2643.

Spectra have been obtained for ^{119}Sn atoms isolated in rare-gas matrices at 4.2 K and are dominated by a single line with an isomer shift of $+3.21 \pm 0.01$ mm s^{-1} relative to BaSnO$_3$ at 300 K (see Figure 27). The shift is independent of rare-gas matrix and tin atomic concentration. The resonance is ascribed to an isolated tin monomer with the atomic

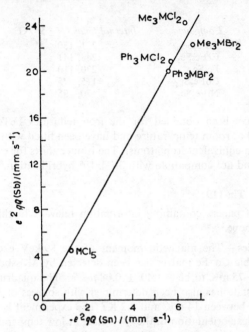

Figure 26 e^2qQ for Sb^V compounds plotted against e^2qQ for isoelectronic Sn^{IV} compounds

configuration $4d^{10}5s^25p^2$. A comparison of the isomer shift data for tin compounds with those for the isolated monomer (see Figure 28) suggests that the electron densities at the tin nuclei in tin compounds are higher than the electron densities in the corresponding free-ion configurations by a factor of 1.25, because of solid-state effects. Detailed analysis of the isomer shift data yields a value of $\Delta R/R = 7.3 \times 10^{-5}$ for the 23.9 keV γ-transition. At higher concentrations a weak quadrupole–doublet resonance, assigned to tin dimers, appears in the spectrum (see Figure 27). The observed quadrupole splitting of 3.5 ± 0.25 mm s^{-1} agrees closely with previous estimates of the quadrupole splitting for a single p_z electron in stannous compounds and it is therefore assumed that the tin dimer contains both 5p electrons in either the p_x or p_y level (i.e. $|q_{\text{dimer}}| = 2|q_{p_x(p_y)}| = |q_{p_z}|$). The data yield a value of $|Q| = 0.065 \pm 0.005$ b for the quadrupole moment of the $I = \frac{3}{2}$ excited state of ^{119}Sn.[356] Studies of ^{119}Sn isolated in solid nitrogen are discussed above.[137]

[356] H. Micklitz and P. H. Barrett, *Phys. Rev. (B)*, 1972, **5**, 1704.

Preliminary spectra have also been given (see Figure 29) for matrix-isolated SnO molecules. The large quadrupole splitting of *ca.* 4.10 mm s^{-1} yields a lower limit of *ca.* 3.0 mm s^{-1} for the quadrupole splitting due to one p_z electron. The more concentrated matrices give spectra with additional doublets, corresponding to simple polymers [*e.g.* Figure 29(c) and

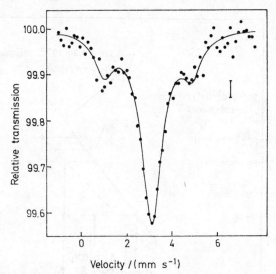

Figure 27 *Mössbauer spectrum of* ^{119}Sn *in an argon matrix at* 4.2 K. Ar/Fe = 60; 205 μg cm^{-2} *tin* (84% *enriched* ^{119}Sn). *The solid curve is a computer fit to the monomer and dimer resonances*
[Reproduced by permission from *Phys. Rev.* (*B*), 1972, **5**, 1704]

(d)]. Similar peaks are also observed in the spectra of matrix b after annealing at 34.5 K.[357]

Accurate values of 1.328 ± 0.003 × 10^{-18} cm^2 for the maximum resonance cross-section of ^{119m}Sn, σ_0, and of 0.63 ± 0.02, 0.45 ± 0.02, and 0.050 ± 0.005 for the *f*-factors of BaSnO$_3$, SnO$_2$, and white tin have been determined by use of a high-resolution silicon–lithium drifted detector.[89]

Mössbauer emission spectra of ^{119}Sb-labelled antimony, Sb$_2$Te$_3$, and Sb$_2$S$_3$ have been obtained (see Figure 30) and indicate the appearance of Sn^0, Sn^{II}, and Sn^{II} + Sn^{IV} states, respectively, in the three solids. The final state of ^{119}Sn therefore depends on the properties of the matrices rather than on the direct effects of the electron-capture decay and the subsequent Auger process. The difference in the distribution of the ^{119}Sn in the

---

[357] A. Bos, A. T. Howe, B. W. Dale, and L. W. Becker, *J.C.S. Chem. Comm.*, 1972, 730.

two chalcogenides of antimony(III) suggests the importance of the electronegativity of the ligand in determining the valence state of the tin. Sb_2S_3 has a relatively low electrical conductivity, in contrast to the other two materials, and because of the low mobility of the electrons in the solid the appearance of more than one valence state of tin is not unexpected.[358] The emission spectra of frozen solutions of $SnCl_2$ in HCl, Me_2CO, and

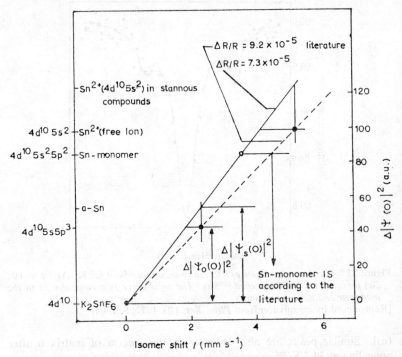

Figure 28 *Correlation between the electron density* $|\psi(0)|^2$ *at the nucleus and the isomer shift for* ^{119}Sn. *The open circle is the measured shift for the tin monomer; the filled circles are taken from the literature*
[Reproduced by permission from *Phys. Rev. (B)*, 1972, **5**, 1704]

MeOH, previously exposed to air, have been measured at 80 K. The systems with HCl and Me_2CO were shown to contain only tin(IV), whereas MeOH stabilized tin in the bivalent state.[359]

In addition to the cubic-to-tetragonal ferroelectric phase transition at 393 K, $BaTiO_3$ also exhibits a tetragonal-to-orthorhombic transition at 278 K and an orthorhombic-to-rhombohedral transition at 183 K. The two lower transitions have now been investigated with the aid of ^{119}Sn

[358] F. Ambe, H. Shoji, S. Ambe, M. Takeda, and N. Saito, *Chem. Phys. Letters*, 1972, **14**, 522.
[359] S. I. Bondarevskii and V. A. Tarasov, *Radiokhimiya*, 1972, **14**, 162.

Mössbauer spectroscopy. The spectra obtained with sources of ^{119}Sn diffused into BaTiO$_3$ show only the presence of tin(IV) in the lattice. The normalized resonance area and the percentage absorption both show minima at the two transition temperatures, consistent with Cochran's

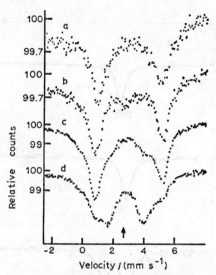

Figure 29 ^{119}Sn *Mössbauer spectra obtained at* 4.2 K *from matrices having* SnO : Ar ca. 1 : 10 000 (a) *and* SnO : N$_2$ ca. 1 : 3000 (b) 1 : 400 (c), *and* 1 : 100 (d). Sn *content* 20—400 μg cm^{-2} (90% ^{119}Sn). *The arrow indicates the position of a tin impurity peak derived from the thermocouple solder*

suggestion that these two phase transitions are caused by lattice instabilities.[360]

Spectra have been obtained for ^{119}Sn impurity atoms in MnO [361] and transferred hyperfine fields have been observed at ^{119}Sn nuclei in the magnetically ordered perovskites Y$_{0.9}$Ca$_{0.1}$Fe$_{0.9}$Sn$_{0.1}$O$_3$,[362, 363] La$_{0.9}$Ca$_{0.1}$-Cr$_{0.9}$Sn$_{0.1}$O$_3$, and La$_{0.7}$Ce$_{0.3}$Mn$_{0.9}$Sn$_{0.1}$O$_3$.[363] At 85 K the field is large (170 kG) in the ferrite, but small (30 kG) in the chromite. These results were explained on the basis of a MO method which indicates that direct transfer of spin density to the 5s orbitals of tin is only possible *via* the e_g orbitals of the paramagnetic constituent and that these are unoccupied in Cr^{3+}(t_{2g}^3). The spectrum of the ferromagnetic manganite indicates a

---

[360] V. G. Bhide and V. V. Durge, *Solid State Comm.*, 1972, **10**, 401.
[361] P. B. Fabrichnyi, E. V. Lamykin, A. M. Babeshkin, and A. N. Nesmeyanov, *Soviet Phys. Solid State*, 1972, **13**, 2874.
[362] I. S. Lyubutin and Yu. S. Vishnyakov, *Kristallografiya*, 1972, **17**, 960.
[363] V. A. Bokov, G. V. Popov, N. N. Parfenova, and G. G. Yushina, *Soviet Phys. Solid State*, 1972, **14**, 83.

large spread of hyperfine fields, ranging from 130 to 200 kG, on account of the distribution of non-equivalent cation-environments.[363] Transferred fields have also been observed in the ferrite garnets $\{Y_{3-x}Ca_x\}[Fe_{2-x}Sn_x](Fe_3)O_{12}$ ($x = 0.1$—0.9) and the existence of a non-collinear spin configuration established for the Fe^{3+} moments in the d-sublattice.[364] The

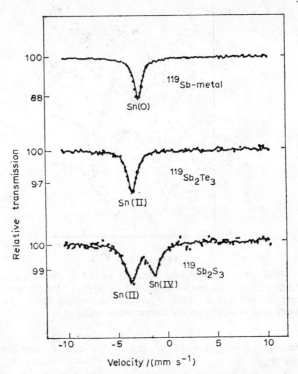

Figure 30 ^{119}Sn *Mössbauer emission spectra of ^{119}Sn-labelled antimony metal, Sb_2Te_3, and Sb_2S_3 at liquid nitrogen temperature versus $BaSnO_3$ at room temperature*
(Reproduced by permission from *Chem. Phys. Letters*, 1972, **14**, 522)

effect of tin impurity atoms on the Morin transition temperature in haematite is described on p. 555.

Spectra have been obtained for ^{119}Sn impurity atoms in gallium in porous glass,[365] and the effects of the glass-crystal transition on the local surroundings of ^{119}Sn impurity atoms in As_2Te_3 and in $As_2Se_3 \cdot As_2Te_3$ have been studied.[366]

[364] I. S. Lyubutin and A. P. Dodokin, *Pis'ma Zhur eksp. i teor. Fiz.*, 1972, **15**, 339.
[365] V. N. Bogomolov, N. A. Klushin, and P. P. Seregin, *Soviet Phys. Solid State*, 1972, **14**, 1729.
[366] P. P. Seregin and L. N. Vasil'ev, *Soviet Phys. Solid State*, 1972, **14**, 1325.

Mössbauer spectra have been presented for palladium-based sols, generated by reduction with Sn^{2+}. From Figure 31 it can be seen that the spectra of the ^{119}Sn atoms in the sol show directly the metal sol core and the inner part of the stabilizing double layer as distinct phases. In Figure 31(a) the Sn^{4+} line is constrained to the position determined in a

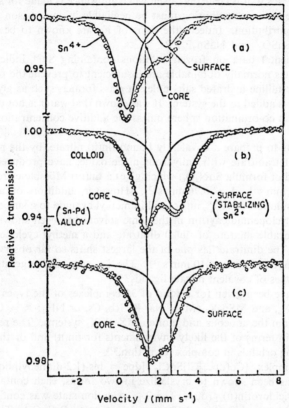

Figure 31 *Mössbauer spectra of frozen Sn–Pd sols. For explanation see text* (Reproduced by permission from *Chem. Phys. Letters*, 1972, **16**, 128)

separate experiment using pure Sn^{4+} dissolved in 4M-HCl. Figure 31(b) shows the spectrum of a centrifugally separated sol in which the Sn^{4+} line is absent and the components of interest are clearly differentiated. Figure 31(c) shows the spectrum of a sample prepared in a similar fashion but having larger particles and therefore a smaller total surface area. These are believed to be the first resonance experiments of any kind to provide information of this type.[367, 368]

[367] R. L. Cohen and K. W. West, *Chem. Phys. Letters*, 1972, **16**, 128.
[368] R. L. Cohen and K. W. West, *J. Electrochem. Soc.*, 1972, **119**, 433.

Tin(II) Compounds.—Estimates of the lattice contribution to the electric-field gradient have been made for the five tin(II) materials SnO, SnSO$_4$, NaSn$_2$F$_5$, CsSnCl$_3$, and SnCl$_2$. The sign of the lattice contribution is positive in all cases except SnCl$_2$, a typical calculated value being 10^{13} e.s.u. Considering that one p_z electron produces a field gradient of about -3×10^{16} e.s.u. it is clear that, even when allowance is made for shielding effects, the lattice contribution must be negligible in comparison with the valence contribution. Indeed, the signs of V_{zz} are known to be negative for SnO, SnSO$_4$, and NaSn$_2$F$_5$.[369]

Experimental data on frozen solutions containing Sn^{2+} indicate that cooling rates normally obtainable are insufficient to prevent the precipitation of crystalline hydrated salts unless a glass former, such as glycerol or methanol, is added to the system. It was shown that water is not displaced from the tin co-ordination sphere unless the additive concentration is very high.[370]

Attempts to prepare a covalently bound tin(II) nitrate, by the reduction of tin(IV) tetranitrate with anhydrous nitric oxide, have produced only a white solid of formula SnN$_2$O$_6$, which gives a tin(IV) Mössbauer resonance ($\delta = 0.29$ mm s^{-1}, $\Delta = 0.96$ mm s^{-1}). However, addition of a freshly prepared solution of 100% nitric acid in methyl cyanide to a suspension of di(methylcyclopentadienyl)tin in the same solvent has been shown to give a non-separable mixture of tin(II) dinitrate and a methyl cyclopentadiene polymer. The dinitrate has one of the largest shifts so far observed for a tin(II) compound ($\delta = 4.10$ mm s^{-1}). The i.r. spectrum suggests that it has two types of covalent nitrate ligands.[371]

Data have been given for a total of 33 complexes of the types MSnX$_3$, MX,MSnX$_3$, and MSn$_2$X$_5$ (M = Na, K, Rb, Cs, or NH$_4$; X = Cl or Br) isolated from the aqueous and molten MX–SnX$_2$ systems. The results are discussed in terms of the likely environments for tin(II) and of the use of tin bonding orbitals in complex formation.[372]

The complex [Co(dpe)$_2$Cl]SnCl$_3$ [dpe = bis-(1,2-diphenylphosphino)-ethane] has been shown to crystallize in two forms, each containing an isolated trichlorotin(II) group. The 2+ oxidation state was confirmed by the large isomer shift ($\delta = 3.10$ mm s^{-1} relative to BaSnO$_3$). This is the first such arrangement found in a transition-metal compound. Normally the SnX$_3^-$ group is linked covalently to the metal and gives an isomer shift in the tin(IV) region.[373] This is found to be the case in the compounds cis- and trans-Fe(SnCl$_3$)$_2$(ArNC)$_4$, cis-FeCl(SnCl$_3$)(ArNC)$_4$, and [Fe-

---

[369] J. D. Donaldson, D. C. Puxley, and M. J. Tricker, *Inorg. Nuclear Chem. Letters*, 1972, **8**, 845.
[370] R. L. Cohen and K. W. West, *Chem. Phys. Letters*, 1972, **13**, 482.
[371] P. G. Harrison, M. I. Khalil, and N. Logan, *Inorg. Nuclear Chem. Letters*, 1972, **8**, 551.
[372] S. R. A. Bird, J. D. Donaldson, and J. Silver, *J.C.S. Dalton*, 1972, 1950.
[373] J. Stalick, D. W. Meek, B. Y. K. Ho, and J. J. Zuckerman, *J.C.S. Chem. Comm.*, 1972, 630.

(ArNC)$_5$SnCl$_3$]ClO$_4$, which have isomer shifts of *ca.* 2 mm s^{-1} relative to BaSnO$_3$. These results show that the formal oxidation state of tin in SnCl$_3$ complexes has significance only when the oxidation state of the other atoms and ligands is well defined. However, the valency of tin in all SnCl$_3$ complexes is four.[236]

The presence of an asymmetric doublet in the Mössbauer spectrum of Sn$_3$BrF$_5$ has now been explained by a crystal-structure determination of this compound by *X*-ray diffraction. Instead of two very distinct tin sites which would be required by a formulation of the type SnF$_2$·SnFBr· SnF$_2$, three very similar tin environments were found in the infinite tin(II) fluoride cationic network, each tin having a pyramidal three-co-ordinated environment.[374]

The first oxidative-addition reactions of stannous halides with carbon–carbon multiply bonded systems to form organotin(II) derivatives have been reported. The i.r. and Mössbauer data for products of the reactions between stannous halides and dimethylacetylene dicarboxylate in dry THF are consistent with the *cis* dimeric structure (12), in which the co-ordination number of each tin atom is raised to six by intermolecular co-ordination *via* the carboxy-groups.[375]

Data have been given for 15 novel tin(II) derivatives obtained by photolysis of dicyclopentadienyltin(II). The 2+ oxidation state was confirmed by the isomer shifts of these compounds which included carboxylates, alkoxides, aryl oxides, oximes, hydroxylamines, metalloxanes, azoles, thiolates, and pseudohalides.[376]

Tin(II) oxalate and phthalate have been studied and are thought to have polymeric structures with bridging carboxylate groups as shown in (13).[377]

The oxidation of tin(II) chalcogenides has been studied.[378]

(12) X = Cl, Br, or I   (13)

Tin(IV) Compounds.—The relationship between ^{119}Sn Mössbauer shifts and atomic parameters, such as Mulliken valence-state electronegativities, has continued to arouse controversy. It appears that equally good correlations are found with both Pauling and Mulliken values.[379, 380]

[374] J. D. Donaldson and D. C. Puxley, *J.C.S. Chem. Comm.*, 1972, 289.
[375] P. G. Harrison, *Inorg. Nuclear Chem. Letters*, 1972, **8**, 555.
[376] P. G. Harrison, *J.C.S. Chem. Comm.*, 1972, 544.
[377] N. W. G. Debye, D. E. Fenton, and J. J. Zuckermann, *J. Inorg. Nuclear Chem.*, 1972, **34**, 352.
[378] P. P. Seregin, S. I. Bondarevskii, V. T. Shipatov, and V. A. Tarasov, *Izvest. Akad. Nauk S.S.S.R., neorg. Materialy*, 1972, **8**, 571.
[379] J. C. Watts and J. E. Huheey, *Chem. Phys. Letters*, 1972, **14**, 89.
[380] R. V. Parish, *Chem. Phys. Letters*, 1972, **14**, 91.

The tin $3d_{\frac{5}{2}}$ electron binding energies for octahedral tin complexes of formula [(MeCH$_2$)$_4$N]$_2$[SnX$_{6-n}$Y$_n$] (X = Y = halogen), determined from X-ray photoelectron spectroscopy (ESCA), have been found to correlate linearly with average ligand electronegativities, Mössbauer isomer shifts, and estimated atomic charges on the tin atom.[381]

The replacement of ethyl groups by halogen atoms in the series Et$_{4-x}$X$_x$Sn (X = halogen) has been shown to increase the isomer shift relative to the tetraethyltin precursor. These results were explained in terms of the relative electronegativities of the ligands, coupled with the rehybridization theory that the s-character of a central atom tends to concentrate in orbitals directed towards electropositive substituents. It was suggested that compounds of the type RSnX$_3$ should show smaller quadrupole splittings than the R$_3$SnX species.[382]

A general MO model for the correlation of Mössbauer quadrupole splitting with stereochemistry has been developed and applied in detail to organotin(IV) compounds. The model, which treats the electric-field gradient at the tin nucleus as a sum of partial field-gradient tensors, was used to discuss the implications of changes in structural type and of distortions from idealized co-ordination geometry. The partial field-gradient associated with a given ligand was shown to be different for tetrahedral, trigonal-bipyramidal-apical, trigonal-bipyramidal-equatorial, and octahedral co-ordination positions, *e.g.* the octahedral value is about 70% of the tetrahedral value. Absolute numerical values for partial field gradient parameters cannot be obtained from experiment and only relative values were discussed. A list of working values for a variety of ligands in tetrahedral or octahedral structures is given in Table 3. The quantity

Table 3 *Working values for partial field-gradient parameters in organotin(IV) compounds*

Tetrahedral structures		Octahedral structures	
Ligand	Value (mm s^{-1})	Ligand	Value (mm s^{-1})
Alkyl	−1.37	Alkyl	−1.03
Ph	−1.26	Ph	−0.95
I	−0.17	I	−0.14
NCS	+0.21	NCS	+0.07
MeCO$_2$	−0.15	½(phen)	−0.04
C$_6$F$_5$	−0.70	½(bipy)	−0.08
C$_6$Cl$_5$	−0.83	DMSO	+0.01
CF$_3$	−0.63	pyO	−0.08
o-CF$_3$C$_6$H$_4$	−1.04	py	−0.10
p-FC$_6$H$_4$	−1.12	½(edt)	−0.56
Co(CO)$_4$	−0.76	CH$_2$CH	−0.96
Mn(CO)$_5$	−0.79	½(dipyam)	−0.17
Re(CO)$_5$	−0.80	½(pic)	+0.06
CpFe(CO)$_2$	−0.91		
HCO$_2$	−0.18		

[381] W. E. Swartz, P. H. Watts, E. R. Lippincott, J. C. Watts, and J. E. Huheey, *Inorg. Chem.*, 1972, **11**, 2632.
[382] N. Watanabe and E. Niki, *Bull. Chem. Soc. Japan*, 1972, **45**, 1.

tabulated is $\frac{1}{2}e^2|Q|([L] - [X])$, where X = F, Cl, or Br, and the sign of Q is negative for the excited state of ^{119}Sn.[383] Quadrupole splittings for fifty compounds calculated by use of these values agree with observed splittings to within 0.4 mm s^{-1} or better.[383]

The signs of the quadrupole coupling constants have been shown to be positive for cis-SnCl$_4$,2MeCN, trans-SnCl$_4$,2PEt$_3$, and trans-SnCl$_4$,2AsEt$_3$, consistent with the order of bond polarity Sn—N (sp-hybridized) > Sn—Cl > Sn—P, Sn—As. The asymmetry parameter, η, is of the order of 0.5 for each compound and suggests a distorted octahedral environment for the central tin atom.[384]

Spectra have been recorded for 25 complexes of the type SnX$_4$L$_2$, where X = Cl, Br, or I; L = R$_3$PO (R = Et, Bu, or Ph), Ph$_3$AsO, Me$_2$SO, Ph$_3$As, R$_3$P (R = Bu, Ph, or C$_8$H$_{17}$), or PhMe$_2$P, or L$_2$ = o-Ph$_2$PC$_6$H$_4$, o-Me$_2$N(C$_6$H$_4$)PPh$_2$, and [o-Me$_2$N(C$_6$H$_4$)]$_2$PPh. In agreement with predictions of the point-charge model it was found that compounds which are known to have trans-structures give quadrupole splittings of approximately 1 mm s^{-1}, whereas those with cis-structures give no resolvable splitting. A cis-configuration is suggested for SnCl$_4$,2PPh$_3$ on the basis of the Mössbauer spectrum, but the i.r. spectrum favours a trans-structure.[385] The 1:1 adducts of SnCl$_4$ with azines (14) and the 2:1 adducts with hydrazones (15) all give zero quadrupole splitting and have similar isomer shifts to each other.[386]

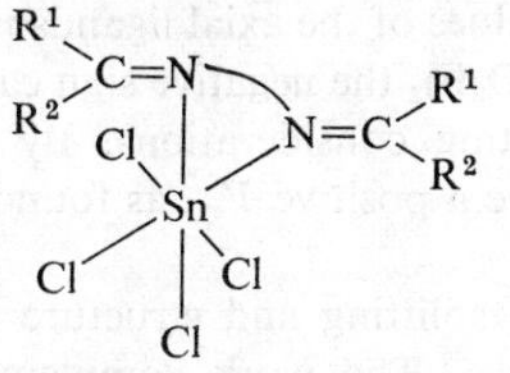

(14) a; R^1 = R^2 = Me
b; R^1 = R^2 = C$_5$H$_{10}$
c; R^1 = Ph, R^2 = H

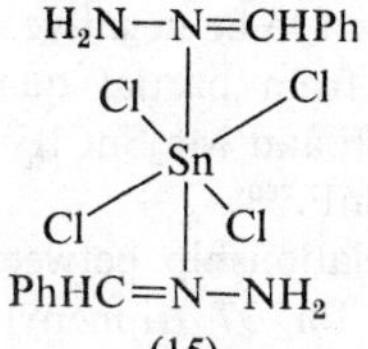

(15)

Data have been given for quick-frozen solutions of SnCl$_4$ and SnI$_4$ in anhydrous (Me$_2$N)$_3$PO, DMF, EtOAc, Me$_2$CO, MeCN, PhNO$_2$, (BuO)$_3$PO, Me$_2$SO, EtOH, and CCl$_4$.[387, 388] The spectrum of solid SnCl$_4$,2DMF is identical with that of a frozen solution of SnCl$_4$ in DMF at 77 K.[389]

Mössbauer and vibrational spectra for the new compounds MeSnF$_3$, MeSnCl$_2$F, Me$_2$SnClF, and SnF$_2$(SO$_3$F)$_2$ indicate that they are all polymeric

---

[383] M. G. Clark, A. G. Maddock, and R. H. Platt, *J.C.S. Dalton*, 1972, 281.
[384] D. Cunningham, M. J. Frazer, and J. D. Donaldson, *J.C.S. Dalton*, 1972, 1647.
[385] P. G. Harrison, B. C. Lane, and J. J. Zuckerman, *Inorg. Chem.*, 1972, **11**, 1537.
[386] C. H. Stapfer, R. W. D'Andrea, and R. H. Herber, *Inorg. Chem.*, 1972, **11**, 204.
[387] A. Vértes, K. Burger, and S. Nagy, *Magyar. Kém. Folyóirat*, 1972, **78**, 476.
[388] A. Vértes and K. Burger, *J. Inorg. Nuclear Chem.*, 1972, **34**, 3665.
[389] W. G. Movius, *J. Inorg. Nuclear Chem.*, 1972, **34**, 3571.

```
    \    Me   /
     F   |   F
      \  |  /
       Sn
      /  |  \
     F   |   F
    /    F    \
         |
        (16)
```

```
         F
     Cl  |
      \  |
       Sn—Me
      /  |
     Cl  F
         |
        (17)
```

```
         F
         |
    Me\  |
       Sn—Cl
    Me/  |
         F
         |
        (18)
```

```
         O
         ‖
      O..S..O
       \ | / F
        \|/ /
      O..Sn..O
        /|\ \
       / | \ O
      O  F
         |
        (19)
```

and have the repeating units shown in structures (16)—(19). There is evidence that $SnCl_2F_2$ does not have a tetrahedrally co-ordinated configuration with C_{2v} symmetry, as suggested earlier, but the true structure of this compound is still in doubt. The signs of the electric-field gradients in these and some related compounds were deduced. For the fluorosulphate compounds $Me_2Sn(SO_3F)_2$, $Me_2Sn(SO_3CF_3)_2$, Me_3SnSO_3F, and $Me_2SnCl_2SO_3F$, and the fluorides Me_2SnF_2, $MeSnF_3$, and SnF_4, a plot of Δ as a function of the sum of the Hammett σ values of the axial ligands is linear if V_{zz} is taken to be negative. For $Cl_2Sn(SO_3F)_2$ the negative sign can also be deduced from partial quadrupole splitting considerations. By contrast, $MeSnCl_2F$ and Me_2SnClF appear to have a positive V_{zz}, as found recently for Me_3SnF.[390]

The relationship between quadrupole splitting and structure has been analysed for 37 triphenyltin compounds. The work complements two recent studies which were discussed in last year's Report. The three

```
       X              Ph              X⎤  Ph
       |  Ph          |               ⎣Sn
    Ph—Sn            Sn                |  Ph
       |  Ph       Ph/ \Ph             |
       X              X               Ph
      (20)          (21)              (22)
```

structures (20)—(22) which are possible for compounds of this type have quadrupole splittings given by the following point-charge expressions:

$$\Delta_{20} \propto -3[R] + 4[X]$$

$$\Delta_{21} \propto -2[R] + 2[X]$$

$$\Delta_{22} \propto (3[R]^2 - 6[R].[X] + 4[X]^2)^{\frac{1}{2}}$$

---

[390] L. E. Levchuk, J. R. Sams, and F. Aubke, *Inorg. Chem.*, 1972, **11**, 43.

It is concluded that triphenyltin nitrate has a structure with bridging nitrate groups and evidence is presented that the oxinate group is chelating in triphenyltin oxinate.[391]

The bis(triorganotin) oxides, $(R_3Sn)_2O$, and the triorganotin hydroxides, R_3SnOH (R = Me, Et, Pr, Bu, Oct, or Ph), have been characterized by i.r. and Mössbauer spectroscopy. The former exhibit quadrupole splittings of 1.18—1.63 mm s^{-1} and are therefore probably tetrahedral monomers, whereas the latter have splittings of 2.78—2.99 mm s^{-1} and are therefore probably co-ordinatively associated into linear polymers containing five-co-ordinate tin, as in (23).[392]

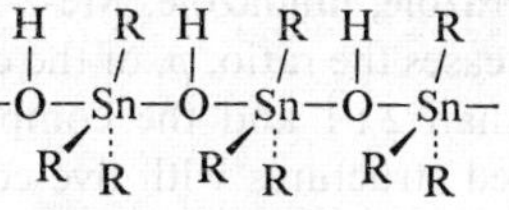

$$\begin{array}{cccccc} & H & & R & & H & & R \\ & | & / & | & / & | & / \\ -O- & Sn & -O- & Sn & -O- & Sn- \\ & / & | & / & | & / & | \\ & R & R & R & R & R & R \end{array}$$

(23)

The tributyltin alkoxides $Bu_3Sn(OR)$ (R = Me or Ph) have been shown to give similar quadrupole splittings and probably have a structure similar to (23) in the solid at 77 K. The compact methyl and planar phenyl groups, like hydrogen, are able to occupy the spaces between the planar R_3Sn moieties. By contrast, the larger organic substituents, R = Et, Pr^n, Bu^t, and CPh$_3$, disrupt this O—Sn co-ordination. The acyclic dialkoxides $Bu_2Sn(OR)_2$ (R = Me or Et) have quadrupole splittings of 2.32 and 2.00 mm s^{-1}, respectively, and are thought to be polymeric with the tin atom occupying a *cis*-R_2SnX_4 configuration (24). The quadrupole splittings for three dibutyltin 1,2-glycoxides are slightly larger (2.72—2.85 mm

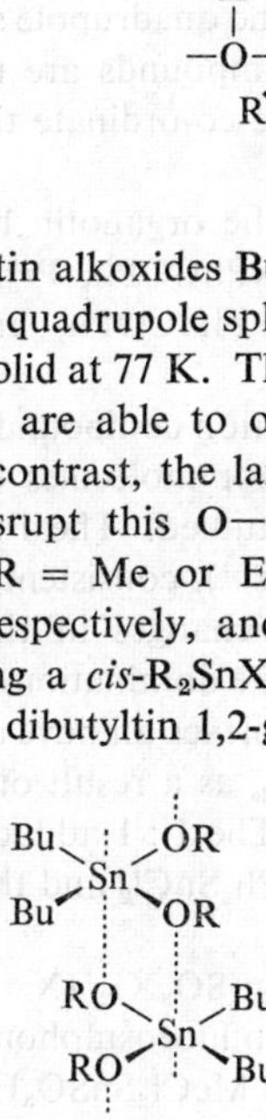

(24) 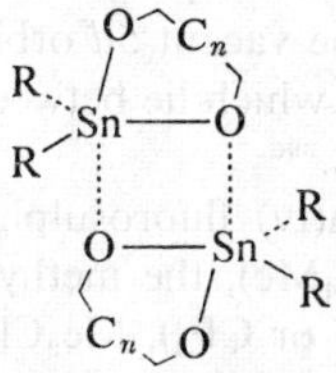   (25) $n = 1$—3

s^{-1}) and are consistent with the presence of the five-co-ordinate dimer (25) in the solid state. The quadrupole splittings for four dialkyltin catechoxides are the largest so far observed for any organotin alkoxide (3.35—3.60 mm s^{-1}) and are consistent with a linear polymeric structure (26) in which the tin atoms are six-co-ordinate and occupy a distorted *trans*-octahedral R_2SnX_4 site.[393]

[391] R. C. Poller and J. N. R. Ruddick, *J. Organometallic Chem.*, 1972, **39**, 121.
[392] J. M. Brown, A. C. Chapman, R. Harper, D. J. Mowthorpe, A. G. Davies, and P. J. Smith, *J.C.S. Dalton*, 1972, 338.
[393] P. J. Smith, R. F. M. White, and L. Smith, *J. Organometallic Chem.*, 1972, **40**, 341.

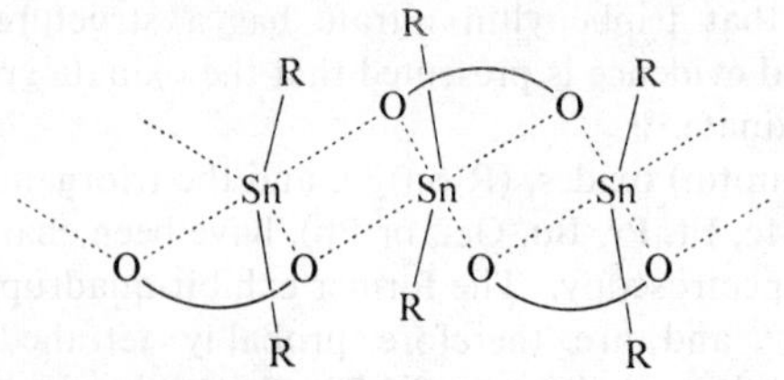

(26) R = Me, Et, Bu, or Oct

Data have been given for eleven organostannylazoles $R_3^1SnR^2$ (R^1 = Me, Et, or Ph; R^2 = pyrazole, imidazole, Me-2-imidazole, benzimidazole, or benzotriazole). In all cases the ratio, ρ, of the quadrupole splitting to the isomer shift is greater than 2 : 1 and the compounds are therefore considered to have associated structures with five-co-ordinate tin in the solid state.[394]

Preliminary data have been recorded for the organotin hydroxylamine derivatives $Me_3Sn-O-NEt_2$, $Me_3Sn-O-NPhCOPh$, $Pr_3^nSn-O-NPh-COPh$, $Ph_3Sn-O-NPhCOPh$, $Me_3Sn-O-NHCOPh$, and $[NEt_3H]^+$-$[Ph_3Sn-O-NCOPh]^-$.[395]

A large number of both mono- and bis-addition compounds of dimethyl- and diphenyl-tin dichloride with oxygen donor molecules of the general type R_nEO (E = C, N, P, or S) have been studied. The 1 : 2 adducts all have large quadrupole splittings (ca. 4.0 mm s⁻¹), consistent with trans-R_2 configurations, but there are no systematic changes in the quadrupole splitting as the ligand is changed, because of the domination of the electric-field gradient by the R groups. The 1 : 2 complexes all have smaller isomer shifts than that of the parent species R_2SnCl_2, as a result of the increased donation into the vacant 5d orbitals of tin. The 1 : 1 adducts have quadrupole splittings which lie between those for Ph_2SnCl_2 and the corresponding 1 : 2 adducts.[396]

The methyltin(IV) fluorosulphonates $Me_2Sn(SO_3X)_2$ (X = F, CF_3, Cl, Me, Et, or C_6H_4Me), the methyltin(IV) chlorofluorosulphonates $MeClSn(SO_3X)_2$ (X = F or CF_3), $Me_2ClSnSO_3F$, and $MeCl_2SnSO_3F$ have all been shown to contain only one type of tin environment and to be polymeric (i.e. they give a room-temperature Mössbauer effect). The quadrupole splittings ($\Delta \sim$ 5.50 mm s⁻¹) for $Me_2Sn(SO_3F)_2$ and $Me_2Sn(SO_3CF_3)_2$ are larger than any values reported previously. To account for these large values it was suggested that the bonding is highly ionic with essentially linear Me_2Sn cations interacting covalently with SO_3X anions. The great electronegativity of the SO_3X groups then leads to a strong withdrawal of p-electron density in the equatorial plane, with concomitant deshielding

---

[394] R. Gassend, M. Delmas, J.-C. Maire, Y. Richard, and C. More, J. Organometallic Chem., 1972, 42, C29.
[395] P. G. Harrison, J. Organometallic Chem., 1972, 38, C5.
[396] B. V. Liengme, R. S. Randall, and J. R. Sams, Canad. J. Chem., 1972, 50, 3212.

of the tin 5s electrons and a resulting positive isomer shift. Simultaneously, a large imbalance develops in the p-orbital charge density on the tin atom, which results in the exceptionally large electric-field gradients. For the compounds $X_2Sn(SO_3F)_2$ (X = Me, F, Cl, Br, or SO_3F), linear correlations were found between the sums of the Pauling electronegativities for the axial ligands and the isomer shifts, and between the sums of the Taft inductive constants (σ^*) for the axial ligands and the quadrupole splittings. The sign of the electric-field gradient is probably negative for the Me_2Sn-$(SO_3X)_2$ derivatives and positive for the Me_3SnSO_3X compounds.[397] In an extension of this work, data were obtained for the compounds R_2SnX_2 (R = Me, Et, Pr^n, Bu^n, or n-C_8H_{17}; X = SO_3F, SO_3CF_3, or PO_2F_2), all of which have quadrupole splittings greater than 4.0 mm s^{-1} and are therefore thought to be octahedral with *trans*-organo groups and bidentate anionic groups bridging through oxygen. Substitution of Et for Me increases the isomer shift but decreases the quadrupole splitting; both parameters then remain almost constant for R = Pr^n and Bu^n within each series. For a given R group, the numerical values of both the isomer shift and the quadrupole splitting decrease in the order X = SO_3F, SO_3CF_3, PO_2F_2, F. The trends can be rationalized in terms of the ligand electronegativities and point-charge considerations.[398]

Mössbauer data for the dimethylchlorotin carboxylates $Me_2ClSnOOCR$ (R = Me, CH_2Cl, $CHCl_2$, CCl_3, CH_2Br, CF_3, C_2F_5, C_3F_7, or CF_2Cl) have been compared with data from compounds of known structure and it is concluded that these materials are polymeric with five-co-ordinate tin atoms. It is interesting to note that the compounds do not display a Mössbauer effect at room temperature, despite the fact that this is usually taken to be diagnostic of polymeric tin compounds.[399]

^{119}Sn Mössbauer spectroscopy and i.r. techniques have been used to distinguish between open-chain and cyclic structures in some organo-tin and -lead esters for which both structures may occur. Dimethyltin(IV)-phthalate (Δ = 3.63 mm s^{-1}) probably has a highly distorted six-co-ordinate octahedral structure with *trans*-methyl groups, but a five-co-ordinate structure with equatorial methyl groups is not ruled out. The five-co-ordinate structure is favoured for the compounds o-phenylenedi-oxydialkyltin(IV) (alkyl = Me, Et, Bu^n, or Oct) and 2,2'-biphenylenedi-oxydimethyltin(IV) (Δ < 4 mm s^{-1}). The species formed from dimethyl-tin(IV) carbonate in a silver chloride matrix is thought to contain equatorial methyl groups and bridging carbonato-groups in a polymeric trigonal-bipyramidal structure. Dimethyltin oxalate monohydrate (Δ = 4.41 mm s^{-1}) is believed to be *trans*-octahedral, involving bridging carboxylate groups. A sixth position is occupied by a water molecule and one

[397] P. A. Yeats, J. R. Samsand F, . Aubke, *Inorg. Chem.*, 1972, **11**, 2634.
[398] T. H. Tan, J. R. Dalziel, P. A. Yeats, J. R. Sams, R. C. Thompson, and F. Aubke, *Canad. J. Chem.*, 1972, **50**, 1843.
[399] C. S. Wang and J. M. Shreeve, *J. Organometallic Chem.*, 1972, **38**, 287.

carboxylate group is bound as an organic ester (27). All of the tin(IV) compounds have $\rho > 2.1$, which is indicative of some intermolecular co-ordination to the tin atom.[377]

From the temperature dependence of the area under the resonance curve for $Me_2Sn(salen)$, it has been concluded that the molecular unit is monomeric with the two oxygen and two nitrogen atoms in the equatorial plane about the metal atom belonging to one $(salen)^{2-}$ moiety (28). The absence of a Goldanskii–Karyagin asymmetry in the spectrum between 78 and 130 K is thought to reflect the severe distortion of the symmetry of the nearest-neighbour environment around the metal atom from an idealized O_h configuration. The observed quadrupole splitting ($\Delta = 3.46$ mm s^{-1}) is rationalized in terms of a modified point-charge formalism which predicts a value of $|3.27|$ mm s^{-1}. The quadrupole splitting of $Ph_2Sn(salen)$ ($\Delta = 2.84$ mm s^{-1}) is intermediate between those normally observed for cis-octahedral and for six-co-ordinate trans-octahedral diphenyltin(IV) compounds, and it is thought that the phenyl groups cause considerable steric distortions of the molecule.[400]

Data have been given for the adducts $RPhSnCl_2,2Me_2SO$ (R = Me, Et, Pr, Bu, or $PhCH_2$).[401]

Mössbauer and i.r. spectra have been obtained for terpyridyl complexes of dimethyl-, di-n-butyl-, and diphenyl-tin di-isothiocyanates and the corresponding $[R_2Sn(NCS)(terpy)]^+[BPh_4]^-$ compounds, and for the 8-(2-pyridylmethyleneamino)quinoline complexes with di-n-butyl- and diphenyl-tin di-isothiocyanates. I.r. spectra of the neutral complexes indicate seven-co-ordination for the tin atoms, and the Mössbauer parameters indicate the presence of axial C—Sn—C bonds with greater tin s-character than in trans-octahedral complexes. The largest quadrupole splitting ($\Delta = 4.73$ mm s^{-1}) is given by the terpyridyl complex with di-n-butyltin di-isothiocyanate.[402]

Data have been given for 31 compounds containing Sn—S bonds.[403] $Sn(tdt)_4$ (tdt = toluene-3,4-dithiolato) gives a large effect at room temperature and is probably polymeric with six-co-ordinate tin, whereas three other spirocyclic bis(dithiolato)tin compounds $Sn(p2dt)_3$, $Sn(p3dt)_3$, and $Sn(tdt)_3$ (p2dt = propane-1,2-dithiolato, p3dt = propane-1,3-dithiolato,

---

[400] R. Barbieri and R. H. Herber, *J. Organometallic Chem.*, 1972, **42**, 65.
[401] K. L. Jaura and V. K. Verma, *Indian J. Chem.*, 1972, **10**, 536.
[402] J. C. May and C. Curran, *J. Organometallic Chem.*, 1972, **39**, 289.
[403] R. C. Poller and J. N. R. Ruddick, *J.C.S. Dalton*, 1972, 555.

tdt = toluene-3,4-dithiolato), are thought to be weakly associated, with five-co-ordinate tin. Sn(SPh)$_4$ does not give a room temperature effect and has zero quadrupole splitting, consistent with a simple tetrahedral structure. These tetrathiolatotin compounds all form adducts of the type SnX$_4$Y$_2$ (X = S, Y = N or O) with uni- and bi-dentate donor molecules and, apart from trimethylamine oxide, the unidentate ligands give *trans*-configurations with quadrupole splittings of 1.77—1.95 mm s^{-1}, whereas the bidentate ligands give *cis*-structures with quadrupole splittings of 0.81—1.38 mm s^{-1}. The relationship $\Delta(trans) \simeq 2\Delta(cis)$ is therefore valid for compounds of the type SnX$_4$Y$_2$ as well as for R$_2$SnX$_4$ complexes. For the latter, the sign of V_{zz} is predicted to be positive for the *trans*-configuration and negative for the *cis*-configuration, because of the greater donor power of the organo R group compared with X. However, as a result of molecular distortions, this sign reversal has not yet been observed. For the SnX$_4$Y$_2$ compounds, the X group is a better donor than Y and the predicted signs of V_{zz} are negative for the *trans* and positive for the *cis* geometry. This prediction was confirmed for the *cis*-bipyridyl and *trans*-diethyl sulphoxide adducts of bis(ethane-1,2-dithiolato)tin by use of a 60 kG applied magnetic field. A number of monothiolatotin compounds were also studied both in the solid and in frozen pyridine solutions. Ph$_3$Sn·S·C$_5$H$_4$N-4, which is five-co-ordinate in the solid, shows no increase in quadrupole splitting (Δ = 2.6 mm s^{-1}) in the pyridine solution, whereas compounds such as Ph$_3$SnSPh, which is approximately tetrahedral in the solid, shows an increase in quadrupole splitting from 1.41 in the solid to 2.39 mm s^{-1} in the frozen solution where interaction with pyridine occurs.[403]

Single-line Mössbauer spectra have been observed for a number of (3-trialkylstannyl)propyl aryl sulphides of the type R$^1_3$SnCH$_2$CH$_2$CH$_2$SR^2 (R^1 = Me, Et, or Bu; R^2 = Ph or *p*-tolyl), which confirms that these compounds are formed by the addition of the arenethiol to the allyltrialkyltin compound in preference to cleavage. Typical compounds which would have been formed by cleavage, *e.g.* Bu$_3$SnSPh and Bu$_3$SnSC$_6$H$_4$Me-*p*, give well-resolved quadrupole-doublet spectra.[404] The spectrum of (BrCH$_2$)$_2$SnO is characteristic of R$_2$SnO-type compounds and rules out alternative formulations such as (HOCH$_2$)$_2$SnBr$_2$. The infusible white solid bis(phenylthiomethyl)tin oxide, (PhSCH$_2$)$_2$SnO, gives a resonance effect at room temperature and is thought to be polymeric. Treatment of the latter with acetic acid does not give the diacetate; instead, the (phenylthiomethyl)stannoxane acetate, [PhSCH$_2$Sn(O)OCOMe]$_x$, is obtained. Data were also given for (PhSCH$_2$)$_4$Sn, (BuSCH$_2$)$_4$Sn, Bu$_3$SnCH$_2$Ph, [PhSCH$_2$Sn(O)OCOMe]$_2$, [PhSn(O)OCOCMe$_3$]$_x$, (PhSCH$_2$)$_2$SnO, and (BrCH$_2$)$_2$SnO.[405]

[404] G. Ayrey, R. D. Brasington, and R. C. Poller, *J. Organometallic Chem.*, 1972, **35**, 105.
[405] R. D. Brasington and R. C. Poller, *J. Organometallic Chem.*, 1972, **40**, 115.

The monoalkyltin orthosulphites (29) have all been shown to give similar spectra and are thought to be isostructural. There are small differences in isomer shift and quadrupole splitting, which are attributed to steric and bonding effects of the groups X, and these indicate that the electron-withdrawing power of the OX groups decreases along the series $SO_3(C_6H_4)Me > OCOMe > OSO_3H > OH$. The results for these compounds were compared with data for the oxy-bridged polymers (30) both

$$\begin{array}{cc}
Bu^n\diagdown\quad O\quad\diagup O\diagdown\quad\diagup X & Bu^n\diagdown\quad O\quad\diagup Bu^n \\
\quad Sn\diagdown\quad\diagup S\diagdown\quad\diagup Sn & -O-Sn\diagup\quad\diagdown Sn-O- \\
X\diagup\quad O\quad\diagdown O\quad\diagdown Bu^n & \diagup\quad\quad\diagdown \\
& X\quad\quad\quad X
\end{array}$$

(29) X = OH, OCOMe, OSO_3H, or $O_3S(C_6H_4)Me$ (30) X = OH or OSO_3H

of which show quadrupole splittings which are significantly smaller than the values for (29). The difference is attributed to a change in stereochemistry due to the fact that the polymers have a less restrictive geometry than that imposed by the four-membered O—Sn—O ring in (29), rather than to differences in electron donor–acceptor properties of the O-bridged moiety compared with the SO_4-bridged group, because the isomer shifts for the two sets of compounds are very similar. Data were also given for (31).[406]

Possible structures have been suggested for various mono-organotin-compounds on the basis of quadrupole splitting data.[407] The sesquisulphides $(RSnS_{1.5})_4$ (R = Me, Et, Bu, $C_8H_{17}{}^n$, or Ph) all give similar data,

$$\begin{array}{c}
Bu^n\quad Bu^n \\
|\quad\quad | \\
-Sn-S-Sn- \\
|\quad\quad | \\
SH\quad SH \\
(31)
\end{array}$$

(32)

suggesting that they are isostructural with the methyl compound which is known to have structure (32). They appear to be one of the few classes of organotin(IV) compounds which are tetrahedral and unassociated in the solid state. The quadrupole splittings are small (1.2—1.5 mm s⁻¹), as expected for a tetrahedral $RSnX_3$ structure. The values observed for the species $RSnCl_3$ lie in the range 1.8—2.0 mm s⁻¹, and are inconclusive in differentiating between tetrahedral and associated five-co-ordinate structures. The organostannoic acids $[RSn(O)OH]_n$ give splittings of 1.29—1.83 mm s⁻¹, which are consistent with structures, of the type (33) and (34), containing tetrahedral tin, but do not exclude the possibility of association.

[406] C. H. Stapfer and R. H. Herber, *J. Organometallic Chem.*, 1972, **35**, 111.
[407] A. G. Davies, L. Smith, and P. J. Smith, *J. Organometallic Chem.*, 1972, **39**, 279.

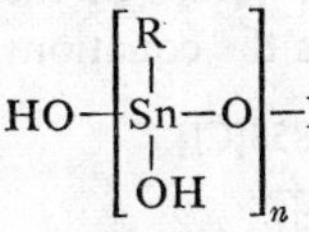

$$\begin{bmatrix} \text{HO} - \underset{\underset{\text{OH}}{|}}{\overset{\overset{\text{R}}{|}}{\text{Sn}}} - \text{O} \end{bmatrix}_n - \text{H}$$
(33)

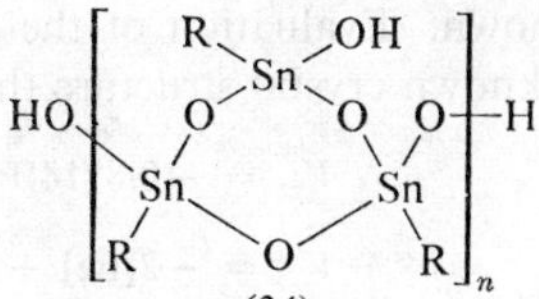

(34)

The quadrupole splitting (1.18 mm s^{-1}) for PhSn(OR^1)$_3$ is very similar to that of the sesquisulphide (32), indicating that the tin atom is probably tetrahedral also. By contrast, the alkyltin trialkoxides R^{1}Sn(OR^2)$_3$ (R^1 = Et, Bu, or n-C$_8$H$_{17}$) have values between 1.9 and 2.0 mm s^{-1} and to account for this increase it is suggested that the tin is five-co-ordinate with a N → Sn interaction of the type shown in (35).[407]

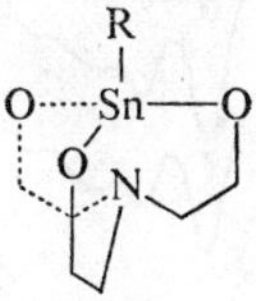

(35)

^{119}Sn spectra have been obtained for the compounds cis-[Fe(CO)$_4$XSnX$_3$] (X = Cl, Br, or I) and for cis- and trans-[Fe(CO)$_4$(SnX$_3$)$_2$] (X = Cl or Br). There is no evidence in the spectra of the latter complexes for intermolecular bridging of the types shown in structures (36) and (37).

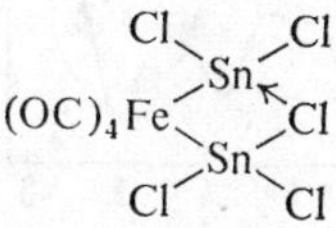

(36) 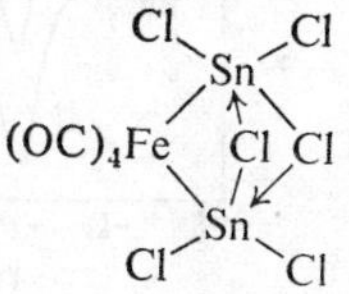   (37)

The electric-field gradient tensor at the tin atom in [{Fe(π-C$_5$H$_5$)(CO)$_2$}$_2$-SnCl$_2$] has been studied in detail by application of a 50 kG magnetic field at 4.2 K and by use of an oriented matrix of single crystals in a zero-field experiment at 78 K.[408] The magnetically perturbed spectrum is shown in Figure 32, together with a series of spectra simulated for a positive e^2qQ (negative V_{zz}) and various values of the asymmetry parameter, η. The result, $\eta = 0.65 \pm 0.05$, constitutes the first determination of the asymmetry parameter at a tin atom. The molecular geometry at the tin is shown in Figure 33: the FeSnFe plane is perpendicular to the ClSnCl plane and consideration of the general electric-field gradient tensor for this geometry reveals that two of the principal axes, arbitrarily labelled i and j, lie in these

[408] T. C. Gibb, R. Greatrex, and N. N. Greenwood, *J.C.S. Dalton*, 1972, 238.

planes as shown. Evaluation of the principal values of the EFG in the light of the known crystal structure then yields the equations:

$$V_{ii} = -0.8714[\text{Fe}] + 0.7858[\text{Cl}]$$

$$V_{jj} = -2[\text{Fe}] + 1.2142[\text{Cl}]$$

$$V_{kk} = +2.8714[\text{Fe}] - 2[\text{Cl}]$$

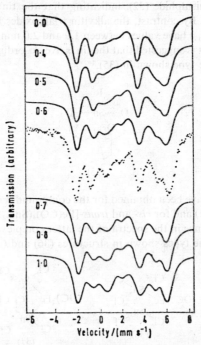

Figure 32 *The Mössbauer spectrum of* $[\{\text{Fe}(\pi\text{-C}_5\text{H}_5)(\text{CO})_2\}_2\text{SnCl}_2]$ *at 4.2 K with a magnetic field of 50 kG applied perpendicular to the direction of observation. The solid curves are computed spectra for a range of values of the asymmetry parameter* η. *The experimentally determined value is* $\eta = 0.65 \pm 0.05$

in which [Fe] and [Cl] refer to the expectation values of $-e\langle\psi|(3\cos^2\theta - 1)r^{-3}|\psi\rangle$ for the Sn—Fe and Sn—Cl bonds, respectively. Numerical evaluation of these equations in the range $0 < R < \infty$, where $R = [\text{Fe}]/[\text{Cl}]$, reveals that the direction of V_{zz} depends on the ratio R as shown in Figure 34 and can adopt any of the three axes. The sign of V_{zz} also alters and shows a complex behaviour. The experimentally determined negative sign of V_{zz} eliminates three of the six possible values for [Fe]/[Cl] and restricts the direction of V_{zz} to either the i or k axis. The results of the single-crystal experiment finally remove this ambiguity and indicate that

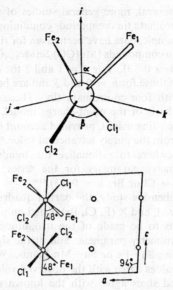

Figure 33 *The molecular geometry of the* $SnFe_2Cl_2$ *unit in* $[\{Fe(\pi-C_5H_5)(CO)_2\}_2\text{-}SnCl_2]$ *and the two possible orientations in the crystal unit cell*

V_{zz} is directed along V_{kk}, at 48° to the *bc* plane of the crystal. It then follows from Figure 34 that [Fe]/[Cl] = 1.2 ± 0.1, which shows conclusively that the value of $\langle r^{-3} \rangle$ is greater for the Sn—Fe bond than for the Sn—Cl bond and that there is therefore less withdrawal of tin 5*p*-electron density into the Sn—Fe bonds than into the Sn—Cl bonds.[408]

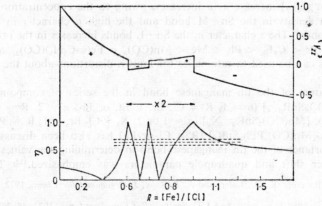

Figure 34 *The dependence of the asymmetry parameter* η *and the arbitrarily normalized value of* V_{zz} *which is denoted by* 'V_{zz}' *on* R = [Fe]/[Cl]. *Note the expanded horizontal scale in the centre*

There have been several, more general studies of the ^{119}Sn quadrupole splitting in four-co-ordinate tin compounds containing tin–transition-metal bonds.[409–414] For example, data have been given for the low-symmetry four co-ordinate organotin compounds [Mn(CO)$_5$SnMe$_{3-n}$Cl$_n$] and [Fe(π-C$_5$H$_5$)-(CO)$_2$SnPh$_{3-n}$Cl$_n$] (n = 0—3). For n = 1 and 2 the quadrupole splittings are much larger than those for n = 0 and 3 and are beyond the range commonly associated with four-co-ordinate tin. These observations can be rationalized on the basis of the point-charge model, using self-consistent partial quadrupole splitting values, provided account is taken of distortions of the bond angles from the purely tetrahedral value.[409] Similar arguments have been used elsewhere to rationalize the trends in the quadrupole splitting and asymmetry parameter for the series [Mn(CO)$_5$SnR$_{3-n}$X$_n$] (R = Me or Ph; X = Cl or Br; n = 0—3).[410]

In a more comprehensive study, the partial quadrupole splitting values for Ph, R (Me, Et, *etc.*), and X (F, Cl, Br), listed on p. 576, have been used to enable predictions to be made of the magnitudes of the quadrupole splitting and the asymmetry parameter and of the signs of e^2qQ for forty compounds containing Sn—Fe or Sn—Mn bonds. With only four exceptions the predicted values agree with the observed values to within 0.4 mm s^{-1} and the predicted signs agree with the known signs and with those determined for Me$_2$ClSnMn(CO)$_5$ (−ve) and MeCl$_2$SnMn(CO)$_5$ (+ve). The predicted values of η for these two complexes are 0.41 and 0.89, respectively, compared with the experimental values of 0.35 and 0.46. The poor agreement in the second case may well reflect the fact that η is very sensitive to variations in the partial quadrupole-splitting values. The quadrupole splitting is much less sensitive. New partial quadrupole splittings were derived for C$_6$F$_5$ (−0.76), Mn(CO)$_5$ (−0.97), and Fe(π-C$_5$H$_5$)(CO)$_2$ (−1.08). The centre shifts for compounds of the type MSnR$_{3-n}$X$_n$ [M = Mn(CO)$_5$ or Fe(π-C$_5$H$_5$)(CO)$_2$; R = Me or Ph; X = Cl, Br, or C$_6$F$_5$] increase as n increases, owing to the concentration of s-electron density in the Sn—M bond and the high p character in the Sn—X bonds. The s character in the Sn—L bonds increases in the order L = Cl, Br < C$_6$F$_5$ < Ph < Me < Mn(CO)$_5$ < Fe(π-C$_5$H$_5$)(CO)$_2$, and this series can be used to rationalize the known distortions about the tin atom.[411]

The nature of the tin–manganese bond in the series of compounds [{Mn(CO)$_5$}$_n$SnR$_{4-n}$] (n = 1, R = Cl, Br, I, Et, or Ph; n = 2, R = Cl, Br, or Ph), [Mn(CO)$_5$SnR$_{3-n}$X$_n$] (n = 1 or 2; X = Cl, Br, or I; R = Ph), and *trans*-[Mn(CO)$_4$PPh$_3$SnR$_3$] (R = Cl or Ph) has also been discussed. The importance of the tin–manganese bond in determining the values of the isomer shift and quadrupole parameters was emphasized.[412] The

---

[409] G. M. Bancroft, K. D. Butler, and A. T. Rake, *J. Organometallic Chem.*, 1972, **34**, 137.
[410] V. B. Liengme, J. R. Sams, and J. C. Scott, *Bull. Chem. Soc. Japan*, 1972, **45**, 2956.
[411] G. M. Bancroft, K. D. Butler, A. T. Rake, and B. Dale, *J.C.S. Dalton*, 1972, 2025.
[412] S. R. A. Bird, J. D. Donaldson, A. F. Le C. Holding, B. Ratcliff, and S. Cenini, *Inorg. Chim. Acta*, 1972, **6**, 379.

quadrupole splitting decreases from 0.61 mm s^{-1} to zero when the carbonyl group *trans-* to the tin atom in [Mn(CO)$_5$SnMe$_3$] is replaced by a ligand of weaker π-accepting ability (*e.g.* PPh$_3$ or AsPh$_3$); this has been attributed to a long-range effect through the π-electrons of the manganese atom and the 5d orbitals of tin. The complete absence of quadrupole splitting in the analogous Ph compounds is attributed to the predominant π-delocalization effect of the phenyl groups attached to the tin atom.[413]

The isomer shifts for the series [Co(CO)$_4$]$_{4-n}$SnX$_n$ (X = Cl, Br, or I, n = 0—4) have been shown to increase non-linearly as the halogens are successively replaced by Co(CO)$_4$ groups, the change in isomer shift becoming smaller as n decreases. The quadrupole splittings are smaller than those of the corresponding organotin halides and are a maximum for n = 2. The results may be summarized by Bent's rule, which suggests that the s-character is concentrated in the orbitals directed toward the more electropositive substituents, in this case in the Sn—Co bonds, while the p-character is concentrated in the Sn—X bonds. This results in Co—Sn—Co angles which are greater than 109° 28' and X—Sn—X angles which are less than this value.[414]

From the temperature dependence of the recoil-free fraction and the Goldanskii–Karyagin asymmetry (zero in this case) of the quadrupole splitting in Me$_2$Sn(S$_2$CNEt$_2$)$_2$, it has been concluded that the compound is monomeric with an essentially isotropic amplitude of vibration for the central tin atom in the temperature range 78—150 K. The latter is perhaps surprising in view of the anisobidentate nature of the dithiocarbamate moieties in this complex. It is also surprising that the tetrakis(dimethyl- and diethyl-dithiocarbamate) complexes give no resolvable quadrupole splitting, despite the fact that they have been shown by X-ray crystallography to contain tin in a distorted octahedral ligand configuration, involving two isobidentate and two unidentate ligands. From isomer shift electronegativity systematics, a group electronegativity of 5.7 (on the Mulliken–Jaffé scale) was deduced for alkyl groups bonded to tin(IV).[415]

Data have been reported for twelve chloro-(NN-dialkyldithiocarbamato)-diorganostannanes R$^1_2$SnCl(S$_2$CNR$^2_2$) [R^1 = Me, Bu, or Ph; R^2 = Me, Et, CH$_2$Ph, or (CH$_2$)$_2$]. The alkyl derivatives have quadrupole splittings in the range 2.72—3.14 mm s^{-1} and the phenyl derivatives, 2.19—2.34 mm s^{-1}. The isomer shifts show less variation and fall in the narrow range 1.08—1.45 mm s^{-1}. Comparison of the above Δ-values with theoretical quadrupole splittings calculated from partial field gradients led to the conclusion that these compounds contain unidentate dithiocarbamato-groups in an essentially tetrahedral stereochemistry.[416]

The ^{119}Sn spectrum of cubanite, Cu$_2$FeSnS$_4$, at 4.2 K is magnetically broadened and indicates a transferred hyperfine field of about 20 kG at the

[413] S. Onaka and H. Sano, *Bull. Chem. Soc. Japan*, 1972, **45**, 1271.
[414] S. Ichiba, M. Katada, and H. Negita, *Bull. Chem. Soc. Japan*, 1972, **45**, 1679.
[415] J. L. K. F. de Vries and R. H. Herber, *Inorg. Chem.*, 1972, **11**, 2458.
[416] B. W. Fitzsimmons and A. C. Sawbridge, *J.C.S. Dalton*, 1972, 1678.

tin nucleus. The ^{57}Fe spectrum of this compound is discussed on p. 566.[349]

Spectra have also been obtained for the compounds SnP, Sn_3P_4, Sn_4P_3, SnAs, Sn_4As_3, Sn_3As_2, and SnSb,[417] and for alloys of the As—S—Sn, P—S—Sn, and P—Se—Sn systems.[418] Glasses in the As—Se—Ge—Sn system [419] and sodium–tin silicate glasses [420] have also been studied.

6 Other Elements

This section covers elements other than iron and tin. In each of the three sub-sections – main group elements, transition elements, and lanthanide and actinide elements – the isotopes are treated in order of increasing atomic number.

It has been shown, by careful analysis of the lineshapes for fourteen different Mössbauer transitions, that dispersion terms are clearly present in absorption spectra of γ-rays with $E2$ or mixed $E2/M1$ character. The dispersion terms are caused by interference effects between conversion electrons, emitted after resonance absorption, and photoelectrons. It was pointed out that this fact has to be taken into account whenever the positions of such absorption lines need to be measured with great accuracy. The fourteen transitions studied were ^{99}Ru (90 keV), ^{166}Er (80.6 keV), ^{170}Yb (84.3 keV), ^{171}Yb (66.7 keV), ^{180}Hf (93.3 keV), ^{182}W (100.01 keV), ^{183}W (46.5 and 99.1 keV), ^{184}W (111.1 keV), ^{186}W (122.5 keV), ^{186}Os (137.2 keV), ^{188}Os (155.0 keV), ^{191}Ir (129.5 keV), and ^{236}U (45.3 keV).[421]

Main Group Elements.—*Germanium* (^{73}Ge). The Mössbauer effect for the 13.3 keV, $\frac{5}{2}+ \to \frac{9}{2}+$, $E2$ transition in ^{73}Ge has been detected for the first time in single crystals of germanium. The source consisted of ^{73}Ga in a germanium lattice, and was produced by the photonuclear reaction ^{74}Ge$(\gamma,p)^{73}$Ga. Scattering geometry was employed to detect the 9.89 keV germanium internal conversion X-rays, thereby avoiding the problem of the very high conversion coefficient. Spectra obtained for an Sb-doped n-type crystal of germanium contain both a singlet and a quadrupole multiplet from ^{73}Ge in tetrahedral and hexagonal interstitial sites, respectively. The best fits to the data yield values of $|Q_{\frac{5}{2}}/Q_{\frac{9}{2}}| = 0.25$ and $|q| = 1.3 \times 10^{16}$ c.g.s. for the quadrupole moment ratio and electric-field gradient, respectively. The latter agrees well with the value of 1.5×10^{16}, estimated by substituting the spectroscopically determined quantity $\langle r^{-3} \rangle = 38.8 \times 10^{24}$ cm^{-3} into the equation $|q| = (\frac{4}{5})e\langle r^{-3}\rangle$.[422, 423]

[417] V. T. Shipatov and P. P. Seregin, *Teor. i eksp. Khim.*, 1972, **8**, 413.
[418] P. P. Seregin, L. N. Vasil'ev, and A. A. Pronkin, *Izvest. Akad. Nauk S.S.S.R., neorg. Materialy*, 1972, **8**, 376.
[419] P. P. Seregin and L. N. Vasil'ev, *Izvest. Akad. Nauk S.S.S.R., neorg. Materialy*, 1972, **8**, 1238.
[420] P. P. Seregin, A. N. Murin, V. G. Bezrodnyi, and V. T. Shipatov, *Vestnik. Leningrad. Univ., Fiz. Khim.*, 1972, **46**, 159.
[421] F. E. Wagner, B. D. Dunlap, G. M. Kalvius, H. Schaller, R. Felscher, and H. Spieler, *Phys. Rev. Letters*, 1972, **28**, 530.
[422] K. Matsui, *Kakuriken Kenkyu Hokoku*, 1972, **4**, 93.
[423] K. Matsui, *Kakuriken Kenkyu Hokoku*, 1972, **4**, 101.

Krypton (^{83}Kr). It has been claimed that the accuracy of measurements with the ^{83}Kr isotope can be nearly doubled by use of proportional counters containing a mixture of argon, neon, and methane.[90] The recoil-free fraction of ^{83}Kr in solid krypton, following the 9.3 keV transition produced by inelastic neutron scattering, has been calculated as a function of temperature in the range 0—90 K. The calculated values are higher than those measured in previous Mössbauer experiments and the differences are attributed to the high vacancy concentration in the source and absorber crystals.[424]

Antimony (^{121}Sb). Data have been reported for the seven antimony(III) halide complexes $Co(NH_3)_6SbX_6$ (X = Cl or Br), $Cs_3Sb_2X_9$ (X = Cl, Br, or I), and $Rb_3Sb_2X_9$ (X = Br or I) and for the three compounds Rb_2SbBr_6, $(NH_4)_2SbBr_6$, and Cs_2SbCl_6, which were shown to contain both antimony(III) and antimony(V) and may perhaps be formulated $M_4Sb^{III}Sb^VX_{12}$. The antimony(III) compounds all have large negative shifts in the range -17.5 to -20.5 mm s^{-1} relative to the $Ba^{121}SnO_3$ source, whereas the antimony(V) sites have shifts in the region -2 to -6 mm s^{-1}. These results are consistent with the known negative sign of $\Delta R/R$ for ^{121}Sb. An increase in the number of Sb—X bonds in these materials, compared with their parent trihalides, leads to a decrease in the isomer shift. This trend is in contrast to that found in tin(II) compounds where addition of halide ions to SnX_2 to form SnX_3^- decreases the s-electron density at the tin nucleus. All of the antimony(III) compounds except $(NH_4)_2SbBr_6$ have small positive quadrupole interactions, e^2qQ ($= eqV_{zz}$; since eQ is negative for ^{121}Sb, V_{zz} in this case is negative also). The positive sign is compatible with a trigonal distortion of the anions from O_h symmetry which results in an excess of $5p_z$ electron character over that of $5p_x$ and $5p_y$ and leads to an overall negative contribution to V_{zz}.[425]

The isomer shifts for the SbX_4^- (X = Cl, Br, or I) anions have been shown to increase in the order Cl < Br < I, *i.e.* with increasing covalent character. This suggests that p-electrons are being withdrawn from antimony and that the contribution from the s-electron character of the lone pair is dominant in determining the shift. The similarity between the parameters for the two complexes Et_4NSbCl_4 and $(pyH)SbBr_4$ and those of $(pyH)SbCl_4$ suggests that the structures of the anions are similar in each case, being based on a trigonal bipyramid with an equatorial lone-pair. This structure is consistent with the positive sign observed for e^2qQ in $SbCl_4^-$ and $SnBr_4^-$ if it is assumed that the contribution from the lone-pair is dominant and that V_{zz} lies in the direction of the lone-pair, as in (38). SbI_4^- has a much smaller, and probably negative, e^2qQ and it is argued that in this anion V_{zz} lies along the axial direction as shown (39). The signs of e^2qQ for the mixed halide complexes $Bu_4^nNSbBr_3I$ (+ve), $Bu_4^nNSbCl_3Br$ (−ve), and $Bu_4^nNSbBr_3Cl$ (−ve) indicate that the principal axis is directed

[424] N. P. Gupta, *Lett. Nuovo Cimento Soc. Ital. Fis.*, 1972, **3**, 599.
[425] J. D. Donaldson, M. J. Tricker, and B. W. Dale, *J.C.S. Dalton*, 1972, 893.

$$\xleftarrow{V_{zz}} \begin{array}{c} X \\ | \diagup X \\ \div Sb \\ | \diagdown X \\ X \end{array} \qquad \begin{array}{c} \uparrow V_{zz} \\ | \\ | \diagup I \\ \div Sb \\ | \diagdown I \\ I \end{array}$$

(38) X = Cl or Br   (39)

through the lone-pair in the first compound but in the axial direction, along a more polar bond, in the other two cases. The fluorides M_2SbF_5 (M = K, Cs, or NH_4) all have similar parameters, consistent with their proposed isostructural nature. The positive values for the quadrupole coupling constants are consistent with a dominant contribution to V_{zz} by the lone pair, which occupies the sixth position in the octahedral co-ordination sphere of these SbF_5^{2-} anions. Essentially the same comments apply to the polymeric compounds $MSbF_4$ (M = Na, K, Cs, or NH_4). In contrast to the data for the other complex halides, the isomer shifts for the fluorides are all greater than those for the parent trihalides, which suggests that there must be extensive mixing of antimony 5s character into the bonding.[426]

A co-operative tunnel effect has been described in some semiconductor hexahalogeno-complexes of antimony.[427]

Data have been given for $SbCl_5$ in quick-frozen solutions of non-aqueous solvents. The isomer shift increases as the donicity of the solvent increases [DMF > $(BuO)_3PO$ > MeCN].[387,388]

Mössbauer spectroscopy, coupled with i.r. and X-ray diffraction techniques, has shown that antimonic acid cannot be dehydrated in air to give products of constant and reproducible weight without simultaneous reduction of some of the antimony(v) to antimony(III). Neither anhydrous Sb_2O_5 nor the hydroxy-oxide Sb_3O_6OH postulated by Dihlström and Westgren can be obtained by this method, but Sb_6O_{13} and β-Sb_2O_4 are well-defined products. Sb_6O_{13} is a cubic defect pyrochlore and probably has the structure $Sb^{III}Sb_2^VO_6O_{0.5}$; it decomposes at 1028 K to β-Sb_2O_4 and not Sb_2O_3. The Mössbauer parameters of β-Sb_2O_4 are similar to those reported for α-Sb_2O_4, but the isomer shifts for antimony(v) in antimonic acid and Sb_6O_{13} are significantly larger than those for α- and β-Sb_2O_4.[428] Stibiconite ($Ca_2Sb_2O_7$) from San Luis Potosi in Mexico contains both antimony(v) and antimony(III) in the approximate ratio of 0.2. It is therefore a 'reduced' non-stoicheiometric pyrochlore rather than a 2–5 pyrochlore. The Mössbauer spectrum provides a direct confirmation of the

---

[426] J. D. Donaldson, J. T. Southern, and M. J. Tricker, *J.C.S. Dalton*, 1972, 2637.
[427] A. Yu. Alexandrov, S. P. Ionov, D. A. Baltrunas, and E. F. Makarov, *Pis'ma Zhur. eskp. i teor. Fiz.*, 1972, **16**, 209.
[428] D. J. Stewart, O. Knop, C. Ayasse, and F. W. D. Woodhams, *Canad. J. Chem.*, 1972, **50**, 690.

correctness of the general formula $(Sb^{III}, Ca)_y Sb^V_{2-z}(O, OH, H_2O)_{6-7}$ which was proposed for stibiconite on the basis of oxygen analysis.[429]

Equations have been given for determining the optimum sample thickness and minimum counting time of Mössbauer samples with nuclei for which the accuracy of measurement under normal conditions is very poor. The method was applied to the determination of the isomer shift of antimony in calcium halogenophosphate luminescent powders.[430, 431]

The sign of the magnetic hyperfine field at an antimony nucleus has been determined for the first time from the ^{121}Sb Mössbauer spectrum of the antimony-substituted nickel ferrite $Ni_{1.2}Fe_{1.7}Sb_{0.1}O_4$. The result, $H_{eff} = -311 \pm 4$ kG at 100 K, was compared with data for tin in yttrium iron garnet and it was suggested that $3d$–$5s$ covalent spin transfer makes the predominant contribution to the hyperfine field.[432]

^{121}Sb spectra have been obtained for two of the iron carbonyl derivatives discussed on p. 539. The increase in isomer shift, which occurs when $SbPh_3$ replaces a carbonyl group in $Fe(CO)_4SbPh_3$ to give $Fe(CO)_3(SbPh_3)_2$, corresponds to a removal of about 0.26 $5s$ electron or, less likely, to an addition of 1.7 $5p$ electron. The fact that e^2qQ is positive in sign for both compounds favours the σ-bonding scheme. A π-back-donation from iron would primarily increase the occupation of the p_x and p_y orbitals and drive e^2qQ negative. However, the reduction in magnitude from +17 to about +10 mm s^{-1} is consistent with a σ-donation along the z-axis from a hybrid orbital with some p character.[245]

The magnetic hyperfine field at ^{121}Sb in Pd_2MnSb has the anomalously large value at 100 K of |579 ± 5| kG (see Figure 35). The field in the closely related compound PdMnSb is |302 ± 5| kG, which is similar to that in other ferromagnetic intermetallics containing manganese and antimony. A possible explanation is that the palladium atoms in Pd_2MnSb attain a small moment which serves to enhance the hyperfine field transfer to the antimony atoms. However, in Co_2MnSn, where the cobalt is known to carry a magnetic moment, there is no such enhancement. A firm explanation is therefore not established, but it might be related to the unusual pressure dependence of T_C reported elsewhere.[433] Magnetic hyperfine fields have also been observed at the antimony nuclei in the Heusler-type alloys Ni_xMnSb (x = 1.0, 1.25, 1.75, 1.9, or 2.0).[434]

Alloys of bismuth and antimony have been shown to give only single-line spectra at 80 K for all compositions between 5 and 100% antimony. The isomer shift becomes more negative with increasing antimony content

---

[429] F. Brisse, D. J. Stewart, V. Seidl, and O. Knop, *Canad. J. Chem.*, 1972, **50**, 3648.
[430] A. Vertes, V. Frakhnoy-Koros, B. Levay, P. A. Gelencser, and M. Ranogajec-Komor, *Acta Chim. Acad. Sci. Hung.*, 1972, **74**, 151.
[431] A. Vertes, V. Frakhnoy-Koros, B. Levay, P. A. Gelencser, and M. Komor, *Magyar. Kém. Folyóirat*, 1972, **78**, 430.
[432] B. J. Evans and L. J. Swartzendruber, *Phys. Rev. (B)*, 1972, **6**, 223.
[433] L. J. Swartzendruber and B. J. Evans, *Phys. Letters*, 1972, **38A**, 511.
[434] L. J. Swartzendruber and B. J. Evans, *Amer. Inst. Phys. Conf. Proc.*, 1972, No. 5 (Pt. 1), 539.

from about 5% antimony, where the alloy makes a transition of the semi-metallic-to-semiconductor type. The isomer shift reaches a maximum negative value at approximately 20—25% antimony which corresponds to the middle of the semiconductor range. The linewidth is also a minimum in this region, but the reason for this is not well understood. There is a

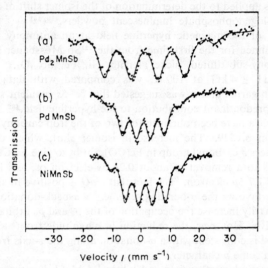

Figure 35 ^{121}Sb *Mössbauer absorption spectra at 100 K. In each case the solid line represents a least-squares fit to a single magnetic hyperfine field. The zero of velocity represents the centre of an* InSb *absorber at 100 K* (Reproduced by permission from *Phys. Letters,* 1972, **38A**, 511)

second semiconductor-to-semimetallic transition at 45—55% antimony and the isomer shift becomes less negative. The s-electron density at the antimony nucleus is therefore greater in the semiconducting alloys.[435]

Tellurium (^{125}Te). New data have been reported for the compounds TeX$_4$, (NH$_4$)$_2$TeX$_6$ (X = Cl, Br, or I), K$_2$TeBr$_6$, and ZnTe and compared with values already in the literature. For the hexahalides the isomer shift increases steadily with increase in electronegativity of the ligand. Since $\Delta R/R$ is positive for ^{125}Te, this trend implies that the bonding is primarily through the 5p orbitals, the 5s^2 electrons on the tellurium being non-bonding and stereochemically non-active.[436] These data have been used in conjunction with known halogen n.q.r. data for the hexahalides to derive an expression between the ^{125}Te isomer shift and the number of effective holes, h_p, in the tellurium 5p orbitals. The relationship was found to be

$$^{125}\delta = (0.44 \pm 0.01) h_p - (0.16 \pm 0.01)$$

[435] J. R. Teague and R. Gerson, *Solid State Comm.*, 1972, **11**, 851.
[436] J. J. Johnstone, C. H. W. Jones, and P. Vasudev, *Canad. J. Chem.*, 1972, **50**, 3037.

with respect to ZnTe, which is remarkably similar to the expression

$$^{125}\delta = (0.44 \pm 0.01)\,h_p - (0.23 \pm 0.01)$$

with respect to I/Cu, obtained previously by Bukshpan et al.[436, 437] This relationship was tested for compounds of the type ^{125}Te(tu)$_4$X$_2$ (tu = thiourea; X = Cl, Br, NO$_3$, or SCN) and was shown to be consistent with the observed isomer shifts and quadrupole couplings.[437, 438] These tellurium(II) complexes give isomer shifts relative to the I/Cu source of 0.74—1.00 mm s^{-1}, and similar shifts are also observed for other tellurium(II) compounds such as cis-Te^{II}(tu)$_2$X and trans-Te^{II}(etu)$_2$X$_2$ (etu = ethylenethiourea; X = Br or I). By contrast, the tellurium(IV) complexes trans-Te^{IV}(tmtu)$_2$X$_4$ (tmtu = tetramethylthiourea; X = Cl or Br) have shifts of about 1.6 mm s^{-1}. These results are consistent with a bonding scheme similar to that described earlier for the hexahalides. Removal of two $5p$ electrons results in a deshielding of the $5s^2$ lone pair from the nucleus and a more positive shift for the higher oxidation state. All of the tellurium(II) compounds give very large quadrupole splittings (15—18 mm s^{-1}), attributable to the imbalance in the occupation of the $5p_x$, $5p_y$, and $5p_z$ orbitals as a result of the square-planar geometry.[438]

The amorphous-to-crystalline phase transition in the Ovonic semiconducting glass Te$_{81}$Ge$_{15}$As$_4$, which occurs on heating at 548 K for one hour, has been observed by the Mössbauer effect in ^{125}Te (and in ^{129}I, see below). The spectrum for the crystalline material is very similar to that for crystalline tellurium metal but differs considerably from that of the amorphous material (see Figure 36) in having a 30% smaller quadrupole splitting and a greater isomer shift (increased s-electron density). The results indicate that the amorphous state has a higher charge asymmetry than the crystalline phase, and this is consistent with the concept of 'dangling' chemical bonds in the amorphous material, which become filled in the crystalline state.[439]

Magnetic fields have been detected at the tellurium nuclei in EuTe, Eu$_3$Te$_4$, Eu$_4$Te$_7$, and Eu$_3$Te$_7$.[440]

Nuclear quadrupole diffraction of resonance γ-radiation has been observed in a tellurium single crystal.[441] The Goldanskii–Karyagin effect could not be detected in polycrystalline tellurium.[442]

Iodine (^{127}I, ^{129}I). The magnetic moment of the 57.6 keV state of ^{127}I has been measured by use of a source of ^{127m}Te, implanted into an iron foil,

---

[437] B. M. Cheyne, J. J. Johnstone, and C. H. W. Jones, *Chem. Phys. Letters*, 1972, **14**, 545.
[438] B. M. Cheyne, C. H. W. Jones, and P. Vasudev, *Canad. J. Chem.*, 1972, **50**, 3677.
[439] D. H. Hafemeister and H. de Waard, *J. Appl. Phys.*, 1972, **43**, 5205.
[440] O. A. Sadovskaya, E. P. Stepanov, V. V. Khrapov, and E. I. Yarembash, *Izvest. Akad. Nauk S.S.S.R., neorg. Materialy.*, 1972, **8**, 815.
[441] V. S. Zasimov, R. N. Kuz'min, A. Yu. Aleksandrov, and A. I. Firov, *Pis'ma Zhur. eksp. i teor. Fiz.*, 1972, **15**, 394.
[442] A. A. Opalenko, I. A. Avenarius, V. P. Gor'kov, B. A. Komissarova, R. N. Kuz'min, and P. N. Zaikin, *Zhur. eksp. i teor. Fiz.*, 1972, **62**, 1037.

together with a single-line $Na_3H_2IO_6$ absorber. The result was $\mu = 2.54 \pm 0.05$ n.m., in good agreement with the calculations of Kisslinger and Sorensen, and of Castel. The hyperfine field at the nucleus of iodine in iron is 1.172 ± 0.016 MG at 4.2 K.[443]

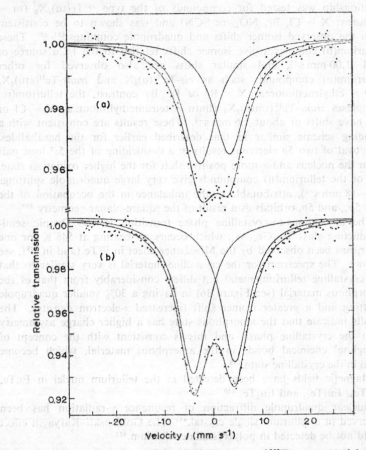

Figure 36 *Mössbauer spectra of the 35.6 keV γ-rays of ^{125}Te transmitted from a $Zn^{125m}Te$ source through $^{125}Te_{81}Ge_{15}As_4$ absorbers in (a) the crystalline and (b) the amorphous phase*
(Reproduced by permission from *J. Appl. Phys.*, 1972, **43**, 5205)

Spectra have been obtained for both single-crystal and polycrystalline $Pb^{127}I_2$. There is a small quadrupole splitting of positive sign, which indicates that the principal electric-field gradient axis is perpendicular to the bonding plane in the sandwich structure. A comparison of the magnitude of the quadrupole coupling in PbI_2 ($+799 \pm 19$ MHz) with that in

[443] N. S. Wolmarans and H. de Waard, *Phys. Rev.* (C), 1972, **6**, 228.

molecular iodine (−2156 ± 10 MHz) indicates a significant amount of covalent bonding. The isomer shift is consistent with pure $p\sigma$-bonding in PbI_2.[444]

An optimal method of extracting hyperfine parameters from complex Mössbauer spectra has been presented for the specific case of ^{129}I, although the method can be adapted for other isotopes. Instead of fitting the data

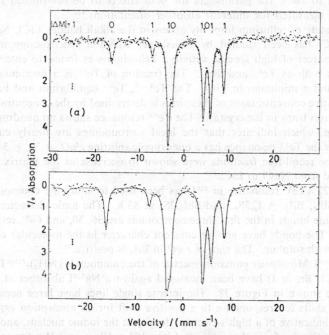

Figure 37 *Mössbauer spectra of HINA. The upper spectrum is a random powder run. The lower spectrum is the same sample aligned by a 4 kG field and quenched* (Reproduced by permission from *Chem. Phys. Letters*, 1972, **15**, 55)

to twelve line positions and then fitting these line positions to theory, the theoretical spectrum is calculated directly from an initial guess of the parameters, which are then optimized simultaneously with a least-squares fit to the data. The required derivatives with respect to each parameter are determined numerically rather than analytically. The results for iodine at 77 K are $e^2qQ = -1526 \pm 4$ MHz, $\delta = +1.019 \pm 0.011$ mm s^{-1}, relative to ZnTe at 77 K, $\eta = 0.180 \pm 0.006$, $\Gamma_{\frac{1}{2}} = 1.385 \pm 0.0022$ mm s^{-1}, and $Q^*/Q = 1.2385 \pm 0.0011$. For $^{127}I_2$, e^2qQ translates to -2180 ± 7 MHz.[75]

The Mössbauer isotope ^{129}I, incorporated in a liquid crystal, has provided a sensitive probe for alignment in the quenched sample. Figure 37

[444] G. Turnbull and E. E. Schneider, *Solid State Comm.*, 1972, **11**, 547.

shows (a) the spectrum of a random powder of the nematic liquid crystal 6-heptyloxy-5-iodo-2-naphthoic acid (HINA) and (b) the spectrum of the same sample after being heated to the nematic phase in a magnetic field of 4 kG and then quenched at 183 K. The alignment causes the $\Delta m = 0$ lines to increase in intensity relative to the $|\Delta m| = 1$ lines. The relative intensities depend also on the Goldanskii–Karyagin asymmetry and expressions were derived to allow the parameters for both effects to be determined from spectra measured for different absorber orientations.[445]

The state of tellurium impurity atoms in the alkali halides LiCl, NaCl, and KCl has been studied by Mössbauer emission spectroscopy using ^{129}Te sources of high specific activity. Tellurium was found to enter the crystals both as Te^{4+} and Te^{6+}. The fraction of Te^{4+} is a maximum in NaCl and a minimum in LiCl. The $Te^{4+} \rightleftharpoons Te^{6+}$ equilibrium and hence the relative concentrations of these ions is determined by the concentration of electron traps in the crystals. The Te^{6+} resonance shows no quadrupole splitting, which indicates that the local surroundings are nearly cubic, whereas the Te^{4+} resonance has a quadrupole splitting $e^2qQ = 30 \pm 3$ mm s^{-1}. The recoil-free fractions were shown to decrease as the matrix was changed from NaCl to LiCl.[446]

The 27.8 keV resonance in ^{129}I has been used to study the compounds AsI_3, SbI_3, BiI_3, $AsI_3,3S_8$, and $SbI_3,3S_8$ at 85 K. The ionic characters of the iodine bonds in the first three compounds are 36, 50, and 69% respectively. The bonds have greater covalent character in the molecular complexes with sulphur. The sign of e^2qQ in BiI_3 is positive.[447]

The ^{129}I Mössbauer emission spectra of the compounds $(NH_4)_2{}^{129m}TeX_6$ (X = Cl, Br, or I) have been measured against a Na^{129}I absorber at 4 K and are shown in Figure 38. The intense single lines have large negative isomer shifts (corresponding to a positive shift for an absorption experiment), indicative of a high s-electron density at the iodine nucleus, and are thought to arise from the octahedral ions $ICl_6{}^-$, $IBr_6{}^-$, and $II_6{}^-$ formed in the radioactive decay. These species have not been observed previously and may only be stable over the lifetime of the Mössbauer transition ($t_{\frac{1}{2}} = 15$ ns). The compounds $^{129}TeX_4$ (X = Cl, Br, or I) were also studied but the emission spectra are much more complex and indicate that the ^{129}I atoms are not found in an environment isostructural and isoelectronic with that of the parent.[436]

The spectrum of $trans$-Te(etu)$_2{}^{129}$I (etu = ethylenethiourea) has been obtained and the isomer shift and quadrupole splitting are consistent with the proposal that the iodine uses only a $5p$ orbital in bonding to tellurium.[438]

On the basis of the magnetic symmetry of antiferromagnetic Fe_2TeO_6 it was correctly predicted that this material should exhibit magnetoelectricity.

---

[445] M. J. Potasek, E. Münck, J. L. Groves, and P. G. Debrunner, *Chem. Phys. Letters*, 1972, **15**, 55.
[446] P. P. Seregin and E. P. Savin, *Soviet Phys. Solid State*, 1972, **13**, 2846.
[447] H. Sakai, *J. Sci. Hiroshima Univ.*, Ser. A-2, 1972, **36**, 47.

It was also predicted that no hyperfine field should exist at the Te^{6+} sites and this prediction has now been confirmed by Mössbauer studies on ^{129}I nuclei produced in Te^{6+} sites by irradiating samples of Fe_2TeO_6 (see also p. 557).[324]

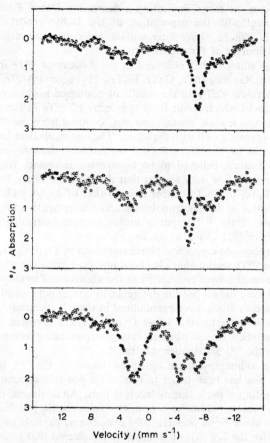

Figure 38 *The ^{129}I Mössbauer emission spectra of the $(NH_4)_2{}^{129m}TeX_6$ compounds (top)* $X = Cl$, *(middle)* $X = Br$, *and (bottom)* $X = I$, *measured at 4 K against a $Na^{129}I$ absorber. The arrows denote the emission lines corresponding to the proposed $^{129}IX_6{}^-$ ions*
(Reproduced by permission from *Canad. J. Chem.*, 1972, **50**, 3037)

Emission spectra have been obtained from sources of ^{129m}Te in GeTe, SnTe, and PbTe against absorbers of $K^{129}I$. Single-line spectra were obtained in each case, the linewidths decreasing along the series. PbTe has the cubic NaCl structure and is therefore expected to give a narrow line. SnTe experiences a phase transition at $T \leqslant 80$ K and the observed broadening may be due to this. The broadening in GeTe is due to a rhombohedral distortion of the NaCl-type lattice. The isomer shift decreases along the series, consistent with reduction of the s-electron density at the iodine nucleus as the $5p$ occupation (and therefore the negative charge on the iodine) increases. The area under the resonance curve decreases along the series, indicating a decreasing Debye θ temperature, in disagreement with the expectation of the Debye model. The solid solutions $Pb_{1-x}Sn_xTe$ show intermediate behaviour, consistent with a random distribution of Pb and Sn atoms.[448]

The charge states of ^{129}I produced in the β-decay of ^{129}Te in the compounds In_2Te_3, Ga_2Te_3, InTe, GaTe, Bi_2Te_3, Sb_2Te_3, and As_2Te_3 have been shown to correlate well with the results of perturbed angular correlations of the ^{127}I nuclei which result from the decay of ^{127}Te in the same compounds. The quadrupole interactions e^2qQ deduced from the ^{129}I emission spectra also correlate well with the latter. The non-equivalent iodine atoms in M_2Te_3 (M = In, Ga, Bi, or Sb) are not resolved. Instead, only a single quadrupole pattern, believed to be an average resonance from the two sites, is observed. The data suggest that the structure of As_2Te_3 is similar to that of Bi_2Te_3 or Sb_2Te_3 rather than to that of As_2Se_3 which is monoclinic. The value of the quadrupole interaction suggests that the octahedral environments of the ^{129}I impurity atoms become increasingly distorted along the series $Bi_2Te_3-Sb_2Te_3-As_2Te_3$.[449]

The amorphous-to-crystalline phase transition in $Te_{81}Ge_{15}As_4$, discussed earlier in the ^{125}Te section, has also been observed using the ^{129m}Te isotope as the Mössbauer source and $Cu^{129}I$ as the absorber. Essentially the same conclusions were drawn but the difference in quadrupole splitting between the two phases is much more pronounced in the ^{129}I spectrum. As a result of this it was suggested that other Ovonic devices prepared with ^{129m}Te should be studied to see whether voltage-dependent phenomena can be observed in the ^{129}I Mössbauer spectra.[439]

Only one iodine resonance has been found in the ^{129}I spectrum of $Fe(CO)_4I_2$; this has been taken to rule out a possible structure involving 'end on' bonding of the iodine molecule to iron. An additional observation, that the asymmetry parameter is zero, also rules out the possibility that the iodine p_x orbitals bond preferentially over the p_y orbitals (in this case z coincides with the I_2 axis). It is therefore considered that iodine bonds as molecular iodine through its p_x and p_y orbitals to give pseudo-eightfold co-ordination at the iron atom as shown in Figure 39. Each bond was estimated to be 45% ionic and 55% covalent, indicating that the iodine has picked up 2×0.45 electrons from the iron atom which is therefore formally in the $+1$ oxidation state.[450]

Caesium (^{133}Cs). The ratio of the g-factor of the 81 keV excited state to that of the ground state has been estimated to be 1.90 ± 0.04 (see also p. 499).[73]

[448] P. P. Seregin and E. P. Savin, *Soviet Phys. Solid State*, 1972, **13**, 2336.
[449] P. P. Seregin and E. P. Savin, *Soviet Phys. Solid State*, 1972, **14**, 1548.
[450] R. Robinette and R. L. Collins, *J. Chem. Phys.*, 1972, **57**, 4319.

Transition Elements.—*Nickel* (^{61}Ni). Spectra have been obtained for the 67.4 keV transition of ^{61}Ni in nickel–palladium alloys throughout the concentration range 0—99.5% Pd. In the ferromagnetic region, 0—98 at.% Pd, the spectra show a partly resolved magnetic hyperfine splitting with a distribution of fields. The average field is negative in pure nickel (-76 kG), changes sign near 50 at.% Pd, and rises to a large positive value

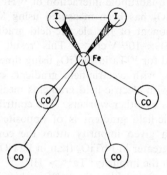

Figure 39 *The structure of* $Fe(CO)_4I_2$
(Reproduced by permission from *J. Chem. Phys.*, 1972, **57**, 4319)

($+173$ kG) at 9 at.% Pd. Qualitative agreement with these results is obtained with a model based on the assumption that $\langle H_{hf} \rangle$ in Ni–Pd has the same contributions from core polarization and bulk conduction-electron polarization, as in other nickel-based alloys, plus a large positive contribution from palladium atoms on neighbouring lattice sites. The nature of the distribution of hyperfine fields at the nickel nuclei was investigated further using applied magnetic fields. From a study of the temperature-dependence of the second-order Doppler shift, ^{61}Ni in palladium was found to have an isomer shift of -23 ± 15 μs relative to ^{61}Ni in nickel, consistent with a slightly greater electron density at the nickel nucleus in palladium ($\Delta R/R$ is negative for ^{61}Ni).[451]

Magnetic fields have also been measured in nickel–platinum alloys as a function of composition and temperature.[452]

Zinc (^{67}Zn). A large anisotropy of the Mössbauer effect has been predicted for the 93.3 keV resonance in ^{67}Zn. Calculations show that in zinc at $T = 0$ K the probability of the effect should be 2.8 times greater in the direction of the *c*-axis than normal to the *c*-axis.[52]

Ruthenium (^{99}Ru). Interference effects have been observed for the 90 keV resonance of ^{99}Ru in ruthenium metal (see p. 590).[421]

Hafnium (^{178}Hf, ^{180}Hf). The isomer shifts observed for the 93.2 keV transition of ^{178}Hf in Hf metal, HfO_2, HfC, and $Hf(C_5H_5)_2Cl_2$ can be

[451] J. E. Tansil, F. E. Obenshain, and G. Czjzek, *Phys. Rev. (B)*, 1972, **6**, 2796.
[452] W. A. Ferrando, R. Segnan, and A. I. Schindler, *Phys. Rev. (B)*, 1972, **5**, 4657.

understood qualitatively in terms of shielding of s-electrons by valence $5d$ electrons. The shift of $+0.19 \pm 0.06$ mm s^{-1} between Hf(C$_5$H$_5$)$_2$Cl$_2$ and the metal was interpreted as evidence of a shrinking of the nuclear charge radius of ^{178}Hf in the first rotational state given by $\Delta\langle r^2\rangle(^{178}\text{Hf}) = -0.37 \times 10^{-3}$ fm^2. This corresponds to a fractional change in the nuclear radius $\Delta\langle r^2\rangle/\langle r^2\rangle(^{178}\text{Hf}) = -0.13 \times 10^{-4}$.[453]

The static electric quadrupole interaction of ^{178}Hf at the titanium site in polycrystalline PbTiO$_3$ has been measured using Mössbauer techniques. The principal component of the electric-field gradient was found to be $V_{zz} = +(10.7 \pm 0.5) \times 10^{17}$ V cm^{-2}. This result was compared with similar data obtained for ^{181}Ta in BaTiO$_3$ using time-differential perturbed angular correlations, with electric-field gradients calculated in a point-charge model, and with electric-field gradients measured at ^{44}Sc and ^{57}Fe in the same titanates by other workers. The contribution of the covalent bonds to the electric-field gradient is of opposite sign from the lattice contribution. For a given impurity atom the covalent contribution is always found to be greater in PbTiO$_3$ than in BaTiO$_3$ and to decrease with decreasing charge on the impurity: Ta^{5+} > Hf^{4+} > Sc^{3+} > Fe^{3+}.[454]

The magnetic hyperfine field at the ^{178}Hf nucleus in iron metal has been found to be 633 ± 40 kG at 4 K and 630 ± 41 kG at 77 K.[73]

Interference effects have been observed for the 93.3 keV resonance of ^{180}Hf in HfZn$_2$ (see p. 590).[421]

Tantalum (^{181}Ta). The high resolution inherent in the small natural linewidth of the 6.2 keV gamma rays ($2\hbar/\tau = 0.0065$ mm s^{-1}) and the large electric quadrupole moments of the ^{181}Ta nuclear levels have been used for the first time in a study of the quadrupole hyperfine interaction of this isotope. Figure 40 shows the completely resolved spectra for ^{181}Ta in rhenium metal, measured at room temperature both parallel and perpendicular to the [0001] axis of the single-crystal ^{181}W(Re) sources. It will be noticed that the experimental linewidths are still *ca.* 500 times greater than the natural linewidth. These spectra can be understood in terms of Figure 41, which shows the expected dependence of the line positions for the $\frac{9}{2} \rightarrow \frac{7}{2}$ $E1$ transition on the ratio of the quadrupole moments. The dotted lines indicate the unique value for this ratio, $Q_{\frac{9}{2}}/Q_{\frac{7}{2}} = 1.133 \pm 0.010$, obtained by least-squares fitting of the data. The sign and magnitude of the nuclear quadrupole interaction were found to be $e^2qQ_{\frac{7}{2}} = -(2.15 \pm 0.02) \times 10^{-6}$ eV (-104 ± 1 mm s^{-1}), and the isomer shift of ^{181}Ta(Re) relative to Ta metal, -14.00 ± 0.10 mm s^{-1}.[455]

The electric quadrupole interaction of ^{181}Ta was subsequently also observed in the hexagonal transition metals osmium ($-2.35 \pm 0.04 \times 10^{-6}$ eV $\equiv -114 \pm 3$ mm s^{-1}), hafnium ($+1.83 \pm 0.10 \times 10^{-6}$ eV $\equiv$

---

[453] P. Boolchand, D. Langhammer, C.-L. Lin, S. Jha, and N. F. Peek, *Phys. Rev.* (C), 1972, **6**, 1093.
[454] G. Schäfer, P. Herzog, and B. Wolbeck, *Z. Physik.*, 1972, **257**, 336.
[455] G. Kaindl, D. Salomon, and G. Wortmann, *Phys. Rev. Letters*, 1972, **28**, 952.

+90 ± 5 mm s⁻¹), and ruthenium (−1.56 ± 0.04 × 10⁻⁶ eV ≡ −76 ± 2 mm s⁻¹). The results show that a single 'point-ion and uniform background model' is unsatisfactory, as this model predicts that the electric-field gradients should be positive in each case. The observed isomer shifts cover a range of 27 mm s⁻¹ and are consistent with the systematics (as yet unpublished) for ^{181}Ta impurities in host metals of the d-transition elements.[456]

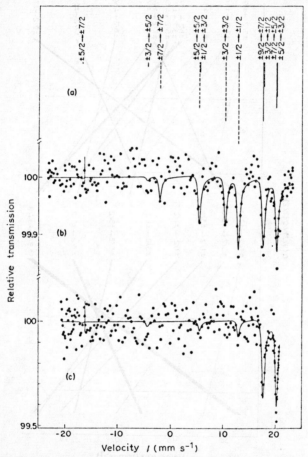

Figure 40 *Velocity spectra of the* 6.2 keV *γ-rays of* ^{181}Ta *in rhenium metal versus a* Ta *metal absorber, with direction of observation* (b) *perpendicular and* (c) *parallel to the* [0001] *axis. Solid lines, the result of a simultaneous least-squares fit of both spectra.* (a) *Positions and intensities of the individual components, solid lines* ($\Delta m = \pm 1$) *and dashed lines* ($\Delta m = 0$, *perpendicular to* [0001] *axis), respectively*
(Reproduced by permission from *Phys. Rev. Letters*, 1972, **28**, 952)

[456] G. Kaindl and D. Salomon, *Phys. Letters*, 1972, **40A**, 179.

Tungsten (^{180}W, ^{182}W, ^{183}W, ^{184}W, ^{186}W). Interference effects have been observed in the spectra of a number of isotopes of tungsten (see p. 590).[421]

The ratio of the quadrupole moments of the first excited states of ^{180}W and ^{182}W has been found to be $^{180}Q : ^{182}Q = 1.031 \pm 0.043$, from measurements following Coulomb excitation in a WO$_3$ absorber.[457]

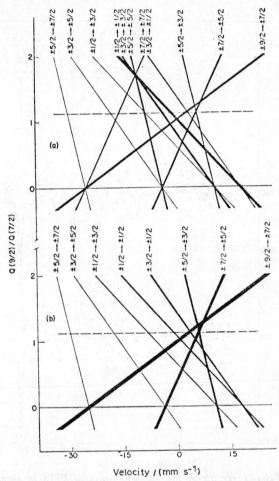

Figure 41 *Dependence of the line positions expected for the $\frac{9}{2} \to \frac{7}{2}$ ground state (E1) transition in ^{181}Ta on the ratio of the quadrupole moments $Q_{\frac{9}{2}}/Q_{\frac{7}{2}}$ for an axially symmetric electric-field gradient:* (a) *perpendicular to the axis of the EFG,* (b) *parallel to it. The widths of the lines are proportional to the intensities of the components*
(Reproduced by permission from *Phys. Rev. Letters*, 1972, **28**, 952)

[457] N. Hershkowitz, S. A. Wender, and A. B. Carpenter, *Phys. Rev.* (C), 1972, **5**, 219.

The results of fitting the transmission integral, describing Mössbauer spectra for five-line hyperfine split absorbers, with sums of single-line Lorentzians have been presented. This procedure can yield fits which are statistically satisfactory but which give incorrect values for the absorber parameters. It was shown how the fitted values depend on the absorber parameters and how they can be used to determine correct values for the absorber parameters. The procedure was demonstrated with data obtained by resonant absorption following Coulomb excitation of ^{182}W using a tungsten metal target and a ^{182}W$_2$B absorber.[72]

The mechanism responsible for the decrease in the saturation magnetization of nickel–tungsten alloys, with increase in the tungsten concentration, has been studied by ^{182}W Mössbauer spectroscopy. It appears that the decrease occurs because the magnetic moments of the tungsten impurities are oriented in the opposite direction to those of the nickel host. An alternative explanation, that the tungsten impurities simply dilute the average magnetic moment, was ruled out.[458]

The 46.5 keV line of ^{183}W has been used to measure the recoil-free fraction, $f_A(T)$, in the temperature range $77 \leqslant T \leqslant 470$ K for a powdered absorber of anhydrous Li$_2$WO$_4$. The analysis of the data was performed with an exact expansion of the transmission integral in powers of the absorber thickness. The effective Debye temperature was found to be $\theta_{\text{eff}} = 172 \pm 9$ K, which is in good agreement with a value of 205 ± 40 K calculated from X-ray diffraction data.[459]

Mössbauer scattering measurements with the 122.6 keV γ-rays of ^{186}W, obtained with a newly developed in-beam spectrometer, have been compared with the results of conventional transmission techniques.[84]

Osmium (^{186}Os, ^{188}Os, ^{189}Os, ^{190}Os). Interference effects have been observed for the ^{186}Os (137.2 keV) and ^{188}Os (155.0 keV) resonances in osmium metal (see p. 590).[421]

Only one other paper has dealt with osmium, but it describes a substantial amount of work and includes a report of the first observation of the 187 keV resonance in ^{190}Os.[460] The 187 keV level is populated in the electron capture of ^{190}Ir ($T_{\frac{1}{2}} = 11$ days), which is itself produced by the cyclotron reaction ^{190}Os($d,2n$)^{190}Ir. It is unlikely that this particular isotope will be used in future chemical studies, because of the very low percentage absorptions which it yields and because its unfavourable nuclear parameters lead to broad unresolved resonances. These points are illustrated in Figure 42, which compares spectra of OsP$_2$ obtained with this and with two other osmium isotopes having $2+ \rightarrow 0+$ spin transitions, namely ^{186}Os (137 keV) and ^{188}Os (155 keV). The latter give better resolution but only marginally improved percentage absorptions.

[458] N. Schibuya, Y. Tsunoda, M. Nishi, and N. Kunitomi, *J. Phys. Soc. Japan*, 1972, **33**, 564.
[459] J. Kaltseis, H. A. Posch, and W. Vogel, *J. Phys. (C)*, 1972, **5**, 2523.
[460] F. E. Wagner, H. Spieler, D. Kucheida, P. Kienle, and R. Wäppling, *Z. Physik*, 1972, **254**, 112.

By contrast, the 69 keV $\frac{5}{2}+ \rightarrow \frac{3}{2}+$ transition in ^{189}Os gives absorptions in the 1—3% range, as shown in Figure 43. This isotope would appear to offer distinct possibilities for future chemical studies. It has a usable source half-life ($T_{\frac{1}{2}} = 13$ days) and gives a single line when incorporated into the cubic iridium lattice. The source preparation is expensive and

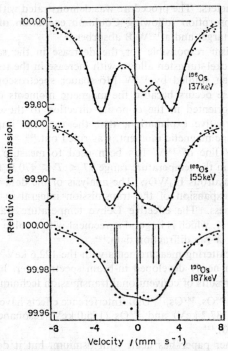

Figure 42 *Transmission Mössbauer spectra of the* 137, 155, *and* 187 keV γ-rays *of* $^{186, 188, 190}$Os *taken with* OsP$_2$ *absorbers and with sources emitting an unsplit resonance line. The curves are the results of least-squares fits. The vertical lines indicate the positions and relative intensities of the individual absorption lines* (Reproduced by permission from *Z. Physik*, 1972, **254**, 112)

involves the cyclotron reaction ^{189}Os$(d,2n)^{189}$Ir, followed by a radiochemical separation. As well as being sensitive to quadrupole interactions, this isotope will also reveal the presence of a magnetic field (see below).

From the spectra of OsP$_2$ discussed earlier, the following ratios of quadrupole moments were derived: $Q_{2+}(^{186}$Os, 137 keV$) : Q_{2+}(^{188}$Os, 155 keV$) : Q_{2+}(^{190}$Os, 187 keV$) : Q_{\frac{3}{2}-}(^{189}$Os, g.s.$) = +1.100 \pm 0.020 : 1.0 : +0.863 \pm 0.051 : -0.586 \pm 0.011$, and $Q_{\frac{5}{2}-}(^{189}$Os, 69 keV$) : Q_{\frac{3}{2}-}(^{189}$Os, g.s.$) = -0.735 \pm 0.012$. From these the following values for the quadrupole moments themselves were obtained: $Q_{2+}(^{186}$Os$) = -1.50 \pm 0.10$ b, $Q_{2+}(^{188}$Os$) = -1.36 \pm 0.09$ b, $Q_{2+}(^{190}$Os$) = -1.18 \pm 0.08$ b, $Q_{\frac{3}{2}-}(^{189}$Os$)$

= $+0.80 \pm 0.06$ b, and $Q_{\frac{5}{2}-}(^{189}Os) = -0.59 \pm 0.05$ b. A measurement with a magnetically split source yielded a value for the $E2/M1$ mixing parameter for the 69 keV transition of ^{189}Os, $\delta = +0.685 \pm 0.025$. The ratio of the g factors of the 69 keV state and the ground state was found to be $g_{\frac{5}{2}-}/g_{\frac{3}{2}-} = 0.895 \pm 0.006$, and the hyperfine field at the osmium nucleus in the magnetic matrix of iron, $H_i = -1135 \pm 20$ kG.[460]

Iridium (^{191}Ir). Interference effects have been observed in the spectra of ^{191}Ir in iridium metal (see p. 590).[421]

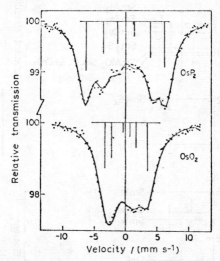

Figure 43 *Transmission Mössbauer spectra of the 69 keV γ-rays of ^{189}Os taken with OsO_2 and OsP_2 absorbers and with a source of ^{189}Ir in Ir metal, which emits an unsplit line. The curves are the results of least-squares fits. The vertical lines indicate the position and relative intensities of the individual absorption lines* (Reproduced by permission from *Z. Physik*, 1972, **254**, 112)

Platinum (^{195}Pt). The g-factor of the first excited $\frac{5}{2}-$ level in ^{195}Pt has been found to be 0.35 ± 0.042 from the magnetic hyperfine splitting in a platinum–iron alloy. Comparison of the isomer shift measurements of the 99 keV and the 130 keV transitions indicate that $\Delta\langle r^2\rangle_{130}/\Delta\langle r^2\rangle_{99} = 1.61 \pm 0.20$.[461]

Gold (^{197}Au). Anomalously narrow Mössbauer linewidths have been reported for the 77.34 keV transition in ^{197}Au in thin absorbers of AuCN, dispersed in an inert powder, and $KAu(CN)_2$ similarly prepared. The average linewidth is 1.54 ± 0.05 mm s^{-1}, which is only 80% of that obtained from lifetime measurements. The latter are in good agreement

[461] B. Wolbeck and K. Zioutas, *Nuclear Phys.* (*A*), 1972, **181**, 289.

with results for absorbers of gold metal obtained previously and corroborated in the present study. Several possible explanations of the effect were suggested.[462]

A large electric hyperfine alignment of ^{197}Au nuclei has been observed by Mössbauer absorption measurements in $KAu(CN)_2$ at temperatures as

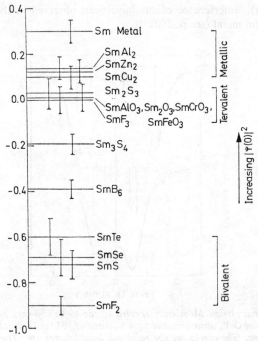

Figure 44 *Isomer shifts in* ^{149}Sm *compounds and alloys at room temperature with respect to a* $Sm:Eu_2O_3$ *source at room temperature. Error flags have been conservatively chosen and include possible impurity effects as well as statistical errors*
[Reproduced by permission from *Phys. Rev. (B)*, 1972, **6**, 18]

low as 36 mK. The nuclear spin–lattice relaxation time was estimated from the time dependence of the asymmetry in the spectra and it was shown that the spectra of oriented samples could be used to determine the recoil-free fraction of the absorber.[463]

Lanthanide and Actinide Elements.—*Samarium* (^{149}Sm). The isomer shift of the 22.5 keV resonance of ^{149}Sm has been studied for various ionic, semi-conducting, and metallic compounds and the results are shown in diagrammatic form in Figure 44. The large shifts between the bivalent and tervalent

[462] W. Potzel and G. J. Perlow, *Phys. Rev. Letters*, 1972, **29**, 910.
[463] T. E. Katila and J. A. Sawicki, *Solid State Comm.*, 1972, **10**, 895.

ionic samarium compounds arise primarily from the change in electronic configuration from $4f^65s^25p^6$ to $4f^55s^25p^6$. The data for the fluorides were used to estimate the difference in mean-square charge radius between the excited and ground state, $\Delta\langle r^2\rangle = +1.2 \times 10^{-3}$ fm². The bivalent compounds show a range of isomer shift due to covalency effects and the intermetallic compounds, which all contain tervalent samarium, exhibit increased isomer shifts due to the contribution of conduction electrons to the s-electron density at the nucleus. This contribution was estimated to be $|\psi|^2(CE) = 0.95 \times 10^{26}$ cm⁻³ in samarium metal. The isomer shifts for the semiconducting SmB_6 and for the chalcogenide Sm_3S_4 are anomalous.[464]

Europium (¹⁵¹Eu, ¹⁵³Eu). The relative transition energies and intensities of the components resulting from quadrupole splitting in ¹⁵¹Eu Mössbauer spectra have been presented as functions of the asymmetry parameter. For $\eta = 1$ the transitions lie symmetrically spaced about the spectrum centroid as expected and the true isomer shift coincides with both the visual estimate and the value calculated from a single Lorentzian fitting. This is not the case for $\eta \neq 1$, as shown in the past. The integrated area of the absorption peak is independent of the electric-field gradient.[465]

The isomer shift of the 21.7 keV transition of ¹⁵¹Eu in EuF_2 and Eu^{2+} : CaF_2 has been measured as a function of temperature between 293 and 873 K. For EuF_2 the temperature dependence of the isomer shift matches that expected from the second-order Doppler effect, whereas for Eu^{2+} : CaF_2 a much larger change is observed, which corresponds to a decrease in electron density at the nucleus. The observed change of electron density is considered to arise from the competition of the volume expansion which reduces $|\psi(0)|^2$ with increasing temperature, and the electron–phonon interaction which has the opposite effect. These two effects cancel one another in FeF_2, whereas in Eu^{2+} : CaF_2 the volume effect dominates.[466]

Data on a number of 1 : 1 amine-polycarboxylate chelates of europium(III) have indicated that the intensity of the Mössbauer effect can be of diagnostic value for detecting polymerization in rare-earth compounds. The chelates $Eu(nta),1.25H_2O$, $H[Eu(edta)]$, $Eu(hedta),H_2O$, and $H_2[Eu(dtpa)]$ have comparatively high resonant effects at room temperature ($\sim 12\%$) and are thought to be polymeric, whereas $Eu(hedta),7H_2O$ and $NH_4[Eu(edta)],8H_2O$ have low ($\sim 3\%$) effects and are monomeric. $Eu(nta),6H_2O$, $H[Eu(edta)],5H_2O$, $(NH_4)_2[Eu(dtpa)],5H_2O$, and $H_2[Eu(dtpa)],5H_2O$ appear to be intermediate between these extremes (nta = nitrilotriacetic acid, hedta = hydroxyethylethylenediaminetriacetic acid, dtpa = diethylenetriaminepenta-acetic acid).[467]

The yellow precipitate from the reaction between europium and NH_4CN has been shown to contain only Eu^{2+} and is therefore $Eu(CN)_2$. The

---

[464] M. Eibschütz, R. L. Cohen, E. Buehler, and J. H. Wernick, *Phys. Rev.* (*B*), 1972, **6**, 18.
[465] A. L. Nichols, N. R. Large, and G. Lang, *Chem. Phys. Letters*, 1972, **15**, 598.
[466] N. Nitsche, J. Pelzl, S. Hufner, and P. Steiner, *Solid State Comm.*, 1972, **10**, 145.
[467] J. L. Mackey and N. N. Greenwood, *J. Inorg. Nuclear Chem.*, 1972, **34**, 1529.

Eu—CN ionicity is estimated to be 37%, using the method of Sanderson which chooses an ionicity of 75% for the bonds in NaF. The dicyanide disproportionates on standing, on warming, or in the presence of water vapour to the tricyanide, in which the Eu—CN bond has an ionic component of 26%. For comparison, the ionicities in EuF_3 and $EuCl_3$ are 41% and 35%, respectively.[468]

The value of ^{151}Eu Mössbauer spectroscopy in detecting new phases in the reactions between europium metal and reagents such as HCN, $(CN)_2$, C_2H_2, and NH_4SCN has been convincingly demonstrated. In liquid ammonia europium reacts with HCN, C_2H_2, and NH_4SCN to produce europium(II) and europium(III) cyanide, europium(III) acetylide, and europium(III) thiocyanate, respectively. Reaction with HCN and $(CN)_2$ at elevated temperatures yields Eu_2C_3, EuC_2, and compounds of the type EuN_xC_y which are thought to be band systems on the basis of their isomer shifts. The isomer shifts for these compounds increase as the linewidths (and hence the quadrupole splittings) increase, which indicates that the population of the 5d orbitals is proportional to the population of the 6s orbital and that the europium orbitals contributing to the band system are hybrids of 6s and $5d_{xz}$, and $5d_{yz}$. In general, ionic europium(II) compounds have shifts in the range -13.9 to -10.9 mm s^{-1} relative to EuF_3, band systems -11.4 to -7.6 mm s^{-1} and ionic europium(III) -0.2 to $+0.9$ mm s^{-1}. EuN was shown to be ionic and to contain Eu^{3+}. For the europium(III) halides and pseudohalides EuX_3 (X = F, Cl, Br, OH, SCN, CN, or I) the isomer shift increases as the bond ionicity decreases in the order listed.[469]

The Mössbauer effect in ^{151}Eu has been observed following Coulomb excitation with 3.3 MeV alpha particles at 85 K and at room temperature. At 85 K the resonance for the Eu_2O_3 target has a more negative isomer shift and a broader linewidth than the room-temperature resonance.[470]

The ratio of the nuclear quadrupole moments of the 21.6 keV excited state and the ground state of ^{151}Eu has been found to be $R = 1.34 \pm 0.03$, and the quadrupole interaction energy -190 ± 7 MHz, from measurements of the hyperfine interaction of ^{151}Eu in Eu_2TiO_4. This material has the K_2NiF_4 structure and orders magnetically at 7.8 K. Unlike other compounds which have been used for determination of R, it contains Eu^{2+} in a site of four-fold symmetry and therefore has an axially symmetric electric-field gradient tensor at the resonant nucleus. Below $T_c = 7.8$ K the magnetic hyperfine field lies very close to the direction of V_{zz}. At 4.2 K and below, the quadrupole interaction energy reduces to 85 MHz when a 2 kG magnetic field is applied. This value is about half of that observed in the absence of a field and implies that the Eu^{2+} magnetic moments are highly susceptible to external fields. This behaviour suggests a ferro-

---

[468] I. J. McColm and S. Thompson, *J. Inorg. Nuclear Chem.*, 1972, **34**, 3801.
[469] I. Colquhoun, N. N. Greenwood, I. J. McColm, and G. E. Turner, *J.C.S. Dalton*, 1972, 1337.
[470] R. L. Lambe and D. Schroeer, *Phys. Letters*, 1972, **41A**, 435.

magnetic ordering in Eu_2TiO_4. The zero-degree value of H_{hf} is estimated to be 305 ± 3 kG.[471]

Spectra of europium iron garnet, EuS, and EuO have yielded the ratios of the g factors and the hyperfine splitting constants for the 21.6 keV state and the ground state in ^{151}Eu and for the 103 keV state and the ground state in ^{153}Eu. From these results values for the hyperfine anomaly $^{\circ}\Delta^{22}_{151} = -0.81\ (8)\%$ and $^{\circ}\Delta^{103}_{153} = +1.8\ (8)\%$ were derived for these two isotopes.[472]

Isomer shifts can be studied both by means of the Mössbauer effect and by nuclear excitation in muonic atoms. Muonic isomer shifts have been reported for five γ-ray transitions in ^{153}Eu and new values of $\Delta\langle r^2\rangle$ have been derived for four europium transitions (see Table 4).

Table 4 Estimates of $\Delta\langle r^2\rangle$ for ^{151}Eu and ^{153}Eu

Isotope	Transition/keV	$\Delta\langle r^2\rangle/\text{fm}^2 \times 10^3$	Ref.
^{153}Eu	103	−101 ± 12	473
^{153}Eu	103	−112.525	475
^{153}Eu	97	−93 ± 11	473
^{153}Eu	97	−86.874	475
^{153}Eu	83	−2.5 ± 0.9	473
^{151}Eu	22	17.9 ± 2.1	473

For the 22 keV transition in ^{151}Eu an isomer shift of 13.1 ± 0.3 mm s^{-1} has been found in the past between Eu^{2+} and Eu^{3+} in chalcogenides by extrapolation to 100% ionicity, and this value can be used to derive the electron-density difference between Eu^{3+} and Eu^{2+}, $\Delta|\psi(0)|^2(Eu^{3+} - Eu^{2+}) = (2.8 \pm 0.3) \times 10^{26}$ cm^{-3}. This calibration can be extrapolated to the other rare earths and, by re-evaluation of Mössbauer data, more accurate values of $\Delta\langle r^2\rangle$ can be obtained for these nuclei.[473, 474] A theoretical study has confirmed the above expressions for $\Delta|\psi(0)|^2$ and has yielded independent estimates of $\Delta\langle r^2\rangle$ for two transitions, which are also listed in Table 4.[475]

Gadolinium (^{155}Gd, ^{157}Gd). The 86.5 keV resonance in ^{155}Gd has been used to measure the hyperfine magnetic and quadrupole interactions in $GdOX_3$ (X = Cl, Br, or I), GdX_3 (X = F, Br, I, or OH), $GdX_3,6H_2O$ (X = Cl, Br, or NO_3), Gd_2O_3, GdOOH, $Gd_2(SO_4)_3$, $Gd_2(SO_4)_3,8H_2O$, $Gd_2(C_2O_4)_3$, and $Gd_2(C_2O_4)_3,10H_2O$. The ratio of the quadrupole moments of the excited state and the ground state of ^{155}Gd was determined to be $R = 0.12 \pm 0.02$ from an analysis of the pure quadrupole spectra of the oxyhalides. The experiments were extended down to 0.04 K and all of the

---

[471] Chia-Ling Chien and F. de S. Barros, *Phys. Letters*, 1972, **38A**, 427.
[472] G. Crecelius, *Z. Physik*, 1972, **256**, 155.
[473] H. K. Walter, H. Backe, R. Engfer, E. Kankeleit, C. Petitjean, H. Schneuwly, and W. U. Schröder, *Phys. Letters*, 1972, **38B**, 64.
[474] H. K. Walter, H. Backe, R. Engfer, E. Kankeleit, C. Petitjean, H. Schneuwly, and W. U. Schröder, *Helv. Phys. Acta*, 1972, **45**, 47.
[475] J. Meyer and J. Speth, *Phys. Letters*, 1972, **39B**, 330.

compounds were found to order magnetically between 0.1 and 10 K. Relaxation effects were observed in some of the spectra close to the magnetic ordering temperature. For the oxyhalides the angle between the principal axis of the electric-field gradient and the magnetic field is $\theta = 90°$, and the magnetic spin is perpendicular to the c-axis. GdF_3 orders at 1.25 ± 0.05 K and has an electric-field gradient, which is not axially symmetric ($\eta = 0.7 \pm 0.1$), ordered at an angle $\theta = 65 \pm 4°$ to the magnetic axis. In view of the fact that the angle between the c-axis and the [111] direction is also ca. 65°, it appears that V_{zz} coincides with the c-axis and that the magnetic axis is along [111].[476]

Experiments with the narrow line of the 64 keV resonance in ^{157}Gd have enabled the isomer shift, magnetic field (-324 ± 11 kG), and the electric-field gradient ($V_{zz} = +0.87 \pm 0.08 \times 10^{17}$ V cm^{-2}) in gadolinium metal to be determined precisely. The ratio of the g-factors of the excited state and the ground state were found to be 0.91 ± 0.01, and $\Delta\langle r^2 \rangle_{64}^{157} = +0.019 \pm 0.005$ fm^2. The latter quantity is related to the change in the mean-square charge radius for the 86.5 keV transition in ^{155}Gd by the expression $\Delta\langle r^2\rangle_{64}^{157}/\Delta\langle r^2\rangle_{86.5}^{155} = 2.45 \pm 0.16$.[477]

Dysprosium (^{161}Dy). The Mössbauer recoil-free fraction for the 25.6 keV resonance in ^{161}Dy in a molybdenum metal host has been found to exhibit an anomalous temperature dependence, decreasing by almost a factor of five as the temperature decreases from 300 to 25 K and then increasing again at 4.2 K. The phenomenon has been reported previously for a ^{161}Dy–Gd_2O_3 source and Dy_2O_3 absorber, for which the recoil-free fraction was less at liquid-air temperature than at room temperature. A minimum in the temperature dependence of the recoil-free fraction could result from a phase transition in the host at the corresponding temperature, but no such transition is known in molybdenum between 4 and 300 K. The effect therefore remains unexplained.[478]

Thermodynamic temperature measurement with a ^{161}Dy Mössbauer thermometer has been described.[98]

Holmium (^{165}Ho). The ratio of the g-factor of the 94.7 keV excited state to that of the ground state in ^{165}Ho has been found to be $g_{ex}/g_{gr} = 0.77 \pm 0.03$.[73]

Erbium (^{166}Er). Interference between conversion electrons, emitted after resonance absorption, and photoelectrons has been observed for the 80.6 keV resonance of ^{166}Er in $ErAl_3$ (see p. 590).[421]

Ytterbium (^{170}Yb, ^{171}Yb). The $0+ \rightarrow 2+$ 84.3 keV transition in ^{170}Yb has been used to study the antiferromagnets Yb_2O_2S and Yb_2O_2Se. The Mössbauer spectra above and below the Néel temperature are unaffected

---

[476] T. E. Katila, V. K. Typpi, G. K. Shenoy, and L. Ninistö, *Solid State Comm.*, 1972, **11**, 1147.
[477] J. Göring, *Z. Physik*, 1972, **251**, 185.
[478] E. W. Rork and P. S. Jastram, *Phys. Rev. Letters*, 1972, **29**, 1297.

Mössbauer Spectroscopy

by slow relaxation phenomena and allow the hyperfine parameters to be accurately determined. The magnetic moments were found to be 1.60 ± 0.05 BM and 1.52 ± 0.05 BM at 1.3 K in Yb^{3+}-containing compounds.[479] The crystalline fields acting on Yb^{3+} in $YbPd_3$, which has the f.c.c. Cu_3Au structure, have been studied in detail by ^{170}Yb Mössbauer spectroscopy. Although ytterbium is tervalent, no magnetic ordering occurs, even at 1.4 K. An external magnetic field induces both a magnetic hyperfine field and an electric-field gradient and the dependence of these on both temperature and applied field shows that the ground state is not an isolated level. All of the experimental observations are reproduced by a cubic-crystalline-field calculation with parameters $A_4 \langle r^4 \rangle = -12 \pm 1$ cm^{-1} and $A_6 \langle r^6 \rangle = 0.6 \pm 0.6$ cm^{-1}. These give a Γ_7 ground state with Γ_8 and Γ_6 lying at 29 and 39 cm^{-1}. The low-field spectra yield an ionic spin-relaxation time of $\tau_R \simeq 4 \times 10^{-11}$ s.[480]

A Mössbauer study of localized moments at ^{170}Yb in gold has yielded the first observation of relaxation effects for a rare-earth ion subjected to an isotropic hyperfine interaction, in a cubic environment, and in the absence of an external magnetic field. The spectra were interpreted in terms of the stochastic model of Clauser and Blume, which in this particular case is equivalent to the perturbation theory of Hirst.[481]

Interference effects have been observed for the ^{170}Yb (84.3 keV) and ^{171}Yb (66.7 keV) resonances in $YbAl_3$ (see p. 590).[421]

Uranium (^{236}U, ^{238}U). Interference effects have been observed for the 45.3 keV resonance of ^{236}U in UO_2 (see p. 590).[421]

Magnetic hyperfine fields of 3.3 ± 0.3 and 3.6 ± 0.3 MG have been shown to exist at the uranium nucleus in ordered US and UP, respectively, by use of the 44.7 keV resonance in ^{238}U. These values, combined with earlier data on UO_2 and USb and with known values of the magnetic moment on the uranium ion, indicate that J-mixing is not dominant in these cubic uranium compounds.[482]

Neptunium (^{237}Np). The compounds $Np(C_5H_5)_4$, $Np(C_5H_5)_3Cl$, and $Np(C_5H_5)_3,3THF$ have been studied at 4.2 K by means of the 59.5 keV resonance in ^{237}Np. The spectra are shown in Figure 45 and are quite different from one another. $Np(C_5H_5)_3,3THF$ gives a sharp line with an isomer shift of 36.4 mm s^{-1} (relative to $NpAl_2$ at 77 K), which is typical of Np^{3+} and indicates that there is very little covalency in the $Np^{3+}-C_5H_5$ bonding. $Np(C_5H_5)_4$ gives a quadrupole-split spectrum ($e^2qQ/4 = 1.66 \pm 0.02$) with an isomer shift of +7.2 mm s^{-1} compared with -5.6 mm s^{-1} for Np^{4+} in NpO_2, which indicates considerable additional shielding of the 6s shell. The spectrum of $Np(C_5H_5)_3Cl$ is a broad, featureless

---

[479] F. Gonzalez Jimenez and P. Imbert, *Solid State Comm.*, 1972, **10**, 9.
[480] I. Nowik, B. D. Dunlap, and G. M. Kalvius, *Phys. Rev. (B)*, 1972, **6**, 1048.
[481] F. Gonzalez Jiminez and P. Imbert, *Solid State Comm.*, 1972, **11**, 861.
[482] G. K. Shenoy, M. Kuznietz, B. D. Dunlap, and G. M. Kalvius, *Phys. Letters*, 1972, **42A**, 61.

resonance spanning 200 mm s⁻¹. If the splitting were due to long-range magnetic ordering the magnetic splitting constant, $g_0\mu_N H_{eff}$, would be about 50 mm s⁻¹. However, magnetic susceptibility results indicate that the compound is paramagnetic at 4.2 K. The broadening is therefore probably caused by paramagnetic relaxation effects.[483]

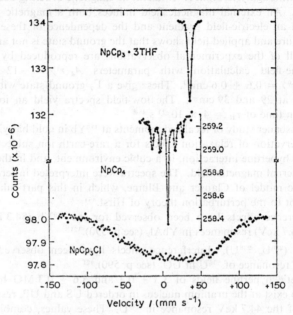

Figure 45 *Mössbauer spectra of some cyclopentadiene compounds of neptunium at 4.2 K*
(Reproduced by permission from *Inorg. Chem.*, 1972, **11**, 1742)

A metamagnetic transition (a field-induced transition from antiferromagnetism to ferromagnetism) has been shown to occur at 11.6 K in neptunium(v) oxalate, $(NpO_2)_2C_2O_4,4H_2O$, by means of magnetic susceptibility measurements. However, the Mössbauer spectrum exhibits magnetic hyperfine splitting which is independent of temperature with $g_0\mu_N H_{eff} = 100$ mm s⁻¹. Splitting observed above the transition temperature is attributed to a long ($\geqslant 10^{-7}$ s) paramagnetic relaxation time.[484]

The four compounds Li_5NpO_6, $[Co(en)_3]NpO_5,xH_2O$, $Ba_3(NpO_5)_2,xH_2O$, and $Ca_3(NpO_5)_2,xH_2O$ have been studied by means of the 59.5 keV resonance in ²³⁷Np, in order to confirm the presence of septavalent neptunium ($5f^06d^07s^2$). The observed isomer shifts are compared with those of neptunium in other oxidation states in Figure 46. In each case the isomer

[483] D. G. Karraker and J. A. Stone, *Inorg. Chem.*, 1972, **11**, 1742.
[484] E. R. Jones and J. A. Stone, *J. Chem. Phys.*, 1972, **56**, 1343.

shift is more negative than that of NpF_6, which is thought to be the most ionic neptunium(VI) compound. Since $\Delta\langle r^2\rangle/\langle r^2\rangle$ for ^{237}Np is negative it follows that the electron density at the neptunium nucleus in each of the four compounds is greater than that in NpF_6. The greater covalency of the

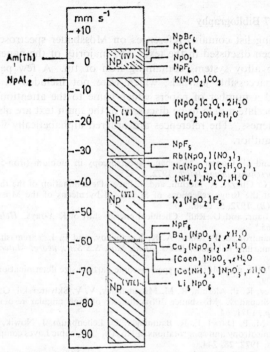

Figure 46 Correlation of isomer shifts of ^{237}Np in different compounds with the oxidation state. The lower ends of the shaded areas are approximate limits predicted by theory

Np—O bonds compared with the Np—F bonds would increase the shielding and decrease the electron density. It therefore follows that the neptunium is septavalent. Li_5NpO_6 exhibits a quadrupole splitting of 15.3 ± 0.5 mm s^{-1}, which disproves the previously assumed O_h symmetry for the NpO_6^{5-} ion. The result is more consistent with a compressed octahedral structure with the neptunyl group lying in the z-direction. The other three compounds have quadrupole splittings which are about twice as large as that for Li_5NpO_6, and this is thought to indicate a reduced bond length for the neptunyl group in these compounds. The more positive isomer shifts for these three compounds are also consistent with this interpretation. The deviation of the asymmetry parameter from zero suggests that the common anion is $[NpO_4(OH)_2]^{3-}$, and not $[NpO_5]^{3-}$, and that this anion adopts a tetragonal-bipyramidal configuration with compression in the z-direction and a rhombic distortion in the xy plane.[485, 486]

Plutonium (^{239}Pu). The recoil-free resonance absorption of the 57.3 keV γ-ray of ^{239}Pu has been observed for the first time and the half-life of the excited state shown to be 101 ± 5 ps.[487]

7 Bibliography

The following list contains references on Mössbauer spectroscopy which have not been discussed in the text. The majority of the papers are concerned with alloy systems containing iron or tin. A few papers, from relatively inaccessible journals, which were not included in last year's Report, and a number of papers which came to the attention of the reviewer at too late a date to be discussed in the main text are also included for completeness. The references are ordered alphabetically with respect to the first author.

M. Ableiter and U. Gonser. Mössbauer spectroscopy in niobium–(iron-57)–hydrogen, *Ber. Kernforschungsanlage Jülich*, 1972, **2**, 727.

G. Albanese, C. Ghezzi, A. Merlini, and S. Pace. Determination of the thermal diffuse scattering at the Bragg reflections of Si and Al by means of the Mössbauer effect, *Phys. Rev. (B)*, 1972, **5**, 1746.

K. Alder, G. Baur, and U. Raff. Chemical energy shift of K X-rays. *Helv. Phys. Acta*, 1972, **45**, 765.

A. Yu. Aleksandrov, S. P. Ionov, A. K. Karabekov, and E. I. Yaremash. Mössbauer effect in samarium tellurides, *Izvest. Akad. Nauk S.S.S.R., neorg. Materialy*, 1971, **7**, 1922.

D. Arnold. E.P.R. and Mössbauer spectroscopic particle size determinations, *Z. Chem.*, 1971, **11**, 409.

A. N. Artem'ev, K. P. Aleshin, R. M. Mirzababaev, V. V. Sklyarevskii, G. V. Smirnov, and E. P. Stepanov. Mössbauer diffractometer of high angular resolution, *Pribory Tekhn. Eksp.*, 1971, 64.

U. Atzmony, M. P. Dariel, E. R. Bauminger, D. Lebenbaum, I. Nowik, and S. Ofer. Magnetic anisotropy and spin rotations in Ho$_x$Tb$_{1-x}$Fe$_2$ cubic Laves compounds, *Phys. Rev. Letters*, 1972, **28**, 244.

I. A. Avenarius, R. N. Kuz'min, and A. A. Opalenko. Mössbauer effect anisotropy in antimony single crystals, *Pis'ma Zhur. eksp. teor. Fiz.*, 1971, **14**, 484.

Yu. F. Babikova, V. P. Filippov, and I. I. Shtan. New intermetallic compound in the zirconium–iron system, *At. Anergy*, 1972, **32**, 484.

D. Barb, S. Constantinescu, L. Diamandescu, and D. Tarina. Contributions to the Mössbauer lineshape in coupled hyperfine interactions, *Rev. Roumaine Phys.*, 1972, **17**, 769.

D. Barb, S. Constantinescu, and D. Tarina. Formulae for the relative γ-transition probabilities of the Mössbauer isotopes, *Rev. Roumaine Phys.*, 1971, **16**, 1005.

D. Barb, S. Constantinescu, and D. Tarina. Mössbauer scattering probability in coupled hyperfine interactions, *Rev. Roumaine Phys.*, 1972, **17**, 3.

D. Barb and M. Morariu. Mössbauer isotope tables, *Inst. Fiz. At. (Rom.)* [*Rep.*], 1972, NR-9.

G. M. Bartenev, A. D. Tsyganov, S. A. Dembovskii, and V. I. Mikailov. Mössbauer effect in tin-containing chalcogenide glasses, *Zhur. struct. Khim.*, 1971, **12**, 926.

Sh. Sh. Bashkirov, I. K. Bel'skaya, N. G. Ivoilov, P. N. Stetsenko, and V. A. Chistyakov. Induced magnetization of nickel in the near-paramagnetic region, *Internat. J. Magn.*, 1972, **2**, 71.

[485] K. Fröhlich, P. Gütlich, and C. Keller, *J.C.S. Dalton*, 1972, 971.
[486] K. Fröhlich, P. Gütlich, and C. Keller, *Angew. Chem.*, 1972, **11**, 57.
[487] J. Gal, Z. Hadari, E. R. Bauminger, and S. Ofer, *Phys. Letters*, 1972, **41B**, 53.

Sh. Sh. Bashkirov, K. V. Ilyatov, V. I. Sinyavskii, and R. Wadas. Magnetic sublattices in hexagonal ferrite $BaFe_{12}O_{19}$, *Electron. Technol.*, 1971, **4**, 3.

Sh. Sh. Bashkirov, S. A. Luchkina, R. A. Manapov, and E. K. Sadykov. Mössbauer spectra of nitrosyl complexes of univalent iron, *Zhur. strukt. Khim.*, 1971, **12**, 1096.

Sh. Sh. Bashkirov, R. A. Manapov, and E. K. Sadykov. Mössbauer spectra of iron bromide crystal hydrates, *Izvest. Vyssh. Ucheb. Zaved., Fiz.*, 1971, **14**, 121.

Sh. Sh. Bashkirov, R. A. Manapov, and E. K. Sadykov. Temperature dependence of the quadrupole splitting of nuclear γ-resonance spectra of an iron(II) iodide hexahydrate crystal, *Teor. i eksp. Khim.*, 1971, **7**, 705.

O. A. Bayukov, V. P. Ikonnikov, and M. I. Petrov. Nuclear γ-resonance spectrometer based on the AI-256 analysers, *Pribory Tekhn. Eksp.*, 1971, 56.

V. F. Belov, P. P. Kirichok, V. Glotov, G. S. Podval'nykh, D. E. Bondarev, and M. N. Shipko. Lithium–manganese ferrites studied by Mössbauer and X-ray methods, *Izvest. Akad. Nauk S.S.S.R., neorg. Materialy*, 1971, **7**, 2106.

V. F. Belov, P. P. Kirichok, and G. S. Podval'nykh. Electronic structure of transition metal ions in scandium-containing ferrites, *Metallofizika*, 1971, 42.

L. H. Bennett, L. J. Swartzendruber, and R. E. Watson. Critical temperatures in irondoped copper-rich copper–nickel alloys, *Amer. Inst. Phys. Conf. Proc.*, 1972, No. 5 (Pt. 2), 1190.

H. C. Benski, R. C. Reno, C. Hohenemser, R. Lyons, and C. Abeledo. New Mössbauereffect measurements on ^{57}Fe in a nickel host: the critical exponent β for Ni, *Phys. Rev. (B)*, 1972, **6**, 4266.

E. Beregi, L. Cser, T. Sterk, A. Vertes, I. Vincze, and B. Zsoldos. Mössbauer study of the iron ions in Ca, V, Bi garnets, *Phys. Status Solidi (A)*, 1971, **8**, K97.

A. I. Beskrovnyi and Yu. M. Ostanevich. Current method for recording Mössbauer spectra, *Pribory Tekhn. Eksp.*, 1971, 54.

M. Boudart, R. L. Garten, and W. N. Delgass. Mössbauer spectroscopy. New tool for the study of catalytic materials, *Mém. Soc. Roy. Sci. Liège, Collect. 8°*, 1971, **1**, 135.

G. Bueche and H. Appel. Angular correlation formalism involving separated nuclear hyperfine transitions, *Z. Physik*, 1972, **250**, 145.

M. A. Butler, G. K. Wertheim, and D. N. E. Buchanan. Domain and wall hyperfine fields in ferromagnetic iron, *Phys. Rev. (B)*, 1972, **5**, 990.

M. A. Butyugin. Automatic Mössbauer spectrometer based on the spectrometric part of a diffractometer, *Zavod. Lab.*, 1972, **38**, 1147.

R. Cammack, C. E. Johnson, D. O. Hall, and K. K. Rao. State of the iron atoms in hydroxylase iron–sulphur proteins, *Biochem. J.*, 1971, **125**, 18p.

R. Cammack, K. K. Rao, D. O. Hall, and C. E. Johnson. Mössbauer studies of adrenoxin. Mechanism of electron transfer in a hydroxylase iron sulphur protein, *Biochem. J.*, 1971, **125**, 849.

S. J. Campbell, P. E. Clark, and P. R. Liddell. Distribution of nearest neighbour atoms in CuFe alloys, *J. Phys. (F)*, 1972, **2**, L114.

J. S. Carlow and R. E. Meads. The iron–palladium–hydrogen alloy system, *J. Phys. (F)*, 1972, **2**, 982.

W. N. Cathey and B. B. Norris. The Mössbauer effect of iron impurities in Cu–Zn alloys, *Phys. Letters*, 1972, **42A**, 331.

A. Chamberod, L. Billard, and H. Rechenberg. Effects of irradiation on the magnetic properties of iron–nickel alloys in the invar region, *Solid State Comm.*, 1972, **10**, 483.

V. V. Chekin, A. I. Velikodnyi, S. N. Glushko, and E. D. Semenova. Dynamics of vibrations of ^{119}Sn nuclei in dilute tin–base alloys, *Soviet Phys. Solid State*, 1972, **14**, 479.

H. S. Cheng, S. J. Yeh, and P. K. Cheng. Mössbauer study of some γ- and neutronirradiated iron carbonyl complexes, *J. Chinese Chem. Soc. (Formosa)*, 1971, **18**, 179.

J. Cirak, J. Lipka, and M. Prejsa. Measurement of the Mössbauer spectra by resonance absorption and resonance scattering, *Fyz. Cas.*, 1971, **21**, 228.

R. L. Collins. Phases and stresses in ferrous metals by Mössbauer spectroscopy, *Internat. J. Nondestruct. Test.*, 1972, **4**, 77.

T. E. Cranshaw. Disturbance produced in an iron lattice by chromium atoms and some other solutes, *J. Phys. (F)*, 1972, **2**, 615.

L. Cser, V. Gorobchenko, and I. Lukashevich. Investigation of hyperfine field distributions in Fe–Sn alloys by the Mössbauer effect, *Phys. Status Solidi. (B)*, 1972, **51**, 339.

L. Dabrowski, J. Piekoszewski, and J. Suwalski. Evaluation of the hyperfine parameters of iron-57 from Mössbauer spectra, *Electron Technol.*, 1971, **4**, 39.

L. M. Dautov, M. M. Kadykenov, D. K. Kaipov, and U. K. Kulbaeva. Calibration of isomeric shifts in Mössbauer spectra based on shifts in X-ray K_α emission lines, *Izvest. Akad. Nauk Kaz. S.S.R., Ser. Fiz.-Mat.*, 1971, **9**, 18.

N. N. Delyagin, E. N. Kornienko, and V. I. Nesterov. Impurity atoms of ^{119}Sn in the helicoidal antiferromagnet Au_2Mn, *Soviet Phys. Solid State*, 1972, **14**, 1590.

I. Dezsi, V. D. Gorobchenko, M. Komor, I. I. Lukashevich, A. Vertes, and K. F. Tsitskishivili. Mössbauer study of frozen ferric perchlorate solutions at different pH, *Acta Chem. Acad. Sci. Hung.*, 1971, **70**, 329.

R. Dicus and M. P. Maley. Investigation of iron local moment behaviour in copper–zinc alloys by Mössbauer effect measurement, *Amer. Inst. Phys. Conf. Proc.*, 1972, No. 5, (Pt. 2), 1179.

J. L. Dormann, L. Brossard, and G. A. Fatseas. Simple analysis of Mössbauer spectra, having many closely spaced internal fields, by line-broadening coefficients, *Phys. Status Solidi (B)*, 1972, **52**, K23.

V. N. Dubinin, V. V. Kuz'movich and V. M. Chegoryan. Use of the nuclear γ-resonance method to study iron and tin in metal–oxygen polymers, *Teor. i eksp. Khim.*, 1971, **7**, 703.

I. A. Dubovtsev, F. A. Sidorenko, A. N. Bortnik, T. S. Shubina, and N. P. Filippova. Mössbauer spectra of ^{57}Fe in $Fe_{1-x}Co_xSi$ and $Fe_{1-x}Ni_xSi$, *Phys. Status Solidi (B)*, 1972, **49**, 405.

J. Dudas, J. Kucera, and T. Zemcik. Mössbauer effect in the surface layer of cobalt-57 diffused into a nickel–aluminium alloy, *Czech. J. Phys.*, 1972, **22**, 332.

B. Dudrewa, S. Grande, K. Melzer, and C. Michalk. Mössbauer investigations of potassium ferrocyanide single crystals, *Wiss. Z. Karl-Marx-Univ. Leipzig, Math.-Naturwiss. Reihe*, 1971, **20**, 565.

H. H. Ettwig and W. Pepperhoff. Mössbauer effect investigations on the σ-phase of binary iron alloys, *Arch. Eisenhuettenwiss.*, 1972, **43**, 271.

P. A. Flinn. Metallurgical applications of the Mössbauer effect, *Appl. Low Energy X-Gamma Rays*, 1971, 123.

F. Y. Fradin, B. D. Dunlap, G. Shenoy, and C. W. Kimball. N.M.R. and Mössbauer effect in Pt–Rh alloys, *Internat. J. Magn.*, 1971, **2**, 415.

E. Frank. Mössbauer spectrometry. Fundamentals, *Cienc. Invest.*, 1971, **27**, 307.

S. N. Glushko, V. V. Chekin, A. I. Velikodnyi, and L. F. Rybal'chenko. Dynamics of tin impurity oscillations in lead, *Zhur. eksp. i teor. Fiz.*, 1972, **62**, 661.

V. I. Goldanskii. Modern trends in nuclear chemistry, *Fiz. Szemle*, 1971, **21**, 329.

M. Goldstein and W. D. Unsworth. Infrared and Raman spectra (3500—70 cm^{-1}) and Mössbauer spectra of some pyrazine complexes of stannic halides, *Spectrochim. Acta*, 1971, **27A**, 1055.

V. A. Golovnin, S. B. Zezin, and R. N. Kuz'min. Effective magnetic fields at the diamagnetic tin atoms in $MnSn_2$, *Soviet Phys. Solid State*, 1972, **14**, 1313.

J. Göring. Nuclear moments of the 67.4 keV level in nickel-61, *Z. Naturforsch.*, 1971, **26a**, 1929.

R. W. Grant. Magnetic structure investigation with polarized recoil-free γ-rays, *Amer. Inst. Phys. Conf. Proc.*, 1972, No. 5 (Pt. 2), 1395.

J. A. R. Griffith, G. R. Isaak, and S. Roman. Observation of polarization transfer in the reaction $^{56}Fe(d,p)^{57}Fe^*$ using the Mössbauer effect, *Phys. Rev. Letters*, 1972, **28**, 375.

A. G. Grigoryan and V. A. Belyakov. Theory of Mössbauer diffraction, *Vestnik. Mosk. Univ. Fiz., Astron.*, 1971, **12**, 668.

G. Grüner, I. Vincze, and L. Cser. Charge- and spin-perturbation around non-magnetic impurities in iron, *Solid State Comm.*, 1972, **10**, 347.

P. L. Gruzin, V. S. Mkrtchyan, Yu. L. Rodionov, Ya. P. Selisskii, and M. Kh. Khachatryan. Effect of germanium on structural changes of iron–aluminium alloys, *Fiz. Metal. Metalloved.*, 1972, **34**, 315.

P. L. Gruzin, Yu. L. Rodionov, Yu. D. Zharov, V. S. Mkrtchyan, A. F. Edneral, and M. D. Perkas. Redistribution of alloying element atoms during the ageing of iron–nickel alloys, *Doklady Akad. Nauk S.S.S.R.*, 1972, **202**, 316.

P. L. Gruzin, Yu. L. Rodionov, V. S. Mkrtchyan, and Yu. A. Li. Effect of cobalt on the redistribution of alloying element atoms in iron-based metals studied by Mössbauer spectroscopy, *Doklady Akad. Nauk S.S.S.R.*, 1972, **204**, 328.

I. G. Gusakovskaya, T. I. Larkina, and V. I. Goldanskii. Effect of glass–crystal phase transition on spin–lattice relaxation time, *Fiz. Tverd. Tela*, 1972, **14**, 2631.

W. C. Harper, C. W. Kimball, and A. T. Aldred. Effect of the Mössbauer lineshape on the determination of hyperfine fields in Fe + Ga solid solution, *Amer. Inst. Phys. Conf. Proc.*, 1972, No. 5 (Pt. 1), 533.

J. Helsen, K. Schmidt, Th. Chakupurakal, R. Coussement, and G. Langouche. Determination of self-duffusion coefficient in a vermiculite by Mössbauer effect, *Bull. Groupe Fr. Argiles*, 1972, **24**, 165.

I. R. Herbert, P. E. Clark, and G. V. H. Wilson. A Mössbauer effect study of dilute Cr–Fe alloys, *J. Phys. and Chem. Solids*, 1972, **33**, 979.

C. L. Herzenberg, D. L. Riley, and R. B. Moler. Applications of Mössbauer spectrometry to lunar and terrestrial rock samples, *Appl. Low Energy X-Gamma Rays*, 1971, 187.

A. W. Hewat. Mössbauer recoil-free fraction and Debye–Waller factors for Nb_3Sn, *Phys. Letters*, 1972, **39A**, 249.

M. C. Hobson. Surface studies by Mössbauer spectroscopy, *Adv. Colloid Interface Sci.*, 1971, **3**, 1.

H. W. Holzinger and H. J. Tiller. Infrared and Mössbauer spectroscopic studies on the incorporation of tin into corona polymers. I. Corona polymerization of tin tetrachloride–benzene gas mixture, *Plaste Kaut.*, 1972, **19**, 656.

H. Horita and E. Hirahara. Mössbauer studies of single crystal iron sulphide, *Sci. Rep. Tohoku Univ., Ser.* 1, 1971, **54**, 127.

G. P. Huffmann and G. R. Dunmyre. Tin-119 and iron-57 Mössbauer study of ordered $FeNi_3$, *Amer. Inst. Phys. Conf. Proc.*, 1972, No. 5 (Pt. 1), 544.

C. Janot and P. Delcroix. Study of mechanically strained beryllium–iron alloys by the Mössbauer effect, *Acta Met.*, 1972, **20**, 637.

C. Janot and B. George. Temperature-dependence study of a vanadium–iron alloy by Mössbauer spectroscopy, *J. Phys. and Chem. Solids*, 1972, **33**, 1023.

C. Janot and H. Gilbert. Mössbauer effect study of the precipitation of iron in beryllium, *Mater. Sci. Eng.*, 1972, **10**, 23.

P. S. Kamenov. Mössbauer experiments on the lifetime of excited nuclei, *Doklady Bolg. Akad. Nauk*, 1972, **25**, 185.

A. A. Kaukis, A. M. Babeshkin, A. A. Bekker, A. N. Nesmeyanov, L. M. Ostrovskaya, and M. I. Tsypin. Mössbauer spectra of $Pd_xSn_{1-x}Te$ alloys, *Izvest. Akad. Nauk S.S.S.R., neorg. Materialy*, 1972, **8**, 1667.

N. Kawamiya, K. Adachi, and Y. Nakamura. Magnetic properties and Mössbauer investigations of Fe–Ga alloys, *J. Phys. Soc. Japan*, 1972, **33**, 1318.

A. Kjekshus and D. Nicholson. Use of indium antimonide as standard for antimony-121 Mössbauer spectroscopy, *Acta Chem. Scand.*, 1971, **25**, 3895.

M. A. Krivoglaz and S. P. Repetskii. Theory of diffusion broadening and weakening of the Mössbauer lines in spectra of impurity atoms in crystals, *Fiz. Metal. Metalloved.*, 1971, **32**, 899.

M. A. Krivoglaz and S. P. Repetskii. Effect of diffusion on Mössbauer spectra with fine structure, *Fiz. Metal. Metalloved.*, 1971, **32**, 1156.

K. Krop. Use of the Mössbauer effect in metal physics, *Hutnik*, 1972, **39**, 314.

L. Kumer, H. Posch, and J. Kaltseis. Mössbauer effect of Fe^0 atoms in cation-exchanged zeolite (type A4), *Phys. Letters*, 1972, **40A**, 59.

R. N. Kuz'min and S. V. Nikitina. Mössbauer effect in a solid solution of tin in germanium, *Soviet Phys. Solid State*, 1972, **13**, 3157.

S. J. Lewis and P. A. Flinn. Mössbauer effect study of diffusion of iron in β-iron–titanium alloys, *Phil. Mag.*, 1972, **26**, 977.

S. Ligenza. A Mössbauer study of hyperfine parameters of ^{119}Sn in antiferromagnetic and paramagnetic states of FeSn, *Phys. Status Solidi (B)*, 1972, **50**, 379.

J. A. Lock and J. F. Reichert. Mössbauer-electronic double resonance in $NH_4(^{57}Fe, Al)(SO_4)_2, 12H_2O$, *J. Magn. Resonance*, 1972, **7**, 74.

E. L. Loh, U. Atzmony, and J. C. Walker. Determination of the Eu–Fe exchange anisotropy in EuIG, *Amer. Inst. Phys. Conf. Proc.*, 1972, No. 5 (Pt. 1), 396.

S. A. Losievskaya and R. N. Kuz'min. Determination of the parameters of long-range order in ordered alloys by the Mössbauer effect, *Izvest. Akad. Nauk S.S.S.R., Metal*, 1972, 179.

J. C. Love and F. E. Obenshain. Hyperfine fields at nickel in Heusler alloys, *Amer. Inst. Phys. Conf. Proc.*, 1972, No. 5 (Pt. 1), 538.

V. A. Makarov, E. B. Granovskii, E. F. Makarov, V. A. Povitskii, and Yu. Ya. Rybakov. Anomalous microregions in ferromagnetic substances studied by γ-resonance spectroscopy, *Zhur. eksp. i Teor. Fiz.*, 1972, **62**, 1827.

Yu. V. Maksimov, I. P. Suzdalev, Yu. P. Yampol'skii, and K. P. Lavrovskii. Effect of carrier on topochemical reactions of iron oxide with acetylene and hydrogen studied by a Mössbauer spectroscopic method, *Kinetika i Kataliz*, 1971, **12**, 1391.

H. Maletta. Giant magnetic moments in $Ni_3Ga(Fe)$, *Z. Physik*, 1972, **250**, 68.

H. Maletta and K. R. P. M. Rao. Discontinuous formation of localized moments in niobium–molybdenum (iron), *Internat. J. Magn.*, 1972, **3**, 5.

H. Maletta, K. R. P. M. Rao, and I. Nowik. Relaxation phenomena due to *s*–*d* exchange interactions of dilute iron in molybdenum, *Z. Physik*, 1972, **249**, 189.

S. S. Malik and S. S. Hanna. Area analysis of neutron and Mössbauer resonances, *Particles Nucl.*, 1971, **1** 268.

T. V. Malysheva. Fundamentals of the nuclear gamma resonance (Mössbauer spectroscopy) method and its applications to the study of isomorphism, *Prob. Izomorfnykh Zameshchenii At. Krist.*, 1971, 231.

A. S. Marfunin, A. R. Mkrtchyan, G. N. Nadzharyan, Ya. M. Nyussik, and A. N. Platnov. Optical and Mössbauer spectra of iron in some layered silicates, *Izvest. Akad. Nauk S.S.S.R., Ser. geol.*, 1971, 87.

G. R. Mather. Anomalous magnetic behaviour of an iron-rich glass, *Amer. Inst. Phys. Conf. Proc.*, 1972, No. 5 (Pt. 2), 821.

J. N. Mathur and H. B. Mathur. Mössbauer study on the nature of iron(II) bonding in bis(dithioacetylacetone) tetrabromoferrate(II) and bis(dithiobenzoylacetone) tetrachloroferrate(II), *Indian J. Chem.*, 1971, **9**, 1348.

L. May, R. Nassif, and M. Sellers. Mössbauer spectroscopy of the undecapeptide of cytochrome *c*, *Appl. Low Energy X-Gamma Rays*, 1971, 257.

A. Mayer, H. Eicher, H. Übelhack, F. Parak, H. Formanek, and K. Winterhalter. Mössbauer spectroscopy of haemoglobins, *European Biophys. Congr. Proc.*, 1st, 1971, **6**, 137.

B. deMayo. Magnetism in gold–iron alloys below 14 at.%Fe, *Amer. Inst. Phys. Conf. Proc.*, 1972, No. 5 (Pt. 1), 492.

W. Meisel, and H. J. Jörg. Studies of the bonding of iron salts to textile fibres by Mössbauer spectroscopy, *Z. Chem.*, 1972, **12**, 302.

T. Mizoguchi, T. Sasaki, and S. Chikazumi. Appearance of ferromagnetism in Fe–Ir–Pt alloys, *Amer. Inst. Phys. Conf. Proc.*, 1972, No. 5 (Pt. 1), 445.

H. Mosbaek and K. G. Poulsen. Mössbauer investigation of the electronic structure of the brown-ring complex, *Acta Chem. Scand.*, 1971, **25**, 2421.

F. Mosora. Mössbauer resonance in methaemoglobin obtained after partial experimental hepatectomy, *Stud. Cercet. Biol.*, Seria Zool., 1971, **23**, 451.

A. N. Murin, V. T. Shipatov, I. P. Polozova, and P. P. Seregin. Tin-containing glasses studied by nuclear γ-resonance effect, *Vestn. Leningrad. Univ. Fiz., Khim.*, 1971, 157.

S. S. Nandwani and S. P. Puri. Anharmonic and magnetic contributions to Mössbauer fraction for 14.4 keV gamma-ray of ^{57}Fe in natural iron, *J. Phys. and Chem. Solids*, 1972, **33**, 973.

S. Nasu and Y. Murakami. Mössbauer isomer shift of metals and alloys, *Nippon Kinzoku Gakkai Kaiho*, 1972, **11**, 267.

D. A. O'Connor, M. W. Reeks, and G. Skyrme. Mössbauer effect study of the lattice dyanmics of iron impurities in Al, Rh, Pd, V, and Mo metals, *J. Phys. (F): Metal Phys.*, 1972, **2**, 1179.

A. Pae, P. Pae, E. Pealo, and L. Uibo. Mechanicochemical reactions and activation of quartz studied by the Mössbauer effect, *Doklady Akad. Nauk S.S.S.R.*, 1971, **200**, 1066.

V. N. Panyushkin and E. N. Yakovlev. Mössbauer effect and superconductivity of tin under pressure, *Soviet Phys. Solid State*, 1972, **14**, 1579.

V. N. Panyushkin and E. N. Yakovlev. Probability of the Mössbauer effect in β-tin at high pressure and the Gruneisen constant, *Zhur. eksp. i teor. Fiz.*, 1972, **62**, 1433.

G. Papadimitriou and J. M. Genin. Mössbauer effect evidence of an ordered $Fe_{15}Si$ solid solution and hyperfine field interpretation of ordered iron-silicon alloys. *Phys. Status Solidi (A)*, 1972, **9**, K19.

R. S. Preston. Mössbauer study of the surface barrier to the diffusion of iron into aluminium, *Met. Trans.*, 1972, **3**, 1831.

D. Raj and S. P. Puri. Re-estimation of impurity–host to host–host coupling constant ratio in iron–molybdenum system, *J. Phys. and Chem. Solids*, 1972, **33**, 2177.

E. Realo, A. Pae, and L. Uibo. Mechanochemical reactions and the activation of quartz studied using the Mössbauer effect, *Eesti NSV Tead. Akad. Toim., Fuus., Mat.*, 1971, **20**, 432.

J. Reedijk. Pyrazoles and imidazoles as ligands. VIII. Mössbauer spectra of octahedral iron(II) solvates containing substituted pyrazoles and imidazoles as ligands. *Rec. Trav. chim.*, 1971, **90**, 1285.

W. M. Reiff, K. S. V. L. Narasimhan, and H. Steinfink. Mössbauer and magnetic investigation of the system $Mn_{5-x}Fe_xGe_3$ ($x = 0.5$, 1.0, or 1.5) *J. Solid State Chem.*, 1972, **4**, 38.

J. A. Sawicki. Electric-field gradients at iron impurities in h.c.p. metals, *Phys. Status Solidi (B)*, 1972, **53**, K103.

N. Schibuya and N. Kunitomi. Magnetic moment on Fe in ternary alloy Fe–Ni–Co, *J. Phys. Soc. Japan*, 1972, **33**, 853.

C. Schwab, B. Meyer, A. Goltzene, and S. Nikitine. Scanning spectrometer with automatic digital data recording, *Rev. Phys. Appl.*, 1971, **6**, 419.

V. A. Semenkin and Yu. A. Shevchenko. Determination of background in Mössbauer experiments based on transmission in iron alloys with 3d transition elements, *Pribory Tekhn. Eksp.*, 1971, 67.

P. P. Seregin, L. N. Vasil'ev, and Z. U. Borisova. Mössbauer effect in semiconductor glasses of the germanium–selenium–tin system, *Izvest. Akad. Nauk S.S.S.R., neorg. Materialy*, 1972, **8**, 567.

T. E. Sharon and C. C. Tsuei. Magnetism in amorphous Fe–Pd–P alloys, *Phys. Rev. (B)*, 1972, **5**, 1047.

M. Shiga, A. Miyoshi, and Y. Nakamura. Mössbauer effect in ordered Fe_3Pt alloy, *Phys. Status Solidi (B)*, 1972, **49**, K195.

M. Shimada, H. Miyamota, F. Kanamaru, and M. Kiozumi. Mössbauer effect and its application to mineralogy, *Kobutsugaku Zasshi*, 1971, **10**, 186.

T. Shinjo and T. Takada. Mössbauer experiments in external magnetic fields, *Bull. Inst. Chem. Res., Kyoto Univ.*, 1971, **49**, 314.

V. T. Shipatov, Yu. P. Kostikov, and P. P. Seregin. Comparison of the chemical shift of $K_{\alpha 1}$ X-ray spectra and the isomeric shift of Mössbauer spectra of tin chalcogenides, *Vestnik Leningrad. Univ. Fiz., Khim.*, 1971, 148.

D. V. Sokol'skii, A. S. Khlystov, A. V. Kuz'minov, and D. K. Kanpov. Iron oxide catalysts on carriers studied by Mössbauer spectroscopy, *Izvest. Akad. Nauk Kaz. S.S.R., Ser. fiz.-mat.*, 1971, **9**, 82.

R. A. Stukan, I. S. Kirin, V. Ya. Mishin, and A. B. Kolyadin. Mössbauer spectroscopic study of the pyrolysis of iron phthalocyanine, *Zhur. neorg. Khim.*, 1972, **17**, 1923.

K. Sumiyama, M. Shiga, and Y. Nakamura. Mössbauer study of metastable f.c.c. iron–rhodium alloy, *Phys. Status Solidi (A)*, 1972, **13**, K75.

I. P. Suzdalev and A. P. Amulyavichus. Observation of electron spin relaxation in antiferromagnets above the Néel point by means of Mössbauer spectroscopy, *Zhur. eksp. i teor. Fiz.*, 1971, **61**, 2354.

L. J. Swartzendruber. Localized moments on iron impurities in niobium–molybdenum alloys. Mössbauer effect absorber study, *Internat. J. Magn.*, 1972, **2**, 129.

L. J. Swartzendruber, L. H. Bennett, and K. R. Kinsman. Crystallographic and magnetic transformations of iron precipitates in copper, *Amer. Inst. Phys. Conf. Proc.*, 1972, No. 5 (Pt. 1), 408.

Y. Tino and J. Arai. Mössbauer effect and the anomalous properties of the invar alloys, *J. Phys. Soc. Japan*, 1972, **32**, 941.

A. M. Van Diepen, K. H. J. Buschow, and J. S. Van Wieringen. Mössbauer effect, magnetization, and crystal structure of the pseudobinary compounds $ThCo_{5-5x}Fe_{5x}$ and $ThNi_{5-5x}Fe_{5x}$, *J. Appl. Phys.*, 1972, **43**, 645.

M. N. Varma and R. W. Hoffmann. Interpretation of Mössbauer spectra in thin iron films, *J. Vac. Sci. Technol.*, 1972, **9**, 177.

L. N. Vasil'ev, P. P. Seregin, and V. T. Shipatov. Mössbauer effect in the tin–phosphorus system, *Izvest. Akad. Nauk S.S.S.R., neorg. Materialy*, 1971, **7**, 2067.

L. N. Vasil'ev, P. P. Seregin, and V. T. Shipatov. Mössbauer effect in the arsenic–selenium–tin system, *Izvest. Akad. Nauk S.S.S.R., neorg. Materialy*, 1971, **7**, 2069.

A. Vertes, M. R. Komor, and M. Suba. Mössbauer study on the reactions occurring in iron-salt solutions, *Ann. Univ. Sci. Budapest Rolando Eotvos Nominatae, Sect. Chim.*, 1971, **12**, 79.

A. Vertes and M. Nagy. Mössbauer study of the interaction of some polymers with iron dichloride, *Radiochem. Radioanalyt. Letters*, 1972, **9**, 221.

A. Vertes, M. Suba, and M. R. Komor. Application of the Mössbauer effect to investigate the structure of iron-salt solutions, *Ann. Univ. Budapest. Rolando Eotvos Nominatae, Sect. Chim.*, 1971, **12**, 69.

I. Vincze. Anomalous temperature dependence of the impurity moment in iron with 4d and 5d impurities, *Solid State Comm.*, 1972, **10**, 341.

I. Vincze and L. Cser. Temperature dependence of the hyperfine field at iron atoms in FeSn, *Phys. Status Solidi (B)*, 1972, **49**, K99.

I. Vincze and L. Cser. Temperature dependence of the hyperfine field at iron atoms around the non-magnetic impurities aluminium and gallium, *Phys. Status Solidi (B)*, 1972, **50**, 709.

I. Vincze and G. Grüner. Temperature dependence of the hyperfine field at iron atoms near 3d impurities, *Phys. Rev. Letters*, 1972, **28**, 178.

B. Window. Mössbauer study of gold–iron alloys, *Phys. Rev. (B)*, 1972 **6**, 2013.

B. Window. Hyperfine field at ^{57}Fe in chromium–manganese alloys, *Amer. Inst. Phys. Conf. Proc.*, 1972, No. 5 (Pt. 1), 522.

V. S. Zavgorodnii, B. I. Rogozev, E. S. Sivenkov, A. A. Petrov, and L. M. Krizhanskii. Unsaturated tin hydrocarbons. XXV. Use of Mössbauer effect to study the electron effects of substituents illustrated by carbofunctional-substituted tin acetylenes, *Zhur. obshchei. Khim.*, 1971, **41**, 2237.

Author Index

Abdel-Gawad, M., 561
Abdullaev, G. K., 260
Abe, S., 158
Abe, T. 122, 257, 384
Abe, Y., 157, 158
Abedini, M., 124
Abel, E. W., 129, 276
Abeledo, C. R., 511, 516, 528, 617
Abeles, F., 197
Abicht, H. P., 378
Ableiter, M., 616
Abley, P., 123
Ablov, A. V., 338, 421, 532
Abraham, M. R., 68
Abramowitz, S., 207, 219, 223
Abu Salah, O. M., 45, 375
Abu-Samn, R. H., 135
Achmatowicz, O., jun., 98
Acquista, N., 207
Acton, N., 41, 58
Adachi, K., 619
Adams, C. J., 209, 216, 222
Adams, D. M., 197, 235, 243, 244, 371, 453
Adams, R. D., 65
Adcock, J. L., 258
Adcock, W., 10, 132, 136
Addison, A. W., 49
Addison, C. C., 438
Adeyemi, S. A., 459
Adijano, G., 260
Adlard, M. W., 52, 341
Adlkofer, J., 61, 131, 173, 245
Adloff, J. P., 495
Advena, J., 139
Afanas'ev, A. M., 495, 525
Afremow, L. C., 99
Agami, C., 379
Agarwal, B. R., 311, 458
Agarwal, N. K., 424, 426, 452, 454
Agarwala, L., 312
Agarwala, U., 312, 341, 463
Agrawal, S. K., 324
Agnihotri, O. P., 233, 326
Agranat, I., 41
Agrell, I., 398
Ahlborn, E., 216, 320, 323
Ahmad, J., 358, 405
Ahmad, N., 31
Ahmed, I. Y., 75
Ahrens, U., 141, 283
Ahuja, I. S., 353, 354, 356
Aida, K., 307, 352

Aikins, J., 111
Aimbinder, N. E., 153
Ainsworth, C., 127, 130, 131
Airey, W., 120, 262
Airoldi, C., 437
Aitken, G. B., 453
Aivazyan, T. M., 512
Ajisaka, K., 97
Akeson, A., 80
Akhtar, M., 67
Aki, O., 98
Akimov, V. K., 263
Akitt, J. W., 80
Aksnes, D. W., 145
Akyüz, S., 344
Alais, L., 44
Alashkevich, V. P., 452
Albanese, G., 499, 616
Albelo, G., 85
Albert, S., 102, 149
Albrand, J. P., 84
Albrecht, H. B., 63
Albrecht, R., 324
Alder, K., 616
Aldred, A. T., 619
Aleeva, G. P., 150
Alefeld, B., 100, 106
Alegranti, C. W., 61
Alei, M., jun., 15
Aleksandrov, A. L., 155
Aleksandrov, A. Yu., 592, 595, 616
Aleksandrova, I. P., 100
Aleksandrova, V. A., 220, 329
Aleksanyan, V. T., 300
Alekseev, V., 515
Aleshin, K. P., 501, 616
Alester, G., 130
Alexander, C. J., 524
Alexander, C. W., 56
Alexander, L. E., 208, 219
Alford, A. L., 97
Alford, K. J., 122, 259
Ali, K. M., 282
Ali, L. H., 41, 371
Ali, M. A., 462
Alichi, A., 81, 385, 388
Al-Kazzaz, Z. M. S., 458
Alkhazov, T. G., 324
Allcock, H. R., 138, 142, 143, 289, 291
Allen, A. D., 331
Allen, C. W., 128, 139, 279, 295
Allen, E. A., 51, 338, 345, 349

Allen, P. E. M., 83
Allenmark, S., 41
Allerhand, A., 12
Allin, E. J., 243
Allison, D. A., 39
Allmann, R., 62
Almqvist, S.-O., 97
Al-Obaichi, K. H., 416
Alonzo, G., 131, 458
Alper, H., 32, 35, 400, 414
Alt, H., 22, 65, 385
Al'tshuler, S. A., 149
Alyaviya, M. K., 405
Alyea, E. C., 18, 405, 456
Alymov, I. M., 168
Amano, T., 181
Ambe, F., 570
Ambe, S., 570
Amirkhanov, B. F., 153
Ammann, D., 77
Amos, L. W., 77
Amulyavichus, A. P., 621
Ananda Murthy, A. S., 463
Anastassiou, A. G., 74
Andersen, R. A., 17, 247
Anderson, C. B., 57, 382
Anderson, D. R., 97
Anderson, G. J., 265
Anderson, J. W., 125, 269, 277, 301
Anderson, N. H., 94
Anderson, R. L., 26
Anderson, S. E., jun., 88
Anderson, S. N., 45
Anderson, W. P., 22, 81, 172, 387, 469, 477
Andersson, R., 97
Ando, I., 98
Andreeff, A., 99, 495
Andreetti, G. D., 456
Andreev, P. P., 150
Andreeva, N. A., 266
Andrew, E. R., 102
Andrews, D. C., 331, 364, 367, 368, 371
Andrews, J. M., 60
Andrews, L., 202, 204, 206, 207
Andrews, S. B., 62
Andriaenssens, G. J., 111
Andrianov, K. A., 12, 87
Andronov, E. A., 491
Andronov, V. F., 9, 10, 12, 13, 126, 131
Andruchow, W., 323
Andzhaparidze, D. I., 263
Anet, F. A. L., 49
Ang, H. G., 35, 139, 475

Author Index

Ang, T. T., 101
Angelici. R. J., 26, 35, 37, 70, 367, 377, 471, 484
Angerer, E. I., 150
Angerman, N. S., 98
Angyal, S. J., 77
Anthonsen, T., 98
Antipin, A. A., 152
Antokol'skaya, I. I., 344
Antokol'skii, G. L., 152
Antonova, N. D., 72
Anxer, M. W., 471
Appec, H., 617
Appel, R., 130, 134, 140
Appelman, E. H., 10, 183, 202
Appelman, M., 70
Appleton, T. G., 54, 55, 58, 86, 346, 338, 379, 403, 404
ApSimon, J. W., 95
Aquista, N., 219
Arai, J., 621
Arai, T., 233
Araneo, A., 43, 335
Arata, Y., 93
Archer, R. D., 323, 458
Ardjomand, S., 164
Aresta, M., 331
Arguello, C. A., 239
Aritomi, M., 84, 278, 433
Arie, G., 232
Aris, K. R., 37
Aris, V., 37, 45
Aritomi, M., 132, 135
Arkangel'skii, I. V., 315
Arkhipenko, D. K., 244
Armbrecht, F. M., jun., 126
Armit, P. W., 69
Armitage, D. A., 130, 294, 406
Armitage, I., 94, 96
Armstrong, R. L., 170, 171
Arnabi, F., 278
Arnau, J. L., 200
Arnett, E. M., 74
Arnold, D., 616
Arnold, D. E. J., 13, 72, 267
Arnold, K., 152
Arnold, V., 123
Arons, R. R., 105, 106
Arro, I., 528
Artemenko, M. V., 343, 410
Artem'ev, A. N., 501, 505, 616
Ashai, Y., 5
Ashby, E. C., 81, 254
Ashe, A. J., tert., 127, 193
Ashley-Smith, J., 47, 70, 381
Ashraf, M., 22
Ashraf, Y., 407
Asker, W. J., 170
Aslanov, K. H. A., 150
Asprey, L. B., 100, 213
Asthana, B. P., 199
Asti, G., 499
Atam, N., 17, 246

Atsarkin, V. A., 150
Attali, S., 69, 488
Atwell, W. H., 136
Atzmony, U., 616, 619
Aubke, F., 261, 273, 445, 578, 581
Aubke, R., 281
Aubord, J., 216
Audibert, P., 152, 353
Auerman, L. N., 259
Aumann, R., 34, 81
Aumont, R., 432
Austin, T. E., 139, 286
Autzen, H., 17, 247
Avasthi, M. N., 222
Avdovich, H. W., 98
Avenarius, I. A., 595, 616
Averill, D. F., 460
Aviram, I., 89
Avkhutsky, L. M., 107
Avramenko, G. I., 83, 135
Axtmann, R. C., 500
Ayant, Y., 163
Ayasse, C., 592
Aymonino, P. J., 318, 319
Ayrey, G., 583
Azikov, B. S., 432

Baalmann, H. H., 290
Baba, S., 53, 342, 345
Babb, D. P., 73, 284
Babeshkin, A. M., 545, 571, 619
Babikova, Yu. F., 616
Babin, V. N., 485
Babior, B. M., 45
Babushkina, T. A., 161
Bacci, M., 440
Bach, R. D., 62
Bachman, D. F., 59
Backe, H., 611
Backes, J., 42, 406
Badalamenti, R., 53
Badalova, R. I., 432
Badand, J. P., 312
Baddley, W. H., 43, 47, 52, 335, 375, 415
Badley, E. M., 55, 377
Badot, D., 567
Baechler, R. D., 72
Baekelmans, P., 128
Bätzel, V., 46, 488
Baggio-Saitovitch, E., 513, 537
Bagnall, K. W., 458
Bailey, P. M., 54
Bailey, R. A., 93, 322, 345, 414
Bairamov, B. K., 244
Baird, M. C., 12, 40, 68, 93, 334, 351, 371
Baisa, D. F., 172
Baiwir, M., 146
Baizer, M. M., 412
Bajaj, P., 131, 277
Bajorek, A., 238
Bajpai, K. K., 257, 398
Bak, B., 223
Baker, B. R., 111
Baker, R. A., 565

Baker, R. W., 58, 346
Bakhadyrkhanov, M. K., 512
Bakhireva, S. I., 452
Balasubrahmanyan, K., 292
Balch, A. L., 32, 47, 55, 375, 380, 414, 435
Baldeschwieler, J. D., 77
Baldo, J., 5
Baldokhin, Yu. V., 499
Balicheva, T. G., 172, 201, 202
Balko, E. N., 414
Ballard, D. H., 45
Ballhausen, C. J., 320
Baltrunas, D. A., 592
Bancroft, G. M., 496, 538, 567, 588
Bandekar, J., 299
Bandler, M., 354
Bandoli, G., 18, 358
Bandy, A. R., 234
Bandyopadhyay, P., 328
Banerjee, A. K., 323
Bánhidai, B., 129
Banks, E., 557
Banister, A. J., 296, 302
Bannaghan, A. J., 510
Banninger, R., 9
Barabash, A. I., 172
Baracco, L., 456
Baraldi, P., 198, 319
Baran, A. A., 110
Baran, E. J., 213, 319
Baranetskaya, N. K., 21, 363
Baranov, S. P., 292, 356
Baranova, E. K., 270
Baranovskii, I. B., 337, 338
Barb, D., 495, 520, 553, 616
Barba, N. A., 532
Barbe, A., 200
Barbe, B., 127
Barbero, C., 501
Barber, P., 234
Barbieri, G., 9, 132, 136
Barbieri, R., 134, 272, 411, 582
Barcelo, J., 231
Barchuk, L. P., 442
Barcza, S., 6
Barefield, E. K., 43, 50, 59, 335, 408
Bargeron, C. B., 506
Barker, G. K., 268
Barlex, D. M., 45, 47, 452, 490
Barna, G., 328
Barnard, R., 445
Barnes, D. J., 51
Barnes, R. G., 105
Barnett, B., 56
Barnett, K. W., 81
Barnhart, D. M., 206
Barraclough, C. G., 220, 329, 333, 528
Barral, J.-C., 198
Barrett, G. J., 120, **260**

Barrett, P. H., 502, 509, 568
Barrie, J. A., 134
Barros, F. de S., 533, 547
Barrow, R. F., 200
Barten, T. J., 300
Bartenev, G. M., 616
Bartex, D. M., 59
Barthels, M. R., 134
Bartish, C. M., 23, 470
Bartlett, N., 209, 222, 351
Barton, D., 125
Barton, T. J., 126, 132, 270
Barvinok, M. S., 402
Bar-Ziv, E., 222
Bascom, W. D., 277
Bashirov, F. I., 149
Bashkirov, Sh. Sh., 616, 617
Basile, L. J., 223
Basolo, F., 49, 403, 466
Basso-Bert, M., 19
Bastow, T. J., 159, 160
Bateman, L. R., 167
Bateman, R. J., 167
Bates, J. B., 210, 211, 218, 230, 247, 263
Batsanov, S. S., 280, 326, 346
Batsanova, L. R., 100
Battey, M. H., 560
Battistuzzi, R., 343, 349, 355
Bauder, A., 177
Baudler, M., 63
Bauer, G., 125, 269
Baugher, J. F., 108
Baumeister, W., 122
Bauminger, E. R., 616
Baur, G., 616
Bayukov, O. A., 547, 617
Bazov, V. P., 273
Beachley, O. T., jun., 113, 258
Beall, T. W., 25, 473
Beam, R. J., 84
Bearden, A. J., 496
Beare, S. D., 97
Beattie, I. R., 139, 208, 219, 227, 285
Beattie, J. K., 42
Beaumont, R. E., 439
Beauté, C., 96
Bebault, G. M., 131
Becher, H. J., 198, 258, 323
Beck, H.-J., 31, 478
Beck, W., 12, 20, 21, 46, 53, 391, 412, 468, 469, 490
Becker, E. D., 3, 11, 64
Becker, H. J., 120
Becker, H. P., 130
Becker, L. W., 569
Becker, R. A., 248
Becker, R. F., 97
Becker, W., 12, 121
Becker, Y., 35, 369
Beckert, D., 152
Beckley, R. S., 63

Bedford, W. M., 51, 415
Bee, M. W., 332
Beech, G., 75
Beecroft, B., 343
Beer, D. C., 23, 467
Beer, H., 292
Beers, Y., 182
Beg, M. A. A., 436
Begalieva, D. U., 319
Bégué, J.-P., 94
Begun, G. M., 355, 363
Behrendt, W., 265
Behrens, H., 331, 337, 366, 367, 371, 399, 491
Behrman, E. J., 43
Beierbeck, H., 95
Beijer, B., 98
Bekker, A. A., 619
Bekturov, A. B., 319
Belaeva, K. F., 431
Belen'kii, M. S., 324
Belfort, G., 102
Belitskii, I. A., 152
Bell, A. P., 19, 315, 362
Bell, B., 26, 42, 65, 302, 321, 334
Bell, C., 197
Bell, L. G., 18, 315
Bell, M. E. B., 301
Bell, N. A., 17, 247
Bell, R. J., 2
Bellama, J. M., 119, 254
Bellet, J., 182
Belluco, U., 54, 55, 377, 415
Belogurov, V. N., 500, 551
Belousov, Yu. A., 485
Belov, V. F., 549, 556, 617
Belov, Yu. V., 148
Belova, V. M., 500
Belozerskii, G. N., 500, 520, 550, 567
Bel'skaya, I. K., 616
Belyaev, L. M., 551
Belyaeva, A. A., 303
Belyakov, V. A., 498, 618
Belyakova, Z. V., 359
Belzile, R., 210
Bénaïm, J., 24, 471
Benczer-Koller, N., 504
Bender, C. F., 200
Bender, C. O., 64
Bender, R., 87
Benedetti, E., 54, 334, 389, 486
Benelli, C., 91
Benfield, F. W. S., 25, 364
Benlian, D., 45, 424
Benkovic, S. J., 77
Benner, L. S., 93, 434
Bennett, L. H., 617, 621
Bennett, M. A., 40, 44, 52, 69, 421, 422, 490
Bennett, M. J., 327
Bennett, R., 51
Bennett, R. D., 97
Bennett, R. L., 28, 475, 476
Benoit, J. P., 244
Ben-Shoshan, R., 35
Benski, H. C., 617

Benson, R. E., 92
Benter, A., 323
Bentham, J. E., 73
Bentley, F. F., 197, 286, 299
Benton-Jones, B., 83, 120
Bentrude, W. G., 97, 144
Beppu, T., 182
Ber, M., 146
Beran, G., 264
Beránek, J., 288
Bercaw, J. E., 18, 315, 415, 489
Beregi, E., 617
Bereman, R. D., 168, 219, 318, 322
Berendsen, H. J. C., 100
Berezina, S. I., 107, 149
Berg, M., 305
Berg, R. W., 337, 404
Berger, R., 436
Berglund, D., 418
Berjot, M., 199
Berke, H., 28, 449
Berlin, K. D., 96
Berman, D. A., 130
Bermann, M., 140, 143, 289, 290
Bermudez, V. M., 297
Bernabei, A., 517
Bernal, I., 536
Bernard, L., 199
Berniaz, A. F., 83, 256
Bernheim, R. A., 103
Berniër, P., 105
Bernot, M., 137
Bernstein, H. J., 199
Bernstein, J. L., 526
Berodias, G., 567
Berringer, B. W., 357, 358
Berry, A., 387
Berry, J. M., 131
Berschied, J. R., jun., 216
Bersuker, I. B., 499
Bertacco, A., 455
Bertan, P. B., 444
Bertazzi, N., 131, 458
Bertini, I., 71, 90, 91
Bertino, R. J., 431
Berwick, M. A., 97
Beskrovnyi, A. I., 617
Beverman, D. R., 152
Beverwijk, C. D. M., 11
Bezdenezhnykh, G. V., 432
Bezman, S. A., 45, 375
Bezrodnyi, V. G., 590
Bezrukov, O. F., 152
Bhacca, N. S., 77, 96
Bhaduri, S., 419
Bhargava, S. C., 550
Bhat, S. N., 198
Bhatnagar, V. M., 403
Bhattacharya, D. L., 497, 502, 508
Bhatti, M., 122
Bhatti, W., 122
Bhayat, I. I., 43
Bhide, V. G., 512, 513, 571
Bhuiyan, A. L., 122, 255
Bianchi, M., 486

Bianco, V. D., 331, 335, 424, 428
Bichlmeir, B., 71
Bickelhaupt, F., 17, 84, 118, 139
Bickley, D. G., 318
Bidzilya, V. A., 75
Bied-Charreton, C., 44
Biedermann, S., 208
Biehl, E. R., 35
Bigam, G., 41
Bigelow, L. A., 139, 286
Bigorgne, M., 342, 344, 392, 422, 475, 480
Bigotto, A., 310
Billard, L., 617
Billups, W. E., 33, 479
Binder, H., 542
Biradar, N. S., 14, 19, 150, 276, 317, 411
Biran, A., 501
Biran, C., 127
Birchall, T., 6, 38, 124, 333, 482, 522, 523, 541
Bird, P. H., 318
Bird, S. R. A., 574, 588
Birkofer, L., 127, 128, 130, 132, 269
Birnbaum, E. R., 9, 78, 335
Biryukov, I. P., 167
Biryukova, T. V., 87
Biscarini, F., 355
Biserni, G. S., 363
Bist, H. D., 243
Bitterwolf, T. E., 31
Bjørseth, A., 191
Bjorkstam, J. L., 100, 111
Blacik, L. J., 367
Black, D. St. C., 409
Black, J. D., 387
Blackmer, G. L., 49, 51
Blackwell, L. F., 91
Blaedel, W. J., 109
Blake, D. M., 44, 53, 335, 430, 431
Blasse, G., 235
Blatz, L. A., 201
Blazer, J., 14
Bleaney, B., 95, 96
Blenderman, W. G., 22, 469
Blicharski, J. S., 3
Blight, D. G., 411
Blinc, R., 111, 157
Blinn, E. L., 320
Blomberg, C., 17, 84
Bloodworth, A. J., 62
Bloomfield, J. J., 132
Bloor, D., 357
Bloor, E. G., 75
Bloor, J. E., 79
Boal, D. H., 281, 287
Boan, E., 70
Boccaccio, G., 97
Boccalon, G., 67
Bochkarev, M. N., 280
Bochkareva, V. A., 26, 323
Bock, J. L., 535
Boddenberg, B., 99, 110
Boden, G., 329

Bodewitz, H. W. H. J., 84
Bodner, G. M., 114, 116
Bodner, R. L., 99
Bodyu, V. G., 532
Böhland, H., 318
Böhler, D., 140, 289
Böhmer, W. H., 530
Boenig, I. A., 142, 263
Boese, R., 474
Bogatov, Yu. E., 263
Bogatyreva, L. V., 25, 26, 364
Bogdanov, A. P., 532
Bogdanov, V. S., 6, 7, 83
Boggs, R. A., 26, 98
Bogomolov, V. N., 572
Bohinc, M., 307
Bohlmann, F., 97
Bohn, H. G , 100, 105, 106
Bohres, E. W., 226
Boian, P., 324
Boicelli, C. A., 132
Bøje, L., 146
Bokov, V. A., 571
Bokshtein, B. S., 545
Bokun, R. Ch., 498
Boldyrev, V. V., 520
Bolesawski, M., 121
Bologa, O. A., 338, 421
Boltaks, B. I., 512
Bolton, K., 189
Bolourtchian, M., 127
Bombieri, G., 358
Bonamartini, A. C., 349
Bonamico, M., 403
Bonardi, M., 564
Bonati, F., 60
Bonazzola, G. C., 501
Bond, A., 33, 35, 482
Bondar, A. M., 152
Bondarenko, G. V., 201
Bondarenko, I. B., 491
Bondarev, D. E., 617
Bondarevskii, S. I., 570, 575
Bonds, W. D., jun., 458
Bonelli, D., 129, 275
Bongaarts, A. L. M., 107
Bonnaire, R., 286, 339, 374
Bonnett, R., 42, 64
Bonnot, A. M., 110
Boolchand, P., 602
Boorman, P. M., 264, 449
Booth, B. C., 490
Booth, B. L., 45, 47, 339, 372, 424
Booth, D. J., 5
Booth, H., 95
Bor, G., 390
Borden, R. S., 81, 149
Borgen, G., 96
Borisenko, A. A., 140
Borisov, A. E., 161
Borisova, Z. U., 621
Borlin, J. J., 113, 151, 249
Borodaevskii, V. E., 326
Borodin, P. M., 80, 84, 149, 150
Borod'ko, Yu. G., 316, 394, 396

Borovskaya, O. R., 549
Borsa, F., 497
Borsdorf, R., 97
Borshagovskii, B. V., 331
Borshch, S. A., 499
Borshon, A. N., 458
Bortnik, A. N., 618
Bos, A., 270, 569
Bos, K. D., 137
Boschi, T., 57, 346, 415
Bose, A. K., 97, 98
Bose, K. S., 59, 462
Botoshanskii, M. M., 338
Bottger, G. L., 222, 228, 338
Botting, B. J., 94
Bottomley, C. G., 92
Bottomley, F., 459
Bottorel, M. P., 432
Bouchal, K., 47
Boucher, D., 186, 330
Boucher, L. J., 51
Boudart, M., 617
Boudjouk, P., 127, 137
Bougon, R., 209
Bouldoukian, A., 382
Boullé, A. L., 293
Bourgeois, P., 127, 128
Bourlas, M. C., 97
Bourne, A. J., 136
Bouquet, G., 475, 480
Bouwma, J., 214, 312, 349
Bouznik, V. M., 13
Bowden, J. A., 26, 386, 420, 471
Bowen, L. H., 539
Bowen, R. E., 114
Bowmaker, G. A., 298
Bowman, R. H., 72
Boyd, G. E., 210, 211, 247, 263
Boyd, R. K., 86
Boyd, W. A., 95
Boyle, L. L., 155
Braca, G., 334, 389
Bradford, C. W., 33
Bradford, R. S., 179
Bradley, C. H., 49
Bradley, C. J., 197
Bradley, D. C., 18, 99, 405
Bradley, E. B., 142, 263
Bradley, J. S., 45
Bradley, R. H., 195
Bradspies, J. I., 227
Brainina, E. M., 65, 362
Braithwaite, M. J., 47, 58, 346
Bramwell, A. F., 98
Brandt, J., 3
Branson, P. R., 46, 413
Brasington, R. D., 583
Brassy, C., 202
Braterman, P. S., 387
Bratton, R. F., 115, 249
Brauer, G., 302
Braun, J., 120
Braunecker, B., 530
Braunstein, P., 314, 343
Bray, P. J., 108, 151, 158
Brec, R., 282

Author Index

Bredereck, H., 130
Breed, L. W., 134
Breeze, P. A., 385
Breitenstein, B., 337
Breitinger, D., 232, 351
Breitmaier, E., 5
Brekhunets, A. G., 103, 152
Brema, J. L., 265
Bresler, L. S., 84
Bressani, T., 501
Bretschneider, E. S., 128, 279
Brewer, D. G., 92
Brewer, L. 200
Březina, F., 439
Brice, M. D., 46
Brice, V. T., 115, 249
Brier, P. N., 195
Brieux de Mandirola, A., 208
Briggs, D., 347
Briggs, J. M., 94
Brigot, N., 199
Brill, P., 107
Brill, T. B., 161, 172, 355, 387, 476
Brinckman, F. E., 27
Brink, G., 202
Brinkhoff, H. C., 64, 355
Brinkman, F. J., 263
Brinkman, M. R., 139, 295
Brinkmann, D., 108
Brintzinger, H. H., 18, 25, 315, 473
Briscoe, G. B., 538
Brisdon, B. J., 441
Brisse, F., 593
Brittain, A. H., 182, 185
Brnjas-Kraljevic, J., 150
Broadhead, P., 260
Brockhaus, M., 21
Brodersen, K., 264
Brodie, A. M., 35, 478
Brodie, J. D., 79
Broida, H. P., 179
Broitman, M. O., 396
Bron, W. E., 227
Brooker, M. H., 210
Brookes, P. R., 52, 339, 373
Brookhart, M., 85, 368
Brossard, L., 618
Brower, L. E., 109
Brown, A. D., jun., 41
Brown, D., 458
Brown, E., 98
Brown, E. S., 135
Brown, F. R., 270
Brown, H. C., 63, 86, 113, 118, 120, 250
Brown, J. D., 241
Brown, J. M., 37, 45, 137, 278, 579
Brown, M. L., 35
Brown, M. P., 114
Brown, N. M. D., 85
Brown, R. A., 322, 385
Brown, R. D., 189
Brown, R. G., 319
Brown, R. J. C., 162

Brown, T. L., 56, 69, 81, 82, 83, 387, 491
Brown, T. M., 317, 319, 322, 448, 454, 455
Browning, I. G., 51
Brownlee, G. S., 438
Brownson, G. W., 307
Brownstein, M., 7, 148, 209, 225, 305
Brubaker, C. H., jun., 323
Brubaker, G. L., 115, 249
Bruce, M. I., 24, 28, 30, 32, 35, 39, 40, 45, 250, 375, 440, 472, 475, 476, 483, 485, 486, 488, 489, 493
Bruecher, E., 78
Brueesch, P., 236
Brüser, W., 320
Brummer, S. B., 277
Brumbach, S. B., 207
Brumnic, A., 151
Brun, E., 101
Brun, G., 293, 303
Brune, H. A., 35, 480
Brunet, J. J., 98
Brunner, H., 21, 25, 37, 370, 414, 468, 472
Brunori, M., 42
Brunvoll, J., 363, 370, 376, 384
Brush, J. R., 465
Brushmiller, J. G., 49, 51
Bryan, G. H., 360
Bryan, J. B., 299
Bryan, P. S., 191
Bryan, R. F., 114, 252
Bryant, R. G., 77, 79
Brynestad, J., 355
Bryukhova, A. N., 168
Bryukhova, E. V., 155, 159, 161, 168, 169, 171
Buchachenko, A. L., 87
Buchanan, D. N. E., 617
Buchler, J. W., 318
Buckingham, A. D., 14, 88
Buckingham, D. A., 49, 85
Buckle, J., 7, 272
Buckley, P. D., 91
Buckley, R. C., 455
Buczkowski, Z., 71
Budarin, L. I., 77
Budzikiewicz, H., 97
Bueche, G., 617
Buehler, E., 609
Bünzli, J. C., 20, 75
Bürger, H., 18, 19, 207, 208, 284, 316
Bürvenich, C., 5
Büssemeier, B., 56
Bues, W., 263
Büttner, H., 40, 419, 480
Buishvili, L. L., 150
Bukin, A. S., 548
Bukshpan, S., 502, 506, 557
Bula, M. J., 121
Bulgakova, G. M., 91
Bulka, G. R., 100
Bullitt, J. G., 67, 101, 370, 384
Bullock, J. I., 262, 445

Bullpitt, M. L., 61, 128
Bulten, E. J., 137
Bulychev, B. M., 253
Bunce, R. J., 62
Bunget, I., 553
Bunzli, J. C., 81
Burchett, S., 423
Burden, F. R., 189
Burdett, J. K., 205
Burg, A. B., 47, 82, 87, 113, 114, 125, 248, 267, 300, 393
Burgada, R., 73
Burgard, M., 162, 169, 440
Burger, K., 496, 537, 539, 555, 577
Burgett, C. A., 95
Burgos, A., 292
Burie, J., 176
Burke, A. R., 6, 67
Burke, J. M., 312, 450, 451
Burke, M. T., 135
Burke, P. L., 113, 120, 250
Burlitch, J. M., 46, 314, 389, 471
Burmeister, J. L., 343, 412, 457
Burnett, L. J., 6
Burnham, R. A., 27, 314
Burns, G., 198, 319
Burns, R. G., 564
Burov, A. I., 127
Burow, D. F., 209
Bursics, L., 352
Busch, D. H., 42, 44, 50, 51, 59, 408, 409
Busch, M. A., 68, 482
Buschow, K. H. J., 621
Busetto, L., 55, 377
Busev, A. I., 263
Bush, M. A., 330
Bush, S. F., 333
Bushnell, G. W., 59, 345
Buslaev, Yu. A., 15, 26, 27, 161, 323
Butchard, C. G., 78
Butcher, A. V., 89, 323
Butler, I. S., 328, 330, 365, 476, 492
Butler, K. D., 538, 567
Butler, M. A., 617
Butler, R. A., 243
Buttery, H. J., 387
Butyugin, M. J., 617
Buzina, N. A., 489, 490, 491
Byers, A. E., 83
Byers, W., 51, 59, 64, 320, 447, 451
Bylinkin, V. A., 500
Byrd, J. E., 123
Byrne, J. E., 86
Bystrov, V. F., 3, 77, 149
Bywater, S., 16

Cabana, A., 210, 211, 216
Cabri, L. J., 566
Cadene, M., 243, 246
Cadraci, G., 38

Cady, G. H., 27, 61, 148, 302, 323, 434, 444, 445
Caglioti, L., 59
Cagnac, B., 110
Cagniant, P., 146
Cain, G., 505
Caira, M. R., 319
Cairns, M. A., 30, 486
Cais, M., 41, 68, 542
Calas, R., 127
Calder, G. V., 204
Calderazzo, F., 28, 434
Calderon, J. L., 65, 395, 472
Calhoun, H. P., 322
Calligaris, M., 50
Caltalini, L., 18
Calvo, C., 56, 220, 329
Cambisi, F., 35, 369
Cameron, T. S., 164
Cammack, R., 536, 617
Campbell, A. J., 101
Campbell, I. D., 159
Campbell, J. R., 96, 148
Campbell, M. J. M., 343, 408, 524
Campbell, S. J., 617
Camus, A., 43, 61, 335, 351, 489
Canepa, P. C., 102
Canters, G. W., 74, 99
Canty, A. J., 30, 67
Canziani, F., 54
Capozzi, G., 121
Capparella, G., 334
Cappuccilli, G., 448
Capwell, R. J., 317
Cara, E., 403
Carabatos, C., 228
Carberry, E., 134, 275
Carbonaro, A., 35, 369, 539
Card, D. W., 408
Card, R. J., 22
Cardacci, G., 368
Cardin, D. J., 46, 339, 373, 490
Cardiot, P., 120
Cardona, R. A., 137, 279
Carey, F. A., 128
Carey, N. A. D., 314
Carey, R. N., 129
Carfagno, P., 436
Cargioli, J. D., 5, 6
Cargioli, P. C., 12
Cariati, F., 61
Carlisle, G. O., 320
Carlow, J. S., 617
Carlson, E. H., 171
Carman, N. J., 131
Carmichael, W. M., 326
Carnevale, A., 106, 152
Caro, P. E., 357
Carpenter, A. B., 499, 604
Carpenter, R. A., 201
Carr, B., 351
Carrea, G., 97
Carroll, M., 12
Carroll, T. X., 524
Carstens, D. H. W., 516

Carter, R. L., 237
Carter, R. O., 193, 271
Cartledge, F. K., 6
Carturan, G., 54
Carty, A. J., 38, 39, 261, 264, 366, 438, 482, 541
Cary, L. W., 26
Casals, P.-F., 97, 98
Casellato, U., 18, 358
Casey, A. T., 320, 448
Casey, C. P., 26, 98
Casey, J. P., 72
Casey, M. E., 314
Casper, J. M., 112, 286, 287
Cassady, J. M., 98
Cassoux, P., 59
Castellano, S. M., 78
Castellucci, E., 277
Cataliotti, R., 376, 387
Cathey, W. N., 617
Cattalini, L., 59, 60, 347, 411, 452
Cattermole, P. E., 391
Caubere, P., 98
Cavalca, L., 349
Cavalieri, A., 493
Cavell, R. G., 8, 15, 19, 25, 59, 64, 72, 84, 132, 141, 279, 292, 296, 298, 320
Cavit, B. E., 30, 423, 486
Cazeau, P., 131
Cazzoli, G., 185
Ceausescu, V., 150
Cebulec, E., 61, 351
Ceccon, A., 67, 363
Ceder, O., 98
Celon, E., 351
Cenini, S., 334, 339, 384, 398, 588
Centofanti, L. F., 143, 248
Cerfontain, H., 97
Cerimele, B. J., 95
Černik, M., 288
Cerruti, L., 325
Cervinez, F., 44
Cervinka, M., 150
Cesaro, S. N., 200, 350
Cesarotti, E., 18, 358
Cessac, G. L., 234
Cetini, G., 30, 32, 35, 483, 486, 487
Çetinkaya, B., 22, 367, 373
Chadaeva, N. A., 141, 153
Chadburn, B. P., 10
Chadha, S. L., 281
Chagin, V. I., 27
Chakdar, N. C., 410
Chakravorty, A., 82, 92
Chakupurakal, Th., 619
Chalmers, A. A., 96
Chamberod, A., 617
Chambers, O. R., 148
Champeney, D. C., 507
Chan, A. W.-L., 70
Chan, C. H., 258
Chan, K. M., 401
Chan, S. C., 401
Chandra, K., 518

Chandra, S., 183, 536
Chang, C. C., 352
Chang, I. F., 228
Chang, J. C., 399
Channing, D. A., 546
Chapelle, J. P., 244
Chapman, A. C., 137, 278, 579
Chappert, J., 511
Charache, S., 42, 151
Charalambous, J., 282
Charkoudian, J. C., 42, 416, 540
Charles, R., 44, 490
Charlton, T. L., 8
Charpin, P., 307
Charrier, C., 24, 73, 471
Chase, L. L., 227
Chatelain, A., 152
Chatrousse, A.-P., 147
Chatt, J., 25, 26, 27, 28, 42, 65, 89, 321, 323, 334, 395, 396, 400
Chatterjee, S., 112, 286
Chau, F. T., 222
Chaudhry, S. C., 435
Chaudhuri, J., 16, 361
Chauvet, G., 231, 427
Chaves, A., 236
Chawla, H. F. S., 97, 98
Checherskaya, L. F., 540
Checherskii, V., 546
Chegoryan, V. M., 618
Chekin, V. V., 617, 618
Chen, D. M., 75
Chen, F., 127, 130
Chen, K.-N., 35
Chen, K. S., 26
Chen, M. M., 190
Chen, T.-M., 77, 89
Chen, Y. T., 140, 441
Cheney, A. J., 53, 75, 347
Cheng, C. Y., 164
Cheng, H. N., 96
Cheng, M. S., 617
Cheng, P. K., 617
Cheng, Y. M., 126
Cheng-Jyisong, 504
Cherepanov, V. M., 525
Cherezov, N. K., 501
Cherkasova, T. G., 490, 491
Chernikov, S. S., 397
Chernov, A. P., 161
Chernov, R. V., 105
Chernyshev, E. A., 127
Chernyshev, Yu. S., 150
Chervonenkis, A. Y. A., 150
Cherwinski, W. J., 58, 347, 413
Cheung, T. T., 358
Chevalier, R., 555
Chew, K. F., 12
Cheyne, B. M., 595
Chia, L. S., 139, 478, 540
Chia-Ling Chien, 611
Chiavassa, E., 501
Chiba, S., 149
Chieh, P. C., 438

Author Index

Chikazumi, S., 620
Chini, P., 46, 54, 372, 493
Chipperfield, J. R., 136
Chirulescu, T., 99
Chisholm, M. H., 11, 53, 54, 379, 380
Chistokletov, V. N., 151
Chistyakov, V. A., 616
Chistyakova, E. A., 343
Chiswell, B., 328, 456, 463
Chivers, T., 140, 200, 290
Chizhik, V. I., 149
Chmelnick, A. M., 88
Chodos, S. L., 234
Chomic, J., 344
Chong, C., 438
Choplin, F., 440
Chorea, M., 438
Chottard, G., 339
Chow, K. K., 343
Chow, S. T., 349
Chow, Y. L., 35
Choy, V. J., 424
Chrisman, B. L., 498
Christe, K. O., 6, 148, 209, 217, 219, 220, 222, 224, 302, 305, 306
Christen, P., 177
Christensen, A. N., 323
Christiaens, L., 146, 147
Christian, D. F., 40, 487
Christiansen, J., 508
Christophliemk, P., 312, 450
Chuiko, A. A., 277, 301
Chujo, R., 98
Chumaevskii, N. A., 197
Chung, C., 127, 148, 202
Churakov, V. G., 491
Churchill, M. R., 45, 375, 450
Churlyaeva, L. A., 11
Chuvaev, V. F., 105, 106, 315
Chvalovský, V., 126, 276
Ciappenelli, D. J., 67, 72
Cichon, J., 208, 284
Cirak, J., 617
Civan, M. M., 103
Claassen, H. H., 202, 224
Clapp, C. H., 61, 447
Clardy, J., 39, 98, 132
Clare, P., 144
Clark, A. C., 341
Clark, A. J., 146, 305
Clark, D. T., 347
Clark, G. M., 122
Clark, G. R., 30, 40, 487
Clark, H. C., 11, 28, 52, 53, 54, 55, 58, 328, 341, 346, 347, 366, 378, 379, 380, 381, 409, 413
Clark, J. W., 297
Clark, M. G., 173, 577
Clark, P. E., 617, 619
Clark, P. W., 52, 422
Clark, R., 35, 483
Clark, R. C., 499
Clark, R. J., 40, 68, 420, 482

Clark, R. J. H., 19, 65, 207, 211, 214, 316, 317, 331
Clarke, A. L., 335
Clarke, E. C. W., 201
Clarke, G. R., 417
Clarke, H. L., 364, 371
Clarke, J. H. R., 355
Clarke, M. R., 306
Claudel, J., 229
Cleare, M. J., 384, 417
Clegg, D. E., 53, 341
Clemens, J., 45, 374
Clemente, D. A., 18, 51, 69, 358
Cleveland, F. F., 426
Cleveland, J. M., 360
Clifford, A. F., 144, 295
Clipsham, R. M., 211
Clough, S., 99
Clouse, A. O., 12
Coakley, M. P., 314
Coates, G. E., 17, 247
Coburn, T., 244
Cocevar, C., 43, 335, 489
Cochoy, R. E., 41
Cock, P. A., 86
Cockerill, A. F., 97
Coe, J. S., 60, 347, 452
Coey, J. M. D., 502, 517, 546, 551
Coffari, E., 200, 350
Coghi, L., 456
Cohen, I. A., 93
Cohen, M. A., 412
Cohen, R. L., 495, 573, 574, 609
Cohen, S. C., 134, 275
Cohn, K., 54, 94, 112, 190, 380, 421
Cohn, M., 80, 152
Coignac, J. P., 234, 242
Coindard, G., 120
Coker, B. M., 9
Colburn, C. B., 37, 335
Cole, T. C., 79
Cole-Hamilton, D. J., 69
Coles, M. A., 65, 316
Coles, R. B., 409
Coletta, F., 71
Collette, J. W., 122
Collier, M. R., 53, 120, 257, 347
Collins, R. L., 500, 504, 600, 617
Collman, J. P., 416, 417, 418, 479
Colman, R. F., 78
Colot, J. L., 99, 157
Colquhoun, I., 356, 610
Colsmann, G., 319
Colton, R., 26, 386, 402, 420, 467, 471, 477
Colts, R. M., 100
Coluccia, S., 325
Commenges, G., 4
Commereuc, D., 24, 471
Commons, C. J., 26, 420
Conard, J., 107
Conder, H. L., 314, 488

Condrate, R. A., 462
Conia, J. M., 97
Conlin, R. T., 129
Connell, P. S., 101
Connelly, N. G., 48, 418, 419
Connick, R. E., 75
Connor, D. E., 45
Connor, J. A., 11, 22, 23, 65, 411, 467
Consiglio, G., 131, 458
Constantinescu, S., 616
Contreras, E., 330
Conway, D. C., 478
Cook, D. F., 402
Cook, K. A., 536
Cook, R. B., 198
Cook, R. L., 182, 183
Cook, R. J., 72
Cook, W. J., 142, 291
Cooke, D. W., 49
Cooke, M. P., 416
Cookson, P. G., 430, 465
Coon, A. D., 120, 260
Cooney, R. P. J., 61, 314, 354, 457
Cooper, C. G., 125, 269
Cooper, D. G., 11
Copin, G., 83
Copperthwaite, R. G., 18, 405
Corain, B., 82, 339, 415
Corbett, J. D., 172
Cord, P. P., 236
Cordes, M., 353
Cordona, R. A., 454
Corey, J. Y., 132, 271
Corfield, P. W. R., 452
Corice, R. J., jun., 211
Cornuel, S., 96
Cornut, J.-C., 203
Corriu, R. J. P., 125, 132, 133, 136, 265
Corset, J., 399
Corsmit, A. F., 235
Cory, R. M., 97
Cosgrove, J. G., 500, 504
Costain, C. C., 189, 190
Costarelli, C., 199
Cot, L., 303
Cottam, G. L., 78
Cotton, F. A., 35, 52, 65, 67, 69, 72, 314, 370, 384, 395, 472
Cotton, S. A., 333
Cottrell, C. E., 86, 101
Couchot, P., 443
Coucouvanis, D., 60
Couret, C., 125, 129, 132, 273
Couret, F., 125
Courrier, W. D., 329
Court, A. S., 128
Courtine, P., 236
Courtois, D., 198
Cousse, H., 136
Coussement, R., 619
Coutts, R. S. P., 316, 362
Couzi, M., 203
Cova, G., 356

Coville, N. J., 330, 476, 492
Covington, A. K., 74, 77
Cowell, R. G., 447, 448, 451
Cowen, J. A., 107
Cowles, R. J. H., 35
Cowley, A. H., 5, 142, 283
Cox, A. P., 41, 182, 185, 371
Cox, D. E., 516, 526
Cox, M., 532
Coyle, T. D., 3, 119, 137, 192, 254, 272
Cozens, R. J., 45, 50, 408, 425
Crabtree, R. H., 27, 395
Cracknell, A. P., 197
Cradock, S., 125, 267
Cragg, R. H., 120
Craig, N. C., 265
Craig, P. G., 25
Craig, P. J., 472
Cramer, J. L., 35
Cramer, R., 47
Crane, R. W., 95
Cranshaw, T. E., 505, 617
Cras, J. A., 60, 345
Crates, P. N., 208
Craven, S. M., 286
Crawford, J. R., 459
Crawforth, C. E., 98
Crea, J., 77
Crease, A. E., 387
Creaser, I. I., 49
Crecelius, G., 611
Creel, R. B., 105
Creemers, H. M. J. C., 314
Creighton, J. A., 265
Cremer, S. E., 94, 141
Cressely, J., 16
Cresswell, P. J., 169, 312
Cretney, W., 315
Crisler, L. R., 432
Crist, J. L. 120, 260
Cristini, A., 354, 403
Cristol, S. J., 63
Croatto, U., 57, 346, 358
Crociani, B., 55, 57, 346, 377, 415
Crombie, L., 98
Cronin, J. T., 434
Cros, G., 83
Cross, R. J., 71
Crossland, R. K., 41, 371
Crow, J. P., 22, 129, 276, 468
Cruset, A., 511, 515, 538
Crutchfield, D. A., 320
Csakvari, B., 546
Csàszar, J., 337, 403
Cser, L., 617, 168, 622
Csontos, G., 430
Cuastalla, G., 415
Cueilleron, J., 119, 258, 262
Cuellar-Ferreira, E., 265
Cuglielminotti, E., 325
Cullen, D., 487

Cullen, W. R., 22, 139, 468, 478, 540
Cummings, S. C., 50, 410, 459
Cummins, R. A., 433
Cundy, C. S., 38, 44, 314
Cunico, R. F., 127
Cunningham, D., 529, 577
Cunningham, P. T., 282
Curl, R. F., 181, 187, 192
Curnette, B., 299
Curran, C., 582
Current, D. H., 171
Curry, J. D., 286
Curtis, D. M., 125
Curtis, E. C., 209, 217, 224, 302
Curtis, M. D., 25, 37, 72, 134, 269, 275
Curtis, N. F., 50, 402, 403
Cushley, R. J., 50, 97
Cusmano, F., 339
Cuthbert, J., 177
Cutler, A., 37, 483
Cuttitta, F., 563
Cyvin, B. N., 269
Cyvin, S. J., 269, 287, 363, 370, 376, 384
Czeskleba, A., 558
Czieslik, G., 120, 289
Czieslik, W., 193
Czjek, G., 601
Czysch, W., 144, 290

Dabard, R., 28, 41, 48, 365
Dabosi, G., 265
Dabrowiak, J. C., 42, 50
Dabrowski, L., 500, 618
Dachlas, B. P., 463
Dämmgen, U., 19
Dagdigian, P. J., 177
Dagys, R., 153
Dahlenburg, L., 45, 374
Dahm, D. J., 523
Dainty, P. J., 296
Dakhis, M. I., 556
Dakkouri, M., 193
Dale, A. J., 98
Dale, B., 588
Dale, B. W., 299, 504, 529, 569, 591
D'Alessio, G. J., 156
Dalgleish, I. G., 71
Dalton, J., 388
Dalziel, J. R., 273, 581
Damasco, M. C., 5
D'Ambrogio, F., 236
Damrauer, R., 135
Damsgard, C. V., 222, 228
D'Andrea, R. W., 399, 577
Danieli, R., 132
Danielsson, I., 75
Danilina, L. I., 418
Danon, J., 495, 513, 537
Danyluk, S. S., 98
Dao, N. Q., 231
Darby, N., 56
Darensbourg, D. J., 28, 369, 394, 395, 467, 471, 478

Darensbourg, M. Y., 37, 471, 480
Dariel, M. P., 616
Darken, J., 532
Darling, J. H., 384, 393
Darnall, D. W., 77
Dartiguenave, M., 344, 422
Dartiguenave, Y., 344, 422
Das, M. K., 7, 272
Das, N. K., 118
Das, R., 16, 245
Dash, K. C., 71, 348
Dash, R. N., 422
Da Silva, E., 232
Dau, E., 329
Dauben, W. G., 56
Daugherty, M., 156
Dautov, L. M., 618
Dautov, R. A., 149
Dautzenberg, J. M. A., 64, 355
Davidenko, N. K., 75
Davidovich, R. L., 280
Davidson, D. W., 111
Davidson, G., 331, 364, 367, 368, 369, 371
Davidson, G. R., 516, 526
Davidson, J. L., 25, 445
Davidson, M. E. A., 83
Davies, A. G., 14, 118, 137, 278, 579, 584
Davies, C. S., 22, 434, 469
Davies, J. D., 48, 418
Davies, J. E. D., 200
Davies, N., 113, 253, 254
Davis, A. R., 359, 438
Davis, B. R., 536
Davis, D. G., 151
Davis, F. A., 118
Davis, R., 416, 437
Davis, R. A., 135
Davis, R. E., 95, 97, 181
Davison, A., 31, 32, 250, 466
Davitashvii, E. G., 357
Dawson, C. R., 131
Dawson, K., 49
Dawson, P., 231, 238
Day, E. D., 59, 64, 320, 447, 451
Day, M. C., 77
Dayal, B., 97, 98
Dazord, J., 249
Deacon, G. B., 63, 139, 261, 285, 427, 430, 431, 442, 465, 466
De Alti, G., 310
Dean, C. R. S., 208
Dean, J. R., 357
Dean, W. K., 21, 24, 39, 364, 420, 471, 479
De Armond, M. K., 326
De'Ath, N. J., 141
Deavenport, D. L., 97
Deb, K. K., 79
Debeau, M., 234, 241
De Beer, J. A., 39, 483, 540
de Bettignies, B., 208
de Bie, M. J. A., 131

Author Index

De Boer, B. G., 450
de Boer, E., 95
De Boer, J. J., 85
De Bolster, M. W. G., 314, 437, 439, 440
Debrunner, P. G., 90, 536, 598
Debye, N. W. G., 273, 575
De Clercq, M., 128, 134
Decroix, B., 147
Dédier, J., 127
Dědina, J., 126
Deeming, A. J., 30, 68
De Filippo, D., 21
Deganello, G., 35
Degetto, S., 60, 347, 411, 452
Dehand, J., 314, 331, 343
Dehnicke, K., 17, 123, 246, 255, 259, 262, 294
Dehouck, C., 134
Deich, A. Y., 167
Deichman, E. N., 333
Deininger, D., 109, 152
Dejak, B., 83
De Jonge, W. J. M., 107
De Jongh, R. O., 19, 361
de Kanter, J. J. P. M., 261
De Kock, C. W., 206, 209
Delcroix, P., 619
Delgass, W. N., 617
de Liefde Meijer, H. J., 361
Delmas, M., 14, 580
Delobel, R., 358
Delorme, P., 241, 315, 427
de Loth, P., 82
Delpuech, J. J., 75
DeLucia, F. C., 182, 183
Deluzarche, A., 16
Delyagin, N. N., 501, 618
Demakov, K. D., 270
Demarco, P. V., 95
Demay, C., 141
de Mayo, A., 620
Dembovskii, S. A., 161, 616
Demco, D., 150
Demerseman, B., 475, 480
Demerseman, P., 98
Demir, T., 164
De Montauzon, D., 46, 489
Demortier, A., 88
Dempster, A. B., 344
Demuth, R., 64
Dem'yanenko, V. P., 316
Demyanko, V. P., 172
Dendoni, A., 347
Den Heijer, J., 437
Deniau, M., 244
Denison, A. B., 517
Denisov, D. A., 359
Denisov, V. M., 77
Dennenberg, R. J., 467
Denney, D. B., 138, 141
Denney, D. Z., 141
Denney, E. J., 177
Denning, J. H., 260, 277, 319

Denniston, M. L., 114, 252
Denoel, J., 146
Dent, A. L., 352
Dent, S. P., 53, 347
Denton, D. L., 115
de Paoli, G. G., 139, 351
de Poorter, B., 128
Derbyshire, W., 12
Dereign, A., 286
de Renzi, A., 54, 55
Dereppe, J. M., 102
Derighetti, B., 101
Dernova, V. S., 212, 266
Derouault, J., 264
Derygin, G., 201
Derzhanski, A. L., 152
de S. Barros, F., 611
Desbat, B., 211
De Sesa, M. A., 317
DeSimone, R. E., 8, 89
Des Marteau, D. D., 146, 148, 301, 308
Desreux, J. F., 94
Dessy, G., 403
Dessy, R. E., 21, 42, 416, 540
Des Tombe, F. J. A., 62, 314
Destombes, J. L., 176, 186, 191
de Trobriand, A., 77
Deutscher, R. L., 411
Dev, R., 27, 61, 323, 434, 444, 445
Devanarayanan, S., 509
Devillanova, F., 21
Devlin, J. P., 241
Devoe, J. R., 495
Devore, E. C., 437
de Vos, D., 137
de Vries, J. L. K. F., 60, 279, 589
de Waard, E. R., 97
de Waard, H., 495, 595, 596
Dexheimer, E. M., 127
Dézsi, I., 495, 496, 517, 525, 618
Dhal, R., 98
Dhingra, M. M., 93, 99
Diamandescu, L., 616
Diaz, A., 403
Dibdin, G. H., 104
Di Bianca, F., 134, 142
Dickson, F. E., 82
Dickson, R. S., 46, 48, 121, 488
Dicus, R., 618
Dieck, R. L., 144, 290
Diemann, E., 216, 318, 320, 323, 325
Diep, L., 45
Dietze, U., 294
Dijkermann, H. A., 176, 177
Dikareva, L. M., 338
Dillon, K. B., 111
Dillon, M. da G. C., 104
Dillon, P. B., 9

Dilworth, J. R., 25, 42, 344, 396
Di Marino, R., 132
Dimitripoulos, C., 163
Dimmel, D. R., 127
Dimsdale, M. J., 42
Dineen, J. A., 48
Dinesh, 166
Dirreen, G. E., 28, 414
Dittmar, E., 524
Dixit, L., 359
Dixneuf, P., 41
Dixon, K. R., 59, 345
Dixon, L. T., 352
Dixon, W. T., 2
Djordjević, C., 65, 454
Dmitriev, I. A., 153
Dmitrikov, V. P., 83
Dobbie, R. C., 29, 37, 332, 478
Dobos, S., 277
Dobson, C. M., 95
Dobson, G. R., 322, 385, 470, 471
Dodgen, H. W., 75, 77
Do-Dinh, C., 555
Dodokin, A. P., 552, 572
Döhler, H. V., 410
Doerffel, K., 95, 96, 301
Dötz, K. H., 125
Dogadina, A. V., 151
Dolby, R., 46
Dolcetti, G., 416, 417, 418
Dollish, F. R., 197
Dolphin, D., 45
Dombek, B. D., 134, 275
Domilideva, I. I., 410
Dominelli, N., 478, 539
Domingos, A. J. P., 67
Domingues, P. H., 537
Domnina, E. S., 89
Donaldson, J. D., 299, 574, 575, 577, 588, 591, 592
Donato, H., jun., 98
Dondoni, A., 60, 452
Dongala, E. B., 98
Doran, M. A., 16
Doretti, L., 285
Dori, Z., 61, 398
Dormann, E., 104
Dormann, J. L., 618
Dorokhov, V. A., 83
Doronina, L. M., 326
Doronzo, S., 331, 335, 424, 428
Doroshev, V. D., 150
Dosser, R. J., 28, 400
Dotzauer, E., 22, 363
Doue, D., 45
Douek, I., 70, 381
Douglas, B. E., 51
Douglas, J. E., 384
Douglas, P. G., 42
Douglas, W. E., 20
Douglas, W. M., 21, 24, 39, 420, 471, 479
Dow, A. W., 128
Dowbor, K., 122
Downs, A. J., 216, 222
Dowty, E., 563

Doyle, M. J., 46, 339, 373, 490
Dräger, M., 146, 300, 352
Drago, R. S., 77, 89, 92, 93
Draguet, C., 73, 146
Drahgicescu, M., 107
Drake, J. E., 121, 125, 268, 269, 277
Dreeskamp, H., 35
Dreizler, H., 176, 193
Dresselhaus, G., 232
Drew, D. A., 191
Drew, M. G. B., 59
Drews, K. A., 22, 469
Drickamer, H. G., 506
Drifford, M., 307
Drobot, D. V., 220, 329
Drokin, A. I., 547
Drozdov, V. A., 9, 10, 12, 13, 126, 131
Druce, P. M., 59, 330, 346
Druck, S. J., 11
Drummond, I., 124, 200
Dryburgh, J. S., 13
Dubac, J., 125, 135, 136
Dubey, B. L., 213, 317
Dubey, S. P., 407
Dubinin, V. N., 618
DuBois, T. D., 59
Dubov, S. S., 414
Dubovitskii, V. A., 10, 151, 168, 316
Dubovtsev, I. A., 505, 618
Dubrulle, A., 186, 191
Ducas, T. W., 181
Duce, D. A., 331, 369
Dudareva, A. G., 263
Dudas, J., 618
Duddell, D. A., 388
Dudley, M. A. H., 200
Dudrewa, B., 618
Dueber, M., 132, 271
Duff, E. J., 339
Duff, J. M., 44, 490
Duffaut, N., 127
Duffy, D. J., 69, 531
Dukes, G. R., 404
Dumas, G. G., 199, 216
Dumler, J. T., 52, 341
Dunaeva, K. M., 359
Duncan, J. L., 453
Dunell, B. A., 101, 111
Dunham, W. R., 496
Dunks, G. B., 117
Dunlap, B. D., 590, 613, 618
Dunmur, R. E., 143
Dunmyre, G. R., 619
Dunn, J. G., 326, 399
Dunn, M. B., 102, 103, 112
Dunn, P., 433
Dunn, T. J., 54
Dunoguès, J., 127
Dunsmore, G., 94
Dunsmuir, J. T. R., 230, 282
Dunstan, P. O., 442
Duong, K. N. V., 45, 424
du Preez, A. L., 37, 38, 485, 487

du Preez, J. G. H., 442
Durand, A., 149
Durge, V. V., 571
Durig, J. R., 112, 190, 192, 193, 221, 265, 268, 271, 286, 287, 297
Durkin, T. R., 58, 122, 282, 422
Durret, D. G., 320
Durst, A., 138
Durst, T., 149
Dustin, D. F., 48
Dutt, N. K., 405, 410
Dutton, G. S., 131
Duxbury, G., 182
Duyckaerts, G., 293, 298, 383
Dvorkin, M. I., 303
Dvornikuv, E. V., 149
Dwek, R. A., 64, 78
Dwivedi, P. C., 198
Dyatkin, B. L., 62, 136
Dyer, D. S., 81
Dyke, T. R., 186
Dymanus, A., 181, 183
Dymock, K., 261, 264
d'Yvoire, F., 242

Eaborn, C., 44, 53, 63, 127, 128, 341, 347, 433
Eachus, S. W., 74
Eady, C. R., 389
Eady, R. R., 536
Eary, J. G., 63, 352
Eastmond, R., 136
Eaton, D. R., 75, 92
Eaton, G. R., 68
Eaton, S. S., 68
Eavenson, C. W., 34
Ebeling, J., 140, 289
Ebert, M., 440
Ebiko, H., 546
Ebina, F., 51
Ebsworth, E. A. V., 13, 72, 73, 125, 267
Edagawa, E., 96
Edgell, W. F., 198
Edmonds, D. T., 157, 162
Edmonson, R. C., 37, 484
Edneral, A. F., 618
Edwards, D. A., 326, 329, 399
Edwards, J., 25, 472
Edwards, P. A., 172, 173, 174
Edwards, T. H., 216
Edzes, H. T., 100
Eeckhaut, Z., 270
Efraty, A., 21, 27, 37, 45, 48, 421, 469
Egiazarov, B. G., 500, 502
Egorochkin, A. N., 124, 127
Egorov, V. N., 357
Egorov, Yu. P., 13, 150, 293
Ehemann, M., 80, 113, 253
Ehntholt, D., 37
Ehntholt, D. J., 35
Ehrig, R., 95
Ehrl, W., 21, 467

Eibschütz, M., 516, 526, 609
Eicher, H., 620
Eichler, B., 443
Eicke, H. F., 123
Einstein, F. W. B., 44, 397, 449
Eisch, J. J., 86, 121, 254
Eisenberg, M., 148, 308
Eisenberg, R., 335, 392
Eisenstadt, A., 34, 35, 41, 369
Eisner, E., 37, 484
Ejchart, A., 98
Ejiri, T., 51
Ekimov, S. P., 501
Ekimovskikh, I. A., 155
Ekong, D. E. U., 97
Elder, P. A., 69, 390
Eley, R. R., 531
Elfimeva, G. I., 458
Elias, J. H., 89
Eliezer, I., 205, 248, 312
Elleman, D. D., 104
Ellenson, W. D., 241
Ellermann, J., 331, 337, 366, 367, 371
Ellert, G. V., 358, 437
Ellestad, O. H., 204, 300
Ellett, J. D., jun., 101
Ellis, J. E., 466
Ellis, P. D., 6, 67
Elrod, C. D., 537
El Saffar, Z. M., 103, 111
Elsbrand, H., 42
Elter, G., 120, 130, 257, 275
Emerson, G. F., 33
Emid, S., 104
Emori, S., 93
Empsall, H. D., 55, 342
Endicott, J. F., 50, 79
Eng-Choon Looi, 199
Engel, R., 90
Engelhardt, G., 12
Engelhardt, U., 86, 298
Engfer, R., 611
Engle, J. L., 77
Engler, R., 146, 284, 300
Englert, M., 55
Engstrom, N., 6
Entine, G., 277
Enwall, E., 449
Epel'baum, M. B., 201
Epps, L. A., 49, 455
Epstein, L. M., 527, 528
Eremenko, V. V., 546
Eremin, Yu. G., 315
Eringis, K., 153
Erlich, K., 33
Ermakov, Yu. A., 149
Ernst, C. R., 40
Ernst, R. R., 3
Eschbach, C. S., 488
Escudié, J., 129, 132, 273
Ettorre, R., 71
Ettwig, H. H., 618
Eujen, R., 207
Evans, B. J., 546, 593
Evans, D. F., 94

Evans, G. O., 22, 389, 468
Evans, J., 85
Evans, J. A., 45, 306
Evans, J. B., 41
Evans, M., 8, 32
Evans, M. C. W., 536
Evans, W. J., 48
Evdokimov, A. M., 87, 125, 152
Everett, G. W., jun., 87, 90
Everitt, G. F., 47, 375
Evstaf'eva, O. M., 358
Evtushenko, N. P., 292
Ewbank, J. D., 376, 384
Ewing, D. F., 136
Ewing, J. J., 181
Ewstaf'eva, O. M., 437
Exarhos, G., 246
Eyring, L., 357
Eysel, H. H., 225, 326
Eysseltova, J., 440

Fabbri, G., 198, 319
Fabrichnyi, B. P., 545, 571
Fachinetti, G., 18, 19, 361
Fackler, J. P., jun., 60, 82, 312, 450, 451
Faffani, A., 38
Falaleeva, L. G., 13
Faleschini, S., 285
Falius, H., 141, 283, 291
Faller, J. W., 56, 70
Faller, P., 145, 146
Fallon, G. D., 466
Fan, H. Y., 231
Fanning, J. C., 537
Fano, V., 567
Fantucci, P., 43, 54, 335, 478
Farach, H. A., 494
Faraglio, G., 142, 285
Farago, M. E., 85
Faraone, F., 59, 339, 347, 465, 490
Fares, V., 403
Farmer, J. B., 129, 275
Farmery, K., 44, 50, 408
Farnell, L. F., 6
Farona, M. F., 330, 454
Farrant, G. C., 34
Farrar, T. C., 3, 11
Farren, J., 177
Fately, W. G., 197
Fatseas, G. A., 618
Favero, G., 415
Favero, P., 185
Fayt, A., 201
Fedarko, M. C., 77
Fedin, E. I., 10, 27, 40, 62, 63, 77
Fedorov, L. A., 31, 61, 62, 82, 370, 485
Fedorov, P. I., 263
Fedoryako, L. I., 105
Fehér, F., 304
Feibush, B., 94
Feizullaeva, Sh. A., 324
Felix, R. A., 132
Felkin, H., 58
Felscher, R., 590

Felten, J. J., 81
Feltham, R. D., 40, 42, 396, 419, 480, 538
Feltri, T., 453
Ferguson, J. A., 35, 37, 414, 484
Ferguson, S. J., 78
Fergusson, J. E., 169, 312
Fernando, W. S., 243, 371
Fernandopulle, M. E., 538
Ferrando, W. A., 601
Ferraro, J. R., 223, 337, 353, 438
Ferrari, R. P., 35, 487
Fenger, J., 515
Fenske, D., 120, 258
Fenske, R. F., 385
Fenster, A. E., 365
Fenton, D. E., 273, 399, 575
Feser, M. F., 130, 295
Fetchin, J. A., 82
Fetter, K., 118, 250
Feuwette, P. H. F. M., 345
Fiat, D., 88, 100
Fichte, B., 131
Field, R. W., 179, 180
Fields, R., 63
Filatova, S. A., 317
Filimonov, V. N., 248
Filippov, N. I., 525
Filippov, V. P., 616
Filippova, N. P., 618
Filleux-Blanchard, M. L., 149
Filoti, G., 520, 530, 548
Finch, A., 208, 304
Findley, D. A. R., 98
Finger, G., 109
Fingh, A., 299
Fink, P., 110
Fink, R., 72
Finn, P., 12, 248
Finnegan, D. J., 185
Firov, A. I., 498, 595
Fischer, E. O., 7, 19, 22, 23, 28, 31, 32, 38, 39, 58, 85, 125, 362, 367, 376, 420, 467, 468, 477, 478, 479, 485, 557
Fischer, H., 85, 468
Fischer, M., 209
Fischer, R. D., 94, 99
Fischer, S., 55
Fischler, I., 35
Fish, R. H., 61, 63, 118, 356
Fish, R. W., 37, 483
Fisher, H. F., 201
Fisher, P. F., 437
Fisher, P. S., 358
Fisher, R. M., 561
Fitjer, L., 146
Fitzgerald, R. J., 397
Fitzky, H. G., 138, 157
Fitzpatrick, N. J., 335, 364, 371
Fitzsimmons, B. W., 10, 532, 589
Flatau, G. N., 141

Flatau, K., 62
Flegel, W., 177
Fleischer, D., 98
Fleischer, E. B., 42
Fleming, S., 113, 142, 259
Fletcher, S. R., 26
Fletcher, W. H., 211, 306
Fletton, R. A., 96
Fleur, A. H. M., 443
Flick, C., 103
Flid, R. M., 344, 382
Flinn, P. A., 546, 618, 619
Flippen, J. L., 35
Flitcroft, N., 449
Flood, T. C., 128
Floriani, C., 18, 19, 361, 434
Florinskaya, V. A., 197
Fluck, E., 3, 13, 140, 141, 144, 542
Fluorescu, V., 553
Flygare, W. H., 181, 191, 193
Foct, J., 547
Foester, R., 112
Foffani, A., 368, 376, 387
Fogarasi, G., 277
Fogelman, J., 75
Foiles, C. L., 171
Fomichev, V. V., 294
Fong, C. W., 136
Fong, D.-W., 79
Fontaine, C., 45, 424
Fontana, S., 19, 467
Forbes, C. E., 60
Ford, P. C., 42, 49, 413
Ford, W. T., 77
Forel, M. T., 198, 261, 264, 276, 442, 451
Foreman, M. I., 33, 85
Forkl, H., 61
Formáček, V., 11
Formanek, H., 42, 620
Forrest, I. W., 230
Forsberg, J. H., 78, 405
Forsellini, E., 358
Forster, D., 392, 523
Forster, W., 329
Fortescue, W. F., 139, 294
Fossey, J., 35
Foster, J., 438
Foster, R. J., 99, 439
Fotiev, A. A., 319
Fouassier, M., 264, 442
Foucaud, A., 97
Fougeroux, P., 339, 374
Fournier, R. P., 210
Foust, R. D., jun., 42, 49, 413
Fowles, G. W. A., 20, 318
Fox, J. P., 9
Fox, K., 211
Fox, L. E., 94
Fox, R. A., 151
Fox, W. B., 301
Foxman, B. M., 85
Fradin, F. Y., 618
Fraenkel, G., 16
Frainnet, F., 131
Frais, P. W., 220, 329

Fraissard, J., 293
Fraknoy-Koros, V., 593
Francis, B. R., 25, 364
Francis, D. J., 49
Francis, H. E., 95
Francis, J., 82
Francis, J. N., 48, 118, 252
Franck, R., 231
Frange, B., 119, 258
Frank, C. W., 506
Frank, E., 526, 618
Franke, B. S., 537
Franke, R., 60, 349
Frankel, L. S., 77
Frankel, R. B., 511, 516, 528
Franz, D. A., 86, 250
Franz, K. D., 468
Franz, M., 127, 132
Fraser, G. W., 147
Fraser, J. W., 406
Fraser, M. S., 43, 47, 335, 375
Fraser, P. J., 46, 488
Fraser, R. R., 97, 149
Fratiello, A., 71, 120
Frauenfelder, H., 495
Frazer, M. J., 282, 529, 577
Frazier, S. E., 139, 143, 289, 293
Frederick, C. G., 508
Frediani, P., 486
Fredrickson, M. J., 134
Freedman, T. B., 204
Freiberg, M., 222
Frembs, D., 466
Freni, M., 478
Freude, D., 108, 109, 110, 152
Freund, S. M., 176, 179
Fricke, B., 497
Fricke, J., 97
Fricke, K., 91
Friedel, R. A., 13
Friedman, M., 63, 356
Friedt, J. M., 495, 513, 515, 538
Fries, D. C., 82
Fringuelli, F., 147
Fripiat, J. J., 109
Fritchie, C. J., jun., 45, 374
Fritz, G., 132, 133, 137, 271
Fritz, H. P., 201, 246, 384, 417, 436
Fritzche, H., 88
Frlec, B., 307
Fröhlich, K., 616
Frohnecke, J., 260
Froix, M. F., 144
Frolov, V. V., 149
Frolov, Yu. L., 271
Frolova, L. A., 260
Fromage, F., 315
Fruchier, A., 95
Frye, C. L., 132, 269
Frye, H., 49, 347, 402
Fuchs, R., 127
Fuger, J., 383
Fuggle, J. C., 318
Fuhr, B. J., 71
Fuhrhop, J. H., 63
Fukai, M., 71
Fukuda, Y., 403, 404
Fukumi, T., 96
Fujii, Y., 51, 51, 131
Fujita, J., 49, 51
Fujiwara, F. Y., 148
Fujiwara, S., 85, 93, 96
Fukumoto, T., 21, 125, 229, 467
Fukushima, E., 100
Fukushima, K., 202
Funabiki, T., 85, 337
Fung, B. M., 12, 91, 107
Fung, K. W., 355
Funnell, N., 131
Furukawa, Y., 51, 162, 171
Fusi, A., 334, 339, 384
Fusii, J. L., 402
Fusina, L., 355
Fyfe, C. A., 101

Gaasch, J. F., 41
Gabes, W., 200, 211
Gabuda, S. P., 110
Gänswein, B., 302
Gager, H. M., 545
Gagliardi, E., 138
Gagnaire, D., 84
Gaidamaka, S. N., 13
Gailar, N. M., 201
Gaines, D. F., 151, 249
Gainsford, A. R., 402
Gainsford, G. J., 33
Gal, J., 616
Galakhov, I. V., 152
Galasso, V., 310
Galindo, A. M. B., 357
Galitskaya, S. M., 453, 457
Galitskii, V. Yu., 110
Galiullin, M. A., 444
Galizzioli, D., 356
Gallagher, T. F., 177
Gallais, F., 82
Gallay, J., 46, 489
Galle, J. E., 62
Galliart, A., 317
Gallo, A. A., 79
Gal'perin, E. L., 414
Galtier, M., 231
Gal'tsev, A. P., 201
Gambaro, A., 71
Gambaryan, N. P., 65
Gambino, O., 30, 32, 35, 483, 486, 487
Gamlitskii, V. Ya., 525
Gănescu, I., 406
Ganiel, U., 566
Ganis, P., 56, 349
Gans, P., 324, 411
Gansow, O. A., 6, 11, 33, 67, 388, 479
Gapeev, A. K., 547, 548
Gardet, J.-J., 242
Gardiner, D. J., 210
Gardner, R. C. F., 39
Garg, A., 353, 354, 356
Garg, S. K., 111
Garg, V. K., 518
Garg, V. N., 427
Gargarinsky, Yu. V., 107
Garlaschelli, G., 46, 372
Garnett, M. W., 79
Garrod, R. E. B., 538
Garrou, P. E., 58, 539
Garten, R. L., 617
Garty, N., 53
Gaspar, P. P., 129
Gasparrini, F., 59
Gassend, R., 580
Gates, P. N., 111, 299, 304
Gatteschi, D., 90, 91, 343
Gatti, G., 10
Gattow, G., 146, 265, 284, 300, 352
Gatzke, A. L., 16
Gaudemar, M., 62, 383
Gaudemer, A., 44, 45, 424
Gaudemer, F., 45
Gauthier, G. J., 41
Gavrilov, Yu. D., 149
Gay, I. D., 107
Gay, R. S., 389
Gaylor, J. R., 44
Gažo, J., 404, 456, 457
Geanangel, R. A., 112, 248
Geary, W. J., 402
Geddes, A. L., 228
Geffarth, U., 263
Gegenheimer, R., 193
Geisel, M., 120
Geisler, T. C., 115, 125, 269
Geissler, H., 271
Gelbaum, L., 90
Gelberg, A., 548
Gelencser, P. A., 593
Gel'fman, M. I., 491
Geller, R., 526
Geller, S., 557
Gendler, T. S., 547
Generalova, N. B., 337
Genin, J. M., 620
Gentile, P. S., 436
Genzel, L., 248
Geoffrion, L. D., 181
George, A. D., 473
George, B., 619
George, R. D., 28, 269, 314
George, T. A., 25, 394, 472, 473
Georgiev, V. I., 152
Gerbeleu, N. V., 532
Gerdau, E., 499
Gerding, H., 200, 211, 277, 278, 299
Gerei, S. V., 382
Gerlach, D. H., 31, 55, 330
Germain, A., 265
Gerry, M. C. L., 183, 184, 187
Gerson, R., 594
Gersonde, K., 79
Gertsev, V. V., 260
Gervais, D., 19
Gervais, F., 243
Geschke, D., 109, 110
Geschke, G., 152
Gevork'yan, S. V., 319

Author Index

Gey, E., 10
Ghedini, M., 59
Ghezzi, C., 616
Ghose, S., 108, 563
Ghosh, S. N., 199, 211
Ghotra, J. S., 99
Giacchero, N., 152, 353
Giacometti, G., 42, 67
Giannoccaro, P., 331, 396, 485
Giannotti, C., 24, 45, 424
Gibb, T. C., 511, 522, 560, 585
Gibb, V. G., 94
Gibbon, G. A., 125, 268
Gibby, M. G., 5, 103
Gibler, D. D., 209
Gibney, K. B., 131
Gibson, A. A. V., 102
Gibson, D. H., 32, 85, 478
Gibson, J. A., 74, 86
Gidney, P. M., 49
Gielen, M., 128, 134, 136
Gielen, P. M., 508
Gielow, P., 141, 295
Giering, W. P., 35, 37, 69, 372, 483
Giesbrecht, E., 442
Giguère, P. A., 200
Gilbert, B., 293, 298, 383
Gilbert, H., 619
Gilchrist, A. B., 44, 397
Gilchrist, T. L., 35
Gilje, J. W., 140
Gill, D. F., 7, 307
Gill, J. B., 216
Gillard, R. D., 49, 51, 428, 457
Gillen, K. T., 15
Gillespie, R. J., 88, 148, 199, 302, 308
Gillies, G. C., 162
Gilman, H., 125
Gil'manov, A. N., 107, 149
Ginzburg, A. G., 27, 28, 365
Giordano, G., 46
Girgis, A. Y., 51
Girling, R. B., 210
Gittsovich, V. N., 500
Giuliani, A. M., 63
Giuliani, R., 199
Giusto, U., 356
Glass, W. K., 21, 25, 386, 469
Glasser, F. P., 552
Glavic, P., 224
Glaze, W. H., 16, 361
Glebov, V. A., 75
Gleiter, R., 542
Gleitzer, C., 547
Glemser, O., 74, 120, 130, 139, 143, 144, 217, 257, 275, 289, 290, 294, 295, 296, 304
Glen, D. N., 201
Glinina, A. G., 453
Glinkina, M. I., 329

Glockling, F., 27, 43, 59, 314, 346, 491
Glonek, T., 138, 144
Glore, J. D., 86, 254
Gloss, G. L., 81
Glotov, V., 617
Gluck, R., 244
Glushko, S. N., 617, 618
Gmehling, J., 87
Goasdoué, N., 62, 383
Gode, H., 260
Godovikov, A. A., 244
Godwin, A. D., 95
Goedken, V. L., 42, 59, 400, 403, 407
Goel, R. G., 101, 142, 285, 294, 434, 439, 443
Göring, J., 612, 618
Goethals, E. J., 146
Goetschel, C. T., 307
Götze, H.-J., 136
Goetze, R., 286
Goffart, J., 383
Gogan, N. J., 22, 434, 464, 469
Goggin, P. L., 58, 59, 63, 352, 404, 452
Goh, L.-Y., 44
Goh, S. H., 44
Goldanskii, V. I., 515, 565, 618, 619
Goldberg, D. E., 405, 455
Goldblatt, M., 226
Golding, B. T., 45
Golding, R. M., 43, 52, 65, 96, 539
Gol'dman, A. M., 338, 421
Goldman, L. M., 199
Goldstein, M., 86, 310, 354, 408, 618
Goleb, J. A., 202
Golen, J., 304
Golino, C. M., 137
Golloch, A., 295
Goloshchapov, M. V., 293
Golovchenko, L. S., 10, 62, 63, 158
Golovnin, V. A., 508, 618
Gol'tyapin, V. V., 171
Goltzene, A., 621
Golubnichaya, M. A., 438
Gombler, W., 80, 141
Gomolea, V., 548
Gomory, P., 546
Gonser, J., 495, 508, 616
Gonzalez Jimenez, F., 613
Gonzalez-Vilchez, F., 213
Good, M. L., 211, 222, 524, 536
Goodall, B. L., 24, 28, 472, 475, 488
Goodall, D. C., 216
Goodfellow, R. J., 58, 59, 63, 352, 404, 452
Goodgame, D. M. L., 311, 407
Goodgame, M., 407
Goodisman, J., 95
Goodman, C. D., 505
Goodman, G. L., 224

Goodman, G. T., 135
Goodman, M., 99
Goodwin, H. A., 436, 530
Gorbatyi, Yu. E., 201
Gorbunov, B. Z., 201
Gordeev, A. D., 155
Gordienko, V. A., 549
Gordy, W., 179, 181, 182, 186
Gore, E. S., 50
Goretzki, H., 105
Gor'kov, V. P., 595
Gorobchenko, V. D., 525, 617, 618
Gorokhvatskii, Ya. B., 382
Goroshchenko, Ya. G., 317
Gosling, K., 122, 255, 259
Gostisa-Mihelcic, B., 152
Gotani, M., 425
Gottardi, W., 210
Goubeau, J., 288, 292, 297
Gouet, F., 432
Gould, G. E., 136
Gouteron-Vaissermann, J., 237, 238
Govil, G., 93
Graddon, D. P., 441
Gräff, G., 177
Graenicher, H., 111, 153
Graff, J., 177
Graham, M. J., 546
Graham, W. A. G., 28, 68, 205, 327, 389, 416, 477
Grande, S., 618
Grandjean, F., 99
Grandjean, J., 94
Grankina, Z. A., 439
Granovskii, E. B., 620
Grant, M. W., 77
Grant, R. W., 550, 557, 561, 618
Grasdalen, H., 77
Grasshof, M., 180
Grassi, B., 334, 389
Graverau, P., 236
Gray, G. A., 141
Gray, H. B., 41, 59, 314, 333, 371, 425, 455, 528
Gray, L. W., 333
Graybeal, J. D., 173
Graziani, M., 54
Graziani, R., 358
Greatrex, R., 540, 560, 585
Grebenshchikov, R. G., 278
Grebenyuk, V. D., 109, 110
Grechishkin, V. S., 155, 172
Greco, A., 25
Grecu, R., 323
Green, B., 144
Green, C. R., 70, 377
Green, G. F. H., 96
Green, M., 28, 30, 33, 45, 46, 47, 55, 85, 116, 252, 342, 374, 413, 475, 486
Green, M. L. H., 20, 25, 318, 364, 443
Green, R. D., 75, 151

Greene, D. L., 355
Greene, P. T., 114, 252
Greenfield, N. J., 78
Greenland, H., 409
Greenler, R. G., 349
Greenwood, N. N., 80, 116, 196, 356, 511, 521, 522, 542, 544, 560, 585, 609, 610
Gregora, I., 230
Grein, F., 92
Greiss, G., 48
Grenoble, D. C., 506
Grey, A. A., 3
Gribov, A. M., 414
Grieb, M. W., 50, 338
Grielen, M., 128
Griffin, R. G., 5, 101
Griffith, J. A. R., 618
Griffith, W. P., 213, 226, 310, 333, 384, 417
Griffiths, D. V., 95
Griffiths, G. H., 440
Griffiths, J. E., 212, 294
Grigg, R., 50
Grigor'ev, A. I., 354
Grigor'ev, V. A., 149
Grigoryan, A. G., 618
Grimaldi, J., 5
Grimes, R. N., 27, 31, 86, 113, 114, 115, 143, 250, 252, 476
Grimm, L. F., 289, 296
Grinvald, A., 121, 264
Grishin, Yu. K., 7, 72
Grizhikhova, R., 551
Grobe, J., 39, 64, 284, 478
Groeneveld, W. L., 314, 434, 437, 439, 440, 443
Grohmann, K., 137
Groll, G., 545
Gromov, B. V., 359
Grootveld, H. H., 17
Grosse, M., 94,
Grote, A., 119
Grotens, A. M., 95
Groves, J. L., 598
Gruber, J. B., 357, 358
Gruen, D. M., 516
Gründemann, E., 131
Grüner, G., 618, 622
Grunberg, P., 236
Grundy, K. R., 30, 417, 423, 486, 487
Grunwald, E., 79
Grutzner, J. B., 77
Gruzin, P. L., 547, 618
Gryff-Keller, A., 71
Grzeskowiak, R., 343, 408
Guastalla, G., 489
Gubin, S. P., 27
Gucwa, K., 405
Günthard, H. H., 177
Günther, H., 3
Guerchais, J. E., 323, 440
Guerts, P. J. M., 349
Guest, A., 214, 329
Gütlich, P., 515, 616
Guggenberger, L. J., 70, 249

Guggenheim, H. J., 526
Guibe, L., 154
Guilhot, B., 242
Gulia, V. G., 314
Guillermet, J., 236, 303
Guillotti, M., 18, 358
Gumprecht, D. F., 567
Gunsalus, I. C., 495, 536
Gunter, J. D., 417
Gupta, G. P., 497
Gupta, N. P., 591
Gupta, R., 10
Gupta, R. K., 79, 151
Gupta, S. K., 63
Gurd, F. R. N., 90
Gurev, V. F., 501
Gurevich, S. Yu., 545
Gurrieri, S., 434
Gusakovskaya, I. G., 619
Gusev, Yu. K., 308
Gushikem, Y., 437, 442
Guss, J. M., 33
Guthrie, D. J. S., 70, 379
Gutowsky, H. S., 96, 102, 149, 151
Gutsze, A., 109
Gutteridge, N. J. A., 97
Gutterman, D. F., 455
Gynane, M. J. S., 122, 255, 256

Haag, A., 119, 121, 139, 141, 144, 146, 283, 295, 296, 304
Haaland, A., 255
Haar, W., 79
Haase, W., 293
Habashy, G. M., 248
Häberlein, M., 121
Hack, F., 24
Hacker, M. J., 45, 306, 490
Hackett, P., 469, 484
Hadari, Z., 616
Haddad, H., 132, 269
Haddad, S., 436
Haddock, S. R., 59, 63, 352, 452
Haddon, W. F., 61
Hadni, A., 229
Hadži, D., D., 104, 224, 229
Häfelinger, G., 24
Hähnke, M., 132, 133, 271
Häni, H., 260
Hafemeister, D. H., 595
Hafner, S. S., 563, 564
Hagen, G., 300
Haggstrom, J., 509
Hagihara, N., 339, 380
Hagnauer, H., 71, 309
Hague, R. H., 205, 207
Hai, Vu., 201
Haiduc, I., 35
Haigh, J. M., 319
Haile, M. A., 399
Haines, L. H., 51
Haines, L. I. B., 86
Haines, L. M., 424
Haines, R. A., 51

Haines, R. J., 37, 38, 39, 481, 483, 484, 485, 487, 540
Hait, D. K., 328
Hájek, B., 293
Hájek, M., 98
Halbert, E. J., 90, 459
Halder, M. C., 323
Hall, C., 73
Hall, D. I., 47, 373
Hall, D. O., 536, 617
Hall, H. K., jun., 62
Hall, J. R., 53, 58, 82, 86, 314, 341, 354, 403, 404
Hall, L. D., 22, 94, 96
Hall, M. B., 385
Hall, M. L., 126
Hall, R. E., 254
Hall, S. R., 566
Hallgren, J. E., 488
Halpern, J., 123, 415, 489
Halstead, T. K., 79
Halton, M. P., 96
Hamada, K., 163
Hambright, P., 320
Hamilton, D. E., 121
Hamilton, J. B., 321, 436
Hamm, D. J., 417
Hammerle, R. H., 180
Hammond, G. S., 425
Han, M. H., 79
Handy, L. B., 27
Handy, P. R., 246
Hanicak, J. E., 16, 361
Hanlan, J. F., 81, 316
Hanna, S. S., 620
Hannan, S. F., 298
Hanousek, F., 250, 251
Hansen, H.-D., 432
Hansen, I. L., 79
Hanslík, T., 251
Hanson, S. W., 94
Hanssgen, D., 134
Hanuza, J., 293, 396, 328
Hanyu, Y., 96
Haony, Y., 516
Haq, M. M., 405, 358
Haque, S. M. E., 411
Hara, T., 17
Harakawa, M., 132, 433
Harber, D., 134
Harbourne, D. A., 478, 540
Harman, J. S., 144, 295
Harmony, M. D., 197
Harney, B. M., 265
Harper, R., 137, 278, 579
Harper, W. C., 619
Harrell, J. W., jun., 103
Harris, C. M., 52, 409, 423
Harris, D. L., 85, 368
Harris, D. O., 179
Harris, J. J., 280
Harris, R. K., 143, 144
Harris, R. O., 322
Harris, W. C., 212
Harrison, B., 438
Harrison, M., 8
Harrison, P. G., 7, 131, 137, 272, 276, 278, 281, 574, 575, 577, 580

Author Index

Harrod, J. F., 351
Hart, F. A., 99
Hart, R. M., 163
Hartley, F. R., 378
Hartley, P. J., 355
Hartman, J. S., 80, 83, 121, 264
Hartmann, M., 194
Hartmann, S. R., 156
Hartmann, V., 80
Hartmanshenn, O., 198
Hartwell, G. E., 58
Hartwig, C. M., 239
Haruda, F., 250
Harvey, A. B., 300
Harvey, K. B., 242
Hathaway, B. J., 456
Hathaway, E. J., 280
Hattori, C., 50
Hatzenbuhler, D. A., 206, 207
Hase, A., 131
Hase, T., 131
Hasegawa, S., 55, 378
Hashi, T., 111
Hashimoto, H., 122
Hashimoto, S., 150
Hashmi, M. A., 436
Hass, Y., 98
Hassall, M. L., 77
Hassner, A., 62, 97
Hastings, R. N., 158
Haszeldine, R. N., 47, 63, 339, 424, 490
Haubold, W., 140, 144
Haul, R., 99, 110
Haupt, H. J., 87, 314, 445
Hauptmann, H., 139
Hausen, H. D., 255, 256
Hauser, A., 109
Hauser, P. J., 417
Hauthal, H. G., 95
Hauw, T. L., 28, 328, 366
Havlicek, M. D., 140
Hawkins, C. J., 2
Haworth, D. T., 120, 260
Hawthorne, M. F., 48, 117, 118, 251, 252
Hay, R. W., 51
Hayashi, J., 46, 372
Hayashi, M., 192
Hayashi, T., 135
Hayes, S. E., 46, 314, 389, 471
Hayman, D. R., 510
Hays, M. J., 25
Hazony, Y., 504
Healy, P. C., 65, 449
Heath, G. A., 27, 395
Heaton, B. T., 49, 56, 347, 381
Heber, R., 447
Heck, R. F., 62
Heckmann, G., 13
Hedgeland, R., 290
Hedwig, G. R., 406
Heess, R., 271
Hegde, M. S., 512
Hegedus, L. S., 416
Heger, G., 526

Heider, R., 248
Heil, B., 430
Heilbronner, E., 128
Heiman, N. D., 498
Heimbach, P., 56
Heinsen, H. H., 325
Heitbaum, J., 177
Heitner, H. I., 69
Helferich, G., 97
Hellams, K. L., 265
Heller, G., 108, 260, 301
Hellwinkel, D., 253
Helminger, P., 182, 183, 186
Helsen, J., 619
Hemmings, J. A. G., 86
Hemmings, R. T., 268
Hencsei, P., 137
Hendra, P. J., 199
Hendricker, D. G., 99, 438, 439, 468
Hendrickson, A. R., 60, 355, 453
Hendrickson, D. N., 41, 314, 371
Henriksen, L., 146
Hendriksma, R. R., 328, 455
Hengge, E., 125, 269, 274, 314
Henneike, H. F., 7, 61, 457
Hennig, H., 349
Hennig, H. J., 150
Henshall, T., 301
Henzi, R., 434
Herber, R. H., 60, 272, 279, 399, 445, 495, 577, 582, 584, 589
Herberhold, M., 22, 65, 385, 416
Herberich, G. E., 48, 481
Herbert, I. R., 619
Herbert, M., 98
Herbeuval, E., 405
Herbst, E., 179
Herbstein, F. H., 68
Herlem, M., 74
Herlinger, A. W., 69
Herlocker, D. W., 92, 440
Hermon, E., 566
Heřmánek, S., 115, 118, 250, 251
Herrington, D. R., 51
Herrmann, P., 152
Herrmann, W. A., 21, 25, 28, 420, 468, 472
Hersey, T. G., 182
Hershkowitz, N., 499, 604
Herskovitz, T., 60
Herz, J. E., 97
Herzenberg, C. L., 619
Herzog, P., 602
Herzog, W., 120, 130, 257, 275
Hess, J., 502
Hesse, G., 119
Hessett, B., 120, 257
Hester, R. E., 210
Hetey, A. H., 350
Hetflejš, J., 276

Hewat, A. W., 619
Hewitt, T. G., 85
Heyde, M. E., 307
Heyding, R. D., 156
Heyns, A. M., 431
Hicks, J. A., 501
Hidai, M., 394
Hiefyje, G. M., 3
Higasi, K., 242
Higginbotham, E., 320
Higson, B. M., 59
Higuchi, T., 131
Hilbers, C. W., 3, 95
Hilborn, R. C., 177
Hild, E., 324
Hildbrand, J., 440
Hill, A. E., 35
Hill, H. A. O., 44, 89
Hill, N. J., 19, 271, 362
Hill, N. L., 451
Hill, R. E. E., 433
Hill, W. E., 37, 335
Hillel, R., 262
Hillgärtner, H., 132
Hillman, M., 9
Hinckley, C. C., 95, 317
Hindman, J. C., 10
Hipp, C. J., 59
Hirabayashi, T., 57, 122
Hirahara, E., 619
Hiraishi, J., 431
Hirashima, T., 61, 382
Hirayama, M., 96
Hirota, E., 181, 182, 183, 191
Hisatsune, I. C., 204, 241
Hitchings, M. R., 200
Ho, B. Y. K., 574
Ho, C., 42, 90, 151
Ho, C. C., 16
Hobday, M. D., 264
Hoberg, H., 123
Hobson, M. C., 496, 545, 619
Hocking, W. H., 183, 184
Hodge, A., 320
Hodges, H. L., 71
Hodgson, P. G., 456
Hoefdraad, H. E., 235
Höfer, R., 144, 294
Höfler, F., 274, 284
Höfler, M., 25, 472, 477
Hoeft, J., 178, 179, 180, 181
Höjgaard, J. J., 517
Hoff, J. T., 236
Hoffman, B. M., 515
Hoffmann, E. G., 3
Hoffmann, H. M. R., 35
Hoffmann, K., 255, 383
Hoffmann, L., 419
Hoffmann, N. W., 418
Hoffmann, R., 73
Hoffmann, R. W., 621
Hofstee, H. K., 361
Hogeveen, H., 97
Hohenemser, C., 617
Hohmann, F., 468
Holah, D. G., 311, 339
Holak, W., 140

Holbrook, K. A., 87
Holding, A. F. le C., 588
Holik, M., 94
Hollaender, J., 130
Hollebone, B. R., 233
Holliday, A. K., 114
Hollingsworth, G., 16
Holloway, C. E., 137
Holly, S., 324
Holm, R., 138, 157
Holm, R. H., 43, 59, 60, 64, 68, 69, 91, 401, 450
Holmes, L., 526
Holmes, R. H., 3
Holmes, R. R., 198, 220
Holste, G., 326
Holt, A., 134
Holt, S. L., 437
Holzinger, H. W., 619
Holtzman, J. L., 496
Homeier, E. H., 317, 427
Homer, J., 93
Homsany, R., 87
Hon, F. H., 41, 97
Hon, J. F., 147, 305
Honeybourne, C. L., 94
Hong, P., 339
Hooper, M. A., 243, 244, 453
Hooz, J., 41
Hopf, F. R., 42
Hopkinson, M. J., 29, 478
Hopous, E. A., 417
Hopper, M. J., 265
Hopper, S. P., 62
Hora, C. J., jun., 220
Horder, J. R., 208
Horita, M., 619
Horlbeck, G., 35, 480
Horn, C. J., 322, 455
Horn, H.-G., 138
Hornreich, R. M., 557
Hornung, V., 128
Horrocks, W. D., jun., 94
Horspool, W. M., 41
Horton, D., 63
Horwood, J. L., 566
Hoskins, L. C., 213
Hosokawa, T., 56
Hosomi, A., 129
Hota, N. K., 322
Hou, F. L., 22, 468
Hough, J. J., 30, 331
Houk, K. N., 97
Houk, L. W., 25, 473
House, D. R., 402
Housley, R. M., 550, 557, 561
Howard, B. J., 186
Howard, G. D., 21, 540
Howard, J., 30, 35, 440, 483
Howard, J. W., 27, 476
Howard-Lock, H. E., 214
Howe, A. T., 542, 544, 569
Howell, F. L., 103
Howell, J. A. S., 22, 35, 479
Howery, D. G., 109
Howie, G. A., 98

Howlett, K. D., 164
Hoxmeier, R. J., 386
Hoy, G. R., 504
Hoyano, J. K., 28, 327, 477
Hoyer, E., 447
Hoyng, P., 176
Hrabák, F., 47
Hrichova, R., 551
Hrung, C. P., 478
Hrynkiewicz, A. Z., 499, 564
Hrynkiewicz, H. U., 544
Hsieh, A. T. T., 25, 40, 388, 471, 475, 492
Hsu, C.-Y., 441
Hubbard, A. F., 63
Huber, F., 87, 445
Huber, H., 94, 95, 206, 246, 342, 345, 398
Huber, R., 42, 137
Hubert, J., 342
Hucl, M. K., 548
Hudson, R. F., 277
Hüfner, S., 495, 609
Huestis, W. H., 42, 78, 401
Hüttel, R., 61
Huffman, G. P., 561, 619
Hughes, A. N., 339
Hughes, M. N., 339, 403, 460
Hughes, R. P., 11, 56
Huggins, R. A., 101
Huheey, J. E., 575, 576
Hui, B. C., 339
Huisman, H. O., 97
Huler, E., 292
Hulett, L. G., 311, 357
Hulscher, J. B., 142
Hunt, D. F., 34, 35, 388, 483
Hunt, J. B., 85
Hunt, J. P., 75, 77
Hunter, B. K., 211
Hunter, G., 297
Hunter, P. W. W., 314
Hunter, R. G., 59, 345
Huong, P. V., 203
Hursthouse, M., 32
Husar, J., 202
Husband, J. P. N., 120
Hussein, M. A., 358
Husson, E., 231
Hutcheon, W. L., 327
Hutchins, J. E. C., 12
Hutchins, R. O., 118, 320
Hutchinson, B., 310, 332
Huttner, G., 31, 32, 467, 479
Huvenne, J. P., 213, 234
Hyams, I. J., 363
Hyman, H. H., 306
Hyne, J. B., 139
Hynes, J. B., 286

Ibbott, D. G., 74
Ibers, J. A., 466
Ibragimov, A. A., 146
Ichiba, S., 589
Ichikawa, K., 123
Ichinose, N., 149

Ignat'ev, I. S., 277, 278
Ignat'ev, Yu. A., 150
Ignatov, B. G., 155
Ihnat, M., 79
Iida, K., 58, 376, 465
Ikeda, F., 98
Ikeda, R., 102
Ikeda, S., 40, 41, 366, 368, 434
Ikeda, Y., 123
Ikonnikov, V. P., 547, 617
Ikram, M., 297
Il'yashenko, V. S., 357
Il'yasova, A. K., 319
Ilyatov, K. V., 617
Imaeda, H., 122
Imai, K., 410
Imai, Y., 352
Imbert, P., 558, 613
Imhoff, D. W., 132, 443
Imshennik, V. K., 521
Incorvia, M. J., 93, 451
Indriksons, A., 134, 274
Ing, S. D., 173
Ingalls, R., 502, 506, 516
Inglis, T., 342, 366
Ingold, K. U., 118
Ingraham, L. L., 9
Ino, H., 496
Inoue, M., 93
Inoue, Y., 98
Intille, G. M., 47, 53, 335, 490
Ionin, B. I., 3, 151
Ionov, S. P., 14, 149, 592, 616
Iordanova, R., 501
Iqbal, M. Z., 28, 45, 475, 489
Iqbal, Z., 239
Ireland, P. R., 33
Irgolic, K. J., 140, 291
Irvine, I., 143, 288
Irving, C. S., 47
Irwin, J. G., 43, 491
Isaak, G. R., 505, 618
Isaev, S. D., 148
Isakov, L. M. I., 547
Ishaq, M., 20, 419
Ishibashi, N., 102
Ishiharo, H., 132
Ishii, Y., 54, 55, 56, 57, 71, 122, 129, 130, 142, 377, 378, 379, 431
Ishikawa, M., 41
Ishikawa, P., 125
Ishimori, T., 428
Ishino, Y., 61, 382
Ishiyama, T., 423
Isida, T., 130
Islip, M., 449
Ismail, Z. K., 207
Isope, K., 311, 425
Issleib, K., 18, 378
Isupov, V. K., 308
Itano, T., 130
Ito, A., 518
Ito, H., 49
Ito, M., 98
Ito, T., 65, 321

Author Index

Ito, Ts., 55, 378, 431
Itoh, K., 71, 122, 129, 130, 229
Ius, A., 97
Ivakin, A. I., 319
Ivanitskii, V. P., 564
Ivannikova, V. N., 419
Ivanov, A. S., 498
Ivanov, E. V., 532
Ivanov, V. T., 77
Ivanova, A. V., 359
Ivanova, N. P., 271
Ivanova, O. M., 359
Ivanov-Emin, B. N., 263, 303, 315, 444
Ivlev, Yu. N., 89
Ivleva, I. N., 394
Ivoilov, N. G., 616
Iwamoto, T., 101, 405
Iwao, T., 55
Iwasaki, M., 359
Iyengar, P. K., 550
Izumi, J., 182

Jack, T., 70
Jackels, S. C., 50, 408
Jackson, P. J., 536
Jackson, W. R., 22, 56, 72
Jacon, M., 199
Jacox, M. E., 200, 204
Jadhao, V. G., 538, 546
Jaenicke, O., 35, 362, 367
Jain, B. D., 383, 458
Jain, M. C., 407
Jain, P. C., 311, 448, 458
Jain, S. C., 324
Jain, Y. S., 243
Jakoubková, M., 276
James, B. D., 358
James, B. R., 44, 334, 338
James, D. W., 241
James, T. L., 77, 109
Jandacek, R. J., 286
Jander, J., 298
Jang, S. D., 462
Janik, J. A., 238, 242
Janik, J. M., 238, 242
Janinický, M., 456
Jankowski, W. C., 2
Janot, C., 495, 619
Jansen, P., 221
Janson, T. R., 81
Jantzen, R., 96
Janz, G. J., 292
Janzen, A. F., 74, 86
Jao, L., 80
Jardine, I., 45
Jarke, F., 426
Jarvie, A. W. P., 134, 136
Jarvis, A. C., 47
Jastram, P. S., 612
Jaura, K. L., 281, 582
Javan, A., 181, 198
Jayaraman, A., 166
Jayne, D., 89
Jean, A., 134
Jeffery, J., 40, 420
Jeffrey, K. R., 163
Jekot, K., 142

Jellinek, F., 48, 214, 312, 336, 349
Jenkinson, M. A., 304
Jensen, F. R., 63
Jensen, K. A., 146
Jeremic, M., 65
Jernigan, R. T., 385, 470
Jessep, H. F., 72, 125, 267
Jesson, J. P., 43, 64, 66, 67
Jessop, K. J., 52
Jetz, W., 26, 35, 37, 471, 484
Jeżowska-Trzebiatowska, B., 293, 310, 328, 332, 396
Jha, S., 602
Jindal, S. L., 98
Jitsugiri, Y., 149
Jitsumori, K., 135
Job, R. C., 37, 134.
Jörg, H. J., 620
Jørgensen, Ch. K., 325
Joesten, M. D., 140, 441
Johannesen, R. B., 3
Johnson, A., 60, 87, 348
Johnson, A. W., 50
Johnson, B. F. G., 22, 30, 33, 35, 38, 47, 67, 68, 70, 85, 96, 374, 381, 389, 416, 419, 478, 479, 481
Johnson, C. E., 534, 536, 617
Johnson, C. S., jun., 99
Johnson, D. A., 423
Johnson, D. H., 146
Johnson, D. R., 192
Johnson, D. W., 437
Johnson, E. H., 61, 425
Johnson, H. D., 115, 249
Johnson, I. K., 139, 285
Johnson, J. S., 127
Johnson, K. A., 343
Johnson, L. F., 2, 69, 77
Johnson, M. D., jun., 45, 94, 95
Johnson, M. F., 77
Johnson, N. P., 329
Johnson, R., 100
Johnson, R. A., 237
Johnson, R. N., 69
Johnson, T. R., 136
Johnson, V., 508
Johnson, W. D., 144
Johnson, W. M., 61, 444, 445
Johnstone, J. J., 594, 595
Jolley, K. W., 91
Jolly, P. W., 55, 56
Jolly, W. L., 12, 112, 113, 248
Joly, M., 136
Jóna, E., 456
Jonas, J., 15
Jonassen, H. B., 410
Jones, C. H. W., 478, 539, 594, 595
Jones, C. J., 48, 118, 252, 416
Jones, D. E. H., 80

Jones, E. M., 11, 22, 23, 411
Jones, E. P., 156
Jones, E. R., 614
Jones, H., 192
Jones, H. V. P., 47, 374
Jones, L. H., 221, 226
Jones, M., jun., 129
Jones, P. J., 219
Jones, R., 78
Jones, R. G., 126, 182
Jones, T. L., 177
Jones, W. E., 40
Joo, W.-Ch., 247
Jordan, A. D., 15, 84, 298
Jordan, R. B., 15, 49, 75, 93
Joseph-Nathan, P., 97
Josey, A. D., 92
Josien, M. L., 198
Josty, P. L., 35, 478, 479
Jotham, R. W., 251
Jouany, C., 112
Jourdan, G., 293
Jouve, P., 198, 200
Jovanović, B., 53, 347
Jozefowicz, J., 405
Jugie, G., 112, 113
Juliano, P. C., 12
Julien, W. M. O., 299
Jumeau, D., 229
Jurczak, J., 98
Jutand, A., 107
Jutzi, P., 118

Kablitz, H.-J., 19
Kachapina, L. M., 394, 396
Kachi, S., 512
Kacmarek, A. J., 146, 301
Kado, T., 152, 156
Kadykenov, M. M., 618
Kaesz, H. D., 3, 386
Kagan, Yu., 495
Kagarakis, C. A., 166, 167
Kagawa, T., 122
Kai, Y., 345
Kaidalova, T. A., 280
Kaindl, G., 602, 603
Kainosho, M., 97
Kaipov, D. K., 618
Kaizu, Y., 17
Kajiwara, M., 50
Kalbfleisch, H., 236
Kalbfus, W., 467
Kalck, P., 485
Kale, A. J., 210
Kalinichenko, A. M., 105
Kalinin, V. N., 115, 152
Kalinnikov, V. T., 320
Kaliraman, S. S., 538
Kaliya, O. L., 344, 382
Kallweit, R., 19
Kal'naya, G. I., 343
Kaloustian, M. K., 48, 117, 251, 252
Kalra, K. L., 450
Kalsotra, B. L., 383, 458
Kaltseis, J., 605, 619
Kalugin, Yu. B., 402

Kalvius, G. M., 495, 590, 613
Kamai, G., 141, 153
Kamata, S., 102
Kamenov, P. S., 501, 619
Kamerling, J. P., 131
Kan, L. S., 78
Kanamaru, F., 553, 621
Kanashiro, T., 149
Kanatomi, H., 59, 460
Kandgetcyan, R. A., 146
Kandil, S. A., 137
Kane, A. R., 70, 249
Kanekar, C. R., 93, 99, 442
Kaneko, M., 150
Kanellakopulos, B., 99, 383, 414, 531
Kane-Maguire, L. A. P., 50, 79
Kaneshima, T., 69
Kang, D. K., 82, 87, 284, 393
Kankeleit, E., 495, 611
Kanohta, K., 150
Kanpov, D. K., 621
Kantlehner, W., 130
Kaplan, M., 524
Kapon, M., 68
Kapoor, P. N., 25, 347, 422, 468
Kapp, W., 126, 272
Karabekov, A. K., 616
Karagounis, G., 103
Karakida, K., 190
Karapet'yants, M. Kh., 304
Karayannis, N. M., 311, 320, 357, 434, 441
Karguppikar, A. M., 240
Karimov, Yu. S., 101
Karitonov, Yu. Ya., 331
Karle, J., 35
Karlow, E. A., 357
Karn, R. A., 135
Karraker, D. G., 614
Karras, M., 148
Karsch, H. H., 53, 341
Kartha, V. B., 442
Kasai, N., 135, 345
Kashin, A. N., 15
Kashiwa, A., 135
Kashyap, S. C., 246
Kasper, H., 95, 96
Katada, M., 589
Katila, T. E., 567, 608, 612
Kato, A., 97
Kato, S., 129, 130, 132
Katochkina, V. S., 315
Katović, V., 409
Katritzky, A. R., 98
Katsuura, T., 129, 130
Kattenburg, H. W., 211
Katyshev, A. N., 152
Katz, B., 98
Katz, J. J., 81, 113, 118, 250
Katz, L., 35, 369
Katz, T. J., 41, 58
Kaufmann, G., 360, 414, 440

Kaukis, A. A., 619
Kawaguchi, S., 53, 311, 342, 345, 400, 425
Kawakami, K., 69, 415
Kawamiya, N., 619
Kawarazaki, S., 107
Kawasaki, K., 51
Kawasaki, Y., 84, 135, 278
Kayel, R. O., 197
Kazachenko, D. V., 427
Kazitsyma, L. A., 2
Keable, H. R., 26, 66, 364, 472
Keat, R., 143, 164, 288
Kebabcioglu, R., 217
Keck, J. M., 123, 139
Kecki, Z., 149
Kedrova, N. S., 253
Kedzia, B. B., 310, 332, 462, 463
Kee, T. G., 72
Keeling, G., 387
Keeling, R. O., 545
Keene, F. R., 50, 401
Keeney, W., 68
Keim, P., 90
Keisch, B., 501
Keiter, R. L., 7, 26, 473
Keith, G. R., 399
Keith, L. H., 97
Keller, C., 616
Keller, G., 75
Keller, H. J., 3, 6, 413
Keller, P. C., 86, 113, 248, 249
Keller, R. M., 42, 89, 90
Keller-Schierlein, W., 97
Kelling, H., 277
Kelly, D. A., 222
Kelsey, D. R., 95
Kemmitt, R. D. W., 45, 47, 55, 59, 306, 452, 490
Kemp, T. J., 41, 371
Kennedy, B. P., 314
Kennelly, W. J., 19, 309
Kenney, M. E., 41
Kent, G. D., 464
Kent, P. W., 78
Kenyon, G. L., 98
Keousim, T., 102
Kepert, D. L., 411
Keramidas, V. G., 229
Kerber, R. C., 35
Kergoat, R., 323, 440
Kerimbekov, A. V., 294
Kerwin, C. M., 59, 461
Keske, R. G., 146
Keskinen, A. E., 70, 341
Kessenikh, A. V., 6, 7, 149
Kettle, S. F. A., 332, 384, 387, 388
Keune, W., 508
Keung, E. C.-H., 35
Kevenaar, P. C. J., 139
Kew, D. J., 220, 329
Khachatryan, M. Kh., 618
Khachaturov, A. S., 84
Khaddar, M. R., 75
Khaibullin, I. B., 270
Khalafalla, D., 546, 556

Khalepp, B. P., 332
Khalil, M. I., 278, 574
Khalilulina, K. K. H., 77
Khamar, M. M., 321
Khan, M. M., 317, 453
Khan, V. P., 153
Khan, W. A., 144
Khandelwal, B. L., 80
Khandros, E. K., 317
Khanna, R. K., 234
Kharboush, M., 382
Khare, G. P., 335, 392
Khare, P. L., 92
Kharitonov, N. P., 266
Kharitonov, Yu. Ya., 333, 337, 359, 458, 462
Kharlamova, E. N., 65
Kharzeeva, S. E., 424, 432
Khashkhozhev, Z. M., 244
Khimich, T. A., 549, 556
Khimich, Yu. P., 500, 550
Khiteeva, V. M., 324
Khlebnikov, V. B., 110
Khlystov, A. S., 621
Khokhlov, R. V., 497
Khopyanova, T. L., 169
Khorshev, S. Ya., 124
Khrapov, V. V., 163, 595
Khudobin, Yu. I., 266
Khvostik, G. M., 11
Kida, S., 410, 461
Kidd, R. G., 14, 75, 145
Kieffer, R., 123
Kiefer, W., 199
Kielbania, A. J., jun., 56
Kiener, V., 367
Kienle, P., 605
Kiennemann, A., 144, 291
Kifer, E. W., 125, 267, 268
Kilby, B. J. L., 55, 377
Kilner, M., 26, 66, 342, 364, 366, 472
Kilponen, R. G., 307
Kim, H., 183, 202, 224
Kimball, C. W., 618, 619
Kimel'fel'd, Ya. M., 344
Kimura, B. Y., 11, 45, 47, 55, 388
King, R. B., 20, 21, 22, 25, 27, 28, 34, 35, 37, 45, 48, 347, 365, 413, 419, 421, 422, 468, 469, 471, 482
King, R. M., 90
Kingston, B. M., 18
Kingston, J. V., 335, 487, 489
Kinnaird, J. K., 93, 459
Kinovalov, L. V., 402
Kinsella, E., 364
Kinsman, K. R., 621
Kinugasa, T., 55
Kipp, E. B., 51
Kira, M., 129
Kirby, R. D., 229
Kirchhoff, W. H., 176
Kiriakidis, T., 131
Kirichok, P. P., 549, 617
Kirilov, M., 17
Kirillov, Yu. B., 357
Kirin, I. S., 308, 410, 621

Author Index

Kiriyama, R., 553
Kirk, R. W., 120, 262
Kirksey, K., 436
Kirmse, R., 349, 447
Kirsch, H. P., 38, 48
Kiselev, A. V., 110
Kiselev, V. G., 83
Kiseleva, N. V., 489
Kiseleva, V. M., 419
Kisin, A. V., 72
Kisleva, L. A., 453, 457
Kiso, Y., 268
Kiss, A. B., 324
Kisyńska, K., 537
Kitagawa, T., 198
Kitamura, T., 131
Kitazawa, H., 150
Kitching, W., 10, 61, 128, 132, 136
Kizhaev, S. A., 106
Kjekshus, A., 565, 619
Kjellstrup, S., 363
Klaeboe, P., 204, 300
Klabunde, U., 84
Klaentschi, N., 315
Klärney, F. G., 98
Klar, G., 123, 139
Klassen, J. J., 83
Kleiman, Yu. L., 148
Klein, G. P., 182
Klein, H.-F., 53, 341
Klein, M. P., 80
Kleinberg, J., 26, 321
Kleinstück, R., 140
Klemenkova, E. S., 366
Klemperer, W. G., 65, 179, 180, 186
Kliegel, W., 119
Kliener, V., 39
Klíma, P., 293
Kline, D., 152
Kline, O., 101
Kline, R. J., 82
Klingebiel, U., 143, 290, 296
Klingen, T. J., 118
Klinger, L. M., 499
Kloker, W., 289
Klopova, Zh. G., 320
Klopsch, A., 294
Klose, G., 152
Klosowski, J. M., 132, 243, 269
Kludt, J. R., 282
Klug, W., 146, 304
Klushin, N. A., 572
Klyuchnikov, N. G., 201
Knapp, J. E., 477
Knauss, L., 7
Knebel, W. J., 55, 58
Knight, J., 30, 486
Knoll, F., 97, 130
Knoll, L., 35, 420
Knop, O., 592, 593
Knoth, W. H., 30, 396
Knowles, P. F., 324
Knowles, P. J., 58
Knox, G. R., 22, 85, 363
Knox, S. A. R., 37, 386, 486

Knunyants, O. L., 152
Knuth, K., 298
Knyazev, A. S., 230
Knyazeva, N. A., 337
Knyazeva, N. N., 452
Ko, H. C., 74
Kobayashi, H., 17, 425, 426, 435
Kobayashi, R., 41
Kobayashi, Y., 98
Kobelt, R., 31, 43, 67, 330
Kobets, L. V., 320
Kobrya, N. V., 549
Koch, K., 140, 297
Kocharyan, L. A., 512
Kocheshkov, K. A., 163, 269
Kochetkova, N. S., 40, 101, 485
Kochi, J. K., 60, 382
Koda, S., 400
Koda, Y., 423
Kodama, G., 86, 115
Kodama, T., 102
Köhler, H., 410, 443
Köhler, K., 232, 351
König, E., 399, 400, 464, 496, 515, 524, 530, 531
Koenig, M. F., 392
Koenig, S. H., 79
Königer, F., 212
Königstein, J. A., 197, 236
Koepke, J. W., 265
Koerner, G. S., 126
Koerner von Gustorf, E., 35, 362, 367, 479
Köttgen, D., 292
Koglin, E., 324
Kogure, T., 132
Kohlrausch, K. W. F., 197
Kohn, B. H., 109
Kohn, H. W., 263
Kohout, J., 457
Koike, Y., 85
Koizumi, M., 553, 621
Kojima, S., 157, 158
Kokes, R. J., 352
Koketsu, J., 142
Kokjma, S., 142
Kokot, S., 423
Kokunov, Yu. V., 26, 323
Kolb, J. R., 19, 65, 309
Kolditz, L., 144
Kolich, C. H., 317
Kollman, P. A., 14
Kollman, V. H., 17
Kolobova, L. V., 248
Kolobova, N. E., 27
Kolopus, J. L., 152
Kolpakov, A. V., 498
Kolta, G. A., 248
Kol'tsov, A. I., 77
Kolyadin, A. B., 410, 621
Komarov, E. V., 280
Komatsu, T., 41
Komeshima, T., 415
Komissarova, B. A., 595
Komissarova, L. N., 105, 259, 314, 315, 438, 458
Komiya, S., 40, 331, 428

Komor, M., 546, 593, 618, 621, 622
Komoriya, A., 71
Komyak, A. I., 320, 359
Kondo, M., 41
Kondo, T., 41
Kondratenkov, G. P., 11, 150
Kondrat'ev, S. N., 121
Kong, P.-C., 43, 52, 335, 342
Konopka, R., 277
Kon Swee Chen, 321
Kontnik-Matecka, B. T., 388
Koola, J., 137, 466
Koonsvitsky, B. P., 40
Kopcewicz, M., 537
Kopvillem, U. Kh., 99
Korablin, L. N., 549
Korcek, S., 118
Korecz, L., 537, 539, 555
Korenevsky, V. A., 72, 135
Koridze, A. A., 27
Kormer, V. A., 11, 56
Kornienko, E. N., 618
Korotaeva, L. G., 303, 315, 444
Korovushkin, V. V., 549
Kosinova, N. M., 105
Kosovtsev, V. V., 151
Kostelnik, R. J., 78
Kostikas, A., 495
Kostikov, Yu. P., 621
Kostiner, E., 509, 557, 564
Kostka, A. G., 81
Kostromina, N. A., 18, 78
Kosuge, K., 512
Kotev, K., 172
Kotlicki, A., 537
Kovács, S., 414
Kovalenko, K. N., 427
Kovalev, I. F., 212, 266, 273
Kovaleva, L. G., 427
Kovar, R. F., 41
Kovtun, N. M., 150
Kow, W. E., 35
Koyama, K., 98
Koyama, T., 397
Kozerski, L., 98
Kozheurov, A., 545
Kozima, S., 130
Kozlov, E. S., 13
Kozlova, N. V., 273
Koz'min, S. A., 77
Kraemer, F., 151
Kraihanzel, C. S., 23, 470
Krakowski, R. A., 499
Kramarenko, F. G., 105
Krannich, L. K., 140, 291
Krasser, W., 226, 324
Kraus, K. F., 330
Krause, L., 158, 163
Krause, V., 420
Krauss, H. L., 325
Krausse, J., 363
Kravchenko, E. A., 161
Kravchenko, O. V., 254
Kravchenko, V. V., 220, 329, 330

Kravtsov, D. N., 10, 62, 63, 82
Krebs, B., 279
Krech, F., 18
Kreevoy, M. M., 12, 112, 202
Kreissl, F. R., 22, 23, 38, 362, 468, 485
Kreiter, C. G., 11, 22, 23, 31, 38, 58, 65, 68, 367, 376, 467, 468, 478, 485
Kremer, S., 530
Kreshkov, A. P., 9, 10, 12, 13, 131
Kreysa, G., 546
Krezerli, E. A., 458
Kricheldorf, H. R., 131
Krieg, B., 467
Kriegsmann, H., 271
Krisher, L. C., 188
Krishnamurthy, N., 229
Krishnan, R. S., 242
Kristoff, J. S., 66, 387
Krivoglaz, M. A., 619
Krivy, I., 242
Krizhanskii, L. M., 501, 622
Krohberger, H., 331, 337, 366, 367, 371
Kronawitter, I., 118, 260
Kroon, J., 142
Krop, K., 619
Kropshofer, H., 147, 303
Kroto, H. W., 189
Kruck, T., 25, 31, 35, 43, 67, 330, 420, 472
Kruczynksi, L., 67, 69
Krüger, C., 56
Kruglaya, O. A., 136
Kruk, C., 97
Krumgal'z, B. S., 74
Krupinskaya, A. V., 308
Krupnick, A. C., 282
Kruse, W., 321
Krylov, E. I., 432
Krylov, V. N., 280
Krynicki, K., 148
Krzhizhanovskaya, E. K., 443
Ksandr, Z., 98
Kubelt, R., 420
Kubik, T. M., 405
Kublashvili, Zh. Sh., 357
Kubo, M., 93, 167
Kubo, Y., 366
Kubota, M., 44, 335, 430, 464
Kuc, T. A., 47
Kucera, J., 618
Kucheida, D., 605
Kuchen, W., 140, 297
Kuchitsu, K., 190
Kuckertz, H., 17, 247
Kuczkowski, R. L., 191, 193
Kudo, K., 233
Kuebel, W. J., 414, 415
Kühnl, H., 263
Kündig, E. P., 206, 342, 393, 398

Kündig, W., 502, 517
Kugel, R. L., 143
Kugel, W., 130
Kuhn, D. E., 33
Kuhtz, B. H., 143, 289
Kukaszewicz, K., 293
Kukushkin, Yu. N., 418, 452, 491
Kul'ba, F. Ya., 264
Kulbaeva, U. K., 618
Kulgawczuk, D. S., 544
Kulkarni, V. H., 19, 276, 317
Kumada, M., 41, 125, 129, 133, 135, 268
Kumer, L., 619
Kumok, V. N., 442, 435
Kunitomi, N., 605, 621
Kunugi, M., 96
Kunze, U., 137, 466
Kuo, M.-C., 45, 374
Kuo, Y.-N., 127, 131
Kupchik, E. J., 137, 279, 454
Kupletskaya, N. B., 2
Kuprii, V. Z., 150
Kupriyanova, G. N., 432
Kurash, V. V., 565
Kurashvili, L. M., 447
Kurbatov, V. P., 427
Kuribayashi, H., 434
Kurilenko, O. D., 109, 110
Kurimura, Y., 150
Kurkin, I. N., 152
Kuroda, K., 51, 428
Kuroda, Y., 355
Kurosawa, A., 341
Kurosawa, H., 52, 380
Kuroya, H., 400
Kursanov, D. N., 21, 27, 28, 363, 365
Kurtz, A. P., 131
Kurucz, I., 555
Kushawaha, V. S., 199
Kuss, M., 295
Kustes, W. A., 84
Kut'ko, V. I., 228, 329
Kuyper, J., 70
Kuzina, A. F., 329
Kuzina, M. G., 338
Kuz'min, I. A., 161
Kuz'min, R. N., 498, 508, 547, 548, 595, 616, 618, 619
Kuz'minov, A. V., 621
Kuz'movich, V. V., 618
Kuznetsov, Yu. S., 545
Kuznietz, M., 613
Kuzyants, G. M., 300
Kvasov, B. A., 10, 26, 27, 62, 63, 115, 151, 152
Kwong, G. Y. W., 282
Kyskin, V. I., 61, 159
Kyuntsel, I. A., 155, 162

Laane, J., 295, 301
Labes, M. M., 311
Labinger, J. A., 45
Labonville, P., 223
Labroue, D., 69, 488

Laces, F., 108
Ladd, J. A., 16
Laffey, K. J., 70
Lafrenz, C., 86
Lagow, R. J., 127, 147
Lagowski, J. J., 282
Lakshmi Agarwala, 341, 463
Lal, K. G., 497
La Mar, G. N., 6, 68, 88, 91
Lambe, R. L., 610
Lambert, J. B., 146
Lambert, L., 211, 216
Lambert, R. L., jun., 62, 63, 126, 132
Lambert-Andron, B., 567
LaMonica, G., 398
Lampe, R. J., 399
Lamykin, E. V., 545, 571
Landa, B., 148, 308
Landa, S., 98
Landesberg, J. M., 35, 369
Landgrebe, J. A., 26, 321
Lane, A. P., 230, 282
Lane, B. C., 49, 281, 403, 577
Lane, J. R., 226
Lane, T. A., 59
Lang, G., 495, 534, 535, 609
Lang, S. A., jun., 98
Lange, D., 433
Lange, V., 443
Langer, M., 304
Langford, C. H., 74
Langhammer, D., 602
Langhout, J. P., 331
Langouche, G., 619
Lanir, A., 89
Lanneau, G. F., 125
Lantzke, I. R., 77
Lapidot, A., 47
Lapp, M., 199
Lappert, M. F., 18, 22, 38, 44, 46, 53, 59, 120, 257, 314, 330, 339, 346, 347, 362, 373, 490
Lapshin, V. D., 149
Lardon, M. A., 87
Large, N. R., 609
Larkin, G. A., 55, 377
Larkina, T. I., 619
Larkworthy, L. F., 321, 445, 532
Larock, R. C., 63
Larsen, B., 98
Larsen, D. W., 74
Larsen, E., 91
Larsen, E. M., 317, 427
Larson, K. W., 49
Larsson, R., 260, 319
Lascombe, J., 198
Lasocki, Z., 83, 151
Latscha, H., 135
Lattes, A., 62
Lau, P. T., 475
Lauer, H. V., 227
Lauer, J. L., 343
Laughlin, D. R., 277

Author Index

Laurent, J.-P., 4, 19, 83, 112
Laurie, S. H., 51
Laurie, V. W., 175
Laussac, J. P., 113
Laval, J. P., 62
Lavalee, D. K., 42
Lavenevskii, O. E., 427
Lavrov, B. B., 201
Lavrovskii, K. P., 620
Lavrukhin, B. D., 27, 87
Lawler, R. G., 3
Laws, E. A., 9
Lazarev, A. N., 197, 277, 278
Lazarev, V. B., 161
Lazareva, N. A., 151
Lazzaroni, R., 55
Leach, J. B., 7, 120, 257
Leach, J. M., 449
Leary, K., 222, 351
Leary, R. D., 72, 132, 279, 292
Lebed, B. M., 551
Lebedev, V. G., 320
Lebedev, V. N., 151
Lebenbaum, D., 616
Lece, A. B., 210
Lechan, R., 528
Lechert, H., 150
Le Corre, C., 547
Lee, A. G., 123, 261, 264
Lee, C. C., 41
Lee, H.-Y., 426
Lee, K. C., 97
Lee, K. W., 328, 456
Lee, L. Y., 505
Lee, M. C., 176
Lee, P. L., 190
Lee, T. H., 244
Leedham, T. J., 373
Leelamani, E. G., 151
Lefelhocz, J. F., 138, 427, 545
Lefevre, F., 94, 97
Lefrant, S., 229
Legg, J. I., 51, 460
Legler, L. E., 98
Legler, Y., 79
Legrand, P., 213, 234
Legros, J.-P., 358
Legros, R., 358
Legzdins, P., 387
Leh, F., 349
Lehman, J. W., 12, 131
Lehmann, H., 42
Lehn, J. M., 75
Lehveld, C. G., 381
Leibfritz, D., 10
Leigh, G. J., 26, 27, 28, 42, 65, 89, 321, 323, 334, 395, 400
Leigh, J. S., jun., 152
Leitereg, T. J., 97
Leites, L. A., 251
Leitich, J., 35
Lelandais, D., 96
Lemerle, C., 276
Lemire, A. E., 73
Lemley, A. T. G., 282, 354
Lenarda, M., 52, 54, 415
Lenkinski, R. E., 95
Lentz, A., 292, 304
Lentzner, H. L., 41
Leonenko, E. P., 452
Leonova, E. V., 40, 101
Leont'ev, V. B., 150
L'Eplatienier, F., 123
Le Plouzennec, M., 28, 365
Lepore, U., 349
Lepoutre, G., 88
Leppard, D. G., 22, 363
Lequan, R.-M., 134, 138, 140
Leroi, G. E., 237
Leroy, J.-M., 358
Leroy, M., 228
Lesbre, M., 137
Leshchenko, A. V., 410
Leshcheva, I. F., 9
Lesiecki, M. L., 208, 209
Lesoille, M., 508
Lester, G. D., 80
Letter, J. E., jun., 93
Letyuk, L. M., 549
Leung, C., 277
Leung, L. M., 53, 431
Leupold, M., 22, 467
Levason, W., 328, 330
Levay, B., 593
Levchuk, L. E., 281, 445, 578
Lever, A. B. P., 233, 314, 348
Levin, I. W., 212, 223, 294
Levin, R. H., 129
Levine, B. A., 95
Levine, L. A., 112
Levisalles, J., 379
Levison, J. J., 450
Levy, G., 82, 144, 291
Levy, G. C., 2, 5, 6, 12, 15
Lewin, A. H., 349
Lewis, D. F., 25, 85, 412, 450
Lewis, J., 22, 30, 33, 35, 47, 67, 70, 85, 96, 374, 381, 389, 478, 479, 481
Lewis, R. A., 69, 450
Lewis, S. J., 619
Ley, R., 180
Li, N. C., 87
Li, Y. S., 190, 192, 193, 271
Li, Yu. A., 618
Liang, C. Y., 357
Liang, Y., 16
Libbey, E. T., 567
Libich, S., 79
Licht, K., 271
Lichtenberg, D. W., 37
Lichtenstein, B., 93
Liddell, P. R., 617
Liddle, K. S., 352
Lide, D. R., 192
Liebig, L., 25, 472
Liegeois-Duyckaerts, M., 237
Liengme, B. V., 281, 580, 588
Ligenza, S., 619
Lillya, C. P., 21, 33
Lim, J. C., 343
Lim, Y.-Y., 92, 93
Limouzi, J., 399
Limouzin, Y., 126, 164
Lin, C.-L., 602
Lin, H. M. W., 397
Lin, L.-P., 33, 479
Lin, S. W., 417
Lin, T. P., 296
Lincoln, S. F., 75, 77
Lind, W., 122, 123, 255
Lindahl, C. B., 148
Lindblom, G., 75
Lindgren, B. J., 349
Lindman, B., 75, 80
Lindner, E., 28, 137, 292, 399, 449, 464, 465, 466, 476, 491, 524
Lindoy, L. F., 59
Lindstrom, T. R., 42, 90
Lines, E. L., 143, 248
Ling, A. C., 31
Ling, D., 127
Lingertat, H., 277
Ling-Fai Wang, J., 200
Linhard, M., 337
Link, R., 515
Linke, S., 41
Linnett, J. W., 548
Lipas, P. O., 567
Lipis, L. V., 357
Lipka, I., 551
Lipka, J., 551, 617
Lipova, I. M., 111
Lipovskii, A. A., 338
Lipowitz, J., 10
Lippard, S. J., 25, 61, 69, 70, 85, 412, 414, 450, 465
Lippincott, E. R., 234, 243, 363, 576
Lipscomb, W. N., 9, 117
Lipsky, S. R., 97
Liquier, J., 436
Lisichkin, I. N., 294
Liskow, D. H., 200
Lisle, J. B., 107
Lister, D. G., 185
Liszi, I., 414
Litchman, W. M., 15
Littlecott, G. W., 55
Liu, C. S., 137
Liu, K.-T., 95
Livingston, K. M. S., 139, 227, 285
Livingstone, S. E., 462
Llabador, Y., 513
Llabrès, G., 146
Llinas, M., 80
Lloyd, A. D., 45, 372
Lloyd, D., 138
Lloyd, J. P., 23, 467
Lloyd, M. K., 23, 411
Lo, G. Y.-S., 323
Lobach, M. I., 11, 56
Lobanova, T. S., 359
Lobkovskii, E. B., 254
Lock, C. J. L., 214, 220, 329
Lock, J. A., 619

Lockman, B., 120, 260
Loeffler, B. M., 464
Loeffler, P. A., 81
Löliger, J., 93
Logan, N., 12, 278, 438, 574
Loginova, E. I., 149
Loh, E. L., 619
Lohmann, B., 81
Loim, N. M., 27
Lokshin, B. V., 21, 28, 31, 65, 362, 363, 365, 366, 370
Long, G. G., 84, 161, 539
Long, J. R., 120, 260
Longato, B., 415
Longoni, G., 54
Loos, K. R., 307
Lopusínski, A., 294
Lorenz, I. P., 464, 524
Lorenz, R., 139, 144, 283, 296, 413
Lorenzelli, V., 241, 315, 427
Losievskaya, S. A., 619
Losilevskii, Ya. A., 497
LoSurdo, A., 77
Lough, R. M., 83
Loutellier, A., 344, 422
Love, J. C., 619
Lovecchio, F. V., 59
Lovejoy, R. W., 255
Low, M. J. D., 282
Lowder, J. E., 204
Lowman, D. W., 6
Lucas, C. R., 20, 318
Lucas, M., 77
Lucas, N. J. D., 186
Lucchini, V., 121
Luchetti, C. A., 353
Luchkina, S. A., 332, 617
Lucken, E. A. C., 154, 162, 164, 169
Łuczak, J., 147
Ludwick, L. M., 82
Lütgemeier, H., 105, 106
Lüttke, W., 146
Lugowski, J. J., 258
Lukas, J., 70, 85
Lukashevich, I. I., 525, 617, 618
Lukeman, B. D., 43
Lunazzi, L., 132
Lundin, R. E., 61
Lunenok-Burmakina, V. A., 150
Lupin, S., 68
Lupis, C. H. P., 546
Lupton, M. K., 142
Lupu, D., 323, 520
Luskus, L. J., 97
Lustig, M., 306
Lutskina, T. K., 520
Lutz, H. D., 248
Lutz, O., 14
Luz, Z., 75, 90, 153
Luzikov, Yu. N., 72
Lyakhova, V. F., 280
Lyapina, S. S., 410
Lygin, V. I., 110
Lykasov, A. A., 545

Lynch, A. H., 443
Lynch, J., 31, 478
Lyons, J. R., 51
Lysy, R., 147
Lyubimov, A. N., 2
Lyubimov, V. S., 14, 149
Lyubutin, I. S., 551, 552, 571, 572

Maass, G., 205
Mabry, T. J., 131
McAlees, A. J., 19, 65, 316, 317
McArdle, P., 33, 35, 96, 481
Macarovici, C. Gh., 317
Macarovici, D., 338
Macášková, L., 404
McAuliffe, C. A., 144, 295, 328, 330, 342, 343, 349, 456
McAvoy, J. S., 251
McBeth, R. L., 516
McCaffrey, D. J. A., 56, 347, 381
McCarley, R. E., 172, 173, 174
McCleverty, J. A., 23, 25, 26, 411, 416
McClory, M. R., 149
McClung, R. E. D., 41
McColm, I. J., 356, 610
McConnell, H. M., 151
MacCordick, J., 169, 247, 360, 414
McCormick, B. J., 345, 355, 465
McCowan, J. D., 81, 316
McDevitt, N. R., 197
MacDiarmid, A. G., 38, 67, 278
McDonald, J., 466
McDonald, J. W., 49
McDonald, R. L., 282, 333
McDonnell, J. J., 41
McDowell, C. A., 102, 112
McDowell, R. S., 213, 226
McElroy, R., 98
McEwen, G. K., 22, 23, 411
McFadyen, W. D., 461
McFarland, C. W., 144
McFarland, J. J., 59, 345
McFarlane, H. C. E., 138
McFarlane, W., 13, 14, 131, 138
McGarvey, B. R., 92
McGaughy, T. W., 91
McGinnety, J. A., 116, 234
McGinnis, J., 41
McGinnis, R. N., 321
McGuire, R. R., 41
Machado, A. A. S. C., 311
Machkhoshvili, R. I., 458, 462
Machmer, P., 154
McIntosh, C. L., 132, 170
Mack, D. P., 142, 291
Mackay, K. M., 28, 269, 314
McKeever, L. D., 3

McKenzie, E. D., 59
Mackey, D. J., 320, 448
Mackey, J. L., 609
Mackie, R. K., 96
McKinney, J. D., 97
Mackor, A., 5, 141, 427
McLauchlan, K. A., 2
MacLaughlin, D. E., 156
Maclean, C., 3
McLean, R. R., 27
McMeeking, J., 373
McMillan, R. S., 334
McMurray, C., 5
McNab, T. K., 516
McNeel, M. L., 64
McNeilly, S. T., 41
McQuaker, N. R., 242
McQuillan, G. P., 453
McQuillin, F. J., 44, 45, 54
Maquire, H. G., 512
McRobbie, D. C., 49, 402
McWhinnie, W. R., 43, 304, 305, 306, 538
Madan, S. F., 444
Madan, S. K., 299, 357
Madden, D. P., 38, 482, 541
Maddock, A. G., 495, 538, 577
Madeja, K., 400, 530, 531
Madl, R., 65
Maeda, S., 208
Maeda, T., 122, 256, 262, 454
Maercker, A., 16, 84
Märkl, G., 127, 139
Mag, P., 537, 539
Magdesieva, N. N., 146, 147, 149, 151
Maggio, F., 53
Magill, J. H., 102
Magnuson, J. A., 17, 80
Magnuson, N. S., 80
Magnuson, R. H., 396
Mague, J. T., 43, 45, 334, 335, 374, 419
Mahadevappa, D. S., 463
Mahale, V. B., 317, 411
Maher, J. P., 8
Mahmond, F. T., 335, 489
Mahmoudi, S., 222, 225
Mahnke, H., 15
Mahoney, J. P., 554
Maier, L., 141
Maigedt, H., 249
Maijs, L., 17
Maiorova, L. P., 280
Maire, J. C., 14, 131, 164, 277, 496, 580
Mairesse, G., 234
Maisel, G., 394
Maitlis, P. M., 54, 56
Maitra, A. N., 260
Majee, B., 10
Majerski, Z., 148
Majoral, J. P., 293
Makarov, E. F., 525, 545, 592, 620
Makarov, V. A., 620
Makarov, Ya. V., 366

Author Index

Makarova, L. G., 9, 25, 26, 37, 364
Makhova, E. T., 253
Makino, F., 193
Makova, M. K., 101
Maksimov, Yu. V., 620
Maksyutin, Yu. K., 163
Malaidza, M., 132, 271
Maletta, H., 620
Maley, M. P., 618
Malhotra, K. C., 305, 435
Malhotra, M. L., 239
Mali, M., 111, 157
Malik, A. U., 453
Malik, S. K., 105
Malik, S. S., 620
Malik, W. V., 407
Malisch, W., 25, 471
Mallmann, A. J., 349
Malone, D., 406
Malone, J. F., 8
Malone, L. J., 249
Maltese, M., 448
Maltina, M. J., 70
Mal'tsev, A. A., 322
Mal'tsev, A. K., 353
Mal'tseva, N. N., 253
Malve, A., 547
Malveger, A. J., 301
Malysheva, T. V., 560, 564, 565, 620
Malyutin, S. A., 304
Mamantov, G., 306, 355
Mambetov, A. A., 319
Manabe, O., 61, 382
Managadze, A. S., 457
Manapov, R. A., 617
Manashirov, O. Ya., 294
Mandache, S., 148
Mande, C., 92
Mandt, J., 279
Manewa, M., 201, 246
Manfredotti, A. G., 349
Mangia, A., 349, 408
Mangini, A., 132
Manhas, M. S., 97, 98
Mani, F., 440
Mank, V. V., 103, 108, 109, 110, 152
Manlief, S. K., 231
Mann, B. E., 8, 11, 12, 44, 339, 490
Mann, J. R., 352
Manni, P. E., 98
Mannik, L., 201
Manning, A. R., 335, 390, 469, 484, 488, 493
Manoharan, P. T., 538
Manojlović-Muir, L., 53, 347
Manoussakis, G. E., 138
Manson, E. L., 181
Mantovani, F., 146
Manzer, L. E., 11, 53, 54, 58, 341, 347, 379, 381, 409, 413
Manzini, G., 50
Maraschin, N. J., 147
Maratke, V. R., 442
March, J. F., 519

Marchand, A., 276
Marchant, L., 515
Marciacq-Rousselot, M.-M., 77
Marcotrigiano, G., 343, 349, 355
Maresca, L., 390
Marfunin, A. S., 620
Margerum, D. W., 404
Margineau, F., 286
Margiolis, K., 123, 259
Margrave, J. L., 137, 205, 207, 272
Mariano, P. S., 98
Maricic, S., 150
Maricondi, C., 527
Marie, J. C., 126
Mariella, R. P., 179
Marino, G., 147
Marino, R. A., 157
Markalous, F., 288
Marketz, H., 125, 269
Marko, L., 430
Marks, J. T., 309
Marks, T. J., 19, 65, 66, 67, 248, 370, 387
Marlière, C., 176
Maroni, V. A., 280
Marquard, D. A., 260
Marquardt, D. N., 416
Marr, G., 41
Marshall, A. G., 94
Marshall, C. J., jun., 134
Marshall, R. C., 90
Marsich, N., 61, 351
Marsmann, H., 146, 304
Marsmann, H. C., 12
Marstokk, K. M., 191
Marston, A. L., 333
Martel, B., 127
Martelli, M., 339
Martens, D., 399
Martgrienko, B. V., 293
Martin, G. J., 97
Martin, H. A., 19, 361
Martin, J. G., 410
Martin, J. S., 74, 148
Martin, M. L., 94, 97
Martin, R. A., 132
Martin, R. B., 59, 95, 98
Martin, R. L., 60, 316, 320, 355, 362, 448, 453
Martin, T. P., 248
Martinengo, S., 46, 372, 493
Martinez, D., 77
Martinez, E. D., 123
Marty, W., 49
Martyn, P. H., 265
Martynenko, L. I., 431, 432
Martynov, B. I., 62, 136
Marvich, R. H., 18, 315
Maryanoff, B. E., 118, 320
Marzilli, L. G., 49, 94, 455
Marzilli, P. A., 94
Masai, H., 380
Masamune, S., 56
Masdupuy, É., 358

Mashchenko, V. M., 277, 301
Maskasky, J. E., 41
Maslennikova, I. S., 401, 402
Maslowsky, E., jun., 101
Mason, P. R., 37, 332
Mason, R., 33, 484
Maspero, F., 47, 373
Masri, F. N., 208, 221
Massé, J., 136
Massé, J. P. R., 132
Massey, A. G., 364
Mast, E., 122
Masters, C., 8, 9, 12, 70, 335, 339
Mastryukov, V. S., 269
Masuda, I., 42, 406, 408, 418
Matejcikova, E., 344
Matejeck, K.-M., 449, 476
Materne, C., 97
Mather, G. R., 620
Mathew, M., 39, 366, 541
Mathieu, J.-P., 238
Mathieu, R., 21, 469
Mathur, H. B., 620
Mathur, J. N., 620
Mathur, S. C., 282
Matic, B., 151
Matinenas, B., 153
Matsubayashi, G., 135
Matsuda, H., 135
Matsuda, I., 129, 130
Matsuda, S., 135
Matsuda, T., 51
Matsui, K., 590
Matsui, M., 98
Matsumara, C., 190, 192
Matsumoto, M., 85
Matsumura, H., 97
Matsumura, Y., 21, 139, 467
Matsuo, M., 62
Mattes, R., 320, 323
Matthews, R. S., 95
Matthews, R. W., 14, 145
Matthias, E., 495
Matuda, K., 151
Matwiyoff, N. A., 17, 88
Matyash, I. V., 105, 564
Maudlin, C. H., 35
Maurer, W., 79
Maurin, M., 303
Mauzerall, D., 63
Mawby, R. J., 40, 420
Maxwell, I. E., 261
May, J. C., 582
May, L., 620
Mayence, G., 134
Mayer, A., 620
Mayer, C., 54
Mayer, T. J., 414
Maylor, R., 216
Mays, M. J., 30, 388, 471, 475, 486, 492, 538
Mays, R., 107
Mazanek, E. S., 544, 546
Mazerolles, P., 125, 135, 136

Mazo, G. Yu., 337, 338
Mazurkiewicz, A., 242
Mazzi, U., 455
Mchedlov-Petrosyan, O. P., 152
Meads, R. E., 499, 552, 617
Meakin, P., 43, 66, 67, 73
Méchin, B., 17
Medvinskii, A. A., 520
Meek, D. W., 342, 350, 418, 452, 456, 574
Meerts, W. L., 181
Mefed, A. F., 150
Mehner, A. W., 88
Mehring, M., 101
Mehrotra, A., 19, 123, 261, 317
Mehrotra, R. C., 19, 123, 131, 144, 273, 277, 317, 463
Mehta, M. L., 222
Meier, S., 134, 275
Meineke, E. W., 23, 468
Meinema, H. A., 5, 73, 141, 142, 285, 427
Meinzer, A. L., 227
Meisel, W., 546, 620
Mekata, M., 508
Melby, E. G., 145
Melcher, L. A., 258
Mellier, A., 202, 236
Mellon, E. K., 9, 119, 258
Mel'nichenko, L. S., 163, 269
Melnik, J. D., 537
Mel'nikov, P. P., 315
Mel'nikova, N. V., 320
Mel'nikova, R. Ya., 338
Mel'nikova, S. I., 121
Meloan, C. E., 423
Melpolder, J. P., 412, 457
Melson, G. A., 59, 315, 461
Melveger, A. J., 203
Melzer, K., 618
Memering, M. N., 342
Menczel, G., 555
Mendelsohn, M. A., 248
Mendelsohn, M. H., 113
Mendicino, F. D., 135
Menge, R., 325
Ménil, F., 521
Mente, P. G., 35
Menzel, H., 128
Mérault, G., 128
Merbach, A., 20, 75, 81
Mercer, E. E., 333
Merenne, C., 424
Mergner, R., 72
Merienne, C., 45
Merle, A., 344, 422
Merlini, A., 616
Mérour, J. Y., 24, 471
Merrell, P. H., 42, 59
Merrithew, P. B., 531
Merryman, D. J., 172
Mertschenk, B., 362, 466
Meshitsuka, S., 242
Mestdagh, H. M., 109
Mestroni, G., 43, 335, 489
Mestyanek, O., 137

Metcalf, S. G., 139, 284
Meth-Cohn, O., 98
Metrschenk, B., 41
Meurier, J., 261
Mews, R., 139, 217, 296
Meyer, B., 621
Meyer, E., jun., 487
Meyer, J., 611
Meyer, P., 58, 376
Meyer, T. J., 35, 37, 459, 484
Meyerstein, D., 49
Michael, C., 131
Michael, G., 131
Michaeli, I., 507
Michalk, C., 618
Michalowski, J. T., 57, 382
Michalska, Z., 151
Michalski, J., 294
Michel, D., 152
Michel, U., 62, 353
Michelsen, K., 49
Michelsen, T. W., 322, 345
Michl, R. J., 349
Michman, M., 53
Micklitz, H., 502, 509, 568
Micoud, M.-H., 59
Mielcarek, J. J., 249
Mihara, N., 130
Mikailov, V. I., 616
Mikaélyan, R. G., 353
Mikhailov, B. M., 83
Mikhailov, G. G., 545
Mikhaleva, I. I., 77
Mikhalevich, K. N., 327
Mikheeva, L. M., 259, 458
Mikheeva, V. I., 253
Miki, M., 58, 444
Mikołajczyk, M., 147
Mikuli, E., 242
Mikulski, C. M., 311, 320, 434
Milburn, R. M., 429
Mildvan, A. S., 77, 89
Mill, B. V., 552
Millar, J. B., 147
Miller, D. A., 318
Miller, E. M., 44, 490
Miller, F. A., 265, 270
Miller, F. J., 459
Miller, J., 32, 414
Miller, J. A., 41
Miller, J. M., 75, 83, 120
Miller, J. R., 63
Miller, J. S., 55, 315, 375
Miller, K., 75
Miller, N. E., 113, 258
Miller, P. J., 243
Miller, R. B., 499
Miller, V. R., 86, 113, 250
Milligan, D. E., 200, 204
Milligan, J. R., 282
Millington, D., 144, 291
Mills, R., 110
Milone, L., 30, 32, 486
Mimura, K., 41
Minamisono, T., 151
Minasz, R. J., 74
Miners, J. O., 409
Minghetti, G., 60

Mink, J., 352, 353
Mink, T. Q., 147
Minkiewicz, V. J., 516
Mirgorodskii, A. P., 277
Mirinskii, D. S., 152
Mironov, K. E., 439
Mirskov, R. G., 271
Mirzababaev, R. M., 498, 616
Misetich, A., 511, 516
Mishchenko, K. P., 74
Mishin, V. Ya., 308, 621
Mishra, I. B., 113, 300
Miskow, M., 97
Mislow, K., 72
Missen, A. W., 89
Mitchell, C. M., 60, 348
Mitchell, P. C. H., 323, 407
Mitchell, P. D., 207, 214
Mitchell, P. R., 51, 120, 428
Mitchell, T. D., 12
Mitchell, T. N., 82, 86
Mitchell, W. J., 326
Mitchener, J. P., 43, 334, 419
Mitchenko, Yu. I., 149
Mitin, A. V., 498
Mitra, S. S., 228
Mitrofanov, K. P., 501
Mitrofanova, N. D., 432
Mitschke, K.-H., 139, 284
Mitsudo, I., 398
Mitsudo, T., 56
Mitsuishi, A., 229
Mitsuishi, T., 150
Mitsuya, M., 401
Mittal, R. L., 453
Mixan, C. E., 146
Miyamota, H., 621
Miyoshi, A., 621
Miyoshi, T., 101
Mizobuchi, A., 151
Mizoguchi, T., 620
Mizukami, F., 49
Mizumoto, S., 182
Mizuta, E., 5, 98
Mkrtchyan, A. R., 512, 620
Mkrtchyan, V. S., 618
Moberg, L. C., 567
Modak, S. G., 92
Mode, V. A., 426
Modena, G., 121
Möbius, R., 324
Möckel, K., 353
Møllendal, H., 191
Möller, K. D., 239
Möller, U., 39, 478
Moeller, T., 144, 290
Mönter, B., 177
Moers, F. G., 331
Mohai, B., 537
Mohan, N., 212, 213
Mohapatra, B. K., 351
Moise, C., 41, 370
Mok, K. F., 35
Mok, K. S., 49
Mokhosoev, M. V., 324
Mokhosoeva, I. V., 106
Moler, R. B., 619

Author Index

Molin, Yu. N., 149
Molinie, P., 282
Molkanov, L. I., 308
Moller, W. M., 557
Molodkin, A. K., 359
Mon, J. P., 227
Monin, J.-P., 41, 370
Montaner, A., 231
Montano, P. A., 501
Montelatica, V., 104
Montenero, A., 349
Mooberry, E. S., 106
Moody, G. J., 440
Moon, R. B., 38, 42
Mooney, J. R., 41
Moore, G. J., 125
Moore, L. F., 302
Moore, M. L., 16, 361
Moore, R. D., 55
Moores, B. M., 171
Morallee, K. G., 44, 89
Morar, Gh., 317
Morariu, M., 520, 553, 616
Morariu, V. V., 110
Moravec, J., 242
Morazzoni, F., 356
More, C., 580
Moreau, J. J. E., 133, 136
Morehouse, R. L., 344
Morehouse, S. M., 70, 465
Morel, G., 97
Morel, J., 147
Moreland, C. G., 84, 115, 249
Moretto, H., 73
Moretto, H. H., 275
Morgan, H. W., 201, 205, 292
Morgan, L. O., 77
Morgan, L. W., 97
Morgunova, M. M., 12
Mori, F., 135
Mori, Y., 98
Moriarty, R. M., 35, 66
Morimoto, A., 98
Morino, Y., 183
Morishima, I., 93
Morissette, S., 200
Morita, H., 311, 425
Moritani, I., 45, 54, 339, 346, 378, 409
Morkovin, N. V., 148, 151
Moroni, V. A., 282
Morosin, B., 231
Moroz, E. M., 280
Moroz, N. K., 152
Morozov, A. I., 320
Morris, D. R., 41, 433, 542
Morris, J. H., 120, 257
Morris, M. F., 333, 522, 523
Morrish, A. H., 502, 556, 565
Morrison, J. A., 188
Morrison, W. H., jun., 41, 314, 371
Morrow, J. S., 90
Morshnev, S. K., 150
Morton, M. J., 88, 199, 302
Morup, S., 526

Mosbaek, H., 620
Mosbo, J. A., 5
Mosel, V. I., 551
Moseley, K., 54
Mosher, M. W., 41
Moskalev, P. N., 410
Moskovits, M., 206, 342, 345, 393, 398
Moskvin, A. I., 359
Mosora, H., 620
Moss, G. P., 94
Moss, J. R., 26, 28, 321, 475
Moss, K. C., 318
Moule, D. C., 265
Mounier, J., 246
Movius, W. G., 138, 577
Mowat, W., 19, 271, 362
Mowthorpe, D. J., 137, 278, 579
Mudgett, M., 144
Mueh, H. J., 28, 414
Müller, A., 212, 213, 214, 216, 223, 312, 318, 319, 320, 323, 325, 349, 450
Müller, B., 363
Müller, D., 108
Mueller, D. C., 63
Müller, G., 323
Müller, H., 17, 246, 481
Müller, J., 22, 28, 31, 41, 58, 123, 259, 362, 376, 466, 467, 477, 478
Müller, P., 480
Müller-Litz, W., 353
Mueller-Warmuth, W., 88, 151
Mueller-Westerhoff, U., 41
Münck, E., 495, 515, 536, 598
Muenter, J. S., 181
Muetterties, E. L., 31, 61, 67, 70, 73, 249, 330
Muir, K. W., 53, 347
Mukaida, M., 428
Mulligan, J. H., 265
Multani, R. K., 383, 458
Munchenbach, B., 331
Muntwyler, R., 97
Murai, S., 131
Murakami, M., 125
Murakami, Y., 620
Murakawa, T., 131
Murase, I., 59, 460
Muraveiskaya, G. S., 344
Murgia, S. M., 38, 368
Muriithi, N., 412
Murin, A. N., 500, 520, 590, 620
Murphy, C. N., 311
Murphy, G. J., 62
Murr, N. U., 48
Murray, B., 353
Murray, B. B., 337
Murray, J. R., 198
Murray, K. S., 45, 50, 323, 407, 408, 425
Murray, M., 86, 143, 144
Murray, M. M., 263
Murray, R. W., 98

Murray, S. G., 330
Murrell, L. L., 83
Murthy, A. S. N., 314
Musco, A., 56
Musgrave, T. R., 464
Musher, J. I., 64
Musker, W. K., 451
Musso, A., 501
Musso, H., 62
Mustacich, R., 89
Muthukrishnan, K., 107, 152
Muto, Y., 410
Myers, T. C., 144
Mysov, E. I., 485

Naberukhin, Yu. I., 201
Nackashi, J. A., 81
Nadzharyan, G. N., 620
Nagai, Y., 6, 132, 335
Nagashima, K., 104
Nagata, C., 150
Nagata, T., 561
Nagibarov, V. R., 99, 149
Nagorny, K., 519
Nagpal, V., 281
Naguyen, K. Z., 80
Nagy, J., 137
Nagy, M., 622
Nagy, S., 577
Nahm, F. C., 120, 255
Najam, A. A., 63, 127
Nakagawa, I., 227
Nakagawa, K., 378
Nakahara, M., 401
Nakamoto, K., 198, 212, 213, 216, 310, 342, 343, 353, 438
Nakamura, A., 25, 47, 52, 58, 321, 342, 372, 397, 424, 444
Nakamura, D., 167
Nakamura, K., 106
Nakamura, T., 232
Nakamura, Y., 107, 311, 400, 425, 619, 621
Nakano, T., 6
Nakao, R., 125
Nakashima, S., 229
Nakhodnova, A. P., 106
Naldini, L., 16
Nandwani, S. S., 620
Nanni, J., 301
Napoletano, T., 43, 335
Narasimhan, K. S. V. L., 621
Nardelli, M., 349, 408, 456
Nardin, G., 50
Naruse, M., 97
Naruto, M., 342
Nasielski, J., 97, 128
Nasledov, D. N., 512
Nassif, R., 620
Nassimbeni, L. R., 319
Nast, R., 45, 374
Nasta, M. A., 38, 278
Nasu, S., 620
Natkaniec, I., 238
Natusch, D. F. S., 89, 151
Naulet, N., 17, 97

Author Index

Naumov, A. D., 83
Nave, C., 399
Navech, J., 293
Navon, G., 49, 89, 98
Néel, J., 405
Neely, J. W., 75, 150
Neese, H.-J., 18, 316
Nefed'ev, A. V., 515
Nefedov, O. M., 353
Negishi, E., 113, 118, 120, 250
Negita, H., 162, 171, 589
Negrebetskii, V. V., 6, 149
Neilands, J. B., 80
Neimysheva, A. A., 152
Nekipelov, V. M., 91
Nelke, J. M., 132
Nelson, A. J., 132
Nelson, G. L., 2
Nelson, N. J., 81, 388
Nelson, S. M., 70, 379
Nemirovskaya, I. B., 28, 365
Nenashev, B. G., 244
Nenasheva, S. N., 244
Nesmeyanov, A. N., 9, 10, 25, 26, 37, 40, 62, 151, 168, 169, 364, 485, 545, 571, 619
Nesterov, V. I., 501, 618
Netterfield, R. P., 176
Netzel, D. A., 3
Neumann, F., 314
Neumann, G., 130, 131
Neumann, W. P., 130, 131, 132
Neuse, E. W., 41, 371
Neuss, G. R. H., 47, 339, 424, 490
Nevett, B. A., 139, 294
Neville, G. A., 98
Newell, I. W., 440
Newlands, M. J., 38, 464, 484
Newman, G. A., 260
Newman, J., 488
Newman, K. E., 74
Newman, P. W. G., 65
Newton, D. C., 197, 235
Newton, M. G., 304
Ng, G., 563
Nguyen, K. Z., 150
Nibler, J. W., 206, 209, 246
Nicholas, K. M., 47
Nicholls, D., 318
Nichols, A. L., 609
Nicholson, D., 619
Nickoloff, P., 35
Nicol, M., 241
Nicolini, M., 415
Nicpon, P. E., 350
Niculescu, V., 148
Niecke, E., 140, 289
Niedenzu, K., 119, 120, 142, 255, 263
Nifant'ev, E. Ye., 140
Nigam, H. L., 448
Nigen, A. M., 90
Nihonyanagi, M., 335
Nijssen, W. P. M., 349

Niki, E., 155, 576
Nikitina, S. A., 338
Nikitina, S. V., 508, 619
Nikitine, S., 621
Nikolaev, V. I., 500, 549
Nikolotova, Z. I., 359
Nikol'skii, G. S., 545
Nikonorova, L. K., 149
Ninistö, L., 612
Nishi, M., 605
Nishii, N., 84, 339, 427
Nishikawa, Y., 400
Nishimura, K., 434
Nitsche, N., 609
Nivellini, G. D., 355
Nixon, E. R., 204
Nixon, J. F., 8, 46, 47, 49, 51, 69, 421, 468
Nizamutdinov, V. M., 100
Noack, F., 105
Noble, P. N., 204
Noda, K., 61, 382
Nöth, H., 12, 21, 22, 80, 113, 118, 121, 253, 260, 286, 467
Nöthe, D., 28, 477
Noggle, J. H., 77, 109
Nogina, O. V., 10, 151, 168
Nolte, C. R., 39, 481, 484
Noltes, J. G., 16, 62, 73, 137, 141, 142, 285, 314, 414, 427
Nomura, T., 428
Nonoyama, K., 349
Nordquest, K., 320
Noren, I. B. E., 42
Norman, A. D., 115, 125, 269
Norman, J. G., jun., 52, 314
Norris, B. D., 617
Norris, C. L., 193
North, B., 69, 372
Northey, H. L., 102
Norton, J. R., 30, 96, 416, 417
Noshiro, M., 149
Nouet, J., 230
Novak, A., 198
Novak, D. P., 81
Novak, L., 227
Novakova, A. A., 547, 548
Novikov, S. S., 462
Novikov, N. V., 322
Novikova, V. N., 161
Novotny, M., 25, 412
Nowak, P., 78
Nowell, I. W., 28, 30, 475
Nowik, I., 613, 616, 620
Nowlin, T., 94, 421
Nozik, A. J., 511
Numata, M., 97, 122
Nunziata, G., 260
Nuretdinov, I. A., 149
Nuzhdina, Yu. A., 293
Nyburg, S. C., 137
Nyholm, R. S., 33, 47, 58, 68, 346, 373
Nyman, C. J., 52, 341
Nyquist, R. A., 197, 352
Nyussik, Ya. M., 620

Oates, G., 74, 148, 304
Obenshain, F. E., 601, 619
Obmoin, B. I., 152
Obozova, L. A., 402
Obradovič, M., 299
Ochiai, E., 59, 334
Ochiai, M., 98
O'Connor, C. H., 424
O'Connor, D. A., 620
O'Connor, T., 39, 366, 541
Odabashyan, G. V., 126
Odehnal, M., 424
Odent, G., 278
Odishaviya, M. A., 201
Odom, J. D., 6, 67, 112, 221, 286
O'Donnell, T. A., 330
Öfele, K., 22, 68, 363, 367
Oertel, R. P., 436
Ofer, S., 616
Offhaus, E., 28, 477
Ogata, R. T., 151
Ogawa, S., 153
Ogden, J. S., 207, 384, 393
Ogino, H., 429
Ogorodnikova, N. A., 27, 251
Ogoshi, H., 310
Ogura, T., 53, 61, 342, 345, 382
Ohkaku, N., 212
Ohkawa, K., 51
Ohneda, Y., 158
Ohnishi, M., 77
Ohno, T., 149
Ohtsuki, M.-A., 6
Ohwada, K., 231, 359
Ohya, T., 525
Oja, T., 157, 158
Ojima, H., 349
Oka, T., 176
Okada, K., 50, 93
Okada, J., 156
Okada, T., 87, 98
Okamoto, N., 28, 416
Okamoto, S., 546
Okamoto, S. I., 546
Okamoto, Y., 87
Okawa, H., 410, 461
Okawara, R., 21, 84, 87, 122, 132, 135, 139, 256, 257, 262, 278, 427, 384, 433, 454, 567
Okhlobystin, O. Yu., 62
Okinaka, H., 512
Okinoshima, H., 133
Okogun, J. I., 97
Okubo, N., 157
Okuda, H., 55
Okuda, T., 162, 171
Okulevich, P. O., 27
Okumoto, T., 51
Olah, G. A., 144, 145
Olander, W. K., 56, 491
Olapinski, H., 122, 256, 261
Oldfield, D., 433
Oleneva, G. I., 462
Olie, K., 211
Oliver, A. J., 339, 490
Oliver, B. G., 359

Author Index

Oliver, J. P., 62, 121
Olsen, C., 35
Olsen, D. N., 358
Olsen, F. P., 258
Olsen, J., 515
Olson, D. C., 400
Olsson, K., 97
Olthof, R., 48, 336
Omaly, J., 312
Omote, Y., 41
Onak, T., 7
Onaka, S., 589
Onderdelinden, A. L., 69, 336
Ondrejović, G., 404
O'Neill, S. R., 142, 287
Ono, K., 518, 525
Onodera, H., 546
Onomichi, M., 233
Onoue, H., 45, 54, 339, 346, 378, 409
Ooi, S., 400
Oosterhuis, W. T., 533, 535, 547
Opalenko, A. A., 595, 616
Oppermann, G., 99, 110
Oram, R. K., 140
Orchard, D. G., 416
Orchin, M., 55
O'Reilly, D. E., 103
Orel, B., 299
Ori, M., 82
Orio, A. A., 455
Orlando, A., 137
Orlova, L. V., 31, 370
Ortalli, I., 567
Ortega, J. D., 435
Orville-Thomas, W. J., 211
Osawa, A., 51
Osborn, J. A., 45
Osborne, A. G., 391
Osgood, R. M., 181
Osipenko, A. N., 153
Oskam, A., 70, 211
Ospici, A., 120
Osredkr, R., 157
Ostanevich, Yu. M., 617
Ostazewski, A. P. P., 250
Ostern, M. J., 396
Ostfeld, D., 93, 478
Ostrovskaya, L. M., 619
Otaguro, W., 239
Otsuka, S., 25, 47, 52, 58, 321, 342, 372, 397, 424, 444
Ouaki, R., 131, 277
Ouellette, R. J., 124, 125
Ouellette, T. J., 73, 284
Ouseph, P. J., 517
Ovcharenko, F. D., 103
Ovchinnikov, A. N., 508
Ovchinnikov, Yu. A., 77
Ovekhov, B. A., 244
Ovenall, D. W., 73
Overend, J., 204, 265
Overend, W. R., 50
Owen, J. D., 116
Owen, P. W., 125, 126, 267
Owens, W. R., 501
Øye, H. A., 263

Oza, C. K., 144
Ozhogin, V. I., 525
Ozin, G. A., 139, 205, 206, 227, 246, 281, 285, 287, 305, 342, 345, 393, 398

Pacansky, J., 204
Pace, E. L., 210
Pace, S., 616
Pachevskaya, V. M., 62
Pachler, K. G. R., 96
Paddock, N. L., 24, 140, 291, 322
Pae, A., 621
Pae, P., 620
Paez, E. A., 533
Page, J. E., 96
Paiaro, G., 55
Paine, R. T., 6, 83, 113, 248, 249
Pakhomova, D. V., 435, 442
Palazzi, A., 55, 377
Palazzotto, M. C., 91
Palenik, G. J., 39, 261, 366, 541
Paliani, G., 322, 376, 387
Palmer, D., 373
Palmer, G., 91, 536
Palmieri, C., 349
Pan, Y.-C. E., 74
Pande, I. M., 399
Pandey, A. V., 453
Pandey, N. K., 213
Pang, J. M., 124
Panigel, R., 49
Pankau, H., 109
Pankowski, M., 342, 422, 475
Pannetier, G., 236, 286, 338, 339, 374
Pant, B. C., 147, 270
Pantzer, R., 288, 292, 297
Panunzi, A., 54, 55
Panyushkin, V. N., 620
Papadimitriou, G., 620
Papko, S. I., 153
Papp, S., 414, 537
Pappas, A. J., 343
Papp-Molnar, E., 555
Paquette, L. A., 98
Parak, F., 151, 509, 525, 620
Parent, C. R., 187
Parfenov, S. V., 111
Parfenova, N. N., 571
Pargamin, L., 546
Parish, R. V., 47, 412, 413, 496, 575
Park, M. J., 108
Parker, D. A., 407
Parker, G. A., 132
Parker, J., 16
Parker, R. S., 103
Parker, W. G., 552
Parkin, C., 352
Parkins, A. W., 35
Parks, J. E., 91, 401
Parnell, D. L., 282
Parnes, Z. N., 27

Parr, W., 137
Parrett, F. W., 262
Parrott, J. C., 261, 442
Parry, R. W., 6, 83, 86, 113, 248, 259
Parshall, G. W., 43, 84, 335
Parshin, A. Ya., 502
Parsons, R. W., 176
Partenheimer, W., 61, 425
Partington, P., 126
Partis, R. A., 32, 414
Pascal, J.-L., 243
Pasdeloup, M., 4
Pasek, E. A., 90, 333, 450, 533
Pasini, A., 18, 339, 358, 384
Pasquier, B., 383
Passaret, M., 432
Pasternak, M., 495
Pasto, D. J., 118
Pastour, P., 147
Pastorek, R., 462
Pasynkiewicz, S., 121, 122
Patel, C. C., 59, 439, 444, 462
Patel, H. A., 264, 541
Patel, K. C., 201, 405, 455
Patel, M. G., 305
Pathak, C. M., 199
Patil, H. R. H., 423
Patmore, D. J., 421
Patoka, G. I., 547
Patrick, P. H., 95
Patrick, T. B., 95
Patterson, G. S., 43
Paugurt, A. P., 106
Paul, J., 384, 387
Paul, J., 97
Paul, K. K., 305
Paul, R. C., 281, 305
Paulmier, C., 147
Pauson, P. L., 22, 48, 85, 363
Pavia, A., 243
Pavlov, B. N., 155
Pavlov, D. V., 502
Pavlov, V. V., 301
Pavlynchenko, V. S., 244
Pavlyukhin, Yu. T., 520
Pawson, D., 332
Paxson, T. E., 117, 251
Payne, D. H., 347
Payne, D. S., 286
Pealo, E., 620
Pearson, E. F., 183
Pearson, P. S., 146, 304, 305
Pearson, R. G., 49
Pearson, R. M., 111
Pechkovskii, V. V., 320, 338
Pecsok, R. L., 111
Peddle, G. J. D., 134
Pedone, C., 54, 56
Peek, N. F., 602
Peercy, P. S., 231
Peerdeman, A. F., 142
Peet, W. G., 31, 330
Peguy, A., 75

Peleg, M., 247
Peleties, N., 147
Pelizzi, C., 349, 408
Pellacani, G. C., 343, 453, 463
Pellizer, G., 44
Pelzl, J., 499, 609
Penfold, B. R., 46, 169, 312
Penland, A. D., 113, 254
Penney, C. M., 199
Pensenstadler, D. F., 286
Pentin, Yu. A., 353
Pepin, C., 211, 216
Pereyre, M., 9
Pepperhoff, W., 618
Percy, G. C., 310, 404
Peregudov, A. S., 82
Pereira, A., 6
Pereyre, M., 125
Perie, J. J., 62
Perkampus, H.-H., 122
Perkas, M. D., 618
Perkins, A., 306
Perkins, H. K., 516
Perkins, P. G., 120, 257
Perlow, G. J., 608
Perov, P. A., 322
Perret, R., 443
Perrett, B. S., 98
Perrin, A., 312, 447
Perrin, C., 312, 447
Perrotti, E., 47, 373
Perry, R. A., 35
Pershikova, N. I., 344
Peruzzo, V., 433
Pesek, J. J., 111
Peshkov, V. P, 502, 552
Peterkin, M. E., 343
Petersen, O., 295
Peterson, E. M., 103
Peterson, G. E., 106, 152
Peterson, L. K., 129, 132, 286
Peterson, M. R., jun., 94
Peterson, R. W., 216
Peterson, R. V., jun., 130
Petit, M. A., 97
Petitjean, C., 611
Petrakis, L., 82
Petráš, P., 449
Petrosyan, V. S., 136
Petrosyants, S. P., 27
Petrov, A. A., 151, 622
Petrov, B. I., 136
Petrov, G., 17
Petrov, K. I., 220, 261, 294, 329, 330, 357
Petrov, M. I., 547, 617
Petrov, M. P., 106
Petrova, G. A., 172, 410
Petrovskii, P. V., 10, 21, 27, 40, 65, 77, 151
Pettit, R., 47
Petushkova, S. M., 357
Petz, W., 32, 261, 432, 482
Petzelt, J., 230
Peuker, C., 271
Peyre, J. J., 500
Peyronel, G., 349, 463

Peytavin, S., 303
Pfeffer, M., 343
Pfeffer, P. E., 97
Pfeifer, H., 99, 109, 110, 148, 150, 152
Pfeiffer, L., 498, 505
Pferrer, S., 193
Phan Dinh Kien, 359
Phillips, C. R., 114
Phillips, D. A., 51
Phillips, D. J., 462
Phillips, J. C., 62
Phillips, K. A., 330
Phillips, L., 6
Phillips, R. P., 37, 486
Piacenti, F., 486
Picard, J.-P., 127
Pick, M. R., 41
Pidcock, A., 53, 341, 347
Piekoszewski, J., 500, 618
Pietropaolo, R., 59, 339, 347, 465, 490
Piette, J. L., 146, 147
Pifat, G., 150
Pignatoro, S., 132
Pignolet, L. H., 69, 91, 450
Pijselman, J., 9
Pilbrow, M. F., 346
Pilipenko, A. T., 20, 439
Pilopovich, D., 146, 148, 219
Pince, R., 385
Pinchas, S., 260
Pine, A. S., 232
Pine, S. H., 87
Pines, A., 5, 103
Pinkerton, A. A., 49, 141, 296
Pinkerton, F. H., 127, 128
Pino, P., 55, 71
Pinter, T. G., 150
Piovesana, O., 448
Pipal, J. R., 5
Piraino, P., 339, 490
Piriou, B., 243
Pirkle, W. H., 97
Pisarev, K. N., 244
Pisciotta, A. V., 90
Pitner, T. P., 59
Pitt, C. G., 129
Pittman, C. U., jun., 22, 468
Pizzino, T., 53
Place, B. M., 499
Plachinda, A. S., 525
Plane, R. A., 354
Platnov, A. N., 620
Platt, R. H., 496, 577
Platzer, N., 98
Plautz, B., 321
Plautz, H., 321, 326
Plazzogna, G., 433
Plešek, J., 115, 118, 250, 251
Pleshkov, A. V., 211
Pletnex, R. N., 153
Plowman, R. A., 463
Plus, M., 247
Plyushchev, V. E., 294
Pochopien, D. J., 41

Podak, M. I., 150
Poddubnyi, I. Ya., 11, 84
Podlaha, J., 449
Podval'nykh, G. S., 549, 617
Poe, M., 79
Pogodilova, E. G., 354
Pohl, L., 97
Poilblanc, M., 385
Poilblanc, R., 21, 46, 69, 469, 485, 488, 489
Pokrovskii, B. I., 547
Poland, J. S., 123, 256, 383
Polansky, O. E., 35, 362, 367
Poldy, F., 99
Poletti, A., 322, 376, 387
Poller, R. C., 55, 346, 378, 378, 579, 582, 583
Pollman, T. G., 81
Pollock, R. J. I., 59, 346
Polovyanyuk, I. V., 9, 37
Polotebneva, N. A., 106
Polozova, I. P., 620
Pol'shin, E. V., 564
Polyakova, V. B., 253, 254
Pomashko, V. P., 502
Pomeroy, R. K., 68, 389
Pomposiello, M. C., 208
Ponticelli, G., 354, 403, 456
Poole, C. P., jun., 104, 494
Pooley, D., 177
Poon, C. K., 49
Pop, I., 148
Popkovich, G. A., 345
Poplavko, Yu. M., 230
Popov, A. I., 74, 246, 306
Popov, A. P., 439
Popov, F. I., 549
Popov, G. V., 571
Popov, Yu. L., 149
Popova, E. A., 242
Popowski, E., 277
Porte, A. L., 164
Porter, E. J., 297
Porter, L. C., 89
Porter, R. J., 37, 332
Porter, S. K., 98
Porto, S. P. S., 236, 239
Posch, H. A., 605, 619
Postgate, J. R., 536
Potakova, V. A., 549
Potasek, M. J., 598
Potemskaya, A. P., 150
Potier, J., 243, 299
Potts, R. A., 349
Potzel, W., 608
Poulet, H., 242
Poulsen, K. G., 620
Poupko, R., 75, 90
Povarennykh, A. S., 319
Povitskii, V. A., 499, 620
Povstyanyi, L. V., 228, 329
Powell, D. B., 292, 297, 332, 373, 382

Author Index

Powell, H. K. J., 406
Powell, J., 11, 56, 70
Powell, K. G., 44, 54
Powles, J. G., 148
Praat, A. P., 70
Prabhakaran, C. P., 439, 444
Pracht, H. J., 71, 83, 130, 137
Prado, J. C., 442
Prakash, H., 275
Prasad, H. S., 294, 443
Prasad, K. G., 556
Prasch, A., 43, 420
Prater, B. E., 334, 414, 538
Pratt, J. R., 127
Pratt, P. L., 510
Praud, J., 83
Preikschat, E., 505
Preiss, H., 299
Prejsa, M., 617
Prelesnik, A., 157
Prescher, G., 122
Pressl, K., 287
Preston, P. N., 136
Preston, R. S., 620
Preti, C., 21, 456
Pretsch, E., 77, 97
Preudhomme, J. M., 236, 278
Prevot, B., 228
Pribush, R. A., 458
Pribytkova, I. M., 58, 72
Price, D. C., 502
Price, E., 144
Price, J. H., 59, 444
Price, J. T., 427
Price, M. G., 428
Prigent, J. P., 312, 447
Principi, G., 500
Pringle, W. C., 227
Prisyazhnyi, V. D., 356
Proctor, D. M., 456
Proffitt, R. T., 9
Prokof'ev, A. K., 62
Prokopenko, V. M., 155
Pronin, I. S., 75
Pronkin, A. A., 590
Protsenko, G. P., 453
Protsenko, P. I., 453
Proulx, T. W., 95
Prout, C. K., 164
Prozorovskaya, Z. N., 105
Prusakov, V. E., 515
Prysyborowski, F., 152
Puchkovskaya, G. A., 172
Puddephatt, R. J., 54, 55, 60, 135, 136, 346, 348, 378
Pudkov, S. D., 525
Pujar, M. A., 14
Pujol, R., 293
Pukhnarevich, V. B., 125
Pulatov, M., 512
Pungor, E., 555
Punjar, M. A., 150
Pupp, M., 106
Purcell, K. F., 216
Puri, S. P., 620

Puşcaşin, M., 324
Pushkina, G. Ya., 438
Pustowka, A. J., 499, 546, 564
Pustowka, A. M., 544
Puxley, D. C., 574, 575
Pytaz, G., 242
Pytlewski, L. L., 311, 320, 357, 434, 441

Quagliano, J. V., 314, 400, 403
Quarterman, L. A., 306
Quastlerová-Hrastijova, M., 457
Quattrochi, A., 353
Quinby, M. S., 396
Quinn, H. A., 56
Quintard, J.-P., 125
Quirk, J. L., 60, 348
Quist, A. S., 210, 211, 230, 247
Qureshi, A. H., 529
Qureshi, A. M., 295

Raab, R. E., 102
Rabenstein, D. L., 3, 71, 79
Rabinovitz, M., 41, 121, 264
Rablen, D. P., 75
Rackham, D. M., 97
Radeglia, R., 10, 96, 98
Rademaker, W. J., 114, 252
Radhakrishna, S., 240
Radheshwar, P. V., 434
Raff, U., 616
Raftery, M. A., 42, 78
Raghu, S., 37
Ragulin, L. I., 152
Rahkamaa, E., 148
Rahman, M. M., 548
Rahman, S. M. F., 358, 405
Rai, A. K., 273
Raichart, D. W., 42, 334
Raj, D., 537, 620
Rajoria, D. S., 513
Rake, A. T., 588
Rakita, P. E., 125
Ramadan, N., 128
Ramadas, S. R., 35
Ramakers, J. E., 55
Ramakers-Blom, J. E., 53, 85, 381
Ramakrishna, J., 107, 152, 166
Ramana Rao, D. V., 422
Ramanujan, P. S., 242
Ramdas, A. K., 244
Ramey, K. C., 66
Ramon, F., 127
Ramos, V. B., 233
Rampone, R., 56
Ramsey, B. G., 118
Ramsey, N. F., 177
Ramshesh, V., 514
Ranade, A., 429
Randaccio, L., 50

Randall, E. W., 6, 11, 32, 94
Randall, G. L. P., 33, 35, 481
Randall, R. S., 281, 580
Raney, J. K., 146, 301
Ranganathan, T. N., 24, 291
Rankin, D. W. H., 13, 72, 73, 143, 267, 283, 287
Ranogajec-Komor, M., 525, 593
Rao, A. P., 407
Rao, C. N. R., 198, 513, 546
Rao, G. R., 513
Rao, G. V. S., 513
Rao, K. K., 536, 617
Rao, K. R. P. M., 620
Rao, N., 204
Rao, V. V. K., 312, 450
Rapp, B., 121, 268
Rasmussen, K., 337, 404
Rasmussen, P. G., 531
Rastogi, D. K., 311, 345, 435, 448, 458
Rastogi, R. P., 213
Ratajczak, H., 294
Ratcliff, B., 53, 341, 347, 384, 588
Ratcliffe, C. T., 301
Rath, J., 264
Rath, W., 499
Rathke, J., 119
Raudsepp, R., 528
Rausch, M. D., 41, 62, 383
Rautenstrauch, V., 127
Rayner-Canham, G. W., 44, 348, 397, 407
Raynes, W. T., 10
Razavi, A., 416
Razumovskii, V. V., 491
Reader, G. W., 50
Realo, E., 500, 621
Rebsch, M., 302
Rechenberg, H., 617
Reddi, S. P., 199
Reddy, G. K. N., 151
Reddy, G. S., 6
Reddy, Y. S., 513
Redfield, A. G., 79, 151
Redhouse, A. D., 488
Redwood, M. E., 139
Reed, C. A., 52, 418
Reed, F. J. S., 58, 59, 404, 452
Reed, G. H., 80, 152
Reed, P. J., 16
Reed, P. R., 255
Reed, T. E., 468
Reedijk, J., 407, 434, 621
Reeks, M. W., 620
Rees, C. W., 35
Rees, L. V. C., 512
Reessing, F., 110
Reetz, T., 120
Reeves, L. W., 64, 75
Reeves, P. C., 35
Reffy, J., 137

Reger, A., 205, 248, 312
Reger, D. L., 32
Regen, S. L., 107
Regis, A., 399
Regler, D., 31, 32, 479
Regnard, J. R., 511
Rehder, D., 14, 19, 385
Rehorek, D., 349
Reich, S., 507
Reichart, U. V., 194
Reichert, B. E., 56
Reichert, J. F., 619
Reichman, S., 218
Reich-Rohrwig, P., 450
Reiff, W. M., 527, 539, 621
Reikhsfel'd, V. O., 125, 152
Reilley, C. N., 94
Reilly, T. J., 113
Reiman, S., 500
Reĭmann, B., 109
Reimann, R. H., 476
Reimer, K. J., 449
Reimer, M. M., 449, 619
Reinheimer, H., 56
Reinhoudt, D. N., 381
Rejhon, J., 276
Relf, J., 61, 457
Remizov, V. G., 444
Rémy, P., 293
Rengaraju, S., 96
Renk, I. W., 24
Rennie, W. J., 126
Reno, R. C., 505, 617
Renoe, B. W., 52, 341
Renson, M., 73, 146, 147
Renwanz, E., 180
Renz, W., 298
Repelin, Y., 231
Repetskii, S. P., 499, 619
Reshamwala, A. S., 166
Resing, H. A., 99, 108
Rest, A. J., 397
Retcofsky, H. L., 13
Rettenmaier, A. J., 471
Rettig, M. F., 85
Reuben, J., 151
Reuter, B., 319
Reutov, O. A., 54, 136
Reuvers, A. J. M., 98
Revitt, D. M., 12, 291
Revokatov, O. P., 111, 152
Rewicki, D., 94
Reynolds, D. J., 227, 251
Reynolds, W. F., 68
Reynoldson, T., 366
Rez, I. S., 172
Reznicek, D. L., 113, 258
Rhee, C., 108, 151
Rhee, S. G., 86, 121, 254
Rheingold, A. L., 21, 42, 416, 540
Rhiger, D. R., 506
Rhim, W.-K., 5, 103, 104
Ribeird Gibran, M. L., 357
Ricci, A., 132
Rice, D. A., 20, 318
Richard, Y., 580
Richards, J. H., 38, 42
Richards, P. L., 89, 323

Richards, R. L., 27, 55, 377, 395
Richardson, M. F., 94
Richelme, S., 137
Richter, R. F., 62
Rick, E. A., 135
Ricks, M. J., 207
Ridley, D. R., 142, 285, 434
Riedel, K. H., 6
Riedl, M. J., 59
Riera, V., 330
Rieskamp, H., 320
Riesner, D., 63
Riess, J. G., 87, 141
Rietz, R. R., 12, 115
Rigamonti, A., 497
Rigatti, G., 67
Rigo, P., 413, 415
Riley, D. L., 619
Riley, P. N. K., 59, 346
Rillema, D. P., 50, 79
Rimai, L., 307
Rinaldi, S., 499
Ring, M. A., 125, 269
Ringel, C., 329
Ripan, R., 324
Ripmeester, J. A., 102, 149
Rippon, D. M., 211
Risen, W. M., jun., 246, 314, 326, 389
Ritter, G., 400, 464, 524, 530, 531
Ritter, J. J., 119, 254
Rivarola, E., 134, 142
Rivière, P., 125, 137
Rix, C. J., 467
Rizvi, S. Q. A., 10, 132, 136
Roark, D. N., 134
Robert, D. U., 141
Robert, R., 202
Roberto, F. Q., 306
Roberts, B. P., 118
Roberts, J. A., 176
Roberts, J. D., 10, 17
Roberts, J. H., 282
Roberts, J. R., 137
Roberts, N. K., 102
Roberts, R. M. G., 26, 126
Robertson, B., 557
Robertson, G. B., 40, 52, 69, 422
Robinette, R., 500, 600
Robinson, B. H., 46, 69, 390
Robinson, E. A., 163
Robinson, S. D., 31, 69, 398, 417, 450
Robinson, W. R., 314, 472, 488
Robinson, W. T., 406, 417
Robson, R., 461
Rocchiccioli-Deltcheff, C., 231
Roche, T. S., 50
Rochon, F. D., 55, 342, 382
Rock, S. L., 193

Rockett, B. W., 5, 41, 433, 542
Rockstroh, C., 25, 122, 257, 405, 471
Rodeheaver, G. T., 34
Rodgers, J., 59
Rodionov, A. N., 269
Rodionov, Yu. L., 618
Rodriguez, E., 131
Rodríguez, V. M., 97
Roe, D. M., 54
Röschenthaler, G. V., 479
Roesky, H. W., 119, 130, 143, 144, 289, 290, 295, 296, 297
Röttelle, H., 480
Rogarski, R., 380
Rogers, D. B., 508
Rogers, K. A., 532
Rogers, M. T., 166, 237
Rogerson, M. J., 90, 459
Rogowski, R., 54
Rogozev, B. I., 622
Rogstad, A., 297
Rohatgi, K. K., 435
Rohbock, K., 318
Rohmer, H. E., 442
Roi, N. I., 202
Rokhlina, E. M., 62, 63
Roland, A., 144
Roland, G., 293
Rolfe, N., 318
Rollmann, L. D., 46, 489
Roman, S., 618
Romanchev, B. P., 564
Romanenko, E. D., 78
Romano, V., 53
Romanov, V. P., 544, 546, 549
Romashkin, I. V., 126
Romiti, P., 478
Ron, M., 508
Ronayne, J., 5
Rondeau, R. E., 97
Roobeek, C. F., 97
Roof, A. A. M., 97
Rooney, J. J., 56
Root, C. A., 93, 434
Roper, W. R., 30, 40, 52, 417, 418, 423, 486, 487
Roques, G., 405
Rork, E. W., 612
Ros, R., 54
Rose, H., 73
Rose, N. J., 50, 408
Rose, P. D., 65
Roseberry, T., 12
Rose-Munch, F., 379
Rosen, A., 120, 255
Rosenberg, E., 11, 32
Rosenberg, M., 548
Rosenberg, R. E., 98
Rosenblatt, G. M., 207
Rosenblum, A., 125
Rosenblum, M., 35, 37, 69, 372, 483
Rosenbuch, P., 40, 482
Rosenkranz, A., 400, 530, 531
Rosenthal, L., 92

Author Index

Roshchupkina, O. S., 316
Rosner, P., 501
Rosolovskii, V. Ya., 261, 292
Ross, D. S., 147, 303
Ross, E. J. F., 196
Ross, M., 563
Ross, S. D., 196, 211, 260, 277, 319
Rossetti, R., 333, 483
Rossi, I., 199
Rossi, M., 331, 396, 428, 485
Rossman, G. R., 333, 425, 528
Rossotti, F. J. C., 149
Roth, K., 9
Roth, R. W., 126, 300
Roth, W. L., 101
Rothbart, H. L., 97
Rothergy, E. F., 120, 255
Roulet, R., 59
Roundhill, D. M., 43, 52, 335, 341
Rouschias, G., 51, 415
Rousseau, M., 230
Roussel, J., 62
Roustan, J. L., 24, 471
Rouxel, J., 282
Rowan, N. S., 429
Rowbottom, J. F., 329
Rowley, R. J., 136
Rowling, P. V., 41
Roy, R., 155
Royo, G. L., 125, 132
Rozanska, H., 232
Rozen, A. M., 359
Rozhdestvenskaya, M. V., 230
Rozhkova, E. V., 382
Rozière, J., 299
Rubin, I. D., 21
Rubinson, K. A., 536
Rubinstein, M., 159
Rubtsov, V. P., 230
Ruby, S. L., 517
Ruddick, J. D., 40
Ruddick, J. N. R., 579, 582
Rudner, B., 280
Rudnitskaya, E. S., 111
Rudolph, R. W., 71, 145, 284
Rüegsegger, P., 517
Rüegsegger, R., 502
Rühlmann, K., 131
Rueterjans, H., 79
Ruff, J. K., 21, 39, 420, 468
Ruitenberg, G., 176
Ruiz-Ramirez, L., 90, 334
Rulmont, A., 228, 231
Rumov, N. N., 436
Ruoff, A., 208
Rupp, H. H., 413
Ruppersberg, H., 508
Ruppert, I., 130
Rupprecht, A., 100
Rush, J. J., 203
Rusnak, L. L., 75, 93
Russ, C. R., 86

Russell, C. A., 98
Russell, D. R., 45, 306
Russell, J. D., 260
Russell, J. W., 35, 265, 388, 483
Rustamov, A. G., 319
Ruth, R. D., 499
Rutt, K. J., 339, 403, 460
Rutt, M. J., 460
Ryadchenko, A. G., 338
Ryan, R. R., 100
Rybakov, Yu. Ya., 620
Rybal'chenko, L. F., 618
Ryan, A. A., 49
Ryan, F. J., 299
Ryan, R. R., 221
Rycroft, D. S., 13
Rys, E. G., 115, 152
Rza-Zade, P. F., 260

Saalfeld, F. E., 38, 278
Sabherwal, I. H., 47
Sabine, R., 139, 285
Sable, H. Z., 79
Saburi, M., 50
Sacco, A., 331, 396, 485
Sacconi, L., 71, 343
Sacher, R. E., 270
Sadovskaya, A. O., 595
Sadykov, A. S., 150
Sadykov, E. K., 617
Säterkvist, C., 102
Sagan, L. S., 140, 291
Sagara, F., 426
Sagdeev, R. Z., 149
Saha, A. K., 155
Saha, H. K., 323
Sahatjian, R. A., 21
Sahni, V. C., 237
Sahoo, B., 359
Saidashev, I. I., 141, 153
Saika, A., 55
Saikin, K. S., 149
Saillant, R. B., 3, 455
St. P. Bunbury, D., 526
St. Denis, J., 62
St. Pyrek, J., 98
Saito, A., 97
Saito, J., 150
Saito, K., 49
Saito, N., 570
Saito, S., 182, 193
Saito, Y., 62, 310, 343, 353
Saji, H., 107
Sakaguchi, U., 93, 96
Sakai, H., 598
Sakai, M., 56
Sakai, S., 54, 56, 377, 379
Sakai, T., 85
Sakakibara, T., 122
Sakash, G. S., 333
Sakellardis, P. U., 166, 167
Sakrikar, S., 45
Sakurai, H., 46, 125, 129, 372
Sala-Pala, J., 440
Salaün, J. R., 97
Salerno, J. G., 49
Sales, K. D., 94
Salienko, S. I., 394

Salih, Z. S., 128
Salmeen, I., 91
Salomon, D., 602, 603
Salomon, M. F., 62
Salomon, R. G., 60
Salov, A. V., 161
Salvador Salvador, P., 151
Salvadori, P., 55
Salvetti, F., 334, 389
Salvi, A., 163
Salwin, A. E., 338
Salzer, A., 58, 376
Sam, D. J., 75
Samedov, F. R., 260
Samitov, Yu. Yu., 148
Samplavskaya, K. K., 304
Sams, J. R., 273, 281, 445, 578, 580, 581, 588
Samuel, E., 62, 383
Sanders, J. R., 31, 330
Sanders, J. K. M., 94, 95
Sandhu, G. K., 440
Sandhu, R. S., 439, 441
Sandhu, S. S., 439, 440, 441
San Filippo, J., jun., 26
Sanger, A. R., 19, 25, 448
Sano, H., 494, 496, 589
Santucci, A., 387
Sanyal, N. K., 359
Sappa, E., 30, 486
Saran, H., 130, 295
Saran, M. S., 25, 28, 35, 413, 471
Saraswati, V., 107
Sarel, S., 35
Sargeson, A. M., 49, 85
Sarma, L. H., 239
Sarnatskii, V. M., 152
Sarneski, J. E., 68, 407
Sartori, P., 141, 293
Saruyama, H., 307
Sas, T. M., 314, 315
Sasaki, T., 620
Sasaki, Y., 101
Sasane, A., 167
Sastri, V. S., 428
Sastry, M. D., 511, 522
Satapathy, K. C., 359
Satgé, J., 125, 129, 132, 137, 273
Sathianandan, K., 210
Sato, A., 131
Sato, F., 58, 82, 376, 451, 465
Sato, K., 429
Sato, M., 58, 82, 376, 451, 465
Sato, S., 139
Sato, S.-I., 132, 433
Sato, T., 75, 261, 441
Satoh, M., 149
Satrawaka, V., 439
Sauer, D. T., 144, 295
Sauer, E. G., 158
Saunders, J. E., 221
Sautrey, D., 370
Sauvage, J. P., 75
Savage, R. O., 554
Savariault, J.-M., 59
Savcchenko, L. T., 105

Savchenkova, A. P., 253
Savenko, N. F., 105
Savin, E. P., 598, 600
Savoie, R., 210
Savory, C. G., 114, 252
Savushkina, V. I., 127
Sawatzky, G. A., 548
Sawbridge, A. C., 10, 589
Sawicka, B. D., 499, 564
Sawicki, J. A., 499, 546, 564, 608, 621
Sawodny, W., 323
Sawyer, D. T., 77
Sawyer, J. O., 357
Sbrana, G., 334, 389
Sbrignadello, G., 390
Scaiano, J. C., 118
Scaife, D. E., 169, 170, 312
Scantlin, W. M., 125
Scaramuzza, L., 403
Scarle, R. D., 323
Schaal, M., 412
Schaal, R., 147
Schaap, H., 461
Schacht, E., 146
Schack, C. J., 146, 147, 148, 209, 219, 224, 305
Schäfer, G., 602
Schäfer, H., 131, 321, 326
Schaefer, H. F., 200
Schaefer, J. P., 62
Schäfer, L., 363, 370, 376, 384
Schaefer, M. A., 327
Schaefer, T., 98
Schaeffer, R., 12, 71, 115, 119
Schaible, B., 122, 255, 261
Schaller, H., 590
Schaper, W., 289
Schardt, K., 137, 465
Scharpen, L. H., 175
Schat, G., 17
Schatz, J., 218
Schauer, W., 180
Schaumann, E., 73
Scheffold, R., 62, 93, 353
Schejter, A., 89
Schenk, M., 99, 495
Scherer, O. J., 72, 140, 295
Scherr, P. A., 62
Schetty, G., 50
Schexnayder, D. A., 11, 388
Schibuya, N., 605, 621
Schiemenz, G. P., 75, 98
Schiller, H. W., 145, 284
Schimitscher, E. J., 357
Schindler, A. I., 601
Schirawski, G., 86
Schirmer, W., 109, 152
Schläpfer, C. W., 213, 216, 342, 343
Schlientz, W. J., 21, 468
Schlodder, R., 391
Schlögl, K., 97
Schmid, A., 474
Schmid, E. R., 439
Schmid, G., 32, 46, 261, 390, 432, 482, 488
Schmid, K. H., 297

Schmidbaur, H., 60, 61, 71, 126, 127, 129, 131, 139, 173, 245, 272, 274, 284, 348, 349
Schmidpeter, A., 140, 289
Schmidt, A., 265, 287
Schmidt, E., 370
Schmidt, H., 325
Schmidt, J., 14
Schmidt, K., 619
Schmidt, K. H., 214, 216, 312, 349
Schmidt, M., 121, 259
Schmidt, P., 47, 275
Schmidt, V. H., 103
Schmidt-Sudhoff, G., 80
Schmiedel, H., 108, 109
Schmolz, A., 105
Schmulbach, C. D., 75, 317
Schmutzler, R., 143, 144
Schnee, W.-D., 241
Schneider, E. E., 597
Schneider, F. M., 318
Schneider, H. J., 131
Schneider, J., 229
Schneider, M. L., 492
Schneuwly, H., 611
Schnitzler, M., 477
Schoenberg, A. R., 172, 387, 477
Schönwasser, R., 177
Scholz, H.-U., 129
Schott, G., 277, 433
Schram, E. P., 58, 86, 122, 254, 282, 422
Schraml, J., 126
Schramm, R. F., 59, 444
Schreiner, A. F., 417
Schrobilgen, G. J., 80, 121, 148, 264, 308
Schroeder, B., 132
Schröder, F. A., 302, 323
Schröder, G., 35
Schröder, H. H. J., 121, 259
Schroeder, L. W., 203
Schröder, W. U., 611
Schroeer, D., 610
Schroth, G., 3
Schué, F., 16, 291
Schürmann, K., 563
Schugar, H. J., 333, 425, 528
Schultz, P., 111
Schulz, R. C., 98
Schulze, H., 214, 325
Schumann, H., 468
Schumann, H.-D., 285
Schumann, W., 320
Schurig, V., 95
Schuster, R. E., 71, 97, 120
Schutte, C. J. H., 237
Schwab, C., 621
Schwab, W., 35
Schwartz, L. D., 113, 249
Schwartz, M. A., 54
Schwartz, P., 126
Schwarz, R., 176
Schwarzhans, K. E., 335
Schweiger, J. R., 142, 283

Schwendemann, R. H., 185, 190
Schwendiman, D., 97
Schwenk, G., 297
Schwenzer, G., 471
Schwering, H. U., 255
Schwirten, K., 61, 173
Schwochau, K., 226
Scibelli, J. V., 269
Sciesinki, J., 242
Sciesinska, E., 242
Scollary, G. R., 335, 487, 489
Scott, A. I., 50
Scott, B. A., 319
Scott, J. C., 588
Scott, J. G. V., 292, 382
Scott, K. L., 428
Scott, P. P., 377
Scott, R. P., 55
Scott, T. A., 156
Scovell, W. M., 457
Scozzafava, A., 90
Scribelli, J. V., 125
Scrocco, M., 199
Sealy, J. M., 131
Searle, G. H., 50, 401
Sebenne, C., 232
Secco, E. A., 201
Secroun, C., 200
Seddon, D., 25, 26, 416
Seddon, K. R., 318
Sedgwick, D. F., 507
Seebach, D., 147
Seeger, R., 542
Seel, F., 80, 141, 297, 479
Seelig, J., 95
Seematter, D. J., 51
Seff, K., 94
Segal, J. A., 38, 68, 419
Segel, S. L., 156
Segnan, R., 601
Sehgal, H. K., 233
Seibold, C. D., 394
Seidel, W. C., 55
Seidl, V., 593
Seifert, H. J., 321, 329
Seitz, L. M., 65
Sekizawa, H., 546
Selbin, J., 96, 320
Seleznev, V. N., 547
Seleznev, V. P., 359
Selig, H., 7, 225
Seliger, J., 157
Selisskii, Ya. P., 618
Sellers, M., 620
Sellmann, D., 28, 394, 395
Semenkin, V. A., 621
Semenenko, K. N., 253, 254
Semenova, E. D., 617
Semenova, G. S., 317, 442
Semin, G. K., 155, 158, 160, 161, 163, 169, 171
Sen, B., 122, 406, 423
Sen, B. K., 328
Sen, D., 260
Seng, N., 141
Sengupta, A., 405
Sengupta, S., 155

Author Index

Sen Gupta, S. K., 435
Senkler, G. H., jun., 72
Senkov, P. E., 551
Senoff, C. V., 44, 70, 101, 341
Senyukova, G. A., 91
Seppelt, K., 144, 147, 225, 302, 303, 308
Septe, B., 45, 424
Serbulenko, M. G., 244
Serebrennikov, V. V., 424, 432, 435, 442
Seregin, P. P., 512, 572, 575, 590, 598, 600, 620, 621
Serfozo, G., 108
Sergeev, N. M., 149
Sergeeva, A. N., 327, 453, 457
Sergeyev, N. M., 7, 14, 15, 72, 83, 135, 140
Sergi, S., 59, 347, 465
Serova, S. A., 345
Serpone, N., 318
Serra, O. A., 357, 442
Serres, B., 125, 135
Servé, M. P., 97
Serve, R. R., 98
Servis, K. L., 80
Setkina, V. N., 21, 27, 28, 363, 365
Seux, R., 97
Sevost'yanova, N. I., 431
Seyferth, D., 62, 63, 126, 128, 132, 383, 488
Seymour, E. F. W., 156
Seymour, S. J., 15
Sgarabotto, P., 456
Shafromskii, V. N., 402
Shah, D. P., 26, 473
Shahab, Y., 97
Shakhnazaryan, A. A., 333
Shamir, J., 148, 209, 260, 305, 306
Shamov, A. I., 500
Shamov, A. M., 502
Shankar, J., 514
Shapiro, B. L., 69, 94, 95
Shapiro, M. J., 95
Shapiro, M. L., 95
Sharma, A. K., 502
Sharma, B. C., 59, 462
Sharma, C. L., 407
Sharma, H. D., 109
Sharma, H. N. K., 497, 502, 508
Sharma, K. K., 281
Sharma, R. K., 123
Sharma, R. P., 545
Sharma, R. R., 502, 550
Sharma, S. K., 246
Sharon, T. E., 621
Sharov, V. A., 432
Sharp, D. W. A., 25, 27, 37, 147, 303, 318, 335, 445
Sharp, K. G., 27, 137, 192, 272
Sharp, R. R., 15
Sharrock, M., 515

Sharrocks, D. N., 24, 472
Shatskii, V. M., 315
Shaw, B. L., 7, 8, 9, 11, 12, 24, 26, 53, 321, 335, 339, 347, 490
Shaw, G., 30, 32, 40, 440, 485, 493
Shaw, H., 49
Shaw, J. H., 200
Shaw, R. A., 164
Shaver, A., 65, 395, 472
Shchegrov, L. N., 338
Shchepkin, V. D., 100
Shcherba, L. D., 303
Shcherbakov, V. N., 110
Shchteinshneider, A. Y. A., 7
Shchukarev, S. A., 201
Shchwerer, F. C., 561
Sheard, B., 152
Shechter, H., 501, 502
Sheka, I. A., 105, 442
Sheldrick, G. M., 297
Sheline, R. K., 106, 386
Shelton, G., 50
Shemyakin, V. N., 401, 402
Shemyakov, A. A., 150
Shenoy, G. K., 612, 613, 618
Shepelev, N. P., 315
Shepherd, I. W., 260
Shepherd, R. E., 45
Shepherd, T. M., 96
Sheppard, N., 75, 382
Sheppard, R., 134
Sheppard, W. A., 73, 146
Sheridan, J., 190
Sherry, A. D., 78
Sherwood, P. M. A., 196
Shevchenko, I. V., 266
Shevchenko, L. L., 20, 439
Shevchenko, Yu. A., 621
Shibata, M., 51
Shiels, T., 21, 25, 386, 469
Shiga, M., 621
Shilkin, S. P., 254
Shilov, A. E., 394, 396
Shilova, A. K., 394, 396
Shimada, M., 553, 621
Shimanouchi, T., 198
Shimizu, F., 185
Shimizu, T., 175
Shimizu, Y., 150
Shimoda, K., 175
Shimony, U., 501
Shimp, L. A., 19, 65, 248, 309
Shimura, Y., 51
Shinjo, T., 621
Shinnik, G. N., 106
Shinno, I., 564
Shinohara, M., 518
Shinra, K., 42, 406, 408, 418
Shiotani, A., 61
Shipatov, V. T., 575, 590, 620, 621
Shipko, M. N., 549, 556, 617
Shirk, J. S., 208

Shirokova, G. N., 261, 292
Shirvinskaya, A. A., 278
Shishkin, E. M., 155
Shishkin, V. A., 155, 162, 172
Shlyapochnikov, V. A., 462
Shmakova, Z. L., 422
Shöllkopf, U., 129
Shoemaker, R. L., 191
Shoji, H., 570
Shok, M., 97
Shore, L., 109
Shore, S. G., 115, 120, 249, 260
Shortland, A., 7, 19, 26, 271, 362
Shoup, R. R., 11, 64
Shporer, M., 90, 103
Shreeve, J. M., 73, 135, 139, 142, 144, 284, 287, 295, 296, 433, 581
Shriver, D. F., 81, 385, 388, 405
Shtan, I. I., 616
Shtrikman, S., 566
Shubina, T. S., 618
Shubochkin, L. K., 438
Shubochkina, E. F., 438
Shuhler, T. A., 103
Shukla, P. R., 401
Shulepov, Yu. V., 108
Shulman, R. G., 152, 153
Shupik, A. N., 91
Shustorovich, E. M., 26
Shutilov, V. A., 152
Shvagerov, V. D., 564, 565
Shvo, Y., 35, 369
Sick, H., 79
Siddiqi, I. W., 211
Siddiqui, Z. U., 22, 469
Sidorenko, F. A., 618
Sidorov, N. K., 280
Siebert, H., 304, 337, 339, 413, 429
Siebert, W., 120
Siedle, A. R., 114, 116
Siefer, G. B., 331
Siegl, W. O., 479
Siekierska, K. E., 515
Sievers, R. E., 50, 94, 459
Sievert, W., 325
Silberman, E., 292
Silhan, W., 97
Silk, C., 177
Sillescu, H., 88
Silva, D. M., 516
Silver, J., 574
Silverman, R. B., 45
Silverthorn, W., 538
Silvestri, A., 131, 458
Silvestro, L., 59, 347, 465
Silvidi, A. A., 77
Simanova, S. A., 452
Simmons, H. E., 75
Simms, P. G., 47, 412, 413
Simon, A., 285
Simon, L., 496
Simon, W., 77, 97
Simonetti, F., 47, 373

Simonnin, M.-P., 73, 138, 140
Simonov, Yu. A., 338
Simons, W. W., 2
Simopoulos, A., 495
Simpson, P., 433
Simpson, J. B., 182
Sinclair, J., 90, 334
Sinden, A. W., 130, 294, 406
Sinegribova, O. A., 317
Singer, M. I. C., 138
Singh, A., 273
Singh, B., 354, 463
Singh, E. B., 343
Singh, G., 6
Singh, G. P., 401
Singh, J., 282
Singh, J. J., 501
Singh, R., 354, 463
Singh, R. A., 343
Singh, R. D., 227
Singh, R. P., 399
Singh, S. K., 564
Singleton, E., 30, 51, 331, 424, 476
Singru, R. M., 538, 546
Sinha, S. P., 151
Sink, C. W., 139, 294, 300
Sinn, E., 43, 65, 409, 423
Sinnema, A., 98
Sinyavskii, V. I., 617
Sipe, J. P., 94
Siracusa, G., 434
Sirmokadam, N. N., 276
Sironen, R. J., 360
Sisido, K., 97, 130
Sisler, H. H., 139, 140, 143, 275, 289, 291, 293
Sisson, D. H., 426
Sivenkov, E. S., 622
Sjöblom, R., 102
Skapski, A. C., 26
Skell, P. S., 35, 125, 126, 267
Skibida, I. I., 91
Sklyarevskii, V. V., 498, 501, 505
Skolozdra, O. E., 327
Skorsepa, J., 344
Škramovská, J., 47
Skvortsov, N. K., 151
Skvortsova, G. G., 89
Sklyarevskii, V. V., 611
Skyrme, G., 620
Sladky, F., 147, 303
Slager, T. L., 349
Slak, J., 111
Slater, J. L., 106
Sledz, J., 16
Sleezer, P. D., 61
Sleight, A. W., 553
Slichter, C. P., 506
Sloane, H. J., 198
Slocum, D. W., 40
Slonim, I. Ya., 2
Slynsarenko, K. F., 410
Smardzewski, R. R., 206, 207
Smart, J. B., 62

Smentowski, F. J., 94
Smetannikova, Yu. S., 512
Smid, J., 64, 95
Smidsrød, O., 98
Smirnov, G. V., 498, 505, 616
Smirnov, S. K., 414
Smirnova, E. M., 344
Smith, A., 98
Smith, A. J., 10, 136
Smith, A. K., 40
Smith, A. W., 532
Smith, B. E., 358, 536
Smith, C. F., 125
Smith, C. W., 97
Smith, D. F., 204
Smith, D. L., 120, 262, 460
Smith, F. E., 436, 530
Smith, G. J., 71
Smith, G. V., 95
Smith, H. D., jun., 121
Smith, J. A. S., 104, 155
Smith, J. D., 122, 259
Smith, J. E., 185
Smith, J. G., 59, 182, 185, 186, 452
Smith, J. N., 319, 448, 454
Smith, L., 14, 579, 584
Smith, M. A. R., 85
Smith, M. R., jun., 125
Smith, P. J., 14, 137, 278, 579, 584
Smith, R. G., 64
Smith, R. V., 536
Smith, S. A., 44, 335, 430
Smith, S. E., 94
Smith, T. D., 264
Smith, V. B., 364
Smith, W. B., 97
Smith, W. H., 218
Smolenskii, G. A., 106
Snaith, R., 120, 129, 257, 275
Sneddon, L. G., 31, 71, 115 252
Snelson, A., 203
Snider, D. E., 200
Snider, R. F., 150
Snow, M. R., 466
So, S. P., 222
Sobata, T., 53, 342
Socrates, G., 52, 341
Sofronova, A. V., 248
Soga, T., 359
Sohn, Y. S., 41, 47, 314, 371, 435
Sokolov, Yu. A., 547
Sokolov, V. I., 54
Sokolov, V. N., 11
Sokol'skii, D. V., 447, 621
Solan, D., 125, 267
Solladié-Cavallo, A., 98
Solladié, G., 98
Šolmajer, T., 299
Solntseva, L. S., 333
Solodar, A. J., 412
Solomon, I. J., 146, 301
Solomon, R. G., 382
Solomon, T. W., 81

Solov'ev, E. E., 150
Solovieva, L. I., 54
Sommer, L. H., 132, 137
Sommer, P., 130
Sommer, U., 108, 301
Songstad, J., 204
Sonoda, N., 146
Sonogashira, K., 339, 380
Soots, V., 229, 243
Sopkova, A., 344
Sorotkin, E. I., 169
Sorokina, L. D., 438
Sorokina, T. A., 54
Soue, K., 403, 404
Soulard, M. H., 276
Sourisseau, C., 383
Soustelle, M., 242
Southern, J. T., 592
Sovocool, G. W., 42
Sowerby, D. B., 12, 144, 291
Soyfer, J. C., 152, 353
Spacu, P., 530
Spagnolo, F., 317
Spanggord, R. J., 62
Spanjaard, D., 151
Sparasci, A. M., 277
Spaulding, L., 55
Speca, A. N., 357, 434, 441
Speight, J. G., 198
Speight, P. A., 157
Spell, H. L., 295
Spencer, A., 43, 51
Spencer, C. T., 403
Spencer, J. L., 116, 252
Spender, M. R., 565
Spendjian, H. K., 330, 476
Speroni, G. P., 71
Speth, J., 611
Spielvogel, B. F., 115, 249
Spiering, H. I., 530, 590, 605
Spiess, H. W., 15, 106
Spiker, R. C., 204, 207
Spindler, H., 349
Spinney, H. G., 14
Spitsyn, V. I., 106, 329, 410, 432, 438
Spoliti, M., 200, 350
Springer, C. S., jun., 94
Šramko, T., 456
Srivastava, A. K., 311, 458
Srivastava, B. N., 506
Srivastava, B. P., 497, 502, 508
Srivastava, J. K., 545, 556
Srivastava, K. P., 424, 452, 454
Srivastava, K. R., 426
Srivastava, R. C., 17, 227, 401
Srivastava, S. P., 227
Srivastava, T. N., 257, 398
Srivastava, T. S., 419, 487
Staab, H. A., 482
Staats, P. A., 201, 205
Stacey, L. M., 104
Stach, H., 109, 152
Stainbank, R. E., 9, 11, 44, 335

Stalick, J. K., 452, 574
Stal'nakhova, L. S., 280
Stamper, P. J., 384, 387
Stanek, T., 238
Stănescu, D., 324
Stanger, C. W., 172, 387
Stanghellini, P. L., 483
Stanke, F., 509
Stanko, V. I., 171
Stapfer, C. H., 399, 445, 577, 584
Stark, Yu. S., 151
Stary, H., 140, 289
Starzewski, K. A. O., 468
Staudacher, F., 21
Stavinin, K. V., 270
Stearns, R. W., 246
Steck, W., 140, 144
Stedronsky, E. R., 41
Steele, D. F., 46, 391
Steenbeckeliers, G., 182, 201
Steenbergen, Chr., 104
Stefani, A., 71
Steger, E., 198, 243
Steiger, H., 335
Stein, H., 243
Steiner, E., 50
Steiner, P., 567, 609
Steinfink, H., 621
Steinkilberg, W., 14
Stejskal, J., 293
Stempfle, W., 3
Stendel, R., 302
Stengle, T. R., 74
Stenhouse, I. A., 98
Stepanov, A. P., 153, 505
Stepanov, E. P., 501, 505, 595, 616
Stephenson, T. A., 46, 69, 90, 334, 391
Stepišnik, J., 104, 111
Steppel, R. N., 97
Sterk, T., 617
Sterlin, S. R., 62, 136
Sterzel, W., 241
Stetger, J., 509
Stetsenko, A. I., 419
Stetsenko, P. N., 616
Steudel, R., 304
Stevens, E. D., 59
Stevens, J. G., 494, 495
Stevens, J. R., 331
Stevens, R. M., 9
Stevens, V. E., 494
Steward, O. W., 127
Stewart, D. J., 592, 593
Stewart, R. P., 28, 416
Stewart, R. S., 82
Štibr, B., 115, 250, 251
Stilbs, P., 145
Stiles, P. J., 88
Stiller, H., 243
Stipanovic, R. D., 94
Stobart, S. R., 27, 28, 37, 83, 269, 314
Stocco, F., 457
Stocco, G. C., 71, 134, 309, 457
Stockmeyer, R., 243

Stocks, J., 306
Stockton, G. W., 74
Stöhr, J., 358
Stojczyk, B., 277
Stole, H., 292
Stolfo, J., 125
Stolyarov, V. L., 264
Stone, F. G. A., 24, 28, 32, 37, 39, 40, 45, 47, 55, 60, 116, 252, 342, 348, 374, 472, 475, 476, 485, 486, 488, 489, 493
Stone, J. A., 614
Stone, N. J., 151
Stone, W. E., 109
Storhoff, B. N., 330, 412
Stormer, B. P., 345, 465
Storr, A., 113, 254, 261
Story, H. S., 101
Stothers, J. B., 11
Stotz, R. W., 315
Stouffs, P., 182
Strait, W. R., 440
Stratford, R., 177
Straub, D. K., 90, 333, 450, 527, 528, 533
Straughan, B. P., 196, 241
Strauss, I., 17
Strel'tsov, L. I., 270
Strimer, P., 229
Stroh, E. G., 143
Strokan, V. N., 20, 439
Strope, D., 81, 388
Strouse, C. E., 17
Stroyer-Hansen, T., 300
Struchkov, Yu. T., 65
Strukov, O. G., 414
Stryland, J. C., 201
Stubbs, M. E., 75
Stucki, H., 67
Studier, M. H., 202
Stühler, H., 127, 139, 284
Stukan, R. A., 515, 532, 621
Stumbreviciute, Z., 62
Stump, N., 100, 106
Stutte, B., 46, 390
Stynes, D. V., 44, 338
Su, A. C. L., 122
Suba, M., 621, 622
Subbaraman, J., 43
Subbaraman, L. R., 43
Subramanian, N., 109
Subramanian, S., 201
Sudarikov, B. N., 359
Sudmeier, J. L., 49, 51
Suenaga, M., 518
Sugawara, F., 232
Sugimoto, K., 151
Sugitani, Y., 104
Sugiyama, N., 41
Sukhova, T. G., 382
Sumida, W. K., 146, 301
Sumiyama, K., 621
Sumodi, A. J., 210
Sun, M. S., 92
Sundberg, R. J., 45
Sunderland, A., 310
Sundermeyer, W., 144
Sunko, D. E., 148

Suprunenko, P. A., 343
Surikov, V. V., 500
Surles, T., 306
Surov, V. N., 253
Sushchinskaya, S. P., 125
Sutcliffe, G. D., 121
Sutcliffe, L. H., 136
Sutherland, R. G., 41
Suthers, R. A., 301
Sutter, D., 193
Sutton, D., 44, 397, 437
Sutula, R. A., 85
Suvorov, A. V., 443
Suwalski, J., 500, 618
Suyunova, Z. E., 103
Suzdalev, I. P, 521, 545, 620, 621
Suzuki, K., 55
Suzuki, S., 201
Suzuki, Y., 376
Svergun, V. I., 161
Sviridov, V. V., 345, 520
Svirmickas, A., 10
Swain, J. R., 8, 46, 421, 468
Swanson, B. I., 226
Swanson, M. E., 472
Swanton, P. F., 56
Swanwick, M. G., 443
Swartz, W. E., 576
Swartzendruber, L. J., 505, 593, 617, 621
Sweeney, A., 50
Swern, D., 98
Swift, J. T., 79
Swile, G. A., 53, 82, 341
Swindell, R. F., 73, 139, 284, 296
Switkes, E. S., 90, 334
Syamal, A., 457
Sych, A. M., 316
Sykes, A. G., 324, 428
Sykes, B. D., 5
Sylvanovich, J. A., jun., 357
Syngh, M. M., 418
Syritso, L. F., 324
Sytsma, L. F., 82
Szabo, A., 495
Szczecinski, P., 71
Szczepański, A., 498
Sze, S. N., 21, 467
Szobyori, L., 406
Szymanski, D., 539
Szymanski, J. T., 137

Tabenko, B. M., 127
Tabereaux, A., 115, 252
Tack, D., 146
Taddei, F., 9, 132, 136
Tadino, A., 146
Tadokoro, S., 96
Taft, R. W., 146
Tagliavini, G., 433
Taillandier, E., 436
Taillandier, M., 436
Taisumov, Kh. A., 253
Takada, T., 512, 621
Takagi, K., 182
Takagi, Y., 95

Takahashi, H., 242
Takahashi, S., 51
Takahashi, Y., 54, 55, 56, 377, 378, 379, 431
Takaki, H., 508
Takano, M., 512, 530
Takaoka, T., 125
Takaya, Y., 135
Takeda, M., 570
Takegami, Y., 56, 398
Takemoto, J., 310
Takemoto, J. H., 332
Takemoto, N., 58, 465
Takeo, H., 181, 183, 187
Taketomi, T., 47
Taki, T., 149
Takizawa, T., 376
Tamaki, A., 60
Tamaki, M., 418, 408
Tamao, K., 268
Tan, B.-Y., 464
Tan, H.-W., 97
Tan, T. H., 273, 581
Tanaka, K., 135
Tanaka, M., 56, 398, 429
Tanaka, S., 150
Tanaka, T., 69, 135, 415, 435
Tananaev, I. V., 261
Tancrede, J., 35
Tandon, S. K., 257
Taneja, A. D., 424, 426
Tanida, H., 97
Taniélian, C., 16, 144, 291
Tanner, D. D., 41, 97
Tansil, J. E., 601
Tarama, K., 85, 337
Taranets, N. A., 324
Tarasevich, A. S., 13, 150
Tarasevich, Yu. I., 103
Tarasov, V. A., 570, 575
Tarasov, V. P., 15
Tarasova, A. I., 259
Taravel, B., 241, 315, 427
Tarelli, J. M., 63
Tarina, D., 520, 553, 616
Tarte, P., 231, 237, 278, 293
Tasaka, S., 129
Tatasevich, Yu. I., 108
Taticchi, A., 147
Tatsuno, Y., 52, 58, 397, 424, 444
Taube, H., 42, 45, 334, 396
Tauber, A., 554, 557
Tauchner, P., 61
Tauchnitz, J., 97
Tayim, H. A., 351, 379, 382
Taylor, B., 136
Taylor, F. B., 310, 431, 529
Taylor, G. A., 125
Taylor, L. T., 409
Taylor, M. D., 358
Taylor, M. J., 282
Taylor, N. J., 262
Taylor, P. C., 159
Taylor, R. C., 97
Taylor, R. S., 324
Taylor, S. H., 85
Tazeeva, N. K., 141, 153

Teague, J. R., 594
Tegenfeldt, J., 102
Telnic, P., 530
Temkin, O. M., 344, 382
Temme, F. P., 155
Templeman, G. J., 74
Tennant, W. C., 52
Tenney, A. S., 260
Teodorescu, M., 530
Tepikin, V. E., 564
Teplov, M. A., 149
Terao, T., 111
Teratani, S., 95
Tereshchenko, G. F., 151
Ternovaya, T. V., 18, 77
Terrier, F., 147
Tertykh, V. A., 277, 301
Teterin, E. G., 315
Tetsu, K., 303, 315
Teuben, J. H., 361
Tew, W. P., 435
Thackeray, J. R., 448
Thakkar, A. L., 95
Thakur, N. V., 99, 442
Thames S. F., 127, 128
Thamm, H., 140, 289
Thavornyutikarn, P., 304, 306
Thé, K.I., 129, 132, 286
The, N. D., 210
Thelen, J., 293
Theophanides, T., 55, 342, 382
Therrell, B. L., jun., 119, 258
Theysohn, W., 16
Theyson, T. W., 401
Thibaut, P., 147
Thibedeau, R. N., 51
Thick, J. A., 61, 398
Thiebeaux, C., 198
Thiele, G., 264
Thiele, K.-H., 320
Thistlethwaite, G. H., 135, 136
Thoai, N., 96
Thomas, B. S., 113, 254
Thomas, J. D. R., 440
Thomas, J. L., 25, 473
Thomas, K., 52, 341
Thomas, K. M., 59, 265
Thomas, M. M., 452
Thomas, P., 2
Thomas, Ph., 349
Thomas, P. M., 517
Thomas, T. E., 211
Thompson, B. C., 78
Thompson, C. L., 536
Thompson, D. W., 138, 318, 427
Thompson, G. L., 98
Thompson, J. C., 73, 79, 137, 272
Thompson, L. K., 37, 343, 484
Thompson, M. L., 114
Thompson, R. C., 273, 581
Thompson, S., 356, 610
Thomson, B. J., 41
Thomson, J., 68

Thomzik, M., 141, 293
Thornhill, D. J., 390
Thornton, D. A., 310, 311, 357, 404
Thornwart, W., 482
Thrane, N., 526
Thyret, H., 32
Tiddy, G. J. T., 75
Tidwell, T. T., 41, 97
Tiemann, E., 178, 179, 180, 181
Tiethof, J. A., 49, 350, 452
Tigelaar, H. L., 181
Tille, D., 407
Tiller, H. J., 619
Timms, P. L., 120, 137, 262
Timofeeva, T. N., 3, 151
Tino, Y., 621
Tirouflet, J., 41, 370
Tisley, D. C., 400
Tkatchenko, I., 56
Tobias, R. S., 71, 233, 309, 457
Toda, S., 150
Todd, K. H., 85
Todd, L. J., 23, 115, 116, 467
Todireanu, G., 107
Todo, I., 152
Törring, T., 178, 179, 180, 181
Tokel, N. E., 59
Tokii, T., 410
Tokuda, A., 54, 377
Tokunan, H., 345
Tolman, C. A., 43, 52, 55
Tom, G. M., 117, 251
tom Dieck, H., 24, 468
Tomala, K., 544
Tomassetti, G., 104
Tomborski, C., 125
Tomic, L., 148
Tomic, M., 148
Tominari, K., 394
Tomlinson, A. A. G., 349, 403
Tomlinson, A. J., 72, 132, 279, 292
Tompa, K., 108
Tong, C. C., 192
Tong, D. A., 164
Tong, H. W., 49
Topart, J., 136
Toren, E. C., jun., 129
Torgeson, D. R., 105, 173
Tori, K., 94, 98
Torocheshnikov, V. N., 14
Torréns, M. A., 527, 528
Tosi, L., 310, 322
Tossidis, I., 450
Tossidis, J. A., 138
Touchard, D., 41
Toužín, J., 288
Tovgashin, Yu. T., 293
Towns, R. L. R., 95
Townsend, C. A., 50
Townsend, M. G., 566
Trabelsi, M., 422, 344
Traber, D. G., 74
Trachevskii, V. V., 20

Author Index

Traficante, D. D., 31, 126, 250
Trahanosky, W. S., 22
Trailina, E. P., 410
Tranquille, M., 451
Trautwein, A., 508
Travis, J. C., 495
Traylor, T. G., 126
Treichel, P. M., 21, 24, 28, 39, 55, 58, 364, 414, 415, 420, 471, 479
Tremmel, G., 337, 429
Tressaud, A., 521
Trias, J. A., 357
Tricker, M. J., 299, 574, 591, 592
Trindle, C., 207
Trippett, S., 140
Trofimenko, S., 66
Trogu, E. F., 21
Troilo, G. G., 57, 346
Troitskaya, A. D., 422
Troitskaya, L. L., 54
Tromme, M., 231
Trommsdorff, K. U., 298
Tronich, W., 126
Trontelj, Z., 100
Trooster, J., 504
Trotter, J., 322
Trotter, P. J., 261
Truelock, M. M., 53, 347
Truex, T. J., 59
Trukhtanov, U. A., 549
Trunov, E. D., 109
Truter, M. R., 399
Tsai, T.-T., 134
Tsang, T., 102, 108
Tsangaris, J. M., 103
Tsapkina, I. V., 358, 437
Tsatsas, A. T., 246
Tsereteli, Yu. I., 74
Tsibris, J. C., 536
Tsintsadze, G. V., 457, 458
Tsitskishivili, K. F., 618
Tsivadze, A. Yu., 457
Tsuboi, M., 241
Tsuchida, E., 150
Tsuei, C. C., 621
Tsukida, K., 98
Tsunekawa, S., 189
Tsunoda, Y., 605
Tsurugi, J., 125
Tsutsui, M., 82, 419, 478, 487
Tsutsumi, S., 131, 146
Tsvetkov, A. A., 359
Tsyganenko, A. A., 248
Tsyganov, A. D., 616
Tsypin, M. I., 619
Tuck, D. G., 83, 123, 256, 383
Tucker, E. E., 204
Tully, M. T., 56, 70
Tupčiauskas, A. P., 14, 15
Turbitt, T. D., 40, 41, 85
Turchi, I. J., 118
Turco, A., 413
Turnblom, E. W., 253
Turnbull, G., 597
Turner, G., 329
Turner, G. E., 356, 610
Turner, J. B., 268
Turner, J. J., 205, 385
Turner, K., 22, 362
Turney, T. W., 69
Turrell, G., 197
Turta, K. I., 532
Tutkunkardes, S., 289
Twigg, M. V., 457
Twist, P. J., 87
Tyagi, R. N., 506
Tyler, J. K., 190
Typpi, V. K., 612
Tyrrell, H. J. V., 211
Tyshchenko, A. A., 150
Tytko, K. H., 214, 312
Tyutev, N. D., 508

Ubozhenko, O. D., 320
Uchida, Y., 394
Udo, F., 152
Udovich, C. A., 40, 420
Uebel, J. J., 98
Übelhack, H., 620
Uemura, S., 123
Uemura, T., 58
Ueno, K., 425, 426
Ugo, R., 339, 398
Uhlenbrock, W., 130, 286
Uhlig, D., 491
Uhrich, D. L., 531
Uibo, L., 620, 621
Ulland, L. A., 132
Ulrich, A., 16
Ulrich, S. E., 111, 131, 276
Underhill, M., 30, 68, 460
Unger, B., 241
Ungermann, C. B., 7
Unsworth, W. D., 310, 354, 618
Uphaus, R. A., 81
Urbach, F. L., 68, 407, 409
Urwin, J. R., 16
Usatenko, Yu. I., 320
Usherov-Marshak, A.-V., 152
Usón, R., 330
Uspenskii, M. N., 547
Ustynyuk, J., 7
Ustynyuk, N. A., 364
Ustynyuk, Yu. A., 9, 14, 15, 25, 26, 58, 72, 83, 135, 140
Utimoto, K., 97
Utrory, K., 290
Uttley, M. F., 31, 69, 417
Utton, D. B., 102
Utvary, K., 144
Uvarova, K. A., 320
Uyemura, M., 208
Uzawa, J., 95

Vaglio, G. A., 32, 35, 486, 487
Vahrenkamp, H., 21, 22, 27, 467
Valenti, V., 478
Vallarino, L. M., 314
Valle, M., 32, 35, 486, 487
Vallorino, L. M., 400, 403
Van Ausdal, R., 180
van Baren, B. A., 104
van Bekkum, H., 98
van Bolhuis, F., 48, 336
Van Bostelen, P., 97
Van Bronswyk, W., 331
van Bruijnsvoort, A., 97
van de Grampel, J. C., 290
Van Den Akker, M., 48, 336
Van Den Bergen, A., 323, 407
van der Ent, A., 69, 336
Van der Kelen, G. P., 14, 271
Van der Kerk, G. J. M., 62, 314
van der Linden, J. G. M., 349
Vander Voet, A., 205, 305
Vanderwielen, A. J., 125, 269
Van der Woude, F., 548
van Diepen, A. M., 565, 621
van Dongen, J. P. C. M., 11
Van Driel, H. M., 170, 171
Van Dyke, C. H., 125, 267, 268
Van Geet, A. L., 74
Van Gorkom, M., 19, 361
Van Hecke, G. R., 88
Van Huong, P., 211
van Ingen Schenau, A. D., 434
van Koten, G., 16, 414
Van Leeuwen, P. W. N. M., 70
Van Leirsburg, D. A., 206
van Oven, H. O., 361
Van Paasschen, J. M., 112, 248
Van Rensburg, D. J. J., 237
van Stapele, R. P., 565
van Tamelen, E. E., 315
Van Thank, N., 199
Van Uitert, L. G., 236
Van Veen, R., 118
van Wageningen, A., 97
Van Wazer, J. R., 138, 140, 143, 144, 289, 290
Van Wieringen, J. S., 621
Vapirev, E., 501
Varache, M., 127
Varghese, G., 199
Varhegyi, G., 108
Várhelyi, C., 406
Varma, M. N., 621
Varret, F., 558
Varsharskii, Yu. S., 489, 490, 491
Vašák, M., 97
Vashman, A. A., 75
Vasil'ev, L. N., 572, 590, 621
Vasil'ev, L. S., 83
Vasilevskis, J., 400

Vasini, E. J., 306
Vassilian, A., 351, 379, 382
Vasudev, P., 478, 539, 594, 595
Vaughan, R. W., 104
Vdovin, V. A., 147, 149, 151
Vear, C. J., 199
Vecchio, G., 97
Vedejs, E., 62
Vedenin, S. V., 100
Vega, A. J., 100
Végh, G., 352, 353
Veksli, Z., 152
Velikodnyi, A. I., 617, 618
Velleman, K.-D., 141, 297
Velthorst, N. H., 3
Velvis, H. P., 290
Venediktov, A. A., 197
Venezky, D. L., 139, 294
Venkataraman, G., 237
Venkateswarlu, K. S., 514
Venugopalan, S., 244
Verdonck, L., 14, 270, 271
Verendyakina, N. A., 331
Vereshchagina, T. Ya., 75
Vergnoux, A. M., 243, 246
Verkade, J. G., 5, 7, 39, 260, 311
Verleur, H. W., 152
Verma, V. K., 582
Vermeer, H., 139
Vernon, W. D., 67
Verrall, K., 463
Vértes, A., 151, 525, 537, 546, 577, 593, 617, 618, 621, 622
Veszpremi, T., 137
Vevere, I., 17
Viau, D., 149
Viccaro, P. J., 535, 547
Vicentini, C., 442
Vickroy, D. G., 306
Victor, R., 35
Victor, T. A., 98
Vidali, M., 18, 358, 411
Viennot, J. P., 199
Vigato, A., 60, 347, 452
Vigato, P. A., 18, 59, 358, 411
Viglino, P., 21
Vijaya, M. S., 166
Vijayaraghavan, R., 105
Vikane, O., 145
Vilkov, L. V., 269
Villa, A. C., 349
Villafranca, J. J., 78, 89
Vincze, I., 617, 618, 622
Vinogradov, I. A., 521
Vinogradov, K. I., 512
Vinogradova, I. S., 108
Vinogradova, L. E., 251
Vinogradova, S. I., 77
Virgo, D., 564
Vishnyakov, Yu. S., 571
Visser, H. D., 62
Visser, J. P., 53, 55, 381
Vitali, D., 28
Vítek, A., 250

Vivien, D., 107
Vladimirova, Z. A., 105
Vlasov, A. G., 197
Vliegenthart, J. F. C., 131
Vodar, B., 201
Vodička, L., 98
Voelter, W., 5
Vogel, M., 37, 414
Vogel, W., 605
Vogl, G., 501
Vogl, W., 501
Vogler, S., 391
Voisin, M., 227
Voitkiv, V. V., 549
Voitländer, J., 107
Vojtech, O., 242
Vokov, S. V., 292
Volf, S., 442
Volger, H. C., 97
Volk, V. I., 357, 444
Vol'kenau, N. A., 169
Volkert, R., 304
Volkov, A. F., 171
Volkov, A. V., 110
Volkov, V. M., 397, 398
Volkova, A. Ya., 565
Volkova, L. S., 397, 398
Volkova, T. A., 326
Volodicheva, M. I., 150
Voloshina, T. D., 106
Volponi, L., 139, 351
von Ammon, R., 94, 99, 414
von Bredow, K., 97
von Deuster, E., 468
von Halasz, S. P. V., 130, 295
Vonnahme, R. L., 32, 85, 478
von Werner, K., 46, 53, 490
Vornberger, W., 127, 129, 274
Vorob'ev, N. I., 320
Vorob'eva, V. Ya., 439
Voronin, A. M., 500
Voronkov, M. G., 212, 266, 273
Voronov, V. K., 89, 149
Vorotilina, T. B., 261
Vorsina, I. A., 294
Vostrikova, L. A., 346
Voyevodskaya, T. I., 58
Vozdvizhenskii, G. S., 149
Vreugdenhil, A. D., 17
Vriesenga, J. R., 75
Vrieze, K., 70
Vuletić, N., 65, 454
Vyas, P. C., 144
Vyayaraghavan, R., 107
Vyazankin, N. S., 124, 127, 136, 280

Waack, R., 16
Waber, J. T., 497
Wada, M., 132, 433
Wadas, R., 617
Waddington, T. C., 111
Wade, K., 120, 129, 257, 275
Wagner, B. E., 91

Wagner, B. O., 10
Wagner, F. E., 590, 605
Wagner, J. L., 378
Wagner, K. P., 58
Wagner, R., 401, 479
Wagner, S., 320
Wagner, W. F., 95
Wahl, G. H., jun., 94
Wahner, E., 144
Wailes, P. C., 19, 316, 362
Wainwright, K. P., 25, 473
Waites, P. C., 315
Waldhör, S., 274
Waldron, B., 403
Waldron, R. H., 61
Waldron, R. W., 447
Walker, D. W., 27
Walker, F. A., 68
Walker, I. M., 92, 137
Walker, I. O., 78
Walker, J. C., 498, 619
Walker, W., 218
Wall, M. C., 223
Wallart, F., 208, 234
Wallbridge, M. G. H., 55, 113, 114, 252, 254, 358, 377
Walrafen, G. E., 201, 294
Walsh, E. J., 143, 289
Walter, H. K., 611
Walter, L. S., 563
Walter, W., 73, 97
Walters, D. B., 95, 97
Walther, B., 25, 122, 257, 471, 405
Walton, D. R. M., 63, 127, 128, 136
Walton, R. A., 400
Wamhoff, H., 97
Wan, J. K. S., 349
Wander, J. D., 63, 77, 96, 122
Wandiga, S. O., 68, 407
Wang, C. H., 237
Wang, C. S.-C., 135, 433, 581
Wang, F. T., 112
Wang, H., 423
Wang, T.-T., 119
Wannagat, U., 17, 86, 134, 247, 275
Wappling, R., 509, 605
Ward, H. R., 3
Ward, J. E. H., 22
Ward, R. T., 329
Wardeska, J. G., 455
Wardle, R., 71
Warner, P., 95
Warren, J. D., 68, 482
Warrier, A. V. R., 324
Washburne, S. S., 130
Wasiutynski, T., 238
Wasser, P., 63
Wasylishen, R., 98
Watanabe, H., 6, 132, 509, 546
Watanabe, I., 172
Watanabe, M., 104
Watanabe, N., 155, 576
Watanabe, Y., 56, 398

Author Index

Waterworth, L. G., 122, 256, 263
Waters, J. M., 30, 40, 417, 487
Watkins, J., 254
Watkins, P. M., 59, 64, 320, 447, 451
Watson, R. E., 617
Watson, W. A., 188
Watt, G. W., 343, 440
Watts, G. B., 118
Watts, J. A., 170
Watts, J. C., 575, 576
Watts, P. H., 576
Watts, W. E., 22, 40, 41, 85, 363
Waugh, F., 128
Waugh, J. S., 5, 101, 103
Way, G. M., 114
Wayland, B. B., 59, 444
Weaver, D. L., 533
Weaver, H. T., 104
Weaver, T. R., 403
Webb, G. A., 314
Weber, A., 98
Weber, J., 145
Weber, K., 84
Weber, P., 125
Weber, W. P., 132
Webster, R. K., 177
Weeks, R. A., 152
Wege, D., 22
Wegener, J., 144, 294
Wehinger, E., 482
Wehrli, F. W., 138, 140
Wei, R. M. C., 410
Weibel, A. T., 121
Weidenbruch, M., 272
Weidlein, J., 122, 139, 255, 256, 261, 284
Weigold, H., 19, 315, 362
Weiher, J. F., 508, 553
Weihofen, U., 537
Weil, J. A., 93, 459
Weimann, B., 3
Weiner, M. A., 126
Weingand, C., 140, 289
Weinstein, G. N., 64
Weinstock, N., 213, 214, 216
Weis, C. D., 97
Weis, J. C., 20, 21, 468, 469
Weise, W., 413
Weiss, A., 99, 154
Weiss, A. J., 9
Weiss, E., 256, 383
Weiss, J., 413
Weiss, S., 222
Weissenhorn, R. G., 24
Weissman, M., 41
Welcker, P. S., 61, 414
Welcman, N., 40, 482
Weller, F., 255, 262
Wells, D., 69, 372
Wells, F. H., 177
Wells, R. D., 98
Welsh, H. L., 201, 243
Wender, S. A., 499, 604
Wendisch, D., 138, 157

Wendling, E., 222, 225
Wennerström, H., 64
Wentworth, R. A. D., 401
Werle, P., 318
Werner, H., 58, 85, 376, 468
Wernick, J. H., 609
Wertheim, G. K., 617
Wesolowski, W., 538
Wessal, N., 272
West, A. R., 317
West, B. O., 45, 56, 139, 323, 407, 408
West, K. W., 573, 574
West, P., 498
West, R., 71, 127, 134, 274
West, R. J., 76
Westermayer, G., 24
Westland, A. D., 412
Westmoreland, T. D., jun., 77
Westrum, E. F., 546
Wetzel, R. B., 98
Wharton, L., 177
Wheatland, D. A., 61, 447
Whiffen, D. H., 182
Whimp, P. O., 52, 69, 422
White, A. H., 65, 320, 448, 449
White, A. I., 6
White, D. A., 412
White, D. L., 128, 383
White, D. M., 6
White, G. L., 122
White, J. F., 454
White, R. F. M., 14, 579
White, W. B., 229
White, W. F., 176
White, W. H., 409
Whitehead, M. A., 158, 163, 211
Whitesides, G. M., 41, 107
Whitfield, H. J., 159, 160
Whiting, D. A., 98
Whitlock, H. W., jun., 67
Whitman, P. J., 50
Whitney, A. G., 107
Whitney, P. M., 93
Whittaker, B., 458
Whittaker, D., 29, 478
Whitten, D. G., 42
Whittingham, M. S., 101
Whittle, K. R., 30, 40, 417, 487
Whittle, M. J., 195
Whyman, R., 391, 392
Wiberg, N., 71, 83, 122, 130, 137, 247, 286, 297
Wicholas, M., 89
Wicke, B. G., 180
Wickman, H. H., 538, 539
Wieckowicz, N. J., 337
Wieghardt, G., 304
Wieghardt, K., 339, 429
Wiegrebe, W., 97
Wiener-Avnear, C. M., 239
Wiener-Avnear, E., 239
Wiersema, R. J., 48, 117, 121, 251, 252
Wies, R., 140, 295

Wigfield, Y. Y., 149
Wignacourt, J. P., 234
Wild, P., 399
Wild, S. B., 25, 473
Wilenzick, R. N., 501
Wiley, J. C., jun., 134
Wilke, G., 19, 55, 56
Wilkie, C. A., 16, 127, 245
Wilkins, B., 47, 69
Wilkins, J. D., 20, 318
Wilkinson, G., 7, 19, 26, 40, 43, 271, 362
Wilkinson, S. P., 22
Wilkinson, W., 51, 338, 345, 349
Willcott, M. R., 95, 97
Wille, G., 18
Willemse, J., 60, 345
Williams, D. E., 95
Williams, D. H., 94, 95
Williams, K. C., 132, 443
Williams, I. R., 151
Williams, J., 464
Williams, M. N., 78
Williams, S. H., 124
Williams, R. J. P., 78, 95, 534
Williams-Smith, D. L., 35
Williamson, A. N., 59, 444
Williamson, D. L., 502, 506
Williamson, D. R., 351
Willis, C. J., 427
Wilson, E. B., 176
Wilson, G. V. H., 619
Wilson, P. W., 359
Wilson, R. D., 146, 147, 148, 305
Wilson, W. W., jun., 59
Wind, R. A., 104
Window, B., 622
Wineburg, J. P., 98
Winfield, J. M., 74, 148, 304, 318
Wing, R. M., 51, 98
Wingfield, J. N., 24, 129, 276, 291, 322
Winkelmann, H., 378
Winkler, E., 31, 32, 38, 467, 478, 479, 485
Winkler, H., 110, 499
Winkler, T., 54
Winnewisser, B. P., 184
Winnewisser, G., 183, 186
Winnewisser, M., 184, 186
Winstead, J. A., 41
Winstein, S., 61
Winter, M. R. C., 534
Winterhalter, K. H., 42, 620
Wintersteiner, P. P., 504
Winterton, N., 35
Wirth, H. E., 77
Wismar, H.-J., 17, 247
Wiszniewska, A., 171
Witiak, J. L., 132
Witschel, J., jun., 107
Wittig, G., 55
Wittstruck, T. A., 95

Author Index

Wojcicki, A., 37, 331, 396, 450
Wojnowska, M., 138, 279
Wojnowski, W., 138, 279
Wojtkowski, P. W., 118
Wolbeck, B., 602, 607
Wolber, P., 275
Wolf, L. R., 35
Wolf, S. N., 188
Wolkowski, Z. W., 96
Wolmarans, N. S., 596
Wolsey, W. C., 401
Wolters, J., 137
Wong, J., 260
Wong, K. S., 138, 427
Wood, E., 478, 539
Wood, E. J., 78
Wood, M., 74
Wood, R. J., 13, 14
Woodhams, F. W. D., 499, 552, 592
Woodruff, R. A., 63
Woods, M., 93, 459
Woodward, L. A., 196
Woodward, P., 30, 35, 440, 483
Woplin, J. R., 143, 144
Work, R. A., 211, 333
Workman, M. O., 342
Worrall, I. J., 122, 123, 255, 256, 263
Worsfold, D. J., 16
Wortmann, G., 602
Wozniak, W. T., 386
Wrackmeyer, B., 12
Wray, V., 6
Wreford, S. S., 31, 250
Wright, A., 127
Wright, J. R., 118
Wright, R. B., 237
Wu, C.-H., 425
Wu, Y., 51
Wüsteneck, A., 321
Wuethrich, K., 42, 89, 90, 153
Wyatt, M., 94
Wyderko, M. E., 544, 546
Wynne, K. J., 146, 304, 305
Wyse, F. C., 179, 181

Xavier, A. V., 78, 95

Yagodin, G. A., 317
Yagupsky, G., 19, 362
Yagupsky, M., 19, 271, 362
Yajima, F., 85
Yakimov, S. S., 525, 549
Yakovlev, E. N., 620
Yakovlev, Yu. M., 550
Yamada, M., 307
Yamada, S., 454
Yamadaya, T., 107
Yamaguchi, V., 509
Yamamoto, A., 40, 331, 366, 368, 428, 434
Yamamoto, G., 146
Yamamoto, H., 509, 546
Yamamoto, K., 41, 56, 129, 133, 135, 398

Yamamoto, T., 368
Yamamoto, Y., 37, 86, 396, 412, 485
Yamamura, H., 553
Yamane, T., 152, 153
Yamanouchi, K., 454
Yamasaki, A., 96
Yamasaki, K., 50
Yamatake, M., 441
Yamazaki, H., 18, 32, 37, 38, 39, 412, 479, 483, 485, 492
Yamazaki, S., 311, 425
Yampol'skii, Yu. P., 620
Yanagida, T., 202
Yang, C.-H. L., 50, 338
Yankelevich, A. Z., 149
Yankovskaya, G. F., 110
Yano, T., 87
Yarembash, E. I., 595, 616
Yarkova, D. G., 427
Yarmarkin, V. K., 512
Yarom, A., 501
Yarrow, D. J., 47, 85
Yarwood, J., 307
Yashina, N. S., 136
Yasufuku, K., 18, 32, 38, 39, 479, 483, 492
Yasuoka, H., 107
Yatsenko, A. P., 319
Yatsimirskii, K. B., 75, 78
Yaymaguchi, K., 509
Yeats, P. A., 261, 273, 445, 581
Yee, K. K., 200
Yeh, C.-L., 35, 66
Yeh, E.-L., 66
Yeh, S. J., 617
Yin, P. K. L., 204
Yoder, C. H., 71, 129, 275
Yoke, J. T., 59
Yokoyama, H., 149
Yonezawa, T., 93
Yoon, N., 93, 451
Yoshida, G., 122, 262, 454
Yoshida, H., 229
Yoshida, I., 425, 435
Yoshida, T., 58, 342, 376
Yoshikawa, Y., 50
Yoshimura, H., 508
Yoshimura, Y., 98
Yoshinaga, A., 181
Young, I. M., 41
Young, J. P., 263
Young, R. T., 282
Young, W. G., 61
Ypenburg, J. W., 277, 278, 300
Yu, S., 16
Yuan, S. T., 299
Yukhin, Yu. M., 294
Yukhnevich, G. V., 357
Yurchenko, A. G., 148
Yushina, G. G., 571
Yutlandov, I. A., 501

Zabransky, B. J., 517
Zacharias, P. S., 82, 92
Zahn, P., 122
Zaikin, P. N., 595

Zaitsev, B. E., 444
Zaitseva, L. L., 444
Zakharchenya, B. P., 244
Zakharkin, L. I., 31, 61, 115, 151, 152, 159, 251, 370
Zakharov, V. P., 230
Zalkin, A., 209
Zamaraev, K. I., 91
Zamojski, A., 98
Zao, N. K., 80, 84
Zanella, A., 85
Zanella, P., 285
Zanger, M., 2
Zannoni, G., 56
Zarkadas, A., 63, 354
Zarli, B., 139, 351, 358
Zarubin, V. N., 549
Zasimov, V. S., 498, 595
Zasyadko, O. A., 271
Zavgorodnii, V. S., 622
Zavin, B. G., 87
Zavorokhina, I. A., 447
Zaw, K., 75
Zayats, M. N., 324
Zdanovich, V. I., 21, 363
Zdero, C., 97
Zecchina, Z., 325
Zeil, W., 154, 193
Zeinalova, Kh. K., 260
Zeldin, M., 120, 255
Zelentsov, V. V., 532
Zelonka, R. A., 12, 40, 93, 334, 371
Zelta, L., 10
Zeltmann, A. H., 6
Zemcik, T., 618
Zemlyanski, N. N., 163, 269
Zengin, N., 344
Zenkin, A. A., 72
Zeppezauer, M., 80
Zerman, J., 288
Zermati, C., 2
Zetlmeisl, M. J., 249
Zetter, M. S., 77
Zezin, S. B., 618
Zhabotinskii, M. E., 150
Zharkov, A. P., 264
Zharov, Yu. D., 618
Zhdanov, A. A., 87
Zhidomirov, F. M., 87
Zhinkin, D. Ya., 12
Zhukov, A. P., 158
Zhukhovitskii, A. A., 545
Ziegler, E., 3
Ziegler, R. J., 314, 326, 389
Ziehn, K.-D., 140
Ziessow, D., 12
Zigan, F., 202
Zimina, G. V., 294
Zimmermann, G., 95
Zimmermann, H., 314
Zingaro, R. A., 140
Zink, J. I., 93, 97, 451
Ziolkowski, J., 310, 332, 396
Ziolo, R. F., 61, 398

Ziomek, J. S., 426
Zioutas, K., 607
Zipperer, W. C., 20, 22, 37, 365, 419
Zlotina, I. B., 27
Zolin, V. F., 358, 437
Zorina, M. L., 324
Zorn, J. C., 180
Zschunke, A., 97

Zsoldos, B., 617
Zubieta, J. A., 85, 450, 484
Zubenko, V. V., 549
Zubkovska, E., 101
Zuckerman, J. J., 131, 273, 276, 281, 574, 575, 577
Züchner, K., 74, 304
Zueva, A. N., 405

Zuika, I., 260
Žumer, S., 111
Zupancic, I., 157
Zurgaro, R. A., 291
Zussman, A., 162
Zverev, N. D., 549
Zvyagin, A. I., 228, 329
Zyuzya, L. A., 20
Zweifel, G., 122